Linear Algebra and Ordinary Differential Equations

Linear Algebra and Ordinary

Advanced
Engineering
Mathematics

Volume 1
Linear Algebra
and Ordinary
Differential
Equations

Volume 2
Complex Analysis
and Applications

Volume 3
Partial
Differential
Equations

Differential Equations

ALAN JEFFREY
Professor of Engineering Mathematics
University of Newcastle upon Tyne, and
Adjunct Professor of Mathematics
University of Delaware

BLACKWELL SCIENTIFIC PUBLICATIONS

BOSTON OXFORD

LONDON EDINBURGH MELBOURNE

© 1990 by
Blackwell Scientific Publications, Inc.
Editorial offices:
3 Cambridge Center, Suite 208
    Cambridge, Massachusetts 02142, USA
Osney Mead, Oxford OX2 0EL, England
8 John Street, London WC1N 2ES
    England
23 Ainslie Place, Edinburgh EH3 6AJ
    Scotland
107 Barry Street, Carlton
    Victoria 3053, Australia

First published 1990

Set by Macmillan India Ltd, Bangalore 560 025
Printed and bound in Great Britain
at the University Press, Cambridge
89 90 91 92 5 4 3 2 1

DISTRIBUTORS

USA
    Blackwell Scientific Publications, Inc.
    Publishers' Business Services
    PO Box 447
    Brookline Village
    Massachusetts 02147
    (*Orders*: Tel: 617 524-7678)

Canada
    Oxford University Press
    70 Wynford Drive
    Don Mills
    Ontario M3C 1J9
    (*Orders*: Tel: 416 441-2941)

Australia
    Blackwell Scientific Publications
    (Australia) Pty Ltd
    107 Barry Street
    Carlton, Victoria 3053
    (*Orders*: Tel: 03 347-0300)

Oustide North America and Australia
    Marston Book Services Ltd
    PO Box 87
    Oxford OX2 0DT
    (*Orders*: Tel: 0865 791155
            FAX: 0865 791927
            Telex: 837515)

Library of Congress
Cataloging-in-Publication Data

Jeffrey, Alan.
    Advanced engineering mathematics/Alan Jeffrey.
        p.   cm.
    Contents: v. 1. Linear algebra and ordinary
differential equations.
        ISBN 0-86542-113-7
        ISBN 0-86542-114-5 (pbk.)
    1. Engineering mathematics.   I. Title.
TA330.J44   1990
620'.001'51—dc20

British Library
Cataloguing in Publication Data

Jeffrey, Alan
    Advanced engineering mathematics.
    Vol. 1: Linear algebra and ordinary differential
    equations
    1. Mathematics
    I. Title
    510

    ISBN 0-86542-113-7
        0-86542-114-5 (pbk)

Contents

Part 3: Ordinary Differential Equations

4   First Order Ordinary Differential Equations, 311

5   Linear Higher Order Ordinary Differential Equations, 382

6   Systems of Linear Differential Equations, 478

7   Laplace Transform and z-Transform, 594

8   Series Solution of Ordinary Differential Equations, 751

Preface

Most of the important physical problems arising in engineering and the applied sciences are either modeled directly by differential equations, or make some use of them in their formulation. As a result a sound understanding of differential equations is an indispensable tool for all who work in these areas. Standing as they do at the frontier between the applied sciences and pure mathematics, differential equations both enrich and are enriched by these different fields. On the one hand techniques for the solution of certain types of equation provide engineers with the means of solving many of their problems, while on the other fresh applications often pose interesting new problems for pure mathematicians, such as the conditions under which a solution exists and is unique. The knowledge that a solution exists but is not unique does not necessarily amount to a mere mathematical abstraction, because although in many situations a unique solution is to be expected, the nonuniqueness of solutions to certain types of two-point boundary value problems provides the basis for the development of Fourier series and more general expansions in terms of orthogonal functions. Consequently, when seeking to formulate a mathematical model of a physical situation, the knowledge that a particular model has a nonunique solution may confirm that the model is in agreement with observation or, conversely, it may be sufficient to invalidate it. Thus, whether a student of engineering, applied mathematics or pure mathematics, the study of differential equations and their applications provides both valuable knowledge and techniques over a wide ranging field and at the same time offers an enriching experience.

This book has been written as a result of many years experience teaching engineers and applied mathematics students. The attention to detail throughout the book reflects the author's familiarity with the difficulties experienced by students encountering their first serious contact with differential equations. In addition to discussing the usual topics to be found in a mainstream text of this type, a serious effort has been made to include some newer material which is important in applications, and so should now find a place in such a book. Students embarking on such a course of differential equations come from diverse backgrounds which cannot always be relied upon to provide them with the necessary prerequisites, so the first chapter has been designed to remedy the most serious of these possible defects.

As linear differential equations and systems play an essential role in many applications, it is essential that the elements of linear algebra are introduced at the outset of any systematic development of the subject. In this book the account of linear algebra takes the form of a detailed development of matrix algebra, preceded by a short discussion of the algebra of vectors which, apart from illustrating important concepts in a concrete manner, is also useful in its own right. When developing the linear algebra of matrices the opportunity has been taken to illustrate a number of applications outside the immediate field of differential equations. These applications are usually to be found in the extensive problem sets at the end of each section, and they have been included to provide motivation and also to help relate the discussion of matrices to different applications which the reader is likely to encounter in other courses.

An examination of the contents listing will show that this book deals mainly with analytical methods of solution, and only briefly with numerical methods for solving initial value problems. This is because although computers enable both linear and nonlinear differential equations and systems to be solved numerically, such an approach is not usually adequate to illustrate the theoretical structure and properties of solutions which should form an essential part of the knowledge of every student. This knowledge provides invaluable help when arriving at solutions and formulating models, and it also helps with the understanding of physical systems which are described by the differential equations involved. Furthermore, the successful numerical solution of initial and boundary value problems for the more difficult differential equations is usually the outcome of a detailed analytical study of the problem followed by careful numerical experimentation, so here also a theoretical understanding of differential equations plays a key role. It is the author's belief that the theoretical development of differential equations greatly benefits when it is supplemented by numerical calculation designed to illustrate key features and orders of magnitude of solutions, but that the theoretical basis of the subject must be properly understood before numerical methods can be fully and efficiently utilized.

In this introductory study of differential equations, computing is important, though it plays a somewhat different role to the one found in more advanced books on the subject and in books specifically devoted to numerical analysis. Its value here is that it enables the student to visualize through numerical and graphical output, based on analytical solutions, the sensitivity of a solution to its initial conditions and to the various parameters which may appear in the equations. The numerical interpretation of analytical solutions also often serves to indicate why numerical methods can fail in certain circumstances. Although access to a computer or a PC is not required in order to use this book, the reader is encouraged to make use of these whenever possible and to experiment numerically with analytical solutions. Mathematical sketch-pad type software for PCs, like MathCAD,™ which is available in a student edition, offers an excellent way of obtaining numerical and graphical output from analytical solutions, even when these involve the summation of series. MathCAD is also suitable for the manipulation of matrix equations and for obtaining purely numerical solutions using a technique like the Runge–Kutta method. This software was, in fact, used to obtain the graphical and numerical results presented throughout the book. Another valuable facility available to users of PCs is the access they offer to various symbolic manipulation packages. These enable a PC to perform a wide range of purely analytical operations in symbolic form, such as integration, Taylor series expansion, matrix and determinantal operations and a variety of other algebraic operations. A simple but useful example of such software now available to the student is the package called DERIVE.™

The conciseness of presentation usual in more advanced accounts of differential equations is inappropriate at this expository level, while the space available for differential equations and examples in standard engineering mathematics texts is sufficient only for the conventional topics to be introduced. The approach adopted in this book is different, because each new idea is introduced with the necessary attention to detail and, in particular, to points students find difficult. Each new idea which is introduced is accompanied by carefully chosen illustrative examples and these, in turn, are reinforced by the problem sets at the end of each section. The problem sets themselves are divided into two parts, the first containing straightforward problems of the type given in the text which are designed to emphasize key concepts and develop manipulative skills, and the second somewhat more difficult group of problems which both extend the text and provide a deeper insight into the subject.

In order to separate examples and theorems from the main body of the text, the end of each example is marked by the symbol ■, while to indicate the end of a theorem and its accompanying proof the symbol □ has been used. So far as possible proofs have been given for all theorems, since they help the reader to appreciate the full implications of the theorem itself, though to keep the level of the book uniform proofs are not always given in their most general form.

As this book is intended for engineers and applied mathematicians, it was considered appropriate that each new mathematical concept should be illustrated first by mathematical examples and then, where desirable, by a suitably chosen application. As this is not intended to be a book on mathematical modeling, the detail of the applications has been kept to a minimum, though where necessary any essential background knowledge has been included in the application itself. It should, however, be stressed that these applications form an important part of the text, providing as they do motivation, useful exercises in manipulation and experience in the interpretation of the underlying mathematical concepts.

The main features of this book are:

1 A detailed explanation of every new concept which is introduced accompanied by carefully worked examples, with special attention being paid to those areas in which students experience difficulty.

2 A broad coverage of the essential standard topics in differential equations, to which has been added new material of importance in the fields of engineering and applied mathematics.

3 Extensive problem sets at the end of most sections, with standard problems to reinforce ideas and develop manipulative skills, and harder problems designed to extend the text and provide a deeper insight into the subject.

4 The use of carefully chosen applications of differential equations to key physical problems to provide motivation and illustrate how the theory can be applied.

5 The provision of an introductory chapter which gathers together the necessary mathematical prerequisites in order to make the book self-contained.

6 Answers to odd-numbered problems which are given in greater than usual detail whenever this is likely to be helpful.

The organization of the book is sequential and in the natural order. If desired, Chapter 2 on the algebra of vectors may be omitted at a first reading, though Section 2.9 must be studied as it contains the essential ideas on which are based the linear algebra of vector spaces. Sections 6.1 through 6.3 of Chapter 6 together with some of the applications of the Laplace transform in Section 7.3 presuppose knowledge of the material of Chapter 3 on matrices. However, the development of the material in Chapters 4, 5, 8 and 9 makes no such demand on Chapter 3.

In the main, the new material presented in this book is confined to Sections 7.4 and 7.5 which deal with the z-transform, to Section 8.8 which introduces asymptotic expansions and proceeds as far as a first encounter with the Laplace and WKBJ methods, and Sections 9.3 and 9.5 through 9.9 which provide a deeper understanding of the properties of Fourier series than is usually offered to engineering students, together with an introduction to Fourier–Bessel expansions.

It now only remains for me to express my gratitude to all who have contributed to the development of this book. Special thanks are due to my past students, both in England and North America, whose questions and reactions to early drafts of the text have helped shape this work, and to my many colleagues whose differing approaches to the teaching of differential equations have influenced my own ideas.

In particular, I wish to acknowledge the special debt of gratitude I owe to my friend and colleague Ivar Stakgold, Chairman of the Department of Mathematical Sciences in the University of Delaware, for making possible my continuing association with his Department. Thanks are also due to David Colton, Richard Weinacht and Robert Gilbert, also of the University of Delaware, the first two for advising on the content and presentation of a differential equations course given there by me, and the third for his friendship and the many stimulating discussions I have enjoyed over the years about most aspects of differential equations.

Grateful thanks must also be given to John Nohel in the University of Madison–Wisconsin, and Joseph Keller in Stanford University who made possible research visits during which valuable insight was also gained into differential equations courses given there for engineers.

An acknowledgement of a different kind is due to the software houses whose packages with the trademarks MathCAD™ and DERIVE™ have already been mentioned and which proved useful during the writing of this book. These are Mathsoft, Inc., Cambridge, MA, who produced MathCAD,™ and Soft Warehouse Inc., Honolulu, HI, who produced DERIVE.™

In conclusion, I must thank Navin Sullivan at Blackwell Scientific Publications for his unfailing patience, encouragement and steady flow of sound advice during successive revisions of the manuscript, David Carpenter for his painstaking editing of the manuscript, and Edward Wates for seeing the book through the press so efficiently.

ALAN JEFFREY
Newcastle upon Tyne
October 1989

Part 1
Mathematical Prerequisites

Chapter 1
Review of Topics from Analysis

The purpose of this chapter is to review some fundamental concepts from analysis along with a few of their most important consequences. After introducing sets, functions and inequalities, the chapter proceeds to a discussion of mathematical induction and polynomials, and then goes on to consider the arithmetic of complex numbers in some detail. Some important properties of integrals not always covered in a first course of calculus are examined and, finally, basic results concerning linear difference equations are developed from first principles. This material forms an essential prerequisite for the development of advanced engineering mathematics. The reader should be thoroughly familiar with these ideas, and with the associated notations and terminology, before proceeding to subsequent chapters in this book.

The concept of a set, together with some of the notation of set theory and the definitions of a function and its inverse are introduced in Sec. 1.1. Ideas involving sets are universal; they occur repeatedly throughout all of mathematics, and at the simplest level the notation of set theory provides a convenient shorthand when writing mathematics. Real numbers and the real line are introduced in Sec. 1.2, along with intervals on a line and some important and useful inequalities.

The basic idea of mathematical induction, which depends for its success on the natural order of the integers, is discussed in Sec. 1.3 and illustrated by a variety of examples. An inductive proof is often used when it provides the quickest derivation of a result. Polynomials and rational functions are defined and discussed in Sec. 1.4, where mention is also made of the fundamental theorem of algebra. Polynomials are introduced at this early stage because of their importance throughout mathematics, and also because the determination of the roots of polynomials leads directly to complex numbers.

In anticipation of the needs of subsequent chapters, the precise meaning of the terms *necessary* and *sufficient* when used to qualify the statement of a theorem are explained and illustrated in Sec. 1.5.

Complex numbers are defined and their properties are developed in a systematic manner throughout the next two sections, with the Cartesian form being considered in Sec. 1.6 and the polar form in Sec. 1.7. In Sec. 1.8 some elementary but important properties of integrals are discussed, ranging from functions defined as integrals and the first form of the fundamental theorem of integral calculus, to improper integrals which form the basis of the Laplace transform which is developed in Chapter 7.

Finally, linear difference equations are introduced in Sec. 1.9, where they are developed from first principles to the point at which simple nonhomogeneous equations can be solved.

1.1 Sets and functions

Set theory[1] forms part of the foundation of mathematics, and due to its generality the concept of a set occurs naturally throughout the mathematics of engineering and science. The basic notion involved is that of the **set** as a collection of distinct objects considered as a single entity. The individual objects in a set are called **elements**, and it is understood that there is an unambiguous membership criterion for elements belonging to a set. Among the sets which frequently arise are those whose elements are real numbers, complex numbers, vectors, matrices and functions.

On occasions it is convenient to define a set by listing its elements, and writing $A = \{a, b, c, d\}$ for the set with elements a, b, c and d. When this is done the order in which the elements are listed is unimportant, so $\{a, d, b, c\}$ and $\{d, c, b, a\}$ both define the *same* set A as above.

If a set is denoted by S, and a is an element of S, this is shown symbolically by writing

$a \in S$.

The set membership symbol $\in$ used in this expression may be read either as 'is an element of', or as 'belongs to'. If b is not an element of set S, this is shown by writing

$b \notin S$.

The negated set membership symbol $\notin$ used here may be read either as 'is not an element of', or as 'does not belong to'. As an example of the use of these symbols, let the set of natural numbers $1, 2, 3, \ldots$ be denoted by $\mathbb{N}$. Then the set comprises all possible integers, and the membership criterion for set $\mathbb{N}$ is 'to be a natural number', so that $17 \in \mathbb{N}$, but $3.2 \notin \mathbb{N}$.

In the chapters which follow, the most commonly occurring set is the set of **real numbers** denoted by $\mathbb{R}$. This comprises the set of **integers** $\mathbb{Z}$ with elements $\ldots -2, -1, 0, 1, 2, 3, \ldots$, the set of **rational numbers** $\mathbb{Q}$ with elements p/q, where p and q are integers having no common factor and $q \neq 0$, and the set $\mathbb{R}^*$ to which belong all the remaining real numbers, called the **irrational numbers**. This last set contains numbers like $\sqrt{2}$, e and π.

If every element of set B is contained in set A, then set B is said to be a **subset** of A, and this is shown by writing

$B \subseteq A$.

The set inclusion symbol $\subseteq$ used in the above expression is to be read 'is a subset of'. The subset relationship between sets A and B leaves open the possibility that in addition to $B \subseteq A$, it is also true that $A \subseteq B$. When this occurs, sets A and B contain the same elements and are said to be **equal**, and we then write $A = B$. Thus $A = B$ if, and only if, $A \subseteq B$ and $B \subseteq A$.

---

[1] Set theory was created by the mathematician GEORG CANTOR (1845–1918) who was born in Russia, but studied in Berlin and subsequently became Professor of Mathematics in Halle in 1879. His work was fundamental to the development of mathematics.

Set B is called a **proper subset** of A, written

$$B \subset A,$$

if both sets A and B have the same membership criterion, but set A contains at least one element which is not in B. The set inclusion symbol $\subset$ used in this expression is to be read 'is a proper subset of'. The **null** or **empty set** $\varnothing$ is the set containing no elements, and it is a proper subset of every set so that, in particular, $\varnothing \in \mathbb{R}$. As another example of a proper subset suppose, as is often done, that the set of rational numbers $\mathbb{Q}$ is defined merely as the set of numbers p/q, with p, q integers and $q \neq 0$. It then follows that $\mathbb{Z} \subset \mathbb{Q}$, because rational numbers like $2/1$, $-6/3$ and $16/2$ are elements of $\mathbb{Z}$, but numbers like p/q, when p, q $(q \neq 0)$ have no common factor, occur only in $\mathbb{Q}$.

Sets may be combined, and the set S to which belong all the elements of two other sets A and B is called the **union** of A and B, and written

$$S = A \cup B,$$

with the symbol $\cup$ being read 'union'. It should be clearly understood that the union operation is not a numerical operation involving addition. Thus if element c occurs in both sets A and B, that same element occurs once in $S = A \cup B$, and there are not two elements c in S as there would be were the elements to be aggregated. For example, if

$$A = \{1, 2, 3, 4\}, \quad B = \{1, 3, 4, 5, 6\} \text{ and}$$
$$S = A \cup B, \quad \text{then } S = \{1, 2, 3, 4, 5, 6\}.$$

Equivalently, $S = \{2, 3, 1, 5, 6, 4\}$, because the order in which elements are listed in a set is unimportant.

Another way of combining sets A and B is to form the set S called their **intersection**, written

$$S = A \cap B,$$

with the symbol $\cap$ being read 'intersection'. The intersection of sets A and B is the set comprising all elements which are common to both A and B. If sets A and B have no common elements, that is if $A \cap B = \varnothing$, they are said to be **disjoint**. By way of illustration, if $A = \{1, 3, 7, 9\}$ and $B = \{1, 6, 9, 2, 27\}$, then $A \cap B = \{1, 9\}$, while if $A = \{1, 4, 9\}$ and $B = \{2, 3, 7\}$, then $A \cap B = \varnothing$, showing that in this case A and B are disjoint.

Relationships often exist between the elements of different sets. One of the most important of these is the relationship known in elementary calculus as a function. Let us now define this concept in general terms using the language and notation of set theory.

We define a **function** f from a set X into a set Y to be a rule that assigns a unique element y of Y to each element x of X. The element $y \in Y$ assigned by the rule to a given element $x \in X$ is called the **value** of f at x, and this value is indicated by writing $f(x)$, so that

$$f(x) = y.$$

The set X containing all elements x is called the **domain of definition** of f (abbreviated to **domain**), and the set of all values of f is called the **range** (or **codomain**) of f, and it is written $f(X)$. In the elementary calculus of functions of a single real variable, x is called the **independent variable** or the **argument** of f, the value $y = f(x)$ is called the **dependent variable**, and the rule defining f is usually an explicit mathematical expression involving x.

In certain contexts geometrical terminology is employed, and the expression **mapping** is used in place of the term **function**. The idea involved here is that if the elements of sets X and Y are represented by points in suitable spaces, the function f can then be thought of as assigning or mapping in a unique manner each point x in space X to a corresponding point y in space Y. When the expression mapping is used, it is usual to refer to $f(x)$ as the **image** of x under f, and to x itself as the **pre-image** of $f(x)$.

When working with functions of a single real variable it is often necessary to interchange the roles of the independent and dependent variables and, for a particular function f, to seek the x corresponding to a given value $y = f(x)$. If all possible y are considered, and to each there corresponds a unique x, this becomes the problem of finding the function inverse to f, which we shall denote by f^{-1}.

The **inverse function** f^{-1} can only be found unambiguously if the function f has the property of being **one-to-one**. That is, if f is such that distinct elements x_1 and x_2 in X always have distinct values $y_1 = f(x_1)$ and $y_2 = f(x_2)$ in Y, so that $f(x_1) \neq f(x_2)$ whenever $x_1, x_2 \in X$ and $x_1 \neq x_2$. In graphical terms this means a function of a single real variable is **one-to-one** on its domain of definition if its graph either increases or decreases steadily. It then follows that for all $x \in X$, one element x corresponds to one element y, and conversely, so that not only is f a function as defined above, but so also is f^{-1}. Clearly, when f^{-1} exists, the domain of f is the range of f^{-1} and the range of f is the domain of f^{-1}.

In general, functions of a single real variable are not one-to-one. They may, however, usually be made so by being regarded as a set of functions with contiguous domains of definition chosen such that each function is one-to-one in its own domain of definition. For example, let the function f be the one illustrated in the graph in Fig. 1.1(a) with domain AD and range PQ. To see this is not one-to-one notice that although the three elements $x_1^{(1)}$, $x_1^{(2)}$ and $x_1^{(3)}$ in the domain each correspond to the unique value y_1 in the range, when the roles of independent and dependent variables are reversed, the single element y_1 corresponds to the three different elements $x_1^{(1)}$, $x_1^{(2)}$ and $x_1^{(3)}$.

If the domain AD is divided up into the three contiguous domains AB, BC and CD shown in Fig. 1.1(b), then taken together the three functions f_1, f_2 and f_3 defined on these

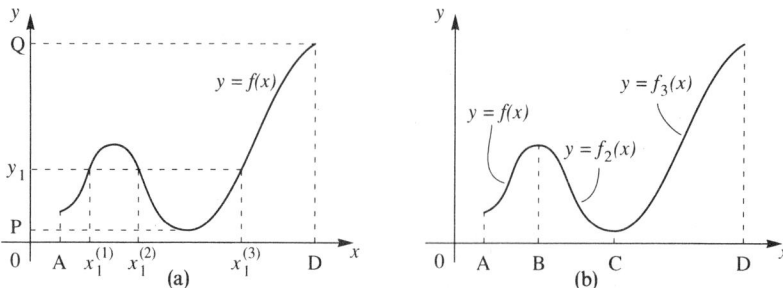

Fig. 1.1   Division of domain of definition to make one-to-one functions

respective domains represent the function f in Fig. 1.1(a), but now f_1, f_2 and f_3 are each one-to-one.

This emphasizes the fact that the specification of the domain forms an integral part of the definition of a one-to-one function. Even when a function is not one-to-one, the specification of the domain still forms part of its definition, and if this is not stated explicitly it is understood to be the largest domain for which the function has meaning. This is sometimes called the **natural domain of definition** of the function.

By way of example, the sine function is defined for all $x \in \mathbb{R}$, so that $\mathbb{R}$ is its natural domain of definition. However an inspection of its graph shows that it is only one-to-one in any interval of length π which has its end points located at any two consecutive points of the sequence $(2n-1)\pi/2$, with $n = 0, 1, 2, \ldots$. By convention, the sine function with its domain of definition lying within the interval with the end points $-\pi/2$ and $\pi/2$, in which it is one-to-one, is called the **principal branch** of the function. The reader will recall from elementary calculus that the principal branch of the tangent function is taken to have this same domain of definition, while the principal branch of the cosine function is taken to have the interval with the end points 0 and π as its domain of definition.

The concept of identity between functions arises frequently and needs to be understood. Two functions f and g, each with the same domain of definition D, are said to be **identical** if

$$f(x) = g(x) \quad \text{for every } x \in D.$$

When f and g are identical we shall write

$$f \equiv g,$$

where the symbol $\equiv$ is to be read 'is identically equal to'.

Identities arise in many different ways and are important. A typical example is the well-known trigonometric identity

$$\cos^2 x \equiv 1 - \sin^2 x,$$

which is true for all $x \in \mathbb{R}$. This identity justifies replacing $1 - \sin^2 x$ by $\cos^2 x$ whenever it occurs, irrespective of the value of x which is involved. Extensive use will be made of identities when partial fractions are developed in Sec. 1.4, and also elsewhere. In practice, the identity symbol is only used when the properties of an identity require emphasis; elsewhere an equality is used in its place.

1.2 Intervals and inequalities

One of the most important properties possessed by the real numbers is their ability to be **ordered**. Their ordering is accomplished by means of the **order relation** 'less than' in conjunction with the set $\mathbb{R}^+$ of positive real numbers. To see how this works, let a, b be real numbers, and suppose $b - a \in \mathbb{R}^+$, then we shall say a is **less** than b and write

$$a < b.$$

If, however, $-(b-a) \in \mathbb{R}^+$, then we shall say a is **greater** than b and write

$a > b$.

The symbol $<$ is to be read 'less than' and the symbol $>$ is to be read 'greater than'.

In particular, the statement $a \in \mathbb{R}^+$ which reads 'a is positive' is equivalent to $a > 0$, and the statement $-a \in \mathbb{R}^+$ which reads 'a is negative' is equivalent to $a < 0$.

The notation is modified in an obvious manner if equality is permitted, and a statement like 'a is less than or equal to b' is written

$a \leq b$.

It is the existence of this order relation between the real numbers comprising $\mathbb{R}$ that allows them to be represented as points on a line. This is usually accomplished by first choosing a point on the line as origin O, a scale by which the magnitude of numbers is to be represented by proportion, and a sense of direction along the line in which the numbers increase. The point representing a number $c \in \mathbb{R}$ is then taken as the point distant from O by an amount proportional to c measured in the appropriate direction along the line. The line itself is called the **real number line**, or simply the **real line**, and there is a one-to-one correspondence between points on the line and numbers. That is, a number corresponds to a unique point on the line, and conversely. For this reason numbers will often be referred to as **points** without risk of ambiguity.

Expressions involving the symbols $<$, $\leq$, $>$ and $\geq$ are called **inequalities**. The simplest use of these is to define **intervals** on a line. Let x be a variable represented by points on a line, and in a diagrammatic representation of an interval as a line, denote included and excluded end points of the interval by a dot and a circle, respectively. The following types of interval arise sufficiently frequently for them to be named:

the **open** interval

$a < x < b$, written (a, b);

the **closed** interval

$a \leq x \leq b$, written $[a, b]$;

the **semi-open** or **semi-closed** intervals

$a \leq x < b$, written $[a, b)$,

$a < x \leq b$, written $(a, b]$;

and the **semi-infinite** intervals

$-\infty < x < a$, written $(-\infty, a)$,

$-\infty < x \leq a$, written $(-\infty, a]$,

$a < x < \infty$, written (a, ∞).

$a \leq x < \infty$, written $[a, \infty)$.

The convention has been used in which an end point that is *excluded* from an interval is enclosed in a round bracket, and one that is *included* is enclosed in a square bracket. Infinity is a limit process and cannot be attained by x, so it is an agreed convention that infinity is enclosed in a round bracket to signify that the interval is open at infinity.

A variant of this notation which is also in use involves denoting a closed interval $a \leq x \leq b$ by $[a, b]$ as above, and an open interval $a < x < b$ by $]a, b[$. The notation extends in an obvious manner and, for example, the interval $a \leq x < b$ becomes $[a, b[$, while the interval $-\infty < x \leq a$ becomes $]-\infty, a]$.

The proof of the following elementary inequalities is left as an exercise for the reader. In their statements use is made of the **implication** symbol $\Rightarrow$ which is to be read 'implies', and of the **two way implication** symbol $\Leftrightarrow$ which is to be read 'implies and is implied by'.

Elementary inequalities

I.1. $a \geq b$ and $c > d$, $\Rightarrow a+c > b+d$,

I.2. $a > b \geq 0$ and $c \geq d > 0 \Rightarrow ac > bd$,

I.3. $k > 0$ and $a > b \Rightarrow ka > kb$, $k < 0$ and $a > b \Rightarrow ka < kb$,

I.4. $a > b \Leftrightarrow -a < -b$,

I.5. $a < 0$, $b > 0 \Rightarrow ab < 0$; $a < 0$, $b < 0 \Rightarrow ab > 0$,

I.6. $a > 0 \Leftrightarrow 1/a > 0$; $a < 0 \Leftrightarrow 1/a < 0$,

I.7. $a > b > 0 \Leftrightarrow 1/b > 1/a > 0$; $a < b < 0 \Leftrightarrow 1/b < 1/a < 0$.

I.8. $0 \leq a < b \Leftrightarrow a^n < b^n$ for n any natural number ($n \in \mathbb{N}$).

The inequality forming the statement of the following theorem is of considerable importance in mathematics. Along with various generalizations, this inequality will arise in subsequent chapters in a variety of different ways.

Theorem 1.1 (Cauchy[2]–Schwarz[3] inequality)

If $a_1, a_2, \ldots, a_n$ and $b_1, b_2, \ldots, b_n$ are any real numbers, then

$$\left(\sum_{k=1}^{n} a_k b_k \right)^2 \leq \left(\sum_{k=1}^{n} a_k^2 \right) \left(\sum_{k=1}^{n} b_k^2 \right). \tag{1}$$

Proof

A sum of squares must be nonnegative, so

$$\sum_{k=1}^{n} (a_k \lambda + b_k)^2 \geq 0,$$

---

[2] AUGUSTIN–LOUIS CAUCHY (1789–1857), a French mathematician of outstanding ability who was Professor of Mathematics in both Turin and Paris. His wide ranging contributions extended over real and complex analysis, differential equations, elasticity and astronomy.

[3] HERMANN AMANDUS SCHWARZ (1845–1921), a German mathematician of distinction who was Professor of Mathematics at Göttingen and Berlin. He made important contributions to geometry, analysis and differential equations.

for all real λ. Expanding the left-hand side of (1) and grouping terms brings the inequality to the form

$$A\lambda^2 + 2B\lambda + C \geq 0,$$

where

$$A = \sum_{k=1}^{n} a_k^2, \quad B = \sum_{k=1}^{n} a_k b_k \quad \text{and} \quad C = \sum_{k=1}^{n} b_k^2.$$

If $A > 0$, setting $\lambda = -B/A$ we obtain $B^2 - AC \leq 0$, which is result (1). If $A = 0$, then $a_1 = a_2 = \ldots = a_n = 0$ and the result is trivial. $\qquad\square$

Definition (Absolute value)

The **absolute value** of a real number a, written $|a|$, is defined as

$$|a| = \begin{cases} a & \text{if } a \geq 0 \\ -a & \text{if } a < 0 \end{cases}. \tag{2}$$

As examples of the absolute value we have $|23| = 23$ and $|-1.9| = 1.9$. The proof of the following elementary inequalities is left as an exercise for the reader. If a and b are real numbers, then:

$$|a| = |-a|, \tag{3}$$

$$|a|^2 = a^2, \tag{4}$$

$$ab \leq |ab| = |a|\,|b|, \tag{5}$$

$$|1/a| = 1/|a| \quad \text{for } a \neq 0, \tag{6}$$

$$|a/b| = |a|/|b| \quad \text{for } b \neq 0. \tag{7}$$

When coupled with inequalities, the absolute value may be used to define intervals on the real line. For example, the inequality $|x| \leq |a|$ comprises the two statements:

$$x \leq |a| \qquad \text{if } x \geq 0,$$

and

$$-x \leq |a| \qquad \text{if } x < 0.$$

Using I.4, this last inequality is seen to be equivalent to $-|a| \leq x$, and so

$$|x| \leq |a| \Leftrightarrow -|a| \leq x \leq |a|. \tag{8}$$

A slightly more complicated example is provided by the inequality $|x - 3| \leq 1$. This comprises the two statements:

$$x - 3 \leq 1 \qquad \text{if } x - 3 \geq 0,$$

and

$$-(x-3)\leq 1 \qquad \text{if } x-3<0.$$

The first inequality shows $x\leq 4$ and the second, after using I. 4, that $2\leq x$. Combining these results yields the inequality $2\leq x\leq 4$. Thus we have proved that

$$|x-3|\leq 1 \Leftrightarrow 2\leq x\leq 4.$$

This result is intuitively obvious, because in words $|x-3|\leq 1$ means 'the set of all points x distant from 3 by an amount less than or equal to 1'. Thus the left-hand end point is $3-1=2$ and the right-hand end point is $3+1=4$, with the end points and all points in between being included in the inequality.

A number of useful inequalities are connected with the absolute value, the most important of which is the following result.

Theorem 1.2 (Triangle inequality)

If a, b are any two real numbers, then

$$|a+b|\leq |a|+|b|. \tag{9}$$

Proof

The proof of this result is simple and proceeds as follows:

$$|a+b|^2=(a+b)^2$$
$$=a^2+2ab+b^2$$
$$=|a|^2+2ab+|b|^2$$
$$\leq |a|^2+2|ab|+|b|^2$$
$$=|a|^2+2|a|\,|b|+|b|^2$$
$$=(|a|+|b|)^2,$$

and because of I.8 this is equivalent to

$$|a+b|\leq |a|+|b|.$$

The reason for the name of this theorem will become clear later when an analogous form of the theorem is considered in Sec. 1.6 in connection with complex numbers.

□

Let $a_1, a_2, \ldots, a_n$ be any real numbers, then successive application of (9) yields the **generalized triangle inequality**

$$|a_1+a_2+\ldots, a_n|\leq |a_1|+|a_2|+\ldots + |a_n|. \tag{10}$$

If in (9) a is replaced by $a+b$ and b by $-b$ we obtain

$$|a|-|b|\leq|a+b|,$$

while replacing a by $-a$ and b by $a+b$ in (9) gives

$$-|a+b|\leq|a|-|b|.$$

We have shown that

$$-|a+b|\leq|a|-|b|\leq|a+b|,$$

and so because of (8) we have

$$|a+b|\geq||a|-|b||. \tag{11}$$

Inequality (9) provides an estimate for the sum $|a+b|$ from above, and inequality (11) an estimate from below. Results of this type will be needed in later chapters.

The manipulation of inequalities and absolute values requires care and a little experience. This following examples, which are typical, illustrate the types of argument that need to be employed.

Example 1

If the real numbers a, b, k are such that $a<b<0$ and $k<0$, prove that

$$\frac{b}{a}<\frac{b+k}{a+k}<1<\frac{a+k}{b+k}<\frac{a}{b}.$$

Solution

To prove that

$$\frac{b}{a}<\frac{b+k}{a+k}$$

we consider the difference

$$\sigma=\frac{b}{a}-\frac{b+k}{a+k}=\frac{k(b-a)}{a(a+k)}.$$

Now as $a<b<0$ it follows that $b-a>0$, but $k<0$, $a<0$ and $a+k<0$, so $\sigma<0$ and thus

$$\frac{b}{a}-\frac{b+k}{a+k}<0, \quad \text{so} \quad \frac{b}{a}<\frac{b+k}{a+k}.$$

Next, to prove that

$$\frac{b+k}{a+k}<1,$$

we consider the difference

$$\Delta = \frac{b+k}{a+k} - 1 = \frac{b-a}{a+k}.$$

However $b-a>0$ and $a+k<0$, so $\Delta<0$ and thus

$$\frac{b+k}{a+k} - 1 < 0, \quad \text{so} \quad \frac{b+k}{a+k} < 1.$$

Thus we have proved that

$$\frac{b}{a} < \frac{b+k}{a+k} < 1.$$

The proof of the remaining inequalities follows in similar fashion. ∎

Example 2

For what values of x is the following equation true

$$|x^2+4x+1| = x+2.$$

Solution

If $x^2+4x+1>0$, then $|x^2+4x+1| = x^2+4x+1$, and the equation becomes

$$x^2+4x+1 = x+2,$$

or

$$x^2+3x-1 = 0.$$

This is true for $x = \frac{1}{2}(-3\pm\sqrt{13})$.

If $x^2+4x+1<0$ then $|x^2+4x+1| = -x^2-4x-1$, and the equation becomes

$$-x^2-4x-1 = x+2,$$

or

$$x^2+5x+3 = 0.$$

This is true for $x = \frac{1}{2}(-5\pm\sqrt{13})$.

Thus the equation is true for the four values of x

$$\tfrac{1}{2}(-3-\sqrt{13}), \quad \tfrac{1}{2}(-3+\sqrt{13}), \quad \tfrac{1}{2}(-5-\sqrt{13}) \quad \text{and} \quad \tfrac{1}{2}(-5+\sqrt{13}). \quad ∎$$

Example 3

Determine the intervals in which the following inequality is valid

$$\frac{x-2}{x+3} > \frac{x-4}{x+2}.$$

Solution

To clear the rational functions, but leave the inequality sign unchanged, we multiply by $(x+2)^2 (x+3)^2$, which is essentially nonnegative (cf. inequality property I.3). This gives the result

$$(x-2)(x+3)(x+2)^2 > (x-4)(x+3)^2(x+2),$$

or

$$(x+2)(x+3)[(x-2)(x+2)-(x-4)(x+3)]>0,$$

which after simplification reduces to

$$(x+2)(x+3)(x+8)>0.$$

Setting

$$P(x)=(x+2)(x+3)(x+8),$$

it follows that

$$P(x)<0 \quad \text{if } x<-8$$

$$P(x)>0 \quad \text{if } -8<x<-3$$

$$P(x)<0 \quad \text{if } -3<x<-2$$

$$P(x)>0 \quad \text{if} -2<x.$$

Thus the inequality in the problem is satisfied if

$$-8<x<-3 \quad \text{or} \quad -2<x. \qquad \blacksquare$$

Problems for Section 1.2

1 Prove inequalities (3) and (4).
2 Prove inequalities (5) and (7).

Find the intervals determined by the following inequalities. Draw them on a diagram using a solid line to represent points of the interval, and indicating an end point belonging to an interval by a dot and an end point excluded from an interval by a circle.

3 $|x-4|<2.$

4 $|x-3|\geq 4.$

5 $|x+6|\leq 6.$

6 $|x+1|>3.$

7 $1<|2x+1|\leq 3.$

8 $0<|x-3|<2.$

9 $\dfrac{x-1}{x+2}>\dfrac{x-3}{x+1}.$

10 $2\leq \left|\dfrac{4x-1}{3}\right|<3.$

11 Prove the inequality

$$(|x|-|y|)^2 \leq (x+y)^2.$$

12 Prove that if x_1 and x_2 are nonnegative numbers, then

$$\frac{x_1+x_2}{2}\geq (x_1 x_2)^{1/2}.$$

This result is called the **arithmetic-mean–geometric-mean inequality.**

13 Use the binomial theorem to prove the **Bernoulli inequality**

$$(1+x)^n \geq 1+nx,$$

for n a positive integer and $x \geq 0$. For a statement of the binomial theorem and an outline of its proof by mathematical induction see Problem 24, Sec. 1.3.

14 Use the binomial theorem to prove that

$$2 < \left(1+\frac{1}{n}\right)^n < 1 + \sum_{r=1}^{n} \frac{1}{r!},$$

with n a positive integer.

15 If $a_1, a_2, \ldots, a_n$ are positive numbers, and $b_k = 1/a_k$ for $k = 1, 2, \ldots, n$, use the Cauchy–Schwarz inequality to prove that

$$\left(\sum_{k=1}^{n} a_k^2\right)\left(\sum_{k=1}^{n} b_k^2\right) \geq n^2.$$

16 If $a_1, a_2, \ldots, a_n$ are positive numbers, use the Cauchy–Schwarz inequality to prove that

$$\left(\sum_{k=1}^{n} a_k\right)^2 \leq n \sum_{k=1}^{n} a_k^2.$$

Harder problems

17 Let $c_0, c_1, \ldots, c_n$ be any real numbers, and set

$$S_n(x) = c_0 + c_1 x + c_2 x^2 + \ldots + c_n x^n,$$

with x a real variable. By setting $a_1 = c_0, a_2 = c_1, \ldots, a_{n+1} = c_n$ and $b_1 = 1, b_2 = x, \ldots, b_{n+1} = x^n$ in the Cauchy–Schwarz inequality with $n+1$ terms, prove that

$$[S_n(x)]^2 \leq \left(\sum_{k=0}^{n} c_k^2\right)\left(\frac{1-x^{2(n+1)}}{1-x^2}\right),$$

for $|x| \neq 1$. Hence deduce that the infinite series $\sum_{k=1}^{\infty} c_k x^k$ is convergent (has a finite sum) for $|x| < 1$ provided $\sum_{k=1}^{\infty} c_k^2$ is convergent.

18 By considering the identity

$$x^n - nx + n - 1 \equiv (x-1)(x^{n-1} + x^{n-2} + \ldots + x - n + 1),$$

prove the inequality

$$x^n + n - 1 \geq nx,$$

for $n \geq 1$ any integer and $x \geq 0$.

[*Hint*: Consider the cases $0 < x < 1$ and $x > 1$ separately, and show the right-hand side of the identity is positive].

1.3 Mathematical induction

A method of proof that will often be used later is **mathematical induction.** An inductive proof depends for its success on the property of the set $\mathbb{N}$ of natural numbers that each number has a unique successor in order of magnitude, so that 1 is followed by a 2, 2 by 3, . . . and so on. This means that if $\mathbb{M}$, a subset of $\mathbb{N}$, contains the number 1, and $n+1 \in \mathbb{M}$ when $n \in \mathbb{M}$, then every number in $\mathbb{M}$ must also be in $\mathbb{N}$ so $\mathbb{M} = \mathbb{N}$. When rephrased, this leads to the following sequence of steps forming the basis of any inductive argument.

The inductive argument

Let $P(n)$ be a *proposition* concerning the positive integers; that is a statement of some property they are believed to possess. Then if,
(i) the proposition is true for $n = 1$, so that $P(1)$ is true,
and
(ii) whenever $P(n)$ is true, so also is $P(n+1)$,
it follows that $P(n)$ is true for all n; otherwise the proposition $P(n)$ is false.

    The way in which proposition $P(n)$ is obtained in individual cases depends on the circumstances. Usually $P(n)$ is found by inspecting the pattern of results obtained by the first few stages of a repetitive process, and then using intuition to suggest the form of the general result at the nth stage. An inductive argument then either confirms or rejects the conjectured form of the general result. In this sense induction leads to what is called a **non-constructive** proof, because although testing the validity of a proposition, it will not help to formulate it in the first place.

    It should be remarked here that on occasions it is necessary to start an inductive argument with some positive integer greater than 1, say with n_0. The same form of inductive argument still applies, but if proposition $P(n)$ is shown to be true, it then follows that this is proved only for $n \geq n_0$.

    The way an inductive proof proceeds is best illustrated by the following examples. It will be seen from these that, apart from the formulation of the original proposition, the difficult stage in an inductive argument is usually showing that $P(n)$ implies $P(n+1)$.

Example 1. Showing the existence of a factor

Prove that for any positive integer n the number $5^{2n+1} + 3^{n+2} \cdot 2^{n-1}$ contains the factor 19.

Solution

This is a problem in which a proposition has been made that must be verified. We set

$$P(n) = 5^{2n+1} + 3^{n+2} \cdot 2^{n-1},$$

and take as our proposition the statement that '$P(n)$ contains the factor 19'.

Step 1

Setting $n=1$ shows

$$P(1)=5^3+3^3 \cdot 2^0 = 152 = 8 \cdot 19.$$

Thus proposition $P(1)$ is true, since it contains the factor 19.

Step 2

We now assume that $P(n)$ contains a factor 19, and try to prove that this implies $P(n+1)$ must also contain the same factor. Replacing n by $n+1$ in $P(n)$ gives

$$
\begin{aligned}
P(n+1) &= 5^{2n+3}+3^{n+3} \cdot 2^n \\
&= 25 \cdot 5^{2n+1} + 6 \cdot 3^{n+2} \cdot 2^{n-1} \\
&= 25 \cdot 5^{2n+1} + 25 \cdot 3^{n+2} \cdot 2^{n-1} - 19 \cdot 3^{n+2} \cdot 2^{n-1} \\
&= 25\,P(n) - 19 \cdot 3^{n+2} \cdot 2^{n-1}.
\end{aligned}
$$

Thus if $P(n)$ contains the factor 19, so also does $P(n+1)$. We have proved that $P(n) \Rightarrow P(n+1)$.

Since $P(1)$ is true it then follows that $P(n)$ is true for all integers $n \geq 1$. Our proof of the existence of the factor 19 in all numbers $P(n)$ for $n \geq 1$ is complete. ∎

Example 2. Checking a proposition

Prove or disprove the proposition that the number $5^n - 3^n$ contains the factor 7 for all integers $n \geq 3$.

Solution

We take as our proposition the statement that '$P(n) = 5^n - 3^n$ contains a factor 7 for $n \geq 3$'.

Step 1

This is a problem in which we must start with $P(3)$ and not $P(1)$. Setting $n=3$ shows $P(3) = 5^3 - 3^3 = 2 \cdot 7^2$. Thus as $P(3)$ contains the factor 7 it follows that $P(3)$ is true.

Step 2

To establish the truth of the proposition for $n \geq 3$ we need to show that $P(n) \Rightarrow P(n+1)$. If this is not the case the proposition is false. However, to show a proposition is false, it is sufficient (and simpler) to show it to be untrue for some choice of $n > 3$. In this case setting $n=4$ gives $P(4) = 5^4 - 3^4 = 2^5 \cdot 17$, which is *not* divisible by 7, so the proposition is false. ∎

Example 3. A problem of differentiation

Prove that

$$\frac{d^n}{dx^n}(\sin x) = \sin\left(x + \frac{n\pi}{2}\right),$$

for n a positive integer.

Solution

Here again a proposition has been made that must be proved. We set

$$P(n) = \sin\left(x + \frac{n\pi}{2}\right),$$

and take as our proposition the statement that

$$\text{'}P(n) = \frac{d^n}{dx^n}(\sin x)\text{'}.$$

Step 1

Setting $n = 1$ gives

$$P(1) = \sin\left(x + \frac{\pi}{2}\right) = \cos x,$$

but $\dfrac{d}{dx}(\sin x) = \cos x,$

so that $P(1)$ is true.

Step 2

Now let us try to prove that $P(n) \Rightarrow P(n+1)$. If $P(n)$ is true then

$$\frac{d^n}{dx^n}(\sin x) = \sin\left(x + \frac{n\pi}{2}\right),$$

so differentiating with respect to x we find

$$\frac{d^{n+1}}{dx^{n+1}}(\sin x) = \cos\left(x + \frac{n\pi}{2}\right). \tag{A}$$

Now replacing n by $n+1$ in $P(n)$ gives

$$P(n+1) = \sin\left[x + \left(\frac{n+1}{2}\right)\pi\right]$$

$$= \sin\left[\left(x + \frac{n\pi}{2}\right) + \frac{\pi}{2}\right]$$

$$= \cos\left(x + \frac{n\pi}{2}\right). \tag{B}$$

As results (A) and (B) are identical we have succeeded in proving that $P(n) \Rightarrow P(n+1)$. The inductive argument has been completed and it follows that

$$\frac{d^n}{dx^n}(\sin x) = \sin\left(x + \frac{n\pi}{2}\right) \qquad \text{for all } n \geq 1.$$

∎

Example 4. A problem concerning coefficients in a series

In the infinite series

$$f(x) = a_0 + a_1 x + a_2 x^2 + \ldots + a_n x^n + \ldots ,$$

it is known that successive coefficients a_r and a_{r+1} are related by the **recursion (recurrence) relation**, or **algorithm**,

$$a_{r+1} = \frac{-a_r}{(r+1)(r+2)},$$

where $a_0 \neq 0$ is given. Find the explicit form of the general coefficient a_n in terms of a_0 and n.

Solution

In this problem a specific proposition $P(n)$ has not been given, so it will be necessary to formulate one before it can be tested. We start by looking at the first few coefficients. Successively setting $r = 0, 1, 2$ and 3 in the recursion relation, and using each result in the next stage of the calculation, we find:

$$a_1 = \frac{-a_0}{1 \cdot 2}, \qquad a_2 = \frac{-a_1}{2 \cdot 3} = \frac{a_0}{1 \cdot 2 \cdot 2 \cdot 3},$$

$$a_3 = \frac{-a_2}{3 \cdot 4} = \frac{-a_0}{1 \cdot 2 \cdot 2 \cdot 3 \cdot 3 \cdot 4}, \qquad a_4 = \frac{-a_3}{4 \cdot 5} = \frac{a_0}{1 \cdot 2 \cdot 2 \cdot 3 \cdot 3 \cdot 4 \cdot 4 \cdot 5}.$$

Inspection of the pattern of results *suggests* that

$$a_n = \frac{(-1)^n a_0}{n!(n+1)!} \qquad \text{for } n \geq 1,$$

but to *prove* this we must use an inductive argument. Accordingly, we take as our proposition $P(n)$ the statement that 'the coefficient a_n in the series has the form

$$P(n) = \frac{(-1)^n a_0}{n!(n+1)!}.$$

Step 1

Setting $n = 1$ gives

$$P(1) = \frac{-a_0}{1.2},$$ which agrees with the result from the recursion relation when $r = 0$, so $P(1)$ is true.

Step 2

Suppose $P(n)$ to be true, then it follows from the recursion relation that

$$P(n+1) = \frac{-P(n)}{(n+1)(n+2)} = \frac{(-1)^{n+1} a_0}{(n+1)(n+2)n!(n+1)!}$$

$$= \frac{(-1)^{n+1} a_0}{(n+1)!(n+2)!}. \tag{C}$$

However, simply replacing n by $n+1$ in $P(n)$ gives

$$P(n+1)=\frac{(-1)^{n+1}a_0}{(n+1)!\,(n+2)!},\tag{D}$$

which is the same result as (C), so we have proved that $P(n)\Rightarrow P(n+1)$. Our inductive argument is complete and we have proved that

$$a_n=\frac{(-1)^n a_0}{n!(n+1)!}\qquad\text{for }n\geq 1.$$

In practice, when problems of this type arise with series solutions of differential equations, once inspection of the pattern of coefficients has suggested the form of a_n, the inductive proof of its correctness is usually omitted. ■

Problems for Section 1.3

Give an inductive proof of the following results concerning series.

1 $1+2+3+\ \ldots\ +n=\dfrac{n(n+1)}{2}$.

2 $a+(a+d)+(a+2d)+\ \ldots\ +[a+(n-1)d]=\dfrac{n[2a+(n-1)d]}{2}$.   (**arithmetic series**)

3 $a+ar+ar^2+\ \ldots\ +ar^{n-1}=\dfrac{a(r^n-1)}{r-1}$, $(r\neq 1)$.   (**geometric series**)

4 $1+3+5+\ \ldots\ +(2n-1)=n^2$.

5 $1+2^2+3^2+\ \ldots\ +n^2=\dfrac{n(n+1)(2n+1)}{6}$.   (**sum of squares**)

6 $1^3+2^3+3^3+\ \ldots\ +n^3=\dfrac{n^2(n+1)^2}{4}$.   (**sum of cubes**)

7 $\dfrac{1}{1\cdot 2}+\dfrac{1}{2\cdot 3}+\ \ldots\ +\dfrac{1}{n(n+1)}=\dfrac{n}{n+1}$.

8 $2+6+18+\ \ldots\ +2\cdot 3^{n-1}=3^n-1$.

Use induction to prove or disprove the following propositions.
9 5^n-2^n is divisible by 3 for all integers $n\geq 1$.
10 7^n-4^n is divisible by 3 for all integers $n\geq 1$.
11 $2^n>n!$ for all integers $n\geq 1$.
12 6^n-2^n is divisible by 4 for all integers $n\geq 1$.

13 $1+\dfrac{1}{2!}+\dfrac{1}{3!}+\ \ldots\ +\dfrac{1}{n!}<2\left[1-\left(\dfrac{1}{2}\right)^n\right]$ for any integer $n\geq 3$.

14 $3^n>2^{n+1}$ for all integers $n\geq 2$.

Give an inductive proof of the following general results.

15 $|a_1 + a_2 + \ldots + a_n| \leq |a_1| + |a_2| + \ldots + |a_n|$. (**generalized triangle inequality**).

16 $\dfrac{d^n}{dx^n}(\sin ax) = a^n \sin\left[a\left(x + \dfrac{n\pi}{2}\right)\right]$, for any real number a and any positive integer n.

17 $\dfrac{d^n}{dx^n}(\cos ax) = a^n \cos\left[a\left(x + \dfrac{n\pi}{2}\right)\right]$, for any real number a and any positive integer n.

18 $x^n - y^n$ is divisible by $x - y$ for any real numbers x, y and all positive integers n.

19 The **Bernoulli inequality**
$$(1+x)^n \geq 1 + nx,$$
for $x \geq -1$ and n a positive integer. This is a stronger form of the result first given in Prob. 13, Sec. 1.2.

In the following problems the successive coefficients a_r and a_{r+1} in a series satisfy the indicated recursion relations, with $a_0 \neq 0$ given and $r \geq 1$ an integer. Deduce the form of the general coefficient a_n, and verify the result by induction.

20 $a_r = \dfrac{-a_{r-1}}{2r(2r+1)}$.

21 $a_r = \dfrac{a_{r-1}}{(2r-1)(2r+1)}$.

22 $a_r = \dfrac{-a_{r-1}}{2^r(r+1)}$.

23 $a_r = \dfrac{-a_{r-1}}{3^r \cdot 5^{r+1}}$.

Harder problems

24 When n is a positive integer and a, b are real numbers, the **binomial theorem** states that the expansion of $(a+b)^n$ is the $n+1$ term expression

$$(a+b)^n \equiv a^n + \binom{n}{1}a^{n-1}b + \binom{n}{2}a^{n-2}b^2 + \ldots + \binom{n}{r}a^{n-r}b^r + \ldots + \binom{n}{n-1}ab^{n-1} + b^n,$$

in which the **binomial coefficient** of the term $a^{n-r}b^r$ is

$$\binom{n}{r} = \frac{n(n-1)(n-2)\ldots(n-r+1)}{r!} = \frac{n!}{r!(n-r)!}, \qquad 0! \equiv 1.$$

Multiply the above statement of the theorem by $(a+b)$, combine terms to find the coefficient of $a^{n-r+1}b^r$ on the right-hand side, and use the result to give an inductive proof of the binomial theorem. The binomial theorem is true when n is replaced by any real number α, though the expansion will contain an infinite number of terms unless α is a positive integer, and the proof will be different. In general, if α and x are real numbers, the **binomial theorem** takes the form

$$(1+x)^{\alpha} = 1 + \alpha x + \frac{\alpha(\alpha-1)}{2!}x^2 + \frac{\alpha(\alpha-1)(\alpha-2)}{3!}x^3 + \ldots$$

$$+ \frac{\alpha(\alpha-1)(\alpha-2)\ldots(\alpha-r+1)}{r!}x^r + \ldots$$

for $|x| < 1$.

25 Make use of the binomial theorem to find (i) the term in the expansion of

$$\left(x^4 - \frac{1}{3x}\right)^{10}$$

which is independent of x and (ii) the coefficient of x^{-5}.

26 Make use of the binomial theorem to find (i) the term in the expansion of

$$\left(x^3 - \frac{1}{2x}\right)^{8}$$

which is independent of x and (ii) the coefficient of x^4.

1.4   Polynomials and partial fractions

A real valued **polynomial** $P_n(x)$ in the real variable x is an algebraic expression of the form

$$P_n(x) \equiv a_0 + a_1 x + \ldots + a_r x^r + \ldots + a_{n-1}x^{n-1} + a_n x^n, \tag{1}$$

in which n is a positive integer and $a_0, a_1, \ldots, a_n$ are real numbers, with $a_n \neq 0$. The numbers $a_0, a_1, \ldots, a_n$ are called the **coefficients** of the polynomial and its **degree** is the number n, its largest exponent. Thus $P_n(x)$ is a polynomial of degree n in x. The term $a_n x^n$ in $P_n(x)$ is the term of greatest degree, and it is called the **leading term** in the polynomial. A polynomial $P_n(x)$ in which the coefficient of the leading terms is unity is called a **monic polynomial**.

The polynomial $P_0(x) = a_0$ is identically constant, while $P_1(x)$ is said to be **linear** in x because the degree of the leading term is unity. All polynomials of degree greater than unity are said to be **nonlinear**. Some low order polynomials are named, and $P_2(x)$, $P_3(x)$, $P_4(x)$ and $P_5(x)$ are called **quadratics**, **cubics**, **quartics** and **quintics**, respectively.

Associated with the polynomial $P_n(x)$ is the polynomial equation

$$P_n(x) = 0. \tag{2}$$

A number ζ is called a **zero** of the polynomial $P_n(x)$, or a **root** of the polynomial equation $P_n(x) = 0$, if $P_n(\zeta) = 0$.

Polynomials arise in the solution of problems in all branches of mathematics, and in a specific application giving rise to a polynomial the nature of its zeros usually has far reaching consequences for the solution. This makes the study of polynomials, and of their zeros, one of considerable practical importance.

As we progress through this text we shall use polynomials in many different ways and, where appropriate, study a number of their properties. At this stage we introduce only the following result, which is certainly the most important property of polynomials.

Theorem 1.3 (Fundamental theorem of algebra)

Every polynomial which is not identically constant has at least one zero. □

A proof of this most important theorem requires the use of analytic functions and so it will not be given here. It is this theorem which justifies the argument that if $P_n(x)$ is any non-constant polynomial, then it always has a zero x_1, and thus a factor $(x-x_1)$, which enables it to be written as

$$P_n(x) \equiv (x-x_1)P_{n-1}(x), \tag{3}$$

where $P_{n-1}(x)$ is the polynomial of degree $n-1$ obtained by dividing $P_n(x)$ by $(x-x_1)$. Proceeding in this manner we conclude that $P_n(x)$ may always be expressed as the product.[4]

$$P_n(x) \equiv a_n(x-x_1)(x-x_2) \ldots (x-x_n), \tag{4}$$

where $x_1, x_2, \ldots, x_n$ are the zeros of $P_n(x)$.

When r of the zeros of $P_n(x)$ are equal, so that $x_{i+1}=x_{i+2}=\ldots=x_{i+r}=\xi$, say, the **zero (root)** ξ is said to be of **order (multiplicity)** r. If the zero ξ is of order r, it follows that $P_n(x)$ contains the factor $(x-\xi)^r$, and we then say the factor $(x-\xi)$ has **multiplicity** r. A zero of order 1 is called a **simple zero**.

For example, if $P_3(x)=x^3-6x^2+12x-8$, then because $P_3(x)=(x-2)^3$ we see this polynomial has a zero 2 of order 3 and a factor $(x-2)$ with multiplicity 3.

The determination of the zeros of a polynomial is a difficult task, and the only explicit formula available is for the quadratic equation

$$ax^2+bx+c=0, \tag{5}$$

whose roots are given by the well-known formula

$$x=\frac{-b\pm\sqrt{b^2-4ac}}{2a}. \tag{6}$$

On occasions this result permits the solution of a quartic as, for example, with the equation

$$x^4-5x^2+4=0.$$

Setting $x^2=u$, and solving the resulting quadratic in u by means of (6), we find $u=1$ or $u=4$. Thus $x=\pm 1$ and $x=\pm 2$ are simple zeros of the quartic, and hence its factors are $(x-1)$, $(x+1)$, $(x-2)$ and $(x+2)$.

---

[4] The factor a_n is included in the right-hand side of (4) so that when expanded the leading term has the coefficient a_n as in (1). If $a_n=1$, the polynomial in (4) is **monic**.

The quotient $P_m(x)/Q_n(x)$ of the two polynomials $P_m(x)$ and $Q_n(x)$ is called a **rational function**, and we shall assume $P_m(x)$ and $Q_n(x)$ have no factors in common. Such a function is defined everywhere except at the n zeros of the divisor $Q_n(x)$.

It follows from elementary division that when $m \geq n$ we may write

$$\frac{P_m(x)}{Q_n(x)} \equiv P^*_{m-n}(x) + \frac{R_s(x)}{Q_n(x)}, \tag{7}$$

where $P^*_{m-n}(x)$ is a polynomial of degree $m-n$ and $R_s(x)$ is a polynomial of degree $s < n$ which is often called the **remainder polynomial**.

Example 1. Rational function

$$\frac{x^4 + 2x^2 + x - 1}{x^2 + x + 1} \equiv x^2 - x + 2 - \frac{3}{x^2 + x + 1}.$$

■

At this point it is appropriate to remind the reader of the distinction between an **equality**, denoted by $=$, and an **identity**, denoted by $\equiv$ which was made clear in Sec. 1.2. Let us illustrate the meaning by using two polynomials $P(x)$ and $Q(x)$.

If $P(x)$ and $Q(x)$ are any two polynomials, the expression $P(x) = Q(x)$ means the *finite* set of values of x which are the *roots* of $P(x) - Q(x) = 0$. However, the expression $P(x) \equiv Q(x)$ means that $P(x)$ and $Q(x)$ are equal for *all* values of x, and so must be identical, both as regards their coefficients and their degree. Thus if $x^2 - 3x + 7 \equiv ax^2 + bx + c$, then $a = 1$, $b = -3$ and $c = 7$. As another illustration, $Q_n(x)$ will be a factor of $P_m(x)$ in (7) if $R_s(x) \equiv 0$.

The name **partial fractions** is given to the process whereby a rational function

$$P_M(x)/Q_N(x) \tag{8}$$

is represented as a sum of simpler rational functions and, when $M \geq N$, also a polynomial of degree $M - N$.

In addition to their use with elementary integration techniques, partial fractions play an important role when inverting Laplace transforms, and in a variety of other ways. For these reasons we now present a summary of this method of reduction, together with two illustrative examples.

Basic rules for partial fraction reduction

(i) If $M \geq N$, divide $P_M(x)$ by $Q_N(x)$ to obtain the representation

$$\frac{P_M(x)}{Q_N(x)} \equiv P^*_{M-N}(x) + \frac{R_S(x)}{Q_N(x)},$$

in which $P^*_{M-N}(x)$ and $R_S(x)$ are polynomials with known coefficients, of degrees $M - N$ and $S(S < N)$, respectively (cf. Ex. 1).

(ii) Factor $Q_N(x)$ into real factors like $(x-a_i)^{m_i}$, if $x=a_i$ is a real root with multiplicity m_i, and into quadratic factors like $(x^2+p_jx+q_j)^{s_j}$ if this factor is repeated s_j times and is such that $p_j^2<4q_j$ (each of the s_j quadratic factors then has a pair of complex conjugate linear factors).

(iii) For each factor $(x-a_i)^{m_i}$ include in the partial fraction representation of $R_S(x)/Q_N(x)$ the terms

$$\frac{A_{i1}}{(x-a_i)}+\frac{A_{i2}}{(x-a_i)^2}+\ldots+\frac{A_{im}}{(x-a_i)^{m_i}},$$

where $A_{i1}, A_{i2}, \ldots, A_{im}$ are unknown constants.

(iv) For each factor like $(x^2+p_jx+q_j)^{s_j}$, include in the partial fraction representation of $R_S(x)/Q_N(x)$ the terms

$$\frac{B_{j1}x+C_{j1}}{(x^2+p_jx+q_j)}+\frac{B_{j2}x+C_{j2}}{(x^2+p_jx+q_j)^2}+\ldots+\frac{B_{js_j}x+C_{js_j}}{(x^2+p_jx+q_j)^{s_j}}$$

where $B_{j1}, \ldots, B_{js_j}, C_{j1}, \ldots, C_{js_j}$ are unknown constants.

(v) To find the so-called **undetermined coefficients** A_i, B_j and C_j in this representation first equate $R_S(x)/Q_N(x)$ to the sum of terms of the form set out in (iii) and (iv) above. Then cross multiply this expression to obtain an identity between a polynomial with known coefficients and one involving the undetermined coefficients. Find the undetermined coefficients by equating the coefficients of corresponding powers of x on each side of this identity, and then solving the resulting simultaneous equations satisfied by the undetermined coefficients. Notice that some of these simultaneous equations may be found very simply if linear factors are present, because if $(x-a_i)^{m_i}$ is a factor, by setting $x=a_i$ every term containing such a factor will vanish.

(vi) The final partial fraction expansion is thus

$$\frac{P_M(x)}{Q_N(x)} \equiv P^*_{M-N}(x)+\text{terms from (iii) and (iv) with the coefficients found in (v).}$$

On occasions quadratic terms like

$$x^2+px+q$$

need to be represented in the form

$$x^2+px+q \equiv (x+\alpha)^2+\beta.$$

This representation is called **completing the square**. Expanding the right-hand side of this identity and equating corresponding powers of x gives

$$\alpha=\tfrac{1}{2}p, \quad \beta=q-\tfrac{1}{4}p^2,$$

so that

$$x^2+px+q \equiv (x+\tfrac{1}{2}p)^2+q-\tfrac{1}{4}p^2.$$

Example 2

Simplify by partial fractions

$$\frac{2x^4+x^3-6x^2-7x-2}{x^3-3x-2}.$$

Solution

We have

$$P_M(x)=2x^4+x^3-6x^2-7x-2 \quad \text{and} \quad Q_N(x)=x^3-3x-2,$$

from which it follows that, $M=4$, $N=3$ and so $M>N$, with $M-N=1$. Thus we must start by dividing $P_M(x)/Q_N(x)$ as in step (i). Routine division followed by factoring $Q_N(x)$ gives

$$\frac{2x^4+x^3-6x^2-7x-2}{x^3-3x-2} \equiv 1+2x+\frac{2x^2+5x}{(x+1)^2(x-2)},$$

and thus in the notation of Rule (i)

$$P_{M-N}^*(x)=P_1^*(x)=1+2x, \quad R_S(x)=R_2(x)=2x^2+5x \quad \text{and} \quad Q_N(x)=Q_3(x)=(x+1)^2(x-2).$$

The factor $(x+1)^2$ has multiplicity 2, and the factor $(x-2)$ multiplicity 1, so following step (iii) we set

$$\frac{R_2(x)}{Q_3(x)} \equiv \frac{2x^2+5x}{(x+1)^2(x-2)} \equiv \frac{A_1}{x+1}+\frac{A_2}{(x+1)^2}+\frac{B_1}{x-2}.$$

Cross multiplication now gives

$$2x^2+5x \equiv A_1(x+1)(x-2)+A_2(x-2)+B_1(x+1)^2, \tag{9}$$
$$\equiv A_1(x^2-x-2)+A_2(x-2)+B_1(x^2+2x+1).$$

Equating corresponding powers of x leads to the three simultaneous equations

(x^2) $2=A_1+B_1$

(x) $5=-A_1+A_2+2B_1,$

(x^0) $0=-2A_1-2A_2+B_1.$

These equations have the solution $A_1=0$, $A_2=1$, and $B_1=2$, so it follows that

$$\frac{2x^2+5x}{(x+1)^2(x-2)} \equiv \frac{1}{(x+1)^2}+\frac{2}{x-2}.$$

Thus adding the polynomial $P_1^*(x)=1+2x$ we arrive at the required partial fraction representation

$$\frac{2x^4+x^3-6x^2-7x-2}{x^3-3x-2} \equiv 1+2x+\frac{1}{(x+1)^2}+\frac{2}{x-2}.$$

Notice that A_2 and B_1 could also have been found directly from (9) by first setting $x=-1$ to cause the factor $(x+1)$ to vanish giving $-3=-3A_2$, so $A_2=1$, and then setting $x=2$ in (9) to cause the factor $(x-2)$ to vanish giving $18=9B_1$, so $B_1=2$. The coefficient A_1 follows from B_1 and A_2 by using them in any equation containing A_1 derived from (9) by equating coefficients of corresponding powers of x. ∎

Example 3

Simplify by partial fractions

$$\frac{3x^4 + 10x^3 + 23x^2 + 23x + 18}{x(x^2 + 2x + 3)^2},$$

and complete the square to simplify the quadratic factors arising in the representation.

Solution

We have

$$P_M(x) \equiv 3x^4 + 10x^3 + 23x^2 + 23x + 18 \quad \text{and} \quad Q_N(x) \equiv x(x^2 + 2x + 3)^2,$$

so $M = 4$ and $N = 5$. As $M < N$ no division is necessary and we proceed to steps (iii) and (iv), for which it follows we must set

$$\frac{3x^4 + 10x^3 + 23x^2 + 23x + 18}{x(x^2 + 2x + 3)^2} \equiv \frac{A}{x} + \frac{Bx + C}{(x^2 + 2x + 3)} + \frac{Dx + E}{(x^2 + 2x + 3)^2}.$$

Cross multiplication gives

$$3x^4 + 10x^3 + 23x^2 + 23x + 18 \equiv A(x^2 + 2x + 3)^2 + (Bx + C)x(x^2 + 2x + 3) + x(Dx + E).$$

Equating coefficients of corresponding powers of x gives the five simultaneous equations

$(x^4) \quad 3 = A + B,$

$(x^3) \quad 10 = 4A + 2B + C,$

$(x^2) \quad 23 = 10A + 3B + 2C + D,$

$(x) \quad 23 = 12A + 3C + E,$

$(x^0) \quad 18 = 9A,$

which have the solution

$$A = 2,\ B - 1,\ C = 0,\ D = 0 \text{ and } E = -1.$$

Thus the partial fraction representation is

$$\frac{3x^4 + 10x^3 + 23x^2 + 23x + 8}{x(x^2 + 2x + 3)^2} \equiv \frac{2}{x} + \frac{x}{(x^2 + 2x + 3)} - \frac{1}{(x^2 + 2x + 3)^2}.$$

It follows by completing the square that

$$x^2 + 2x + 3 \equiv (x + 1)^2 + 2,$$

so the representation after completing the square becomes

$$\frac{3x^4 + 10x^3 + 23x^2 + 23x + 18}{x(x^2 + 2x + 3)^2} \equiv \frac{2}{x} + \frac{x}{(x + 1)^2 + 2} - \frac{1}{[(x + 1)^2 + 2]^2}.$$

∎

Problems for Section 1.4

Use partial fractions to simplify the following expressions and when a quadratic factor occurs in the denominator simplify it by completing the square.

1 $\dfrac{x^3 + x^2 - x + 3}{x^2 + x - 2}$

2 $\dfrac{3x^3 + 4x^2 - 2x + 1}{x^2 + x - 2}$

3 $\dfrac{2x^2 + 9x + 8}{x^3 + 5x^2 + 8x + 4}$

4 $\dfrac{x^5 + 2x^3 + 2x^2 + 3x + 4}{(x^2 - 2x + 2)(x^2 - x + 1)}$

5 $\dfrac{x - 2}{(x^2 - 2x + 2)(x^2 - x + 1)}$

6 $\dfrac{3x^2 - 4x + 7}{x^3 - 3x + 2}$

7 $\dfrac{2x^3 - 2x^2 - 2x - 3}{x^2(x^2 + x + 1)}$

8 $\dfrac{6x^3 - x^2 + 8x + 24}{x(x + 3)(x^2 - 2x + 4)}$

9 $\dfrac{1}{x^3 + 2x^2 + 5x}$

10 $\dfrac{1}{x^4 + 3x^3 + 2x^2}$

1.5   Remarks on the statement of theorems

This is a suitable point at which to draw the attention of the reader to the way in which theorems are formulated. An important part of every theorem is a statement (sometimes implied) of the conditions under which it is true. Thus in Pythagoras' theorem the condition is a restriction to *right angled* triangles, while in Theorem 1.3 the restriction is to *non-constant* polynomials. However, in subsequent chapters, different types of restriction will arise in which the terms *necessary* and *sufficient* are needed to qualify conditions in theorems. It is important that the meanings of these terms should be clearly understood, so we now define them and illustrate their use.

A condition in a theorem is **necessary** if without it the statement of the theorem is false. A condition in a theorem is **sufficient** if it ensures the truth of the theorem, though a weaker condition might also ensure its truth. If conditions in a theorem are both **necessary and sufficient**, this means no weaker set of conditions may be found which will ensure the truth of the theorem.

The distinction between necessary and sufficient conditions is best illustrated by example. Consider the requirement that a polynomial contains the factor $(x - 1)$. Then it is sufficient to ensure this by requiring the polynomial to contain the factor $(x - 1)^2$, though this is not a necessary condition.

A different example is provided by considering a differentiable function $f(x)$ defined for $-\infty < x < \infty$. Then for $f(x)$ to have a local extremum (max or min) at $x = a$ it is necessary that $f'(a) = 0$. However, this is not a sufficient condition because it is also true that $f'(a) = 0$ when $x = a$ is a point of inflection of the graph of $y = f(x)$.

As a final and slightly more complicated example consider the infinite series $\sum\limits_{n=1}^{\infty} a_n$.

Then for this series to converge it is necessary that $\lim\limits_{n \to \infty} a_n = 0$, but this is *not* a sufficient condition, because the **harmonic series** $\sum\limits_{n=1}^{\infty} \dfrac{1}{n}$ is divergent, yet its nth term $a_n = 1/n$ is such that $\lim\limits_{n \to \infty} a_n = 0$.

It is usually difficult to find necessary and sufficient conditions for all theorems, so frequently the stated conditions are merely sufficient for the use that will be made of the theorem.

1.6 Complex numbers in Cartesian form

The inadequacy of the real number system for carrying out all arithmetic calculations was noticed early on in the development of mathematics. This discovery was made as a result of attempts to find a general method of solution for the quadratic equation, which is the simplest nonlinear polynomial equation. For example, when using (6) in Sec. 1.4 to find the roots of either of the equations

$$x^2 + 3x + 10 = 0 \quad \text{or} \quad x^2 + 2 = 0,$$

the square root of a negative number arises. This is a new form of number, and it is not one which belongs to the real number system.

Similarly, if the zeros of the cubic $P_3(x) = x^3 - 1$ are required, then by rewriting it as $P_3(x) = (x-1)(x^2 + x + 1)$ we see at once that $x = 1$ is a zero, but the zeros of the remaining quadratic factor $(x^2 + x + 1)$ again involve the square root of a negative number. Thus seeking to perform ordinary arithmetic operations on equations with real numbers as coefficients can lead to numbers outside the real number system.

This limitation of the real number system was overcome by the introduction of **complex numbers**[5], and these in turn led to the development of the theory of analytic functions, also called the theory of functions of a complex variable, which is now so important in engineering and science. The extension of the real number system to overcome these difficulties is accomplished by introducing a new quantity i, called the **unit imaginary element**. The quantity i is defined as the solution to the equation $x^2 + 1 = 0$, so that

$$i^2 = -1. \tag{1}$$

---

[5] An early example involving the square root of a negative integer was given in 1545 by the Italian mathematician JEROME CARDAN (GEROLAMO CARDANO) (1501–1576) in his major work on algebra *Ars Magna*. In this he gave the roots of $x^2 - 10x + 40 = 0$ as $5 \pm \sqrt{-15}$, but dismissed them as being useless. He encountered complex numbers again when solving cubic equations by the method due to NICOLO OF BRESCIA (TARTAGLIA) (1499–1557), but again rejected them as meaningless. The term complex number itself was introduced by CARL FRIEDRICH GAUSS (cf. the footnote 1 in Sec. 3.4).

A more general form of number called a **complex number** is defined as the combination

$$x+iy,$$

where x and y are real numbers, and y is regarded as the multiplier of the unit imaginary element i. If $y=0$ the complex number $x+iy$ reduces to the real number x. It is understood that iy and yi have the same meaning[6].

The real numbers x and y appearing in the complex number $x+iy$ are called the **real part** and the **imaginary part** of $x+iy$, respectively. This relationship between x, y and $x+iy$ is shown symbolically by writing

$$x=\operatorname{Re}(x+iy) \quad \text{and} \quad y=\operatorname{Im}(x+iy), \tag{2}$$

with Re being read 'real part of' and Im being read 'imaginary part of'. It must be emphasized that the real and imaginary parts of a complex number are both *real* numbers.

When a complex number is considered as a single entity it is usual to denote it by z, so that

$$z=x+iy, \quad \operatorname{Re}z=x \quad \text{and} \quad \operatorname{Im}z=y. \tag{3}$$

Two complex numbers $z_1=x_1+iy_1$ and $z_2=x_2+iy_2$ are said to be **equal** if, and only if, there is equality between their respective real and imaginary parts, so that $x_1=x_2$ and $y_1=y_2$.

Example 1. Equality

If $z_1=z_2$, with $z_1=-7+ib$ and $z_2=a+11i$, it follows that $a=-7$ and $b=11$. We then have $\operatorname{Re}z_1=\operatorname{Re}z_2=-7$, and $\operatorname{Im}z_1=\operatorname{Im}z_2=11$.   ■

If $\operatorname{Im}z=0$ we shall say z is a **purely real** number, whereas if $\operatorname{Re}z=0$ we shall say it is a **purely imaginary** number. It is an accepted abbreviation to write $z=0$ to denote the zero complex number $z=0+0i$.

In the chapters which follow it will often be necessary to refer to the set of all possible complex numbers. This set will be denoted by the symbol $\mathbb{C}$, so a statement like 'z is an arbitrary complex number' may be contracted to $z\in\mathbb{C}$.

The arithmetic operations of addition, subtraction, multiplication and division of the complex numbers $z_1=x_1+iy_1$ and $z_2=x_2+iy_2$ are defined as follows:

Addition of complex numbers

The **sum**

$$z_1+z_2=(x_1+iy_1)+(x_2+iy_2)$$
$$=(x_1+x_2)+i(y_1+y_2), \tag{4}$$

---

[6] When y has a specific numerical value, say 11, it is customary to write 11i, but when y is a variable it is usual to write iy, though there is no strict observance of these conventions.

and so

$$\text{Re}(z_1 + z_2) = \text{Re}\, z_1 + \text{Re}\, z_2, \quad \text{and} \quad \text{Im}(z_1 + z_2) = \text{Im}\, z_1 + \text{Im}\, z_2.$$

Subtraction of complex numbers

The subtraction operation is defined as the inverse of the operation of addition, so the **difference**

$$z_1 - z_2 = (x_1 + iy_1) - (x_2 + iy_2)$$
$$= (x_1 - x_2) + i(y_1 - y_2), \tag{5}$$

showing that

$$\text{Re}(z_1 - z_2) = \text{Re}\, z_1 - \text{Re}\, z_2, \quad \text{and} \quad \text{Im}(z_1 - z_2) = \text{Im}\, z_1 - \text{Im}\, z_2.$$

Example 2. Addition and subtraction

If $z_1 = 2 + 5i$ and $z_2 = 7 - 2i$, then $z_1 + z_2 = 9 + 3i$ and $z_1 - z_2 = -5 + 7i$. ∎

Multiplication of complex numbers

The **product**

$$z_1 z_2 = (x_1 + iy_1)(x_2 + iy_2)$$
$$= (x_1 x_2 - y_1 y_2) + i(x_1 y_2 + x_2 y_1), \tag{6}$$

showing that

$$\text{Re}(z_1 z_2) = x_1 x_2 - y_1 y_2, \quad \text{and} \quad \text{Im}(z_1 z_2) = x_1 y_2 + x_2 y_1.$$

In particular, if a is purely real and $z = x + iy$ is an arbitrary complex number, it follows from (6) that

$$az = ax + iay,$$

or

$$\text{Re}(az) = ax \quad \text{and} \quad \text{Im}(az) = ay.$$

Thus multiplication of a complex number by a real number a involves multiplying both the real and the imaginary parts of the complex number by a.

Replacing a by $1/b$, where $b \neq 0$, shows that division of a complex number by a nonzero real number b involves dividing both the real and imaginary parts of the complex number by b.

The form of the product $z_1 z_2$ given in (6) follows directly by expanding the expression $(x_1 + iy_1)(x_2 + iy_2)$ as though it were an ordinary product, using the result $i^2 = -1$, and then collecting together the real and imaginary parts. In practice, multiplication is always performed in this manner rather than by using definition (6).

Example 3. Multiplication

If $z_1 = 1 - i$ and $z_2 = 2 + 3i$, then $3z_1 = 3 - 3i, \frac{1}{4}z_2 = \frac{1}{2} + \frac{3}{4}i$ and

$z_1 z_2 = (1 - i)(2 + 3i) = 2 + 3i - 2i - 3(i^2)$

$\qquad = 2 + i - 3(-1) = 5 + i.$ ■

When performing repeated multiplications it is necessary to use the following results which follow directly from the definition $i^2 = -1$;

$i^3 = i(i^2) = -i, \quad i^4 = (i^2)(i^2) = (-1)(-1) = 1,$

$i^5 = i(i^4) = i \quad$ and, in general,

$i^{2n} = (-1)^n, \quad i^{2n+1} = (-1)^n i \quad$ for $n = 0, 1, 2, \ldots.$ $\qquad$ (7)

Division of complex numbers

Division is defined as the inverse of the operation of multiplication. When $z_2 \neq 0$, the **quotient**

$$\frac{z_1}{z_2} = \frac{x_1 + iy_1}{x_2 + iy_2} = \left(\frac{x_1 x_2 + y_1 y_2}{x_2^2 + y_2^2} \right) + i \left(\frac{x_2 y_1 - x_1 y_2}{x_2^2 + y_2^2} \right), \qquad (8)$$

showing that

$$\mathrm{Re}(z_1/z_2) = \frac{x_1 x_2 + y_1 y_2}{x_2^2 + y_2^2}, \quad \text{and} \quad \mathrm{Im}(z_1/z_2) = \frac{x_2 y_1 - x_1 y_2}{x_2^2 + y_2^2}.$$

This formal definition of the quotient z_1/z_2 is difficult to remember, so let us arrive at this result in a different way. We now show how division may be accomplished very simply by using the *complex conjugate* of a complex number.

Complex conjugate of a complex number

Let $z = x + iy$ be an arbitrary complex number. Then the complex number $x - iy$, denoted by $\bar{z}$, is called the **complex conjugate** of z. The operation of forming the complex conjugate of an expression is called **complex conjugation**.

The rule for forming the complex conjugate of a complex quantity simply involves reversing the sign of the imaginary part.

Example 4. Complex conjugate

If $z = 3 + 4i$, then $\bar{z} = 3 - 4i$;   if $z = 6i$, then $\bar{z} = -6i$,   while if $z = 7$, then $\bar{z} = 7$.   ■

It follows directly from the definitions of z and $\bar{z}$ that

$$\bar{\bar{z}} = z,$$ (9)

and

$$z\bar{z} = x^2 + y^2 \, (\geq 0).$$ (10)

We also have

$$z + \bar{z} = 2x \quad , \quad z - \bar{z} = 2iy$$

and so

$$\operatorname{Re} z = x = \frac{1}{2}(z + \bar{z}) \quad , \quad \operatorname{Im} z = y = \frac{1}{2i}(z - \bar{z}).$$ (11)

Notice that $z = \bar{z}$ if, and only if, z is purely real, and that $z = -\bar{z}$ if, and only if, z is purely imaginary.

Let us now determine the quotient

$$\frac{z_1}{z_2} = \frac{x_1 + iy_1}{x_2 + iy_2}$$

when $z_2 \neq 0$. The difficulty in interpreting z_1/z_2 arises because, as yet, we are uncertain how to divide by a complex number. However, if both numerator and denominator of this quotient are multiplied by $\bar{z}_2$, the complex conjugate of z_2, it follows from (10) that the denominator becomes purely real and we obtain

$$\frac{z_1}{z_2} = \frac{(x_1 + iy_1)(x_2 - iy_2)}{x_2^2 + y_2^2}.$$

Expanding the numerator, collecting its real and imaginary parts and dividing each by the real number $x_2^2 + y_2^2$, we arrive at result (8). In practice, division is always performed in this manner rather than by using definition (8). This method is illustrated in the following example.

Example 5. Division

If $z_1 = 4 - 2i$ and $z_2 = 3 + 4i$, then

$$\frac{z_1}{z_2} = \frac{4 - 2i}{3 + 4i} = \frac{(4 - 2i)(3 - 4i)}{(3 + 4i)(3 - 4i)} = \frac{(4 - 2i)(3 - 4i)}{25}$$

$$= \frac{12 - 16i - 6i + 8i^2}{25} = \frac{4}{25} - \frac{22}{25}i.$$

■

Taking the complex conjugate of the arithmetic operations of addition, subtraction, multiplication and division we obtain the following useful results:

$$\overline{(z_1+z_2)}=\bar{z}_1+\bar{z}_2 \ , \quad \overline{(z_1-z_2)}=\bar{z}_1-\bar{z}_2,$$

$$\overline{z_1 z_2}=\bar{z}_1\bar{z}_2 \ , \quad \overline{\left(\frac{z_1}{z_2}\right)}=\frac{\bar{z}_1}{\bar{z}_2}. \tag{12}$$

Repeated use of the third result in (12) with $z_1=z_2=z$ shows

$$\overline{(z^n)}=(\bar{z})^n, \quad \text{for } n \text{ a positive integer.} \tag{13}$$

Representation of numbers in the complex plane – Cartesian form

As with real numbers, a geometrical interpretation of a complex number is often extremely useful. The approach used is to represent $z=x+iy$ as a point P in a plane, using rectangular Cartesian axes Ox and Oy with origin O, and (x, y) as the Cartesian coordinates of P. The length scales are always chosen to be the same on both the x- and y-axes which are called the **real axis** and the **imaginary axis**, respectively. The plane in which this representation is used is called the **complex plane** (Fig. 1.2). On account of the notation $z=x+iy$, the complex plane is also called the **z-plane**[7]. For obvious reasons the complex number $x+iy$ is said to be expressed in **Cartesian form**, or sometimes **real and imaginary form**.

Purely real numbers lie on the real axis and purely imaginary numbers lie on the imaginary axis. Since the correspondence between complex numbers and their represen-

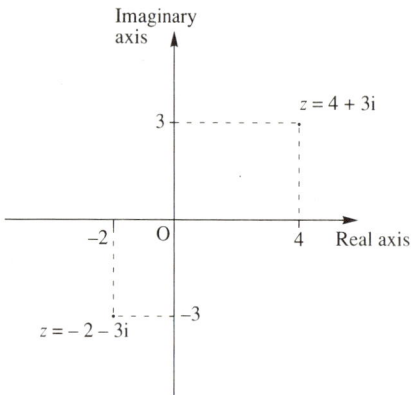

Fig. 1.2 Representation of numbers in the complex plane

---

[7] The original name for the complex plane, or z-plane, was the **Argand diagram**. This name derives from JEAN ROBERT ARGAND (1768–1822), the self-taught Swiss mathematician in whose work (1806) this representation appeared. This same idea was anticipated by some nine years in a little known paper by the Norwegian surveyor CASPAR WESSEL (1745–1818).

tation as points in the complex plane is unique, and conversely, it is customary to refer either to points z in the complex plane, or to complex numbers z, depending on whichever is the more convenient.

The operations of addition, subtraction and complex conjugation have simple and useful geometrical interpretations in the complex plane. To make these clear we consider first the case of addition, and draw lines from the origin to the points z_1 and z_2 as shown in Fig. 1.3(a).

By completing the parallelogram in that diagram, and appealing to the definition of addition which involves adding the real and imaginary parts separately, the sum $z_1 + z_2$ is seen to be the complex number represented by the end of the diagonal of the

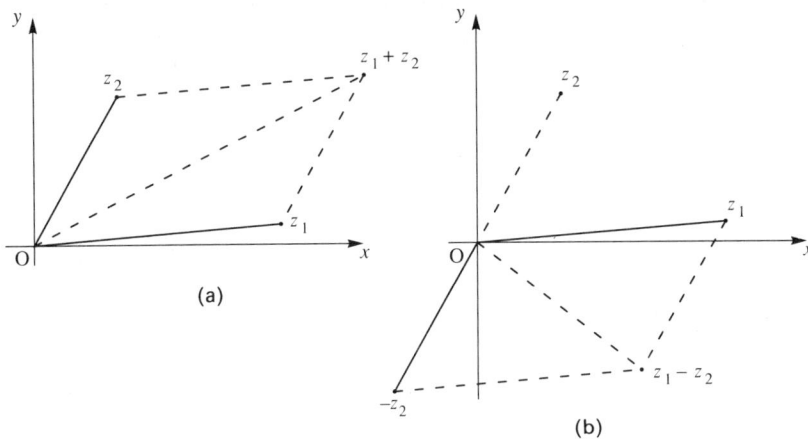

(a)

(b)

Fig. 1.3 Addition and subtraction of complex numbers. (a) Addition in the complex plane (b) Subtraction in the complex plane

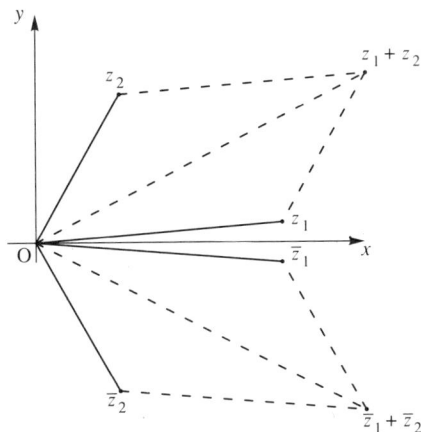

Fig. 1.4 Complex conjugation in the complex plane applied to the sum $z_1 + z_2$

parallelogram drawn from the origin. This is called the **parallelogram law** for addition of complex numbers, and it is the same law by which forces are added in mechanics.

Subtraction of complex numbers follows in similar fashion by adding the complex numbers z_1 and $-z_2$ as shown in Fig. 1.3(b), where $-z_2 = (-1)z_2$.

If z is an arbitrary point in the complex plane, the operation of complex conjugation represents a **reflection** of that point in the real axis. That is, in the operation of complex conjugation, the real axis acts as though it were a mirror (Fig. 1.4), producing an image below the real axis of the representation above, and conversely.

Commutative, associative and distributive laws

The commutative, associative and distributive laws for real numbers, and the laws for arithmetic operations on complex numbers, may be combined to establish the following properties for arithmetic operations on any arbitrary complex numbers z_1, z_2 and z_3.

Commutative law for addition
$$z_1 + z_2 = z_2 + z_1.$$

Commutative law for multiplication
$$z_1 z_2 = z_2 z_1.$$

Associative law for addition
$$(z_1 + z_2) + z_3 = z_1 + (z_2 + z_3).$$

Associative law for multiplication
$$z_1(z_2 z_3) = (z_1 z_2)z_3.$$

Distributive law of multiplication with respect to addition
$$z_1(z_2 + z_3) = z_1 z_2 + z_1 z_3.$$

The unit element 1 and zero element 0 in the complex number system are unique and such that

$$z + 0 = 0 + z, \quad z + (-z) = (-z) + z = 0, \quad z \cdot 1 = z.$$

These laws assert that the algebraic rules for the combination of complex numbers are the same as those for real numbers, to which complex numbers reduce when their imaginary parts are zero.

The modulus and the triangle inequality

The **modulus** (also **absolute value** or **magnitude**) of an arbitrary complex number $z = x + iy$, denoted by $|z|$, is defined as the real nonnegative number

$$|z| = \sqrt{x^2 + y^2}. \tag{14}$$

In geometrical terms this is the distance from the origin to the point representing z in the complex plane.

Example 6. Modulus

If $z = 3 + 7i$, $|z| = \sqrt{58}$ and if $z = -3 + 4i$, $|z| = 5$, while if $z = 7$, $|z| = 7$ and if $z = -2i$, $|z| = 2$.

■

A number of useful results follow directly from (14), the simplest of which are

$$|z| = |-z| = |\bar{z}| = |-\bar{z}| = \sqrt{z\bar{z}} = (x^2 + y^2)^{1/2}, \tag{15}$$

$$|z^2| = |z|^2 \quad \text{and} \quad |z_1 z_2| = |z_1||z_2|,$$

together with the inequalities

$$|x| = |\operatorname{Re} z| \le |z| \quad \text{and} \quad |y| = |\operatorname{Im} z| \le |z|. \tag{16}$$

If z_1 and z_2 are arbitrary complex numbers, setting $z = z_1 - z_2$ and using the first equality in (15) shows that

$$|z_1 - z_2| = |z_2 - z_1|. \tag{17}$$

Geometrically, this asserts that if z_1 and z_2 are represented by the points P_1 and P_2 in the complex plane (Fig. 1.5), the distance from P_1 to P_2 is the same as the distance from P_2 to P_1 (distance is symmetric).

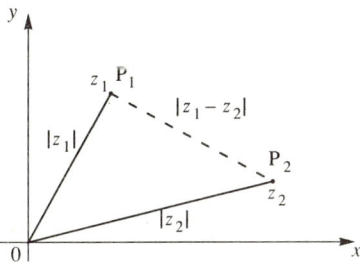

Fig. 1.5 Distance between points P_1 and P_2

The order relation $<$ (or $>$) which enables real numbers to be arranged in a natural order relative to each other has no meaning with respect to complex numbers. This is because real numbers can be made to correspond to points on a line and, relative to a given point (number) on the line, any other point (number) lies either to its left (is less than) or to its right (is greater than), whereas complex numbers correspond to points in a plane, so no such ordering exists. For example, -3 is less than 7, but we can make no corresponding statement about $1 + i$ and $1 - i$. We can, however, order the moduli of complex numbers since these are real nonnegative numbers.

The most important inequality involving the modulus is the **triangle inequality**

$$|z_1 + z_2| \leq |z_1| + |z_2|, \tag{18}$$

which has already been encountered in Sec. 1.2 in terms of real numbers.

As z_1 and z_2 are complex numbers, the proof of (18) must be different. We begin by establishing another inequality involving the modulus which will be needed in the proof. Our starting point for this is the observation that combining the first result of (11) with the first result of (16), and then setting $z = z_1 \bar{z}_2$, gives

$$|z_1 \bar{z}_2 + \bar{z}_1 z_2| = 2\mathrm{Re}(z_1 \bar{z}_2) \leq 2|z_1 \bar{z}_2|.$$

Then, as (15) implies $|z_1 \bar{z}_2| = |z_1||z_2|$, we have proved that

$$|z_1 \bar{z}_2 + \bar{z}_1 z_2| \leq 2|z_1||z_2|. \tag{19}$$

Our proof of (18) starts from the result

$$
\begin{aligned}
|z_1 + z_2|^2 &= (z_1 + z_2)\overline{(z_1 + z_2)} \\
&= z_1 \bar{z}_1 + z_1 \bar{z}_2 + \bar{z}_1 z_2 + z_2 \bar{z}_2 \\
&= |z_1|^2 + z_1 \bar{z}_2 + \bar{z}_1 z_2 + |z_2|^2.
\end{aligned}
$$

Then, since $z_1 \bar{z}_2 + \bar{z}_1 z_2$ is a real number (possibly negative), replacing it by its absolute value will not decrease the right-hand side of this last result, and employing (19) will further strengthen the inequality to give

$$
\begin{aligned}
|z_1 + z_2|^2 &\leq |z_1|^2 + 2|z_1||z_2| + |z_2|^2 \\
&= (|z_1| + |z_2|)^2.
\end{aligned}
$$

Taking the positive square root of both sides of this expression then establishes the triangle inequality (18).

This result is so called because of its representation in the complex plane (Fig. 1.6). It asserts in algebraic form the result due to Euclid that the length of any side of a triangle is less than or equal to the sum of the lengths of the other two sides (equality is possible only if the vertices are collinear). The application of inequalities such as the triangle inequality

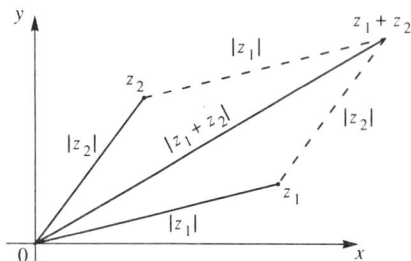

Fig. 1.6   Triangle inequality in the complex plane

to algebraic identities gives rise to other useful inequalities. For example, starting from the fact that $z_1 \equiv (z_1 - z_2) + z_2$ and applying (18), we find

$$|z_1| \le |z_1 - z_2| + |z_2| \quad \text{or} \quad |z_1| - |z_2| \le |z_1 - z_2|.$$

A similar argument applied to $z_2 \equiv (z_2 - z_1) + z_1$ shows

$$-|z_2 - z_1| \le |z_1| - |z_2|.$$

However $|z_2 - z_1| = |z_1 - z_2|$, so taken together these two inequalities show that

$$-|z_1 - z_2| \le |z_1| - |z_2| \le |z_1 - z_2|,$$

which by (8) of Sec. 1.2 may be written

$$||z_1| - |z_2|| \le |z_1 - z_2|. \tag{20}$$

Disks in the complex plane

If z is an arbitrary complex number, $\rho > 0$ a given real number and a is a given complex number, the inequality

$$|z - a| < \rho \tag{21}$$

defines the set of all points in the complex plane distant from the point a by an amount which is less than ρ. Thus, in geometrical terms, (21) represents all points z belonging to the interior of a **circular disk** of radius ρ with its center at $z = a$ in the complex plane. Such a disk from which the boundary is excluded is called an **open disk**, and is shown as the shaded region in Fig. 1.7 in which the excluded boundary is indicated by a dotted line. This geometrical interpretation of inequality (21) is often used when working with analytic functions, as is the inequality

$$|z - a| \le \rho,$$

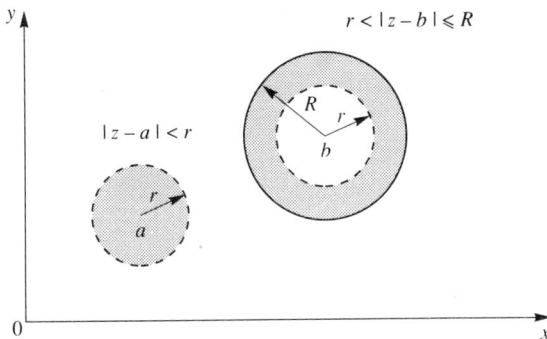

Fig. 1.7   Open disk of radius ρ centered on $z = a$ and an annulus centered on $z = b$

which represents all points z belonging to that same disk together with all the points z which lie on the boundary. Such a disk which contains all the boundary points is called a **closed disk**. The circular boundary itself is defined by the expression

$$|z - a| = \rho.$$

When $\rho = 1$ such a circle is called a **unit circle**.

The disk

$$0 < |z - a| < \rho,$$

from which only the central point $z = a$ is omitted, is often useful in complex analysis and it is called a **punctured disk.**

Annular regions are also of importance, and the points forming their boundaries may, or may not, be included in the region. A typical annulus centered on $z = b$, with the points on the inner boundary of radius r excluded and the points on the outer boundary of radius R included is shown in Fig. 1.7. The annulus itself is shown as a shaded region, with a dotted line indicating an excluded boundary and a full line an included one. The points forming this annulus may be identified symbolically by writing $r < |z - b| \leq R$.

Complex polynomials and roots of polynomials with real coefficients

The importance to engineering and applied mathematics of polynomials in the real variable x extends also to polynomials in the complex variable z. In view of this, and in preparation for later use, we now define a complex polynomial and use the elementary properties of complex numbers to deduce a number of very useful properties. The definition of a complex polynomial is a direct generalization of the definition given in Sec. 1.4.

If n is a positive integer, and $a_0, a_1, \ldots, a_n$ are arbitrary complex numbers $(a_n \neq 0)$, the expression

$$P_n(z) \equiv a_0 + a_1 z + a_2 z^2 + \ldots + a_n z^n \tag{22}$$

is called a **complex polynomial** (or simply a **polynomial**) in z of **degree** n. The numbers a_0, $a_1, \ldots, a_n$ are called the **coefficients** of the polynomial. For any given z the polynomial $P_n(z)$, defines a complex number. Any number ζ such that $P_n(\zeta) = 0$ is called a **zero** of $P_n(z)$, and $z = \zeta$ is called a **root** of the equation $P_n(z) = 0$. The number ζ may be either real or complex.

We are now in a position to determine the roots of all quadratic equations for which the coefficients are real. The most general case in which the coefficients are complex must be postponed until we have examined how to find the roots of an arbitrary complex number.

We start from the general quadratic equation with real coefficients

$$az^2 + bz + c = 0, \quad (a \neq 0), \tag{23}$$

which we write in the standard form

$$z^2 + \frac{b}{a}z + \frac{c}{a} = 0. \tag{24}$$

Our objective is to express (24) in the form

$$(z+\alpha)^2 = \beta, \tag{25}$$

by **completing the square**, for then z follows once the square root of β has been found. We achieve our purpose, which involves finding α and β, by expanding (25) and equating the coefficient of each power of z in (25) to the corresponding coefficient in (24), since the result must be an identity in z. This leads to the results

$$\alpha = b/(2a) \quad \text{and} \quad \beta = (b^2 - 4ac)/4a^2.$$

Solving (25) for z and using these expressions for α and β brings us to the familiar expression for the roots of a quadratic equation

$$z = \frac{-b \pm \sqrt{b^2 - 4ac}}{2a},$$

where now we can permit $b^2 - 4ac$, called the **discriminant** of (23) (or (24)), to be either positive or negative. If $b^2 - 4ac \geq 0$ way may take the square root in the usual manner, but if $b^2 - 4ac < 0$ we use the result

$$\sqrt{b^2 - 4ac} = \sqrt{(-1)(4ac - b^2)} = \sqrt{-1}\sqrt{4ac - b^2} = i\sqrt{4ac - b^2}.$$

Thus we have shown that

$$z = \frac{-b \pm \sqrt{b^2 - 4ac}}{2a} \quad \text{if } b^2 \geq 4ac \tag{26}$$

while

$$z = \frac{-b \pm i\sqrt{4ac - b^2}}{2a} \quad \text{if } b^2 < 4ac.$$

Example 7. Complex roots

Find the roots of

$$z^2 + 3z + 10 = 0.$$

Solution

Using (26) the roots are found to be $z = (-3 + i\sqrt{31})/2$ and $z = (-3 - i\sqrt{31})/2$, and the corresponding factors of the equation are

$$\left(z + \frac{3}{2} - i\frac{\sqrt{31}}{2}\right) \quad \text{and} \quad \left(z + \frac{3}{2} + i\frac{\sqrt{31}}{2}\right).$$

■

We shall return to the general problem of finding the roots when the cofficients are complex, so let us now consider some other properties of polynomial equations with real coefficients. Suppose that in (22) the coefficients $a_0, a_1, \ldots, a_n$ of $P_n(z)$ are arbitrary *real* numbers $(a_n \neq 0)$. Then from (12) and (13), and the fact that $\bar{a}_r = a_r, r = 0, 1, \ldots, n$ (because the a_r are real), we conclude that

$$\overline{P_n(z)} \equiv a_0 + a_1 \bar{z} + a_2 (\bar{z})^2 + \ldots + a_n (\bar{z})^n \equiv P_n(\bar{z}).$$

Consequently, if ζ is a zero of $P_n(z)$, this shows $\bar{\zeta}$ must also be a zero of $P_n(z)$. So, if a polynomial equation with real coefficients has a complex root $z = \zeta$, it must also have a corresponding complex root $z = \bar{\zeta}$. Hence $(z - \zeta)$ and $(z - \bar{\zeta})$ must be factors of $P_n(z)$ giving rise to the quadratic factor

$$(z - \zeta)(z - \bar{\zeta}) \equiv [z^2 - (\zeta + \bar{\zeta})z + \zeta\bar{\zeta}].$$

Since both $\zeta + \bar{\zeta}$ and $\zeta\bar{\zeta}$ are always real this quadratic factor has real coefficients.

Any real roots of $P_n(z) = 0$ will, of course, correspond to real factors. So, using the observation that the product of linear and quadratic factors with real coefficients will generate a polynomial with real coefficients we arrive at the following result.

Theorem 1.4 (Roots of polynomials with real coefficients)

Let the arbitrary real numbers $a_0, a_1, \ldots, a_n$ $(a_n \neq 0)$ be the coefficients of the polynomial of degree n

$$P_n(z) \equiv a_0 + a_1 z + a_2 z^2 \ldots + a_n z^n.$$

Then the n roots of $P_n(z) = 0$ are such that either they are real or, if complex, they occur in complex conjugate pairs.

☐

Although this theorem provides information about the structure of the roots it gives no indication of how they may be found. It must be stressed that the theorem provides *no* information about the roots of polynomial equations with complex coefficients.

One use of this theorem arises if, by some means, one complex root $z = \zeta$ can be found. For then another is $z = \bar{\zeta}$, and we may immediately remove the quadratic factor $[z^2 - (\zeta + \bar{\zeta})z + \zeta\bar{\zeta}]$ from $P_n(z)$, thereby reducing its degree by 2 and so simplifying the task of finding the other roots.

Example 8. Roots of a quadratic

Find the roots of $P_4(z) = 0$ when

$$P_4(z) \equiv z^4 + 7z^3 + 29z^2 + 61z + 70,$$

given $z = -2 + i\sqrt{3}$ is a root.

Solution

The conditions of Theorem 1.4 apply showing $z = -2 - i\sqrt{3}$ is also a root. So $(z + 2 - i\sqrt{3})$ and $(z + 2 + i\sqrt{3})$ must be factors, leading to the quadratic factor

$$(z + 2 - i\sqrt{3})(z + 2 + i\sqrt{3}) = z^2 + 4z + 7.$$

Division of $P_4(z)$ by $z^2 + 4z + 7$ gives $z^2 + 3z + 10$, so that

$$P(z) \equiv (z^2 + 4z + 7)(z^2 + 3z + 10).$$

Equating the last factor to zero to determine the remaining two roots gives, (see Ex. 7), $z = (-3 + i\sqrt{31})/2$ and $z = (-3 - i\sqrt{31})/2$. The four roots of $P_4(z) = 0$ are thus $-2 + i\sqrt{3}$, $-2 - i\sqrt{3}$, $(-3 + i\sqrt{31})/2$ and $(-3 - i\sqrt{31})/2$.

■

Problems for Section 1.6

1 Prove result (7).

2 If $z_1 = (\sqrt{2} + i\sqrt{2})/2$ and $z_2 = -(\sqrt{2} + i\sqrt{2})/2$ find

$$z_1 + z_2, \quad z_1 - z_2, \quad z_1/i, \quad z_1^2 \quad \text{and} \quad z_2^2.$$

If $z_1 = 2 + 5i$, $z_2 = 1 - i$ and $z_3 = 2 + 3i$, find:

3 $z_1 + 2z_2 - z_3$ **4** $(z_1 + z_2)^2$ **5** $z_1 z_2 / z_3$

6 $\text{Re} \dfrac{1}{z_1 + z_2}$ **7** $\text{Im} \dfrac{z_1}{z_2 - z_3}$ **8** $\text{Re}(z_1 \bar{z}_2)$

9 $\text{Im}(z_1 \bar{z}_1)$ **10** $\text{Re}\left(\dfrac{1}{\bar{z}_2}\right)$

11 Prove the results in (12) for arbitrary z_1 and z_2.

12 Prove result (13) by mathematical induction.

13 If $z_1 = 3 + 2i$ and $z_2 = -1 + i$ verify that $z_1 + z_2$ and $z_1 - z_2$ obey the parallelogram rule for addition.

14 Prove the commutative laws for addition and subtraction of complex numbers.

15 Prove the associative laws for addition and subtraction of complex numbers.

16 Verify inequalities (18) and (20) if $z_1 = 2 + 3i$ and $z_2 = -1 + 4i$.

17 Prove that $|z^n| = |z|^n$ for n a positive integer.

18 Prove the **generalized triangle inequality**

$$|z_1 + z_2 + \ldots + z_n| \leq |z_1| + |z_2| + \ldots + |z_n|.$$

19 By setting $z_1 = x_1 + iy_1$ and $z_2 = z_2 + iy_2$, and using the definition of the modulus, prove

$$|z_1 + z_2|^2 + |z_1 - z_2|^2 = 2(|z_1|^2 + |z_2|^2).$$

20 Use result (19) to prove that

$$(1 + z_1 \bar{z}_2)(1 + \bar{z}_1 z_2) \leq 1 + 2|z_1||z_2| + |z_1|^2 |z_2|^2 \leq (1 + |z_1|^2)(1 + |z_2|^2).$$

21 Use the generalized triangle inequality of Prob. 1.18 to prove

$$|z_0+z_1+ \ldots +z_n| \geq |z_0|-|z_1|-|z_2|- \ldots -|z_n|.$$

22 Sketch the following disks in the complex plane, indicating the boundary of an open disk by a dotted line and the boundary of a closed disk by a solid line:

$$|z-3| \leq 1, \quad |z+2-i|<2, \quad |z-2i| \leq 3.$$

23 Which of the points $1+\frac{1}{2}i$, $1-\frac{1}{2}i$, $2+i$ and $2+i\dfrac{\sqrt{3}}{2}$ lie either in or on the boundary of the closed annulus

$$\tfrac{1}{4} \leq |z-1-i| \leq 1?$$

Find the roots, and hence the factors, of:

24 $z^2+4=0$ **25** $z^2+9z+25=0$

26 $z^2+6z+12=0$ **27** $2z^2+6z+9=0$

28 $z^4-16=0$

29 Prove that if the real numbers a_0, a_1, $\ldots$, a_n $(a_n \neq 0)$ are the coefficients of

$$P_n(z) \equiv a_0+a_1z+a_2z^2+\ldots+a_nz^n,$$

then $P_n(z)=0$ must have at least one real root if n is odd.

30 Use the result of Prob. 29 to find the roots of the cubic equation

$$z^3+3z^2+2z-6=0.$$

[*Hint*: First find the real root by inspection (trial and error) and after removing the corresponding factor from the cubic equation find the two remaining roots.]

31 By setting $w=z^2$ in the quartic equation

$$z^4+13z^2+36=0,$$

solve the resulting quadratic equation for w and hence find the four roots z.

32 Find all the roots of

$$z^4+2z^3+21z^2+32z+80=0,$$

given that $4i$ is a root.

33 Find all of the roots of

$$z^4+7z^3+25z^2+45z+42=0,$$

given that $-2+i\sqrt{3}$ is a root.

34 Given $-i$ is a root of

$$z^4+(i-1)z^3+(4-i)z^2+4(i-1)z-4i=0,$$

remove the corresponding factor and use the result of Prob. 29 to find the remaining three roots. Why would it be incorrect to argue that if $-i$ is a root then so also is i, and thus that (z^2+1) is a factor?

Harder problems

35 If $z_1, z_2, \ldots, z_n$ and $w_1, w_2, \ldots, w_n$ are arbitrary complex numbers the **Cauchy–Schwarz inequality** for complex numbers asserts that

$$\left| \sum_{k=1}^{n} z_k w_k \right|^2 \le \left(\sum_{k=1}^{n} |z_k|^2 \right) \left(\sum_{k=1}^{n} |w_k|^2 \right).$$

Use this inequality to show that if

$$P_m(z) = 1 + z + z^2 + \ldots + z^m,$$

then for $|z| \ne 1$

$$|P_m(z)|^2 \le (m+1) \left(\frac{1 - |z|^{2m+2}}{1 - |z|^2} \right).$$

In practical problems giving rise to polynomials, the location of the zeros in the complex plane is important since this determines the nature of the solution. In control theory, for example, a system will only be stable (have a response which does not increase without bound) if all the zeros of a certain polynomial lie to the left of the imaginary axis. The following two problems illustrate the type of information concerning the location of zeros which may be deduced from the coefficients of a polynomial.

36 Use the result of Prob. 21 to prove that if $a_0, a_1, \ldots, a_n$ are arbitrary complex numbers with $a_0 \ne 0$ and $a_n \ne 0$, the polynomial

$$P_n(z) \equiv a_0 + a_1 z + \ldots + a_n z^n$$

does not vanish (has no zero) in the open disk $|z| < \rho$, where ρ is the single positive root of

$$|a_0| - |a_1||z| - |a_2||z|^2 - \ldots - |a_n||z|^n = 0.$$

37 Use the result of Prob. 36 to prove that all the zeros of

$$P_n(z) \equiv a_0 + a_1 z + \ldots + a_n z^n$$

lie in the open annulus $\rho < |z| < R$, where ρ is as defined in that problem, and $R = 1/r$ with r the single positive root of

$$|a_n| - |a_{n-1}||w| - |a_{n-2}||w|^2 - \ldots - |a_0||w|^n = 0.$$

Illustrate the result by applying it to a quadratic polynomial with real coefficients of your own choice and plotting the location of the zeros inside the annulus. [*Hint:* Set $z = 1/w$.]

1.7 Complex numbers in polar form. Roots

The $x + iy$ representation of complex numbers introduced in the previous section came from the use of Cartesian coordinates in the complex plane. An alternative and equally important representation involves the use of the polar coordinates r, θ to identify points in the complex plane. The origin O from which the radial distance r is measured is the same as for Cartesian coordinates, while the x-axis is taken as the reference line from which the angle θ is measured in radians, with the *positive sense* being taken **counterclockwise**.

Provided $r \neq 0$, any point P in the complex plane with Cartesian coordinates (x, y) is related to the polar coordinates (r, θ) of that same point (Fig. 1.8) by

$$x = r\cos\theta, \quad y = r\sin\theta. \tag{1}$$

In polar form the complex number $z = x + iy$ becomes

$$z = r(\cos\theta + i\sin\theta), \tag{2}$$

and in this representation the nonnegative number r is the modulus of z, for

$$|z| = \sqrt{x^2 + y^2} = \sqrt{r^2(\cos^2\theta + \sin^2\theta)} = r \geq 0. \tag{3}$$

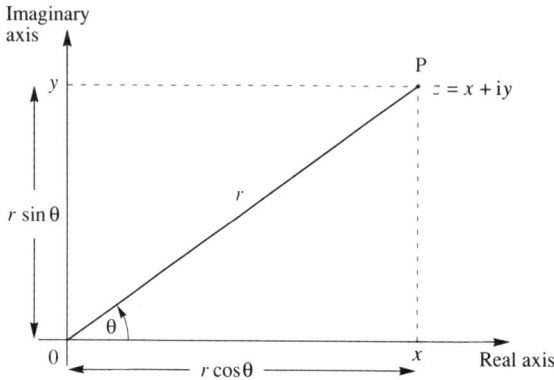

Fig. 1.8 Cartesian and polar representations of a complex number

The angle θ measured counterclockwise from the reference line Ox to OP is called the **argument** of z, written arg z, and it is always defined provided $z \neq 0$. Since the trigonometric functions sine and cosine are periodic with period 2π, the argument is indeterminate to within an arbitrary integral multiple of 2π. To avoid ambiguity it is necessary that an interval of length 2π be selected within which θ is required to lie. The choice of this interval is arbitrary, but by convention it is taken to be

$$-\pi < \theta \leq \pi. \tag{4}$$

The value of θ satisfying this inequality will be called the **principal value** of the argument of z. It is usual to denote the principal value by Arg z, so that

$$\arg z = \operatorname{Arg} z + 2k\pi \quad (k = 0, \pm 1, \pm 2, \ldots). \tag{5}$$

The process of adding or subtracting integral multiples of 2π from an argument until the result satisfies inequality (4) is called determining the argument **modulo** 2π.

On account of this definition of the argument, the **equality** $z_1 = z_2$ is to be interpreted in the polar representation as meaning the two conditions

$$|z_1| = |z_2| \quad \text{and} \quad \arg z_1 = \text{Arg } z_2 + 2k\pi \quad (k = 0, \pm 1, \pm 2, \ldots). \tag{6}$$

In terms of (r, θ) we have the obvious results

$$\sin \theta = \frac{y}{r}, \quad \cos \theta = \frac{x}{r}, \quad \tan \theta = \frac{y}{x} \quad \text{with } r^2 = x^2 + y^2, \tag{7}$$

so that

$$\arg z = \theta = \arc \sin \frac{y}{r} = \arc \cos \frac{x}{r} = \arc \tan \frac{y}{x}. \tag{8}$$

When determining the principal value of an argument from (8) it is necessary to consider both arc sin (y/r) and arc cos (x/r), rather than just arc tan (y/x), in order that the quadrant in which θ is located is correctly identified.

Two results which will often be needed are the polar representations of the real number 1 and the complex number i. As $1 = (\cos 0 + i \sin 0)$ and $i = \left(\cos \frac{\pi}{2} + i \sin \frac{\pi}{2} \right)$, we see that:

if $z = 1$, then $r = 1$ and Arg $1 = 0$,

while

if $z = i$, then $r = 1$ and Arg $i = \pi/2$.

Example 1. Polar representation

Express z in polar form if (i) $z = 2\sqrt{3} + 2i$ and (ii) $z = -2 - 2i$.

Solution

(i) As $z = 2\sqrt{3} + 2i$, $r = |z| = 4$. Removing the factor r to express z in form (2) gives

$$z = 4 \left(\frac{\sqrt{3}}{2} + \frac{i}{2} \right).$$

Thus $\cos \theta = \sqrt{3}/2$, and $\sin \theta = 1/2$, and so $\theta = \text{Arg } z = \pi/6$. It then follows that

$$z = 4 \left(\cos \frac{\pi}{6} + i \sin \frac{\pi}{6} \right).$$

We have shown

$$r = |z| = 4 \text{ and } \arg z = \frac{\pi}{6} + 2k\pi \quad (k = 0, \pm 1, \pm 2, \ldots).$$

(ii) As $z = -2 - 2i$, $r = |z| = 2\sqrt{2}$. Proceeding as before gives

$$z = 2\sqrt{2} \left(\frac{-1}{\sqrt{2}} - \frac{i}{\sqrt{2}} \right).$$

Thus $\cos\theta = -1/\sqrt{2}$, $\sin\theta = -1/\sqrt{2}$, and so $\theta = \text{Arg}\, z = -3\pi/4$. Hence we may write

$$z = 2\sqrt{2}\left[\cos\left(\frac{-3\pi}{4}\right) + i\sin\left(\frac{-3\pi}{4}\right)\right],$$

and we have shown

$$r = |z| = 2\sqrt{2} \text{ and } \arg z = \frac{-3\pi}{4} + 2k\pi \qquad (k = 0,\ \pm 1,\ \pm 2, \ldots).$$

These results are illustrated in Fig. 1.9.

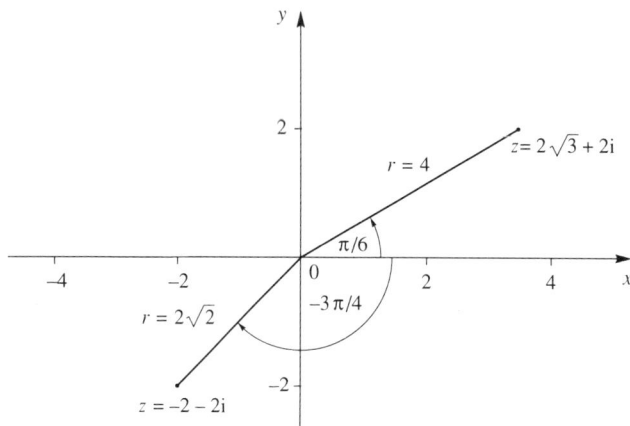

Fig. 1.9   The complex numbers and principal values of the arguments in Example 1

When expressed in polar form, multiplication and division of complex numbers is particularly simple and has a useful geometrical interpretation. To see this we consider two arbitrary complex numbers z_1 and z_2 where

$$z_1 = r_1(\cos\theta_1 + i\sin\theta_1) \quad \text{and} \quad z_2 = r_2(\cos\theta_2 + i\sin\theta_2). \tag{9}$$

Direct multiplication gives

$$z_1 z_2 = r_1 r_2[(\cos\theta_1\cos\theta_2 - \sin\theta_1\sin\theta_2) + i(\sin\theta_1\cos\theta_2 + \cos\theta_1\sin\theta_2)], \tag{10}$$

or

$$z_1 z_2 = r_1 r_2[\cos(\theta_1 + \theta_2) + i\sin(\theta_1 + \theta_2)]. \tag{11}$$

Inspection of (11) shows that when multiplying two complex numbers in polar form the modulus of the product is the *product* of their respective moduli, while the principal value of the argument of the product is the *sum* of the respective arguments (modulo 2π). Symbolically these results become

$$|z_1 z_2| = |z_1||z_2| = r_1 r_2 \quad \text{and} \quad \arg(z_1 z_2) = \arg z_1 + \arg z_2,$$

or

$$\text{Arg}(z_1 z_2) = \arg z_1 + \arg z_2 \text{ (modulo } 2\pi). \tag{12}$$

Notice that as $i = \left(\cos\dfrac{\pi}{2} + i\sin\dfrac{\pi}{2}\right)$, result (12) shows that multiplication of a complex number z by i leaves its modulus unchanged but increases its argument by $\pi/2$, with $|z|$ remaining unchanged.

Similar reasoning applied to division shows that

$$\frac{z_1}{z_2} = \left(\frac{r_1}{r_2}\right)[\cos(\theta_1 - \theta_2) + i\sin(\theta_1 - \theta_2)] \qquad (z_2 \neq 0). \tag{13}$$

That is, the modulus of a quotient is the quotient of the respective moduli, while the argument of a quotient is the difference of the respective arguments (modulo 2π). In symbolic form these results become

$$\left|\frac{z_1}{z_2}\right| = \frac{|z_1|}{|z_2|} = \frac{r_1}{r_2}(r_2 \neq 0), \quad \text{and} \quad \arg(z_1/z_2) = \arg z_1 - \arg z_2,$$

or

$$\mathrm{Arg}(z_1/z_2) = \arg z_1 - \arg z_2 \qquad (\text{modulo } 2\pi). \tag{14}$$

Example 2. Multiplication and division in polar form

If $z_1 = 4\left[\cos\left(\dfrac{\pi}{4}\right) + i\sin\left(\dfrac{\pi}{4}\right)\right]$ and $z_2 = 3\left[\cos\left(\dfrac{\pi}{6}\right) + i\sin\left(\dfrac{\pi}{6}\right)\right]$,

then $r_1 = |z_1| = 4$, $\theta_1 = \mathrm{Arg}\,z_1 = \pi/4$; $r_2 = |z_2| = 3$, $\theta_2 = \mathrm{Arg}\,z_2 = \pi/6$.

Thus $r_1 r_2 = |z_1 z_2| = 12$, $\mathrm{Arg}(z_1 z_2) = \theta_1 + \theta_2 = \pi/4 + \pi/6 = 5\pi/12$ and

$r_1/r_2 = |z_1/z_2| = 4/3$, $\mathrm{Arg}(z_1/z_2) = \theta_1 - \theta_2 = \pi/4 - \pi/6 = \pi/12$,

leading to

$$z_1 z_2 = 12\left(\cos\frac{5\pi}{12} + i\sin\frac{5\pi}{12}\right) \quad \text{and} \quad \frac{z_1}{z_2} = \frac{4}{3}\left(\cos\frac{\pi}{12} + i\sin\frac{\pi}{12}\right).$$

∎

Theorem 1.5 (De Moivre's theorem[8])

De Moivre's theorem states that for any integer n

$$(\cos\theta + i\sin\theta)^n = \cos n\theta + i\sin n\theta. \tag{15}$$

Proof

The result follows by induction as an immediate consequence of (12) when n is a positive integer. It is trivial when $n = 0$, and it is a consequence of the result for positive integral n and of (13) when n is a negative integer.

---

[8] Named after the French mathematician ABRAHAM DE MOIRE (1667–1754) who settled in London as a young man, became a respected contemporary of ISAAC NEWTON, and made contributions to the study of trigonometric functions and the theory of mathematical probability. This theorem was implicit in a note dated 1722, though this used a result published in his earlier 1707 paper in *Phil. Trans.*, London. The form of the theorem as stated here was due to LEONHARD EULER (cf. next footnote).

Combining (12), (14) and (15) gives the useful results that if $z = r(\cos\theta + i\sin\theta)$, then for positive integral n

$$z^n = r^n(\cos n\theta + i\sin n\theta) \tag{16}$$

and

$$\frac{1}{z^n} = \frac{1}{r^n}(\cos n\theta - i\sin n\theta) \qquad (z \neq 0). \tag{17}$$

$\square$

The polar representation is particularly convenient when raising a complex number to a high power, as illustrated by the next example.

Example 3. Raising a complex number to a power

Find $(2 + i2\sqrt{3})^{16}$.

Solution

Setting $z = 2 + i2\sqrt{3}$ we see $r = |z| = 4$ and $\theta = \text{Arg } z = \pi/3$. Now $r^{16} = 4^{16}$ and $\arg(z^{16})$ $= 16 \text{ Arg } z = 16\pi/3$, so $\text{Arg}(z^{16}) = 16\pi/3$ (modulo 2π) $= -2\pi/3$ showing that

$$z^{16} = 4^{16}\left[\cos\left(\frac{-2\pi}{3}\right) + i\sin\left(\frac{-2\pi}{3}\right)\right] = 4^{16}\left(-\frac{1}{2} - i\frac{\sqrt{3}}{2}\right).$$

∎

If $z = r(\cos\theta + i\sin\theta)$, then its complex conjugate

$$\bar{z} = r(\cos\theta - i\sin\theta) = r[\cos(-\theta) + i\sin(-\theta)]. \tag{18}$$

This shows very clearly that the geometrical interpretation of complex conjugation is a reflection in the real axis, since the modulus of z is unchanged, but the sign of its argument is reversed.

Representations for $\sin\theta$ and $\cos\theta$ follow by combining the definition of z with result (18) in the case $|z| = r = 1$ to obtain

$$\cos\theta = \frac{z + \bar{z}}{2} \quad \text{and} \quad \sin\theta = \frac{z - \bar{z}}{2i}. \tag{19}$$

Euler's formula[9]

When functions of a complex variable are considered it is usual to define e^z as

$$e^z = e^{x+iy} = e^x(\cos y + i\sin y). \tag{20}$$

[9] LEONHARD EULER (1707–1783), as Swiss mathematician of truly outstanding ability who was born in Basel and studied there under Johann Bernoulli. In 1727 he was appointed Professor of Mathematics in St Petersburg where he spent the remainder of his life, apart from the period 1741–1766 which was spent in Berlin. He made contributions to all branches of mathematics and also to the physical sciences. Euler was probably the most prolific contributor to mathematical literature of all time, and part of his work resulted in the founding of the theory of differential equations. He was totally blind during the last seventeen years of his life, yet his memory was such that his output of mathematical publications was in no way diminished either in quantity or quality.

It is then shown that the function e^z is in some ways similar to the real variable exponential function e^x. If we now adopt this definition, set $x=0$ and replace y by θ, we arrive at the important **Euler formula**

$$e^{i\theta} = \cos\theta + i\sin\theta, \tag{21}$$

in which θ is in radians.

A comparison of (2) and (21) then shows we have obtained the alternative representation for z in **polar form**

$$z = r(\cos\theta + i\sin\theta) = re^{i\theta}. \tag{22}$$

The complex number $e^{i\theta}$ has a unit modulus, since

$$|e^{i\theta}| = \sqrt{\cos^2\theta + \sin^2\theta} = 1, \tag{23}$$

and so the line drawn from the origin to the point representing $e^{i\theta}$ is of unit length and makes an angle θ with the positive real axis Ox (Fig. 1.8).

Taking special choices of θ in (21) gives us the useful representations

$$e^{2\pi i} = 1,$$

$$e^{\pi i} = e^{-\pi i} = -1.$$

$$e^{i\pi/2} = i,$$

$$e^{-i\pi/2} = -i,$$

$$e^{i(\theta + 2k\pi)} = e^{i\theta}, \quad (k = 0, \pm 1, \pm 2, \ldots). \tag{24}$$

The arguments used to arrive at results (11) and (13) for multiplication and division, respectively, also show that

$$e^{i\theta_1} \cdot e^{i\theta_2} = e^{i(\theta_1 + \theta_2)} \tag{25}$$

$$\frac{e^{i\theta_1}}{e^{i\theta_2}} = e^{i(\theta_1 - \theta_2)}, \tag{26}$$

from which follow the results

$$(e^{i\theta})^{-1} = e^{-i\theta} = \overline{(e^{i\theta})}. \tag{27}$$

More generally, it also follows that if $z_1 = x_1 + iy$ and $z_2 = x_2 + iy_2$,

$$e^{z_1 + z_2} = e^{z_1} e^{z_2} \tag{28}$$

and, as a special case,

$$e^{x+iy} = e^x e^{iy}. \tag{29}$$

Euler's formula (21) also implies the important definitions of the *real variable trigonometric functions sine and cosine*:

$$\cos\theta = \frac{e^{i\theta} + e^{-i\theta}}{2} \quad \text{and} \quad \sin\theta = \frac{e^{i\theta} - e^{-i\theta}}{2i}. \tag{30}$$

These results are, of course, equivalent to (1) because of (27).

Roots of complex numbers

The task of finding the roots of a given complex number z arises when considering equations of the form

$$w^n = z \qquad (n=1, 2, \ldots). \tag{31}$$

From the fundamental theorem of algebra it follows there will be n values of w satisfying this equation, and it will be seen shortly that these values are all distinct. They are called the **nth roots** of z and are denoted collectively by writing

$$w = \sqrt[n]{z} \text{ or, equivalently, } w = z^{1/n}.$$

From now on, as is customary when working in the complex plane, the symbols $\sqrt[n]{}$ and $()^{1/n}$ will be interpreted as being **n-valued (many-valued)**. To determine these n distinct values we start from (31) and set

$$z = r(\cos\theta + i\sin\theta) \quad \text{and} \quad w = \rho(\cos\phi + i\sin\phi) \tag{32}$$

to obtain

$$\rho^n(\cos n\phi + i\sin n\phi) = r(\cos\theta + i\sin\theta). \tag{33}$$

The definition of equality of complex numbers given in (6) now requires that

$$\rho^n = r \quad \text{and} \quad n\phi = \theta + 2k\pi, \quad (k=0, \pm1, \pm2, \ldots). \tag{34}$$

The required nth roots of z now follow by combining w in (32) and (34) to give

$$\rho = r^{1/n} \quad \text{and} \quad \phi = \frac{\theta + 2k\pi}{n}, \qquad (k=0, \pm1, \pm2, \ldots). \tag{35}$$

The form of ϕ in (35) shows there are only n distinct values of ϕ (modulo 2π), so there are only n distinct nth roots of z. The parameter k in (35) may be allocated any n consecutive integral values, but for convenience we set $k=0, 1, \ldots, n-1$ to obtain

$$z^{1/n} = r^{1/n}\left[\cos\left(\frac{\theta + 2k\pi}{n}\right) + i\sin\left(\frac{\theta + 2k\pi}{n}\right)\right], \qquad (k=0, 1, 2, \ldots, n-1).$$

In terms of the Euler notation this may be expressed more concisely as

$$z^{1/n} = r^{1/n}\exp[i(\theta + 2k\pi)/n], \qquad (k=0, 1, 2, \ldots, n-1). \tag{36}$$

From among the n values of $z^{1/n}$ in (36), the one corresponding to $k=0$ and $\theta = \operatorname{Arg} z$ is called the **principal value** of $w = z^{1/n}$.

Inspection of (36) shows all the nth roots of z lie equally spaced around a circle of radius $r^{1/n}$ centered on the origin. The angle between the lines drawn from the origin to any two consecutive roots is $2\pi/n$, so if one root can be found by inspection all the rest follow without further computation. Suppose $w^3 = 1$, then inspection shows $w=1$ is a root. The three roots of this equation thus lie equally spaced at angles $2\pi/3$ around the unit circle centered on the origin (the circle $|z|=1$). Their arguments are 0, $2\pi/3$, $4\pi/3$ so the three **cube roots of unity**, as they are called, are

$$w_0 = e^{0i} = 1, \quad w_1 = \exp[(2\pi i)/3] = \left(\cos\frac{2\pi}{3} + i\sin\frac{2\pi}{3} \right) = \left(-\frac{1}{2} + i\frac{\sqrt{3}}{2} \right) \quad \text{and}$$

$$w_2 = \exp[(4\pi i)/3] = \left(\cos\frac{4\pi}{3} + i\sin\frac{4\pi}{3} \right) = \left(-\frac{1}{2} - i\frac{\sqrt{3}}{2} \right).$$

Example 4. *n*th roots of unity

Find the n complex numbers called the *n*th **roots of unity** which are the solutions of the equation

$$z^n = 1.$$

Solution

If $z=1$, then $r=|z|=1$ and $\theta = \text{Arg }z = 0$, so result (36) shows that

$$(1)^{1/n} = \exp[(2k\pi i)/n], \quad (k=0, 1, 2, \ldots, n-1). \tag{37}$$

The n complex numbers given by (37) are the required *n*th **roots of unity**. ∎

The roots in this last example are equally spaced around the unit circle $|z|=1$ at angles $2\pi/n$, with one root located at $z=1$ (Fig. 1.10). Let us define ω as $\omega = \exp[(2\pi i)/n]$, corresponding to $k=1$ in (37). Then because $|\omega|=1$, multiplication of a complex number by ω leaves its modulus unchanged, but increases its argument by $2\pi/n$. Consequently, as 1 is an *n*th root of unity, the set of roots (37) may be written $1, \omega, \omega^2, \ldots, \omega^{n-1}$.

Example 5. Fourth roots of a complex number

Find $(8 + i8\sqrt{3})^{1/4}$.

Solution

Setting $z = 8 + i\,8\sqrt{3}$, we have $r = |z| = 16$ and $\theta = \text{Arg }z = \pi/3$. Putting $n=4$ in (36), and using $r=16$, $\theta = \pi/3$, gives for the fourth roots w_0, w_1, w_2 and w_3 of z

$$w_k = 16^{1/4}\exp\{[(1+6k)\pi i]/12\}, \quad (k=0, 1, 2, 3).$$

The location of these roots is shown in Fig. 1.11. ∎

We are now in a position to complete our discussion of quadratic equations by considering the most general form

$$az^2 + bz + c = 0, \tag{38}$$

in which the coefficients a, b and c are arbitrary complex numbers with $a \neq 0$.

However, we first observe that result (36) makes the $\pm$ sign unnecessary when considering the square root of a complex number. Both $\sqrt{z}$ and $-\sqrt{z}$ are contained in

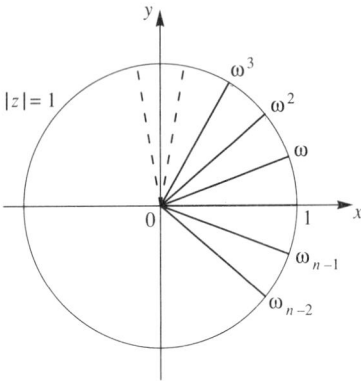

Fig. 1.10   *n*th roots of unity

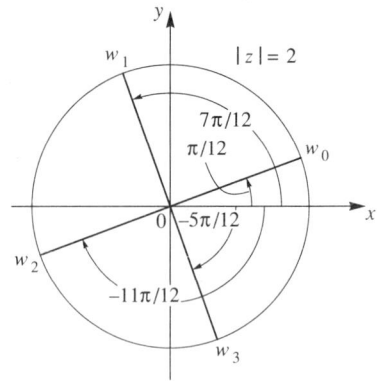

Fig. 1.11 $w=(8+i\,8\,\sqrt{3})^{1/4}$

(36) where they correspond, respectively, to $k=0$ and $k=1$. Arguing as with (24) and completing the square again brings us to the result $(z+\alpha)^2 = \beta$ given in (25). However β is now complex, so $\sqrt{\beta}$ will be two-valued as just discussed, while as before $\alpha=b/2a$ and $\beta=(b^2-4ac)/(4a^2)$, so the roots of (38) become

$$z = \frac{-b}{2a} + \sqrt{\frac{b^2-4ac}{4a^2}}. \tag{39}$$

The explicit form of the roots of (38) follows directly from (39) by identifying z in (36) with $(b^2-4ac)/4a^2$, and then setting $n=2$ to find $\sqrt{(b^2-4ac)/4a^2}$.

Example 6. Roots of a quadratic equation with complex coefficients

Find the roots of $z^2 + 4z - 4i = 0$.

Solution

Setting $a=1$, $b=4$, $c=-4i$ in (39) gives

$$z = -2 + 2\sqrt{1+i}.$$

We have $r=|1+i| = \sqrt{2}$, $\theta=\mathrm{Arg}\,(1+i)=\pi/4$ so from (36) the two square roots of $1+i$ are

$$2^{1/4}\exp\{[1+8k)\pi i]/8\}, \qquad (k=0,\,1).$$

The roots z_1 and z_2 of the quadratic equation are thus

$$z_{k+1} = -2 + 2^{5/4}\exp\{[1+8k)\pi i]/8\}, \qquad (k=0,\,1). \qquad\blacksquare$$

Example 7. Summing two useful trigonometric series

Find closed form expressions for the sums

$$\sum_{k=1}^{n} \cos kx \quad \text{and} \quad \sum_{k=1}^{n} \sin kx.$$

Solution

The two series are seen to be, respectively, the real and imaginary parts of the complex series

$$\sum_{k=1}^{n} e^{ikx}.$$

This is a geometric series, and its closed form sum follows by identifying it with the standard elementary result

$$a + ar + ar^2 + \ldots + ar^{n-1} = \frac{a(r^n - 1)}{r - 1}.$$

Setting $a = e^{ix}$ and $r = e^{ix}$ gives

$$\sum_{k=1}^{n} e^{ikx} = \frac{e^{i(n+1)x} - e^{ix}}{e^{ix} - 1}.$$

Multiplying the numerator and denominator of the right-hand side by $e^{-\frac{1}{2}ix}$ leads to

$$\sum_{k=1}^{n} e^{ikx} = \frac{e^{i(n+\frac{1}{2})x} - e^{\frac{1}{2}ix}}{e^{\frac{1}{2}ix} - e^{-\frac{1}{2}ix}}$$

$$= \frac{e^{i(n+\frac{1}{2})x} - e^{\frac{1}{2}ix}}{2i \sin \frac{1}{2}x}.$$

Equating the real parts of this identity gives as the closed form expression for the first sum

$$\sum_{k=1}^{n} \cos kx = \frac{\sin(n+\frac{1}{2})x}{2 \sin \frac{1}{2}x} - \frac{1}{2}. \tag{40}$$

Similarly, equating the imaginary parts of the identity leads to the other closed form expression which is required

$$\sum_{k=1}^{n} \sin kx = \frac{1}{2} \cot \frac{1}{2}x - \frac{\cos(n+\frac{1}{2})x}{2 \sin \frac{1}{2}x}. \tag{41}$$

Results (40) and (41) are called the **Lagrange trigonometric identities.** ∎

Problems for Section 1.7

Plot each of the following numbers in the complex plane and obtain r, Arg z and arg z.

1 $1 + i\sqrt{3}$ **2** $1 - i$ **3** -4 **4** $2i$ **5** $3 + 4i$

Express in polar form

6 $(1+i)/\sqrt{2}$ **7** $-\sqrt{3} - i$ **8** $3i$ **9** -7

Express in polar form $z_1 z_2$ and z_1/z_2 when

10 $z_1 = 1 + i$ and $z_2 = 1 - i$

11 $z_1 = \sqrt{3} + i$ and $z_2 = -3i$

12 Let z_1 and z_2 be any two distinct points such that $|z_1| = |z_2|$. Find the angle between the straight line through z_1 and z_2 and the straight line through i z_1 and i z_2.

13 Give an inductive proof of de Moivre's theorem.

14 Set $n=4$ in de Moivre's theorem and expand the left-hand side. By equating the respective real and imaginary parts of the resulting equation find expressions for $\sin 4\theta$ and $\cos 4\theta$ in terms of powers of $\sin\theta$ and $\cos\theta$.

15 Repeat Prob. 14 with $n=5$ and hence find expressions for $\cos 5\theta$ and $\sin 5\theta$ in terms of powers of $\sin\theta$ and $\cos\theta$.

16 Find $(-1+i)^{25}$.

17 Express $1-i\sqrt{3}$ and $2-2i$ in polar form and then by means of de Moivre's theorem find $[(1-i\sqrt{3})/(2-2i)]^{11}$.

18 Find the polar forms of $\bar{z}$ and $1/z$ given $z=3+4i$, and hence find $\bar{z}^2$ and $1/z^3$.

19 Use definition (20) to prove that if $z_1 = x_1 + iy_1$ and $z_2 = x_2 + iy_2$, then $e^{z_1 + z_2} = e^{z_1} e^{z_2}$. Deduce that $e^{x+iy} = e^x e^{iy}$.

20 Expand $\left(\dfrac{e^{i\theta} + e^{-i\theta}}{2}\right)^4$, group terms and use result (30) to show

$$\cos^4\theta = \tfrac{1}{8}(\cos 4\theta + 4 \quad \cos \quad 2\theta + 3).$$

21 Expand $\left(\dfrac{e^{i\theta} - e^{-i\theta}}{2i}\right)^5$, group terms and use result (30) to show

$$\sin^5\theta = \frac{1}{16}(\sin 5\theta - 5\sin 3\theta + 10\sin\theta).$$

22 Find $(64 + 64i)^{1/3}$.

23 Find $\left(4\sqrt{2} - i\dfrac{8}{\sqrt{2}}\right)^{1/5}$.

24 If $w^n = z^m$ with m, n positive integers and $z=r(\cos\theta+i\sin\theta)$ prove

$$z^{m/n} = r^{m/n}\left[\cos\left(\frac{m(\theta+2k\pi)}{n}\right) + i\sin\left(\frac{m(\theta+2k\pi)}{n}\right)\right], \qquad (k=0,\,1,\,2,\,\ldots,\,n-1)$$

25 Use the result of Prob. 24 to find $(1+i\sqrt{3})^{3/4}$.

26 Use the result of Prob. 24 to find $(-8\sqrt{3} + 8i)^{2/5}$.

27 Prove that

$$1 + \omega + \omega^2 + \ldots + \omega^{n-1} = 0,$$

where $\omega \neq 1$ is any nth root of unity.

28 If ζ is any nth root of the arbitrary complex number z, and ω is the nth root of unity given by $\omega = \exp[(2\pi i)/n]$, prove that the n values of $z^{1/n}$ are

$$\zeta,\ \zeta\omega,\ \zeta\omega^2,\ \ldots,\ \zeta\omega^{n-1}.$$

29 Show by the method of Ex. 6 that the equation

$$z^2 + (3+i)z + 3i = 0$$

has the roots $z = -3$ and $z = -i$.

30 By setting $\omega = z^2$, use the method of Ex. 6 to find the roots of

$$z^4 + 6iz^2 - 9 = 0.$$

Harder problems

31 Give a geometrical proof that

$$|z-1| \leq ||z|-1| + |z||\text{Arg } z|.$$

32 If ω is the nth root of unity given by $\omega = \exp[(2\pi i)/n]$, prove by considering the geometry of the nth roots of unity $1, \omega, \omega^2, \ldots, \omega^{n-1}$ that

(a) $1 + \cos\left(\dfrac{2\pi}{n}\right) + \cos\left(\dfrac{4\pi}{n}\right) + \ldots + \cos\left(\dfrac{n}{2} \cdot \dfrac{2\pi}{n}\right) = 0$ for n even,

(b) $1 + \cos\left(\dfrac{2\pi}{n}\right) + \cos\left(\dfrac{4\pi}{n}\right) + \ldots + \cos\left(\dfrac{(n-1)\pi}{n}\right) > 0$ for n odd.

33 This problem provides an alternative way of establishing the first Lagrange trigonometric identity

$$\tfrac{1}{2} + \cos\theta + \cos 2\theta + \ldots + \cos n\theta = \frac{\sin[(n+\tfrac{1}{2})\theta]}{2\sin(\tfrac{1}{2}\theta)}.$$

Multiply the expression

$$S = \tfrac{1}{2} + \sum_{r=1}^{n} \cos rx$$

by $2\sin\tfrac{1}{2}x$, replace the products $2\sin\tfrac{1}{2}x \cos rx$ in the summation by the difference of sines and cancel terms to arrive at the required result.

34 Consider the quadratic equation

$$az^2 + bz + c = 0$$

where a, b and c are arbitrary complex numbers with $a \neq 0$, and set $\zeta = (b^2 - 4ac)/4a^2$. If $\alpha = \text{Re } \zeta$ and $\beta = \text{Im } \zeta$, prove that the roots of the equation are

$$z_1 = -\frac{b}{2a} + [\sqrt{(|\zeta|+\alpha)/2} + i(\text{sgn }\beta)\sqrt{(|\zeta|-\alpha)/2}\,],$$

and

$$z_2 = -\frac{b}{2a} - [\sqrt{(|\zeta|+\alpha)/2} + i(\text{sgn }\beta)\sqrt{(|\zeta|-\alpha)/2}\,],$$

where sgn $\beta = 1$ for $\beta > 0$, sgn $\beta = 0$ for $\beta = 0$ and sgn $\beta = -1$ for $\beta < 0$ is the **signum function.** [*Hint*: Complete the square, set $\sqrt{\zeta} = p + iq$ and express p^2 and q^2 in terms of α and β. Use the equation for β to justify the elimination of two of the four equations which arise.]

1.8 Some properties of integrals

This section reviews a few important but less familiar topics from the calculus which will be required in later chapters. Although the material is elementary, and so is suitable for inclusion in a first course on the calculus, much of it is often omitted.

A function defined by an integral

The definite integral (**Riemann integral**) I of a function $f(x)$ over the interval $[a, b]$ is a number. In any first course on the calculus this number is defined as

$$I = \lim_{n \to \infty} \sum_{i=1}^{n} f(\xi_i) \Delta_i, \tag{1}$$

where $[a, b]$ is divided into n intervals $\Delta_i = x_i - x_{i-1}$ with $a = x_0 < x_1 < \ldots < x_n = b$, ξ_i is any point in Δ_i, and when it exists the limit is understood to be taken so that all $\Delta_i \to 0$ as $n \to \infty$. In the usual notation for this definite integral

$$I = \int_a^b f(x) \, dx, \tag{2}$$

the number a is called the **lower limit of integration**, the number b the **upper limit of integration**, $[a, b]$ the **interval of integration**, x the **variable of integration** and $f(x)$ the **integrand**. The sole purpose of displaying the variable x in (2) is to emphasize that f is to be integrated with respect to the variable of integration over the interval $[a, b]$, so that $[a, b]$ is the domain of f. Any other symbol would serve equally well in place of x, since the choice of symbol has no effect on the value of the limit I when it exists. For this reason, the variable of integration appearing in a definite integral is called a **dummy variable**, and the expressions

$$\int_a^b f(x) \, dx, \quad \int_a^b f(t) \, dt, \quad \int_a^b f(\theta) \, d\theta$$

with the dummy variables x, t and θ are all equal to I.

A function may be defined in terms of an integral by replacing a limit of integration by a variable, by using an integrand in which a variable is included as a parameter and, indeed, in many other ways. Thus a function $F(x)$ may be defined as

$$F(x) = \int_a^x f(t) \, dt, \tag{3}$$

where the integrand $f(t)$ is defined on as interval $a \le t \le b$, and x lies within this same interval. Notice the use here of the dummy variable t to avoid confusion with the upper limit x. In expression (3), x is to be regarded as a constant upper limit when evaluating the integral defining $F(x)$, in which $a \le t \le x$. If an **indefinite integral**, or **primitive**, $G(t)$ of $f(t)$ can be found, so that $dG/dt = f(t)$ for $a \le t \le b$, it follows from the second fundamental theorem of calculus that

$$F(x) = G(t)|_a^x = G(x) - G(a), \qquad \text{for } a \le x \le b. \tag{4}$$

An important example of a function defined as an integral is provided by the **error function**

$$\text{erf} \, x = \frac{2}{\sqrt{\pi}} \int_0^x e^{-t^2} \, dt. \tag{5}$$

We remark that the indefinite integral of e^{-t^2} is not an elementary function, so $\operatorname{erf} x$ cannot be evaluated in terms of simpler functions as in (4).

An example of a function defined by an integral in which a variable appears as a parameter is provided by the **gamma function**

$$\Gamma(x) = \int_0^\infty e^{-t} t^{x-1} \, dt. \tag{6}$$

This definite integral, with a semi-infinite interval of integration, is an example of the class of integrals known as **improper integrals** which will be considered later in this section.

More general than functions of the form given in (3) are those functions $G(x)$ which may be defined as

$$G(x) = \int_{\mu(x)}^{\psi(x)} g(t, x) \, dt, \tag{7}$$

where the functions $\mu(x)$, $\psi(x)$ are continuous and defined on some interval $a \le x \le b$, and $g(t, x)$ is integrable with respect to t for $\mu(x) \le t \le \psi(x)$.

Example 1

Evaluate

$$G(x) = \int_x^{x^2} (x + 3t)^2 \, dt.$$

Solution

This simple example of the type given in (7) can be evaluated without difficulty. Expanding the integrand, regarding x as a constant and integrating with respect to t gives

$$G(x) = \int_x^{x^2} (x^2 + 6xt + 9t^2) \, dt$$
$$= (x^2 t + 3xt^2 + 3t^3)|_x^{x^2}$$
$$= 3x^6 + 3x^5 + x^4 - 7x^3. \qquad \blacksquare$$

Differentiation of an integral with respect to a parameter

It will suffice if we consider the derivative of the function $G(x)$ defined in (7) in which x enters the integral as a parameter.

Theorem 1.6 (Differentiation of an integral with respect to a parameter)

Let $\phi(x)$, $\psi(x)$ be differentiable functions in some interval $a \le x \le b$, and let $g(t, x)$ be both integrable with respect to t over the interval $\phi(x) \le t \le \psi(x)$ and differentiable with respect to x.

Then

$$\frac{d}{dx}\int_{\phi(x)}^{\psi(x)} g(t, x)\,dt = \int_{\phi(x)}^{\psi(x)} \frac{\partial g}{\partial x}\,dt + \frac{d\psi}{dx} g[\psi(x), x] - \frac{d\phi}{dx} g[\phi(x), x]. \tag{8}$$

Proof

The proof of this result involves expressing the left-hand side dG/dx in terms of the definition of a derivative as

$$\frac{dG}{dx} = \lim_{h \to 0} \left(\frac{G(x+h) - G(x)}{h} \right),$$

approximating $\phi(x+h)$, $\psi(x+h)$ and $g(t, x+h)$ by the linear terms in their Taylor series expansions, and then taking the limit as $h \to 0$. The details are left as an exercise for the interested reader. $\square$

Example 2

Use Theorem 1.6 to determine dG/dt given

$$G(x) = \int_{x}^{x^2} (x+3t)^2\,dt.$$

Solution

Making the identifications $\phi(x) = x$, $\psi(x) = x^2$, $g(t, x) = (x+3t)^2$ we see these satisfy the conditions of the theorem so it may be used to find dG/dx. We obtain

$$\frac{dG}{dx} = \int_{x}^{x^2} 2(x+3t)\,dt + 2x(x+3x^2)^2 - (x+3x)^2$$

$$= 18x^5 + 15x^4 + 4x^3 - 21x^2,$$

which agrees with the result obtained by differentiating the result of Ex. 1. ■

Two special cases of Theorem 1.6 are of importance. The first is when $f(t, x) = f(t)$ is a function only of t, $\phi(x) = a$ (const.) and $\psi(x) = x$, for then (8) reduces to

$$\frac{d}{dx}\int_{a}^{x} f(t)\,dt = f(x), \tag{9}$$

which is the **first form of the fundamental theorem of calculus**. The other special case arises when $\phi(x) = a$, $\psi(x) = b$, with $a < b$ constants, for then (8) reduces to

$$\frac{d}{dx}\int_{a}^{b} g(t, x)\,dt = \int_{a}^{b} \frac{\partial g}{\partial x}\,dt, \tag{10}$$

which is the result usually known as **differentiation under the integral sign**.

Example 3

Show that

$$G(x) = \int_1^2 \frac{dt}{x+3t} = \frac{1}{3}\ln\left(\frac{x+6}{x+3}\right).$$

Use this result together with (9) to evaluate without further integration the integral

$$\int_1^2 \frac{dt}{(x+3t)^2}.$$

Solution

$$\int_1^2 \frac{dt}{x+3t} = \frac{1}{3}\ln(x+3t)\big|_1^2 = \frac{1}{3}[\ln(x+6) - \ln(x+3)] = \frac{1}{3}\ln\left(\frac{x+6}{x+3}\right).$$

Differentiating this result with respect to x gives

$$\frac{d}{dx}\int_1^2 \frac{dt}{x+3t} = \frac{d}{dx}\left[\frac{1}{3}\ln(x+6) - \frac{1}{3}\ln(x+3)\right].$$

An application of (9) to the left-hand side followed by simplification then shows that

$$\int_1^2 \frac{dt}{(x+3t)^2} = \frac{1}{x^2+9x+18}.$$

■

Integral inequalities and mean value theorems

Inequalities involving integrals have many uses, and collectively they are called **integral inequalities**.

Theorem 1.7 (Integral inequalities)

Let $f(x)$, $g(x)$ be continuous functions over the interval $[a, b]$. Then,

    (i) if $0 \le f(x) \le g(x)$,

$$0 \le \int_a^b f(x)\,dx \le \int_a^b g(x)\,dx; \tag{11}$$

    (ii) if $g(x) > 0$ and m, M are such that

$$m \le f(x) \le M \quad \text{for } a \le x \le b,$$

$$m \int_a^b g(x)\,dx \le \int_a^b f(x)g(x)\,dx \le M \int_a^b g(x)\,dx; \tag{12}$$

    (iii) $$\left| \int_a^b f(x)\,dx \right| \le \int_a^b |f(x)|\,dx. \tag{13}$$

Proof

The proofs of these results follow directly from the definition of a definite integral as the limit of a sum and they are left as exercises for the reader. A geometrical illustration of the correctness of these results is easily given if particular functions are considered. For example, Fig. 1.12(a) shows a typical function $f(x)$ which is both positive and negative over an interval $[a, b]$. When a definite integral is interpreted in terms of areas, the sign of the integrand causes areas above the x-axis to be counted as positive and those below to be counted as negative. Figure 1.12(b) shows the graph of $|f(x)|$ in which all the areas are positive, so a comparison of these two diagrams justifies inequality (13) with respect to this function. Clearly, the result must be true in general.

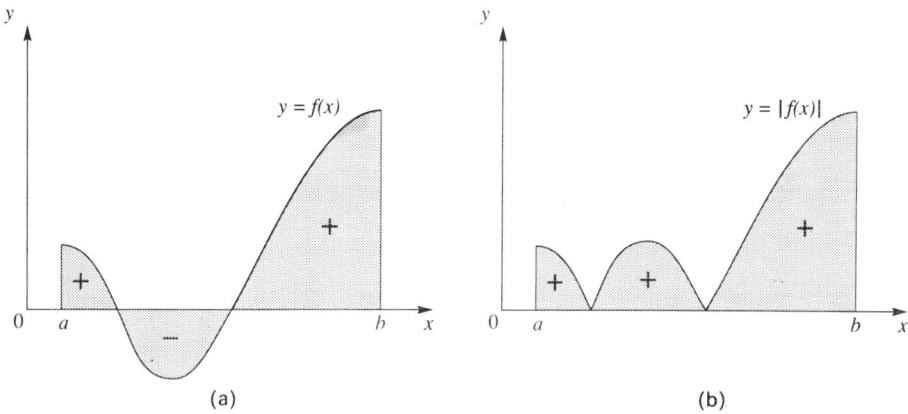

Fig. 1.12   Geometrical justification of $\left| \int_a^b f(x)\,dx \right| \le \int_a^b |f(x)|\,dx$

□

Result (12) may be re-expressed in a useful way if it is first noticed that a number μ exists, with $m \le \mu \le M$, such that

$$\int_a^b f(x)g(x)\,dx = \mu \int_a^b g(x)\,dx. \tag{14}$$

For setting $g(x) \equiv 1$, and using the fact that as $f(x)$ is continuous there must be at least one number ξ in the interval $[a, b]$ for which $\mu = f(\xi)$ (the *intermediate value theorem*), brings (14) to the form

$$\int_a^b f(x)\,dx = (b-a)f(\xi), \quad a \le \xi \le b. \tag{15}$$

This result is called the **first mean value theorem for integrals**.

The second mean value theorem for integrals is expressible in several different forms, of which the following is perhaps the most common.

Theorem 1.8 (Second mean value theorem for integrals)

Let $f(x)$ be continuous, and $g(x)$ bounded, monotonic increasing and differentiable with a continuous derivative in the interval $[a, b]$. Then there is a number ξ in $[a, b]$ such that

$$\int_a^b f(x) g(x) \, dx = g(a) \int_a^\xi f(x) \, dx + g(b) \int_\xi^b f(x) \, dx. \tag{16}$$

Proof

Let $F(x)$ be an indefinite integral of $f(x)$ $[F'(x) = f(x)]$. Integration by parts gives

$$\int_a^b f(x) g(x) \, dx = F(b) g(b) - F(a) g(a) - \int_a^b F(x) g'(x) \, dx. \tag{17}$$

As $g(x)$ is monotonic increasing $g'(x) \geq 0$. Thus applying the first mean value theorem for integrals as given in (14) to the integral on the right-hand side of (17) we find

$$\int_a^b F(x) g'(x) \, dx = F(\xi) \int_a^b g'(x) dx = F(\xi)[g(b) - g(a)]. \tag{18}$$

Using (18) in (17) and rearranging terms gives

$$\int_a^b f(x) g(x) \, dx = g(a)[F(\xi) - F(a)] + g(b)[F(b) - F(\xi)]. \tag{19}$$

As $F(x)$ is an indefinite integral of $f(x)$, use of the first fundamental theorem of calculus (result (9)), then gives

$$\int_a^b f(x) g(x) \, dx = g(a) \int_a^\xi f(x) \, dx + g(b) \int_\xi^b f(x) \, dx, \tag{20}$$

and the proof is complete. □

The final inequality to be discussed is the **Cauchy–Schwarz integral inequality**, which is the direct analog of Theorem 1.1.

Theorem 1.9 (Cauchy–Schwarz integral inequality)

Let $f(x)$, $g(x)$ be bounded piecewise continuous functions on the interval $[a, b]$. Then

$$\left| \int_a^b f(x) g(x) \, dx \right| \leq \sqrt{\int_a^b [f(x)]^2 \, dx \int_a^b [g(x)]^2 \, dx}.$$

Proof

Consider the function

$$F(x) = [\lambda f(x) + g(x)]^2,$$

where λ is a parameter. Then as $F(x) \geq 0$, it follows that

$$\int_a^b [\lambda f(x) + g(x)]^2 \, dx \geq 0,$$

or

$$A\lambda^2 + 2B\lambda + C \geq 0,$$

for all real λ, where

$$A = \int_a^b [f(x)]^2 \, dx, \quad B = \int_a^b f(x) g(x) \, dx \quad \text{and} \quad C = \int_a^b [g(x)]^2 \, dx.$$

The quantity A is nonnegative and will only vanish if $f(x) \equiv 0$, in which case the theorem is trivial. Assuming $A > 0$, and setting $\lambda = -B/A$, we obtain $B^2 - AC \leq 0$, or

$$\left(\int_a^b f(x) g(x) \, dx \right)^2 - \int_a^b [f(x)]^2 \, dx \int_a^b [g(x)]^2 \, dx \leq 0.$$

Taking the square root of this expression yields

$$\left| \int_a^b f(x) g(x) \, dx \right| \leq \sqrt{\int_a^b [f(x)]^2 \, dx \int_a^b [g(x)]^2 \, dx},$$

which is the required result.

$\square$

Example 4

Estimate the integral

$$\int_0^1 \sqrt{1+x} \, dx$$

using the Cauchy–Schwarz integral inequality and compare the result with the exact value.

Solution

Setting $f(x) = \sqrt{1+x}$, $g(x) \equiv 1$ and using Theorem 1.9 gives

$$\left| \int_0^1 \sqrt{1+x} \, dx \right| = \int_0^1 \sqrt{1+x} \, dx \leq \sqrt{\int_0^1 (1+x) \, dx \int_0^1 1 . \, dx} = \sqrt{3/2},$$

and so

$$\int_0^1 \sqrt{1+x} \, dx < 1.2247.$$

The change of variable $u = 1 + x$ shows the exact result to be

$$\int_0^1 \sqrt{1+x} \, dx = \int_1^2 u^{1/2} \, du = \frac{2}{3}(2\sqrt{2} - 1) = 1.2190.$$

■

Improper integrals

Definite integrals in which the integrand becomes infinite within or at the end of a finite interval of integration, or in which the interval of integration is either semi-infinite or infinite are called **improper integrals**. These integrals arise naturally in engineering and science, but they are only of physical and practical interest when they are finite. As improper integrals may be finite (**convergent**), oscillatory or infinite (**divergent**), it is necessary to devise tests for their convergence in order to know when such integrals are useful. First we define these integrals which are of two basic types.

An **improper integral of the first kind** is of the form

$$\int_a^b f(x)\,dx, \tag{21}$$

where the limits of integration $a < b$ are both finite, but the integrand $f(x)$ is infinite at a point c of the interval $[a, b]$.

A typical example of a function $f(x)$ which would produce such an improper integral is shown in Fig. 1.13. The improper integral of the first kind is defined as

$$\int_a^b f(x)\,dx = \lim_{\varepsilon \to 0} \int_a^{c-\varepsilon} f(x)\,dx + \lim_{\delta \to 0} \int_{c+\delta}^d f(x)\,dx, \tag{22}$$

where $\varepsilon, \delta > 0$. If both limits on the right-hand side of (22) exist and are finite the integral is said to **converge** to the sum of the limits. If either or both of these limits is infinite, or is not defined due to oscillation, the integral will be said to **diverge**. For simple functions $f(x)$ the integrals in (22) may be evaluated by routine calculation, after which proceeding to the limit as $\varepsilon, \delta \to 0$ establishes the value of the integral.

Definition (22) needs to be modified in an obvious manner when $f(x)$ is infinite at an end point of $[a, b]$. Thus, for example, if $f(x)$ is infinite at a, we define the integral as

$$\int_a^b f(x)\,dx = \lim_{\varepsilon \to 0} \int_{a+\varepsilon}^b f(x)\,dx, \quad \varepsilon > 0, \tag{23}$$

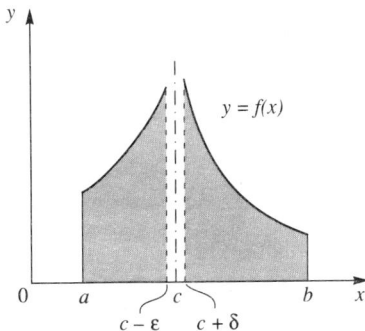

Fig. 1.13 A function $f(x)$, infinite at $x = c$, leading to an improper integral of the first kind

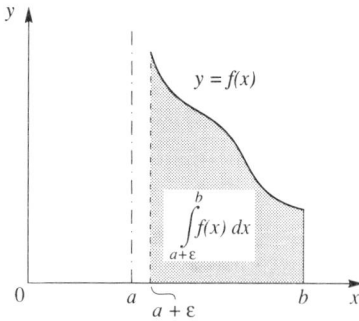

Fig. 1.14 Function $f(x)$ with infinite value at $x = a$.

and the integral will be said to **converge** to this limit when it has a finite value. If the limit is infinite, or does not exist due to oscillation, the integral will be said to **diverge**. The geometrical interpretation of (23) is shown in Fig. 1.14.

Example 5

Investigate the convergence of

$$\int_0^a \frac{dx}{x^\mu}, \text{ with } \mu > 0.$$

Solution

The integrand has an infinity at the origin, so following (23) we write

$$\int_0^a \frac{dx}{x^\mu} = \lim_{\varepsilon \to 0} \int_\varepsilon^a \frac{dx}{x^\mu}.$$

Provided $\mu \neq 1$, we have

$$\int_0^a \frac{dx}{x^\mu} = \lim_{\varepsilon \to 0} \left(\frac{x^{1-\mu}}{1-\mu} \right) \Big|_\varepsilon^a = \frac{a^{1-\mu}}{1-\mu} - \lim_{\varepsilon \to 0} \left(\frac{\varepsilon^{1-\mu}}{1-\mu} \right).$$

If $0 < \mu < 1$ the limit in the second term is zero and the improper integral converges to the value $a^{1-\mu}/(1-\mu)$. If $\mu > 1$ the limit is infinite, and so the integral diverges. When $\mu = 1$ we have

$$\int_0^a \frac{dx}{x} = \lim_{\varepsilon \to 0} \int_\varepsilon^a \frac{dx}{x} = \ln a - \lim_{\varepsilon \to 0} (\ln \varepsilon) = \infty,$$

so the integral again diverges. Thus this improper integral is convergent to $a^{1-\mu}/(1-\mu)$ for $0 < \mu < 1$ and it is divergent when $\mu \geq 1$. ∎

Many functions leading to improper integrals cannot be integrated in terms of known functions, so their convergence must be established in some other way. The simplest of these involves using one of the integral inequalities of Theorem 1.7. The underlying idea is simple, as it merely involves using an integral inequality to compare the required improper integral with a known integral of similar but simpler structure. The following example illustrates this approach.

Example 6

Use an integral inequality approach to investigate the convergence of

$$\int_0^{1/2} \frac{dx}{x(1-x)^2}.$$

Solution

The integrand is infinite at the origin, so this is an improper integral, and we have the inequality

$$0 < \frac{1}{x} < \frac{1}{x(1-x^2)}, \quad \text{for } 0 \le x \le \tfrac{1}{2}.$$

Setting $f(x) = 1/x$, $g(x) = 1/[x(1-x^2)]$ and using (11), we find

$$0 < \int_0^{1/2} \frac{dx}{x} < \int_0^{1/2} \frac{1}{x(1-x^2)} dx. \cdot$$

However, from Example 4 the middle integral is known to be divergent, so these inequalities imply the integral under investigation must also be divergent.

In this case the integral can be evaluated by means of partial fractions, from which it follows that

$$\int_0^{1/2} \frac{dx}{x(1-x^2)} = \lim_{\varepsilon \to 0} \int_\varepsilon^{1/2} \frac{dx}{x(1-x^2)} = \lim_{\varepsilon \to 0} \left[\ln\left(\frac{-x}{(1-x^2)^{1/2}} \right) \right]_\varepsilon^{1/2} = \infty,$$

thereby confirming our previous conclusion. ∎

An **improper integral of the second kind** has one of the following two forms

$$\int_a^\infty f(x)\,dx \quad \text{or} \quad \int_{-\infty}^\infty f(x)\,dx, \tag{24}$$

where $f(x)$ is everywhere finite.

These improper integrals are defined as follows:

$$\int_a^\infty f(x)\,dx = \lim_{R \to +\infty} \int_a^R f(x)\,dx, \tag{25}$$

and

$$\int_{-\infty}^\infty f(x)\,dx = \lim_{R_1, R_2 \to +\infty} \int_{-R_1}^{R_2} f(x)\,dx. \tag{26}$$

The integrals in (25) and (26) will be said to **converge** to the values of these limits when they are finite, and to **diverge** when the limits are either infinite or indeterminate. Geometrical interpretations of the definitions in (25) and (26) are given in Figs 1.15(a), (b), respectively.

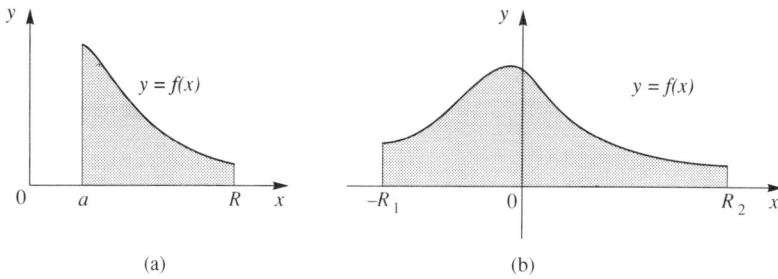

Fig. 1.15 Functions leading to improper integrals of the second kind

Example 7

Investigate the convergence of the improper integral

$$\int_a^\infty \frac{dx}{1+x^2} \cdot$$

Solution

$$\int_a^\infty \frac{dx}{1+x^2} = \lim_{R\to+\infty} \int_a^R \frac{dx}{1+x^2} = \lim_{R\to+\infty} (\text{arc tan } x) \Big|_a^R$$

$$= \lim_{R\to+\infty} (\text{arc tan } R) - \text{arc tan } a = \frac{\pi}{2} - \text{arc tan } a,$$

so the integral is convergent.   ∎

Example 8

The improper integral $\int_a^\infty \cos x \, dx$ is divergent, because

$$\int_a^\infty \cos x \, dx = \lim_{R\to+\infty} \int_a^\infty \cos x \, dx = \lim_{R\to+\infty} (\sin R) - \sin a,$$

and the limit in the last expression is not defined because $\sin R$ oscillates between ± 1 as $R+\infty$.   ∎

The integral inequalities in Theorem 1.7 often help determine convergence or divergence of an improper integral of the second kind when no indefinite integral can be found in terms of known functions. This is illustrated in the next example.

Example 9

Investigate the convergence of

$$\int_1^\infty \frac{x \, dx}{(x^5 + \sin^2 x)^{1/2}}$$

Solution

The integrand is nowhere infinite in the semi-infinite interval of integration, so this is an improper integral of the second kind. We have the inequality

$$0 < \frac{x}{(x^5 + \sin^2 x)^{1/2}} = \frac{1}{\left(x^3 + \left(\frac{\sin x}{x}\right)^2\right)^{1/2}} < \frac{1}{x^{3/2}}, \qquad 1 \le x < \infty.$$

It follows from (11) that

$$0 < \int_1^\infty \frac{x \, dx}{(x^5 + \sin^2 x)^{1/2}} < \int_1^\infty \frac{dx}{x^{3/2}}.$$

Now

$$\int_1^\infty \frac{dx}{x^{3/2}} = \lim_{R \to +\infty} \int_1^R \frac{dx}{x^{3/2}}$$

$$= \lim_{R \to +\infty} (-2x^{-1/2}) \Big|_1^R = -2 \lim_{R \to +\infty} (R^{-1/2}) + 2 = 2$$

Thus the last improper integral in the inequality is finite (converges to 2), so the integral under investigation must also be convergent. Its precise value is unknown, though the integral inequality shows that

$$0 < \int_1^\infty \frac{x \, dx}{(x^5 + \sin^2 x)^{1/2}} < 2. \qquad \blacksquare$$

Definite integrals also arise with the features of improper integrals of both kinds. When this occurs their interval of integration must be divided up into subintervals so that an improper integral of either the first or second kind arises in each subinterval. Thus, for example, the definite integral

$$\int_{1/2}^\infty \frac{dx}{x(1 - x^2)}$$

could be represented as the sum

$$\int_{1/2}^\infty \frac{dx}{x(1 - x^2)} = \int_{1/2}^1 \frac{dx}{x(1 - x^2)} + \int_1^a \frac{dx}{x(1 - x^2)} + \int_a^\infty \frac{dx}{x(1 - x^2)},$$

with $a > 1$ any number. The first two integrals on the right-hand side are improper integrals of the first kind, and the third is an improper integral of the second kind.

Differentiation under the integral sign may be applied to improper integrals containing a parameter, provided the integral is convergent and the integrand is differentiable with respect to the parameter. The following example illustrates an application of this process to an improper integral of the second kind.

Example 10

Evaluate the improper integral

$$\int_0^\infty e^{-st} \sin at \, dt.$$

Use the result together with differentiation under the integral sign to evaluate

$$\int_0^\infty e^{-st} t \sin at \, dt.$$

Solution

The first result is obtained by using integration by parts, for

$$\int_0^\infty e^{-st} \sin at \, dt = \lim_{R \to +\infty} \int_0^R e^{-st} \sin at \, dt$$

$$= \lim_{R \to +\infty} \left(\frac{e^{-st}(-s \sin at - a \cos at)}{s^2 + a^2} \right)_0^R$$

$$= \frac{a}{s^2 + a^2}.$$

Assuming that differentiation under the integral sign with respect to the parameter s is permissible for this improper integral, we have

$$\frac{d}{ds} \int_0^\infty e^{-st} \sin at \, dt = \frac{d}{ds} \left(\frac{a}{s^2 + a^2} \right),$$

or

$$\int_0^\infty \frac{\partial}{\partial s} (e^{-st} \sin at) \, dt = \frac{-2as}{(s^2 + a^2)^2},$$

and thus

$$\int_0^\infty e^{-st} t \sin at \, dt = \frac{2as}{(s^2 + a^2)^2}.$$

Integrals of this type will be encountered again in Chapter 7 when the Laplace transform is discussed. We shall see that $a(s^2 + a^2)^{-1}$ is the Laplace transform of $\sin at$, and $2as(s^2 + a^2)^{-2}$ is the Laplace transform of $t \sin at$. Thus one Laplace transform has been deduced from the other by differentiation under the integral sign.

Differentiation of improper integrals of the second kind with respect to a parameter is a widely used technique for evaluating integrals involving a parameter. It is also used to evaluate integrals in which no parameter is involved, but into which one may be appropriately introduced. Thus it is necessary to know when such an operation is permissible on an improper integral of the second kind.

The theorem we shall use, but not prove, is the following one.

Theorem 1.10 (Differentiation under the integral sign of an improper integral of the second kind)

Let $f(t, x)$ and $\partial f/\partial x$ be continuous functions and let

$$F(x) = \int_a^\infty f(t, x)\,dt$$

be convergent. Then a sufficient condition that

$$\frac{dF}{dx} = \int_a^\infty \frac{\partial f}{\partial x}\,dt,$$

is that a function $g(t)$ can be found such that $|\partial f/\partial x| \le g(t)$ for $a \le t < \infty$ and $\int_a^\infty g(t)\,dt$ is convergent. □

It is possible to formulate this theorem more generally using a less restrictive condition on $\partial f/\partial x$, but the stated condition will suffice for our needs and it is straightforward to use. Notice that since the condition for differentiating under the integral sign is only a *sufficient* condition, this operation may still be possible even when a suitable function $g(t)$ cannot be found.

Example 11

Use Theorem 1.10 to justify differentiating under the integral sign in Ex. 10.

Solution

$$f(t, s) = e^{-st} \sin at \quad \text{and} \quad \partial f/\partial s = -e^{-st} t \sin at$$

are continuous functions, and

$$\int_0^\infty e^{-st} \sin at\,dt$$

is convergent (it equals $a/(s^2 + a^2)$), so the first part of Theorem 1.10 is satisfied.

In this case, in order to show that a function $f(t)$ exists with the property required by Theorem 1.10, it is first necessary to introduce two compensating exponential factors and to rewrite $\partial f/\partial s$ as

$$\frac{\partial f}{\partial s} = -e^{-st} e^{\beta t} e^{-\beta t} t \sin at = -e^{-(s-\beta)t} e^{-\beta t} t \sin at,$$

for any $\beta > 0$.

It follows from this that

$$|\partial f/\partial s| = e^{-(s-\beta)t} |\sin at| t e^{-\beta t} < t e^{-\beta t},$$

provided $s > 2\beta > 0$. We now set $g(t) = t e^{-\beta t}$, and it is a simple matter to check that $\int_0^\infty g(t)\,dt$ is convergent.

Thus the conditions of Theorem 1.10 are satisfied so that differentiation under the integral sign is permissible for $s > 0$, since $\beta > 0$ was arbitrary. This justifies the last calculation in Ex. 10.

■

Problems for Section 1.8

Determine the following functions defined in terms of integrals.

1 $F(x) = \displaystyle\int_0^x (1 + \sin t) \, dt.$

2 $F(x) = \displaystyle\int_2^x t \, e^t dt.$

3 $F(x) = \displaystyle\int_1^x t \ln t \, dt.$

4 $F(x) = \displaystyle\int_3^x a^t dt.$

5 $F(x) = \displaystyle\int_x^{\ln x} (1 + e^{2t}) \, dt.$

6 $F(x) = \displaystyle\int_1^{\tan x} \frac{dt}{1 + t^2}, \quad (\pi/4 \le x < \pi/2).$

7 $F(x) = \displaystyle\int_x^{x^2} (x^2 + 3t) \, dt.$

8 $F(x) = \displaystyle\int_x^{1+x^2} \sin(xt) \, dt, \quad (x > 0).$

Use Theorem 1.6 to find $F'(x)$ when $F(x)$ is defined as follows:

9 $F(x) = \displaystyle\int_x^{x^2} e^{-t^2} dt.$

10 $F(x) = \displaystyle\int_x^0 \sqrt{1 + t^6} \, dt.$

11 $F(x) = \displaystyle\int_{1/x}^{\sqrt{x}} \cos(t^2) \, dt, \ (x > 0).$

12 $F(x) = \displaystyle\int_x^{x^2} \left(\frac{x+1}{t} \right) dt, \quad (x > 0).$

13 $F(x) = \displaystyle\int_{x^2}^x \frac{dt}{x + t}.$

14 $F(x) = \displaystyle\int_x^{x^2} \sinh(x + 2t) \, dt.$

15 Use geometrical illustrations of functions $f(x)$ and $g(x)$ to illustrate the validity of Theorem 1.7 (i) and (ii).

16 Verify Theorem 1.7 (iii) by direct calculation when (a) $f(x) = \sin x$, $a = 0$, $b = 2\pi$, (b) $f(x) = x \cos x$, $a = 0$, $b = \pi$.

In the following questions verify the first mean value theorem for integrals given in (15) by finding a number ξ for which the result is true for the given functions and intervals. Is the number ξ unique, and if not for what other values of ξ in the given interval is the theorem true?

17 $f(x) = \cos x$, $a = -\pi$, $b = \pi$.

18 $f(x) = \sin x$, $a = -\pi/2$, $b = \pi/2$.

19 $f(x) = 1/x$, $a = 2$, $b = 4$.

20 $f(x) = |x|$, $a = -1$, $b = 1$.

In the following questions verify Theorem 1.8 by finding a number ξ for which the result is true for the given functions and intervals. Is the number ξ unique, and if not for what other values of ξ in the given interval is the theorem true?

21 (a) $f(x) \equiv 1$, $g(x) = x$ for any real $a \le x \le b$,
(b) $f(x) = \cos x$, $g(x) = x$, $a = 0$, $b = 2\pi$.

22 (a) $f(x) \equiv 1$, $g(x) = x^2$ for any real $a \le x \le b$,
(b) $f(x) = x$, $g(x) = \sin x$, $a = 0$, $b = \pi/2$.

Use the Theorem 1.9 to obtain an upper estimate of the following definite integrals.

23 $I = \int_0^1 \sqrt{(1+x^4)(1+\sin x)}\, dx$ **24** $\int_0^1 \sqrt{\dfrac{\cos x}{1+x^2}}\, dx$

Investigate the convergence of each of the following improper integrals of the first kind using integration, and find the value of any integral which is convergent.

25 $\int_0^1 \dfrac{dx}{\sqrt{1-x^2}}$ **26** $\int_{-1}^1 \sqrt{\dfrac{1-x}{1+x}}\, dx$

27 $\int_1^2 \dfrac{x+1}{x^2-1}\, dx$ **28** $\int_1^2 \dfrac{dx}{\sqrt{x-1}}$

29 $\int_0^{\pi/2} \cot x \, dx$ **30** $\int_0^{1/2} \dfrac{dx}{x \ln x}$

31 $\int_0^{1/2} \dfrac{dx}{x (\ln)^2}$ **32** $\int_0^1 \dfrac{dx}{x^3-5x^2}$

Investigate the convergence of each of the following integrals of the first kind using inequalities, or by any other appropriate means.

33 $\int_1^2 \dfrac{2x+1}{x^2-1}\, dx$ **34** $\int_0^1 \dfrac{dx}{x^{7/2}-4x^2}$

35 $\int_0^1 \dfrac{dx}{x^{1/2}+x^3}$ **36** $\int_0^{\pi/2} \dfrac{1+\sin x}{x}\, dx$

37 $\int_{\pi/2}^{3\pi/2} \dfrac{1-\sin x}{x-2}\, dx$ **38** $\int_{-\pi/2}^0 \dfrac{\sin x}{x^{2/3}}\, dx$

Investigate the convergence of each of the following improper integrals of the second kind using integration, and find the value of any integral which is convergent.

39 $\int_1^\infty \dfrac{dx}{x^\mu}$ **40** $\int_0^\infty \dfrac{dx}{x\sqrt{x^2-1}}$

41 $\int_{-\infty}^\infty \dfrac{dt}{\cosh t}$ **42** $\int_{-\infty}^\infty \dfrac{d\theta}{1+\theta^2}$

43 $\int_0^\infty \dfrac{dx}{x(1+x^2)}$ **44** $\int_a^\infty \dfrac{dx}{x \ln x}$

45 $\int_a^\infty \dfrac{dx}{x (\ln x)^2}$ **46** $\int_{-\infty}^\infty \dfrac{dx}{x^2+4x+9}$

47 $\Gamma(n) = \int_0^\infty x^{n-1} e^{-x}\, dx$ (integral n)

48 $\int_2^\infty \dfrac{dx}{(x^2-1)^2}$

Investigate the convergence of each of the following integrals of the second kind using inequalities, or by any other appropriate means.

49 $\displaystyle\int_1^\infty \frac{dx}{\sqrt{x^3+1}}$ **50** $\displaystyle\int_2^\infty \frac{dx}{\sqrt{x^5+x^2+3}}$

51 $\displaystyle\int_1^\infty \frac{dx}{\sqrt{\sqrt{x+1}}}$ **52** $\displaystyle\int_3^\infty \frac{dx}{3x+2x^{2/3}+4}$

53 $\displaystyle\int_{\pi/4}^\infty \frac{\sin x}{x^2}\, dx$ **54** $\displaystyle\int_1^\infty \frac{1+x}{x^\mu(1+2x)}\, dx$

Harder problems

55 The effect of discharging an electric current is sometimes measured by using a ballistic device whose response depends either on the integral of the current I while it is discharging, or on the integral of I^2 over this same period. As, in theory, the discharge takes an infinitely long time, the quantities of interest are

$$F=\int_0^\infty I\, dt \quad\text{and}\quad G=\int_0^\infty I^2\, dt,$$

rather than the instantaneous values of I or I^2. Find F and G when the discharge is a damped oscillatory process described by $I=I_0 e^{-kt}\sin \omega t$, with $k>0$, $\omega>0$ constants (see Sec. 5.7).

56 A particle of mass m moves along a line in such a way that its energy is conserved. Thus if x is its displacement along the line, its equation of motion is

$$\frac{1}{2}m\left(\frac{dx}{dt}\right)^2 + \phi(x)=E,$$

where $\frac{1}{2}m(dx/dt)^2$ is the kinetic energy, $\phi(x)$ is the potential energy, and $E\geq 0$ is the total energy of the particle. If $E-\phi(x)=\frac{1}{2}m\omega^2(a^2-x^2)$, it follows from the equation of motion that the particle must oscillate between $x=\pm a$. Integrate the equation of motion to determine the time $\frac{1}{2}T$ taken by the particle to move from $x=-a$ to $x=a$, and hence show its period T is given by

$$T=\frac{2}{\omega}\int_{-a}^a \frac{dx}{(a^2-x^2)^{1/2}}.$$

Evaluate the period T determined by this improper integral.

57 Prove Theorem 1.6 by using the definition

$$\frac{dG}{dx}=\lim_{h\to 0}\left(\frac{G(x+h)-G(x)}{h}\right),$$

approximating $g(t, x+h)$, $\phi(x+h)$ and $\psi(x+h)$ by

$$g(t, x+h)=g(t, x) + h\left(\frac{\partial g}{\partial x}\right),$$

$$\phi(x+h)=\phi(x)+h\left(\frac{\mathrm{d}\phi}{\mathrm{d}x}\right),$$

$$\psi(x+h)=\psi(x)+h\left(\frac{\mathrm{d}\psi}{\mathrm{d}x}\right),$$

and proceeding to the limit.

58 Use the definition of a definite integral given in (1) to prove the results of Theorem 1.7.

59 Prove that if in Theorem 1.8 the conditions imposed on $g(x)$ are replaced by

(i) $g(x) \geq 0$, bounded, monotonic increasing and differentiable with a continuous derivative in the interval $[a,b]$, then for some ξ in $[a,b]$

$$\int_a^b f(x)\,g(x)\mathrm{d}x = g(b)\int_\xi^b f(x)\,\mathrm{d}x;$$

while if they are replaced by

(ii) $g(x) \leq 0$, bounded, monotonic decreasing and differentiable with a continuous derivative in the interval $[a,b]$, then for some ξ in $[a,b]$,

$$\int_a^b f(x)g(x)\,\mathrm{d}x = g(a)\int_a^\xi f(x)\,\mathrm{d}x.$$

60 A continuous function $f(x)$ is said to be **convex** on $[a,b]$ if for any two points P, Q on the graph of $y=f(x)$, the points on the chord PQ all lie above the curve $y=f(x)$. Two examples of convex functions are shown in Fig. 1.16. The curve in Fig. 1.16(a) is continuous, but not differentiable at the point S, whereas the function in Fig. 1.16(b) is both continuous and differentiable at all points in $[a,b]$. It is easily seen that if $f(x)$ is twice differentiable on $[a,b]$, it will be convex on this interval if $f''(x)>0$ for $a \leq x \leq b$.

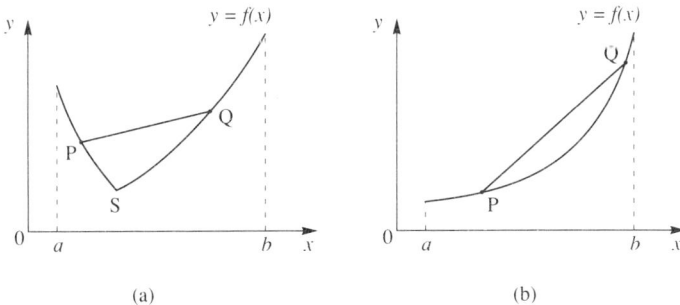

Fig. 1.16 Convex functions (a) continuous on $[a, b]$ but not differentiable at S (b) continuous and differentiable on $[a, b]$

Show the gamma function

$$\Gamma(x)=\int_0^\infty t^{x-1}\mathrm{e}^{-t}\mathrm{d}t$$

is defined for any real $x>0$, and that it is convex for $x>0$. A graph of $\Gamma(x)$ for $x>0$ is shown in Fig. 1.17.

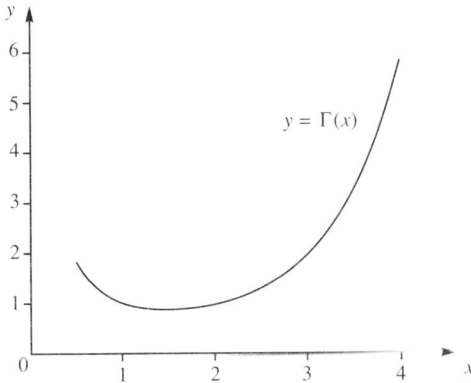

Fig. 1.17 Graph of $\Gamma(x)$

The improper integral $\displaystyle\int_a^\infty f(x)\,dx$ is said to be **absolutely convergent** if both the improper integrals

$$\int_a^\infty f(x)\,dx \quad \text{and} \quad \int_a^\infty |f(x)|dx$$

are convergent. Absolute convergence implies convergence, because by Theorem 1.7(iii)

$$\left|\int_a^b f(x)\,dx\right| \le \int_a^b |f(x)|\,dx, \quad \text{for any } b>a.$$

Letting $b \to +\infty$, the convergence of the integral on the right then implies the convergence of the integral on the left.

Absolute convergence of improper integrals is useful in many ways, but mainly because the most powerful tests for convergence are usually tests for absolute convergence. Notice that although the absolute convergence of an improper integral implies the convergence of the integral, the converse is not necessarily true. See, for example, Prob. 64, which shows that $\displaystyle\int_a^\infty (\sin x/x)\,dx$ is convergent, but that $\displaystyle\int_a^\infty |\sin x/x|\,dx$ is divergent. An improper integral such as this which is convergent, but not absolutely convergent, is said to be **conditionally convergent.**

61 Prove that if $f(x)$ is bounded for $x \ge a>0$ and a number $\mu>1$ can be found such that

$$\lim_{x \to +\infty} |x^\mu f(x)| < A$$

with $A>0$ a finite constant, then $\displaystyle\int_a^\infty f(x)\,dx$ is absolutely convergent.

62 Prove that if $f(x)$ is bounded for $x \ge a>0$ and a number $\mu \le 1$ can be found such that $x^\mu f(x)$ is bounded below by a positive constant, that is

$$x^\mu f(x) \ge A, \ x \ge a$$

then $\displaystyle\int_a^\infty f(x)\,dx$ is divergent.

63 Test the following improper integrals for absolute convergence or divergence using the results of Probs 61 and 62.

(i) $\displaystyle\int_a^\infty \frac{\sin x}{x^{1+n}}\,dx$

(ii) $\displaystyle\int_0^\infty \frac{\sin mx}{a^2+x^2}\,dx$

(iii) $\displaystyle\int_0^\infty \frac{x^5}{(a^2+x^2)^3}\,dx$

(iv) $\displaystyle\int_0^\infty \frac{\sin x \tanh x}{a^2+x^2}\,dx$

64 The purpose of this problem is to establish the conditional convergence of the so-called **Dirichlet integral**

$$I = \int_0^\infty \frac{\sin x}{x}\,dx.$$

It will be sufficient to consider the integral

$$I_1 = \int_{\pi/2}^\infty \frac{\sin x}{x}\,dx.$$

This follows because when I is written in the form

$$I = \int_0^\infty \frac{\sin x}{x}\,dx = \int_0^{\pi/2} \frac{\sin x}{x}\,dx + \int_{\pi/2}^\infty \frac{\sin x}{x}\,dx,$$

the first integral on the right-hand side is a proper integral which is finite, since $\lim\limits_{x\to 0}(\sin x/x) = 1$.

(i) Prove I_1 converges by integrating by parts and comparing the result with $\displaystyle\int_{\pi/2}^\infty \frac{dx}{x^\mu}$. This then establishes the convergence of I as it is the sum of two finite quantities.

(ii) The Dirichlet integral will be conditionally convergent if

$$I_2 = \int_0^\infty \left|\frac{\sin x}{x}\right|\,dx = \int_0^{\pi/2} \frac{|\sin x|}{x}\,dx + \int_{\pi/2}^\infty \frac{|\sin x|}{x}\,dx$$

is divergent. That is, if

$$I_3 = \int_{\pi/2}^\infty \frac{|\sin x|}{x}\,dx$$

is divergent, because the first of the two integrals on the right of I_2 is the same as the corresponding integral in I and so is finite. Prove the divergence of I_3 by using the inequality

$$\frac{|\sin x|}{x} \ge \frac{\sin^2 x}{x} = \frac{1-\cos^2 x}{2x}$$

together with the fact that

$$\int_a^\infty \frac{\cos x}{x}\,dx \quad (a>0)$$

converges, which may be proved in the same way as the convergence of I_1. This completes the proof of the conditional convergence of I.

65 Show by means of Theorem 1.10 that

$$F(\beta) = \int_0^\infty \frac{e^{-\alpha t} \sin \beta t}{t} \, dt, \quad (\alpha > 0)$$

may be differentiated under the integral sign with respect to β. Integrate the expression obtained for $F'(\beta)$, and use the fact that $F(0) = 0$ to show

$$F(\beta) = \arctan (\beta/\alpha).$$

For any fixed β, allow $\alpha \to 0$ and thus deduce the value of

$$\int_0^\infty \frac{\sin \beta t}{t} \, dt,$$

and hence the value of the Dirichlet integral (Prob. 64).

66 (i) Show by example that if $\int_a^\infty f(x) \, dx$ converges and the function $g(x)$ is bounded, the integral $\int_a^\infty f(x) \, g(x) \, dx$ need not necessarily converge.

(ii) Prove that if $\int_a^\infty f(x) \, dx$ is absolutely convergent, then $\int_a^\infty f(x) \, g(x) \, dx$ is also absolutely convergent.

1.9   Linear difference equations

To discuss difference equations it is necessary to use the notion of a **sequence** of numbers, by which we mean a set of numbers enumerated in a particular order. More precisely, the numbers $u_0, u_1, u_2, \ldots$, form a **sequence,** with the understanding that u_0 is the zeroth term of the sequence, u_1 the first term, u_2 the second term, and so on. In this notation the suffix indicates the position of the associated number in the sequence. The nth term u_n of the sequence is called the **general term** of the sequence, and it is convenient to use it to denote the sequence $u_0, u_1, \ldots, u_N$ more concisely by writing $\{u_n\}_{n=0}^N$. Here the lower limit indicates the first term and the upper limit the last term of the sequence. The sequence will be called a **finite sequence** if N is finite and **infinite sequence** if N is infinite. On occasion, when the value of N is immaterial, the limits will be omitted.

Two examples of second order constant coefficient linear difference equations satisfied by a sequence of real numbers $\{u_n\}$ are

$$u_{n+2} + 4u_{n+1} + 2u_n = 2^n, \tag{1}$$

and

$$u_{n+2} + 2u_{n+1} + 4u_n = 0. \tag{2}$$

These equations are **linear** because the elements of the sequence $\{u_n\}$ occurring in the equations only appear to degree one. The **coefficients** of a difference equation are the multipliers of the elements of the sequence $\{u_n\}$ appearing in the equation, and in the examples given above the coefficients are all constants. The **order** of a difference equation is defined as the difference between the greatest and smallest suffixes appearing in the equation. Thus both difference equations given above are second order, because $(n+2) - n = 2$.

Difference equation (1) is said to be **nonhomogeneous** because it contains a nonzero term (in this case 2^n) independent of u_n. Difference equation (2) is said to be **homogeneous** because it contains no such term.

Clearly, if the real numbers u_n and u_{n+1} are specified, each difference equation will determine the real number u_{n+2} uniquely, and hence also $u_{n+3}, u_{n+4}, \ldots$. When working with second order difference equations it is usual to specify u_0 and u_1, called the **initial values** for the difference equation, and then to use them with the equation itself to find the general term u_n as a function of n. This process is known as solving the difference equation, and it may be accomplished in a variety of ways. In this section we shall only outline a straightforward algebraic approach. An alternative approach in terms of the z-transform, which has certain advantages, will be given in Sec. 7.5.

It will suffice for us to discuss second order constant coefficient linear difference equations because this is the most important case and the arguments involved generalize to higher order difference equations in an obvious manner.

(i) Homogeneous difference equations

We shall consider first the second order **homogeneous linear difference equation** with real coefficients a_1, a_2

$$u_{n+2} + a_1 u_{n+1} + a_2 u_n = 0, \quad (a_2 \neq 0). \tag{3}$$

Having determined its general solution, we shall then see how to find its solution subject to the initial conditions $u_0 = \alpha$, $u_1 = \beta$, where α, β are arbitrary real numbers.

The structure of the suffixes in (3) is suggestive of the rule for the combination of indices, so we shall try to find a solution of the form

$$u_n = \lambda^n, \tag{4}$$

where λ is a constant to be determined.

Direct substitution of (4) into (3) shows such a solution is possible provided λ satisfies the equation

$$\lambda^{n+2} + a_1 \lambda^{n+1} + a_2 \lambda^n = 0.$$

Unless $\lambda = 0$, corresponding to a **trivial solution**, we may cancel the factor λ^n and arrive at the following quadratic equation satisfied by λ

$$\lambda^2 + a_1 \lambda + a_2 = 0. \tag{5}$$

This equation is called the **characteristic equation** of difference equation (3). The associated polynomial

$$P(\lambda) = \lambda^2 + a_1 \lambda + a_2 \tag{6}$$

is called the **characteristic polynomial** of difference equation (3).

Three possibilities now arise, according as:

(i) $a_1^2 - 4a_2 > 0$; the roots λ_1, λ_2 of (5) are real and distinct,

(ii) $a_1^2 - 4a_2 < 0$; the roots λ_1, λ_2 of (5) are complex conjugates,

(iii) $a_1^2 - 4a_2 = 0$; the roots λ_1, λ_2 of (5) are real and equal.

Case (i)

When $a_1^2 - 4a_2 > 0$ the roots λ_1, λ_2 of (5) are real and distinct, so it follows from (4) that λ_1^n and λ_2^n are two different solutions of (3). It is easily verified by direct substitution that

$$u_n = A\lambda_1^n + B\lambda_2^n \tag{7}$$

is also a solution of (3) where A, B are arbitrary constants. As only two distinct values of λ are possible, we conclude that (7) is the most general form of solution possible for (3). The solution (7) is called the **general solution** of difference equation (3).

If a solution of (3) is to be found subject to the initial conditions $u_0 = \alpha$, $u_1 = \beta$ the arbitrary constants A, B in (7) must be matched to these conditions, Setting $n = 0, 1$ in (7) we then find

$$(n=0) \quad \alpha = A + B$$
$$(n=1) \quad \beta = A\lambda_1 + B\lambda_2,$$

so

$$A = \frac{\beta - \alpha\lambda_2}{\lambda_1 - \lambda_2}, \quad B = \frac{\alpha\lambda_1 - \beta}{\lambda_1 - \lambda_2}. \tag{8}$$

Thus the solution of the difference equation subject to the stated initial conditions is

$$u_n = \left(\frac{\beta - \alpha\lambda_2}{\lambda_1 - \lambda_2}\right)\lambda_1^n + \left(\frac{\alpha\lambda_1 - \beta}{\lambda_1 - \lambda_2}\right)\lambda_2^n, \tag{9}$$

with $n = 0, 1, 2, \ldots$.

Case (ii)

Here $a_1^2 - 4a_2 < 0$, so the roots λ_1, λ_2 of (5) are complex conjugates. The arguments in case (i) above leading to the general solution (7) and to the solution subject to the initial conditions (9) still apply, but the results may be more conveniently expressed as follows.

Let the complex conjugate roots λ_1, λ_2 be expressed in the polar form

$$\lambda_1 = \rho e^{i\omega} \quad \text{and} \quad \lambda_2 = \rho e^{-i\omega}. \tag{10}$$

Then using de Moivre's theorem the general solution (7) is expressible in the form

$$u_n = \rho^n (K e^{in\omega} + L e^{-in\omega}) \tag{11}$$

where now, as complex numbers are involved, we must allow K and L to be complex. However u_n is to be real, but this will only be possible if L and K are complex conjugates, so $L = \bar{K}$. Setting $L = \frac{1}{2}(A + iB)$ with A, B real arbitrary constants (the factor $\frac{1}{2}$ is for convenience), substitution into (11) shows the general solution to be

$$u_n = \rho^n(A \cos n\omega + B \sin n\omega), \tag{12}$$

with $n = 0, 1, 2, \ldots$.

The solution corresponding to the stated initial conditions follows as in case (i) by using (11) and matching the values of A and B to $u_0 = \alpha$ and $u_1 = \beta$. We find that

$(n = 0)$ $\quad \alpha = A$

$(n = 1)$ $\quad \beta = \rho(A \cos \omega + B \sin \omega)$,

so

$$A = \alpha, \quad B = \frac{\beta - \rho\alpha \cos \omega}{\rho \sin \omega} \tag{13}$$

The solution of the difference equation subject to the stated initial conditions is thus

$$u_n = \rho^n \left[\alpha \cos n\omega + \left(\frac{\beta - \rho\alpha \cos \omega}{\rho \sin \omega} \right) \sin n\omega \right], \tag{14}$$

with $n = 0, 1, \ldots$.

Case (iii)

Here $a_1^2 - 4a_2 = 0$, so the roots λ_1, λ_2 of (5) are equal and $\lambda_1 = \lambda_2 = \mu$ (say). The general solution (7) ceases to be valid because both terms now combine leaving only one arbitrary constant instead of two. To find another solution which may be combined with μ^n to form the general solution in this case we proceed as follows.

Substituting $u_n = \mu^n$ into the left-hand side of (3) and using (6) leads to the identity

$$\mu^{n+2} + a_1 \mu^{u+1} + a_2 \mu^n \equiv \mu^n P(\mu).$$

Differentiation with respect to μ then gives

$$(n+2)\mu^{n+1} + (n+1)a_1\mu^u + na_2\mu^{n-1} \equiv n\mu^{n-1} P(\mu) + \mu^n P'(\mu). \tag{15}$$

However $P(\mu) = 0$, and since in this case we may write $P(\mu)$ as

$$P(\mu) = (\lambda - \mu)^2,$$

it follows by differentiation with respect to μ that

$$P'(\lambda) = 2(\lambda - \mu),$$

which shows that $P'(\mu) = 0$. Thus the right-hand side of (15) vanishes, and the resulting expression then shows $u_n = n\mu^{u-1}$ is also a solution of (3).

Thus in this case $n\mu^{n-1}$ is another solution of (3), so the general solution when there are equal real roots becomes

$$u_n = A\mu^n + Bn\mu^{n-1}. \tag{16}$$

The solution of the difference equation subject to the stated initial conditions follows as before by matching the arbitrary constants A, B to the conditions $u_0 = \alpha$, $u_1 = \beta$. We find

$(n=0)$ $\alpha = A,$

$(n=1)$ $\beta = A\mu + B,$

so

$$A = \alpha, \quad B = \beta - \alpha\mu. \tag{17}$$

Thus the solution of the difference equation subject to the stated initial conditions becomes

$$u_n = \alpha(1-n)\mu^n + \beta n\mu^{n-1}, \tag{18}$$

with $n = 0, 1, 2, \ldots$.
Table 1.1 presents a summary of the results obtained so far.

Table 1.1   General solution u_n of difference equation $u_{n+2} + a_1 u_{n+1} + a_2 u_n = 0$.

Condition on coefficients	Nature of roots	General solution
$a_1^2 - 4a_2 > 0$	real and distinct roots λ_1, λ_2	$u_n = A\lambda_1^n + B\lambda_2^n$
$a_1^2 - 4a_2 < 0$	complex conjugate roots $\lambda_1 = \rho e^{i\omega}, \lambda_2 = \rho e^{-i\omega}$	$u_n = \rho^n(A \cos n\omega + B \sin n\omega)$
$a_1^2 - 4a_2 = 0$	equal real roots $\lambda_1 = \lambda_2 = \mu$	$u_n = A\mu^n + Bn\mu^{n-1}$

Example 1. Homogeneous difference equation

Find the general solution of the difference equation

$$u_{n+2} - u_{n+1} + \tfrac{1}{4}u_n = 0,$$

and hence obtain the solution satisfying the initial conditions $u_0 = 1$, $u_1 = 2$.

Solution

The characteristic equation is

$$\lambda^2 - \lambda + \tfrac{1}{4} = 0,$$

and it has the repeated real root $\lambda = \frac{1}{2}$. This corresponds to case (iii) above with $\mu = \frac{1}{2}$, so the general solution has the form

$$u_n = A(\tfrac{1}{2})^n + Bn(\tfrac{1}{2})^{n-1}.$$

To satisfy the initial conditions A and B must be chosen such that

$$(n=0) \quad 1 = A$$
$$(n=1) \quad 2 = \tfrac{1}{2}A + B.$$

Thus $A = 1$, $B = 3/2$ and the solution of the difference equation satisfying the given initial conditions is

$$u_n = (\tfrac{1}{2})^n + \tfrac{3}{2}n(\tfrac{1}{2})^{n-1}, \quad n = 0, 1, 2, \ldots \qquad \blacksquare$$

(ii) Nonhomogeneous difference equations

We shall only consider the second order constant coefficient nonhomogeneous linear difference equation

$$u_{n+2} + a_1 u_{n+1} + a_2 u_n = f(n), \tag{19}$$

in which $f(n)$ is one of the functions a, ak^n, $a \sin n\omega$ or $a \cos n\omega$, with a, k, ω constants.

Associated with (19) is the homogeneous equation obtained by setting $f(n) \equiv 0$. This has a general solution we shall denote by $\Phi_c(n)$ which may be determined from Table 1.1 according to the nature of the discriminant $a_1^2 - 4a_2$. We will now say that function $u_n = \Phi_p(n)$ is a **particular solution** of the nonhomogeneous difference equation (19) if u_n satisfies (19), which is equivalent to requiring that

$$\Phi_p(n+2) + a_1 \Phi_p(n+1) + a_2 \Phi_p(n) = f(n). \tag{20}$$

Direct substitution then shows

$$u_n = \Phi_c(n) + \Phi_p(n) \tag{21}$$

also to be a solution of (19). It is not difficult to prove that this is the most general form of solution possible for (19), though we shall not bother to establish this here. The solution in (21) comprising the sum of the general solution of the associated homogeneous equation and a particular solution of the nonhomogeneous equation (19) is called the **complete solution** of (19).

It now only remains for us to discover how to find particular solutions when $f(n)$ is one of the functions a, ak^n, $a \sin n\omega$ or $a \cos n\omega$ mentioned above. A different and more powerful form of approach to difference equations which allows a wider choice of functions $f(n)$ will be given in Sec. 7.5.

Inspection of difference equation (19) and of the functions $f(n)$ listed above shows that, provided $f(n)$ does not occur in $\Phi_c(n)$, the only possible choice of function $\Phi_p(n)$ corresponding to a given $f(n)$ must be of the general form given in Table 1.2. The constants C, D are undetermined and must be chosen so that when $\Phi_p(n)$ is substituted

Table 1.2 Trial functions $\Phi_p(n)$ for $u_{n+2}+a_1u_{n+1}+a_2u_n=f(n)$

Nonhomogeneous term $f(n)$	Trial function $\Phi_p(n)$ corresponding to $f(n)$ provided $f(n)$ is not contained in $\Phi_c(n)$
c	C
dk^n	Dk^n
$c\sin n\omega$	$C\cos n\omega + D\sin n\omega$
$c\cos n\omega$	$C\cos n\omega + D\sin n\omega$

into (19) it becomes an identity, for only then will $\Phi_p(n)$ be a solution of (19). This method, in which a trial function with undetermined coefficients is used and the coefficients are then determined by substitution, is called the **method of undetermined coefficients**. The same form of approach will be encountered again in Chapter 5 when we come to discuss linear differential equations.

If initial conditions are given for the nonhomogeneous difference equation (19), then they must be used in conjunction with the complete solution (21) in order to determine the unknown constants which occur in it.

Example 2. Nonhomogeneous difference equation

Find the complete solution of

$$u_{n+2}-u_{n+1}+\frac{5}{2}u_n=5\sin\frac{n\pi}{4},$$

and hence find the solution subject to the initial conditions $u_0=0$, $u_1=1$.

Solution

The characteristic equation of the associated homogeneous difference equation is

$$\lambda^2-\lambda+\tfrac{5}{2}=0.$$

This has the complex conjugate roots

$$\lambda_1=\tfrac{1}{2}(1+3i)\quad\text{and}\quad\lambda_2=\tfrac{1}{2}(1-3i),$$

which in polar form may be written as

$$\lambda_1=(\sqrt{10}/2)e^{i\theta}\quad\text{and}\quad\lambda_2=(\sqrt{10}/2)e^{-i\theta},\text{ with }\theta=\arctan 3.$$

Thus in the notation of case (ii) $\rho=\sqrt{10}/2$, $\omega=\theta$. Table 1.1 now shows that the general solution $\Phi_c(n)$ of the associated homogeneous equation is

$$\Phi_c(n)=(\sqrt{10}/2)^n(A\cos n\theta+B\sin n\theta).$$

To find a particular solution $\Phi_p(n)$ we see from Table 1.2 that an appropriate form of trial function corresponding to $f(n) = 5 \sin(n\pi/4)$ is

$$\Phi_p(n) = C \cos\frac{n\pi}{4} + D \sin\frac{n\pi}{4}.$$

Substituting $u_n = \Phi_p(n)$ into the difference equation gives

$$C \cos(n+2)\frac{\pi}{4} + D \sin(n+2)\frac{\pi}{2} - \left[C \cos(n+1)\frac{\pi}{4} + D \sin(n+1)\frac{\pi}{4} \right]$$

$$+ \frac{5}{2}\left[C \cos\frac{n\pi}{4} + D \sin\frac{n\pi}{4} \right] = 5 \sin\frac{n\pi}{4}.$$

Expanding the sines and cosines where necessary and grouping terms involving $\sin(n\pi/4)$ and $\cos(n\pi/4)$ this becomes

$$\left[\left(\frac{5}{2} - \frac{1}{\sqrt{2}}\right)C + \left(1 - \frac{1}{\sqrt{2}}\right)D \right]\cos\frac{n\pi}{4} + \left[\left(\frac{1}{\sqrt{2}} - 1\right)C + \left(\frac{5}{2} - \frac{1}{\sqrt{2}}\right)D \right]\sin\frac{n\pi}{4} = 5 \sin\frac{n\pi}{4}.$$

If $\Phi_p(n)$ is to be a particular solution this expression must be an identity for all n. However, this is only possible if the coefficients of $\sin(n\pi/4)$ and $\cos(n\pi/4)$ on both sides of the equality are identical. Equating the corresponding coefficients of $\sin(n\pi/4)$ we find

$$\left(\frac{1}{\sqrt{2}} - 1\right)C + \left(\frac{5}{2} - \frac{1}{\sqrt{2}}\right)D = 5.$$

Similarly, equating the corresponding coefficients of $\cos(n\pi/4)$ gives

$$\left(\frac{5}{2} - \frac{1}{\sqrt{2}}\right)C + \left(1 - \frac{1}{\sqrt{2}}\right)D = 0,$$

because there is no term $\cos(n\pi/4)$ on the right-hand side. This pair of equations has the solution

$$C = -0.4437, \qquad D = 2.7163,$$

so the required particular solution is

$$\Phi_p(n) = -0.4437 \cos\frac{n\pi}{4} + 2.7163 \sin\frac{n\pi}{4}.$$

The complete solution u_n, which is the sum of $\Phi_c(n)$ and $\Phi_p(n)$ is thus

$$u_n = (\sqrt{10}/2)^n (A \cos n\theta + B \sin n\theta) - 0.4437 \cos\frac{n\pi}{4} + 2.7163 \sin\frac{n\pi}{4},$$

with $n = 0, 1, 2, \ldots$.

To find the solution subject to the initial conditions $u_0 = 0$, $u_1 = 1$ we must now match the constants A and B occurring in the complete solution u_n so that $u_0 = 0$ and $u_1 = 1$. Thus A and B must be determined from the two equations

$$(n=0) \qquad 0 = A - 0.4437$$

$$(n=1) \qquad 1 = \frac{1}{2}A + \frac{3}{2}B - 0.4437\left(\frac{1}{\sqrt{2}}\right) + 2.7163\left(\frac{1}{\sqrt{2}}\right).$$

These have the solution

$$A = 0.4437, \qquad B = -0.5526,$$

so the required solution of the difference equation subject to the given initial condition is

$$u_n = (\sqrt{10}/2)^n (0.4437 \cos n\theta - 0.5526 \sin n\theta) - 0.4437 \cos \frac{n\pi}{4} + 2.7163 \sin \frac{n\pi}{4},$$

with $n = 0, 1, \ldots$. ◾

Before leaving the subject of difference equations we remark in passing that they need not necessarily have constant coefficients, nor need they be linear. Furthermore, their elements u_n may be numbers, as above, functions, matrices, vectors or many other mathematical entities.

The following is an example of a first order **variable coefficient** linear difference equation whose elements u_n are numbers

$$u_n + n u_{n-1} = 1. \tag{22}$$

This is linear because u_n and u_{n-1} only occur to degree one, but it has the variable coefficient n as a multiplier for u_{n-1}. By using integration by parts it is easily shown that this difference equation is satisfied by the definite integral

$$u_n = \int_0^1 x^n e^{x-1} \, dx,$$

in which the integer n occurs as a parameter. In this context (22) is usually called a **recursion relation**. Successive use of (22) enables u_n to be evaluated for large n in terms of the simple result $u_0 = 1 - e^{-1}$ which follows from elementary integration.

The first order difference equation

$$u_{n+1} = 5 - \left(\frac{2u_n + 1}{u_n^2} \right) \tag{23}$$

whose elements u_n are numbers is **nonlinear**, because the bracketed term on the right-hand side is nonlinear. This recursion relation arises when solving for the largest root of the cubic equation

$$x^3 - 5x^2 + 2x + 1 = 0,$$

when u_n and u_{n+1} are successive approximations to the root. For example, starting with the approximation $u_0 = 5$, some simple calculations show $u_1 = 4.56$, $u_2 = 4.5133$, $u_3 = 4.5078$, $u_4 = 4.5071$ and $u_5 = 4.5070$, which is the value of the root correct to four places of decimals.

As a final example, the second order variable coefficient linear difference equation

$$(n-1)u_n + (2-n)u_{n-2} = \sec^{n-2} x \tan x \tag{24}$$

has *functions* for its elements u_n. This is easily seen because a routine application of integration by parts shows the difference equation (recursion relation) is satisfied by the indefinite integral

$$u_n = \int \sec^n x \, dx.$$

Problems for Section 1.9

In each of the following problems find the roots of the characteristic equation, the general solution of the difference equation and the solution satisfying the stated initial conditions.

1 $u_{n+2} - 5u_{n+1} + 6u_n = 0;$ $\qquad u_0 = 1, \, u_2 = 0.$
2 $u_{n+2} - 3u_{n+1} + 2u_n = 0;$ $\qquad u_0 = 1, \, u_1 = 3.$
3 $u_{n+2} + 2u_{n+1} + u_n = 0;$ $\qquad u_0 = 1, \, u_1 = 2.$
4 $u_{n+2} - 4u_{n+1} + 4u_n = 0;$ $\qquad u_0 = u_1 = 1.$
5 $u_{n+2} + 2u_{n+1} + 2u_n = 0;$ $\qquad u_0 = 1, \, u_1 = \sqrt{2}.$
6 $u_{n+2} - 2u_{n+1} + 2u_n = 0;$ $\qquad u_0 = 0, \, u_1 = 1/\sqrt{2}.$

In each of the following problems find the roots of the characteristic equation, the complete solution of the difference equation and the solution satisfying the stated initial conditions.

7 $u_{n+2} + u_{n+1} - 6u_n = 4^n;$ $\qquad u_0 = 0, \, u_1 = 1.$
8 $u_{n+2} - 9u_n = 7;$ $\qquad u_0 = 1, \, u_1 = -1.$
9 $u_{n+2} + u_n = \sin n;$ $\qquad u_0 = 0, \, u_1 = 1.$
10 $u_{n+2} + 4u_n = 2 + 3^n;$ $\qquad u_0 = 0, \, u_1 = 1.$

(*Hint*: Construct $\Phi_p(n)$ by combining entries from Table 1.2)

Part 2
Vectors and Linear Algebra

Linear algebra is one of the most important and versatile branches of mathematics. For the purposes of this text it has been divided into two parts, the first of which we call vector algebra and the second matrices. It will be seen later that both vectors and matrices are special cases of the unifying concept of a general vector space. Vectors and their associated algebra provide the natural mathematical setting for the discussion of physical problems involving directed quantities such as force, velocity and magnetic field. The reason for including the introduction to vector algebra contained in Chapter 2 is that geometrical vectors provide an important concrete example of a vector space. Thus their study offers a natural introduction to the vector space concepts underlying matrices and their application to systems of differential equations. In Chapter 2 the essentials of vector algebra are developed to the point at which the scalar product is seen to play a fundamental role after which, for the sake of completeness, a brief discussion of the vector product and various geometrical applications of vectors has been included. The straightforward but important extension of geometrical vectors to n-dimensional euclidean space in order to arrive at the notion of a general linear vector space is made at the end of the chapter.

It will be seen from the account of matrices given in Chapter 3, and from the problem sets in particular, that matrices have applications which are both numerous and very diverse in nature. Just as the main purpose of Chapter 2 is to develop vector algebra against a background of physical applications, with an emphasis on the underlying vector space, so Chapter 3 aims to achieve the same result for matrices. The matrix algebra developed in Chapter 3 will be used freely throughout the remainder of this text. In particular, it will be seen to play an essential part in the study of differential equations.

Chapter 2
Algebra of Vectors

Preparatory to the discussion of vectors, ordered number pairs and triples are introduced in Sec. 2.1 along with the notion of a directed line segment. In Sec. 2.2 a scalar is first defined, and then vectors are defined in terms of directed line segments. A geometrical approach to vectors in $\mathbb{R}^3$ and to their addition and multiplication by a scalar is presented in Sec. 2.3. The fundamental ideas of linear independence and a basis for geometrical vectors in $\mathbb{R}^3$ are also introduced in that section. The algebraic approach to vectors in component form is the subject of Sec. 2.4, in which coordinate systems other than Cartesian coordinates are also discussed. The scalar product is defined and applied in Sec. 2.5, and the vector product in Sec. 2.6, while combinations of products are considered in Sec. 2.7. Some geometrical applications of vectors are made in Sec. 2.8 concerning the geometry of lines, planes and their mutual intersection. Finally, in Sec. 2.9. the concept of a general vector space is introduced and related to geometrical vectors in $\mathbb{R}^3$ and to vectors in $\mathbb{R}^n$.

2.1   Preliminaries

Most of this chapter will be concerned with the development of a geometrical approach to the algebra of vectors in three-dimensional euclidean space, also called **euclidean 3-space**. This space, which is easily visualized, is the one which arises most frequently in applications of mathematics to engineering and science. We shall adopt a widely used convention and denote it symbolically by $\mathbb{R}^3$, which in words is pronounced 'R three'. In this notation the exponent 3 indicates the number of spatial dimensions involved, and $\mathbb{R}$ denotes the real number line introduced in Sec. 1.2.

When required, points P in $\mathbb{R}^3$ may be identified uniquely in terms of their Cartesian coordinates (a_1, a_2, a_3) relative to three mutually orthogonal (perpendicular) axes $L_1, L_2,$ and L_3 with a common origin and scale of measurement. Here the representation (a_1, a_2, a_3) of a point P in $\mathbb{R}^3$ is called an **ordered number triple**. In general, changing the order of the three real numbers a_1, a_2 and a_3 in an ordered number triple will change the point which is represented.

It is understood that the first number a_1 in an ordered number triple is the number corresponding to the point at which the L_1-axis is cut by the plane containing P drawn parallel to the L_2 and L_3-axes as shown in Fig. 2.1(a). The second and third numbers a_2 and a_3 are defined in similar fashion as the points at which the L_2 and L_3-axes are cut by the planes containing P drawn parallel to the L_1 and L_3-axes, and the L_1 and L_2-axes, respectively. These planes, together with a_2 and a_3, are illustrated in Figs 2.1 (b, c).

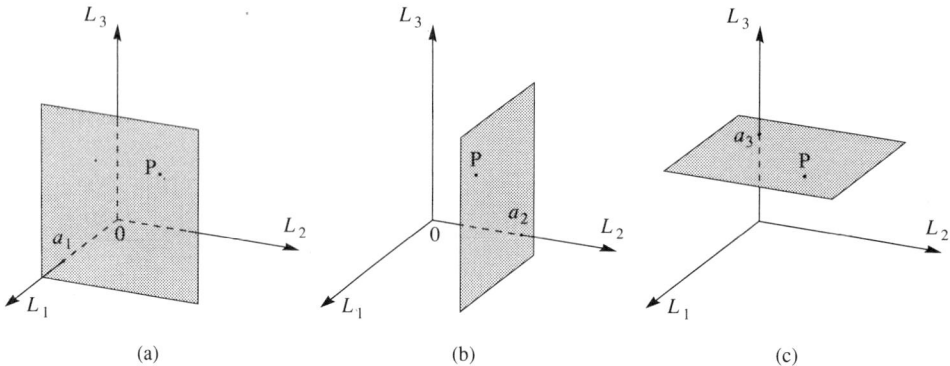

(a)   (b)   (c)

Fig. 2.1   Point $P(a_1, a_2, a_3)$ defined in terms of mutually orthogonal axes L_1, L_2 and L_3 in $\mathbb{R}^3$

Contained within euclidean 3-space are two-dimensional spaces in the form of planes (euclidean 2-spaces) and one-dimensional spaces in the form of straight lines (euclidean 1-spaces, or axes). These lower dimensional spaces are called **subspaces** of $\mathbb{R}^3$. In terms of the convention just introduced, euclidean 2-space is denoted by $\mathbb{R}^2$ and euclidean 1-space by $\mathbb{R}$, which in words are pronounced 'R two' and 'R', respectively.

When required, points P in $\mathbb{R}^2$ may be identified uniquely in terms of their Cartesian coordinates $(a_1, a_2,)$ relative to two mutually orthogonal axes L_1 and L_2 lying within the plane in question and sharing a common origin and scale of measurement. The

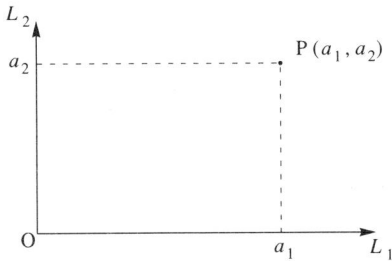

Fig. 2.2   Point $P(a_1, a_2)$ defined in terms of mutually orthogonal axes L_1 and L_2 in $\mathbb{R}^2$

representation (a_1, a_2) of a point P in $\mathbb{R}^2$ is called an **ordered number pair** and changing the order of the two real numbers a_1 and a_2 will, in general, change the point which is represented. In $\mathbb{R}^2$, a_1 is the number corresponding to the point at which the L_1-axis is cut by the line through P parallel to the L_2-axis, and a_2 is defined in similar fashion. This familiar convention is illustrated in Fig. 2.2.

These ideas and notations generalize immediately to higher dimensional space called **euclidean n-space**. This space, in which it is understood there are $n > 3$ dimensions, is denoted by $\mathbb{R}^n$ and pronounced 'R n'. Although $\mathbb{R}^n$ is not easy to visualize, it will prove to be of importance to us later. In such a space there are considered to be n mutually orthogonal axes $L_1, L_2, \ldots, L_n$, all with a common origin and scale of measurement. Relative to these axes, a point P is represented by its coordinates in the form of the **ordered n-tuple** $(a_1, a_2, \ldots, a_n)$ of real numbers. By analogy with $\mathbb{R}^3$, the ith number a_i in the ordered n-tuple $(a_1, a_2, \ldots, a_n)$ defining a point P in $\mathbb{R}^n$ is the point at which the L_i-axis is cut by the generalized plane (**hyperplane**) containing P drawn parallel to the remaining $n - 1$ axis. As with ordered number pairs and triples, the order in which the n real numbers $a_1, a_2, \ldots, a_n$ appear in an ordered n-tuple is important, and changing it will, in general, change the point which is so represented.

Of special significance in any account of vectors in $\mathbb{R}^3$ are the related concepts of a line segment and a directed line segment. We define a **line segment** to be the portion of the straight line that lies between and joins two distinct points P and Q in $\mathbb{R}^3$. A **directed line segment** is then defined to be a line segment in $\mathbb{R}^3$ to which has been added a sense of direction. The **sense** of a directed line segment joining two points P and Q is defined to be a direction along the line either from P to Q or from Q to P. In a graphical representation the sense of a directed line segment is indicated by the addition of an arrow. This convention is illustrated in Fig. 2.3 in which the directed line segment from P to Q is

Fig. 2.3   Directed line segment from P to Q

represented. The **magnitude** (also **modulus**) of a directed line segment is defined to be equal to its length, and so is nonnegative.

2.2   Scalars and vectors

The introduction of vectors into mathematics arose mainly because the quantitative description of many of the physical properties encountered in engineering and science often requires more than a single real number. At the simplest level some quantities are completely specified once their magnitude is given. That is, they are characterized by a single real number representing their size in terms of some appropriate units, though the size may vary with position in $\mathbb{R}^3$ and with time. Since this number measures the scale of the quantity relative to the system of units being employed, quantities of this type are called **scalars**. Typical examples of scalars are mass, volume, density, pressure, electric charge, chemical concentration, work and temperature.

At the next level there are quantities which are described completely only when both their magnitude and direction in $\mathbb{R}^3$ are given, though as with scalars these quantities may also vary with position in $\mathbb{R}^3$ and with time. Such quantities are called **vectors**. Here, when defining a vector, the term **direction** refers not only to the orientation of the line of the vector in $\mathbb{R}^3$, but also to the sense in which it acts along this line. Typical examples of vectors in $\mathbb{R}^3$ are the position of a point relative to a given set of axes, the displacement of a point, velocity, force, momentum, acceleration, angular velocity and magnetic field.

The directed line segment defined in Section 2.1 provides a natural geometrical means of representing a vector quantity. To accomplish this the orientation of the directed line segment is taken parallel to the vector, the sense along the line segment is chosen to be the same as that of the vector, and the length of the directed line segment is set equal to the magnitude of the vector quantity measured according to some scale.

The scale used for geometrical vectors throughout this text is a universally accepted one. In it a directed line segment of unit length in $\mathbb{R}^3$ is taken to represent a vector of unit magnitude in whatever units are appropriate. Thus, for example, a directed line segment of unit length could represent a velocity of 1 km/hour, a force of 1N (Newton) or an angular velocity of 1 radian/second. The advantage of this convention is that the length of a line segment is the magnitude of the associated vector quantity in the appropriate units. The equivalence between a vector quantity and its directed line segment representation leads to directed line segments being regarded as though they are the vectors themselves. In future we shall not usually distinguish between them.

2.3   Vectors – a geometrical approach in $\mathbb{R}^3$

To proceed further a suitable notation is needed for vector quantities. The simplest notation involves labeling the ends of a directed line segment representing a vector by letters, say P and Q, when the vector with its sense directed from P to Q is then denoted either by $\overrightarrow{PQ}$ or by $\underline{PQ}$. Similarly, the vector along the same line segment but with its sense reversed, so it is directed from Q and P, would be written either $\overrightarrow{QP}$ or by $\underline{QP}$. Now although this simple notation is helpful when developing geometrical arguments, it becomes inconvenient when performing general vector algebra. For such purposes we shall use an alternative notation in which a vector is denoted by a single bold-faced letter such as **a**, **b**, **ω**, or **H**. The magnitude of a vector will be denoted by enclosing the vector symbol between two pairs of vertical rules thus $\| \cdot \|$, so the magnitudes of **a**, **ω** and **H** are $\|\mathbf{a}\|$, $\|\mathbf{\omega}\|$ and $\|\mathbf{H}\|$, respectively. When a general algebraic approach to vectors in $\mathbb{R}^n$ is developed it is customary to refer to the magnitude $\|\mathbf{a}\|$ of **a** as the **euclidean norm** of **a**.

In handwriting, vector quantities are usually either underlined or written with an arrow above them, so that **a**, **ω** and **H** would either be written as $\underline{a}$, $\underline{\omega}$ and $\underline{H}$ or as $\vec{a}$, $\vec{\omega}$ and $\vec{H}$. We remark in passing that other notations in use denote the magnitudes of the vectors **a**, **ω**, and **H** in $\mathbb{R}^3$ either by $|\mathbf{a}|$, $|\mathbf{\omega}|$ and $|\mathbf{H}|$, or by a, ω and H, respectively.

It is customary to refer to the end of a directed line segment indicated by the positive sense along it (arrow) as the terminal point (tip) of the vector and to the other end as the initial point (base) of the vector. So, in Fig. 2.4(a), A is the initial point and B the terminal point of the vector $\overrightarrow{AB}$.

Usually it is unnecessary to consider that the initial point of a vector is located at a particular position. This being so, it is convenient to regard vectors represented by directed line segments which are parallel and have the same length and sense as being **equivalent**. When two vectors **a** and **b** are equivalent we shall say they are equal and write **a** = **b**. Thus *equality* means equivalence under a parallel displacement without change of length. Displacements of this sort are called **translations** or, in the context of mechanics, **rigid body translations**. Figure 2.4(b) illustrates a set of equivalent vectors, each of which is equal to $\overrightarrow{AB}$.

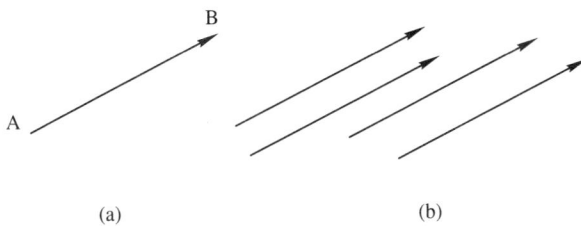

(a) (b)

Fig. 2.4 (a) Vector $\overrightarrow{AB}$ with initial point A and terminal point B. (b) Equivalent vectors

To develop the algebra of vectors in $\mathbb{R}^3$ we must first define two algebraic operations on vectors called addition of vectors, and multiplication of vectors by scalars. Here **addition** means a rule which associates with any two vectors **a** and **b** in $\mathbb{R}^3$ a combination **a** + **b** called the **sum** of **a** and **b**; and **multiplication** of a vector by a scalar means a rule which associates with any scalar λ and vector **a** a combination $\lambda\mathbf{a}$ called the **scalar multiple** of **a** by λ.

Definition (Vector addition – triangle law)

To find the sum $\mathbf{a} + \mathbf{b}$ of any two vectors $\mathbf{a}$ and $\mathbf{b}$ in $\mathbb{R}^3$ the vector $\mathbf{b}$ is first translated until its initial point coincides with the terminal point of $\mathbf{a}$. The sum $\mathbf{a} + \mathbf{b}$ is then defined to be the vector whose initial point is the initial point of $\mathbf{a}$ and whose terminal point is the terminal point of $\mathbf{b}$.

Figures 2.5(a, b) illustrate the use of this definition to arrive at the vector sums $\mathbf{a} + \mathbf{b}$ and $\mathbf{b} + \mathbf{a}$, respectively. These diagrams also make clear why vector addition is often said to be performed by the **triangle law**. Since the triangles in Figs 2.5(a, b) are similar they may be combined to give the parallelogram in Fig. 2.5(c). This last diagram illustrates the fact that vector addition is **commutative**, because

$$\mathbf{a} + \mathbf{b} = \mathbf{b} + \mathbf{a}. \tag{1}$$

It also explains why vector addition is often said to obey the **parallelogram law**. Clearly, the triangle law and the parallelogram law for vector addition are equivalent.

The **parallelogram law** derived from Fig. 2.5(c) for the addition of any two vectors $\mathbf{a}$ and $\mathbf{b}$ in $\mathbb{R}^3$ involves first translating vector $\mathbf{b}$ until its initial point coincides with the initial point of $\mathbf{a}$. The parallelogram is completed, and a diagonal is drawn from the initial point of $\mathbf{a}$ to the opposite corner of the parallelogram. The vector sum $\mathbf{a} + \mathbf{b}$ is then the vector represented by this diagonal with its initial point coincident with the initial point of $\mathbf{a}$.

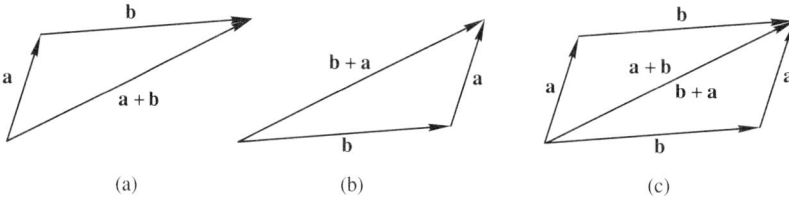

Fig. 2.5 Triangle and parallelogram laws for vector addition   (a) sum $\mathbf{a} + \mathbf{b}$ (b) sum $\mathbf{b} + \mathbf{a}$ (c) vector addition is commutative because $\mathbf{a} + \mathbf{b} = \mathbf{b} + \mathbf{a}$

The **zero** or **null vector 0** in $\mathbb{R}^3$ is a vector of zero magnitude (length) and undefined direction. It is defined by the requirement that if $\mathbf{a}$ is any vector in $\mathbb{R}^3$, then

$$\mathbf{a} + \mathbf{0} = \mathbf{0} + \mathbf{a} = \mathbf{a}. \tag{2}$$

In agreement with the properties of the real number zero, we define $-\mathbf{0} = \mathbf{0}$.

Let us now consider the vector equation $\mathbf{a} + \mathbf{b} = \mathbf{0}$, where $\mathbf{a}$ and $\mathbf{b}$ are vectors in $\mathbb{R}^3$. Then it is apparent from the definition of vector addition that the directed line segments representing $\mathbf{a}$ and $\mathbf{b}$ must be parallel, their lengths must be equal, but they must have opposite senses. Vector $\mathbf{b}$ is called the **negative** of vector $\mathbf{a}$, and is written $-\mathbf{a}$. To obtain $-\mathbf{a}$ from $\mathbf{a}$ the sense of the directed line segment representing $\mathbf{a}$ is merely reversed. The relationship between $\mathbf{a}$ and $-\mathbf{a}$ is shown in Fig. 2.6.

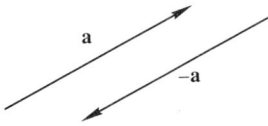

Fig. 2.6   Vector **a** and its negative −**a**

Definition (Vector subtraction)

Let **a** and **b** be any two vectors in $\mathbb{R}^3$, then the operation of subtracting **b** from **a** is defined by

$$\mathbf{a} - \mathbf{b} = \mathbf{a} + (-\mathbf{b}).$$

To use this definition to determine the difference **a** − **b**, vector **b** is first reversed to obtain −**b**. The difference **a** − **b** is then obtained by vector addition of **a** and −**b**. These three steps are illustrated in Fig. 2.7 (a, b, c).

An alternative approach to the determination of the difference **a** − **b** is suggested by examination of Fig. 2.7(c). If the terminal points of **a** and **b** are brought into coincidence by means of a translation, the vector **a** − **b** has its initial point at the initial point of **a** and its terminal point at the initial point of **b** (Fig. 2.8).

Let **a**, **b** and **c** be any three vectors in $\mathbb{R}^3$, and consider the vector sums $(\mathbf{a} + \mathbf{b}) + \mathbf{c}$ and $\mathbf{a} + (\mathbf{b} + \mathbf{c})$ illustrated in Fig. 2.9(a, b). The points P and Q are the same in both diagrams so

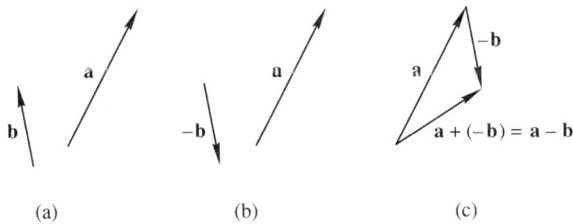

(a)   (b)   (c)

Fig. 2.7   (a) Vectors **a** and **b**   (b) Vectors **a** and −**b**   (c) Difference **a** − **b**

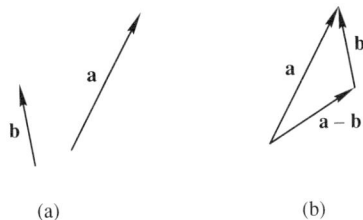

(a)   (b)

Fig. 2.8   (a) Vectors **a** and **b**   (b) Difference **a** − **b**

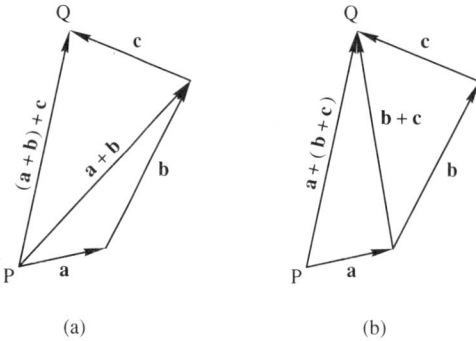

(a)   (b)

Fig. 2.9   (a) $\vec{PQ} = (a+b)+c$ (b) $\vec{PQ} = a+(b+c)$

this establishes that vector addition is **associative**, because from (a) $\vec{PQ} = (a+b)+c$, while from (b) $\vec{PQ} = a+(b+c)$, showing

$$(a+b)+c = a+(b+c). \tag{3}$$

Setting $d = b+c$, the commutative property of vector addition shows $a+d = d+a$, or $a+(b+c) = (b+c)+a$. Thus the order in which vector addition is performed does not affect the result, so brackets may be omitted without ambiguity and, for example, $a+b+c = b+a+c = a+c+b$.

Definition (Scalar multiple of a vector)

Let a be any vector in $\mathbb{R}^3$ and λ be any scalar (real number). Then the **product** λa, called the scalar multiple of a vector, is defined to be a vector of length $|\lambda| \, \|a\|$ in the same direction as that of a when $\lambda > 0$, and in the direction opposite to that of a when $\lambda < 0$. The product λa is defined to be 0 if either $\lambda = 0$ or $a = 0$.

Several different scalar multiples of a are shown in Fig. 2.10 in which $(-1)a$ is simply the negative of a, and so may be written $-a$.

Any vector in $\mathbb{R}^3$ which is of unit magnitude is called a **unit vector**. We shall adopt the useful convention in which an arbitrary vector and a unit vector in the same direction are denoted by the same bold-faced symbol, with the caret symbol ˆ (usually called 'hat') inserted above the unit vector. Thus $\hat{a}$, $\hat{w}$ and $\hat{F}$ are unit vectors in the direction of a, w and F, respectively, with $\|\hat{a}\| = \|\hat{w}\| = \|\hat{F}\| = 1$. It follows from the definition of the scalar

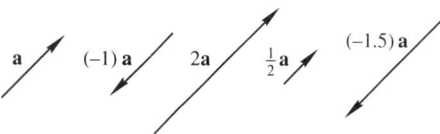

Fig. 2.10   Scalar multiples of a

multiple of a vector that if $\mathbf{u} \neq 0$ is an arbitrary vector, then

$$\mathbf{u} = \|\mathbf{u}\| \, \hat{\mathbf{u}}. \tag{4}$$

The scalar multiplication operation has the following properties, all of which are almost immediate consequences of the definition of the operation. If λ, μ are any two scalars, and $\mathbf{a}$, $\mathbf{b}$ are any two vectors in $\mathbb{R}^3$, then

scalar multiplication is distributive

$$\lambda(\mathbf{a} + \mathbf{b}) = \lambda\mathbf{a} + \lambda\mathbf{b}, \tag{5}$$

$$(\lambda + \mu)\mathbf{a} = \lambda\mathbf{a} + \mu\mathbf{a}, \tag{6}$$

scalar multiplication is associative

$$\lambda(\mu\mathbf{a}) = \mu(\lambda\mathbf{a}) = \lambda\mu\mathbf{a}, \tag{7}$$

scaling by unity

$$1\mathbf{a} = \mathbf{a}. \tag{8}$$

We shall prove property (5), and leave the proof of the other properties as an exercise for the reader. In Fig. 2.11, diagram (a) shows the vector sum $\mathbf{a} + \mathbf{b}$. Scaling the vectors $\mathbf{a}$, $\mathbf{b}$ and $\mathbf{a} + \mathbf{b}$ by $\lambda > 0$ then leads to the triangle in diagram (b) which is similar to that in (a), showing that $\lambda\mathbf{a} + \lambda\mathbf{b} = \lambda(\mathbf{a} + \mathbf{b})$. If $\lambda < 0$ the triangle in diagram (c) is obtained, and again $\lambda\mathbf{a} + \lambda\mathbf{b} = \lambda(\mathbf{a} + \mathbf{b})$. This completes the proof, because if $\lambda = 0$ then by definition $\lambda\mathbf{a} = \lambda\mathbf{b} = \lambda(\mathbf{a} + \mathbf{b}) = 0$ and the result is still true.

The set of all geometrical vectors in $\mathbb{R}^3$ together with the operations of addition and multiplication by a scalar as defined above is said to constitute a **vector space**. The following summary of the algebraic properties of this vector space will serve later as the basis for the axioms used to define a general vector space.

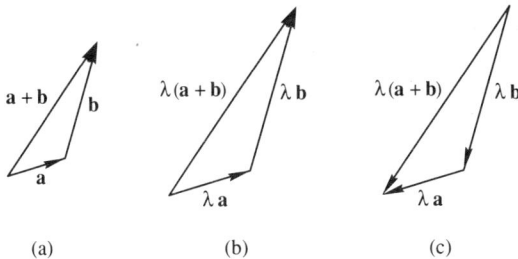

(a)   (b)   (c)

Fig. 2.11   (a) Sum $\mathbf{a} + \mathbf{b}$ (b) Scaling by $\lambda > 0$ (c) Scaling by $\lambda < 0$

Properties of the vector space of geometrical vectors in $\mathbb{R}^3$

If **a**, **b** and **c** are any three vectors in $\mathbb{R}^3$ and λ, μ are any two scalars, then

Addition is commutative

P.1 $\mathbf{a}+\mathbf{b}=\mathbf{b}+\mathbf{a}$,

Addition is associative

P.2 $(\mathbf{a}+\mathbf{b})+\mathbf{c}=\mathbf{a}+(\mathbf{b}+\mathbf{c})$,

Zero vector

P.3 $\mathbf{a}+\mathbf{0}=\mathbf{0}+\mathbf{a}=\mathbf{a}$,

Negative vector (additive inverse)

P.4 $\mathbf{a}+(-\mathbf{a})=\mathbf{0}$,

Scalar multiplication is distributive

P.5 $\lambda(\mathbf{a}+\mathbf{b})=\lambda\mathbf{a}+\lambda\mathbf{b}$,

P.6 $(\lambda+\mu)\mathbf{a}=\lambda\mathbf{a}+\mu\mathbf{a}$,

Scalar multiplication is associative

P.7 $\lambda(\mu\mathbf{a})=(\lambda\mu)\mathbf{a}$,

Multiplication by unity

P.8 $1\mathbf{a}=\mathbf{a}$.

Since all geometrical vectors in both a line and a plane are also geometrical vectors in $\mathbb{R}^3$, they obey the same rules for addition and scalar multiplication as the geometrical vectors in $\mathbb{R}^3$. Thus properties P.1 to P.8 are also true of geometrical vectors in $\mathbb{R}^1$ and $\mathbb{R}^2$.

Linear dependence and independence in $\mathbb{R}^3$

Let $\mu_1, \mu_2, \ldots, \mu_n$ be any n scalars and $\mathbf{a}_1, \mathbf{a}_2, \ldots, \mathbf{a}_n$ be any n vectors in $\mathbb{R}^3$. Then the vector sum

$$\mathbf{l}_n = \mu_1\mathbf{a}_1 + \mu_2\mathbf{a}_2 + \ldots + \mu\,\mathbf{a}_n \tag{9}$$

is called a **linear combination** of the n vectors.

If $\mathbf{a}_1 \neq \mathbf{0}$ is fixed, and μ_1 is allowed to be arbitrary, the set of all vectors generated by $\mathbf{l}_1 = \mu_1\mathbf{a}_1$ will be **collinear** (lie on a line). If $\mathbf{a}_1$ and $\mathbf{a}_2$ are any two fixed nonzero vectors in $\mathbb{R}^3$ with line segments which are not parallel, and μ_1, μ_2 are allowed to be arbitrary, the set of all vectors generated by $l_2 = \mu_1 \mathbf{a}_1 + \mu_2 \mathbf{a}_2$ will be **coplanar** (lie in a plane). Finally, let $\mathbf{a}_1$, $\mathbf{a}_2$ and $\mathbf{a}_3$ be any three fixed nonzero vectors in $\mathbb{R}^3$ with line segments which when taken pairwise lie in three non-parallel planes. Then, whenever $\mu_3 \neq 0$, for any given μ_1, μ_2 and μ_3, $l_3 = \mu_1 \mathbf{a}_1 + \mu_2 \mathbf{a}_2 + \mu_3 \mathbf{a}_3$ will be a vector in $\mathbb{R}^3$ which will lie outside the plane containing $\mathbf{a}_1$ and $\mathbf{a}_2$. By allowing μ_1, μ_2 and μ_3 to arbitrary it is possible to generate all possible vectors in $\mathbb{R}^3$ from the general representation $l_3 = \mu_1 \mathbf{a}_1 + \mu_2 \mathbf{a}_2 + \mu_3 \mathbf{a}_3$. Thus the linear combination of vectors in l_3 provides the most general representation of an arbitrary geometrical vector in $\mathbb{R}^3$.

Any three vectors in $\mathbb{R}^3$ which used in the linear combination l_3 enable the generation of any one of the complete set of vectors in $\mathbb{R}^3$ are said to form a **basis** for the vector space of geometrical vectors in $\mathbb{R}^3$. The vectors in a basis are said to **span** their associated vector space, and the number of vectors in the basis is called the **dimension** of the vector space. It is clear that a basis for the vector space of geometrical vectors in $\mathbb{R}^3$ is not unique, that any basis in the space is equivalent to any other basis, that $\mathbf{a}_1$, $\mathbf{a}_2$ and $\mathbf{a}_3$ as just defined form a basis, and that the dimension of the vector space of geometrical vectors in $\mathbb{R}^3$ is three. The space of collinear vectors is a one-dimensional vector space and the space of coplanar vectors is a two-dimensional vector space. These lower dimensional vector spaces are said to be **subspaces** of the space of geometrical vectors in $\mathbb{R}^3$.

The fact that if $\mathbf{a}_1$, $\mathbf{a}_2$ and $\mathbf{a}_3$ form a basis for the vector space of geometrical vectors in $\mathbb{R}^3$, then all vectors in that space can be generated from the linear combination

$$\mu_1 \mathbf{a}_1 + \mu_2 \mathbf{a}_2 + \mu_3 \mathbf{a}_3,$$

will be used in Sec. 2.4 to express vectors in terms of components.

The choice of vectors for a basis $\mathbf{a}_1$, $\mathbf{a}_2$, $\mathbf{a}_3$ for the vector space of geometrical vectors in $\mathbb{R}^3$ is such that no basis vector can be expressed as a linear combination of the other two. This idea provides the motivation for the formal definition of linear independence that now follows.

Let $\mathbf{a}_1$, $\mathbf{a}_2$ and $\mathbf{a}_3$ be any three vectors in $\mathbb{R}^3$ and μ_1, μ_2 and μ_3 scalars. Then $\mathbf{a}_1$, $\mathbf{a}_2$ and $\mathbf{a}_3$ are said to be **linearly independent** if the only solution to the equation

$$\mu_1 \mathbf{a}_1 + \mu_2 \mathbf{a}_2 + \mu_3 \mathbf{a}_3 = \mathbf{0} \tag{10}$$

is $\mu_1 = \mu_2 = \mu_3 = 0$. If, however, (10) is true for some μ_1, μ_2 and μ_3 which are not all zero, then $\mathbf{a}_1$, $\mathbf{a}_2$ and $\mathbf{a}_3$ are said to be **linearly dependent**. In general, the vectors $\mathbf{a}_1, \mathbf{a}_2, \ldots, \mathbf{a}_n$ in $\mathbb{R}^n$ are linearly dependent if the equation

$$\mu_1 \mathbf{a}_1 + \mu_2 \mathbf{a}_2 + \ldots + \mu_n \mathbf{a}_n = \mathbf{0} \tag{11}$$

is true for some $\mu_1, \mu_2, \ldots, \mu_n$ which are not all zero.

It follows from the definition of a basis that any four or more nonzero geometrical vectors in $\mathbb{R}^3$ must be linearly dependent. To see this it will suffice to consider the case of any four vectors $\mathbf{a}_1$, $\mathbf{a}_2$, $\mathbf{a}_3$ and $\mathbf{a}_4$ in $\mathbb{R}^3$, since the argument is similar when more vectors are involved. If all the vectors are collinear we have $\mathbf{a}_1 = \mu_2 \mathbf{a}_2 = \mu_3 \mathbf{a}_3 = \mu_4 \mathbf{a}_4$ for some nonzero μ_2, μ_3 and μ_4. It thus follows that

$$3\mathbf{a}_1 - \mu_2 \mathbf{a}_2 - \mu_3 \mathbf{a}_3 - \mu_4 \mathbf{a}_4 = \mathbf{0},$$

and as 3, $-\mu_2$, $-\mu_3$ and $-\mu_4$ are not all zero we see $\mathbf{a}_1$, $\mathbf{a}_2$, $\mathbf{a}_3$ and $\mathbf{a}_4$ are linearly dependent.

If all the vectors are coplanar, then at least two of them, say $\mathbf{a}_1$ and $\mathbf{a}_2$, must have line segments which are not parallel and serve to define the plane. Both $\mathbf{a}_3$ and $\mathbf{a}_4$ may then be expressed as linear combinations of $\mathbf{a}_1$ and $\mathbf{a}_2$, so that again the four vectors are linearly dependent. Finally, if three of the vectors are linearly independent, say $\mathbf{a}_1$, $\mathbf{a}_2$ and $\mathbf{a}_3$, they may serve as a basis. Then for some μ_1, μ_2 and μ_3 it follows that

$$\mathbf{a}_4 = \mu_1 \mathbf{a}_1 + \mu_2 \mathbf{a}_2 + \mu_3 \mathbf{a}_3,$$

showing that again the four vectors are linearly dependent. Our proof is complete.

We conclude this section with two examples involving simple applications of vectors. The first is a geometrical problem in which vectors determine the location of certain points in space relative to a given origin O. Vectors of this type are called **position vectors**, and it is understood that the initial point of a position vector is located at the origin O. When vectors are localized in this manner the initial point of the vector is always regarded as being at the origin.

The second example is taken from mechanics and involves forces. Experiment shows forces are vector quantities, and that when the lines of action of all the forces acting on a body pass through a single point P (are **concurrent**), their combined effect is equivalent to that of a single force **R**, equal to the vector sum of the forces, whose line of action also passes through P. Here the **line of action** of a force (vector) means the line in $\mathbb{R}^3$ to which its directed line segment belongs. The force **R** itself is called the **resultant** of the forces acting through P.

Example 1. Division of a line in a given ratio

Given the position vectors of points A and B relative to an origin O are $\overrightarrow{OA} = \mathbf{a}$, $\overrightarrow{OB} = \mathbf{b}$, find $\overrightarrow{AB}$. Determine the position vector $\mathbf{r} = \overrightarrow{OR}$ of the point R on the line segment AB which divides it such that the ratio of the lengths $AR/RB = p/q$.

Solution

Vector addition shows (Fig. 2.12) that $\mathbf{a} + \overrightarrow{AB} = \mathbf{b}$, so adding $-\mathbf{a}$ to both sides of the equation gives $\overrightarrow{AB} = \mathbf{b} - \mathbf{a}$.

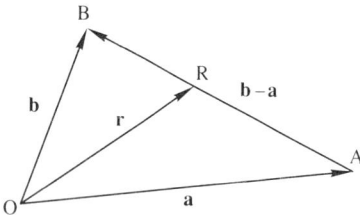

Fig. 2.12   Position vectors **a**, **b** and **r** on line segment AB

The conditions of the problem require

$$\frac{AR}{RB} = \frac{p}{q} \quad \text{or, equivalently,} \quad \frac{AR}{AB-AR} = \frac{p}{q},$$

from which we see

$$AR = \left(\frac{p}{p+q}\right)AB.$$

Since R is a point on the line of the vector $\overrightarrow{AB}$, we deduce from this result that

$$\overrightarrow{AR} = \left(\frac{p}{p+q}\right)\overrightarrow{AB}.$$

Now $\mathbf{r} = \mathbf{a} + \overrightarrow{AR}$, and we have already seen $\overrightarrow{AB} = \mathbf{b} - \mathbf{a}$, so combining results gives

$$\mathbf{r} = \mathbf{a} + \left(\frac{p}{p+q}\right)(\mathbf{b} - \mathbf{a}),$$

or

$$\mathbf{r} = \frac{q\mathbf{a} + p\mathbf{b}}{p+q}. \qquad \blacksquare$$

Example 2. Resultant force

Forces $\mathbf{F}_1$ and $\mathbf{F}_2$ of magnitudes $7N$ and $3N$, respectively, are applied to point P of a body in equilibrium in such a manner that they are directed away from P with their lines of action inclined to one another at an angle $\pi/6$. Find the vector sum of these forces, and hence the force which must be applied to point P if the body is to remain in equilibrium.

Solution

The configuration is shown in Fig. 2.13(a), in which the directed line segments representing the forces have been drawn to scale with $\|\mathbf{F}_1\| = 7$ and $\|\mathbf{F}_2\| = 3$. The vector sum of the concurrent forces acting through P, that is the resultant force $\mathbf{R} = \mathbf{F}_1 + \mathbf{F}_2$, is shown in Fig. 2.13(b).

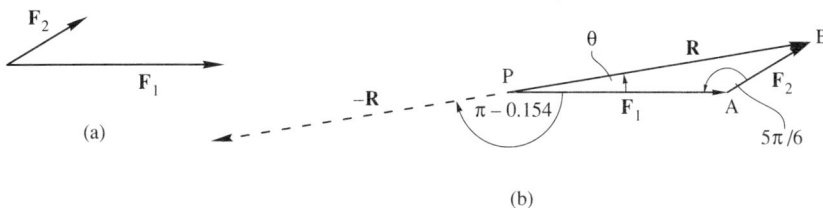

Fig. 2.13 (a) Forces acting on P   (b) Resultant force $\mathbf{R}$ acting on P

Applying the cosine law to triangle PAB gives

$$\|\mathbf{R}\|^2 = \|\mathbf{F}_1\|^2 + \|\mathbf{F}_2\|^2 - 2\|\mathbf{F}_1\|\,\|\mathbf{F}_2\|\,\cos\frac{5\pi}{6},$$

from which we find

$\|\mathbf{R}\| = 9.715\,N$. From the sine law

$$\frac{\|\mathbf{F}_2\|}{\sin\theta} = \frac{\|\mathbf{R}\|}{\sin\dfrac{5\pi}{6}},$$

from which we find $\theta = 0.154$ radians. The resultant force $\mathbf{R}$ acting at P has thus been determined, for $\|\mathbf{R}\|$ and θ are known, and the sense of $\mathbf{R}$ is indicated by the arrow in Fig. 2.13(b).

The concurrent forces $\mathbf{F}_1$ and $\mathbf{F}_2$ acting through P are equivalent to their resultant $\mathbf{R}$ also acting through P. Thus if a further force $-\mathbf{R}$ is applied through P, the total force acting on the body will be $\mathbf{R} + (-\mathbf{R}) = \mathbf{0}$, so the body will remain in equilibrium. The force $-\mathbf{R}$ necessary to maintain equilibrium is shown as the dotted vector in Fig. 2.13(b). ∎

Problems for Section 2.3

1 Using an appropriate scale, draw any two vectors $\mathbf{a}$ and $\mathbf{b}$ of magnitudes 2 and 3, respectively. Construct the vectors $\mathbf{c} = 2\mathbf{a} + \mathbf{b}$ and $\mathbf{d} = 2\mathbf{a} - \mathbf{b}$ and determine their magnitudes by measurement. Use the vectors $\mathbf{c}$ and $\mathbf{d}$ constructed in the first part of the problem to show by further construction and measurement that $\mathbf{c} + \mathbf{d} = 4\mathbf{a}$ and $\mathbf{d} - \mathbf{c} = -2\mathbf{b}$.

2 Using an appropriate scale, draw any three vectors $\mathbf{a}$, $\mathbf{b}$ and $\mathbf{c}$ in a plane with magnitude 2, 5 and 4, respectively. Verify by construction and measurement that if $\mathbf{f} = \mathbf{a} + \mathbf{b}$ and $\mathbf{g} = \mathbf{c} - \mathbf{b}$, then $\mathbf{f} + \mathbf{g} = \mathbf{a} + \mathbf{c}$.

3 If $\mathbf{a}$ and $\mathbf{b}$ are perpendicular, find the magnitudes of the vectors $\mathbf{a} + \mathbf{b}$, $\mathbf{a} - \mathbf{b}$, $\mathbf{a} + 2\mathbf{b}$ and $4\mathbf{a} - 4\mathbf{b}$, in terms of $a = \|\mathbf{a}\|$ and $b = \|\mathbf{b}\|$.

4 Give a geometrical reason why if $\mathbf{a}$ and $\mathbf{b}$ are any two vectors, then

$$\|\mathbf{a} + \mathbf{b}\| \le \|\mathbf{a}\| + \|\mathbf{b}\|.$$

When is the equality possible?

5 Express the following statements in vector notation.
   (i) P is the midpoint of OA, where O is the origin.
   (ii) CD is parallel to AB, three times as long as AB, and oppositely directed to AB.
   (iii) P is the point on the line OA distant n times as far from O as from A, and on the opposite side of O to A, where O is the origin. Also express PA in terms of OA.

6 Show by geometrical arguments that the sum and difference of the vectors $\mathbf{a}$ and $\mathbf{b}$ represented by the sides of an equilateral triangle meeting at a vertex V, and directed away from it, are at right angles to one another.

7 Let $P_1, P_2, \ldots, P_n$ be n points equally spaced around a circle with its center at the origin O. Give a geometrical reason why $\overrightarrow{OP}_1 + \overrightarrow{OP}_2 + \ldots + \overrightarrow{OP}_n = \mathbf{0}$.

8 If $\mathbf{a}$, $\mathbf{b}$ and $\mathbf{c}$ are any three nonzero vectors, no two of which have parallel line segments, show the condition that they are all parallel to a plane is $\mathbf{a} + \mathbf{b} + \mathbf{c} = \mathbf{0}$.

9 Let the diagonals of three adjacent faces of a cube drawn from their common corner represent three vectors whose magnitude equals the length of the diagonal, and let them all be directed away from the corner. Show the sum of these three vectors is a vector along the diagonal of the cube drawn from that corner, and that its magnitude is twice the length of this diagonal.

10 Use the result of Ex. 1 to prove that the medians of a triangle are concurrent at a point P which is two thirds of the distance along each median from its vertex.

The next three problems involve **velocity** which is a vector quantity with a magnitude which is a **speed** (scalar). If the velocity of an object relative to some reference point is $\mathbf{v}$, and the velocity of an observer relative to that same reference point is $\mathbf{u}$, the velocity $\mathbf{v} - \mathbf{u}$ is defined to be the **relative velocity** of the object with respect to the observer.

11 The velocity of a boat relative to the water is 6m/s in a direction due north and the velocity of the water relative to the earth is 2m/s due east. Find the speed and direction of the boat relative to the earth.

12 The velocity of a boat relative to the earth is 5m/s northeast. Relative to the water the boat appears to move northwest at 5m/s. Find the speed and direction of the water.

13 A cyclist travelling east at 5m/s finds a steady wind appears to blow from the north. On doubling his speed he finds it appears to blow from the northeast. Find the speed and direction of the wind.

14 Let the lines of action of n arbitrary forces $\mathbf{F}_1, \mathbf{F}_2, \ldots, \mathbf{F}_n$ in space pass through a point P. Use vector addition to combine these n vectors to form an open **polygon of forces** in space, in such a way that the terminal point of $\mathbf{F}_i$ is placed at the initial point of $\mathbf{F}_{i+1}$, for $i = 1, 2, \ldots, n-1$. Explain why the resultant force $\mathbf{R}$ is equivalent to a vector drawn from the initial point of $\mathbf{F}_1$ to the terminal point of $\mathbf{F}_n$, and that its line of action also passes through P (see Fig. 2.14).

Deduce from the polygon of forces in Fig. 2.14 that:

(a) The system of forces acting on P will be in equilibrium if the polygon is closed (the terminal point of $\mathbf{F}_n$ and the initial point of $\mathbf{F}_1$ coincide;

(b) The additional force acting through P necessary to bring the system of forces into equilibrium is $-\mathbf{R}$.

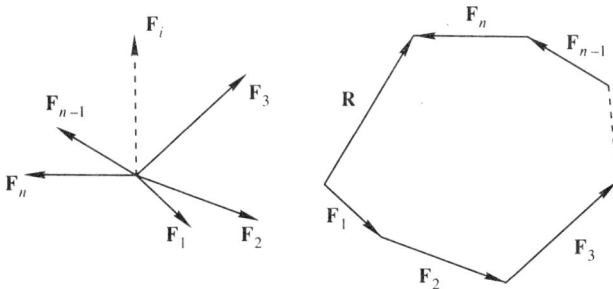

Fig. 2.14   Polygon of forces

2.4   Vectors in component form

A geometrical vector is determined once its magnitude and direction in $\mathbb{R}^3$ have been specified. The analytical determination of such a vector may be accomplished in a

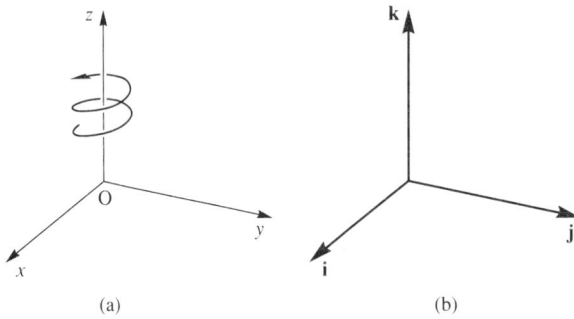

Fig. 2.15   (a) Right-handed system of Cartesian axes $O(x, y, z)$ (b) Triad of unit vectors

particularly simple manner in $\mathbb{R}^3$ if three mutually orthogonal (perpendicular) Cartesian axes $O(x, y, z)$ are introduced, and a unit vector is then associated with each of the three axes. Specifically, three mutually orthogonal unit vectors (a **triad**) $\mathbf{i}, \mathbf{j}$ and $\mathbf{k}$ are introduced which are parallel to and directed in the positive sense along the x-, y- and z-axes, respectively, the axes themselves being oriented as shown in Fig. 2.15.

The caret symbol ^ signifying a unit vector is unnecessary in conjunction with these vectors, since by convention the symbols $\mathbf{i}, \mathbf{j}$ and $\mathbf{k}$ are reserved exclusively for this purpose.

When working with rectangular Cartesian axes we shall always assume they form a right-handed set. That is, they are chosen such that the positive sense along the z-axis is determined by the way in which a right-handed screw would advance were it to be rotated about the z-axis from the x-direction to the y-direction. This is illustrated in Fig. 2.15. In a left-handed set of axes a left-handed screw would replace the right-handed screw, so the sense of the z-axis would be reversed. Left-handed sets will *not* be employed in what follows.

By means of vector addition and the scaling of vectors it is possible to represent any geometrical vector $\mathbf{v}$ in $\mathbb{R}^3$ in terms of $\mathbf{i}, \mathbf{j}$ and $\mathbf{k}$ as the vector sum

$$\mathbf{v} = v_1\mathbf{i} + v_2\mathbf{j} + v_3\mathbf{k}. \tag{1}$$

Result (1) involves using $\mathbf{i}, \mathbf{j}$ and $\mathbf{k}$ as a basis for the vector space of geometrical vectors in $\mathbb{R}^3$. This is justified because $\mathbf{i}, \mathbf{j}$ and $\mathbf{k}$ are linearly independent. When expressed in the form shown in (1), the scalars v_1, v_2 and v_3 are called the x, y and z-**components** of $\mathbf{v}$, respectively. It is usual to write the components of $\mathbf{v}$ as the ordered triple (v_1, v_2, v_3). As a vector is completely determined once its components are known we shall consider $\mathbf{v}$ and the ordered triple (v_1, v_2, v_3) to be interchangeable.

Since vectors are invariant under a translation, the initial point of $\mathbf{v}$ may be located at the origin O, when (v_1, v_2, v_3) become the Cartesian coordinates of the terminal point P of $\mathbf{v}$. This interpretation of $\mathbf{v}$ as a position vector is illustrated in Fig. 2.16 for the vector with components $(3, -2, 4)$. It follows directly from Pythagoras's theorem that

$$\|\mathbf{v}\| = \|\overrightarrow{OP}\| = (3^2 + (-2)^2 + 4^2)^{1/2} = \sqrt{29}$$

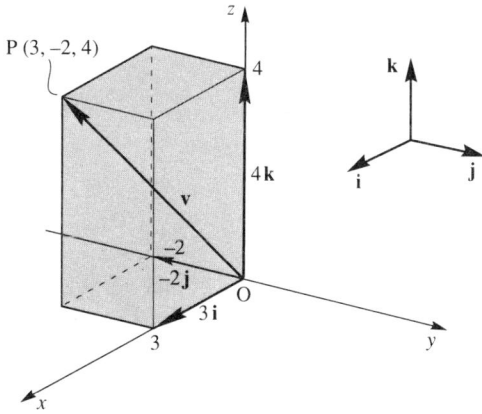

Fig. 2.16 Vector $\mathbf{v} = 3\mathbf{i} - 2\mathbf{j} + 4\mathbf{k}$ with components $(3, -2, 4)$

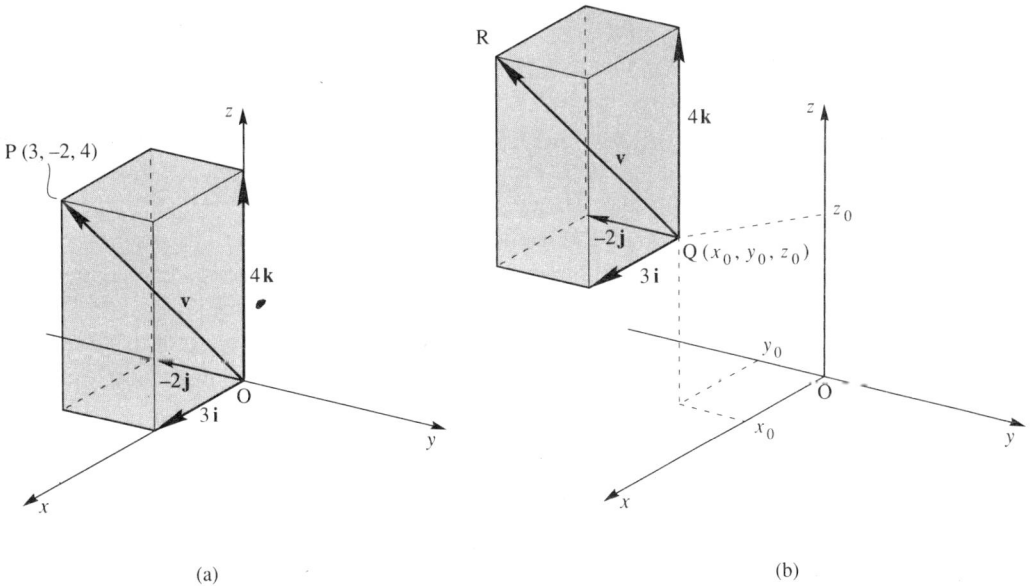

(a)

(b)

Fig. 2.17 (a) Initial point of $\mathbf{v} = 3\mathbf{i} - 2\mathbf{j} + 4\mathbf{k}$ located at origin O. (b) Initial point of $\mathbf{v} = 3\mathbf{i} - 2\mathbf{j} + 4\mathbf{k}$ located at (x_0, y_0, z_0)

If the vector $\mathbf{v}$ with components $(3, -2, 4)$ is translated so its initial point lies at the arbitrary point Q with coordinates (x_0, y_0, z_0), its terminal point will then lie at the point R with the coordinates $(x_0 + 3, y_0 - 2, z_0 + 4)$, as shown in the right-hand diagram in Fig. 2.17. The direction of $\mathbf{v}$ and its magnitude remain the same as they were in the left-hand diagram in Fig. 2.17. The algebraic proof of the invariance of the length follows from the fact that we have already seen $\| \overrightarrow{OP} \| = \sqrt{29}$, while shifting the origin to Q we have from

Pythagoras' theorem that

$$\|\overrightarrow{QR}\| = \{[(x_0+3)-x_0]^2 + [(y_0-2)-y_0]^2 + [(z_0+4)-z_0]^2\}^{1/2} = \sqrt{29}.$$

This example demonstrates that if a general vector has its initial point P_1 at (x_1, y_1, z_1), and its terminal point P_2 at (x_2, y_2, z_2), then its components (v_1, v_2, v_3) must be defined to be

$$v_1 = x_2 - x_1, \quad v_2 = y_2 - y_1, \quad v_3 = z_2 - z_1. \tag{2}$$

Translating $\mathbf{v}$ is equivalent to moving its initial and terminal points so that the components defined in (2) remain invariant. Shifting the origin to P_1 in Fig. 2.17 and using Pythagoras' theorem gives the general result

$$\|\mathbf{v}\| = (v_1^2 + v_2^2 + v_3^2)^{1/2}. \tag{3}$$

In terms of the coordinates of the initial and terminal points of $\mathbf{v}$, the magnitude $\|\mathbf{v}\|$ in (3) becomes

$$\|\mathbf{v}\| = [(x_2-x_1)^2 + (y_2-y_1)^2 + (z_2-z_1)^2]^{1/2}. \tag{4}$$

Notice that in (2) each component of $\mathbf{v}$ is obtained by subtracting the appropriate coordinate of the initial point of the vector from the corresponding coordinate of its terminal point. Thus a component will be positive if when viewed from its coordinate axis the vector points in the direction in which that coordinate increases; otherwise the component will be negative. This may be seen clearly in Fig. 2.16 representing the vector $\mathbf{v}$ with components $(3, -2, 4)$ in which the second component is negative. The geometrical interpretation of the components of the vector $\mathbf{v}$ represented by the directed line segment $\overrightarrow{PQ}$ is shown in Fig. 2.18.

In terms of components, the **zero** or **null vector 0** is defined to be the vector $(0, 0, 0)$ in which every component is zero. This then agrees with the previous definition of $\mathbf{0}$, for it has zero magnitude and its direction is unspecified.

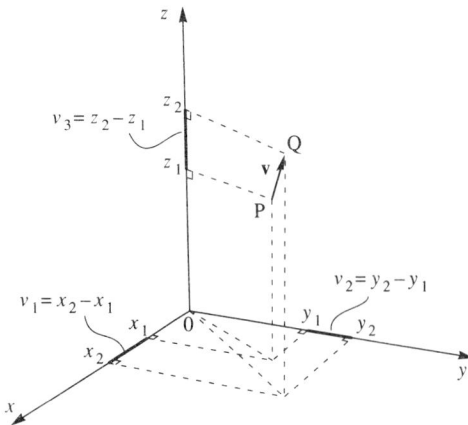

Fig. 2.18   Geometrical interpretation of the components of $\mathbf{v} = \overrightarrow{PQ}$

By now it should be clear that any three vectors **a**, **b** and **c** may be used as a basis in $\mathbb{R}^3$, provided only that when taken pairwise their line segments lie in three non-parallel planes. If such an arbitrary nonorthogonal basis is used, the components of **v** are defined as the scalar coefficients v_1, v_2 and v_3 of **a**, **b** and **c** in the vector representation

$$\mathbf{v} = v_1\mathbf{a} + v_2\mathbf{b} + v_3\mathbf{c},$$

as illustrated in Fig. 2.19.

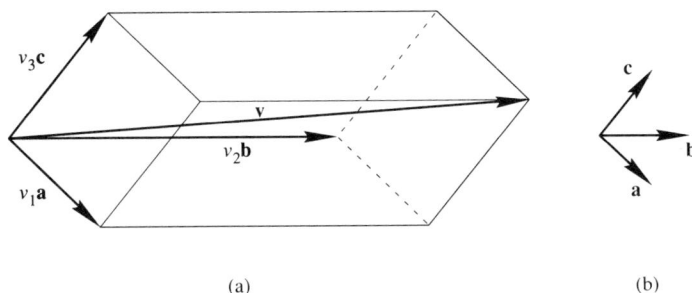

(a)   (b)

Fig. 2.19 (a) Vector $\mathbf{v} = v_1\mathbf{a} + v_2\mathbf{b} + v_3\mathbf{c}$ (b) Nonorthogonal basis vectors

However if **a**, **b** and **c** are not mutually orthogonal the task of finding $\|\mathbf{v}\|$ and angles between lines becomes much harder. It is for these and related reasons that hereafter we shall confine attention to a basis of mutually orthogonal unit vectors.

Equality of vectors

The invariance of a vector under a translation (parallel displacement) means that its components remain unchanged. Thus, if

$$\mathbf{u} = u_1\mathbf{i} + u_2\mathbf{j} + u_3\mathbf{k}, \quad \mathbf{v} = v_1i + v_2j + v_3k \quad \text{and} \quad \mathbf{u} = \mathbf{v},$$

then

$$u_1 = v_1, \; u_2 = v_2, \; u_3 = v_3. \tag{5}$$

Negative of a vector

By definition, the negative of the vector **u**, written $-\mathbf{u}$ is obtained by reversing the sense of **u**. When expressed in component form this involves reversing the sign of each component. Thus if $\mathbf{u} = u_1\mathbf{i} + u_2\mathbf{j} + u_3\mathbf{k}$, then

$$-\mathbf{u} = -u_1\mathbf{i} - u_2\mathbf{j} - u_3\mathbf{k}. \tag{6}$$

Vector addition and subtraction

The triangle law for the addition of the vectors $\mathbf{u}$ and $\mathbf{v}$ is readily interpreted in terms of components. If $\mathbf{u} = u_1\mathbf{i} + u_2\mathbf{j} + u_3\mathbf{k}$, $\mathbf{v} = v_1\mathbf{i} + v_2\mathbf{j} + v_3\mathbf{k}$, then

$$\mathbf{u} + \mathbf{v} = u_1\mathbf{i} + u_2\mathbf{j} + u_3\mathbf{k} + v_1\mathbf{i} + v_2\mathbf{j} + v_2\mathbf{k} = (u_1 + v_1)\mathbf{i} + (u_2 + v_2)\mathbf{j} + (u_3 + v_3)\mathbf{k}. \tag{7}$$

When vectors are defined in component form the **commutative** property of vector addition $\mathbf{u} + \mathbf{v} = \mathbf{v} + \mathbf{u}$ is a consequence of the fact that the addition of real numbers is commutative. The **associative** property

$$(\mathbf{u} + \mathbf{v}) + \mathbf{w} = \mathbf{u} + (\mathbf{v} + \mathbf{w}) \tag{8}$$

for the addition of the three vectors $\mathbf{u}$, $\mathbf{v}$ and $\mathbf{w}$ when defined in component form follows in similar fashion.

The vector subtraction $\mathbf{u} - \mathbf{v}$ is defined as the vector addition $\mathbf{u} + (-\mathbf{v})$, so that when expressed in component form it follows from (6) and (7) that

$$\mathbf{u} - \mathbf{v} = (u_1 - v_1)\mathbf{i} + (u_2 - v_2)\mathbf{j} + (u_3 - v_3)\mathbf{k}. \tag{9}$$

Scaling a vector

We know from Sec. 2.3 that multiplying a vector $\mathbf{u}$ by a real scalar multiplier λ (**scaling** it by λ) changes its length by $|\lambda|$, and that its sense is unchanged if $\lambda > 0$, but is reversed if $\lambda < 0$. It follows from this that in component form such scaling merely involves multiplying each component by λ. Thus for any real λ, if $\mathbf{u} = u_1\mathbf{i} + u_2\mathbf{j} + u_3\mathbf{k}$, we have

$$\lambda\mathbf{u} = \lambda u_1\mathbf{i} + \lambda u_2\mathbf{j} + \lambda u_3\mathbf{k}. \tag{10}$$

Result (3) shows, as required, that

$$\|\lambda\mathbf{u}\| = (\lambda^2 u_1^2 + \lambda^2 u_2^2 + \lambda^2 u_3^2)^{1/2}$$

$$= |\lambda| \, \|\mathbf{u}\|. \tag{11}$$

As a special case of scaling we remark that the negative of a vector involves scaling by $\lambda = -1$. Setting $\lambda = -1$ in (11) then gives the familiar result

$$\| -\mathbf{u} \| = \| \mathbf{u} \|.$$

Unit vectors

If $\mathbf{u} = u_1\mathbf{i} + u_2\mathbf{j} + u_3\mathbf{k}$ is an arbitrary vector, it follows directly from (4), Sec. 2.3, that in component form the unit vector $\hat{\mathbf{u}}$ in the direction of $\mathbf{u}$ is

$$\hat{\mathbf{u}} = \frac{1}{\|\mathbf{u}\|}\mathbf{u} = \frac{u_1\mathbf{i} + u_2\mathbf{j} + u_3\mathbf{k}}{(u_1^2 + u_2^2 + u_3^2)^{1/2}}. \tag{12}$$

Since $\hat{\mathbf{u}}$ is scaled by $1/\|\mathbf{u}\|$ in (12), this may be written

$$\hat{\mathbf{u}} = \frac{u_1}{\|\mathbf{u}\|}\mathbf{i} + \frac{u_2}{\|\mathbf{u}\|}\mathbf{j} + \frac{u_3}{\|\mathbf{u}\|}\mathbf{k}. \tag{13}$$

Cylindrical and spherical polar coordinates

Two other orthogonal coordinate systems which are frequently used are those called **cylindrical** and **spherical polar coordinates**. These are illustrated in Figs 2.20 and 2.21, together with the associated triads of unit vectors. The positive sense in which an angle is to be measured is indicated by an arrow. It should be noticed that the triads of unit vectors used in these coordinate systems depend on the point in question, and are not the same for all points in space as is the basis $\mathbf{i}$, $\mathbf{j}$, and $\mathbf{k}$ used in Cartesian coordinates.

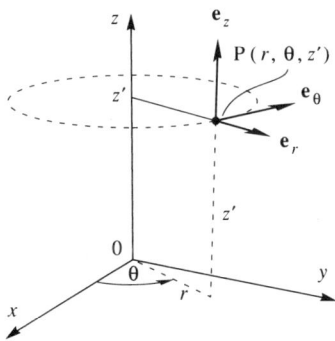

Fig. 2.20 Cylindrical polar coordinates

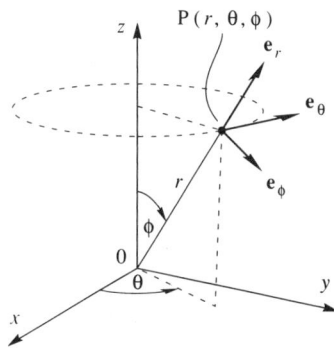

Fig. 2.21 Spherical polar coordinates

In the cylindrical polar coordinates shown in Fig. 2.20 the unit vector $\mathbf{e}_r$ is drawn parallel to the (x, y)-plane in the direction of increasing r, $\mathbf{e}_\theta$ parallel to the (x, y)-plane in the direction of increasing *azimuthal angle* θ and $\mathbf{e}_z$ parallel to the z-axis in the direction of increasing z. It is easily established that if P has the Cartesian coordinates (x, y, z), then its corresponding cylindrical polar coordinates (r, θ, z') are related to x, y and z by

$$x = r\cos\theta, \quad y = r\sin\theta, \quad z = z'. \tag{14}$$

Solving for r, θ and z', these are seen to be equivalent to

$$r = (x^2 + y^2)^{1/2}, \quad \theta = \arctan\frac{y}{x}, \quad z' = z, \tag{15}$$

where $0 \le r < \infty$.

Although the coordinate z' in cylindrical polars is the same as the coordinate z in Cartesian coordinates, it is important to distinguish between the two when such systems are used in vector field theory.

The quadrant in which the azimuthal angle θ is located must be determined from the expression in (15) taken in conjunction with the signs of x and y.[1] Since θ is indeterminate up to a multiple of 2π it is customary to restrict it to $-\pi < \theta \le \pi$.

In the spherical polar coordinates shown in Fig. 2.21 the unit vector e_r is drawn in the OP direction in the sense of increasing r, e_θ is drawn tangential to the dotted circle parallel to the (x, y)-plane in the sense of increasing θ, and e_ϕ lies in the plane containing OP and the z-axis, at right angles to OP and drawn in the sense of increasing ϕ. If point P has the Cartesian coordinates (x, y, z), then its corresponding polar coordinates (r, θ, ϕ) are related to x, y and z by

$$x = r\cos\theta\sin\phi, \quad y = r\sin\theta\sin\phi, \quad z = r\cos\phi. \tag{16}$$

Solving for r, θ and ϕ, these are seen to be equivalent to

$$r = (x^2 + y^2 + z^2)^{1/2}, \quad \theta = \arctan\frac{y}{x}, \quad \phi = \arccos\frac{z}{(x^2 + y^2 + z^2)^{1/2}}, \tag{17}$$

where $0 \le r < \infty$.

The quadrants in which θ and ϕ are located must be determined from the expressions in (17) taken in conjunction with the signs of x, y and z. It is now necessary that both θ and ϕ be restricted to avoid ambiguity, so we shall require that $-\pi < \theta \le \pi$ and $0 \le \phi < \pi$.

Example 1. Cartesian coordinates

Let the vectors **a**, **b** and **c** be defined as

$$\mathbf{a} = \mathbf{i} + 2\mathbf{j} - \mathbf{k}, \ \mathbf{b} = 2\mathbf{i} - 3\mathbf{j} + \mathbf{k}, \ \mathbf{c} = -\mathbf{i} - 3\mathbf{k}.$$

Then

$$\|\mathbf{a}\| = \sqrt{6}, \ \|\mathbf{b}\| = \sqrt{14}, \ \|\mathbf{c}\| = \sqrt{10},$$

$$2\mathbf{a} = 2\mathbf{i} + 4\mathbf{j} - 2\mathbf{k}, \ 0.5\mathbf{b} = \mathbf{i} - 1.5\mathbf{j} + 0.5\mathbf{k}, \ 3\mathbf{c} = -3\mathbf{i} - 9\mathbf{k}.$$

$$2\mathbf{a} + 3\mathbf{c} = -\mathbf{i} + 4\mathbf{j} - 11\mathbf{k}, \ 2\mathbf{a} - 3\mathbf{c} = 5\mathbf{i} + 4\mathbf{j} + 7\mathbf{k},$$

$$\hat{\mathbf{a}} = \frac{1}{\sqrt{6}}(\mathbf{i} + 2\mathbf{j} - \mathbf{k}) = \frac{1}{\sqrt{6}}\mathbf{i} + \frac{2}{\sqrt{6}}\mathbf{j} - \frac{1}{\sqrt{6}}\mathbf{k},$$

$$\hat{\mathbf{b}} = \frac{1}{\sqrt{14}}(2\mathbf{i} - 3\mathbf{j} + \mathbf{k}) = \frac{2}{\sqrt{14}}\mathbf{i} - \frac{3}{\sqrt{14}}\mathbf{j} + \frac{1}{\sqrt{14}}\mathbf{k},$$

$$\hat{\mathbf{c}} = \frac{1}{\sqrt{10}}(-\mathbf{i} - 3\mathbf{k}) = -\frac{1}{\sqrt{10}}\mathbf{i} - \frac{3}{\sqrt{10}}\mathbf{k}.$$

■

---

[1] An alternative definition of θ which takes explicit account of the signs of x and y is obtained by requiring that θ satisfies both of the following expressions

$$\theta = \arcsin\frac{y}{(x^2 + y^2)^{1/2}} \quad \text{and} \quad \theta = \arccos\frac{x}{(x^2 + y^2)^{1/2}},$$

where again $-\pi < \theta \le \pi$.

Example 2. Cylindrical and spherical coordinates

If a point P has the Cartesian coordinates $\left(\dfrac{-7}{2}, \dfrac{-7\sqrt{3}}{2}, \sqrt{15}\right)$, find (a) its cylindrical polar coordinates and (b) its spherical polar coordinates.

Solution

(a) $r = \left[\left(\dfrac{-7}{2}\right)^2 + \left(\dfrac{-7\sqrt{3}}{2}\right)^2\right]^{1/2} = 7,$

$\theta = \arctan\left[\left(\dfrac{-7\sqrt{3}}{2}\right)\bigg/\left(\dfrac{-7}{2}\right)\right] = -2\pi/3(-120°),$

$z' = \sqrt{15}.$

(b) $r = \left[\left(\dfrac{-7}{2}\right)^2 + \left(\dfrac{-7\sqrt{3}}{2}\right)^2 + \left(\sqrt{15}\right)^2\right]^{1/2} = 8,$

$\theta = \arctan\left[\left(\dfrac{-7\sqrt{3}}{2}\right)\bigg/\left(\dfrac{-7}{2}\right)\right] = -2\pi/3\ (-120°),$

$\phi = \arctan\left(\dfrac{\sqrt{15}}{8}\right) = 0.339$ radians (approx 61°).

■

Problems for Section 2.4

1 Find the components and magnitude of each of the following vectors with the given initial point P and terminal point Q.
  (a) P(1, 0, −3), Q(2, 1, 7)
  (b) P(2, −1, 2), Q(4, −1, 6)
  (c) P(−1, −2, −6), Q(1, 4, 3)
  (d) P(3, 7, −4), Q(3, 7, −5).

2 Find either the initial point P or the terminal point Q of each of the following vectors $\overrightarrow{PQ}$ using the given end point of the vector
  (a) P(1, 2, 1), $\overrightarrow{PQ} = 2\mathbf{i} - \mathbf{j} + \mathbf{k}$
  (b) P(−1, 3, −6), $\overrightarrow{PQ} = -\mathbf{i} + \mathbf{j} + 3\mathbf{k}$
  (c) Q(2, 1, 6), $\overrightarrow{PQ} = \mathbf{i} + \mathbf{j} + \mathbf{k}$
  (d) Q(3, 1, 4), $\overrightarrow{PQ} = \mathbf{i} - 4\mathbf{j} - 4\mathbf{k}.$

If $\mathbf{a} = \mathbf{i} + 2\mathbf{j} + 3\mathbf{k}$, $\mathbf{b} = -\mathbf{i} + \mathbf{j}$ and $\mathbf{c} = 3\mathbf{i} + \mathbf{j} + 2\mathbf{k}$ and $\mathbf{d} = -2\mathbf{i} + 6\mathbf{j} - \mathbf{k}$, find the following expressions:
3 $\|\mathbf{a}\|, \|\mathbf{b}\|, \|\mathbf{c}\|, \|\mathbf{d}\|$ **4** $4\mathbf{a}, \tfrac{1}{3}\mathbf{c}, -2\mathbf{d}$ **5** $\mathbf{a} + \mathbf{b} + \mathbf{c}, \mathbf{c} + \mathbf{a} + \mathbf{b}$
6 $\mathbf{a} + 2\mathbf{b}, \mathbf{a} - 2\mathbf{b}$ **7** $\mathbf{a} + 2\mathbf{b} - \mathbf{c} + \mathbf{d}, 2\mathbf{a} + \mathbf{b} + \mathbf{c} - 2\mathbf{d}, \mathbf{b} + \mathbf{c} + 2(\mathbf{a} - \mathbf{d})$

8 $\|\mathbf{a}+\mathbf{b}\|$, $\|\mathbf{a}\|+\|\mathbf{b}\|$, $\|\mathbf{b}\|-\|\mathbf{a}\|$, $|\|\mathbf{b}\|-\|\mathbf{a}\||$, $\|\mathbf{a}-\mathbf{c}\|$, $\|2(\mathbf{a}+\mathbf{b})\|$, $2\|\mathbf{a}+\mathbf{b}\|$,

9 $\hat{\mathbf{a}}$, $\hat{\mathbf{b}}$, $\hat{\mathbf{c}}$, $\hat{\mathbf{d}}$.

10 Solve the following vector equations for the unknown scalars, a, b and c given
   (i) $3\mathbf{i}+b\mathbf{j}+3\mathbf{k}=a\mathbf{i}+7\mathbf{j}-6c\mathbf{k}$
   (ii) $a\mathbf{i}+2\mathbf{j}+c\mathbf{k}=3\mathbf{i}+2\mathbf{j}-7\mathbf{k}$
   (iii) $a\mathbf{i}+3\mathbf{j}+c\mathbf{k}=2\mathbf{i}+b\mathbf{j}+2c\mathbf{k}$
   (iv) $7\mathbf{i}+b\mathbf{j}+3\mathbf{k}=a\mathbf{i}+c\mathbf{j}+3\mathbf{k}$.

Let a unit vector represent a force of unit magnitude. What are the magnitudes of the forces represented by the following vectors
11 $2\mathbf{i}-\mathbf{j}+\mathbf{k}$, $-\mathbf{i}-3\mathbf{j}-\mathbf{k}$, $\mathbf{i}+\mathbf{j}+\mathbf{k}$?
12 What vectors represent a force of:
   (a) magnitude 9 in the direction of the vector $\mathbf{i}+2\mathbf{j}+\mathbf{k}$
   (b) magnitude 3 in the opposite direction to the vector $-\mathbf{i}+\mathbf{j}+3\mathbf{k}$?

Let a unit vector represent a velocity of 1 km/hour and take the x- and y-axes such that they point due east and due north, respectively.
13 What vector represents a velocity:
   (a) 4 km/hour due south,
   (b) 9 km/hour in a northeasterly direction,
   (c) 5 km/hour 30° south of east?
14 A bus moves northwest at 40 km/hour and a cyclist moves due east at 20 km/hour.
   (a) What is the speed and direction of the cyclist relative to the bus?
   (b) What is the speed and direction of the bus relative to the cyclist?
15 Find the resultant of a force of magnitude 6 units in the direction $\mathbf{i}+3\mathbf{j}-2\mathbf{k}$ and a force of magnitude 3 units in the direction $2\mathbf{i}-\mathbf{j}+3\mathbf{k}$. What is its magnitude?
16 A rectangular parallelepiped is such that a corner is located at the origin with the three edges which meet at the origin lying along the x-, y- and z-axes. If the lengths of the edges along these respective axes are 3, 2 and 3 find in terms of the unit vectors $\mathbf{i}$, $\mathbf{j}$ and $\mathbf{k}$ the vector forming the diagonal drawn from the origin to the opposite corner of the parallelepiped. If the lines of action forces of magnitudes 3, 2 and 3 units act through the origin in the $\mathbf{i}$, $\mathbf{j}$ and $\mathbf{k}$ directions, respectively, what is the magnitude of the resultant force?
17 Prove by means of vectors that the sum of the vectors drawn from the center of a square to each of its corners is the null vector.

Find (a) the cylindrical polar coordinates and (b) the spherical polar coordinates of the point with Cartesian coordinates:
18 $(1, 1, 1)$ 19 $(1, -1, -2)$ 20 $(-1, -1, -1)$
21 Find the spherical polar coordinates of the point with the cylindrical polar coordinates
$$\left(2, \frac{\pi}{3}, 3\right).$$

22 Find the cylindrical polar coordinates of the point with the spherical polar coordinates
$$\left(6, \frac{-\pi}{6}, \frac{\pi}{3}\right).$$

2.5   Scalar product (dot product)

The idea of the projection of the length (magnitude) of a vector in $\mathbb{R}^3$ onto a given line leads naturally to its generalization in the form of the **scalar product** of two vectors, also called the **inner product** or **dot product** of the two vectors. If **a** and **b** are any two vectors, their **scalar product**, written **a**·**b**, is defined as

$$\mathbf{a}\cdot\mathbf{b}=\begin{cases}\|\mathbf{a}\|\,\|\mathbf{b}\|\cos\theta & \text{when } \mathbf{a}\neq\mathbf{0},\ \mathbf{b}\neq\mathbf{0},\\ 0 & \text{when } \mathbf{a}=\mathbf{0} \text{ or } \mathbf{b}=\mathbf{0},\end{cases} \tag{1}$$

where θ, measured counterclockwise and chosen so $0\leq\theta\leq\pi$, is the angle between the lines of action of the two vectors when their initial points are brought into coincidence by a translation. Here the dot between **a** and **b** in (1) signifies that a scalar product is involved and leads to the alternative term *dot product* for (1). The angle θ between typical pairs of vectors is illustrated in Fig. 2.22.

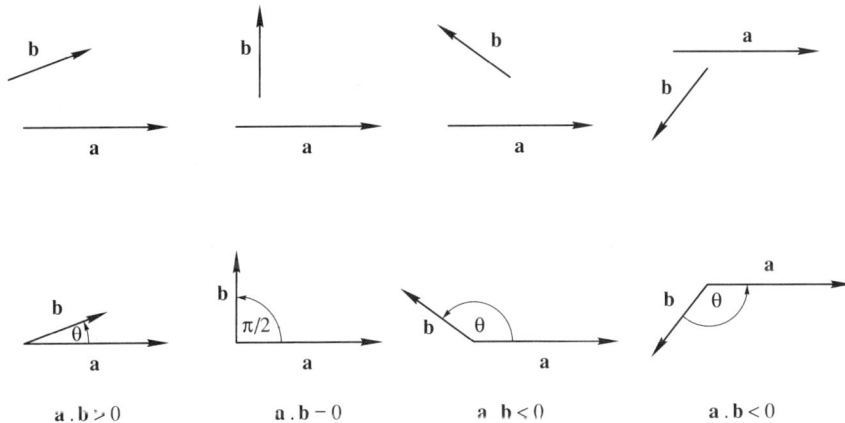

Fig. 2.22   Angle between two vectors

By definition, the scalar product of two vectors is a real number (a scalar) which may be positive, negative or zero. It is for this reason that the name 'scalar product' has been given to this type of product, and it should not be confused with the scalar multiple of a vector introduced in Sec. 2.3, which is a vector. The scalar product will be negative when the angle θ between the vectors is obtuse.

Now suppose that **a** and **b** are nonzero vectors. Then it follows from (1) and the condition $0\leq\theta\leq\pi$ that $\mathbf{a}\cdot\mathbf{b}=0$ if, and only if, $\theta=\pi/2$. That is, $\mathbf{a}\cdot\mathbf{b}=0$ if the vectors **a** and **b** are orthogonal. This result is of sufficient importance for it to be stated formally.

Theorem 2.1 (Orthogonality of vectors)

Two nonzero vectors **a** and **b** will be orthogonal if, and only, if $\mathbf{a}\cdot\mathbf{b}=0$.   □

The following useful results are all direct consequences of the definition given in (1). If we set $\mathbf{b}=\mathbf{a}$, then $\mathbf{a}\cdot\mathbf{a}=\|\mathbf{a}\|^2$. Thus the magnitude $\|\mathbf{a}\|$ of the vector **a** is given in terms of the scalar product as

$$\|\mathbf{a}\| = \sqrt{\mathbf{a} \cdot \mathbf{a}} \geq 0. \tag{2}$$

We conclude that

$$\mathbf{a} \cdot \mathbf{a} = 0 \text{ if, and only if, } \mathbf{a} = \mathbf{0}. \tag{3}$$

If λ, μ are any two nonzero real numbers, we find
Multiplication by a scalar is associative

$$\lambda(\mathbf{a}) \cdot (\mu \mathbf{b}) = \lambda \mu \mathbf{a} \cdot \mathbf{b}. \tag{4}$$

Setting $\lambda = 1/\|\mathbf{a}\|$, $\mu = 1/\|\mathbf{b}\|$ and using (1), (2) and (4) we obtain

$$\cos\theta = \frac{\mathbf{a} \cdot \mathbf{b}}{\|\mathbf{a}\| \|\mathbf{b}\|} = \frac{\mathbf{a} \cdot \mathbf{b}}{\sqrt{\mathbf{a} \cdot \mathbf{a}} \sqrt{\mathbf{b} \cdot \mathbf{b}}} = \hat{\mathbf{a}} \cdot \hat{\mathbf{b}}. \tag{5}$$

This result is useful for finding the angle θ between two vectors.
We also have the result

$$\mathbf{a} \cdot \mathbf{b} = \|\mathbf{a}\| \|\mathbf{b}\| \cos\theta = \|\mathbf{b}\| \|\mathbf{a}\| \cos\theta = \mathbf{b} \cdot \mathbf{a},$$

showing the
Commutativity of a scalar product

$$\mathbf{a} \cdot \mathbf{b} = \mathbf{b} \cdot \mathbf{a}. \tag{6}$$

Now let λ, μ be any two real numbers and consider the behavior of the scalar product with respect to vector addition, where $\mathbf{a}$, $\mathbf{b}$ and $\mathbf{c}$ are arbitrary vectors in $\mathbb{R}^3$. We then find, with respect to vector addition, the
Linearity of the scalar product

$$(\lambda \mathbf{a} + \mu \mathbf{b}) \cdot \mathbf{c} = \lambda \mathbf{a} \cdot \mathbf{c} + \mu \mathbf{b} \cdot \mathbf{c}, \tag{7}$$

and with $\lambda = \mu = 1$, the
Distributivity of the scalar product

$$(\mathbf{a} + \mathbf{b}) \cdot \mathbf{c} = \mathbf{a} \cdot \mathbf{c} + \mathbf{b} \cdot \mathbf{c}. \tag{8}$$

Taking the absolute value of (1) gives

$$|\mathbf{a} \cdot \mathbf{b}| = \|\mathbf{a}\| \|\mathbf{b}\| |\cos\theta|,$$

but since $|\cos\theta| \leq 1$ we arrive at the **Schwarz inequality**

$$|\mathbf{a} \cdot \mathbf{b}| \leq \|\mathbf{a}\| \|\mathbf{b}\|. \tag{9}$$

This result may be used to derive a further inequality involving the magnitudes of vectors. From (2), (8) with $\mathbf{c} = \mathbf{a} + \mathbf{b}$ and (6) we have

$$\|\mathbf{a} + \mathbf{b}\|^2 = (\mathbf{a} + \mathbf{b}) \cdot (\mathbf{a} + \mathbf{b}) = \mathbf{a} \cdot \mathbf{a} + 2\mathbf{a} \cdot \mathbf{b} + \mathbf{b} \cdot \mathbf{b}$$
$$= \|\mathbf{a}\|^2 + 2\mathbf{a} \cdot \mathbf{b} + \|\mathbf{b}\|^2.$$

However, since $\mathbf{a}\cdot\mathbf{b}\leq|\mathbf{a}\cdot\mathbf{b}|$, using (9) this leads to the inequality

$$\|\mathbf{a}+\mathbf{b}\|^2 \leq \|\mathbf{a}\|^2 + 2\|\mathbf{a}\|\,\|\mathbf{b}\| + \|\mathbf{b}\|^2,$$

and so, by taking the positive square root, to the **triangle inequality**

$$\|\mathbf{a}+\mathbf{b}\| \leq \|\mathbf{a}\| + \|\mathbf{b}\|. \tag{10}$$

The reason for this name follows from the triangle law for the vector addition of $\mathbf{a}$ and $\mathbf{b}$. This same inequality has already been encountered with complex numbers (Sec. 1.6) which may be regarded as two-dimensional vectors. It is a familiar result from euclidean geometry that the sum of the lengths of any two sides of a triangle ($\|\mathbf{a}\| + \|\mathbf{b}\|$) is greater than, or equal to, the length of the third side ($\|\mathbf{a}+\mathbf{b}\|$). Equality only occurs when the vertices are collinear, that is when the line segments representing $\mathbf{a}$ and $\mathbf{b}$ are parallel. The general situation is illustrated in Fig. 2.23.

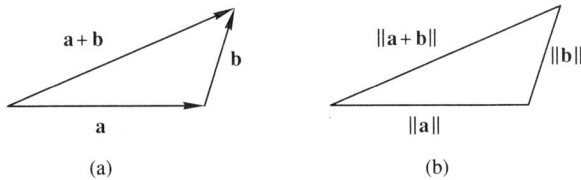

Fig. 2.23 (a) Triangle law for addition (b) Triangle inequality $\|\mathbf{a}+\mathbf{b}\| \leq \|\mathbf{a}\| + \|\mathbf{b}\|$

The scalar product $\mathbf{a}\cdot\mathbf{b}$ takes on a simple and convenient form when the vectors $\mathbf{a}$ and $\mathbf{b}$ are expressed in terms of a basis comprising orthogonal unit vectors. Notice first that if we work with the unit vectors $\mathbf{i}$, $\mathbf{j}$ and $\mathbf{k}$, it follows from (1) that

$$\mathbf{i}\cdot\mathbf{i}=\mathbf{j}\cdot\mathbf{j}=\mathbf{k}\cdot\mathbf{k}=1, \tag{11}$$

but that

$$\mathbf{i}\cdot\mathbf{j}=\mathbf{j}\cdot\mathbf{i}=0, \quad \mathbf{j}\cdot\mathbf{k}=\mathbf{k}\cdot\mathbf{j}=0, \quad \mathbf{k}\cdot\mathbf{i}=\mathbf{i}\cdot\mathbf{k}=0. \tag{12}$$

Then if the vectors $\mathbf{a}$ and $\mathbf{b}$ are written in the component form

$$\mathbf{a}=a_1\mathbf{i}+a_2\mathbf{j}+a_3\mathbf{k} \quad \text{and} \quad \mathbf{b}=b_1\mathbf{i}+b_2\mathbf{j}+b_3\mathbf{k}, \tag{13}$$

after expanding $\mathbf{a}\cdot\mathbf{b}$ term by term and using (11) and (12) we find the basic result

$$\mathbf{a}\cdot\mathbf{b}=a_1 b_1 + a_2 b_2 + a_3 b_3. \tag{14}$$

To show the scalar product is a generalization of the notion of a projection, consider the scalar product $\mathbf{a}\cdot\hat{\mathbf{b}}$, where $\mathbf{a}, \hat{\mathbf{b}}$ are vectors in $\mathbb{R}^3$. Then, since $\|\hat{\mathbf{b}}\|=1$, it follows from the definition of a scalar product that

$$\mathbf{a}\cdot\hat{\mathbf{b}}=\|\mathbf{a}\|\,\|\hat{\mathbf{b}}\|\cos\theta = \|\mathbf{a}\|\cos\theta, \tag{15}$$

where θ is the angle between $\mathbf{a}$ and $\hat{\mathbf{b}}$. The two representative diagrams in Fig. 2.24 show that $\mathbf{a}\cdot\hat{\mathbf{b}}$ may be regarded as the component of $\mathbf{a}$ in the direction of $\hat{\mathbf{b}}$. This component will

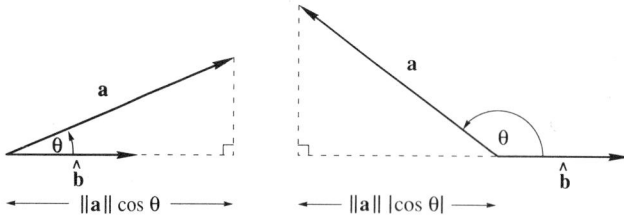

Fig. 2.24

be positive when θ is acute, and negative when θ is obtuse. Thus the scalar product $\mathbf{a} \cdot \hat{\mathbf{b}}$ is the length of the projection of a line of length $\|\mathbf{a}\|$ onto a line parallel to $\hat{\mathbf{b}}$ multiplied by ± 1, according as θ is acute or obtuse. If only the length of the projection of $\mathbf{a}$ in the direction of $\hat{\mathbf{b}}$ is required it is necessary to compute $|\mathbf{a} \cdot \hat{\mathbf{b}}|$. As an arbitrary vector $\mathbf{b}$ and its associated unit vector $\hat{\mathbf{b}}$ both have the same direction, it follows directly that the length of the projection of $\mathbf{a}$ in the direction of $\mathbf{b}$ is again $|\mathbf{a} \cdot \hat{\mathbf{b}}|$.

Example 1. Work done by a force

In mechanics, the **work** W done by a constant force $\mathbf{F}$ in moving its point of application through a displacement $\mathbf{d}$ is defined as the product of $\|\mathbf{d}\|$ and the component of $\mathbf{F}$ in the direction of $\mathbf{d}$. Considering Fig. 2.25 we see that

$$W = \|\mathbf{d}\| (\|\mathbf{F}\| \cos \theta) = \mathbf{d} \cdot \mathbf{F}, \tag{16}$$

where $0 \leq \theta \leq \pi$ is measured as shown in the diagram.

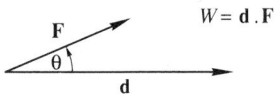

Fig. 2.25   Work done by constant force $\mathbf{F}$

This expression for the work done by $\mathbf{F}$ is true irrespective of whether or not other forces are acting on that same point of application. If they are, and their combined effect is to make the point of application move in a direction opposite to that of the component of $\mathbf{F}$ in the direction of motion, then the work W done by $\mathbf{F}$ will be negative (θ will be obtuse).   ∎

Example 2. Elementary calculations with vectors

Let $\mathbf{a} = 2\mathbf{i} + \mathbf{j} - \mathbf{k}$, $\mathbf{b} = \mathbf{i} - 3\mathbf{j} + \mathbf{k}$. Find

(i) $\mathbf{a} . \mathbf{b}$ (ii) the angle between $\mathbf{a}$ and $\mathbf{b}$ (iii) the length of the projection of $\mathbf{a}$ in the direction of $\mathbf{b}$ (iv) the length of the projection of $\mathbf{b}$ in the direction of $\mathbf{a}$ (v) the work done by a constant force of 7 newtons in the direction $\mathbf{a}$ which displaces its point of application by 3 meters in the direction opposite to $\mathbf{b}$.

Solution

(i) To determine $\mathbf{a} \cdot \mathbf{b}$ we use (14) to find

$\mathbf{a} \cdot \mathbf{b} = (2\mathbf{i} + \mathbf{j} - \mathbf{k}) \cdot (\mathbf{i} - 3\mathbf{j} + \mathbf{k}) = 2 \times 1 + 1 \times (-3) + (-1) \times 1 = -2.$

(ii) To determine the angle θ between $\mathbf{a}$ and $\mathbf{b}$ we use (5). Since $\|\mathbf{a}\| = \sqrt{6}$ and $\|\mathbf{b}\| = \sqrt{11}$, we have

$$\frac{\mathbf{a} \cdot \mathbf{b}}{\|\mathbf{a}\| \|\mathbf{b}\|} = \frac{-2}{\sqrt{66}}, \quad \text{so } \theta = 104.3°.$$

(iii) To find the length of the projection of $\mathbf{a}$ in the direction of $\mathbf{b}$ we must first construct $\hat{\mathbf{b}} = \mathbf{b}/\|\mathbf{b}\|$ and then take the absolute value of (15). We find

$$\mathbf{a} \cdot \hat{\mathbf{b}} = \frac{\mathbf{a} \cdot \mathbf{b}}{\|\mathbf{b}\|} = \frac{-2}{\sqrt{11}}.$$

The length of the required projection is thus $|\mathbf{a} \cdot \hat{\mathbf{b}}| = 2/\sqrt{11}$.

(iv) We proceed as in (iii) above, but now we must find $|\hat{\mathbf{a}} \cdot \mathbf{b}|$. We have

$$\hat{\mathbf{a}} \cdot \mathbf{b} = \frac{\mathbf{a} \cdot \mathbf{b}}{\|\mathbf{a}\|} = \frac{-2}{\sqrt{6}},$$

so the length of the required projection is $|\hat{\mathbf{a}} \cdot \mathbf{b}| = 2/\sqrt{6}$.

(v) We shall use the convention that the unit vector $\hat{\mathbf{a}}$ denotes a force of 1 newton in the direction $\mathbf{a}$, and the unit vector $\hat{\mathbf{b}}$ a displacement of 1 meter in the direction $\mathbf{b}$. Then $\mathbf{F} = 7\hat{\mathbf{a}}$, $\mathbf{d} = -3\hat{\mathbf{b}}$ so

$$W = \mathbf{F} \cdot \mathbf{d} = -21\hat{\mathbf{a}} \cdot \hat{\mathbf{b}} = \frac{-21\mathbf{a} \cdot \mathbf{b}}{\|\mathbf{a}\| \|\mathbf{b}\|} = \frac{(-21) \times (-2)}{\sqrt{66}} =$$

$$= \frac{42}{\sqrt{66}} \mathrm{J} \text{ (joules)}.$$

∎

Direction cosines

The orientation of a vector $\mathbf{v}$ relative to a given system of Cartesian axes $O(x, y, z)$ may be specified by giving the angles α, β and γ between $\mathbf{v}$ and the unit vectors $\mathbf{i}$, $\mathbf{j}$ and $\mathbf{k}$, respectively, the angles being measured as between the vectors in a scalar product. Consider first the case of an arbitrary vector $\mathbf{v} = v_i \mathbf{i} + v_2 \mathbf{j} + v_3 \mathbf{k}$ in $\mathbb{R}^3$. Translate $\mathbf{v}$ so that its initial point coincides with the origin O, and let the angles α, β and γ be as shown in Fig. 2.26(a). Then since $OA = \|\mathbf{v}\|$,

$$\cos \alpha = \frac{v_1}{\|\mathbf{v}\|}, \quad \cos \beta = \frac{v_2}{\|\mathbf{v}\|}, \quad \cos \gamma = \frac{v_3}{\|\mathbf{v}\|}, \tag{17}$$

where $0 \leq \alpha \leq \pi$, $0 \leq \beta \leq \pi$ and $0 \leq \gamma \leq \pi$. The three numbers $\cos \alpha$, $\cos \beta$ and $\cos \gamma$ are called the **direction cosines** of the vector $\mathbf{v}$. They are usually denoted by the ordered triple (l, m, n) where

$$l = \cos \alpha, \quad m = \cos \beta, \quad n = \cos \gamma. \tag{18}$$

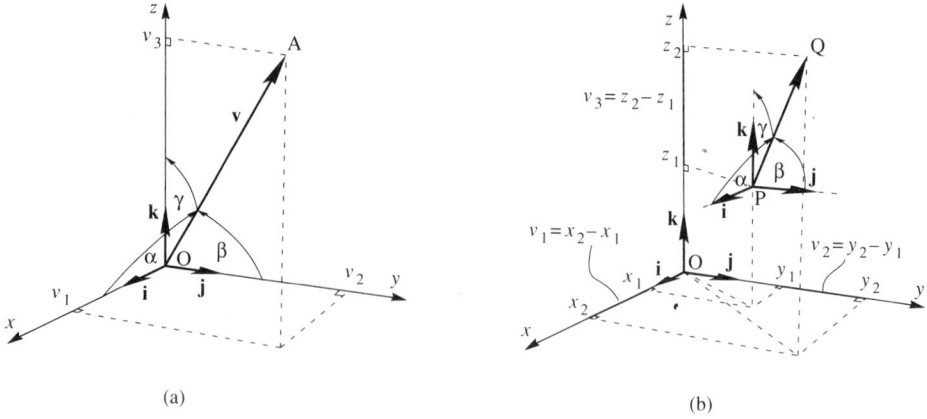

(a)   (b)

Fig. 2.26   (a) Vector drawn from origin   (b) General vector

As $\|\mathbf{v}\|^2 = v_1^2 + v_2^2 + v_3^2$, it follows from (17) that the direction cosines satisfy the identity

$$l^2 + m^2 + n^2 = 1. \tag{19}$$

The usefulness of this identity is such that it proves more convenient to work with the direction cosines l, m and n than with the angles α, β and γ themselves. Any set of numbers r_1, r_2 and r_3 which are proportional to l, m and n, respectively, are called the **direction ratios** of a vector $\mathbf{v}$. Setting $r_1 = kl$, $r_2 = km$ and $r_3 = kn$, with k a constant of proportionality, and substituting into (19) shows $k^2 = r_1^2 + r_2^2 + r_3^2$. It then follows that

$$l = \frac{r_1}{(r_1^2 + r_2^2 + r_3^2)^{1/2}}, \quad m = \frac{r_2}{(r_1^2 + r_2^2 + r_3^2)^{1/2}}, \quad n = \frac{r_3}{(r_1^2 + r_2^2 + r_3^2)^{1/2}}. \tag{20}$$

Thus any vector $\mathbf{v} = r_1\mathbf{i} + r_2\mathbf{j} + r_3\mathbf{k}$ in $\mathbb{R}^3$ will have direction cosines given by the expressions in (20).

Let us now consider the general vector $\mathbf{v} = v_1\mathbf{i} + v_2\mathbf{j} + v_3\mathbf{k}$ and the angles α, β and γ as shown in Fig. 2.26(b), where the initial point P of $\mathbf{v}$ is at (x_1, y_1, z_1) and the terminal point Q is at (x_2, y_2, z_2). In terms of the coordinates of P and Q the components of $\mathbf{v}$ are

$$v_1 = x_2 - x_1, \quad v_2 = y_2 - y_1, \quad v_3 = z_2 - z_1. \tag{21}$$

In this case

$$l = \frac{x_2 - x_1}{\|\mathbf{v}\|}, \quad m = \frac{y_2 - y_1}{\|\mathbf{v}\|}, \quad n = \frac{z_2 - z_1}{\|\mathbf{v}\|}, \tag{22}$$

when we again have the result $l^2 + m^2 + n^2 = 1$, but now

$$\|\mathbf{v}\| = [(x_2 - x_1)^2 + (y_2 - y_1)^2 + (z_2 - z_1)^2]^{1/2}. \tag{23}$$

It follows directly from this that $\hat{\mathbf{v}}$, the unit vector in the direction of $\mathbf{v}$, is

$$\hat{\mathbf{v}} = l\mathbf{i} + m\mathbf{j} + n\mathbf{k}. \tag{24}$$

For any real $k \neq 0$ the numbers kv_1, kv_2 and kv_3 will be the direction ratios of $\mathbf{v}$.

Example 3. Direction cosines

In cases (a) to (c) find both the direction cosines of the given vector and the angles it makes with the positive x-, y- and z-axes, respectively.

(a) $\mathbf{a} = \mathbf{i} + 2\mathbf{j} - 2\mathbf{k}$;

(b) the vector $\mathbf{a}$ with the direction ratios $2, -1, 2$;

(c) the vector $\mathbf{a}$ with its initial point at $(2, -1, -2)$ and its terminal point at $(1, 2, 4)$.

(d) find the terminal point of the vector $\mathbf{a}$ of magnitude 4 which has its initial point at $(1, -3, 2)$ and the direction ratios $-2, 2, 1$.

Solution

(a) $\|\mathbf{a}\| = [1^2 + 2^2 + (-2)^2]^{1/2} = 3$, and so from (17)

$$l = \tfrac{1}{3}, \quad m = \tfrac{2}{3}, \quad n = \tfrac{-2}{3},$$

giving $\alpha = 70.5°$, $\beta = 48.2°$, $\gamma = 131.8°$.

(b) Here $r_1 = 2$, $r_2 = -1$, $\gamma_3 = 2$ and $(r_1^2 + r_2^2 + r_3^2)^{1/2} = 3$, so from (20) we have

$$l = \tfrac{2}{3}, \quad m = -\tfrac{1}{3}, \quad n = \tfrac{2}{3},$$

giving $\alpha = 48.2°$, $\beta = 109.5°$, $\gamma = 48.2°$.

(c) Setting $\mathbf{a} = a_1\mathbf{i} + a_2\mathbf{j} + a_3\mathbf{k}$ we find $a_1 = 1 - 2 = -1$, $a_2 = 2 - (-1) = 3$, $a_3 = 4 - (-2) = 6$, so $\|\mathbf{a}\| = [(-1)^2 + 3^2 + 6^2]^{1/2} = \sqrt{46}$. From (22) we have

$$l = \frac{-1}{\sqrt{46}}, \quad m = \frac{3}{\sqrt{46}}, \quad n = \frac{6}{\sqrt{46}},$$

giving $\alpha = 98.5°$, $\beta = 63.7°$, $\gamma = 27.8°$.

(d) As $r_1 = -2$, $r_2 = 2$, $r_3 = 1$, result (20) becomes

$$l = \tfrac{-2}{3}, \quad m = \tfrac{2}{3}, \quad n = \tfrac{1}{3}.$$

From (24), the unit vector $\hat{\mathbf{a}}$ with these direction cosines is

$$\mathbf{a} = \tfrac{-2}{3}\mathbf{i} + \tfrac{2}{3}\mathbf{j} + \tfrac{1}{3}\mathbf{k},$$

so the vector $\mathbf{a}$ in this direction which is of magnitude 4 will be

$$\mathbf{a} = 4\hat{\mathbf{a}} = \tfrac{-8}{3}\mathbf{i} + \tfrac{8}{3}\mathbf{j} + \tfrac{4}{3}\mathbf{k}.$$

Since the initial point of the required vector has the position vector $\mathbf{i} - 3\mathbf{j} + 2\mathbf{k}$ we see that the terminal point will have the position vector

$$\mathbf{r} = (\mathbf{i} - 3\mathbf{j} + 2\mathbf{k}) + (-\tfrac{8}{3}\mathbf{i} + \tfrac{8}{3}\mathbf{j} + \tfrac{4}{3}\mathbf{k}),$$

giving

$$\mathbf{r} = \tfrac{-5}{3}\mathbf{i} - \tfrac{1}{3}\mathbf{j} + \tfrac{10}{3}\mathbf{k}.$$

In terms of Cartesian coordinates this terminal point will be located at the point $(-\tfrac{5}{3}, -\tfrac{1}{3}, \tfrac{10}{3})$.

■

Problems for Section 2.5

Use the following vectors to find (i) $\mathbf{a} \cdot \mathbf{b}$, (ii) the angle θ between $\mathbf{a}$ and $\mathbf{b}$, (iii) the length of the projection of $\mathbf{a}$ in the direction of $\mathbf{b}$ (iv) the length of the projection of $\mathbf{b}$ in the direction of $\mathbf{a}$:

1 $\mathbf{a} = \mathbf{i} - \mathbf{j} + 2\mathbf{k}$, $\mathbf{b} = 2\mathbf{i} - 2\mathbf{j} + \mathbf{k}$,
2 $\mathbf{a} = 2\mathbf{i} - 3\mathbf{j} + \mathbf{k}$, $\mathbf{b} = \mathbf{i} - \mathbf{j} - \mathbf{k}$,
3 $\mathbf{a} = \mathbf{i} + 2\mathbf{j} + \mathbf{k}$, $\mathbf{b} = -\mathbf{i} - 2\mathbf{j} + \mathbf{k}$,
4 $\mathbf{a} = \mathbf{i} + 7\mathbf{k}$, $\mathbf{b} = -2\mathbf{i} - 14\mathbf{k}$.
5 Given $\mathbf{a} = \mathbf{i} + 2\mathbf{j} + 3\mathbf{k}$, $\mathbf{b} = 2\mathbf{i} - \mathbf{j} - \mathbf{k}$, $\mathbf{c} = -3\mathbf{i} - \mathbf{j} + \mathbf{k}$, $\lambda = -2$ and $\mu = 3$ verify (7) and (8).
6 Verify (9) given that $\mathbf{a} = \mathbf{i} + 2\mathbf{j} + 2\mathbf{k}$, $\mathbf{b} = \mathbf{i} - \mathbf{j} + 3\mathbf{k}$.
7 Verify (9) given that $\mathbf{a} = 2\mathbf{i} - \mathbf{j} + \mathbf{k}$, $\mathbf{b} = 6\mathbf{i} - 3\mathbf{j} + 3\mathbf{k}$.
8 Verify (10) given that $\mathbf{a} = \mathbf{i} + \mathbf{j} + \mathbf{k}$, $\mathbf{b} = 2\mathbf{i} - 3\mathbf{j} + \mathbf{k}$.
9 Verify (10) given that $\mathbf{a} = -\mathbf{i} + 3\mathbf{j} + 2\mathbf{k}$, $\mathbf{b} = 3\mathbf{i} - 9\mathbf{j} - 6\mathbf{k}$.
10 In which of the following three cases are the vectors $\mathbf{a}$ and $\mathbf{b}$ orthogonal:
    (i) $\mathbf{a} = \mathbf{i} + \mathbf{j} + \mathbf{k}$, $\mathbf{b} = -\mathbf{i} + \mathbf{j} + \mathbf{k}$,
    (ii) $\mathbf{a} = 3\mathbf{i} + \mathbf{j} + 2\mathbf{k}$, $\mathbf{b} = -\mathbf{i} + \mathbf{j} + \mathbf{k}$,
    (iii) $\mathbf{a} = 4\mathbf{i} + 3\mathbf{k}$, $\mathbf{b} = 6\mathbf{j}$?
11 Use vector methods to prove that the diagonals of a rhombus are orthogonal.
12 Use vector methods to show that the angle between the diagonals drawn from the same corner of two adjacent faces of a unit cube is $\pi/3$.
13 Consider the triangle shown in Fig. 2.27 whose sides are the vectors $\mathbf{a}$, $\mathbf{b}$ and $\mathbf{c}$. Starting from the law for vector addition $\mathbf{c} = \mathbf{a} + \mathbf{b}$, form a suitable scalar product and hence prove the cosine law for triangles

$$c^2 = a^2 + b^2 - 2ab \cos C,$$

where $a = \|\mathbf{a}\|$, $b = \|\mathbf{b}\|$, $c = \|\mathbf{c}\|$ and C is the angle between $\mathbf{a}$ and $\mathbf{b}$ measured as shown in the diagram.

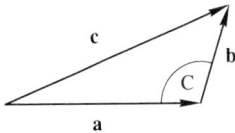

Fig. 2.27

14 A constant force $\mathbf{F}$ moves its point of application through a displacement $\mathbf{d}$. Prove by vector methods that the work done by $\mathbf{F}$ is independent of any component of $\mathbf{F}$ which is orthogonal to $\mathbf{d}$.
15 Find the work done by a constant force of 9 newtons in the direction $3\mathbf{i} + \mathbf{j} - \mathbf{k}$ which displaces its point of application (i) by 4 meters in the direction $\mathbf{i} - 6\mathbf{j} - \mathbf{k}$ and (ii) by 3 meters in the direction $2\mathbf{i} + \mathbf{j} - 2\mathbf{k}$. Comment on the sign of the result in each case.
16 A constant force $\mathbf{F}$ moves its point of application from the point with position vector $\mathbf{a}$ to the point with position vector $\mathbf{b}$. Write down the vector expression for the work W done by the force and find W given that $\mathbf{F}$ is 5 newtons in the direction $\mathbf{i} + \mathbf{j} + \mathbf{k}$ while $\mathbf{a} = 2\mathbf{i} + \mathbf{j} + \mathbf{k}$ and $\mathbf{b} = 3\mathbf{i} + 2\mathbf{j} + 4\mathbf{k}$, with a unit vector representing a displacement of one meter.
17 In each of the following cases find the direction ratios, direction cosines and the angles made by the vector with the x-, y- and z-axes given that the initial point A and terminal point B of the

vector is
(a) $A(2, 2, 3),$ $B(1, 4, -2),$
(b) $A(-1, 2, -3),$ $B(1, 1, -1),$
(c) $A(0, 1, 2),$ $B(0, -1, -2),$
(d) $A(0, 0, 0),$ $B(2, 1, -3).$

18 In each of the following cases find the angles made by the vector with the x-, y- and z-axes given that the direction ratios of the vector are:
(a) $1, 1, 1$
(b) $1, -1, 2$
(c) $3, 1, 2$
(d) $1, 0, -\sqrt{3}.$

19 Find the vector of magnitude 5 with the direction ratios $1, 1, \sqrt{2}$ which has:
(a) its terminal point at $(1, 4, 2),$
(b) its initial point at $(1, 1, 1).$

20 What is the magnitude and what are the direction cosines of the force $\mathbf{F}$ needed at P to bring into equilibrium the forces $\mathbf{i}+\mathbf{j}+\mathbf{k}$, $2\mathbf{i}-\mathbf{j}+3\mathbf{k}$ and $2\mathbf{i}+\mathbf{j}+2\mathbf{k}$ acting through P, where a unit vector represents a unit force?

2.6 Vector product (cross product)

It is useful to introduce another form of product involving any two vectors $\mathbf{a}$ and $\mathbf{b}$ in $\mathbb{R}^3$ called their **vector product**, or **cross product**. This is written

$$\mathbf{a} \times \mathbf{b},$$

also sometimes $\mathbf{a} \wedge \mathbf{b}$, and it is a vector which we now define.

Let θ, measured counterclockwise and chosen such that $0 \le \theta \le \pi$, be the angle between the lines of action of the two vectors $\mathbf{a}$ and $\mathbf{b}$ when their initial points have been brought into coincidence by a translation. Denote by $\hat{\mathbf{n}}$ the unit vector normal to the plane containing $\mathbf{a}$ and $\mathbf{b}$, chosen such that $\mathbf{a}$, $\mathbf{b}$ and $\hat{\mathbf{n}}$ in this order form a right-handed set of vectors (see Fig. 2.28). Then the **vector product** $\mathbf{a} \times \mathbf{b}$ is defined as

$$\mathbf{a} \times \mathbf{b} = \begin{cases} \|\mathbf{a}\| \, \|\mathbf{b}\| \sin \theta \, \hat{\mathbf{n}} & \text{when } \mathbf{a} \ne \mathbf{0}, \, \mathbf{b} \ne \mathbf{0} \\ \mathbf{0} & \text{when } \mathbf{a} = \mathbf{0} \text{ or } \mathbf{b} = \mathbf{0}. \end{cases} \tag{1}$$

Now $\sin \theta \ge 0$ for $0 \le \theta \le \pi$, so it follows directly from (1) that

$$\|\mathbf{a} \times \mathbf{b}\| = \|\mathbf{a}\| \, \|\mathbf{b}\| \sin \theta. \tag{2}$$

Inspection of Fig. 2.28 shows that in geometrical terms $\|\mathbf{a} \times \mathbf{b}\|$ is the area of the parallelogram OACB.

Definition (1) shows $\mathbf{a} \times \mathbf{b} = \mathbf{0}$ if $\theta = 0$ or $\theta = \pi$. Thus, if neither $\mathbf{a}$ nor $\mathbf{b}$ is a null vector, the vector product $\mathbf{a} \times \mathbf{b} = \mathbf{0}$ only when the line segments of the vectors $\mathbf{a}$ and $\mathbf{b}$ are parallel. In particular, it follows that $\mathbf{a} \times \mathbf{a} = \mathbf{0}$ for any vector $\mathbf{a}$.

Applying the definition of a vector product to the product $\mathbf{b} \times \mathbf{a}$ gives

$$\mathbf{b} \times \mathbf{a} = \|\mathbf{b}\| \, \|\mathbf{a}\| \sin \theta \, \hat{\mathbf{m}}, \tag{3}$$

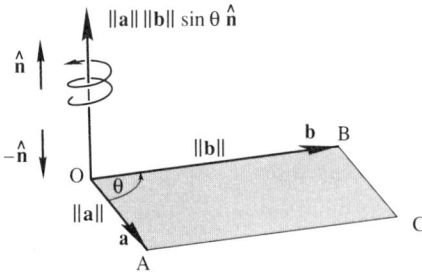

Fig. 2.28   Vector product $\mathbf{a} \times \mathbf{b}$

where from (1) the vectors $\mathbf{b}$, $\mathbf{a}$ and $\hat{\mathbf{m}}$ must form a right-handed set. As in this system a right-handed screw advances in the opposite sense to that of $\hat{\mathbf{n}}$, we have $\hat{\mathbf{m}} = -\hat{\mathbf{n}}$, so (3) becomes

$$\mathbf{b} \times \mathbf{a} = -\|\mathbf{b}\|\,\|\mathbf{a}\| \sin \theta \, \hat{\mathbf{n}}. \tag{4}$$

Comparing (1) and (4) gives the important result for vector products that

$$\mathbf{b} \times \mathbf{a} = -(\mathbf{a} \times \mathbf{b}). \tag{5}$$

Result (5) shows the vector product is *not* commutative, and demonstrates the need to preserve the order of the factors in such a product. Taking the magnitude of both sides of (5) gives the result

$$\|\mathbf{a} \times \mathbf{b}\| = \|\mathbf{b} \times \mathbf{a}\|. \tag{6}$$

Important special cases of vector products arise when only the mutually orthogonal unit vectors $\mathbf{i}$, $\mathbf{j}$ and $\mathbf{k}$ are involved. It is easily seen from (1) that

$$\mathbf{i} \times \mathbf{j} = \mathbf{k}, \quad \mathbf{j} \times \mathbf{k} = \mathbf{i}, \quad \mathbf{k} \times \mathbf{i} = \mathbf{j}. \tag{7}$$

Result (5) then gives the three further results

$$\mathbf{j} \times \mathbf{i} = -\mathbf{k}, \quad \mathbf{k} \times \mathbf{j} = -\mathbf{i}, \quad \mathbf{i} \times \mathbf{k} = -\mathbf{j}, \tag{8}$$

Definition (1) also shows that

$$\mathbf{i} \times \mathbf{i} = \mathbf{j} \times \mathbf{j} = \mathbf{k} \times \mathbf{k} = 0. \tag{9}$$

These results are extremely useful and they are often required when performing vector algebra. Results (7) tell us that $\mathbf{i}, \mathbf{j}$ and $\mathbf{k}$ form a right-handed mutually orthogonal system of unit vectors.

The form of (7) and (8) may be remembered algebraically by noticing that each result is obtained as a cyclic permutation of the vectors $\mathbf{i}, \mathbf{j}$ and $\mathbf{k}$ when written in this order. That is, we may move the first vector to appear to the last position while moving each of the other two vectors one place to the left but leaving their order unchanged, as indicated symbolically in the table below. There are only three different permutations of this type which are possible, after which the same ones are regenerated, as may be seen from the last three entries in the table. If the order of two vectors in such a permutation is inverted, the sign of the associated vector product must be changed. For example, a permutation like $\mathbf{k}$, $\mathbf{j}$, $\mathbf{i}$ will yield $\mathbf{k} \times \mathbf{j} = -\mathbf{i}$, which is one of the three entries in (8).

	$\times$ =		
	$\mathbf{i}$	$\mathbf{j}$	$\mathbf{k}$
Cycle of three permutations	$\mathbf{j}$	$\mathbf{k}$	$\mathbf{i}$
	$\mathbf{k}$	$\mathbf{i}$	$\mathbf{j}$
	$\mathbf{i}$	$\mathbf{j}$	$\mathbf{k}$
Repeat of cycle	$\mathbf{j}$	$\mathbf{k}$	$\mathbf{i}$
	$\mathbf{k}$	$\mathbf{i}$	$\mathbf{j}$

Let λ, μ be any two real numbers and $\mathbf{a}$, $\mathbf{b}$ any two vectors in $\mathbb{R}^3$. Then definition (1) shows

$$(\lambda\mathbf{a}) \times (\mu\mathbf{b}) = \lambda\mu(\mathbf{a} \times \mathbf{b}) = (\mu\mathbf{a}) \times (\lambda\mathbf{b}), \tag{10}$$

so that scaling the factors in a vector product simply scales the product itself. This establishes the fact that the vector product is **associative** with respect to multiplication by a scalar.

We now derive an alternative and extremely useful definition of the vector product in terms of the components of $\mathbf{a}$ and $\mathbf{b}$. Squaring (2), using the identity $\cos^2\theta = 1 - \sin^2\theta$, and result (5) from Sec. 2.6, we obtain

$$\|\mathbf{a} \times \mathbf{b}\|^2 = \|\mathbf{a}\|^2 \|\mathbf{b}\|^2 - (\mathbf{a} \cdot \mathbf{b})^2, \tag{11}$$

which is known as **Lagrange's identity**[2].

Expanding the right-hand side of (11) and rearranging terms it is possible to express it in determinantal form[3] as

$$\|\mathbf{a} \times \mathbf{b}\|^2 = \begin{vmatrix} a_2 & a_3 \\ b_2 & b_3 \end{vmatrix}^2 + \begin{vmatrix} a_1 & a_3 \\ b_1 & b_3 \end{vmatrix}^2 + \begin{vmatrix} a_1 & a_2 \\ b_1 & b_2 \end{vmatrix}^2. \tag{12}$$

Now $\mathbf{a} \times \mathbf{b}$ is a vector in $\mathbb{R}^3$ so, apart from a choice of sign, the determinants in (12) must be its x-, y-, and z-components, though as yet it is unclear which these are. To resolve the matter let $\hat{\mathbf{e}}_1$, $\hat{\mathbf{e}}_2$ and $\hat{\mathbf{e}}_3$ be the permutation of $\mathbf{i}$, $\mathbf{j}$ and $\mathbf{k}$ for which

$$\mathbf{a} \times \mathbf{b} = \begin{vmatrix} a_2 & a_3 \\ b_2 & b_3 \end{vmatrix} \hat{\mathbf{e}}_1 + \begin{vmatrix} a_1 & a_3 \\ b_1 & b_3 \end{vmatrix} \hat{\mathbf{e}}_2 + \begin{vmatrix} a_1 & a_2 \\ b_1 & b_2 \end{vmatrix} \hat{\mathbf{e}}_3. \tag{13}$$

[2] JOSEPH LOUIS LAGRANGE (1736–1813), born in Turin of French extraction, worked for twenty years in Berlin and then in Paris. He was one of most outstanding mathematicians for all time and made contributions to algebra, the calculus, differential equations, the calculus of variations and mechanics.
[3] The reader unfamiliar with the definition of second and third order determinants is referred to Sec. 3.7.

Considering the vector product $i \times j = k$ in (13) shows $k = \hat{e}_3$, so the unit vector $\hat{e}_3 = k$. Repeating this same argument with the vector products $j \times k = i$ and $k \times i = j$ shows, respectively, that $\hat{e}_1 = i$ and $\hat{e}_2 = -j$. Incorporating these results into (13) we obtain

$$a \times b = \begin{vmatrix} a_2 & a_3 \\ b_2 & b_3 \end{vmatrix} i - \begin{vmatrix} a_1 & a_3 \\ b_1 & b_3 \end{vmatrix} j + \begin{vmatrix} a_1 & a_2 \\ b_1 & b_2 \end{vmatrix} k, \tag{14}$$

which is one form of the definition of the vector product $a \times b$ in terms of components.

A concise form for this vector product becomes possible once it is recognized that (14) may be regarded as the expansion of the following symbolic third order determinant in terms of the elements of its first row

$$a \times b = \begin{vmatrix} i & j & k \\ a_1 & a_2 & a_3 \\ b_1 & b_2 & b_3 \end{vmatrix}. \tag{15}$$

This is to be regarded as a symbolic determinant because although the entries in the last two rows are numbers, those in the first row are vectors. Notice that interchanging the order of the vectors in the vector product interchanges two rows, and thus changes the sign of $a \times b$ in agreement with (5).

We have established the equivalence of the geometrical definition of the vector product given in (1) and the algebraic definition in terms of coordinates given in (15).

Using (15) it is a straightforward matter to prove that if a, b and c are any three vectors in $\mathbb{R}^3$, the vector product is **distributive** with respect to addition, so that

$$a \times (b+c) = a \times b + a \times c,$$

and

$$(a+b) \times c = a \times c + b \times c. \tag{16}$$

Example 1. Vector product

If $a = i + 3j - k$ and $b = -3i + k$ find (i) $a \times b$, and (ii) two unit vectors which are orthogonal to the plane containing a and b.

Solution

(i) From (15)

$$a \times b = \begin{vmatrix} i & j & k \\ 1 & 3 & -1 \\ -3 & 0 & 1 \end{vmatrix} = 3i + 2j + 9k.$$

(ii) By definition $a \times b$ is normal to the plane containing a and b, so

$$\hat{n} = \frac{a \times b}{\|a \times b\|} = \frac{1}{\sqrt{94}}(3i + 2j + 9k)$$

must be one of the required unit vectors. The other is the vector

$$-\hat{\mathbf{n}}=\frac{-1}{\sqrt{94}}(3\mathbf{i}+2\mathbf{j}+9\mathbf{k}).$$

∎

Example 2. Test for linear independence

Two arbitrary nonzero vectors **a** and **b** will form a basis for the space of geometrical vectors in any plane parallel to them (the vector space $\mathbb{R}^2$) if they are linearly independent. Now from Sec. 2.3 we know **a** and **b** will be linearly independent if the only solution to the vector equation

$$\lambda_1\mathbf{a}+\lambda_2\mathbf{b}=\mathbf{0} \tag{17}$$

is $\lambda_1=\lambda_2=0$. Forming the vector product of (17) from the left with **a** gives

$$\lambda_1(\mathbf{a}\times\mathbf{a})+\lambda_2(\mathbf{a}\times\mathbf{b})=\mathbf{a}\times\mathbf{0}$$

or, as $\mathbf{a}\times\mathbf{a}=\mathbf{0}$,

$$\lambda_2(\mathbf{a}\times\mathbf{b})=\mathbf{0},$$

which will have the solution $\lambda_2=0$ provided $\mathbf{a}\times\mathbf{b}\neq\mathbf{0}$. Similarly, forming the vector product of (17) from the right with **b** gives

$$\lambda_1(\mathbf{a}\times\mathbf{b})+\lambda_2(\mathbf{b}\times\mathbf{b})=\mathbf{0}\times\mathbf{b},$$

or, as $\mathbf{b}\times\mathbf{b}=\mathbf{0}$,

$$\lambda_1(\mathbf{a}\times\mathbf{b})=\mathbf{0},$$

which will have the solution $\lambda_1=0$ provided $\mathbf{a}\times\mathbf{b}\neq\mathbf{0}$. Thus a·and **b** will be **linearly independent** if, and only if, $\mathbf{a}\times\mathbf{b}\neq\mathbf{0}$.

If $\mathbf{a}\times\mathbf{b}=\mathbf{0}$, λ_1 and λ_2 may be chosen arbitrarily, showing that then **a** and **b** are linearly dependent. This means **a** is a multiple of **b**, so no unique plane can exist parallel to **a** and **b**.

We have proved that if **a** and **b** are any two nonzero vectors in $\mathbb{R}^3$, then they are linearly independent if, and only if, $\mathbf{a}\times\mathbf{b}\neq\mathbf{0}$, otherwise they are linearly dependent.

∎

Example 3. Moment of a force and moment of momentum

In mechanics the **moment**, or **torque**, produced by a force about a point is important because it represents the turning effect produced by the force. By definition, the moment **M** of force **F** about a point P in space is the vector product

$$\mathbf{M}=\mathbf{r}\times\mathbf{F},$$

where **r** is the position vector, relative to P, of any point Q on the line of action of the force **F**.

The situation is illustrated in Fig. 2.29 in which l is the perpendicular distance of P from the line of action Λ of **F**, and θ is the angle between **r** and **F** as shown. The line of action L of the moment **M** passes through P, is normal to the plane containing **r** and **F**, and from the definition of a vector product, **r**, **F** and **M**, in this order, form a right-handed set.

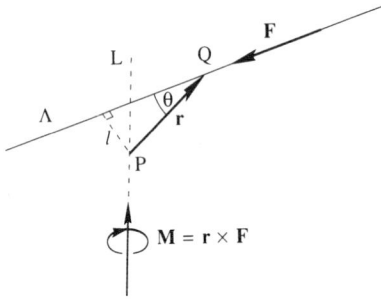

Fig. 2.29 Moment of force **F** about P

As $l = PQ \sin \theta = \|\mathbf{r}\| \sin \theta$, we see from definition (1) of a vector product that

$$\|\mathbf{M}\| = l\|\mathbf{F}\|.$$

Thus the magnitude of the moment **M** of **F** about P is the product of the perpendicular distance of P from the line of action of the force, and the magnitude $\|\mathbf{F}\|$ of the force itself. ∎

Consider next a particle of mass m and velocity **v** located at a point with position vector **r** relative to some reference point. By definition the linear momentum of the particle is $m\mathbf{v}$. Then by analogy with the definition of a moment, the vector

$$\mathbf{h} = \mathbf{r} \times (m\mathbf{v}) = m(\mathbf{r} \times \mathbf{v}) \tag{18}$$

is defined to be the **moment of momentum** or **angular momentum** of the particle about this point. This idea enters into mechanics when examining the dynamical effect of a moment applied to a rigid body.

Example 4. Angular velocity

Consider the rotation of a rigid body about a fixed axis L passing through a point Q. Suppose its angular speed about L is ω, and let a point P in the body have the position vector **r** with respect to O

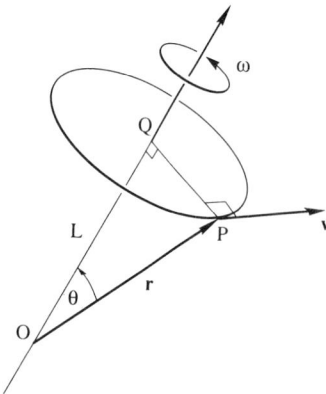

Fig. 2.30 Angular velocity

(Fig. 2.30). Then P moves around a circle with center Q and radius $\|\mathbf{r}\|\sin\theta$, so its tangential speed $v = \omega\|\mathbf{r}\|\sin\theta$. The instantaneous velocity $\mathbf{v}$ of P has magnitude v and is directed along the tangent to the circle passing through P in the direction of rotation. It then follows from the definition of a vector product that we may write

$$\mathbf{v} = \boldsymbol{\omega} \times \mathbf{r}, \tag{19}$$

where the vector $\boldsymbol{\omega}$ with magnitude ω is directed along L in the way in which a right-handed screw would advance due to the rotation. The vectors $\boldsymbol{\omega}$, $\mathbf{r}$ and $\mathbf{v}$, in this order, form a right-handed set. The vector $\boldsymbol{\omega}$ is called the **angular velocity** of the body, while the vector $\mathbf{v}$ is called the **linear velocity** of P. ∎

Problems for Section 2.6

In Problems 1 to 11 set

$$\mathbf{a} = \mathbf{i} + \mathbf{j} - \mathbf{k}, \quad \mathbf{b} = \mathbf{j} + 2\mathbf{k} \quad \text{and} \quad \mathbf{c} = 2\mathbf{i} - \mathbf{j} + \mathbf{k}.$$

1 Find $\mathbf{a} \times \mathbf{b}$, $\mathbf{a} \times \mathbf{c}$, $\mathbf{c} \times \mathbf{a}$.

2 Find $\mathbf{b} \times \mathbf{c}$, $\mathbf{a} \times (2\mathbf{b})$, $\|\mathbf{b} \times \mathbf{c}\|$, $\|\mathbf{c} \times (2\mathbf{b})\|$.

Find the area of the parallelogram with the following adjacent sides:

3 $\mathbf{a}$, $\mathbf{a} + 2\mathbf{b}$,   **4** $\mathbf{b} + \mathbf{c}$, $\mathbf{a} + \mathbf{b} + \mathbf{c}$.

Find the area of the triangle with the following adjacent sides:

5 $\mathbf{a} + \mathbf{b}$, $\mathbf{a} - \mathbf{c}$.   **6** $2\mathbf{a} - \mathbf{c}$, $2\mathbf{a} + \mathbf{c}$.

Evaluate

7 $\mathbf{a} \times (\mathbf{b} + \mathbf{c})$, $\mathbf{a} \times \mathbf{b} + \mathbf{a} \times \mathbf{c}$.

8 $\mathbf{a} \times (\mathbf{b} + 3\mathbf{c})$, $\mathbf{a} \times (2\mathbf{a} + \mathbf{b} - \mathbf{c})$.

9 $\mathbf{a} \cdot (\mathbf{b} \times \mathbf{c})$, $(\mathbf{a} \times \mathbf{b}) \cdot \mathbf{c}$.

10 $(\mathbf{a} \cdot \mathbf{b})(\mathbf{a} \times \mathbf{b})$, $(\mathbf{a} \times \mathbf{b}) \cdot (\mathbf{a} \times \mathbf{c})$.

11 $\mathbf{a} \times (\mathbf{b} \times \mathbf{c})$, $(\mathbf{a} \times \mathbf{b}) \times \mathbf{c}$.

12 Derive result (14) for the vector product $\mathbf{a} \times \mathbf{b}$ of two arbitrary vectors $\mathbf{a}$ and $\mathbf{b}$ by considering $(a_1\mathbf{i} + a_2\mathbf{j} + a_3\mathbf{k}) \times (b_1\mathbf{i} + b_2\mathbf{j} + b_3\mathbf{k})$ and using results (7) to (9).

13 Find the two unit vectors which are orthogonal to both $-\mathbf{i} + 2\mathbf{j} - \mathbf{k}$ and $2\mathbf{i} + \mathbf{k}$.

14 Find a vector which is orthogonal to the vectors $\overrightarrow{AB}$ and $\overrightarrow{AC}$ given that A, B and C are the points (1, 2, 1), (2, 1, 2) and (1, 0, 3), respectively.

15 Verify by direct calculation that
$$\|\mathbf{a}\|^2\|\mathbf{b}\|^2 - (\mathbf{a} \cdot \mathbf{b})^2 = (a_2 b_3 - a_3 b_2)^2 + (a_1 b_3 - a_3 b_1)^2 + (a_1 b_2 - a_2 b_1)^2.$$

16 Use the Lagrange identity to prove that if $\mathbf{a} = \lambda\mathbf{b}$ with $\lambda \neq 0$ an arbitrary scalar, then $\mathbf{a} \times \mathbf{b} = \mathbf{0}$.

17 Let $\mathbf{a}$, $\mathbf{b}$ and $\mathbf{c}$ be the vectors forming the sides of the triangle shown in Fig. 2.31.

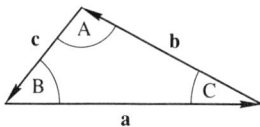

Fig. 2.31

Use the fact that $\mathbf{a}+\mathbf{b}+\mathbf{c}=\mathbf{0}$ together with the results of forming successive vector products of this equation with $\mathbf{a}$, $\mathbf{b}$ and $\mathbf{c}$ to prove the **sine law** for triangles

$$\frac{a}{\sin A}=\frac{b}{\sin B}=\frac{c}{\sin C},$$

where $a=\|\mathbf{a}\|$, $b=\|\mathbf{b}\|$ and $c=\|\mathbf{c}\|$.

18 By using the properties of a determinant, or otherwise, prove that if $\mathbf{a}\times\mathbf{b}=\mathbf{0}$, the corresponding components of $\mathbf{a}$ and $\mathbf{b}$ are proportional.

19 Are the following pairs of vectors linearly dependent or orthogonal?
(a) $3\mathbf{i}+3\mathbf{j}$, $2\mathbf{i}-3\mathbf{j}$
(b) $2\mathbf{i}-\mathbf{j}+3\mathbf{k}$, $-4\mathbf{i}+\mathbf{j}+3\mathbf{k}$ (c) $-\mathbf{i}+4\mathbf{j}+2\mathbf{k}$, $3\mathbf{i}-12\mathbf{j}-6\mathbf{k}$
(d) $\mathbf{i}+4\mathbf{k}$, $-4\mathbf{i}+6\mathbf{j}+\mathbf{k}$ (e) $4\mathbf{i}+3\mathbf{j}-2\mathbf{k}$, $-12\mathbf{i}-9\mathbf{j}+6\mathbf{k}$.

20 Find the moment of the force $2\mathbf{i}-\mathbf{j}+\mathbf{k}$ about the point $(2, 3, 4)$ given that the line of action of the force passes through the point $(1, 2, 1)$.

21 A force $3\mathbf{i}+2\mathbf{j}-\mathbf{k}$ acts through a point $(1, 1, 1)$ in a rigid body. What moment must be applied about the point $(2, 1, 3)$ in the body in order that the moment due to the force is neutralized?

22 A particle of mass 3 located at the point $(1, 3, 2)$ moves with the velocity $\mathbf{i}+\mathbf{j}+3\mathbf{k}$. Find the moment of momentum of this particle about the point $(1, 0, 1)$.

23 Find the linear velocity of the point $(2, 4, 3)$ in a rigid body given that the axis of rotation passes through the origin and the angular velocity of the body is $\mathbf{i}+\mathbf{j}-2\mathbf{k}$.

2.7 Combinations of scalar and vector products

Scalar and vector products may be combined to yield more complicated products than those discussed in the two previous sections. Not all such combinations are meaningful, and even when they are it is usually necessary to insert parentheses to avoid ambiguity. These are necessary, for example, in an expression like $\mathbf{a}\times\mathbf{b}\times\mathbf{c}$ which could mean either $\mathbf{a}\times(\mathbf{b}\times\mathbf{c})$ or $(\mathbf{a}\times\mathbf{b})\times\mathbf{c}$, because $\mathbf{a}\times(\mathbf{b}\times\mathbf{c})\neq(\mathbf{a}\times\mathbf{b})\times\mathbf{c}$ (cf. Prob. 11, Sec. 2.6). In an expression like $(\mathbf{a}\cdot\mathbf{b})\mathbf{c}$ they are helpful, though not essential, because $(\mathbf{a}\cdot\mathbf{b})$ is a scalar. However an expression like $\mathbf{a}\cdot\mathbf{b}\cdot\mathbf{c}$ is meaningless with or without parentheses. This is so because although $\mathbf{a}\cdot\mathbf{b}$ and $\mathbf{b}\cdot\mathbf{c}$ are well-defined scalars, neither $(\mathbf{a}\cdot\mathbf{b})\cdot\mathbf{c}$ nor $\mathbf{a}\cdot(\mathbf{b}\cdot\mathbf{c})$ has any meaning, for the scalar product of a vector and a scalar is not defined.

The two most frequently encountered expressions involving three arbitrary vectors $\mathbf{a}$, $\mathbf{b}$ and $\mathbf{c}$ are

$$\mathbf{a}\cdot(\mathbf{b}\times\mathbf{c})\qquad\qquad\textbf{(scalar triple product)}\qquad\qquad\qquad(1)$$

and

$$\mathbf{a}\times(\mathbf{b}\times\mathbf{c})\qquad\qquad\textbf{(vector triple product)}.\qquad\qquad\qquad(2)$$

The first of these is a scalar and the second is a vector.

The scalar triple product is easily evaluated in terms of the components of $\mathbf{a}$, $\mathbf{b}$ and $\mathbf{c}$. Using (14) from Sec. 2.6 to evaluate $\mathbf{b}\times\mathbf{c}$ we find

$$\mathbf{b}\times\mathbf{c}=\mathbf{i}\begin{vmatrix}b_2 & b_3\\c_2 & c_3\end{vmatrix}-\mathbf{j}\begin{vmatrix}b_1 & b_3\\c_1 & c_3\end{vmatrix}+\mathbf{k}\begin{vmatrix}b_1 & b_2\\c_1 & c_2\end{vmatrix}.$$

Forming the scalar product with **a** this becomes

$$\mathbf{a}\cdot(\mathbf{b}\times\mathbf{c})=a_1\begin{vmatrix}b_2 & b_3\\c_2 & c_3\end{vmatrix}-a_2\begin{vmatrix}b_1 & b_3\\c_1 & c_3\end{vmatrix}+a_3\begin{vmatrix}b_1 & b_2\\c_1 & c_2\end{vmatrix},$$

showing that in terms of a third order determinant

$$\mathbf{a}\cdot(\mathbf{b}\times\mathbf{c})=\begin{vmatrix}a_1 & a_2 & a_3\\b_1 & b_2 & b_3\\c_1 & c_2 & c_3\end{vmatrix},\qquad(3)$$

The scalar product is commutative, so this may also be written

$$\mathbf{a}\cdot(\mathbf{b}\times\mathbf{c})=(\mathbf{b}\times\mathbf{c})\cdot\mathbf{a}.\qquad(4)$$

Because interchanging two rows in a determinant changes its sign, it follows from (3) that

$$\mathbf{a}\cdot(\mathbf{b}\times\mathbf{c})=-\mathbf{b}\cdot(\mathbf{a}\times\mathbf{c})=\mathbf{b}\cdot(\mathbf{c}\times\mathbf{a}).$$

A repetition of this form of argument leads to the useful results

$$\mathbf{a}\cdot(\mathbf{b}\times\mathbf{c})=\mathbf{b}\cdot(\mathbf{c}\times\mathbf{a})=\mathbf{c}\cdot(\mathbf{a}\times\mathbf{b}).\qquad(5)$$

These results are best remembered by noticing that the dot and cross in the scalar triple product may always be interchanged without altering the product.

A frequently used notation for the scalar triple product $\mathbf{a}\cdot(\mathbf{b}\times\mathbf{c})$ is [**a b c**], with the order of the vectors in the square brackets indicating their cyclic order in the product. In this notation (5) becomes

$$[\mathbf{a}\ \mathbf{b}\ \mathbf{c}]=[\mathbf{b}\ \mathbf{c}\ \mathbf{a}]=[\mathbf{c}\ \mathbf{a}\ \mathbf{b}].\qquad(6)$$

A convenient geometrical interpretation of the scalar triple product is in terms of the volume V of a parallelepiped which has the vectors **a**, **b** and **c** as its adjacent edges (Fig. 2.32).

To see this, recall that $\|\mathbf{a}\times\mathbf{b}\|$ is the area A of the shaded parallelogram forming the base of the parallelepiped while $|\mathbf{c}\cdot\hat{\mathbf{n}}|$, with $\hat{\mathbf{n}}$ the unit vector normal to A in the direction of $\mathbf{a}\times\mathbf{b}$, is its height $H=\|\mathbf{c}\|\,|\cos\theta|$, so that

$$V=AH=\|\mathbf{a}\times\mathbf{b}\|\,\|\mathbf{c}\|\,|\cos\theta|=|(\mathbf{a}\times\mathbf{b})\cdot\mathbf{c}|.\qquad(7)$$

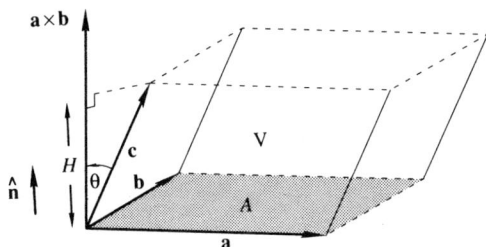

Fig. 2.32   Volume of parallelepiped $V=|(\mathbf{a}\times\mathbf{b})\cdot\mathbf{c}|$

The volume is a geometrical invariant of the parallelepiped, and so will be the same whichever pair of vectors is taken to define the base (**a** and **b** in Fig. 2.32). Thus the results in (5) also follow from (7) by taking either **b** and **c** or **c** and **a** as the pair of vectors defining the base.

Suppose three arbitrary vectors **a**, **b** and **c** are considered as a basis for the vector space $\mathbb{R}^3$. Then they must be linearly independent, which in geometrical terms means that taken pairwise they must define three distinct planes. This is equivalent to requiring the volume of the parallelepiped with adjacent sides **a**, **b** and **c** to be nonzero. Consequently, the three vectors **a**, **b** and **c** will be **linearly independent** if, and only if, $\mathbf{a} \cdot (\mathbf{b} \times \mathbf{c}) \neq 0$, and they will be **linearly dependent** if, and only if, $\mathbf{a} \cdot (\mathbf{b} \times \mathbf{c}) = 0$. Thus the value of the scalar triple product provides a simple test for the linear independence of its constituent vectors.

The vector triple products $\mathbf{a} \times (\mathbf{b} \times \mathbf{c})$ and $(\mathbf{a} \times \mathbf{b}) \times \mathbf{c}$ may be expanded by means of the following identities

$$\mathbf{a} \times (\mathbf{b} \times \mathbf{c}) = (\mathbf{a} \cdot \mathbf{c})\mathbf{b} - (\mathbf{a} \cdot \mathbf{b})\mathbf{c}, \tag{8}$$

$$(\mathbf{a} \times \mathbf{b}) \times \mathbf{c} = (\mathbf{a} \cdot \mathbf{c})\mathbf{b} - (\mathbf{b} \cdot \mathbf{c})\mathbf{a}. \tag{9}$$

In general $\mathbf{a} \times (\mathbf{b} \times \mathbf{c}) \neq (\mathbf{a} \times \mathbf{b}) \times \mathbf{c}$ so that the vector product is **not associative**.

We prove only (9). Take the x-axis parallel to **a** and the y-axis such that **b** is parallel to the (x, y)-plane. It then follows that in component form **a**, **b** and **c** may be written

$$\mathbf{a} = a_1 \mathbf{i}, \quad \mathbf{b} = b_1 \mathbf{i} + b_2 \mathbf{j}, \quad \mathbf{c} = c_1 \mathbf{i} + c_2 \mathbf{j} + c_3 \mathbf{k}.$$

As a result

$$\mathbf{a} \times \mathbf{b} = a_1 b_2 \mathbf{k},$$

and so

$$(\mathbf{a} \times \mathbf{b}) \times \mathbf{c} = -a_1 b_2 c_2 \mathbf{i} + a_1 b_2 c_1 \mathbf{j}.$$

Substituting the same expressions for **a**, **b** and **c** in (9) also gives this result. Identity (9) is thus proved, for the x-, y- and z-axes may always be chosen in this manner. Identity (8) follows from (9) as a result of a suitable permutation of **a**, **b** and **c** and an interchange of the order of the factors in the vector products.

More complicated forms of product can be simplified by use of the above identities. As the only illustration of this consider the product $\mathbf{a} \cdot (\mathbf{b} \times (\mathbf{c} \times \mathbf{d}))$. Setting $\mathbf{h} = \mathbf{c} \times \mathbf{d}$ this becomes the scalar triple product $\mathbf{a} \cdot (\mathbf{b} \times \mathbf{h})$ which, because of (5), is equal to $(\mathbf{a} \times \mathbf{b}) \cdot \mathbf{h}$. Thus we have shown that

$$\mathbf{a} \cdot (\mathbf{b} \times (\mathbf{c} \times \mathbf{d})) = (\mathbf{a} \times \mathbf{b}) \cdot (\mathbf{c} \times \mathbf{d}). \tag{10}$$

A more interesting form of this same result is obtained if we first expand the vector triple product $\mathbf{b} \times (\mathbf{c} \times \mathbf{d})$ using (8), and then form the scalar product of the result with **a**. After using (10) in the left-hand side we find that

$$(\mathbf{a} \times \mathbf{b}) \cdot (\mathbf{c} \times \mathbf{d}) = (\mathbf{a} \cdot \mathbf{c})(\mathbf{b} \cdot \mathbf{d}) - (\mathbf{a} \cdot \mathbf{d})(\mathbf{b} \cdot \mathbf{c}) = \begin{vmatrix} \mathbf{a} \cdot \mathbf{c} & \mathbf{a} \cdot \mathbf{d} \\ \mathbf{b} \cdot \mathbf{c} & \mathbf{b} \cdot \mathbf{d} \end{vmatrix}, \tag{11}$$

which is known as the **generalized Lagrange identity**. This reduces to the ordinary Lagrange identity (11) of Sec. 2.6 when $\mathbf{c}=\mathbf{a}$ and $\mathbf{d}=\mathbf{c}$.

Summary of properties of scalar and vector products

Commutativity of scalar product

$$\mathbf{a}\cdot\mathbf{b}=\mathbf{b}\cdot\mathbf{a}.$$

Linearity of scalar product with respect to addition

$$(\lambda\mathbf{a}+\mu\mathbf{b})\cdot\mathbf{c}=\lambda\mathbf{a}\cdot\mathbf{c}+\mu\mathbf{b}\cdot\mathbf{c}$$

Distributivity of scalar product with respect to addition

$$(\mathbf{a}+\mathbf{b})\cdot\mathbf{c}=\mathbf{a}\cdot\mathbf{c}+\mathbf{b}\cdot\mathbf{c}$$

Schwarz inequality

$$|\mathbf{a}\cdot\mathbf{b}|\le\|\mathbf{a}\|\ \|\mathbf{b}\|$$

Triangle inequality

$$\|\mathbf{a}+\mathbf{b}\|\le\|\mathbf{a}\|+\|\mathbf{b}\|.$$

For a mutually orthogonal set of unit vectors $\mathbf{i}, \mathbf{j}, \mathbf{k}$

$$\mathbf{i}\cdot\mathbf{i}=\mathbf{j}\cdot\mathbf{j}=\mathbf{k}\cdot\mathbf{k}=1$$
$$\mathbf{i}\cdot\mathbf{j}=\mathbf{j}\cdot\mathbf{i}-0,\quad \mathbf{j}\cdot\mathbf{k}=\mathbf{k}\cdot\mathbf{j}=0,\quad \mathbf{k}\cdot\mathbf{i}=\mathbf{i}\cdot\mathbf{k}=0$$
$$\mathbf{i}\times\mathbf{j}=\mathbf{k},\quad \mathbf{j}\times\mathbf{k}=\mathbf{i},\quad \mathbf{k}\times\mathbf{i}=\mathbf{j}$$
$$\mathbf{j}\times\mathbf{i}=-\mathbf{k},\quad \mathbf{k}\times\mathbf{j}=-\mathbf{i},\quad \mathbf{i}\times\mathbf{k}=-\mathbf{j}$$
$$\mathbf{i}\times\mathbf{i}=\mathbf{j}\times\mathbf{j}=\mathbf{k}\times\mathbf{k}=\mathbf{0}.$$

Let $\mathbf{a}=a_1\mathbf{i}+a_2\mathbf{j}+a_3\mathbf{k}, \mathbf{b}=b_1\mathbf{i}+b_2\mathbf{j}+b_3\mathbf{k}.$

Scalar product in terms of components
$$\mathbf{a}\cdot\mathbf{b}=a_1b_1+a_2b_2+a_3b_3$$

Magnitude in terms of the scalar product
$$\|\mathbf{a}\|=\sqrt{\mathbf{a}\cdot\mathbf{a}}=(a_1^2+a_2^2+a_3^2)^{1/2}\ge 0$$

Vector product in terms of components

$$\mathbf{a}\times\mathbf{b}=\begin{vmatrix} \mathbf{i} & \mathbf{j} & \mathbf{k} \\ a_1 & a_2 & a_3 \\ b_1 & b_2 & b_3 \end{vmatrix}.$$

The vector product is *not* commutative, but

$$\mathbf{a} \times \mathbf{b} = -\mathbf{b} \times \mathbf{a}.$$

The vector product is *not* associative so, in general,

$$\mathbf{a} \times (\mathbf{b} \times \mathbf{c}) \neq (\mathbf{a} \times \mathbf{b}) \times \mathbf{c}.$$

Distributivity of vector product with respect to addition

$$\mathbf{a} \times (\mathbf{b} + \mathbf{c}) = \mathbf{a} \times \mathbf{b} + \mathbf{a} \times \mathbf{c}$$
$$(\mathbf{a} + \mathbf{b}) \times \mathbf{c} = \mathbf{a} \times \mathbf{c} + \mathbf{b} \times \mathbf{c}.$$

Scalar triple product

$$\mathbf{a} \cdot (\mathbf{b} \times \mathbf{c}) = \begin{vmatrix} a_1 & a_2 & a_3 \\ b_1 & b_2 & b_3 \\ c_1 & c_2 & c_3 \end{vmatrix},$$

$$\mathbf{a} \cdot (\mathbf{b} \times \mathbf{c}) = \mathbf{b} \cdot (\mathbf{c} \times \mathbf{a}) = \mathbf{c} \cdot (\mathbf{a} \times \mathbf{b})$$
$$= -\mathbf{a} \cdot (\mathbf{c} \times \mathbf{b}) = -\mathbf{b} \cdot (\mathbf{a} \times \mathbf{c}) = -\mathbf{c} \cdot (\mathbf{b} \times \mathbf{a}).$$

Vector triple product

$$\mathbf{a} \times (\mathbf{b} \times \mathbf{c}) = (\mathbf{a} \cdot \mathbf{c})\mathbf{b} - (\mathbf{a} \cdot \mathbf{b})\mathbf{c}$$
$$(\mathbf{a} \times \mathbf{b}) \times \mathbf{c} = (\mathbf{a} \cdot \mathbf{c})\mathbf{b} - (\mathbf{b} \cdot \mathbf{c})\mathbf{a}.$$

Generalized Lagrange identity

$$(\mathbf{a} \times \mathbf{b}) \cdot (\mathbf{c} \times \mathbf{d}) = (\mathbf{a} \cdot \mathbf{c})(\mathbf{b} \cdot \mathbf{d}) - (\mathbf{a} \cdot \mathbf{d})(\mathbf{b} \cdot \mathbf{c})$$

$$= \begin{vmatrix} \mathbf{a} \cdot \mathbf{c} & \mathbf{a} \cdot \mathbf{d} \\ \mathbf{b} \cdot \mathbf{c} & \mathbf{b} \cdot \mathbf{d} \end{vmatrix}.$$

Linear independence

Vectors $\mathbf{a}$ and $\mathbf{b}$ will be linearly independent if, and only if, $\mathbf{a} \times \mathbf{b} \neq \mathbf{0}$, and they will be linearly dependent if, and only if, $\mathbf{a} \times \mathbf{b} = \mathbf{0}$. Vectors $\mathbf{a}$, $\mathbf{b}$ and $\mathbf{c}$ will be linearly dependent if, and only if, $\mathbf{a} \cdot (\mathbf{b} \times \mathbf{c}) \neq \mathbf{0}$, and they will be linearly dependent if, and only if, $\mathbf{a} \cdot (\mathbf{b} \times \mathbf{c}) = \mathbf{0}$.

Problems for Section 2.7

Find the scalar triple product $\mathbf{a} \cdot (\mathbf{b} \times \mathbf{c})$ if the respective vectors $\mathbf{a}$, $\mathbf{b}$ and $\mathbf{c}$ are as follows:

1 $\mathbf{i}$, $3\mathbf{j}$, $\mathbf{k}$ 2 $\mathbf{i}+\mathbf{j}$, $\mathbf{j}+\mathbf{k}$, $\mathbf{i}+\mathbf{k}$
3 $\mathbf{i}$, $2\mathbf{i}-\mathbf{j}$, $\mathbf{k}$ 4 $2\mathbf{i}+\mathbf{j}$, $3\mathbf{k}$, $2\mathbf{j}+\mathbf{k}$
5 $\mathbf{i}+\mathbf{j}$, $\mathbf{i}-\mathbf{j}$, $3\mathbf{k}$ 6 $2\mathbf{i}-\mathbf{j}+\mathbf{k}$, $2\mathbf{i}-\mathbf{j}$, $-4\mathbf{i}+2\mathbf{j}$

Find the volume of the parallelepiped with the following adjacent edges:

7 $i-j, i+j, -2k$ **8** $2i-j+k, 2j, i+k$

9 $i+j+k, j, -k$ **10** $2i, i+2j, i+j+k$

11 $i+2j+3k, 2i+j-k, -3j-7k$.

12 Prove the volume V of the tetrahedron with adjacent edges a, b and c as shown in Fig. 2.33 is

$$V = \left| \frac{1}{6} a \cdot (b \times c) \right|.$$

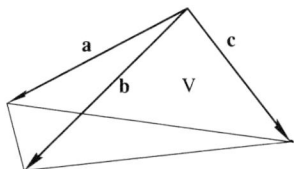

Fig. 2.33

13 What condition on λ will ensure the following three vectors will be linearly independent:

$$i+3j+2k, \quad 4i+5j-\frac{2\lambda}{3}k, \quad 3i+2j-4k?$$

14 Prove $a \cdot (b \times c) = 0$ if, and only if, the vectors a, b and c are all parallel to a single plane.

15 Which of the following sets of three vectors is linearly dependent and which is linearly independent?

(i) $i+2j+k, 2i-k, 4i+4j+k$

(ii) $i+j+k, i-j+k, i+j$

(iii) $i+j, i+k, j$

(iv) $2i-j+3k, -i+2j+k, 3j+5k$

(v) $3i+3j, -2j, 4k$.

16 If $\|a\| = 2$ and $\|b\| = 3$ show

$$(a \times b) \cdot (a+2b) + (a \cdot b)^2 + (a \times b)^2 = 36.$$

17 Prove

$$(a \times b) \cdot [(a \times c) \times d] = (a \cdot d)[a \cdot (b \times c)].$$

18 Prove

$$(a \times b) \cdot [(b \times c) \times (c \times a)] = [a \cdot (b \times c)]^2.$$

19 Prove

$$[a \cdot (b \times c)]^2 = \begin{vmatrix} a \cdot a & a \cdot b & a \cdot c \\ b \cdot a & b \cdot b & b \cdot c \\ c \cdot a & c \cdot b & c \cdot c \end{vmatrix},$$

in which the expression on the right is known as **Gram's determinant.**

20 Let a, b, c and d be vectors and x, y and z scalars satisfying the equation

$$(b \times c)x + (c \times a)y + (a \times b)z + d = 0.$$

Show that when **a**, **b** and **c** are linearly independent

$$x = -\frac{\mathbf{a} \cdot \mathbf{d}}{\mathbf{a} \cdot (\mathbf{b} \times \mathbf{c})}, \quad y = -\frac{\mathbf{b} \cdot \mathbf{d}}{\mathbf{a} \cdot (\mathbf{b} \times \mathbf{c})}, \quad z = -\frac{\mathbf{c} \cdot \mathbf{d}}{\mathbf{a} \cdot (\mathbf{b} \times \mathbf{c})}.$$

21 Let **a**, **b**, **c** and **d** be vectors and x, y and z scalars satisfying the equation

$$\mathbf{a}x + \mathbf{b}y + \mathbf{c}z + \mathbf{d} = \mathbf{0}.$$

By forming the scalar product of this equation with $(\mathbf{b} \times \mathbf{c})$, $(\mathbf{a} \times \mathbf{c})$ and $(\mathbf{a} \times \mathbf{b})$, respectively, show that when **a**, **b** and **c** are linearly independent

$$x = -\frac{\mathbf{d} \cdot (\mathbf{b} \times \mathbf{c})}{\mathbf{a} \cdot (\mathbf{b} \times \mathbf{c})}, \quad y = -\frac{\mathbf{d} \cdot (\mathbf{c} \times \mathbf{a})}{\mathbf{a} \cdot (\mathbf{b} \times \mathbf{c})}, \quad z = -\frac{\mathbf{d} \cdot (\mathbf{a} \times \mathbf{b})}{\mathbf{a} \cdot (\mathbf{b} \times \mathbf{c})}.$$

Find the vector triple products $\mathbf{a} \times (\mathbf{b} \times \mathbf{c})$ and $(\mathbf{a} \times \mathbf{b}) \times \mathbf{c}$ given the respective vectors **a**, **b** and **c** are:

22 **i**, **j**, **k**   **23** **i**, **i**+**j**, **j**+**k**   **24** **i**−**j**, **i**+**j**, **i**+**k**
25 **i**+**j**, 2**i**+2**j**, **i**+**k**   **26** **i**+**j**, **j**+**k**, **i**+**k**.
27 Prove that for arbitrary vectors **a** and **b**

$$(\mathbf{a} \times (\mathbf{b} \times \mathbf{a})) \cdot \mathbf{b} \geq 0 \quad \text{and} \quad ((\mathbf{a} \times \mathbf{b}) \times \mathbf{a}) \cdot \mathbf{b} \geq 0.$$

28 Prove that for arbitrary vectors **a**, **b** and **c**,

$$\mathbf{a} \times (\mathbf{b} \times \mathbf{c}) + \mathbf{b} \times (\mathbf{c} \times \mathbf{a}) + \mathbf{c} \times (\mathbf{a} \times \mathbf{b}) = \mathbf{0}.$$

29 If **a** is an arbitrary vector and **n̂** is a given unit vector prove

$$\mathbf{a} = (\mathbf{\hat{n}} \cdot \mathbf{a})\mathbf{\hat{n}} + \mathbf{\hat{n}} \times (\mathbf{a} \times \mathbf{\hat{n}}).$$

This shows that vector **a** may always be represented as the sum of a vector parallel to a given unit vector **n̂** and a vector orthogonal to **n̂**, and also how this representation may be obtained from **a** and **n̂**. Find the form taken by this representation when $\mathbf{a} = \mathbf{i} + 2\mathbf{j} + \mathbf{k}$ and **n̂** is in the direction **i**−**j**.

30 Prove that for arbitrary vectors **a**, **b**, **c** and **d**,

$$(\mathbf{a} \times \mathbf{b}) \times (\mathbf{c} \times \mathbf{d}) = [\mathbf{a} \cdot (\mathbf{c} \times \mathbf{d})]\mathbf{b} - [\mathbf{b} \cdot (\mathbf{c} \times \mathbf{d})]\mathbf{a} = [\mathbf{a} \cdot (\mathbf{b} \times \mathbf{d})]\mathbf{c} - [\mathbf{a} \cdot (\mathbf{b} \times \mathbf{c})]\mathbf{d}.$$

31 Prove that for arbitrary vectors **a**, **b** and **c**,

$$(\mathbf{a} \times \mathbf{b}) \cdot [\mathbf{c} \times (\mathbf{c} \times \mathbf{a})] = (\mathbf{a} \cdot \mathbf{c})[\mathbf{a} \cdot (\mathbf{b} \times \mathbf{c})].$$

32 By considering the product $(\mathbf{a} \times \mathbf{b}) \times (\mathbf{c} \times \mathbf{d})$, with **a**, **b**, **c** and **d** arbitrary vectors, prove the identity

$$[\mathbf{a} \cdot (\mathbf{b} \times \mathbf{c})]\mathbf{d} = [\mathbf{d} \cdot (\mathbf{b} \times \mathbf{c})]\mathbf{a} + [\mathbf{a} \cdot (\mathbf{d} \times \mathbf{c})]\mathbf{b} + [\mathbf{a} \cdot (\mathbf{b} \times \mathbf{d})].$$

2.8   Geometrical applications of scalar and vector products

Many instructive and useful applications of scalar and vector products are to be found in a vector approach to coordinate geometry. The few cases which are discussed in this section have been chosen because they arise frequently as part of more general problems, and also because of the vector techniques which are necessary to resolve them.

Straight line

Figure 2.34 shows a straight line L in $\mathbb{R}^3$ through the point A with position vector **a** in the direction of a nonzero arbitrary vector **b**. If **r** is the position vector of a representative point P on the line it is always possible to choose the scalar λ so that $\overrightarrow{AP} = \lambda\mathbf{b}$. To allow P to be any point on L it is necessary that the parameter λ be such that $-\infty < \lambda < \infty$. The law for vector addition applied to the triangle OAP then gives as the vector equation for the straight line L with parameter λ

$$\mathbf{r} = \mathbf{a} + \lambda\mathbf{b}, \tag{1}$$

where $-\infty < \lambda < \infty$. Since **a** is the position vector of any point on L and **b** is any vector along the line, the appearance of (1) when expressed with **a** and **b** in component form will depend on the choice of **a** and **b**.

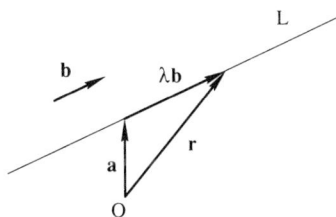

Fig. 2.34 Vector equation of straight line L.

The more familiar Cartesian equations for L follow from (1) by equating the **i**, **j** and **k** components and eliminating λ. Setting $\mathbf{r} = x\mathbf{i} + y\mathbf{j} + z\mathbf{k}$, $\mathbf{a} = a_1\mathbf{i} + a_2\mathbf{j} + a_3\mathbf{k}$ and $\mathbf{b} = b_1\mathbf{i} + b_2\mathbf{j} + b_3\mathbf{k}$ and equating components of (1) gives

$x = a_1 + \lambda b_1$ (equality of **i** components)

$y = a_2 + \lambda b_2$ (equality of **j** components)

$z = a_3 + \lambda b_3$ (equality of **k** components)

Finally, eliminating λ gives[4]

$$\frac{x - a_1}{b_1} = \frac{y - a_2}{b_2} = \frac{z - a_3}{b_3} (= \lambda), \tag{2}$$

which is the familiar Cartesian form of the equations of a straight line in $\mathbb{R}^3$.

[4] If, say, $b_2 = 0$ in (2), the corresponding numerator must vanish identically giving $y \equiv a_2$, otherwise this term would be infinite. The equations then reduce to

$$\frac{x - a_1}{b_1} = \frac{z - a_3}{b_3} \text{ with } y = a_2,$$

so the line lies in the plane $y = a_2$.

Notice that when written in the **standard form** (2), in which the coefficients of x, y and z are all equal to unity, (a_1, a_2, a_3) is a point on the line L while b_1, b_2, b_3, are the direction ratios of L.

Example 1

(i) Find the vector equation of the straight line through point A at $(1, 2, 1)$ and point B at $(2, 4, 3)$.

(ii) Express in vector form the straight line $\dfrac{2x-1}{3} = \dfrac{-(y+2)}{4} = \dfrac{3z+2}{6} = \lambda.$

Solution

(i) A vector $\mathbf{b}$ along L is the vector $\overrightarrow{AB}$, so $\mathbf{b} = \overrightarrow{AB} = \mathbf{i} + 2\mathbf{j} + 2\mathbf{k}$. Point A lies on L so its position vector $\mathbf{a} = \mathbf{i} + 2\mathbf{j} + \mathbf{k}$ will serve as the vector $\mathbf{a}$ in (1). The vector equation of L then becomes

$$\mathbf{r} = (\mathbf{i} + 2\mathbf{j} + \mathbf{k}) + \lambda(\mathbf{i} + 2\mathbf{j} + 2\mathbf{k}),$$

for $-\infty < \lambda < \infty$.

(ii) Expressing the equations in standard form by making the coefficients of x, y and z unity gives

$$\frac{x - 1/2}{3/2} = \frac{y + 2}{-4} = \frac{z + 2/3}{2}.$$

The position vector $\mathbf{a}$ of a point on the line is thus $\mathbf{a} = \frac{1}{2}\mathbf{i} - 2\mathbf{j} - \frac{2}{3}\mathbf{k}$, while a vector $\mathbf{b}$ along the line is $\mathbf{b} = \frac{3}{2}\mathbf{i} - 4\mathbf{j} + 2\mathbf{k}$. The required vector equation of the line becomes

$$\mathbf{r} = \tfrac{1}{2}\mathbf{i} - 2\mathbf{j} - \tfrac{2}{3}\mathbf{k} + \lambda(\tfrac{3}{2}\mathbf{i} - 4\mathbf{j} + 2\mathbf{k}),$$

for $-\infty < \lambda < \infty$.

It should again be emphasized that the form of this equation depends on the choice of $\mathbf{a}$ and $\mathbf{b}$ in (1). Setting $\lambda = 2$ in the Cartesian equations to find another point on the line gives $x = \frac{7}{2}$, $y = -10$, $z = \frac{10}{3}$, while the direction ratios of the line may equally well be taken as 3, -8, 4 giving $\mathbf{b} = 3\mathbf{i} - 8\mathbf{j} + 4\mathbf{k}$. So another form of the vector equation of the same line is

$$\mathbf{r} = (\tfrac{7}{2}\mathbf{i} - 10\mathbf{j} + \tfrac{10}{3}\mathbf{k}) + \lambda(3\mathbf{i} - 8\mathbf{j} + 4\mathbf{k}),$$

for $-\infty < \lambda < \infty$. ∎

Perpendicular distance between two skew lines

Let the straight lines L_1 and L_2 in $\mathbb{R}^3$ be skew (not parallel) and have the vector equations

$$L_1: \mathbf{r} = \mathbf{a}_1 + \lambda\mathbf{b}_1$$
$$L_2: \mathbf{r} = \mathbf{a}_2 + \mu\mathbf{b}_2. \tag{3}$$

We shall determine the perpendicular distance d between the two lines as shown in Fig. 2.35.

By definition, the perpendicular distance d between the lines L_1 and L_2 is the length of the projection of any straight line segment AB joining a point A on L_1 and a point B on L_2

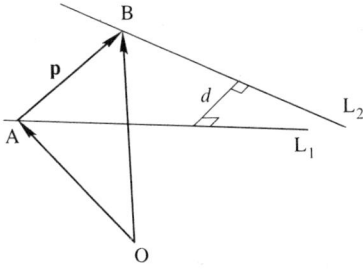

Fig. 2.35 Skew lines a perpendicular distance d apart

onto a line which is perpendicular to both L_1 and L_2. For A we take the point with position vector $\mathbf{a}_1$ and for B the point with position vector $\mathbf{a}_2$, so $\mathbf{p} = \overrightarrow{AB} = \mathbf{a}_2 - \mathbf{a}_1$. A vector $\mathbf{N}$ normal to both L_1 and L_2 is $\mathbf{N} = \mathbf{b}_1 \times \mathbf{b}_2$, so a unit vector normal to these lines is

$$\hat{\mathbf{N}} = \frac{\mathbf{b}_1 \times \mathbf{b}_2}{\|\mathbf{b}_1 \times \mathbf{b}_2\|}. \tag{4}$$

The length of the projection is $|\mathbf{p} \cdot \hat{\mathbf{N}}|$, so that

$$d = \frac{1}{\|\mathbf{b}_1 \times \mathbf{b}_2\|} |\mathbf{p} \cdot (\mathbf{b}_1 \times \mathbf{b}_2)|. \tag{5}$$

Notice that d is expressible in terms of the magnitude of the scalar triple product $\mathbf{p} \cdot (\hat{\mathbf{b}}_1 \times \hat{\mathbf{b}}_2)$.

An immediate consequence of this result is that for the lines L_1 and L_2 to intersect it is necessary that

$$\mathbf{p} \cdot (\mathbf{b}_1 \times \mathbf{b}_2) = 0, \tag{6}$$

for then $d = 0$. When intersection occurs, the point of intersection $\mathbf{r} = \mathbf{c}$ may be found from (3) by first solving the vector equation

$$\mathbf{c} = \mathbf{a}_1 + \lambda \mathbf{b}_1 = \mathbf{a}_2 + \mu \mathbf{b}_2$$

for λ and μ, and then substituting either λ or μ into equations (3) to find $\mathbf{r}$.

Example 2

Find the perpendicular distance between line L_1 with direction ratios 1, 0, -2 through the point $(2, 1, -1)$, and line L_2 with direction ratios 1, 1, -1, through the point $(2, 2, 0)$. If the lines intersect find their point of intersection.

Solution

Here $\mathbf{a}_1 = 2\mathbf{i} + \mathbf{j} - \mathbf{k}$, $\mathbf{a}_2 = 2\mathbf{i} + 2\mathbf{j}$, $\mathbf{b}_1 = \mathbf{i} - 2\mathbf{k}$ and $\mathbf{b}_2 = \mathbf{i} + \mathbf{j} - \mathbf{k}$. So $\mathbf{p} = \mathbf{a}_2 - \mathbf{a}_1 = \mathbf{j} + \mathbf{k}$ and $\mathbf{N} = \mathbf{b}_1 \times \mathbf{b}_2$ $= 2\mathbf{i} - \mathbf{j} + \mathbf{k}$, when $\hat{\mathbf{N}} = \dfrac{1}{\sqrt{6}}(2\mathbf{i} - \mathbf{j} + \mathbf{k})$. The perpendicular distance d between the lines is

$$d = \frac{1}{\sqrt{6}} |(\mathbf{j} + \mathbf{k}) \cdot (2\mathbf{i} - \mathbf{j} + \mathbf{k})| = 0,$$

so the lines intersect.

The position vector $\mathbf{r} = \mathbf{c}$ of the point of intersection is given by $\mathbf{c} = 2\mathbf{i} + \mathbf{j} - \mathbf{k} + \lambda(\mathbf{i} - 2\mathbf{k}) = 2\mathbf{i} + 2\mathbf{j} + \mu(\mathbf{i} + \mathbf{j} - \mathbf{k})$. This has the solution $\lambda = 1$, $\mu = -1$, so $\mathbf{c} = 3\mathbf{i} + \mathbf{j} - 3\mathbf{k}$. ∎

Perpendicular distance of a point from a straight line

It is required to find the perpendicular distance d of a point P with position vector $\mathbf{c}$ in $\mathbb{R}^3$ from the straight line L with the vector equation $\mathbf{r} = \mathbf{a} + \lambda\mathbf{b}$ (Fig. 2.36).

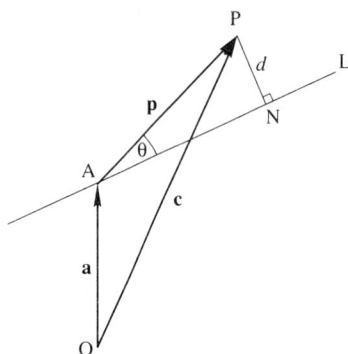

Fig. 2.36   Perpendicular distance d of P from L

By definition, d is the length of the projection of the line segment AP in the direction of the normal to L through P. Now a vector normal to both the line L and $\mathbf{p} = \mathbf{c} - \mathbf{a}$ is the vector $\mathbf{b} \times \mathbf{p}$, so a vector parallel to NP is $(\mathbf{b} \times \mathbf{p}) \times \mathbf{b}$. A unit vector $\hat{\mathbf{N}}$ parallel to NP is thus

$$\hat{\mathbf{N}} = \frac{1}{\|(\mathbf{b} \times \mathbf{p}) \times \mathbf{b}\|} (\mathbf{b} \times \mathbf{p}) \times \mathbf{b}. \tag{7}$$

The length of the projection of $\mathbf{p}$ in the direction $\hat{\mathbf{N}}$ is

$$|\hat{\mathbf{N}} \cdot \mathbf{p}| = \frac{1}{\|(\mathbf{b} \times \mathbf{p}) \times \mathbf{b}\|} |((\mathbf{b} \times \mathbf{p}) \times \mathbf{b}) \cdot \mathbf{p}|$$

$$= \frac{1}{\|(\mathbf{b} \times \mathbf{p}) \times \mathbf{b}\|} |[(\mathbf{b} \cdot \mathbf{b})(\mathbf{p} \cdot \mathbf{p}) - (\mathbf{p} \cdot \mathbf{b})^2]|.$$

However, the expression in parentheses is merely $\|\mathbf{b}\|^2 \|\mathbf{p}\|^2 (1 - \cos^2 \theta)$, which is nonnegative, so the absolute value is unnecessary and length d is given by

$$d = \frac{1}{\|(\mathbf{b} \times \mathbf{p}) \times \mathbf{b}\|} [(\mathbf{b} \cdot \mathbf{b})(\mathbf{p} \cdot \mathbf{p}) - (\mathbf{p} \cdot \mathbf{b})^2]. \tag{8}$$

Example 3

Find the perpendicular distance d of the point $(1, 0, 1)$ from the straight line L with direction ratios 1, 1, 1 which passes through the point $(1, 3, 2)$.

Solution

Here $\mathbf{p} = \mathbf{i} + \mathbf{k} - (\mathbf{i} + 3\mathbf{j} + 2\mathbf{k}) = -3\mathbf{j} - \mathbf{k}$ and $\mathbf{b} = \mathbf{i} + \mathbf{j} + \mathbf{k}$. Thus as $(\mathbf{b} \times \mathbf{p}) \times \mathbf{b} = 4\mathbf{i} - 5\mathbf{j} + \mathbf{k}$, $\|(\mathbf{b} \times \mathbf{p}) \times \mathbf{b}\| = \sqrt{42}$ and $(\mathbf{b} \cdot \mathbf{b})(\mathbf{p} \cdot \mathbf{p}) - (\mathbf{p} \cdot \mathbf{b})^2 = 14$ so result (8) shows $d = 14/\sqrt{42}$.

■

The plane

Let a plane Π be determined by the requirements that it is normal to the vector $\mathbf{n}$ and it contains the point P with position vector $\mathbf{a}$ (Fig. 2.37). Suppose Q is any point on Π distinct from P, and let its position vector be $\mathbf{r}$. Then $\overrightarrow{PQ} = \mathbf{r} - \mathbf{a}$ lies in the plane Π and so is orthogonal to $\mathbf{n}$. It thus follows that

$$\mathbf{n} \cdot (\mathbf{r} - \mathbf{a}) = 0$$

or, equivalently,

$$\mathbf{n} \cdot \mathbf{r} = \mathbf{n} \cdot \mathbf{a}, \tag{9}$$

which is the vector equation of the plane Π.

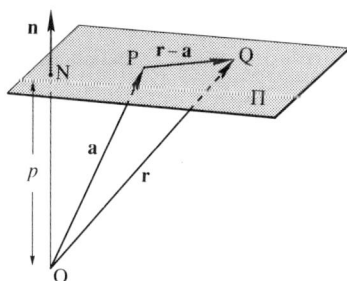

Fig. 2.37 Plane Π with normal $\mathbf{n}$ and containing point P

If $\mathbf{n}$ is normalized to the unit vector $\hat{\mathbf{n}}$, result (9) becomes

$$\hat{\mathbf{n}} \cdot \mathbf{r} = \hat{\mathbf{n}} \cdot \mathbf{a},$$

or

$$\hat{\mathbf{n}} \cdot \mathbf{r} = p, \tag{10}$$

where $|\mathbf{p}| = ON$ is the perpendicular distance of the plane from the origin. As $\overrightarrow{ON} = p\hat{\mathbf{n}}$, we see the foot N of the perpendicular from the origin to plane Π has the position vector $p\hat{\mathbf{n}}$. Equation (10) is sometimes called the **normal** or **standard form** of the equation of a plane.

If $\mathbf{r} = x\mathbf{i} + y\mathbf{j} + z\mathbf{k}$, $\mathbf{n} = n_1\mathbf{i} + n_2\mathbf{j} + n_3\mathbf{k}$ and $\mathbf{a} = a_1\mathbf{i} + a_2\mathbf{j} + a_3\mathbf{k}$ the Cartesian form of (9) becomes

$$n_1 x + n_2 y + n_3 z = n_1 a_1 + n_2 a_2 + n_3 a_3. \tag{11}$$

Similarly, the Cartesian form of (10) becomes

$$lx + my + nz = p, \tag{12}$$

with l, m, n the direction cosines (components) of $\hat{\mathbf{n}}$

$$l = \frac{n_1}{||\mathbf{n}||}, \quad m = \frac{n_2}{||\mathbf{n}||}, \quad n = \frac{n_3}{||\mathbf{n}||}, \tag{13}$$

and

$$p = la_1 + ma_2 + na_3. \tag{14}$$

Examination of (11) shows that a general Cartesian equation of the form

$$ax + by + cz = f \tag{15}$$

represents a plane with normal $\mathbf{n}$ having the direction ratios a, b, c, located at a perpendicular distance (regarded as nonnegative)

$$d = \frac{|f|}{(a^2 + b^2 + c^2)^{1/2}} \tag{16}$$

from the origin. The foot of the perpendicular from the origin to the plane has the position vector $\mathbf{r} = (\operatorname{sgn} f)\, d\hat{\mathbf{n}}$, where sgn denotes the signum function.

The angle θ between two planes with normals $\mathbf{m}$ and $\mathbf{n}$ is the angle between their normals, so from their scalar product we have

$$\cos\theta = \frac{\mathbf{m}\cdot\mathbf{n}}{||\mathbf{m}||\,||\mathbf{n}||}. \tag{17}$$

As an illustration of the use of (10) we now determine the perpendicular distance d of a

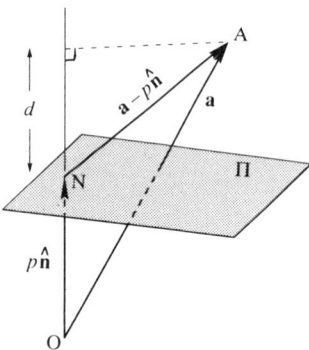

Fig. 2.38 Perpendicular distance d of A from plane Π

point A with position vector **a** from the plane Π whose normal form is

$$\hat{\mathbf{n}} \cdot \mathbf{r} = p. \tag{18}$$

The situation is illustrated in Fig. 2.38. By definition, d is the length of the projection of any line drawn from A to a point on the plane in the direction of the normal to the plane. Point N lies on the plane, so NA is such a line. The position vector of N is $p\hat{\mathbf{n}}$, and the position vector of A is **a**, so

$$\overrightarrow{\text{NA}} = \mathbf{a} - p\hat{\mathbf{n}}, \tag{19}$$

from which it follows that $d = |(\overrightarrow{\text{NA}}) \cdot \hat{\mathbf{n}}|$, or

$$d = |\mathbf{a} \cdot \hat{\mathbf{n}} - p|. \tag{20}$$

Example 4

Find (i) the equation of the plane having a normal with direction ratios 1, -1, 2 and containing the point (1, 0, 1), (ii) the perpendicular distance d of the point $(-2, -3, -1)$ from the plane $x + 2y + z = -1$, (iii) the angle between the planes $x + y + z = 3$ and $x - y + 2z = 1$.

Solution

(i) Here the normal $\mathbf{n} = \mathbf{i} - \mathbf{j} + 2\mathbf{k}$ and $\mathbf{a} = \mathbf{i} + \mathbf{k}$, so setting $\mathbf{r} = x\mathbf{i} + y\mathbf{j} + z\mathbf{k}$, (9) becomes

$$x - y + 2z = 3.$$

(ii) Expressing the equation of the plane in normal form gives

$$\frac{1}{\sqrt{6}}x + \frac{2}{\sqrt{6}}y + \frac{1}{\sqrt{6}}z = \frac{-1}{\sqrt{6}}.$$

The unit normal to this plane $\hat{\mathbf{n}} = \frac{1}{\sqrt{6}}\mathbf{i} + \frac{2}{\sqrt{6}}\mathbf{j} + \frac{1}{\sqrt{6}}\mathbf{k}$, and thus the foot of the perpendicular from the origin to the plane has position vector $-\frac{1}{\sqrt{6}}\hat{\mathbf{n}}$. Point $(-2, -3, -1)$ has position vector $\mathbf{a} = -2\mathbf{i} - 3\mathbf{j} - \mathbf{k}$, so $\mathbf{a} \cdot \hat{\mathbf{n}} = -9/\sqrt{6}$. The perpendicular distance of the point from the plane is thus $d = |-9/\sqrt{6} - (-1)/\sqrt{6}| = 8/\sqrt{6}$.

(iii) The normals to the two planes are $\mathbf{m} = \mathbf{i} + \mathbf{j} + \mathbf{k}$ and $\mathbf{n} = \mathbf{i} - \mathbf{j} + 2\mathbf{k}$, so from (17) the angle θ between the planes is determined by

$$\cos\theta = \frac{2}{\sqrt{3} \cdot \sqrt{6}}, \text{ so that } \theta = 61.9°. \qquad \blacksquare$$

Line of intersection of two nonparallel planes

The planes Π_1 and Π_2 in $\mathbb{R}^3$ with the equations

$$m_1 x + m_2 y + m_3 z = q,$$
$$n_1 x + n_2 y + n_3 z = s, \tag{21}$$

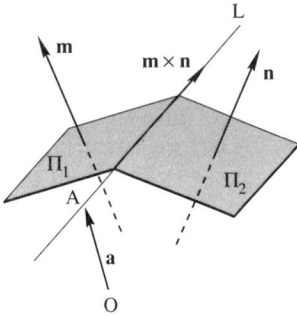

Fig. 2.39   Line L of intersection of two planes

will intersect in a straight line L provided they are not parallel (Fig. 2.29). Their line of intersection lies in each plane and so will be orthogonal to both the normal $\mathbf{m} = m_1\mathbf{i} + m_2\mathbf{j} + m_3\mathbf{k}$ to Π_1 and the normal $\mathbf{n} = n_1\mathbf{i} + n_2\mathbf{j} + n_3\mathbf{k}$ to Π_2, showing its direction must be along $\mathbf{m} \times \mathbf{n}$.

The general equation of a line in this direction must be

$$\mathbf{r} = \mathbf{a} + \lambda \mathbf{m} \times \mathbf{n}, \tag{22}$$

where $\mathbf{a} = a_1\mathbf{i} + a_2\mathbf{j} + a_3\mathbf{k}$ is the position vector of any point A on the line L and $-\infty < \lambda < \infty$. As line L is common to both planes, the vector $\mathbf{r} = \mathbf{a}$ must satisfy each equation in (21). That is, we must set $x = a_1$, $y = a_2$ and $z = a_3$ in each equation and, as a result, obtain the two conditions

$$m_1 a_1 + m_2 a_2 + m_3 a_3 = q,$$
$$n_1 a_1 + n_2 a_2 + n_3 a_3 = s. \tag{23}$$

Assigning one of three components a_1, a_2, a_3 arbitrarily, for $\mathbf{a}$ is an arbitrary point on L, enables the other two components to be found uniquely from these two simultaneous equations. The position vector $\mathbf{a}$ of a point on the line of intersection can thus be found, after which the equation of the line L itself follows from (22) with this choice of $\mathbf{a}$.

Example 5

Find the line of intersection of the planes with the equations $x - y + 3z = 1$ and $2x + y - z = 3$.

Solution

A vector $\mathbf{m}$ normal to the first plane is $\mathbf{m} = \mathbf{i} - \mathbf{j} + 3\mathbf{k}$ and a vector $\mathbf{n}$ normal to the second plane is $\mathbf{n} = 2\mathbf{i} + \mathbf{j} - \mathbf{k}$, so the direction of the line of intersection is $\mathbf{m} \times \mathbf{n} = -2\mathbf{i} + 7\mathbf{j} + 3\mathbf{k}$. From (21) the components a_1, a_2 and a_3 of a point $\mathbf{r} = \mathbf{a}$ on the line of intersection must satisfy the equations

$$a_1 - a_2 + 3a_3 = 1 \quad \text{and} \quad 2a_1 + a_2 - a_3 = 3.$$

Setting $a_3 = 0$ (this can be done arbitrarily, so we choose a convenient value) and solving for a_1 and a_2 gives $a_1 = 4/3$, $a_2 = 1/3$. Thus $\mathbf{a} = \frac{4}{3}\mathbf{i}, +\frac{1}{3}\mathbf{j}$, and from (22) the vector equation of the line of intersection is seen to be

$$\mathbf{r} = \tfrac{4}{3}\mathbf{i} + \tfrac{1}{3}\mathbf{j} + \lambda(-2\mathbf{i} + 7\mathbf{j} + 3\mathbf{k}),$$

for $-\infty < \lambda < \infty$. ∎

Problems for Section 2.8

1 Find the two unit vectors parallel to the line

$$\mathbf{r} = \mathbf{i} - \mathbf{j} + 3\mathbf{k} + \lambda(\mathbf{i} + 4\mathbf{j} + 2\mathbf{k}).$$

2 Find the position vectors of three distinct points on the straight line

$$\mathbf{r} = \mathbf{i} + \mathbf{j} + 3\mathbf{k} + \lambda(2\mathbf{i} - \mathbf{j} - \mathbf{k}).$$

In each of the following problems find the equation of the straight line through the point P in the given direction:

3 P(1, 2, 1) , direction ratios, 2, 0, 3.
4 P(2, 4, −1) , direction ratios, 1, 2, 1.
5 P(1, −1, −2) , direction ratios, 1, 0, 1.
6 P(0, 2, 0) , direction ratios, 1, 1, 2.

In each of the following problems find the equation of the straight line through the points P and Q:

7 P(2, −3, 1) , Q(1, 2, −3).
8 P(1, 4, 1) , Q(2, −2, 3).
9 P(1, 1, 1) , Q(2, 2, 3).
10 P(1, −1, 3) , Q(1, 0, 1).
11 Express the following straight lines in the vector form $\mathbf{r} = \mathbf{a} + \lambda\mathbf{b}$.

(i) $\dfrac{x+2}{3} = \dfrac{y+1}{2} = \dfrac{z-1}{-1}$ (ii) $\dfrac{2x-1}{2} = \dfrac{3y+4}{-2} = \dfrac{-2z+3}{5}$.

12 Re-express the following vector equations of a straight line in Cartesian form:

(i) $\mathbf{r} = 3\mathbf{i} + \mathbf{k} + \lambda(2\mathbf{i} - \mathbf{j} + \mathbf{k})$,
(ii) $\mathbf{r} = \mathbf{i} - 2\mathbf{j} - 3\mathbf{k} + \lambda(\mathbf{i} + 4\mathbf{j} - \mathbf{k})$,
(iii) $\mathbf{r} = -\mathbf{i} + \mathbf{j} - 3\mathbf{k} + \lambda(\mathbf{i} + 2\mathbf{k})$.

13 Which of the following pairs of lines intersect, and when they do what is their point of intersection:

(i) $-x = y - 3 = \dfrac{z-7}{4}$ and $\dfrac{x+2}{-3} = \dfrac{y-2}{0} = \dfrac{z-5}{2}$,

(ii) $\mathbf{r} = -\mathbf{i} + \mathbf{j} + 6\mathbf{k} + \lambda(-2\mathbf{i} + \mathbf{j} + 5\mathbf{k})$
and $\mathbf{r} = \mathbf{i} + 2\mathbf{k} + \mu(2\mathbf{i} - \mathbf{j} - 5\mathbf{k})$,
(iii) $\mathbf{r} = 2\mathbf{i} - \mathbf{j} + \mathbf{k} + \lambda(\mathbf{i} + 2\mathbf{j} + 3\mathbf{k})$
and $\mathbf{r} = 4\mathbf{i} + \mathbf{j} + 2\mathbf{k} + \mu(2\mathbf{i} + 2\mathbf{j} + \mathbf{k})$?

14 Use result (5) to prove by vector methods that any two lines in a plane will always intersect provided they are not parallel.

15 Find the perpendicular distance between the line with direction ratios 1, 2, 1 through the point (1, 0, 1) and the line with direction ratios 1, 3, 1 through the point (1, 1, 2).

16 Find the perpendicular distance between the line with direction ratios 2, 0, 3 through the point (1, 1, 1) and the line with direction ratios 1, 1, 2 through the point (2, 1, 2).

17 Find the perpendicular distance of the point (2, 3, 1) from the straight line with direction ratios 1, −1, 1 through the point (1, 2, 1).

18 Find the perpendicular distance of the point (1, 2, 1) from the straight line through the points (1, 1, 1) and (2, 2, 1).

Find the vector and Cartesian forms of the equation of the plane with the given normal **n** containing the given point **a**:

19 $\mathbf{n} - \mathbf{i} + 3\mathbf{j} - 4\mathbf{k}$, $\mathbf{a} = 2\mathbf{i} - \mathbf{j} - \mathbf{k}$.

20 $\mathbf{n} = 2\mathbf{i} - \mathbf{j} + 2\mathbf{k}$, $\mathbf{a} = \mathbf{i} - \mathbf{k}$.

21 $\mathbf{n} = 3\mathbf{i} - 3\mathbf{j} + \mathbf{k}$, $\mathbf{a} = 2\mathbf{i} + 2\mathbf{j}$.

22 $\mathbf{n} = \mathbf{i} + \mathbf{j} + 2\mathbf{k}$, $\mathbf{a} = -\mathbf{j} + \mathbf{k}$.

23 Find the equation of the plane containing the points (1, 0, 1), (2, 1, 1) and (3, 2, 4).

24 Show the equation of the plane containing the two parallel straight lines $\mathbf{r} = \mathbf{a} + \lambda\mathbf{b}$ and $\mathbf{r} = \mathbf{c} + \mu\mathbf{b}$ may be written

$$[(\mathbf{c} - \mathbf{a}) \times \mathbf{b}] \cdot \mathbf{r} = [(\mathbf{c} - \mathbf{a}) \times \mathbf{b}] \cdot \mathbf{a}.$$

25 Find the angles between:

(i)   two planes with normals $\mathbf{m} = \mathbf{i} - \mathbf{j} + 2\mathbf{k}$, $\mathbf{n} = \mathbf{i} + 2\mathbf{j} + \mathbf{k}$,

(ii) the two planes $3x + y - z = 3$ and $x + y - 2z = -1$,

(iii) the plane $x + y + z = -1$ and the plane with a normal having the direction ratios 1, 2, 0 and containing the point (1, 1, 3).

26 Show the condition that the two planes

$$a_1 x + a_2 y + a_3 z = p$$

and

$$b_1 x + b_2 y + b_3 z = q$$

are orthogonal is $a_1 b_1 + a_2 b_2 + a_3 b_3 = 0$.

27 Find the perpendicular distance of the point (2, 1, 4) from the plane $2x + 3y - z = 1$.

28 Find the perpendicular distance of the point (1, 2, 4) from the plane whose normal has direction ratios 1, 1, 1 and which contains the point (1, 1, 2).

Find the line of intersection of the following pairs of planes:

29 $x + 2y - z = 1$, $2x + y + z = 2$.

30 $2x - y - z = 2$, $x + 2y + 2z = 3$.

31 $x - y + 2z = 3$, $x + y + z = 1$.

32 $x - y + z = 1$, $x + y - z = 2$.

33 Derive result (5) by considering parallel planes Π_1 and Π_2 containing the lines L_1 and L_2, respectively.

2.9 Vector spaces

A **vector space** was first introduced in Sec. 2.3 using the concrete example of the vector space of geometrical vectors in $\mathbb{R}^3$. In that section the laws of addition and multiplication by a scalar were defined in such a way that geometrical vectors exhibit the properties of vector quantities in the physical world. It turns out that if the idea of a vector space is generalized, by basing its defining axioms on the properties of geometrical vectors in $\mathbb{R}^3$, the resulting concept has wide application throughout mathematics.

Although abstract, the study of vector spaces is of real importance to engineers and applied mathematicians because of the central role it plays in fields such as numerical analysis, ordinary and partial differential equations and functional analysis, all of which are used extensively. It is the purpose of this section to provide a brief introduction to vector spaces in which only the essential ideas will be introduced. The following axioms used to define a general vector space are based on properties P.1 to P.8 possessed by geometrical vectors and listed in Sec. 2.3.

Definition (Vector space)

A **vector space** V (also called a **linear space**) is a nonempty set of elements **x, y, z**, . . . called **vectors** for which there are defined two algebraic operations. One of these, called **vector addition**, is a rule for associating with each pair of vectors **x** and **y** in V another vector **x + y** called the **sum** of **x** and **y**. The other, called **multiplication** of a vector by a scalar, is a rule for associating with each vector **x** and each scalar λ a vector λ**x** called the **scalar multiple** of **x**. The following axioms are assumed to be true, with **x, y, z** representing any vectors in V and λ, μ any scalars, either real or complex.

Addition
A.1 **x + y** is a vector in V (**closure** under addition)
A.2 **x + y = y + x**   (**commutativity**)
A.3 **x + (y + z) = (x + y) + z**   (**associativity**)
A.4 There is a unique **zero** or **null vector 0** in V with the property that

$$\mathbf{x + 0 = x}.$$

A.5 Each vector **x** in V has associated with it a unique element $-$**x** called the **negative** (**additive inverse**) of **x** with the property that

$$\mathbf{x + (-x) = 0}.$$

Multiplication
M.1 λ**x** is in V (**closure** under scalar multiplication)
M.2 $\lambda(\mathbf{x + y}) = \lambda\mathbf{x} + \lambda\mathbf{y}$ (**distributivity**)

M.3 $(\lambda + \mu)\mathbf{x} = \lambda\mathbf{x} + \mu\mathbf{x}$ **(distributivity)**
M.4 $\lambda(\mu\mathbf{x}) = (\lambda\mu)\mathbf{x}$ **(associativity)**
M.5   If $\lambda = 1$ then

$$1\mathbf{x} = \mathbf{x}.$$ **(scaling by unity)**.

If in the above definition the scalars $\lambda, \mu, \ldots$ are taken to be real numbers V is called a **real vector space**, while if the scalars $\lambda, \mu, \ldots$ are taken to be complex numbers it is called a **complex vector space.**

A comparison of these ten axioms with properties P.1 to P.8 possessed by geometrical vectors in $\mathbb{R}^3$ shows the inclusion of the **closure axioms** A.1 and M.1. These ensure that in a general vector space the result of addition and multiplication is to produce vectors which themselves belong to the same vector space, and do not lie outside it. A formal statement of these properties was unnecessary in the case of the vector space of geometrical vectors in $\mathbb{R}^3$, since they were included in the definitions of addition and scalar multiplication of geometrical vectors.

Let $\lambda_1, \lambda_2, \ldots, \lambda_m$ be any m scalars and $\mathbf{x}_1, \mathbf{x}_2, \ldots, \mathbf{x}_m$ any m vectors belonging to V. Then the sum

$$\lambda_1\mathbf{x}_1 + \lambda_2\mathbf{x}_2 + \ldots + \lambda_m\mathbf{x}_m, \tag{1}$$

which by A.1 also belongs to V, is called a **linear combination** of the vectors $\mathbf{x}_1, \mathbf{x}_2, \ldots, \mathbf{x}_m$. If the set of vectors $\mathbf{x}_1, \mathbf{x}_2, \ldots, \mathbf{x}_m$ is fixed, but the scalars $\lambda_1, \lambda_2, \ldots, \lambda_m$ are allowed to take all possible values, a set U of vectors is generated containing all the resulting linear combinations (1). The set U, itself a vector space, is said to be **spanned** by the vectors $\mathbf{x}_1, \mathbf{x}_2, \ldots, \mathbf{x}_m$, which are themselves called the **span** of U and written span $\{\mathbf{x}_1, \mathbf{x}_2, \ldots, \mathbf{x}_m\}$.

A set of vectors $\mathbf{x}_1, \mathbf{x}_2, \ldots, \mathbf{x}_m$ will be said to be **linearly independent** if, and only if, the vector equation

$$\lambda_1\mathbf{x}_1 + \lambda_2\mathbf{x}_2 + \ldots + \lambda_m\mathbf{x}_m = \mathbf{0} \tag{2}$$

is only true when the scalars $\lambda_1 = \lambda_2 = \ldots = \lambda_m = 0$. If the vectors $\mathbf{x}_1, \mathbf{x}_2, \ldots, \mathbf{x}_m$ are such that (2) is true for some set of scalars $\lambda_1, \lambda_2, \ldots, \lambda_m$ which are not all zero then the vectors $\mathbf{x}_1, \mathbf{x}_2, \ldots, \mathbf{x}_m$ will be said to be **linearly dependent**.

A set of vectors $\mathbf{e}_1, \mathbf{e}_2, \ldots, \mathbf{e}_n$ belonging to V will be called a **basis** for the vector space V when it spans V and is linearly independent. The **dimension** n of a vector space V is the number of vectors in a basis for V. For a vector space V of dimension n, this definition implies every set of m vectors in V will be linearly dependent if $m > n$.

If the dimension of a vector space is finite the space is called a **finite dimensional vector space**. If, however, the vectors needed to form a basis are infinite in number the associated vector space is called an **infinite dimensional vector space.**

The importance of the dimension of a space lies in the fact that it may be proved that all finite dimensional vector spaces with the same dimension have identical algebraic properties. As a result the study of the simplest space with dimension n provides information about the structure of *all* other spaces of equal dimension. The preservation of algebraic properties in this way among all vector spaces of the same dimension is called an **isomorphism**. This means, for example, that all vector spaces of dimension 3 are **isomorphic** to the vector space of all geometrical vectors in $\mathbb{R}^3$.

A set of vectors U will be called a **subset** of the vector space V if every element of U belongs to V. It will be called a **proper subset** of the vector space V if every element of U belongs to V, but V contains at least one element which does not belong to U. Thus V is a subset of itself, but it is not a proper subset of itself.

A nonempty subset U of a vector space V is called a **subspace** of V, and is itself a vector space with the same algebraic operations as in V if the linear combination $\lambda \mathbf{x} + \mu \mathbf{y}$ is in U whenever $\mathbf{x}$, $\mathbf{y}$ are in U and λ, μ are scalars. This is equivalent to requiring the elements of U to satisfy the closure properties A.1 and M.1 of the axioms defining V. We see from this that V is a subspace of the vector space V, though if U is a proper subset of V, then span $\{U\} < $ span $\{V\}$.

Example 1. Examples of finite dimensional vector spaces

(i) The space V containing all geometrical vectors in $\mathbb{R}^3$ is a real vector space with vector addition and the multiplication of a vector by a scalar defined as in Sections 2.3 or 2.4. The space is of dimension 3 because $\mathbf{i}$, $\mathbf{j}$ and $\mathbf{k}$ form a basis for all such vectors. The vectors $\mathbf{a} = \mathbf{i}$, $\mathbf{b} = \mathbf{i} + 2\mathbf{j}$ and $\mathbf{c} = \mathbf{i} + \mathbf{j} + 3\mathbf{k}$ are linearly independent, and so also form a basis for $\mathbb{R}^3$, because the only solution to

$$\lambda_1 \mathbf{a} + \lambda_2 \mathbf{b} + \lambda_3 \mathbf{c} = 0$$

is $\lambda_1 = \lambda_2 = \lambda_3 = 0$. However, the vectors $\mathbf{a} = \mathbf{i} + 5\mathbf{j} - 5\mathbf{k}$, $\mathbf{b} = \mathbf{i} + \mathbf{j} + \mathbf{k}$, $\mathbf{c} = 2\mathbf{j} - 3\mathbf{k}$ are linearly dependent because $\mathbf{a} = \mathbf{b} + 2\mathbf{c}$. Vectors $\mathbf{b}$ and $\mathbf{c}$ thus span a subspace of $\mathbb{R}^3$ of dimension 2 (a plane in $\mathbb{R}^3$).

(ii) Let V be the set of vectors $\mathbf{x}$ comprising all ordered number pairs $\mathbf{x} = (a, b)$ with a, b real numbers. Let $\mathbf{x} = (a, b)$, $\mathbf{y} = (c, d)$ be any two vectors in V, and define vector addition by the rule

$$\mathbf{x} + \mathbf{y} = (a + c, b + d),$$

and the product $\lambda \mathbf{x}$, with λ any real number, by the rule

$$\lambda \mathbf{x} = (\lambda a, \lambda b).$$

It is easily checked that the vectors in V satisfy axioms A and M, which shows that V is a vector space. We may take as a basis for this vector space the vectors $\mathbf{e}_1 = (1, 0)$ and $\mathbf{e}_2 = (0, 1)$, because every vector $\mathbf{x} = (a, b)$ may be written $\mathbf{x} = a\mathbf{e}_1 + b\mathbf{e}_2$, showing the dimension of this space is 2.

(iii) As a final example, let the space V be the set of vectors $\mathbf{x}$ comprising all ordered n-tuples of real numbers $\mathbf{x} = (x_1, x_2, \ldots, x_n)$. Let $\mathbf{x} = (x_1, x_2, \ldots, x_n)$ and $\mathbf{y} = (y_1, y_2, \ldots, y_n)$ be any two vectors in V, and define vector addition by the rule

$$\mathbf{x} + \mathbf{y} = (x_1 + y_1, x_2 + y_2, \ldots, x_n + y_n),$$

and the product $\lambda \mathbf{x}$, with λ any real number, by the rule

$$\lambda \mathbf{x} = (\lambda x_1, \lambda x_2, \ldots, \lambda x_n). \qquad \blacksquare$$

It is a straightforward matter to verify that the vectors in V satisfy axioms A and M, which shows that V is a vector space. We may take as a basis for this n-dimensional space the vectors $e_1 = (1, 0, 0, \ldots, 0)$, $e_2 = (0, 1, 0, \ldots, 0)$, $\ldots$, $e_n = (0, 0, 0, \ldots, 1)$, since every vector x in V may be expressed as a linear combination of $e_1, e_2, \ldots, e_n$.

The space V involved in this example is a direct generalization of the vector space of geometrical vectors in $\mathbb{R}^3$ to an analogous space of vectors in $\mathbb{R}^n$. In the next chapter we shall discover that this space is fundamental to the study of matrices.

A specially useful class of vector spaces is that in which it is possible to associate with each pair of vectors x, y in the space V a real number, written (x, y), called the **inner product**[5] of x and y. A way of doing this has already been encountered in the case of geometrical vectors in $\mathbb{R}^3$ through the use of the ordinary scalar product of two vectors. As a direct generalization of this idea, when the inner product satisfies the following axioms, the vector space V is said to be a **real inner product space**:

P.1   For all x, y, z in V and all scalars λ, μ,

$(\lambda x + \mu y, z) = \lambda(x, z) + \mu(y, z)$ (**Linearity** of inner product)

P.2   For all x, y in V,

$(x, y) = (y, x)$ (**Symmetry** of inner product)

P.3   For every x in V,

$(x, x) \geq 0$, with   (**Positive definiteness** of inner product)
$(x, x) = 0$ if, and only if, $x = 0$,

By direct analogy with the scalar product of two geometrical vectors in $\mathbb{R}^3$, two vectors x, y in an inner product space V are said to be **orthogonal** when

$(x, y) = 0$.

Example 2. Finite dimensional inner product space

The vector space V with vectors x comprising all ordered n-tuples of real numbers $x = (x_1, x_2, \ldots, x_n)$, with the operations of addition and scalar multiplication defined as in Ex. 1 (iii), becomes an inner product space when the inner product (x, y) is defined as

$(x, y) = x_1 y_1 + x_2 y_2 + \ldots + x_n y_n$.

To establish this assertion it is necessary to verify that the inner product just defined possesses properties P.1 to P.3.

P.1   Linearity of the inner product

By definition

$\lambda x + \mu y = (\lambda x_1 + \mu y_1, \lambda x_2 + \mu y_2, \ldots, \lambda x_n + \mu y_n)$,

and so from the definition of the inner product

$(\lambda x + \mu y, z) = (\lambda x_1 z_1 + \mu y_1 z_1, \lambda x_2 z_2 + \mu y_2 z_2, \ldots, \lambda x_n z_n + \mu y_n z_n)$
$= \lambda(x_1 z_1 + x_2 z_2 + \ldots + x_n z_n) + \mu(y_1 z_1 + y_2 z_2 + \ldots + y_n z_n)$
$= \lambda(x, z) + \mu(y, z)$.

[5] The inner product is also called the **scalar** or **dot product** of x and y, and it is then written $x \cdot y$.

P.2 Symmetry of inner product

From the definition of an inner product,

$$(\mathbf{x}, \mathbf{y}) = x_1 y_1 + x_2 y_2 + \ldots + x_n y_n$$
$$= y_1 x_1 + y_2 x_2 + \ldots + y_n x_n = (\mathbf{y}, \mathbf{z}).$$

P.3 Positive definiteness

From the definition of an inner product,

$$(\mathbf{x}, \mathbf{x}) = x_1^2 + x_2^2 + \ldots + x_n^2,$$

showing $(\mathbf{x}, \mathbf{x})$ is a sum of squares. Thus $(\mathbf{x}, \mathbf{x}) > 0$ if $\mathbf{x} \neq \mathbf{0}$, and $(\mathbf{x}, \mathbf{x}) = 0$ if, and only if, $\mathbf{x} = \mathbf{0}$, showing $(\mathbf{x}, \mathbf{x})$ is **positive definite** (see Sec. 3.12). ■

The inner product may be used to introduce the notion of a length into a general inner product space V by defining the **norm** $\|\mathbf{x}\|$ of a vector $\mathbf{x}$ in V as

$$\|\mathbf{x}\| = \sqrt{(\mathbf{x}, \mathbf{x})},$$

which from P.3 is seen to be a nonnegative number. Only the zero vector $\mathbf{x} = \mathbf{0}$ has a zero norm.

To see that the norm exhibits the properties of a distance we first recall that if $d(\mathbf{x}, \mathbf{y})$ is the euclidean distance between points with geometrical position vectors $\mathbf{x}$ and $\mathbf{y}$ in $\mathbb{R}^3$, then d satisfies the axioms:

D.1 $d(\mathbf{x}, \mathbf{y}) > 0$ for $\mathbf{x} \neq \mathbf{y}$

      $d(\mathbf{x}, \mathbf{y}) = 0$ if, and only if, $\mathbf{x} = \mathbf{y}$,   (distance is strictly **positive**)

D.2 $d(\mathbf{x}, \mathbf{y}) = d(\mathbf{y}, \mathbf{x})$,   (distance is **symmetric**)

D.3 $d(\mathbf{x}, \mathbf{z}) \leq d(\mathbf{x}, \mathbf{y}) + d(\mathbf{y}, \mathbf{z})$ (distance satisfies **triangle inequality**).

The norm already exhibits properties D.1 and D.2 by virtue of its definition. Property D.3 follows by using the definition of a norm and the axioms of an inner product space to prove that

$$\|\mathbf{x} + \mathbf{y}\| \leq \|\mathbf{x}\| + \|\mathbf{y}\| \qquad \text{(triangle inequality)}.$$

The **norm** $\|\mathbf{x}\| = (x_1^2 + x_2^2 + \ldots + x_n^2)^{1/2}$ associated with the inner product space of Ex. 2 is called the **euclidean norm**, because when $n = 3$ it reduces to an ordinary distance in $\mathbb{R}^3$.

Another inequality involving norms which follows directly from the definition of the norm and the axioms of an inner product space is

$$|(\mathbf{x}, \mathbf{y})| \leq \|\mathbf{x}\| \|\mathbf{y}\| \qquad \text{(Cauchy–Schwarz inequality)}.$$

Example 3. Infinite dimensional inner product space

Finally we mention the inner product space of continuous functions $f(x), g(x), \ldots$, defined on an interval $a \leq x \leq b$ with the ordinary definitions of addition and multiplication by a scalar, in which the inner product is defined by the definite integral

$$(f, g) = \int_a^b w(x) f(x) g(x) \, dx,$$

with $w(x) \geq 0$ a given **weight function**. Properties P.1 to P.3 required of an inner product follow directly from this definition and from the properties of the definite integral. This form of inner product will arise later in connection with both Fourier series and eigenfunction expansions, when the spaces involved will be infinite dimensional. In the case of the half-range Fourier sine expansion on the interval $0 \leq x \leq L$ to be encountered later, the weight function $w(x) \equiv 1$, and $f(x) = \sin(m\pi x/L)$, $g(x) = \sin(n\pi x/L)$ for $m, n = 1, 2, \ldots$.

The elements, or vectors, in this inner product space are the functions $\sin(\pi x/L)$, $\sin(2\pi x/L)$, $\sin(3\pi x/L), \ldots$. These vectors are mutually **orthogonal** over the interval $0 \leq x \leq L$, in the sense that the inner product of two different vectors is zero, because of the elementary result

$$\left(\sin\frac{m\pi x}{L}, \sin\frac{n\pi x}{L} \right) = \begin{cases} \displaystyle\int_0^L \sin\frac{m\pi x}{L} \sin\frac{n\pi x}{L} \, dx = 0 & \text{for } m \neq n, \\[2mm] \dfrac{L}{2} & \text{for } m = n. \end{cases}$$

If we take $w(x) \equiv 1$, and consider the more general case in which $f(x)$, $g(x)$ and $h(x)$ are any continuous functions defined on the interval $a \leq x \leq b$, then by setting $x = f(x) - h(x)$ and $y = h(x) - g(x)$ the **triangle inequality** takes the form

$$\left(\int_a^b [f(x) - g(x)]^2 \, dx \right)^{1/2} \leq \left(\int_a^b [f(x) - h(x)]^2 \, dx \right)^{1/2} + \left(\int_a^b [h(x) - g(x)]^2 \, dx \right)^{1/2},$$

and the **Cauchy–Schwarz inequality** bceomes

$$\left| \int_a^b f(x) g(x) \, dx \right| \leq \left(\int_a^b [f(x)]^2 \, dx \right)^{1/2} \left(\int_a^b [g(x)]^2 \, dx \right)^{1/2}. \qquad \blacksquare$$

Problems for Section 2.9

Show that the following sets of elements with the given algebraic operations constitute a vector space. When a space is finite dimensional find a basis for it, and hence determine its dimension.

1 All geometrical vectors of the form $\lambda \mathbf{j} + \mu \mathbf{k}$, with λ, μ any real numbers and vector addition and multiplication by a scalar defined as in $\mathbb{R}^3$.

2 All geometrical vectors of the form $\lambda(\mathbf{i} + \mathbf{j}) + \mu(2\mathbf{i} + \mathbf{k})$, with λ, μ any real numbers and vector addition and multiplication by a scalar defined as in $\mathbb{R}^3$.

3 All geometrical vectors of the form $\lambda(\mathbf{i} + 2\mathbf{j} + 7\mathbf{k})$, with λ any real number and vector addition and multiplication by a scalar defined as in $\mathbb{R}^3$.

4 The set of all functions $y(x) = \lambda \sinh x + \mu \cosh x$, with λ, μ any real numbers and with the usual definitions of addition and multiplication by a real number.

5 The set of all complex numbers, with the usual definitions of addition and multiplication.

6 The set of all real polynomials of degree not exceeding 3 with the usual definitions of addition and multiplication by a real number.

7 The set of all complex polynomials with the usual definitions of addition and multiplication by a complex number.

8 The set of all continuous functions of x defined on the closed real interval $a \leq x \leq b$ with the usual definitions of addition and multiplication by a real number.

9 The set of all real numbers x, y, z, ... in the closed interval $[0, 1]$, with addition defined by $x + y =]x + y[$ and multiplication by a real number λ defined by $\lambda x =]|\lambda|x|[$ where $]k[$ denotes the decimal fractional part of any real nonnegative number k; thus $]2.7[= 0.7$.

In problems 10 to 16 determine which of the following sets of elements with the given algebraic operations constitutes a vector space and give your reasons.

10 The set of all irrational numbers, both positive and negative, with the usual definitions of addition and multiplication.

11 The set of rational numbers with the usual definitions of addition and multiplication by any real number λ.

12 The set $\mathbb{N}$ of all natural numbers, with the usual definitions of addition and multiplication.

13 The set of exponential functions defined on the interval $[1, 4]$ with the usual definitions of addition and multiplication by a real number.

14 The set of all polynomials of degree 5 with the usual definitions of addition and multiplication by a real number.

15 The set of all ordered triples of real numbers in which the last number is nonnegative, with the definitions of addition and multiplication by an arbitrary real scalar as in $\mathbb{R}^3$.

16 The set of all sequences $X = \{x_n\}$, $Y = \{y_n\}$, ... such that their nth terms tend to zero, with addition defined by

$$X + Y = \{x_n + y_n\},$$

and multiplication by a scalar λ defined by

$$\lambda X = \{\lambda x_n\}.$$

Problems 17 to 22 concern the vector space V with vectors $\mathbf{x} = (x_1, x_2, \dots, x_n)$, $\mathbf{y} = (y_1, y_2, \dots, y_n)$, ... in which the x_i, y_i, ... are real numbers. Addition of vectors is defined by

$$\mathbf{x} + \mathbf{y} = (x_1 + y_1, x_2 + y_2, \dots, x_n + y_n)$$

and multiplication of a vector by any real scalar λ is defined by

$$\lambda \mathbf{x} = (\lambda x_1, \lambda x_2, \dots, \lambda x_n).$$

Determine which of the following sets constitutes a vector space, giving reasons for your answer and a basis for the space.

17 The set of all vectors in $\mathbb{R}^n$ $(n > 2)$ with the second component equal to 0.

18 The set of all vectors in $\mathbb{R}^n$ $(n \geq 2)$ with the first and last components rational numbers.

19 The set of all vectors in $\mathbb{R}^n$ $(n \geq 3)$ where the first and third components x_1, x_3 satisfy $4x_1 - x_3 = 5$.

20 The set of all vectors in $\mathbb{R}^n$ $(n \geq 2)$ where the first two components x_1, x_2 satisfy $3x_1 - 4x_2 = 0$.

21 The set of all vectors in $\mathbb{R}^n$ $(n \geq 3)$ in each of which at least one of the first and third components is 0.

22 The set of all vectors in $\mathbb{R}^n$ for which the sum of the squares of the components is positive.

23 The set of all vectors in $\mathbb{R}^n$ for which the sum of the squares of the components is nonnegative.

24 Let a set V have vectors in $\mathbb{R}^n$ $(n>3)$ satisfying the usual definitions of vector addition and multiplication by a scalar. Take the set U to be all vectors of the form

$$\mathbf{x} = (\alpha, \beta, \gamma, x_4, x_5, \ldots, x_n),$$

with α, β, γ given real numbers. Find a basis for U and show it is a subspace of V of dimension $n-3$.

25 In $\mathbb{R}^4$ let the space V be spanned by the four vectors
$$(1, 3, 2, 4), (1, 2, 0, 2), (2, 5, 2, 6), (1, 0, 1, -1).$$

Determine the dimension of V.

26 Derive the triangle inequality for norms.

27 Prove the norm satisfies property D.3.

28 Derive the Cauchy–Schwarz inequality for norms.

Chapter 3
Matrices

After introducing matrices together with their notation in Sec. 3.1, the operations of matrix addition, multiplication by a scalar and transposition are defined and illustrated in Sec. 3.2. General matrix multiplication is discussed in Sec. 3.3. In Sec. 3.4 the relationship between matrices and systems of linear equations is introduced and their solution by elimination described. Section 3.5 develops the notions of linear dependence, rank and the reduced echelon form of a matrix. The ideas of Sec. 3.5 are developed further when the solution of systems of equations is re-examined in greater detail in Sec. 3.6. Determinants are defined and their properties deduced in systematic fashion in Sec. 3.7. The relationship of determinants to rank and to Cramer's rule is explained in Sec. 3.8. The existence of the inverse matrix is considered in Sec. 3.9, together with its properties and method of computation. A simple physical problem is used to motivate and introduce the fundamental algebraic eigenvalue problem in Sec. 3.10. The determination of the eigenvalues and associated eigenvectors of a matrix is then discussed together with the Gerschgorin circle theorem, similarity transformations, and orthogonal matrices. The important and useful problem of the diagonalization of a matrix is considered in Sec. 3.11. In Sec. 3.12 the diagonalization process is related to quadratic forms. Finally, in Sec. 3.13, the **LU** and Cholesky factorizations of a nonsingular matrix are derived and discussed.

3.1   Introductory ideas

Many physical problems need to be described in terms of sets of numbers, either real or complex, which are arranged in an array of m rows and n columns. Such an array is called a **matrix** and it contains mn separate **elements** (entries). It is the purpose of this chapter to explain how algebraic operations may be performed on matrices as single entities, rather than stage by stage on groups of elements.

A **matrix** will be denoted by a single bold-faced letter like $\mathbf{A}, \mathbf{B}, \mathbf{x}, \mathbf{y}, \mathbf{d}, \ldots$. When it is sufficient to describe only the shape of a matrix, without the need to refer to its detailed structure, we shall speak simply of an $m \times n$ matrix if it has m rows and n columns. Here the symbols $m \times n$ are to be read 'm by n'. It is to be understood that by the **shape** of a matrix we mean the number m of its rows and the number n of its columns. If the detailed structure needs to be displayed, this is accomplished by enclosing the elements of the matrix, arranged in the appropriate order, within rectangular brackets, as follows

$$\mathbf{A} = \begin{bmatrix} 1 & 4 & 0 \\ 2 & 1 & -4 \\ 1 & 2 & 1 \end{bmatrix}, \quad \mathbf{B} = \begin{bmatrix} 1 & 0 & 2 & 1 \\ 2 & 1 & -3 & 2 \\ 4 & -7 & 6 & 1 \end{bmatrix}, \quad \mathbf{d} = \begin{bmatrix} 1 \\ -2 \\ 3 \end{bmatrix}.$$

Here $\mathbf{A}$ is a 3×3, $\mathbf{B}$ a 3×4 and $\mathbf{d}$ a 3×1 matrix.

The matrix $\mathbf{A}$ may, for example, represent the matrix of coefficients of the unknowns x_1, x_2 and x_3 in the following system of linear equations

$$\begin{aligned} x_1 + 4x_2 \quad &= 1 \\ 2x_1 + x_2 - 4x_3 &= -2 \\ x_1 + 2x_2 + x_3 &= 3. \end{aligned}$$

The matrix $\mathbf{d}$ could then represent the terms on the right-hand side of these equations (nonhomogeneous terms).

An arbitrary $m \times n$ matrix is written

$$\mathbf{A} = \begin{bmatrix} a_{11} & a_{12} & \cdots & a_{1n} \\ a_{21} & a_{22} & \cdots & a_{2n} \\ . & . & \cdots & . \\ a_{m1} & a_{m2} & \cdots & a_{mn} \end{bmatrix},$$

where a_{ij} denotes the **general** or representative element of $\mathbf{A}$ in row i and column j. Here the double subscript notation tells us the location within the matrix of the entry in question. Thus, if $a_{23} = 17$, we know the entry 17 must appear in row 2 and column 3 of the matrix. The full matrix display may be contracted to $\mathbf{A} = [a_{ij}]$ once the shape of the matrix is known, for then it is understood that $i = 1, 2, \ldots, m;\ j = 1, 2, \ldots, n$.

Matrices comprising either a single row or a single column are called **row vectors** and **column vectors**, respectively. More precisely, a $1 \times n$ matrix is called an n element row vector and an $m \times 1$ matrix is called an m element column vector. Row and column vectors will be denoted by lower case letters.

In addition to row and column vectors, we also single out as being of special importance matrices having the same number of rows as columns. Matrices of this type are called **square matrices**, and they are said to be of order n when their shape is $n \times n$.

If $\mathbf{A}$ is a square matrix of order n the elements $a_{11}, a_{22}, \ldots, a_{nn}$ are said to form the **leading diagonal**, also called the **principal diagonal**, of $\mathbf{A}$. The leading diagonal runs from top left to bottom right and comprises the entries within the shaded strip in the following matrix

$$
\begin{bmatrix}
a_{11} & a_{12} & \cdot & \cdot & \cdot & a_{1n} \\
a_{21} & a_{22} & \cdot & \cdot & \cdot & a_{2n} \\
\cdot & \cdot & \cdot & \cdot & \cdot & \cdot \\
\cdot & \cdot & \cdot & \cdot & \cdot & \cdot \\
\cdot & \cdot & \cdot & \cdot & \cdot & \cdot \\
a_{n1} & a_{n2} & \cdot & \cdot & \cdot & a_{nn}
\end{bmatrix}
$$

Elements not on the leading diagonal of a square matrix are called **off-diagonal** elements. An important special case of a square matrix is one in which all the off-diagonal elements are zero. This is called a **diagonal matrix**, and some typical examples are given below.

$$
\begin{bmatrix} 1 & 0 \\ 0 & 1 \end{bmatrix}, \quad
\begin{bmatrix} 1 & 0 & 0 \\ 0 & 3 & 0 \\ 0 & 0 & 5 \end{bmatrix}, \quad
\begin{bmatrix} 2 & 0 & 0 & 0 \\ 0 & 7 & 0 & 0 \\ 0 & 0 & 0 & 0 \\ 0 & 0 & 0 & 4 \end{bmatrix}
$$

One special class of diagonal matrices is the set of 1×1 square matrices $\mathbf{A} = [a]$ containing only a single element $a_{11} = a$ which may be considered either as the 1×1 matrix $[a]$, or as the number a. Another is the set of diagonal matrices of order n ($n = 1, 2, \ldots$) in which each entry in the leading diagonal is unity. These are called **unit matrices** or **identity matrices**. In what follows, all such unit matrices will be denoted by the same symbol $\mathbf{I}$, irrespective of their order. Unit matrices of order 2, 3, and 4, respectively, have the form

$$
\mathbf{I} = \begin{bmatrix} 1 & 0 \\ 0 & 1 \end{bmatrix}, \quad
\mathbf{I} = \begin{bmatrix} 1 & 0 & 0 \\ 0 & 1 & 0 \\ 0 & 0 & 1 \end{bmatrix}, \quad
\mathbf{I} = \begin{bmatrix} 1 & 0 & 0 & 0 \\ 0 & 1 & 0 & 0 \\ 0 & 0 & 1 & 0 \\ 0 & 0 & 0 & 1 \end{bmatrix}.
$$

If for any reason the order n of a unit matrix needs to be indicated, this notation is easily modified by the inclusion of a subscript n. In this notation the first of the matrices above would be written $\mathbf{I}_2$ and the last $\mathbf{I}_4$.

The natural definition of **equality** of two matrices requires corresponding entries to be equal and, by implication, the matrices to be of the same shape. Thus, if $\mathbf{A} = [a_{ij}]$,

$\mathbf{B} = [b_{ij}]$, we shall say 'A equals B', and write $\mathbf{A} = \mathbf{B}$ if $\mathbf{A}$ and $\mathbf{B}$ are both $m \times n$ matrices, and

$$a_{ij} = b_{ij} \qquad \text{for } i = 1, 2, \ldots, m; \quad j = 1, 2, \ldots, n.$$

For example, equality of the following 2×3 matrices

$$\begin{bmatrix} 1 & a_{12} & 3 \\ 2 & a_{22} & a_{23} \end{bmatrix} = \begin{bmatrix} 1 & 4 & b_{13} \\ 2 & 3 & 6 \end{bmatrix}$$

implies $a_{12} = 4$, $b_{13} = 3$, $a_{22} = 3$ and $a_{23} = 6$.

A **zero** or **null matrix** is a matrix in which every element is equal to zero. We shall denote all zero matrices by the same symbol $\mathbf{0}$, irrespective of their shape. Thus the 3×1 and 2×2 zero matrices have the form

$$\mathbf{0} = \begin{bmatrix} 0 \\ 0 \\ 0 \end{bmatrix}, \qquad \mathbf{0} = \begin{bmatrix} 0 & 0 \\ 0 & 0 \end{bmatrix}.$$

3.2 Addition of matrices, multiplication by a number and the transposition operation

Matrix addition

To develop the algebra of matrices we start by defining the operation of matrix addition and then the scaling of a matrix by a number (real or complex). These, together with the transposition operation which will also be introduced, are the simplest matrix operations. The two matrices $\mathbf{A}$ and $\mathbf{B}$ are said to be **conformable for addition** only if they each have the same shape. Addition will be signified by the usual $+$ sign. Let $\mathbf{A} = [a_{ij}]$, $\mathbf{B} = [b_{ij}]$ both be $m \times n$ matrices. Then their **sum** is the $m \times n$ matrix $\mathbf{C} = [c_{ij}]$, where

$$c_{ij} = a_{ij} + b_{ij} \qquad \text{for } i = 1, 2, \ldots, m; \quad j = 1, 2, \ldots, n.$$

Thus the sum $\mathbf{C}$ of two $m \times n$ matrices $\mathbf{A}$ and $\mathbf{B}$ is itself an $m \times n$ matrix, each of whose elements is obtained as the sum of the corresponding elements of $\mathbf{A}$ and $\mathbf{B}$.

Example 1. Matrix addition

If

$$\mathbf{A} = \begin{bmatrix} 2 & 1 \\ 0 & 2 \\ -1 & 3 \end{bmatrix} \quad \text{and} \quad \mathbf{B} = \begin{bmatrix} 1 & 3 \\ -1 & 4 \\ 2 & 1 \end{bmatrix}, \quad \text{then } \mathbf{A} + \mathbf{B} = \begin{bmatrix} 3 & 4 \\ -1 & 6 \\ 1 & 4 \end{bmatrix}. \quad \blacksquare$$

It follows immediately from the definition of matrix addition that if **A**, **B** and **C** are conformable for addition, then:

A.1 if **A** and **B** are $m \times n$ matrices, so also is **A** + **B** (**closure** under addition)
A.2 **A** + **B** = **B** + **A** (matrix addition is **commutative**)
A.3 **A** + (**B** + **C**) = (**A** + **B**) + **C** (matrix addition is **associative**)
A.4 **A** + **0** = **A** (**zero** matrix)
A.5 **A** + (−**A**) = **0**, where if $A = [a_{ij}]$, $-A = [-a_{ij}]$ (**additive inverse**)

The matrix −**A** which is obtained from **A** by multiplying each of its elements by −1 is called the **negative** of **A**.

The **difference** of two conformable matrices **A** and **B** then follows as the sum **A** + (−**B**). We may, without ambiguity, write the difference of **A** and **B** as **A** − **B**.

On account of A.3 the sum **A** + **B** + **C** is unambiguous and may be written in place of either **A** + (**B** + **C**) or (**A** + **B**) + **C**.

Example 2. Matrix subtraction

If

$$A = \begin{bmatrix} 1 & 2 & 3 \\ 1 & 1 & 0 \end{bmatrix} \quad \text{and} \quad B = \begin{bmatrix} 1 & -4 & 7 \\ 2 & 1 & -4 \end{bmatrix}, \quad \text{then } A - B = \begin{bmatrix} 0 & 6 & -4 \\ -1 & 0 & 4 \end{bmatrix}.$$

■

Multiplication of matrices by a number

The product of an $m \times n$ matrix $A = [a_{ij}]$ by the number λ (real or complex), written either as λA or $A\lambda$, is defined as the $m \times n$ matrix $\lambda A = [\lambda a_{ij}]$. Thus

$$\lambda A = A\lambda = \begin{bmatrix} \lambda a_{11} & \lambda a_{12} & \cdots & \lambda a_{1n} \\ \lambda a_{21} & \lambda a_{22} & \cdots & \lambda a_{2n} \\ . & . & \cdots & . \\ \lambda a_{m1} & \lambda a_{m2} & \cdots & \lambda a_{mn} \end{bmatrix}. \tag{1}$$

On account of this definition, when n is integral we have

$$\underbrace{A + A + \ldots + A}_{n \text{ times}} = nA. \tag{2}$$

Example 3. Multiplying a matrix by a number

If

$$A = \begin{bmatrix} 1.5 & -3 \\ 9 & 4.5 \end{bmatrix}, \quad \text{then } 3A = \begin{bmatrix} 4.5 & -9 \\ 27 & 13.5 \end{bmatrix} \quad \text{and} \quad -2A = \begin{bmatrix} -3 & 6 \\ -18 & -9 \end{bmatrix}.$$

■

It follows immediately from the definition of multiplication of a matrix by a number that if **A** and **B** are conformable for addition and λ, μ are arbitrary numbers, then:

M.1   if **A** is an $m \times n$ matrix, so also is $k\mathbf{A}$ (**closure** under multiplication)
M.2 $\lambda(\mathbf{A} + \mathbf{B}) = \lambda\mathbf{A} + \lambda\mathbf{B}$ (**distributivity** of multiplication)
M.3 $(\lambda + \mu)\mathbf{A} = \lambda\mathbf{A} + \mu\mathbf{A}$ (**distributivity** of multiplication)
M.4 $\lambda(\mu\mathbf{A}) = \mu(\lambda\mathbf{A})$ (**associativity** of multiplication)
M.5 $1\mathbf{A} = \mathbf{A}$ (**scaling by unity**)

On account of M.4 we may, without ambiguity, write $\lambda\mu\mathbf{A}$ in place of either $\lambda(\mu\mathbf{A})$ or $\mu(\lambda\mathbf{A})$. The negative of **A** introduced in A.5 follows from (1) as $-\mathbf{A} = (-1)\mathbf{A}$.

A comparison of A.1 to A.5 and M.1 to M.5 above with the axioms of a vector space set out in Chapter 2, Sec. 2.9 shows that, for fixed m and n, the set of all $m \times n$ matrices constitutes a vector space. The vector space will be of dimension mn, since there are mn elements in an $m \times n$ matrix. It will be a real vector space when the matrix elements are all taken to be real numbers, and a complex vector space when they belong to the field of complex numbers.

A **basis** for this vector space of matrices is provided by the set of mn matrices $\mathbf{E}_{pq}$ ($p = 1, 2, \ldots, m; q = 1, 2, \ldots, n$), where $\mathbf{E}_{pq}$ is the $m \times n$ matrix with zero elements everywhere except for the element in row p and column q which is equal to unity. Thus for the real vector space of 2×2 matrices, a basis is provided by

$$\mathbf{E}_{11} = \begin{bmatrix} 1 & 0 \\ 0 & 0 \end{bmatrix}, \quad \mathbf{E}_{12} = \begin{bmatrix} 0 & 1 \\ 0 & 0 \end{bmatrix}, \quad \mathbf{E}_{21} = \begin{bmatrix} 0 & 0 \\ 1 & 0 \end{bmatrix}, \quad \mathbf{E}_{22} = \begin{bmatrix} 0 & 0 \\ 0 & 1 \end{bmatrix}.$$

Transposition of a matrix

The **transpose** $\mathbf{A}^\mathsf{T}$ of an $m \times n$ matrix $\mathbf{A} = [a_{ij}]$ is the $n \times m$ matrix obtained from **A** by interchanging the rows and columns of **A**. That is, the ith row in **A** becomes the ith column in $\mathbf{A}^\mathsf{T}$ and the jth column in **A** becomes the jth row in $\mathbf{A}^\mathsf{T}$. If $\mathbf{A} = [a_{ij}]$, then $\mathbf{A}^\mathsf{T} = [a_{ji}]$, with $i = 1, 2, \ldots, m; j = 1, 2, \ldots, n$. Thus, if

$$\mathbf{A} = \begin{bmatrix} a_{11} & a_{12} & \cdots & a_{1n} \\ a_{21} & a_{22} & \cdots & a_{2n} \\ \cdot & \cdot & \cdots & \cdot \\ a_{m1} & a_{m2} & \cdots & a_{mn} \end{bmatrix}, \quad \text{then } \mathbf{A}^\mathsf{T} = \begin{bmatrix} a_{11} & a_{21} & \cdots & a_{m1} \\ a_{12} & a_{22} & \cdots & a_{m2} \\ \cdot & \cdot & \cdots & \cdot \\ a_{1n} & a_{2n} & \cdots & a_{mn} \end{bmatrix}. \tag{3}$$

Example 4. Transposition of matrices

If

$$\mathbf{a} = \begin{bmatrix} 1 \\ 3 \\ 2 \end{bmatrix} \quad \text{and} \quad \mathbf{B} = \begin{bmatrix} 1 & 2 & 3 \\ 2 & 1 & 4 \\ 1 & 0 & 1 \\ 2 & -1 & -3 \end{bmatrix}, \quad \text{then}$$

$$\mathbf{a}^\mathsf{T} = [1 \quad 3 \quad 2] \quad \text{and} \quad \mathbf{B}^\mathsf{T} = \begin{bmatrix} 1 & 2 & 1 & 2 \\ 2 & 1 & 0 & -1 \\ 3 & 4 & 1 & -3 \end{bmatrix}.$$

■

It is a trivial consequence of definition (3) that if $\mathbf{A}$ and $\mathbf{B}$ are conformable for addition and λ is any number, then

$$(\mathbf{A} + \mathbf{B})^\mathsf{T} = \mathbf{A}^\mathsf{T} + \mathbf{B}^\mathsf{T}, \tag{4}$$

$$(\mathbf{A}^\mathsf{T})^\mathsf{T} = \mathbf{A}, \tag{5}$$

$$(\lambda \mathbf{A})^\mathsf{T} = \lambda \mathbf{A}^\mathsf{T}. \tag{6}$$

Special square matrices

The square matrix $\mathbf{A} = [a_{ij}]$ of order n with real elements a_{ij} is said to be **symmetric** if $a_{ij} = a_{ji}$ $(i, j = 1, 2, \ldots, n)$. That is, $\mathbf{A}$ is symmetric if

$$\mathbf{A} = \mathbf{A}^\mathsf{T}. \tag{7}$$

The square matrix $\mathbf{A} = [a_{ij}]$ of order n with real elements a_{ij} is said to be **skew-symmetric** if $a_{ij} = -a_{ij}$ $(i, j = 1, 2, \ldots, n)$. Thus $\mathbf{A}$ is skew-symmetric if

$$\mathbf{A} = -\mathbf{A}^\mathsf{T}. \tag{8}$$

The elements in the leading diagonal of a skew-symmetric matrix $\mathbf{A} = [a_{ij}]$ must all be zero, for on the leading diagonal we have $a_{ii} = -a_{ii}$ which is only possible if $a_{ii} = 0$ $(i = 1, 2, \ldots, n)$.

Let $\mathbf{A} = [a_{ij}]$ be an arbitrary square matrix of order n with the real elements a_{ij}. Then, starting from the identity

$$a_{ij} = \tfrac{1}{2}(a_{ij} + a_{ji}) + \tfrac{1}{2}(a_{ij} - a_{ji})$$

for $i, j = 1, 2, \ldots, n$, we may write $\mathbf{A}$ in the form

$$\mathbf{A} = [\tfrac{1}{2}(a_{ij} + a_{ji}) + \tfrac{1}{2}(a_{ij} - a_{ji})]$$
$$= \tfrac{1}{2}[a_{ij} + a_{ji}] + \tfrac{1}{2}[a_{ij} - a_{ji}]$$
$$= \tfrac{1}{2}(\mathbf{A} + \mathbf{A}^\mathsf{T}) + \tfrac{1}{2}(\mathbf{A} - \mathbf{A}^\mathsf{T}).$$

The matrix $\mathbf{S} = \tfrac{1}{2}(\mathbf{A} + \mathbf{A}^\mathsf{T})$ is symmetric and the matrix $\mathbf{K} = \tfrac{1}{2}(\mathbf{A} - \mathbf{A}^\mathsf{T})$ is skew-symmetric. We have thus established that any square matrix $\mathbf{A}$ with real elements may be written as the sum $\mathbf{S} + \mathbf{K}$ of a unique symmetric matrix $\mathbf{S} = \tfrac{1}{2}(\mathbf{A} + \mathbf{A}^\mathsf{T})$ and a unique skew-symmetric matrix $\mathbf{K} = \tfrac{1}{2}(\mathbf{A} - \mathbf{A}^\mathsf{T})$.

Example 5. Decomposition into symmetric and skew-symmetric matrices

The matrices

$$A = \begin{bmatrix} 4 & 5 & 1 \\ 5 & 3 & -2 \\ 1 & -2 & 0 \end{bmatrix} \quad \text{and} \quad B = \begin{bmatrix} 1 & 0 & 0 \\ 0 & 7 & 0 \\ 0 & 0 & 9 \end{bmatrix}$$

are symmetric, whereas the matrix

$$C = \begin{bmatrix} 0 & 3 & -4 \\ -3 & 0 & 3 \\ 4 & -3 & 0 \end{bmatrix}$$

is skew-symmetric. The matrix

$$D = \begin{bmatrix} 1 & 4 & 3 \\ -1 & 0 & 2 \\ 2 & 1 & 4 \end{bmatrix},$$

which is neither symmetric nor skew-symmetric, may be written as the sum $D = S + K$ of a symmetric matrix S and a skew-symmetric matrix K with

$$S = \tfrac{1}{2}(D + D^T) = \begin{bmatrix} 1 & \tfrac{3}{2} & \tfrac{5}{2} \\ \tfrac{3}{2} & 0 & \tfrac{3}{2} \\ \tfrac{5}{2} & \tfrac{3}{2} & 4 \end{bmatrix} \quad \text{and} \quad K = \tfrac{1}{2}(D - D^T) = \begin{bmatrix} 0 & \tfrac{5}{2} & \tfrac{1}{2} \\ -\tfrac{5}{2} & 0 & \tfrac{1}{2} \\ -\tfrac{1}{2} & -\tfrac{1}{2} & 0 \end{bmatrix}. \quad \blacksquare$$

Two types of square matrix which are specially useful in computations are the **upper triangular matrix U**, in which all elements below the leading diagonal are zero, and the **lower triangular matrix L** in which all elements above the leading diagonal are zero. In general, these are called **triangular matrices**. Typical examples of such matrices are

$$L = \begin{bmatrix} 1 & 0 & 0 & 0 \\ 2 & 0 & 0 & 0 \\ 3 & 4 & 1 & 0 \\ -1 & 3 & 7 & 3 \end{bmatrix}, \quad U = \begin{bmatrix} 3 & 4 & 0 \\ 0 & 7 & 2 \\ 0 & 0 & -1 \end{bmatrix}.$$

Problems for Section 3.2

Let $A = \begin{bmatrix} 1 & -1 & 3 \\ 2 & 0 & 4 \\ 1 & -1 & 0 \end{bmatrix}$, $B = \begin{bmatrix} -2 & 3 & 1 \\ 0 & 1 & 2 \\ 4 & 0 & -3 \end{bmatrix}$, $C = \begin{bmatrix} 1 & 0 \\ 2 & 3 \\ 1 & -1 \end{bmatrix}$.

Find the following matrices when the indicated operations are defined.

1 $A + B$ 2 $B - A$ 3 $2A - B$ 4 $B - 3A$
5 $A + C$ 6 $A + A^T$ 7 $A - A^T$ 8 $A + B^T$
9 $C + C^T$ 10 $[(C^T)^T]^T$ 11 $A^T + (B^T)^T$

12 Express **A** in the form **A**=**S**+**K** where **S** is a symmetric matrix and **K** is a skew-symmetric matrix.

13 Express **B** in the form **B**=**S**+**K** where **S** is a symmetric matrix and **K** is a skew-symmetric matrix.

14 Express **B**^{T} in the form **B**^{T}=**S**+**K** where **S** is a symmetric matrix and **K** is a skew-symmetric matrix. How could you deduce this result from the answer to Prob. 13 without further calculation?

15 Does the set of all 3×3 symmetric matrices form a vector space? If so find a basis for this space and hence its dimension.

16 Does the set of all 3×3 skew-symmetric matrices form a vector space? If so find a basis for this space and hence its dimension.

17 Let **B**$_{pq}$ ($p=1, 2, \ldots, m$; $q=1, 2, \ldots, n$) be an $m \times n$ matrix in which the element in row p and column q is an arbitrary real number, while all its other elements are zero. Does such a set of mn matrices always form a basis for the vector space of all $m \times n$ real matrices?

18 If **B**$_{pq}$ is a basis for the vector space of all real square matrices of order n prove that **B**$_{pq}^{\mathsf{T}}$ is also a basis.

19 Prove that the set of all real lower triangular matrices of order n is a vector space. What is the dimension of this vector space? Find a basis for the case $n=3$.

20 A *strictly lower (upper) triangular matrix* is one in which all elements on and above (below) the leading diagonal are zero. What is the dimension of this vector space? Find a basis for the vector space of strictly lower triangular matrices of order 3.

Harder problems

The following problems touch upon some of the applications of matrices to **combinatorics**; that is to the study of arrangements of objects. The nature of these objects is unimportant and they may, for example, be semiconductor circuit elements on a patch board, the routing of traffic in a one-way system or the testing of components after each successive stage in a production line, with the possibility of acceptance, rejection as scrap, or return for a repeat of the previous operation. The problems are straightforward. They have only been separated from the preceding ones because study of them is not essential for what is to follow.

21 A **directed graph**, or as it is often known, a **digraph**, is a finite set of elements V called **vertices** and an associated finite set of **arcs (edges)** along which a direction is defined (**ordered arcs**). If the vertices are numbered, an arc from vertex V$_i$ to vertex V$_j$ is denoted by (V$_i$, V$_j$), and is considered to be different from an arc from V$_j$ to V$_i$, denoted by (V$_j$, V$_i$). Digraphs are most conveniently represented graphically by a numbered set of points representing the vertices joined by arcs along which an arrow indicates the ordering of the connection. A **loop** in which an arc returns to its point of origin indicates a connection of a vertex with itself as, for example, in the case of a turning circle at the end of a street. In a digraph the mere crossing of arcs does not indicate a connection between them. It is to be regarded as being similar to the situation in which roads pass over or under one another.

Two vertices are said to be **adjacent** if joined by an arc, and two arcs are said to be **adjacent** if they have at least one vertex in common. An **adjacency matrix** may be defined for a digraph with vertices V$_1$, V$_2$, $\ldots$, V$_n$ as the matrix **A**=$[a_{ij}]$, in which a_{ij} is the number of arcs of the form

(V_i, V_j). Verify that the digraph in Fig. 3.1 has the adjacency matrix

$$A = \begin{bmatrix} 0 & 1 & 1 & 0 \\ 0 & 1 & 1 & 0 \\ 2 & 1 & 0 & 0 \\ 0 & 0 & 1 & 1 \end{bmatrix}.$$

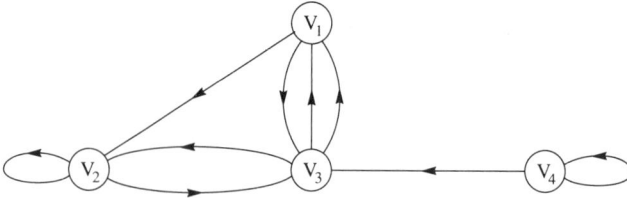

Fig. 3.1 General diagraph

22 Draw the digraph corresponding to the adjacency matrix

$$A = \begin{bmatrix} 0 & 1 & 0 & 1 \\ 1 & 0 & 2 & 0 \\ 1 & 1 & 0 & 1 \\ 0 & 0 & 1 & 1 \end{bmatrix}.$$

23 Find the adjacency matrix for the digraph in Fig. 3.2 in which any two vertices are joined by precisely one arc. Such a digraph is called a **tournament** because it enables the representation of any competition in which a draw is not allowed. An arrow directed away from vertex V_i to V_j may be regarded as the player V_i winning the match between V_i and V_j, and conversely if the arrow is directed into V_i. An entry 1 in the adjacency matrix then represents a 'win' and a zero a 'loss' in the match represented by that element.

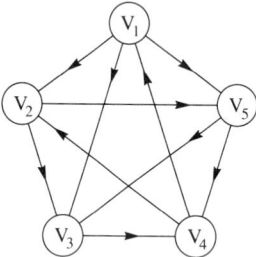

Fig. 3.2 Tournament diagraph

24 A **nodal incidence matrix**, or **vertex incidence matrix** (a **node** is an alternative name for a **vertex**) may be constructed from a digraph if the arcs are numbered as well as the vertices. It is the matrix $A = [a_{ij}]$ where

$$a_{ij} = \begin{cases} +1 \text{ if arc } j \text{ leaves vertex } V_i \\ -1 \text{ if arc } j \text{ enters vertex } V_i \\ 0 \text{ if arc } j \text{ does not join vertex } V_i. \end{cases}$$

Verify that the following nodal incidence matrix describes the digraph in Fig. 3.3

$$A = \begin{bmatrix} 1 & 0 & 0 & 1 & 0 & 1 \\ -1 & 1 & -1 & 0 & 0 & 0 \\ 0 & -1 & 0 & 0 & -1 & -1 \\ 0 & 0 & 1 & -1 & 1 & 0 \end{bmatrix}.$$

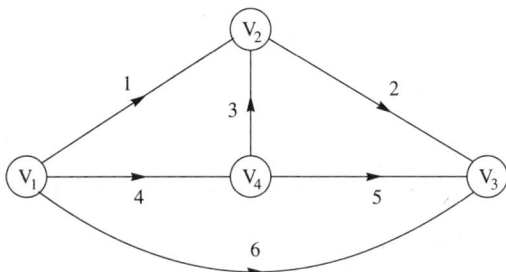

Fig. 3.3

25 Draw the digraph for which the nodal incidence matrix is

$$A = \begin{bmatrix} 1 & 0 & 0 & 0 & 0 & 1 \\ -1 & 1 & 0 & 1 & 0 & 0 \\ 0 & 0 & -1 & -1 & 1 & 0 \\ 0 & 0 & 0 & 0 & -1 & -1 \\ 0 & -1 & 1 & 0 & 0 & 0 \end{bmatrix}.$$

26 Every digraph has associated with it an **underlying graph** which is obtained from the digraph by the removal of the arrows. This graph indicates associations of vertices without regard to order (direction). An **adjacency matrix for a graph** depends on the numbering of the vertices and is the *symmetric* matrix $A = [a_{ij}]$ in which a_{ij} is the number of arcs joining the vertices V_i and V_j. Verify that the adjacency matrix for the underlying graph in Fig. 3.1 is given by

$$A = \begin{bmatrix} 0 & 1 & 3 & 0 \\ 1 & 1 & 2 & 0 \\ 3 & 2 & 0 & 1 \\ 0 & 0 & 1 & 1 \end{bmatrix}.$$

27 Find the adjacency matrices for the underlying graphs in Figs 3.2 and 3.3.

28 Many fundamental processes in engineering and science involve **discrete random processes**, in the sense that they may be described by a sequence of discrete random variables $\{X_n\} = \{X_0, X_1, \ldots\}$. The set of possible values attainable by $\{X_n\}$ is called the **state space** of the process, and when there is only a finite number of states these may be indexed by the integers $1, 2, \ldots$. With this notation in mind we see that $X_n = r$ means that after n steps the process is in state r. By way of example, the state r could represent the number remaining in a waiting line after the nth person has been served, the next measurement of the water level in a reservoir (to the nearest inch) for which n successive weekly measurements have already been made, or the number of encoding errors in a data transmission system after n transmissions.

Let there be n states and suppose p_{ij} is the probability that the process moves from state i to state j. Then, for any given i, since the system must be in one of the n possible states, we have the result

$$\sum_{j=1}^{n} p_{ij} = 1 \quad (i = 1, 2, \ldots, n),$$

with $p_{ij} \geq 0$ because these are probabilities. As a special case, if when the process is in state i it can move only to state j, then $p_{ij} = 1$, because a probability of 1 is a 'certainty' (a probability of 0 is an 'impossibility').

Each p_{ij} is called a **transition probability** for the system, and the $n \times n$ matrix $\mathbf{P} = [p_{ij}]$ is called the **transition matrix** for the process. When, given the transition matrix $\mathbf{P}$ and the state of the system at a given stage, its probabilistic behavior at the next stage is determined solely by $\mathbf{P}$ and the knowledge of its *present state*, the process is called a **Markov process** or a **Markov chain**. Thus, in a Markov process, the history of the process up to the nth stage does not influence the probabilistic behavior at the $(n+1)$th stage. A typical transition matrix has the form

$$\mathbf{P} = \begin{bmatrix} \frac{3}{4} & \frac{1}{4} & 0 & 0 & 0 \\ \frac{1}{6} & \frac{1}{3} & \frac{1}{2} & 0 & 0 \\ 0 & 0 & \frac{1}{2} & \frac{1}{4} & \frac{1}{4} \\ 0 & \frac{1}{6} & 0 & \frac{1}{2} & \frac{1}{3} \\ 0 & 0 & 0 & \frac{3}{4} & \frac{1}{4} \end{bmatrix}.$$

Prove that when $\mathbf{P}$ and $\mathbf{Q}$ are transition matrices, then so also is $\alpha \mathbf{P} + (1-\alpha)\mathbf{Q}$ for $0 \leq \alpha \leq 1$.

29 A production line involves three successive operations, each of which is followed by an inspection. At any stage an article may be accepted and passed on to the next stage, recycled through the same stage to correct a fault, or rejected as scrap. Suppose the probability of acceptance at each stage is p_1, the probability of recycling at each stage is p_2, and the probability of rejection at each stage is p_3, where p_1, p_2 and p_3 are all nonnegative and $p_1 + p_2 + p_3 = 1$. Suppose also that finished and scrap articles do not re-enter the production line. Then the production line may be described by the five states listed and indexed below:

State

1	article is in the first production stage
2	article is in the second production stage
3	article is in the third production stage
4	article is finished
5	article is scrapped.

Verify that the transition matrix for the production line is

$$\mathbf{P} = \begin{bmatrix} 0 & 0 & 0 & 0 & 1 \\ 0 & 0 & 0 & 1 & 0 \\ p_2 & p_3 & 0 & 0 & p_1 \\ 0 & p_2 & p_3 & 0 & p_1 \\ 0 & 0 & p_2 & p_3 & p_1 \end{bmatrix}.$$

The movement of an article along the production line is called a one-dimensional **random walk**. A generalization of a random walk may be made to two and three dimensions and has many applications. Typical of these is the Brownian motion of particles suspended in a liquid, the movement of an atom within a defective crystal lattice in which some sites (possible locations for atoms) are vacant, the motion of a neutron in the moderator of a nuclear reactor, and the microscopic study of diffusion processes in general.

30 How would the transition matrix in Prob. 29 need to be modified if the probabilities of acceptance, recycling and rejection as scrap are different at each production stage? You may assume they are, respectively, p_1, p_2, p_3 at the first stage, q_1, q_2, q_3 at the second stage and r_1, r_2, r_3 at the third stage.

3.3 Matrix multiplication. Linear transformations. Differentiation

The fundamental matrix operation which must now be introduced is the multiplication of a matrix by a matrix. We shall see that not all pairs of matrices may be multiplied together, and that when matrix multiplication is possible the order in which it is performed is important.

The most elementary type of matrix multiplication involves the product **ab** of an n element row vector **a** by an n element column vector **b**. This product, in which the order of the vectors is important, yields a single number (1×1 matrix). If

$$\mathbf{a} = [a_1 \ a_2 \ \ldots \ a_n] \quad \text{and} \quad \mathbf{b} = \begin{bmatrix} b_1 \\ b_2 \\ \vdots \\ b_n \end{bmatrix}, \tag{1}$$

the product **ab** is defined as the number

$$\mathbf{ab} = a_1 b_1 + a_2 b_2 + \ldots + a_n b_n = \sum_{s=1}^{n} a_s b_s. \tag{2}$$

Notice that this product is defined only when the number of elements (columns) in vector **a** is equal to the number of elements (rows) in vector **b**. When this condition is satisfied, **a** and **b** are said to be **conformable** for the product **ab**.

Because of the obvious analogy with geometrical vectors the product in (2) is often called the **dot product** of **a** and **b**, when it is written $\mathbf{a} \cdot \mathbf{b}$. Again by analogy with geometrical vectors, the row vector **a** and column vector **b** are said to be **orthogonal** if $\mathbf{ab} = 0$.

Example 1. Product of a row and a column vector

If $\mathbf{a} = [1, \ -3, \ 4]$, $\mathbf{b} = \begin{bmatrix} 2 \\ 1 \\ -6 \end{bmatrix}$ and $\mathbf{c} = \begin{bmatrix} -1 \\ 1 \\ 1 \end{bmatrix}$, then

$$\mathbf{ab} = 1 \cdot 2 + (-3) \cdot 1 + 4 \cdot (-6) = -25 \quad \text{and} \quad \mathbf{ac} = 1 \cdot (-1) + (-3) \cdot 1 + 4 \cdot 1 = 0.$$

The second product **ac** shows that row vector **a** and column vector **c** are orthogonal. ∎

The part played in general matrix multiplication by the elementary operation involving the product of a row and a column vector introduced in (2) is best seen by considering the effect of two successive linear transformations. At this stage we take a linear transformation to be a system of simultaneous equations of degree one relating a set of m variables to a set of n variables. To illustrate matters, let the variables $(x_1, x_2), (y_1, y_2)$ and (z_1, z_2, z_3) be related by the two linear transformations

$$y_1 = a_{11}x_1 + a_{12}x_2$$
$$y_2 = a_{21}x_1 + a_{22}x_2 \tag{3}$$

and

$$x_1 = b_{11}z_1 + b_{12}z_2 + b_{13}z_3$$
$$x_2 = b_{21}z_1 + b_{22}z_2 + b_{23}z_3. \tag{4}$$

Typically, (z_1, z_2, z_3) might be the coordinates of a point within a chemical reactor, with x_1 the temperature and x_2 the flow rate at that point. Then y_1 and y_2 could describe the local rates at which two reaction products are produced.

Substituting (4) into (3) gives

$$y_1 = c_{11}z_1 + c_{12}z_2 + c_{13}z_3$$
$$y_2 = c_{21}z_1 + c_{22}z_2 + c_{23}z_3, \tag{5}$$

where the general coefficient

$$c_{ij} = a_{i1}b_{1j} + a_{i2}b_{2j} = \sum_{s=1}^{2} a_{is} b_{sj}, \tag{6}$$

for $i = 1, 2; j = 1, 2, 3$. Thus, for example, $c_{11} = a_{11}b_{11} + a_{12}b_{21}$, $c_{13} = a_{11}b_{13} + a_{12}b_{23}$ and $c_{21} = a_{21}b_{11} + a_{22}b_{21}$.

The coefficients of the variables on the right-hand sides of (3), (4) and (5) may be used to define **coefficient matrices** for the respective systems of equations. We shall set

$$A = \begin{bmatrix} a_{11} & a_{12} \\ a_{21} & a_{22} \end{bmatrix}, \quad B = \begin{bmatrix} b_{11} & b_{12} & b_{13} \\ b_{21} & b_{22} & b_{23} \end{bmatrix}, \quad C = \begin{bmatrix} c_{11} & c_{12} & c_{13} \\ c_{21} & c_{22} & c_{23} \end{bmatrix}. \tag{7}$$

The fact that (4) was substituted into (3) to obtain (5) suggests that we define the matrix product AB, performed in this order, to be

$$AB = C, \tag{8}$$

with the elements c_{ij} of C being given by (6).

Notice that A is a 2×2 matrix and B a 2×3 matrix, and the matrix product AB in that order yields C which is a 2×3 matrix. Symbolically, the shape of the matrix product

AB=**C** may be recorded by writing

Number of
columns in product

$(2 \times 2) \qquad (2 \times 3) \quad = \quad (2 \times 3).$

Same
number

Number of rows in product

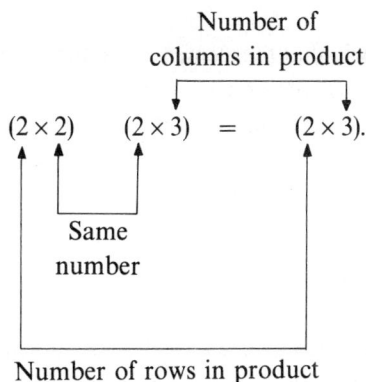

We shall see that this pattern of result applies to all matrix products.

To remove ambiguity about the order in which matrix multiplication is performed we shall say **A premultiplies B** when **A** appears as the first factor in the product **AB**. We shall say **B postmultiplies A** when **B** appears as the last factor in the product **AB**.

Example 2. Matrix multiplication

If

$$A = \begin{bmatrix} 1 & 4 \\ 2 & 1 \end{bmatrix} \quad \text{and} \quad B = \begin{bmatrix} 1 & 2 & 1 \\ 0 & 2 & -3 \end{bmatrix},$$

then from (6)

$$AB = \begin{bmatrix} 1 & 4 \\ 2 & 1 \end{bmatrix}\begin{bmatrix} 1 & 2 & 1 \\ 0 & 2 & -3 \end{bmatrix} = \begin{bmatrix} 1 \cdot 1 + 4 \cdot 0 & 1 \cdot 2 + 4 \cdot 2 & 1 \cdot 1 + 4 \cdot (-3) \\ 2 \cdot 1 + 1 \cdot 0 & 2 \cdot 2 + 1 \cdot 2 & 2 \cdot 1 + 1 \cdot (-3) \end{bmatrix} = \begin{bmatrix} 1 & 10 & -11 \\ 2 & 6 & -1 \end{bmatrix}.$$

∎

Now we may write **A** and **B** in (7) as

$$A = \begin{bmatrix} \mathbf{a}_1 \\ \mathbf{a}_2 \end{bmatrix} \quad \text{and} \quad B = [\mathbf{b}_1 \quad \mathbf{b}_2 \quad \mathbf{b}_3], \tag{9}$$

where $\mathbf{a}_i$ is the row vector formed from the ith row of **A** and $\mathbf{b}_j$ is the column vector formed from the jth column of **B**, so

$$\mathbf{a}_i = [a_{i1} a_{i2}] \quad \text{and} \quad \mathbf{b}_j = \begin{bmatrix} b_{1j} \\ b_{2j} \end{bmatrix}.$$

Then, in terms of result (2) (with $n = 2$), we see that **AB**=**C** is given by

$$C = \begin{bmatrix} \mathbf{a}_1\mathbf{b}_1 & \mathbf{a}_1\mathbf{b}_2 & \mathbf{a}_1\mathbf{b}_3 \\ \mathbf{a}_2\mathbf{b}_1 & \mathbf{a}_2\mathbf{b}_2 & \mathbf{a}_2\mathbf{b}_3 \end{bmatrix}, \tag{10}$$

showing that

$$c_{ij} = \mathbf{a}_i \mathbf{b}_j, \tag{11}$$

for $i = 1, 2; j = 1, 2, 3$. This shows clearly that the element c_{ij} in the matrix product $\mathbf{AB} = \mathbf{C}$ is obtained as the product of the ith row vector $\mathbf{a}_i$ of $\mathbf{A}$ and the jth column vector $\mathbf{b}_j$ of $\mathbf{B}$.

The general rule for matrix multiplication now follows as a direct generalization of (10), once it has been noticed that there must be as many columns in $\mathbf{A}$ as there are rows in $\mathbf{B}$ in order that the products in (11) should be defined.

Definition. Product of matrices

The $m \times n$ matrix $\mathbf{A} = [a_{ij}]$ and the $p \times q$ matrix $\mathbf{B} = [b_{ij}]$ are conformable for the matrix product $\mathbf{AB}$ only when $n = p$. The result is then an $m \times q$ matrix $\mathbf{C} = [c_{ij}]$ with the elements

$$c_{ij} = a_{i1}b_{1j} + a_{i2}b_{2j} + \ldots + a_{in}b_{nj} = \sum_{s=1}^{n} a_{is}b_{sj}, \tag{12}$$

for $i = 1, 2, \ldots, m; j = 1, 2, \ldots, q$.

To relate this definition to the elementary product defined in (2), let $\mathbf{a}_i$ be the ith row of $\mathbf{A}$ and $\mathbf{b}_j$ the jth column of $\mathbf{B}$, so that

$$\mathbf{A} = \begin{bmatrix} \mathbf{a}_1 \\ \mathbf{a}_2 \\ \vdots \\ \mathbf{a}_m \end{bmatrix} \quad \text{and} \quad \mathbf{B} = [\mathbf{b}_1 \quad \mathbf{b}_2 \quad \ldots \quad \mathbf{b}_q], \tag{13}$$

Then (12) shows that when the product $\mathbf{AB} = \mathbf{C}$ is defined,

$$\mathbf{C} = \begin{bmatrix} \mathbf{a}_1\mathbf{b}_1 & \mathbf{a}_1\mathbf{b}_2 & \cdots & \mathbf{a}_1\mathbf{b}_q \\ \mathbf{a}_2\mathbf{b}_1 & \mathbf{a}_2\mathbf{b}_2 & \cdots & \mathbf{a}_2\mathbf{b}_q \\ \cdot & \cdot & \cdots & \cdot \\ \mathbf{a}_m\mathbf{b}_1 & \mathbf{a}_m\mathbf{b}_2 & \cdots & \mathbf{a}_m\mathbf{b}_q \end{bmatrix} \tag{14}$$

It is worth emphasizing the following points illustrated by (14):

(i) the product of an $m \times n$ matrix and an $n \times q$ matrix is an $m \times q$ matrix;

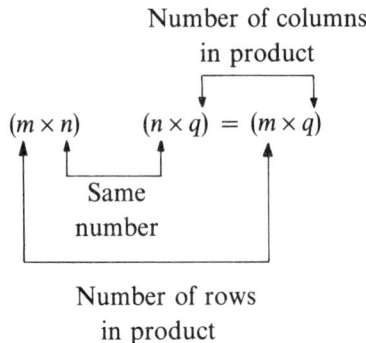

Number of columns
in product

$$(m \times n) \qquad (n \times q) = (m \times q)$$

Same
number

Number of rows
in product

(ii) the element c_{ij} in the product $\mathbf{AB}=\mathbf{C}$ is obtained as the product of the ith row vector of $\mathbf{A}$ and the jth column vector of $\mathbf{B}$ as defined in (12). This is often described as the process of '*multiplying a row into a column*';

(iii) pre- or postmultiplication of matrix $\mathbf{A}$ by a conformable unit matrix $\mathbf{I}$ leaves matrix $\mathbf{A}$ unchanged, so $\mathbf{IA}=\mathbf{AI}=\mathbf{A}$;

(iv) although the matrix product $\mathbf{AB}$ may be defined, it does not necessarily follow that the product $\mathbf{BA}$ is also defined nor, if it is, that $\mathbf{AB}$ and $\mathbf{BA}$ are equal;

(v) it is not necessary for $\mathbf{A}$ or $\mathbf{B}$ to be the zero matrix in order that $\mathbf{AB}=\mathbf{0}$;

(vi) whereas when $\mathbf{a}$ is a $1 \times n$ matrix (row vector) and $\mathbf{b}$ is an $n \times 1$ matrix (column vector) the product $\mathbf{ab}$ is a scalar (1×1 matrix), the product $\mathbf{ba}$ is an $n \times n$ matrix. The columns of $\mathbf{a}$ and the rows of $\mathbf{b}$ are single elements so the element c_{ij} in the product $\mathbf{ba}=\mathbf{C}$ is the single product $c_{ij}=b_i a_j$, where b_j is the ith element of $\mathbf{b}$ and a_j is the jth element of $\mathbf{a}$;

Example 3. General matrix multiplication

Let

$$\mathbf{a}=[1 \quad 2 \quad 3], \quad \mathbf{b}=\begin{bmatrix} 1 \\ -1 \\ 2 \end{bmatrix}, \quad \mathbf{A}=\begin{bmatrix} 2 & 1 & 0 \\ 3 & 2 & 1 \end{bmatrix}, \quad \mathbf{B}=\begin{bmatrix} 1 & 2 \\ 3 & 1 \\ 0 & 2 \end{bmatrix}, \quad \mathbf{C}=\begin{bmatrix} -1 & 4 \\ 2 & 1 \end{bmatrix},$$

$$\mathbf{D}=\begin{bmatrix} 6 & -5 \\ -2 & 1 \end{bmatrix}, \quad \mathbf{E}=\begin{bmatrix} 1 & 5 \\ 2 & 6 \end{bmatrix}, \quad \mathbf{F}=\begin{bmatrix} 0 & 4 & 5 \\ 0 & 0 & -2 \\ 0 & 0 & 0 \end{bmatrix},$$

$$\mathbf{G}=\begin{bmatrix} 0 & 0 & 7 \\ 0 & 0 & 0 \\ 0 & 0 & 0 \end{bmatrix} \quad \text{and} \quad \mathbf{I}=\begin{bmatrix} 1 & 0 & 0 \\ 0 & 1 & 0 \\ 0 & 0 & 1 \end{bmatrix}.$$

Then

$$\mathbf{ab}=5, \quad \mathbf{ba}=\begin{bmatrix} 1 & 2 & 3 \\ -1 & -2 & -3 \\ 2 & 4 & 6 \end{bmatrix}, \quad \mathbf{aB}=[7 \quad 10], \quad \mathbf{aI}=\mathbf{a}, \quad \mathbf{Ib}=\mathbf{b},$$

$$\mathbf{AB}=\begin{bmatrix} 5 & 5 \\ 9 & 10 \end{bmatrix}, \quad \mathbf{BA}=\begin{bmatrix} 8 & 5 & 2 \\ 9 & 5 & 1 \\ 6 & 4 & 2 \end{bmatrix}, \quad (2\mathbf{A})\mathbf{B}=\mathbf{A}(2\mathbf{B})=2\mathbf{AB}=\begin{bmatrix} 10 & 10 \\ 18 & 20 \end{bmatrix},$$

$$\mathbf{CD}=\begin{bmatrix} -14 & 9 \\ 10 & -9 \end{bmatrix}, \quad \mathbf{DC}=\begin{bmatrix} -16 & 19 \\ 4 & -7 \end{bmatrix},$$

$$DE = ED = \begin{bmatrix} -4 & 0 \\ 0 & -4 \end{bmatrix}, \quad FG = 0, \quad Fb = \begin{bmatrix} 6 \\ -4 \\ 0 \end{bmatrix} \quad \text{and} \quad IF = FI = F.$$

Products like $\mathbf{b}\mathbf{A}$, $\mathbf{A}\mathbf{C}$ and $\mathbf{b}\mathbf{F}$ are not defined. ∎

The results illustrated in Ex. 3, which are all direct consequences of the definition of a matrix product, may be formulated as the next main result. Their proof will be left as an exercise.

Theorem 3.1 (Matrix operations involving multiplication and addition)

Let $\mathbf{A}$, $\mathbf{B}$ and $\mathbf{C}$ be matrices which are conformable for the following operations of matrix multiplication and addition, and let λ be an arbitrary scalar. Then:

matrix multiplication is associative

$$(\lambda\mathbf{A})\mathbf{B} = \mathbf{A}(\lambda\mathbf{B}) = \lambda\mathbf{A}\mathbf{B},$$

$$\mathbf{A}(\mathbf{B}\mathbf{C}) = (\mathbf{A}\mathbf{B})\mathbf{C} = \mathbf{A}\mathbf{B}\mathbf{C},$$

matrix multiplication is distributive with respect to addition

$$\mathbf{A}(\mathbf{B}+\mathbf{C}) = \mathbf{A}\mathbf{B}+\mathbf{A}\mathbf{C},$$

$$(\mathbf{A}+\mathbf{B})\mathbf{C} = \mathbf{A}\mathbf{C}+\mathbf{B}\mathbf{C},$$

matrix multiplication is noncommutative

If $\mathbf{A}\mathbf{B}$ and $\mathbf{B}\mathbf{A}$ are both defined then, in general, $\mathbf{A}\mathbf{B} \neq \mathbf{B}\mathbf{A}$. □

On account of the associative property of matrix multiplication, for any square matrix $\mathbf{A}$ we may, without ambiguity, set $\mathbf{A}\mathbf{A} = \mathbf{A}^2$, $\mathbf{A}\mathbf{A}\mathbf{A} = \mathbf{A}^3, \ldots$.

The operation of matrix transposition plays an important role in matrix algebra and this next result shows how it relates to the product of matrices.

Theorem 3.2 (Transposition of a product)

If the matrix product $\mathbf{A}\mathbf{B}$ is defined, then

$$(\mathbf{A}\mathbf{B})^\mathsf{T} = \mathbf{B}^\mathsf{T}\mathbf{A}^\mathsf{T}.$$

Proof

Let $\mathbf{A} = [a_{ij}]$ and $\mathbf{B} = [b_{ij}]$ be $m \times n$ and $n \times p$ matrices, respectively, so the product $\mathbf{A}\mathbf{B} = \mathbf{C}$ is an $m \times p$ matrix and $\mathbf{C}^\mathsf{T}$ is a $p \times m$ matrix. Writing $\mathbf{C} = [c_{ij}]$ we have $c_{ij} = \mathbf{a}_i \mathbf{b}_j$, so that if

$$\mathbf{C}^\mathsf{T} = [\tilde{c}_{ij}], \text{ then by definition}$$

$$\tilde{c}_{ij} = \mathbf{a}_j \mathbf{b}_i = a_{j1}b_{1i} + a_{j2}b_{2i} + \ldots + a_{jn}b_{ni}.$$

Set $\mathbf{B}^{\mathsf{T}}\mathbf{A}^{\mathsf{T}} = \mathbf{D} = [d_{ij}]$, then

$$d_{ij} = \mathbf{b}_i^{\mathsf{T}}\mathbf{a}_j^{\mathsf{T}} = [b_{1i} \ b_{2i} \ \cdots \ b_{ni}] \begin{bmatrix} a_{j1} \\ a_{j2} \\ \vdots \\ a_{jn} \end{bmatrix}$$

$$= b_{1i}a_{j1} + b_{2i}a_{j2} + \ldots + b_{ni}a_{jn}$$
$$= a_{j1}b_{1i} + a_{j2}b_{2i} + \ldots + a_{jn}b_{ni}$$
$$= \mathbf{a}_j\mathbf{b}_i \text{ for } i = 1, 2, \ldots p; \ j = 1, 2, \ldots, m.$$

Thus $c_{ij} = d_{ij}$ from which it follows at once that

$$(\mathbf{AB})^{\mathsf{T}} = \mathbf{B}^{\mathsf{T}}\mathbf{A}^{\mathsf{T}}. \qquad \qquad \square$$

Systems of equations and linear transformations

In terms of matrix multiplication the system of equations

$$y_1 = a_{11}x_1 + a_{12}x_2 + \ldots + a_{1n}x_n$$
$$y_2 = a_{21}x_1 + a_{22}x_2 + \ldots + a_{2n}x_n$$

$$\qquad \cdot \qquad \cdot \qquad \cdot \qquad \cdot \qquad \cdot$$

$$y_m = a_{m1}x_1 + a_{m2}x_2 + \ldots + a_{mn}x_n \qquad \qquad (15)$$

may be written concisely as

$$\mathbf{y} = \mathbf{Ax}, \qquad \qquad (16)$$

$$\text{with } \mathbf{y} = \begin{bmatrix} y_1 \\ y_2 \\ \vdots \\ y_m \end{bmatrix} \quad \mathbf{A} = \begin{bmatrix} a_{11} & a_{12} & \cdots & a_{1n} \\ a_{21} & a_{22} & \cdots & a_{2n} \\ \cdot & \cdot & \cdots & \cdot \\ a_{m1} & a_{m2} & \cdots & a_{mn} \end{bmatrix} \quad \text{and} \quad \mathbf{x} = \begin{bmatrix} x_1 \\ x_2 \\ \vdots \\ x_n \end{bmatrix} \qquad (17)$$

Similarly, the system

$$x_1 = b_{11}z_1 + b_{12}z_2 + \cdots + b_{1p}z_p$$
$$x_2 = b_{21}z_1 + b_{22}z_2 + \cdots + b_{2p}z_p$$

$$\qquad \cdot \qquad \cdot \qquad \cdot \qquad \cdots \qquad \cdot$$

$$x_n = b_{n1}z_1 + b_{n2}z_2 + \cdots + b_{np}z_p \qquad \qquad (18)$$

can be written

$$\mathbf{x} = \mathbf{Bz}, \qquad \qquad (19)$$

with

$$\mathbf{z} = \begin{bmatrix} z_1 \\ z_2 \\ \vdots \\ z_p \end{bmatrix} \quad \text{and} \quad \mathbf{B} = \begin{bmatrix} b_{11} & b_{12} & \cdots & b_{1p} \\ b_{21} & b_{22} & \cdots & b_{2p} \\ \cdot & \cdot & \cdots & \cdot \\ b_{n1} & b_{n2} & \cdots & b_{np} \end{bmatrix} \tag{20}$$

Combining (16) and (19) gives

$$\mathbf{y} = \mathbf{ABz}, \tag{21}$$

which shows how $\mathbf{y}$ may be related directly to $\mathbf{z}$ through the new coefficient matrix $\mathbf{AB}$, which is the product of the coefficient matrices in (17) and (20). Expression (21) is, of course, a direct generalization of the result obtained previously in (5).

Example 4. Linear transformation

If

$$y_1 = 2x_1 + x_2 \qquad x_1 = 3z_1 + 2z_2$$

$$\text{and}$$

$$y_2 = x_1 - 3x_2 \qquad x_2 = 2z_1 - z_2,$$

use matrices to express y_1 and y_2 in terms of z_1 and z_2.

Solution

Set $\mathbf{y} = \begin{bmatrix} y_1 \\ y_2 \end{bmatrix}$, $\mathbf{x} = \begin{bmatrix} x_1 \\ x_2 \end{bmatrix}$, $\mathbf{z} = \begin{bmatrix} z_1 \\ z_2 \end{bmatrix}$, $\mathbf{A} = \begin{bmatrix} 2 & 1 \\ 1 & -3 \end{bmatrix}$ and $\mathbf{B} = \begin{bmatrix} 3 & 2 \\ 2 & -1 \end{bmatrix}$.

Then $\mathbf{y} = \mathbf{Ax}$, $\mathbf{x} = \mathbf{Bz}$ so $\mathbf{y} = \mathbf{ABz}$.

$$\mathbf{AB} = \begin{bmatrix} 2 & 1 \\ 1 & -3 \end{bmatrix} \begin{bmatrix} 3 & 2 \\ 2 & -1 \end{bmatrix} = \begin{bmatrix} 8 & 3 \\ -3 & 5 \end{bmatrix} \quad \text{so}$$

$$\begin{bmatrix} y_1 \\ y_2 \end{bmatrix} = \begin{bmatrix} 8 & 3 \\ -3 & 5 \end{bmatrix} \begin{bmatrix} z_1 \\ z_2 \end{bmatrix} \quad \text{giving}$$

$$y_1 = 8z_1 + 3z_2$$

$$y_2 = -3z_1 + 5z_2. \qquad \blacksquare$$

Example 5. Rotational transformation

Two rectangular coordinate systems Oxy and $Ox'y'$ have a common origin, but their corresponding axes are inclined at an angle θ to each other. Find the relationship between (x, y) and (x', y'), express it in matrix form and discuss the result.

Solution

The situation is illustrated in Fig. 3.4 where the arbitrary point P has the coordinates (ξ, η) in the unprimed coordinate system and the coordinates (ξ', η') in the primed system

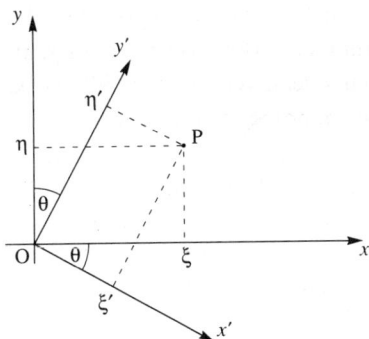

Fig. 3.4 Rotation of coordinates

A routine calculation shows that

$$\xi' = \xi \cos\theta - \eta \sin\theta$$
$$\eta' = \xi \sin\theta + \eta \cos\theta.$$

Replacing the coordinates (ξ, η) of the specific point P by the general coordinates (x, y), and the corresponding primed coordinates (ξ', η') by (x', y'), we arrive at the required result

$$\mathbf{x}' = \mathbf{A}\mathbf{x},$$

where

$$\mathbf{x}' = \begin{bmatrix} x' \\ y' \end{bmatrix}, \quad \mathbf{x} = \begin{bmatrix} x \\ y \end{bmatrix} \quad \text{and} \quad \mathbf{A} = \begin{bmatrix} \cos\theta & -\sin\theta \\ \sin\theta & \cos\theta \end{bmatrix}.$$

Matrix $\mathbf{A}$ is an example of an **orthogonal matrix**; that is a matrix with the property that if $\mathbf{a}_i$ and $\mathbf{a}_j$ are any two column vectors of $\mathbf{A}$ then

$$\mathbf{a}_i^T \mathbf{a}_j = \begin{cases} 1 \text{ for } i=j \\ 0 \text{ for } i \neq j. \end{cases}$$

Giving specific values to θ in $\mathbf{A}$ shows the effect of a rotation on the coordinate system. Thus if $\theta = \pi/2$, corresponding to

$$\mathbf{A} = \mathbf{A}_1 = \begin{bmatrix} 0 & -1 \\ 1 & 0 \end{bmatrix},$$

we see that $x' = -y$ and $y' = x$. Similarly, if $\theta = \pi$, corresponding to

$$\mathbf{A} = \mathbf{A}_2 = \begin{bmatrix} -1 & 0 \\ 0 & -1 \end{bmatrix},$$

we see that $x' = -x$ and $y' = -y$. Since this last result may also be obtained as a result of two successive rotations of $\pi/2$, it should be produced from two successive applications of the orthogonal matrix $\mathbf{A}_1$; that is from $\mathbf{A}_1^2$. This is confirmed by the following calculation

$$\mathbf{A}_1^2 = \begin{bmatrix} 0 & -1 \\ 1 & 0 \end{bmatrix}\begin{bmatrix} 0 & -1 \\ 1 & 0 \end{bmatrix} = \begin{bmatrix} -1 & 0 \\ 0 & -1 \end{bmatrix} = \mathbf{A}_2. \qquad \blacksquare$$

If we regard a matrix vector as an ordered set of numbers it may be taken to define the coordinates of a point in a space with the appropriate number of dimensions. Thus $\mathbf{y}$, an ordered m-tuple of real numbers, defines a point in the euclidean vector space $\mathbb{R}^m$ and $\mathbf{x}$, an ordered n-tuple of real numbers, a point in the corresponding vector space $\mathbb{R}^n$. The operation in (16) may thus be considered as a **transformation** of points in $\mathbb{R}^n$ into points in $\mathbb{R}^m$. An obvious geometrical analogy suggests that the operation defined in (16) is called a **mapping** of $\mathbf{x}$ into $\mathbf{y}$. In this context the vector $\mathbf{y}$ is then said to be the **image** of $\mathbf{x}$ under the mapping.

A basis $\mathbf{e}_1, \mathbf{e}_2, \ldots, \mathbf{e}_n$ for the space of vectors $\mathbf{x}$ is seen to be provided by the n vectors

$$
\mathbf{e}_1 = \begin{bmatrix} 1 \\ 0 \\ 0 \\ \vdots \\ 0 \end{bmatrix}, \quad
\mathbf{e}_2 = \begin{bmatrix} 0 \\ 1 \\ 0 \\ \vdots \\ 0 \end{bmatrix}, \quad \ldots, \quad
\mathbf{e}_n = \begin{bmatrix} 0 \\ 0 \\ 0 \\ \vdots \\ 1 \end{bmatrix},
$$

because any vector $\mathbf{x}$ with elements $x_1, x_2, \ldots, x_n$ can be represented uniquely as the linear combination

$$
\mathbf{x} = x_1 \mathbf{e}_1 + x_2 \mathbf{e}_2 + \ldots + x_n \mathbf{e}_n.
$$

An **inner product** $(\mathbf{x}_1, \mathbf{x}_2)$ may be introduced into this n dimensional vector space through the definition

$$
(\mathbf{x}_1, \mathbf{x}_2) = \mathbf{x}_1^T \mathbf{x}_2. \tag{22}
$$

It then follows from (2) that $(\mathbf{x}_1, \mathbf{x}_2)$ satisfies properties P.1 to P.3 in Sec. 2.9. The norm $\|\mathbf{x}\|$ follows from (22) as

$$
\|\mathbf{x}\| = \sqrt{\mathbf{x}^T \mathbf{x}}, \tag{23}
$$

and it introduces the notion of *length* into this n dimensional euclidean vector space.

Differentiation of a matrix

Matrices often arise with elements which are functions of some variable, as in the case of Example 5. When the elements of an $m \times n$ matrix $\mathbf{A}$ depend on x, say, so that $\mathbf{A} = [a_{ij}(x)]$, we define the derivative of $\mathbf{A}$ with respect to x, written $d\mathbf{A}/dx$, to be

$$
\frac{d\mathbf{A}}{dx} = \left[\frac{da_{ij}}{dx}\right] = \begin{bmatrix}
\dfrac{da_{11}}{dx} & \dfrac{da_{12}}{dx} & \cdots & \dfrac{da_{1n}}{dx} \\[2ex]
\dfrac{da_{21}}{dx} & \dfrac{da_{22}}{dx} & \cdots & \dfrac{da_{2n}}{dx} \\[2ex]
\cdot & \cdot & \cdots & \cdot \\[2ex]
\dfrac{da_{m1}}{dx} & \dfrac{da_{m2}}{dx} & \cdots & \dfrac{da_{mn}}{dx}
\end{bmatrix}. \tag{24}
$$

It follows directly from the properties of matrices and from the rules for differentiation that when $\mathbf{A}+\mathbf{B}$ and $\mathbf{AB}$ are defined and have elements depending on x,

$$\frac{\mathrm{d}}{\mathrm{d}x}(\mathbf{A}+\mathbf{B}) = \frac{\mathrm{d}\mathbf{A}}{\mathrm{d}x} + \frac{\mathrm{d}\mathbf{B}}{\mathrm{d}x}, \tag{25}$$

$$\frac{\mathrm{d}}{\mathrm{d}x}(\mathbf{AB}) = \frac{\mathrm{d}\mathbf{A}}{\mathrm{d}x}\mathbf{B} + \mathbf{A}\frac{\mathrm{d}\mathbf{B}}{\mathrm{d}x}. \tag{26}$$

An immediate consequence of (24) is that if the elements a_{ij} of $\mathbf{A}$ are all constant, then $\mathrm{d}\mathbf{A}/\mathrm{d}x = \mathbf{0}$. Higher derivatives follow by repeated application of (24) so, for example,

$$\frac{\mathrm{d}^2\mathbf{A}}{\mathrm{d}x^2} = \frac{\mathrm{d}}{\mathrm{d}x}\left[\frac{\mathrm{d}\mathbf{A}}{\mathrm{d}x}\right].$$

If the elements of $\mathbf{A}$ involve several variables $x, y, \ldots$, then the partial derivatives $\partial\mathbf{A}/\partial x$, $\partial\mathbf{A}/\partial y$, $\partial^2\mathbf{A}/\partial x\partial y, \ldots$, of $\mathbf{A}$ may be defined in an analogous fashion.

Example 6. Differentiation of a matrix

Given

$$\mathbf{A} = \begin{bmatrix} \cos x & 0 \\ 0 & \sin x \end{bmatrix}, \quad \mathbf{B} = \begin{bmatrix} 2 & x \\ x^3 & 0 \end{bmatrix}, \quad \mathbf{C} = \begin{bmatrix} \sinh(xy) & x^2+y^2 \\ 3x^2 & \ln(xy) \end{bmatrix},$$

find

$$\frac{\mathrm{d}}{\mathrm{d}x}(\mathbf{A}+\mathbf{B}), \quad \frac{\mathrm{d}}{\mathrm{d}x}(\mathbf{AB}), \quad \frac{\mathrm{d}^2\mathbf{B}}{\mathrm{d}x^2}, \quad \frac{\partial\mathbf{C}}{\partial x} \quad \text{and} \quad \frac{\partial^2\mathbf{C}}{\partial y\partial x}.$$

Solution

$$\frac{\mathrm{d}}{\mathrm{d}x}(\mathbf{A}+\mathbf{B}) = \begin{bmatrix} -\sin x & 0 \\ 0 & \cos x \end{bmatrix} + \begin{bmatrix} 0 & 1 \\ 3x^2 & 0 \end{bmatrix} = \begin{bmatrix} -\sin x & 1 \\ 3x^2 & \cos x \end{bmatrix},$$

$$\frac{\mathrm{d}}{\mathrm{d}x}(\mathbf{AB}) = \begin{bmatrix} -\sin x & 0 \\ 0 & \cos x \end{bmatrix}\begin{bmatrix} 2 & x \\ x^3 & 0 \end{bmatrix} + \begin{bmatrix} \cos x & 0 \\ 0 & \sin x \end{bmatrix}\begin{bmatrix} 0 & 1 \\ 3x^2 & 0 \end{bmatrix}$$

$$= \begin{bmatrix} -2\sin x & \cos x - x\sin x \\ x^3\cos x + 3x^2\sin x & 0 \end{bmatrix},$$

$$\frac{\mathrm{d}^2\mathbf{B}}{\mathrm{d}x^2} = \frac{\mathrm{d}}{\mathrm{d}x}\begin{bmatrix} 0 & 1 \\ 3x^2 & 0 \end{bmatrix} = \begin{bmatrix} 0 & 0 \\ 6x & 0 \end{bmatrix},$$

$$\frac{\partial\mathbf{C}}{\partial x} = \begin{bmatrix} y\cosh(xy) & 2x \\ 6x & 1/x \end{bmatrix},$$

$$\frac{\partial^2\mathbf{C}}{\partial y\partial x} = \begin{bmatrix} \cosh(xy)+xy\sinh(xy) & 0 \\ 0 & 0 \end{bmatrix}. \quad \blacksquare$$

Row permutation matrix

An $n \times n$ matrix $\mathbf{P}$ derived from the $n \times n$ unit matrix $\mathbf{I}$ by interchanging its rows in some specific order (permuting them) is called a **row permutation matrix**. Thus only a single 1 occurs in each row and column of $\mathbf{P}$, with all its other elements being zeros. The effect of premultiplying an $n \times m$ matrix $\mathbf{A}$ with elements a_{ij} by $\mathbf{P}$ is to interchange the rows of $\mathbf{A}$ in the same order as that used to derive $\mathbf{P}$ from $\mathbf{I}$.

The proof of this result follows from the definition of matrix multiplication. Let $s_1 \, s_2, \ldots, s_n$ be some permutation of the numbers $1, 2, \ldots, n$, and denote by $\mathbf{P}_i$ the ith row of $\mathbf{P}$ in which the s_ith element is 1 and all the others are zero. Then $\mathbf{P}$ is obtained from $\mathbf{I}$ by permuting its rows in the order $s_1, s_2, \ldots, s_n$. Now let $\mathbf{a}_j$ be the jth column of $\mathbf{A}$.

Then it follows from the definition of matrix multiplication given in (14) that

$$\mathbf{PA} = \begin{bmatrix} \mathbf{p}_1\mathbf{a}_1 & \mathbf{p}_1\mathbf{a}_2 & \cdots & \mathbf{p}_1\mathbf{a}_m \\ \mathbf{p}_2\mathbf{a}_1 & \mathbf{p}_2\mathbf{a}_2 & \cdots & \mathbf{p}_2\mathbf{a}_m \\ . & . & \cdots & . \\ \mathbf{p}_n\mathbf{a}_1 & \mathbf{p}_n\mathbf{a}_2 & \cdots & \mathbf{p}_n\mathbf{a}_m \end{bmatrix}. \tag{27}$$

However, $\mathbf{p}_i\,\mathbf{a}_j = a_{s_ij}$, so that

$$\mathbf{PA} = \begin{bmatrix} a_{s_11} & a_{s_11} & \cdots & a_{s_1m} \\ a_{s_21} & a_{s_22} & \cdots & a_{s_2m} \\ . & . & \cdots & . \\ a_{s_n1} & a_{s_n2} & \cdots & a_{s_nm} \end{bmatrix},$$

which is simply the matrix obtained from $\mathbf{A}$ by permuting its rows in the order $s_1, s_2, \ldots, s_n$ used to derive $\mathbf{P}$ from $\mathbf{I}$.

An important result follows from (27) by setting $\mathbf{A} = \mathbf{P}^\mathsf{T}$, the transpose of $\mathbf{P}$. The element in row i and column j then becomes

$$\mathbf{p}_i\mathbf{p}_j^\mathsf{T} = \begin{cases} 0, & i \neq j \\ 1, & i = j \end{cases},$$

for $i, j = 1, 2, \ldots, n$. Thus we arrive at the result

$$\mathbf{PP}^\mathsf{T} = \mathbf{I}, \tag{28}$$

which by taking the transpose and using Theorem 3.2 shows that

$$\mathbf{PP}^\mathsf{T} = \mathbf{P}^\mathsf{T}\mathbf{P} = \mathbf{I}. \tag{29}$$

If a sequence of row permutations is carried out on a conformable matrix in the order $\mathbf{P}_1, \mathbf{P}_2, \ldots, \mathbf{P}_m$, it follows by matrix multiplication that the final order in which the rows are arranged is represented by the row permutation matrix

$$\mathbf{P} = \mathbf{P}_m\mathbf{P}_{m-1} \cdots \mathbf{P}_2\mathbf{P}_1. \tag{30}$$

Notice that the order in which the matrices P_r appear in the product (30) must be that in which the row operations corresponding to them are performed on the conformable matrix in question.

Example 7. Row permutation by matrix multiplication

Verify the permutation property of the row permutation matrix

$$P = \begin{bmatrix} 0 & 1 & 0 & 0 \\ 0 & 0 & 1 & 0 \\ 1 & 0 & 0 & 0 \\ 0 & 0 & 0 & 1 \end{bmatrix} \quad \text{using } A = \begin{bmatrix} 1 & 9 \\ 7 & 0 \\ 2 & 4 \\ 3 & 5 \end{bmatrix}.$$

Solution

$$PA = \begin{bmatrix} 7 & 0 \\ 2 & 4 \\ 1 & 9 \\ 3 & 5 \end{bmatrix}.$$

The row permutation matrix P is obtained from I by permuting its rows into the order row 2, row 3, row 1 and row 4, which is seen from the above result to be the same as the order in which the rows of PA are obtained from A. ∎

Example 8. Row permutation operations

(i) Write down the row permutation matrix that permutes rows 1, 2, 3 and 4 of a conformable matrix A into the order 3, 1, 4, 2.

(ii) Let P_1 be the row permutation matrix that interchanges rows 3 and 4 of a conformable matrix A, and P_2 the matrix that interchanges rows 1 and 3. Verify that the effect on A of first interchanging rows 3 and 4, and then rows 1 and 3, is described by the row permutation matrix

$$P = P_2 P_1.$$

Solution

$$\text{(i) } P = \begin{bmatrix} 0 & 0 & 1 & 0 \\ 1 & 0 & 0 & 0 \\ 0 & 0 & 0 & 1 \\ 0 & 1 & 0 & 0 \end{bmatrix}.$$

(ii) $P_1 = \begin{bmatrix} 1 & 0 & 0 & 0 \\ 0 & 1 & 0 & 0 \\ 0 & 0 & 0 & 1 \\ 0 & 0 & 1 & 0 \end{bmatrix}$ $P_2 = \begin{bmatrix} 0 & 0 & 1 & 0 \\ 0 & 1 & 0 & 0 \\ 1 & 0 & 0 & 0 \\ 0 & 0 & 0 & 1 \end{bmatrix}$

$P = P_2 P_1 = \begin{bmatrix} 0 & 0 & 0 & 1 \\ 0 & 1 & 0 & 0 \\ 1 & 0 & 0 & 0 \\ 0 & 0 & 1 & 0 \end{bmatrix}.$

P rearranges the rows of A in the order 4, 2, 1, 3. This is the same as first interchanging rows 3 and 4 of A to obtain the order 1, 2, 4, 3, and then interchanging rows 1 and 3 of this new arrangement to obtain the order 4, 2, 1, 3. ■

Problems for Section 3.3

Let

$$a = [1 \quad -1 \quad 2], \quad b = \begin{bmatrix} 3 \\ 2 \\ 1 \end{bmatrix}, \quad B = \begin{bmatrix} 1 & -2 & 4 \\ 2 & 0 & 3 \\ 1 & 2 & 1 \end{bmatrix}, \quad C = \begin{bmatrix} 2 & 1 & 1 \\ -1 & 2 & -1 \end{bmatrix}.$$

Where defined, find the following matrix products.

1 ab	2 ba	3 aB	4 Ba	5 aC
6 Cb	7 Ba$^\mathsf{T}$	8 b$^\mathsf{T}$B	9 BC	10 CC$^\mathsf{T}$
11 B^2	12 BB$^\mathsf{T}$	13 aB$^\mathsf{T}$b		
14 If				

$$A = \begin{bmatrix} 1 & 2 & 0 \\ 1 & 1 & 0 \\ 2 & 1 & 2 \end{bmatrix} \quad \text{and} \quad B = \begin{bmatrix} 2 & 1 & 1 \\ 1 & 2 & 2 \\ 1 & 2 & 1 \end{bmatrix}$$

find BA and AB.

15 Find an example of 2×2 matrix A such that

$$A^2 = -I.$$

16 Prove from the definitions of matrix addition and multiplication that

$$A(B+C) = AB + AC$$

and

$$(A+B)C = AC + BC.$$

17 Use Theorem 3.2 to prove that if A, B and C are $n \times n$ matrices, then

$$(ABC)^\mathsf{T} = C^\mathsf{T} B^\mathsf{T} A^\mathsf{T}, \text{ and that } (A^m)^\mathsf{T} = (A^\mathsf{T})^m \text{ (}m\text{ integral)}.$$

18 Expand the following expressions in which **A** and **B** are $n \times n$ matrices:

(i) $(\mathbf{AB})^2$ (ii) $[(\mathbf{AB})^\mathsf{T}]^2$ (iii) $[(\mathbf{AB}^\mathsf{T})^2]^\mathsf{T}$ (iv) $[(\mathbf{B}^\mathsf{T})^2(\mathbf{A}^\mathsf{T})^2]^\mathsf{T}$.

19 Express the following system in matrix form

$$3x_1 + 2x_2 + x_3 = 7$$
$$4x_1 - x_2 + x_3 = 1$$
$$x_1 - 6x_2 + 7x_3 = 4.$$

20 Given

$$y_1 = 3x_1 + 2x_2 \qquad\qquad x_1 = 4z_1 + z_2$$

$$\text{and}$$

$$y_2 = 2x_1 - x_2 \qquad\qquad x_2 = z_1 - 2z_2$$

use matrix methods to express y_1 and y_2 directly in terms of z_1 and z_2.

21 What is the geometrical interpretation of the transformation of points in the plane by the linear transformation of the form

$$\mathbf{y} = \mathbf{Ax} + \mathbf{b},$$

where

(i) $\mathbf{A} = \begin{bmatrix} 1 & 0 \\ 0 & 1 \end{bmatrix}$, $\mathbf{b} = 0$ (ii) $\mathbf{A} = \begin{bmatrix} 2 & 0 \\ 0 & 2 \end{bmatrix}$, $\mathbf{b} = 0$

(iii) $\mathbf{A} = \begin{bmatrix} 3 & 0 \\ 0 & 3 \end{bmatrix}$, $\mathbf{b} = \begin{bmatrix} 1 \\ 3 \end{bmatrix}$ (iv) $\mathbf{A} = \begin{bmatrix} 0 & 1 \\ 1 & 0 \end{bmatrix}$, $\mathbf{b} = 0$

(v) $\mathbf{A} = \begin{bmatrix} \lambda & 0 \\ 0 & \mu \end{bmatrix}$, $\mathbf{b} = 0$ (vi) $\mathbf{A} = \begin{bmatrix} -1 & 0 \\ 0 & 1 \end{bmatrix}$, $\mathbf{b} = \begin{bmatrix} 3 \\ -2 \end{bmatrix}$.

22 An $n \times n$ matrix **A** is **orthogonal** when for $i = 1, 2, \ldots, n; \, j = 1, 2, \ldots, n$.

$$\mathbf{a}_i^\mathsf{T} \mathbf{a}_j = \begin{cases} 1 \text{ for } i = j \\ 0 \text{ for } i \neq j, \end{cases}$$

where $\mathbf{a}_i$ and $\mathbf{a}_j$ are any two column vectors belonging to **A**. Use this definition to prove that **A** is orthogonal if

$$\mathbf{AA}^\mathsf{T} = \mathbf{I} \text{ or, equivalently, if } \mathbf{A}^\mathsf{T}\mathbf{A} = \mathbf{I},$$

where **I** is the $n \times n$ unit matrix. Use this result to verify that

(a) the following matrices are orthogonal for all θ, ϕ and ψ:

$$\mathbf{A}_1 = \begin{bmatrix} \cos\theta & -\sin\theta & 0 \\ \sin\theta & \cos\theta & 0 \\ 0 & 0 & 1 \end{bmatrix}, \quad \mathbf{A}_2 = \begin{bmatrix} \cos\phi & 0 & -\sin\phi \\ 0 & 1 & 0 \\ \sin\phi & 0 & \cos\phi \end{bmatrix},$$

$$\mathbf{A}_3 = \begin{bmatrix} 1 & 0 & 0 \\ 0 & \cos\psi & -\sin\psi \\ 0 & \sin\psi & \cos\psi \end{bmatrix};$$

(b) to determine which of the following matrices is orthogonal:

(i) $\begin{bmatrix} \dfrac{\sqrt{2}}{5} & \dfrac{1}{\sqrt{5}} \\ \dfrac{1}{\sqrt{5}} & \dfrac{-2}{\sqrt{5}} \end{bmatrix}$
(ii) $\begin{bmatrix} \dfrac{\sqrt{3}}{2} & \dfrac{1}{2} \\ \dfrac{-1}{2} & \dfrac{\sqrt{3}}{2} \end{bmatrix}$
(iii) $\begin{bmatrix} \dfrac{1}{2} & \dfrac{-\sqrt{3}}{2} \\ \dfrac{\sqrt{3}}{2} & \dfrac{1}{2} \end{bmatrix}$

(iv) $\begin{bmatrix} \dfrac{1}{\sqrt{2}} & 0 & -\dfrac{1}{\sqrt{2}} \\ 0 & 1 & 0 \\ \dfrac{1}{\sqrt{2}} & 0 & \dfrac{1}{\sqrt{2}} \end{bmatrix}$
(v) $\begin{bmatrix} \dfrac{1}{\sqrt{2}} & 0 & \dfrac{-1}{\sqrt{2}} \\ 0 & 1 & 0 \\ \dfrac{-1}{\sqrt{2}} & 0 & \dfrac{1}{\sqrt{2}} \end{bmatrix}$
(vi) $\begin{bmatrix} \dfrac{1}{2} & -\dfrac{\sqrt{3}}{2} & 0 \\ \dfrac{\sqrt{3}}{2} & \dfrac{1}{2} & 0 \\ 0 & 0 & 1 \end{bmatrix}$.

23 Prove results (25) and (26).

Harder problems

24 The square matrices **L** and **U** known, respectively, as **lower** and **upper triangular matrices**, are of considerable importance in numerical analysis. Use the definition of matrix multiplication to prove that the product of two lower (upper) triangular matrices is a lower (upper) triangular matrix.

25 The **trace** of an $n \times n$ matrix $\mathbf{A} = [a_{ij}]$, written $\mathrm{tr}\,(\mathbf{A})$ is the sum of the elements on its leading diagonal, so

$$\mathrm{tr}\,(\mathbf{A}) = \sum_{i=1}^{n} a_{ii}.$$

If $\mathbf{A}$, $\mathbf{B}$ and $\mathbf{C}$ are $n \times n$ matrices and λ is an arbitrary scalar, prove that the operation tr has the following properties:
  (i) $\mathrm{tr}\,(\mathbf{A} + \mathbf{B}) = \mathrm{tr}\,(\mathbf{A}) + \mathrm{tr}\,(\mathbf{B})$
  (ii) $\mathrm{tr}\,(\lambda \mathbf{A}) = \lambda \,\mathrm{tr}\,(\mathbf{A})$
  (iii) $\mathrm{tr}\,(\mathbf{A}^T) = \mathrm{tr}\,(\mathbf{A})$
  (iv) $\mathrm{tr}\,(\mathbf{AB}) = \mathrm{tr}\,(\mathbf{BA})$

26 Prove that if $\mathbf{A}$ and $\mathbf{B}$ are $n \times n$ orthogonal matrices, then so also is the product $\mathbf{AB}$. Verify this using the matrices from Prob. 22(a).

27 The length of a vector $\mathbf{x}$ in the n-dimensional euclidean vector space $\mathbb{R}^n$ is given by the norm

$$\|\mathbf{x}\| = \sqrt{\mathbf{x}^T \mathbf{x}}.$$

Prove that if $\mathbf{A}$ is an $n \times n$ orthogonal matrix, then the linear transformation

$$\mathbf{y} = \mathbf{Ax}$$

preserves length. That is, the length of $\mathbf{x}$ is the same as the length of its image $\mathbf{y}$ under the given linear transformation. Is this still true for the linear transformation

$$y = Ax + b,$$

where b is an arbitrary n element column vector?

28 The **Fibonacci sequence** is the sequence of numbers

$$0, 1, 1, 2, 3, 5, 8, 13, 21, \ldots,$$

in which each member of the sequence is the sum of two which precede it. The sequence arises in many parts of mathematics, the physical sciences and throughout biology. The **algorithm** for generating the member u_{n+2} from u_{n+1} and u_n is

$$u_{n+2} = u_{n+1} + u_n \quad \text{with} \quad u_0 = 0 \quad \text{and} \quad u_1 = 1.$$

Show the sequence may be generated from the linear transformation

$$a_{n+1} = Aa_n,$$

where

$$a_n = \begin{bmatrix} u_{n+1} \\ u_n \end{bmatrix} \quad \text{and} \quad A = \begin{bmatrix} 1 & 1 \\ 1 & 0 \end{bmatrix},$$

and find a_{n+1} in terms of a_0.

29 A **transition matrix**, also called a **stochastic matrix**, is an $n \times n$ matrix $P = [p_{ij}]$ with nonnegative elements such that the sum of the elements in each row is unity. That is, $p_{ij} \geq 0$ and $\sum_{j=1}^{n} p_{ij} = 1$ (see Prob. 28, Sec. 3.2). Prove that P^2 is a transition matrix, and hence that so also is P^n for $n > 2$.

30 A block of p adjacent rows and q adjacent columns of an $m \times n$ matrix A is said to form a **submatrix** of A. The division of A into submatrices by drawing lines between rows and columns of A is called the **partitioning** of A. Let $A_{ij}^{(rs)}$ denote the $r \times s$ submatrix in the ith row and jth column of the partitioned matrix A with submatrices as elements. If A is an $m \times n$ matrix and B an $n \times r$ matrix, verify that the matrix product AB may be represented as follows in terms of products of submatrices

$$\left[\begin{array}{c|c} A_{11}^{(m_1 n_1)} & A_{12}^{(m_1 n_2)} \\ \hline A_{21}^{(m_2 n_1)} & A_{22}^{(m_2 n_2)} \end{array} \right] \left[\begin{array}{c|c} B_{11}^{(n_1 r_1)} & B_{12}^{(n_1 r_2)} \\ \hline B_{21}^{(n_2 r_1)} & B_{22}^{(n_2 r_2)} \end{array} \right]$$

$$\left[\begin{array}{c|c} A_{11}^{(m_1 n_1)} B_{11}^{(n_1 r_1)} + A_{12}^{(m_1 n_2)} + B_{21}^{(n_2 r_1)} & A_{11}^{(m_1 n_1)} B_{12}^{(n_1 r_2)} + A_{12}^{(m_1 n_2)} B_{22}^{(n_2 r_2)} \\ \hline A_{21}^{(m_2 n_1)} B_{11}^{(n_1 r_1)} + A_{22}^{(m_2 n_2)} B_{21}^{(n_2 r_1)} & A_{21}^{(m_2 n_1)} B_{12}^{(n_1 r_2)} + A_{22}^{(m_2 n_2)} B_{22}^{(n_2 r_2)} \end{array} \right],$$

where $m_1 + m_2 = m$ and $n_1 + n_2 = n$.

31 If $A = \begin{bmatrix} 0 & 3 & 0 \\ 3 & 0 & 4 \\ 0 & 4 & 0 \end{bmatrix}$, show that (a) $A^{2r-1} = 5^{2r-2} A$, $r = 1, 2, \ldots,$

(b) $A^{2r} = 5^{2r-2} \begin{bmatrix} 9 & 0 & 12 \\ 0 & 25 & 0 \\ 12 & 0 & 16 \end{bmatrix}$, $r = 1, 2, \ldots.$

32 If $A = \begin{bmatrix} 1 & 1 & 0 \\ 0 & 1 & 1 \\ 0 & 0 & 1 \end{bmatrix}$, show that $A^n = \begin{bmatrix} 1 & n & \frac{1}{2}n(n-1) \\ 0 & 1 & n \\ 0 & 0 & 1 \end{bmatrix}$, and hence that

$$\sum_{n=0}^{\infty} \frac{t^n}{n!} A^n = \begin{bmatrix} e^t & te^t & \frac{1}{2}t^2 e^t \\ 0 & e^t & te^t \\ 0 & 0 & e^t \end{bmatrix}.$$

33 Let a rod of length l, cross-sectional area A and Young's modulus of elasticity E be subjected to a force F along its length which causes a small change of length d $(d/l \ll 1)$. Then in the theory of linear elasticity the **stress** τ (force per unit area) is proportional to the **strain** ε (extension per unit length), with Young's modulus as the constant of proportionality. Thus $\tau = F/A$ and $\varepsilon = d/l$, so that

$$F = kd,$$

with $k = EA/l$ being called the **stiffness** of the rod.

Figure 3.5(a) shows a rod of length l, cross-sectional area A and Young's modulus of elasticity E with its left-hand end located at (x_1, y_1) and its right-hand end at (x_2, y_2). Suppose, as illustrated in Fig. 3.5(b), that as a result of a force F acting along the rod with horizontal and vertical components F_{1x}, F_{1y} at the left-hand end, and F_{2x}, F_{2y} at the right-hand end, respectively, the left-hand end moves to $(x_1 + \alpha_1, y_1 + \beta_1)$ and the right-hand end to $(x_2 + \alpha_2, y_2 + \beta_2)$. Assume the rod to be in equilibrium, and $\alpha_1, \beta_1, \alpha_2, \beta_2$ to be sufficiently small that the resulting change of inclination θ of the rod to the x-axis may be neglected.

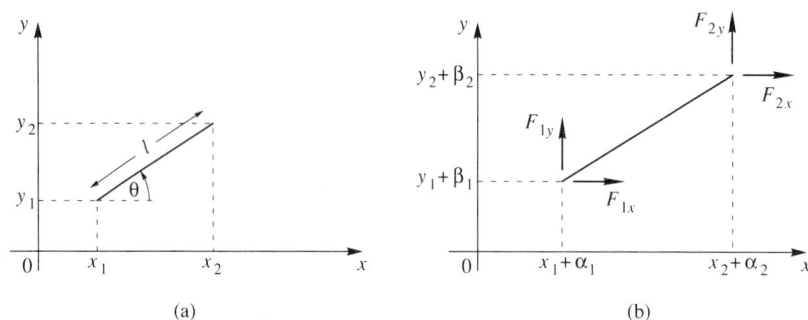

(a)   (b)

Fig. 3.5

Show that if $p = \cos \theta$, $q = \sin \theta$, then

$$F = kd,$$

where

$$F = \begin{bmatrix} F_{1x} \\ F_{1y} \\ F_{2x} \\ F_{2y} \end{bmatrix}, \quad d = \begin{bmatrix} \alpha_1 \\ \beta_1 \\ \alpha_2 \\ \beta_2 \end{bmatrix} \quad \text{and} \quad k = \begin{bmatrix} p^2 & pq & -p^2 & -pq \\ pq & q^2 & -pq & -q^2 \\ -p^2 & -pq & p^2 & pq \\ -pq & -q^2 & pq & q^2 \end{bmatrix}.$$

The matrix **k** occurring here is called the **stiffness matrix** for the rod, and it is a generalization of the stiffness k defined earlier for a one-dimensional problem. The stiffness matrix plays an important part in the analysis of deformable structures.

34 Let

$$A = \begin{bmatrix} 0 & -c & b \\ c & 0 & -a \\ -b & a & 0 \end{bmatrix} \quad \text{and} \quad x = [a, b, c],$$

where $a^2 + b^2 + c^2 = 1$.

(a) Show $A^2 = x^T x - I$.

(b) Prove that $A^3 = -A$.

(c) Find A^4 in terms of a, b and c.

3.4 Systems of linear equations. Solution by elimination

General considerations

Systems of linear simultaneous equations arise frequently; so a study of their properties and methods of solution is of fundamental importance. At this stage our attention must be confined to the basic matrix algebra of the problem, and to the way this relates to the nature of a solution when one exists. The question of how numerical answers are best obtained will be deferred until later.

The most general **system of linear simultaneous equations** involves m equations in the n unknowns $x_1, x_2, \ldots, x_n$ and has the form

$$a_{11}x_1 + a_{12}x_2 + \cdots + a_{1n}x_n = b_1$$
$$a_{21}x_1 + a_{22}x_2 + \cdots + a_{2n}x_n = b_2$$

$$a_{m1}x_1 + a_{m2}x_2 + \cdots + a_{mn}x_n = b_m, \tag{1}$$

where the coefficients a_{ij} and the b_i are given numbers. This may be written in matrix form as

$$Ax = b, \tag{2}$$

where

$$A = \begin{bmatrix} a_{11} & a_{12} & \cdots & a_{1n} \\ a_{21} & a_{22} & \cdots & a_{2n} \\ \vdots & \vdots & \cdots & \vdots \\ a_{m1} & a_{m2} & \cdots & a_{mn} \end{bmatrix}, \quad x = \begin{bmatrix} x_1 \\ x_2 \\ \vdots \\ x_n \end{bmatrix} \quad \text{and} \quad b = \begin{bmatrix} b_1 \\ b_2 \\ \vdots \\ b_m \end{bmatrix}. \tag{3}$$

Here, the matrix **A** containing the coefficients a_{ij} is called the **coefficient matrix** of (1), and

x is called the **solution vector.** When every element b_i of **b** is zero, equations (1) are said to form a **homogeneous system.** The system is said to be **nonhomogeneous** if at least one element of **b** is nonzero. Depending on m, n, **A** and **b**, we shall see there may be a unique solution to (1), an infinite number of solutions or no solution at all. The following simple examples illustrate this behavior.

Example 1. Uniqueness, nonuniqueness and nonexistence of solutions

(i) **Nonhomogeneous system with $m=n=2$. Unique solution**
The system

$$x_1 + 2x_2 = 4$$
$$x_1 - x_2 = 1,$$

has the unique solution $x_1 = 2$, $x_2 = 1$. In graphical terms this corresponds to the point of intersection P of the two straight lines shown in Fig. 3.6.

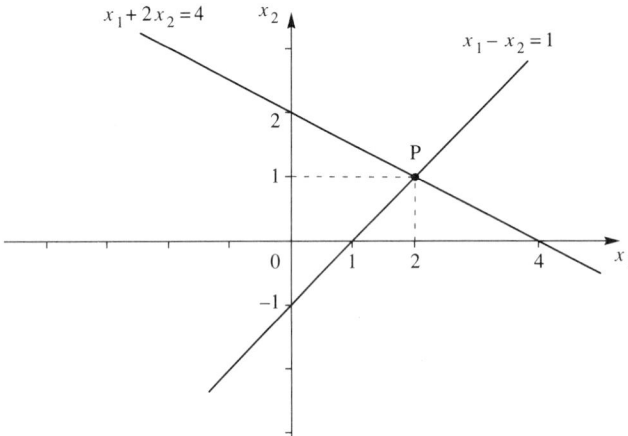

Fig. 3.6   Unique solution

(ii) **Nonhomogeneous system with $m=n=2$. Nonunique solution**
The system

$$x_1 - 2x_2 = 4$$
$$2x_1 - 4x_2 = 8,$$

is such that the second equation is twice the first one. Solving for x_1 and setting $x_2 = k$, a parameter, we find the solution $x_1 = 4 + 2k$, $x_2 = k$ for $-\infty < k < \infty$. Thus this system has a **one-parameter family** of solutions (a single infinity), and so its solution is not unique. The infinity of solutions arises because there is in fact only one equation for the two unknowns x_1 and x_2. The equations imply no contradiction, and so are said to be **consistent.** Although this fact is obvious in this simple case it ceases to become so when more equations and unknowns are involved, and one or more of the equations is obtained as the sum of multiples of others. In graphical terms the two straight lines involved in this problem coincide, and any point P on the common line shown in Fig. 3.7 is then a solution.

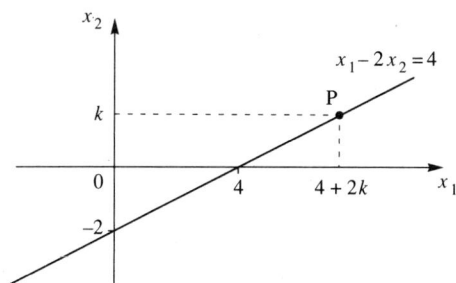

Fig. 3.7 Nonunique solution

(iii) **Nonhomogeneous system with $m=n=2$. No solution**

The system

$$2x_1 + x_2 = 4$$

$$2x_1 + x_2 = 8$$

has no solution, since the second equation contradicts the first. That is, the left-hand sides are identical but the right-hand sides are different. This is readily seen in graphical terms, because the lines are parallel straight lines and so have no point of intersection (Fig. 3.8). Contradictory systems of this type are said to be **inconsistent**.

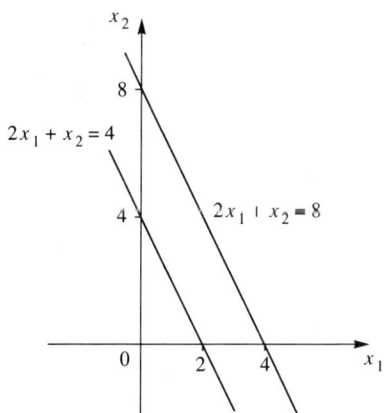

Fig. 3.8 No point of intersection. No solution

(iv) **Homogeneous system with $m=n=2$**

The system

$$3x_1 - x_2 = 0$$

$$x_1 + px_2 = 0$$

is homogeneous and it has the **null** or **trivial** solution $x_1 = x_2 = 0$. To obtain another solution we first combine the equations to obtain

$$x_1(1 + 3p) = 0.$$

If $1+3p \neq 0$ this yields $x_1 = 0$, and hence $x_2 = 0$, which is just the trivial solution. However, if $p = -1/3$, x_1 becomes arbitrary, and from the first equation $x_2 = 3x_1$. So, setting $x_1 = k$, a parameter, we obtain the **one-parameter** family of solutions

$$x_1 = k, \quad x_2 = 3k \qquad \text{for } -\infty < k < \infty.$$

Thus there is only the unique trivial solution if $p \neq -\frac{1}{3}$, and a nonunique family of solutions if $p = -\frac{1}{3}$. In graphical terms, the situation may be illustrated as in Fig. 3.9. Both lines pass through the origin which is always a solution (trivial). For $p \neq -\frac{1}{3}$ the lines are distinct and the trivial solution is the *only* solution (unique). If $p = -\frac{1}{3}$ the lines coincide and any point P on the common line is a solution (non-unique).

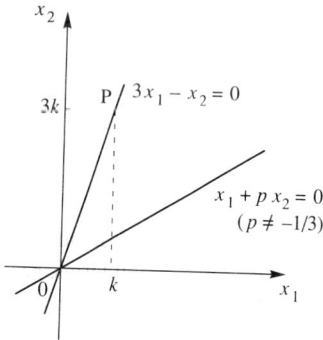

Fig. 3.9   Homogeneous system

(v) **Nonhomogeneous system $m = 3$, $n = 2$**

The system

$$x_1 + 2x_2 = 4$$
$$x_1 - x_2 = 1$$
$$2x_1 - x_2 = p$$

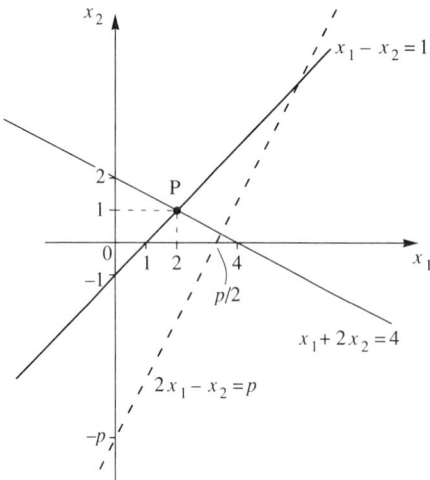

Fig. 3.10   Unique solution when $p = 3$; otherwise no solution

is the same as in (i) except for the third equation. The first two equations may be solved to give $x_1 = 2, x_2 = 1$, but the third will only be consistent if $p = 3$. There will thus be a unique solution if $p = 3$, and no solution for any other value of p. The situation is illustrated in Fig. 3.10. This shows that only when $p = 3$ will the dotted line representing the third equation pass through the unique point of intersection P of the first two equations allowing all three equations to be satisfied. ∎

Before proceeding to a more detailed discussion of these matters it is useful to find the relationship between system (1), and the linear transformation

$$\mathbf{y} = \mathbf{A}\mathbf{x} \qquad (4)$$

introduced in Eq. (16), Sec. 3.3. This was used to relate an n-component vector $\mathbf{x}$ to an m-component vector $\mathbf{y}$ called its **image** under the transformation. The implication was that when $\mathbf{x}$ is given, (4) determines its image $\mathbf{y}$. If the roles of $\mathbf{x}$ and $\mathbf{y}$ are reversed, a natural question to ask is: for what $\mathbf{x}$ will a particular image $\mathbf{y} = \mathbf{b}$ be obtained? Letting $\mathbf{b}$ have the elements $b_1, b_2, \ldots, b_m$ (given numbers) we arrive at the system of simultaneous equations (1).

Gaussian elimination[1]

Any direct approach to the solution of a system of equations is always based on the process of elimination of variables. The method we now describe is known as **Gaussian elimination**, and it involves a systematic elimination procedure to reduce the original system to one of triangular form, which is then easily solved by back-substitution.

The process is best described by means of a numerical example. Let us solve the system

$$
\begin{aligned}
x_1 + 2x_2 - x_3 + x_4 &= 0 \\
2x_1 + 3x_2 + x_3 + 2x_4 &= 7 \\
x_1 + x_2 + 3x_3 - x_4 &= 14 \\
-x_1 + x_2 + 2x_3 + x_4 &= 5.
\end{aligned}
\qquad (5)
$$

Rather than use the full system of equations in (5), we shall work instead with what is called the **augmented matrix** associated with a system of simultaneous equations. For a system such as (1), this is just the coefficient matrix $\mathbf{A}$ of the system, *augmented* on the right-hand side by the inclusion of the constants contained in the vector $\mathbf{b}$, and usually written $\mathbf{A} | \mathbf{b}$.

---

[1] CARL FRIEDRICH GAUSS (1777–1855), a German mathematician who is universally regarded as the greatest mathematician of the nineteenth century. He ranks with Isaac Newton as one of the greatest mathematicians of all time. He made fundamental contributions to all aspects of mathematics and also to science, including astronomy and electricity.

In the case of system (5), the augmented matrix is as shown below where, for the sake of emphasis, the constant terms have been separated from the coefficient matrix by a dotted line

$$
A|b = \left[\begin{array}{cccc|c}
1 & 2 & -1 & 1 & 0 \\
2 & 3 & 1 & 2 & 7 \\
1 & 1 & 3 & -1 & 14 \\
-1 & 1 & 2 & 1 & 5
\end{array}\right]. \tag{6}
$$

The augmented matrix is equivalent to the full system of equations, because entries in its ith row define the ith equation, while an entry in its jth column refers to the variable x_j. The last column contains the constant terms on the right-hand side, so the dotted line may be regarded as taking the place of the equality signs in the equations.

Now a system of simultaneous linear equations will not be altered as a result of any one of the following elementary algebraic operations being performed on the equations:

(i) the interchange of any two equations,
(ii) the multiplication of an equation by a nonzero constant,
(iii) the addition to any equation of a multiple of another equation.

In terms of the augmented matrix, these become the following **elementary row operations** on a matrix

(I) the interchange of any two rows,
(II) the multiplication of elements in a row by a nonzero constant,
(III) the addition to the elements in a row of a multiple of the corresponding elements in another row.

For what is to follow we now adopt the notation that R_1, R_2, ... , refers to row 1, row 2, ... , and $3R_2$ to three times row 2. Then $R_2 - 3R_1$ is the operation of subtracting from the elements of row 2, three times the corresponding elements of row 1.

Gaussian elimination applied to the four equations in four unknowns represented by (6) amounts to using elementary row operations to reduce the augmented matrix shown in (6) to one in which the coefficient matrix is an upper triangular matrix. That is, a square matrix with zeros below the leading diagonal.

The elimination process proceeds as follows, with the row operations showing how each transformation of the augmented matrix has been obtained from the previous one:

Step 1

$$
A|b = \left[\begin{array}{cccc|c}
1 & 2 & -1 & 1 & 0 \\
2 & 3 & 1 & 2 & 7 \\
1 & 1 & 3 & -1 & 14 \\
-1 & 1 & 2 & 1 & 5
\end{array}\right],
\quad
\begin{array}{l}
R_1 \\
R_2 - 2R_1 \\
R_3 - R_1 \\
R_4 + R_1
\end{array}
\left[\begin{array}{cccc|c}
1 & 2 & -1 & 1 & 0 \\
0 & -1 & 3 & 0 & 7 \\
0 & -1 & 4 & -2 & 14 \\
0 & 3 & 1 & 2 & 5
\end{array}\right],
$$

$$
\begin{array}{c}
\text{Step 2} \\
\begin{array}{c}
R_1 \\
R_2 \\
R_3 - R_2 \\
R_4 + 3R_2
\end{array}
\left[
\begin{array}{cccc|c}
1 & 2 & -1 & 1 & 0 \\
0 & -1 & 3 & 0 & 7 \\
0 & 0 & 1 & -2 & 7 \\
0 & 0 & 10 & 2 & 26
\end{array}
\right],
\end{array}
\qquad
\begin{array}{c}
\text{Step 3} \\
\begin{array}{c}
R_1 \\
R_2 \\
R_3 \\
R_4 - 10R_3
\end{array}
\left[
\begin{array}{cccc|c}
1 & 2 & -1 & 1 & 0 \\
0 & -1 & 3 & 0 & 7 \\
0 & 0 & 1 & -2 & 7 \\
0 & 0 & 0 & 22 & -44
\end{array}
\right].
\end{array}
\qquad (7)
$$

Interpreting this last transformation of the augmented matrix gives the following system which, because it has been obtained by elementary row operations, is equivalent to (5):

$$
\begin{aligned}
x_1 + 2x_2 - x_3 + x_4 &= 0 \\
-x_2 + 3x_3 &= 7 \\
x_3 - 2x_4 &= 7 \\
22x_4 &= -44.
\end{aligned}
\qquad (8)
$$

The final equation in (8) gives $x_4 = -2$. Using this result in the third equation then gives $x_3 = 3$, and proceeding upwards sequentially in this fashion gives $x_2 = 2$ and, finally, $x_1 = 1$. So system (5) has the solution

$$
x_1 = 1, \quad x_2 = 2, \quad x_3 = 3, \quad x_4 = -2. \qquad (9)
$$

This general process of finding an unknown in an equation from the unknowns determined previously is called **back-substitution**.

Before proceeding to a brief discussion of the general system of simultaneous linear equations given in (1), two remarks should be made in preparation for later use. The first is that any two matrices which have been obtained one from the other by means of elementary row operations are said to be **row-equivalent**. If matrices A and B are row-equivalent this is indicated by writing $A \sim B$. The matrices in (7) are all row-equivalent.

The second remark is that in Gaussian elimination it may happen that the method terminates prematurely due to the generation of a zero on the leading diagonal of one of the transformed coefficient matrices. Suppose this occurs at the kth stage of the calculation so that $a_{kk}^* = 0$. Then to overcome the difficulty the kth row must be interchanged with any one of rows $k+1, k+2, \ldots, n$ below it which has a nonzero element in its kth column. Thereafter the elimination process may be continued as before. In Gaussian elimination and its variants the number a_{kk}^* used to reduce to zero the elements occurring below it in the kth column is called the kth **pivotal element, or simply the kth pivot**.

In practice, to retain accuracy should a pivot be very small, and also to overcome the possible occurrence of a zero pivot at any stage of the calculation, the above method is usually modified as follows to become what is called **Gaussian elimination with partial pivoting**. In this method, at each stage k a search is made of the elements occurring in rows $k, k+1, \ldots, n$ of the kth column, to find the element with the largest absolute value. This row and the kth are then interchanged, when necessary, before proceeding with the elimination of the remaining nonzero entries in the kth column. The method of Gaussian elimination with total pivoting will not be discussed, because although it is of some theoretical interest it is less efficient than the method just described.

Example 2.   Gaussian elimination with partial pivoting

Use Gaussian elimination with partial pivoting to solve

$$2x_1 + x_2 - x_3 + x_4 = 2$$
$$4x_1 + 2x_2 + 2x_3 - x_4 = 2$$
$$x_1 + x_2 - x_3 + x_4 = 1$$
$$3x_1 + 2x_2 + 2x_3 - 2x_4 = -1.$$

Solution

$$\mathbf{A}|\mathbf{b} = \begin{bmatrix} 2 & 1 & -1 & 1 & | & 2 \\ 4 & 2 & 2 & -1 & | & 2 \\ 1 & 1 & -1 & 1 & | & 1 \\ 3 & 2 & 2 & -2 & | & -1 \end{bmatrix} \quad \begin{matrix} \text{Interchange} \\ R_1 \text{ and } R_2 \\ R_3 \\ R_4 \end{matrix} \sim \begin{bmatrix} 4 & 2 & 2 & -1 & | & 2 \\ 2 & 1 & -1 & 1 & | & 2 \\ 1 & 1 & -1 & 1 & | & 1 \\ 3 & 2 & 2 & -2 & | & -1 \end{bmatrix}$$

$$\begin{matrix} R_1 \\ R_2 - \frac{1}{2}R_1 \\ R_3 - \frac{1}{4}R_1 \\ R_4 - \frac{3}{4}R_1 \end{matrix} \begin{bmatrix} 4 & 2 & 2 & -1 & | & 2 \\ 0 & 0 & -2 & \frac{3}{2} & | & 1 \\ 0 & \frac{1}{2} & -\frac{3}{2} & \frac{5}{4} & | & \frac{1}{2} \\ 0 & \frac{1}{2} & \frac{1}{2} & -\frac{5}{4} & | & -\frac{5}{2} \end{bmatrix} \sim \begin{matrix} R_1 \\ \text{Interchange} \\ R_2 \text{ and } R_3 \\ R_4 \end{matrix} \begin{bmatrix} 4 & 2 & 2 & -1 & | & 2 \\ 0 & \frac{1}{2} & -\frac{3}{2} & \frac{5}{4} & | & \frac{1}{2} \\ 0 & 0 & -2 & \frac{3}{2} & | & 1 \\ 0 & \frac{1}{2} & \frac{1}{2} & -\frac{5}{4} & | & -\frac{5}{2} \end{bmatrix}$$

$$\begin{matrix} R_1 \\ R_2 \\ R_3 \\ R_4 - R_2 \end{matrix} \begin{bmatrix} 4 & 2 & 2 & -1 & | & 2 \\ 0 & \frac{1}{2} & -\frac{3}{2} & \frac{5}{4} & | & \frac{1}{2} \\ 0 & 0 & -2 & \frac{3}{2} & | & 1 \\ 0 & 0 & 2 & -\frac{5}{2} & | & -3 \end{bmatrix} \sim \begin{matrix} R_1 \\ R_2 \\ R_3 \\ R_4 + R_3 \end{matrix} \begin{bmatrix} 4 & 2 & 2 & -1 & | & 2 \\ 0 & \frac{1}{2} & -\frac{3}{2} & \frac{5}{4} & | & \frac{1}{2} \\ 0 & 0 & -2 & \frac{3}{2} & | & 1 \\ 0 & 0 & 0 & -1 & | & -2 \end{bmatrix}.$$

Back-substitution then shows

$$x_4 = 2, \ x_3 = 1, \ x_2 = -1 \text{ and } x_1 = 1.$$

Notice that in the calculation, as the element with the largest absolute value in column 1 occurred in row 2 (the element 4), it was necessary to start by interchanging rows 1 and 2 to bring this element into the pivotal position (1, 1). The occurrence of a zero in position (2, 2) of the leading diagonal at the next stage necessitated a further interchange of rows. As both elements below this zero in column 2 had the same magnitude (in fact they were equal to $\frac{1}{2}$) either row to which they belonged could have been interchanged with row 2. In the calculation rows 2 and 3 were interchanged. Thereafter the elimination process was continued without a further interchange of rows being necessary. ∎

General case *m = n*

The numerical examples just discussed are typical of system (1) when $m = n$ and the coefficient matrix $\mathbf{A}$ in the augmented matrix $\mathbf{A}|\mathbf{b}$ is row-similar to an upper triangular matrix in which every element on the leading diagonal is nonzero. In such a case the solution is unique and follows as in the examples.

However, for some systems in which $m = n$, elementary operations may show that the augmented matrix is row-similar to

$$
\left[
\begin{array}{ccccccc:c}
a_{11} & a_{12} & a_{13} & \cdots & a_{1r} & a_{1r+1} & \cdots & a_{1n} & b_1 \\
0 & a_{22}^* & a_{23}^* & \cdots & a_{2r}^* & a_{2r+1}^* & \cdots & a_{2n}^* & b_2^* \\
0 & 0 & a_{33}^* & \cdots & a_{3r}^* & a_{3r+1}^* & \cdots & a_{3n}^* & b_3^* \\
\cdot & \cdot & \cdot & \cdots & \cdot & \cdot & \cdots & \cdot & \cdot \\
0 & 0 & 0 & \cdots & a_{rr}^* & a_{rr+1}^* & \cdots & a_{rn}^* & b_r^* \\
0 & 0 & 0 & \cdots & 0 & 0 & \cdots & 0 & b_{r+1}^* \\
0 & 0 & 0 & \cdots & 0 & 0 & \cdots & 0 & b_{r+2}^* \\
\cdot & \cdot & \cdot & \cdots & \cdot & \cdot & \cdots & \cdot & \cdot \\
0 & 0 & 0 & \cdots & 0 & 0 & \cdots & 0 & b_n^*
\end{array}
\right],
\tag{10}
$$

with $r < n$ and all the elements $a_{11}, a_{22}^*, \ldots, a_{rr}^*$ nonzero. Then, since all elements in the last $n-r$ rows of the coefficient matrix are zero, the corresponding $n-r$ equations may lead to a contradiction, for the equations are inconsistent if any one of the numbers $b_{r+1}^*, b_{r+2}^*, \ldots, b_n^*$ is nonzero. However, if $b_{r+1}^* = b_{r+2}^* = \ldots = b_n^* = 0$ a solution is possible, though it will not be unique, for then r suitably chosen unknowns will depend on the values of the remaining $n-r$ unknowns which may be assigned arbitrarily as parameters.

In Ex. 1 (iii), where $m = n = 2$ we have seen by graphical means that no solution exists. To see this by Gaussian elimination, notice that

$$
A|b = \begin{bmatrix} 2 & 1 & 4 \\ 2 & 1 & 8 \end{bmatrix} \quad \text{and} \quad \begin{matrix} R_1 \\ R_2 - R_1 \end{matrix} \begin{bmatrix} 2 & 1 & 4 \\ 0 & 0 & 4 \end{bmatrix}
$$

are row equivalent. The last row in the transformed matrix implies the contradiction '$0 = 4$', and so the equations are inconsistent.

In Ex. 1 (ii) we have that

$$
A|b = \begin{bmatrix} 1 & -2 & 4 \\ 2 & -4 & 8 \end{bmatrix} \quad \text{and} \quad \begin{matrix} R_1 \\ R_2 - 2R_1 \end{matrix} \begin{bmatrix} 1 & -2 & 4 \\ 0 & 0 & 0 \end{bmatrix}
$$

are row equivalent. There is no contradiction implied by row 2, and row 1 of the transformed matrix leads to the one-parameter family of solutions $x_1 = 4 + 2k$, $x_2 = k$ for $-\infty < k < \infty$ as the equations are consistent.

General case $m > n$

Elementary row operations will always reduce the augmented matrix of a system in which $m > n$ to a row equivalent one of the form

$$
\begin{bmatrix}
a_{11} & a_{12} & a_{13} & \cdots & a_{1r} & a_{1r+1} & \cdots & a_{1n} & \vline & b_1 \\
0 & a_{22}^* & a_{23}^* & \cdots & a_{2r}^* & a_{2r+1}^* & \cdots & a_{2n}^* & \vline & b_2^* \\
0 & 0 & a_{33}^* & \cdots & a_{3r}^* & a_{3r+1}^* & \cdots & a_{3n}^* & \vline & b_3^* \\
\cdot & \cdot & \cdot & \cdots & \cdot & \cdot & \cdots & \cdot & \vline & \cdot \\
0 & 0 & 0 & \cdots & a_{rr}^* & a_{rr+1}^* & \cdots & a_{rn}^* & \vline & b_r^* \\
0 & 0 & 0 & \cdots & 0 & 0 & \cdots & 0 & \vline & b_{r+1}^* \\
0 & 0 & 0 & \cdots & 0 & 0 & \cdots & 0 & \vline & b_{r+2}^* \\
\cdot & \cdot & \cdot & \cdots & \cdot & \cdot & \cdots & \cdot & \vline & \cdot \\
0 & 0 & 0 & \cdots & 0 & 0 & \cdots & 0 & \vline & b_m^*
\end{bmatrix},
\tag{11}
$$

in which $r \leq n$, and the elements $a_{11}, a_{22}^*, \ldots, a_{rr}^*$ are all nonzero. The situation is then similar to the one already discussed. No solution will exist if any one of the numbers $b_{r+1}^*, b_{r+2}^*, \ldots, b_m^*$ is nonzero for the equations will be inconsistent. However, if $b_{r+1}^* = b_{r+2}^* = \ldots = b_m^* = 0$, an $(n-r)$ parameter solution will exist. The results for the homogeneous case follow from the above reasoning by setting $\mathbf{b} = \mathbf{0}$.

We remark here that a matrix in which the first nonzero entry in any row is a 1 and this appears to the right of the first nonzero entry in the preceding row is said to be in **echelon form**. Thus matrices (10) and (11) can be brought into echelon form by dividing the first row by a_{11}, the second row by $a_{22}^*, \ldots$, and the rth row by a_{rr}^*.

In Ex. 1 (v), where $m = 3$, $n = 2$, we have seen by graphical means that no solution exists unless $p = 3$. To show this by Gaussian elimination, notice that

$$
\mathbf{A}|\mathbf{b} =
\begin{bmatrix}
1 & 2 & 4 \\
1 & -1 & 1 \\
2 & -1 & p
\end{bmatrix}
\quad \sim \quad
\begin{matrix} R_1 \\ R_2 - R_1 \\ R_3 - 2R_1 \end{matrix}
\begin{bmatrix}
1 & 2 & 4 \\
0 & -3 & -3 \\
0 & -5 & p-8
\end{bmatrix}
$$

$$
\sim \quad
\begin{matrix} R_1 \\ R_2 \\ R_3 - \frac{5}{3}R_2 \end{matrix}
\begin{bmatrix}
1 & 2 & 4 \\
0 & -3 & -3 \\
0 & 0 & p-3
\end{bmatrix}
$$

are all row-equivalent. Consequently, only when $p = 3$ will the equations be consistent and a solution exist. This solution is unique, because $r = n = 2$ when $p = 3$. Back-substitution then shows $x_2 = 1$ and $x_1 = 2$.

Problems for Section 3.4

Use Gaussian elimination to solve the following systems of linear simultaneous equations.

1
$$
\begin{aligned}
x_1 + x_2 + x_3 &= 3 \\
2x_1 - x_2 - x_3 &= 6 \\
x_1 + 2x_2 + 3x_3 &= 2
\end{aligned}
$$

2
$$
\begin{aligned}
2x_1 - 2x_2 + x_3 &= -4 \\
3x_1 - x_2 + 2x_3 &= -3 \\
x_1 + x_2 - 3x_3 &= 9
\end{aligned}
$$

3     $x_1 + 2x_2 + x_3 = 1$
         $x_1 - x_2 + 2x_3 = 10$
         $2x_1 + x_3 = 7$

4    $3x_1 + x_3 = 11$
      $2x_1 + 3x_2 + x_3 = 5$
      $x_1 + 2x_2 - 2x_3 = -3$

5    $2x_1 + x_2 - 4x_3 = 0$
     $-4x_1 + 5x_2 - 6x_3 = 0$
      $x_1 + 2x_2 - 7x_3 = 0$

6    $x_1 + 2x_2 + 5x_3 = 0$
     $2x_1 + x_2 - 4x_3 = 0$
    $-x_1 + 3x_2 - 5x_3 = 0$

7    $4x_1 + 9x_2 + x_3 = 0$
     $2x_1 + 13x_2 + 9x_3 = 0$
     $2x_1 - 5x_2 + x_3 = 0$

8    $x_1 + x_2 + x_3 = 0$
     $2x_1 + 3x_2 + x_3 = 0$
     $3x_1 + 4x_2 + x_3 = 0$

9    $x_1 + 3x_2 - x_3 = 4$
     $x_1 - x_2 + 2x_3 = 1$

10 $5x_1 + x_2 + 2x_3 = 4$
     $3x_1 + 2x_2 - 3x_3 = 1$

11    $x_1 + 3x_2 + 2x_3 + x_4 = 4$
     $2x_1 + 2x_2 + x_3 + x_4 = 3$
     $3x_1 + x_2 - x_3 - 2x_4 = -1$

12 $2x_1 + 2x_2 - 2x_3 + x_4 = 5$
   $-x_1 + 2x_2 + 5x_3 + x_4 = 5$
     $x_1 + 2x_2 + 5x_3 + 2x_4 = 8$

13    $x_1 + x_2 - x_3 + x_4 = 5$
     $2x_1 - x_2 + 2x_3 - x_4 = 1$

14 $x_1 + 2x_2 - x_3 + 2x_4 = 1$
    $3x_1 + 5x_2 + 4x_3 + 5x_4 = 6$

15    $x_1 + 3x_2 - x_3 - x_4 = 10$
     $2x_1 + x_2 - 4x_3 + 2x_4 = 4$
     $3x_1 - x_2 + 2x_3 + 5x_4 = -11$
   $-x_1 + 2x_2 + 2x_3 - 3x_4 = 7$
     $2x_1 + x_2 - x_3 + 2x_4 = 1 + p$
     $x_1 + 3x_2 + 4x_3 + 3x_4 = q$

16 $x_1 + 2x_2 + 4x_3 - x_4 = 2$
   $2x_1 + 5x_2 + 6x_3 - 3x_4 = 5$
     $x_1 + 3x_2 + 2x_2 - 2x_4 = 3$
   $4x_1 + 9x_2 + 14x_3 - 5x_3 = 9$

Harder problems

The following problems all lead to simultaneous equations. In the formulation of the first four it is necessary to use **Kirchhoff's laws** governing the instantaneous voltages and currents flowing in an electrical network, together with **Ohm's law**. These may be stated as follows:

Kirchhoff's current law
The sum of the currents entering a junction equals the sum of the currents leaving it.

Kirchhoff's voltage law
The algebraic sum of the voltage drops around any closed loop is zero, where a voltage increase is regarded as a negative voltage drop.

Ohm's law
The current I flowing through a resistance of magnitude R across which is applied a voltage V is given by the result $I = E/R$.

17 The diagram in Fig. 3.1 represents a **Wheatstone bridge**. This apparatus is used to measure an unknown resistance R_x by comparison with the known resistances R_1, R_2 and R_3 which are adjusted so that no current from the cell with voltage E_0 passes through the branch BD in which there is an ammeter with resistance R_g. Use Kirchhoff's laws to show the unknown currents

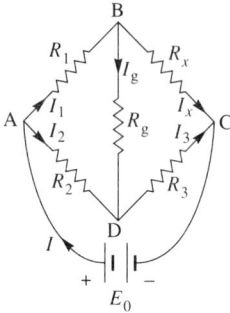

Fig. 3.11   Wheatstone bridge

I, I_1, I_2, I_3 and I_x satisfy the equations

$$I_1 + I_2 = I, \quad I_x + I_3 = I$$
$$I_g + I_x = I_1, \quad I_2 + I_g = I_3$$
$$R_x I_x - R_3 I_3 - R_g I_g = 0$$
$$R_1 I_1 + R_g I_g - R_2 I_2 = 0$$
$$R_1 I_1 + R_x I_x = E_0.$$

Deduce that the current $I_g = 0$ (the bridge is **balanced**) when

$$R_1 R_3 = R_2 R_x.$$

18 Six resistors of equal magnitude R are joined to form the edges of a tetrahedron as shown in Fig. 3.12. If a voltage E_0 is applied across the resistor AD, use Kirchhoff's laws to find the currents flowing through each resistor. What would be the effect on the currents flowing in the network if the resistor BC were to be replaced by one of magnitude $3R$? Show that if the network of resistors is replaced by a single resistor of magnitude $\frac{1}{2}R$ across AD the current drawn from the cell is unchanged. The resistance $\frac{1}{2}R$ is called the **equivalent resistance** of the network. An equivalent resistance is often used when a subcircuit is disconnected for repair, since it leaves unaltered the current flow in the rest of the circuit which may still be in use.

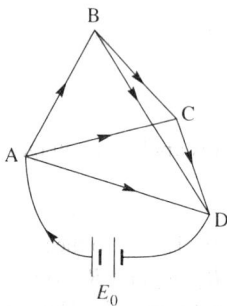

Fig. 3.12   Tetrahedron of resistors of equal magnitude R

19 Twelve resistors of equal magnitude R are joined to form the edges of a cube as shown in Fig. 3.13. If a voltage E_0 is applied across one of the resistors, use Kirchhoff's laws to show the currents I_1, I_2 and I_3 are

$$I_1 = \frac{5E_0}{14R}, \quad I_2 = \frac{E_0}{R}, \quad I_3 = \frac{2E_0}{7R}.$$

What is the equivalent resistance of the network?

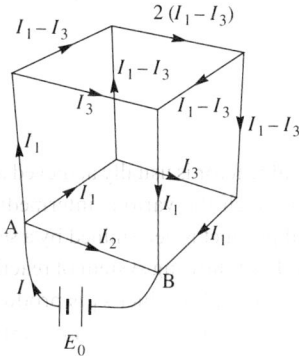

Fig. 3.13 Cube formed by twelve identical resistors

20 Use Kirchhoff's laws to find the equations governing the currents I_1, I_2 and I_3 which flow in the circuit illustrated in Fig. 3.14 in which E_1 and E_3 are the voltages produced by the two cells. Find the currents I_1, I_2 and I_3. A circuit of this type is sometimes used as a simple regulator to keep the voltage across a device with resistance R_2 close to the value E_2, when the device itself draws a large current from a source whose voltage E_1 may fluctuate.

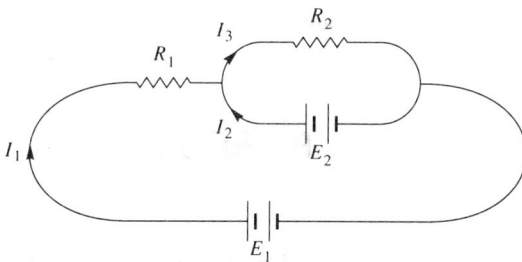

Fig. 3.14 Voltage regulator

21 In a telephone network, cells received by exchanges A, B and C are routed to exchange D for onward transmission. Calls entering through exchange A may either go directly to D or, if the lines are full, via exchanges B or C, while calls entering through exchanges B and C go directly to D. The network is shown in Fig. 3.15 together with the number of calls per hour entering or leaving the four exchanges. If all calls are connected, find the number of calls per hour passing through the different lines. Show the solution is not unique and find any constraints that are placed on the number of calls passing through the lines.

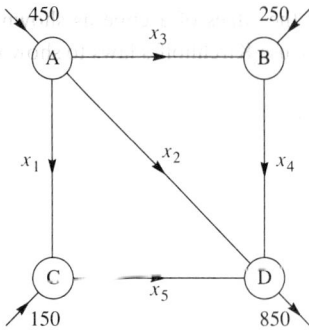

Fig. 3.15   Telephone network

22 In industrial chemical processes the end product from a chemical reactor is usually achieved as a result of several intermediate chemical reactions. The rate at which the various intermediate products are produced and then recombined to produce the end product is determined by a set of first order differential equations called **reaction-rate equations**. The following system of reaction rate equations is typical. The equations relate the rate at which the end product x_1 is produced to the rates at which the two intermediate products x_2 and x_3 are produced. All other quantities in the equations may be regarded as constants describing the reactions.

$$\frac{dx_1}{dt} = 6x_1 - x_2 - 59.5$$

$$\frac{dx_2}{dt} = 3x_2 + 2x_3 - 4.5$$

$$\frac{dx_3}{dt} = 0.1x_1 + 2x_2 - x_3 - 0.5.$$

Write this system in matrix form, giving a suitable definition for the derivative of a matrix vector whose elements are functions of t. Find the rates of production of x_1, x_2 and x_3 when the chemical reactor is operating under equilibrium conditions.

3.5   Linear independence. Rank. Reduced echelon form

To proceed further it is necessary that we develop certain ideas which are basic to linear algebra. Some of these have already been encountered in Sec. 2.9 in connection with geometrical vectors and so will be familiar. Let $\mathbf{a}_1, \mathbf{a}_2, \ldots, \mathbf{a}_m$ be n element row vectors. Then if $\alpha_1, \alpha_2, \ldots, \alpha_m$ are any scalars, an expression of the form

$$\alpha_1\mathbf{a}_1 + \alpha_2\mathbf{a}_2 + \ldots + \alpha_m\mathbf{a}_m \tag{1}$$

is called a **linear combination** of the m vectors.

If the equation

$$\alpha_1\mathbf{a}_1 + \alpha_2\mathbf{a}_2 + \ldots + \alpha_m\mathbf{a}_m = \mathbf{0} \tag{2}$$

is only true when $\alpha_1 = \alpha_2 = \ldots = \alpha_m = 0$, the vectors $\mathbf{a}_1, \mathbf{a}_2, \ldots \mathbf{a}_m$ are said to be **linearly independent**. When equation (2) is true for some constants $\alpha_1, \alpha_2, \ldots, \alpha_m$ which are not all zero, the vectors $\mathbf{a}_1, \mathbf{a}_2, \ldots, \mathbf{a}_m$ are said to be **linearly dependent**.

For example, $\mathbf{a}_1 = [1 \ 0 \ 0]$, $\mathbf{a}_2 = [0 \ 1 \ 0]$ and $\mathbf{a}_3 = [0 \ 0 \ 1]$ are linearly independent, because

$$\alpha_1 \mathbf{a}_1 + \alpha_2 \mathbf{a}_2 + \alpha_3 \mathbf{a}_3 = [\alpha_1 \quad \alpha_2 \quad \alpha_3] = \mathbf{0}$$

only when $\alpha_1 = \alpha_2 = \alpha_3 = 0$. However, the vector $\mathbf{a} = [3 \ 0 \ -7]$ is linearly dependent on these vectors because $\mathbf{a} = 3\mathbf{a}_1 - 7\mathbf{a}_3$ or, equivalently, $\mathbf{a} - 3\mathbf{a}_1 + 7\mathbf{a}_3 = \mathbf{0}$.

These ideas are needed to define an important number called the **rank** of a matrix. The rank of matrix $\mathbf{A}$, written rank $\mathbf{A}$, is the greatest number of linearly independent row vectors belonging to $\mathbf{A}$. So, by definition, the rank of a zero matrix is 0.

Inspection of the matrix

$$\mathbf{A} = \begin{bmatrix} 1 & 3 & 0 & 2 \\ 1 & -3 & 1 & -1 \\ 1 & 9 & -1 & 5 \end{bmatrix} \tag{3}$$

shows rank $\mathbf{A} = 2$, because $R_3 = 2R_1 - R_2$ and rows 1 and 2 are linearly independent. We shall discuss how the rank may be determined once the next result has been established.

Theorem 3.3 (Row operations and rank)

The following statements are equivalent:

(i) Elementary row operations do not alter the rank of a matrix.
(ii) Matrices which are row-equivalent have the same rank.

Proof

The result is almost immediate. Firstly, the interchange of two rows of matrix $\mathbf{A}$ (elementary row operation I) and the multiplication of the elements of a row of $\mathbf{A}$ by a nonzero constant (elementary row operation II) cannot alter the number of linearly independent rows of $\mathbf{A}$, and hence the rank of $\mathbf{A}$. Thus it only remains to consider the effect of the addition to a row of $\mathbf{A}$ of a multiple of another row (elementary row operation III). Each of the two rows involved is expressible as a linear combination of the linearly independent rows of $\mathbf{A}$, so the new row is also expressible as a linear combination of the same set of linearly independent rows of $\mathbf{A}$. Thus elementary row operation III cannot alter the number of linearly independent rows of $\mathbf{A}$, and hence the rank of $\mathbf{A}$. The result is proved. $\qquad\square$

To find the rank of an arbitrary $m \times n$ matrix $\mathbf{A}$ we first need to describe how elementary row operations may be used to bring $\mathbf{A}$ to what is called the **row-equivalent reduced echelon form** of $\mathbf{A}$ or, more simply, the **reduced echelon form** of $\mathbf{A}$. The reduced echelon form $\mathbf{A}_E$ of $\mathbf{A}$ is a matrix which in addition to being in echelon form is such that the first nonzero entry in a given row is the only nonzero entry in its column (cf. **echelon matrix** at the end of Sec. 3.4). The transformation is accomplished using only elementary row operations in a manner similar to Gaussian elimination, so that by Theorem 3.3 A_E has the same rank as A. The reduction of matrix A to A_E is accomplished by what is called the **Gauss–Jordan method**, and it produces a matrix with the following general form. In a sense, this reduction amounts to Gaussian elimination with simultaneous back-substitution.

$$A_E = \begin{bmatrix} \text{col·}j_1 & \text{col·}j_2 & \text{col·}j_3 & \cdots & \text{col·}j_r \end{bmatrix}$$

(matrix diagram, reduced echelon form, r rows, m rows, n columns) $\qquad$ (4)

The essential features of A_E may be summarized as follows:

(a) the r separate columns located between the pairs of dotted lines each contain only the single nonzero element 1, the first nonzero element of the row in which it is located;

(b) the number r of rows comprising elements which are not all zero (**nonzero rows**) is such that $r \le \min(m, n)$;

(c) the first nonzero element in a row always lies to the right of the first nonzero element in the row above;

(d) the symbol * signifies an element which is not, in general, zero;

(e) all elements outside the shaded region are zero.

Bringing a matrix to its reduced echelon form (Gauss–Jordan method)

The following steps describe the operations which must be performed to bring an arbitrary $m \times n$ matrix A to its reduced echelon form A_E.

Step 1. Counting from the left, find the first column of **A** whose elements are not all zeros. Let this be column j_1, say.

Step 2. Transfer to the first row of the matrix the row containing the first nonzero element in column j_1 found in Step 1. Call this element the unnormalized pivot for row 1.

Step 3. Divide row 1 of the transformed matrix of Step 2 by the nonzero pivot for row 1. This will produce a new row 1 with a pivot equal to 1 in column j_1.

Step.4. Subtract suitable multiples of the new row 1 of Step 3 from all rows below it to reduce to zero all the elements in column j_1 below the pivot. This will generate a new column j_1 with a pivot equal to 1 as its first element and zeros elsewhere.

Step 5. Define a submatrix with $m-1$ rows and $n-j_1$ columns whose elements are obtained from the matrix in Step 4 by deleting the first row and the first j_1 columns.

Step 6. Apply to the submatrix of Step 5 the operations in steps numbered 1 to 4. This will generate a new column j_2 containing a pivot equal to 1 as its second entry with zeros below it.

Step 7. Subtract a suitable multiple of the new row 2 of the transformed full matrix from row 1 to reduce to zero the element above the pivot. This will generate a new column j_2 with zero elements everywhere except at the pivot which is equal to 1 and located in the second row.

Step 8. Define a new submatrix with $m-2$ rows and $n-j_2$ columns by deleting from the full matrix of Step 7 the first two rows and the first j_2 columns.

Step 9. Keep repeating the cycle of operations in Steps 1 through 8. After the kth cycle this will result in a new column j_k being produced with a pivot equal to 1 in its kth row and zeros elsewhere. Stop either when a submatrix is generated containing only zero elements, or when no further submatrices exist. The final full matrix obtained by this process is the required reduced echelon form $\mathbf{A}_E$ of matrix **A**.

The method by which the echelon form $\mathbf{A}_E$ is obtained from **A** automatically replaces n element row vectors $\mathbf{a}_1, \mathbf{a}_2, \ldots, \mathbf{a}_m$ in **A** between which there is linear dependence by zero row vectors in $\mathbf{A}_E$. Consequently, the nonzero rows $\hat{\mathbf{a}}_1, \hat{\mathbf{a}}_2, \ldots, \hat{\mathbf{a}}_r$ that remain in $\mathbf{A}_E$ are all linearly independent. From the definition of rank and from Theorem 3.3 it then follows immediately that the rank of **A** is equal to the number of nonzero rows in $\mathbf{A}_E$.

If, now, we consider the m element column vectors of $\mathbf{A}_E$, we see by considering the structure of (4) that all columns of $\mathbf{A}_E$ may be obtained as linear combinations of the r column vectors $\hat{\mathbf{c}}_1, \hat{\mathbf{c}}_2, \ldots, \hat{\mathbf{c}}_r$ containing the pivots. Inspection shows these r column vectors to be linearly independent; so we have shown the rather surprising result that the number of linearly independent column vectors in $\mathbf{A}_E$, hence also in **A**, is the same as the number of linearly independent row vectors in **A**.

Suppose that the n element row vectors $\mathbf{a}_1, \mathbf{a}_2, \ldots, \mathbf{a}_m$ of **A** are considered to belong to a vector space $\mathbb{R}^n$, the set of all ordered n-tuples of numbers, and the m element column vectors $\mathbf{c}_1, \mathbf{c}_2, \ldots, \mathbf{c}_n$ of **A** to a vector space $\mathbb{R}^m$, the set of all ordered m-tuples of numbers.

Then the last result implies that if rank $A = r$, the vectors

$$\hat{a}_1, \hat{a}_2, \ldots, \hat{a}_r$$

form a basis for what will be called the vector **row space** of A. This is the space to which the particular rows of A belong, and as it is r-dimensional it will, in general, be a subspace of the vector space $\mathbb{R}^n$ to which all n element rows belong. Similarly, the vectors

$$\hat{c}_1, \hat{c}_2, \ldots, \hat{c}_r$$

form a basis for what will be called the vector **column space** of A. This is the space to which the particular columns of A belong, and since it also is r-dimensional it will, in general, be a subspace of the vector space $\mathbb{R}^m$ to which all m element columns belong.

Consider next an arbitrary set of s row vectors $k_1, k_2, \ldots, k_s$ each containing n elements, and let K be the matrix with these vectors as its rows. Then from the definition of rank, the vectors will be linearly independent if rank $K = s$, as they will be linearly dependent if rank $K < s$. An immediate and important consequence of this is that if $n < s$, then rank $K \leq n < s$, so the s vectors must be **linearly dependent**.

It is for this reason that the unit vectors i, j and k may be used to form a basis in three-dimensional euclidean space. Representing them in the form $k_1 = [1 \quad 0 \quad 0]$, $k_2 = [0 \quad 1 \quad 0]$ and $k_3 = [0 \quad 0 \quad 1]$ we see rank $K = 3$, showing that they form a basis in the vector space $\mathbb{R}^3$. Fewer than three vectors could not span the vector space $\mathbb{R}^3$, while more than three must always be linearly dependent in that space.

We now reformulate the above results as two theorems which are far-reaching in their consequences.

Theorem 3.4 (Rank, and row and column vector spaces)

Let A be an arbitrary $m \times n$ matrix with reduced echelon form A_E. Then,

(i) rank A is equal to the number of nonzero rows in A_E;
(ii) the number of linearly independent rows in A is equal to the number of linearly independent columns in A.

Suppose rank $A = r$, and denote the nonzero rows of A_E by $\hat{a}_1, \hat{a}_2, \ldots, \hat{a}_r$, and the columns of A_E containing the pivots by $\hat{c}_1, \hat{c}_2, \ldots, \hat{c}_r$. Then,

(iii) the vectors $\hat{a}_1, \hat{a}_2, \ldots, \hat{a}_r$ form a basis for the r-dimensional row vector space to which the rows of A belong;
(iv) the vectors $\hat{c}_1, \hat{c}_2, \ldots, \hat{c}_r$ form a basis for the r-dimensional column vector space to which the columns of A belong. $\square$

Theorem 3.5 (Linear dependence and rank)

Let $k_1, k_2, \ldots, k_s$ be an arbitrary set of n element row vectors, and K be the matrix with these vectors as its elements. Then the s vectors will be,

(i) linearly independent if rank $K = s$;
(ii) linearly dependent if rank $K < s$;
(iii) linearly dependent if $n < s$. $\square$

Example 1. Reduced echelon form

Bring to reduced echelon form

$$A = \begin{bmatrix} 0 & 0 & 1 & 3 & 1 & 3 \\ 2 & 0 & 4 & 2 & 0 & 4 \\ 1 & 0 & 3 & 0 & 0 & -1 \\ 1 & 0 & 1 & 0 & 0 & -1 \end{bmatrix}.$$

Find rank A, a basis for the row vector space and a basis for the column vector space of A.

Solution

The notation of Sec. 3.4 will be used to describe in a concise manner the row operations which are necessary to bring A to reduced echelon form. Wherever possible steps will be combined. The unnormalized pivot in a column will be enclosed in a dotted circle while the normalized pivot will be enclosed in a full circle.

$$A = \begin{bmatrix} 0 & 0 & 1 & 3 & 1 & 3 \\ 2 & 0 & 4 & 2 & 0 & 4 \\ 1 & 0 & 3 & 0 & 0 & -1 \\ 1 & 0 & 1 & 0 & 0 & -1 \end{bmatrix} \sim \begin{array}{l} R_1 = R_2 \\ R_2 = R_1 \\ R_3 \\ R_4 \end{array} \begin{bmatrix} ② & 0 & 4 & 2 & 0 & 4 \\ 0 & 0 & 1 & 3 & 1 & 3 \\ 1 & 0 & 3 & 0 & 0 & -1 \\ 1 & 0 & 1 & 0 & 0 & -1 \end{bmatrix}$$

$$\sim \begin{array}{l} \tfrac{1}{2}R_1 \\ R_2 \\ R_3 \\ R_4 \end{array} \begin{bmatrix} ① & 0 & 2 & 1 & 0 & 2 \\ 0 & 0 & 1 & 3 & 1 & 3 \\ 1 & 0 & 3 & 0 & 0 & -1 \\ 1 & 0 & 1 & 0 & 0 & -1 \end{bmatrix} \sim \begin{array}{l} R_1 \\ R_2 \\ R_3 - R_1 \\ R_4 - R_1 \end{array} \begin{bmatrix} ① & 0 & 2 & 1 & 0 & 2 \\ 0 & 0 & 1 & 3 & 1 & 3 \\ 0 & 0 & 1 & -1 & 0 & -3 \\ 0 & 0 & -1 & -1 & 0 & -3 \end{bmatrix}$$

$$\sim \begin{array}{l} R_1 \\ R_2 \\ R_3 \\ R_4 \end{array} \begin{bmatrix} ① & 0 & 2 & 1 & 0 & 2 \\ 0 & 0 & ① & 3 & 1 & 3 \\ 0 & 0 & 1 & -1 & 0 & -3 \\ 0 & 0 & -1 & -1 & 0 & -3 \end{bmatrix} \sim \begin{array}{l} R_1 - 2R_2 \\ R_2 \\ R_3 - R_2 \\ R_4 + R_2 \end{array} \begin{bmatrix} ① & 0 & 0 & -5 & -2 & -4 \\ 0 & 0 & ① & 3 & 1 & 3 \\ 0 & 0 & 0 & -4 & -1 & -6 \\ 0 & 0 & 0 & 2 & 1 & 0 \end{bmatrix}$$

$$\sim \begin{array}{l} R_1 \\ R_2 \\ R_3 \\ R_4 \end{array} \begin{bmatrix} ① & 0 & 0 & -5 & -2 & -4 \\ 0 & 0 & ① & 3 & 1 & 3 \\ 0 & 0 & 0 & ④ & -1 & -6 \\ 0 & 0 & 0 & 2 & 1 & 0 \end{bmatrix} \sim \begin{array}{l} R_1 \\ R_2 \\ -\tfrac{1}{4}R_3 \\ R_4 \end{array} \begin{bmatrix} ① & 0 & 0 & -5 & -2 & -4 \\ 0 & 0 & ① & 3 & 1 & 3 \\ 0 & 0 & 0 & ① & \tfrac{1}{4} & \tfrac{3}{2} \\ 0 & 0 & 0 & 2 & 1 & 0 \end{bmatrix}$$

$$\sim \begin{array}{l} R_1 + 5R_3 \\ R_2 - 3R_3 \\ R_3 \\ R_4 - 2R_3 \end{array} \begin{bmatrix} ① & 0 & 0 & 0 & -\tfrac{3}{4} & \tfrac{7}{2} \\ 0 & 0 & ① & 0 & \tfrac{1}{4} & -\tfrac{3}{2} \\ 0 & 0 & 0 & ① & \tfrac{1}{4} & \tfrac{3}{2} \\ 0 & 0 & 0 & 0 & ⑴{2} & -3 \end{bmatrix} \sim \begin{array}{l} R_1 \\ R_2 \\ R_3 \\ 2R_4 \end{array} \begin{bmatrix} ① & 0 & 0 & 0 & -\tfrac{3}{4} & \tfrac{7}{2} \\ 0 & 0 & ① & 0 & \tfrac{1}{4} & -\tfrac{3}{2} \\ 0 & 0 & 0 & ① & \tfrac{1}{4} & \tfrac{3}{2} \\ 0 & 0 & 0 & 0 & ① & -6 \end{bmatrix}$$

$$\sim \begin{array}{l} R_1 + \tfrac{3}{4}R_4 \\ R_2 - \tfrac{1}{4}R_4 \\ R_3 - \tfrac{1}{4}R_4 \\ R_4 \end{array} \begin{bmatrix} ① & 0 & 0 & 0 & 0 & -1 \\ 0 & 0 & ① & 0 & 0 & 0 \\ 0 & 0 & 0 & ① & 0 & 3 \\ 0 & 0 & 0 & 0 & ① & -6 \end{bmatrix} = A_E.$$

There are four pivots in this reduction, and in the notation of Steps 1 through 9 they correspond to the column numbers $j_1 = 1$, $j_2 = 3$, $j_3 = 4$ and $j_4 = 5$. The reduced echelon form A_E of A contains four nonzero rows; so by Theorem 3.4 rank $A = 4$. Thus the row vector space has dimension 4. This is a subspace of the vector space $\mathbb{R}^5$ to which five element row vectors belong; that is the vector space $\mathbb{R}^5$ comprising all ordered quintuples of numbers. A basis for the row vector space is

$$\hat{a}_1 = [1 \ \ 0 \ \ 0 \ \ 0 \ \ 0 \ \ -1],$$
$$\hat{a}_2 = [0 \ \ 0 \ \ 1 \ \ 0 \ \ 0 \ \ \ \ 0],$$
$$\hat{a}_3 = [0 \ \ 0 \ \ 0 \ \ 1 \ \ 0 \ \ \ \ 3],$$
$$\hat{a}_4 = [0 \ \ 0 \ \ 0 \ \ 0 \ \ 1 \ \ -6].$$

By Theorem 3.4 the column vector space also has dimension 4. This same theorem shows that a basis for the vector space $\mathbb{R}^4$ of all ordered quadruples of numbers to which the column vectors belong is

$$\hat{c}_1 = \begin{bmatrix} 1 \\ 0 \\ 0 \\ 0 \end{bmatrix}, \quad \hat{c}_2 = \begin{bmatrix} 0 \\ 1 \\ 0 \\ 0 \end{bmatrix}, \quad \hat{c}_3 = \begin{bmatrix} 0 \\ 0 \\ 1 \\ 0 \end{bmatrix} \quad \text{and} \quad \hat{c}_4 = \begin{bmatrix} 0 \\ 0 \\ 0 \\ 1 \end{bmatrix}.$$

Notice that in this example the vector space to which all four element column vectors belong, and the column vector space for this matrix A both happen to have the same dimension. ∎

Example 2. Application of rank to dimensional analysis

When conducting scale model experiments to assist with a design project it is essential to know how any conclusions which are reached may be scaled to relate to the finished design. The present example shows how the notion of rank may be used to determine the *similarity law* by which such scaling may be accomplished. The particular situation to be considered concerns experiments involving the steady incompressible flow of a viscous liquid past a model of given shape.

As the scaling of a particular model leads to geometrically similar shapes, the type of model being tested will be characterized by a representative length l (say the length of the model). It is reasonable to expect the fluid flow to depend on the fluid density ρ, the dynamic viscosity of the fluid η, the steady fluid speed far from the model U, and the acceleration due to gravity g.

The approach used is to try to find combinations of l, ρ, η, U and g of the form

$$l^{m_1} \, \rho^{m_2} \, \eta^{m_3} \, U^{m_4} \, g^{m_5} = k, \tag{A}$$

so that k is simply a real number. Such combinations are said to be **dimensionless**. If the number k is determined for a given model and fluid, the experimental results which give rise to k will then relate to any other similar model and fluid flow for which k has the same value.

Denoting the fundamental physical quantities of length, mass and time by L, M and T, respectively, the dimensions of the parameters involved are

Physical quantity	l	ρ	η	U	g
Dimensions	L	ML^{-3}	$ML^{-1}T^{-1}$	LT^{-1}	LT^{-2}

If k is to be dimensionless we must have

$$L^{m_1}(ML^{-3})^{m_2} \, (ML^{-1}T^{-1})^{m_3} \, (LT^{-1})^{m_4} \, (LT^{-2})^{m_5} = L^0 M^0 T^0,$$

so equating the exponents of L, M and T on either side of this expression gives

(exponents of L) $m_1 - 3m_2 - m_3 + m_4 + m_5 = 0$

(exponents of M) $m + m_3 = 0$

(exponents of T) $-m_3 - m_4 - 2m_5 = 0.$

The coefficient matrix $\mathbf{A}$ for these three equations in five unknowns is

$$\mathbf{A} = \begin{bmatrix} 1 & -3 & -1 & 1 & 1 \\ 0 & 1 & 1 & 0 & 0 \\ 0 & 0 & -1 & -1 & -2 \end{bmatrix}.$$

When $\mathbf{A}$ is brought into its reduced echelon from $\mathbf{A}_E$ we find

$$\mathbf{A}_E = \begin{bmatrix} 1 & 0 & 0 & -1 & -3 \\ 0 & 1 & 0 & -1 & -2 \\ 0 & 0 & 1 & 1 & 2 \end{bmatrix},$$

and because the first three columns are linearly independent rank $\mathbf{A}_E = $ rank $\mathbf{A} = 3$. Thus from the five unknowns m_1 to m_5, $5 - 3 = 2$ of them may be assigned arbitrarily. When deciding which these two unknowns are to be, it is necessary to remember that the 3×3 matrix remaining after the columns corresponding to these two unknowns have been deleted must still be of rank 3.

Inspection of $\mathbf{A}_E$ shows the variables m_2 and m_3 cannot be chosen arbitrarily, because when columns 2 and 3 are deleted from $\mathbf{A}_E$ the resulting 3×3 matrix has rank 2 because its last two rows are proportional. Let us then decide to assign m_1 and m_4 arbitrarily. Matrix $\mathbf{A}_E$ corresponds to the equations

$$m_1 - m_4 - 3m_5 = 0$$
$$m_2 - m_4 - 2m_5 = 0$$
$$m_3 + m_4 + 2m_5 = 0,$$

which in terms of $m_1 = \alpha$, $m_4 = \beta$ have the solution

$$m_1 = \alpha, \quad m_2 = \tfrac{1}{3}(2\alpha + \beta), \quad m_3 = -\tfrac{1}{3}(2\alpha + \beta), \quad m_4 = \beta \quad \text{and} \quad m_5 = \tfrac{1}{3}(\alpha - \beta).$$

The choice $\alpha = -1$, $\beta = 2$ gives $m_1 = -1$, $m_2 = m_3 = 0$, $m_4 = 2$ and $m_5 = -1$, and (A) becomes

$$k = F = \frac{U^2}{lg},$$

which is called the **Froude number** for the flow.

The alternative choice $\alpha = \beta = 1$ gives $m_1 = m_2 = m_4 = 1$, $m_3 = -1$ and $m_5 = 0$, and this time (A) becomes

$$k = R = \frac{l\rho U}{\eta},$$

which is called the **Reynolds number** for the flow. Any other choices of α and β will merely give rise to dimensionless numbers k of the form $R^p F^q$. For example, setting $\alpha = 1$, $\beta = -2$ leads to $k = F^{-1}$, while setting $\alpha = 1$, $\beta = -1$ leads to $k = R^{1/3} F^{-2/3}$.

In this rather simple case the identification of R and F could have been obtained either by inspection or by direct solution of the original system, without appeal to the notion of rank.

However, this form of reasoning leads to a systematic approach, and so is essential for more complicated situations when the number and form of the dimensionless combinations is not obvious. In such cases it is also helpful to use the connection between rank and determinants (Theorem 3.19) when deciding which unknowns may be chosen arbitrarily. ∎

Example 3. Statical determinacy of pin jointed plane frames

This example illustrates the part played by rank in the statical analysis of a special class of structures called *plane frames*. A system of rigid rods connected together at their ends by pin joints in such a manner that they form a rigid structure in a plane is called a **plane frame** or **truss**. If a rod may be removed from a plane frame without altering its rigidity within the plane the rod is said to be **redundant**. A typical plane frame is shown in Fig. 3.16(a), in which any one of the eight rods connected together at five joints is redundant.

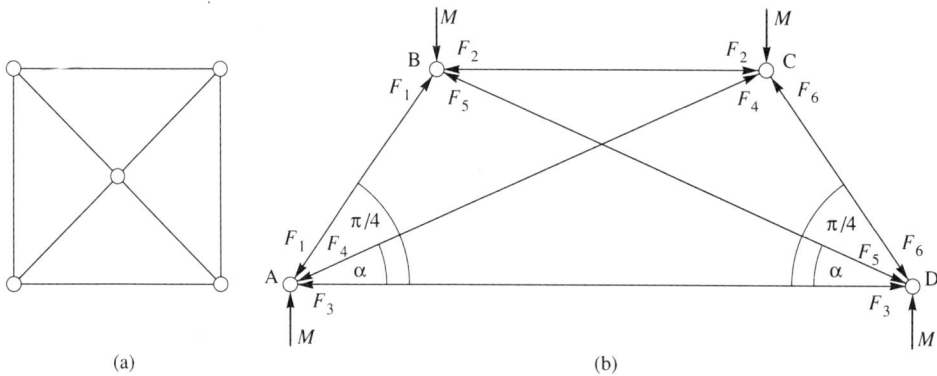

(a) (b)

Fig. 3.16 (a) Eight element plane frame (b) Loaded six element plane frame

It is known from mechanics that a rigid body will be in equilibrium if

(i) the resultant of the external forces acting on the body is zero, and
(ii) there is no moment acting on the body.

For a plane frame, condition (i) gives rise to two linear equations which must be satisfied by the external forces (balancing forces in two mutually perpendicular directions in the plane of the frame). Condition (ii) gives rise to one further linear equation which must be satisfied by the external forces. There are thus three linear constraints imposed on the external forces.

Now suppose there are n bars and m joints, then there will be n stresses to be determined (one for each rod). Balancing all stresses and external forces in mutually perpendicular directions within the plane of the frame at each joint will lead to $2m$ linear equations for the stresses, but these are subject to the three constraints just mentioned. Thus, in all, there will be $2m - 3$ equations to be satisfied by the n stresses.

There will be as many equations as there are unknown stresses if $n = 2m - 3$. A plane frame will be said to be **statically determinate** if $n = 2m - 3$ and the equations determine a *unique* stress in each of the n rods. If $n > 2m - 3$ there will be more unknown stresses than there are equations, so some stresses will then be indeterminate. When this occurs the plane frame is said to be **statically indeterminate**, and a rod in which the stress is indeterminate corresponds to a *redundant* rod.

A necessary and sufficient condition for a plane frame to be statically indeterminate (to contain redundant rods) is seen to be that $n > 2m - 3$. A necessary condition for a plane frame to be statically determinate (there is a unique stress in each rod) is that $n=2m-3$, though this is not a sufficient condition because linear dependence may exist between the equations. A necessary and sufficient condition for statical determinacy is that rank $\mathbf{A}=\text{rank } \mathbf{A}|\mathbf{b}=n$, where $\mathbf{A}$ and $\mathbf{A}|\mathbf{b}$ are the coefficient matrix and the augmented matrix, respectively, of the equations determining the stresses.

These ideas are illustrated by the example shown in Fig. 3.16(b) involving a loaded frame with six rods and four joints. Here $n=6$, $m=4$ and $2m-3=5<n=6$, showing redundancy is present in the plane frame.

To relate this problem to a specific matrix and its rank we resolve forces horizontally and vertically at each of the joints A, B, C and D, where F_i is the magnitude of the stress in the ith rod.

Joint A

$$F_1 \cos \frac{\pi}{4} + F_4 \cos \alpha + F_3 = 0$$

$$F_1 \sin \frac{\pi}{4} + F_4 \sin \alpha = M,$$

Joint B

$$F_1 \cos \frac{\pi}{4} - F_2 - F_5 \cos \alpha = 0$$

$$F_1 \sin \frac{\pi}{4} + F_5 \sin \alpha = M,$$

Joint C

$$F_2 + F_4 \cos \alpha - F_6 \cos \frac{\pi}{4} = 0$$

$$F_4 \sin \alpha + F_6 \sin \frac{\pi}{4} = M,$$

Joint D

$$F_3 + F_5 \cos \alpha + F_6 \cos \frac{\pi}{4} = 0$$

$$F_5 \sin \alpha + F_6 \sin \frac{\pi}{4} = M.$$

The augmented coefficient matrix $\mathbf{A}|\mathbf{b}$ for this system is

$$\mathbf{A}|\mathbf{b} = \left[\begin{array}{cccccc|c}
\cos \frac{\pi}{4} & 0 & 1 & \cos \alpha & 0 & 0 & 0 \\
\sin \frac{\pi}{4} & 0 & 0 & \sin \alpha & 0 & 0 & M \\
\cos \frac{\pi}{4} & -1 & 0 & 0 & -\cos \alpha & 0 & 0 \\
\sin \frac{\pi}{4} & 0 & 0 & 0 & \sin \alpha & 0 & M \\
0 & 1 & 0 & \cos \alpha & 0 & -\cos \frac{\pi}{4} & 0 \\
0 & 0 & 0 & \sin \alpha & 0 & \sin \frac{\pi}{4} & M \\
0 & 0 & 1 & 0 & \cos \alpha & \cos \frac{\pi}{4} & 0 \\
0 & 0 & 0 & 0 & \sin \alpha & \sin \frac{\pi}{4} & M
\end{array}\right]$$

$$\underbrace{\phantom{\cos \frac{\pi}{4} \quad 0 \quad 1 \quad \cos \alpha \quad 0 \quad 0}}_{\mathbf{A}} \quad \underbrace{}_{\mathbf{b}}$$

A routine calculation shows this is row equivalent to

$$
A_r | b_r = \underbrace{\begin{bmatrix}
\cos\frac{\pi}{4} & 0 & 1 & \cos\alpha & 0 & 0 \\
0 & 1 & 1 & \cos\alpha & \cos\alpha & 0 \\
0 & 0 & 1 & (\cos\alpha - \sin\alpha) & 0 & 0 \\
0 & 0 & 0 & (\cos\alpha - \sin\alpha) & -\cos\alpha & -\cos\frac{\pi}{4} \\
0 & 0 & 0 & 0 & \sin\alpha & \cos\frac{\pi}{4} \\
0 & 0 & 0 & 0 & 0 & 0 \\
0 & 0 & 0 & 0 & 0 & 0 \\
0 & 0 & 0 & 0 & 0 & 0
\end{bmatrix}}_{A_r} \underbrace{\left.\begin{matrix}
0 \\ 0 \\ -M \\ -M \\ M \\ 0 \\ 0 \\ 0
\end{matrix}\right]}_{b_r}
$$

Both A_r and $A_r | b_r$ have rank 5, so the six stress magnitudes F_1 to F_6 may be determined in terms of $6-5=1$ arbitrarily assigned stress. The rod with an arbitrarily assigned stress is a redundant rod. In the plane frame shown in Fig. 3.16(b) any one of the rods is redundant.

If required, the stress magnitudes may be determined from this last result by back substitution once either F_5 or F_6 has been assigned arbitrarily. Thus if F_6 is chosen arbitrarily, it follows from the fifth row of $A_r | b_r$ that

$$
F_5 = \frac{1}{\sin\alpha}\left(M - \frac{1}{\sqrt{2}}F_6\right),
$$

from which F_4 to F_1 then follow in similar fashion.

■

Problems for Section 3.5

Find which of the following systems of vectors is linearly independent and which is linearly dependent.

1 [1, 0], [0, 1]   **2** [1, 0, 0], [2, 0, 1], [1, 0, 1]
3 [7, 0, 14], [1, 3, 0], [0, 0, 1]   **4** [1, 2], [2, 3], [1, 3]
5 [1, 3, 2], [−3, 11, 4], [−2, 4, 1]   **6** [1, 1, 0, 0], [1, 0, 1, 0], [1, 0, 0, 1]

Find the rank of the following matrices

7 $\begin{bmatrix} 2 & 1 \\ 0 & 0 \\ 7 & 4 \end{bmatrix}$ **8** $\begin{bmatrix} 1 & 3 & 1 \\ 0 & 0 & 0 \\ 2 & 6 & 2 \end{bmatrix}$ **9** $\begin{bmatrix} 1 & 4 & 2 \\ 3 & -4 & 1 \\ 5 & -12 & 0 \end{bmatrix}$

10 $\begin{bmatrix} 1 & 2 & 1 & 3 \\ 2 & 0 & -1 & 2 \\ 1 & -2 & -2 & -1 \end{bmatrix}$ **11** $\begin{bmatrix} 1 & 4 & 7 & 2 \\ 0 & 0 & 0 & 0 \\ 2 & 8 & -14 & 4 \\ 1 & 1 & 1 & 1 \end{bmatrix}$

12 $\begin{bmatrix} 0 & 3 & 1 \\ 2 & 4 & 2 \\ 1 & 1 & 1 \\ 0 & 1 & 0 \end{bmatrix}$

13 Let $\mathbf{A}$ be an $m \times n$ matrix with rank $\mathbf{A} = r$, and $\mathbf{\Lambda}$ be an $m \times m$ diagonal matrix with nonzero elements in its leading diagonal. Find, stating your reasons,

   (i) rank $\mathbf{\Lambda}$ (ii) rank $(3\mathbf{A})$,   (iii) rank $(\mathbf{\Lambda A})$

   (iv) rank $\mathbf{\Lambda}^{\mathsf{T}}$ (v) rank $\mathbf{A}^{\mathsf{T}}$ (vi) rank $(\mathbf{\Lambda A})^{\mathsf{T}}$

14 Show by example that if the product $\mathbf{AB}$ is defined and rank $\mathbf{A} \neq 0$, rank $\mathbf{B} \neq 0$, it is not necessarily true that

      (rank $\mathbf{A}$) $\cdot$ (rank $\mathbf{B}$) and rank $(\mathbf{AB})$ are the same.

In each of the following examples bring the matrix to its reduced echelon form and hence find (i) its rank, (ii) a basis for its row space and (iii) a basis for its column space.

15 $\begin{bmatrix} 2 & 2 & 6 \\ 1 & 0 & 2 \\ 1 & 1 & 1 \end{bmatrix}$ **16** $\begin{bmatrix} 0 & 4 & 2 & 1 \\ 3 & 6 & 3 & 3 \\ 3 & 2 & 1 & 2 \end{bmatrix}$

17 $\begin{bmatrix} 0 & 1 & -1 & -4 & 1 \\ 2 & 4 & 2 & 0 & 2 \\ 1 & 2 & 1 & 3 & 2 \\ 1 & 1 & 2 & 1 & 1 \end{bmatrix}$ **18** $\begin{bmatrix} 1 & 3 & 1 & 2 \\ 1 & 1 & 3 & 0 \\ 2 & 4 & 1 & 2 \\ 3 & 2 & 3 & 1 \\ 1 & 0 & 1 & 2 \end{bmatrix}$

19 $\begin{bmatrix} 1 & 3 & 2 & 1 & 1 \\ 1 & 0 & -1 & 1 & 1 \\ 2 & 1 & 1 & 1 & 2 \\ 0 & 1 & -2 & 3 & 0 \end{bmatrix}$ **20** $\begin{bmatrix} 1 & 2 & 1 & 3 & 3 & 2 \\ 2 & 4 & 3 & 2 & 1 & 4 \\ 1 & 3 & 2 & 1 & 1 & 1 \end{bmatrix}$

3.6 Systems of linear equations: existence and form of solution in terms of rank

The general question of the existence and nature of the solution of a system of simultaneous equations has already been discussed in Sec. 3.4 in connection with Gaussian elimination. Now that we have available the notion of rank, and a computational procedure for its determination from the reduced echelon form of a matrix, our earlier results may be re-expressed in the following more concise and convenient form. We leave as an exercise for the reader the task of verifying that the conclusions of Sec. 3.4 are contained in the following.

Theorem 3.6 (Dependence of the solution of a system of linear equations on rank)

Associate with the system of m linear equations

$$a_{11}x_1+a_{12}x_2+ \ldots +a_{1n}x_n=b_1$$
$$a_{21}x_1+a_{22}x_2+ \ldots +a_{2n}x_n=b_2$$
$$\cdots\cdots\cdots\cdots\cdots\cdots\cdots\cdots$$
$$a_{m1}x_1+a_{m2}x_2+ \ldots +a_{mn}x_n=b_m, \tag{1}$$

involving the n unknowns $x_1, x_2, \ldots, x_n$, the coefficient matrix $\mathbf{A}$ and the augmented matrix $\mathbf{A}|\mathbf{b}$,

$$
\mathbf{A} =
\begin{bmatrix}
a_{11} & a_{12} & \cdots & a_{1n} \\
a_{21} & a_{22} & \cdots & a_{2n} \\
\cdot & \cdot & \cdots & \cdot \\
a_{m1} & a_{m2} & \cdots & a_{mn}
\end{bmatrix}
\quad \text{and} \quad
\mathbf{A}|\mathbf{b} =
\begin{bmatrix}
a_{11} & a_{12} & \cdots & a_{1n} & b_1 \\
a_{21} & a_{22} & \cdots & a_{2n} & b_2 \\
\cdot & \cdot & \cdots & \cdot & \cdot \\
a_{m1} & a_{m2} & \cdots & a_{mn} & b_m
\end{bmatrix}. \tag{2}
$$

Then system (1) is such that:

(i) it has a unique solution when rank $\mathbf{A}=$ rank $\mathbf{A}|\mathbf{b}=n$;

(ii) it has a nonunique solution when rank $\mathbf{A}=$ rank $\mathbf{A}|\mathbf{b}=r<n$, for then r suitably chosen unknowns will depend on the values of the remaining $n-r$ unknowns which may be assigned arbitrary values in the consistent system of equations;

(iii) when a solution exists (unique or nonunique) it may be found from the reduced echelon form $\mathbf{A}|\mathbf{b}_E$ of the augmented matrix by means of back-substitution;

(iv) no solution exists when rank $\mathbf{A} \neq$ rank $\mathbf{A}|\mathbf{b}$ for the system is inconsistent.

$\square$

In Sec. 3.4 we saw that system (1) is called **homogeneous** when the constants $b_1, b_2, \ldots, b_m$ are such that $b_1=b_2= \ldots =b_m=0$. The nature of the solution of a homogeneous system now follows directly from the results of Theorem 3.6.

It is clear that a homogeneous system always has the solution $x_1=x_2= \ldots =x_n=0$ or, equivalently, the solution $\mathbf{x}=\mathbf{0}$. This is called the **trivial solution** of the system. To discover if more interesting **nontrivial solutions** are possible, that is solutions which are not identically zero, we need to make use of Theorem 3.6. To do this we notice first that as $\mathbf{b}=\mathbf{0}$, both the coefficient matrix $\mathbf{A}$ and the augmented matrix $\mathbf{A}|\mathbf{0}$ must have the same rank.

Let us suppose rank $\mathbf{A}=$ rank $\mathbf{A}|\mathbf{0}=r$. Then, from Theorem 3.6, if $r=n$ a unique solution exists. However, $r=n$ implies the existence of n linearly independent homogeneous equations, and in this case if they are to be consistent the only possible solution is the trivial solution $\mathbf{x}=\mathbf{0}$.

Next let us suppose that $r<n$. Then it follows from Theorem 3.6(ii) that nontrivial solutions exist, though they are not unique. In this case the equations are consistent, though since there are only r which are linearly independent, it follows that r suitably

chosen unknowns will depend on the values of the remaining $n-r$ unknowns, which may be assigned arbitrary values. This shows that in this case the solutions belong to a vector space of dimension $n-r$.

At this point it is appropriate to remark that in more abstract accounts of linear algebra the space to which the solutions of $\mathbf{Ax}=\mathbf{0}$ belong is called either the **null space** or the **kernel** of $\mathbf{A}$. The dimension of the null space (kernel) is then called the **nullity** of $\mathbf{A}$, written $N(\mathbf{A})$, so we have the result

$$\text{rank } \mathbf{A} + N(\mathbf{A}) = n.$$

We mention that homogeneous systems have the useful and important special property that a linear combination of two nontrivial solutions is itself a solution. This is easily seen, because if $\mathbf{x}_1$ and $\mathbf{x}_2$ are any two solutions, and α, β are arbitrary constants, then

$$\alpha \mathbf{Ax}_1 = \mathbf{0} \quad \text{and} \quad \beta \mathbf{Ax}_2 = \mathbf{0}.$$

Hence

$$\mathbf{A}(\alpha \mathbf{x}_1) = \mathbf{0} \quad \text{and} \quad \mathbf{A}(\beta \mathbf{x}_2) = \mathbf{0},$$

so that

$$\mathbf{A}(\alpha \mathbf{x}_1 + \beta \mathbf{x}_2) = \mathbf{0},$$

showing $\alpha \mathbf{x}_1 + \beta \mathbf{x}_2$ is also a solution. This property is *not*, of course, true of non-homogeneous systems.

Finally, from Theorem 3.6 (ii), we see that if $m<n$ it follows automatically that rank $\mathbf{A}<n$; so in this case nontrivial solutions will always exist. We have thus established the following result.

Theorem 3.7 (Dependence of the solution of a homogeneous system on rank)

Consider the homogeneous system of m linear equations

$$a_{11}x_1 + a_{12}x_2 + \ldots + a_{1n}x_n = 0$$

$$a_{21}x_1 + a_{22}x_2 + \ldots + a_{2n}x_n = 0$$

$$\cdots \cdots \cdots \cdots \cdots \cdots \cdots \cdots$$

$$a_{m1}x_1 + a_{m2}x_2 + \ldots + a_{mn}x_n = 0 \tag{3}$$

involving the n unknowns $x_1, x_2, \ldots, x_n$ and the coefficient matrix

$$\mathbf{A} = \begin{bmatrix} a_{11} & a_{12} & \cdots & a_{1n} \\ a_{21} & a_{22} & \cdots & a_{2n} \\ \cdots & \cdots & \cdots & \cdots \\ a_{m1} & a_{m2} & \cdots & a_{mn} \end{bmatrix}. \tag{4}$$

Then:

(i) If rank $A = n$, system (3) only has the trivial solution
$$x_1 = x_2 = \ldots = x_n = 0;$$

(ii) if rank $A = r < n$, system (3) has the trivial solution and also nontrivial solutions which are not unique, with the property that r suitably chosen unknowns will depend on the values of the remaining $n - r$ unknowns, which may be assigned arbitrary values;

(iii) if $\mathbf{x}_1$ and $\mathbf{x}_2$ are any two solutions of system (3), and α, β are arbitrary constants, $\alpha \mathbf{x}_1 + \beta \mathbf{x}_2$ is also a solution of system (3);

(iv) if $m < n$, in addition to the trivial solution, system (3) also has nontrivial solutions. □

Example 1. Homogeneous equations $m = n$

Find the solution of

$$x_1 + 2x_2 - x_3 + x_4 = 0$$
$$2x_1 - x_2 - 4x_3 + 2x_4 = 0$$
$$x_1 - 3x_2 - 3x_3 + x_4 = 0$$
$$x_1 - 8x_2 - 5x_3 + x_4 = 0.$$

Solution

We shall solve this system by bringing it to its reduced echelon form, though Gaussian elimination would work equally well.

$$A|0 = \begin{bmatrix} 1 & 2 & -1 & 1 & 0 \\ 2 & -1 & -4 & 2 & 0 \\ 1 & -3 & -3 & 1 & 0 \\ 1 & -8 & -5 & 1 & 0 \end{bmatrix} \begin{matrix} R_1 \\ R_2 - 2R_1 \\ R_3 - R_1 \\ R_4 - R_1 \end{matrix} \sim \begin{bmatrix} 1 & 2 & -1 & 1 & 0 \\ 0 & -5 & -2 & 0 & 0 \\ 0 & -5 & -2 & 0 & 0 \\ 0 & -10 & -4 & 0 & 0 \end{bmatrix}$$

$$\begin{matrix} R_1 \\ -\frac{1}{5}R_2 \\ R_3 \\ R_4 \end{matrix} \sim \begin{bmatrix} 1 & 2 & -1 & 1 & 0 \\ 0 & 1 & \frac{2}{5} & 0 & 0 \\ 0 & -5 & -2 & 0 & 0 \\ 0 & -10 & -4 & 0 & 0 \end{bmatrix} \begin{matrix} R_1 - 2R_2 \\ R_2 \\ R_3 + 5R_2 \\ R_4 + 10R_2 \end{matrix} \sim \begin{bmatrix} 1 & 0 & -\frac{9}{5} & 1 & 0 \\ 0 & 1 & \frac{2}{5} & 0 & 0 \\ 0 & 0 & 0 & 0 & 0 \\ 0 & 0 & 0 & 0 & 0 \end{bmatrix}.$$

The last result implies the two equations

$$x_1 - \tfrac{9}{5}x_3 + x_4 = 0 \quad \text{and} \quad x_2 + \tfrac{2}{5}x_3 = 0.$$

A suitable choice for the two unknowns which are to be expressed in terms of the remaining two is x_1 and x_2, for then we may assign x_3 and x_4 arbitrarily. Thus a two-parameter family of solutions of this homogeneous system is

$$x_1 = \tfrac{9}{5}\alpha - \beta, \quad x_2 = -\tfrac{2}{5}\alpha, \quad x_3 = \alpha \quad \text{and} \quad x_4 = \beta,$$

with α, β arbitrary. Included in this family of solutions is the trivial solution corresponding to $\alpha = \beta = 0$.

We could not, for example, have allocated x_2 and x_3 arbitrarily because of the second equation relating only x_2 and x_3. Alternative choices for the two unknowns could equally well have been x_1 and x_2, or x_2 and x_4, for neither of these choices places any constraint on the corresponding pairs of unknowns x_3 and x_4, and x_1 and x_3 which can then be allocated arbitrarily.

■

Example 2. Homogeneous equations $m < n$.

Find the solution of

$$x_1 + 3x_2 - x_3 = 0$$
$$x_1 + 4x_2 - 2x_3 = 0.$$

Solution

In this case $m = 2$ and $n = 3$, so by Theorem 3.6(ii) nontrivial solutions will exist.

$$\mathbf{A} | \mathbf{0} = \begin{bmatrix} 1 & 3 & -1 & 0 \\ 1 & 4 & -2 & 0 \end{bmatrix} \underset{R_2 - R_1}{\overset{R_1}{\sim}} \begin{bmatrix} 1 & 3 & -1 & 0 \\ 0 & 1 & -1 & 0 \end{bmatrix}$$

$$\underset{R_2}{\overset{R_1 - 3R_2}{\sim}} \begin{bmatrix} 1 & 0 & 2 & 0 \\ 0 & 1 & -1 & 0 \end{bmatrix}.$$

This last result implies the two equations

$$x_1 + 2x_3 = 0 \quad \text{and} \quad x_2 - x_3 = 0.$$

Setting $x_3 = \alpha$, with α arbitrary, we then find the one-parameter family of solutions

$$x_1 = -2\alpha, \quad x_2 = \alpha, \quad x_3 = \alpha.$$

■

3.7 Determinants

This section, which is largely independent of the remainder of the chapter, is concerned with the definition and algebraic properties of determinants. Although their theory belongs to algebra, it is not necessary to present it as part of an account of matrices. Determinants arise throughout most of mathematics, and their important properties are used in many different ways, so in a text such as this there is no ideal place at which to introduce them. The present section has been chosen because of their relevance to rank, systems of linear equations, and to inverse matrices which have still to be introduced. For the purposes of the remainder of this chapter only the definition of a determinant and some of their simplest properties will be needed.

A **determinant** is a number which may be associated in a unique way with any square matrix whose n^2 elements are themselves numbers. If, however, the elements of the matrix are functions, the associated determinant will itself also be a function. Let $\mathbf{A}$ be an $n \times n$ matrix. Then its determinant is indicated either by writing $\det \mathbf{A}$ or $|\mathbf{A}|$.

Determinants are classified according to the number of elements they contain in their leading diagonal, and if $\mathbf{A}$ is an $n \times n$ matrix its determinant $\det \mathbf{A}$ is called an **nth order determinant**.

Inductive definition of a determinant

There are several different ways in which a determinant may be defined. We shall employ an inductive definition with respect to its order n as being the most direct for our purposes.

For $n=1$, when $\mathbf{A}=[a_{11}]$, we define

$\det \mathbf{A}=a_{11}$.

Then, assuming determinants of all orders up to and including $n-1$ have been defined, we take as our definition of an nth order determinant the expression

$$
\det \mathbf{A} =
\begin{vmatrix}
a_{11} & a_{12} & \cdots & a_{1n} \\
a_{21} & a_{22} & \cdots & a_{2n} \\
\cdots & \cdots & \cdots & \cdots \\
a_{n1} & a_{n2} & \cdots & a_{nn}
\end{vmatrix}
= \sum_{j=1}^{n} (-1)^{1+j} a_{1j} M_{1j} \,,
\tag{1}
$$

where M_{1j} is called the **minor** of the element a_{1j} and is the $n-1$th order determinant derived from $\mathbf{A}$ by deleting the elements in its first row and jth columns. Specifically, we have

$$
M_{1j}=
\begin{vmatrix}
a_{21} & a_{22} & \cdots & a_{2\,j-1}\,a_{2\,j+1} & \cdots & a_{2n} \\
a_{31} & a_{32} & \cdots & a_{3\,j-1}\,a_{3\,j+1} & \cdots & a_{3n} \\
\cdots & \cdots & \cdots & \cdots & \cdots & \cdots \\
a_{n1} & a_{n+} & \cdots & a_{n\,j-1}\,a_{n\,j-1} & \cdots & a_{nn}
\end{vmatrix}
\quad n-1 \text{ rows .}
\tag{2}
$$

$$n-1 \text{ columns}$$

The simplest nontrivial example of a determinant arises when $n=2$. In this case (1) shows

$$
\begin{vmatrix} a_{11} & a_{12} \\ a_{21} & a_{22} \end{vmatrix}
=(-1)^{1+1} a_{11} a_{22} + (-1)^{1+2} a_{12} a_{21} = a_{11} a_{22} - a_{12} a_{21} .
$$

Similarly, when $n=3$,

$$
\begin{vmatrix}
a_{11} & a_{12} & a_{13} \\
a_{21} & a_{22} & a_{23} \\
a_{31} & a_{32} & a_{33}
\end{vmatrix}
=(-1)^{1+1} a_{11}
\begin{vmatrix} a_{22} & a_{23} \\ a_{32} & a_{33} \end{vmatrix}
+(-1)^{1+2} a_{12}
\begin{vmatrix} a_{21} & a_{23} \\ a_{31} & a_{33} \end{vmatrix}
$$

$$
+(-1)^{1+3} a_{13}
\begin{vmatrix} a_{21} & a_{22} \\ a_{31} & a_{32} \end{vmatrix}
$$

$$
= a_{11}(a_{22} a_{33} - a_{23} a_{32}) - a_{12}(a_{21} a_{33} - a_{23} a_{31})
$$

$$
+ a_{13}(a_{21} a_{32} - a_{22} a_{31}).
$$

Example 1

Evaluate the determinants

(i) $\begin{vmatrix} 3 & 1 \\ 4 & 7 \end{vmatrix}$, (ii) $\begin{vmatrix} 1 & -4 & 2 \\ 1 & 3 & 1 \\ 2 & 0 & 3 \end{vmatrix}$ and (iii) $\begin{vmatrix} \sin x & \cos x \\ \sinh x & \cosh x \end{vmatrix}$

Solution

(i) $\begin{vmatrix} 3 & 1 \\ 4 & 7 \end{vmatrix} = 3 \cdot 7 - 1 \cdot 4 = 17,$

(ii) $\begin{vmatrix} 1 & -4 & 2 \\ 1 & 3 & 1 \\ 2 & 0 & 3 \end{vmatrix} = (1) \begin{vmatrix} 3 & 1 \\ 0 & 3 \end{vmatrix} - (-4) \begin{vmatrix} 1 & 1 \\ 2 & 3 \end{vmatrix} + (2) \begin{vmatrix} 1 & 3 \\ 2 & 0 \end{vmatrix}$

$= 1(9-0) + 4(3-2) + 2(0-6) = 1.$

(iii) $\begin{vmatrix} \sin x & \cos x \\ \sinh x & \cosh x \end{vmatrix} = \sin x \cosh x - \cos x \sinh x.$ ∎

It is convenient to incorporate the alternation of sign produced by the factor $(-1)^{1+j}$ in (1) into what is called the **cofactor** of the element a_{1j}. This is accomplished by defining the **cofactor** C_{1j} of the element a_{1j} to be the **signed minor** (minor with an appropriate sign)

$$C_{1j} = (-1)^{1+j} M_{1j}, \qquad (3)$$

so that result (1) becomes

$$\det \mathbf{A} = \sum_{j=1}^{n} a_{1j} C_{1j}. \qquad (4)$$

The notion of a minor and its associated cofactor extends to every element of $\mathbf{A}$ in a natural manner. We define the minor M_{ij} of the element a_{ij} to be the value of the determinant of order $n-1$ derived from $\mathbf{A}$ by deleting the elements of the ith row and jth column. The **cofactor** C_{ij} of element a_{ij} is then defined to be

$$C_{ij} = (-1)^{i+j} M_{ij}. \qquad (5)$$

Inspection of (4) shows that $\det \mathbf{A}$ has been expanded in terms of the elements and cofactors of the *first row* of $\mathbf{A}$. It is a remarkable fact that this form of expansion of a determinant may be generalized, and that the same value ($\det \mathbf{A}$) is obtained if the determinant is expanded in terms of the elements and cofactors of *any* row or column of $\mathbf{A}$. This fundamental result is known as the **Laplace**[2] **expansion** of a determinant in terms of the elements of a row or column. It forms the basic result upon which all our

---

[2] PIERRE SIMON LAPLACE (1749–1827), French mathematician of outstanding ability who made fundamental contributions to analysis, differential equations, probability and celestial mechanics. He regarded mathematics as a tool with which to investigate physical phenomena and in addition to celestial mechanics he studied hydrodynamics, sound propagation, surface tension and many other topics.

subsequent results will be based, though its proof will be deferred until the end of this section.

Theorem 3.8 (Fundamental Laplace expansion theorem)

Let A be an $n \times n$ matrix. Then,

(i) det A may be expanded in terms of the elements of the ith row as

$$\det A = a_{i1} C_{i1} + a_{i2} C_{i2} + \ldots + a_{in} C_{in} = \sum_{j=1}^{n} a_{ij} C_{ij}$$

for any fixed i such that $1 \le i \le n$;

(ii) det A may be expanded in terms of the elements of the jth column as

$$\det A = a_{1j} C_{ij} + a_{2j} C_{2j} + \ldots + a_{nj} C_{nj} = \sum_{i=1}^{n} a_{ij} C_{ij}$$

for any fixed j such that $1 \le j \le n$.   □

This theorem provides the key to the derivation of many important and useful algebraic properties of determinants. These may either be used to simplify the evaluation of a determinant or to derive algebraic properties of expressions which may be represented in determinantal form.

Theorem 3.9 (Transposition of a determinant)

If A is an $n \times n$ matrix and A^T its transpose, then

$$\det A = \det A^T.$$

Proof

The result follows directly from Theorem 3.8, because expanding det A in terms of the elements of the ith row (column) is the same as expanding det A^T in terms of elements of the ith column (row).   □

Theorem 3.10 (Multiplication by a constant)

If the elements of the ith row (column) of a determinant are all multiplied by a constant k, then the value of the determinant is multiplied by k.

Proof

The result follows directly by expanding det A in terms of the elements of the ith row (column) and removing the factor k.   □

Theorem 3.11 (Zero row or column)

A determinant in which all the elements of a row or column are zero has the value zero.

Proof

The result follows directly by expanding the determinant in terms of the row or column in which all the elements are zero. $\square$

Theorem 3.12 (Elements in row or column of the form $a_{ij}^{(1)} + a_{ij}^{(2)}$)

Let $A = [a_{ij}]$ be an $n \times n$ matrix such that each element a_{ij} of the ith row (jth column) is of the form $a_{ij} = a_{ij}^{(1)} + a_{ij}^{(2)}$. Then

$$\det A = \det A_1 + \det A_2,$$

where A_1 and A_2 are derived from A by replacing the elements of the ith row (jth column) by $a_{ij}^{(1)}$ and $a_{ij}^{(2)}$, respectively.

Proof

The result follows directly by expanding the terms of the elements of the ith row (jth column). $\square$

Theorem 3.13 (Interchange of rows or columns)

If any two rows or two columns of det A are interchanged, then the sign of det A is reversed.

Proof

The proof is by induction, and starts from the fact that the result is clearly true for $n = 2$, when

$$\begin{vmatrix} a_{11} & a_{12} \\ a_{21} & a_{22} \end{vmatrix} = a_{11} a_{22} - a_{12} a_{21}.$$

We now assume it to be true for determinants of order $n - 1$. For a determinant of order n we select a row not involved in the interchange and expand det A in terms of its elements. Suppose it is the ith row, then each cofactor will be a determinant of order $n - 1$ in which two rows have been interchanged. Then, from our induction hypothesis, each cofactor must have its sign reversed. Thus the result is proved, because it is true for $n = 2$ and so by induction it is also true for $n > 2$. A similar proof establishes the result when columns are interchanged. $\square$

Theorem 3.14 (Proportionality of rows or columns)

If the elements of any two rows or two columns of det A are proportional, then det $A=0$.

Proof

Let the constant of proportionality between the two rows (columns) be k. Then from Theorem 3.10 det $A=k$ det A_1, where det A_1 now has two identical rows (columns). By Theorem 3.13, interchanging the two identical rows (columns) in det A_1 will reverse its sign, yet the determinant itself is unchanged, so that det $A_1=0$, and hence det $A=0$. $\square$

Theorem 3.15 (Addition of a multiple of a row (column) to a row (column))

The value of a determinant remains unaltered if a multiple of a row (column) is added to another row (column).

Proof

Suppose the elements a_{ij} of the ith row (jth column) of A are replaced by $a_{ij}+ka_{rj}$, to produce a matrix A_1. Then by Theorems 3.10 and 3.12,

$$\det A_1 = \det A + k \det A_2,$$

where A_2 is obtained from A by replacing the ith row (jth column) by the elements a_{rj}. Thus det A_2 has two identical rows (columns), and so vanishes by Theorem 3.14. Thus det $A_1 = $ det A, and the theorem is proved. $\square$

Theorem 3.16 (Multiplication of determinants)

Let A and B be two $n \times n$ matrices. Then

$$\det(AB) = \det A \det B.$$

Proof

We start with the case $n=2$. Set

$$D_1 = \begin{vmatrix} a_{11} & a_{12} \\ a_{21} & a_{22} \end{vmatrix}, \qquad D_2 = \begin{vmatrix} b_{11} & b_{12} \\ b_{21} & b_{22} \end{vmatrix}$$

and consider the determinant

$$D = \begin{vmatrix} a_{11} & a_{12} & 0 & 0 \\ a_{21} & a_{22} & 0 & 0 \\ -1 & 0 & b_{11} & b_{12} \\ 0 & -1 & b_{21} & b_{22} \end{vmatrix}.$$

Expanding by elements of the last column, and then again expanding the two third order determinants that result by the elements of their last columns gives

$$D = -b_{12}\begin{vmatrix} a_{11} & a_{12} & 0 \\ a_{21} & a_{22} & 0 \\ 0 & -1 & b_{21} \end{vmatrix} + b_{22}\begin{vmatrix} a_{11} & a_{12} & 0 \\ a_{21} & a_{22} & 0 \\ -1 & 0 & b_{11} \end{vmatrix}$$

$$= -b_{12}b_{21}\begin{vmatrix} a_{11} & a_{12} \\ a_{21} & a_{22} \end{vmatrix} + b_{11}b_{22}\begin{vmatrix} a_{11} & a_{12} \\ a_{21} & a_{22} \end{vmatrix}$$

$$= \begin{vmatrix} a_{11} & a_{12} \\ a_{21} & a_{22} \end{vmatrix}\begin{vmatrix} b_{21} & b_{12} \\ b_{21} & b_{22} \end{vmatrix} = D_1 D_2.$$

If, now, we appeal to Theorem 3.15 and replace the first row of D by $R_1 + a_{11}R_3 + a_{12}R_4$ and the second row by $R_2 + a_{21}R_3 + a_{22}R_4$, we have

$$D = D_1 D_2 = \begin{vmatrix} 0 & 0 & a_{11}b_{11}+a_{12}b_{21} & a_{11}b_{12}+a_{12}b_{22} \\ 0 & 0 & a_{21}b_{11}+a_{22}b_{21} & a_{21}b_{12}+a_{22}b_{22} \\ -1 & 0 & b_{11} & b_{12} \\ 0 & -1 & b_{21} & b_{22} \end{vmatrix}.$$

When expanded by its first two columns this gives the result

$$\begin{vmatrix} a_{11} & a_{12} \\ a_{21} & a_{22} \end{vmatrix}\begin{vmatrix} b_{11} & b_{12} \\ b_{21} & b_{22} \end{vmatrix} = \begin{vmatrix} a_{11}b_{11}+a_{12}b_{21} & a_{11}b_{12}+a_{12}b_{22} \\ a_{21}b_{11}+a_{22}b_{21} & a_{21}b_{12}+a_{22}b_{22} \end{vmatrix}.$$

The terms of the determinant on the right-hand side are the terms resulting from the matrix product **AB**, so the result is true when $n=2$.

This process generalizes in an obvious manner. Let $A=[a_{ij}]$, $B=[b_{ij}]$ be $n\times n$ matrices and consider the determinant

$$D = \begin{vmatrix} a_{11} & a_{12} & \cdots & a_{1n} & 0 & \cdots & 0 \\ & & & & & & \\ & & & & & & \\ a_{n1} & a_{n2} & \cdots & a_{nn} & 0 & \cdots & 0 \\ -1 & 0 & \cdots & 0 & b_{11} & \cdots & b_{1n} \\ 0 & -1 & \cdots & 0 & & & \\ & & & & & & \\ 0 & 0 & \cdots & -1 & b_{n1} & \cdots & b_{nn} \end{vmatrix}.$$

Then, arguing as before, it follows that $D=D_1 D_2$, where $D_1 = \det A$ and $D_2 = \det B$. We now make the following replacements in D:

new row $1 = R_1 + a_{11}R_{n+1}+a_{12}R_{n+2}+\ \ldots\ +a_{1n}R_{2n}$

new row $2 = R_2 + a_{21}R_{n+1}+a_{22}R_{n+2}+\ \ldots\ +a_{2n}R_{2n}$

$\cdots\cdots\cdots\cdots\cdots\cdots\cdots\cdots\cdots\cdots$

new row $n = R_n + a_{n1}R_{n+1}+a_{n2}R_{n+2}+\ldots+a_{nn}R_{2n}.$

Expanding the resulting determinant by its first n columns gives

$$D = (-1)^n \begin{vmatrix} -1 & 0 & \cdots & 0 \\ 0 & -1 & \cdots & 0 \\ \cdot & \cdot & \cdots & \cdot \\ 0 & 0 & \cdots & -1 \end{vmatrix} \begin{vmatrix} c_{11} & \cdot & \cdots & c_{1n} \\ c_{21} & \cdot & \cdots & c_{2n} \\ \cdot & \cdot & \cdots & \cdot \\ c_{n1} & \cdot & \cdots & c_{nn} \end{vmatrix},$$

where

$$c_{ij} = a_{i1}b_{1j} + a_{i2}b_{2j} + \cdots + a_{in}b_{nj}.$$

However, the first determinant of order n on the right-hand side has the value $(-1)^n$ so that

$$D = D_1 D_2 = \begin{vmatrix} c_{11} & \cdot & \cdots & c_{1n} \\ \cdot & \cdot & \cdots & \cdot \\ \cdot & \cdot & \cdots & \cdot \\ c_{n1} & \cdot & \cdots & c_{nn} \end{vmatrix}.$$

Since the c_{ij} are the terms in the matrix product $\mathbf{AB}$ the result is proved.

$\square$

Theorem 3.17 (Generalized Laplace expansion)

Let $\mathbf{A} = [a_{ij}]$ be an $n \times n$ matrix with C_{ij} the cofactor of the element a_{ij} in det $\mathbf{A}$, then

(i) the sum of the products of the elements of the ith row with the corresponding cofactors of the kth row is

$$a_{i1}C_{k1} + a_{i2}C_{k2} + \cdots + a_{in}C_{kn} = \begin{cases} 0 & \text{for } i \neq k \\ \det \mathbf{A} & \text{for } i = k; \end{cases}$$

(ii) the sum of the products of the elements of the jth column with the corresponding cofactors of the kth column is

$$a_{1j}C_{1k} + a_{2j}C_{2k} + \cdots + a_{nj}C_{nk} = \begin{cases} 0 & \text{for } j \neq k \\ \det \mathbf{A} & \text{for } j = k. \end{cases}$$

Proof

When in (i) $i = k$, or in (ii) $j = k$, this is a restatement of the Laplace expansion theorem. Consider case (i) when $i \neq k$, so that the right-hand side vanishes. The left-hand side is seen to be the expansion of a determinant obtained from $\mathbf{A}$ by replacing the kth row by the jth row. This determinant vanishes since it has two identical rows and result (i) is proved. Result (ii) follows in similar fashion.

$\square$

Theorem 3.18 (Differentiation of a determinant)

Let $\mathbf{A} = [a_{ij}]$ be an $n \times n$ matrix whose elements a_{ij} are differentiable functions of x. Then,

$$\frac{d}{dx}(\det \mathbf{A}) = \det \mathbf{A}_1 + \det \mathbf{A}_2 + \ldots + \det \mathbf{A}_n,$$

where $\mathbf{A}_i$ is the determinant derived from $\mathbf{A}$ by differentiation of the elements of its ith row with respect to x.

Proof

When expanded, a determinant of order n contains products ($n!$ in all) of the form

$$a_{1\alpha} a_{2\beta} \ldots a_{n\nu},$$

where $\alpha, \beta, \ldots \nu$ is some permutation of the integers $1, 2, \ldots, n$. Differentiation of any such term gives

$$\frac{d}{dx}(a_{1\alpha} a_{2\beta} \ldots a_{n\nu}) = \left(\frac{da_{1\alpha}}{dx}\right) a_{2\beta} \ldots a_{n\nu} + a_{1\alpha}\left(\frac{da_{2\beta}}{dx}\right) \ldots a_{n\nu}$$

$$+ \ldots + a_{1\alpha} a_{2\beta} \ldots \left(\frac{da_{n\nu}}{dx}\right).$$

The theorem then follows by differentiating each such product, and recombining the terms into the sum of n determinants in such a way that the ith determinant differs only from $\det \mathbf{A}$ in that the elements of its ith row have been differentiated. $\square$

We conclude this section by providing the promised proof of the fundamental Laplace expansion theorem (Theorem 3.8). The proof, which is included only for the sake of completeness, may be omitted by any reader who feels it to be unnecessary.

The inductive definition of a determinant used so far is inconvenient for a proof of Laplace's equation as it leads to awkward notation. Accordingly, we shall proceed from an alternative though equivalent definition of a determinant, which allows a much simpler proof.

Let us start by noticing that $n!$ terms will arise in the expansion of an nth order determinant, each of the form $a_{1\alpha} a_{2\beta} \ldots a_{n\nu}$, where $(\alpha, \beta, \ldots, \nu)$ is some permutation of the integers $1, 2, \ldots, n$. When any two of the numbers in the permutation are out of their natural numerical order they are said to be **inverted**. For any given permutation, the number of such inversions required to bring it to its natural numerical order is unique. We shall denote this number by p. A permutation is said to be *even* when p is even and *odd* when it is odd. As the number of inversions will be needed shortly, it is necessary to make quite clear how it may be determined.

Inversions are most easily counted with the help of a simple diagram, a specific example of which is shown in Fig. 3.17 for the permutation $(4, 5, 3, 2, 1)$. In the top line the

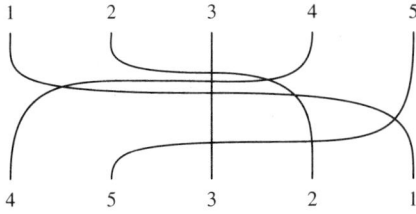

Fig. 3.17

integers 1 to 5 are arranged in their natural order, while in the bottom line they are displayed in the order of the permutation.

A little thought then shows that the number of inversions p is equal to the number of intersections of the lines joining each number in the top line to the corresponding number in the bottom line. In the case of the permutation $(4, 5, 3, 2, 1)$ we have $p = 9$.

With these preliminaries out of the way, an alternative but equivalent definition of the determinant of order n associated with the $n \times n$ matrix $\mathbf{A} = [a_{ij}]$ can now be formulated. We set

$$\det \mathbf{A} = \Sigma \pm a_{1\alpha} a_{2\beta} \ldots a_{n\nu}, \tag{6}$$

where the summation is over all $n!$ products of this type, and the positive sign is taken when a permutation is even and the negative sign when it is odd. The reader should check this expansion against the explicit expressions for $\det \mathbf{A}$ given earlier for $n = 2$ and $n = 3$ to see that they are the same.

Now the coefficient in $\det \mathbf{A}$ of the element a_{11} is simply

$$\Sigma \pm a_{2\beta} a_{3\gamma} \ldots a_{n\nu}, \tag{7}$$

and because the row and column suffixes in a_{11} are in their correct order, they will make no contribution to the sign of (7). Notice that in each term in (7) the number of inversions involved must now be counted as the number required to bring $(\beta, \gamma, \ldots, \nu)$ to its natural order $(2, 3, \ldots, n)$.

There are $(n-1)!$ terms in (7), which together comprise the definition of the determinant derived from $\mathbf{A}$ by deleting the elements of the first row and first column and multiplying the result by $+1$. This is just the cofactor C_{11} of $\det \mathbf{A}$. Let us now find the coefficient of a_{ij} in $\det \mathbf{A}$. First we bring the ith row into the position of the first row, and then the jth column into the position of the first column. Then a_{ij} becomes the first element of the leading diagonal, though because of the transposition of rows and columns $\det \mathbf{A}$ will have undergone $(i-1)+(j-1)$ changes of sign. The coefficient of a_{ij} in $\det \mathbf{A}$ then follows as before. However, now it will be the determinant derived from $\mathbf{A}$ by deleting the elements of the ith row and jth column, which is then multiplied by the factor $(-1)^{i+j-2} = (-1)^{i+j}$. This is just the cofactor C_{ij} of a_{ij} in $\det \mathbf{A}$.

In the expression

$$a_{11}C_{11} + a_{12}C_{12} + \ldots + a_{1n}C_{1n} \tag{8}$$

there are $n(n-1)!$ terms; a total of $n!$ in all. They are all different and all occur with their correct sign in det $\mathbf{A}$, so since det $\mathbf{A}$ has only $n!$ terms, (8) is the Laplace expansion of det $\mathbf{A}$ in terms of the elements of its first row. By bringing a different row into the position of the first row, the same argument applies and so establishes Theorem 3.8 (i). The second part of the theorem follows directly by using the result det $\mathbf{A} = \det \mathbf{A}^T$. $\square$

Example 2. Basic manipulation of determinants

Evaluate the following determinants after simplification:

(i) $\begin{vmatrix} 1 & 1 & 1 \\ 1 & 7 & 0 \\ -2 & 4 & -3 \end{vmatrix}$
(ii) $\begin{vmatrix} 2 & 4 & 6 \\ 1 & 0 & 1 \\ 2 & 0 & 4 \end{vmatrix}$
(iii) $\begin{vmatrix} 1 & 1 & 1 \\ a & b & c \\ a^2 & b^2 & c^2 \end{vmatrix}$

(iv) $\begin{vmatrix} 1 & a & a^2 \\ 1 & b & b^2 \\ 1 & c & c^2 \end{vmatrix}$
(v) $\begin{vmatrix} 2 & 2 & -2 \\ 1 & 7 & 4 \\ 3 & 6 & 4 \end{vmatrix}$ given that $\begin{vmatrix} 2 & 1 & -1 \\ 1 & 4 & 2 \\ 3 & 1 & 2 \end{vmatrix} = 27,$

(vi) $\begin{vmatrix} 1 & a+2b & 0 \\ 1 & a+3b & 0 \\ 0 & 0 & c \end{vmatrix}$
(vii) $\dfrac{d}{dx} \begin{vmatrix} 1 & x \\ e^{-x} & e^x \end{vmatrix}.$

Solution

In arriving at the solutions use will be made of most theorems in this section. However, there is usually more than one way of simplifying a determinant, and the reader is encouraged to obtain these results by alternative means.

(i) $R_2 - 3R_1 - R_3 = 0$ so the determinant is equivalent to one in which the elements of row two are all zeros. Thus the value of the determinant is zero.

(ii) $\det \mathbf{A} = \begin{vmatrix} 2 & 4 & 6 \\ 1 & 0 & 1 \\ 2 & 0 & 4 \end{vmatrix} = 2 \begin{vmatrix} 1 & 2 & 3 \\ 1 & 0 & 1 \\ 2 & 0 & 4 \end{vmatrix} = 2^2 \begin{vmatrix} 1 & 2 & 3 \\ 1 & 0 & 1 \\ 1 & 0 & 2 \end{vmatrix}.$

Expanding in terms of elements of the first row then gives $\det \mathbf{A} = 2^2(-2) = -8.$

(iii)

$\begin{vmatrix} 1 & 1 & 1 \\ a & b & c \\ a^2 & b^2 & c^2 \end{vmatrix} = \begin{vmatrix} 1 & 0 & 0 \\ a & (b-a) & (c-a) \\ a^2 & (b^2-a^2) & (c^2-a^2) \end{vmatrix} = \begin{vmatrix} 1 & 0 & 0 \\ a & (b-a) & (c-a) \\ a^2 & (b+a)(b-a) & (c+a)(c-a) \end{vmatrix}$

$= (b-a)(c-a) \begin{vmatrix} 1 & 0 & 0 \\ a & 1 & 1 \\ a^2 & (b+a) & (c+a) \end{vmatrix} = (b-a)(c-a) \begin{vmatrix} 1 & 0 & 0 \\ a & 1 & 0 \\ a^2 & (b+a) & (c-b) \end{vmatrix}$

$= (b-a)(c-a)(c-b) \begin{vmatrix} 1 & 0 & 0 \\ a & 1 & 0 \\ a^2 & (b+a) & 1 \end{vmatrix} = (b-a)(c-a)(c-b).$

A determinant of this form is called an **alternant**.

(iv) The determinant is the transpose of the determinant in (iii), so the result is the same, giving

$$\begin{vmatrix} 1 & a & a^2 \\ 1 & b & b^2 \\ 1 & c & c^2 \end{vmatrix} = (b-a)(c-a)(c-b).$$

(v)

$$\det A = \begin{vmatrix} 2 & 2 & -2 \\ 1 & 7 & 4 \\ 3 & 6 & 4 \end{vmatrix} = 2 \begin{vmatrix} 2 & 2 & -1 \\ 1 & 7 & 2 \\ 3 & 6 & 2 \end{vmatrix}.$$

Now replace C_2 by $C_2 - C_1 - C_3$ to find

$$\det A = 2 \begin{vmatrix} 2 & 1 & -1 \\ 1 & 4 & 2 \\ 3 & 1 & 2 \end{vmatrix}.$$

Using the quoted value of this last determinant shows $\det A = 2 \times 27 = 54$.

(vi) By Theorem 2.12,

$$\begin{vmatrix} 1 & a+2b & 0 \\ 1 & a+3b & 0 \\ 0 & 0 & c \end{vmatrix} = \begin{vmatrix} 1 & a & 0 \\ 1 & a & 0 \\ 0 & 0 & c \end{vmatrix} + \begin{vmatrix} 1 & 2b & 0 \\ 1 & 3b & 0 \\ 0 & 0 & c \end{vmatrix}$$

$$= 0 + bc \begin{vmatrix} 1 & 2 & 0 \\ 1 & 3 & 0 \\ 0 & 0 & 1 \end{vmatrix} = bc \begin{vmatrix} 1 & 0 & 0 \\ 0 & 1 & 0 \\ 0 & 0 & 1 \end{vmatrix} = bc.$$

(vii) By Theorem 2.18

$$\frac{d}{dx} \begin{vmatrix} 1 & x \\ e^{-x} & e^x \end{vmatrix} = \begin{vmatrix} 0 & 1 \\ e^{-x} & e^x \end{vmatrix} + \begin{vmatrix} 1 & x \\ -e^{-x} & e^x \end{vmatrix} = -e^{-x} + e^x + xe^{-x} = (x-1)e^{-x} + e^x.$$

∎

The routine evaluation of a determinant with arbitrary numerical elements is best carried out by applying the Gaussian elimination process to the rows of the determinant (cf. Sec. 3.4). This reduces the determinant to one of upper triangular form, the value of which is equal to the product of the elements on its leading diagonal.

When using this method it should be remembered that, as a determinant is involved, whenever two rows are interchanged due to the occurrence of a zero pivot (or a very small one) a change of sign of the determinant will result. Thus, if during the elimination process r row interchanges are involved, the value of the determinant will be the product of the elements in the leading diagonal of the upper triangular form times the factor $(-1)^r$. The value of the determinant will be zero if an identically zero pivot arises during the elimination procedure which cannot be removed by means of a row interchange.

For ease of illustration, in the following example this method is applied to a determinant with integer elements. In practice the elements are usually decimal numbers. Row interchanges are introduced into the elimination process when either an identically zero pivot occurs, or a pivot is so small that using it would introduce large errors in the subsequent elimination process.

Example 3. Evaluation of a determinant by Gaussian elimination

Use Gaussian elimination to evaluate

$$|A| = \begin{vmatrix} 3 & 1 & 4 & 2 \\ 9 & 3 & 4 & 4 \\ 1 & 2 & 1 & 1 \\ 1 & 1 & 1 & 2 \end{vmatrix}.$$

Solution

$$|A| = \begin{vmatrix} 3 & 1 & 4 & 2 \\ 9 & 3 & 4 & 4 \\ 1 & 2 & 1 & 1 \\ 1 & 1 & 1 & 2 \end{vmatrix} \quad \begin{matrix} R_2 - 3R_1 \\ R_3 - \frac{1}{3}R_1 \\ R_4 - \frac{1}{3}R_1 \end{matrix} \quad \begin{vmatrix} 3 & 1 & 4 & 2 \\ 0 & 0 & -8 & -2 \\ 0 & \frac{5}{3} & -\frac{1}{3} & \frac{1}{3} \\ 0 & \frac{2}{3} & -\frac{1}{3} & \frac{4}{3} \end{vmatrix} \quad \begin{matrix} \text{Interchange} \\ R_2 \text{ and } R_3 \end{matrix} \quad \begin{vmatrix} 3 & 1 & 4 & 2 \\ 0 & \frac{5}{3} & -\frac{1}{3} & \frac{1}{3} \\ 0 & 0 & -8 & -2 \\ 0 & \frac{2}{3} & -\frac{1}{3} & \frac{4}{3} \end{vmatrix}$$

$$\begin{matrix} R_4 - \frac{2}{5}R_2 \end{matrix} \quad \begin{vmatrix} 3 & 1 & 4 & 2 \\ 0 & \frac{5}{3} & -\frac{1}{3} & \frac{1}{3} \\ 0 & 0 & -8 & -2 \\ 0 & 0 & -\frac{1}{5} & \frac{6}{5} \end{vmatrix} \quad \begin{matrix} R_4 - \frac{1}{40}R_3 \end{matrix} \quad \begin{vmatrix} 3 & 1 & 4 & 2 \\ 0 & \frac{5}{3} & -\frac{1}{3} & \frac{1}{3} \\ 0 & 0 & -8 & -2 \\ 0 & 0 & 0 & \frac{5}{4} \end{vmatrix}.$$

Thus

$$|A| = (-1)^r \, 3 \cdot \tfrac{5}{3} \cdot (-8) \cdot \tfrac{5}{4} = (-1)^{r+1} \, 50,$$

where r is the number of row interchanges involved. In this case $r = 1$, so we have $|A| = 50$. ■

Problems for Section 3.7

Evaluate the following determinants

1 (i) $\begin{vmatrix} 1 & 3 \\ 6 & 2 \end{vmatrix}$ (ii) $\begin{vmatrix} 2 & 4 \\ 4 & 8 \end{vmatrix}$ (iii) $\begin{vmatrix} 1 & b \\ a & ab \end{vmatrix}$ (iv) $\begin{vmatrix} 3 & 9 \\ 9 & 3 \end{vmatrix}$

(v) $\begin{vmatrix} 4 & 2 \\ 1 & 7 \end{vmatrix}$ (vi) $\begin{vmatrix} \cos x & \sin x \\ -\sin x & \cos x \end{vmatrix}$ (vii) $\begin{vmatrix} x^3 & x^2 \\ 1 & x \end{vmatrix}$.

2 (i) $\begin{vmatrix} 1 & 2 & -2 \\ 3 & 2 & 0 \\ 1 & 2 & -8 \end{vmatrix}$ (ii) $\begin{vmatrix} a & 0 & 0 \\ 0 & b & 0 \\ 0 & 0 & c \end{vmatrix}$.

3 (i) $\begin{vmatrix} 2 & -1 & 6 \\ 3 & 2 & 4 \\ 1 & -1 & 2 \end{vmatrix}$, (ii) $\begin{vmatrix} 1 & 3 & 6 \\ 0 & -1 & 4 \\ 0 & 0 & -3 \end{vmatrix}$.

4 (i) $\begin{vmatrix} 1 & 2 & 3 \\ 5 & -1 & 4 \\ 3 & -5 & -2 \end{vmatrix}$, (ii) $\begin{vmatrix} 2 & 0 & 0 \\ 3 & -7 & 0 \\ 1 & 3 & 2 \end{vmatrix}$.

5 What are the signs of the terms bfg, cdh, and bdk in the determinant

$$\begin{vmatrix} a & b & c \\ d & e & f \\ g & h & k \end{vmatrix} ?$$

Write down the cofactors of a, f and k in the above determinant and the minors of b and d.

6 Verify the Laplace expansion theorem for determinants (Theorem 3.8) by expanding the following determinant in six different ways in terms of the elements of each row and then the elements of each column

$$\begin{vmatrix} 1 & 3 & -2 \\ 2 & 0 & 1 \\ 4 & 2 & 6 \end{vmatrix}.$$

Use the fact that

$$\begin{vmatrix} 3 & 1 & -2 \\ 8 & -5 & 7 \\ 4 & 0 & 1 \end{vmatrix} = -35$$

to find without expansion the value of each of the following determinants.

7 $\begin{vmatrix} 3 & 8 & 4 \\ 1 & -5 & 0 \\ -2 & 7 & 1 \end{vmatrix}$ **8** $\begin{vmatrix} 3 & -2 & -2 \\ 8 & 7 & 10 \\ 4 & 1 & 0 \end{vmatrix}$

9 $\begin{vmatrix} 3 & 3 & -2 \\ 8 & -12 & 7 \\ 4 & -1 & 1 \end{vmatrix}$ **10** $\begin{vmatrix} 11 & 1 & -2 \\ -20 & -5 & 7 \\ 0 & 0 & 1 \end{vmatrix}$

11 $\begin{vmatrix} 4 & 1 & -2 \\ 3 & -5 & 7 \\ 4 & 0 & 1 \end{vmatrix}$ **12** $\begin{vmatrix} 6 & 2 & -4 \\ 8 & -5 & 7 \\ 4 & 0 & 1 \end{vmatrix}$ **13** $\begin{vmatrix} 3 & 3 & 10 \\ 8 & -15 & -35 \\ 4 & 0 & -5 \end{vmatrix}$

14 Prove that if $\mathbf{A}$ is the $n \times n$ lower-triangular matrix

$$\mathbf{A} = \begin{bmatrix} a_{11} & 0 & 0 & \cdots & 0 \\ a_{21} & a_{22} & 0 & \cdots & 0 \\ a_{31} & a_{32} & a_{33} & \cdots & 0 \\ \cdot & \cdot & & \cdots & 0 \\ a_{n1} & a_{n2} & a_{n3} & \cdots & a_{nn} \end{bmatrix},$$

then $\det \mathbf{A} = a_{11} a_{22} a_{33} \ldots a_{nn}$.

15 Prove that if $\mathbf{A}$ is the $n \times n$ upper-triangular matrix

$$\mathbf{A} = \begin{bmatrix} a_{11} & a_{12} & a_{13} & \cdots & a_{1n} \\ 0 & a_{22} & a_{23} & \cdots & a_{2n} \\ 0 & 0 & a_{33} & \cdots & a_{3n} \\ \cdot & \cdot & \cdot & \cdots & \cdot \\ 0 & 0 & 0 & \cdots & a_{nn} \end{bmatrix},$$

then $\det \mathbf{A} = a_{11} a_{22} a_{33} \ldots a_{nn}$.

In each of the following determinants establish the stated result without expansion.

16
$$\begin{vmatrix} 1 & a & b+c \\ 1 & b & c+a \\ 1 & c & a+b \end{vmatrix} = 0$$

17
$$\begin{vmatrix} 1 & a & a^3 \\ 1 & b & b^3 \\ 1 & c & c^3 \end{vmatrix} = (a+b+c)(b-c)(c-a)(a-b)$$

18
$$\begin{vmatrix} 1 & a & a^2 - bc \\ 1 & b & b^2 - ca \\ 1 & c & c^2 - ab \end{vmatrix} = 0$$

19
$$\begin{vmatrix} 1 & a & b \\ a & 1 & b \\ a & b & 1 \end{vmatrix} = (a+b+1)(a-1)(b-1)$$

20
$$\begin{vmatrix} b+c & c+a & a+b \\ q+r & r+q & p+q \\ y+z & z+x & x+y \end{vmatrix} = 2 \begin{vmatrix} a & b & c \\ p & q & r \\ x & y & z \end{vmatrix}$$

21
$$\begin{vmatrix} 1 & 1 & 1 & 1 \\ 1 & 2 & 2^2 & 2^3 \\ 1 & 3 & 3^2 & 3^3 \\ 1 & 4 & 4^2 & 4^3 \end{vmatrix} = 2!\,3!$$

22 Use the determinant in Prob. 6 to verify Theorem 3.17 (ii).

23 Evaluate in determinantal form

$$\frac{\mathrm{d}}{\mathrm{d}x} \begin{vmatrix} x^2 & \sin x & \cosh x \\ \cos x & \cosh x & \sinh x \\ 1 & e^x & e^{-x} \end{vmatrix}.$$

24 Evaluate

$$\frac{\mathrm{d}}{\mathrm{d}x} \begin{vmatrix} x & x^3 & 0 \\ 1 & x^2 & 0 \\ 0 & 0 & x \end{vmatrix}$$

25 Verify the expansion of a general determinant of order three using Eq. (6).

26 Prove that the alternant

$$\begin{vmatrix} 1 & 1 & 1 & \cdots & 1 & 1 \\ a_1 & a_2 & a_3 & \cdots & a_{n-1} & a_n \\ a_1^2 & a_2^2 & a_3^2 & \cdots & a_{n-1}^2 & a_n^2 \\ \cdot & \cdot & \cdot & \cdots & \cdot & \cdot \\ a_1^{n-1} & a_2^{n-1} & a_3^{n-1} & \cdots & a_{n-1}^{n-1} & a_n^{n-1} \end{vmatrix}$$

is the product of all terms of the form $(a_i - a_j)$, with $i \neq j$, and a factor $\varepsilon = \pm 1$ chosen such that the sign of the leading term $a_2 a_3^2 a_4^3 \ldots a_n^{n-1}$ in the product is $+1$.

27 Prove without expanding the nth order determinant that

$$\begin{vmatrix} k & 1 & 1 & \ldots & 1 \\ 1 & k & 1 & \ldots & 1 \\ 1 & 1 & k & \ldots & 1 \\ . & . & . & \ldots & 1 \\ 1 & 1 & 1 & \ldots & k \end{vmatrix} = (k+n-1)(k-1)^{n-1}$$

3.8 Determinants and rank. Cramer's rule

We now establish the connection between the rank of an $m \times n$ matrix $\mathbf{A}$, and the determinants of all the square submatrices $\mathbf{B}$ which may be formed from $\mathbf{A}$ by deleting rows and columns. First, though, let us recall that Theorems 3.3 and 3.4 showed;

(i) rank $\mathbf{A}$, namely the largest number of linearly independent row vectors belonging to $\mathbf{A}$, is also the largest number of linearly independent column vectors belonging to $\mathbf{A}$, and

(ii) row-equivalent matrices have the same rank, so that transforming a matrix by any sequence of elementary row operations will not alter its rank.

With these ideas in mind, let $\mathbf{B}$ be any $p \times p$ submatrix which may be formed from $\mathbf{A}$, and suppose rank $\mathbf{B} = r$, with $1 \leq r < p$. Then a Gaussian-type reduction of $\mathbf{B}$ will always show it is row-equivalent to a $p \times p$ upper-triangular matrix $\mathbf{U}$, with nonzero elements in the first r positions of its leading diagonal, and zeros elsewhere along the leading diagonal.

Now we have seen that elementary row operations will not alter the number of linearly independent rows in a determinant, though they may lead to its multiplication by a nonzero scalar (positive or negative). Thus, since the determinant of an upper-triangular matrix is equal to the product of the elements on its leading diagonal, we conclude that det $\mathbf{B} = 0$ if $1 < r < p$, but that det $\mathbf{B} \neq 0$ if $r = p$. Consequently, rank $\mathbf{B} = p$ if, and only if, det $\mathbf{B} \neq 0$, while rank $\mathbf{B} < p$ if det $\mathbf{B} = 0$.

This last result shows that if a general $m \times n$ matrix $\mathbf{A}$ has at least one non-vanishing determinant of order p, then rank $\mathbf{A} \geq p$. By considering all $p \times p$ matrices which may be formed from $\mathbf{A}$ we arrive at the following result.

Theorem 3.19 (Test for rank using determinants)

Let $\mathbf{A}$ be an $m \times n$ matrix. Then rank $\mathbf{A} = r$ if, and only if, the determinants of all the square submatrices of order $r+1$ which may be formed from $\mathbf{A}$ vanish, but there is at least one square submatrix of $\mathbf{A}$ of order r with a nonzero determinant. In particular, a square matrix $\mathbf{A}$ of order n will have rank n if, and only if, det $\mathbf{A} \neq 0$. ☐

Although this theorem is of theoretical importance, it does not provide a practical method for the determination of rank. Should this be required, it is most easily found by reducing the matrix in question to its echelon form as described in Ex. 1, Sec. 3.5. The real significance of Theorem 3.19 lies in its last statement, which gives a necessary and sufficient condition for a square matrix to have a multiplicative inverse.

Before discussing the inverse of a matrix, we first derive Cramer's rule[3] for the solution of a system of n linear equations in the n unknowns $x_1, x_2, \ldots x_n$ in terms of determinants.

Let the system be written

$$a_{11}x_1 + a_{12}x_2 + \ldots + a_{1j}x_j + \ldots + a_{kn}x_n = b_1$$
$$a_{21}x_1 + a_{22}x_2 + \ldots + a_{2j}x_j + \ldots + a_{2n}x_n = b_2$$

$$\begin{matrix} . & . & . & . \\ . & . & \cdots & . \end{matrix}$$

$$a_{n1}x_1 + a_{n2}x_2 + \ldots + a_{nj}x_j + \ldots + a_{nn}x_n = b_n, \tag{1}$$

which in terms of matrices becomes

$$\mathbf{Ax} = \mathbf{b}, \tag{2}$$

where

$$\mathbf{A} = \begin{bmatrix} a_{11} & \cdots & a_{1n} \\ . & \cdots & . \\ . & \cdots & . \\ . & \cdots & . \\ a_{n1} & \cdots & a_{nn} \end{bmatrix}, \quad \mathbf{x} = \begin{bmatrix} x_1 \\ . \\ . \\ . \\ x_n \end{bmatrix} \quad \text{and} \quad \mathbf{b} = \begin{bmatrix} b_1 \\ . \\ . \\ . \\ b_n \end{bmatrix}. \tag{3}$$

Now associate with each element a_{ij} of $\mathbf{A}$ its cofactor C_{ij}. Then for some fixed j form the sum of C_{1j} times the first equation in (1), C_{2j} times the second equation in (1), $\ldots$, and C_{nj} times the last equation in (1) to obtain

$$C_{1j}(a_{11}x_1 + a_{12}x_2 + \ldots a_{1j}x_j + \ldots + a_{1n}x_n)$$
$$+ C_{2j}(a_{21}x_1 + a_{22}x_2 + \ldots + a_{2j}x_j + \ldots + a_{2n}x_n)$$
$$+ \ldots + C_{ij}(a_{i1}x_1 + a_{i2}x_2 + \ldots + a_{ij}x_j + \ldots + a_{ij}x_n)$$
$$+ \ldots + C_{nj}(a_{n1}x_1 + a_{n2}x_2 + \ldots + a_{nj}x_j + \ldots + a_{nn}x_n)$$
$$= b_1 C_{1j} + b_2 C_{2j} + \ldots + b_i C_{ij} + \ldots + b_n C_{nj}. \tag{4}$$

Grouping the coefficients of $x_1, x_2, \ldots, x_n$ this becomes

$$x_1 \sum_{i=1}^{n} a_{i1} C_{ij} + x_2 \sum_{i=1}^{n} a_{i2} C_{ij} + \ldots + x_j \sum_{i=1}^{n} a_{ij} C_{ij}$$
$$+ \ldots + x_n \sum_{i=1}^{n} a_{in} C_{ij} = \sum_{i=1}^{n} b_i C_{ij}. \tag{5}$$

The coefficients of all but the unknown x_j in (5) are seen to be the sum of the products of the elements of one column of $\mathbf{A}$ and the cofactors of another column of $\mathbf{A}$. Thus, by Theorem 3.17(ii), these all vanish, while the coefficient of x_j in (5) is seen to be det $\mathbf{A}$.

[3] GABRIEL CRAMER (1704–1752), a Swiss mathematician who made contributions to algebra and geometry. The result now known as Cramer's rule was first formulated by Maclaurin around 1729 and published posthumously in his *Treatise on Algebra* (1748). Cramer's form of the result appeared in his book *Traité des courbes algébriques* (1750) which was to become a standard reference work. It was so well written and frequently quoted that after his death Cramer was, on occasions, credited with results not original with him.

Thus (5) reduces to the result

$$x_j \det \mathbf{A} = \sum_{i=1}^{n} b_i C_{ij}, \qquad (6)$$

which has a unique solution for x_j provided $\det \mathbf{A} \neq 0$. Since j was any one of the integers 1, 2, . . . , n result (6) is true for $j = 1, 2, . . . , n$, and so provides the required solution for the unknowns $x_1, x_2, . . . , x_n$ in (1).

The result may be simplified still further by noticing that the right-hand side of (6) is simply the expansion by elements of the jth column of $\det \mathbf{A}_j$, where matrix $\mathbf{A}_j$ is derived from $\mathbf{A}$ by replacing the elements of the jth column by the elements $b_1, b_2, . . . , b_m$. It is this form of the result which is attributed to Cramer, and which we have now proved.

Theorem 3.20 (Cramer's rule)

Let $D = \det \mathbf{A}$ be the determinant of the coefficients of the system of n linear equations

$$a_{11}x_1 + a_{12}x_2 + \ldots + a_{1n}x_n = b_1$$
$$a_{21}x_1 + a_{22}x_2 + \ldots + a_{2n}x_n = b_2$$

$$\qquad \cdot \qquad \cdot \qquad \cdots \qquad \cdot \qquad \cdot$$

$$a_{n1}x_1 + a_{n2}x_2 + \ldots + a_{nn}x_n = b_n$$

in the n unknowns $x_1, x_2, . . . , x_n$. Then, provided $D \neq 0$, the system has the unique solution

$$x_1 = \frac{D_1}{D}, \; x_2 = \frac{D_2}{D}, \; . . . , \; x_n = \frac{D_n}{D},$$

where $D_i = \det \mathbf{A}_i$, with $\mathbf{A}_i$ the matrix derived from $\mathbf{A}$ by replacing the elements of the ith column by the elements $b_1, b_2, . . . , b_n$.

If the system is homogeneous and $D \neq 0$ it has only the trivial solution $x_1 = x_2 = \ldots = x_n = 0$. If the system is homogeneous and $D = 0$ it also has nontrivial solutions.

□

It should be emphasized that, due to the magnitude of the task of evaluating large order determinants, this method is usually only useful for obtaining numerical solutions to systems when n is small (say $n \leq 4$). Nevertheless, despite this limitation, it is still of considerable theoretical importance, for we shall see that it provides an explicit representation for the inverse of an $n \times n$ matrix $\mathbf{A}$ for arbitrary n.

Example 1. Cramer's rule

Solve by Cramer's rule

$$x_1 + x_2 + x_3 = 0$$
$$2x_1 - 5x_2 - 3x_3 = 10$$
$$4x_1 + 8x_2 + 2x_3 = 4.$$

Solution

$$D = \begin{vmatrix} 1 & 1 & 1 \\ 2 & -5 & -3 \\ 4 & 8 & 2 \end{vmatrix} = 34, \qquad D_1 = \begin{vmatrix} 0 & 1 & 1 \\ 10 & -5 & -3 \\ 4 & 8 & 2 \end{vmatrix} = 68,$$

$$D_2 = \begin{vmatrix} 1 & 0 & 1 \\ 2 & 10 & -3 \\ 4 & 4 & 2 \end{vmatrix} = 0, \qquad D_3 = \begin{vmatrix} 1 & 1 & 0 \\ 2 & -5 & 10 \\ 4 & 8 & 4 \end{vmatrix} = -68,$$

so $x_1 = 2$, $x_2 = 0$ and $x_3 = -2$. ∎

Problems for Section 3.8

Where appropriate, use Cramer's rule to solve the following systems of equations. In each case compare the task of using this method of solution with that of obtaining the solution by Gaussian elimination.

1
$$2x_1 - x_2 = -1$$
$$x_1 + 4x_2 = 13$$

2
$$2x_1 + 3x_2 = 4$$
$$3x_1 - 2x_2 = -7$$

3
$$3x_1 + 4x_2 = 0$$
$$x_1 - 3x_2 = 0$$

4
$$-7x_1 + 4x_2 = 2$$
$$14x_1 - 8x_2 = -4$$

5
$$x_1 + x_2 + 2x_3 = 4$$
$$x_1 - x_2 + x_3 = 4$$
$$2x_1 + x_2 - x_3 = -1$$

6
$$x_1 + 2x_2 - x_3 = -1$$
$$2x_1 + x_2 + 2x_3 = 0$$
$$x_1 + 3x_3 + 3x_3 = -7$$

7
$$x_1 + 3x_2 + x_3 = 2$$
$$x_2 - 2x_3 = -3$$
$$2x_1 + 5x_2 + 4x_3 = 7$$

8
$$x_1 + 3x_2 + x_3 - 2x_4 = 0$$
$$2x_1 - 2x_3 - 4x_4 = 0$$
$$x_1 + x_2 + x_4 = 0$$
$$2x_1 + 5x_2 + 3x_3 + 6x_4 = 0$$

9
$$x_1 + x_2 + x_3 + x_4 = 1$$
$$x_1 + 2x_2 + x_3 - x_4 = 2$$
$$x_1 + 3x_2 - 2x_3 + 2x_4 = -8$$
$$2x_1 + x_2 + 3x_3 - 3x_4 = 10$$

10
$$x_1 - x_2 - x_3 - x_4 = 4$$
$$2x_1 + x_2 + x_3 + x_4 = -1$$
$$x_1 + 3x_2 + 2x_3 + 2x_4 = -5$$
$$x_1 - x_2 + x_3 - x_4 = 2$$

3.9 Inverse matrices

An $n \times n$ matrix $A = [a_{ij}]$ is said to have an **inverse** or, more precisely, a **multiplicative inverse**, denoted by A^{-1}, if

$$AA^{-1} = A^{-1}A = I, \tag{1}$$

where I is the $n \times n$ unit matrix. The matrix A is said to be **nonsingular** when such an inverse exists, and to be **singular** when it has no inverse.

To see that when an inverse exists it is unique, let us suppose the $n \times n$ matrices P and Q are two *different* inverses of A, then

$$AP = I \quad \text{and} \quad QA = I.$$

Consequently, from the first expression,

$$Q(AP) = QI, \quad \text{or} \quad (QA)P = Q.$$

However, $QA = I$, showing that $P = Q$, so that when it exists the inverse is *unique*. This result also implies that if A and B are $n \times n$ matrices, and $AB = I$ or $BA = I$, then B is the inverse of A, and conversely.

An immediate use for the inverse A^{-1} of A is provided by the system of n linear equations in the n unknowns $x_1, x_2, \ldots, x_n$ when written in the matrix form

$$Ax = b, \tag{2}$$

where A, x and b are as defined in Eq. (2), Sec. 3.8. If A is nonsingular, we may pre-multiply (2) by A^{-1} to obtain

$$A^{-1}Ax = A^{-1}b,$$

and hence the solution

$$x = A^{-1}b. \tag{3}$$

Similarly, consider a **linear transformation**

$$y = Ax \tag{4}$$

relating the vectors x and y each with the n elements $x_1, x_2, \ldots, x_n$ and $y_1, y_2, \ldots y_n$, respectively, with an $n \times n$ matrix A. Then, if A is nonsingular, x is related to y by the inverse transformation

$$x = A^{-1}y. \tag{5}$$

To find an explicit representation for A^{-1} when A is nonsingular we use result (6) of Sec. 3.8 and represent x in the form

$$x = \frac{1}{\det A} \begin{bmatrix} \sum_{i=1}^{n} b_i C_{i1} \\ \sum_{i=1}^{n} b_i C_{i2} \\ \vdots \\ \sum_{i=1}^{n} b_i C_{in} \end{bmatrix},$$

which is equivalent to

$$x = \frac{1}{\det A} \begin{bmatrix} C_{11} & C_{21} & \cdots & C_{n1} \\ C_{12} & C_{22} & \cdots & C_{n2} \\ \cdot & \cdot & \cdots & \cdot \\ C_{1n} & C_{2n} & \cdots & C_{nn} \end{bmatrix} \begin{bmatrix} b_1 \\ b_2 \\ \cdot \\ b_n \end{bmatrix}. \tag{6}$$

Identifying (3) and (6), and using the fact that when it exists the inverse is unique, shows that if $\det A \neq 0$,

$$A^{-1} = \frac{1}{\det A} \begin{bmatrix} C_{11} & C_{21} & \cdots & C_{n1} \\ C_{12} & C_{22} & \cdots & C_{n2} \\ . & . & \cdots & . \\ C_{1n} & C_{2n} & \cdots & C_{nn} \end{bmatrix} = \frac{C}{\det A}, \tag{7}$$

where $C = [C_{ij}]$ is the transpose of the matrix of cofactors of A. Matrix C is called the **adjoint** of A and written adj A. This has proved the next result which is of considerable theoretical importance.

Theorem 3.21 (Inverse of a matrix)

The inverse of a nonsingular $n \times n$ matrix $A = [a_{ij}]$ is given by

$$A^{-1} = \frac{\text{adj } A}{\det A},$$

where adj A is the transpose of the matrix of cofactors of A.

Theorem 3.19 may be used to interpret the existence of an inverse matrix in terms of rank. It is equivalent to the statement that the $n \times n$ matrix A is nonsingular if rank $A = n$, and it is singular if rank $A < n$.

The following simple result, which is often useful, is an immediate consequence of Theorem 3.21. Let A be the diagonal matrix

$$A = \begin{bmatrix} a_{11} & 0 & 0 & \cdots & 0 \\ 0 & a_{22} & 0 & \cdots & 0 \\ 0 & 0 & a_{33} & \cdots & 0 \\ . & . & . & \cdots & . \\ 0 & 0 & 0 & \cdots & a_{nn} \end{bmatrix}.$$

Then, if each entry on the leading diagonal of A is nonzero,

$$A^{-1} = \begin{bmatrix} 1/a_{11} & 0 & 0 & \cdots & 0 \\ 0 & 1/a_{22} & 0 & \cdots & 0 \\ 0 & 0 & 1/a_{33} & \cdots & 0 \\ . & . & . & \cdots & . \\ 0 & 0 & 0 & \cdots & 1/a_{nn} \end{bmatrix}. \tag{8}$$

As with Cramer's rule, because of the magnitude of the task of evaluating determinants, Theorem 3.21 only provides a practical method for the determination of an inverse matrix when n is small (say $n \leq 4$).

Example 1. Inverse of a matrix

Find A^{-1} given

$$A = \begin{bmatrix} 1 & 2 & 1 \\ 2 & 1 & 0 \\ 0 & 1 & 3 \end{bmatrix}.$$

Solution

If **C** is the matrix of cofactors of **A**, a simple calculation establishes that

$$\mathbf{C} = \begin{bmatrix} 3 & -6 & 2 \\ -5 & 3 & -1 \\ -1 & 2 & -3 \end{bmatrix}.$$

Expanding det **A** in terms of the elements of any row or column, using the corresponding cofactors from **C**, shows det $\mathbf{A} = -7$. Thus

$$\text{adj } \mathbf{A} = \mathbf{C}^\mathsf{T} = \begin{bmatrix} 3 & -5 & -1 \\ -6 & 3 & 2 \\ 2 & -1 & -3 \end{bmatrix},$$

and so

$$\mathbf{A}^{-1} = \frac{\text{adj } \mathbf{A}}{\det \mathbf{A}} = \begin{bmatrix} -3/7 & 5/7 & 1/7 \\ 6/7 & -3/7 & -2/7 \\ -2/7 & 1/7 & 3/7 \end{bmatrix}.$$

Theorem 3.22 (Properties of the inverse matrix)

Let **A** and **B** be any two nonsingular $n \times n$ matrices and m be any integer. Then,

(i) $(\mathbf{A}^{-1})^{-1} = \mathbf{A}$,

(ii) $(\mathbf{AB})^{-1} = \mathbf{B}^{-1}\mathbf{A}^{-1}$,

(iii) $(\mathbf{A}^{-1})^m = (\mathbf{A}^m)^{-1}$,

(iv) $(\mathbf{A}^{-1})^\mathsf{T} = (\mathbf{A}^\mathsf{T})^{-1}$,

(v) If **A** is nonsingular and its elements are functions of x, then

$$\frac{d\mathbf{A}^{-1}}{dx} = -\mathbf{A}^{-1}\frac{d\mathbf{A}}{dx}\mathbf{A}^{-1},$$

(vi) $\det \mathbf{A}^{-1} = 1/(\det \mathbf{A})$.

Proof

(i) When **A** is nonsingular $\mathbf{A}^{-1}$ exists, and so $\mathbf{A}^{-1}$ has as its inverse the matrix **A**. In terms of our notation we thus have

$$(\mathbf{A}^{-1})^{-1} = \mathbf{A}.$$

(ii) By definition, if **AB** is nonsingular, then

$$(\mathbf{AB})(\mathbf{AB})^{-1} = \mathbf{I}.$$

Pre-multiplication by $\mathbf{A}^{-1}$ gives

$$\mathbf{A}^{-1}\mathbf{A}\,\mathbf{B}(\mathbf{AB})^{-1} = \mathbf{A}^{-1}.$$

but since $\mathbf{A}^{-1}\mathbf{A} = \mathbf{I}$ this becomes

$$\mathbf{B}(\mathbf{AB})^{-1} = \mathbf{A}^{-1}.$$

Pre-multiplication by $\mathbf{B}^{-1}$ together with the result $\mathbf{B}^{-1}\mathbf{B}=\mathbf{I}$ then gives the required result

$$(\mathbf{AB})^{-1}=\mathbf{B}^{-1}\mathbf{A}^{-1}.$$

(iii) Let $\mathbf{A}_1, \mathbf{A}_2, \ldots, \mathbf{A}_m$ be any m nonsingular $n \times n$ matrices. Then, by repeated application of (ii), it follows that

$$(\mathbf{A}_1 \mathbf{A}_2 \ldots \mathbf{A}_{m-1} \mathbf{A}_m)^{-1} = \mathbf{A}_m^{-1} \mathbf{A}_{m-1}^{-1} \ldots \mathbf{A}_2^{-1} \mathbf{A}_1^{-1},$$

so setting $\mathbf{A}_1 = \mathbf{A}_2 = \ldots = \mathbf{A}_m = \mathbf{A}$ we obtain

$$(\mathbf{A}^m)^{-1} = (\mathbf{A}^{-1})^m.$$

(iv) This result follows immediately from Theorem 3.2 by setting $\mathbf{B} = \mathbf{A}^{-1}$. For we then have

$$(\mathbf{A}\mathbf{A}^{-1})^{\mathsf{T}} = (\mathbf{A}^{-1})^{\mathsf{T}} \mathbf{A}^{\mathsf{T}},$$

but since $\mathbf{A}\mathbf{A}^{-1} = \mathbf{I}$ and $\mathbf{I}^{\mathsf{T}} = \mathbf{I}$ this is equivalent to

$$(\mathbf{A}^{\mathsf{T}})^{-1} = (\mathbf{A}^{-1})^{\mathsf{T}}.$$

(v) As $\mathbf{A}$ is nonsingular $\mathbf{A}\mathbf{A}^{-1} = \mathbf{I}$, so that

$$\frac{d}{dx}(\mathbf{A}\mathbf{A}^{-1}) = \frac{d\mathbf{I}}{dx} = \mathbf{0}.$$

Thus

$$\frac{d\mathbf{A}}{dx}\mathbf{A}^{-1} + \mathbf{A}\frac{d\mathbf{A}^{-1}}{dx} = 0,$$

from which it follows by pre-multiplication by $\mathbf{A}^{-1}$ that

$$\frac{d\mathbf{A}^{-1}}{dx} = -\mathbf{A}^{-1}\frac{d\mathbf{A}}{dx}\mathbf{A}^{-1}.$$

(vi) The result follows at once from $\mathbf{A}\mathbf{A}^{-1} = \mathbf{I}$ and Theorem 3.16. ☐

The elementary division law of algebra states that if a and b are scalars such that $ab = 0$, then either $a = 0$ or $b = 0$. Rather surprisingly this is not true in matrix algebra (see the product $\mathbf{FG}$ in Ex. 3, Sec. 3.3) where the corresponding result is as follows.

Theorem 3.23 (Matrix analog of the division law of algebra)

If $\mathbf{A}$ and $\mathbf{B}$ are $n \times n$ matrices such that $\mathbf{AB} = \mathbf{0}$, then $\mathbf{A} = \mathbf{0}$, or $\mathbf{B} = \mathbf{0}$, or $\mathbf{A}$ and $\mathbf{B}$ are both singular.

Proof

The result is obvious if $\mathbf{A} = \mathbf{0}$, or $\mathbf{B} = \mathbf{0}$. Consider the remaining case $\mathbf{A} \neq \mathbf{0}$ and $\mathbf{B} \neq \mathbf{0}$ and suppose, if possible, that $\det \mathbf{A} \neq 0$. Then $\mathbf{A}^{-1}$ exists, and since $\mathbf{AB} = \mathbf{0}$, we have

$A^{-1}AB = 0$, showing that $B = 0$, which is a contradiction. A similar argument applies if $\det B \neq 0$, and so the result is proved. $\qquad\qquad\qquad\qquad\qquad\qquad\qquad\qquad\qquad\square$

The elementary row operations which reduce a matrix to its echelon form as described in Sec. 3.5 provide a useful alternative to Theorem 3.21 for finding the inverse of a matrix. We take as our starting point the observation that pre-multiplication of an $n \times n$ matrix A by an $n \times n$ matrix T is equivalent to performing elementary row operations on A.

For example, if A is a 2×2 matrix and

$$T = \begin{bmatrix} 3 & 0 \\ 1 & -7 \end{bmatrix},$$

then TA is the matrix obtained from A by replacing row 1 by $3 \times$ row 1 and row 2 by row $1 - 7$ row 2.

Now let the matrices $T_1, T_2, \ldots, T_m$ be any sequence of $n \times n$ matrices whose combined effect as successive pre-multipliers of A, in this order, will be to bring A to its reduced echelon form. Then it follows from Theorem 3.19 that when $\det A \neq 0$ this reduced echelon form will be the $n \times n$ matrix I, so in that case

$$T_m T_{m-1} \cdots T_2 T_1 A = I. \tag{9}$$

Post-multiplication by A^{-1} then gives the result

$$T_m T_{m-1} \cdots T_2 T_1 = A^{-1},$$

which may be rewritten as

$$A^{-1} = T_m T_{m-1} \cdots T_2 T_1 I. \tag{10}$$

This shows that the sequence of row operations performed on a nonsingular matrix A to reduce it to the unit matrix I will simultaneously transform the unit matrix I into A^{-1}. The nonexistence of an inverse will correspond to the reduction of A to a matrix whose rank is less than n.

When using this method it is best to replace the matrices $T_1, T_2, \ldots$ by their equivalent row operations, and to transform A and I simultaneously, step by step, as shown in the following examples.

Example 2. Inverse of a matrix by row transformations

If possible, find A^{-1}, given that

$$A = \begin{bmatrix} 1 & 2 & 1 \\ 2 & 1 & 0 \\ 1 & 2 & 4 \end{bmatrix}.$$

Solution

We shall use the notation that R_i denotes ith row and R_i^* the new ith row. Thus $R_2^* = R_1 + 3R_2$ signifies that the new second row is obtained as the sum of the old first row and three times the old second row.

$$A = \begin{bmatrix} 1 & 2 & 1 \\ 2 & 1 & 0 \\ 1 & 2 & 4 \end{bmatrix} \qquad I = \begin{bmatrix} 1 & 0 & 0 \\ 0 & 1 & 0 \\ 0 & 0 & 1 \end{bmatrix}$$

$$\begin{array}{c} R_1^* = R_1 \\ R_2^* = R_2 - 2R_1 \\ R_3^* = R_3 - R_1 \end{array} \begin{bmatrix} 1 & 2 & 1 \\ 0 & -3 & -2 \\ 0 & 0 & 3 \end{bmatrix} \qquad \begin{bmatrix} 1 & 0 & 0 \\ -2 & 1 & 0 \\ -1 & 0 & 1 \end{bmatrix}$$

$$\begin{array}{c} R_1^* = R_1 + \frac{2}{3}R_2 \\ R_2^* = R_2 \\ R_3^* = \frac{1}{3}R_3 \end{array} \begin{bmatrix} 1 & 0 & -\frac{1}{3} \\ 0 & -3 & -2 \\ 0 & 0 & 1 \end{bmatrix} \qquad \begin{bmatrix} -\frac{1}{3} & \frac{2}{3} & 0 \\ 2 & 1 & 0 \\ -\frac{1}{3} & 0 & \frac{1}{3} \end{bmatrix}$$

$$\begin{array}{c} R_1^* = R_1 + \frac{1}{3}R_3 \\ R_2^* = R_2 + 2R_3 \\ R_3^* = R_3 \end{array} \begin{bmatrix} 1 & 0 & 0 \\ 0 & -3 & 0 \\ 0 & 0 & 1 \end{bmatrix} \qquad \begin{bmatrix} -\frac{4}{9} & \frac{2}{3} & \frac{1}{9} \\ -\frac{8}{3} & 1 & \frac{2}{3} \\ -\frac{1}{3} & 0 & \frac{1}{3} \end{bmatrix}$$

$$\begin{array}{c} R_1^* = R_1 \\ R_2^* = -\frac{1}{3}R_2 \\ R_3^* = R_3 \end{array} \begin{bmatrix} 1 & 0 & 0 \\ 0 & 1 & 0 \\ 0 & 0 & 1 \end{bmatrix} \qquad \begin{bmatrix} -\frac{4}{9} & \frac{2}{3} & \frac{1}{9} \\ \frac{8}{9} & -\frac{1}{3} & -\frac{2}{9} \\ -\frac{1}{3} & 0 & \frac{1}{3} \end{bmatrix}.$$

Since **A** has been reduced to **I** it is nonsingular, and so has an inverse. This inverse is given by

$$A^{-1} = \begin{bmatrix} -\frac{4}{9} & \frac{2}{3} & \frac{1}{9} \\ \frac{8}{9} & -\frac{1}{3} & -\frac{2}{9} \\ -\frac{1}{3} & 0 & \frac{1}{3} \end{bmatrix}.$$

∎

Example 3. Row transformations with a singular matrix

If possible, find A^{-1} given that

$$A = \begin{bmatrix} 2 & 1 & 3 \\ 4 & -3 & -1 \\ 1 & -2 & -2 \end{bmatrix}.$$

Solution

$$A = \begin{bmatrix} 2 & 1 & 3 \\ 4 & -3 & -1 \\ 1 & -2 & -2 \end{bmatrix} \qquad I = \begin{bmatrix} 1 & 0 & 0 \\ 0 & 1 & 0 \\ 0 & 0 & 1 \end{bmatrix}$$

$$\begin{array}{c} R_1^* = R_1 \\ R_2^* = R_2 - 2R_1 \\ R_3^* = R_3 - \frac{1}{2}R_1 \end{array} \begin{bmatrix} 2 & 1 & 3 \\ 0 & -5 & -7 \\ 0 & -\frac{5}{2} & -\frac{7}{2} \end{bmatrix} \qquad \begin{bmatrix} 1 & 0 & 0 \\ -2 & 1 & 0 \\ -\frac{1}{2} & 0 & 1 \end{bmatrix}$$

$$\begin{array}{c} R_1^* = R_1 \\ R_2^* = R_2 - 2R_3 \\ R_3^* = R_3 \end{array} \begin{bmatrix} 2 & 1 & 3 \\ 0 & 0 & 0 \\ 0 & -\frac{5}{2} & -\frac{7}{2} \end{bmatrix} \qquad \begin{bmatrix} 1 & 0 & 0 \\ -1 & 1 & -2 \\ -\frac{1}{2} & 0 & 1 \end{bmatrix}.$$

The last matrix on the left-hand side has a zero row, so its rank is less than 3 and it cannot be reduced to the 3×3 unit matrix $\mathbf{I}$. Thus matrix $\mathbf{A}$ is singular, and so has *no* inverse.

∎

This **Gauss–Jordan** type method for the determination of an inverse matrix is well suited to digital computation. Most computer subroutine libraries contain a program based on this approach. They will either find an inverse matrix, or determine that the matrix is singular to within some predetermined error.

Example 4. Determinants involving matrix algebra

Given that $\mathbf{A}$ and $\mathbf{B}$ are $n \times n$ matrices such that $\det \mathbf{A} = 2$ and $\det \mathbf{B} = 5$, find (i) $\det(\mathbf{A}^3)$ (ii) $\det(3\mathbf{A})^{-1}$ (iii) $\det(3\mathbf{A}^{-1})$ (iv) $\det(\mathbf{A}^\mathsf{T})^{-1}$ (v) $\det(\mathbf{B}^{-3}\mathbf{A}^\mathsf{T})$ (vi) $\det(\text{adj } \mathbf{A} \text{ adj } \mathbf{B})$.

Solution

(i) As $\mathbf{A}^3 = \mathbf{A}^2\mathbf{A}$, it follows from Theorem 31.6 that
$$\det(\mathbf{A}^3) = \det(\mathbf{A}^2)\det(\mathbf{A}).$$
A further application of the theorem gives
$$\det(\mathbf{A}^3) = (\det \mathbf{A})^3 = 2^3.$$

(ii) It follows from the definition of the matrix $3\mathbf{A}$ and Theorem 3.10 that
$$\det(3\mathbf{A}) = 3^n \det \mathbf{A} = 3^n \cdot 2.$$
Now
$$(3\mathbf{A})(3\mathbf{A})^{-1} = \mathbf{I},$$
so applying Theorem 3.16 we obtain
$$\det(3\mathbf{A}) \cdot \det(3\mathbf{A})^{-1} = 1,$$
and consequently
$$\det(3\mathbf{A})^{-1} = \frac{1}{\det(3\mathbf{A})} = \frac{1}{3^n \cdot 2}.$$

(iii) We start from the fact that
$$3\mathbf{A}^{-1}\mathbf{A} = 3\mathbf{I}.$$
An application of Theorem 3.16 yields
$$\det(3\mathbf{A}^{-1}) \cdot \det \mathbf{A} = 3^n,$$
and thus
$$\det(3\mathbf{A}^{-1}) = \frac{3^n}{\det \mathbf{A}} = \frac{3^n}{2}.$$

(iv) We begin with the fact that
$$\mathbf{A}^\mathsf{T}(\mathbf{A}^\mathsf{T})^{-1} = \mathbf{I},$$
from which it follows after applying Theorem 3.16 that
$$\det(\mathbf{A}^\mathsf{T}) \cdot \det(\mathbf{A}^\mathsf{T})^{-1} = 1,$$

and so

$$\det (\mathbf{A}^\mathsf{T})^{-1} = \frac{1}{\det (\mathbf{A}^\mathsf{T})}.$$

Note $\det (\mathbf{A}^\mathsf{T}) = \det \mathbf{A}$ (Theorem 3.9), so we see that

$$\det (\mathbf{A}^\mathsf{T})^{-1} = \frac{1}{\det \mathbf{A}} = \frac{1}{2}.$$

(v) Applying Theorem 3.16, and then Theorem 3.9, gives

$$\det (\mathbf{B}^{-3} \mathbf{A}^\mathsf{T}) = \det (\mathbf{B}^{-3}) \det (\mathbf{A}^\mathsf{T})$$

$$= \frac{\det (\mathbf{A}^\mathsf{T})}{(\det \mathbf{B})^3} = \frac{\det \mathbf{A}}{(\det \mathbf{B})^3} = \frac{2}{5^3}.$$

(vi) We begin from the result

$$\mathbf{B} \mathbf{A} \mathbf{A}^{-1} \mathbf{B}^{-1} = \mathbf{I},$$

which because of the definitions of $\mathbf{A}^{-1}$ and $\mathbf{B}^{-1}$ may be written as

$$\mathbf{B} \mathbf{A} \operatorname{adj} \mathbf{A} \operatorname{adj} \mathbf{B} = (\det \mathbf{A})(\det \mathbf{B}) \, \mathbf{I}.$$

Applying Theorem 3.16 leads to the result

$$(\det \mathbf{B}) (\det \mathbf{A}) \det (\operatorname{adj} \mathbf{A} \operatorname{adj} \mathbf{B}) = (\det \mathbf{A})^n (\det \mathbf{B})^n,$$

and hence to

$$\det (\operatorname{adj} \mathbf{A} \operatorname{adj} \mathbf{B}) = (\det \mathbf{A})^{n-1} (\det \mathbf{B})^{n-1} = 2^{n-1} \cdot 5^{n-1} = 10^{n-1}. \qquad \blacksquare$$

Problems for Section 3.9

1 If

$$\mathbf{A} = \begin{bmatrix} a_{11} & a_{12} \\ a_{21} & a_{22} \end{bmatrix} \quad \text{find } \mathbf{A}^{-1} \text{ and}$$

verify that $\mathbf{A} \mathbf{A}^{-1} = \mathbf{A}^{-1} \mathbf{A} = \mathbf{I}$.

2 Prove (8).

Find the inverse when it exists and verify the result by multiplication.

3 $\begin{bmatrix} 1 & 3 \\ 2 & -4 \end{bmatrix}$ **4** $\begin{bmatrix} 2 & 4 \\ 1 & 2 \end{bmatrix}$ **5** $\begin{bmatrix} 2 & 4 \\ 1 & 3 \end{bmatrix}$

6 $\begin{bmatrix} \cos \theta & \sin \theta \\ \sin \theta & -\cos \theta \end{bmatrix}$ **7** $\begin{bmatrix} \cos \theta & 0 & -\sin \theta \\ 0 & 1 & 0 \\ \sin \theta & 0 & \cos \theta \end{bmatrix}$

8 $\begin{bmatrix} 3.2 & 0 & 0 \\ 0 & 0.4 & 0 \\ 0 & 0 & 1.6 \end{bmatrix}$ **9** $\begin{bmatrix} 2 & 2 & 0 \\ 0 & 3 & 1 \\ 1 & 0 & 1 \end{bmatrix}$

10 $\begin{bmatrix} 2 & 1 & -1 \\ -1 & 2 & 0 \\ 0 & 1 & 3 \end{bmatrix}$ **11** $\begin{bmatrix} 1 & -4 & 2 \\ 5 & -6 & 11 \\ 3 & 2 & 7 \end{bmatrix}$

Solve the following simultaneous equations by means of a suitable inverse matrix.

12 $\begin{aligned} x_1 + x_2 - x_3 &= 6 \\ 2x_1 - x_2 + 3x_3 &= -9 \\ x_1 - 2x_2 - 2x_3 &= 3 \end{aligned}$ **13** $\begin{aligned} x_1 - 2x_2 + x_3 &= -5 \\ -2x_1 + 3x_2 + 2x_3 &= 1 \\ -x_1 + 4x_2 + 3x_3 &= -1 \end{aligned}$

14 $\begin{aligned} 2x_1 \phantom{{}+3x_2} + x_3 &= 0 \\ x_1 + 3x_2 - 4x_3 &= -15 \\ x_1 + x_2 + 2x_3 &= 7 \end{aligned}$

Find the inverse of the given linear transformation.

15 $\begin{aligned} y_1 &= x_1 + x_2 - 3x_3 \\ y_2 &= -x_1 \phantom{{}+x_2} + 2x_3 \\ y_3 &= -3x_1 + 5x_2 \end{aligned}$ **16** $\begin{aligned} y_1 &= 9x_1 - 4x_2 + x_3 \\ y &= x_1 + 8x_2 - 2x_3 \\ y_3 &= -2x_1 + 3x_2 + 4x_3 \end{aligned}$

17 Given

$$A = \begin{bmatrix} 1 & 2 \\ 4 & 1 \end{bmatrix}, \qquad B = \begin{bmatrix} 2 & 4 \\ 1 & 3 \end{bmatrix} \quad \text{and} \quad m = 2,$$

verify Theorem 3.22 parts (ii), (iii) and (iv).

18 Given

$$A = \begin{bmatrix} \cos x & \sin x \\ \sin x & -\cos x \end{bmatrix}, \text{ verify Theorem 3.22(v).}$$

19 Prove that if the $n \times n$ matrix A is symmetric, then adj A is symmetric. Hence show that if it is also true that det $A \neq 0$, then A^{-1} is symmetric.

20 Prlve that if A is an $n \times n$ matrix, then

    (i) A adj $A = (\text{adj } A)A = (\det A)I$,
    (ii) $\det(\text{adj } A) = (\det A)^{n-1}$.

21 Verify results (i) and (ii) of Prob. 20 using

$$A = \begin{bmatrix} -3 & 0 & 2 \\ 0 & 1 & 1 \\ -2 & 2 & 2 \end{bmatrix}.$$

22 Use the result of Prob. 20(i) with A replaced by adj A, and then the result of Prob. 20(ii), to prove that

$$\text{adj(adj } A) = (\det A)^{n-2} A.$$

23 Verify the result of Prob. 22 when

    (i) $A = \begin{bmatrix} 3 & 0 & -1 \\ 2 & 1 & 0 \\ 4 & 2 & 1 \end{bmatrix}$,   (ii) $A = \begin{bmatrix} 1 & 0 & 2 \\ 3 & 1 & 4 \\ 0 & 1 & -2 \end{bmatrix}$.

Where possible, use row transformations to find the inverse A^{-1} of the given matrix A.

24 $A = \begin{bmatrix} 1 & 0 & 0 \\ 0 & \dfrac{1}{\sqrt{2}} & -\dfrac{1}{\sqrt{2}} \\ 0 & \dfrac{1}{\sqrt{2}} & \dfrac{1}{\sqrt{2}} \end{bmatrix}$
25 $A = \begin{bmatrix} 1 & 2 & 3 \\ 4 & 5 & 6 \\ 2 & -1 & 3 \end{bmatrix}$

26 $A = \begin{bmatrix} 2 & 1 & 0 \\ 3 & -2 & 1 \\ 4 & 0 & -1 \end{bmatrix}$
27 $A = \begin{bmatrix} 2 & -1 & 6 \\ 1 & 10 & -9 \\ -1 & 4 & -7 \end{bmatrix}$

28 $A = \begin{bmatrix} 1 & 2 & -1 & 0 \\ 2 & 1 & 1 & 2 \\ 0 & 3 & 1 & 0 \\ 0 & 3 & 5 & 2 \end{bmatrix}$
29 $A = \begin{bmatrix} 1 & 0 & 0 & -\frac{2}{3} \\ 0 & 0 & 0 & \frac{1}{3} \\ 1 & 0 & 1 & -\frac{2}{3} \\ -\frac{1}{2} & \frac{1}{2} & -\frac{1}{2} & \frac{1}{3} \end{bmatrix}$

30 Find a vector x such that

$$\begin{bmatrix} 1 & 4 & -1 \\ 2 & 1 & 3 \\ 1 & -8 & 8 \end{bmatrix} x = \begin{bmatrix} -6.5 \\ 10 \\ 34 \end{bmatrix}$$

31 Find a matrix B such that

$$\begin{bmatrix} 3 & 2 & 1 \\ -1 & 1 & -3 \\ 2 & 1 & 0 \end{bmatrix} B = \begin{bmatrix} 6 & 0 & 12 \\ 0 & 3 & 0 \\ 12 & 0 & 60 \end{bmatrix}$$

32 A **four-terminal network** is a general linear electrical circuit element found in transmission lines. It may contain resistance, capacitance, self-inductance and mutual-inductance, but no internal current source. Any such network has one pair of input terminals and one pair of output terminals across each of which voltage may be applied. A network is represented in the manner shown in Fig. 3.18, in which v_1 and i_1 are the input voltage and current, respectively, and v_2 and i_2 are the corresponding output quantities. The arrows indicate the sign conventions for voltage and current.

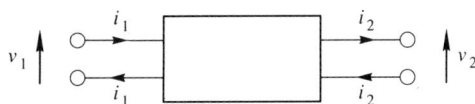

Fig. 3.18 Four-terminal network

The input and output voltages may be described in terms of the currents by the equations

$$v_1 = z_{11} i_1 - z_{12} i_2$$
$$v_2 = z_{12} i_1 - z_{22} i_2, \tag{A}$$

in which the z_{ij} are constants describing the particular four-terminal network involved.
   If

$$\mathbf{v} = \begin{bmatrix} v_1 \\ v_2 \end{bmatrix}, \quad \mathbf{i} = \begin{bmatrix} i_1 \\ i_2 \end{bmatrix} \quad \text{and} \quad \mathbf{Z} = \begin{bmatrix} z_{11} & -z_{12} \\ z_{12} & -z_{22} \end{bmatrix},$$

show

$\mathbf{i} = \mathbf{Z}^{-1}\mathbf{v}$, provided $z_{11}z_{22} - z_{12}^2 \neq 0$.

33 In a four-terminal network it is more useful to compare input and output quantities than currents and voltages. Show from Eqns (A) that if input and output vectors $\mathbf{x}_1$ and $\mathbf{x}_2$ are defined as

$$\mathbf{x}_1 = \begin{bmatrix} v_1 \\ i_1 \end{bmatrix}, \quad \mathbf{x}_2 = \begin{bmatrix} v_2 \\ i_2 \end{bmatrix}, \quad \text{then}$$

$$\mathbf{x}_1 = \mathbf{T}\mathbf{x}_2,$$

with

$$\mathbf{T} = \begin{bmatrix} \dfrac{z_{11}}{z_{12}} & \dfrac{\det \mathbf{Z}}{z_{12}} \\ \dfrac{1}{z_{12}} & \dfrac{z_{22}}{z_{12}} \end{bmatrix}.$$

The matrix $\mathbf{T}$ is called the **transmission matrix** for the network.

34 Prove det $\mathbf{T} = 1$, and hence that

$$\mathbf{x}_2 = \mathbf{T}^{-1}\mathbf{x}_1.$$

Deduce that $\mathbf{T}$ describes transmission through the four-terminal network from terminals 1 to terminals 2, while $\mathbf{T}^{-1}$ describes transmission in the reverse direction.

35 Prove that if n four-terminal networks are connected in **cascade** as shown in Fig. 3.19, then

$$\mathbf{x}_1 = \mathbf{T}_1\mathbf{T}_2 \ldots \mathbf{T}_n\mathbf{x}_{n+1},$$

where $\mathbf{x}_k$ and $\mathbf{T}_k$ are the input vector and transmission matrix, respectively, at the kth stage. Deduce that when joined in cascade the equivalent four-terminal network relating the input $\mathbf{x}_1$ and output $\mathbf{x}_{n+1}$ has the transmission matrix $\mathbf{T} = \mathbf{T}_1\mathbf{T}_2 \ldots \mathbf{T}_n$. Express $\mathbf{x}_{n+1}$ in terms of $\mathbf{x}_1$ and $\mathbf{T}_1, \mathbf{T}_2, \ldots, \mathbf{T}_n$.

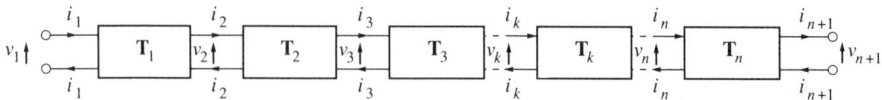

Fig. 3.19 n four-terminal networks in cascade

36 Find the inverse of the matrix

$$\mathbf{B} = \begin{bmatrix} 1 & 2 & 3 \\ 1 & 3 & 3 \\ 1 & 2 & 4 \end{bmatrix}.$$

If $A = \begin{bmatrix} 0 & 2 & 3 \\ 1 & 2 & 3 \\ 1 & 2 & 3 \end{bmatrix}$, and $(A+I)X = A - I$,

find X.

37 Find $(B-2I)^{-1}$ given

$$B = \begin{bmatrix} 5 & -2 & -1 \\ -4 & 3 & -1 \\ 2 & 0 & 3 \end{bmatrix}.$$

Hence find the vector x satisfying

$$ABx = 2Ax + AC, \quad \text{where} \quad A = \begin{bmatrix} 1 & a & b \\ 0 & 1 & c \\ 0 & 0 & 1 \end{bmatrix} \quad \text{and} \quad C = \begin{bmatrix} -1 \\ 1 \\ 3 \end{bmatrix}.$$

3.10 Algebraic eigenvalue problems. Eigenvalues

Throughout mathematics the **algebraic eigenvalue problem** associated with matrices occurs naturally and is of fundamental importance, though for simplicity the name is usually shortened to the **eigenvalue problem**. The full name is necessary only when a distinction needs to be made between the matrix eigenvalue problem and other nonalgebraic eigenvalue problems that arise in connection with differential equations and elsewhere. We shall see that the algebraic eigenvalue problem is only defined for square matrices so, with the exception of column vectors, all matrices which occur in the remainder of this section will be $n \times n$ matrices.

To introduce matters, let us consider the linear transformation

$$Ax = y, \tag{1}$$

where $A = [a_{ij}]$ is an $n \times n$ matrix and

$$x = \begin{bmatrix} x_1 \\ x_2 \\ \vdots \\ x_n \end{bmatrix}, \qquad y = \begin{bmatrix} y_1 \\ y_2 \\ \vdots \\ y_n \end{bmatrix}. \tag{2}$$

This has the property that it transforms any n element vector x into an n element vector y. We now ask a question which at first sight may seem a little strange, and it is this: when is a vector x transformed by the linear transformation (1) into a vector y which is proportional to x? The reason for this question will become apparent later when we come to study applications. The question is equivalent to asking for what vectors x does (1) produce a vector y of the form $y = \lambda x$, with λ a scalar. Setting $y = \lambda x$ in (1) then leads to the matrix equation

$$\mathbf{A}\mathbf{x} = \lambda\mathbf{x}, \tag{3}$$

which is the starting point of the eigenvalue problem.

Equation (3) always has the trivial vector solution $\mathbf{x} = \mathbf{0}$ for any λ, though this is not an interesting solution. If, however, the number λ is equal to one of a finite set of special numerical values called the **eigenvalues** or **characteristic values** of matrix $\mathbf{A}$, a corresponding set of nontrivial solutions $\mathbf{x} \neq \mathbf{0}$ exists. These are called the **eigenvectors** or **characteristic vectors** of matrix $\mathbf{A}$ corresponding to the eigenvalues[4].

The set of eigenvalues, which may be complex, is called the **spectrum** of $\mathbf{A}$, and contains useful information about the algebraic properties of $\mathbf{A}$. Let the eigenvalues be plotted as points in the complex plane. Then the radius ρ of the smallest circle centered on the origin which contains all such points is called the **spectral radius** of $\mathbf{A}$. If the eigenvalues are all real, ρ is equal to the largest of the set of absolute values of the eigenvalues, and if they are complex to the largest of the set of moduli of the eigenvalues.

To discover how the eigenvalues and eigenvectors of a matrix $\mathbf{A}$ may be determined from (3) we must use the fact that $\mathbf{x} = \mathbf{I}\mathbf{x}$, to rewrite (3) as

$$(\mathbf{A} - \lambda\mathbf{I})\mathbf{x} = \mathbf{0}. \tag{4}$$

Notice that the unit matrix is necessary in this expression, because $\mathbf{A}$ is an $n \times n$ matrix and λ is simply a number (scalar). Expression (4) is a homogeneous system of linear equations for the n elements $x_1, x_2, \ldots, x_n$ of the vector $\mathbf{x}$, in which the coefficient matrix is $\mathbf{A} - \lambda\mathbf{I}$. It follows from Theorem 3.20 that this system will only have nontrivial vector solutions $\mathbf{x} \neq \mathbf{0}$ when

$$\det(\mathbf{A} - \lambda\mathbf{I}) = 0, \tag{5}$$

which imposes an algebraic condition on the permissible values of λ (eigenvalues).

To examine the nature of the eigenvalues, set

$$P(\lambda) \equiv \det(\mathbf{A} - \lambda\mathbf{I}) = \begin{vmatrix} a_{11} - \lambda & a_{12} & \cdots & a_{1n} \\ a_{21} & a_{22} - \lambda & \cdots & a_{2n} \\ \cdot & \cdot & \cdots & \cdot \\ a_{n1} & a_{n2} & \cdots & a_{nn} - \lambda \end{vmatrix}, \tag{6}$$

when the eigenvalues λ determined by (5) are seen to be the *roots* of

$$P(\lambda) = 0, \tag{7}$$

that is they are the *zeros* of $P(\lambda)$. As λ appears only once in each row (column) of (6), it follows from the expansion of a determinant that $P(\lambda)$ is a polynomial of degree n in λ. This is called the **characteristic polynomial** of $\mathbf{A}$, and Eq. (7) determining the eigenvalues is called the **characteristic equation** of $\mathbf{A}$. The fundamental theorem of algebra then asserts

---

[4] In older texts the names **latent root** and **proper value** are used in place of eigenvalue, with the corresponding terms **latent vector** and **proper vector** being used in place of eigenvector. The term eigenvalue is derived from the German word *Eigenwert* meaning intrinsic value.

that $P(\lambda)=0$ will have n roots, though these may be complex even if the elements of **A** are real. When any root $\lambda=\mu$ of the characteristic equation is repeated r times (has **algebraic multiplicity** r) the eigenvalue μ is said to be of the rth **order**. An eigenvalue is said to have **geometric multiplicity** k if it has associated with it k linearly independent eigenvectors. In general, when the order of the eigenvalue exceeds 1, the algebraic and geometric multiplicities are not the same.

If the n eigenvalues of **A** are denoted by $\lambda_1, \lambda_2, \ldots, \lambda_n$, the corresponding eigenvectors $\mathbf{x}_1, \mathbf{x}_2, \ldots, \mathbf{x}_n$ follow by solving the homogeneous system derived from (4) by setting $\lambda=\lambda_k, \mathbf{x}=\mathbf{x}_k$ to give

$$(\mathbf{A} - \lambda_k\mathbf{I})\mathbf{x}_k=0, \tag{8}$$

with $k=1, 2, \ldots, n$.

Notice that an eigenvector is indeterminate up to a scalar multiplier, because if $\mathbf{x}$ is an eigenvector, then so also is $\alpha\mathbf{x}$ for $\alpha\neq0$ any scalar. This fact is often used to advantage to *normalize* eigenvectors when performing numerical calculations. If $\mathbf{x}$ is an eigenvector with real elements $x_1, x_2, \ldots, x_n$, then the corresponding **normalized eigenvector** $\hat{\mathbf{x}}$ is defined as

$$\hat{\mathbf{x}} = \frac{1}{\|\mathbf{x}\|}\begin{bmatrix} x_1 \\ x_2 \\ \vdots \\ x_n \end{bmatrix}, \tag{9}$$

where

$$\|\mathbf{x}\| = (\mathbf{x}^T\mathbf{x})^{1/2} = (x_1^2+x_2^2+ \ldots +x_n^2)^{1/2} \tag{10}$$

is the **norm** or euclidean length of $\mathbf{x}$ in $\mathbb{R}^n$.

The **algebraic eigenvalue problem** mentioned at the start of this section is the problem of determining the eigenvalues and eigenvectors of a given $n \times n$ matrix **A**.

The linearly independent eigenvectors of this problem form a basis for the space to which all solutions of the problem belong. That is, the space comprising all possible linear combinations of these eigenvectors. This is called the **solution space** or **eigenspace** of the eigenvalue problem.

When n is large the numerical solution of eigenvalue problems poses major difficulties and special methods are necessary. The discussion which now follows will be confined to calculations involving small n and to some important theoretical aspects of the eigenvalue problem.

Example 1. Real eigenvalue problem

Find the eigenvalues, eigenvectors, normalized eigenvectors and spectral radius of

$$\mathbf{A} = \begin{bmatrix} -5 & 3 \\ 10 & -4 \end{bmatrix}.$$

Determine a basis for the solution space of the eigenvalue problems.

Solution

$$P(\lambda) = \begin{vmatrix} -(5+\lambda) & 3 \\ 10 & -(4+\lambda) \end{vmatrix} = \lambda^2 + 9\lambda - 10 = (\lambda - 1)(\lambda + 10).$$

The eigenvalues determined by the characteristic equation $P(\lambda) = 0$ are $\lambda_1 = 1$ and $\lambda_2 = -10$. When $\lambda = \lambda_1$ the matrix equation (8) becomes the pair of consistent equations

$$-6x_1 + 3x_2 = 0$$
$$10x_1 - 5x_2 = 0,$$

so $x_2 = 2x_1$. Setting $x_1 = \alpha$, with $\alpha \neq 0$ arbitrary, then shows that an eigenvector $\mathbf{x}_1$ of $\mathbf{A}$ corresponding to the eigenvalue λ_1 is

$$\mathbf{x}_1 = \alpha \begin{bmatrix} 1 \\ 2 \end{bmatrix}.$$

Since $\|\mathbf{x}_1\| = \alpha(1^2 + 2^2)^{1/2} = \alpha\sqrt{5}$, the normalized eigenvector $\hat{\mathbf{x}}_1$ corresponding to the eigenvalue λ_1 is

$$\hat{\mathbf{x}}_1 = \begin{bmatrix} \dfrac{1}{\sqrt{5}} \\ \dfrac{2}{\sqrt{5}} \end{bmatrix}.$$

A similar argument with $\lambda = \lambda_2$ reduces (8) to the pair of consistent equations

$$5x_1 + 3x_2 = 0$$
$$10x_1 + 6x_2 = 0,$$

showing $x_2 = -\frac{5}{3}x_1$. So an eigenvector $\mathbf{x}_2$ of $\mathbf{A}$ corresponding to the eigenvalue λ_2 is

$$\mathbf{x}_2 = \beta \begin{bmatrix} 1 \\ -\frac{5}{3} \end{bmatrix},$$

with $\beta \neq 0$ arbitrary. Since $\|\mathbf{x}_2\| = \beta[1 + (-\frac{5}{3})^2]^{1/2} = \frac{1}{3}\beta\sqrt{34}$, the normalized eigenvector $\hat{\mathbf{x}}_2$ corresponding to the eigenvalue λ_2 is

$$\hat{\mathbf{x}}_2 = \begin{bmatrix} \dfrac{3}{\sqrt{34}} \\ -\dfrac{5}{\sqrt{34}} \end{bmatrix}.$$

The eigenvalues $\lambda_1 = 1$ and $\lambda_2 = -10$ are both real, so the spectral radius of $\mathbf{A}$ is $\rho = \max\{|\lambda_1|, |\lambda_2|\}$ $= 10$. As $\mathbf{x}_1$ and $\mathbf{x}_2$ are linearly independent they form a basis for the solution space, which in this case is a real two-dimensional vector space. ∎

Example 2. Complex eigenvalue problem

Find the eigenvalues, eigenvectors and spectral radius of

$$A = \begin{bmatrix} 1 & 1 \\ -1 & 2 \end{bmatrix}.$$

Determine a basis of the solution space for the eigenvalue problem.

Solution

$$P(\lambda) = \begin{vmatrix} 1-\lambda & 1 \\ -1 & 2-\lambda \end{vmatrix} = \lambda^2 - 3\lambda + 3.$$

The eigenvalues determined by the characteristic equation $P(\lambda)=0$ are the complex conjugate numbers $\lambda_1 = \frac{1}{2}(3 - i\sqrt{3})$ and $\lambda_2 = \frac{1}{2}(3 + i\sqrt{3})$, each with modulus $\sqrt{3}$. When $\lambda = \lambda_1$ the matrix equation (8) becomes the pair of consistent equations

$$\left(-\frac{1}{2} + i\frac{\sqrt{3}}{2}\right)x_1 + x_2 = 0$$

$$-x_1 + \left(\frac{1}{2} + i\frac{\sqrt{3}}{2}\right)x_2 = 0,$$

showing $x_2 = \frac{1}{2}(1 - i\sqrt{3})x_1$. Thus setting $x_1 = \alpha$, with $\alpha \neq 0$ an arbitrary complex constant, shows that an eigenvector $\mathbf{x}_1$ of A corresponding to the eigenvalue λ_1 is

$$\mathbf{x}_1 = \alpha \begin{bmatrix} 1 \\ \frac{1}{2} - i\frac{\sqrt{3}}{2} \end{bmatrix}.$$

Repeating this argument with $\lambda = \lambda_2$ shows that an eigenvector $\mathbf{x}_2$ of A corresponding to the eigenvalue λ_2 is

$$\mathbf{x}_2 = \beta \begin{bmatrix} 1 \\ \frac{1}{2} + i\frac{\sqrt{3}}{2} \end{bmatrix},$$

with $\beta \neq 0$ an arbitrary complex constant. The eigenvalues λ_1 and λ_2 each have modulus $\sqrt{3}$, so the spectral radius of A is $\rho = \sqrt{3}$.

Here also $\mathbf{x}_1$ and $\mathbf{x}_2$ are linearly independent and so form a basis for a two-dimensional solution space. This time, however, since the elements are complex, it is a complex vector space. ∎

Motivation for the eigenvalue problem formulated in (3) is most easily provided by making an application of matrices to linear constant coefficient differential equations. All that will be required for this application is the knowledge that a general solution of such an equation for $y(t)$ is an arbitrary linear combination of elementary solutions of the form $e^{\lambda t}$. For example, the general solution of the simple scalar equation

$$\frac{dy}{dt} = ky \tag{11}$$

is of the form $y = Ce^{\lambda t}$, where C is an arbitrary constant and λ is a parameter. To determine λ we substitute y into the differential equation and, after cancellation of the factor $Ce^{\lambda t}$, find $\lambda = k$. Thus the general solution of (11) is $y = Ce^{kt}$.

Now consider the coupled system of constant coefficient linear first order differential equations

$$\frac{dy_1}{dt} = a_{11}y_1 + a_{12}y_2$$

$$\frac{dy_2}{dt} = a_{21}y_1 + a_{22}y_2. \tag{12}$$

This system might, for example, determine the concentrations y_1 and y_2 of two constituents in a chemical process as a function of the time t. By defining

$$y = \begin{bmatrix} y_1 \\ y_2 \end{bmatrix}, \qquad A = \begin{bmatrix} a_{11} & a_{12} \\ a_{21} & a_{22} \end{bmatrix}, \tag{13}$$

and using the definition of differentiation of a matrix whose elements are functions of t (Sec. 3.3), system (12) may be written as the matrix differential equation

$$\frac{dy}{dt} = Ay. \tag{14}$$

By analogy with the method of solution of the single differential equation (11) just considered, let us try to find a solution of (14) of the form

$$y = ae^{\lambda t}, \tag{15}$$

where a is a constant two element column vector and λ is a parameter to be determined from (14). Substituting (15) into (14) gives

$$\lambda ae^{\lambda t} = Aae^{\lambda t},$$

which after cancellation of the nonzero scalar factor $e^{\lambda t}$ shows λ and a are determined by the eigenvalue problem

$$Aa = \lambda a. \tag{16}$$

Thus the permissible values of λ in (15) are the eigenvalues λ_1 and λ_2 of matrix A, and the constant vectors a are the corresponding eigenvectors of A.

The general solution of (14) is thus

$$y = a_1 e^{\lambda_1 t} + a_2 e^{\lambda_2 t}, \tag{17}$$

where a_1 and a_2 are eigenvectors of A corresponding to the eigenvalues λ_1 and λ_2. This solution shows quite clearly how the basis a_1, a_2 of the solution space of the eigenvalue problem determines the solution space of the system of differential equations in (12).

Example 3. 3×3 matrix with distinct eigenvalues

Find the eigenvalues, and normalized eigenvectors of

$$\mathbf{A} = \begin{bmatrix} 0 & 2 & 4 \\ 1 & 1 & -2 \\ -2 & 0 & 5 \end{bmatrix}.$$

Determine a basis for the solution space of the eigenvalue problem.

Solution

$$P(\lambda) = \begin{vmatrix} -\lambda & 2 & 4 \\ 1 & 1-\lambda & -2 \\ -2 & 0 & 5-\lambda \end{vmatrix} = (1-\lambda)(\lambda-2)(\lambda-3).$$

The eigenvalues determined by the characteristic equation $P(\lambda) = 0$ are $\lambda_1 = 1$, $\lambda_2 = 2$ and $\lambda_3 = 3$, so the eigenvalues are all **distinct** (different).

When $\lambda = \lambda_1$ the matrix equation (8) becomes the set of three consistent equations

$$\begin{aligned} -x_1 + 2x_2 + 4x_3 &= 0 \\ x_1 \qquad\quad -2x_3 &= 0 \\ -2x_1 \qquad\quad +4x_3 &= 0, \end{aligned}$$

showing $x_3 = \frac{1}{2}x_1$ and $x_2 = -\frac{1}{2}x_1$. So setting $x_1 = \alpha$, with $\alpha \neq 0$ arbitrary, we find the eigenvector $\mathbf{x}_1$ of $\mathbf{A}$ corresponding to λ_1 is

$$\mathbf{x}_1 = \alpha \begin{bmatrix} 1 \\ -\frac{1}{2} \\ \frac{1}{2} \end{bmatrix}.$$

When $\lambda = \lambda_2$ a similar argument shows the eigenvector $\mathbf{x}_2$ of $\mathbf{A}$ corresponding to λ_2 is

$$\mathbf{x}_2 = \beta \begin{bmatrix} 1 \\ -\frac{1}{3} \\ \frac{2}{3} \end{bmatrix}, \quad \text{with } \beta \neq 0 \text{ arbitrary,}$$

and when $\lambda = \lambda_3$ the eigenvector $\mathbf{x}_3$ of $\mathbf{A}$ corresponding to λ_3 is found to be

$$\mathbf{x}_3 = \gamma \begin{bmatrix} 1 \\ -\frac{1}{2} \\ 1 \end{bmatrix}, \quad \text{with } \gamma \neq 0 \text{ arbitrary.}$$

The normalized eigenvectors are

$$\hat{\mathbf{x}}_1 = \frac{1}{\sqrt{6}} \begin{bmatrix} 2 \\ -1 \\ 1 \end{bmatrix}, \quad \hat{\mathbf{x}}_2 = \frac{1}{\sqrt{14}} \begin{bmatrix} 3 \\ -1 \\ 2 \end{bmatrix} \quad \text{and} \quad \hat{\mathbf{x}}_3 = \frac{1}{3} \begin{bmatrix} 2 \\ -1 \\ 2 \end{bmatrix}.$$

The eigenvectors x_1, x_2 and x_3 are all linearly independent, so they form a basis for the solution space of the eigenvalue problem. As the elements of the eigenvectors are real and there are three linearly independent eigenvectors the solution space is a real three-dimensional vector space.

∎

Example 4. 3×3 matrix with repeated eigenvalue and three eigenvectors

Find the eigenvalues and normalized eigenvectors of

$$A = \begin{bmatrix} 1 & 0 & 0 \\ 1 & 3 & 2 \\ 1 & 2 & 3 \end{bmatrix}.$$

Determine a basis for the solution space of the eigenvalue problem.

Solution

$$P(\lambda) = \begin{vmatrix} 1-\lambda & 0 & 0 \\ 1 & 3-\lambda & 2 \\ 1 & 2 & 3-\lambda \end{vmatrix} = (5-\lambda)(\lambda-1)^2.$$

The eigenvalues determined by the characteristic equation $P(\lambda)=0$ are $\lambda_1=5$, $\lambda_2=1$ and $\lambda_3=1$, so $\lambda=1$ is a twice repeated eigenvalue (algebraic multiplicity 2). When $\lambda=\lambda_1$ the matrix equation (8) becomes the set of three consistent equations

$$-4x_1 \qquad\qquad = 0$$
$$x_1 - 2x_2 + 2x_3 = 0$$
$$x_1 + 2x_2 - 2x_3 = 0,$$

showing $x_1=0$ and $x_2=x_3$. So setting $x_2=\alpha$, with $\alpha \neq 0$ arbitrary, shows an eigenvector x_1 of A corresponding to λ_1 is

$$x_1 = \begin{bmatrix} 0 \\ \alpha \\ \alpha \end{bmatrix}.$$

When $\lambda=\lambda_2$ or $\lambda=\lambda_3$ the matrix equation (8) becomes the set of three consistent equations

$$0x_1 + 0x_2 + 0x_3 = 0$$
$$x_1 + 2x_2 + 2x_3 = 0$$
$$x_1 + 2x_2 + 2x_3 = 0. \qquad\qquad (A)$$

These show that any two of x_1, x_2 and x_3 may be specified arbitrarily. So, setting $x_2=\beta$ and $x_3=\gamma$, with β, $\gamma \neq 0$ arbitrary, we have $x_1 = -2(\beta+\gamma)$. The eigenvector corresponding to $\lambda=1$ then becomes

$$x = \begin{bmatrix} -2(\beta+\gamma) \\ \beta \\ \gamma \end{bmatrix} = \begin{bmatrix} -2\beta \\ \beta \\ 0 \end{bmatrix} + \begin{bmatrix} -2\gamma \\ 0 \\ \gamma \end{bmatrix}.$$

As β, γ are arbitrary and independent this shows that there are two eigenvectors $\mathbf{x}_2$ and $\mathbf{x}_3$ corresponding to $\lambda=1$, are

$$\mathbf{x}_2 = \begin{bmatrix} -2\beta \\ \beta \\ 0 \end{bmatrix} \quad \text{and} \quad \mathbf{x}_3 = \begin{bmatrix} -2\gamma \\ 0 \\ \gamma \end{bmatrix}.$$

The eigenvectors $\mathbf{x}_2$ and $\mathbf{x}_3$ are linearly independent, so the eigenvalue $\lambda=1$ has geometric multiplicity 2.

This result might have been anticipated from Theorem 3.7 (ii), because $\mathbf{A}$ is a 3×3 matrix and rank $(\mathbf{A}-\lambda\mathbf{I})=1$ for $\lambda=\lambda_2$ and $\lambda=\lambda_3$, so there must be $3-1=2$ linearly independent solutions.

In this case, although an eigenvalue was repeated, each eigenvalue had associated with it as many linearly independent eigenvectors as the order (algebraic multiplicity) of the eigenvalue. The elements of the linearly independent eigenvectors $\mathbf{x}_1$, $\mathbf{x}_2$ and $\mathbf{x}_3$ are real, so these eigenvectors form a basis for a real three-dimensional solution space. The corresponding normalized eigenvectors are

$$\hat{\mathbf{x}}_1 = \frac{1}{\sqrt{2}} \begin{bmatrix} 0 \\ 1 \\ 1 \end{bmatrix}, \quad \hat{\mathbf{x}}_2 = \frac{1}{\sqrt{5}} \begin{bmatrix} -2 \\ 1 \\ 0 \end{bmatrix}, \quad \hat{\mathbf{x}}_3 = \frac{1}{\sqrt{5}} \begin{bmatrix} -2 \\ 0 \\ 1 \end{bmatrix}. \qquad \blacksquare$$

Example 5. 3×3 matrix with repeated eigenvalue and only two eigenvectors

Find the eigenvalues and eigenvectors of

$$\mathbf{A} = \begin{bmatrix} 5 & 10 & 7 \\ 0 & -3 & -3 \\ 0 & 3 & 3 \end{bmatrix}.$$

Determine a basis for the solution space of the eigenvalue problem.

Solution

$$P(\lambda) = \begin{vmatrix} 5-\lambda & 10 & 7 \\ 0 & -(3+\lambda) & -3 \\ 0 & 3 & 3-\lambda \end{vmatrix} = (5-\lambda)\lambda^2.$$

The eigenvalues determined by the characteristic equation $P(\lambda)=0$ are $\lambda_1=5$, $\lambda_2=0$ and $\lambda_3=0$, with $\lambda=0$ an eigenvalue of degree two (algebraic multiplicity 2). When $\lambda=\lambda_1$ the matrix equation (8) becomes the set of three consistent equations

$$10x_2 + 7x_3 = 0$$
$$-8x_2 - 3x_3 = 0$$
$$3x_3 - 2x_3 = 0,$$

which imposes no condition on x_1, but can only be satisfied if $x_2=x_3=0$. Thus assigning x_1 arbitrarily by setting $x_1=\alpha$, with $\alpha \neq 0$ arbitrary, we see that an eigenvector $\mathbf{x}_1$ of $\mathbf{A}$ corresponding to λ_1 is

$$\mathbf{x}_1 = \alpha \begin{bmatrix} 1 \\ 0 \\ 0 \end{bmatrix}.$$

When $\lambda = \lambda_2$ or $\lambda = \lambda_3$ the matrix equation (8) becomes the set of three consistent equations

$$5x_1 + 10x_2 + 7x_3 = 0$$
$$-3x_2 - 3x_3 = 0$$
$$3x_2 + 3x_3 = 0,$$

with the single solution $x_2 = -5/3\, x_1$ and $x_3 = -x_2$ where x_1 is arbitrary. So setting $x_1 = \beta$, with $\beta \neq 0$ arbitrary, an eigenvector $\mathbf{x}_2$ of $\mathbf{A}$ corresponding to λ_2 and to λ_3 is seen to be

$$\mathbf{x}_2 = \beta \begin{bmatrix} 1 \\ -\frac{5}{3} \\ \frac{5}{3} \end{bmatrix}.$$

Here $\lambda = 0$ has algebraic multiplicity 2, but geometric multiplicity 1, so $\mathbf{x}_1$ and $\mathbf{x}_2$ form a basis for a real two-dimensional solution space.

The existence of only one eigenvector corresponding to the twice repeated eigenvalue could have been anticipated from Theorem 3.7(ii). As $\mathbf{A}$ is a 3×3 matrix, and rank $(\mathbf{A} - \lambda\mathbf{I}) = 2$ for $\lambda = \lambda_2$ or $\lambda = \lambda_3$, the number of linearly independent solutions must be $3 - 2 = 1$.   ∎

Two useful computational checks on the eigenvalues of a matrix may be deduced from (6). First we use the fact that $\lambda_1, \lambda_2, \ldots, \lambda_n$ are the zeros of $P(\lambda)$ to write it in the form

$$(\lambda_1 - \lambda)(\lambda_2 - \lambda) \ldots (\lambda_n - \lambda) \equiv \det(\mathbf{A} - \lambda\mathbf{I}). \tag{18}$$

Setting $\lambda = 0$ in this identity we obtain the result

$$\lambda_1 \lambda_2 \ldots \lambda_n = \det \mathbf{A}, \tag{19}$$

which provides the first check.

The other check involves equating the coefficients of λ^{n-1} on either side of (18). On the left-hand side this coefficient is easily seen to be $(-1)^{n-1}(\lambda_1 + \lambda_2 + \ldots + \lambda_n)$. Theorem 3.12 may be used to prove that the corresponding coefficient on the right-hand side is $(-1)^{n-1}(a_{11} + a_{22} + \ldots + a_{nn}) = (-1)^{n-1}\operatorname{tr}(\mathbf{A})$. Equating these two results then gives the second check

$$\lambda_1 + \lambda_2 \ldots + \lambda_n = \operatorname{tr}(\mathbf{A}). \tag{20}$$

Taken together, (19) and (20) form the next result.

Theorem 3.24 (Sum and product of eigenvalues)

If $\mathbf{A}$ is an $n \times n$ matrix with the eigenvalues $\lambda_1, \lambda_2, \ldots, \lambda_n$, then

(i) $\lambda_1 + \lambda_2 + \ldots + \lambda_n = \operatorname{tr}(\mathbf{A})$,

(ii) $\lambda_1 \lambda_2 \ldots \lambda_n = \det \mathbf{A}$.   □

The reader should verify that the eigenvalues found in the last five examples satisfy (i) and (ii) of Theorem 3.24, irrespective of whether they are real or complex.

Theorem 3.25 (Eigenvalues of special matrices)

Let A be an $n \times n$ matrix with the eigenvalues $\lambda_1, \lambda_2, \ldots, \lambda_n$ and the corresponding eigenvectors $x_1, x_2, \ldots, x_n$. Then

  (i) if A is a diagonal matrix the diagonal elements are the eigenvalues;

  (ii) if A is either an upper or lower triangular matrix the diagonal elements are the eigenvalues;

  (iii) for $r > 1$ an integer, the eigenvalues of A^r are $\lambda_1^r, \lambda_2^r, \ldots, \lambda_n^r$;

  (iv) if $\det A \neq 0$, the eigenvalues of A^{-1} are $1/\lambda_1, 1/\lambda_2, \ldots, 1/\lambda_n$.

Proof

Result (i) is immediate, while result (ii) follows by expanding $\det |A - \lambda I|$ in terms of the elements of its first column. Result (iii) is proved by induction. First, since for $r = 1, 2, \ldots, n$,

$$A x_r = \lambda_r x_r,$$

it follows that

$$A^2 x_r = \lambda_r A x_r = \lambda_r^2 x_r,$$

showing that if A has the eigenvalues $\lambda_1, \lambda_2, \ldots, \lambda_n$, then A^2 has the eigenvalues $\lambda_1^2, \lambda_2^2, \ldots, \lambda_n^2$. Now suppose the result to be true for A^n with $n > 2$ an integer, so that

$$A^n x_r = \lambda_r^n x_r.$$

Then

$$A^{n+1} x_r = \lambda_r^n A x_r = \lambda_r^{n+1} x_r,$$

showing the result is also true for A^{n+1}. Since it is true for A^2 it follows by induction that it is true for all integral $n \geq 2$.

Result (iv) is proved as follows. For $r = 1, 2, \ldots, n$

$$A x_r = \lambda_r x_r,$$

so when $\det A \neq 0$, A^{-1} exists and

$$A^{-1} A x_r = \lambda_r A^{-1} x_r,$$

or

$$x_r = \lambda_r A^{-1} x_r.$$

This is equivalent to

$$A^{-1}x_r = \frac{1}{\lambda_r} x_r,$$

which establishes the required result.   □

Let $P(t)$ be a polynomial of degree n in the real variable t (n is the highest power of t) and take A to be an $n \times n$ matrix. Then a **matrix polynomial** $P(A)$ of degree n may be derived from $P(t)$ by replacing each power of t by the corresponding power of A, with t^0 being replaced by the $n \times n$ unit matrix I. Thus if

$$P(t) = 4t^3 - t^2 + 2t + 7,$$

it is of degree 3, and the corresponding matrix polynomial of degree 3 is

$$P(A) = 4A^3 - A^2 + 2A + 7I.$$

When $N(A)$ and $D(A)$ are matrix polynomials with det $D(A) \neq 0$, an expression of the form $R(A) = N(A) [D(A)]^{-1}$ is called a **rational function** of A.

Theorem 3.26 (Eigenvalues and eigenvectors of rational functions)

Let A be an $n \times n$ matrix and $R(A) = N(A)[D(A)]^{-1}$, where $N(A)$ and $D(A)$ are matrix polynomials with det $D(A) \neq 0$. Then if A has the eigenvalues $\lambda_1, \lambda_2, \ldots, \lambda_n$, and the corresponding eigenvectors $x_1, x_2, \ldots, x_n$, the rational function $R(A)$ has the same eigenvectors as A and the eigenvalues $R(\lambda_1), R(\lambda_2), \ldots, R(\lambda_n)$ corresponding to $x_1, x_2, \ldots, x_n$, respectively.

Proof

From Theorem 3.25 it follows that for $r = 1, 2, \ldots, n$,

$$N(A)x_r = N(\lambda_r)x_r \quad \text{and} \quad D(A)x_r = D(\lambda_r)x_r.$$

Now $R(A)$ is an $n \times n$ matrix, so post-multiplying it by $D(A)$ we find

$$R(A)D(A) = N(A).$$

Hence for $r = 1, 2, \ldots, n$,

$$R(A)D(A)x_r = N(A)x_r,$$

or

$$R(A)D(\lambda_r)x_r = N(\lambda_r)x_r.$$

Since $N(\lambda_r)$ and $D(\lambda_r)$ are scalars this may be written

$$R(A)x_r = \frac{N(\lambda_r)}{D(\lambda_r)} x_r,$$

which is equivalent to

$$R(\mathbf{A})\mathbf{x}_r = R(\lambda_r)\mathbf{x}_r,$$

thereby establishing the result that $R(\mathbf{A})$ has the eigenvalue $R(\lambda_r)$ corresponding to the eigenvector $\mathbf{x}_r$, for $r = 1, 2, \ldots, n$. ☐

This theorem has many applications and greatly simplifies the task of calculating the eigenvalues of a rational function $R(\mathbf{A})$ once the eigenvalues of $\mathbf{A}$ are known. They are simply the numbers $R(\lambda_r) = N(\lambda_r)/D(\lambda_r)$ for $r = 1, 2, \ldots, n$.

Example 6. Eigenvalues and eigenvectors of a rational function

Given

$$\mathbf{A} = \begin{bmatrix} -5 & 3 \\ 10 & -4 \end{bmatrix},$$

find the eigenvalues and eigenvectors of $R(\mathbf{A}) = N(\mathbf{A})[D(\mathbf{A})]^{-1}$, when $N(\mathbf{A}) = \mathbf{A}^2 - 2\mathbf{A} + \mathbf{I}$ and $D(\mathbf{A}) = \mathbf{A}^3 + \mathbf{A} + 4\mathbf{I}$.

Solution

From Ex. 1, the eigenvalues and eigenvectors of $\mathbf{A}$ are

$$\lambda_1 = 1, \quad \lambda_2 = -10 \quad \text{and} \quad \mathbf{x}_1 = \alpha \begin{bmatrix} 1 \\ 2 \end{bmatrix}, \quad \mathbf{x}_2 = \beta \begin{bmatrix} 1 \\ -\frac{5}{3} \end{bmatrix},$$

respectively. From Theorem 3.26 it follows that the eigenvectors of $R(\mathbf{A})$ are also $\mathbf{x}_1$ and $\mathbf{x}_2$. The corresponding eigenvalues are $R(\lambda_r) = N(\lambda_r)/D(\lambda_r)$, with $r = 1, 2$. Now

$$R(\lambda) = \frac{\lambda^2 - 2\lambda + 1}{\lambda^3 + \lambda + 4},$$

so the eigenvalues of $R(\mathbf{A})$ are $R(\lambda_1) = R(1) = 0$ and $R(\lambda_2) = R(-10) = -121/1006$. ∎

Specific examples of this next important result may be seen in the worked examples already discussed.

Theorem 3.27 (Distinct eigenvalues and their eigenvectors)

The eigenvectors corresponding to distinct eigenvalues of $\mathbf{A}$ are linearly independent. In particular, if all the eigenvalues are distinct then all the eigenvectors are linearly independent.

Proof

Let $\lambda_1, \lambda_2, \ldots, \lambda_r$ be distinct eigenvalues of an $n \times n$ matrix $\mathbf{A}$ $(0 < r \le n)$ with the corresponding eigenvectors $\mathbf{x}_1, \mathbf{x}_2, \ldots, \mathbf{x}_r$ and suppose, if possible, that these eigenvectors are linearly dependent. This means that for some $k < r$ we may write

$$\mathbf{x}_{k+1} = \alpha_1 \mathbf{x}_1 + \alpha_2 \mathbf{x}_2 + \ldots + \alpha_k \mathbf{x}_k, \tag{21}$$

where the constants α_s are unique and such that not every one is zero.

Pre-multiplying (21) by $\mathbf{A}$ and using the result $\mathbf{A}\mathbf{x}_s = \lambda_s \mathbf{x}_s$ we find

$$\lambda_{k+1} \mathbf{x}_{k+1} = \alpha_1 \lambda_1 \mathbf{x}_1 + \alpha_2 \lambda_2 \mathbf{x}_2 + \ldots + \alpha_k \lambda_k \mathbf{x}_k. \tag{22}$$

Now $\lambda_{k+1} \neq 0$, since this would imply all the $\lambda_1, \lambda_2, \ldots, \lambda_k$ are zero, so we may divide by λ_{k+1} to obtain

$$\mathbf{x}_{k+1} = \frac{\alpha_1 \lambda_1}{\lambda_{k+1}} \mathbf{x}_1 + \frac{\alpha_2 \lambda_2}{\lambda_{k+1}} \mathbf{x}_2 + \ldots + \frac{\alpha_k \lambda_k}{\lambda_{k+1}} \mathbf{x}_k. \tag{23}$$

Comparing (21) and (23) gives

$$\alpha_s = \frac{\alpha_s \lambda_s}{\lambda_{k+1}}, \qquad s = 1, 2, \ldots, k. \tag{24}$$

Because not every α_s is zero, this implies that λ_{k+1} is contained in the set of eigenvalues $\lambda_1, \lambda_2, \ldots, \lambda_k$. This is a contradiction, so we have proved that distinct eigenvalues have eigenvectors which are linearly independent. The final statement of the theorem follows at once by setting $r = n$. □

Examples 4 and 5 show that when an eigenvalue has order (multiplicity) m there may, or may not, be a corresponding set of m linearly independent eigenvectors, though Theorem 3.27 offers no information in such circumstances.

Let $\mathbf{A}$, $\mathbf{B}$ be two $n \times n$ matrices. Then $\mathbf{B}$ is said to be **similar** to $\mathbf{A}$ if a nonsingular $n \times n$ matrix $\mathbf{M}$ exists such that

$$\mathbf{B} = \mathbf{M}^{-1} \mathbf{A} \mathbf{M}. \tag{25}$$

Matrix $\mathbf{B}$ is then said to be related to $\mathbf{A}$ by a **similarity transformation**. As result (25) is equivalent to

$$\mathbf{A} = \mathbf{N}^{-1} \mathbf{B} \mathbf{N},$$

where $\mathbf{N} = \mathbf{M}^{-1}$, so we see that $\mathbf{A}$ is also similar to $\mathbf{B}$. Thus when (25) is true it suffices to say merely that $\mathbf{A}$ and $\mathbf{B}$ are **similar**. The most important class of similar matrices which will be considered later arises when $\mathbf{A}$ is similar to a diagonal matrix $\mathbf{\Lambda}$.

Theorem 3.28 (Properties of similar matrices)

(i) If $\mathbf{A}$ and $\mathbf{B}$ are similar, then so also are $\mathbf{A}^k$ and $\mathbf{B}^k$ with $k > 1$ an integer:

(ii) if $\mathbf{A}$ and $\mathbf{B}$ are similar, then $\det \mathbf{A} = \det \mathbf{B}$;

(iii) similar matrices have the same eigenvalues.

Proof

To see (i) is true when $k = 2$, notice that if $\mathbf{B} = \mathbf{M}^{-1}\mathbf{AM}$, then $\mathbf{B}^2 = \mathbf{M}^{-1}\mathbf{AMM}^{-1}\mathbf{AM}$ $= \mathbf{M}^{-1}\mathbf{A}^2\mathbf{M}$, which shows $\mathbf{B}^2$ and $\mathbf{A}^2$ are similar. The proof for arbitrary integral k follows by induction and is left as an exercise for the reader.

To establish (ii) suppose $\mathbf{A}$ and $\mathbf{B}$ are similar, with $\mathbf{B} = \mathbf{M}^{-1}\mathbf{AM}$. Then from Theorems 2.16 and 2.22 (vi) we have

$$\det \mathbf{B} = \det(\mathbf{M}^{-1}\mathbf{AM}) = \det \mathbf{M}^{-1} \det \mathbf{A} \det \mathbf{M}$$
$$= \det \mathbf{M}^{-1} \det \mathbf{M} \det \mathbf{A} = \det \mathbf{A}.$$

Result (iii) follows from

$$\det(\mathbf{B} - \lambda\mathbf{I}) = \det(\mathbf{M}^{-1}\mathbf{AM} - \lambda\mathbf{M}^{-1}\mathbf{M})$$
$$= \det[\mathbf{M}^{-1}(\mathbf{A} - \lambda\mathbf{I})\mathbf{M}],$$

because using (ii) in the right-hand side gives

$$\det(\mathbf{B} - \lambda\mathbf{I}) = \det(\mathbf{A} - \lambda\mathbf{I}).$$

This shows $\mathbf{A}$ and $\mathbf{B}$ have the same characteristic polynomials, and thus the same eigenvalues. $\qquad\square$

Many eigenvalue problems involve symmetric matrices. These have very convenient mathematical properties, knowledge of which can greatly simplify the subsequent analysis. The next result lists the two most important of these properties.

Theorem 3.29 (Eigenvalues and eigenvectors of symmetric matrices)

In an $n \times n$ matrix $\mathbf{A}$ is real and symmetric, then

(i) its eigenvalues are all real;

and

(ii) it has n eigenvectors which are mutually orthogonal;

Proof

The proof will use the fact that if $\mathbf{x}, \mathbf{y}$ are n element column vectors and $\mathbf{A}$ is an $n \times n$ symmetric matrix, $\mathbf{y}^\mathsf{T}\mathbf{Ax}$ is a scalar and so is equal to its transpose. Taking the transpose of $\mathbf{y}^\mathsf{T}\mathbf{Ax}$ and using Theorem 3.2 gives

$$\mathbf{y}^\mathsf{T}\mathbf{Ax} = \mathbf{x}^\mathsf{T}\mathbf{Ay}, \tag{26}$$

where use has been made of the fact that for a symmetric matrix $\mathbf{A}^\mathsf{T} = \mathbf{A}$.

To prove (i) we need the result that if λ is a complex number and $\bar{\lambda}$ is its complex conjugate, then λ is real if $\lambda = \bar{\lambda}$ (see Sec. 1.6). Let λ be an eigenvalue of $\mathbf{A}$ and $\mathbf{x}$ the corresponding eigenvector, then

$$\mathbf{A}\mathbf{x} = \lambda \mathbf{x}. \tag{27}$$

Taking the complex conjugate, and using the fact that $\bar{\mathbf{A}} = \mathbf{A}$ because $\mathbf{A}$ is real, gives

$$\mathbf{A}\bar{\mathbf{x}} = \bar{\lambda}\bar{\mathbf{x}}, \tag{28}$$

so that $\bar{\mathbf{x}}$ is an eigenvector corresponding to the eigenvalue $\bar{\lambda}$.

Pre-multiplication of (27) by $\bar{\mathbf{x}}^{\mathsf{T}}$ and (28) by $\mathbf{x}^{\mathsf{T}}$ gives

$$\bar{\mathbf{x}}^{\mathsf{T}}\mathbf{A}\mathbf{x} = \lambda\bar{\mathbf{x}}^{\mathsf{T}}\mathbf{x} \quad \text{and} \quad \mathbf{x}^{\mathsf{T}}\mathbf{A}\bar{\mathbf{x}} = \bar{\lambda}\mathbf{x}^{\mathsf{T}}\bar{\mathbf{x}}. \tag{29}$$

Using (29) in conjunction with (26) in which we set $y = \bar{\mathbf{x}}$ shows

$$\lambda\bar{\mathbf{x}}^{\mathsf{T}}\mathbf{x} = \bar{\lambda}\mathbf{x}^{\mathsf{T}}\bar{\mathbf{x}}.$$

However $\bar{\mathbf{x}}^{\mathsf{T}}\mathbf{x} = \mathbf{x}^{\mathsf{T}}\bar{\mathbf{x}}$, so $\lambda = \bar{\lambda}$, which completes the proof.

To prove (ii) we need to show that if $\mathbf{x}_r$, $\mathbf{x}_s$ are eigenvectors of $\mathbf{A}$ corresponding to the eigenvalues λ_r, λ_s, then $\mathbf{x}_r^{\mathsf{T}}\mathbf{x}^s = 0$ for $r \neq s$. We have

$$\mathbf{A}\mathbf{x}_r = \lambda_r\mathbf{x}_r \quad \text{and} \quad \mathbf{A}\mathbf{x}_s = \lambda_s\mathbf{x}_s,$$

so

$$\mathbf{x}_s^{\mathsf{T}}\mathbf{A}\mathbf{x}_r = \lambda_r\mathbf{x}_s^{\mathsf{T}}\mathbf{x}_r \quad \text{and} \quad \mathbf{x}_r^{\mathsf{T}}\mathbf{A}\mathbf{x}_s = \lambda_s\mathbf{x}_r^{\mathsf{T}}\mathbf{x}_s. \tag{30}$$

It follows from (26) with $\mathbf{x} = \mathbf{x}_r$ and $\mathbf{y} = \mathbf{x}_s$ that $\mathbf{x}_s^{\mathsf{T}}\mathbf{A}\mathbf{x}_r = \mathbf{x}_r^{\mathsf{T}}\mathbf{A}\mathbf{x}_s$, so as $\mathbf{x}_s^{\mathsf{T}}\mathbf{x}_r = \mathbf{x}_r^{\mathsf{T}}\mathbf{x}_s$, result (30) implies

$$(\lambda_r - \lambda_s)\mathbf{x}_r^{\mathsf{T}}\mathbf{x}_s = 0. \tag{31}$$

So, for $\lambda_r \neq \lambda_s$, it follows that $\mathbf{x}_r^{\mathsf{T}}\mathbf{x}_s = 0$, which was to be shown. We see that $\mathbf{x}_r^{\mathsf{T}}\mathbf{x}_r = \|\mathbf{x}_r\|^2 \neq 0$ for all r, since not every element of $\mathbf{x}_r$ can be zero.

We have proved the orthogonality of the eigenvectors of real symmetric matrices corresponding to distinct eigenvalues. Although we shall not establish the result here, it can be shown that this is still true when the eigenvalues of a real symmetric matrix are repeated. □

In physical situations giving rise to eigenvalue problems it is often important to know if the eigenvalues are positive or negative or, should they be complex, if they have positive or negative real parts. The consequences of this qualitative information can be seen in solution (17) to the system of differential equations given in (12). If $\lambda_1 < 0$, $\lambda_2 < 0$, the elements of the solution vector $\mathbf{y}$ have the qualitative property that they will decay exponentially with time. In the context of physical applications such behavior is important and a solution of this type is said to be **stable**. If, however, an eigenvalue is positive, the element of the solution vector associated with it will have the qualitative property that it will grow exponentially with time. This property is also important in physical applications and a solution which exhibits such behavior is said to be **unstable**. Similar behavior of the solution occurs if the eigenvalues are complex, and their real parts are either negative (stable) or positive (unstable).

The next result provides information about the boundary of a region in the complex plane within which the eigenvalues of an arbitrary matrix are located. If this region lies entirely to the left or right of the imaginary axis the theorem provides information about the sign of the real parts of all the eigenvalues. This information is based solely on the coefficients of the matrix, and *not* on the coefficients of the characteristic polynomial. It has many uses, amongst which are to be found applications to control theory (stability/instability) and to numerical analysis (stability of difference schemes).

Both the statement of this theorem and its proof involve elementary properties of complex numbers.

Theorem 3.30 (Gerschgorin's circle theorem)

Each of the eigenvalues of an arbitrary $n \times n$ matrix **A** lies in at least one of the circles $C_1, C_2, \ldots, C_n$ in the complex plane, where circle C_r with radius ρ_r has its center at a_{rr}, where a_{rr} is the rth element of the leading diagonal of **A**, and

$$\rho_r = \sum_{\substack{j=1 \\ j \neq r}}^{n} |a_{rj}| = |a_{r1}| + |a_{r2}| + \ldots + |a_{r,r-1}| + |a_{r,r+1}| + \ldots + |a_{rn}|.$$

Proof

The rth equation of the eigenvalue problem $\mathbf{Ax} = \lambda \mathbf{x}$ may be written

$$(\lambda - a_{rr})x_r = \sum_{\substack{j=1 \\ j \neq r}}^{n} a_{rj}x_j, \qquad \text{for } r = 1, 2, \ldots, n.$$

Taking the modulus of the equation and using the generalized triangle inequality we find

$$|\lambda - a_{rr}| < \sum_{\substack{j=1 \\ j \neq r}}^{n} |a_{rj}| \frac{|x_j|}{|x_r|}, \qquad \text{for } r = 1, 2, \ldots, n.$$

Thus if x_r is the element of **x** with the greatest modulus, $|x_j|/|x_r| \leq 1$ for $r = 1, 2, \ldots, n$, and the result of the theorem follows directly. $\square$

Notice that circle C_r will have zero radius (contract to the point a_{rr}) when the off-diagonal elements in the rth row are all zero.

Example 7. Location of eigenvalues

Use Gerschgorin's circle theorem to identify the region in the complex plane containing the eigenvalues of

$$\mathbf{A} = \begin{bmatrix} 0 & 2 & 4 \\ 1 & 1 & -2 \\ -2 & 0 & 5 \end{bmatrix}.$$

Solution

From Theorem 3.30, the eigenvalues of **A** must lie in the region comprising the **union** (sum of areas) of the three circles C_1, C_2 and C_3, where

C_1 with its center at 0 has radius $\rho_1 = |2| + |4| = 6$;

C_2 with its center at 1 has radius $\rho_2 = |1| + |-2| = 3$;

C_3 with its center at 5 has radius $\rho_3 = |-2| + 0 = 2$.

These circles are illustrated in Fig. 3.20, with their union being shown as the shaded region. Also shown in the figure as the points A, B and C are the eigenvalues 1, 2 and 3 found in Ex. 3.

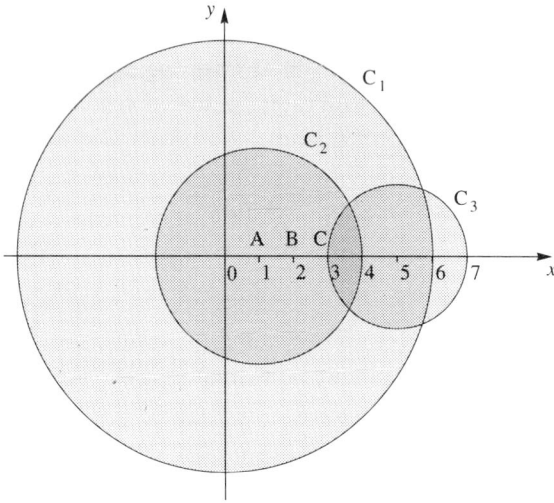

Fig. 3.20   Gerschgorin circles for Ex. 7.

In this case the imaginary axis (y-axis) intersects the union of the circles, so Theorem 3.30 provides no information about the location of the eigenvalues relative to the imaginary axis.

∎

Orthogonal matrices

To provide motivation for the discussion of orthogonal matrices which is to follow, we select an example from euclidean geometry. Let O (x_1, x_2, x_3) and O (y_1, y_2, y_3) be two orthogonal coordinate systems with a common origin. Then either of these systems may be transformed into the other by a suitable rotation of coordinates about O.

If a point P not at the origin has the coordinates (x_1, x_2, x_3) and (y_1, y_2, y_3) in the respective systems, it follows from Pythagoras' theorem that

$$x_1^2 + x_2^2 + x_3^2 = y_1^2 + y_2^2 + y_3^2,$$

because length OP is the same in each system. It is this preservation of the algebraic form determining length when an orthogonal coordinate system is rotated about its origin that underlies all considerations concerning orthogonal matrices.

We now generalize this property to n dimensions and express the result in terms of matrices. Let $\mathbf{A}$ be an $n \times n$ matrix and $\mathbf{x}$, $\mathbf{y}$ and n element column vectors, all with real elements. Then the linear transformation

$$\mathbf{y} = \mathbf{A}\mathbf{x} \tag{32}$$

maps any vector $\mathbf{x}$ in $\mathbb{R}^n$ into another vector $\mathbf{y}$ also in $\mathbb{R}^n$. The squares $\|\mathbf{x}\|^2$, $\|\mathbf{y}\|^2$ of the euclidean lengths of $\mathbf{x}$ and $\mathbf{y}$ are

$$\|\mathbf{x}\|^2 = x_1^2 + x_2^2 + \ldots x_n^2 = \mathbf{x}^T\mathbf{x}$$

and

$$\|\mathbf{y}\|^2 = y_1^2 + y_2^2 + \ldots + y_n^2 = \mathbf{y}^T\mathbf{y}.$$

Let us discover how these are related by (32).

Taking the transpose of (32) and forming the product $\mathbf{y}^T\mathbf{y}$ we obtain

$$\mathbf{y}^T\mathbf{y} = \mathbf{x}^T\mathbf{A}^T\mathbf{A}\mathbf{x},$$

or

$$\|\mathbf{y}\|^2 = \mathbf{x}^T\mathbf{A}^T\mathbf{A}\mathbf{x}. \tag{33}$$

If instead of being arbitrary $\mathbf{A}$ has the special property that

$$\mathbf{A}^T\mathbf{A} = \mathbf{I}, \tag{34}$$

then (33) becomes

$$\|\mathbf{y}\|^2 = \|\mathbf{x}\|^2. \tag{35}$$

This establishes that for this class of matrices transformation (32) preserves length. This is the very property with which we began our discussion. It is immaterial whether we consider that (32) rotates $\mathbf{x}$ to become $\mathbf{y}$, or whether it expresses the relationship between the coordinates of the same vector in different orthogonal coordinate systems obtained by rotation about the origin.

An $n \times n$ matrix $\mathbf{A}$ with the property

$$\mathbf{A}^T\mathbf{A} = \mathbf{I}, \tag{36}$$

is called an **orthogonal** matrix. Hereafter, we shall denote an orthogonal matrix by the symbol $\mathbf{Q}$. The following properties of an orthogonal matrix are a direct consequence of (36):

$$\mathbf{Q}^T = \mathbf{Q}^{-1}, \quad \text{so} \quad \mathbf{Q}^T\mathbf{Q} = \mathbf{I} \text{ and } \mathbf{Q}\mathbf{Q}^T = \mathbf{I}. \tag{37}$$

The row permutation matrix $\mathbf{P}$ introduced at the end of Sec. 3.3 is a simple example of an orthogonal matrix.

Example 8. Orthogonal matrix

Show that

$$Q = \begin{bmatrix} \cos\theta & \sin\theta \\ -\sin\theta & \cos\theta \end{bmatrix}$$

is an orthogonal matrix.

Solution

To show Q is orthogonal it is only necessary to prove that $Q^T Q = I$, and this follows by direct calculation.   ∎

If the columns of Q are represented by the vectors $q_1, q_2, \ldots, q_n$, matrix Q may be expressed in the partitioned form $Q = [q_1 \; q_2 \; \cdots \; q_n]$, and so

$$Q^T = \begin{bmatrix} q_1^T \\ q_2^T \\ \vdots \\ q_n^T \end{bmatrix}.$$

Then since $Q^T Q = I$,

$$Q^T Q = \begin{bmatrix} q_1^T q_1 & q_1^T q_2 & q_1^T q_3 & \cdots & q_1^T q_n \\ q_2^T q_1 & q_2^T q_2 & q_2^T q_3 & \cdots & q_2^T q_n \\ q_3^T q_1 & q_3^T q_2 & q_3^T q_3 & \cdots & q_3^T q_n \\ \cdot & \cdot & \cdot & \cdots & \cdot \\ q_n^T q_1 & q_n^T q_2 & q_n^T q_3 & \cdots & q_n^T q_n \end{bmatrix} = \begin{bmatrix} 1 & 0 & 0 & \cdots & 0 \\ 0 & 1 & 0 & \cdots & 0 \\ 0 & 0 & 1 & \cdots & 0 \\ \cdot & \cdot & \cdot & \cdots & \cdot \\ 0 & 0 & 0 & \cdots & 1 \end{bmatrix},$$

showing that

$$q_i^T q_j = \begin{cases} 0, & i \neq j \\ 1, & i = j \end{cases} \quad \text{for } i, j = 1, 2, \ldots, n. \tag{38}$$

In terms of matrix vectors, result (38) asserts that the columns of an orthogonal matrix are **orthonormal**; that is the columns are *mutually orthogonal* and such that the vectors comprising the columns are all of *unit length* (their norm is 1).

As might be anticipated, the eigenvalue problem for orthogonal matrices is simpler than for other matrices. The main property of the eigenvalues is contained in (iv) of the following theorem which also lists other useful properties of orthogonal matrices. Result (v) will be needed in Sec. 3.12.

Theorem 3.31 (Orthogonal matrices)

(i) If Q is an orthogonal matrix, then det $Q = \pm 1$;

(ii) If Q_1 and Q_2 are two orthogonal $n \times n$ matrices, then their product $Q_1 Q_2$ is an orthogonal matrix;

(iii) If the rows or columns of an orthogonal matrix are permuted, the resulting matrix is also on orthogonal matrix;

(iv) The eigenvalues of an orthogonal matrix are all of unit modulus;

(v) A matrix whose columns are the normalized eigenvectors of a symmetric matrix with distinct eigenvalues is orthogonal. This result is still true if a symmetric matrix has a repeated eigenvalue, provided the normalized eigenvectors corresponding to the repeated eigenvalue are chosen so they are mutually orthogonal.

Proof

Result (i) follows from the fact that $\det \mathbf{Q}^T = \det \mathbf{Q}$ and $\det(\mathbf{Q}^{-1}) = 1/\det \mathbf{Q}$, showing that $(\det \mathbf{Q})^2 = 1$.

To establish (ii) we start from $\mathbf{Q}_1^T \mathbf{Q}_1 = \mathbf{I}$ and $\mathbf{Q}_2^T \mathbf{Q}_2 = \mathbf{I}$. We have $(\mathbf{Q}_1 \mathbf{Q}_2)^T \mathbf{Q}_1 \mathbf{Q}_2 = \mathbf{Q}_2^T \mathbf{Q}_1^T \mathbf{Q}_1 \mathbf{Q}_2 = \mathbf{Q}_2^T \mathbf{Q}_2 = \mathbf{I}$, and the result is proved.

In the case of column permutations, result (iii) follows from the fact that the columns of $\mathbf{Q}$ are orthonormal. Permuting the columns of $\mathbf{Q}$ to produce a matrix $\mathbf{Q}_1$ causes a corresponding permutation in the rows of $\mathbf{Q}_1^T$ so it is still true that $\mathbf{Q}_1^T \mathbf{Q}_1 = \mathbf{I}$. A similar proof applies to row permutations.

To establish (iv) we must again appeal to the elementary properties of complex numbers. Consider the eigenvalue problem

$$\mathbf{Q}\mathbf{x} = \lambda \mathbf{x}.$$

Taking the complex conjugate of this result and using the fact that $\mathbf{Q}$ is real gives

$$\mathbf{Q}\bar{\mathbf{x}} = \bar{\lambda}\bar{\mathbf{x}}, \quad \text{so} \quad \bar{\mathbf{x}}^T \mathbf{Q}^T = \bar{\lambda}\bar{\mathbf{x}}^T.$$

Thus

$$\bar{\mathbf{x}}^T \mathbf{Q}^T \mathbf{Q}\mathbf{x} = \lambda\bar{\lambda}\mathbf{x}^T \mathbf{x},$$

but as $\mathbf{Q}$ is orthogonal $\mathbf{Q}^T \mathbf{Q} = \mathbf{I}$, causing the result to simplify to

$$\bar{\mathbf{x}}^T \mathbf{x} = \lambda\bar{\lambda}\bar{\mathbf{x}}^T \mathbf{x}.$$

Hence, since $\bar{\mathbf{x}}^T \mathbf{x} = \|\mathbf{x}\|^2 = 1$, it follows that $\lambda\bar{\lambda} = |\lambda|^2 = 1$, and we have proved that the eigenvalues λ have unit modulus.

Result (v) is an immediate consequence of Theorem 3.29 (ii). The normalization of the eigenvectors is necessary because the columns (rows) of an orthogonal matrix must be orthonormal. If when they are determined the normalized eigenvectors corresponding to a repeated eigenvalue are not orthogonal they are unsuitable as columns of $\mathbf{Q}$ and must be replaced by an equivalent orthonormal set. This may be constructed by means of the **Gram–Schmidt orthogonalization process** (Prob. 36, Sec. 3.10). ☐

Problems for Section 3.10

Find the eigenvalues and spectral radius for each of the following matrices. In addition, when the eigenvalues are real, find both the eigenvectors and normalized eigenvectors of the matrix and thus determine a basis for the solution space.

1 $\begin{bmatrix} 7 & 0 \\ 0 & -2 \end{bmatrix}$ **2** $\begin{bmatrix} 1 & 3 \\ 0 & -11 \end{bmatrix}$ **3** $\begin{bmatrix} 5 & 0 \\ 1 & -9 \end{bmatrix}$

4 $\begin{bmatrix} 2 & 1 \\ 1 & 2 \end{bmatrix}$ **5** $\begin{bmatrix} 1 & 3 \\ 1 & 2 \end{bmatrix}$ **6** $\begin{bmatrix} 5 & 4 \\ 1 & 2 \end{bmatrix}$

7 $\begin{bmatrix} 2 & 4 & 1 \\ 1 & -2 & -1 \\ 0 & 0 & 0 \end{bmatrix}$ **8** $\begin{bmatrix} 2 & -\frac{5}{2} & 1 \\ 2 & -2 & 2 \\ 0 & 0 & 0 \end{bmatrix}$ **9** $\begin{bmatrix} 4 & 1 & 0 \\ 0 & 4 & 0 \\ 0 & 0 & 4 \end{bmatrix}$

10 $\begin{bmatrix} 7 & 0 & 0 \\ 3 & 7 & 0 \\ 1 & 0 & 1 \end{bmatrix}$ **11** $\begin{bmatrix} 4 & 0 & 0 \\ 0 & 1 & 3 \\ 0 & 3 & 1 \end{bmatrix}$ **12** $\begin{bmatrix} -11 & 0 & 0 \\ 0 & 2 & 5 \\ 0 & 0 & 2 \end{bmatrix}$

Verify Theorem 3.24 in the case of the following matrices.

13 $\begin{bmatrix} 2 & 2 & 0 \\ 1 & 2 & 0 \\ 2 & -2 & 3 \end{bmatrix}$ **14** $\begin{bmatrix} 3 & 2 & 2 \\ 1 & 4 & 1 \\ -2 & -4 & -1 \end{bmatrix}$

Find the eigenvalues and eigenvectors of the rational function $R(A) = N(A)[D(A)]^{-1}$ when **A** and the matrix polynomials $N(A)$ and $D(A)$ are defined as follows.

15 $A = \begin{bmatrix} 3 & 0 \\ 3 & 4 \end{bmatrix}$, $N(A) = A + 2I$, $D(A) = A - I$.

16 $A = \begin{bmatrix} 1 & 5 \\ 2 & 4 \end{bmatrix}$, $N(A) = A^2 - 3A + 5I$, $D(A) = 2A^2 - I$.

17 $A = \begin{bmatrix} 1 & 4 \\ 0 & -4 \end{bmatrix}$, $N(A) = A^2 + A - 3I$, $D(A) = A^3$.

18 $A = \begin{bmatrix} 1 & 8 \\ -2 & 11 \end{bmatrix}$, $N(A) = A^3 - 3I$, $D(A) = A^3 + 3I$.

19 Verify Theorem 3.27 when
   (i) **A** is the matrix in Prob. 7;
   (ii) **A** is the matrix in Prob. 10;
   (iii) **A** is the matrix in Prob. 11.
20 Complete the proof of Theorem 3.28 (i).

Using a nonsingular 3×3 matrix **M** of your own choice, verify by direct calculation that the eigenvalues of the following matrices **A** and those of the similar matrices $M^{-1}AM$ are identical.

21 $A = \begin{bmatrix} 2 & -2 & 3 \\ 1 & 1 & 1 \\ 1 & 3 & -1 \end{bmatrix}$ **22** $A = \begin{bmatrix} 1 & -1 & 1 \\ 4 & 0 & -1 \\ 4 & -2 & 1 \end{bmatrix}$

23 Verify Theorem 3.29 when

(i) $A = \begin{bmatrix} 1 & 2 & 0 \\ 2 & 2 & -2 \\ 0 & -2 & 3 \end{bmatrix}$ (ii) $A = \begin{bmatrix} 1 & -2 & 0 \\ -2 & 1 & 0 \\ 0 & 0 & -1 \end{bmatrix}$.

Sketch the Gerschgorin circles for each of the following matrices, marking on the sketch the location of the eigenvalues. When the eigenvalues lie either to the left or right of the imaginary axis (y-axis), find whether the union of the circles also lies to the left of right of the imaginary axis, respectively.

24 $A = \begin{bmatrix} 2 & 0 & 0 \\ -\frac{1}{2} & 2 & 0 \\ \frac{1}{4} & 0 & 1 \end{bmatrix}$ **25** $A = \begin{bmatrix} 5 & -2 & 0 \\ -2 & 6 & 2 \\ 0 & 2 & 7 \end{bmatrix}$

26 $A = \begin{bmatrix} 1 & 2 & 4 \\ 2 & -2 & 2 \\ 4 & 2 & 1 \end{bmatrix}$ **27** $A = \begin{bmatrix} 1 & 0 & -1 \\ 1 & 2 & 1 \\ 2 & 2 & 3 \end{bmatrix}$

28 Show that

$$\begin{bmatrix} -\frac{2}{3} & \frac{1}{3} & \frac{2}{3} \\ \frac{1}{3} & -\frac{2}{3} & \frac{2}{3} \\ -\frac{2}{3} & -\frac{2}{3} & -\frac{1}{3} \end{bmatrix}$$

is orthogonal.

29 Which of the following matrices are orthogonal?

(i) $\begin{bmatrix} 1 & 0 & 0 \\ 0 & -\sin\theta & \cos\theta \\ 0 & -\cos\theta & \sin\theta \end{bmatrix}$ (ii) $\begin{bmatrix} 0 & \cos\theta & -\sin\theta \\ 1 & 0 & 0 \\ 0 & \sin\theta & \cos\theta \end{bmatrix}$

(iii) $\begin{bmatrix} 0 & \dfrac{1}{\sqrt{2}} & -\dfrac{1}{\sqrt{2}} \\ -\dfrac{1}{\sqrt{2}} & \dfrac{1}{2} & \dfrac{1}{2} \\ -\dfrac{1}{\sqrt{2}} & -\dfrac{1}{2} & -\dfrac{1}{2} \end{bmatrix}$.

30 Show that the eigenvalues of

$$\begin{bmatrix} 1 & 0 & 0 \\ 0 & \cos\theta & -\sin\theta \\ 0 & \sin\theta & \cos\theta \end{bmatrix}$$ are $\lambda = 1$ and $\lambda = \cos\theta \pm i\sin\theta$.

Harder problems

31 Complete the proof of Theorem 3.24 (i).

32 Prove that the spectral radius $\rho(A)$ of an $n \times n$ matrix A satisfies the inequality

$$\rho(A) \leq \max_i \sum_{j=1}^{n} |a_{ij}|.$$

33 Use Theorem 3.24 (ii) together with Theorem 3.30 to prove that if A is an $n \times n$ matrix for which the radius ρ_r of the rth Gerschgorin circle C_r satisfies the inequality

$$\rho_r < |a_{rr}| \qquad \text{for } r = 1, 2, \ldots, n,$$

then $\det A \neq 0$. Give a geometrical interpretation of the inequality.

34 Prove that if λ is an eigenvalue of an orthogonal matrix Q, then so also is $1/\lambda$.

35 If Q is an orthogonal matrix, then $\det Q = \pm 1$. Q is said to be **proper** if $\det Q = 1$ and **improper** if $\det Q = -1$. Prove that the product of two proper or two improper orthogonal matrices is a proper orthogonal matrix.

36 The **Gram–Schmidt orthogonalization process** replaces a set of linearly independent column vectors $a_1, a_2, \ldots, a_n$ by a set of mutually orthogonal column vectors $k_1, k_2, \ldots, k_n$. The *algorithm* by which they are generated is

$$k_1 = a_1,$$

and

$$k_i = a_i - \frac{k_1^T a_i}{k_1^T k_1} k_1 - \frac{k_2^T a_i}{k_2^T k_2} k_2 - \cdots - \frac{k_{i-1}^T a_i}{k_{i-1}^T k_{i-1}} k_{i-1},$$

for $i = 2, 3, \ldots, n$.

Give an inductive proof that $k_i^T k_j = 0$ for $i \neq j$, and hence that the set of column vectors $\hat{k}_1, \hat{k}_2, \ldots, \hat{k}_n$ is orthonormal, where $\hat{k}_i = k_i / (k_i^T k_i)$, for $i = 1, 2, \ldots n$.

When $a_1, a_2, \ldots, a_n$ are n element column vectors forming a basis for a space $\mathbb{R}^n$, the Gram–Schmidt orthogonalization process shows that this basis may always be replaced by an equivalent orthogonal basis $k_1, k_2, \ldots, k_n$.

By means of the Gram–Schmidt process, find an orthonormal basis for each of the following systems of vectors. Use the normalized vectors in each orthogonal basis as the columns in a matrix to construct an orthogonal matrix Q.

37 $a_1 = \begin{bmatrix} 1 \\ 2 \end{bmatrix}, \qquad a_2 = \begin{bmatrix} -1 \\ 1 \end{bmatrix}.$ **38** $a_1 = \begin{bmatrix} 1 \\ -1 \end{bmatrix}, \qquad a_2 = \begin{bmatrix} 0 \\ 1 \end{bmatrix}.$

39 $a_1 = \begin{bmatrix} 1 \\ 1 \\ 1 \end{bmatrix}, \qquad a_2 = \begin{bmatrix} 0 \\ 1 \\ 0 \end{bmatrix}, \qquad a_3 = \begin{bmatrix} -1 \\ 0 \\ 1 \end{bmatrix}.$

40 $a_1 = \begin{bmatrix} 1 \\ 0 \\ 0 \end{bmatrix}, \qquad a_2 = \begin{bmatrix} 0 \\ -1 \\ 0 \end{bmatrix}, \qquad a_3 = \begin{bmatrix} 1 \\ 0 \\ 2 \end{bmatrix}.$

41 Use the Gram–Schmidt process to find a basis for the vectors

$$\mathbf{a}_1 = \begin{bmatrix} 1 \\ -1 \\ 2 \end{bmatrix}, \qquad \mathbf{a}_2 = \begin{bmatrix} 2 \\ 1 \\ 1 \end{bmatrix}, \qquad \mathbf{a}_3 = \begin{bmatrix} 4 \\ -1 \\ 5 \end{bmatrix}.$$

For what type of space is this a basis?

42 When $\mathbf{a}_1, \mathbf{a}_2$ and $\mathbf{a}_3$ are geometrical vectors in $\mathbb{R}^3$, the Gram–Schmidt orthogonalization process takes the form

$$\mathbf{k}_1 = \mathbf{a}_1, \quad \mathbf{k}_2 = \mathbf{a}_2 - \frac{\mathbf{k}_1 \cdot \mathbf{a}_2}{\|\mathbf{k}_1\|^2} \mathbf{k}_1$$

and

$$\mathbf{k}_3 = \mathbf{a}_3 - \frac{\mathbf{k}_1 \cdot \mathbf{a}_3}{\|\mathbf{k}_1\|^2} \mathbf{k}_1 - \frac{\mathbf{k}_2 \cdot \mathbf{a}_3}{\|\mathbf{k}_2\|^2} \mathbf{k}_2,$$

where $\mathbf{k}_1, \mathbf{k}_2$ and $\mathbf{k}_3$ is a triad of orthogonal geometrical vectors. Interpret this construction of $\mathbf{k}_1, \mathbf{k}_2$ and $\mathbf{k}_3$ in terms of geometry.

Use this result to find $\mathbf{k}_1, \mathbf{k}_2$ and $\mathbf{k}_3$ when $\mathbf{a}_1 = \mathbf{i} + \mathbf{j} + 2\mathbf{k}, \mathbf{a}_2 = \mathbf{i} + \mathbf{j}$ and $\mathbf{a}_3 = \mathbf{i} + \mathbf{k}$. Obtain the same triad of orthogonal vectors using geometrical arguments based on the vector product.

43 Find the eigenvalues and eigenvectors of the $n \times n$ matrix

$$\mathbf{A} = \begin{bmatrix} a & b & b & b & \dots & b \\ b & a & b & b & \dots & b \\ b & b & a & b & \dots & b \\ \cdot & \cdot & \cdot & \cdot & \cdot & \cdot \\ b & b & b & b & \dots & a \end{bmatrix}.$$

3.11 Diagonalizability of matrices. The Cayley–Hamilton theorem

Many problems leading to matrices are better understood if a transformation can be found which simplifies the structure of a matrix while leaving its eigenvalues unchanged. As it was established in Theorem 3.28 (ii) that similar matrices have the same eigenvalues, this suggests that simplification of the structure of an arbitrary $n \times n$ matrix $\mathbf{A}$ should be attempted by means of a similarity transformation.

The greatest possible simplification of structure arises when $\mathbf{A}$ is similar to a diagonal matrix

$$\Lambda = \begin{bmatrix} d_1 & 0 & 0 & \cdots & 0 \\ 0 & d_2 & 0 & \cdots & 0 \\ 0 & 0 & d_3 & \cdots & 0 \\ \cdot & \cdot & \cdot & \cdots & 0 \\ 0 & 0 & 0 & \cdots & d_n \end{bmatrix}. \tag{1}$$

In terms of a similarity transformation this is equivalent to requiring a nonsingular matrix **P** to exist with the property that

$$\mathbf{P}^{-1}\mathbf{A}\mathbf{P}=\Lambda. \tag{2}$$

When such a similarity transformation can be found, the matrix **A** is said to be **diagonalizable**, and **P** is said to be a **diagonalizing matrix** for **A**.

Thus we need to find necessary and sufficient conditions for **A** to be diagonalizable and, when it is, a method by which the diagonalizing matrix **P** may be constructed. As a preliminary to this task we make the following observations about the diagonal matrix Λ;

(i) if **P** is an arbitrary $n \times n$ matrix with columns denoted by the vectors $\mathbf{p}_1, \mathbf{p}_2, \ldots \mathbf{p}_n$, then the vector comprising the ith column of the matrix product $\mathbf{P}\Lambda$ is $d_i\mathbf{p}_i$;
(ii) the n eigenvalues of Λ are the elements $d_1, d_2, \ldots, d_n$ of the leading diagonal;
(iii) the n linearity independent normalized eigenvectors $l_1, l_2, \ldots, l_n$ of Λ corresponding to the eigenvalues $d_1, d_2, \ldots, d_n$ are

$$l_1 = \begin{bmatrix} 1 \\ 0 \\ 0 \\ \vdots \\ 0 \end{bmatrix}, \quad l_2 = \begin{bmatrix} 0 \\ 1 \\ 0 \\ \vdots \\ 0 \end{bmatrix}, \quad \ldots, \quad l_n = \begin{bmatrix} 0 \\ 0 \\ 0 \\ \vdots \\ 1 \end{bmatrix}.$$

Notice that a diagonal matrix always has n linearly independent eigenvectors, because to an eigenvalue with algebraic multiplicity r there correspond r linearly independent eigenvectors. This means that in a diagonal matrix the algebraic and geometric multiplicity of an eigenvalue is always the same.

Let **A** be a matrix with n linearly independent eigenvectors corresponding to the eigenvalues $\lambda_1, \lambda_2, \ldots, \lambda_n$, respectively, some of which may be repeated. Then

$$\mathbf{A}\mathbf{p}_i = \lambda_i\mathbf{p}_i \quad \text{for } i = 1, 2, \ldots, n. \tag{3}$$

Define **P** to be the $n \times n$ matrix whose columns are the eigenvector of **A**; that is, the ith column of **P** is the eigenvectors $\mathbf{p}_i$, for $i = 1, 2, \ldots, n$. The columns of **P** are linearly independent, so rank $\mathbf{P} = n$ and thus det $\mathbf{P} \neq 0$, showing that $\mathbf{P}^{-1}$ exists. We now prove that this choice of **P** in similarity transformation (2) will diagonalize **A**.

Pre-multiply (2) by **P** to obtain

$$\mathbf{A}\mathbf{P} = \mathbf{P}\Lambda. \tag{4}$$

Writing **P** in the partitioned form $\mathbf{P} = \begin{bmatrix} \mathbf{p}_1 & \mathbf{p}_2 & \ldots & \mathbf{p}_n \end{bmatrix}$, where vector $\mathbf{p}_i$ represents the ith column of **P**, substituting into (4), and using property (i) to interpret the product $\mathbf{P}\Lambda$, leads to the matrix equation

$$\begin{bmatrix} \mathbf{A}\mathbf{p}_1 & \mathbf{A}\mathbf{p}_2 & \ldots & \mathbf{A}\mathbf{p}_n \end{bmatrix} = \begin{bmatrix} d_1\mathbf{p}_1 & d_2\mathbf{p}_2 & \ldots & d_n\mathbf{p}_n \end{bmatrix}. \tag{5}$$

Equating corresponding columns on either side of (5) gives the n matrix equations

$$\mathbf{A}\mathbf{p}_i = d_i \mathbf{p}_i, \qquad \text{for } i = 1, 2, \ldots, n. \tag{6}$$

Comparison of (3) and (6) shows this choice of $\mathbf{P}$ will indeed diagonalize $\mathbf{A}$, and that $\lambda_i = d_i$, for $i = 1, 2, \ldots, n$. The method of construction of $\mathbf{P}$ ensures that the *order* in which the *eigenvectors* occur in the matrix is the same as the *order* in which the corresponding *eigenvalues* occur along the diagonal of Λ.

Diagonalization of $\mathbf{A}$ is only possible if it has n linearly independent eigenvectors, for if this is not the case $\mathbf{P}$ cannot be constructed. Consequently, if $\mathbf{A}$ does not have n linearly independent eigenvectors if cannot be diagonalized. We have proved the following result.

Theorem 3.32 (Diagonalizability of matrices)

(i) The necessary and sufficient condition that an $n \times n$ matrix $\mathbf{A}$ should be diagonalizable is that it has n linearly independent eigenvectors;

(ii) A diagonalizable matrix $\mathbf{A}$ will be diagonalized by a similarity transformation involving the matrix $\mathbf{P}$ whose columns are the eigenvectors of $\mathbf{A}$;

(iii) If a diagonalizable matrix $\mathbf{A}$ has eigenvalues $\lambda_1, \lambda_2, \ldots, \lambda_n$ with the corresponding eigenvectors $\mathbf{p}_1, \mathbf{p}_2, \ldots, \mathbf{p}_n$, respectively, and the ith column of $\mathbf{P}$ is $\mathbf{p}_i$, for $i = 1, 2, \ldots, n$, then

$$\mathbf{P}^{-1}\mathbf{A}\mathbf{P} = \Lambda,$$

where

$$\Lambda = \begin{bmatrix} \lambda_1 & 0 & 0 & \cdots & 0 \\ 0 & \lambda_2 & 0 & \cdot & 0 \\ 0 & 0 & \lambda_3 & \cdots & 0 \\ \cdot & \cdot & \cdot & \cdots & \cdot \\ 0 & 0 & 0 & \cdots & \lambda_n \end{bmatrix}$$

(iv) When it exists, the diagonalizing matrix $\mathbf{P}$ is not unique. If the ith and jth columns of $\mathbf{P}$ are interchanged, then so also are the entries λ_i and λ_j on the leading diagonal of Λ.

(v) An $n \times n$ matrix $\mathbf{A}$ with fewer than n linearly independent eigenvectors cannot be diagonalized. □

Three immediate and useful deductions from Theorem 3.32 are the following:

(i) a real symmetric matrix may always be diagonalized;

(ii) a matrix with distinct eigenvalues may always be diagonalized;

(iii) a diagonalizable matrix $\mathbf{A}$ with real elements may give rise to a diagonalizing matrix $\mathbf{P}$ with complex elements.

Results (i) and (ii) follow from the fact that real symmetric matrices and matrices with distinct eigenvalues always have a full set of linearly independent eigenvectors. Result (iii) follows because the zeros of a characteristic polynomial with real coefficients may be complex (Theorem 1.4).

Example 1. Diagonalization of matrices

Determine if the following matrices can be diagonalized and, where possible, find a diagonalizing matrix **P**.

$$\text{(i) } \mathbf{A} = \begin{bmatrix} 1 & 0 & 0 \\ 1 & 3 & 2 \\ 1 & 2 & 3 \end{bmatrix}, \quad \text{(ii) } \mathbf{A} = \begin{bmatrix} 5 & 10 & 7 \\ 0 & -3 & -3 \\ 0 & 3 & 3 \end{bmatrix}, \quad \text{(iii) } \mathbf{A} = \begin{bmatrix} 1 & 1 \\ -1 & 2 \end{bmatrix}.$$

Solution

(i) This matrix was consider in Ex. 4, Sec. 3.10. **A** has the eigenvalues $\lambda_1 = 5$, $\lambda_2 = 1$ and $\lambda_3 = 1$, so an eigenvalue is repeated. However **A** has the three linearly independent eigenvectors

$$x_1 = \begin{bmatrix} 0 \\ 1 \\ 1 \end{bmatrix}, \quad x_2 = \begin{bmatrix} -2 \\ 1 \\ 0 \end{bmatrix} \quad \text{and} \quad x_3 = \begin{bmatrix} -2 \\ 0 \\ 1 \end{bmatrix},$$

and so it is diagonalizable. Constructing **P** gives

$$\mathbf{P} = \begin{bmatrix} 0 & -2 & -2 \\ 1 & 1 & 0 \\ 1 & 0 & 1 \end{bmatrix},$$

from which it follows that

$$\mathbf{P}^{-1}\mathbf{AP} = \begin{bmatrix} 5 & 0 & 0 \\ 0 & 1 & 0 \\ 0 & 0 & 1 \end{bmatrix}.$$

Had the eigenvectors been ordered differently in **P** so that, say,

$$\mathbf{P} = \begin{bmatrix} -2 & 0 & -2 \\ 1 & 1 & 0 \\ 0 & 1 & 1 \end{bmatrix},$$

then the diagonalized matrix would become

$$\mathbf{P}^{-1}\mathbf{AP} = \begin{bmatrix} 1 & 0 & 0 \\ 0 & 5 & 0 \\ 0 & 0 & 1 \end{bmatrix}.$$

(ii) This matrix was considered in Ex. 5, Sec. 3.10. **A** has the eigenvalues $\lambda_1 = 5$, $\lambda_2 = 0$ and $\lambda_3 = 0$, so an eigenvalue is repeated. However, **A** has only two linearly independent eigenvectors and so is not diagonalizable.

(iii) This matrix was considered in Ex. 2, Sec. 3.10. $\mathbf{A}$ has the complex eigenvalues $\lambda_1 = \frac{1}{2}(3 - i\sqrt{3})$ and $\lambda_2 = \frac{1}{2}(3 + i\sqrt{3})$, and the complex linearly independent eigenvectors

$$\mathbf{x}_1 = \begin{bmatrix} 1 \\ \frac{1}{2}(1 - i\sqrt{3}) \end{bmatrix} \quad \text{and} \quad \mathbf{x}_2 = \begin{bmatrix} 1 \\ \frac{1}{2}(1 + i\sqrt{3}) \end{bmatrix}$$

and so is diagonalizable. Constructing $\mathbf{P}$ gives

$$\mathbf{P} = \begin{bmatrix} 1 & 1 \\ \frac{1}{2}(1 - i\sqrt{3}) & \frac{1}{2}(1 + i\sqrt{3}) \end{bmatrix},$$

from which is follows that

$$\mathbf{P}^{-1}\mathbf{A}\mathbf{P} = \begin{bmatrix} \frac{1}{2}(3 - i\sqrt{3}) & 0 \\ 0 & \frac{1}{2}(3 + i\sqrt{3}) \end{bmatrix}.$$

Thus, in this case, a real matrix is similar to a complex diagonal matrix. ∎

We now make use of the process of diagonalization to prove a special case of the *Cayley–Hamilton theorem*.

Consider an arbitrary $n \times n$ matrix $\mathbf{A}$ and denote its characteristic polynomial by $P_n(\lambda)$, so that

$$P_n(\lambda) \equiv \det|\mathbf{A} - \lambda\mathbf{I}|. \tag{7}$$

When expanded, $P_n(\lambda)$ may always be written in the form

$$P_n(\lambda) \equiv (-1)^n \lambda^n \mid c_{n-1}\lambda^{n-1} + \ldots + c_1\lambda + c_0. \tag{8}$$

The eigenvalues $\lambda_1, \lambda_2, \ldots, \lambda_n$ of $\mathbf{A}$, which may be multiple, are the zeros of $P_n(\lambda)$, so that

$$P_n(\lambda_i) = 0, \qquad i = 1, 2, \ldots, n. \tag{9}$$

Now suppose $\mathbf{A}$ is diagonalizable by means of the matrix $\mathbf{P}$ defined in Theorem 3.32, and that its diagonal form is the matrix Λ in that same theorem. Then

$$\mathbf{P}^{-1}\mathbf{A}\mathbf{P} = \Lambda,$$

and so

$$\mathbf{A} = \mathbf{P}\Lambda\mathbf{P}^{-1}. \tag{10}$$

It then follows that

$$\mathbf{A}^2 = \mathbf{P}\Lambda\mathbf{P}^{-1}\mathbf{P}\Lambda\mathbf{P}^{-1} = \mathbf{P}\Lambda^2\mathbf{P}^{-1},$$
$$\mathbf{A}^3 = \mathbf{A}\mathbf{A}^2 - \mathbf{P}\Lambda\mathbf{P}^{-1}\mathbf{P}\Lambda^2\mathbf{P}^{-1} = \mathbf{P}\Lambda^3\mathbf{P}^{-1}$$

and, in general,

$$\mathbf{A}^m = \mathbf{P}\Lambda^m\mathbf{P}^{-1}, \qquad m = 1, 2, \ldots. \tag{11}$$

Now, if

$$\boldsymbol{\Lambda} = \begin{bmatrix} \lambda_1 & & & & 0 \\ & \lambda_2 & & & \\ & & \ddots & & \\ 0 & & & & \lambda_n \end{bmatrix}, \quad \text{then} \quad \boldsymbol{\Lambda}^m = \begin{bmatrix} \lambda_1^m & & & & 0 \\ & \lambda_2^m & & & \\ & & \ddots & & \\ 0 & & & & \lambda_n^m \end{bmatrix}$$

and so

$$\boldsymbol{A}^m = \boldsymbol{P} \begin{bmatrix} \lambda_1^m & & & & 0 \\ & \lambda_2^m & & & \\ & & \ddots & & \\ 0 & & & & \lambda_n^m \end{bmatrix} \boldsymbol{P}^{-1}, \qquad m = 1, 2, \ldots. \tag{12}$$

Let us now examine the form of the matrix polynomial $P_n(\boldsymbol{A})$, that is the matrix polynomial of degree n derived from the characteristic polynomial by replacing λ by $\boldsymbol{A}$ and λ^0 by $\boldsymbol{I}$. We see that

$$P_n(\boldsymbol{A}) = (-1)^n \boldsymbol{A}^n + c_{n-1} \boldsymbol{A}^{n-1} + \ldots + c_1 \boldsymbol{A} + c_0 \boldsymbol{I},$$

but because of (12) this may be written as

$$P_n(\boldsymbol{A}) = \boldsymbol{P} \{ -1)^n \boldsymbol{\Lambda}^n + c_{n-1} \boldsymbol{\Lambda}^{n-1} + \ldots + c_1 \boldsymbol{\Lambda} + c_0 \boldsymbol{I} \} \boldsymbol{P}^{-1}. \tag{13}$$

The bracketed matrix polynomial in (13) is the sum of diagonal matrices, and so it is itself a diagonal matrix. The ith element in the leading diagonal is simply $P_n(\lambda_i)$, but λ_i is an eigenvalue of $\boldsymbol{A}$ so that $P_n(\lambda_i) = 0$. This is true for each element in the leading diagonal, so the bracketed matrix polynomial in (13) is the zero matrix and we have proved that

$$P_n(\boldsymbol{A}) = 0. \tag{14}$$

Expressed in words, this result asserts that $\boldsymbol{A}$ satisfies its own characteristic equation. Although we shall not prove it, this result is in fact true for an arbitrary $n \times n$ matrix $\boldsymbol{A}$ which may, or may not, be diagonalizable. This general result is known as the **Cayley–Hamilton theorem**.

Theorem 3.33 (Cayley–Hamilton theorem)

Let $\boldsymbol{A}$ be an arbitrary $n \times n$ matrix $\boldsymbol{A}$ with the characteristic polynomial

$$P_n(\lambda) = (-1)^n \lambda^n + c_{n-1} \lambda^{n-1} + \ldots + c_1 \lambda + c_0.$$

Then

$$P_n(\boldsymbol{A}) \equiv (-1)^n \boldsymbol{A}^n + c_{n-1} \boldsymbol{A}^{n-1} + \ldots + c_1 \boldsymbol{A} + c_0 \boldsymbol{I} = \boldsymbol{0}. \qquad \square$$

This theorem finds many applications in the theory of matrices, and it is also of some practical value when performing computations. Perhaps its most important consequence is that it shows that an arbitrary matrix polynomial of degree N involving the $n \times n$ matrix $\mathbf{A}$ may always be reduced to a matrix polynomial in $\mathbf{A}$ of degree less than n.

A simple application of the Cayley–Hamilton theorem is to the determination of $\mathbf{A}^{-1}$ when $\det \mathbf{A} \neq 0$. This will be the case if $c_0 \neq 0$, because $\det \mathbf{A} = P_n(0)$ and from (8) we see that $P_n(0) = c_0$. Thus, when $c_0 \neq 0$, pre-multiplication of the result of Theorem 3.33 by $\mathbf{A}^{-1}$ followed by a rearrangement of the terms gives

$$\mathbf{A}^{-1} = \frac{-1}{c_0}[(-1)^n \mathbf{A}^{n-1} + c_{n-1}\mathbf{A}^{n-2} + \ldots + c_1\mathbf{I}]. \tag{15}$$

Thus $\mathbf{A}^{-1}$ is expressible as a linear combination of $\mathbf{I}, \mathbf{A}, \mathbf{A}^2, \ldots, \mathbf{A}^{n-1}$.

Example 2. Applications of the Cayley–Hamilton Theorem

Given that

$$\mathbf{A} = \begin{bmatrix} -2 & 2 & 3 \\ -2 & 3 & 2 \\ -4 & 2 & 5 \end{bmatrix},$$

use the Cayley–Hamilton theorem to find (i) $\mathbf{A}^{-1}$, (ii) $\mathbf{A}^4$.

Solution

A routine calculation establishes that the characteristic polynomial

$$P_3(\lambda) = 6 - 11\lambda + 6\lambda^2 - \lambda^3,$$

and as $P_3(0) = 6$, $\mathbf{A}$ is nonsingular. Thus, by the Cayley–Hamilton theorem,

$$6\mathbf{I} - 11\mathbf{A} + 6\mathbf{A}^2 - \mathbf{A}^3 = 0.$$

Pre-multiplication by $\mathbf{A}^{-1}$ followed by a rearrangement of the terms then gives via result (15)

$$\mathbf{A}^{-1} = \tfrac{1}{6}\{11\mathbf{I} - 6\mathbf{A} + \mathbf{A}^2\}.$$

Now

$$\mathbf{A}^2 = \begin{bmatrix} -12 & 8 & 13 \\ -10 & 9 & 10 \\ -16 & 8 & 17 \end{bmatrix},$$

so using this in the expression for $\mathbf{A}^{-1}$ we find

$$\mathbf{A}^{-1} = \tfrac{1}{6}\begin{bmatrix} 11 & -4 & -5 \\ 2 & 2 & -2 \\ 8 & -4 & -2 \end{bmatrix}.$$

To determine $\mathbf{A}^4$ we start from the result that

$$\mathbf{A}^3 = 6\mathbf{I} - 11\mathbf{A} + 6\mathbf{A}^2.$$

Pre-multiplication by $\mathbf{A}$ gives

$$\mathbf{A}^4 = 6\mathbf{A} - 11\mathbf{A}^2 + 6\mathbf{A}^3,$$

but $\mathbf{A}^3 = 6\mathbf{I} - 11\mathbf{A} + 6\mathbf{A}^2$,

and so

$$\mathbf{A}^4 = 36\mathbf{I} - 60\mathbf{A} + 25\mathbf{A}^2.$$

Summing the matrices we find that

$$\mathbf{A}^4 = \begin{bmatrix} -144 & 80 & 145 \\ -130 & 81 & 130 \\ -160 & 80 & 161 \end{bmatrix}.$$

■

Low powers of $\mathbf{A}$ may be found as indicated in Example 2, but if high powers are required a better method is needed. The derivation and application of such a method forms the subject of the last part of this section. Let $\mathbf{A}$ be an $n \times n$ matrix with the nth degree characteristic polynomial $P_n(\lambda)$ and the n distinct eigenvalues $\lambda_1, \lambda_2, \dots, \lambda_n$, so that

$$P_n(\lambda_i) = 0, \qquad i = 1, 2, \dots, n. \tag{16}$$

Let $r \geq n$ be an integer. Then it follows by elementary division that

$$\frac{\lambda^r}{P_n(\lambda)} \equiv Q_{r-n}(\lambda) + \frac{b_{n-1}\lambda^{n-1} + b_{n-2}\lambda^{n-2} + \dots + b_1\lambda + b_0}{P_n(\lambda)}, \tag{17}$$

where $Q_{r-n}(\lambda)$ is a polynomial in λ of degree $r - n$, and $b_0, b_1, \dots, b_{n-1}$ are constants. This is equivalent to the identity

$$\lambda^r \equiv Q_{r-n}(\lambda)P_n(\lambda) + b_{n-1}\lambda^{n-1} + b_{n-2}\lambda^{n-2} + \dots + b_1\lambda + b_0. \tag{18}$$

Result (18) enables powers of $\mathbf{A}$ to be evaluated very simply. Replacing λ by $\mathbf{A}$ in (18), we obtain the matrix polynomial identity

$$\mathbf{A}^r \equiv Q_{r-n}(\mathbf{A})P_n(\mathbf{A}) + b_{n-1}\mathbf{A}^{n-1} + b_{n-2}\mathbf{A}^{n-2} + \dots + b_1\mathbf{A} + b_0\mathbf{I}, \qquad r \geq n. \tag{19}$$

However by the Cayley–Hamilton theorem $P_n(\mathbf{A}) = \mathbf{0}$, so (19) reduces to the identity

$$\mathbf{A}^r \equiv b_{n-1}\mathbf{A}^{n-1} + b_{n-2}\mathbf{A}^{n-2} + \dots + b_1\mathbf{A} + b_0\mathbf{I}. \tag{20}$$

To determine the coefficients $b_0, b_1, \dots, b_{n-1}$ we now use the fact that $P_n(\lambda_i) = 0$, and set $\lambda = \lambda_i$ in (18) to obtain the n simultaneous equations

$$(\lambda_i)^r = b_{n-1}(\lambda_i)^{n-1} + b_{n-2}(\lambda_i)^{n-2} + \dots + b_1\lambda_i + b_0, \qquad i = 1, 2, \dots, n. \tag{21}$$

When the $b_0, b_1, \dots, b_{n-1}$ have been determined from (21), $\mathbf{A}^r$ then follows from (20).

Example 3. Powers of A without repeated eigenvalues

Find $\mathbf{A}^r$ in terms of $\mathbf{A}$ and $\mathbf{I}$, given that

$$\mathbf{A} = \begin{bmatrix} 1 & 2 \\ 3 & 2 \end{bmatrix}.$$

Solution

$\mathbf{A}$ is a 2×2 matrix, so we see from (20) that we must seek a result of the form

$$\mathbf{A}^r = b_1 \mathbf{A} + b_0 \mathbf{I}, \qquad r \geq 2.$$

$P_2(\lambda) = \det|\mathbf{A} - \lambda \mathbf{I}| = \lambda^2 - 3\lambda - 4$, so the eigenvalues of $\mathbf{A}$ are distinct, with $\lambda_1 = 4$ and $\lambda_2 = -1$. Using these results in (21) we have

$$4^r = 4b_1 + b_0, \quad (-1)^r = -b_1 + b_0, \qquad r \geq 2,$$

and thus

$$b_0 = \frac{4^r + 4(-1)^r}{5}, \quad b_1 = \frac{4^r - (-1)^r}{5}, \qquad r \geq 1,$$

The required result is thus

$$\mathbf{A}^r = \frac{4^r - (-1)^r}{5} \mathbf{A} + \frac{4^r + 4(-1)^r}{5} \mathbf{I}, \qquad r \geq 2.$$

It is easily verified that this result is, in fact, also true for $r=0$ and $r=1$. ∎

In the event that eigenvalues with multiplicity greater than 1 arise the argument must be modified, because then (21) yields fewer than n equations from which to determine the n constants $b_0, b_1, \ldots, b_{n-1}$.

It will suffice to consider in detail only the case of an eigenvalue with multiplicity 2, because if a higher multiplicity occurs it may be treated in similar fashion. Suppose $\lambda_{n-1} = \lambda_n = \mu$ is the repeated eigenvalue, so we may write

$$P_n(\lambda) = (\lambda - \mu)^2 R_{n-2}(\lambda), \tag{22}$$

with $R_{n-2}(\lambda)$ is a polynomial in λ of degree $n-2$ for which $\lambda = \mu$ is not a zero. Then

$$P_n(\mu) = 0, \tag{23}$$

and because

$$P'_n(\lambda) = \frac{\mathrm{d}P_n}{\mathrm{d}\lambda} = 2(\lambda - \mu)R_{n-2}(\lambda) + (\lambda - \mu)^2 \frac{\mathrm{d}R_{n-2}}{\mathrm{d}\lambda},$$

it also follows that

$$P'_n(\mu) = 0. \tag{24}$$

The $n-1$ distinct eigenvalues $\lambda_1, \lambda_2, \ldots, \lambda_{n-2}$ and μ used in (21) yield only $n-1$ different equations. The remaining equation follows by differentiating (18) with respect to

λ, setting $\lambda = \mu$, and using (23) and (24) to obtain

$$r\mu^{r-1} = (n-1)b_{n-1}\mu^{n-2} + (n-2)b_{n-2}\mu^{n-3} + \ldots + 2b_2\mu + b_1. \tag{25}$$

We now determine the constants $b_0, b_1, \ldots, b_{n-1}$ from the $n-1$ equations (21) and the single equation (25).

If a triple eigenvalue occurs, say $\lambda_{n-2} = \lambda_{n-1} = \lambda_n = \mu$, then (21) yields only $n-2$ different equations. The argument then proceeds as follows. Setting

$$P_n(\lambda) = (\lambda - \mu)^3 R_{n-3}(\lambda), \tag{26}$$

repeated differentiation shows that

$$P_n(\mu) = P'_n(\mu) = P''_n(\mu) = 0. \tag{27}$$

Two successive differentiations of (18), followed by setting $\lambda = \mu$ and making use of (27) gives the two further equations

$$r\mu^{r-1} = (n-1)b_{n-1}\mu^{n-2} + (n-2)b_{n-2}\mu^{n-3} + \ldots + 2b_2\mu + b_1, \tag{28}$$

$$r(r-1)\mu^{r-2} = (n-1)(n-2)b_{n-1}\mu^{n-3} + (n-2)(n-3)b_{n-2}\mu^{n-4} + \ldots$$
$$+ 3\cdot 2b_3\mu + 2b_2. \tag{29}$$

The constants then follow by solving the $n-2$ equations (21) with (28) and (29).

Example 4. Powers of A when repeated eigenvalues are involved

Find A^r given that

$$A = \begin{bmatrix} 1 & 2 \\ 0 & 1 \end{bmatrix}.$$

Solution

A is a 2×2 matrix, so we see from (20) that we must seek a result of the form

$$A^r = b_1 A + b_2 I, \qquad r \geq 2.$$

$P_2(\lambda) = \det |A - \lambda I| = (1 - \lambda)^2$, so $\lambda = 1$ is a double eigenvalue of A. Setting $\lambda_2 = \lambda_1 = \mu$, with $\mu = 1$, (21) becomes

$$1 = b_1 + b_0$$

and (25) becomes

$$r = b_1.$$

Thus

$$b_0 = 1 - r, \quad b_1 = r,$$

and so A^r may be expressed in terms of A and I by

$$A^r = rA + (1-r)I, \qquad r \geq 2.$$

Inspection shows this is also true for $r=0$ and $r=1$. Combining the matrices on the right-hand side shows that

$$\mathbf{A}' = \begin{bmatrix} 1 & 2r \\ 0 & 1 \end{bmatrix}, \qquad r = 0, 1, 2, \ldots.$$

■

Problems for Section 3.11

Determine if the following matrices can be diagonalized and, where this is possible, find a diagonalizing matrix **P**.

1 $\begin{bmatrix} -2 & 1 \\ 1 & -2 \end{bmatrix}$ **2** $\begin{bmatrix} -1 & 2 & 4 \\ 2 & 2 & -2 \\ 4 & -2 & -1 \end{bmatrix}$ **3** $\begin{bmatrix} 5 & -2 & 0 \\ -2 & 6 & 2 \\ 0 & 2 & 7 \end{bmatrix}$

4 $\begin{bmatrix} 4 & 0 & 0 \\ 0 & 3 & 1 \\ 0 & 1 & 3 \end{bmatrix}$ **5** $\begin{bmatrix} -1 & 1 & 0 \\ -4 & 3 & 0 \\ 1 & 0 & 2 \end{bmatrix}$ **6** $\begin{bmatrix} 1 & 2 & 4 \\ 2 & -2 & 2 \\ 4 & 2 & 1 \end{bmatrix}$

7 $\begin{bmatrix} 5 & 0 & 0 \\ 0 & 0 & -1 \\ 0 & 1 & 0 \end{bmatrix}$ **8** $\begin{bmatrix} 1 & 0 & -3 \\ 0 & -2 & 0 \\ -3 & 0 & 1 \end{bmatrix}$

Use the Cayley–Hamilton theorem to find $\mathbf{A}^{-1}$ and $\mathbf{A}^3$ in each of the following problems.

9 $\mathbf{A} = \begin{bmatrix} 4 & 1 \\ 2 & 3 \end{bmatrix}$ **10** $\mathbf{A} = \begin{bmatrix} -1 & 1 \\ 11 & 1 \end{bmatrix}$ **11** $\mathbf{A} = \begin{bmatrix} 4 & 6 \\ 1 & -3 \end{bmatrix}$

Use the Cayley–Hamilton theorem to find $\mathbf{A}^{-1}$ and $\mathbf{A}^4$ in each of the following problems.

12 $\mathbf{A} = \begin{bmatrix} 0 & 1 & 0 \\ 1 & 0 & 1 \\ 1 & 0 & -1 \end{bmatrix}$ **13** $\mathbf{A} = \begin{bmatrix} 1 & 0 & 2 \\ 0 & 1 & 1 \\ 0 & 0 & 1 \end{bmatrix}$ **14** $\mathbf{A} = \begin{bmatrix} 1 & 1 & 1 \\ 1 & 0 & 1 \\ 0 & 0 & 1 \end{bmatrix}$

15 Prove that if the $n \times n$ matrix **A** has the characteristic polynomial

$$P_n(\lambda) = (-1)^n \lambda^n + c_{n-1} \lambda^{n-1} + c_{n-2} \lambda^{n-2} + \ldots + c_1 \lambda + c_0,$$

then

$$\text{adj } \mathbf{A} = -[(-1)^n \mathbf{A}^{n-1} + c_{n-1} \mathbf{A}^{n-2} + \ldots + c_1 \mathbf{I}].$$

16 Prove that provided $\mathbf{A}^{-1}$ exists, $\mathbf{A}^{-2}$ is a linear combination of

$$\mathbf{I}, \mathbf{A}, \mathbf{A}^2, \ldots, \mathbf{A}^{n-1}.$$

17 Find $\mathbf{A}'$ in terms of **A** and **I** given

$$\mathbf{A} = \begin{bmatrix} -1 & 3 \\ 0 & 2 \end{bmatrix},$$

and hence find $\mathbf{A}^{100}$.

18 Find A^r in terms of A and I given

$$A = \begin{bmatrix} 1 & 0 \\ 4 & -1 \end{bmatrix},$$

and hence show $A^r = I$ when r is even and $A^r = A$ when r is odd.

19 Find A^r in terms of A^2, A, I given

$$A = \begin{bmatrix} 1 & 0 & 1 \\ 0 & 0 & 2 \\ 0 & 0 & -1 \end{bmatrix},$$

and hence show $A^r = A^2$ when r is even and $A^r = A$ when r is odd.

20 Find A^r in terms of A^2, A and I given

$$A = \begin{bmatrix} 1 & 1 & 1 \\ 0 & 1 & 2 \\ 0 & 0 & 2 \end{bmatrix}.$$

21 Show that if

$$A = \begin{bmatrix} 0 & 3 & 0 \\ 3 & 0 & 4 \\ 0 & 4 & 0 \end{bmatrix}, \quad \text{then } A^{2r-1} = 5^{2r-2}A \quad \text{and} \quad A^{2r} = 5^{2r-2}A^2,$$

for $r = 1, 2, \dots$.

22 Show that if

$$A = \begin{bmatrix} 1 & 1 & 0 \\ 0 & 1 & 1 \\ 0 & 0 & 1 \end{bmatrix}, \quad \text{then} \quad A^r = \begin{bmatrix} 1 & r & \frac{1}{2}r(r-1) \\ 0 & 1 & r \\ 0 & 0 & 1 \end{bmatrix}, \quad r = 1, 2, \dots.$$

23 Find A^r in terms of A^2, A and I given

$$A = \begin{bmatrix} 2 & 0 & 0 \\ 1 & 2 & 0 \\ 0 & 1 & 2 \end{bmatrix}.$$

24 Find A^r in terms of A^2, A and I given

$$A = \begin{bmatrix} 2 & 0 & 0 \\ 1 & 1 & 0 \\ 0 & 1 & 2 \end{bmatrix}.$$

Harder problems

25 Let A be a 3×3 matrix with the characteristic polynomial

$$P_3(\lambda) = c_0 + c_1\lambda + c_2\lambda^2 + c_3\lambda^3,$$

and suppose $\det(A-I)\neq 0$. By writing $(A-I)^{-1}=b_0 I+b_1 A+b_2 A^2$, and determining the coefficients b_0, b_1 and b_2 by multiplying by $A-I$ and using the Cayley–Hamilton theorem, show that

$$(A-I)^{-1}=\frac{-1}{\det(A-I)}[c_1 I+c_2(A+I)+c_3(A^2+A+I)].$$

26 Let A be a 3×3 matrix with the characteristic polynomial

$$P_3(\lambda)=c_0+c_1\lambda+c_2\lambda^2+c_3\lambda^3,$$

and suppose $\det(A+I)\neq 0$. By writing $(A+I)^{-1}=b_0 I+b_1 A+b_2 A^2$, and determining the coefficients b_0, b_1 and b_2 by multiplying by $A+I$ and using the Cayley–Hamilton theorem, show that

$$(A+I)^{-1}=\frac{-1}{\det(A+I)}[c_1 I+c_2(A-I)+c_3(A^2-A+I)].$$

27 Let A be an $n\times n$ matrix with the characteristic polynomial

$$P_n(\lambda)=c_0+c_1\lambda+ \ldots +c_n\lambda^n,$$

and suppose $\det(A-I)\neq 0$. Show that

$$(A-I)^{-1}=\frac{-1}{\det(A-I)}[c_1 I+c_2(A+I)+c_3(A^2+A+I)+ \ldots$$

$$+c_n(A^{n-1}+A^{n-2}+ \ldots +A+I)].$$

28 Let A be an $n\times n$ matrix with the characteristic polynomial

$$P_n(\lambda)=c_0+c_1\lambda+ \ldots +c_n\lambda^n,$$

and suppose $\det(A+I)\neq 0$. Show that

$$(A+I)^{-1}=\frac{-1}{\det(A+I)}[c_1 I+c_2(A-I)+ \ldots +c_n(A^{n-1}-A^{n-2}$$

$$+A^{n-3}+ \ldots +(-1)^{n-2}A+(-1)^{n-1}I)].$$

29 Show that if

$$A=\begin{bmatrix} \frac{3}{2} & \frac{3}{4} \\ -\frac{5}{4} & -\frac{1}{2} \end{bmatrix},$$

then

$$A^r=\tfrac{1}{2}[-(\tfrac{3}{4})^r+3(\tfrac{1}{4})^r]I+2[(\tfrac{3}{4})^r-(\tfrac{1}{4})^r]A,$$

for $r=1, 2, \ldots$.

Set

$$S_n=I+A+A^2+ \ldots +A^n=\sum_{r=0}^{n} A^r,$$

where by definition $A^0\equiv I$. Then if as $n\to\infty$ the sum of the series forming the element in row i and column j of S_n converges to the finite limit s_{ij}, and this is true for $i, j=1, 2, \ldots , n$, the resulting **infinite matrix series** is said to have the sum **S**, where $S = [s_{ij}]$, and this is written

$$S = \sum_{r=0}^{\infty} A^r.$$

Use the matrix **A** and the expression for A^r in the first part of this problem to show that

$$\sum_{r=0}^{\infty} A^r = (16/3) A.$$

[*Hint*: Make use of the sum of a geometric series. $\displaystyle\sum_{r=0}^{\infty} a^r = \frac{1}{1-a}$ for $|a| < 1$.]

30 Show that if

$$A = \begin{bmatrix} 0 & -k \\ k & 0 \end{bmatrix},$$

then

$$A^r = (i)^{r-1} k^{r-1} \left[\frac{1-(-1)^r}{2} \right] A + (i)^r k^r \left[\frac{1+(-1)^r}{2} \right] I,$$

and hence that

$$A^{2m} = (-1)^m k^{2m} I \qquad \text{for } m = 0, 1, \ldots,$$

and

$$A^{2m-1} = (-1)^{m-1} k^{2m-2} A \qquad \text{for } m = 1, 2, \ldots.$$

Use the interpretation of the sum of an infinite matrix series in Problem 29 to show that with this choice of **A**,

$$\sum_{r=0}^{\infty} \frac{t^r}{r!} A^r = \frac{1}{k} \sin kt \, A + \cos kt \, I,$$

and hence that

$$\sum_{r=0}^{\infty} \frac{t^r}{r!} A^r = \begin{bmatrix} \cos kt & -\sin kt \\ \sin kt & \cos kt \end{bmatrix}.$$

Notice that this result gives meaning to the formal expansion of the matrix function

$$e^{tA} = I + tA + \frac{t^2}{2!} A^2 + \frac{t^3}{3!} A^3 + \ldots + \frac{t^n}{n!} A^n + \ldots.$$

3.12   Quadratic forms

Special polynomial expressions called bilinear forms and quadratic forms arise naturally in mathematics and in its application to subjects such as mechanics, statistics, optimization and control. Each is a polynomial in many variables, with bilinear forms being of degree 1 and quadratic forms of degree 2 in the variables concerned.

The combination of the $2n$ variables $x_1, x_2, \ldots, x_n$ and $y_1, y_2, \ldots, y_n$ into an expression of the form

$$B = \sum_{i=1}^{n} \sum_{j=1}^{n} a_{ij} x_i y_j = a_{11} x_1 y_1 + a_{12} x_1 x_2 + \ldots + a_{nn} x_n y_n, \tag{1}$$

in which the coefficients a_{ij} are numbers, is called a **bilinear form** in the $2n$ variables. The name derives from the fact that B is a linear expression in each x_i and y_j, and so *bilinear* in the x_i and y_j.

Bilinear form B in (1) may be written as the matrix expression

$$B = B(\mathbf{x}, \mathbf{y}) = \mathbf{x}^T \mathbf{A} \mathbf{y}, \tag{2}$$

if we set

$$\mathbf{x} = \begin{bmatrix} x_1 \\ x_2 \\ \vdots \\ x_n \end{bmatrix}, \qquad \mathbf{y} = \begin{bmatrix} y_1 \\ y_2 \\ \vdots \\ y_n \end{bmatrix} \quad \text{and} \quad \mathbf{A} = \begin{bmatrix} a_{11} & a_{12} & \cdots & a_{1n} \\ a_{21} & a_{22} & \cdots & a_{2n} \\ \cdot & \cdot & \cdots & \cdot \\ a_{n1} & a_{n2} & \cdots & a_{nn} \end{bmatrix}, \tag{3}$$

when $\mathbf{A}$ is then called the **coefficient matrix** of the form B.

Example 1. Bilinear form

Express the bilinear form

$$B = 2x_1 y_1 + 7x_1 y_2 - 3x_2 y_1 + 11 x_2 y_2$$

in terms of matrices.

Solution

Setting $\mathbf{x} = \begin{bmatrix} x_1 \\ x_2 \end{bmatrix}$, $\mathbf{y} = \begin{bmatrix} y_1 \\ y_2 \end{bmatrix}$ inspection shows that $\mathbf{A} = \begin{bmatrix} 2 & 7 \\ -3 & 11 \end{bmatrix}$. ∎

The simplest example of a bilinear form has already been encountered in Sec. 3.3, Eq. (22) where the inner product $(\mathbf{x}, \mathbf{y})$ of $\mathbf{x}$ and $\mathbf{y}$ was defined as

$$(\mathbf{x}, \mathbf{y}) = x_1 y_1 + x_2 y_2 + \ldots + x_n y_n. \tag{4}$$

The expression on the right-hand side is simply $\mathbf{x}^T \mathbf{y}$, and it follows from (2) by setting $\mathbf{A} = \mathbf{I}$. Thus as a special case of a bilinear form we have

$$(\mathbf{x}, \mathbf{y}) = \mathbf{x}^T \mathbf{y}. \tag{5}$$

When $\mathbf{y} = \mathbf{x}$ in (2) the expression B becomes quadratic in the variables $x_1, x_2, \ldots, x_n$, and is then called a *quadratic form*. We shall denote a quadratic form by Φ. Thus a **quadratic form** Φ may be written as

$$\Phi = \mathbf{x}^T \mathbf{A} \mathbf{x} = \sum_{i=1}^{n} \sum_{j=1}^{n} a_{ij} x_i x_j = a_{11} x_1^2 + a_{12} x_1 x_2 + a_{21} x_2 x_1 + \ldots + a_{nn} x_n^2. \tag{6}$$

Example 2. Quadratic forms and moments of inertia

A quadratic form arises in mechanics when expressing the moment of inertia of a body about an arbitrary axis passing through an origin O in a rigid body. Let I_{xx}, I_{yy}, I_{zz} and I_{xy}, I_{xz}, I_{yz} be the principal moments and products of inertia, respectively, of the body relative to a set of orthogonal axes $O(x, y, z)$ fixed in the body. Then the moment of inertia I_L of the body about an axis L through O with direction cosines l, m, n relative to $O(x, y, z)$ can be shown to be

$$I_L = I_{xx}l^2 + I_{yy}m^2 + I_{zz}n^2 - 2I_{xy}lm - 2I_{xz}ln - 2I_{yz}mn.$$

In this case I_L is a quadratic form in the variables l, m, n.   ∎

Inspection of (6) shows that if α is the coefficient of the product term $x_i x_j$ in Φ, then $\alpha = a_{ij} + a_{ji}$. It is convenient to use this fact in the matrix representation of Φ by setting $a_{ij} = a_{ji} = \frac{1}{2}\alpha$, thus causing $\mathbf{A}$ to become symmetric. This involves no loss of generality and greatly simplifies all subsequent manipulation. Henceforth, the coefficient matrix $\mathbf{A}$ in a quadratic form will always be assumed to be a symmetric matrix.

Example 3. Quadratic form

Express in matrix notation the quadratic form

$$\Phi = 3x_1^2 - x_2^2 + 5x_3^2 + 4x_1x_2 - 7x_1x_3 + 9x_2x_3.$$

Solution

If $\mathbf{A} = [a_{ij}]$, then by inspection $a_{11} = 3$, $a_{22} = -1$, $a_{33} = 5$. Because $\mathbf{A}$ is symmetric, $a_{12} = a_{21} = \frac{1}{2} \times 4 = 2$, $a_{13} = a_{31} = \frac{1}{2} \times (-7) = -3.5$ and $a_{23} = a_{32} = \frac{1}{2} \times 9 = 4.5$. Thus $\Phi = \mathbf{x}^\mathsf{T}\mathbf{A}\mathbf{x}$, with

$$\mathbf{x} = \begin{bmatrix} x_1 \\ x_2 \\ x_3 \end{bmatrix} \quad \text{and} \quad \mathbf{A} = \begin{bmatrix} 3 & 2 & -3.5 \\ 2 & -1 & 4.5 \\ -3.5 & 4.5 & 5 \end{bmatrix}.$$

∎

We now come to the real objective of this section, which is to show how the structure of a quadratic form Φ may be simplified by means of a change of variable. To be specific, we shall show how to find a new set of variables $z_1, z_2, \ldots, z_n$ depending linearly on $x_1, x_2, \ldots, x_n$ so that Φ depends only on the squares $z_1^2, z_2^2, \ldots, z_n^2$, and not on products like $z_1z_2, z_2z_3, \ldots$, etc.

This simplification may be achieved by introducing an orthogonal matrix $\mathbf{Q}$ which diagonalizes $\mathbf{A}$. Such a matrix $\mathbf{Q}$ always exists because $\mathbf{A}$ is symmetric (Theorem 2.31 (v)). When the eigenvalues of $\mathbf{A}$ are distinct, $\mathbf{Q}$ is constructed by taking the normalized eigenvectors of $\mathbf{A}$ as the columns of $\mathbf{Q}$. If an eigenvalue is repeated, then for the corresponding columns of $\mathbf{Q}$ it is necessary to take a set of orthonormal eigenvectors corresponding to that eigenvalue. If such a set cannot be found by inspection it may always be constructed from any set of eigenvectors corresponding to the repeated eigenvalue by means of the Gram–Schmidt orthogonalization process (Prob. 36, Sec. 3.10).

When $\mathbf{Q}$ has been constructed we set

$$\mathbf{x} = \mathbf{Qz} \tag{7}$$

in (6), with $\mathbf{z} = [z_1, z_2, \ldots, z_n]^\mathsf{T}$, and as a result obtain

$$\Phi = \mathbf{x}^\mathsf{T} \mathbf{Ax} = \mathbf{z}^\mathsf{T} \mathbf{Q}^\mathsf{T} \mathbf{AQz}. \tag{8}$$

Now $\mathbf{Q}$ diagonalizes $\mathbf{A}$ to give

$$\mathbf{Q}^\mathsf{T} \mathbf{AQ} = \Lambda, \tag{9}$$

so that (8) becomes

$$\Phi = \mathbf{x}^\mathsf{T} \mathbf{Ax} = \mathbf{z}^\mathsf{T} \Lambda \mathbf{z},$$

$$= \lambda_1 z_1^2 + \lambda_2 z_2^2 + \ldots + \lambda_n z_n^2, \tag{10}$$

where $\lambda_1, \lambda_2, \ldots, \lambda_n$ are the eigenvalues of $\mathbf{A}$. Expression (10), which is called the **canonical form** of the quadratic form Φ, provides the desired simplification. Thus the required change of variables to reduce Φ to its canonical form is given by (7). Quadratic form Φ is said to be **diagonalized** by the orthogonal matrix $\mathbf{Q}$.

It should be remembered that Λ is a diagonal matrix in which the eigenvalues of $\mathbf{A}$ appear as the elements of the leading diagonal in the same order as that in which the corresponding normalized eigenvectors of $\mathbf{A}$ appear as the columns of $\mathbf{Q}$.

A quadratic form is said to be **positive definite** if it is strictly positive (never zero) for all vectors $\mathbf{x} \neq \mathbf{0}$. Similarly, it is said to be **negative definite** if it is strictly negative (never zero) for all vectors $\mathbf{x} \neq \mathbf{0}$. A quadratic form which may assume both positive and negative values for vectors $\mathbf{x} \neq \mathbf{0}$ is said to be **indefinite**. Inspection of (10) shows a quadratic form is positive definite if the eigenvalues are all nonnegative and negative definite if they are all nonpositive. A quadratic form is indefinite if the eigenvalues of $\mathbf{A}$ are not all of the same sign. An $n \times n$ matrix $\mathbf{A}$ is said to be **positive definite**, **negative definite** or **indefinite** if its associated quadratic form is so classified.

Example 4. Diagonalization and canonical form when A has distinct eigenvalues

Find the canonical form of

$$\Phi = 5x_1^2 + 6x_2^2 + 7x_3^2 - 4x_1 x_2 + 4x_2 x_3 \text{ and identify its type.}$$

Solution

$\Phi = \mathbf{x}^\mathsf{T} \mathbf{Ax}$, with

$$\mathbf{x} = \begin{bmatrix} x_1 \\ x_2 \\ x_3 \end{bmatrix} \quad \text{and} \quad \mathbf{A} = \begin{bmatrix} 5 & -2 & 0 \\ -2 & 6 & 2 \\ 0 & 2 & 7 \end{bmatrix}.$$

The eigenvalues of $\mathbf{A}$ are $\lambda_1 = 3$, $\lambda_2 = 6$, $\lambda_3 = 9$ and the corresponding normalized eigenvectors are

$$\hat{\mathbf{x}}_1 = \begin{bmatrix} \frac{2}{3} \\ \frac{2}{3} \\ -\frac{1}{3} \end{bmatrix}, \quad \hat{\mathbf{x}}_2 = \begin{bmatrix} \frac{2}{3} \\ -\frac{1}{3} \\ \frac{2}{3} \end{bmatrix} \quad \text{and} \quad \hat{\mathbf{x}}_3 = \begin{bmatrix} -\frac{1}{3} \\ \frac{2}{3} \\ \frac{2}{3} \end{bmatrix}.$$

Thus the orthogonal matrix $\mathbf{Q}$ which will diagonalize Φ, and the corresponding diagonal matrix Λ, are

$$\mathbf{Q} = \begin{bmatrix} \frac{2}{3} & \frac{2}{3} & -\frac{1}{3} \\ \frac{2}{3} & -\frac{1}{3} & \frac{2}{3} \\ -\frac{1}{3} & \frac{2}{3} & \frac{2}{3} \end{bmatrix} \quad \text{and} \quad \Lambda = \begin{bmatrix} 3 & 0 & 0 \\ 0 & 6 & 0 \\ 0 & 0 & 9 \end{bmatrix}.$$

The canonical form of Φ is thus

$$\Phi = \mathbf{z}^T \Lambda \mathbf{z} = 3z_1^2 + 6z_2^2 + 9z_3^2,$$

which is seen to be positive definite. Since $\mathbf{x} = \mathbf{Qz}$, the explicit change of variables involved is

$$x_1 = \tfrac{2}{3}z_1 + \tfrac{2}{3}z_2 - \tfrac{1}{3}z_3$$
$$x_2 = \tfrac{2}{3}z_1 - \tfrac{1}{3}z_2 + \tfrac{2}{3}z_3$$
$$x_3 = -\tfrac{1}{3}z_1 + \tfrac{2}{3}z_2 + \tfrac{2}{3}z_3. \qquad\qquad\qquad\blacksquare$$

Example 5. Diagonalization and canonical form when A has repeated eigenvalues

Find the canonical form of

$$\Phi = 2x_1 x_2 + 2x_1 x_3 - 2x_1 x_4 - 2x_2 x_3 + 2x_2 x_4 + 2x_3 x_4 \text{ and identify its type.}$$

Solution

In matrix form

$$\Phi = \mathbf{x}^T \mathbf{A} \mathbf{x} \text{ with } \mathbf{x} = \begin{bmatrix} x_1 \\ x_2 \\ x_3 \\ x_4 \end{bmatrix} \quad \text{and} \quad \mathbf{A} = \begin{bmatrix} 0 & 1 & 1 & -1 \\ 1 & 0 & -1 & 1 \\ 1 & -1 & 0 & 1 \\ -1 & 1 & 1 & 0 \end{bmatrix}.$$

The eigenvalues of $\mathbf{A}$ are $\lambda_1 = -3$ and $\lambda_2 = \lambda_3 = \lambda_4 = 1$, so the eigenvalue 1 has algebraic multiplicity 3. The normalized eigenvector corresponding to $\lambda_1 = -3$ is

$$\hat{\mathbf{x}}_1 = \begin{bmatrix} \frac{1}{2} \\ -\frac{1}{2} \\ -\frac{1}{2} \\ \frac{1}{2} \end{bmatrix}.$$

The remaining three eigenvectors must be determined from the conditions which follow from $(\mathbf{A} - \lambda\mathbf{I})\mathbf{x} = \mathbf{0}$ with $\lambda = 1$; that is from the single condition $x_1 - x_2 - x_3 + x_4 = 0$. Assigning $x_2 = \alpha$, $x_3 = \beta$, $x_4 = \gamma$ arbitrarily, we find

$$\mathbf{x} = \begin{bmatrix} \alpha + \beta - \gamma \\ \alpha \\ \beta \\ \gamma \end{bmatrix} = \alpha \begin{bmatrix} 1 \\ 1 \\ 0 \\ 0 \end{bmatrix} + \beta \begin{bmatrix} 1 \\ 0 \\ 1 \\ 0 \end{bmatrix} + \gamma \begin{bmatrix} -1 \\ 0 \\ 0 \\ 1 \end{bmatrix}. \tag{A}$$

The three vectors on the right-hand side are linearly independent, so they are the eigenvectors corresponding to $\lambda_2 = \lambda_3 = \lambda_4 = 1$.

Now by means of suitable choices of α, β, γ we need to construct three mutually orthogonal vectors $\mathbf{x}_2$, $\mathbf{x}_3$ and $\mathbf{x}_4$ from this general expression for the eigenvectors which correspond to $\lambda_2 = \lambda_3 = \lambda_4 = 1$. One possible choice which may be found by inspection is vector $\mathbf{x}_2$ with $\alpha = 1$, $\beta = \gamma = 0$, vector $\mathbf{x}_3$ with $\alpha = 0$, $\beta = \gamma = 1$ and vector $\mathbf{x}_4$ with $\alpha = -1$, $\beta = 1$, $\gamma = -1$, when

$$\mathbf{x}_2 = \begin{bmatrix} 1 \\ 1 \\ 0 \\ 0 \end{bmatrix}, \quad \mathbf{x}_3 = \begin{bmatrix} 0 \\ 0 \\ 1 \\ 1 \end{bmatrix}, \quad \mathbf{x}_4 = \begin{bmatrix} 1 \\ -1 \\ 1 \\ -1 \end{bmatrix}.$$

Thus the corresponding three normalized eigenvectors are

$$\hat{\mathbf{x}}_2 = \begin{bmatrix} \dfrac{1}{\sqrt{2}} \\ \dfrac{1}{\sqrt{2}} \\ 0 \\ 0 \end{bmatrix}, \quad \hat{\mathbf{x}}_3 = \begin{bmatrix} 0 \\ 0 \\ \dfrac{1}{\sqrt{2}} \\ \dfrac{1}{\sqrt{2}} \end{bmatrix}, \quad \hat{\mathbf{x}}_4 = \begin{bmatrix} \dfrac{1}{2} \\ -\dfrac{1}{2} \\ \dfrac{1}{2} \\ -\dfrac{1}{2} \end{bmatrix},$$

so that a diagonalizing matrix is

$$\mathbf{Q} = \begin{bmatrix} \dfrac{1}{2} & \dfrac{1}{\sqrt{2}} & 0 & \dfrac{1}{2} \\ -\dfrac{1}{2} & \dfrac{1}{\sqrt{2}} & 0 & -\dfrac{1}{2} \\ -\dfrac{1}{2} & 0 & \dfrac{1}{\sqrt{2}} & \dfrac{1}{2} \\ \dfrac{1}{2} & 0 & \dfrac{1}{\sqrt{2}} & -\dfrac{1}{2} \end{bmatrix} \quad \text{corresponding to } \mathbf{\Lambda} = \begin{bmatrix} -3 & 0 & 0 & 0 \\ 0 & 1 & 0 & 0 \\ 0 & 0 & 1 & 0 \\ 0 & 0 & 0 & 1 \end{bmatrix}.$$

The canonical form becomes

$$\Phi = z^T A z = -3z_1^2 + z_2^2 + z_3^2 + z_4^2,$$

which is seen to be an indefinite form since the eigenvalues are not all of the same sign. The explicit change of variables to accomplish this reduction is $x = Qz$, or

$$x_1 = \frac{1}{2}z_1 + \frac{1}{\sqrt{2}}z_2 + \frac{1}{2}z_4$$

$$x_2 = -\frac{1}{2}z_1 + \frac{1}{\sqrt{2}}z_2 - \frac{1}{2}z_4$$

$$x_3 = -\frac{1}{2}z_1 + \frac{1}{\sqrt{2}}z_3 + \frac{1}{2}z_4$$

$$x_4 = \frac{1}{2}z_1 + \frac{1}{\sqrt{2}}z_3 - \frac{1}{2}z_4.$$

If a mutually orthogonal set of eigenvectors corresponding to $\lambda_2 = \lambda_3 = \lambda_4 = 1$ had not been found by inspection, and this is usually difficult to do, the Gram–Schmidt orthogonalization process could have been used. This involves a certain amount of calculation but always provides a mutually orthogonal set of vectors.

If from (A) we had taken the three linearly independent but nonorthogonal vectors

$$a_1 = \begin{bmatrix} 1 \\ 1 \\ 0 \\ 0 \end{bmatrix}, \quad a_2 = \begin{bmatrix} 1 \\ 0 \\ 1 \\ 0 \end{bmatrix} \quad \text{and} \quad a_3 = \begin{bmatrix} -1 \\ 0 \\ 0 \\ 1 \end{bmatrix}$$

as the vectors to be used in the Gram–Schmidt orthogonalization process, we would have found the equivalent orthonormal system

$$\hat{x}_2 = \begin{bmatrix} \dfrac{1}{\sqrt{2}} \\ \dfrac{1}{\sqrt{2}} \\ 0 \\ 0 \end{bmatrix}, \quad \hat{x}_3 = \begin{bmatrix} \dfrac{1}{\sqrt{6}} \\ -\dfrac{1}{\sqrt{6}} \\ \sqrt{\dfrac{2}{3}} \\ 0 \end{bmatrix} \quad \text{and} \quad \hat{x}_4 = \begin{bmatrix} \dfrac{-1}{2\sqrt{3}} \\ \dfrac{1}{2\sqrt{3}} \\ \dfrac{1}{2\sqrt{3}} \\ \dfrac{\sqrt{3}}{2} \end{bmatrix}. \tag{B}$$

Thus in this case the diagonalizing matrix would be

$$
\mathbf{A} =
\begin{bmatrix}
\dfrac{1}{2} & \dfrac{1}{\sqrt{2}} & \dfrac{1}{\sqrt{6}} & -\dfrac{1}{2\sqrt{3}} \\[2ex]
-\dfrac{1}{2} & \dfrac{1}{\sqrt{2}} & -\dfrac{1}{\sqrt{6}} & \dfrac{1}{2\sqrt{3}} \\[2ex]
-\dfrac{1}{2} & 0 & \sqrt{\dfrac{2}{3}} & \dfrac{1}{2\sqrt{3}} \\[2ex]
\dfrac{1}{2} & 0 & 0 & \dfrac{\sqrt{3}}{2}
\end{bmatrix}
, \quad \text{and again } \mathbf{\Lambda} =
\begin{bmatrix}
-3 & 0 & 0 & 0 \\
0 & 1 & 0 & 0 \\
0 & 0 & 1 & 0 \\
0 & 0 & 0 & 1
\end{bmatrix}.
$$

The canonical form is unchanged with

$$
\Phi = -3z_1^2 + z_2^2 + z_3^2 + z_4^2,
$$

but this time the explicit change of variables $\mathbf{x} = \mathbf{Q}\mathbf{z}$ to accomplish this reduction is different, with

$$
x_1 = \frac{1}{2}z_1 + \frac{1}{\sqrt{2}}z_2 + \frac{1}{\sqrt{6}}z_3 - \frac{1}{2\sqrt{3}}z_4
$$

$$
x_2 = -\frac{1}{2}z_1 + \frac{1}{\sqrt{2}}z_2 - \frac{1}{\sqrt{6}}z_3 + \frac{1}{2\sqrt{3}}z_4
$$

$$
x_3 = -\frac{1}{2}z_1 + \sqrt{\frac{2}{3}}z_3 + \frac{1}{2\sqrt{3}}z_4
$$

$$
x_4 = \frac{1}{2}z_1 + \frac{\sqrt{3}}{2}z_4.
$$

This example illustrates how quite different orthogonalizing matrices, and hence changes of variable, can achieve a reduction to the same canonical form when eigenvalues are repeated.

Since $\mathbf{Q}$ is orthogonal $\mathbf{Q}^{\mathsf{T}} = \mathbf{Q}^{-1}$, and so should it be needed the variable change from $\mathbf{x}$ to $\mathbf{z}$ is always given by the result $\mathbf{z} = \mathbf{Q}^{\mathsf{T}}\mathbf{x}$. Thus this case

$$
z_1 = \frac{1}{2}x_1 - \frac{1}{2}x_2 - \frac{1}{2}x_3 + \frac{1}{2}x_4
$$

$$
z_2 = \frac{1}{\sqrt{2}}x_1 + \frac{1}{\sqrt{2}}x_2
$$

$$
z_3 = \frac{1}{\sqrt{6}}x_1 - \frac{1}{\sqrt{6}}x_2 + \sqrt{\frac{2}{3}}x_3
$$

$$
z_4 = -\frac{1}{2\sqrt{3}}x_1 + \frac{1}{2\sqrt{3}}x_2 + \frac{1}{2\sqrt{3}}x_3 + \frac{\sqrt{3}}{2}x_4. \qquad \blacksquare
$$

The identification of a quadratic form (and also of its associated matrix **A**) as being positive definite, negative definite, or indefinite, is determined by the signs of the eigenvalues of **A**. These are not always easy to find, and in many applications it is necessary to be able to identify a quadratic form without using the eigenvalues of **A** at all. The next theorem tells us how to do this by using determinants derived from **A**.

Theorem 3.34 (Identification of quadratic forms)

Let Φ be the quadratic form

$$\Phi = x^T A x$$

in the n variables $x_1, x_2, \ldots, x_n$. Then the criteria given in (i) and (ii) below are equivalent, and are also both necessary and sufficient to identify Φ.

(i) Let the eigenvalues of **A** be $\lambda_1, \lambda_2, \ldots, \lambda_m$ then

   (a) Φ will be positive definite if $\lambda_i > 0$ for $i = 1, 2, \ldots, n$;
   (b) Φ will be negative definite if $\lambda_i < 0$ for $i = 1, 2, \ldots, n$;
   (c) Φ will be indefinite if the λ_i are of both signs.

(ii) Let Δ_r be the determinant derived from **A** by retaining only the first r rows and columns, so that

$$\Delta_r = \begin{vmatrix} a_{11} & a_{12} & \cdots & a_{1r} \\ a_{21} & a_{22} & \cdots & a_{2r} \\ \cdot & \cdot & \cdots & \cdot \\ a_{r1} & a_{r2} & \cdots & a_{rr} \end{vmatrix}.$$

Then,

   (a) Φ will be positive definite if $\Delta_r > 0$ for $r = 1, 2, \ldots, n$;
   (b) Φ will be negative definite if $(-1)^r \Delta_r > 0$ for $r = 1, 2, \ldots, n$;
   (c) Φ will be indefinite if neither conditions (a) nor (b) are satisfied.

Proof

The proof of (i) has already been given when showing how Φ may be reduced to canonical form. The proof of (ii) is harder and involves showing the stated conditions are both necessary and sufficient.

Necessity of conditions (ii)(a)

The characteristic polynomial $\det(A - \lambda I)$ of matrix **A** is of degree n, so if the eigenvalues of **A** are $\lambda_1, \lambda_2, \ldots \lambda_n$, it follows that we may write

$$\det(A - \lambda I) \equiv (\lambda_1 - \lambda)(\lambda_2 - \lambda) \ldots (\lambda_n - \lambda). \tag{11}$$

This is an identity in λ, so setting $\lambda=0$ we obtain

$$\det \mathbf{A} = \lambda_1 \lambda_2 \ldots \lambda_n. \tag{12}$$

Thus if Φ is positive definite it must follow that

$$\det \mathbf{A} > 0, \tag{13}$$

because in that case $\lambda_r > 0$ for $r = 1, 2, \ldots, n$.

Now if Φ is positive definite it will remain so for any choice of $x_1, x_2, \ldots, x_n$ and, in particular, when $x_{r+1} = x_{r+2} = \ldots = x_n = 0$. This reduces Φ to a quadratic form Φ_r in the r variables $x_1, x_2, \ldots, x_r$ whose $r \times r$ coefficient matrix $\mathbf{A}_r$ is derived from $\mathbf{A}$ by retaining only the first r columns and rows. As Φ_r is positive definite we see from (13) that we must have

$$\Delta_r > 0. \tag{14}$$

However condition (14) must be true for $r = 1, 2, \ldots, n$, thereby showing Φ will be positive definite if $\Delta_r > 0$ for $r = 1, 2, \ldots, n$. Conditions (ii) (a) are thus necessary.

Sufficiency of conditions (ii)(a)

Consider the quadratic form

$$\Phi = \sum_{i,j=1}^{n} a_{ij} x_i x_j \quad \text{with } a_{11} \neq 0 \text{ and } a_{ij} = a_{ji}.$$

Then setting $x_2 = x_3 = \ldots = x_n = 0$ reduces Φ to the form $\Phi_1 = a_{11} x_1^2$, so as $\Delta_1 > 0$ form Φ_1 will be positive definite, because $\Delta_1 = a_{11}$.

Next, setting $x_3 = x_4 = \ldots = x_n = 0$ reduces Φ to the form

$$\Phi_2 = a_{11} x_1^2 + 2a_{12} x_1 x_2 + a_{22} x_2^2,$$

or to

$$\Phi_2 = a_{11}\left(x_1 + \frac{a_{12}}{a_{11}} x_2\right)^2 + \left(a_{22} - \frac{a_{12}^2}{a_{11}}\right) x_2^2.$$

As $a_{11} > 0$ and $\Delta_2 > 0$ it follows that Φ_2 is positive definite, because

$$a_{22} - \frac{a_{12}^2}{a_{11}} = \frac{1}{a_{11}} \Delta_2 > 0.$$

Proceeding in this fashion until all $x_1, x_2, \ldots, x_n$ have been considered enables us to show conditions (ii) (a) are also sufficient for Φ to be positive definite, and our proof is complete.

The proof of conditions (ii) (b) follows in similar fashion when use is made of the fact that when Φ is negative definite all the eigenvalues must be negative, giving rise to the condition $(-1)^n \det \mathbf{A} > 0$ in place of (13). Condition (ii)(c) is the only remaining possibility once conditions (ii) (a), (b) have been excluded. ☐

Example 6

Identify the nature of the quadratic forms

(i) $\Phi = 2x_1^2 - 8x_1 x_2 + x_2^2$ (ii) $\Phi = 2x_1^2 + 2x_1 x_2 + 2x_1 x_3 + 4x_2^2 + 2x_2 x_3 + x_3^2$.

Solution

(i) $\mathbf{A} = \begin{bmatrix} 2 & -4 \\ -4 & 2 \end{bmatrix}$, so $\Delta_1 = 2 > 0$ and $\Delta_2 = \begin{vmatrix} 2 & -4 \\ -4 & 2 \end{vmatrix} = -12 < 0.$

Thus Φ is indefinite by Theorem 3.34 (ii) (c). ∎

(ii) $\mathbf{A} = \begin{bmatrix} 2 & 1 & 1 \\ 1 & 4 & 1 \\ 1 & 1 & 1 \end{bmatrix}$, so $\Delta_1 = 2 > 0$, $\Delta_2 = \begin{vmatrix} 2 & 1 \\ 1 & 4 \end{vmatrix} = 7 > 0$ and $\Delta_3 = \begin{vmatrix} 2 & 1 & 1 \\ 1 & 4 & 1 \\ 1 & 1 & 1 \end{vmatrix} = 3 > 0.$

Thus Φ is positive definite by Theorem 3.34(ii)(a).

Example 7. Quadratic forms and stationary points

This example illustrates an application of quadratic forms to the identification of the **stationary points** of a function of two variables. Let $f(x, y)$ be a function capable of expansion by Taylor's theorem about the point (a, b) to give

$$f(a+h, b+k) = f(a, b) + hf_x(a, b) + kf_y(a, b)$$
$$+ \frac{1}{2!}[h^2 f_{xx}(a, b) + 2hk f_{xy}(a, b) + k^2 f_{yy}(a, b)] + R(a, b, h, k),$$

where suffixes denote partial differentiation, $R(a, b, h, k)$ is the remainder term and use has been made of the equality of mixed derivatives $f_{xy} = f_{yx}$.

If (a, b) is a stationary point of $f(x, y)$ it follows that $f_x(a, b) = f_y(a, b) = 0$, and so at such a point

$$f(a+h, b+k) - f(a, b) = \frac{1}{2!}[h^2 f_{xx}(a, b) + 2hk f_{xy}(a, b) + k^2 f_{yy}(a, b)] + R(a, b, h, k).$$

In the neighborhood of (a, b) when h, k are sufficiently small $R(a, b, h, k) \rightarrow 0$ and the sign of the left-hand side will be determined by the quadratic form

$$\Phi = h^2 f_{xx}(a, b) + 2hk f_{xy}(a, b) + k^2 f_{yy}(a, b),$$

for which

$$\mathbf{A} = \begin{bmatrix} f_{xx}(a, b) & f_{xy}(a, b) \\ f_{xy}(a, b) & f_{yy}(a, b) \end{bmatrix}.$$

If Φ is positive definite, $f(a+h, b+k)-f(a, b)>0$ for all suitably small h, k, showing (a, b) must be a **local minimum**. If Φ is negative definite it follows that (a, b) must be a **local maximum**. When Φ is indefinite (a, b) will be a **saddle point**. The conditions of Theorem 3.34 (ii) applied to det $\mathbf{A}$ at stationary points of $f(x, y)$ thus identify local maxima, minima and saddle points of $f(x, y)$.

If, for example, (a, b) is a local maximum we see from Theorem 3.34 (ii) (b) that $\Delta_1<0$ and $\Delta_2>0$ or, equivalently, $f_{xx}(a, b)<0$ and $f_{xx}(a, b)f_{yy}(a, b)-f_{xy}^2(a, b)>0$, which are the familiar conditions derived in any first course in calculus. The conditions for a local minimum and a saddle point follow directly from the other conditions in Theorem 3.34 (ii).

The matrix $\mathbf{A}$ associated with this quadratic form is called a **Hessian** matrix. Both the Hessian matrix and this form of analysis extend immediately to allow the identification of stationary points of functions of many variables. ∎

Problems for Section 3.12

Find the coefficient matrices for the following bilinear and quadratic forms.

1 $B=5x_1y_1+3x_1y_2-x_1y_3+x_2y_1-6x_2y_3+x_3y_1+7x_3y_3$

2 $B=-3x_1y_2+4x_1y_3+5x_2y_1-11x_2y_3+x_3y_1-3x_3y_2$

3 $\Phi=4x_1^2+9x_2^2-5x_3^2+4x_1x_2+8x_1x_3+6x_2x_3$

4 $\Phi=7x_1^2+4x_3^2+10x_1x_2+6x_1x_3-2x_2x_3$

5 $\Phi=x_1^2-2x_2^2-x_3^2+4x_1x_2-6x_1x_3+x_2x_3$

6 Use the definition of a bilinear form to prove the following properties:
   (i) $B(\mathbf{x}, \mathbf{y})=B(\mathbf{y}, \mathbf{x})$,
   (ii) $B(\lambda\mathbf{x}, \mathbf{y})=\lambda B(\mathbf{x}, \mathbf{y})$ for λ a scalar,
   (iii) $B(\mathbf{x}+\mathbf{y}, \mathbf{z})=B(\mathbf{x}, \mathbf{z})+B(\mathbf{y}, \mathbf{z})$.

Express each quadratic form Φ in its canonical form, stating the transformation which will lead to this reduction and whether the form is positive definite, negative definite or indefinite.

7 $\Phi=x_1^2+3x_2^2+3x_3^2-2x_2x_3$

8 $\Phi=\frac{2}{9}x_1^2+\frac{5}{9}x_2^2+\frac{11}{9}x_3^2-\frac{20}{9}x_1x_2+\frac{4}{9}x_1x_3+\frac{16}{9}x_2x_3$

9 $\Phi=3x_1^2-x_2^2+3x_3^2-2x_1x_3$

10 $\Phi=\frac{3}{49}x_1^2-\frac{9}{49}x_2^2+\frac{104}{49}x_3^2-\frac{72}{49}x_1x_2+\frac{60}{49}x_1x_3-\frac{132}{49}x_2x_3$

11 $\Phi=-2x_1^2-\frac{7}{3}x_2^2-\frac{5}{3}x_3^2-\frac{4}{3}x_1x_2+\frac{4}{3}x_1x_3$

12 $\Phi=\frac{20}{9}x_1^2+\frac{14}{9}x_2^2+\frac{11}{9}x_3^2+\frac{16}{9}x_1x_2+\frac{4}{9}x_1x_3-\frac{20}{9}x_2x_3$

13 Find the canonical form of

$$\Phi=x_2^2+2x_1x_3,$$

and give two different diagonalizing matrices which will lead to such a reduction when the eigenvalues in Λ are arranged in increasing numerical order along the leading diagonal. Are there any other diagonalizing matrices?

14 Find the symmetric matrix $\mathbf{A}$ with eigenvalues $\lambda_1 = -1$, $\lambda_2 = 0$ and $\lambda_3 = 1$ and the corresponding eigenvectors

$$\mathbf{x}_1 = \begin{bmatrix} -2 \\ -1 \\ 2 \end{bmatrix}, \quad \mathbf{x}_2 = \begin{bmatrix} 2 \\ -2 \\ 1 \end{bmatrix} \quad \text{and} \quad \mathbf{x}_3 = \begin{bmatrix} 1 \\ 2 \\ 2 \end{bmatrix}.$$

15 Find the symmetric matrix $\mathbf{A}$ with the eigenvalues $\lambda_1 = 1$, $\lambda_2 = 1$ and $\lambda_3 = 2$ and the corresponding eigenvectors

$$\mathbf{x}_1 = \begin{bmatrix} 1 \\ 0 \\ 0 \end{bmatrix}, \quad \mathbf{x}_2 = \begin{bmatrix} 0 \\ 1 \\ 1 \end{bmatrix} \quad \text{and} \quad \mathbf{x}_3 = \begin{bmatrix} 0 \\ 1 \\ -1 \end{bmatrix}.$$

Use Theorem 3.33 to identify the nature of the following quadratic forms:

16 $\Phi = 3x_1^2 - 6x_1x_2 + 9x_2^2$.

17 $\Phi = -4x_1^2 + 4x_1x_2 - 4x_2^2 - 6x_1x_3 + 6x_2x_3 + x_3^2$.

18 The form Φ in Problem 11.

19 The form Φ in Problem 12.

20 $\Phi = -2x_1^2 - 3x_2^2 - 3x_3^2 + 2x_1x_2 + 2x_1x_3 + 4x_2x_3$.

Harder problems

21 Let $f(x_1, x_2, x_3)$ be a function capable of expansion by Taylor's theorem about the point (a, b, c) as

$$f(a+h_1, b+h_2, c+h_3) = f(a, b, c) + h_1 f_{x_1}(a, b, c) + h_2 f_{x_2}(a, b, c) + h_3 f_{x_3}(a, b, c)$$

$$+ \frac{1}{2!} \left[\left(h_1 \frac{\partial}{\partial x_1} + h_2 \frac{\partial}{\partial x_2} + h_3 \frac{\partial}{\partial x_3} \right) \left(h_1 f_{x_1} + h_2 f_{x_2} + h_3 f_{x_3} \right) \right]_{(a, b, c)}$$

$$+ R(a, b, c, h_1, h_2, h_3),$$

with $R(a, b, c, h_1, h_2, h_3)$ the remainder term and $[\cdot]_{(a, b, c)}$ indicating that the quantity $[\cdot]$ is to be evaluated at (a, b, c).

Arguing as in Example 6, show that when (a, b, c) is a stationary point of $f(x_1, x_2, x_3)$, so $f_{x_1}(a, b, c) = f_{x_2}(a, b, c) = f_{x_3}(a, b, c) = 0$, its nature is determined by the quadratic form

$$\Phi = \mathbf{d}^T \mathbf{H} \mathbf{d},$$

where

$$\mathbf{d} = \begin{bmatrix} h_1 \\ h_2 \\ h_3 \end{bmatrix} \quad \text{and} \quad \mathbf{H} = \begin{bmatrix} f_{x_1 x_1} & f_{x_1 x_2} & f_{x_1 x_3} \\ f_{x_2 x_1} & f_{x_2 x_2} & f_{x_2 x_3} \\ f_{x_3 x_1} & f_{x_3 x_2} & f_{x_3 x_3} \end{bmatrix}_{(a, b, c)}.$$

The matrix $\mathbf{H}$ is the **Hessian matrix** for $f(x_1, x_2, x_3)$. Deduce that (a, b, c) is a local maximum when Φ is negative definite, a local minimum when it is positive definite and a saddle point when it is indefinite. What are the equivalent conditions for the identification of stationary points in terms of the Δ_r of Theorem 3.33 (ii)?

22 Show $f(x_1, x_2, x_3)=(x_1+x_2+x_3)\exp[-(x_1^2+x_2^2+x_3^2)]$ has stationary points at $(\pm 1/\sqrt{6}, \pm 1/\sqrt{6}, \pm 1/\sqrt{6})$. Use these results to show that the corresponding Hessian matrices are

$$\mathbf{H}_\pm = \mp\frac{2}{\sqrt{6}}e^{-1/2}\begin{bmatrix} 4 & 1 & 1 \\ 1 & 4 & 1 \\ 1 & 1 & 4 \end{bmatrix},$$

and hence that $(1/\sqrt{6}, 1/\sqrt{6}, 1/\sqrt{6})$ is a maximum and $(-1/\sqrt{6}, -1/\sqrt{6}, -1/\sqrt{6})$ a minimum of $f(x_1, x_2, x_3)$.

23 Let $p>0, q>0$ be arbitrary real numbers, and define the $n \times n$ matrix $\mathbf{A}$ and the $n \times 1$ vector $\mathbf{x}$ as

$$\mathbf{A}=\begin{bmatrix} p & q & q & \cdot & \cdot & \cdot & q \\ q & p & q & \cdot & \cdot & \cdot & q \\ q & q & p & \cdot & \cdot & \cdot & q \\ \cdot & \cdot & \cdot & & & & \cdot \\ \cdot & \cdot & \cdot & & & & \cdot \\ \cdot & \cdot & \cdot & & & & \cdot \\ q & q & q & \cdot & \cdot & \cdot & p \end{bmatrix}, \mathbf{x}=\begin{bmatrix} x_1 \\ x_2 \\ x_3 \\ \cdot \\ \cdot \\ \cdot \\ x_n \end{bmatrix}.$$

(i) How can Gerschgorin's theorem (Theorem 3.30) be used to deduce a sufficient condition for the positive definiteness of $\mathbf{x}^T\mathbf{A}\mathbf{x}$?

(ii) Use Theorem 3.30(i) to find the necessary and sufficient conditions on p and q such that the quadratic form $\mathbf{x}^T\mathbf{A}\mathbf{x}$ is (a) positive definite and (b) indefinite.

(iii) Show that the quadratic form $\mathbf{x}^T\mathbf{A}\mathbf{x}$ can never be negative definite.

3.13 The LU and Cholesky factorization methods

This section describes two special methods for the numerical solution of systems of simultaneous equations. Both methods are computationally efficient, and each is based on a factorization of the coefficient matrix of the system. The first method, called the **LU factorization method,** is the more general of the two. It may be applied to any system with a nonsingular coefficient matrix, and it is specially well suited to the repeated solution of a system of equations in which only the nonhomogeneous terms differ from one computation to the next. The other method, called the **Cholesky factorization method,** applies to systems with symmetric coefficient matrices which are positive definite.

The LU factorization method

Many classes of problem arising in engineering, physics and numerical analysis necessitate the solution of a matrix equation

$$\mathbf{A}\mathbf{x} = \mathbf{b} \tag{1}$$

in which the $n \times n$ matrix $\mathbf{A}$ is fixed, but the $n \times 1$ vector $\mathbf{b}$ may be any one of a sequence of vectors $\mathbf{b}_1, \mathbf{b}_2, \ldots, \mathbf{b}_m$. This happens, for example, in the analysis of structural problems where the $n \times n$ matrix $\mathbf{A}$ describes a given structure, and so is fixed, while the vectors $\mathbf{b}_1, \mathbf{b}_2, \ldots, \mathbf{b}_m$ describe the various different loadings to which the structure is to be subjected. The elements of the vector $\mathbf{x}$ then describe the forces acting at the joints of the structure. The use of Gaussian elimination to solve for the elements $x_1, x_2, \ldots, x_n$ of $\mathbf{x}$ for each choice of $\mathbf{b}$ becomes very time-consuming when n is large, so an alternative approach is desirable.

The method we now describe overcomes the problem by expressing $\mathbf{A}$ as the product of two special matrix factors $\mathbf{L}$ and $\mathbf{U}$ by using a method closely related to Gaussian elimination. Once these factors have been obtained we shall see that the determination of $\mathbf{x}$ for a given vector $\mathbf{b}$ becomes a simple matter in which only $\mathbf{b}$ is changed from one calculation to another.

The factorization we shall seek is

$$\mathbf{A} = \mathbf{LU}, \tag{2}$$

in which $\mathbf{A}$ is a nonsingular $n \times n$ matrix, $\mathbf{L}$ is a *lower triangular* $n \times n$ matrix with 1's on its leading diagonal, and $\mathbf{U}$ is a nonsingular *upper triangular* $n \times n$ matrix. Matrix $\mathbf{L}$ is nonsingular because det $\mathbf{L} \neq 0$, as may be seen from the fact that det $\mathbf{A} = (\det \mathbf{L})(\det \mathbf{U})$, taken together with the result det $\mathbf{L} = 1$ and det $\mathbf{A} \neq 0$. Here the symbols and $\mathbf{L}$ and $\mathbf{U}$ are used as mnemonics for a *lower* triangular and an *upper* triangular matrix, respectively. For the moment let us suppose $\mathbf{L}$ and $\mathbf{U}$ have been found, and see how $\mathbf{x}$ may be determined from (1). It is easily seen that (1) is equivalent to first solving the matrix equation

$$\mathbf{Ly} = \mathbf{b} \tag{3}$$

for $\mathbf{y}$, and then with this $\mathbf{y}$ solving the matrix equation

$$\mathbf{Ux} = \mathbf{y} \tag{4}$$

for $\mathbf{x}$. This follows because substitution of (4) into (3) gives $\mathbf{LUx} = \mathbf{b}$, and $\mathbf{LU} = \mathbf{A}$.

The advantage offered by this approach is that $\mathbf{L}$ and $\mathbf{U}$ need be determined only once from a given matrix $\mathbf{A}$, after which the pair of equations (3) and (4) are solved for each choice of $\mathbf{b}$. On account of the lower triangular form of $\mathbf{L}$, the solution of (3) is obtained by *forward* substitution. The value of y_1 is determined immediately, after which y_2 follows from y_1, y_3 follows from y_2 and y_1, and so on, until finally y_n is obtained. Using these $y_1, y_2, \ldots, y_n$ in (4), the required elements of the solution vector $\mathbf{x}$ follow by *back*-substitution in the order $x_n, x_{n-1}, \ldots, x_1$, as in Gaussian elimination.

We must now discover how to reconstruct matrix $\mathbf{A}$ from the matrix $\mathbf{U}$ obtained by the Gaussian elimination process used to convert $\mathbf{A}$ to upper triangular echelon form. To accomplish this it is necessary to perform on $\mathbf{U}$, in reverse order, the converse of each operation which generated $\mathbf{U}$ from $\mathbf{A}$.

At each stage of the Gaussian elimination process leading to $\mathbf{U}$ from $\mathbf{A}$ let us record the multipliers used in the successive columns of a lower triangular matrix $\mathbf{L}$, and in doing so we will assume no zero pivot occurs. The matrix **L** thus has the form

$$
\mathbf{L} = \begin{bmatrix}
1 & 0 & 0 & . & . & . & 0 \\
m_{21} & 1 & 0 & . & . & . & 0 \\
m_{31} & m_{32} & 1 & . & . & . & 0 \\
. & . & . & . & . & . & . \\
. & . & . & . & . & . & . \\
m_{n1} & m_{n2} & m_{n3} & . & . & . & 1
\end{bmatrix}. \tag{5}
$$

In (5) the numbers $m_{21}, m_{31}, \ldots, m_{n1}$ in column 1 are the respective multiples of row 1 of **A** which must be subtracted from rows 2 to n to reduce to zero the elements $a_{21}, a_{31}, \ldots, a_{n1}$ of column 1 of **A**. Similarly, $m_{32}, m_{42}, \ldots, m_{n2}$ are the respective multiples of the new row 2 of **A** which must be subtracted from the new rows 3 to n to reduce to zero the elements $a_{32}^*, a_{42}^*, \ldots, a_{n2}^*$ of column 2 of the modified matrix **A**, and so on.

To recover **A** from **U**, row 1 of **U** must be left unchanged, as it is simply row 1 of the original matrix **A** which was used at the start of the elimination process. To row 2 of **U** must be added m_{21} times row 1 of **U** (row 1 of **A**). Similarly, to row 3 of **U** must be added m_{31} times row 1 of **U** (row 1 of **A**) and m_{32} times row 2 of **U**, and so on. Notice that the presence of zeros below the leading diagonal of **U** prevents the inclusion of unwanted terms at each stage which would otherwise result from the addition to a row of multiples of rows occurring above it.

Now the effect of pre-multiplication of a matrix by a lower triangular matrix with 1's on its leading diagonal is simply to add to each row of the matrix multiples of rows of the matrix which occur above it. Inspection shows the elements of the rows of the matrix **L** in (5) to be precisely the multipliers needed to reconstruct **A** from **U** and, furthermore, they occur in the right order. Hence the lower triangular matrix **L** in (5) is the matrix required in the factorization of **A** given in (2). Thus, to arrive at the **LU** factorization, matrix **U** is obtained by the Gaussian elimination process applied to **A**, and matrix **L** is constructed simultaneously from the successive steps used in the elimination process.

We remark that not every nonsingular matrix **A** has an **LU** factorization, though it can be shown that such a factorization can always be found if the order of the rows of **A** is altered. A simple case illustrating these remarks is provided by the nonsingular matrix

$$
\begin{bmatrix}
0 & 0 & 1 \\
0 & 1 & 0 \\
1 & 0 & 0
\end{bmatrix}.
$$

If for any reason one or more row interchanges occur in the Gaussian elimination process used to arrive at the **LU** factorization of **A**, the method described above must be modified.

To see the effect on the **LU** factorization process when row interchanges of the $n \times n$ matrix **A** are involved we need to use the notion of a *row permutation matrix* **P** introduced at the end of Sec. 3.3. It will be recalled that **P** is a matrix obtained from the $n \times n$ unit matrix **I** by interchanging the order of its rows in some specific manner. The matrix **P** has the property that the matrix product **PA** is obtained from the $n \times n$ matrix **A** by making the same interchange of rows in **A** that was used to define **P** in terms of **I**.

Suppose that the elimination process involves m row interchanges, with the first being represented by the row permutation matrix $\mathbf{P}_1$, the second by $\mathbf{P}_2$, and so on, up to the last which is represented by $\mathbf{P}_m$. Then the final permutation of the rows of **A** will be described by the row permutation matrix

$$\mathbf{P} = \mathbf{P}_m \mathbf{P}_{m-1} \cdots \mathbf{P}_2 \mathbf{P}_1. \tag{6}$$

Notice that the order in which the row permutation matrices are combined in **P** must be same as that in which the row permutations themselves are made to **A**. Thus, in terms of **P**, it follows directly that the row interchanges used in the elimination process cause the **LU** factorization to be performed on **PA** instead of **A**, so that

$$\mathbf{PA} = \mathbf{LU}, \tag{7}$$

or

$$\mathbf{A} = \mathbf{P}^{-1}\mathbf{LU}. \tag{8}$$

Let us now consider the solution of

$$\mathbf{Ax} = \mathbf{b} \tag{9}$$

when the above row interchanges are involved. Pre-multiplying (9) by **P** and using (7) we obtain the result

$$\mathbf{LUx} = \mathbf{Pb}, \tag{10}$$

which is analogous to (1). The solution of this system in terms of the triangular matrices **L** and **U** then proceeds as before, except for the fact that (10) is regarded as the two simultaneous triangular systems (compare with (3) and (4))

$$\mathbf{Ly} = \mathbf{Pb} \tag{11}$$

and

$$\mathbf{Ux} = \mathbf{y}. \tag{12}$$

Should it be required, $\mathbf{P}^{-1}$ is easily determined from (28) in Sec. 3.3. We saw there that every row permutation matrix **P** has the property that $\mathbf{PP}^\mathsf{T} = \mathbf{I}$, so it follows at once that $\mathbf{P}^{-1} = \mathbf{P}^\mathsf{T}$ (**P** is an *orthogonal* matrix).

Reversing the order of the operations in the **LU** factorization processes described above brings us to the following algorithm.[5]

---

[5] An **algorithm** is a computational rule detailing the sequence of mathematical operations to be performed in order to arrive at a specific result.

Algorithm for the LU factorization of A

Case (i) No row interchanges necessary

1. Apply Gaussian elimination without partial pivoting to the $n \times n$ coefficient matrix **A**. The upper triangular matrix obtained will then be the matrix **U**.
2. Record the multipliers m_{ij} used in the Gaussian elimination process in a lower triangular matrix as indicated in (5). This is the matrix **L**.
3. The required factorization is then

$$A = LU$$

Case (ii) Row interchange necessary

4. Apply Gaussian elimination with partial pivoting to the $n \times n$ coefficinet matrix **A**. The upper triangular matrix obtained will then be the matrix **U**.
5. As in (5), after each successive stage of the elimination process record the multipliers m_{ij} used in the reduction process in successive columns of a lower triangular matrix **L*** which has 1's on its leading diagonal. Whenever an interchange of rows occurs during the elimination process, interchange the elements in the corresponding rows of **L*** which have been determined up to that stage, and which lie to the left of the leading diagonal. When the elimination process has been completed the matrix determined in this fashion is the required matrix **L**.
6. Let **P** be the row permutation matrix describing the combined effect of all the row interchanges performed during the elimination process. The required factorization is then

$$PA = LU$$

or, equivalently,

$$A = P^T LU$$

Example 1. LU factorization with no row interchanges

Obtain the **LU** factorization of the matrix

$$A = \begin{bmatrix} 2 & 1 & 4 & 1 \\ 1 & 2 & 1 & 2 \\ 3 & 1 & 2 & 1 \\ 4 & 1 & 0 & 4 \end{bmatrix}.$$

Use the result to solve the matrix equation

$$Ax = b,$$

when $b = [1, 1, 1, -1]^T$.

Solution

$$A = \begin{bmatrix} 2 & 1 & 4 & 1 \\ 1 & 2 & 1 & 2 \\ 3 & 1 & 2 & 1 \\ 4 & 1 & 0 & 4 \end{bmatrix} \begin{array}{l} R_1 \\ R_2 - \frac{1}{2}R_1 (m_{21} = \frac{1}{2}) \\ R_3 - \frac{3}{2}R_2 (m_{31} = \frac{3}{2}) \\ R_4 - 2R_1 (m_{41} = 2) \end{array} \begin{bmatrix} 2 & 1 & 4 & 1 \\ 0 & \frac{3}{2} & -1 & \frac{3}{2} \\ 0 & -\frac{1}{2} & -4 & -\frac{1}{2} \\ 0 & -1 & -8 & 2 \end{bmatrix}$$

$$\begin{array}{l} R_1 \\ R_2 \\ R_3 - (-\frac{1}{3}) R_2 (m_{32} = -\frac{1}{3}) \\ R_4 - (-\frac{2}{3}) R_2 (m_{42} = -\frac{2}{3}) \end{array} \begin{bmatrix} 2 & 1 & 4 & 1 \\ 0 & \frac{3}{2} & -1 & \frac{3}{2} \\ 0 & 0 & -\frac{13}{3} & 0 \\ 0 & 0 & -\frac{26}{3} & 3 \end{bmatrix} \begin{array}{l} R_1 \\ R_2 \\ R_3 \\ R_4 - 2R_3 (m_{43} = 2) \end{array} \begin{bmatrix} 2 & 1 & 4 & 1 \\ 0 & \frac{3}{2} & -1 & \frac{3}{2} \\ 0 & 0 & -\frac{13}{3} & 0 \\ 0 & 0 & 0 & 3 \end{bmatrix}.$$

Thus the matrix factors **L** and **U** are

$$L = \begin{bmatrix} 1 & 0 & 0 & 0 \\ \frac{1}{2} & 1 & 0 & 0 \\ \frac{3}{2} & -\frac{1}{3} & 1 & 0 \\ 2 & -\frac{2}{3} & 2 & 1 \end{bmatrix}, \quad U = \begin{bmatrix} 2 & 1 & 4 & 1 \\ 0 & \frac{3}{2} & -1 & \frac{3}{2} \\ 0 & 0 & -\frac{13}{3} & 0 \\ 0 & 0 & 0 & 3 \end{bmatrix}.$$

Equation (3) becomes

$$\begin{aligned} y_1 &= 1 \\ \tfrac{1}{2}y_1 + y_2 &= 1 \\ \tfrac{3}{2}y_1 - \tfrac{1}{3}y_2 + y_3 &= 1 \\ 2y_1 - \tfrac{2}{3}y_2 + 2y_3 + y_3 + y_4 &= -1 \end{aligned}$$

with the solution obtained by *forward* substitution $y_1 = 1$, $y_2 = \frac{1}{2}$, $y_3 = -\frac{1}{3}$, $y_4 = -2$. Using these results in (4) gives

$$\begin{aligned} 2x_1 + x_2 + 4x_3 + x_4 &= 1 \\ \tfrac{3}{2}x_2 - x_3 + \tfrac{3}{2}x_4 &= \tfrac{1}{2} \\ -\tfrac{13}{3}x_3 &= -\tfrac{1}{3} \\ 3x_4 &= -2. \end{aligned}$$

which has the solution obtained by *backward* substitution $x_4 = -\frac{2}{3}$, $x_3 = \frac{1}{13}$, $x_2 = \frac{41}{39}$ and $x_1 = \frac{2}{13}$. ∎

The next example illustrates the effect of two row interchanges. The first of these is essential because of the occurrence of a zero pivot, while the second is introduced merely to demonstrate its effect on the construction of **L** and **P**.

Example 2. LU factorization with row interchanges

Obtain the **LU** factorization of the matrix

$$A = \begin{bmatrix} 2 & 1 & 4 & 1 \\ 2 & 1 & 3 & 2 \\ 4 & 3 & 9 & 3 \\ 6 & 1 & 0 & 4 \end{bmatrix},$$

and use it to solve $Ax = b$ given that $b^{\mathsf{T}} = [-2, 0, -5, 9]$.

Solution

$$A = \begin{bmatrix} 2 & 1 & 4 & 1 \\ 2 & 1 & 3 & 2 \\ 4 & 3 & 9 & 3 \\ 6 & 1 & 0 & 4 \end{bmatrix} \begin{array}{l} R_1 \\ R_2 - R_1 \\ R_3 - 2R_1 \\ R_4 - 3R_1 \end{array} \begin{bmatrix} 2 & 1 & 4 & 1 \\ 0 & 0 & -1 & 1 \\ 0 & 1 & 1 & 1 \\ 0 & -2 & -12 & 1 \end{bmatrix}.$$

A row interchange is now essential to remove the zero pivot in row 2 column 2 so the elimination process can proceed. Let us interchange rows 2 and 4.

$$\begin{array}{l} \text{Interchange} \\ R_2 \text{ and } R_4 \end{array} \begin{bmatrix} 2 & 1 & 4 & 1 \\ 0 & -2 & -12 & 1 \\ 0 & 1 & 1 & 1 \\ 0 & 0 & -1 & 1 \end{bmatrix} \begin{array}{l} R_1 \\ R_2 \\ R_3 - (-\frac{1}{2})R_2 \\ R_4 \end{array} \begin{bmatrix} 2 & 1 & 4 & 1 \\ 0 & -2 & -12 & 1 \\ 0 & 0 & -5 & \frac{3}{2} \\ 0 & 0 & -1 & 1 \end{bmatrix}.$$

Although no further interchange of rows is necessary, in order to demonstrate the effect of more than one row interchange on **L** and **U** we now interchange rows 3 and 4.

$$\begin{array}{l} \text{Interchange} \\ R_3 \text{ and } R_4 \end{array} \begin{bmatrix} 2 & 1 & 4 & 1 \\ 0 & -2 & -12 & 1 \\ 0 & 0 & -1 & 1 \\ 0 & 0 & -5 & \frac{3}{2} \end{bmatrix} \begin{array}{l} R_1 \\ R_2 \\ R_3 \\ R_4 - 5R_3 \end{array} \begin{bmatrix} 2 & 1 & 4 & 1 \\ 0 & -2 & -12 & 1 \\ 0 & 0 & -1 & 1 \\ 0 & 0 & 0 & -\frac{7}{2} \end{bmatrix}.$$

Thus, after an elimination process involving two row interchanges, we find

$$\mathbf{U} = \begin{bmatrix} 2 & 1 & 4 & 1 \\ 0 & -2 & -12 & -1 \\ 0 & 0 & -1 & 1 \\ 0 & 0 & 0 & -\frac{7}{2} \end{bmatrix}.$$

We now construct **L** sequentially to take account of row interchanges. At each step corresponding to an elimination the entries shown as dots have still to be determined.

$$\mathbf{L}^* = \begin{bmatrix} 1 & 0 & 0 & 0 \\ 1 & 1 & 0 & 0 \\ 2 & . & 1 & 0 \\ 3 & . & . & 1 \end{bmatrix} \begin{array}{l} \text{Interchange elements} \\ \text{determined so far in} \\ R_2 \text{ and } R_4 \text{ to left of} \\ \text{leading diagonal} \end{array} \begin{bmatrix} 1 & 0 & 0 & 0 \\ 3 & 1 & 0 & 0 \\ 2 & . & 1 & 0 \\ 1 & . & . & 1 \end{bmatrix}$$

Step 1

$$\begin{bmatrix} 1 & 0 & 0 & 0 \\ 3 & 1 & 0 & 0 \\ 2 & -\frac{1}{2} & 1 & 0 \\ 1 & 0 & . & 1 \end{bmatrix} \begin{array}{l} \text{Interchange elements} \\ \text{determined so far in} \\ R_3 \text{ and } R_4 \text{ to left of} \\ \text{leading diagonal} \end{array} \begin{bmatrix} 1 & 0 & 0 & 0 \\ 3 & 1 & 0 & 0 \\ 1 & 0 & 1 & 0 \\ 2 & -\frac{1}{2} & . & 1 \end{bmatrix}$$

Step 2

$$\mathbf{L} = \begin{bmatrix} 1 & 0 & 0 & 0 \\ 3 & 1 & 0 & 0 \\ 1 & 0 & 1 & 0 \\ 2 & -\frac{1}{2} & 5 & 1 \end{bmatrix}$$

Step 3

The first row interchange brought rows R_1, R_2, R_3, R_4 of A into the order R_1, R_4, R_3, R_2, while the second row interchange permuted them into the final order R_1, R_4, R_2, R_3. Thus the permutation matrix involved here is

$$P = \begin{bmatrix} 1 & 0 & 0 & 0 \\ 0 & 0 & 0 & 1 \\ 0 & 1 & 0 & 0 \\ 0 & 0 & 1 & 0 \end{bmatrix},$$

so we have arrived at the factorization

$$PA = LU.$$

It is left to the reader to verify that this result is, in fact, true.

For $b = [-2, 0, -5, 9]^T$ system (11) becomes

$$
\begin{aligned}
y_1 && = -2 \\
3y_1 + y_2 && = 9 \\
y_1 + y_3 && = 0 \\
2y_1 - \tfrac{1}{2}y_2 + 5y_3 + y_4 && = -5,
\end{aligned}
$$

which when solved by *forward* substitution gives

$$y_1 = -2, \quad y_2 = 15, \quad y_3 = 2, \quad y_4 = -\tfrac{7}{2}$$

Then, finally, system (12) becomes

$$
\begin{aligned}
2x_1 + x_2 + 4x_3 + x_4 &= -2 \\
-2x_2 - 12x_3 + x_4 &= 15 \\
-x_3 + x_4 &= 2 \\
-\tfrac{7}{2}x_4 &= -\tfrac{7}{2},
\end{aligned}
$$

which when solved by backward substitution gives $x_4 = 1$, $x_3 = -1$, $x_2 = -1$ and $x_1 = 1$, which is the required solution. ∎

The Cholesky factorization method

The **Cholesky factorization method** applies to the solution of systems of simultaneous equations

$$Ax = b, \tag{13}$$

in which the $n \times n$ matrix A is *symmetric* and *positive definite*. It involves factorizing A in the form

$$A = QQ^T, \tag{14}$$

where Q is a real nonsingular lower triangular matrix.

To see how the solution of (13) is obtained once **L** is known we first write (13) as

$$\mathbf{QQ^T x = b}. \tag{15}$$

Then, as in the **LU** factorization method, the solution of (8) is equivalent to the solution of

$$\mathbf{Qy = b} \tag{16}$$

for **y**, followed by the solution of

$$\mathbf{Q^T x = y} \tag{17}$$

for **x**.

System (16) is *lower triangular* in form, and so is solved for the elements of **y** by *forward* substitution, while system (17) is *upper triangular* in form, and so when **y** is known, is solved by *back* substitution for the elements of **x**.

It now remains for us to prove that when **A** is symmetric and positive definite it may be expressed as in (14), and then to find the elements q_{ij} of **Q**. The reader who does not need a detailed justification of representation (14) should proceed directly to the Cholesky factorization algorithm and Example 3. We already know that when **A** is nonsingular, and no zero pivots occur, a factorization exists of the form

$$\mathbf{A = LU}, \tag{18}$$

where **L** is a lower triangular matrix with 1's on its leading diagonal and **U** is a nonsingular upper triangular matrix with elements u_{ij}. Let us now define the nonsingular diagonal matrix **D** whose elements $d_{ii} = u_{ii}$ for $i = 1, 2, \ldots, n$. Then $\mathbf{R^T = D^{-1}U}$ is an upper triangular matrix with 1's on its leading diagonal. Thus, as

$$\mathbf{A = LU = LD(D^{-1}U) = LDR^T},$$

this shows **A** may always be expressed as the product of factors

$$\mathbf{A = LDR^T}. \tag{19}$$

Next we show that if **A** is symmetric then **L = R**. Consider the matrix $\mathbf{M = R^{-1}A(R^{-1})^T}$. Then **M** is symmetric, because as $\mathbf{A^T = A}$ it follows that $\mathbf{M^T = M}$. Matrix **M** is also lower triangular, because as $\mathbf{(R^{-1})^T = (R^T)^{-1}}$ we have from (19) that $\mathbf{A(R^{-1})^T = LD}$ which is a lower triangular matrix. Consequently as **M** is the product of the two lower triangular matrices $\mathbf{R^{-1}}$ and $\mathbf{A(R^{-1})^T}$ it must also be lower triangular. Thus as **M** is both symmetric and lower triangular it must be *diagonal*. Now

$$\mathbf{M = R^{-1}A(R^{-1})^T = R^{-1}LD}$$

can only be diagonal if $\mathbf{R^{-1}}$ is diagonal. However $\mathbf{R^{-1}L}$ is a lower triangular matrix with 1's on its leading diagonal, so $\mathbf{R^{-1}L = I}$ and thus **L = R**. Thus we conclude from (12) that if

A is symmetric it has the representation

$$\mathbf{A} = \mathbf{L}\mathbf{D}\mathbf{L}^\mathsf{T}. \tag{20}$$

Finally, let us now suppose **A** to be both symmetric and positive definite. Then if **x** is an $n \times 1$ vector, $\mathbf{x}^\mathsf{T}\mathbf{A}\mathbf{x}$ will be a positive definite quadratic form. Setting $\mathbf{X} = \mathbf{L}^\mathsf{T}\mathbf{x}$ it follows that

$$\mathbf{x}^\mathsf{T}\mathbf{A}\mathbf{x} = \mathbf{X}^\mathsf{T}\mathbf{D}\mathbf{X}$$

is a positive definite quadratic form, which is only possible if the elements d_{ii} of the leading diagonal of the diagonal matrix **D** are all positive.

Thus we may set $\mathbf{D} = \mathbf{D}_1^2$ where $\mathbf{D}_1$ is a diagonal matrix with the positive real elements $\sqrt{d_{11}}, \sqrt{d_{22}}, \ldots, \sqrt{d_{nn}}$ on its leading diagonal. Then, setting $\mathbf{Q} = \mathbf{L}\mathbf{D}_1$, it follows directly from (20) that

$$\mathbf{A} = \mathbf{Q}\mathbf{Q}^\mathsf{T},$$

as was to be proved.

The elements q_{ij} of the lower triangular matrix **Q** now follow from (14) when it is written in the form

$$\begin{bmatrix} a_{11} & a_{21} & \cdots & a_{n1} \\ a_{21} & a_{22} & \cdots & a_{n2} \\ . & . & \cdots & . \\ a_{n1} & a_{n2} & \cdots & a_{nn} \end{bmatrix} = \begin{bmatrix} q_{11} & 0 & \cdots & 0 \\ q_{21} & q_{22} & \cdots & 0 \\ . & . & \cdots & . \\ q_{n1} & q_{n2} & \cdots & q_{nn} \end{bmatrix} \begin{bmatrix} q_{11} & q_{21} & \cdots & q_{n1} \\ 0 & q_{22} & \cdots & q_{n2} \\ . & . & \cdots & . \\ 0 & 0 & \cdots & q_{nn} \end{bmatrix},$$

by expanding the matrix product and equating corresponding elements in the lower triangular arrays on each side of the equality (in representing **A** use has been made of the fact that $a_{ij} = a_{ji}$).

The explicit representation for the q_{ij} obtained in this manner becomes

$$q_{ii} = \begin{cases} \sqrt{a_{11}} & \text{for } i = 1, \\ \left(a_{ii} - \sum_{k=1}^{i-1} q_{ik}^2 \right)^{1/2} & \text{for } i = 2, 3, \ldots, n \end{cases}$$

and   (21)

$$q_{ij} = \begin{cases} a_{i1}/q_{11} & \text{for } j = 1, \, i = 2, 3, \ldots, n \\ \left(a_{ij} - \sum_{k=1}^{j-1} q_{ik}q_{jk} \right)\Big/ q_{jj} & \text{for } j = 2, 3, \ldots, n, \, i = j+1, \ldots, n. \end{cases}$$

The quantities under the square root in (14) will all be positive provided **A** is positive definite. If a diagonal element q_{ii} in the matrix **Q** involves the square root of a negative number matrix **A** will not be positive definite and the Cholesky factorization method will fail. Thus the Cholesky factorization of **A** provides a test for its positive definiteness.

Algorithm for the Cholesky factorization of A

Let A be a symmetric $n \times n$ matrix with elements a_{ij}.

1. Determine the elements q_{ij} of the lower triangular matrix

$$Q = \begin{bmatrix} q_{11} & 0 & 0 & \cdots & 0 \\ q_{21} & q_{22} & 0 & \cdots & 0 \\ q_{31} & q_{33} & q_{33} & \cdots & 0 \\ \cdot & \cdot & \cdot\cdot & \cdots & \cdot \\ q_{n1} & q_{n2} & q_{n3} & \cdots & q_{nn} \end{bmatrix}$$

using the expressions

$$q_{ii} = \begin{cases} \sqrt{a_{11}} & \text{for } i=1, \\ \left(a_{ii} - \displaystyle\sum_{k=1}^{i-1} q_{ik}^2 \right)^{1/2} & \text{for } i=2, 3, \ldots, n \end{cases}$$

and

$$q_{ij} = \begin{cases} a_{i1}/q_{11} & \text{for } j=1, \quad i=2, \ldots, n \\ \left(a_{ij} - \displaystyle\sum_{k=1}^{j-1} q_{ik} q_{jk} \right) \bigg/ q_{jj} & \text{for } j=2, 3, \ldots, n, \quad i=j+1, \ldots, n. \end{cases}$$

2. If all the q_{ii} are positive the matrix A is positive definite and may be represented in the form

$$A = QQ^{\mathsf{l}}.$$

3. If a diagonal element q_{ii} in matrix Q is complex the matrix A is not positive definite and the factorization fails.

It is clear that symmetric negative definite matrices A may also be factorized by the Cholesky method by considering the factorization of $-A$ which is then a symmetric positive definite matrix.

Example 3. Cholesky method

Use the Cholesky method to show the symmetric matrix

$$A = \begin{bmatrix} 5 & -2 & 0 \\ -2 & 6 & 2 \\ 0 & 2 & 7 \end{bmatrix}$$

is positive definite and to obtain its representation in the form

$$A = QQ^{\mathsf{T}}.$$

Use the matrix $\mathbf{Q}$ to solve the system of equations

$\mathbf{Ax} = \mathbf{b}$

when $\mathbf{b} = [1, -1, 1]^{\mathsf{T}}$.

Solution

$$q_{11} = \sqrt{5} = 2.2361, \qquad q_{21} = a_{21}/q_{11} = -0.8944, \qquad q_{31} = 0$$
$$q_{22} = (a_{22} - q_{21}^2)^{1/2} = 2.2804, \qquad q_{32} = (a_{32} - q_{31}\, q_{21})/q_{22} = 0.8770$$
$$q_{33} = (a_{33} - q_{31}^2 - q_{32}^2)^{1/2} = 2.4962.$$

As all the diagonal elements are positive the matrix $\mathbf{A}$ is positive definite and

$$\mathbf{Q} = \begin{bmatrix} 2.2361 & 0 & 0 \\ -0.8944 & 2.2804 & 0 \\ 0 & 0.8770 & 2.4962 \end{bmatrix}.$$

It is easily verified that $\mathbf{Q}\mathbf{Q}^{\mathsf{T}} = \mathbf{A}$.

Using results (16) and (17) to solve the system

$\mathbf{Ax} = \mathbf{b}$

shows we must first solve the system $\mathbf{Qy} = \mathbf{b}$ and then the system $\mathbf{Q}^{\mathsf{T}}\mathbf{x} = \mathbf{y}$. The first of these is the lower triangular system

$$\begin{aligned} 2.2361\,y_1 &= 1 \\ -0.8944\,y_1 + 2.2804\,y_2 &= -1 \\ 0.8770\,y_2 + 2.4962\,y_3 &= 1 \end{aligned}$$

with the solution $y_1 = 0.4472$, $y_2 = -0.2631$ and $y_3 = 0.4931$. Finally, solving $\mathbf{Q}^{\mathsf{T}}\mathbf{x} = \mathbf{y}$ for $\mathbf{x}$, we obtain $x_1 = 0.1235$, $x_2 = -0.1913$ and $x_3 = 0.1975$. ∎

Before leaving this topic it will be instructive to compare the computational effort required by some of the different methods described in this chapter for the solution of linear systems of n simultaneous equations. The criterion employed as the basis for comparison will be the number of multiplications involved. This is because the operation of multiplication on a digital computer is far more time-consuming than that for addition; so the number of multiplications involved provides a good measure of the time taken.

A solution obtained by Cramer's rule requires the evaluation of $n+1$ determinants, and as each involves $(n-1)n!$ multiplications we see that an application of Cramer's rule requires $(n^2 - 1)n!$ multiplications.

A solution by Gaussian elimination requires approximately $\frac{1}{3}n^3 + n^2$ multiplications. Thus if the nonhomogeneous term $\mathbf{b}$ in a system of equations is changed m times, Gaussian elimination will require approximately $m(\frac{1}{3}n^3 + n^2)$ multiplications.

When the **LU** factorization method is used with m vectors $\mathbf{b}$, approximately $\frac{1}{3}n^3 + mn^3$ multiplications are needed, which represents a considerable saving over the successive use of Gaussian elimination m times.

These few results give some idea of the relative efficiency of the methods described so far. The results demonstrate clearly why the use of Cramer's rule is to be avoided for the numerical solution of systems involving more than about four equations. The real importance of Cramer's rule lies in the theoretical insight it provides in the many branches of mathematics giving rise to linear systems of simultaneous equations.

It should be recalled here that det **A** may be deduced directly from the upper triangular array produced by Gaussian elimination when applied to a determinant (cf. Ex. 3, Sec. 3.4 and the remarks preceding it). The above comparison of computational effort shows that the evaluation of determinants by Gaussian elimination is certainly the most efficient method in terms of the number of multiplications involved.

Problems for Section 3.13

The first two problems involve determining the **LU** factorization of a 3×3 matrix **A** from first principles. Assuming no zero divisors occur, the elements l_{ij} and u_{ij} of **L** and **U** may be found by solving the following nine simultaneous equations

$$u_{11} = a_{11}, u_{12} = a_{12}, u_{13} = a_{13}$$

$$l_{21}u_{11} = a_{21}, l_{21}u_{12} + u_{22} = a_{22}, l_{21}u_{13} + u_{23} = a_{23}$$

$$l_{31}u_{11} = a_{31}, l_{31}u_{12} + l_{32}u_{22} = a_{22}, l_{31}u_{13} + l_{32}u_{23} + l_{33}u_{33} = a_{33}$$

obtained by equating corresponding elements in the matrix equation $\mathbf{A} = \mathbf{LU}$ when written in the form

$$\begin{bmatrix} a_{11} & a_{12} & a_{13} \\ a_{21} & a_{22} & a_{23} \\ a_{31} & a_{32} & a_{33} \end{bmatrix} = \begin{bmatrix} 1 & 0 & 0 \\ l_{21} & 1 & 0 \\ l_{31} & l_{32} & 1 \end{bmatrix} \begin{bmatrix} u_{11} & u_{12} & u_{13} \\ 0 & u_{22} & u_{23} \\ 0 & 0 & u_{33} \end{bmatrix}.$$

$$\mathbf{1 \ A} = \begin{bmatrix} 2 & -1 & 3 \\ 8 & -1 & 13 \\ 2 & -10 & -2 \end{bmatrix} \qquad \mathbf{2 \ A} = \begin{bmatrix} 3 & -2 & 1 \\ 6 & -3 & 4 \\ 3 & -3 & 2 \end{bmatrix}$$

Apply case (i) of the **LU** factorization algorithm to each of the following problems to find the factorization for the given matrix **A**, and then use the factors **L** and **U** to solve the matrix equation

$$\mathbf{Ax = b}$$

for the stated vector **b**. In each case verify the solution **x** by repeating the calculation using case (ii) of the algorithm with two row interchanges of your own choice introduced into the elimination process.

$$\mathbf{3 \ A} = \begin{bmatrix} 2 & -1 & 1 \\ 6 & -1 & 4 \\ 4 & -4 & 4 \end{bmatrix}; \qquad \mathbf{b} = \begin{bmatrix} 5 \\ 15 \\ 16 \end{bmatrix}.$$

4 $A = \begin{bmatrix} 5 & 1 & 2 \\ -10 & -1 & -3 \\ -5 & 0 & -2 \end{bmatrix}$; $b = \begin{bmatrix} 10 \\ -21 \\ -12 \end{bmatrix}$.

5 $A = \begin{bmatrix} -1 & 0 & 1 & 2 \\ -1 & 1 & 3 & 3 \\ 1 & 0 & 1 & 1 \\ -2 & 1 & 6 & 9 \end{bmatrix}$; $b = \begin{bmatrix} 2 \\ 4 \\ 3 \\ 12 \end{bmatrix}$.

6 $A = \begin{bmatrix} 3 & 0 & 1 & 0 \\ 6 & 1 & 4 & 0 \\ 0 & -1 & -1 & 4 \\ 0 & 2 & 5 & 5 \end{bmatrix}$; $b = \begin{bmatrix} -1 \\ 3 \\ -11 \\ 2 \end{bmatrix}$.

7 $A = \begin{bmatrix} 1 & 0 & 0 & 1 \\ 0 & 0 & 1 & 2 \\ 0 & 1 & 0 & 1 \\ 1 & 2 & 1 & 0 \end{bmatrix}$; $b = \begin{bmatrix} -1 \\ -2 \\ -3 \\ 1 \end{bmatrix}$.

8 $A = \begin{bmatrix} 1 & 1 & 0 & 1 \\ 1 & 0 & 0 & 2 \\ 0 & 0 & 3 & -1 \\ 1 & 2 & -1 & 0 \end{bmatrix}$; $b = \begin{bmatrix} 0 \\ 0 \\ 4 \\ -1 \end{bmatrix}$.

Use the Cholesky factorization algorithm to determine which of the following matrices is positive definite, and when A is positive definite find the matrix Q in the representation $A = QQ^T$.

9 $A = \begin{bmatrix} 2 & -4 \\ -4 & 1 \end{bmatrix}$ **10** $A = \begin{bmatrix} 3 & -3 \\ -3 & 9 \end{bmatrix}$

11 $A = \begin{bmatrix} 2 & 1 & 0 \\ 1 & 1 & 1 \\ 0 & 1 & 4 \end{bmatrix}$ **12** $A = \begin{bmatrix} 1 & 0 & 0 \\ 0 & \frac{3}{2} & -\frac{1}{2} \\ 0 & -\frac{1}{2} & \frac{3}{2} \end{bmatrix}$

Harder problems

It was proved in the argument leading to (13) that provided a *symmetric* matrix A has an LU factorization it can be represented in the form $A = LDL^T$, where L is a lower triangular matrix with 1's on its leading diagonal and D is a diagonal matrix. A representation of this type is often more useful than the Cholesky factorization when dealing with symmetric matrices which are not necessarily positive definite. The two problems which follow show how this representation may be determined from first principles provided no zero divisor occurs.

13 By considering a symmetric 3×3 matrix $\mathbf{A}$ and writing

$$\mathbf{A} = \begin{bmatrix} a_{11} & a_{21} & a_{31} \\ a_{21} & a_{22} & a_{32} \\ a_{31} & a_{32} & a_{33} \end{bmatrix}, \quad \mathbf{L} = \begin{bmatrix} 1 & 0 & 0 \\ l_{21} & 1 & 0 \\ l_{31} & l_{32} & 1 \end{bmatrix} \quad \text{and} \quad \mathbf{D} = \begin{bmatrix} d_1 & 0 & 0 \\ 0 & d_2 & 0 \\ 0 & 0 & d_3 \end{bmatrix},$$

show by equating corresponding elements in the representation $\mathbf{A} = \mathbf{LDL}^{\mathsf{T}}$, that provided no zero divisor occurs the elements of $\mathbf{L}$ and $\mathbf{D}$ are determined by the expressions

$d_1 = a_{11}, \quad l_{21} = a_{21}/d_1, \quad d_2 = a_{22} - d_1 l_{21}^2$

$l_{31} = a_{31}/d_1, \qquad l_{32} = (a_{32} - d_1 l_{31} l_{21})/d_2$

$d_3 = a_{33} - d_1 l_{31}^2 - d_2 l_{32}^2.$

Find $\mathbf{L}$ and $\mathbf{D}$ when

$$\mathbf{A} = \begin{bmatrix} -1 & 1 & -1 \\ 1 & 1 & 1 \\ -1 & 1 & 2 \end{bmatrix}.$$

14 Using the same form of argument as in Problem 13 show that if $\mathbf{A}$ is a symmetric 4×4 matrix with the representation $\mathbf{A} = \mathbf{LDL}^{\mathsf{T}}$ and no zero divisor occurs, the elements of $\mathbf{L}$ and $\mathbf{D}$ are determined by the expressions given in Problem 13 together with the additional expressions

$l_{41} = a_{41}/d_1, \quad l_{42} = (a_{42} - d_1 l_{41} l_{21})/d_2,$

$l_{43} = (a_{43} - d_1 l_{41} l_{31} - d_2 l_{42} l_{32})/d_3 \quad \text{and}$

$d_4 = a_{44} - d_1 l_{41}^2 - d_2 l_{42}^2 - d_3 l_{43}^2.$

How may the nature of the quadratic form associated with $\mathbf{A}$ be deduced from the $\mathbf{LDL}^{\mathsf{T}}$ factorization of $\mathbf{A}$?

Part 3
Ordinary Differential Equations

The importance of differential equations is such that a large part of any text on engineering or applied mathematics must be devoted to their study. The central position occupied by such equations arises mainly because they offer the natural description of many physical problems, but also because of the part they play in defining special functions of use throughout virtually all of engineering and science. On account of this, the physical origins of differential equations will be used to provide motivation for the development of the underlying theory of the subject itself. Whenever useful and appropriate, special functions defined in terms of differential equations will be introduced and their properties developed to the point at which they may be used in applications.

Ordinary differential equations involve the study of equations in which an unknown function of a single independent variable appears together with some of its derivatives, and from which it must be determined by integration. When considering differential equations in general, a simple form of classification is necessary so that their study may be organized and developed in a systematic manner. In its simplest form this is accomplished by using a number called the order of a differential equation, which is simply the order of the highest derivative to appear in the equation. In general it follows that the higher the order of a differential equation, the more complicated is the task of finding its solution, but the richer are the properties exhibited by the solution.

A further and even more important classification of differential equations is provided by their separation into one or other of the two mutually exclusive classes called linear and nonlinear differential equations. The many applications of linear differential equations which arise, coupled with the simple structure of their solutions, makes them the most important single class to be studied. Within the class of linear differential equations, the order of an equation provides a very convenient and natural sub-classification. Knowledge of the structure of the solution of a linear differential equation will be seen to be valuable when considering initial value problems for differential equations of any order. In such problems all the conditions which must be specified to identify a unique solution are given at a single point on the real line (initial conditions). It will be seen that such knowledge is equally important when considering two-point boundary value problems for second order linear differential equations. In such problems a solution must be found in the interval between two distinct points on the real line, at each of which conditions on the solution are specified (boundary values). This class of problem will be found to provide examples of how differential equations can have a unique solution, a nonunique solution, or even no solution at all. The significance of nonunique solutions to boundary value problems will become apparent in Chapter 9 when eigenfunctions are introduced and used to expand arbitrary functions.

The study of nonlinear ordinary differential equations is difficult, but as many important applications involve special equations belonging to this class, a brief discussion of some of the simplest types of nonlinear equation will be presented. In the main, attention will be confined to first order nonlinear differential equations and to simple systems. The numerical solution of differential equations of both linear and nonlinear type forms a major part of numerical analysis, though only the numerical solution of initial value problems will be discussed here.

The last three chapters introduce the Laplace transform method for solving initial value problems for constant coefficient differential equations, the z-transform, the series solution of linear second order variable coefficient equations with and without regular singular points, and the theory of Fourier series coupled with an introduction to boundary value problems.

Chapter 4
First Order Ordinary
Differential Equations

The fundamental notions of an ordinary differential equation, its solution, and an initial value problem are introduced in Sec. 4.1, in which some of the physical origins of first order ordinary differential equations are also described. The concept of isoclines together with their geometrical interpretation is introduced in Sec. 4.2. This idea is then used to illustrate the qualitative behavior of solutions in the plane and also, through the Euler polygon, to introduce the most elementary numerical method of solution for an initial value problem. The simplest class of nonlinear ordinary differential equations to solve, the separable equations, are considered in Sec. 4.3. Exact equations are discussed in Sec. 4.4 along with their relationship to separable equations, after which the basic idea of an integrating factor is introduced and developed in a systematic manner. This approach is continued in Sec. 4.5, in which an integrating factor is used to obtain the general solution of the most fundamental of all linear ordinary differential equations, the linear first order equation. Section 4.6 develops the related geometrical concepts of orthogonal and isogonal trajectories which find many physical applications, the most important being those arising in connection with the study of potential theory in two dimensions. In Sec. 4.7 fundamental questions concerning the existence and uniqueness of solutions are considered. The section concludes with a brief discussion of an iterative method of solution due to E. Picard, which is of considerable theoretical importance and also of some practical value. Finally, Sec. 4.8 describes the accurate and straightforward Runge–Kutta method for numerically integrating first order differential equations.

4.1   Differential equations and their origins

A **differential equation** is an equation relating the derivatives of an unknown function, the function itself, the variables in terms of which the function is defined and constants. When functions of several real variables are involved, such equations are called **partial differential equations**, and when only functions of a single real variable are involved they are called **ordinary differential equations**. The chapters which follow will be devoted exclusively to ordinary differential equations, so we will be able to use the abbreviation *differential equations* without fear of ambiguity.

Typical examples of ordinary differential equations are

$$y' + 4y = 2 + e^{-x}, \tag{1}$$

$$y'' + 7xy' - 4(1 - x^2)y = 0, \tag{2}$$

$$y'(x + 2y) + y = -2x. \tag{3}$$

The **diffusion (heat) equation**

$$\kappa \frac{\partial u}{\partial t} = \frac{\partial^2 u}{\partial x^2} + \frac{\partial^2 u}{\partial y^2} \tag{4}$$

with $\kappa > 0$ constant is an example of a partial differential equation for the function $u(x, y, t)$ of the three independent variables x, y and t.

Ordinary differential equations are classified according to their order. The **order** n of a differential equation is the order of the highest derivative to appear in the equation.

Some of the most important and useful differential equations arising in applications belong to the class of linear differential equations. The general **linear** nth order ordinary differential equation is of the form

$$a_0(x) \frac{d^n y}{dx^n} + a_1(x) \frac{d^{n-1} y}{dx^{n-1}} + \ldots + a_n(x)y = f(x), \tag{5}$$

where $a_0(x), a_1(x), \ldots, a_n(x)$ and $f(x)$ are given functions of x. When, as is often the case, $a_0(x), a_1(x), \ldots, a_n(x)$ are absolute constants, (5) is called a **constant coefficient** nth order ordinary differential equation. Any differential equation not of the form shown in (1) is said to be **nonlinear**.

The nth order linear differential equation (5) will be said to be **nonhomogeneous** when $f(x) \neq 0$, and to be **homogeneous** when $f(x) \equiv 0$. When equation (5) is nonhomogeneous the function $f(x)$ is called the **nonhomogeneous term**.

The examples in (1), (2) and (3) are, respectively, a first order linear constant coefficient differential equation with nonhomogeneous term $2 + e^{-x}$, a second order linear homogeneous differential equation and a first order nonlinear differential equation. The diffusion equation shown in (4) is a second order linear partial differential equation.

A **solution** of an nth order differential equation is an n times differentiable function

$$y = g(x), \tag{6}$$

which when substituted into the equation satisfies it identically in some interval $a < x < b$.

The function

$$y = g(x) = 6e^{-4x} + \tfrac{1}{2} + \tfrac{1}{3}e^{-x}$$

is seen to be a solution of (1), because replacing y and y' in (1) by g and g', respectively, leads to the identity

$$2 + e^{-x} \equiv 2 + e^{-x},$$

for all x.

The solution $y = g(x)$ introduced in (6) is called an **explicit solution**, because it determines y directly in terms of x. On occasions solutions arise in the less convenient form

$$F(x, y) = 0,$$

from which it is impossible to deduce an explicit representation for y in terms of x. Solutions of this type are called **implicit solutions**. For example,

$$\ln(x^2 + y^2) + \text{arc tan}\left(\frac{y}{x}\right) = 1$$

is an implicit solution of (3), as may be verified by differentiating it with respect to x to find the equation satisfied by x, y and y'.

Each integration leading to the solution $y = g(x)$ of a differential equation involves the introduction of an arbitrary constant. Thus when solving an nth order differential equation, n arbitrary constants will enter into the solution. Such a solution of an nth order ordinary differential equation which contains n arbitrary constants will be called a **general solution**. Sometimes the arbitrary constants arising in a general solution are subject to constraints. For example, if the function $\ln(A - x^2)$ were to arise in a solution, with A an arbitrary constant, it would be necessary to restrict A to arbitrary positive values, for only then would $\ln(A - x^2)$ be a real valued function for $x^2 < A$. When the arbitrary constants in a general solution are assigned specific values, the solution which results is called a **particular solution**. Thus $y = 6e^{-4x} + \tfrac{1}{2} + \tfrac{1}{3}e^{-x}$ is a particular solution of (1), which has the general solution $y = Ae^{-4x} + \tfrac{1}{2} + \tfrac{1}{3}e^{-x}$, with A an arbitrary constant.

Sometimes special solutions of ordinary differential equations exist which cannot be obtained from the general solution by means of a suitable choice of arbitrary constants. These degenerate solutions are called **singular solutions**. The first order nonlinear differential equation

$$(y')^2 + 4y^3(y - 1) = 0$$

has the general solution

$$y = \frac{1}{1 + (x + A)^2},$$

but inspection shows it also has the two special solutions $y \equiv 0$ and $y \equiv 1$. Neither of the two special solutions may be obtained from the general solution by means of a particular choice of A, so they are singular solutions of the differential equation.

Example 1

It is easily verified that

$$y = \tfrac{1}{3}x^3, \qquad y = \tfrac{1}{3}x^3 + 1, \qquad y = \tfrac{1}{3}x^3 - 4$$

are all solutions of the differential equation

$$y' = x^2$$

which has the general solution

$$y = \tfrac{1}{3}x^3 + A,$$

with A an arbitrary constant. Representative solutions corresponding to different values of A are shown in Fig. 4.1.

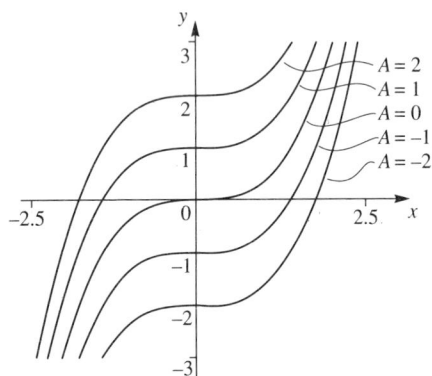

Fig. 4.1   Solutions of $y' = x^2$

Differential equations are of great importance in most branches of engineering and science. This is because, when expressed in mathematical terms, physical laws and their consequences usually give rise to differential equations. The process of deriving a mathematical formulation of a physical situation is called **mathematical modeling**, or simply **modeling**.

Example 2. Newton's law of cooling

It is known from experiment that when an object is heated to a moderate temperature (say less than 200 °C), and placed in air at around room temperature, it will give up its heat to the atmosphere according to Newton's law of cooling[1]. This law, which is approximate, states that the rate of

---

[1] ISSAC NEWTON (1642–1727), born in Lincolnshire, England, and appointed Lucasian professor of mathematics at Trinity College, Cambridge in 1663. One of the most outstanding mathematicians and scientists of all time who in his great work *Principia Mathematica* contributed to the founding of the calculus, to algebra and geometry. He made significant contributions to all of the physical sciences, the most fundamental of which included his formulation of the universal law of gravitation and the laws of motion in mechanics.

cooling is proportional to the difference between the temperature of the surface of the body and the ambient air temperature.

To formulate this in terms of a differential equation, let t be the time, T the surface temperature at time t and T_0 the ambient air temperature. The rate of cooling is thus dT/dt, while the difference in temperature between the surface of the body and the air at time t is $T - T_0$. If $\lambda > 0$ is a constant of proportionality depending on the body in question, and determined by experiment, Newton's law of cooling takes the mathematical form

$$\frac{dT}{dt} = -\lambda(T - T_0),$$

where the negative sign is necessary because the surface temperature of the body *decreases* with time.

Rewriting (7) we obtain

$$\frac{dT}{dt} + \lambda T = \lambda T_0, \tag{7}$$

showing the temperature $T(t)$ obeys a first order linear constant coefficient nonhomogeneous differential equation. The nonhomogeneous term in (7) is λT_0.

We shall see shortly that the general solution of (7) is

$$T = A e^{-\lambda t} + T_0,$$

where A is an arbitrary constant. To select an appropriate numerical value for A in a given physical situation it is necessary to know the temperature of the surface of the body when it was first placed in the air. If this is T_1, say, and time is measured from the moment the body is placed in the cool air, it follows that $T(0) = T_1$. A condition of this type, which is imposed on a solution of a differential equation at the *start*, is called an **initial condition**. Combining this initial condition with the general solution gives the particular solution

$$T = (T_1 - T_0)e^{-\lambda t} + T_0 \quad \text{for } t \geq 0. \tag{8}$$

∎

Example 3. Resisted motion

It is known from experiment that the resistance experienced by a body of constant mass when falling vertically under gravity in a resisting medium is μv^α, where v is its speed and $\alpha > 0$ and $\mu > 0$ are constants depending on the body and the resisting medium.

To formulate the differential equation governing the motion we use Newton's second law of motion. This asserts that, in the appropriate units, the force acting on a body is equal to the rate of change of its momentum.

If the body has mass m, its momentum is mv, and the rate of change of its momentum is thus mdv/dt, where t is the time. Denoting the acceleration due to gravity by g, the forces acting on the body are the downward force mg due to gravity, and the upward resisting force μv^α opposing motion. The resultant force acting on the body is thus $mg - \mu v^\alpha$, so equating this to the rate of change of momentum we arrive at the differential equation of motion

$$m\frac{dv}{dt} = mg - \mu v^\alpha,$$

or equivalently

$$m\frac{dv}{dt} + \mu v^{\alpha} = mg. \tag{9}$$

This is a first order nonlinear nonhomogeneous differential equation. The nonhomogeneous term in (9) is mg. ∎

Example 4. Xenon poisoning in a nuclear reactor after shutdown

Radioactive material decays spontaneously by the emission of various elementary particles and radiation to form atoms of a new element, which might also be radioactive. The radioactivity of an element is known to be directly proportional to the number of atoms of the element present. Thus if $M(t)$ denotes the mass of a specific radioactive element present at time t, it follows that the rate of decay dM/dt must be proportional to M. If $\lambda > 0$ is the constant of proportionality for the given radioactive element, it follows that the mass M must obey the differential equation

$$\frac{dM}{dt} = -\lambda M. \tag{10}$$

The negative sign is necessary because M is decaying. The constant λ, which is specific to the radioactive element involved, is called its **decay constant**.

The general solution of this differential equation is easily seen to be

$$M = Ae^{-\lambda t},$$

with A an arbitrary constant. So if $M = M_0$ at time $t = 0$, that is if the initial condition $M(0) = M_0$ is imposed on the solution, it follows at once that

$$M = M_0 e^{-\lambda t}$$

is the decay law for the mass of the radioactive element.

It is convenient to characterize the decay of a radioactive element in terms of its **half-life**, which is the time taken for half of a given mass of the element to decay. For example, the half-life of the isotope of uranium ^{235}U used as fuel in a nuclear reactor is 7.13×10^8 years, whereas the half-lives of the radioactive isotope of iodine ^{135}I and the radioactive isotope of xenon ^{135}Xe produced by the fission decay chain within a reactor are 6.68 hours and 9.13 hours, respectively.

^{135}Xe is produced by radioactive decay of ^{135}I, and when a nuclear reactor is operating at power both the iodine and xenon are destroyed (burnt out) by the high neutron flux in the reactor.

If the reactor is shut down after a period of steady operation, the flux falls to a very low level, but the isotope of iodine continues to feed the isotope of xenon into the reactor core by decay. ^{135}Xe is a strong absorber of neutrons, and its buildup could prevent startup of the reactor until the xenon has decayed sufficiently. The transient behavior of the xenon within the reactor is seen to be described by the differential equation

$$\frac{dM_{Xe}}{dt} = -\lambda_{Xe} M_{Xe} + \lambda_I (M_I)_0 e^{-\lambda_I t}.$$

In this equation t is the time, M_{Xe} is the mass of ^{135}Xe present at time t, $(M_I)_0$ is the mass of ^{135}I present at shutdown and λ_{Xe}, λ_I are the decay constants for xenon and iodine, respectively. The first term on the right-hand side describes the rate of decay of the xenon, while the second term describes its rate of replenishment by the decay of the iodine. The buildup of ^{135}Xe in a reactor is called **xenon poisoning**.

The differential equation for the xenon is a first order linear constant coefficient non-homogeneous differential equation, with $\lambda_1(M_1)_0 e^{-\lambda_1 t}$ as its nonhomogeneous term. ■

Example 5. The *R–L–C* circuit

A voltage $v(t)$ which varies with time t is applied to the circuit shown in Fig. 4.2 containing an inductance L, a resistor of resistance R and a capacitor of capacitance C, all in series.

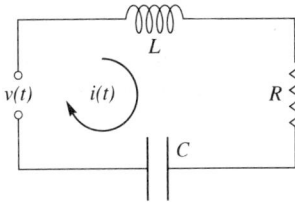

Fig. 4.2 The *R–L–C* circuit

Let the current flowing in the circuit at time t be $i(t)$. Then it is known from the elementary theory of electric circuits that the voltage drop across the inductance is $L di/dt$, while from Ohm's law the voltage drop across the resistor is Ri. If q is the charge on a plate of the capacitor, it follows from the definition of the capacitance C that the voltage drop across the capacitor is q/C. Equating the sum of the voltage drops around the circuit to the imposed voltage $v(t)$ (Kirchhoff's voltage law) leads to the differential equation

$$L\frac{di}{dt} + Ri + \frac{q}{C} = v(t).$$

The current i is defined as the rate of change of charge, so $i = dq/dt$. In terms of q the differential equation becomes

$$L\frac{d^2q}{dt^2} + R\frac{dq}{dt} + \frac{q}{C} = v(t). \tag{11}$$

This is a second order linear constant coefficient nonhomogeneous differential equation, with nonhomogeneous term $v(t)$. ■

Problems for Section 4.1

Determine the order of each of the following differential equations. Verify that the given function is a solution and state whether it is a general or a particular solution.

1 $y' - 4y = 0$, $\qquad y = Ae^{4x}$,

2 $y'' - 2y = 0$, $\qquad y = \sinh x\sqrt{2} + 3\cosh x\sqrt{2}$.

3 $y' + y \tan x = \sec x$, $y = \sin x + 7 \cos x$.
4 $xy' - xy = e^x$, $y = e^x(A + \ln|x|)$.
5 $y'' - 2yy' = 0$, $y = A \tan(Ax + B)$
6 $y'''' - 4y'' + 4y = 1$, $y = \frac{1}{4} + (A + Bx)e^{x\sqrt{2}} + (C + Dx)e^{-x\sqrt{2}}$.
7 $y'' + 4y' + 5y = 15$, $y = 3 + e^{-2x}(4 \cos x - 3 \sin x)$.

Find the general solution of each of the following differential equations by integrating the equation as often as necessary to find y.

8 $y' = 4 + e^{-3x}$ **9** $y'' = \sin 3x$ **10** $y''' = x^2$

11 $y'' = 2 \cosh 3x + \sinh x$ **12** $(x - 1)y' - 2x = 0$.

In each of the following differential equations integrate as often as necessary to find the form of the general solution for y, and then find the particular solution satisfying the stated initial conditions.

13 $y' = x^2 - 2x + 1$, with $y = 1$ when $x = 0$.
14 $y' = e^{2x} + 2 \sinh 3x$, with $y = 2$ when $x = 1$.

15 $y'' = x^2 - \sin\left(\dfrac{\pi x}{2}\right)$, with $y = 0$ and $y' = 2$ when $x = 1$.

16 $y'' = \dfrac{1}{x - 2}$, with $y = 1$ and $y' = 5$ when $x = 3$.

17 $y''' = x + 1$, with $y = 0$, $y' = 1$, $y'' = 2$ when $x = 2$.
18 $y'' = xe^x$, with $y = 3$, $y' = -1$ when $x = 0$.
19 The tangent to a curve at P meets the x-axis at Q. Find the differential equation of the curve for which at all such points P the gradient of the curve equals k times the length of the projection of PQ onto the y-axis.
20 The tangent to a curve at P meets the x-axis at Q. Find the differential equation of the curve for which at all such points P the gradient of the curve at P equals k times the length of the projection of PQ onto the x-axis.
21 Find the differential equation of the curve at each point of which the radius of curvature equals the rate of change of the gradient.
22 Use the arguments of Ex. 5 to find the differential equation for the charge q on the plate of a capacitor in an electrical circuit containing an inductance L and a capacitor of capacity C in series with a voltage source $v(t)$.
23 Let the maximum value of the nonnegative quantity $y(x)$ be $L > 0$. Find the differential equation governing the growth of $y(x)$ given that for every x its rate of growth is proportional both to $y(x)$ and to the growth which is still possible.
24 Find the differential equation governing the temperature $T(t)$ of the object in Ex. 2 if the ambient air temperature $T_0 = h(t)$ is time dependent.
25 A rocket and its fuel of combined initial mass M kg moves vertically upwards into the atmosphere under a constant thrust F. The fuel burns steadily at a rate of m kg/second, and the air resistance on the rocket is kv^α, where k, α are constants and v is the speed of the rocket at time t after launching. Find the differential equation governing the speed v while the rocket is still in the atmosphere and before the fuel is exhausted.
26 A spherical ball of ice melts at a rate proportional to its surface area. Find the differential equation governing its volume V as a function of its radius r.

27 A radioactive isotope of mass $M(t)$ at time t has a decay constant λ_1. It is also produced as a decay product of two other radioactive elements which are present, with initial masses M_2 and M_3 and decay constants λ_2 and λ_3, respectively. If the respective proportions of M_2 and M_3 which decay into this isotope are k_2 and k_3, find the differential equation for M.

4.2 First order differential equations and isoclines

The definition of an ordinary differential equation given in Sec. 4.1 implies that the most general first order equation may be written in the *implicit* form

$$F(x, y, y') = 0, \tag{1}$$

or if it can be solved for y', as the *explicit* equation

$$y' = f(x, y). \tag{2}$$

A **solution** of (1) or (2) is a differentiable function $y = g(x)$ with the property that these equations become identities in x when y and y' are replaced by $g(x)$ and $g'(x)$, respectively. The graph of $y = g(x)$ corresponding to a specific solution of (1) or (2) is called a **solution curve**, or an **integral curve** of the differential equation.

It is sometimes convenient to classify a first order ordinary differential equation according to its **degree**. When (1) can be re-expressed as a polynomial equation in y' (not necessarily in x and y as well), its **degree** is defined to be the degree m of that polynomial. Thus the differential equation

$$(y')^{3/2} = 1 + x^2 + \sin^2 y$$

is of degree 3, because after clearing the radical the equation becomes

$$(y')^3 - (1 + x^2 + \sin^2 y)^2 = 0,$$

which is a cubic in y'.

It follows from the fundamental theorem of algebra (Theorem 1.3) that a polynomial of degree m in y' can be solved to find m values of y', each of which depends on x and y, though not all of which are necessarily different. Thus, since solutions are obtained by integrating y', it then follows that an equation of the form (1), which can be re-expressed as a polynomial in y' of degree m, can have at most m different solutions passing through any point at which its solutions exist. Clearly, such an equation can only have a unique solution when it is of degree 1.

By analogy, any first order ordinary differential equation (1) which is not expressible as a polynomial in y' will be said to be of **degree** m if it has at most m different values of y' at points where its solutions exist.

Let us now turn to some important and useful geometrical considerations concerning first order ordinary differential equations. When discussing these matters we shall work with the general differential equation (1), since this contains (2) as a special case. Suppose (1) is defined at an arbitrary point P_0 located at (x_0, y_0) in the (x, y)-plane, and take the number k_0 such that

$$F(x_0, y_0, k_0) = 0. \tag{3}$$

Then a comparison of (1) and (3) shows $y' = k_0$ is the numerical value of the gradient of a solution curve of (1) at the point (x_0, y_0). It follows directly from this that the tangent line to a solution curve passing through (x_0, y_0) makes an angle $\alpha = \arctan k_0$ with the x-axis.

This suggests how an approximate graphical solution of (1) may be constructed satisfying the initial condition $y(x_0) = y_0$. To arrive at this solution first draw through P_0 in the sense of increasing x a small straight line segment with gradient k_0. Let its end P_1 be at (x_1, y_1). Then this segment provides a tangent line approximation to a solution curve through P_0 in the interval $x_0 \leq x \leq x_1$. Next, take k_1 such that

$$F(x_1, y_1, k_1) = 0,$$

and draw through P_1 in the sense of increasing x a small straight line segment with gradient k_1. Let its end P_2 be at (x_2, y_2). This segment provides a tangent line approximation to a solution curve through P_1 in the interval $x_1 \leq x \leq x_2$.

Proceeding in this fashion generates a polygonal line connecting the points $P_0, P_1, P_2, \ldots$, which represents an approximation to an exact solution of (1) passing through P_0. This is called an **Euler polygonal approximation**, and it is illustrated in Fig. 4.3.

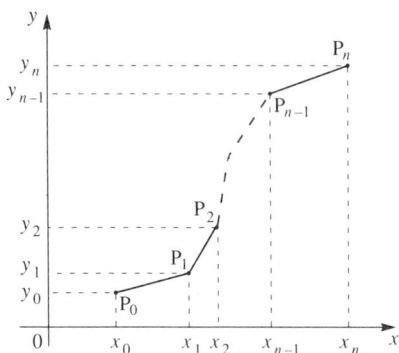

Fig. 4.3   Euler polygonal approximation to the solution of $F(x, y, y') = 0$ with $y(x_0) = y_0$

If the Euler polygonal approximation is developed from P_0 in the sense of decreasing x an approximate solution is obtained for $x < x_0$. This is often called a **backwards solution** of (1) relative to the condition $y(x_0) = y_0$. It is reasonable to expect an Euler polygonal approximation to become a better approximation to an exact solution as the lengths of the segments $P_0 P_1$, $P_1 P_2$, $\ldots$ are reduced. That this is in fact true will be seen later.

Let us now change our viewpoint. Rather than seeking a specific Euler polygonal approximation through some point P_0, we shall try to discover the general behavior of all solution curves of (1). To do this we shall set $y'=k$, with k an arbitrary constant of our choosing, and consider

$$F(x,\ y,\ k)=0. \tag{4}$$

This is now an equation, and *not* a differential equation, for by keeping k constant equation (4) becomes an implicit equation for y in terms of x. If the graph of y as a function of x defined by (4) is drawn, it will have the property that wherever it intersects a solution curve of (1), the tangent line to the solution curve will make an angle $\alpha=\arc\tan k$ with the x-axis. Curves of this type, at every point of which a solution curve has a tangent line making the same angle with the x-axis, are called **isoclines** (lines of constant inclination of the tangent lines). It must be emphasized that in general an isocline is *not* a solution curve, but a line giving information about the set of all solution curves.

Drawing small straight line segments at an inclination $\alpha=\arc\tan k$ to the x-axis through points spaced along an isocline defined by $k=$ const. indicates the direction of all solution curves which intersect that isocline. Repeating this process for different isoclines associates a tangent line direction with all points in the $(x,\ y)$-plane at which differential equation (1) is defined. This specification of a direction at points of the $(x,\ y)$-plane is said to define a **direction field** in the $(x,\ y)$-plane. If representative isoclines and their associated straight line elements are drawn for a specific differential equation, the direction field indicated by the straight line elements shows the qualitative behavior of all solutions of the equation.

Isoclines and their associated tangent line segments may be used to construct Euler polygonal approximations in an obvious manner when an initial condition $y(x_0)=y_0$ is specified. This is accomplished by starting from the point $(x_0,\ y_0)$ and tracing a polygonal line in the sense of increasing x in such a manner that each segment of it between successive isoclines is a continuation of the tangent line segment at the point on the isocline from which the segment originated. If accurate numerical solutions are required, rather than a qualitative general picture of the behavior of solutions, methods like the one described in Sec. 4.8 must be used.

Example 1. Isoclines

Find and graph the isoclines and the direction field of the differential equation

$$y'+y=3-\tfrac{1}{2}x^2. \tag{5}$$

Use the result to determine the approximate numerical solution of this differential equation subject to

(a) the initial condition $y(-1)=-1.5$, and
(b) the initial condition $y(-2.5)=3$.

Solution

Setting $y'=k$ and solving for y shows the equation of the isoclines is given in explicit form by

$$y=(3-k)-\tfrac{1}{2}x^2.$$

This equation represents a family of parabolas with parameter k. These parabolas are shown as the solid curves in Fig. 4.4 and they correspond to values of k from -4 to 5, at increments of 1. This information is sometimes written more concisely as $k = -4(1)5$. The appropriate line segments are drawn at intervals along each isocline and, for example, the ones on the isocline corresponding to $k = 2$ are inclined at an angle arc tan $2 \approx 63°$ while those corresponding to $k = -4$ are inclined at an angle arc tan $(-4) \approx -76°$ to the x-axis.

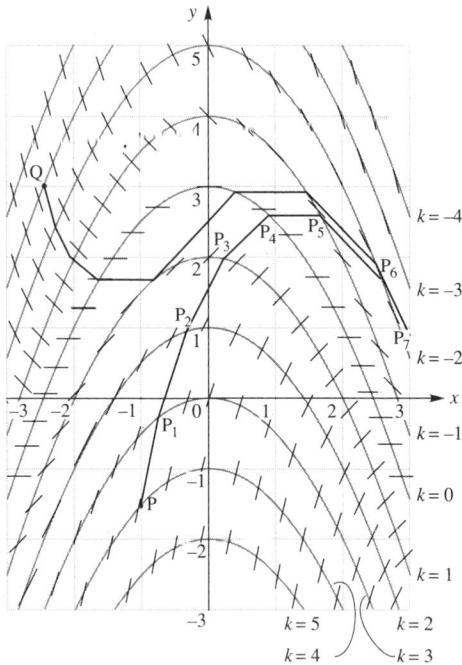

Fig. 4.4 Isoclines and approximate solution curves of $y' + y = 3 - \frac{1}{2}x^2$

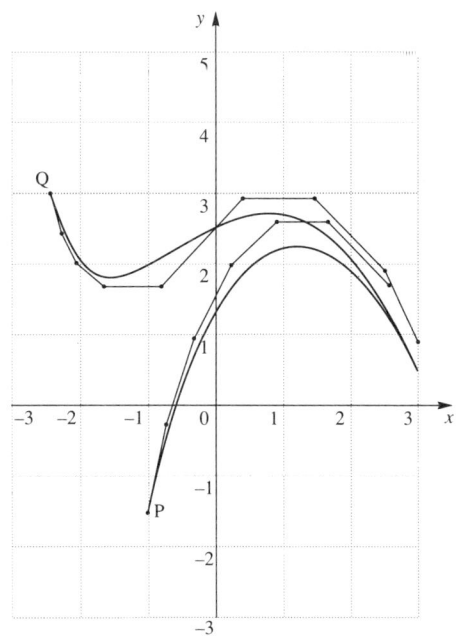

Fig. 4.5 Comparison of approximate and exact solutions

Let us now use these isoclines to obtain an approximate numerical solution of (5) subject to initial condition (a) in which $y(-1) = -1.5$; that is $y = -1.5$ when $x = -1$. Starting at P, the point $(-1, -1.5)$, and moving in the sense of increasing x, we draw a polygonal line whose segments join successive isoclines in such a way that each segment is an extension of the tangent line segment at the point of the isocline from which it originates. In Fig. 4.4 the process just described generates the polygonal line through P joining the points P_1 to P_7. This line is the required Euler polygonal approximation to an exact solution of (5) with initial condition (a).

Repeating the process, but this time starting with initial condition (b) in which $y(-2.5) = 3$, or what is equivalent $y = 3$ when $x = -2.5$, generates the Euler polygonal approximation shown in Fig. 4.4 as the polygonal line originating from Q which is the point $(-2.5, 3)$. To improve the accuracy of this simple graphical method of solution it would be necessary to construct isoclines at smaller increments of k to reduce the error of extrapolation between isoclines.

Equation (5) is a first order linear differential equation, and using the method to be described in Sec. 4.5 it can be shown to have the general solution

$$y = Ae^{-x} + 2 + x - \tfrac{1}{2}x^2, \tag{6}$$

where A is an arbitrary constant. Using initial condition (a) in (6) corresponding to $x = -1$, $y = -1.5$ and solving for A shows $A = -2/e$. The particular solution corresponding to initial condition (a) is thus

$$y = -2e^{-(1+x)} + 2 + x - \tfrac{1}{2}x^2.$$

A similar argument shows the particular solution corresponding to initial condition (b) is

$$y = 6.625e^{-(2.5+x)} + 2 + x - \tfrac{1}{2}x^2.$$

A comparison of these exact solutions and the Euler polygonal approximations obtained in Fig. 4.4 is made in Fig. 4.5. The exact solutions are shown as full lines and the approximate solutions are dotted polygonal lines. ∎

The reader should appreciate that isoclines are not always easy to construct, because y frequently cannot be obtained explicitly from the expression $F(x, y, k) = 0$. When y is determined implicitly it may be necessary to employ a combination of numerical computation and curve sketching techniques (location of asymptotes and turning points etc.) to construct the isoclines.

It should be noticed that sometimes the arbitrary values of k in (4) are restricted by the functions which are involved. For example, if the function arc cos k arises in the equation for an isocline, it follows that although k is arbitrary it must be restricted so that $|k| \le 1$. Similarly, if $\sqrt{k}$ arises, it is necessary to restrict k to positive values.

Finally, as indefinite integrals will be used repeatedly in what follows, we take this opportunity to recall their definition. For any continuous function f defined on the interval I, let the function F be such that $F'(x) = f(x)$ for $x \in I$. The F is called a **primitive** of f, and if c is an arbitrary constant, the **indefinite integral** notation $\int f(x)\,dx$ is used to mean

$$\int f(x)\,dx = F(x) + c, \quad \text{for } x \in I. \tag{7}$$

Thus the indefinite integral of f in (7) is uniquely determined only as far as an additive arbitrary constant of integration c.

A representation for F follows from the fundamental theorem of calculus, which asserts that if a is any point of the interval I on which f is defined, then

$$F(x) = \int_a^x f(t)\,dt, \quad \text{for } x(\ge a) \in I. \tag{8}$$

It follows from this that the **indefinite integral** (7) may also be written as

$$\int f(x)\,dx = \int_a^x f(t)\,dt + c, \quad \text{for } x(\ge a) \in I. \tag{9}$$

Although the existence of an additive arbitrary constant is implied by the indefinite integral sign $\int$, the importance of such constants in the solution of a differential equation makes it desirable to display them explicitly when working with general results. Thus, when discussing the method of separation of variables (Sec. 4.3), when arriving at the expression

$$\int f(x)\,dx = \int g(y)\,dy, \qquad\qquad (10)$$

we choose to write it as

$$\int f(x)\,d(x) = \int g(y)\,dy + c, \qquad\qquad (11)$$

in order to emphasize the need for the integration constant c. Expression (11) is, of course, equivalent to (10) because of (7). It should be understood that in (11) the two arbitrary constants arising from $\int f(x)dx$ and $\int g(y)dy$ have been combined to form the single arbitrary constant c.

Problems for Section 4.2

In each of the following problems graph the isoclines and the direction fields of the given differential equation. Draw polygonal approximations corresponding to several different initial conditions and compare the results with the exact solution found by integration.

1 $y' = \frac{1}{2}x^2$ **2** $y' = 1 + x$ **3** $y' = \dfrac{1}{x-1}$

4 $y' = \sin x$ **5** $y' = \tanh x$ **6** $y' = e^{(\frac{1}{2}x)}$

In each of the following problems graph the isoclines and the direction fields of the given differential equation. Select an initial condition, and after drawing the corresponding polygonal approximation, compare it with the exact solution obtained by using a suitable value of the arbitrary constant in the stated general solution.

7 $y' = x + 2y$;   general solution $y = Ae^{2x} - \frac{1}{2}(x + \frac{1}{2})e^{-2x}$

8 $y' = -2y/x$;   general solution $y = A/x^2$

9 $y' = 3x^2/y^2$;   general solution $y^3 = 3x^3 + A$

10 $y' = 2y^2/x^2$;   general solution $y = x/(2 + Ax)$

11 $y' = 3y/x^2$;   general solution $y = Ae^{-3/x}$

12 $y' = (1 + y^2)/x^2$ general solution $y = \tan(\frac{1}{3}x^3 + A)$

13 $y' = \dfrac{x^2 - y}{x}$;   general solution $y = \dfrac{A}{x} + \dfrac{x^3}{4}$

14 $y' = xy^2/(1 + y^2)$;   general solution $(y^2 - 1) = y(\frac{1}{2}x^2 + A)$

Harder problems

15 The drag on a sphere falling slowly under gravity in a viscous fluid is given by the expression (Stokes' formula)

$$F = 6\pi R \eta u,$$

where R is the radius of the sphere, η is the dynamic viscosity of the fluid and u is the speed of fall. The equation of motion is thus

$$m \frac{du}{dt} = mg - 6\pi R \eta u,$$

where m is the mass of the sphere and g is the acceleration due to gravity. The sphere will reach its terminal speed of fall U when $du/dt \equiv 0$, so from the equation of motion it follows that $U = mg/6\pi R \eta$. This result may be used to determine the dynamic viscosity η in terms of the measured value of U.

In suitably scaled units, the equation of motion for a particular sphere and fluid takes the form

$$\frac{du}{dt} = 2 - 0.5u,$$

Use the method of isoclines to find the approximate solution for u corresponding to $u = 0$ when $t = 0$ (the sphere starts from rest). Use your result to estimate the value of t at which u attains 90 per cent of its terminal speed. Compare your approximate solution with the exact solution $u = 4(1 - e^{-0.5t})$.

16 If there is no capacitor in the electric circuit shown in Fig. 4.2, the current i obeys the differential equation

$$L \frac{di}{dt} + Ri = v(t),$$

where $v(t)$ is the applied voltage. Use the method of isoclines to determine the approximate behavior of the current in the interval $0 \le t \le 2$, given that $R/L = \frac{1}{2}$, $v(t)/L = \sin 2t$ and $i = 0$ when $t = 0$. Compare your answer with the exact solution $i = \frac{1}{17}(8e^{-(\frac{1}{2}t)} + 2 \sin 2t - 8 \cos 2t)$.

17 The transient behavior of a mass M of poison in a nuclear reactor obeys the differential equation (see Ex. 4, Sec. 4.1)

$$\frac{dM}{dt} = -M + 3e^{-t}.$$

Use the method of isoclines to determine the approximate behavior of M in the interval $0 \le t \le 2$, given that $M = 1$ when $t = 0$. Compare your approximate solution with the exact solution $M = 4e^{-t} - 3e^{-2t}$, and use your approximate solution to find when the mass M of poison attains its maximum value.

18 The **error function** erf(x) often arises in engineering problems involving heat transfer, in diffusion problems in general and in statistics. It is defined as the integral

$$\mathrm{erf}(x) = \frac{2}{\sqrt{\pi}} \int_0^x e^{-t^2} \, dt.$$

Setting $y = \mathrm{erf}(x)$, it follows by differentiation of this integral with respect to x that $\mathrm{erf}(a)$ is the solution of the differential equation

$$\frac{\mathrm{d}y}{\mathrm{d}x} = \frac{2}{\sqrt{\pi}} e^{-x^2}$$

at $x = a$, corresponding to the initial condition $y(0) = 0$. Use the method of isoclines to determine the approximate value of $\mathrm{erf}(0.5)$. Compare your result with the exact value to four places of decimals $\mathrm{erf}(0.5) = 0.5205$.

The **Fresnel integrals** $C(x)$ and $S(x)$ occur in diffraction problems involving elasticity, in the diffraction of water waves, in optics and elsewhere. They are defined as the integrals

$$C(x) = \int_0^x \cos\left(\frac{\pi}{2} t^2\right) \mathrm{d}t \quad \text{and} \quad S(x) = \int_0^x \sin\left(\frac{\pi}{2} t^2\right) \mathrm{d}t.$$

19 Set $y = C(x)$ and consider the differential equation

$$\frac{\mathrm{d}y}{\mathrm{d}x} = \cos\left(\frac{\pi}{2} x^2\right),$$

subject to the initial condition $y(0) = 0$. Use the method of isoclines to determine the approximate value of $C(1)$ from this equation. Compare your results with the exact value to four places of decimals $C(1) = 0.7799$.

20 Set $y = S(x)$ and consider the differential equation

$$\frac{\mathrm{d}y}{\mathrm{d}x} = \sin\left(\frac{\pi}{2} x^2\right),$$

subject to the initial condition $y(0) = 0$. Use the method of isoclines to determine the approximate value of $S(1.5)$ from this equation. Compare your result with the exact value to four places of decimals $S(1.5) = 0.6975$.

21 The **Clairaut equation**[2] has the form

$$y = xp + f(p),$$

where $p = y'$ and f is an arbitrary function of its argument. Differentiating with respect to x leads to the result

$$[x + f'(p)]\frac{\mathrm{d}p}{\mathrm{d}x} = 0.$$

This equation is satisfied if either $\mathrm{d}p/\mathrm{d}x = 0$, or if $x + f'(p) = 0$. The condition $\mathrm{d}p/\mathrm{d}x = 0$ implies $p = c$, and so from the original equation

$$y = cx + f(c), \tag{A}$$

with c an arbitrary constant. This solution of the original equation represents a family of straight lines with c as a parameter.

---

[2] ALEXIS CLAUDE CLAIRAUT (1713–1765), a French mathematician who contributed to the study of space curves and through them to differential geometry. The differential equation bearing his name was part of this work. He also made important contributions to the study of the shape of the earth and the motion of the moon.

Combining the second condition

$$x = -f'(p) \tag{B}$$

with the original equation leads to the result

$$y = -pf'(p) + f(p). \tag{C}$$

Taken together, (B) and (C) define a **singular solution** of the original equation with p as a parameter.

Show that this singular solution is the envelope of the family of straight line solutions (A). Find the singular solution of the Clairaut equation

$$y = px + a\sqrt{-2p}.$$

Sketch it together with some representative straight line solutions of the form (A).

4.3 Separable equations

First order differential equations capable of being written in either of the forms

$$y' = f(x)\,g(y) \tag{1}$$

or

$$F(x)\,G(y)\,y' + f(x)\,g(y) = 0 \tag{2}$$

are said to have **separable variables,** or simply to be **separable equations.** This is because the variables x and y occur in such a way that these differential equations can be rewritten with only a function of x appearing on one side of the equation and a function of y on the other. Thus (1) and (2) may be re-expressed as

$$\frac{1}{g(y)}y' = f(x), \tag{3}$$

and

$$\frac{G(y)}{g(y)}y' = -\frac{f(x)}{F(x)}, \tag{4}$$

in which the variables have been separated.

The significance of this separation of variables follows at once, because integrating both sides of these equations with respect to x, and remembering that $y' = dy/dx$, gives

$$\int \frac{1}{g(y)}\frac{dy}{dx}\,dx = \int f(x)\,dx \tag{5}$$

and

$$\int \frac{G(y)}{g(y)} \frac{dy}{dx} dx = - \int \frac{f(x)}{F(x)} dx. \tag{6}$$

The rule for integration by substitution then permits the left-hand sides of each of these results to be rewritten as an integral with respect to y, leading to the results

$$\int \frac{1}{g(y)} dy = \int f(x) dx + c \tag{7}$$

and

$$\int \frac{G(y)}{g(y)} dy = - \int \frac{f(x)}{F(x)} dx + c, \tag{8}$$

where for the reason given in connection with (11) of Sec. 4.2 the integration constant c is shown explicitly.

It is essential the arbitrary constant c in (7) and (8) is introduced immediately the indicated integrations have been performed, and before the expressions are simplified to determine y explicitly. The constant c arises as a result of the integration, and only by including it at this stage can it enter into the solution in the correct manner.

Example 1

Solve the differential equation

$$y' = x^2(1 + y^2).$$

Solution

Separating the variables and integrating both sides as in (7) gives

$$\int \frac{1}{1 + y^2} dy = \int x^2 dx,$$

which after the integrations have been performed shows

$$\text{arc tan } y = \tfrac{1}{3}x^3 + c$$

where c is an arbitrary constant. Thus the explicit general solution is

$$y = \tan(\tfrac{1}{3}x^3 + c).$$

■

Example 2

Solve the differential equation

$$x^2 y^2 y' - (1 + x)(1 + y) = 0.$$

Solution

Separating the variables and integrating both sides as in (8) gives

$$\int \frac{y^2}{1+y} dy = \int \frac{1+x}{x^2} dx,$$

which after simplification becomes

$$\int \left(y - 1 + \frac{1}{1+y} \right) dy = \int \left(\frac{1}{x^2} + \frac{1}{x} \right) dx.$$

Performing the indicated integrations gives

$$\frac{1}{2} y^2 - y + \ln|1+y| = -\frac{1}{x} + \ln|x| + c,$$

with c an arbitrary constant. This is an implicit equation for y and cannot be further simplified.

■

When a first order differential equation is required to satisfy an initial condition $y = y_0$ when $x = x_0$, written

$$y(x_0) = y_0, \tag{9}$$

the combination of the equation and the initial condition is called an **initial value problem**. This name arises because in many physical situations the independent variable is the time, and a condition of this type specifies how a solution is to start. The following examples of physical problems leading to separable equations illustrate how initial conditions arise. They also show how the initial condition must be used to determine the appropriate value of the arbitrary constant in the general solution if it is to satisfy the initial value problem.

Example 3. Newton's law of cooling

It was shown in Ex. 2, Sec. 4.1, that when a body cools in air subject to Newton's law of cooling, its surface temperature $T(t)$ at time t satisfies the differential equation

$$\frac{dT}{dt} = -\lambda(T - T_0), \tag{10}$$

where λ is a constant depending on the body and T_0 is the constant ambient air temperature. We shall now solve an initial value problem for this equation in which the initial condition is $T(0) = T_1$. Separating the variables in (10) and integrating both sides gives

$$\int \frac{1}{T - T_0} dT = -\int \lambda dt,$$

or

$$\ln(T - T_0) = -\lambda t + c,$$

where c is an arbitrary constant.

Taking the exponential of both sides of this equation then shows

$$T - T_0 = A e^{-\lambda t},$$

where $A = e^c$. As c was an arbitrary constant, it follows that A must be an arbitrary positive constant. No useful purpose is served by performing arithmetic on arbitrary constants, so from this point onwards we shall work with A rather than c. The general solution of (10) is thus

$$T = A e^{-\lambda t} + T_0.$$

To solve the initial value problem it is necessary to choose A so that $T(0) = T_1$. Setting $T = T_1$ and $t = 0$ in the general solution shows

$$T_1 = A + T_0 \quad \text{or} \quad A = T_1 - T_0.$$

The required solution of the initial value problem is thus

$$T = (T_1 - T_0) e^{-\lambda t} + T_0 \quad \text{for} \quad t \geq 0. \tag{11}$$

This is the solution quoted in (8), Sec. 4.1, and it shows that with the passage of time the body surface temperature decays exponentially to the ambient air temperature T_0.

■

Example 4. The logistic equation

The study of population growth, the rate of learning, the spread of information, the acceptance of newly marketed products and various other analogous situations is often based on the mathematical model called the **logistic equation**, or the **equation of inhibited growth**

$$\frac{dP}{dt} = aP \left(1 - \frac{P}{L} \right), \tag{12}$$

where a and L are positive constants.

This equation is derived on the assumption that a population P cannot grow beyond some known positive value L, and that the rate of growth of P is proportional both to P and to $\left(1 - \frac{P}{L} \right)$, the fraction of growth which remains possible. The positive number a in (12) is a proportionality constant. Let us solve an initial value problem for this equation in which the initial condition $P(0) = P_0$, with $0 < P_0 < L$.

Separating the variables and integrating both sides of (12) gives

$$\int \frac{1}{P \left(1 - \dfrac{P}{L} \right)} \, dP = \int a \, dt,$$

which after simplification by means of partial fractions becomes

$$\int \left(\frac{1}{P} + \frac{1}{L - P} \right) dP = \int a \, dt. \tag{13}$$

Carrying out the indicated integrations we find

$$\ln\left(\frac{P}{L-P}\right)=at+c,$$

where c is an arbitrary constant. Taking the exponential of both sides of this equation shows

$$\frac{P}{L-P}=Ae^{at}.$$

Here again we have set $A=e^c$, with A an arbitrary positive constant. Rearranging terms brings the general solution of the logistic equation to the form

$$P=\frac{L}{1+Ae^{-at}}\qquad\text{for } t\geq0.\tag{14}$$

To solve the stated initial value problem it is now necessary to choose A so that $P(0)=P_0$. Setting $P=P_0$ and $t=0$ in (14) gives

$$A=\frac{P_0}{L-P_0}.$$

Combining this result with (14) finally gives for the solution of the initial value problem

$$P=\frac{L}{1+\left(\dfrac{L}{P_0}-1\right)e^{-at}}\qquad\text{for } t\geq0.\tag{15}$$

The behavior of this solution is shown in Fig. 4.6. A simple calculation establishes that the curve has a point of inflection at Q, where $P=\frac{1}{2}L$ and $t=\frac{1}{a}\ln\left(\dfrac{L}{P_0}-1\right)$.

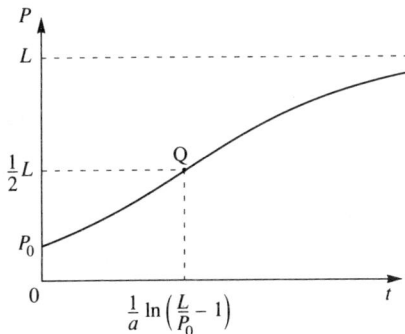

Fig. 4.6 Logistic curve with point of inflection at Q

Example 5. Vertical resisted motion

In this example we determine the speed v at time t of a particle of mass m falling vertically downward from rest in a resisting medium in which the resistance is kv^2, with $k>0$

a constant. The equation of motion obtained by equating the rate of change of momentum d(mv)/dt to the algebraic sum of the gravitational force mg acting downward on the particle and the resistance kv^2 acting upward is

$$m\frac{dv}{dt}=mg-kv^2,\qquad(16)$$

Setting $V^2=mg/k$, separating the variables and integrating both sides gives

$$\int\frac{V^2}{V^2-v^2}\,dv=\int g\,dt,$$

which is equivalent to

$$\int\left(\frac{1}{V-v}+\frac{1}{V+v}\right)dv=\int\frac{2g}{V}\,dt.$$

Carrying out the indicated integrations we obtain

$$\ln\left|\frac{V+v}{V-v}\right|=\frac{2gt}{V}+A,\qquad(17)$$

with A an arbitrary constant.

Using the initial condition $v=0$ when $t=0$ in the general solution (17) then shows $A=0$. Making use of this result in (17), taking the exponential of both sides and simplifying gives

$$v=V\left(\frac{e^{2gt/V}-1}{e^{2gt/V}+1}\right)$$

$$=V\tanh\,(gt/V).\qquad(18)$$

As $t\rightarrow\infty$ so $\tanh\,(gt/V)\rightarrow1$, showing that as $t\rightarrow\infty$ so $v\rightarrow V=(mg/k)^{1/2}$, which is the **terminal speed** of fall of the body. Some typical terminal speeds are listed below:

beach ball   19 meters/second
small pebble   21 meters/second
large pebble   40 meters/second

■

Example 6. Nonuniqueness of an initial value problem

This example demonstrates that not all initial value problems are unique, and so shows the need for conditions which guarantee the uniqueness that is usually expected of solutions of physical problems. It will suffice for us to consider the initial value problem

$$y'=2y^{1/2}\quad\text{with}\quad y(0)=0.\qquad(19)$$

Separating variables and integrating both sides this becomes

$$\int\frac{1}{y^{1/2}}\,dy=\int2dx,\quad\text{or}\quad2y^{1/2}=2x+c,$$

where c is an arbitrary constant. Carrying out the indicated integrations, but following the accepted convention and omitting arithmetic operations on arbitrary constants (not writing $c/2$ in place of c), we obtain the general solution

$$y^{1/2} = x + c$$

or

$$y = (x + c)^2.$$

The initial condition $y(0) = 0$ is satisfied if we set $c = 0$, so $y = x^2$ is a solution of the initial value problem (19). However this is not the only solution of (19) with this initial condition. To see this take an arbitrary number $a > 0$ and define $g_a(x)$ as

$$g_a(x) = \begin{cases} 0 & \text{for } -\infty < x \leq a \\ (x - a)^2 & \text{for } a < x < \infty \end{cases}$$

It is easily verified that $g(x)$ is a solution of the differential equation in (19), but $g_a(0) = 0$ so that $g_a(x)$ also satisfies the stated initial condition. Thus, as $a > 0$ was arbitrary, we see that in this case initial value problem (19) possesses an infinity of solutions, four of which are shown in Fig. 4.7.

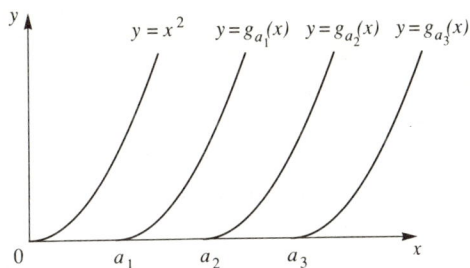

Fig. 4.7 Nonuniqueness of solution of the initial value problem $y' = 2y^{1/2}$, $y(0) = 0$.

If the differential equation

$$P(x, y) + Q(x, y)y' = 0 \tag{20}$$

can be reduced to either of the forms

$$y' = f\left(\frac{x}{y}\right) \tag{21}$$

or

$$y' = f\left(\frac{a_1 x + b_1 y + c_1}{a_2 x + b_2 y + c_2}\right), \tag{22}$$

a simple change of variables will reduce it to separable form.

We preface our remarks concerning (21) by defining an algebraic function $P(x, y)$ to be **algebraically homogeneous** of degree n if $P(\lambda x, \lambda y) = \lambda^n P(x, y)$. That is, if λx and λy are written in place of x and y in $P(x, y)$, and if λ^n is factored out of the resulting expression, the function which remains is simply $P(x, y)$. For example, if $P(x, y) = x^3 - 2xy^2 + 4x^2 y$ $- y^3$, then $P(\lambda x, \lambda y) = (\lambda x)^3 - 2(\lambda x)(\lambda y)^2 + 4(\lambda x)^2(\lambda y) - (\lambda y)^3 = \lambda^3 P(x, y)$, showing $P(x, y)$ to be homogeneous of degree three.

The form (21) follows from form (20) if $P(x, y)$ and $Q(x, y)$ are both algebraically homogeneous functions with the same degree n. The variable change $y = xu$, which is equivalent to

$$\frac{dy}{dx} = u + x\frac{du}{dx},\tag{23}$$

then reduces (21), and thus (20), to the separable form

$$u + x\frac{du}{dx} = f(u).\tag{24}$$

Separating variables and integrating both sides of (24) shows the general solution of (21) in terms of x and u is

$$\int \frac{1}{f(u) - u}\,du = \int \frac{1}{x}\,dx = 1n|x| + c,\tag{25}$$

with c an arbitrary constant. The solution in terms of y follows by setting $u = y/x$.

Differential equations of the form shown in (20) which are reducible to form (21) are called **homogeneous equations.** In this context the term homogeneous refers only to the special algebraic structure of $P(x, y)$ and $Q(x, y)$ in (20). It is important that the use of the expression 'homogeneous', which is historical, should not be confused with its modern and more fundamental meaning described in Sec. 4.1, where it refers to a differential equation (usually linear) with a zero right-hand side.

Equations of the form shown in (22) are called **near-homogeneous.** Setting $x = X + \alpha$ and $y = Y + \beta$, with α, β constants to be determined, it follows that

$$\frac{dY}{dX} = \frac{a_1 X + b_1 Y + a_1 \alpha + b_1 \beta + c_1}{a_2 X + b_2 Y + a_2 \alpha + b_2 \beta + c_2}.$$

This becomes the *homogeneous* differential equation

$$\frac{dY}{dX} = \frac{a_1 X + b_1 Y}{a_2 X + b_2 Y},\tag{26}$$

provided α and β are chosen such that

$$a_1 \alpha + b_1 \beta + c_1 = 0 \quad \text{and} \quad a_2 \alpha + b_2 \beta + c_2 = 0.\tag{27}$$

The constants α and β will be determined if

$$\begin{vmatrix} a_1 & b_1 \\ a_2 & b_2 \end{vmatrix} \neq 0.$$

In this case the solution of (27) follows by setting $Y=Xu$ and solving the resulting equation by separating the variables. If, however,

$$\begin{vmatrix} a_1 & b_1 \\ a_2 & b_2 \end{vmatrix} = 0,$$

it follows that $a_2x+b_2y=k(a_1x+a_2y)$, with $k=$ const. The solution of (22) then follows by setting $u=a_1x+b_1y$ and separating variables in the transformed differential equation

$$\frac{du}{dx} = a_1 + b_1 \left(\frac{u+c_1}{ku+c_2} \right). \tag{28}$$

As a last special case we mention that any equation of the form

$$y' = f(ax+by+c), \tag{29}$$

may be transformed into the separable equation

$$u' = a + bf(u) \tag{30}$$

by means of the change of variable

$$u = ax+by+c. \tag{31}$$

Example 7. Homogeneous differential equation

Find the general solution of

$$y' = 2e^{y/x} + \frac{y}{x}.$$

Solution

Setting $y=xu$ reduces the differential equation to

$$x\frac{du}{dx} = 2e^u.$$

Separating variables and integrating gives

$$\int e^{-u} du = \int \frac{2}{x} dx = 2\ln|x| + c,$$

where c is an arbitrary constant. To simplify the manipulation we set $c=\ln(A^2)$, with A an arbitrary constant, when it follows that

$$-e^{-u} = 2\ln|x| + \ln A^2,$$

or

$$u = -\ln \ln \left(\frac{1}{Ax} \right)^2.$$

As $u=y/x$ it follows that the general solution is

$$y=-x \ln \ln\left(\frac{1}{Ax}\right)^2.$$

∎

We conclude this section by introducing the *differential form* of (20) which is often to be found in textbooks, and which will be used later in our discussion of exact differential equations.

If x and y are related by the functional relationship $y=f(x)$, the *differentials* dx and dy are defined such that $dy=f'(x)dx$, with dx arbitrary. As $y'=f'(x)$, it then follows that

$$dy=y'\,dx. \tag{32}$$

Multiplication of (20) by the differential dx gives

$$P(x,\ y)dx+Q(x,\ y)y'\,dx=0,$$

which because of (32) may be written

$$P(x,\ y)dx+Q(x,\ y)dy=0. \tag{33}$$

This last result is called the **differential form** of (20), and it should be remembered it is simply an alternate way of expressing (20).

Problems for Section 4.3

Find the general solution of each of the following separable equations.

1 $x+yy'=0$ **2** $y'=-y/x$

3 $(1+x)y+(1-y)xy'=0$ **4** $y'=y/2x$

5 $y'=x^2/y^2$ **6** $y'=y^2/x^2$

7 $y'=x(y+1)$ **8** $y'=1/[x(y+1)]$

9 $y'+\dfrac{x(1+y^2)}{y(1+x^2)}=0$ **10** $y'\cot x+y=2$

11 $(y\sec^2 x+\sec x\tan x)+(\tan x+2y)y'=0$

12 $xyy'-\sqrt{y^2+1}=0$

Solve the initial value problem for each of the following separable equations.

13 $(1+e^x)yy'-e^x=0$ with $y(0)=1$.

14 $2x^2yy'+y^2=2$ with $y(-1)=2$.

15 $(xy^2+x)+(x^2y-y)y'=0$ with $y(0)=1$.

16 $y'\sin x-y \ln y=0$ with $y(\pi/2)=1$.

17 $(x^2-1)y'+2xy^2=0$ with $y(\sqrt{2})=1/2$.

18 $\left(\dfrac{2x-1}{y}\right)+\left(\dfrac{x-x^2}{y^2}\right)y'=0$ with $y(2)=6$.

19 $y=\exp(x+2y)$ with $y(1)=-1/2$.

20 $3e^x \tan y + (1-e^x) \sec^2 y\, y' = 0$ with $y(1) = \pi/4$.

21 $y' = 3.10^{x+y}$ with $y(1) = -1$.

Find the general solution of each of the following differential equations.

22 $y' = xy/(x^2 - y^2)$.

23 $y' = (x^2 + y^2)/x^2$.

24 $y' = \dfrac{x+y-3}{x-y-1}$.

25 $y' = (x^2 - y^2)/x^2$.

26 $y' = \dfrac{2x+y-1}{4x+2y+5}$.

27 $y' = \dfrac{y}{x}\left(1 + \dfrac{y}{x}\right)$.

28 $y' = \dfrac{x+2y+1}{2x+4y+3}$.

29 $y' = \dfrac{1-3x-3y}{1+x+y}$.

30 $y' = \sqrt{4x+2y-1}$.

Harder problems

31 A particle of mass m falls from rest under gravity in a resisting medium in which the resistance is mkv, where v is the particle speed at time t. Show the equation of motion is

$$m\frac{dv}{dt} = mg - mkv,$$

and hence that the particle speed is given by

$$v = \frac{g}{k}(1 - e^{-kt}).$$

If x is the distance fallen by the particle at time t, show by using the result $dv/dt = v\, dv/dx$ in the equation of motion that

$$x = \frac{g}{k^2}\ln\left(\frac{g}{g-kv}\right) - \frac{v}{k}.$$

32 An electrical circuit contains an inductance L, a resistor of resistance R and a constant voltage source V_0, all connected in series. Derive the differential equation determining the current i flowing in the circuit at time t. Solve this equation by separation of variables to show that if $i = i_0$ at time $t = 0$, then

$$i(t) = \frac{V_0}{R} - \left(\frac{V_0}{R} - i_0\right)e^{-(R/L)t}.$$

33 A particle of mass m is projected vertically upward with initial speed U in a medium in which the resistance is mkv, where v is the particle speed at time t. Show that the equation of motion is

$$m\frac{dv}{dt} = -m(g+kv),$$

and hence that the particle speed is given by

$$v = \left(U + \frac{g}{k}\right)e^{-kt} - \frac{g}{k}.$$

Let x be the distance moved by the particle from its starting point. Use the result $dv/dt = v\,dv/dx$ in the equation of motion to show that when $v = V$ the particle has risen to the height h, where

$$h = \frac{U - V}{k} - \frac{g}{k^2}\ln\left(\frac{g + kU}{g + kV}\right), \qquad 0 \le V < U.$$

34 A particle P moves along a curve in the (x, y)-plane in such a way that the angle between the line joining P to the origin O and the x-axis is half of the angle made by the tangent line to the curve at P and the x-axis. Show that if P has coordinates (x, y), the curve followed by P is determined by the differential equation

$$y' = \frac{2xy}{x^2 - y^2}.$$

Find the equation of the path of P given that it passes through (x_0, y_0).

35 A particle of mass m is projected vertically upward with initial speed U in a medium in which the resistance is mkv^2, where v is the particle speed at time t. Show the equation of motion is

$$m\frac{dv}{dt} = -m(g + kv^2),$$

and hence that the particle speed is given by

$$v = c \tan\left[\arc\tan\left(\frac{U}{c}\right) - \frac{gt}{c}\right],$$

where $c = \sqrt{(g/k)}$.

Let x be the distance moved by the particle from its starting point. Use the result $dv/dt = v\,dv/dx$ in the equation of motion to show that when $v = V$ the particle has risen to the height h, where

$$h = \frac{c^2}{2g}\ln\left(\frac{c^2 + U^2}{c^2 + V^2}\right),$$

where $c = \sqrt{(g/h)}$.

36 A tank contains V gal of concentrated liquid fertilizer obtained by dissolving M lb of solid fertilizer in water. To dilute it for use water is pumped into the tank at the rate of Q gal/min. If mixing is considered to be instantaneous and diluted liquid fertilizer is drawn off at the rate of Q gal/min, show the differential equation for the mass m lb of solid fertilizer remaining dissolved at time t after the start of the process is

$$\frac{dm}{dt} + \frac{Q}{V}m = 0,$$

and hence that

$$m = Me^{-(Q/V)t}.$$

4.4 Exact differential equations and integrating factors

The first order differential equation

$$M(x, y)\,dx + N(x, y)\,dy = 0 \tag{1}$$

is said to be **exact** if a function $F(x, y)$ exists for which

$$\frac{\partial F}{\partial x} = M(x, y) \quad \text{and} \quad \frac{\partial F}{\partial y} = N(x, y). \tag{2}$$

In this case (1) may be replaced by

$$\frac{\partial F}{\partial x}\,dx + \frac{\partial F}{\partial y}\,dy = 0. \tag{3}$$

Now the total differential dF of $F(x, y)$ is

$$dF = \frac{\partial F}{\partial x}\,dx + \frac{\partial F}{\partial y}\,dy, \tag{4}$$

so a comparison of (3) and (4) shows that (1) implies

$$dF = 0. \tag{5}$$

Integration of (5) gives

$$F(x, y) = c, \tag{6}$$

with c an arbitrary constant. The *solution curves*, or *integral curves*, of (1) are the curves defined by (6), from which must be excluded curves passing through points at which $M(x, y) = N(x, y) = 0$. The exclusion of curves through these points is necessary because at such points (1) imposes no restriction on dx and dy.

Let us now derive a simple test for the exactness of a differential equation. To do this we appeal to the result from elementary calculus that if $\partial F/\partial x$, $\partial F/\partial y$ and $\partial^2 F/\partial x\partial y$ exist and are continuous in some region in the (x, y)-plane, then $\partial^2 F/\partial y\partial x$ also exists and is continuous in the same region and, furthermore,

$$\frac{\partial^2 F}{\partial x\partial y} = \frac{\partial^2 F}{\partial y\partial x} \tag{7}$$

throughout the region. This means that if (1) is exact, it follows from (2) and (7) that a *necessary* condition for (1) to have a solution of the form of (6) which is continuous and twice differentiable throughout some region is

$$\frac{\partial M}{\partial y} = \frac{\partial N}{\partial x}. \tag{8}$$

It is a straightforward matter to prove that (8) is also a *sufficient* condition, though the proof is left as an exercise for the reader. Condition (8) thus provides a simple test which establishes whether or not differential equation (1) is exact.

Sometimes the structure of an exact equation is sufficiently simple that its solution may be found by inspection. However, if the solution is not obvious, it can be found by integration using equations (2) as follows. Integrating the first of equations (2) with respect to x, and regarding y as a constant, gives

$$F = \int M(x,\ y)\,dx + h(y) + c, \tag{9}$$

where $h(y)$ is an arbitrary function of y, and c is an arbitrary constant. In (9) the function $h(y) + c$ appears as the 'constant' of integration when integrating with respect to x, because (9) is the most general expression for which $\partial F/\partial x = M(x, y)$. The unknown function $h(y)$ follows by integration of the expression for dh/dy obtained by differentiating (9) partially with respect to y and combining the result with the second of equations (2).

Alternatively, we may proceed the other way round, by starting with the result

$$F = \int N(x,\ y)\,dy + k(x) + d, \tag{10}$$

obtained by integrating the second of equations (2) with respect to y regarding x as a constant, where $k(x)$ is an arbitrary function of x and d is an arbitrary constant. The unknown function $k(x)$ then follows by integration of the expression for dk/dx obtained by differentiating (10) partially with respect to x and combining the result with the first of equations (2).

Example 1

Solve

$$(2xy + 4)dx + x^2\,dy = 0.$$

Solution

We use (8) to test for exactness, with $M = 2xy + 4$ and $N = x^2$. As $\partial M/\partial y = 2x$ and $\partial N/\partial x = 2x$, (8) is satisfied identically, showing the equation to be exact. From (9) we find

$$F = \int (2xy + 4)\,dx + h(y) + c,$$

giving

$$F = x^2 y + 4x + h(y) + c.$$

Now $\partial F/\partial y = x^2 + dh/dy$, but from (2) $\partial F/\partial y = N = x^2$, showing $dh/dy \equiv 0$, and hence that $h = \text{const}$. The solution of the exact equation is thus

$$x^2 y + 4x = d, \quad \text{or} \quad y = \frac{d - 4x}{x^2},$$

where d is an arbitrary constant (it combines c and h).   ∎

Example 2

Solve

$$e^x \sin y \, dx + (2y + e^x \cos y) \, dy = 0.$$

Solution

Setting $M = e^x \sin y$ and $N = 2y + e^x \cos y$ we find (8) is satisfied identically, showing the equation to be exact. From (9) we find

$$F = \int e^x \sin y \, dx + h(y) + c,$$

giving

$$F = e^x \sin y + h(y) + c.$$

From this result and (2) we have

$$\frac{\partial F}{\partial y} = e^x \cos y + \frac{dh}{dy} = N = 2y + e^x \cos y,$$

showing

$$\frac{dh}{dy} = 2y \quad \text{and so} \quad h = y^2 + \text{const.}$$

The solution is thus

$$e^x \sin y + y^2 = c,$$

with c an arbitrary constant.

∎

As a general rule, equations of the form

$$P(x, y) \, dx + Q(x, y) \, dy = 0 \tag{11}$$

are not exact. They may, however, be made exact by multiplying them by a suitable function $\mu(x, y)$ called an **integrating factor**. It is a simple matter to show an integrating factor μ exists, but in general such a factor is difficult to find.

Let us first establish the existence of an integrating factor for (11) when the equation is not exact. If $F(x, y) = \text{const.}$ is a solution of (11), then

$$dF = \frac{\partial F}{\partial x} \, dx + \frac{\partial F}{\partial y} \, dy = 0. \tag{12}$$

Now let μ be a function such that the equation

$$\mu P(x, y) \, dx + \mu Q(x, y) \, dy = 0 \tag{13}$$

is exact. Then a comparison of (12) and (13) shows

$$\frac{\partial F}{\partial x} = \mu P \quad \text{and} \quad \frac{\partial F}{\partial y} = \mu Q. \tag{14}$$

Using (14) in (12) we find

$$dF = \mu(P\,dx + Q\,dy). \tag{15}$$

This shows (11) becomes a total differential after multiplication by μ, thereby proving the existence of an integrating factor for (11).

As the solution curves of (11) are $F(x, y) = $ const., it follows that $G(F) = $ const., where $G(u)$ is an arbitrary function of its argument u. Thus when μ is an integrating factor for (11), so also is $G(F)\mu$, from which we see that integrating factors are not unique.

Example 3. Nonuniqueness of integrating factor

The differential equation

$$dx + (x/y)\,dy = 0$$

is not exact, but inspection shows $\mu = y$ is an integrating factor, because

$$y[dx + (x/y)dy] = y\,dx + x\,dy = 0$$

is exact, and it has the solution $F_1(x, y) \equiv xy = $ const. The integrating factor vanishes when $y = 0$, but inspection of the equation shows $y \equiv 0$ is in fact a solution.

To find another integrating factor, notice that if $F_2(x, y) \equiv 1 + x^2 y^2$, then the total differential of $F_2(x, y) = $ const. is

$$2xy^2\,dx + 2x^2 y\,dy = 0,$$

or

$$2xy^2[dx + (x/y)\,dy] = 0.$$

Thus $\mu = 2xy^2$ is seen to be another integrating factor of the original differential equation corresponding to the solution $F_2(x, y) = $ const. This integrating factor vanishes when $x = 0$ or $y = 0$, but inspection of the equation shows $x \equiv 0$ and $y \equiv 0$ are in fact both solutions.

A third integrating factor follows from the fact that if $F_3(x, y) \equiv \tanh(xy)$, the total differential of $F_3(x, y) = $ const. is

$$\text{sech}^2(xy)(y\,dx + x\,dy) = 0,$$

or

$$y\,\text{sech}^2(xy)[dx + (x/y)\,dy] = 0.$$

Thus $\mu = y\,\text{sech}^2(xy)$ is yet another integrating factor of the original differential equation, this time corresponding to the solution $F_3(x, t) = $ const. This integrating factor vanishes when $y = 0$, but it again follows from the differential equation that $y \equiv 0$ is a solution.

Although at first sight the three solutions $F_1 = $ const., $F_2 = $ const., and $F_3 = $ const. appear to be different, it is easy to see that they are in fact all equivalent to the simplest solution $xy = c$. It is clear that more integrating factors could be found, though these would also give rise to solutions equivalent to $xy = c$. ∎

When the structure of a nonexact equation is simple an integrating factor may often be found by inspection. If this approach fails there are two special cases in which an integrating factor may be obtained by integration. To see how these cases arise, suppose (11) is not exact and let μ be an integrating factor, so that

$$\mu P(x, y)\,dx + \mu Q(x, y)\,dy = 0 \tag{16}$$

is exact. Then from (8) it must follow that

$$\frac{\partial}{\partial y}(\mu P) = \frac{\partial}{\partial x}(\mu Q). \tag{17}$$

Performing the indicated differentiations and simplifying gives

$$\mu\left(\frac{\partial P}{\partial y} - \frac{\partial Q}{\partial x}\right) = Q\frac{\partial \mu}{\partial x} - P\frac{\partial \mu}{\partial y}. \tag{18}$$

Now suppose μ is a function of x only, so that $\partial\mu/\partial y \equiv 0$. Then an inspection of (18) shows the expression

$$\frac{1}{Q}\left(\frac{\partial P}{\partial y} - \frac{\partial Q}{\partial x}\right) \equiv g(x) \tag{19}$$

can only be a function of x, so (18) reduces to the separable equation for μ

$$\frac{d\mu}{dx} = g(x)\mu. \tag{20}$$

Separating variables and integrating gives

$$\mu = A\,\exp\left[\int g(x)\,dx\right], \tag{21}$$

where A is an arbitrary constant.

If, instead, μ is a function of y only, so that $\partial\mu/\partial x \equiv 0$, it follows in similar fashion that the expression

$$\frac{1}{P}\left(\frac{\partial P}{\partial y} - \frac{\partial Q}{\partial x}\right) \equiv h(y) \tag{22}$$

can only be a function of y. In this case (18) reduces to the separable equation

$$\frac{d\mu}{dy} = -h(y)\mu, \tag{23}$$

with the solution

$$\mu = A\,\exp\left[-\int h(y)\,dy\right]. \tag{24}$$

in which A is an arbitrary constant. As μ multiplies the entire differential equation, there is no loss of generality if we set $A = 1$ in both (21) and (24).

These arguments have established the following useful theorem on integrating factors.

Theorem 4.1 (Integrating factors)

Let

$$P(x, y)\,dx + Q(x, y)\,dy = 0 \tag{25}$$

be an arbitrary first order differential equation. Then,

(i) if the expression

$$\frac{1}{Q}\left(\frac{\partial P}{\partial y} - \frac{\partial Q}{\partial x}\right) \equiv g(x)$$

is a function of x only, differential equation (25) has an integrating factor

$$\mu(x) = \exp\left[\int g(x)\,dx\right]$$

which is only a function of x;

(ii) if the expression

$$\frac{1}{P}\left(\frac{\partial P}{\partial y} - \frac{\partial Q}{\partial x}\right) \equiv h(y)$$

is a function of y only, differential equation (25) has an integrating factor

$$\mu(y) = \exp\left[-\int h(y)\,dy\right]$$

which is only a function of y. $\qquad\square$

Example 4. An important integrating factor

Solve

$$p(x)y\,dx + dy = 0.$$

Solution

This equation in which $P = p(x)y$ and $Q = 1$ is not exact. It is easily seen that $(\partial P/\partial y - \partial Q/\partial x)/Q = p(x)$, so that Theorem 4.1(i) applies with $g(x) = p(x)$. Thus an integrating factor $\mu = \mu(x)$ exists, with

$$\mu = \exp\left[\int p(x)\,dx\right].$$

Multiplying the differential equation by μ we arrive at

$$\exp\left[\int p(x)\,dx\right]p(x)y\,dx + \exp\left[\int p(x)\,dx\right]dy = 0,$$

which is now exact. Inspection of this shows it may be written as

$$\frac{d}{dx}\left\{\exp\left[\int p(x)\,dx\right]y\right\} = 0,$$

and after integration this becomes

$$\exp\left[\int p(x)\,dx\right]y = c,$$

with c an arbitrary constant.

⋮

The general solution is thus

$$y = c \exp\left[-\int p(x)\,dx\right].$$

This result could, of course, have been obtained by separating the variables and integrating, but it was instructive to obtain it via Theorem 4.1(i). For future reference we notice the original differential equation may also be written in the form

$$y' + p(x)y = 0.$$

■

Example 5. Finding an integrating factor

Solve

$$(2xy + x^2y + \tfrac{1}{3}y^3)\,dx + (x^2 + y^2)\,dy = 0.$$

Solution

This equation in which $P = 2xy + x^2y + \tfrac{1}{3}y^2$ and $Q = x^2 + y^2$ is not exact. A simple calculation shows $(\partial P/\partial y - \partial Q/\partial x)/Q \equiv 1$, so that Theorem 4.1 (i) applies with $g(x) \equiv 1$. It follows from this that an integrating factor $\mu = \mu(x)$ exists, with

$$\mu = \exp\left(\int 1.dx\right) = \exp(x).$$

Thus the differential equation

$$e^x(2xy + x^2y + \tfrac{1}{3}y^2)\,dx + e^x(x^2 + y^2)\,dy = 0$$

is exact, and proceeding as in Ex. 1 a routine calculation shows the solution to be

$$ye^x(x^2 + \tfrac{1}{3}y^3) = c,$$

with c an arbitrary constant.

■

Problems for Section 4.4

In each of the following problems find the total differential dF of the given function $F(x, y)$ and hence find the exact differential equation satisfied by $F(x, y) = \text{const.}$

1 $F(x, y) = x^2y + 3x - y.$

2 $F(x, y) = xy \sin x + x^2 - y^2.$

3 $F(x, y) = \sin x \cosh y + \cos x \sinh y.$

4 $F(x, y) = x \ln(1 + y^2) + x^2y^2.$

5 $F(x, y) = x^2 e^{x/y}.$

6 $F(x, y) = \cosh\left(x + \dfrac{1}{y}\right) + 2x + y^3.$

7 $F(x, y) = \sinh\left(\dfrac{x}{y} + \dfrac{2y}{x}\right).$

8 $F(x, y) = x \tanh xy.$

Determine whether or not the following differential equations are exact and solve them using an appropriate integrating factor determined from Theorem 4.1 where necessary.

9 $\dfrac{2x}{y^3}dx+\left(\dfrac{y^2-3x^2}{y^4}\right)dy=0.$

10 $(x+2y+4)dx+(2x+y+1)dy=0.$

11 $(\cosh y+y\cosh x)dx+(x\sinh y+\sinh x)dy=0.$

12 $\left(\dfrac{2x}{x^2-3y^2}+y\right)dx+\left(x-\dfrac{6y}{x^2-3y^2}\right)dy=0.$

13 $(e^{x+y}+x\,e^{x+y})dx+(x\,e^{x+y}+1)dy=0.$

14 $(x+y)dx+(x+2y)dy=0.$

15 $(y^3\cos xy+2xy)dx+(xy^2\cos xy-y\sin xy-2x^2)dy=0.$

16 $(x^3-3xy^2+2)dx+(y^2-3x^2y)dy=0.$

17 $\dfrac{2x}{y^3}dx+\left(\dfrac{y^2-3x^2}{y^4}\right)dy=0.$

18 $(x^2-3y^2)dx+2xy\,dy=0.$

19 $(x^2+2x+y^2)dx+2xy\,dy=0.$

20 $(x+y^2)dx-2xy\,dy=0.$

21 $(y+xy^2)dx-x\,dy=0.$

22 $\dfrac{y}{x}dx+(y^3-\ln|x|)dy=0.$

23 $(x\sin y+y\cos y)dx+(x\cos y-y\sin y)dy=0.$

24 $(x+e^{x/y})dx+e^{x/y}\left(1-\dfrac{x}{y}\right)dy=0$ with $y(1)=3.$

25 $2x\,dx-\left(\dfrac{2x^2}{y}+1\right)dy=0$ with $y(0)=\tfrac14.$

26 $(2y^3-x)dx-3xy^2\,dy=0$ with $y(1)=2.$

27 $dx-x\cot y\,dy=0$ with $y(3)=\pi/3.$

Harder problems

28 Show that if $Q\neq0$ and P and Q have continuous second order partial derivatives, an integrating factor $\mu\neq0$ of

$$P(x,\,y)dx+Q(x,\,y)dy=0$$

can be found which depends only on x provided

$$Q\left(\dfrac{\partial^2 Q}{\partial x\partial y}-\dfrac{\partial^2 P}{\partial y^2}\right)=\dfrac{\partial Q}{\partial y}\left(\dfrac{\partial Q}{\partial x}-\dfrac{\partial P}{\partial y}\right).$$

Determine the corresponding condition for an integrating factor $\mu\neq0$ to exist which depends only on y.

29 Let μ_1 and μ_2 be any two integrating factors of

$$P(x, y)\,dx + Q(x, y)\,dy = 0$$

which are not proportional. Show that the function

$$\mu_1 + k\mu_2 \qquad (k = \text{const.})$$

is a solution of the differential equation. Verify this result by appeal to the three integrating factors found in Ex. 3.

4.5 Linear first order differential equations

The most general **first order linear differential equation** with continuous coefficients may be written in the form

$$h(x)y' + k(x)y = m(x). \tag{1}$$

For solutions on intervals on which $h(x) \neq 0$ this equation can be divided by $h(x)$ to obtain the simpler form

$$y' + p(x)y = q(x), \tag{2}$$

with $p(x)$, $q(x)$ continuous functions. Hereafter we shall work only with form (2), and unless stated to the contrary we shall assume $p(x)$ and $q(x)$ are continuous.

Since the solution of (2) is fundamental to the study of linear differential equations of all orders, we will derive it in two quite different ways. The first method will display the structure of the solution of (2), and the second will introduce a method by which the solution of a homogeneous equation may be used to find the solution of a non-homogeneous equation; both methods will be encountered again in connection with higher order linear differential equations.

Method 1. (Integrating factor)

In this method the solution of (2) will be represented as the sum of two distinct parts by setting

$$y = y_c + \tilde{y}_p, \tag{3}$$

where the nature of y_c and $\tilde{y}_p$ will become clear later. Using (3) in (2) we obtain

$$y_c' + \tilde{y}_p' + p(x)y_c + p(x)\tilde{y}_p = q(x). \tag{4}$$

We now define y_c to be the solution of the *homogeneous* equation

$$y_c' + p(x)y_c = 0, \tag{5}$$

and $\tilde{y}_p$ to be *any* solution of the *nonhomogeneous* equation

$$\tilde{y}_p' + p(x)\tilde{y}_p = q(x). \tag{6}$$

Then from Ex. 4, Sec. 4.4, it follows that $\mu = \exp[\int p(x)\,dx]$ is an integrating factor of (5), and

$$y_c = c\exp[-\int p(x)\,dx] \tag{7}$$

is its general solution, with c an arbitrary constant. Multiplying (6) by μ enables the equation to be written

$$\frac{d}{dx}\left(\exp[\int p(x)\,dx]\tilde{y}_p\right) = \exp[\int p(x)\,dx]q(x).$$

Integration then shows

$$\exp[\int p(x)\,dx]\tilde{y}_p = \int \exp[\int p(x)\,dx]q(x)\,dx + c^*,$$

leading to

$$\tilde{y}_p = c^*\exp[-\int p(x)\,dx] + \exp[-\int p(x)\,dx]\int \exp[\int p(x)\,dx]\,q(x)\,dx, \tag{8}$$

where c^* is an arbitrary constant.

At first sight there would seem to be the two arbitrary constants c and c^* in the solution of this first order equation, which is impossible. However, the complete solution is $y = y_c + \tilde{y}_p$, and the part of the solution in (8) depending on c^* is already contained in y_c, so without loss of generality we may always set $c^* = 0$ thereby making $\tilde{y}_p$ unique. From now on, with this choice of c^*, we shall write y_p in place of $\tilde{y}_p$.

Thus (2) has the general solution

$$y = c\exp[-\int p(x)\,dx] + \exp[-\int p(x)\,dx]\int \exp[\int p(x)\,dx]q(x)\,dx. \tag{9}$$

The importance of this form of derivation of general result (9) is that it shows the solution of (2) may always be represented in the form $y = y_c + y_p$; that is, as the sum of

$$y_c = c\exp[-\int p(x)\,dx] \tag{10}$$

which contains the arbitrary constant and is the solution of the homogeneous equation (5), and

$$y_p = \exp[-\int p(x)\,dx]\int \exp[\int p(x)\,dx]q(x)\,dx, \tag{11}$$

a unique function containing no additive multiple of y_c, with the property that when it is substituted into the left-hand side of (2) it gives rise to the particular function $f(x)$ that appears on the right-hand side.

Equation (5) leading to y_c is called the **associated homogeneous equation** (for (2)), or sometimes the **reduced equation**, and its solution y_c is called the **complementary function** of (2). The function $\tilde{y}_p$ is called a **particular solution** of (2), and we shall call the function y_p in (11), which is free from any additive multiple of y_c, and so is unique, the **particular integral** of (2). The term particular integral is often used in other texts to refer to any particular solution $\tilde{y}_p$, but here we find it convenient to confine its use to the unique function y_p just defined. The representation of a solution as the sum of the complementary function and the particular integral will be seen later to be true for linear differential equations of any order.

When seeking the solution of a specific linear first order differential equation, the reader is advised not to use result (9), which is mainly of theoretical importance. In practice, (2) is multiplied by the integrating factor $\exp\left[\int p(x)\,dx\right]$ to allow it to be written in the form

$$\frac{d}{dx}\left\{\exp\left[\int p(x)\,dx\right]y\right\} = \exp\left[\int p(x)\,dx\right]q(x), \tag{12}$$

and this result is then integrated to find y.

Before proceeding to discuss a number of examples, let us first use this result to prove a very important property of solutions of initial value problems for linear first order differential equation; namely that they are **unique**. The existence of a solution has, of course, already been established in arriving at the construction given in (9).

Suppose, if possible, that two different solutions exist for the nonhomogeneous differential equation (2), each corresponding to the same initial condition $y(a)=k$. Let these solutions be y_1 and y_2. Then it follows that

$$y_1' + p(x)y_1 = q(x), \quad \text{with } y_1(a)=k,$$

and

$$y_2' + p(x)y_2 = q(x), \quad \text{with } y_2(a)=k.$$

Setting $w = y_1 - y_2$, and subtracting the second of these equations from the first, shows that

$$w' + p(x)w = 0, \text{ with } w(a) = 0.$$

Thus the difference w between these two solutions satisfies a homogeneous linear first order differential equation with the homogeneous initial condition $w(a)=0$. It thus follows from (7) that

$$w = c \exp\left[-\int p(x)\,dx\right], \text{ with } w(a) = 0.$$

Since $\exp\left[-\int p(x)dx\right]$ can never be zero, the initial condition $w(a)=0$ can only be satisfied if $c = 0$. This implies $w \equiv 0$, so that $y_1(x) \equiv y_2(x)$, for $x \geq a$, thereby establishing the uniqueness of the solution of the initial value problem for (2). We have proved the following result.

Theorem 4.2 (Existence and uniqueness theorem)

The first order linear differential equation with continuous coefficients

$$y' + p(x)y = q(x)$$

which is defined on some interval I and subject to the initial condition $y(a) = k$ has a unique solution.   □

Example 1

Solve

$$y' - (\tan x)y = \cos x.$$

Solution

Here $p(x) = -\tan x$, so the integrating factor

$$\mu = \exp(-\int \tan x \, dx) = \exp(\ln|\cos x|) = \cos x.$$

Multiplication of the differential equation by $\mu = \cos x$ allows it to be written (see (12))

$$\frac{d}{dx}(y \cos x) = \cos^2 x.$$

Integrating this gives

$$y \cos x = \int \cos^2 x \, dx = \int \tfrac{1}{2}(1 + \cos^2 x) \, dx$$
$$= \tfrac{1}{2}x + \tfrac{1}{4}\sin 2x + c.$$

Dividing by $\cos x$ and using the trigonometric identity $\sin 2x = 2 \sin x \cos x$ shows the general solution to be

$$y = c \sec x + \tfrac{1}{2}(x \sec x + \sin x).$$ ■

The complementary function is $y_c = c \sec x$ and the particular integral is $y_p = \tfrac{1}{2}(x \sec x + \sin x)$.

Example 2

Solve

$$y' + y = 3 - \tfrac{1}{2}x^2$$

Solution

This is the differential equation encountered in Ex. 1., Sec. 4.2. We have $p(x) = 1$ so the integrating factor

$$\mu = \exp[\int p(x) \, dx] = \exp(\int dx) = \exp(x).$$

Multiplying the equation by $\mu = e^x$ allows it to be written

$$\frac{d}{dx}(e^x y) = 3e^x - \tfrac{1}{2}x^2 e^x.$$

Integrating we find

$$e^x y = \int 3e^x \, dx - \tfrac{1}{2}\int x^2 e^x \, dx$$

$$= 3e^x - \tfrac{1}{2}(x^2 - 2x + 2)e^x + c,$$

so that

$$y = c e^{-x} + 2 + x - \tfrac{1}{2}x^2.$$

This is the general solution quoted in (6) in Sec. 4.2. The complementary function $y_c = c e^{-x}$ and the particular integral $y_p = 2 + x - \tfrac{1}{2}x^2$. ∎

Example 3

Solve

$$y' - \frac{4}{(x+2)}y = (x+2)^5,$$

subject to the initial condition $y(0) = 8$.

Solution

In this case $p(x) = -4/(x+2)$, so the integrating factor

$$\mu = \exp\left(-\int \frac{4}{x+2}\, dx\right) = \exp[-4\ln|x+2|] = \frac{1}{(x+2)^4}.$$

Multiplication of the differential equation by $\mu = (x+2)^{-4}$ allows it to be written

$$\frac{d}{dx}[(x+2)^{-4} y] = x + 2.$$

Integrating this equation and solving for y gives

$$y = (\tfrac{1}{2}x^2 + 2x + c)(x+2)^4.$$

This is an initial value problem, so c must be determined such that $y(0) = 8$. Setting $x = 0$, $y = 8$ gives $c = \tfrac{1}{2}$, so the required solution is

$$y = \tfrac{1}{2}(x^2 + 4x + 1)(x+2)^4.$$ ∎

Method 2. (Variation of parameters)

The starting point of this approach, called the method of **variation of parameters** (sometimes **variation of constants**), is the result of Ex. 4, Sec. 4.4, that the general solution of

$$y' + p(x)y = 0 \tag{13}$$

is

$$y = c \exp\left[-\int p(x)\,dx\right],$$ (14)

with c an arbitrary constant.

This suggests that the solution of the nonhomogeneous equation

$$y' + p(x)y = q(x)$$ (15)

could be represented in the form

$$y = v(x)\exp\left[-\int p(x)\,dx\right],$$ (16)

where the unknown function $v(x)$ has now replaced the constant c in (14).

To show this is possible, and to find the function v, we now substitute (16) into (15). The result is the equation

$$v' \exp\left[-\int p(x)\,dx\right] = q(x),$$

where use has been made of the elementary result

$$\frac{d}{dx}\left\{\exp\left[-\int p(x)\,dx\right]\right\} = -p(x)\exp\left[-\int p(x)\,dx\right].$$

Thus

$$v' = \exp\left[\int p(x)\,dx\right]q(x),$$

which is a separable equation with the solution

$$v = \int \exp\left[\int p(x)\,dx\right] q(x)\,dx + c.$$ (17)

Combining (16) and (17) then gives for the general solution of (15)

$$y = c\exp\left[-\int p(x)\,dx\right] + \exp\left[-\int p(x)\,dx\right]\int \exp\left[\int p(x)\,dx\right] q(x)\,dx,$$ (18)

which is identical with (9).

In practice, when variation of parameters is used to obtain the solution of a specific differential equation, the functions, $\exp\left[\int p(x)\,dx\right]$ and $\exp\left[-\int p(x)\,dx\right]$ are determined first, and then v is found from (17) and combined with (16) to give the solution.

This method of solution was first introduced by Lagrange, and its name 'variation of parameters' (or constants) comes from the fact that the constant c in (14) is replaced by the function (parameter) $v(x)$.

Example 4

Solve

$$y' + 4y = x.$$

Solution

As $p(x) = 4$ it follows that

$$\exp\left[\int p(x)\,dx\right] = \exp\left(\int 4dx\right) = \exp(4x),$$

and similarly that

$$\exp\left[-\int p(x)\,dx\right] = \exp(-4x).$$

Now $q(x) = x$, so

$$v(x) = \int \exp\left[\int p(x)\,dx\right] q(x)\,dx + c = \int x\exp(4x)dx + c = \tfrac{1}{4}(x - \tfrac{1}{4})\exp(4x) + c.$$

Thus the required general solution is

$$y = v(x)\exp\left[-\int p(x)\,dx\right] = \left[\tfrac{1}{4}(x - \tfrac{1}{4})\exp(4x) + c\right]\exp(-4x)$$

giving

$$y = c\exp(-4x) + \tfrac{1}{4}(x - \tfrac{1}{4}).$$

The complementary function $y_c = c\,e^{-4x}$ and the particular integral $y_p = \tfrac{1}{4}(x - \tfrac{1}{4})$. ∎

Example 5

Solve the initial value problem

$$y' + (2\cot x)y = \operatorname{cosec} x, \quad \text{with } y(\pi/4) = 2 - \sqrt{2}.$$

Solution

We have $p(x) = 2\cot x$, so

$$\exp\left[\int p(x)dx\right] = \exp\left(\int 2\cot x\,dx\right) - \exp\left[\ln(\sin^2 x)\right] = \sin^2 x,$$

and similarly,

$$\exp\left[-\int p(x)dx\right] = 1/\sin^2 x.$$

As $q(x) = \operatorname{cosec} x$ it follows that

$$v(x) = \int \exp\left[\int p(x)dx\right] q(x)dx + c$$

$$= \int \sin^2 x\,\operatorname{cosec} x\,dx + c = \int \sin x\,dx + c = -\cos x + c.$$

The required general solution is thus

$$y = v(x)\exp\left[-\int p(x)dx\right] = (c - \cos x)\frac{1}{\sin^2 x} = c\operatorname{cosec}^2 x - \operatorname{cosec} x\cot x.$$

Using the initial condition $y(\pi/4) = 2 - \sqrt{2}$ in this result shows $c = 1$, so the solution of the initial value problem is

$$y = \operatorname{cosec} x(\operatorname{cosec} x - \cot x). \quad ∎$$

Sometimes the reversal of the roles of x as independent variable and y as dependent variable will convert a nonlinear first order differential equation into a linear one which can be solved by either of the above methods. Although this situation does not arise often it is worth illustrating it by means of an example.

Example 6. Linearization by interchange of variables

Solve

$$\frac{dy}{dx} = \frac{1}{3x + \sin 3y}.$$

Solution

This first order equation, which is nonlinear when y is regarded as the dependent variable, becomes linear when x is regarded as the dependent variable. Rewriting the equation with the roles of x and y interchanged, which is accomplished by taking the reciprocal of both sides of the equation, yields

$$\frac{dx}{dy} - 3x = \sin 3y.$$

This equation has the integrating factor

$$\exp[\int -3dy] = e^{-3y},$$

and so by Method 1 may be written

$$\frac{d}{dy}(e^{-3y}x) = e^{-3y}\sin 3y.$$

Integration gives

$$e^{-3y}x = -\tfrac{1}{6}(\sin 3y + \cos 3y)e^{-3y} + c,$$

so that finally the general solution is seen to be

$$x = ce^{3y} - \tfrac{1}{6}(\sin 3y + \cos 3y). \qquad\blacksquare$$

Example 7. An application involving the hardening of an adhesive

The hardness $H(t)$ of an epoxy resin adhesive bonding two sample plates together at time t after its application is defined as

$$H(t) = \frac{\text{shear strength at time } t}{\text{ultimate shear strength when fully cured}}.$$

Find the shear strength $H(t)$, given the initial shear strength $H(0) = 0$, and that $H(t)$ satisfies the differential equation

$$\frac{dH}{dt} + \frac{2t}{k}H = \frac{2t}{k},$$

with $k > 0$ a constant depending on the adhesive. How may the hardening law be checked and k found by experiment?

Solution

This is an initial value problem for a nonhomogeneous linear equation, in which $p(t)=q(t)=2t/k$, and $H(0)=0$. Now

$$\exp[-\int p(t)dt]=\exp(-t^2/k), \quad \text{so that} \quad \exp[\int p(t)dt]=\exp(t^2/k).$$

Thus with the obvious notational changes it follows at once from (17) that

$$v(t)=\int \frac{2t}{k}\exp(t^2/k)dt$$

$$=\exp(t^2/k)+c.$$

Using (16) we find

$$H(t)=1+c\exp(-t^2/k),$$

but $H(0)=0$ so we must set $c=-1$, and thus

$$H(t)=1-\exp(-t^2/k).$$

It follows from the form of $H(t)$ that

$$k=-t^2/\{\ln[1-H(t)]\}.$$

Thus the validity of the hardening law can be checked for a given adhesive, and the appropriate value of k determined, by graphing the quantity $-t^2/\{\ln[1-H(t)]\}$ determined by experiment for different values of t. If this quantity remains substantially constant the law represents a good approximation, and the value of the constant is the desired value of k. ∎

We conclude by mentioning the **Bernoulli**[3] **equation**, which by means of a simple transformation may be changed from a nonlinear equation to a linear one capable of solution by the methods of this section. It has the form

$$y'+p(x)y=q(x)y^\alpha, \tag{19}$$

with $\alpha \neq 0$ and $\alpha \neq 1$.

The transformation

$$z=y^{1-\alpha} \tag{20}$$

converts (19) to a linear first order differential equation for z. This follows because differentiating (20) with respect to x gives

$$\frac{dz}{dx}=(1-\alpha)y^{-\alpha}\frac{dy}{dx}, \tag{21}$$

and when this is used in (19) together with (20) it yields the equation

$$z'+(1-\alpha)p(x)z=(1-\alpha)q(x). \tag{22}$$

---

[3] JAKOB (JAMES) BERNOULLI (1654–1705), Swiss mathematician born in Basel where he held the chair of mathematics until his death. He belonged to one of the most distinguished families in the history of mathematics. His main contributions were to the theory of probability, the theory of elasticity and the calculus.

The cases $\alpha = 0$ and $\alpha = 1$ are excluded, because (19) reduces to (2) when $\alpha = 0$, and to a homogeneous equation when $\alpha = 1$.

Example 8. Bernoulli equation

Solve

$$y' - \frac{2}{x} y = xy^{1/2}.$$

Solution

This is a Bernoulli equation with $\alpha = \frac{1}{2}$, $p(x) = 2/x$ and $q(x) = x$. Using the transformation $z = y^{1/2}$, or making direct use of (20), we find

$$z' - \frac{1}{x} z = \frac{1}{2} x.$$

Solving this by either method 1 or 2 leads to

$$z = \frac{1}{2} x^2 + cx,$$

or as $z = y^{1/2}$, to the general solution

$$y = (\tfrac{1}{2} x^2 + cx)^2. \qquad\qquad \blacksquare$$

Problems for Section 4.5

1 Show that if y_1 and y_2 are any two solutions of

$$y' + p(x) y = 0,$$

then $y = c_1 y_1 + c_2 y_2$ is also a solution for any two arbitrary constants c_1 and c_2.

2 Show that if y_1 and y_2 are any two solutions of

$$y' + p(x) y = q(x),$$

then $y = y_1 - y_2$ is a solution of the associated homogeneous equation

$$y' + p(x) y = 0.$$

3 Explain why, when evaluating the integrating factor $\mu = \exp[\int p(x) dx]$, the constant of integration in the antiderivative $\int p(x) dx$ may always be set equal to zero.

4 Explain why although the antiderivatives $\int p(x) dx$ and $\int -p(x) dx$ appear in the particular integral (11), and each has an associated constant of integration, there is no constant of integration present in y_p.

Use either of Methods 1 or 2 to find the general solution for each of the following differential equations.

5 $y' + ky = e^{-kx}.$ **6** $y' + y = 2 - x^2.$

7 $y' + \dfrac{1}{x}y = x^2.$

8 $y' + y = 2 + x^2.$

9 $y' + \dfrac{1}{x-2}y = x - 1.$

10 $y' - \dfrac{2}{x+1}y = (x+1)^3.$

11 $y' + y \tan x = \sec x.$

12 $y' + y \cos x = \tfrac{1}{2}\sin 2x.$

13 $y' + \dfrac{n}{x}y = \dfrac{3}{x^n}.$

14 $y' - 2xy = 2x^3.$

15 $y' - \dfrac{2}{x \ln|x|}y = \dfrac{1}{x}.$

16 $y' - \dfrac{1}{x}y = \dfrac{1}{\ln|x|}.$

Use either of Methods 1 or 2 to solve the following initial value problems.

17 $y' - 2y = e^x$ with $y(0) = 1.$

18 $y' + \dfrac{3}{x}y = \dfrac{\ln|x|}{x^4}$ with $y(1) = 1.$

19 $y' - \dfrac{1}{1-x^2}y = 1 + x$ with $y(0) = 0.$

20 $y' - y \tan x = \sec x$ with $y(0) = 0.$

21 $y' + \dfrac{1-2x}{x^2}y = 1$ with $y(-1) = 2.$

22 $y' - \dfrac{a}{x}y = \dfrac{x+1}{x}$ with $y(1) = 1.$

Use either of Methods 1 or 2 coupled with an interchange of dependent and independent variables to solve the following problems.

23 $y' + \dfrac{2}{x-y} = 0.$

24 $y' - \dfrac{1}{x \cot y + \sin^2 y} = 0.$

25 $y' - \dfrac{1}{x \tan y + \sec y} = 0,$ with $y\left(\dfrac{4\pi}{3}\right) = \dfrac{\pi}{3}.$

26 $y' - \dfrac{1}{x(3x-1)\sin y} = 0,$ with $y(1/4) = \dfrac{\pi}{2}.$

27 $y' + \dfrac{1}{2x}y = \dfrac{1}{y^3},$ with $y(1) - 1.$

28 $y' - \dfrac{y}{3x+y^5} = 0,$ with $y(0) = 1.$

Solve the following Bernoulli equations.

29 $y' - 4y = -xy^2.$

30 $y' - \dfrac{1}{x}y = \dfrac{4y^2}{x^2}.$

31 $y' - \dfrac{4}{x}y = x\sqrt{y}.$

32 $y' + xy = x^3 y^3.$

33 Show Method 1 may be used to solve the Bernoulli equation

$$y' + p(x)y = q(x)y^\alpha$$

by setting $y = uv$ and requiring u to satisfy the homogeneous equation

$$u' + p(x)u = 0.$$

Use the method outlined in Prob. 33 to solve the initial value problems for the following Bernoulli equations.

34 $y' - \dfrac{1}{x}y = -\dfrac{3x^2}{y^2}$ with $y(1) = 2.$

35 $y' + \dfrac{1}{x}y = -xy^2$ with $y(1) = 4.$

36 $y' - \dfrac{x}{1-x^2}y = \dfrac{axy^2}{1-x^2}$ with $y(0)=1$.   **37** $y' + \dfrac{2}{x}y = y^2 \dfrac{\ln|x|}{x}$ with $y(1)=2$.

Solve the Bernoulli equation obtained by interchanging dependent and independent variables in each of the following equations.

38 $y' = \dfrac{1}{xy + x^2 y^3}$.   **39** $y' = \dfrac{3x^2}{ax^3 + y + 1}$.

40 Solve the initial value problem for the linear first order differential equation arising from **Newton's law of cooling** encountered in Ex. 2, Sec. 4.1, in which

$$\frac{dT}{dt} + \lambda T = \lambda T_0$$

with $\lambda = $ const., and $T(0) = T_1$.

41 Solve the initial value problem for the first order differential equation arising from the **drag on a sphere** encountered in Prob. 15, Sec. 4.2, in which

$$\frac{du}{dt} + 0.5u = 2 \quad \text{with } u(0)=0.$$

42 Solve the initial value problem for the first order differential equation arising from the **R–L circuit** encountered in Prob. 16, Sec. 4.2, in which

$$L\frac{di}{dt} + Ri = L \sin 2t \quad \text{with } i(0)=0 \text{ and } R/L = 1/2.$$

43 Solve the initial value problem for the first order differential equation arising from the **xenon poisoning** problem encountered in Prob. 17, Sec. 4.2, in which

$$\frac{dM}{dt} + M = 3e^{-t} \quad \text{with } M(0)=1.$$

44 Solve the initial value problem for the Bernoulli equation arising from the **logistic equation** encountered in Ex. 4, Sec. 4.3, in which

$$\frac{dP}{dt} - aP = -\frac{a}{L}P^2 \quad \text{with } P(0)=P_0.$$

45 Solve the initial value problem for the differential equation arising from the **R–L circuit** encountered in Prob. 32, Sec. 4.3, in which

$$L\frac{di}{dt} + Ri = V_0 \quad \text{with } i(0)=i_0.$$

46 Solve the initial value problem for the differential equation arising from the **resisted motion** problem encountered in Prob. 33, Sec. 4.3, in which

$$v' + kv = -g \quad \text{with } v(0)=U.$$

Harder problems

47 Write the linear first order differential equation (2) in the form

$$dy+[p(x)y-q(x)]\,dx=0,$$

and apply Theorem 4.1(i) to find an integrating factor. Hence derive the form of solution given in (9).

48 Show the solution of the initial value problem $y'+ay=b\sin\omega t$, with $y(0)=0$ and a, b, ω constants can be written in the form

$$y=\frac{\omega b}{a^2+\omega^2}e^{-at}+\frac{b}{(a^2+\omega^2)^{1/2}}\sin(\omega t-\varepsilon),$$

where

$$\sin\varepsilon=\frac{\omega}{(a^2+\omega^2)^{1/2}}\quad\text{and}\quad\cos\varepsilon=\frac{a}{(a^2+\omega^2)^{1/2}}.$$

Notice that the solution is the sum of an exponential term and an oscillatory term. If $a>0$, the first term decays as $t\rightarrow+\infty$, leaving only the oscillatory term. The solution y is then said to be **stable**. However, if $a<0$, the first term grows without bound as $t\rightarrow+\infty$ causing the solution to become unbounded. The solution y is then said to be **unstable**.

When $a>0$, the first term is called the **transient term**, the second is called the **steady state solution** and ε is called the **phase angle**. Solutions of this sort arise in many engineering problems in which a system described by a linear first order differential equation starts from rest under the excitation of an external disturbance $b\sin\omega t$. When $a>0$ the transient term describes the effect of the startup of the system on the complete solution. The steady state term describes the long-term behavior of the solution once the manner of startup has ben 'forgotten' by the complete solution. Compare this result with the solution of Prob. 42.

49 Let a **particular solution** of (2) be any solution of the form (9) in which the arbitrary constant c is given a specific value. Show that if y_1 and y_2 are any two different particular solutions of (2), its general solution is given by

$$\frac{y-y_1}{y_2-y_1}=k,$$

with $k\neq0$ an arbitrary constant.

Linear first order differential equations with coefficients defined in piecewise fashion. It may happen that the coefficient $p(x)$ in the linear first order differential equation

$$y'+p(x)y=q(x) \tag{A}$$

is defined in a piecewise fashion, such that

$$p(x)=\begin{cases}p_1(x), & x\leq a,\\ p_2(x), & x>a.\end{cases}$$

This happens, for example, when $p(x)$ is discontinuous at $x=a$, and also when it is everywhere continuous but defined differently according as $x<a$ or $x>a$.

To solve an initial value problem for such an equation subject to the initial condition

$$y(x_0) = b, \quad \text{with } x_0 < a, \tag{B}$$

it is necessary to recognize that (A) is, in fact, two separate differential equations; one for $x_0 \le x \le a$, and the other for $x > a$. Although the derivative y' may be discontinuous at $x = a$, the solution y itself must be continuous there, since it is obtained by the process of integration. Thus the connection between the differential equations defined on adjacent intervals is provided by the *continuity* of the solution at $x = a$.

The initial value problem (A) and (B) is thus solved by setting

$$y = \begin{cases} y_1(x), & x_0 \le x \le a, \\ y_2(x), & x \ge a, \end{cases}$$

and then finding $y_1(x)$ in the interval $x_0 \le x \le a$ from

$$y_2' + p_1(x) y_1 = q(x), \quad \text{with } y_1(x_0) = b,$$

and $y_2(x)$ in the interval $x \ge a$ from

$$y_2' + p_2(x) y = q(x), \quad \text{with } y_2(a) = y_1(a).$$

50 Solve

$$y' + p(x)y = 2, \quad \text{with } y(1) = 2,$$

given that

$$p(x) = \begin{cases} 1/x, & 1 \le x \le 2 \\ 1/2, & x \ge 2. \end{cases}$$

Is the derivative of the solution continuous at $x = 2$?

51 Solve

$$y' + p(x)y = 1, \quad \text{with } y(1) = 1,$$

given that

$$p(x) = \begin{cases} 1/x, & 1 \le x \le 2 \\ -1/x, & x > 2. \end{cases}$$

Is the derivative of the solution continuous at $x = 2$?

52 Solve

$$y' + p(x)y = e^{-x}, \quad \text{with } y(0) = 3,$$

given that

$$p(x) = \begin{cases} 1, & 0 \le x \le 1 \\ -2, & x > 1. \end{cases}$$

Is the derivative of the solution continuous at $x = 1$?

53 Solve

$$y' + p(x)y = x, \quad \text{with } y(0) = 2,$$

given that

$$p(x) = \begin{cases} x, & 0 \leq x \leq 1 \\ 1, & x \geq 1. \end{cases}$$

Is the derivative of the solution continuous at $x = 1$?

54 An equation of the form

$$y' = p(x)y^2 + q(x)y + r(x)$$

is called a **Riccati equation**[4], and it contains as special cases the linear first order equation ($p(x) \equiv 0$) and the Bernoulli equation ($r(x) \equiv 0$). This equation cannot in general be solved by integration as this leads to transcendental functions whose properties are unknown. However, if a particular solution can be found by any means (usually by inspection), this can be used to find the general solution in terms of an integral. Show that if y_1 is any particular solution, then

(i) the substitution $y = y_1 + u$ leads to the equation

$$u' - (2py_1 + q)u = pu^2,$$

which is a Bernoulli equation for u;

(ii) the subsitution $y = y_1 + \dfrac{1}{v}$ leads to the equation

$$v' + (2py_1 + q)v + p = 0,$$

which is a linear first order equation for v.

Thus once either of these equations has been solved, the general solution of the Riccati equation is known either as $y = y_1 + u$ or as $y = y_1 + \dfrac{1}{v}$.

Use either of the methods of Prob. 54 to solve the following Riccati equations.

55 $y' = y^2 + \dfrac{1}{x}y - \dfrac{3}{x^2}.$ $\qquad \left(\text{Set } y_1 = \dfrac{1}{x} \right).$

56 $y' = xy^2 - \dfrac{2}{x}y - \dfrac{1}{x^3}$ $\qquad \left(\text{Set } y_1 = \dfrac{1}{x^2} \right).$

57 $y' = \left(\dfrac{1-x}{2x^2} \right)y^2 + \dfrac{1}{x}y + \dfrac{x-1}{2}$ $\qquad (\text{Set } y_1 = x).$

58 $y' = \dfrac{1}{2}y^2 - \dfrac{2}{x}y + \dfrac{1}{2x^2}$ $\qquad \left(\text{Set } y_1 = \dfrac{1}{x} \right).$

---

[4] JACOPO FRANCESCO, COUNT RICCATI of Venice (1676–1754), an Italian mathematician whose contribution to mathematics was mainly concerned with differential equations, but who also worked in geometry and acoustics.

4.6   Orthogonal and isogonal trajectories

A **one-parameter family** of plane curves is a set of curves capable of expression in the form

$$F(x, y, c) = 0, \tag{1}$$

where the parameter c is usually restricted to a given interval. Individual curves correspond to specific values of c. For example, the equation

$$x^2 + y^2 = c^2 \tag{2}$$

defines a family of concentric circles with radius $|c|$ which are centered on the origin, as shown by the full lines in Fig. 4.8(a). Similarly, the equation

$$y = cx^2 \tag{3}$$

defines a family of parabolas, as shown by the full lines in Fig. 4.8(b). The parabolas lie in the upper half plane for $c > 0$ and the lower half plane for $c < 0$.

(a)   (b)

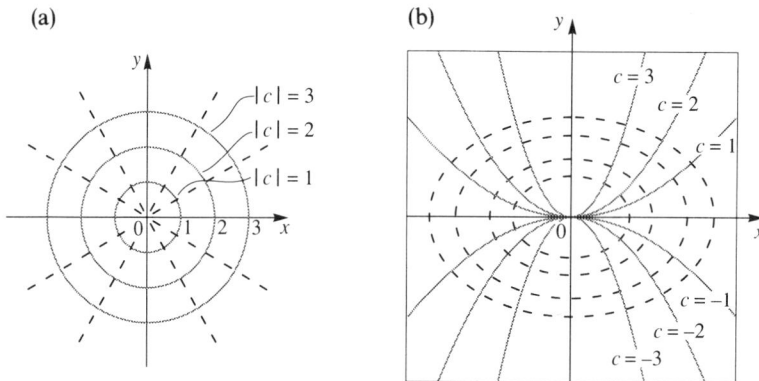

Fig. 4.8   (a) Family of concentric circles   (b) Family of parabolas

Such families of curves arise frequently in applications in which the parameter c usually represents a quantity of physical interest in a two-dimensional problem. Typically, c might be a temperature, an electrostatic potential, a gravitational potential, the value of an elastic strain function, or a velocity potential in a fluid flow. Graphing the curves $c = $ const. for different values of c helps to show visually how c varies throughout some region of the (x, y)-plane.

A curve which intersects each member of the family of curves (1) at right angles is called an **orthogonal trajectory** of the family. The set of all possible orthogonal trajectories corresponding to the family of curves in (1) itself forms a family of curves.

A very simple example of orthogonal trajectories is provided by the dashed radial lines in Fig. 4.8(a) which intersect each circle at right angles. In the case of the parabolas in Fig. 4.8(b), the orthogonal trajectories are the ellipses shown as dashed lines. The presence of one or more singular points at which one family of curves becomes degenerate (the circles at the origin in Fig. 4.8(a)) is a common feature of orthogonal trajectories. A more interesting example of orthogonal trajectories is shown in Fig. 4.9 in which two families of coaxial circles are involved, with one being shown as the full lines and the other as the dashed lines. In this case both P and Q are singular points at which the dashed circles degenerate to points.

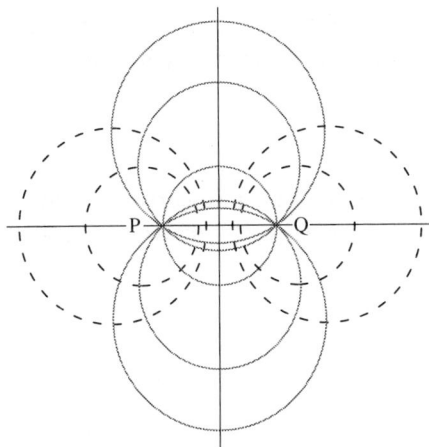

Fig. 4.9 Orthogonal trajectories involving two families of coaxial circles

The full significance of orthogonal trajectories will become clear later, but for the moment something of their importance can be appreciated by studying Table 4.1. This shows some of the possible relationships which exist between the families of curves $c = $ const. and their corresponding orthogonal trajectories.

Table 4.1 Physical processes and orthogonal trajectories

Physical process	Family of curves $c = $const.	Orthogonal trajectories
Heat conduction	Isothermals	Lines of heat flow
Electrostatics	Equipotentials	Lines of force (field lines)
Gravitation	Gravitational equipotentials	Lines of force
Elasticity	Lines of constant strain	Lines of stress
Fluid flow	Fluid equipotentials	Streamlines

Thus the dashed lines in Fig. 4.9 could, for example, be interpreted as the equipotentials due to two line changes of equal intensity but opposite sign positioned normal to the plane of the paper, one through P and the other through Q. The full lines in Fig. 4.9 would then correspond to the lines of force (field lines).

In order to determine the differential equation satisfied by the orthogonal trajectories of (1) it is necessary that we first find the differential equation satisfied by the family of curves belonging to (1). This is accomplished by differentiating (1) implicitly with respect to x to obtain

$$\frac{\partial F}{\partial x} + \frac{\partial F}{\partial y}\frac{dy}{dx} = 0, \tag{4}$$

and then eliminating the parameter c between (1) and (4). When this is done we arrive at a differential equation satisfied by the family of curves in (1) which we shall denote by

$$\frac{dy}{dx} = f(x,y). \tag{5}$$

Let us now turn our attention to the orthogonal trajectories. Two curves C_1 and C_2 will intersect at an angle α at an arbitrary point P if, when measured as shown in Fig. 4.10, the angle between their tangents at P is α.

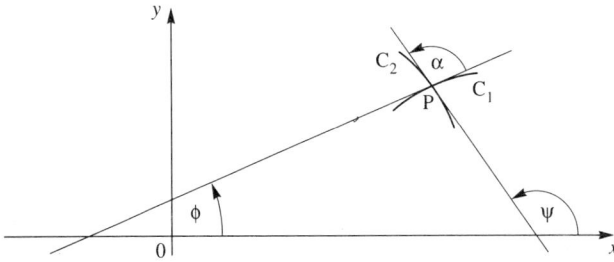

Fig. 4.10   Curves intersecting at an angle α

To relate this to orthogonal trajectories we let C_1 be a member of the family of curves belonging to (1). Then the curve C_2 will be the orthogonal trajectory to C_1 at P if $\alpha = \pi/2$. To connect the angles ϕ, ψ and α in Fig. 4.10 with gradients we use the fact that $\alpha = \psi - \phi$ together with the elementary trigonometric identity

$$\tan\alpha = \tan(\psi - \phi) = \frac{\tan\psi - \tan\phi}{1 + \tan\psi \tan\phi}. \tag{6}$$

Then if $\alpha = \pi/2$, as $\tan\pi/2$ is infinite, but the numerator of (6) is finite, it must follow from (6) that

$$1 + \tan\psi \tan\phi = 0,$$

showing that

$$\tan\psi = -\frac{1}{\tan\phi}. \tag{7}$$

Now if dy/dx is the gradient of curve C_1 belonging to (1) at P, $dy/dx = \tan\phi$, and if

dy_0/dx is the gradient of its orthogonal trajectory at P, $dy_0/dx = \tan\psi$. Thus from (7) we see these two gradients are related by the expression

$$\frac{dy}{dx} = -1 \bigg/ \left(\frac{dy_0}{dx}\right). \tag{8}$$

Substituting for dy/dx in (8) from (5), and omitting the suffix zero, it follows that the differential equation satisfied by the orthogonal trajectories of (1) is

$$\frac{dy}{dx} = -\frac{1}{f(x, y)}. \tag{9}$$

The arbitrary constant of integration which enters into the integral of (9) is analogous to the parameter c in (1), and when a value is assigned to it this determines a specific orthogonal trajectory.

Example 1. Parabolas and ellipses

Find the orthogonal trajectories of

$$y = cx^2$$

Solution

This is the family of parabolas shown in Fig. 4.8(b), and differentiation of the equation gives $y' = 2cx$. Elimination of c between this result and $y = cx^2$ shows the differential equation for these curves is

$$\frac{dy}{dx} = \frac{2y}{x}.$$

This corresponds to equation (5), so from (9) the differential equation of the orthogonal trajectories of the family of parabolas must be

$$\frac{dy}{dx} = -\frac{x}{2y}.$$

Integration leads to the family of orthogonal trajectories

$$2y^2 + x^2 = a,$$

where $a > 0$ is an arbitrary constant. This is a family of ellipses, and they are shown in Fig. 4.8(b) as the dashed lines. ∎

Example 2. Two families of coaxial circles

Find the orthogonal trajectories of

$$\frac{x}{x^2 + y^2} = c.$$

Solution

A simple manipulation of this equation puts it into the form

$$\left(x-\frac{1}{2c}\right)^2+y^2=\left(\frac{1}{2c}\right)^2.$$

This is a family of coaxial circles with radius $|1/2c|$ and their center at $\left(\dfrac{1}{2c},0\right)$ on the x-axis, and they are all tangent to the y-axis at the origin. These are shown as the solid curves in Fig. 4.11.

To find the differential equation satisfied by them we first differentiate their equation to obtain

$$1=2c(x+yy').$$

The elimination of c between this equation and the equation of the family $x=c(x^2+y^2)$ then shows the required differential equation corresponding to (5) is

$$\frac{dy}{dx}=\frac{y^2-x^2}{2xy}.$$

Consequently the differential equation corresponding to (9) determining the orthogonal trajectories is

$$\frac{dy}{dx}=\frac{2xy}{x^2-y^2}.$$

This equation is homogeneous of degree 1, and so it may be solved by setting $u=y/x$ as in Sec. 4.3, when we find

$$\frac{y}{x^2+y^2}=a,$$

where a is an arbitrary constant of integration. This is easily seen to be equivalent to

$$x^2+\left(y-\frac{1}{2a}\right)^2=\left(\frac{1}{2a}\right)^2,$$

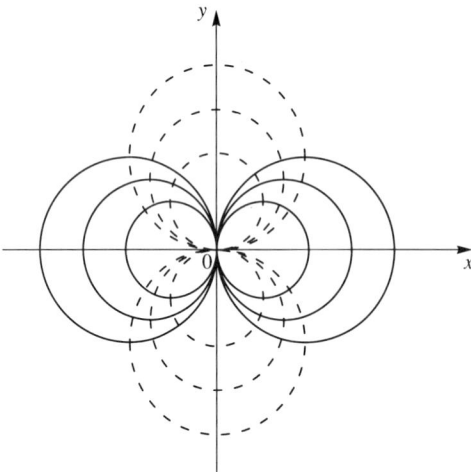

Fig. 4.11   Two families of coaxial circles

which is a family of coaxial circles of radius $|1/2a|$ with their center at $\left(0, \dfrac{1}{2a}\right)$ on the y-axis, and they are all tangent to the x-axis at the origin. These are shown as the dashed curves in Fig. 4.11. ∎

An **isogonal trajectory** is a curve which intersects each member of a family of curves (1) at a constant angle α. Thus an orthogonal trajectory is a special case of an isogonal trajectory in which $\alpha = \pi/2$. To determine the isogonal trajectories for (1) we use (5) and (6).

Let C_1 in Fig. 4.10 be a member of family (1), so that $dy/dx = \tan\phi$ at an arbitrary point P, and let C_2 be an isogonal trajectory with gradient $dy_1/dx = \tan\psi$. Then setting $\tan\alpha = k$, result (6) relates dy/dx and dy_1/dx as follows

$$k = \frac{\dfrac{dy_1}{dx} - \dfrac{dy}{dx}}{1 + \dfrac{dy_1}{dx}\dfrac{dy}{dx}}. \tag{10}$$

Substituting for dy/dx in (10) from (5), and omitting the suffix I, we find the differential equation satisfied by the isogonal trajectories of (1) is

$$\frac{dy}{dx} = \frac{k + f(x, y)}{1 - kf(x, y)}. \tag{11}$$

Here again, the constant of integration which enters into the integral of (11) is analogous to the parameter c in (1).

Example 3. Isogonal trajectories – equiangular spirals

Find the isogonal trajectories of the family of radial lines

$$y = cx$$

for the case $\alpha = \pi/3$.

Solution

The differential equation satisfied by this family of lines is easily seen to be

$$\frac{dy}{dx} = \frac{y}{x}.$$

Thus $f(x, y)$ in differential equation (5) is seen to be $f(x, y) = y/x$, and $k = \tan\alpha = \tan\pi/3 = \sqrt{3}$. The differential equation of the isogonal trajectories corresponding to (11) is thus

$$\frac{dy}{dx} = \frac{\sqrt{3} + (y/x)}{1 - \sqrt{3}(y/x)}.$$

This is a homogeneous equation of degree 1 and its solution obtained by means of the change of variable $u=y/x$ is easily shown to be

$$\ln(x^2+y^2)^{1/2} = \frac{1}{\sqrt{3}}\operatorname{arc\,tan}(y/x)+\ln a,$$

with a an arbitrary constant. These curves are called **equiangular spirals** (also **logarithmic spirals**), and some representative members of the family are shown in Fig. 4.12. The equation of the equiangular spiral takes on a simpler form when polar coordinates are used, for setting

$$r=(x^2+y^2)^{1/2} \quad \text{and} \quad \theta = \operatorname{arc\,tan}(y/x)$$

it becomes

$$r=a\exp[\theta/\sqrt{3}].$$

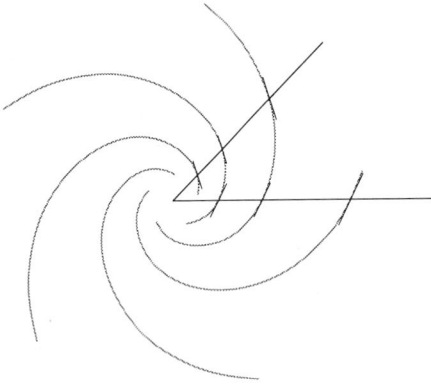

Fig. 4.12   Equiangular spiral

The last example suggests it would be useful to know the form taken by orthogonal trajectories in polar coordinates. To derive this result consider Fig. 4.13(a) in which Q is a point adjacent to P on a curve C belonging to the family

$$F(r, \theta, c)=0, \tag{12}$$

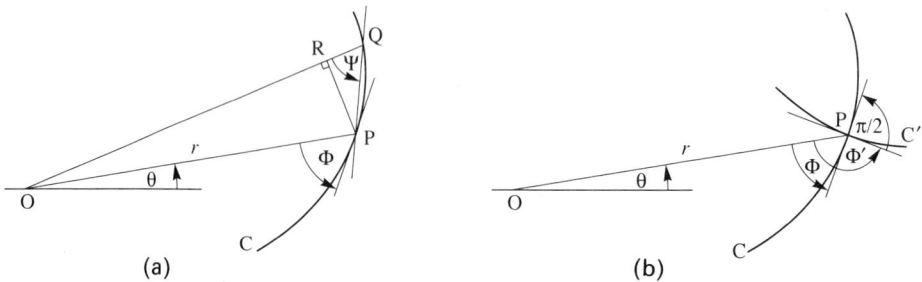

Fig. 4.13   Angles between tangent to curve C, the radius vector and the orthogonal trajectory

and PR is at right angles to OQ. Inspection of Fig. 4.13(a) shows $\Psi \to \Phi$ as $Q \to P$. Since

$$\tan \Psi = \frac{RP}{RQ},$$

and in terms of differentials $RQ \approx dr$ and $RP \approx r\, d\theta$, we have in the limit as $Q \to P$

$$\tan \Phi = r\frac{d\theta}{dr}. \tag{13}$$

If C' in Fig. 4.13(b) is the orthogonal trajectory to C at P, then $\Phi' - \Phi = \pi/2$, and so

$$\tan \Phi = -\cot \Phi',$$

or

$$\tan \Phi \tan \Phi' = -1. \tag{14}$$

Combining this with (13) we find the gradient of the orthogonal trajectory C' to C at P is

$$\tan \Phi' = -\frac{1}{r}\frac{dr}{d\theta}. \tag{15}$$

Differentiating (12) implicitly gives

$$\frac{\partial F}{\partial r} + \frac{\partial F}{\partial \theta}\frac{d\theta}{dr} = 0,$$

which can be rewritten as

$$\frac{\partial F}{\partial r} + \frac{1}{r}\frac{\partial F}{\partial \theta}r\frac{d\theta}{dr} = 0. \tag{16}$$

Thus from (15), replacing $r(d\theta/dr)$ in this equation by $-(1/r)(dr/d\theta)$, the differential equation of the orthogonal trajectories of (12) will be obtained by eliminating c between (12) and

$$\frac{\partial F}{\partial r}r\, d\theta - \frac{1}{r}\frac{\partial F}{\partial \theta}dr = 0. \tag{17}$$

Example 4. Cardioids

Find the orthogonal trajectories of the cardioids

$$r = c(1 - \cos \theta) \qquad (c > 0).$$

Solution

Setting $F \equiv r + c \cos \theta - c$ brings the equation of the cardioids into the form $F(r, \theta, c) = 0$, as in (12), so that (17) becomes

$$r\, d\theta + \frac{1}{r}c \sin \theta\, dr = 0.$$

The equation of the orthogonal trajectories now follows by eliminating c between this equation and $r = c(1 - \cos \theta)$ to obtain

$$\left(\frac{1 - \cos \theta}{\sin \theta} \right) d\theta + \frac{1}{r} dr = 0.$$

This has separable variables and is best integrated in the form

$$\int \frac{1}{\sin \theta} d\theta - \int \frac{\cos \theta}{\sin \theta} d\theta = - \int \frac{1}{r} dr.$$

Integration then shows that

$$-\frac{1}{2} \ln \left| \frac{1 + \cos \theta}{1 - \cos \theta} \right| - \ln |\sin \theta| = -\ln r + \ln a,$$

where for convenience the arbitrary constant is written $\ln a$. This is equivalent to

$$\left(\frac{1 - \cos \theta}{1 + \cos \theta} \right) \frac{1}{\sin^2 \theta} = \frac{a^2}{r^2}.$$

Using the identity $\sin^2 \theta = 1 - \cos^2 \theta$ and taking the square root reduces this to

$$r = a(1 + \cos \theta).$$

This family of orthogonal trajectories is another family of cardioids, rotated through an angle π about the origin relative to the original family. A representative member of the original family of cardioids is shown as the full line in Fig. 4.14, and one of its orthogonal trajectories is shown as the dashed line.

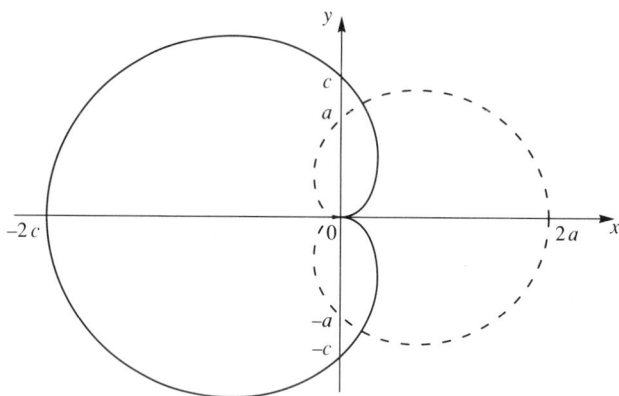

Fig. 4.14   Orthogonal families of cardioids   ∎

Problems for Section 4.6

Find the families of orthogonal trajectories corresponding to each of the following families of curves. Sketch a representative curve from each family together with an orthogonal trajectory.

1 $xy = c.$ **2** $x^2 + y^2 = c^2.$
3 $y^2 = cx.$ **4** $y = cx^\alpha$, with $\alpha \neq 0$ a constant.

5 $y = ce^{-x/\alpha}$, with $\alpha > 0$ constant.   **6** $x^2 + (y-c)^2 = 1 + c^2$.
7 $(x^2 + y^2)^2 = cxy$.   **8** $cx^2 + y^2 = 1$.
9 $y^2(2c - x) = x^3$.

Find the families of isogonal trajectories with the stated angle of intersection α corresponding to each of the following families of curves.

10 $y = cx$, with $\alpha = \pi/6$.   **11** $x^2 = 2a(y - x\sqrt{3})$, with $\alpha = \pi/3$.
12 $y^2 = 4cx$, with $\alpha = \pi/4$.

Find the families of orthogonal trajectories corresponding to each of the following families of curves.

13 $r = c \cos\theta$.   **14** $r = c(1 + \cos\theta)$, with $c > 0$.
15 $r^2 = c \cos\theta$, with $c > 0$.   **16** $r^m = c^m \cos m\theta$.

Harder problems

17 Let $u(x, y)$ and $v(x, y)$ be two functions which satisfy the (partial differential) equations

$$\frac{\partial u}{\partial x} = \frac{\partial v}{\partial y} \quad \text{and} \quad \frac{\partial u}{\partial y} = -\frac{\partial u}{\partial x}.$$

Show that if $u(x, y) = c = \text{const.}$ defines a family of curves, then its family of orthogonal trajectories is given by $v(x, y) = a = \text{const.}$ The two stated equations are called the **Cauchy–Riemann equations** and they arise in connection with complex analysis.

18 Find the differential equation satisfied by the isogonal trajectories of $F(r, \theta, c) = 0$ when the angle of intersection is α.

4.7 Existence, uniqueness and an iterative method of solution

The first six sections of this chapter have been concerned with the mathematical modeling of physical situations giving rise to first order differential equations, and the development of some methods for their solution. The tacit assumption has been made that a solution exists, and apart from Ex. 6 of Sec. 4.3 and Sec. 4.5, no discussion has been offered about its uniqueness.

The two fundamental questions which must now be considered are the following:

1. **Existence of solutions.** What conditions will ensure the existence of a solution of an initial value problem for a general first order differential equation?
2. **Uniqueness of solutions.** What conditions will ensure the uniqueness of a solution of an initial value problem for a general first order differential equation?

These questions are of considerable practical importance, because in physical situations modeled by initial value problems it is normally expected that a solution can be found (it **exists**) and, furthermore, that there is only one solution (it is **unique**). If no solution exists where one is expected this will indicate a failure of the mathematical model used to derive the differential equation. Should more than one solution exist this will either indicate an important feature of the physical problem, or a failure of the model. In either event such a problem will require further investigation, and this could lead to some modification of the mathematical model.

The following nonphysical examples illustrate the different types of situation which may arise. The equations involved are sufficiently simple that when solutions exist they can be found by inspection.

(a) The initial value problem

$$(y')^2 = -1 \quad \text{with} \quad y(0) = 1$$

has no solution for the given initial condition, or indeed for any other, since the left-hand side is essentially nonnegative.

(b) The initial value problem

$$|y'| = \sin x \quad \text{with} \quad y(\tfrac{3}{2}\pi) = 1$$

has no solution in the interval $\frac{3}{2}\pi < x < 2\pi$, because the left-hand side is essentially nonnegative while $\sin x < 0$ for $\frac{3}{2}\pi < x < 2\pi$. However, if the initial condition is changed to $y(\frac{1}{2}\pi) = 1$ then a solution certainly exists in the interval $\frac{1}{2}\pi \le x < \pi$, but it is not unique. To understand the nonuniqueness, notice that this initial value problem has the solution

$$y = 1 - \cos x \text{ when } y' > 0 \text{ and } \tfrac{1}{2}\pi \le x < \pi,$$

and the solution

$$y = 1 + \cos x \text{ when } y' < 0 \text{ and } \tfrac{1}{2}\pi \le x < \pi.$$

(c) The initial value problem

$$y' = \cos x \quad \text{with} \quad y(\tfrac{1}{2}\pi) = 2$$

has the unique solution (by Theorem 4.2)

$$y = 1 + \sin x \qquad \text{for } x \ge \tfrac{1}{2}\pi.$$

To determine the existence and uniqueness of solutions in more complicated cases it is necessary to appeal to a suitable general theorem. The proofs of such theorems are beyond the techniques we have at our disposal at present so we shall merely state two key theorems and then illustrate their use.

Theorem 4.3 (Peano existence theorem[5])

Let $f(x, y)$ be continuous at all points (x, y) of the rectangular region R defined by

$$|x - x_0| < \alpha, \quad |y - y_0| < \beta,$$

[5] GIUSEPPE PEANO (1858–1932), an Italian mathematician who made significant contributions to symbolic logic and differential equations.

and bounded within R with

$$|f(x, y)| \le M.$$

Then there is at least one solution of the differential equation

$$y' = f(x, y)$$

which satisfies the initial condition $y(x_0) = y_0$, for x in the interval $|x - x_0| < \delta$, where $\delta = \min \{\alpha, \alpha/M\}$. □

The meaning of this existence theorem becomes clearer once reference is made to the diagrams in Fig. 4.15. To understand them notice first that the boundedness of $f(x, y)$ in R, expressed by requiring $|f(x, y)| \le M$ throughout R, places bounds of $\pm M$ on the derivative y' of the solution $y(x)$ in R. This means that if a solution curve passes through (x_0, y_0), then it must lie between the two straight lines passing through that point with gradients $\pm M$.

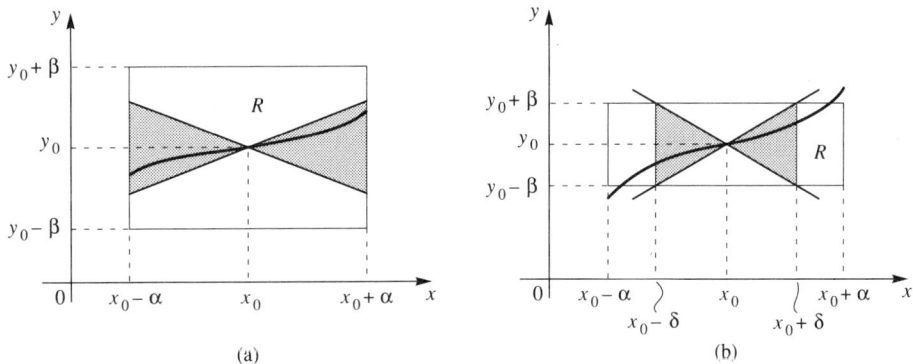

Fig. 4.15 Solutions exist in shaded regions defined by Peano existence theorem

The cases $\beta \ge \alpha M$ and $\beta < \alpha M$ are shown in Figs 4.15(a) and (b), respectively. In (a) the solution curve traverses the full width of R, while in (b) it traverses only the interval $x_0 - \delta \le x \le x_0 + \delta$, after which it *may* leave R. As the statement of the theorem depends on conditions prescribed within R it cannot provide information about the solution curve once it leaves R.

Example 1. Existence

Consider the initial value problem

$$y' = y^2 \text{ with } y(0) = 1,$$

and take region R to be $|x| < 2$, $|y - 1| < \frac{1}{2}$. It then follows that $\alpha = 2$, $\beta = \frac{1}{2}$, $|f| = |y^2| \le \frac{9}{4}$, so $\delta = \min \{2, \frac{8}{9}\}$ which corresponds to the situation in Fig. 4.15(b). Thus there is at least one solution of the stated initial value problem in the region R' defined by $|x| < \frac{8}{9}$, $|y - 1| < \frac{1}{2}$.

In fact the solution is $y = 1/(1-x)$, and it is unique in R', though this fact does not follow from Theorem 4.2, but by solving the problem directly. This example also serves to illustrate that even if $f(x, y)$ is continuous everywhere, it does not necessarily follow that the solution of an initial value problem is continuous for all x. In this case $f = y^2$ is continuous everywhere yet the solution $y(x)$ is discontinuous at $x = 1$. ∎

The next theorem utilizes a condition on $f(x, y)$ called a **Lipschitz condition**[6]. The function $f(x, y)$ is said to satisfy a Lipschitz condition with respect to y in R with constant m if for any (x, y_1) and (x, y_2) in R it is true that

$$|f(x, y_2) - f(x, y_1)| \le m|y_2 - y_1|. \tag{1}$$

Theorem 4.4 (Uniqueness)

Let R be the region defined in Theorem 4.3, and suppose the function $f(x, y)$ is such that

(i) it is bounded in R with $|f(x, y)| \le M$,
(ii) it satisfies a Lipschitz condition with respect to y at all points in R.

Then the initial value problem

$$y' = f(x, y), \text{ with } y(x_0) = y_0,$$

has a unique solution for all x such that

$$|x - x_0| < \delta, \text{ where } \delta = \min\{\alpha, \alpha/M\}.$$

☐

The stated conditions in these last two theorems are some of the simplest under which existence and uniqueness may be proved. They are thus only *sufficient* conditions, and may be weakened in different ways to produce more general theorems. Notice that if condition (ii) is omitted, Theorem 4.4 reduces to Theorem 4.3.

Example 2. Nonuniqueness

Consider the initial value problem

$$y' = y^{1/3}, \text{ with } y(0) = 0,$$

and take the region R to be $|x| < 1$, $|y| < 2$. Then $f(x, y) = y^{1/3}$ is continuous throughout R, and such that $|f(x, y)| \le 2^{1/3}$ in R, so condition (i) of Theorem 4.3 is satisfied. However the solution is *not* unique, because condition (ii) does not hold at the origin. To see this notice that

$$\left| \frac{f(x, y_2) - f(x, y_1)}{y_2 - y_1} \right| = \left| \frac{y_2^{1/3} - y_1^{1/3}}{y_2 - y_1} \right| = \frac{1}{|y_1^{2/3} + y_1^{1/3} y_2^{1/3} + y_2^{2/3}|},$$

which is unbounded for arbitrarily small y_1 and y_2.

---

[6] RUDOLF OTTO LIPSCHITZ (1832–1903), a German mathematician whose contributions to mathematics were in differential equations, differential geometry, algebra and number theory.

This example is simple enough for nonuniqueness to be established directly without appeal to Theorem 4.3. It is clear that $y \equiv 0$ is a degenerate solution, and a routine calculation shows that another solution is

$$y = \begin{cases} 0 & \text{for } x < 0 \\ (\tfrac{2}{3}x)^{3/2} & \text{for } x \geq 0, \end{cases}$$

so at least two different solutions exist for the same initial value problem. ∎

When a solution of an initial value problem exists and is unique it may be constructed by means of an iterative technique due to Picard[7]. This involves writing the initial value problem of Theorem 4.3 in the form

$$\frac{dy}{dx} = f[x, y(x)], \text{ with } y(x_0) = y_0, \tag{2}$$

and integrating from x_0 to x to obtain

$$y(x) = y_0 + \int_{x_0}^{x} f[t, y(t)]\, dt. \tag{3}$$

Now let $y_0(x)$ be an arbitrary differentiable function defined in R with a continuous first order derivative, and define the function $y_n(x)$ by the requirement that

$$y_n(x) = y_0 + \int_{x_0}^{x} f[t, y_{n-1}(t)]\, dt, \tag{4}$$

for $n = 1, 2, \ldots$. Result (4) defines an **iterative process** which generates an infinite sequence of functions $y_1(x), y_2(x), \ldots$, called **iterates**, starting from the arbitrary function $y_0(x)$ which for convenience is usually defined as $y_0(x) \equiv y_0$. The idea is that each successive iterate is a better approximation to the solution than the previous iterate. As the solution is unknown, by starting with $y_0(x) \equiv y_0$ then at least the starting approximation $y_0(x)$ satisfies the initial condition $y(x_0) = y_0$. The convergence of the sequence of iterates to an exact solution can be shown to follow if $y_0(x)$ is chosen as described and the conditions of Theorem 4.3 are satisfied.

This method can be used to obtain approximate solutions, but the calculation is usually tedious, and sometimes becomes impossible due to the generation of an integrand during the iteration process which cannot be integrated analytically. Its main use is in establishing existence and uniqueness theorems, though a Picard iterate is sometimes used to start a numerical computation.

Example 3. Picard iteration

Find the first two Picard iterates $y_1(x)$ and $y_2(x)$ for the initial value problem

$$y' = x^2 + y^2, \text{ with } y(1) = 2.$$

---

[7] EMILE PICARD (1856–1941), a leading French mathematician whose work in differential equations was of fundamental importance to the development of the subject. He also made many contributions to analysis and algebra.

We set $f(x, y) = x^2 + y^2$ in (4) and start with $y_0(x) \equiv 2$. This gives for the first iterate

$$y_1(x) = 2 + \int_1^x \{t^2 + [y_0(t)]^2\}\,dt$$

$$= 2 + \int_1^x (t^2 + 4)\,dt = -\frac{7}{3} + 4x + \frac{x^3}{3}.$$

The second iterate is then

$$y_2(x) = 2 + \int_0^x \left[t^2 + \left(-\frac{7}{3} + 4t + \frac{t^3}{3} \right)^2 \right] dt$$

$$= \frac{39}{630} + \frac{49}{9}x - \frac{28}{3}x^2 + \frac{17}{3}x^3 - \frac{7}{18}x^4 + \frac{8}{15}x^5 + \frac{1}{63}x^7.$$

This example illustrates the rapid increase in complexity of the integrations needed to determine successive iterates which is typical of this method. ■

Example 4. Problem of integration caused by Picard method

The problems of integration which often arise when determining Picard iterates are well illustrated by the initial value problem

$$y' = x + \cos y, \quad \text{with} \quad y(0) = \pi/2.$$

Setting $y_0(x) \equiv \pi/2$ we find the first iterate $y_1(x) = \int_0^x \left(t + \cos\frac{\pi}{2} \right) dt = \frac{1}{2}x^2.$

The second iterate is

$$y_2(x) = \int_0^x [t + \cos(\tfrac{1}{2}t^2)]\,dt$$

$$= \tfrac{1}{2}x^2 + \int_0^x \cos(\tfrac{1}{2}t^2)\,dt,$$

and this involves an integral which is not capable of evaluation in terms of elementary functions. In fact the integral in $y_2(x)$ is the *Fresnel integral* $C(x)$ (see the remark preceding Prob. 19 in Sec. 4.2). ■

Problems for Section 4.7

1 Give reasons why the initial value problem

$$(y')^2 + |y| = 0 \text{ with } y(0) = 1$$

has no solution.

2 Find all the solutions of the initial value problem

$$y' = y^{2/3} \text{ with } y(0) = 0.$$

3 Find all the solutions of the initial value problem

$$y' = 2(y - 1)^{1/2} \text{ with } y(0) = 1.$$

4 Show the initial value problem

$$y' = (y-3)^{2/3} \text{ with } y(1) = 5$$

has a unique solution for $|x-1| < 2$, $|y-5| < 1$.

5 Show the initial value problem

$$y' = x^2 + 2y^2 \text{ with } y(1) = 2$$

has a unique solution for $|x-1| < 1$, $|y-2| < 1$.

6 Show the initial value problem

$$y' = x + \sin y \text{ with } y(0) = \pi/2$$

has a unique solution for $|x| < 1$, $|y - \frac{1}{2}\pi| < \frac{1}{4}\pi$.

7 Show the solution of the initial value problem

$$y' = (1+x^2)\sqrt{|y|} \text{ with } y(0) = 0$$

is not unique in any region containing the origin.

8 Show $f(x, y) = x|y|$ satisfies a Lipschitz condition in y within the region R defined by $|x| < a$, $|y| < b$, but that $\partial f/\partial y$ is undefined except at the origin. Does the initial value problem

$$y' = x|y| \text{ with } y(0) = 0$$

have a solution in region R and is it unique?

9 Define the function f as

$$f(x, y) = \begin{cases} \dfrac{xy}{x^2 + y^2} & \text{for } (x, y) \neq (0, 0) \\ 0 & \text{for } (x, y) = (0, 0) \end{cases}$$

Determine the nature of the solution of the initial value problem

$$y' = f(x, y) \text{ with } y(0) = 0$$

in the region R defined by $|x| < 1$, $|y| < 1$.

10 Show that if $f(x, y)$ is Lipschitz continuous in y, then f is a continuous function of y alone.

11 Show that the existence and boundedness of $\partial f/\partial y$ in a region R implies that f is Lipschitz continuous in y in region R.

12 Continue the calculation in Ex. 3 and find the third iterate $y_3(x)$.

13 Find the Picard iterates $y_1(x)$ to $y_4(x)$ for the initial value problem

$$y' = xy \text{ with } y(0) = 1,$$

starting with $y_0(x) = 1$.

14 Find the Picard iterates $y_1(x)$ to $y_3(x)$ for the initial value problem

$$y' = y - \sin x \quad \text{with} \quad y(0) = 1,$$

starting with $y_0(x) = 1$.

The following problems illustrate the effect of using different starting approximations.

15 Find the third Picard iterate $y_3(x)$ for the initial value problem

$$y' = y \text{ with } y(0) = 1,$$

starting with (a) $y_0(x) = 1$, (b) $y_0(x) = \cos x$ and (c) $y_0(x) = \cosh x$. Graph these iterates for $0 \leq x \leq 2$ and compare the results with the exact solution $y = e^x$.

16 Find the third Picard iterate $y_3(x)$ for the initial value problem

$$y' = x + y \text{ with } y(0) = 1,$$

starting with (a) $y_0(x) = 1$, (b) $y_0(x) = \cos x$ and (c) $y_0(x) = \cosh x$. Graph these iterates for $0 \leq x \leq 2$ and compare the results with the exact solution $y = 2e^x - x - 1$.

4.8 Numerical solution of first order equations by the Runge–Kutta method

An accurate numerical solution is often required to an initial value problem for which there is no known analytical solution. In such circumstances it becomes necessary to use one of a number of special methods for the numerical solution of differential equations. These methods determine the solution $y(x)$ of the differential equation

$$\frac{dy}{dx} = f(x, y), \tag{1}$$

subject to the initial condition

$$y(x_0) = y_0, \tag{2}$$

at discrete intervals of x separated one from the other by an interval h called the integration **step length**. Thus, if N integration steps are involved, numerical methods generate the solution in the form

$$y_n = y(x_n), \quad \text{where} \quad x_n = x_0 + nh \tag{3}$$

and $n = 1, 2, \ldots, N$.

One of the most useful and frequently used methods is the **Runge–Kutta**[8] **fourth order method**, usually known simply as the **Runge–Kutta method**. The popularity of the method arises from the fact that:

(i) the method is accurate, with the error involved when an integration step of length h is used being approximately h^5,
(ii) the method is straightforward to use,
(iii) the method allows the adjustment of the step length h at any stage of the calculation.

---

[8] CARL DAVID TOLMÉ RUNGE (1856–1927), a German applied mathematician who was professor of applied mathematics at Göttingen, and WILHELM KUTTA (1867–1944) a German aerodynamicist who contributed to the study of fluid mechanics.

As the name implies, the Runge–Kutta fourth order method is one of a family of methods of similar type, in which the accuracy and complexity increases with the order of the method. All Runge–Kutta methods are based on a Taylor series type approach, similar in some ways to the Taylor series derivation of Simpson's rule for numerical integration.

It is difficult to estimate the error involved more accurately than h^5 when an integration step of length h is taken. In practice, when this information is important, usually due to uncertainty about the choice of step length, a calculation is repeated with the step length doubled and then compared with the results of the previous calculation. We describe the method in the form of an algorithm.

Runge–Kutta algorithm for a first order equation

Let the differential equation to be integrated be

$$\frac{dy}{dx} = f(x, y),$$

subject to the initial condition

$$y(x_0) = y_0.$$

Set $x_n = x_0 + nh$ and $y_n = y(x_n)$, where h is the step length. The algorithm then determines the value y_{n+1} at x_{n+1} from y_n at x_n by means of the following steps.

Step 1
Compute

$$k_{1n} = hf(x_n, y_n)$$
$$k_{2n} = hf(x_n + \tfrac{1}{2}h, y_n + \tfrac{1}{2}k_1)$$
$$k_{3n} = hf(x_n + \tfrac{1}{2}h, y_n + \tfrac{1}{2}k_2)$$
$$k_{4n} = hf(x_n + h, y_n + k_3).$$

Step 2
Compute

$$k_n = \tfrac{1}{6}(k_{1n} + 2k_{2n} + 2k_{3n} + k_{4n}).$$

Step 3
The numerical estimate of y_{n+1} at x_{n+1} is then given by

$$y_{n+1} = y_n + k_n.$$

Example 1. Runge–Kutta calculation

Use the Runge–Kutta method with an integration step $h=0.2$ to determine y_1 to y_5, given that

$$\frac{dy}{dx} = 2\cos\left(\frac{\pi x^2}{2}\right) + 0.5\,y^2,$$

subject to the initial condition $x_0 = 0$, $y_0 = 0$ (i.e. $y(0) = 0$).

Solution

n	x_n	k_{1n}	k_{2n}	k_{3n}	k_{4n}
0	0	0.400000	0.403951	0.404030	0.415535
1	0.2	0.415633	0.433594	0.434704	0.457983
2	0.4	0.458041	0.483893	0.486674	0.513813
3	0.6	0.513501	0.537692	0.541535	0.563018
4	0.8	0.562133	0.578153	0.581597	0.598557
5	1.0	0.597792	0.623317	0.630337	0.690787

n	x_n	k_n	y_n
0	0	0.405249	0
1	0.2	0.435036	0.405249
2	0.4	0.485498	0.840285
3	0.6	0.539162	1.325784
4	0.8	0.580032	1.864946
5	1.0	0.632648	2.444978

Problems for Section 4.8

In the following problems use the Runge–Kutta method together with the given step length h and initial condition to determine y_1 to y_5.

1 $\dfrac{dy}{dx} = \dfrac{x^2 + 2\tan(xy)}{1+x^2}$, $h=0.1$, $y(0)=1$.

2 $\dfrac{dy}{dx} = \dfrac{2x^2 - y^2 + 1}{1+x^2}$, $h=0.3$, $y(0)=1$.

3 $\dfrac{dy}{dx} = \dfrac{3\sin(x+y^2)}{\sin(xy)}$, $h=0.3$, $y(0.5)=1$.

4 $\dfrac{dy}{dx} = \dfrac{x^2 - 2y^2 + 1}{1+x^2}$, $h=0.3$, $y(0)=1$.

5 $\dfrac{dy}{dx} = 1 + \tanh(xy^2 + 1)$, $h = 0.4$, $y(0) = -1$.

6 $\dfrac{dy}{dx} = \dfrac{x^2 - 2y^2 + 1}{1 + x^2}$, $h = 0.5$, $y(-1) = -0$.

7 $\dfrac{dy}{dx} = \sin\left(\dfrac{3}{\sqrt{x^2 + 2y^2}}\right)$, $h = 0.2$, $y(0) = 0.5$.

8 $\dfrac{dy}{dx} = \cos(\pi x^2/2) + 0.5y^2$, $h = 0.2$, $y(0) = 0$.

Chapter 5
Linear Higher Order Ordinary Differential Equations

This chapter deals mainly with linear higher order constant coefficient differential equations. However it also lays the foundations for the discussion of variable coefficient linear second order differential equations which follows later.

The general linear higher order differential equation is introduced in Sec. 5.1 along with its associated reduced equation. The properties possessed by solutions of all linear differential equations are then identified. After introducing the idea of the linear dependence and independence of a set of functions, these same ideas are related to functions which are solutions of a linear differential equation and to Wronskians. These ideas are then used to establish a fundamental connection with linear algebra through the introduction of the concept of a basis for the solution of a differential equation.

Second order constant coefficient equations of homogeneous type are considered in Sec. 5.2, and the form of their solution (the complementary function) is derived. Initial value problems are considered along with some simple boundary value problems, which also serve to illustrate one of the ways in which a differential equation may have a unique solution, a nonunique solution, or no solution at all. Higher order differential equations of homogeneous type form the subject of Sec. 5.3, in which the structure of their general solution is examined.

A brief introduction to the differential operator D is offered in Sec. 5.4, mainly because of its notational convenience. The operator D method is not developed beyond an elementary stage as the discussion of the powerful Laplace transform method which follows later renders this unnecessary. Nonhomogeneous linear differential equations and particular integrals form the topic of discussion in Sec. 5.5, in which both the elementary method of undetermined coefficients and the powerful general method of variation of parameters are developed at length. Section 5.6 describes the method by which the order of a linear differential equation may be reduced once a solution is known, thereby simplifying the task of finding the remaining linearly independent solutions.

The very detailed discussion of oscillatory behavior presented in Sec. 5.7 is based on the work of both Sec. 5.2 and Sec. 5.5, which it reinterprets in terms of key physical examples of different types. In particular, the solution of a certain type of nonhomogeneous linear second order differential equation is discussed in terms of free undamped, free damped and forced oscillatory behavior. In each case the detailed structure of the solution is derived and its physical interpretation is discussed. Included amongst the physical phenomena considered is the behavior of a system at resonance, the effect of beats, the Q factor, the phase angle, and the power necessary to sustain forced

382

damped oscillations in a mechanical system. A brief discussion of the complex method for the determination of particular integrals and its application to the complex impedance of the R–L–C circuit is also included.

The reduction of a general linear second order differential equation to its normal form, in which the first derivative term is absent, forms the subject of Sec. 5.8. On occasions, such a reduction enables the solution of a variable coefficient linear second order differential equation to be derived, but its main use will come later when asymptotic solutions by the WKBJ method are discussed. Finally, in Sec. 5.9, a more advanced topic is considered when the Green's functions for initial value problems and two-point boundary value problems are introduced through the general solution found by the method of variation of parameters.

5.1 Linear higher order ordinary differential equations

In this first section we establish some fundamental properties of all linear higher order differential equations. The study of specific types of equations and their manner of solution will follow later. It will be recalled that when introducing differential equations in Sec. 4.1, the general linear nth order ordinary differential equation was defined as an equation of the form

$$a_0(x)\frac{d^n y}{dx^n}+a_1(x)\frac{d^{n-1} y}{dx^{n-1}}+ \ldots +a_n(x)y=f(x). \tag{1}$$

The functions $a_0(x), a_1(x), \ldots, a_n(x)$ are called the **coefficients** of the differential equation, and $f(x)$ is known as the **nonhomogeneous term**. Unless otherwise stated, we shall assume the coefficients of (1) are all continuous functions, and that $a_0(x) \neq 0$ in any interval in which a solution is required. The effect of the vanishing of $a_0(x)$ at certain points will be considered when series solutions are discussed.

Higher order differential equations which cannot be expressed as in (1) will be said to be **nonlinear.** Thus the equations

$$y''' \sin y+xy'+(1-x^2)y=\cos x$$

and

$$y'''+3xy''+(\sin x)y'+y^3=4$$

are both nonlinear; the first because of the term $y''' \sin y$ and the second because of the term y^3.

A solution of (1) is an n times differentiable function

$$y=\phi(x), \tag{2}$$

defined in some suitable interval $a<x<b$, which when substituted into (1) reduces it to an identity in x.

For example, the linear second order differential equation

$$x^2y''-4xy'+6y=x$$

has as a solution the function

$$y=c_1 x^3+c_2 x^2+\tfrac{1}{2}x,$$

valid for $-\infty<x<\infty$, with c_1 and c_2 two arbitrary constants. This is a **variable coefficient** differential equation, and it is representative of the type shown in (1).

Most of this chapter will be concerned with higher order **constant coefficient** differential equations, corresponding to the case in which the coefficients $a_i(x)$ in (1) are all absolute constants. Although certain aspects of the variable coefficient case will be considered here, the main discussion of such equations will be postponed until later.

A typical example of a second order constant coefficient differential equation is

$$y'' - 5y' + 6y = \sin 2x.$$

This has as a solution the function

$$y = c_1 e^{2x} + c_2 e^{3x} + \tfrac{1}{52}(\sin 2x + 5 \cos 2x),$$

valid for all x, where again c_1 and c_2 are arbitrary constants.

Before proceeding with a detailed discussion of the constant coefficient case, we shall establish some important properties possessed by all linear differential equations. We begin by considering the homogeneous form of (1)

$$a_0(x)\frac{d^n y}{dx^n} + a_1(x)\frac{d^{n-1}y}{dx^{n-1}} + \ \ldots \ + a_n(x)y = 0, \tag{3}$$

which we call the **associated**, or **reduced**, homogeneous differential equation corresponding to (1). Letting $y = \phi(x)$ be a solution, direct substitution into (3) shows that $y = c\phi(x)$ is also a solution, with c an arbitrary constant. This property may be expressed by saying that solutions of a homogeneous differential equation may be scaled and still remain solutions. Notice that (3) always has the **trivial** solution $y \equiv 0$.

Direct substitution into (3) also shows that if $\phi_1(x)$, $\phi_2(x)$, ..., $\phi_m(x)$ are any m solutions of (3), then so also is the linear combination

$$y = c_1\phi_1(x) + c_2\phi_2(x) + \ \ldots \ + c_m\phi_m(x), \tag{4}$$

with $c_1, c_2, \ldots, c_m$ arbitrary constants. The property exhibited in (4) may be expressed by saying that a linear combination of solutions of a homogeneous differential equation is also a solution. This very important result is usually called the **linear superposition principle** or **linearity principle** for solutions of linear homogeneous differential equations. We have established the following theorem.

Theorem 5.1 (Scaling and linear superposition of solutions)

(i) If $y = \phi(x)$ is a solution of the associated homogeneous form of (1) defined on some interval $a < x < b$, then so also is $y = c\phi(x)$, with c an arbitrary constant (scaling of solutions).

(ii) If $\phi_1(x)$, ..., $\phi_m(x)$ are any m solutions of the associated homogeneous form of (1) defined on some interval $a < x < b$, then so also is $y = c_1\phi_1(x) + c_2\phi_2(x) + \ldots + c_m\phi_m(x)$, with $c_1, c_2, \ldots, c_m$ arbitrary constants (linear superposition). $\qquad\Box$

As with first order differential equations, initial value problems play an important part in both the theory and application of higher order differential equations. For an nth order differential equation (linear or nonlinear), an **initial value problem** is the determination of a solution which satisfies arbitrarily prescribed values of y, dy/dx, ..., $d^{n-1}y/dx^{n-1}$ at some point $x = x_0$. The **initial conditions** for such an equation involve the specification of the n values $y(x_0) = y_0$, $y^{(1)}(x_0) = y_1$, ..., $y^{(n-1)}(x_0) = y_{n-1}$. Only the n

quantities $y_0, y_1, \ldots, y_{n-1}$ may be specified arbitrarily, because the equation itself then determines $y^{(n)}(x_0)$ and all higher order derivatives.

The solution of an initial value problem for a linear differential equation of any order has the important property that when it exists it is unique. This has already been proved for a first order equation using a very simple argument, but the proof for a general nth order linear differential equation is harder and longer, so it will be omitted. An outline of the proof for a general second order linear differential equation is given at the end of this section in Prob. 39. The general result may be stated as follows.

Theorem 5.2 (Uniqueness of solution for nth order equation)

Let the linear nth order differential equation

$$a_0(x)y^{(n)} + a_1(x)y^{(n-1)} + \ldots + a_n(x)y = f(x)$$

have coefficients which are continuous on an interval I on which $a_0(x) \neq 0$. Then the solution satisfying the initial conditions $y(x_0) = y_0, y^{(1)}(x_0) = y_1, \ldots, y^{(n-1)}(x_0) = y_{n-1}$ is unique for $x \geq x_0$ in I. $\qquad\qquad\square$

To proceed further with our discussion of linear differential equations we need to use the notion of the linear independence of a set of functions. This is a generalization of the corresponding concepts introduced in Chapters 2 and 3 when defining a basis for a vector space.

Definition (Linear dependence and independence of functions)

The n functions $\phi_1(x), \phi_2(x), \ldots, \phi_n(x)$ are said to be **linearly dependent** on the interval $a \leq x \leq b$ if constants $c_1, c_2, \ldots, c_n$ can be found, not all zero, such that

$$c_1\phi_1(x) + c_2\phi_2(x) + \ldots + c_n\phi_n(x) \equiv 0 \tag{5}$$

for all x in the interval. If (5) is only true for all x in the interval when $c_1 = c_2 = \ldots = c_n = 0$, the n functions are said to be **linearly independent** on the interval $a \leq x \leq b$.

This definition simplifies if only the two functions $\phi_1(x)$ and $\phi_2(x)$ are involved. For we see that in this case $\phi_1(x)$ and $\phi_2(x)$ will be linearly dependent if they are proportional, and they will be linearly independent if $\phi_1(x)/\phi_2(x)$ is a nonconstant function of x.

It is an immediate consequence of the definition that if two or more of the set of n functions $\phi_1(x), \phi_2(x), \ldots, \phi_n(x)$ are linearly dependent, then the complete set must be linearly dependent. For example, if $\phi_1(x)$ and $\phi_2(x)$ are linearly dependent, nonzero constants c_1 and c_2 exist such that $c_1\phi_1(x) + c_2\phi_2(x) = 0$ for $a \leq x \leq b$. Consequently, by setting $c_3 = c_4 = \ldots = c_n = 0$, result (5) is automatically satisfied, thus proving the linear dependence of the n functions $\phi_i(x)$.

Example 1. Linear dependence and independence

(a) The functions 1, x, x^2 are linearly independent on the interval $-\infty < x < \infty$. This follows because

$$c_1 + c_2 x + c_3 x^2 = 0$$

is a quadratic equation and so can only be true for two values of x and not for all x in $-\infty < x < \infty$. Thus if this is to be true for all x it must follow that $c_1 = c_2 = c_3 = 0$.

However, the functions 1, x, x^2 and $(2+3x)^2$ are linearly dependent on the interval $-\infty < x < \infty$, because $(2+3x)^2$ is a linear combination of the first three functions.

(b) The functions $\sin^2 \theta$ and $\cos^2 \theta$ are linearly independent on the interval $0 \le \theta \le 2\pi$ because they are not proportional.

The functions 1, $\sin^2 \theta$ and $\cos^2 \theta$ are linearly dependent on the interval $0 \le \theta \le 2\pi$ because they are related linearly by the trigonometric identity $\sin^2 \theta + \cos^2 \theta = 1$.

(c) The functions e^x, e^{2x}, e^{3x} and e^{4x} are linearly independent for $-\infty < x < \infty$, because $e^{2x} = (e^x)^2$, $e^{3x} = (e^x)^3$ and $e^{4x} = (e^x)^4$. ∎

Rather than determining linear dependence or independence case by case, as in Ex. 1, it is desirable to find a test to determine when an arbitrary set of functions $g_1(x)$, $g_2(x)$, ..., $g_n(x)$ is linearly dependent or independent in some interval $a \le x \le b$. The test we now derive assumes the functions all to be $(n-1)$ times differentiable in the interval.

We start from the fact that the n functions $g_1(x)$, $g_2(x)$, ..., $g_n(x)$ will be linearly independent in the interval $a \le x \le b$ if

$$c_1 g_1(x) + c_2 g_2(x) + \ldots + c_n g_n(x) \equiv 0 \tag{6}$$

implies $c_1 = c_2 = \ldots = c_n = 0$. Successive differentiation of (6) leads to the n equations

$$c_1 g_1(x) + c_2 g_2(x) + \ldots + c_n g_n(x) = 0$$
$$c_1 g_1^{(1)}(x) + c_2 g_2^{(1)}(x) + \ldots + c_n g^{(1)}(x) = 0$$
$$\vdots \qquad \vdots \qquad \vdots \qquad \vdots$$
$$c_1 g^{(n-1)}(x) + c_2 g_2^{(n-1)}(x) + \ldots + c_n g^{(n-1)}(x) = 0, \tag{7}$$

where we have denoted $d^r g / dx^r$ by $g^{(r)}(x)$.

It now follows from Theorem 3.20 that this homogeneous system of equations for $c_1, c_2, \ldots, c_n$ can only have the trivial solution $c_1 = c_2 = \ldots = c_n = 0$ if for some x in the interval $a \le x \le b$ the determinant $W \neq 0$, where

$$W = \begin{vmatrix} g_1(x) & g_2(x) & \cdots & g_n(x) \\ g_1^{(1)}(x) & g_2^{(1)}(x) & \cdots & g_n^{(1)}(x) \\ \cdot & \cdot & \cdots & \cdot \\ g_1^{(n-1)}(x) & g_2^{(n-1)}(x) & \cdots & g_n^{(n-1)}(x) \end{vmatrix}. \tag{8}$$

Thus $g_1(x)$, $g_2(x)$, ..., $g_n(x)$ will be linearly independent for $a \le x \le b$ if $W \neq 0$. The functional determinant W in (8), sometimes written $W(g_1, g_2, \ldots, g_n)$ to make explicit the functions involved, is called the **Wronskian**[1] of the n functions $g_1(x)$, $g_2(x)$, ..., $g_n(x)$.

---

[1] JOZEF MARIA HOENE WRONSKI (1778–1853), a Polish philosopher and mathematician now remembered only because of the Wronskian.

Observe that this same argument establishes that the vanishing of the Wronskian is a *necessary* condition for the n functions $g_1(x), g_2(x), \ldots, g_n(x)$ to be linearly dependent on $a \le x \le b$, though it is *not* a sufficient condition. That this is so may be seen by considering the functions

$$g_1(x) = \begin{cases} x^4, & x \ge 0 \\ 0, & x < 0 \end{cases} \quad \text{and} \quad g_2(x) = \begin{cases} 0, & x \ge 0 \\ x^4, & x < 0. \end{cases}$$

The Wronskian of $g_1(x)$ and $g_2(x)$ is identically zero, because

$$W = \begin{cases} \begin{vmatrix} x^4 & 0 \\ 4x^3 & 0 \end{vmatrix} = 0 & \text{for } x \ge 0, \\[4mm] \begin{vmatrix} 0 & x^4 \\ 0 & 4x^3 \end{vmatrix} = 0 & \text{for } x < 0, \end{cases}$$

yet in fact the functions are linearly independent because they are not proportional.

If, however, the functions $g_1(x), g_2(x), \ldots, g_n(x)$ are solutions of a homogeneous differential equation with continuous coefficients, it is easy to show that the vanishing of their Wronskian is both a necessary and a sufficient condition for the functions to be linearly dependent.

To see this, suppose the functions are solutions of (3), and notice that if $W(\alpha) = 0$ for $a \le \alpha \le b$, then we know from Theorem 3.7 that the equations

$$c_1 g_1(\alpha) + c_2 g_2(\alpha) + \ldots + c_n g_n(\alpha) = 0$$
$$c_1 g_1^{(1)}(\alpha) + c_2 g_2^{(1)}(\alpha) + \ldots + c_n g_n^{(1)}(\alpha) = 0$$

$$\cdot \qquad \cdot \qquad \cdots \qquad \cdot$$

$$c_1 g_1^{(n-1)}(\alpha) + c_2 g_2^{(n-1)}(\alpha) + \ldots + c_n g_n^{(n-1)}(\alpha) = 0$$

have a solution $c_1^*, c_2^*, \ldots, c_n^*$ in which not all the c_i^* vanish (a nontrivial solution). Setting

$$\Phi(x) = c_1^* g_1(x) + c_2^* g_2(x) + \ldots + c_n^* g_n(x),$$

we know $\Phi(x)$ is a solution of (3), and that it satisfies the initial conditions $\Phi(\alpha) = \Phi^{(1)}(\alpha) = \ldots = \Phi^{(n-1)}(\alpha) = 0$. However $y \equiv 0$ is a solution of (3) satisfying the same initial conditions, so by the uniqueness theorem it follows that $\Phi(x)$ must be the trivial solution $\Phi(x) \equiv 0$. This implies linear dependence between the n functions $g_1(x), g_2(x), \ldots, g_n(x)$. We have thus established our next important result.

Theorem 5.3 (Wronskian test for linear dependence and independence)

Let the homogeneous differential equation (3) with continuous coefficients defined for $a \le x \le b$ have solutions $\phi_1(x), \phi_2(x), \ldots, \phi_n(x)$. Then a necessary and sufficient condition that they will be linearly independent is that their Wronskian is nonvanishing for $a \le x \le b$. If, however, the Wronskian vanishes identically, the solutions are linearly dependent. $\square$

Although we shall not prove it, the vanishing of the Wronskian of n functions is also a necessary and sufficient condition for their linear dependence if they are **analytic functions** (capable of expansion in convergent power series on the interval $a \leq x \leq b$).

Example 2. Wronskian test for linear independence

The functions $\cos(\ln x^2)$ and $\sin(\ln x^2)$ are solutions of

$$x^2 y'' + xy' + 4y = 0,$$

as may be verified by direct substitution. They are linearly independent solutions on any interval not containing the origin because the Wronskian

$$W = \begin{vmatrix} \cos(\ln x^2) & \sin(\ln x^2) \\ -\dfrac{2}{x}\sin(\ln x^2) & \dfrac{2}{x}\cos(\ln x^2) \end{vmatrix} = \frac{2}{x} \neq 0. \qquad \blacksquare$$

Sometimes it is useful to have a representation for the Wronskian of the linearly independent solutions of a homogeneous equation in terms of the coefficients of the equation. Such a result is provided by **Abel's formula**[2].

$$W(x) = A \exp\left(-\int \frac{a_1(x)}{a_0(x)}\, dx\right) \quad (A = \text{const.}), \tag{9}$$

for the Wronskian $W(x)$ of the n linearly independent solutions $\phi_1, \phi_2, \ldots, \phi_n$ of $a_0(x)y^{(n)} + a_1(x)y^{(n-1)} + \ldots + a_n(x)y = 0$, in which the coefficients $a_0(x), \ldots, a_n(x)$ are continuous and $a_0(x) \neq 0$.

The proof of this result follows most easily by deriving the differential equation satisfied by W. Differentiating W by means of Theorem 3.18 produces

$$W' = \begin{vmatrix} \phi_1 & \phi_2 & \cdots & \phi_n \\ \phi_1' & \phi_2' & \cdots & \phi_n' \\ \vdots & \vdots & & \vdots \\ \phi_1^{(n-2)} & \phi_2^{(n-2)} & \cdots & \phi_n^{(n-2)} \\ \phi_1^{(n)} & \phi_2^{(n)} & \cdots & \phi_n^{(n)} \end{vmatrix},$$

because each of the other $n-1$ determinants which result from the differentiation contains two identical rows, and so vanishes.

Now ϕ_r is a solution of

$$a_0(x)\phi_r^{(n)} + a_1(x)\phi_r^{(n-1)} + \ldots + a_n(x)\phi_r^{(n)} = 0,$$

---

[2] NIELS HENRIK ABEL (1802–1829), a Norwegian mathematician born of a poor family who due to his outstanding ability gained entry to the University of Oslo in 1821. During his short life he made contributions to algebra, convergence, the theory of functions and integral equations.

for $r = 1, 2, \ldots, n$, so it follows that

$$\phi_r^{(n)} = -\frac{1}{a_0}\left(a_1 \phi_r^{(n-1)} + a_2 \phi_r^{(n-2)} + \ldots + a_0 \phi_r \right).$$

Replacing each element of the last row of W' by the corresponding form of the this result, and then subtracting appropriate multiples of the first $n-1$ rows from this new nth row, produces the result

$$W' = \begin{vmatrix} \phi_1 & \phi_2 & \cdots & \phi_n \\ \phi_1' & \phi_2' & \cdots & \phi_n' \\ \vdots & \vdots & & \vdots \\ \phi_1^{(n-2)} & \phi_2^{(n-2)} & \cdots & \phi_n^{(n-2)} \\ -\dfrac{a_1}{a_0}\phi_1^{(n-1)} & -\dfrac{a_1}{a_0}\phi_2^{(n-1)} & \cdots & -\dfrac{a_1}{a_0}\phi_n^{(n-1)} \end{vmatrix}.$$

Removing the factor $-(a_1/a_0)$ from the last row we obtain

$$W' = -\frac{a_1}{a_0} W.$$

Thus W satisfies the first order equation

$$W' + \frac{a_1(x)}{a_0(x)} W = 0,$$

and from (7) of Sec. 4.5 this has the general solution

$$W(x) = A \exp\left(-\int \frac{a_1(x)}{a_0(x)}\, dx \right),$$

which is **Abel's formula**. The advantage of this result is that it provides the functional form of the Wronskian immediately, though the scaling is indeterminate.

To determine the scale factor A, let the indefinite integral in W above be written in the alternative form

$$W(x) = A \exp\left(-\int_{x_0}^{x} \frac{a_1(s)}{a_0(s)}\, ds \right),$$

where x_0 is an arbitrary point at which the n solutions ϕ_r are defined. Then setting $x = x_0$ shows $A = W(x_0)$, and we arrive at the more precise statement of Abel's formula

$$W(x) = W(x_0) \exp\left(-\int_{x_0}^{x} \frac{a_1(s)}{a_0(s)}\, ds \right). \tag{10}$$

This form of the result is mainly of theoretical interest, since the Wronskian must be evaluated at $x = x_0$ in order to determine $W(x_0)$.

As an application of Abel's formula, consider the differential equation

$$x^2y'' + xy' - y = 0.$$

Making the identifications $a_0(x) = x^2$ and $a_1(x) = x$, we see from (9) that

$$W = A \exp\left(-\int \frac{1}{x}\, dx\right) = \frac{A}{x}.$$

Thus the functional form of the Wronskian has been determined without any knowledge of two linearly independent solutions of the differential equation. In point of fact these solutions are $\phi_1 = x$ and $\phi_2 = 1/x$, so the precise expression for the Wronskian is

$$W = \begin{vmatrix} x & \dfrac{1}{x} \\ 1 & -\dfrac{1}{x^2} \end{vmatrix} = -\frac{2}{x},$$

showing that $A = -2$.

Taken together, the results of Theorems 5.1(ii) and 5.3 tell us that the homogeneous equation (3) can have only n linearly independent solutions $\phi_1(x), \phi_2(x), \ldots, \phi_n(x)$. Thus all possible solutions of (3) must be capable of representation in the form of a linear combination of these n solutions. For this reason, the linear combination

$$y = c_1\phi_1(x) + c_2\phi_2(x) + \ldots + c_n\phi_n(x), \tag{11}$$

with $c_1, c_2, \ldots, c_n$ arbitrary constants, is called the **general solution** of (3). The set of n functions $\phi_1, \phi_2, \ldots, \phi_n$ forms a **basis** for the solutions of (3), sometimes called a **fundamental system** for (3). We now restate these results formally.

Theorem 5.4 (Structure of the general solution – homogeneous case)

The general linear nth order homogeneous differential equation

$$a_0(x)y^{(n)} + a_1(x)y^{(n-1)} + \ldots + a_n(x)y = 0$$

possesses n linearly independent solutions $\phi_1(x), \phi_2(x), \ldots, \phi_n(x)$ on some interval I, and its general solution on I is of the form

$$y = c_1\phi_1(x) + c_2\phi_2(x) + \ldots + c_n\phi_n(x),$$

with $c_1, c_2, \ldots, c_n$ arbitrary constants.

Proof

Suppose there are $m > n$ solutions $\phi_1, \phi_2, \ldots, \phi_m$ of (3), and let α be any point of I. Then, as in the proof of Theorem 5.3, we know there is always a nontrivial solution

$c_1^*, c_2^*, \ldots, c_n^*$ of the system

$$c_1\phi_1(\alpha)+c_2\phi_2(\alpha)+ \ldots +c_m\phi_m(\alpha)=0$$
$$c_1\phi_1^{(1)}(\alpha)+c_2\phi_2^{(1)}(\alpha)+ \ldots +c_m\phi_m^{(1)}(\alpha)=0$$

$$\begin{array}{ccccccc} . & & . & & . & & . \\ . & & . & . & . & & . \\ . & & . & & . & & . \end{array}$$

$$c_1\phi_1^{(n-1)}(\alpha)+c_2\phi_2^{(n-1)}(\alpha)+ \ldots +c_m\phi_m^{(n-1)}(\alpha)=0.$$

With such a choice of constants c_i^*, form the linear combination

$$\Phi(x)=c_1^*\phi_1(x)+c_2^*\phi_2(x)+ \ldots +c_m^*\phi_m(x).$$

Then $\Phi(x)$ is a solution of (3), and it follows from the above equations that

$$\Phi(\alpha)=\Phi^{(1)}(\alpha)= \ldots =\Phi^{(n-1)}(\alpha)=0.$$

However, the trivial solution $y\equiv0$ is a solution of (3) and, together with all its derivatives, it vanishes at $x=\alpha$. The uniqueness theorem then asserts that $\Phi(x)$ must be this trivial solution, and so $\Phi(x)\equiv0$. This establishes that the m solutions must be linearly dependent when $m>n$.

To complete the proof, observe that any function of the form (11) must satisfy (3), while at $x=\alpha$ it follows from Cramer's rule that unique constants $c_1, c_2, \ldots, c_n$ can be found such that

$$c_1\phi_1(\alpha)+c_2\phi_2(\alpha)+ \ldots +c_n\phi_n(\alpha)=y(\alpha)$$
$$c_1\phi_1^{(1)}(\alpha)+c_2\phi_2^{(1)}(\alpha)+ \ldots +c_n\phi_n^{(1)}(\alpha)=y^{(1)}(\alpha)$$

$$\begin{array}{ccccccc} . & & . & & . & & . \\ . & & . & . & . & & . \end{array}$$

$$c_1\phi_1^{(n-1)}(\alpha)+c_2\phi_2^{(n-1)}(\alpha)+ \ldots +c_n\phi_n^{(n-1)}(\alpha)=y^{(n-1)}(\alpha).$$

This is so because the denominator in Cramer's rule is simply the Wronskian, and as the functions form a basis for the solutions of (3) it must follow that $W(\alpha)\neq0$. The function

$$\Psi(x)=y(x)-c_1\phi_1(x)-c_2\phi_2(x)- \ldots -c_n\phi_n(x)$$

satisfies (3) and the homogeneous initial conditions

$$\Psi(\alpha)=\Psi^{(1)}(\alpha)= \ldots =\Psi^{(n-1)}(\alpha)=0.$$

Thus by the uniqueness theorem Ψ must be the trivial solution, $\Psi(x)\equiv0$, from which it follows that (11) is the general solution of (3). $\square$

Finally, we establish a fundamental relationship between solutions of the non-homogeneous equation (1) and the associated homogeneous equation (3). Any solution $y=\tilde{y}_p(x)$ which satisfies the nonhomogeneous equation (1), and which does not contain arbitrary constants, is said to be a **particular solution** of (1).

If we denote by $y_c(x)$ the linear combination

$$y_c(x)=c_1\phi_1(x)+c_2\phi_2(x)+ \ldots +c_n\phi_n(x), \tag{12}$$

where $\phi_1, \phi_2, \ldots, \phi_n$ are linearly independent solutions of the homogeneous equation (3), it follows at once that

$$y = y_c(x) + \tilde{y}_p(x) \tag{13}$$

must also be a solution. The function $y_c(x)$ in (12) is called the **complementary function** of the nonhomogeneous differential equation (1); it is the general solution of (3).

Expression (13) is the **general solution** of the nonhomogeneous equation (1), and it is a consequence of the uniqueness theorem that all solutions of (1) may be expressed in the form shown in (13). In agreement with Sec. 4.4, if we remove from $\tilde{y}_p(x)$ any terms in the form of a linear combination of $\phi_1, \phi_2, \ldots, \phi_n$ with specific numerical coefficients, and absorb them into $y_c(x)$ which has arbitrary coefficients, there remains a function $y_p(x)$ which we shall call the **particular integral** of (1). This definition differs slightly from the one used in some other accounts of differential equations in so far as there the name particular integral is usually given to $\tilde{y}_p(x)$, rather than to the *unique* function $y_p(x)$ just defined.

Thus *a* particular solution is a general solution of (1) in which the arbitrary constants have been given specific numerical values, while *the* particular integral of (1) is the general solution in which the arbitrary constants have been set equal to zero.

Unlike solutions of (3), the particular integral $y_p(x)$ (and also particular solutions $\tilde{y}_p(x)$) cannot be scaled to yield another particular integral. We have arrived at our last fundamental result in this introductory section.

Theorem 5.5 (Structure of the general solution–nonhomogeneous case)

The general solution of (1) may always be written in the form

$$y = y_c(x) + y_p(x),$$

where the complementary function $y_c(x)$ is the general solution of (3) and $y_p(x)$ is the particular integral of (1). □

Example 3. Constant coefficient equation

The differential equation

$$y''' + 2y'' - y' - 2y = x + 1$$

has the associated homogeneous equation

$$y''' + 2y'' - y' - 2y = 0.$$

The functions $\phi_1(x) = e^{-x}$, $\phi_2(x) = e^x$ and $\phi_3(x) = e^{-2x}$ are three linearly independent solutions of the homogeneous equation. The function $\tilde{y}_p(x) = 2e^{-x} + e^x - 3e^{-2x} - \frac{1}{2}x - \frac{1}{4}$ is a particular solution of the nonhomogeneous equation, but $y_p(x) = -\frac{1}{2}x - \frac{1}{4}$ is the particular integral. The general solution is

$$y = c_1 e^{-x} + c_2 e^x + c_3 e^{-2x} - \frac{1}{2}x - \frac{1}{4}. \qquad ■$$

Example 4. Variable coefficient equation

The differential equation

$$x^2 y'' - 4xy' + 6y = x$$

has the associated homogeneous equation

$$x^2 y'' - 4xy' + 6y = 0.$$

The function $\phi_1(x) = x^3$ is a solution of the homogeneous differential equation, as are $4\phi_1(x), \frac{1}{2}\phi_1(x)$ and $c_1\phi_1(x)$, with c_1 an arbitrary constant. Similarly, $\phi_2(x) = x^2$ is a linearly independent solution of the homogeneous differential equation, as are $-3\phi_2(x)$ and $c_2\phi_2(x)$, with c_2 an arbitrary constant. The particular integral $y_p(x)$ of the nonhomogeneous differential equation is $y_p(x) = \frac{1}{2}x$. The general solution of the equation is

$$y = c_1 x^3 + c_2 x^2 + \tfrac{1}{2}x.$$ ∎

Problems for Section 5.1

In each of the following problems verify that the given functions are solutions of the stated differential equation.

1   $\sin 2x$, $\cos 2x$, $7 \sin 2x$, $3 \sin 2x - 4 \cos 2x$; $y'' + 4y = 0$.
2   $\sinh 3x$, $\cosh 3x$, $5 \cosh 3x$, $6 \sinh 3x - 2 \cosh 3x$; $y'' - 9y = 0$.
3   e^{7x}, e^{-4x}, $2e^{7x} + 3e^{-4x}$; $y'' - 3y' - 28y = 0$.
4   e^{-3x}, $4e^{-3x}$, xe^{-3x}, $(1 + 2x)e^{-3x}$; $y'' + 6y' + 9y = 0$.
5   $e^{-x} \sin 3x$, $e^{-x} \cos 3x$, $2e^{-x} \sin 3x - 4e^{-x} \cos 3x$; $y'' + 2y' + 10y = 0$.
6   $\sin(\ln x)$, $\cos(\ln x)$, $2 \sin(\ln x) + 3 \cos(\ln x)$; $x^2 y'' + xy' + y = 0$.

In each of the following problems verify that $y_p(x)$ is the particular integral of the stated differential equation, and that $y_c(x)$ is a solution of the associated homogeneous differential equation.

7   $y_p(x) = -xe^{2x} \cos x$, $y_c(x) = c_1 e^{2x} \sin x + c_2 e^{2x} \cos x$; $y'' - 4y' + 5y = 2e^{2x} \sin x$.
8   $y_p(x) = -e^{-2x} \cos 2x$, $y_c(x) = c_1 e^{-2x} \sin x + c_2 e^{-2x} \cos x$; $y'' + 4y' + 5y = 3e^{-2x} \cos 2x$.
9   $y_p(x) = -(2x \ln x + 4)$, $y_c(x) = c_1 x + c_2 x^2$; $x^2 y'' - 2xy' + 2y = 2x - 8$.
10   $y_p(x) = \frac{1}{2}(\ln x)^2$, $y_c(x) = (c_1 + c_2 \ln x)x$; $x^2 y'' - xy' + y = x$.

Reduction of order of linear second order differential equations with y absent

In the differential equation

$$a_0(x)\frac{d^2 y}{dx^2} + a_1(x)\frac{dy}{dx} = f(x),$$

the variable y is only present implicitly, as there is no term containing the undifferentiated variable y. The *order* of this differential equation may be *reduced* by means of the substitution $u = dy/dx$, for it can then be brought into the form

$$\frac{du}{dx} + p(x)u = q(x).$$

This is a first order linear differential equation which may be solved by the method of Sec. 4.5. The general solution of the original second order differential equation follows by integration of $dy/dx = u$. One integration constant will enter when determining u and another when determining y from u, so the general solution will contain two arbitrary constants. An **initial value problem** for this second order differential equation will involve specifying y and y' at some given value $x = x_0$, say $y(x_0) = y_0$ and $y'(x_0) = y_1$.

Use this method to solve the following problems, finding either the general solution or, if initial conditions are given, the solution of the initial value problem.

11 $3y'' - y' = 0$.

12 $xy'' - y' = 0$.

13 $(1 + x^2)y'' + xy' = 0$.

14 $(1 + x^2)y'' + 2xy' = 0$.

15 $(1 - x^2)y'' - xy' = 2$.

16 $xy'' + y' = x^2$.

17 $xy'' - y' = 3x^2$.

18 $y'' - y' = e^x$, with $y(0) = 3$ and $y'(0) = 2$.

19 $y'' - y' = \sin x$, with $y(0) = 1$ and $y'(0) = 1$.

20 $y'' - y' = 2 \cosh x$, with $y(0) = 2$ and $y'(0) = 0$.

21 $y'' \sin x + 2y' \cos x = \cos x$, with $y(\pi/4) = 0$ and $y'(\pi/4) = 0$.

22 $xy'' + y' = 8x^3$, with $y(1) = 5/2$ and $y'(1) = 4$.

Reduction of order of nonlinear and higher order differential equations with y absent

Provided the variable y is only present implicitly in a differential equation, it is still possible to reduce its order even though it may be nonlinear, or of order greater than 2. This is accomplished by means of the substitution $u = d^m y/dx^m$, where $d^m y/dx^m$ is the lowest order derivative in the differential equation. Once u has been determined the general solution follows by m successive integrations of $d^m y/dx^m = u$.

Use the method outlined above to find the general solution of each of the following differential equations or, if initial conditions are given, the solution of the initial value problem.

23 $xy'' + (y')^2 = 1$, with $y(2) = -\pi$ and $y'(2) = 0$.

24 $y'' + (y')^2 = 1$.

25 $xy''' - 4y'' = 0$, with $y(-1) = 2$, $y'(-1) = -3$ and $y''(-1) = 10$.

26 $ky'' = y'[1 + (y')^2]$.

27 $y''' - 4y'' = 0$, with $y(1) = 1$, $y'(1) = 0$ and $y''(1) = -1$.

Reduction of order when x is absent

In a differential equation of the form $F(y, y', y'') = 0$ the independent variable x is not present explicitly, though it is present in implicit form through the dependence of y on x. The change of variable $u = dy/dx$, together with the result $y'' = (du/dy)u$ which follows from the chain rule, reduces the differential equation to a first order equation in which y appears as the independent variable.

Use the method outlined above to find the general solution of each of the following differential equations or, if initial conditions are given, the solution of the initial value problem.

28 $yy'' + 2(y')^2 - 2yy' = 0$.

29 $[1 + (y')^2]^3 = a^2(y'')^2$. [*Hint*: in first integration use substitution $u = 1/t$.]

30 $2yy'' + 2(y')^2 - 3yy' = 0$, with $y(0) = 1$ and $y'(0) = 2$.

31 $2yy'' - (y')^2 = 0$, with $y(-6) = 2$, $y'(-6) = 4$.

32 $yy'' + (y')^2 + 1 = 0$.

33 $y'' + \alpha y = 0$, considering the two cases

(i) $\alpha = k^2$ and (ii) $\alpha = -k^2$.

In each of the following problems determine whether the given functions are linearly dependent or independent on the associated interval.

34 x, x^2, x^3, $(1+x)^3$ for $-1 \leq x \leq 1$.

35 1, $\sinh^2 3x$, $\cosh^2 3x$,   for $-\infty < x < \infty$.

36 1, $\sinh^2 x$, $\cos^2 x$,   for $-4 \leq x \leq 4$.

37 x, $|x|$, $x/|x|$ for $-\infty < x < \infty$, with $x/|x|$ defined to be zero when $x = 0$.

38 1, $\ln(1+x)$, $\ln[3(1+x)^2]$ for $x > 0$.

Harder problems

39 (**Uniqueness of solution of a general second order equation**) Starting from the equation

$$y'' + a_1(x)y' + a_2(x)y = f(x),$$

subject to the initial conditions $y(x_0) = y_0$ and $y'(x_0) = y_1$, show that if two different solutions y_1 and y_2 exist satisfying the same initial condition and $u = y_1 - y_2$, then

$$u'' + a_1(x)u' + a_2(x)u = 0, \text{ with } u(x_0) = u'(x_0) = 0.$$

Define $w = u^2 + (u')^2$, and show that

$$w' = 2[1 - a_2(x)]uu'' - 2a_1(x)(u_1')^2.$$

By substituting for u'' from the differential equation, and making use of the fact that since

$$(u \pm u')^2 \geq 0 \text{ it follows that } 2|uu'| \leq u^2 + (u')^2,$$

show that

$$w' \leq [1 + 2|a_1(x)| + |a_2(x)|]w.$$

Letting $M = \max[1 + 2|a_1(x)| + |a_2(x)|]$ on an interval $x_0 \leq x \leq b$ on which the solution is defined, integrate this last inequality to show

$$w(x) \leq w(x_0) \exp[M(x - x_0)].$$

Use the initial condition on $w(x)$ together with the essentially nonnegative nature of w to conclude that $w(x) \equiv 0$, and hence that $y_1(x) \equiv y_2(x)$ for $x_0 \leq x \leq b$.

40 Let ϕ_1 be a known solution of

$$a_0(x)y'' + a_1(x)y' + a_2(x) = 0$$

in an interval I on which $a_0(x) \neq 0$. By writing Abel's formula in the form

$$\phi_1 \phi_2' - \phi_1' \phi_2 = A \exp\left[-\int \frac{a_1(x)}{a_0(x)} dx\right],$$

and solving this nonhomogeneous linear first order equation for ϕ_2, show that a basis for the solutions of the homogeneous second order equation is provided by the functions

$$\phi_1(x) \quad \text{and} \quad \phi_2(x) = \phi_1(x) \int \frac{\exp\left(-\int \frac{a_1(x)}{a_0(x)} dx\right)}{[\phi_1(x)]^2} dx.$$

41 Use the method of Prob. 40 to find the general solution of each of the following differential equations.

(a) $x^2 y'' + 3xy' + y = 0$, given $1/x$ is a solution;
(b) $x^2 y'' - 2xy' + 2y = 0$, given x is a solution;
(c) $(1+x)^2 y'' - 3(1+x)y' + 4y = 0$, given $(1+x)^2$ is a solution.

5.2 Second order constant coefficient equations – homogeneous case

The most general homogeneous linear second order constant coefficient differential equation may always be written in the standard form

$$y'' + ay' + by = 0, \tag{1}$$

where a and b are arbitrary real constants. To find its general solution, we start from the observation that (7), Sec. 4.5, shows the general solution of a homogeneous first order constant coefficient equation to be of the form $y = ce^{\lambda x}$, with c an arbitrary constant, and λ a constant depending on the differential equation.

Let us now extend this idea and try to find two linearly independent solutions of (1) of this same form. It follows at once from Theorem 5.4 that if such solutions can be found, the general solution of (1) must be of the form

$$y = c_1 \exp(\lambda_1 x) + c_2 \exp(\lambda_2 x). \tag{2}$$

The initial conditions for (1) are the specification of $y(x_0) = y_0$ and $y'(x_0) = y_1$, which will then determine the values of the arbitrary constants c_1, c_2, and so make solution (2) unique.

To justify this conjecture about the form of the solution, and to determine λ_1, λ_2, substitute

$$y = ce^{\lambda x} \tag{3}$$

into (1) to obtain

$$(\lambda^2 + a\lambda + b)ce^{\lambda x} = 0, \tag{4}$$

which must be true for all x. Thus either $c = 0$, corresponding to the **trivial solution** $y \equiv 0$, or

$$\lambda^2 + a\lambda + b = 0. \tag{5}$$

This equation, called the **characteristic equation** (sometimes the **auxiliary equation**) of (1), determines the permissible values λ_1, λ_2 of λ in (3) in order that there should be nontrivial solutions. The nature of the roots of (5) will depend on the discriminant $a^2 - 4b$ of the quadratic equation. There are three cases:

(I) the roots of (5) will be real and distinct if $a^2 - 4b > 0$; when

$$\lambda_1 = \tfrac{1}{2}(-a + \sqrt{a^2 - 4b}), \quad \lambda_2 = \tfrac{1}{2}(-a - \sqrt{a^2 - ab}),$$

(II) the roots of (5) will be complex conjugates if $a^2 - 4b < 0$; when

$$\lambda_1 = \tfrac{1}{2}(-a + i\sqrt{4b - a^2}), \quad \lambda_2 = \tfrac{1}{2}(-a - i\sqrt{4b - a^2}), \quad \text{and}$$

(III) the roots of (5) will be real and equal (double root) if $a^2 - 4b = 0$; when

$$\lambda_1 = \lambda_2 = -\tfrac{1}{2}a.$$

In case (I), the two functions $\exp(\lambda_1 x)$ and $\exp(\lambda_2 x)$ represent two linearly independent solutions of (1), because as $\lambda_1 \neq \lambda_2$, $\exp(\lambda_1 x)/\exp(\lambda_2 x) = \exp[(\lambda_1 - \lambda_2)x]$ is a function of x (see remark following (5) in Sec. 5.1). Thus in this case the general solution is

$$y = c_1 \exp(\lambda_1 x) + c_2 \exp(\lambda_2 x), \tag{6}$$

with c_1, c_2 two arbitrary real constants.

In case (II) a similar argument shows that

$$\exp[\tfrac{1}{2}(-a + i\sqrt{4b - a^2})x] \quad \text{and} \quad \exp[\tfrac{1}{2}(-a - i\sqrt{4b - a^2})x]$$

represent two linearly independent solutions of (1), and so the general solution must be an arbitrary linear combination of these functions. Setting $p = -\tfrac{1}{2}a$, $q = \tfrac{1}{2}\sqrt{4b - a^2}$, the general solution becomes

$$y = A \exp[(p + iq)x] + B \exp[(p - iq)x], \tag{7}$$

where now the arbitrary constants A and B must be complex. However, as the solution y of (1) is required to be real, it follows that the two terms in (7) must be complex conjugates. This is only possible if $A = \bar{B}$, so setting $B = \tfrac{1}{2}(c_1 + ic_2)$ with c_1, c_2 arbitrary real constants, (7) takes the form

$$y = \exp(px)(c_1 \cos qx + c_2 \sin qx). \tag{8}$$

Finally, in case (III) there is only the single function $\exp(\lambda_1 x)$ corresponding to the double root $\lambda = \lambda_1$. To find a second solution notice that in this case $x \exp(\lambda_1 x)$ is also a solution[3] of (1). As $\exp(\lambda_1 x)$ and $x \exp(\lambda_1 x)$ are linearly independent (their ratio is a function of x), it follows that in this last case the general solution has the form

$$y = (c_1 + xc_2)\exp(\lambda_1 x). \tag{9}$$

We have established the following important result.

---

[3] When the method of **reduction of order** is discussed later, an explicit argument will be developed showing how this other solution may be found from $\exp(\lambda_1 x)$.

Theorem 5.6 (Solution of homogeneous equation)

The form of the general solution of the homogeneous linear second order constant coefficient differential equation

$$y'' + ay' + by = 0, \tag{10}$$

in which a, b are real constants, depends on the discriminant $a^2 - 4b$. If c_1, c_2 are arbitrary real constants, then:

(i) if $a^2 - 4b > 0$, the general solution is

$$y = c_1 \exp(\lambda_1 x) + c_2 \exp(\lambda_2 x), \tag{11}$$

with $\lambda_1 = \frac{1}{2}(-a + \sqrt{a^2 - 4b})$, $\lambda_2 = \frac{1}{2}(-a - \sqrt{a^2 - 4b})$;

(ii) if $a^2 - 4b < 0$, the general solution is

$$y = \exp(px)(c_1 \cos qx + c_2 \sin qx), \tag{12}$$

with $p = -\frac{1}{2}a$, $q = \frac{1}{2}\sqrt{4b - a^2}$; and,

(iii) if $a^2 - 4b = 0$, the general solution is

$$y = (c_1 + c_2 x)\exp(\lambda_1 x), \tag{13}$$

with $\lambda_1 = -\frac{1}{2}a$. ☐

In terms of the notion of an abstract vector space introduced in Sec. 2.9, which occurred again in Chapter 3, we see that in each case the solutions of (10) form a **two-dimensional vector space** over $\mathbb{R}$. The two linearly independent functions which are combined with constants c_1, c_2 to form $y(x)$ thus form a **basis** for the solutions of (10) in each of the three cases.

The functions forming a basis in each case are best summarized as in Table 5.1.

Table 5.1   Functions forming a basis for the differential equation $y'' + ay' + by = 0$

Discriminant	Roots	Basis functions	Interval
(i) $a^2 - 4b > 0$	Real and distinct $\lambda_1 \neq \lambda_2$	$\exp(\lambda_1 x)$, $\exp(\lambda_2 x)$	all x
(ii) $a^2 - 4b < 0$	Complex conjugates $\lambda_1 = p + iq, \lambda_2 = p - iq$	$e^{px}\cos qx$, $e^{px}\sin qx$	all x
(iii) $a^2 - 4b = 0$	Real and double λ_1 (twice)	$\exp(\lambda_1 x)$, $x\exp(\lambda_1 x)$	all x

Example 1

Solve the initial value problem

$$y'' + 4y' = 0,$$

with $y(1) = 1$ and $y'(1) = 3$.

Solution

In the notation of (10), $a = 4$, $b = 0$ and so $a^2 - 4b = 16$. This is case (i) of Theorem 5.6 corresponding to

$$\lambda_1 = \tfrac{1}{2}(-a + \sqrt{a^2 - 4b}) = 0, \qquad \lambda_2 = \tfrac{1}{2}(-a - \sqrt{a^2 - 4b}) = -4.$$

The functions forming a basis for the solutions of this differential equation are thus $e^{0x} \equiv 1$ and e^{-4x}, so the general solution is

$$y = c_1 + c_2 e^{-4x}.$$

Using the first initial condition $y(1) = 1$ we find

$$1 = c_1 + e^{-4}c_2.$$

Differentiating the general solution gives

$$y' = -4c_2 e^{-4x},$$

which taken with the second initial condition $y'(1) = 3$ gives

$$3 = -4e^{-4}c_2.$$

Thus $c_2 = -(3/4)e^4$ and $c_1 = 7/4$, and so the solution of the initial value problem is

$$y = \tfrac{1}{4}(7 - 3e^{4(1-x)}). \qquad \blacksquare$$

Example 2

Determine the general solution of

$$y'' + 2y' + 7y = 0,$$

and hence find the solution of the initial value problem for this equation in which $y(0) = 0$ and $y'(0) = 1$.

Solution

In the notation of (10), $a = 2$, $b = 7$ and so $a^2 - 4b = -24$. This is case (ii) of Theorem 5.6 corresponding to $p = -\tfrac{1}{2}a = -1$, $q = \tfrac{1}{2}\sqrt{4b - a^2} = \sqrt{6}$. The functions forming a basis for solutions of this differential equation are thus $e^{-x}\cos\sqrt{6}x$ and $e^{-x}\sin\sqrt{6}x$, so the general solution is

$$y = e^{-x}(c_1 \cos\sqrt{6}x + c_2 \sin\sqrt{6}x).$$

To find the solution of the initial value problem, we first use the condition $y(0) = 0$ in the general solution. This yields

$$0 = c_1,$$

which reduces the solution to

$$y = c_2 e^{-x}\sin\sqrt{6}x.$$

Differentiation of this solution gives

$$y' = -c_2 e^{-x} \sin\sqrt{6}x + c_2\sqrt{6}e^{-x}\cos\sqrt{6}x,$$

which used with the second condition $y'(0) = 1$ shows

$$1 = c_2\sqrt{6} \quad \text{or} \quad c_2 = 1/\sqrt{6}.$$

The solution of the initial value problem is thus

$$y = \frac{1}{\sqrt{6}}e^{-x}\sin\sqrt{6}x.$$

∎

Example 3

Find the homogeneous differential equation which has as a basis for its solutions the functions e^{-2x} and xe^{-2x}

Solution

This is case (iii) of Theorem 5.6 in which $\lambda_1 = -2$. Now $\lambda_1 = -\frac{1}{2}a$, so we see $a = 4$, and because $a^2 = 4b$ it then follows that $b = 4$. Thus the equation must be

$$y'' + 4y' + 4y = 0.$$

∎

When working with higher order differential equations another type of problem often arises called a **two-point boundary value problem (boundary value problem** for short). In such problems, which are of considerable importance in engineering and science, a solution of a differential equation is sought in which $y(x)$ (or $y'(x)$) assumes prescribed values at two different points, say at $x = a$ and at $x = b$ $(a < b)$. Typically the solution $y(x)$ might be required to satisfy the conditions $y(a) = y_0$ and $y(b) = y_1$, called the **boundary conditions** (or **boundary data**) for the differential equation, with y_0 and y_1 arbitrary. The resulting solution, when it exists, is then defined for x in the interval $a \le x \le b$. In the physical problems giving rise to such conditions the points $x = a$ and $x = b$ usually correspond to spatial boundaries (hence the name boundary value problem).

Boundary value problems are more complicated than initial value problems, because their solutions are not always uniquely determined by the boundary data and, indeed, sometimes do not even exist. These features are illustrated in the following example.

Example 4. Boundary value problems

Solve the differential equation

$$y'' + y = 0$$

subject to the boundary conditions (i) $y(0) = 0$, $y(\pi/2) = 3$, (ii) $y(0) = 0$, $y(\pi) = 0$ and (iii) $y(0) = 1$, $y(\pi) = 1$.

Solution

In the notation of (10), $a = 0$, $b = 1$ and so $a^2 - 4b = -4$. This is case (ii) of Theorem 5.6 corresponding to $p = 0$, $q = 1$. The functions forming a basis for the solutions of this differential equation are thus $\cos x$ and $\sin x$, so the general solution is

$$y = c_1 \cos x + c_2 \sin x.$$

Case (i)

The first boundary condition $y(0) = 0$ shows $c_1 = 0$, which reduces the general solution to

$$y = c_2 \sin x.$$

The second boundary condition $y(\pi/2) = 3$ then shows $c_2 = 3$; so the solution of the boundary value problem in case (i) exists and is unique, and is given by

$$y = 3 \sin x \qquad \text{for } 0 \le x \le \pi/2.$$

Case (ii)

The first boundary condition $y(0) = 0$ again shows $c_1 = 0$, thereby reducing the general solution to

$$y = c_2 \sin x.$$

However, in this case the second boundary condition $y(\pi) = 0$ is automatically satisfied for any value of c_2. Thus in this case the solution of the boundary value problem exists but is not unique, for it is

$$y = c_2 \sin x \qquad \text{for } 0 \le x \le \pi,$$

with c_2 arbitrary. This family of solutions is seen to contain the trivial solution $y \equiv 0$, corresponding to $c_2 = 0$.

Case (iii)

The first boundary condition $y(0) = 1$ shows $c_1 = 1$, which reduces the general solution to

$$y = \cos x + c_2 \sin x.$$

The second boundary condition $y(\pi) = 1$ is incompatible with this solution for it leads to the contradiction '$1 = -1$'. Thus there is no solution to this last boundary value problem. ∎

The next example shows how the determination of the load at which the theory of linear elasticity predicts a simple strut will *buckle* is determined by the solution of a boundary value problem. This is known as the **Euler strut problem**, after Leonhard Euler who was the first person to study this phenomenon.

Example 5. The Euler strut

It has been found by experiment that when a strut of uniform construction is subjected to a compressive load F it exhibits no transverse displacement until F exceeds some critical value F_1.

Once this load has been exceeded buckling occurs and large deflections are produced as a result of small changes in the load. Thereafter, prior to the collapse of the strut due to excessive buckling, the shape of the strut undergoes rapid and irregular change.

This phenomenon is illustrated in Fig. 5.1 for a rod of length a, where it is assumed that $F > F_1$ (the effect of gravity is ignored). Using the linear theory of elasticity it is possible to show that the transverse displacement y as a function of the length x along the strut satisfies the differential equation

$$\frac{d^2 y}{dx^2} = -\frac{Fy}{EI}.$$

In this equation E is Young's modulus of elasticity for the material of the strut, and I is the moment of inertia of the cross-section of the strut about its axis.

Fig. 5.1   Buckling of an Euler strut

The boundary conditions to be imposed at the ends of the strut are that the transverse displacement $y=0$, so we must have

$$y(0)=0 \quad \text{and} \quad y(a)=0.$$

Setting $k^2 = F/EI$ enables the general solution of the differential equation to be written as

$$y = c_1 \cos kx + c_2 \sin kx.$$

The first boundary condition $y(0) = 0$ shows $c_1 = 0$, while the second then shows we must have

$$c_2 \sin ka = 0.$$

This condition is satisfied if $c_2 = 0$, or if $ka = n\pi$, with $n = 0, 1, 2, \ldots$. The condition $c_2 = 0$ (an $n = 0$) corresponds to $y \equiv 0$, in which case the strut is unbuckled. The first nontrivial solution occurs when $n = 1$, so that $k = \pi/a$ and the deflection is

$$y = c_2 \sin \frac{\pi}{a},$$

with c_2 arbitrary.

Remembering $F = EIk^2$, it follows that $n = 1$ corresponds to the critical compressive load

$$F_1 = EI \frac{\pi^2}{a^2}.$$

This corresponds to the **buckling load** for the strut, because the cases $n > 1$ simply give loads in excess of F_1. The indeterminacy of the amplitude c_2 of the maximum displacement when buckled indicates a failure of the linear theory of elasticity to model the situation.   ■

On occasions a boundary value problem is associated with a semi-infinite interval, so that one condition is specified at a finite point and the other at infinity. In such cases the **boundary condition at infinity** often takes the form of a statement to the effect that the solution is *finite* at infinity. This means that if the general solution of a differential equation is

$$y = c_1 \phi_1(x) + c_2 \phi_2(x),$$

where, say, $\phi_1(\infty)$ is infinite and $\phi_2(\infty)$ is finite, then we must set $c_1 = 0$ in order to obtain the correct form of the solution.

Example 6. Boundary condition at infinity

Solve the boundary value problem

$$y'' + y' - 6y = 0,$$

with $y(0) = 4$ and $y(\infty) = 0$.

Solution

In the notation of (10), $a = 1, b = -6$, so $a^2 - 4b = 25$. This is case (i) of Theorem 5.6 corresponding to

$$\lambda_1 = \tfrac{1}{2}(-a + \sqrt{a^2 - 4b}) = 2, \quad \lambda_2 = \tfrac{1}{2}(-a - \sqrt{a^2 - 4b}) = -3.$$

The functions forming a basis for the solutions of this differential equation are thus e^{2x} and e^{-3x}; so the general solution is

$$y = c_1 e^{2x} + c_2 e^{-3x}.$$

Now the boundary condition $y(0) = 4$ shows

$$4 = c_1 + c_2.$$

However the boundary condition at infinity is $y(\infty) = 0$, and this can only be satisfied if we set $c_1 = 0$ in order to remove the term e^{2x} which becomes infinite at infinity. Thus it follows that $c_2 = 4$, and so the solution of the boundary value problem on this semi-infinite interval is

$$y = 4e^{-3x}. \qquad \blacksquare$$

When working with solutions of variable coefficient linear second order equations, it can also happen that the linearly independent solutions ϕ_1 and ϕ_2 in the general solution

$$y = c_1 \phi_1(x) + c_2 \phi_2(x)$$

have the property that one of them, say ϕ_2, becomes infinite at a finite point $x = a$. Should this occur, and a boundary value problem arises on the interval $[a, b]$, then the boundary condition most frequently imposed at $x = a$ is the requirement that the solution remains finite (bounded) at $x = a$. Thus if $\phi_2(a)$ is infinite, it is necessary to set $c_2 = 0$ for the solution y to remain bounded on $[a, b]$.

Example 7. Boundedness condition at a finite point

The variable coefficient equation

$$x^2 y'' + xy' - y = 0$$

has the two linearly independent solutions $\phi_1 = x$ and $\phi_2 = 1/x$, and thus the general solution

$$y = c_1 x + \frac{c_2}{x}.$$

If a boundary value problem on $[0, 1]$ is to be considered, then although the boundary condition imposed at $x = 1$ may be arbitrary (the specification of $y(1)$, $y'(1)$ or even of $\alpha y(1) + \beta y'(1)$), for the solution to remain bounded it is necessary to set $c_2 = 0$. Thus the solution to be used to match the boundary condition at $x = 1$ must be $y = c_1 x$. ■

Problems for Section 5.2

Find the general solution for each of the following differential equations.

1 $y'' - 3y' + 2y = 0.$
2 $y'' - y' - 6y = 0.$
3 $y'' + 4y' + 13y = 0.$
4 $y'' + 6y' + 9y = 0.$
5 $y'' - 4y' + y = 0.$
6 $2y'' - y' + y = 0.$
7 $y'' - 3y' = 0.$
8 $y'' - 4y' + 4y = 0.$
9 $y'' + ky = 0.$
10 $2y'' + 5y' - 3y = 0.$

Solve the following initial value problems.

11 $y'' - 4y' + 3y = 0,$ with $y(0) = 1,$ $y'(0) = 2.$
12 $y'' + 3y' + 2y = 0,$ with $y(0) = 1,$ $y'(0) = -1.$
13 $y'' + 3y' = 0,$ with $y(1) = 0,$ $y'(1) = -1.$
14 $y'' + 2y' = 0,$ with $y(0) = 1,$ $y'(0) = 0.$
15 $y'' - 4y = 0,$ with $y(2) = 0,$ $y'(2) = 1.$
16 $y'' + 16y = 0,$ with $y(\pi) = -3,$ $y'(\pi) = 0.$
17 $y'' + 2y' + y = 0,$ with $y(0) = 0,$ $y'(0) = 2.$
18 $y'' + 2y' + 5y = 0,$ with $y(\pi/2) = 0, y'(\pi/2) = 1.$

In each of the following problems find the homogeneous differential equation which has as a basis for its solutions the two given functions.

19 $e^x, e^{-4x}.$
20 $1, e^{2x}.$
21 $\sin 4x, \cos 4x.$
22 $e^{-x} \cos \sqrt{2} x, e^{-x} \sin \sqrt{2} x.$

23 e^{3x}, e^{-3x}.

24 $\sinh 3x, \cosh 3x$.

25 $\exp(3x/2)\cos\dfrac{\sqrt{11}}{2}x, \exp(3x/2)\sin\dfrac{\sqrt{11}}{2}x$.

Solve the following boundary value problems.

26 $y'' + 3y' = 0$,   with $y(0) = 0$, $y(1) = 1$.

27 $y'' + \pi^2 y = 0$,   with $y(0) = 0$, $y(1) = 0$.

28 $y'' - 2y' + 2y = 0$,   with $y(0) = 0$, $y\left(\dfrac{\pi}{2}\right) = 1$.

29 $y'' - 4y' + 2y = 0$,   with $y(0) = 0$, $y(1) = 1$.

30 $y'' - 4y' + 2y = 0$,   with $y(0) = 0$, $y'(1) = 1$.

31 $y'' + 4y = 0$,   with $y'(0) = 0$, $y\left(\dfrac{\pi}{2}\right) = 1$.

In each of the following problems, find the value of k for which the stated boundary value problem has a solution, and give the form of the solution.

32 $y'' + k^2 y = 0$,   with $y'(0) = 0$, $y(a) = 0$.

33 $y'' + k^2 y = 0$,   with $y'(0) = 0$, $y'(a) = 0$.

34 $y'' + k^2 y = 0$,   with $y(0) = 0$, $y(a) = 0$.

35 $y'' + k^2 y = 0$,   with $y(0) = 0$, $y'(a) = 0$.

36 Show that the boundary value problem

$$y'' + k^2 y = 0, \quad \text{with} \quad y(0) + \pi y'(0) = 0 \quad \text{and} \quad y(\pi) = 0,$$

will only have a solution if k is a solution of the equation $\tan k\pi = k\pi$. Determine the approximate values of the first two positive numbers k, by sketching the graphs of $y = \tan \pi x$ and $y = \pi x$, and finding where they intersect. Is the number of values of k satisfying this equation finite or infinite?

Solve the following boundary value problems on semi-infinite intervals.

37 $y'' - 2y' - 2y = 0$,   with $y(1) = 2e^3$ and $y(-\infty) = 0$.

38 $y'' - 4y = 0$,   with $y(0) = 6$, $y(\infty) = 0$.

39 $y'' - 2y' - 8y = 0$,   with $y(-\tfrac{1}{2}) = 3e$, $y(\infty)$ finite.

40 $y'' + 2y' - 3y = 0$,   with $y(-1) = 1$, $y(-\infty)$ finite.

5.3 Higher order constant coefficient equations – homogeneous case

The method of Sec. 5.2 extends at once to homogeneous linear higher order constant coefficient equations which may always be written in the form

$$y^{(n)} + a_1 y^{(n-1)} + \ldots + a_n y = 0, \tag{1}$$

where $a_1, a_2, \ldots, a_n$ are real numbers. Setting

$$y = c\,e^{\lambda x}, \tag{2}$$

a similar argument to that of Sec. 5.2 shows the characteristic equation of (1) is

$$\lambda^n + a_1 \lambda^{n-1} + \ldots + a_n = 0. \tag{3}$$

This nth degree polynomial will have n roots $\lambda_1, \lambda_2, \ldots, \lambda_n$. It follows from Theorem 1.4 that as the coefficients of (3) are real, the roots will either be real or, if complex, will occur in complex conjugate pairs.

Before analyzing the nature of the possible forms of the general solution of (1), let us first agree to call a root of polynomial (3) which occurs only once a **simple root**, and one which is repeated r times a root with **multiplicity** r (**r-fold** root). Thus a simple root has multiplicity one. The roots of the characteristic equation (3) will then satisfy one or more of the following criteria:

(i) the roots will all be real and distinct (all have multiplicity one);
(ii) one or more real roots may have multiplicity greater than one;
(iii) one or more pairs of complex conjugate roots may have multiplicity greater than one.

A direct extension of the arguments used to arrive at the general solution of the second order equation in Sec. 5.3 brings us to the following theorem which tells us how to obtain the general solution of a homogeneous linear nth order constant coefficient differential equation.

Theorem 5.7 (The structure of the general solution)

Let y be the general solution of the homogeneous linear nth order differential equation (1). Then,

(i) corresponding to each real and distinct root $\lambda = \lambda_m$ of (3), the general solution y of (1) will contain a term

$$c_m \exp(\lambda_m x),$$

with c_m an arbitrary real constant;
(ii) corresponding to each real root $\lambda = \lambda_m$ of (3) with multiplicity r, the general solution y of (1) will contain the r terms

$$c_1 \exp(\lambda_m x) + c_2 x \exp(\lambda_m x) + c_3 x^2 \exp(\lambda_m x) + \ldots + c_r x^{r-1} \exp(\lambda_m x),$$

with $c_1, c_2, \ldots, c_r$ arbitrary real constants, and;
(iii) corresponding of each pair of complex conjugate roots $\lambda = p_m + iq_m$ and $\lambda = p_m - iq_m$ of (3), each with multiplicity s, the general solution y of (1) will contain the $2s$ terms

$$\exp(p_m x)(a_1 \cos q_m x + b_1 \sin q_m x) + x \exp(p_m x)(a_2 \cos q_m x + b_2 \sin q_m x) + \ldots$$
$$+ x^{s-1} \exp(p_m x)(a_s \cos q_m x + b_s \sin q_m x),$$

with $a_1, a_2, \ldots, a_s, b_1, b_2, \ldots, b_s$ arbitrary real constants.

□

The reader is left the task of verifying by differentiation that each of the terms in (ii) and (iii) is, in fact, a linearly independent solution of (1).

To demonstrate one form of argument which may be used for this purpose we now prove that if the root $\lambda = \mu$ (real or complex) has multiplicity 2 (is a double root), then $e^{\mu x}$ and $xe^{\mu x}$ are both solutions of (1). It is obvious that $e^{\mu x}$ is a solution, so it is only necessary to prove that $xe^{\mu x}$ is a solution.

Factor out the double root from the characteristic polynomial, and write the polynomial in the form

$$\lambda^n + a_1\lambda^{n-1} + \ldots + a_{n-1}\lambda + a_n \equiv (\lambda - \mu)^2 P(\lambda),$$

where $P(\lambda)$ is a polynomial of degree $n-2$. Differentiating this expression with respect to λ and setting $\lambda = \mu$ in the result causes the right-hand side to vanish, and we obtain the result

$$n\mu^{n-1} + (n-1)a_1\mu^{n-2} + \ldots + a_{n-1} = 0. \tag{4}$$

Now if $y = xe^{\mu x}$, then $y^{(r)} = r\mu^{r-1}e^{\mu x} + \mu^r xe^{\mu x}$. Substituting this into the left-hand side of (1) and using the fact that $\lambda = \mu$ is a root of the characteristic equation reduces the result to the expression

$$(n\mu^{n-1} + (n-1)a_1\mu^{n-1} + \ldots + a_{n-1})e^{\mu x}.$$

However this expression is zero because of (4), so we have proved $xe^{\mu x}$ is a solution. The form of argument extends in an obvious manner to roots with higher multiplicity.

Let us use Theorem 5.3 to prove the exponential solutions in Theorem 5.7 form a basis for the solutions of the homogeneous equation (1). Setting $\phi_r(x) = \exp(\lambda_r x)$, the Wronskian becomes

$$W = \begin{vmatrix} \exp(\lambda_1 x) & \exp(\lambda_2 x) & \ldots & \exp(\lambda_n x) \\ \lambda_1\exp(\lambda_1 x) & \lambda_2\exp(\lambda_2 x) & \ldots & \lambda_n\exp(\lambda_n x) \\ . & . & \ldots & . \\ \lambda_1^{n-1}\exp(\lambda_1 x) & \lambda_2^{n-1}\exp(\lambda_2 x) & \ldots & \lambda_n^{n-1}\exp(\lambda_n x) \end{vmatrix},$$

so after removing the n factors $\exp(\lambda_r x)$, with $r = 1, 2, \ldots, n$, we have

$$W = \exp[(\lambda_1 + \lambda_2 + \ldots + \lambda_n)x] \begin{vmatrix} 1 & 1 & \ldots & 1 \\ \lambda_1 & \lambda_2 & \ldots & \lambda_n \\ . & . & \ldots & . \\ \lambda_1^{n-1} & \lambda_2^{n-1} & \ldots & \lambda_n^{n-1} \end{vmatrix}.$$

The exponential factor in W can never be zero, and the determinant is an alternant (Sec. 3.7, Ex. 2(iii) and Prob. 26) and so, apart from its sign, is equal to the product of all possible factors of the form $(\lambda_i - \lambda_j)$ with $i \neq j$. As the roots are distinct none of these factors can vanish, so $W \neq 0$. This proves the linear independence of the n different exponential solutions $\exp(\lambda_r x)$ in case (i), for all x.

The linear independence of solutions like

$$\exp(\lambda_m x), \ x\exp(\lambda_m x), \ \ldots, \ x^{r-1}\exp(\lambda_m x),$$

for all x, where λ_m is a root (real or complex) with multiplicity r, will be proved if we can show the r functions $1, x, x^2, \ldots, x^{r-1}$ are linearly independent for all x. Defining $\phi_s(x) = x^s$, with $s = 0, 1, \ldots, r-1$, the Wronskian becomes

$$W(1, x, x^2, \ldots, x^{r-1}) = \begin{vmatrix} 1 & x & x^2 & \ldots & x^{r-1} \\ 0 & 1 & 2x & \ldots & (r-1)x^{r-2} \\ 0 & 0 & 2! & \ldots & (r-1)(r-2)x^{r-3} \\ . & . & . & \ldots & . \\ 0 & 0 & 0 & \ldots & (r-1)! \end{vmatrix}.$$

This is an upper triangular determinant (Sec. 3.7, Prob. 15), so its value is the product of the elements on the leading diagonal, none of which is zero. Thus $W \neq 0$, and so the functions are linearly independent for all x. This completes the proof of Theorem 5.7(ii), and 5.7 (iii) follows in similar fashion.

It follows at once from Theorem 5.7 that the solutions of (1) form an n-dimensional vector space over $\mathbb{R}$. The n functions corresponding to the roots of the characteristic equation (multiplicities counted) form a basis for this space.

Example 1

Find the general solution of

$$y^{(4)} - y = 0,$$

and hence determine a basis for its solutions.

Solution

The characteristic equation is

$$\lambda^4 - 1 = 0,$$

which has the roots $\lambda_1 = 1$, $\lambda_2 = -1$, $\lambda_3 = i$ and $\lambda_4 = -i$. These are all simple roots, so it follows from Theorem 5.7 (i) and (iii) that the general solution is

$$y = c_1 e^x + c_2 e^{-x} + c_3 \cos x + c_4 \sin x,$$

for all x. A basis for the solutions for all x is provided by the set of functions e^x, e^{-x}, $\cos x$ and $\sin x$.

∎

Example 2

Find the general solution of

$$y^{(3)} - 3ay^{(2)} + 3a^2 y^{(1)} - a^3 y = 0,$$

and hence determine a basis for its solutions.

Solution

The characteristic equation is

$$\lambda^3 - 3a\lambda^2 + 3a^2\lambda - a^3 = 0,$$

which has the triple root (multiplicity 3) $\lambda = a$. It then follows from Theorem 5.7 (ii) that the general solution is

$$y = (c_1 + c_2 x + c_3 x^2)e^{ax},$$

for all x. A basis for the solutions for all x is provided by the set of functions e^{ax}, xe^{ax} and $x^2 e^{ax}$.

■

Example 3

Find the general solution of

$$y^{(4)} - 8y^{(3)} + 42y^{(2)} - 104y^{(1)} + 169y = 0,$$

and hence determine a basis for its solutions.

Solution

The characteristic equation is

$$\lambda^4 - 8\lambda^3 + 42\lambda^2 - 104\lambda + 169 = 0,$$

which can be factored to give

$$(\lambda^2 - 4\lambda + 13)^2 = 0.$$

The roots of $\lambda^2 - 4\lambda + 13 = 0$ are $\lambda = 2 \pm 3i$, so the characteristic equation has the four roots $2 + 3i$, $2 + 3i$, $2 - 3i$ and $2 - 3i$. There is thus a pair of complex conjugate roots with multiplicity 2. It then follows from Theorem 5.7 (iii) that the general solution is

$$y = e^{2x}(a_1 \cos 3x + b_1 \sin 3x) + xe^{2x}(a_2 \cos 3x + b_2 \sin 2x),$$

for all x. A basis for the solutions for all x is provided by the set of functions $e^{2x}\cos 3x$, $e^{2x}\sin 3x$, $xe^{2x}\cos 3x$ and $xe^{2x}\sin 3x$.

■

Example 4

Determine the homogeneous differential equation which has as a basis for its solutions defined on $\mathbb{R}$ the functions e^x, $e^{2x}\cos 2x$ and $e^{2x}\sin 2x$.

Solution

Inspection of the form of the functions comprising a basis for solutions of the differential equation shows the roots of its characteristic equation are $\lambda = 1$, $\lambda = 2 + 2i$ and $\lambda = 2 - 2i$. The characteristic equation itself is thus

$$(\lambda - 1)(\lambda - 2 - 2i)(\lambda - 2 + 2i) = 0,$$

or

$$\lambda^3 - 5\lambda^2 + 12\lambda - 8 = 0.$$

This corresponds to the homogeneous differential equation

$$y^{(3)} - 5y^{(2)} + 12y^{(1)} - 8y = 0.$$ ∎

Problems for Section 5.3

In each of the following problems determine whether the given set of functions is linearly dependent or linearly independent on the stated interval.

1 $1, x, x^3$,   for $0 \leq x \leq 1$.

2 $x, \sin x, \cos x$,   for all x.

3 $1, \sin^2 x, \cos 2x$,   for $0 \leq x \leq 2\pi$.

4 $1, e^x, e^{2x}$,   for all x.

5 $e^x, \cosh x, \sinh x$,   for $x \geq 0$.

6 $e^x, e^{2x}, \sinh 2x, \cosh 2x$,   for all x.

In each of the following problems, find the general solution of the differential equation, and hence determine a basis for its solutions.

7 $y'' - 3y' + 2y = 0.$

8 $y'' - y' - 6y = 0.$

9 $y''' - 3y' + 2y = 0.$

10 $y''' + 5y'' + 7y' + 3y = 0.$

11 $y'' - 6y' + 18y = 0.$

12 $y'' + 4y' + 5y = 0.$

13 $y''' - 6y'' + 12y' - 8y = 0.$

14 $y^{(4)} - 16y = 0.$

15 $y^{(4)} + 18y^{(2)} + 81y - 0.$

16 $y^{(4)} + y = 0.$

17 $y^{(4)} - 3y^{(3)} + 3y^{(2)} - y^{(1)} = 0.$

18 $y^{(4)} - 18y^{(2)} + 81y = 0.$

Solve the following initial value problems.

19 $y''' - y'' + y' - y = 0$, with $y(0) = 1$, $y'(0) = -2$, $y''(0) = 1$.

20 $y''' - y' = 0$, with $y(0) = 1$, $y'(0) = 0$, $y''(0) = 1$.

21 $y''' - y'' - 4y' = 0$, with $y(0) = 0$, $y'(0) = 0$, $y''(0) = 1$.

22 $y''' + y'' - y' - y = 0$, with $y(0) = 1$, $y'(0) = y''(0) = 0$.

23 $y^{(4)} + 8y^{(2)} + 16y = 0$, with $y(0) = 1$, $y^{(1)}(0) = y^{(2)}(0) = y^{(3)}(0) = 0$.

24 $y^{(4)} + 5y^{(2)} + 4y = 0$,   with $y(0) = 0$, $y^{(1)}(0) = 1$, $y^{(2)}(0) = y^{(3)}(0) = 0$.

25 The functions $1, e^{-x}$ and e^x form a basis for the solutions of a homogeneous differential equation defined for all x. Deduce the form of the differential equation, write down its general solution and hence find the solution satisfying the initial conditions $y(0) = 2$, $y'(0) = 0$, $y''(0) = -1$.

26 The functions e^{-x}, $\sin 3x$ and $\cos 3x$ form a basis for the solutions of a homogeneous differential equation defined for all x. Deduce the form of the differential equation, write down its general solution and hence find the solution satisfying the initial conditions $y(\pi/2) = 0$, $y'(\pi/2) = 1$ and $y''(\pi/2) = 1$.

5.4   Differential operators

A notation which is sometimes used to signify the operation of differentiation is $D \equiv d/dx$. Thus it is appropriate to call D the **differentiation operator**, and when it acts on a differentiable function y we have the results

$$Dy = \frac{dy}{dx}, \quad D^2 y = D\left(\frac{dy}{dx}\right) = \frac{d}{dx}\left(\frac{dy}{dx}\right) = \frac{d^2 y}{dx^2}, \tag{1}$$

and in general

$$D^n y = \frac{d^n y}{dx^n}, \tag{2}$$

for integral n, with the understanding that $D^0 y \equiv y$ so that $D^0 \equiv 1$. On account of the linearity of the differentiation operation it follows that if f and g are suitably differentiable functions, and α and β are constants, then

$$D^n[\alpha f + \beta g] = \alpha D^n f + \beta D^n g. \tag{3}$$

In terms of the operator D, the homogeneous linear nth order constant coefficient differential equation (1) in Sec. 5.3 becomes

$$D^n y + a_1 D^{n-1} y + \ldots + a_n y = 0. \tag{4}$$

A concise notation for this equation is

$$L[y] = 0, \tag{5}$$

where

$$L \equiv D^n + a_1 D^{n-1} + \ldots + a_n \tag{6}$$

is called an **nth order linear differential operator**; it is a generalization of the elementary operation of differentiation as it represents a sum of scaled differentiation operations of different orders.

The operator L is called a **linear differential operator** because if f and g are solutions of $L[y] = 0$, then so also is $\alpha f + \beta g$, with α, β arbitrary constants.

The choice of the integer n and of the numerical coefficient $a_1, a_2, \ldots, a_n$ completely determines L. In this case the operator L has the form of a 'polynomial' in D, so it is appropriate to write $L \equiv P(D)$, when because of (6) differential equation (1) of Sec. 5.3 becomes

$$P(D)[y] = 0. \tag{7}$$

For example, if $P(D) = D^3 + 3D^2 + 4D + 1$, then

$$L[y] = P(D)[y] = 0$$

becomes

$$(D^3 + 3D^2 + 4D + 1)y = y''' + 3y'' + 4y' + y = 0.$$

As we have the results

$$D[e^{\lambda x}] = \lambda e^{\lambda x}, \quad D^2[e^{\lambda x}] = \lambda^2 e^{\lambda x}, \ldots, \quad D^n[e^{\lambda x}] = \lambda^n e^{\lambda x}, \tag{8}$$

it follows that if $P(D)$ operates on the function $e^{\lambda x}$ it generates the result

$$P(D)[e^{\lambda x}] = (\lambda^n + a_1\lambda^{n-1} + \ldots + a_n)e^{\lambda x} = P(\lambda)e^{\lambda x}, \tag{9}$$

where now $P(\lambda)$ is an ordinary algebraic polynomial in λ. Thus if $e^{\lambda x}$ is a solution of (1) in Sec. 5.3, $P(D)[e^{\lambda x}] = 0$, and so it must follow that

$$P(\lambda)e^{\lambda x} = 0. \tag{10}$$

This can only be true for all x if λ is a root (real or complex) of the polynomial equation

$$P(\lambda) = 0, \tag{11}$$

which we recognize as the characteristic equation already obtained in (3) of Sec. 5.3. Thus to each distinct root λ_i of (11) there will correspond a solution of (1) of Sec. 5.3 of the form $\exp(\lambda_i x)$.

If the roots of (11) are $\lambda_1, \lambda_2, \ldots, \lambda_n$, $P(\lambda)$ may be factored and written

$$P(\lambda) \equiv (\lambda - \lambda_1)(\lambda - \lambda_2) \ldots (\lambda - \lambda_n).$$

It then follows that because the order of these n factors may be permuted without altering $P(\lambda)$, so also may be the corresponding factors of $P(D)$ without altering the differential operator. For example, if $P(D) = D^3 - 2D^2 - D + 2$, then

$$P(D) = (D-1)(D+1)(D-2) = (D+1)(D-1)(D-2) = (D-2)(D-1)(D+1).$$

We conclude from this that if the differential equation $P(D)[y] = 0$ is considered, then

$$(D-1)(D+1)(D-2)y = (D+1)(D-1)(D-2)y = (D-2)(D-1)(D+1)y$$

$$= y''' - 2y'' - y' + 2y = 0.$$

Thus the operator $P(D)$ may be factored in any order without changing the associated differential equation.

To see how this approach extends to multiple roots suppose, for example, that $\lambda = \mu$ is a double root of (11), and write the characteristic equation in the form

$$P(\lambda) = (\lambda - \mu)^2 Q(\lambda) = 0, \tag{12}$$

where $Q(\lambda)$ is a polynomial in λ of degree $n-2$. Then $P(\mu) = 0$, and as

$$\frac{dP}{d\lambda} = P'(\lambda) = 2(\lambda - \mu)Q(\lambda) + (\lambda - \mu)^2 Q'(\lambda),$$

it also follows that $P'(\mu) = 0$. Thus, if $\lambda = \mu$ is a double root of $P(\mu) = 0$, not only is it true that $P(\mu) = 0$, but also that $P'(\mu) = 0$. Now from (9), if $e^{\lambda x}$ is a solution of (1) of Sec. 5.3,

$$P(D)[e^{\lambda x}] = P(\lambda)e^{\lambda x} = 0,$$

Differentiating this result with respect to λ gives

$$P(D)[xe^{\lambda x}] = P'(\lambda)e^{\lambda x} + P(\lambda)xe^{\lambda x} = 0. \tag{13}$$

However, if $\lambda = \mu$ is a double root of $P(\lambda) = 0$ we have seen that both $P(\mu) = 0$ and $P'(\mu) = 0$, so that (13) shows $xe^{\mu x}$ is also a solution of (1) of Sec. 5.3.

Thus, as in Sec. 5.3, if $\lambda = \mu$ is a double root of (11), the two solutions of (1) of Sec. 5.3 corresponding to this double root are $e^{\mu x}$ and $xe^{\mu x}$. The linear independence of these two solutions has already been established in Sec. 5.3. This form of argument extends in an obvious manner to real or complex roots of any multiplicity, and may also be used to show that the associated linearly independent solutions are those listed in Theorem 5.7.

Apart from providing a convenient alternative notation for differential equations, the differential operator method is often useful when proving special results like the one connected with (13). In fact we shall have occasion to use the method again for such a purpose in Sec. 5.5 when discussing the method of undetermined coefficients.

Let us close this section by providing a simple example of a typical differential operator result which is often useful. We shall prove that if $Q(D)$ and $R(D)$ are two such operators, then

$$Q(D)R(D)[e^{\alpha x}] = Q(\alpha)R(\alpha)e^{\alpha x}. \tag{14}$$

The result is almost immediate and relies on (9). We have $R(D)[e^{\alpha x}] = R(\alpha)e^{\alpha x}$, with $R(\alpha)$ an ordinary polynomial in α, so that

$$Q(D)R(D)[e^{\alpha x}] = Q(D)[R(\alpha)e^{\alpha x}] = R(\alpha)Q(D)[e^{\alpha x}] = R(\alpha)Q(\alpha)e^{\alpha x} = Q(\alpha)R(\alpha)e^{\alpha x}. \tag{15}$$

By now it will have become clear that the differential operator method described above, which applies only to constant coefficient equations, offers no real advantages over the approach of Sec. 5.3 when applied to homogeneous equations. Such advantages as it has only become apparent when nonhomogeneous equations are considered, and then only for rather restricted forms of the nonhomogeneous term. For this reason we shall not develop this particular operator method any further. We shall, however, return to the topic of linear operators, though of a different form, when we come to discuss the powerful Laplace transform method for the solution of initial value problems.

Problems for Section 5.4

In each of the following problems apply the given differential operator to the stated functions.
1 $2D + 3$; e^{2x}, xe^x, $\cos 3x$
2 $D^2 - 1$; $\sin x + 2 \cos x$, $\cosh 3x$, $4 - \sinh 2x$
3 $(D+1)(D+2)$; e^{3x}, $\sinh 2x$, $1 + \sin x + \cosh x$
4 $(D+2)^2$; e^{-x}, $1 + \cosh 2x$, $x^2 + \cos x$.

Find the general solution of each of the following equations.

5 $(D^3-3D-3)y=0$.

6 $(D^3-D^2-D+1)y=0$.

7 $(D^3+6D^2+12D+8)y=0$.

8 $(D^4-2D^2+1)y=0$.

Harder problems

9 Consider the differential equation

$$(D^2+aD+b)y=0.$$

and suppose its characteristic equation

$$\lambda^2+a\lambda+b=0$$

has the two real roots λ_1 and λ_2 (which may be equal).
Write the equation in the form

$$(D-\lambda_1)(D-\lambda_2)y=0,$$

set $u=(D-\lambda_2)y$, and find u which is the solution of a first order homogeneous equation. Then solve the linear first order differential equation

$$y'-\lambda_2 y=u$$

to determine the solution y of the original second order equation. Hence show that when $\lambda_1 \neq \lambda_2$ the general solution is

$$y=A\exp(\lambda_1 x)+B\exp(\lambda_2 x).$$

and when $\lambda_1=\lambda_2=\alpha$ it is

$$y=Ae^{\alpha x}+Bxe^{\alpha x}.$$

10 Show that if the characteristic equation $P(\lambda)=0$ of the differential equation $P(D)[y]=0$ has a triple root $\lambda=\mu$, then the corresponding solutions are $e^{\mu x}$, $xe^{\mu x}$ and $x^2 e^{\mu x}$.

11 Extend the method of Problem 9 to find the general solution of a third order homogeneous constant coefficient equation whose characteristic equation has the real roots λ_1, λ_2 and λ_3 (not necessarily distinct) by writing it in the form

$$(D-\lambda_1)(D-\lambda_2)(D-\lambda_3)y=0.$$

12 Show by substitution of $y_p=Ae^{\alpha x}$, with A a constant to be determined, that if α is not a root of the characteristic equation of the nonhomogeneous constant coefficient equation

$$P(D)[y]=ke^{\alpha x},$$

the particular integral $y_p(x)$ is obtained by setting $A=k/P(\alpha)$.

5.5   Nonhomogeneous linear differential equations

The purpose of this section will be to describe two methods by which particular integrals may be determined. When working with the first of these methods it will be useful to use the concise notation introduced in Sec. 5.4 in which a general linear nth order differential equation is written

$$L[y] = f(x), \tag{1}$$

where L denotes the **differential operator**

$$L \equiv a_0(x) D^n + a_1(x) D^{n-1} + \ldots + a_n(x). \tag{2}$$

If the solution of (1) is expressed in the form

$$y = y_c(x) + y_p(x), \tag{3}$$

the **complementary function** $y_c(x)$ will be the general solution of

$$L[y] = 0.$$

The determination of $y_c(x)$ has already been discussed in some detail for the case of constant coefficient equations, and this knowledge will prove valuable in what is to follow.

The first method to be considered is the most elementary one, and is known as the method of **undetermined coefficients**. It is straightforward to use, but applies only to constant coefficient equations with nonhomogeneous terms containing polynomials, exponentials or trigonometric functions. When the nonhomogeneous term is not one of these elementary functions, or when the differential equation has variable coefficients, the more powerful method of **variation of parameters** must be used. This second method starts from a knowledge of the n linearly independent solutions of (4) which enter into the complementary function, and then constructs the particular integral by means of integration.

Depending on the functions involved, the integrals generated by the method of variation of parameters may be straightforward, difficult to evaluate, or even incapable of expression in terms of elementary functions. The latter situation reflects the fact that an elementary differential equation with a comparatively simple nonhomogeneous term may give rise to a solution involving integrals which cannot be evaluated analytically. It is necessary to solve initial value problems for such equations by means of numerical analysis.

Method of undetermined coefficients

When only constant coefficient equations are involved, and the nonhomogeneous term can be represented as a linear combination of a polynomial, exponentials, sines and cosines and products of exponentials and sines and cosines, the most elementary way of finding a particular integral (and often the simplest) is by means of the so-called method of **undetermined coefficients**. This method has as its basis the observation that repeated differentiation of such functions merely gives rise to similar functions.

Thus inspection of a nonhomogeneous term of this form shows the general form to be expected of a particular integral, but not the numerical coefficients of the functions which are involved. The coefficients are found by substituting the general form for $y_p(x)$ into the left-hand side of the differential equation $L[y]=f(x)$ and equating coefficients of corresponding terms on either side of the equation to make the result an identity.

These simple observations, taken together with the way in which linearly independent solutions of the homogeneous equation are to be associated with repeated roots (Theorem 5.7), lead to the following rules which form the basis of the method of undetermined coefficients.

The coefficients $A_0, A_1, \ldots, A_m, B, C, D$ and F occurring in the right-hand column of Table 5.2 are the undetermined coefficients which must be found by substitution of the appropriate form of $y_p(x)$ into $L[y]=f(x)$. If the nonhomogeneous term $f(x)$ in (1) is representable as a linear combination of the terms found in Column 1 of Table 5.2, then the general form of $y_p(x)$ is obtained by adding together the corresponding terms found in Column 2 of the table.

Rules 1(i), 2(i), 3(i) and 4(i) are self-evident from what has already been said about differentiation of the simple functions involved. Rule 1(ii) follows because when an undifferentiated term y is absent, the lowest order derivative term in $L[y]$ can, in principle, give rise to all of the powers of x in the nonhomogeneous term. Thus the lowest degree term to be included in $y_p(x)$ must be x^r, and as the nonhomogeneous term is of degree m in x, the highest degree term to be included must be x^{m+r}.

Let us prove rule 3(ii) for the case $m=2$. The case in which m is an arbitrary integer can be proved in an analogous fashion. Suppose $\lambda=\mu$ is a triple root of the characteristic equation $P(\lambda)=0$ of the associated homogeneous equation, and let $P(\lambda)$ be of degree n in λ. Then it follows that

$$P(\lambda)=(\lambda-\mu)^3 Q(\lambda), \tag{5}$$

with $Q(\lambda)$ a polynomial of degree $n-3$ in λ. Two successive differentiations of (3) shows that in addition to the result $P(\mu)=0$, we also have $P'(\mu)=0$ and $P''(\mu)=0$.

Now let the differential operator $L \equiv P(D)$ giving rise to the characteristic polynomial $P(\lambda)$ operate on $e^{\alpha x}$. We have

$$P(D)[e^{\alpha x}]=P(\alpha)e^{\alpha x}, \tag{6}$$

which when differentiated three times with respect ot α gives

$$P(D)[xe^{\alpha x}]=P'(\alpha)e^{\alpha x}+P(\alpha)xe^{\alpha x}, \tag{7}$$

$$P(D)[x^2 e^{\alpha x}]=P''(\alpha)e^{\alpha x}+2P'(\alpha)xe^{\alpha x}+P(\alpha)x^2 e^{\alpha x}, \tag{8}$$

$$P(D)[x^3 e^{\alpha x}]=P'''(\alpha)e^{\alpha x}+3P''(\alpha)xe^{\alpha x}+3P'(\alpha)x^2 e^{\alpha x}+P(\alpha)x^3 e^{\alpha x}. \tag{9}$$

Table 5.2   Rules for the method of undetermined coefficients

Terms occurring in $f(x)$	Corresponding terms to be included in the particular integral $y_p(x)$
1. A polynomial in x of degree m.	(i) If $L[y]$ contains an undifferentiated term y, include in $y_p(x)$ terms of the form $$A_0 x^m + A_1 x^{m-1} + \ldots + A_m$$
	(ii) If $L[y]$ does not contain an undifferentiated term y, and r is the lowest order of derivative occurring in $L[y]$, include in $y_p(x)$ terms of the form $$A_0 x^{m+r} + A_1 x^{m+r-1} + \ldots + A_m x^r$$
2. $\sin qx$, $\cos qx$	(i) If $\sin qx$ and/or $\cos qx$ are not contained in the complementary function, include in $y_p(x)$ terms of the form $$B \sin qx + C \cos qx$$
	(ii) If the complementary function contains the terms $x^r \sin qx$ and $x^r \cos qx$ for $r = 0, 1, \ldots, m$, include in $y_p(x)$ terms of the form $$Bx^{m+1} \sin qx + Cx^{m+1} \cos qx$$
3. e^{ax}	(i) If e^{ax} is not contained in the complementary function, include in $y_p(x)$ the term $$De^{ax}$$
	(ii) If the complementary function contains the terms e^{ax}, xe^{ax}, ..., $x^m e^{ax}$, include in $y_p(x)$ the term $$Dx^{m+1} e^{ax}$$
4. $e^{px} \sin qx$, $e^{px} \cos qx$	(i) If $e^{px} \sin qx$ and/or $e^{px} \cos qx$ are not cotained in the complementary function, include in $y_p(x)$ terms of the form $$Ee^{px} \sin qx + Fe^{px} \cos qx$$
	(ii) If the complementary function contains the terms $x^r e^{px} \sin qx$ and $x^r e^{px} \cos qx$ for $r = 0, 1, \ldots, m$, include in $y_p(x)$ terms of the form $$Ex^{m+1} e^{px} \sin qx + Fx^{m+1} e^{px} \cos qx.$$

Setting $\alpha = \mu$ and using $P(\mu) = P'(\mu) = P''(\mu) = 0$ we see from (6), (7) and (8) (as was already known) that $e^{\mu x}$, $xe^{\mu x}$ and $x^2 e^{\mu x}$ are all solutions of the homogeneous equation, while (9) shows that

$$P(D)[x^3 e^{\mu x}] = P'''(\mu)e^{\mu x}. \tag{10}$$

This is the required result, because it shows that setting $y_p(x) = Cx^3 e^{\mu x}$ will generate the function $e^{\mu x}$ appearing in the nonhomogeneous term.

Rule 2(ii) is a special case of rule 4(ii) in which $p = 0$. Rule 4(ii) may be proved using this same form of argument by allowing μ to be complex, setting $\mu = p + iq$, and combining the results corresponding to μ and $\bar{\mu}$ to obtain the terms $e^{px} \sin qx$ and $e^{px} \cos qx$.

Example 1

Find the particular integral of

$$y'' + 3y' + 2y = 3 \sin 2x.$$

Solution

The characteristic equation of the associated homogeneous equation is

$$\lambda^2 + 3\lambda + 2 = 0.$$

This has the roots $\lambda = -1$ and $\lambda = -2$, so the functions forming a basis for the solutions of this equation are e^{-x} and e^{-2x}, and so the complementary function is

$$y_c(x) = c_1 e^{-x} + c_2 e^{-2x}.$$

The nonhomogeneous term $3 \sin 2x$ can be obtained from entry 2 of Table 5.2 by multiplication by 3, so the method of undetermined coefficients may be used. As $\sin 2x$ is not contained in the complementary function, it follows from entry 2(i) of Table 5.2 that corresponding to the term $3 \sin 2x$ we must take for the particular integral a function of the form

$$y_p(x) = B \sin 2x + C \cos 2x.$$

Substituting $y = y_p(x)$ in the differential equation

$$y'' + 3y' + 2y = 3 \sin 2x,$$

we find after simplification that

$$-(2B + 6C) \sin 2x + (6B - 2C) \cos 2x = 3 \sin 2x.$$

This must be an identity in x, so equating corresponding coefficients on each side of the expression we arrive at the simultaneous equations

$$-2B - 6C = 3 \qquad \text{(coefficients of } \sin 2x)$$
$$6B - 2C = 0 \qquad \text{(coefficients of } \cos 2x).$$

These have the solution $B = -3/20$, $C = -9/20$, so the required particular integral is

$$y_p(x) = -\tfrac{3}{20} (\sin 2x + 3 \cos 2x).$$

■

Example 2

Find the general solution of

$$y'' - 2y' + 5y = 4 + x^2 + 5 \cos 3x - 2e^{2x}.$$

Solution

The characteristic equation of the associated homogeneous differential equation is

$$\lambda^2 - 2\lambda + 5 = 0,$$

which has the roots $\lambda = 1 \pm 2i$. The functions forming a basis for the solutions of this equation are thus $e^x \sin 2x$ and $e^x \cos 2x$, so the complementary function is

$$y_c(x) = c_1 e^x \sin 2x + c_2 e^x \cos 2x.$$

The functions in the nonhomogeneous term represent a linear combination of the functions found in the left-hand column of Table 5.2, so the method of undetermined coefficients may be used. It follows from entry 1(i) of the table that the terms to be included in $y_p(x)$ corresponding to the polynomial $4 + x^2$ are

$$A_0 x^2 + A_1 x + A_2.$$

Since the term $\cos 3x$ does not occur in the complementary function, it follows from entry 2(i) that the terms to be included in $y_p(x)$ corresponding to the term $5 \cos 3x$ in the nonhomogeneous term are

$$B \sin 3x + C \cos 3x.$$

Finally, since the term e^{2x} does not occur in the complementary function, it follows from entry 3(i) that the term to be included in $y_p(x)$ corresponding to the term $-2e^{2x}$ in the nonhomogeneous term is

$$De^{2x}.$$

Thus for $y_p(x)$ we take the function

$$y_p(x) = A_0 x^2 + A_1 x + A_2 + B \sin 3x + C \cos 3x + De^{2x}.$$

Substituting $y = y_p(x)$ into the differential equation

$$y'' - 2y' + 5y = 4 + x^2 + 5 \cos 3x - 2e^{2x},$$

and simplifying by combining corresponding terms, leads to the result

$$(2A_0 - 2A_1 + 5A_2) + (5A_1 - 4A_0)x + 5A_0 x^2 + (6C - 4B) \sin 3x$$
$$-(6B + 4C) \cos 3x + 5De^{2x} = 4x^2 + 5 \cos 3x - 2e^{2x}.$$

This must be an identity in x, so equating the coefficients of corresponding terms on each side of this expression we obtain the simultaneous equations

$2A_0 - 2A_1 + 5A_2 = 4$,	(coefficients of x^0),
$5A_1 - 4A_0 = 0$,	(coefficients of x),
$5A_0 = 1$,	(coefficients of x^2),
$6C - 4B = 0$,	(coefficients of $\sin 3x$),
$-6B - 4C = 5$,	(coefficients of $\cos 3x$),
$5D = -2$.	(coefficients of e^{2x}).

These have the solution

$$A_0 = \frac{1}{5}, \quad A_1 = \frac{4}{25}, \quad A_2 = \frac{98}{125}, \quad B = -\frac{15}{26}, \quad C = -\frac{5}{13}, \quad D = -\frac{2}{5},$$

and so the required particular integral is

$$y_p(x)=\frac{1}{5}\left(x^2+\frac{4}{5}x+\frac{98}{25}\right)-\frac{15}{13}\left(\frac{1}{2}\sin 3x+\cos 3x\right)-\frac{2}{5}e^{2x}.$$

The general solution is $y=y_c(x)+y_p(x)$, with $y_c(x)$ and $y_p(x)$ defined as above.

∎

Example 3

Find the particular integral of

$$y''-4y'+13y=3e^{2x}\sin 3x.$$

Solution

The characteristic equation of the associated homogeneous equation is

$$\lambda^2-4\lambda+13=0,$$

with the roots $\lambda=2\pm3i$. The functions forming a basis for the solutions of this differential equation are thus $e^{2x}\sin 3x$ and $e^{2x}\cos 3x$; so the complementary function

$$y_c(x)=c_1e^{2x}\sin 3x+c_2e^{2x}\cos 3x.$$

Now in this case the nonhomogeneous term $3e^x\sin 3x$ is contained in the complementary function. Thus we must use entry 4(ii) of Table 5.2 with $m=0$, $p=2$ and $q=3$, and take for the particular integral a function of the form

$$y_p(x)=Exe^{2x}\sin 3x+Fxe^{2x}\cos 3x.$$

Substitution of $y=y_p(x)$ into the original differential equation followed by simplification then leads to the result

$$6Ee^{2x}\cos 3x-6Fe^{2x}\sin 3x=3e^{2x}\sin 3x.$$

Equating the coefficients of corresponding terms in this expression to make it an identity we find

$$6E=0 \qquad \text{(coefficients of } e^{2x}\cos 3x),$$
$$-6F=3 \qquad \text{(coefficients of } e^{2x}\sin 3x).$$

Thus $E=0$, $F=-\frac{1}{2}$, and so the required particular integral is

$$y_p(x)=-\tfrac{1}{2}xe^{2x}\cos 3x.$$

∎

Example 4

Find the particular integral of

$$y'''+3y''=x^2+1.$$

Solution

This differential equation does not contain an undifferentiated term y, so as the nonhomogeneous term is a polynomial of degree 2, and the lowest order of derivative is 2, it follows from entry 1(ii) of

Table 5.2 that we must take for the particular integral an expression of the form

$$y_p(x) = b_0 x^4 + b_1 x^3 + b_2 x^2.$$

Substitution of $y = y_p(x)$ into the differential equation then leads to the result

$$36 b_0 x^2 + (24 b_0 + 18 b_1)x + 6(b_1 + b_2) = x^2 + 1.$$

Equating the coefficients of corresponding powers of x on each side of this expression we obtain the simultaneous equations

$$
\begin{array}{ll}
36 b_0 = 1 & \text{(coefficients of } x^2), \\
24 b_0 + 18 b_1 = 0 & \text{(coefficients of } x), \\
6 b_1 + 6 b_2 = 1 & \text{(coefficients of } x^0).
\end{array}
$$

These have the solution $b_0 = 1/36$, $b_1 = -1/27$ and $b_2 = 11/9$, so the required particular integral is

$$y_p(x) = \frac{1}{36} x^4 - \frac{1}{27} x^3 + \frac{11}{9} x^2.$$

∎

Example 5

Find the particular integral of

$$y''' - 3y'' + 3y' - y = 3e^x.$$

Solution

The associated homogeneous differential equation has the characteristic equation

$$\lambda^3 - 3\lambda^2 + 3\lambda - 1 = 0,$$

which may be written

$$(\lambda - 1)^3 = 0.$$

Thus $\lambda = 1$ is a triple root, and so the functions forming a basis for the solutions of this equation are e^x, xe^x and $x^2 e^x$. The complementary function is thus

$$y_c(x) = c_1 e^x + c_2 x e^x + c_3 x^2 e^x.$$

The nonhomogeneous term $3e^x$ is contained in the complementary function, so we must use entry 3(ii) of Table 5.2 with $\alpha = 1$ and $r = 2$. Thus for the particular integral we must take an expression of the form

$$y_p(x) = D x^3 e^x.$$

Proceeding as before, and substituting $y = y_p(x)$ in the differential equation, we arrive at the result

$$6D = 3.$$

Thus $D = \frac{1}{2}$, and the required particular integral is

$$y_p(x) = \frac{1}{2} x^3 e^x.$$

∎

Example 6

Find the complementary function, particular integral and general solution of

$$y^{(4)} + 8y^{(2)} + 16y = 4 \sin 2x.$$

Solution

The characteristic equation of the associated homogeneous equation is

$$\lambda^4 + 8\lambda^2 + 16 = 0,$$

which may be written

$$(\lambda^2 + 4)^2 = 0.$$

Thus $\lambda = \pm 2i$ are each double roots of this equation. The functions forming a basis for the solutions of this equation are thus $\sin 2x$, $x \sin 2x$, $\cos 2x$ and $x \cos 2x$, and so the complementary function is

$$y_c(x) = c_1 \sin 2x + c_2 x \sin 2x + c_3 \cos 2x + c_3 x \cos 2x.$$

The nonhomogeneous term $4 \sin 2x$ is contained in the complementary function, so we must use entry 2(ii) of Table 5.2 with $q = 2$ and $m = 1$. Thus for the particular integral we must take an expression of the form

$$y_p(x) = Bx^2 \sin 2x + Cx^2 \cos 2x.$$

Proceeding as before, and substituting $y = y_p(x)$ into the differential equation, we arrive at the result

$$-32(B \sin 2x + C \cos 2x) = 4 \sin 2x.$$

Equating the coefficients of corresponding terms on each side of this expression we find

$$-32B = 4 \quad \text{(coefficients of } \sin 2x),$$
$$-32C = 0 \quad \text{(coefficients of } \cos 2x).$$

Thus $B = -\frac{1}{8}$, $C = 0$, and so the particular integral is

$$y_p(x) = -\frac{1}{8}x^2 \sin 2x.$$

The general solution is $y = y_c(x) + y_p(x)$, with $y_c(x)$ and $y_p(x)$ defined as above. ∎

Method of variation of parameters

In this subsection a general method for the determination of particular integrals is developed called the **method of variation of parameters**. It uses as its starting point the complementary function; that is, the solution of the associated homogeneous equation. This method applies to all linear differential equations, irrespective of whether or not they have constant coefficients, and it presupposes that the complementary function is known. Although we have so far only discussed the determination of the complementary function belonging to constant coefficient equations, for the sake of generality we shall develop the method using the general second order linear differential equation with continuous coefficients and nonhomogeneous term

$$a(x)y'' + b(x)y' + c(x)y = f(x),$$ (11)

in which $a(x) \neq 0$ in the interval in which the solution is defined. Once the method is understood it is an easy matter to extend it to linear differential equations of any order.

The basic idea is to seek the particular integral $y_p(x)$ of (11) in the form

$$y_p(x) = u_1(x)\phi_1(x) + u_2(x)\phi_2(x),$$ (12)

where $\phi_1(x)$ and $\phi_2(x)$ are assumed to be two known linearly independent solutions of the associated homogeneous equation

$$a(x)y'' + b(x)y' + c(x)y = 0$$ (13)

(they form a *basis* for the solutions), and $u_1(x)$, $u_2(x)$ are functions which must be determined. (Compare this with Method 2, Sec. 4.5.)

The name 'variation of parameters' comes from the fact that the complementary function of (11) is

$$y_c(x) = c_1\phi_1(x) + c_2\phi_2(x),$$ (14)

so that if c_1 and c_2 are allowed to vary with x (are *parameters*), it is reasonable to seek $y_p(x)$ in the form given in (12).

Starting from (12) it follows that

$$y_p' = u_1\phi_1' + u_2\phi_2' + u_1'\phi_1 + u_2'\phi_2,$$

so if we now *choose* u_1, u_2 to be such that

$$u_1'\phi_1 + u_2'\phi_2 = 0$$ (15)

it follows that

$$y_p' = u_1\phi_1' + u_2\phi_2'.$$ (16)

Notice that the effect of condition (15) which has been imposed on u_1 and u_2 is to make y_p' in (16) have the same form as if u_1 and u_2 had been constants. Differentiation of (16) leads to the result

$$y_p'' = u_1\phi_1'' + u_2\phi_2'' + u_1'\phi_1' + u_2'\phi_2'.$$ (17)

We now set $y = y_p(x)$ in (11), use results (12), (16) and (17), and group the terms to obtain

$$u_1[a(x)\phi_1'' + b(x)\phi_1' + c(x)\phi_1] + u_2[a(x)\phi_2'' + b(x)\phi_2' + c(x)\phi_1] +$$
$$+ a(x)u_1'\phi_1' + a(x)u_2'\phi_2' = f(x).$$ (18)

The bracketed terms vanish identically because ϕ_1 and ϕ_2 are solutions of (13), so that

(18) reduces to

$$a(x)u_1' \phi_1' + a(x)u_2' \phi_2' = f(x). \tag{19}$$

We now solve (15) and (19) for u_1' and u_2' and integrate to find $u_1(x)$ and $u_2(x)$, after which $y_p(x)$ follows from (12). A simple calculation shows that since $a(x) \neq 0$,

$$u_1' = -\frac{\phi_2 f}{Wa}, \quad u_2' = \frac{\phi_1 f}{Wa}, \tag{20}$$

where

$$W = \phi_1 \phi_2' - \phi_1' \phi_2$$

is the Wronskian of ϕ_1 and ϕ_2. It follows that $W \neq 0$ because ϕ_1 and ϕ_2 are two linearly independent solutions of (13) (Theorem 5.3).

Integrating results (20) and setting the arbitrary constants of integration equal to zero we find

$$u_1 = -\int \frac{\phi_2 f}{Wa} dx, \quad u_2 = \int \frac{\phi_1 f}{Wa} dx. \tag{21}$$

The required particular integral (12) is thus

$$y_p(x) = -\phi_1(x) \int \frac{\phi_2 f}{Wa} dx + \phi_2(x) \int \frac{\phi_1 f}{Wa} dx. \tag{22}$$

Had the constants of integration c_1 and c_2 been included in (21), the result corresponding to (22) would have been

$$y = c_1 \phi_1(x) + c_2 \phi_2(x) - \phi_1(x) \int \frac{\phi_2 f}{Wa} dx + \phi_2(x) \int \frac{\phi_1 f}{Wa} dx, \tag{23}$$

which will be recognized as the general solution of the original nonhomogeneous equation (11). Thus by leaving c_1 and c_2 arbitrary, this method generates the general solution of the nonhomogeneous equation (11) in terms of the general solution of the associated homogeneous equation (13).

Notice that this method would still have worked had a condition other than (15) been imposed on u_1 and u_2, provided only that such a condition does not prevent (12) being a solution. The condition chosen here, which is the standard one, was selected because in general it brings about the greatest simplification. The apparently arbitrary nature of this condition arises because the *two* functions u_1 and u_2 which were introduced at the start are only required to satisfy the *single* condition that (12) must be a solution, so an additional condition is necessary to determine u_1 and u_2.

Example 7

Find the particular integral of

$$y'' - 2y' + y = 4xe^x.$$

Solution

The characteristic equation of the associated homogeneous equation is

$$\lambda^2 - 2\lambda + 1 = 0, \quad \text{or} \quad (\lambda - 1)^2 = 0.$$

As $\lambda = 1$ is a double root of this equation, the functions forming a basis for the solutions are e^x and xe^x (Theorem 5.7(ii)), so the complementary function is

$$y_c(x) = c_1 e^x + c_2 xe^x.$$

The nonhomogeneous term is contained in the complementary function, but it is more complicated than the cases listed in entry 3, Table 5.2; so we shall find the particular integral by means of the method of variation of parameters.

Setting $\phi_1(x) = e^x$ and $\phi_2(x) = xe^x$, the Wronskian $W(\phi_1, \phi_2)$ becomes

$$W(\phi_1, \phi_2) = \begin{vmatrix} e^x & xe^x \\ e^x & e^x + xe^x \end{vmatrix} = e^{2x}.$$

The particular integral $y_p(x)$ now follows from (22) after making the identifications $a(x) = 1$, $f(x) = 4xe^x$ when we find

$$y_p(x) = -e^x \int \frac{xe^x \cdot 4xe^x}{e^{2x}} dx + xe^x \int \frac{e^x \cdot 4xe^x}{e^{2x}} dx.$$

Integrating, and remembering that to find $y_p(x)$ we must set the constants of integration equal to zero, we obtain

$$y_p(x) = \tfrac{2}{3} x^3 e^x.$$

The general form of this result might have been anticipated in view of the arguments which led to Table 5.2. However, although the entries in Table 5.2 could be extended to allow for the nonhomogeneous term being any function in the basis for the solutions, it is not worth while doing so because of other methods which will be developed later. ∎

Example 8

Find the general solution of

$$y'' + y = \tan x, \quad \text{for } -\pi/2 < x < \pi/2.$$

Solution

The nonhomogeneous term involved here is more complicated than those for which the method of undetermined coefficients may be used; so we must use the method of variation of parameters. The associated homogeneous equation is a constant coefficient equation which is easily seen to have for its solution

$$y_c(x) = c_1 \cos x + c_2 \sin x,$$

so we shall set $\phi_1(x) = \cos x$ and $\phi_2(x) = \sin x$. Then the Wronskian

$$W(\phi_1, \phi_2) = \begin{vmatrix} \cos x & \sin x \\ -\sin x & \cos x \end{vmatrix} \equiv 1.$$

The general solution now follows from (23), where from (21)

$$u_1(x) = -\int \sin x \tan x \, dx \quad \text{and} \quad u_2(x) = \int \cos x \tan x \, dx.$$

Routine integration establishes

$$u_1(x) = \sin x - \ln|\sec x + \tan x|, \quad u_2(x) = -\cos x \qquad \text{for } -\pi/2 < x < \pi/2.$$

Substitution of these results into the expression for y shows the general solution to be

$$y = c_1 \cos x + c_2 \sin x - \cos x \ln|\sec x + \tan x| \qquad \text{for } -\pi/2 < x < \pi/2.$$

The restriction on x to the interval $-\pi/2 < x < \pi/2$ is necessary in order to confine $\tan x$ to an interval in which it is differentiable.

Not only does this example illustrate the use of the method of variation of parameters, but it also illustrates the complexity of particular integrals. This is seen to be so even in the case of constant coefficient equations when the nonhomogeneous term is anything other than one of the simple functions listed in Table 5.2. In this case the nonhomogeneous term $\tan x$ gave rise to the particular integral

$$y_p(x) = -\cos x \ln|\sec x + \tan x|. \qquad \blacksquare$$

Example 9

Find the general solution of

$$y'' - 4y = \frac{\sin x}{1 + x^2}.$$

Solution

It is at once apparent that

$$y_c(x) = c_1 e^{-2x} + c_2 e^{2x},$$

so we shall set $\phi_1(x) = e^{-2x}$ and $\phi_2(x) = e^{2x}$. The Wronskian is thus

$$W(\phi_1, \phi_2) = \begin{vmatrix} e^{-2x} & e^{2x} \\ -2e^{-2x} & 2e^{2x} \end{vmatrix} \equiv 4.$$

The general solution then follows from (23) in the form

$$y = c_1 e^{-2x} + c_2 e^{2x} - \frac{e^{-2x}}{4} \int \frac{e^{2x} \sin x}{1 + x^2} \, dx + \frac{e^{2x}}{4} \int \frac{e^{-2x} \sin x}{1 + x^2} \, dx,$$

though the two integrals involved cannot be evaluated analytically. If they are to be determined numerically they need to be expressed in the form

$$\int_{x_0}^{x} \frac{e^{2t} \sin t}{1 + t^2} \, dt \quad \text{and} \quad \int_{x_0}^{x} \frac{e^{-2t} \sin t}{1 + t^2} \, dt,$$

with x_0 chosen to be some convenient number. $\qquad \blacksquare$

Problems for Section 5.5

Use the method of undetermined coefficients to find the general solution of each of the following differential equations.

1 $y'' - 4y' + 4y = x^2$.

2 $y'' - 8y' + 7y = 6$.

3 $y'' - 2y' - 3y = e^{4x}$.

4 $y'' + 2y' + y = e^{2x}$.

5 $y'' - 3y' + 2y = \sin x$.

6 $y'' + y = 4 \sin x$.

7 $y'' + y = \cos x$.

8 $y'' - 4y' + 8y = 2e^{2x} + \sin 2x$.

9 $y'' - 2y' + y = \sin x + \sinh x$.

10 $y'' + y' = 4 + 2 \sin^2 x$.

11 $y'' - 2y' + 5y = e^x \cos 2x$.

12 $y'' - 4y' + 5y = 2e^{2x} \sin x$.

13 $y'' - 2y' - 8y = 3e^x - 8 \cos 2x$.

14 $y'' - 2y' + 10y = \sin 3x + 2e^{2x}$.

15 $y'' - 2y' = e^{3x} + 4$.

16 $y''' - 9y' = 8e^x - 90x$.

17 $y''' + 4y' = 20 + 16 \sin 2x$.

18 $y'''' - 2y''' + y'' = e^x + x^3$.

19 $y''' - y = x^3 + 2$.

20 $y'''' + y''' = 2 \cos 4x$.

Use the method of undetermined coefficients to find the general solution of each of the following differential equations, and hence the solution of the stated initial value problem.

21 $y'' + 4y = \sin x$,   with $y(0) = 1$, $y'(0) = 1$.

22 $y'' + n^2 y = k \sin px (p \neq n)$,   with $y(0) = a$, $y'(0) = b$.

23 $y''' + y'' = 4 - 12e^{2x}$,   with $y(0) = y'(0) = y''(0) = 0$.

24 $y'''' - y = 5 \cos x$,   with $y(0) = y'(0) = y''(0) = y'''(0) = 0$.

25 $y''' - 4y'' + 5y' - 2y = 2x + 3$,   with $y(0) = 1$, $y'(0) = y''(0) = 0$.

26 $y'' + 4y' + 5y = 3e^{-2x} \cos 2x$,   with $y\left(\dfrac{\pi}{2}\right) = 1$, $y'\left(\dfrac{\pi}{2}\right) = 0$.

Use the method of variation of parameters to find the general solution of each of the following differential equations.

27 $y'' - 2y' + y = 4e^x$.

28 $y'' - y = 2x \sin x$.

29 $y'' - 2y' + 2y = 4e^x \sin x$.

30 $y'' - 2y' = 3x + 2xe^x$.

31 $y'' - 4y' + 4y = xe^{2x}$.

32 $y'' - 2y = 2xe^x(\cos x - \sin x)$.

33 $y'' + y = \sec x$.

34 $y'' + y = \operatorname{cosec} x$.

35 $y'' + y = \cot x$.

36 $y'' - 2y' + y = e^x/x.$

37 $y'' - \dfrac{1}{x}y' = x.$

38 $y'' + 2y' + y = e^{-x}/x.$

Harder problems

Generalization of the method of variation of parameters

The method of variation of parameters may be extended to the general linear nth order differential equation

$$y^{(n)} + a_1(x)y^{(n-1)} + \ldots + a_n(x)y = f(x). \tag{A}$$

Let $\phi_1(x)$, $\phi_2(x)$, $\ldots$, $\phi_n(x)$ be a basis for the solutions of the equation, and seek a particular integral $y_p(x)$ in the form

$$y_p(x) = u_1(x)\phi_1(x) + \ldots + u_n(x)\phi_n(x), \tag{B}$$

with $u_1(x)$, $u_2(x)$, $\ldots$, $u_n(x)$ functions which are to be determined. Then proceeding as in Sec. 5.5, the functions $u_1(x)$, $\ldots$, $u_n(x)$ follow by solving the system of equations

$$u_1'\phi_1 + u_2'\phi_2 + \ldots + u_n'\phi_n = 0,$$
$$u_1'\phi_1' + u_2'\phi_2' + \ldots + u_n'\phi_n' = 0,$$
$$\ldots \ldots \ldots \ldots \ldots$$
$$u_1'\phi_1^{(n-2)} + u_2'\phi_2^{(n-2)} + \ldots + u_n'\phi_n^{(n-2)} = 0,$$
$$u_1'\phi_1^{(n-1)} + u_2'\phi_2^{(n-1)} + \ldots + u_n'\phi_n^{(n-1)} = f(x), \tag{C}$$

for $u_1'(x)$, $u_2'(x)$, $\ldots$, $u_n'(x)$.

Using Cramer's rule, the solution of (C) may be written in the form

$$u_1'(x) = \frac{U_i(x)f(x)}{W(\phi_1, \phi_2, \ldots, \phi_n)}, \qquad \text{for } i = 1, 2, \ldots, n, \tag{D}$$

where $W(\phi_1, \phi_2, \ldots, \phi_n)$ is the Wronskian of the functions forming a basis for solutions of (A) (and so cannot vanish), and $U_i(x)$ is the determinant obtained from $W(\phi_1, \phi_2, \ldots, \phi_n)$ by replacing the ith column by $(0, 0, \ldots, 1)^{\mathsf{T}}$.

Thus integration of (D) gives

$$u_i(x) = \int \frac{U_i(x)f(x)\,dx}{W(\phi_1, \phi_2, \ldots, \phi_n)}, \qquad \text{for } i = 1, 2, \ldots, n. \tag{E}$$

Setting the constants of integration equal to zero in (E) and substituting in (B) gives the required particular integral. If the constants of integration are left as arbitrary constants, expression (B) becomes the general solution of (A).

Use this generalization of the method of variation of parameters to find the general solution of each of the following differential equations and, where appropriate, the solution of the stated initial value problem.

39 $y''' - 5y'' + 8y' - 4y = e^{2x}.$

40 $y'''' - 3y'' - 4y = \cosh x.$

41 $y''' + y'' + y' + y = xe^x$.

42 $y''' + y' = \sec x$.

43 $y''' + 2y'' + 2y' + y = x$, with $y(0) = y'(0) = y''(0) = 0$.

Cauchy–Euler equation

A linear variable coefficient differential equation of the form

$$x^n y^{(n)} + a_1 x^{n-1} y^{(n-1)} + \ldots + a_{n-1} xy' + a_n y = f(x),$$

in which $a_1, a_2, \ldots, a_n$ are constants, is called a Cauchy–Euler equation. Such a differential equation may be solved by purely algebraic means when it is homogeneous ($f(x) \equiv 0$) by setting

$$y = Ax^m.$$

Substituting the result

$$\frac{d^r y}{dx^r} = m(m-1)(m-2)\ldots(m-r+1)Ax^{m-r}$$

into the original differential equation leads to a polynomial equation for m of degree n. If it has purely real roots $m_1, m_2, \ldots, m_n$ which are all distinct, the corresponding functions forming a basis for the solutions are

$$\phi_1 = x^{m_1}, \phi_2 = x^{m_2}, \ldots, \phi_n = x^{m_n}.$$

If a real root $m_1 = \alpha$ has multiplicity s, the method of variation of parameters may be used to show that the corresponding s linearly independent solutions are

$$\phi_1 = x^\alpha, \phi_2 = x^\alpha \ln x, \ldots, \phi_s = x^\alpha (\ln x)^{s-1}.$$

If m_1 and $\bar{m}_1$ are complex conjugate roots, with $m_1 = \alpha + i\beta$, the corresponding pair of linearly independent solutions is

$$\phi_1 = x^\alpha \cos(\beta \ln x) \quad \text{and} \quad \phi_2 = x^\alpha \sin(\beta \ln x).$$

If m_1 and $\bar{m}_1$ are complex roots each with multiplicity s, and $m_1 = \alpha + i\beta$, the corresponding $2s$ linearly independent solutions which may be found by the method of variation of parameters are

$$\phi_1 = x^\alpha \cos(\beta \ln x), \qquad \phi_2 = x^\alpha \sin(\beta \ln x),$$
$$\phi_3 = x^\alpha \ln x \cos(\beta \ln x), \qquad \phi_4 = x^\alpha \ln x \sin(\beta \ln x),$$

$$\phi_{2s-1} = x^\alpha (\ln x)^{s-1} \cos(\beta \ln x), \qquad \phi_{2s} = x^\alpha (\ln x)^{s-1} \sin(\beta \ln x).$$

If the Cauchy–Euler equation is nonhomogeneous ($f(x) \neq 0$), its general solution may always be found by first constructing the complementary function, y_c, and then using the method of variation of parameters to find y_p.

An alternative approach to the solution of the nonhomogeneous Cauchy–Euler equation is to use the substitution

$$x = e^t$$

which reduces the equation to a constant coefficient equation. However, unless $f(x)$ is very simple in form, this is likely to lead to a complicated nonhomogeneous term involving the variable t, so the previous approach is often simpler.

44 Verify that with the change of variable $x = e^t$,

$$\frac{dy}{dx} = \frac{1}{x}\frac{dy}{dt}, \quad \frac{d^2y}{dt^2} = \frac{1}{x^2}\left(\frac{d^2y}{dt^2} - \frac{dy}{dt}\right),$$

$$\frac{d^3y}{dt^3} = \frac{1}{x^3}\left(\frac{d^3y}{dt^3} - \frac{3d^2y}{dt^2} + \frac{2dy}{dt}\right).$$

Find the general solution of each of the following differential equations.

45 $x^2 y'' + 3xy' + y = 0$.

46 $x^2 y'' - 2y = 0$.

47 $x^2 y'' + xy' + 4y = 0$.

48 $x^2 y'' + xy' + y = 0$.

49 $x^2 y'' - 3xy' + 4y = 0$.

50 $x^2 y'' + xy' + y = 1$.

51 $x^2 y'' - 4xy' + 6y = x$.

52 $x^2 y'' - 2y = \sin(\ln x)$.

53 The Lagrange equation

$(ax+b)^n y^{(n)} + A_1(ax+b)^{n-1} y^{(n-1)} + \ldots + A_{n-1}(ax+b)y' + A_n y = f(x)$, in which a, b, A_1, $A_2, \ldots, A_n$ are constants, is a generalization of the Cauchy–Euler equation.

Show that the substitution

$$ax + b = e^t$$

leads to the results

$$\frac{dy}{dx} = \frac{a}{(ax+b)}\frac{dy}{dt}, \quad \frac{d^2y}{dx^2} = \frac{a^2}{(ax+b)^2}\left(\frac{d^2y}{dt^2} - \frac{dy}{dt}\right),$$

$$\frac{d^3y}{dx^3} = \frac{a^3}{(ax+b)^3}\left(\frac{d^3y}{dt^3} - \frac{3d^2y}{dt^2} + \frac{2dy}{dt}\right), \ldots,$$

and this reduces the Lagrange equation to a constant coefficient equation. Use this approach to find the general solution of

$$(1+x)^2 y'' - 3(1+x)y' + 4y = (1+x)^3.$$

5.6 General reduction of the order of a linear differential equation. Integral method

Once a solution ϕ_1 of a homogeneous linear nth order differential equation is known, it may be used to transform the equation to one of order $n-1$, thereby simplifying the task of determining the other $n-1$ linearly independent solutions which appear in its general solution. When the general solution itself has been found, it can then be used with the method of variation of parameters, or with another method, to find the general solution of the corresponding nonhomogeneous equation.

The **reduction of order** is achieved by seeking a solution for either constant or variable coefficient homogeneous linear differential equations in the form

$$y = \phi_1 v, \tag{1}$$

where v is a function to be determined. We illustrate the method in the case of second and third order variable coefficient equations, though the method applies to equations of any order. The most general homogeneous second order equation may be written

$$a_0(x)y'' + a_1(x)y' + a_2(x)y = 0, \tag{2}$$

where the coefficient functions a_0, a_1 and a_2 are continuous functions of x in some interval in which $a_0(x) \neq 0$.

Differentiation of (1) yields

$$y' = \phi_1 v' + \phi_1' v \quad \text{and} \quad y'' = \phi_1 v'' + 2\phi_1' v' + \phi_1'' v. \tag{3}$$

Substituting results (3) into (2) and grouping terms then gives

$$a_0 \phi_1 v'' + (2a_0 \phi_1' + a_1 \phi_1)v' + (a_0 \phi_1'' + a_1 \phi_1' + a_2 \phi_1)v = 0. \tag{4}$$

The last group of bracketed terms in (4) vanishes, because ϕ_1 is a solution of (2). Thus by setting $w = v'$ we are left with the linear first order equation for w

$$a_0 \phi_1 w' + (2a_0 \phi_1' + a_1 \phi_1)w = 0,$$

which because $a_0(x) \neq 0$, may be rewritten in the form

$$\frac{dw}{dx} + \left(\frac{2\phi_1'}{\phi_1} + \frac{a_1}{a_0} \right) w = 0. \tag{5}$$

The equation in (5) is a homogeneous linear first order differential equation with the integrating factor (Sec. 4.5)

$$\mu = \exp\left[\int \left(\frac{2\phi_1'}{\phi_1} + \frac{a_1}{a_0} \right) dx \right] = \exp(2 \ln \phi_1) \exp\left[\int \left(\frac{a_1}{a_0} \right) dx \right]$$

$$= \phi_1^2 \exp\left[\int \left(\frac{a_1}{a_0} \right) dx \right].$$

Making use of this integrating factor, it then follows directly from Sec. 4.5 (see also Ex. 4, Sec. 4.4) that (5) has the solution

$$w = \frac{B}{\phi_1^2} \exp\left[-\int \left(\frac{a_1}{a_0} \right) dx \right], \tag{6}$$

where B is an arbitrary constant of integration. However, as $w = v'$, a further integration of (6) shows that

$$v = B \int \frac{\exp[-\int (a_1/a_0)dx]}{\phi_1^2} dx. \tag{7}$$

Combining (1) and (7) we thus arrive at the following expression for the second linearly independent solution of (2)

$$y = B\phi_1 \int \frac{\exp[-\int(a_1/a_0)dx]}{\phi_1^2} dx. \tag{8}$$

As ϕ_1 is also a solution of (2), an arbitrary linear combination of it with the function in (8) must be the general solution of (2). We thus arrive at the following form of the general solution of (2)

$$y = A\phi_1 + B\phi_1 \int \frac{\exp[-\int(a_1/a_0)dx]}{\phi_1^2} dx, \tag{9}$$

in which both A and B are arbitrary constants. Notice that the constant of integration may be omitted when evaluating (8), because when (8) is used in (9) any such constant will merely introduce a term already contained in $A\phi_1$. Because of the form of (8), this method of finding a second linearly independent solution of (2) is also called the **integral method**.

Results (8) and (9) have considerable theoretical interest but are difficult to remember. In practice, when finding the second linearly independent solution $\phi_1 v$ from ϕ_1, it is easier to find v directly by substituting (1) into (2). This approach is used in the example which follows.

Example 1. Reduction of order in second order equation

Find two linearly independent solutions of

$$x^2 y'' + xy' - y = 0,$$

given that $\phi_1 = x$ is a solution. Use these two linearly independent solutions to find the general solution of

$$x^2 y'' + xy' - y = 2x^2 + 1.$$

Solution

In this case (1) takes the form

$$y = xv.$$

Substituting this expression into the given homogeneous differential equation reduces it to

$$v'' + \frac{3}{x}v' = 0, \quad \text{or to} \quad w' + \frac{3}{x}w = 0,$$

where $w = v'$. The last equation has the solution $w = c_1/x^3$. Thus

$$w = \frac{dv}{dx} = \frac{c_1}{x^3},$$

so a further integration shows that

$$v = \frac{c_1}{x^2} + c_2,$$

where the numerical factor arising from the integration of $1/x^3$ has been absorbed into the arbitrary constant c_1. The general solution of the homogeneous equation (the complementary function) is thus $y_c = xv$, or

$$y_c = \frac{c_1}{x} + c_2 x.$$

This shows that two linearly independent solutions of the homogeneous equation are x and $1/x$.

The solution of the nonhomogeneous equation may be found either by the method of variation of parameters, or by finding the particular integral y_p by the method of undetermined coefficients. If the method of undetermined coefficients is used, an appropriate trial function for y_p is

$$y_p = ax^2 + b.$$

The term in x which might at first sight be expected to occur in y_p has been omitted because x occurs as a function in the complementary function y_c.

A simple calculation shows $a = \frac{2}{3}$ and $b = -1$, so the general solution of the nonhomogeneous equation $y = y_c + y_p$ becomes

$$y = \frac{c_1}{x} + c_2 x + \frac{2}{3} x^2 - 1.$$

■

The approach just used for a second order equation may also be applied to the homogeneous third order linear variable coefficient equation

$$a_0(x) y''' + a_1(x) y'' + a_2(x) y' + a_3(x) = 0, \tag{10}$$

in which a_0, a_1, a_2 and a_3 are continuous functions of x in some interval in which $a_0(x) \neq 0$. Setting $y = \phi_1 v$, where now ϕ_1 is a solution of (10), and proceeding as before, we arrive at the result

$$a_0 \phi_1 v''' + (3a_0 \phi_1' + a_1 \phi_1) v'' + (3a_0 \phi_1'' + 2a_1 \phi_1' + a_2 \phi_1) v'$$

$$+ (a_0 \phi_1''' + a_1 \phi_1'' + a_2 \phi_1' + a_3 \phi_1) v = 0.$$

Once again the last group of bracketed terms vanishes, because now ϕ_1 is a solution of (10), and by setting $w = v'$ we are left with the second order linear differential equation for w

$$a_0 \phi_1 w'' + (3a_0 \phi_1' + a_1 \phi_1) w' + (3a_0 \phi_1'' + 2a_1 \phi_1' + a_2 \phi_1) w = 0. \tag{11}$$

Thus the order of the equation to be solved has been reduced from three to two.

Two arbitrary constants will be introduced when finding w from (11), and a further one will arise when integrating $v' = w$ to find v. Substitution of this function v into $y = \phi_1 v$ will then give the general solution of (10). This general solution will comprise the sum of three functions, each multiplied by an arbitrary constant of integration, with one of the functions being ϕ_1. Both of the other two functions arising in the general solution will be linearly independent solutions of (10), and each will be different from ϕ_1.

Example 2. Reduction of order in a third order equation

Find three linearly independent solutions of

$$y''' - 3ay'' + 3a^2y' - a^3y = 0,$$

given that $\phi_1 = e^{ax}$ is a solution.

Solution

In this case (1) takes the form

$$y = e^{ax}v,$$

and substitution into the given differential equation reduces it to

$$v''' = 0,$$

so that $v = c_1 + c_2x + c_3x^2$.

The general solution y is thus given by

$$y = e^{ax}(c_0 + c_1x + c_2x^2).$$

Inspection of this result shows that three linearly independent solutions of the equation are e^{ax}, xe^{ax} and x^2e^{ax} (compare this with Ex. 2 of Sec. 5.3). ■

Problems for Section 5.6

Find the general solution of each of the following homogeneous linear differential equations, given that the indicated function ϕ_1 is a solution.

1 $y'' - 4y' + 5y = 0$, with $\phi_1 = e^{2x}\sin x$.
2 $y''' - 9y = 0$, with $\phi_1 = e^{-3x}$.
3 $y''' + y'' = 0$, with $\phi_1 = c_1$ (const).
4 $y^{(4)} - y = 0$, with $\phi_1 = e^x$.
5 $(x^2 + 1)y'' + 2xy' = 0$, with $\phi_1 = c_1$(const).
6 $xy'' - 2y' + 9x^2y = 0$, with $\phi_1 = \sin(x^3)$.
7 $x^3y''' + 3x^2y'' + xy' - 8y = 0$, with $\phi_1 = x^2$.
8 $x^3y''' - x^2y'' - 3xy' = 0$, with $\phi_1 = x^4$.

5.7 Oscillatory behavior

One of the most fundamental forms of time-dependent behavior of physical systems is that involving oscillation. This phenomenon may arise in the form of mechanical vibrations, the fluctuation of electric charge on the plates of a capacitor in an electric circuit, the pitching and yawing motion of ships and airplanes, the motion of the top of a fluid column in a manometer tube, and in many other ways.

Oscillations in linear systems may be divided into the following three categories, which are arranged in order of increasing complexity:

(i) free undamped oscillations,
(ii) free damped oscillations,
(iii) forced oscillations.

In their simplest form, oscillations in all three categories are governed by second order constant coefficient linear differential equations. When such equations are homogeneous the oscillations are said to be **free** (or **natural**), but when the equations are non-homogeneous, and the nonhomogeneous term is a sinusoidal function, the oscillations are said to be **forced**. This latter situation arises in many important applications. The initial conditions for the second order differential equations describing oscillations will determine the precise way in which the oscillations are started.

All possible forms of solution of the general homogeneous second order constant coefficient linear differential equation have already been derived in Sec. 5.2. Thus it only remains for these solutions to be interpreted in the context of oscillatory behavior. To provide motivation for this interpretation we first discuss some of the ways in which oscillatory phenomena arise in engineering and science.

(i) Free undamped oscillations

The basic differential equation governing **free undamped oscillations** is

$$\frac{d^2 y}{dt^2} + \omega_0^2 y = 0, \tag{1}$$

in which ω_0 is real. From Section 5.2 this equation is known to have the general solution

$$y = A \cos \omega_0 t + B \sin \omega_0 t, \tag{2}$$

in which the numbers A and B are arbitrary real constants.

A more convenient form of solution follows from (2) by first writing it in the form

$$y = \sqrt{A^2 + B^2} \left(\frac{A}{\sqrt{A^2 + B^2}} \cos \omega_0 t + \frac{B}{\sqrt{A^2 + B^2}} \sin \omega_0 t \right), \tag{3}$$

setting $b = \sqrt{A^2 + B^2}$, and then choosing ε such that

$$\sin \varepsilon = \frac{A}{\sqrt{A^2 + B^2}} \quad \text{and} \quad \cos \varepsilon = \frac{B}{\sqrt{A^2 + B^2}}$$

or, equivalently,

$$\varepsilon = \arctan(A/B).$$

For when this is done (3) becomes

$$y = b(\sin \varepsilon \cos \omega_0 t + \cos \varepsilon \sin \omega_0 t),$$

or

$$y = b \sin (\omega_0 t + \varepsilon). \tag{4}$$

This version of (2) shows that the solution of (1) is always expressible as a sinusoid with **amplitude** b and argument $\omega_0 t + \varepsilon$. As the sine function is defined for all values of its argument, this sinusoidal oscillation will continue with the same amplitude for all time. In physics, the motion of any particle which can be described by (4) is called **simple harmonic motion**.

Simple harmonic motion corresponds to the motion of point Q in Fig. 5.2, the projection of P onto the y-axis, as P moves around a circle of radius b at an angular rate ω_0 radians per unit of time. The angle ε determines the point R on the circle at which P is located at time $t = 0$.

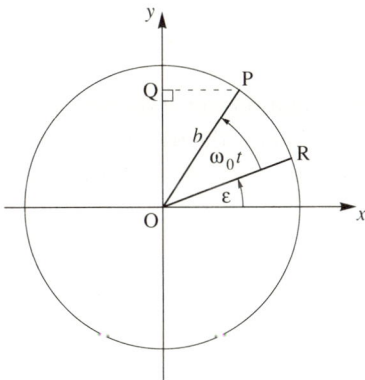

Fig. 5.2   Simple harmonic motion with $y = OQ = b \sin (\omega_0 t + \varepsilon)$

The amplitude of the oscillations described by (4) does not decay with time, and so they are said to be **undamped** oscillations. In the representation of the solution of (1) given in (4), it is usual to regard the numbers b and ε as the two arbitrary constants which must appear in the general solution of (1). Thus when working with oscillatory problems, b and ε take the place of the arbitrary constants A and B introduced in (2). The quantity ε (an angle in radians) in (4) is called the **phase angle** (or constant) for the oscillation, and by convention this angle is taken to lie in the interval $-\pi < \varepsilon \le \pi$. Inspection of (4) shows that the phase angle represents the argument of the solution at time $t = 0$. When $\varepsilon > 0$, the angle ε is called a **phase advance**, and when $\varepsilon < 0$ a **phase lag**.

A function $f(x)$ is said to be **periodic** with **period** X if

$$f(x) = f(x + X) \tag{5}$$

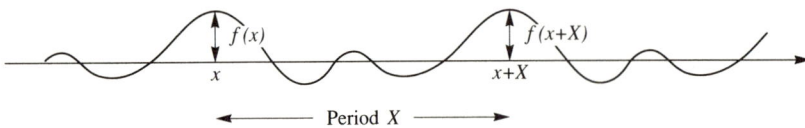

Fig. 5.3 Periodic function $f(x)$ with period X

for all x, and there is no smaller value of X for which this result is always true. This situation is illustrated in Fig. 5.3.

The sine function is an example of a periodic function with period 2π. It follows that the sine function in (4) must be periodic with period 2π with respect to its argument $\omega_0 t + \varepsilon$. Thus if the **time period** of the oscillation is T, increasing t by an amount T will increase the argument $\omega_0 t + \varepsilon$ of the sine function by 2π, so that

$$\omega_0 T = 2\pi. \tag{6}$$

In terms of ω_0, the **time period** T of the oscillation described by (4) is thus

$$T = \frac{2\pi}{\omega_0}. \tag{7}$$

The quantity ω_0 is called the **angular frequency** of the oscillation, and if the period T is measured in seconds, ω_0 is measured in radians second^{-1}. Finally, the **frequency** f of the oscillation is defined as the number of oscillations in a unit of time (usually one second), so

$$f = \frac{1}{T}. \tag{8}$$

When combined with (6), result (8) becomes

$$f = \frac{\omega_0}{2\pi}. \tag{9}$$

Thus all information about the periodicity of the solutions of (1) is contained in the single parameter ω_0. The amplitude b and the phase angle ε are determined by the initial conditions for (1) in conjunction with (4).

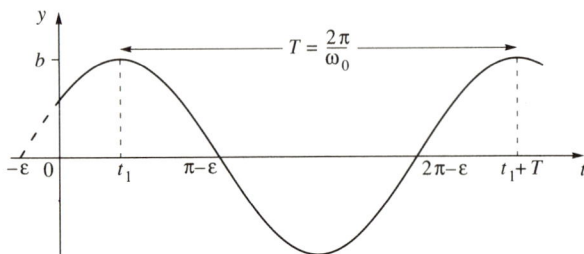

Fig. 5.4 The oscillation $y = b \sin(\omega_0 t + \varepsilon)$

It is usual for the period T to be measured in seconds, when the corresponding unit for the frequency f is called the **hertz** (Hz). Consequently 60Hz is a frequency of 60 cycles a second.

The oscillation represented by (4) is interpreted in graphical terms in Fig. 5.4, which also makes clear the meaning of the amplitude b of the oscillation and the phase angle ε.

The two examples which follow illustrate ways in which the differential equation in (1) may arise as the mathematical model for simple physical problems.

Example 1. Free undamped oscillations of a mass-spring system

Figure 5.5(a) represents the equilibrium configuration of a mechanical system comprising two identical springs of negligible mass, each with unstretched length l, resting on a smooth horizontal table and fastened rigidly to the table at points A and B distant $2l$ apart, but joined in the center to a mass m which is free to move. The system is set in motion by displacing m a distance $d\,(<l)$ along the line AB to point P as shown in Fig. 5.5(b), and then releasing the mass from rest.

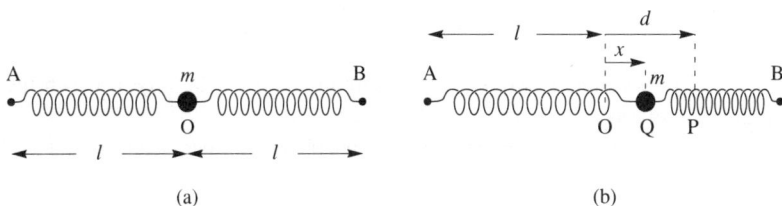

Fig. 5.5 Mass-spring system on a smooth horizontal table

To arrive at the equation of motion of the mass we must make use of experimental information about the behavior of springs. It is known from **Hooke's law**[4], which was deduced from a series of experiments, that the restoring force exerted by most elastic materials when displaced by stretching or compressing them a small amount from their equilibrium length is proportional to the displacement. The constant of proportionality itself must be obtained by experiment in each individual case.

Denoting the spring constant of proportionality by K, it follows that when the mass m is distant x from O as shown in Fig. 5.5(b), each spring will have had its length changed by the same amount x, one having been stretched and the other compressed. Thus each spring will exert a force Kx on mass m. The spring in position AQ is in tension, and the one in position QB is in compression, so the total restoring force on the mass will be of magnitude $2Kx$ in the negative x-direction.

Newton's second law asserts that, in suitable units, the rate of change of momentum of a body is equal to the applied force. Thus as the momentum of the constant mass m is $m(dx/dt)$, its rate of change of momentum must be

$$\frac{d}{dt}\left(m\frac{dx}{dt}\right)=m\frac{d^2x}{dt^2}.$$

---

[4] ROBERT HOOKE (1635–1703), an English physicist and contemporary of Isaac Newton, who made contributions to astronomy, mechanics and cartography.

Equating this to the force $-2Kx$, in which the negative sign arises because the force acts in the opposite sense to the rate of change of momentum, we arrive at the equation of motion

$$m\frac{d^2x}{dt^2} = -2Kx,$$

or

$$\frac{d^2x}{dt^2} + \frac{2K}{m}x = 0. \tag{10}$$

Comparing this result with (1) shows

$$\omega_0^2 = \frac{2K}{m},$$

so from (4) the general solution is

$$x = b \sin\left(\sqrt{\frac{2K}{m}}\,t + \varepsilon\right), \tag{11}$$

where the amplitude b and phase angle ε have still to be determined from the way in which the motion is started (initial conditions).

For convenience we take $t=0$ at the start of the motion, when the mass is released from rest at a distance d to the right of O. Hence at $t=0$ result (11) becomes

$$d = b \sin \varepsilon. \tag{12}$$

As m is released from rest we must have $dx/dt = 0$ when $t=0$, so differentiating (11) and then setting $t=0$ gives

$$0 = b\sqrt{\frac{2K}{m}}\,\cos \varepsilon. \tag{13}$$

Thus since ε is such that $-\pi < \varepsilon < \pi$, and b and d in (12) are both positive, it follows that the phase angle $\varepsilon = \pi/2$. Result (12) then shows $b = d$, so the solution becomes

$$x = d \sin\left(\sqrt{\frac{2K}{m}}\,t + \frac{\pi}{2}\right),$$

or what is equivalent,

$$x = d \cos\left(\sqrt{\frac{2K}{m}}\,t\right). \tag{14}$$

The simple mass-spring system thus has an angular frequency $\omega_0 = \sqrt{2K/m}$, a time period of oscillation $T = 2\pi/\omega_0 = \pi\sqrt{2m/K}$, and a frequency of oscillation $f = 1/T = (1/\pi)\sqrt{K/(2m)}$. Physical intuition suggests that for a given mass, the stiffer the springs (the larger K), the faster will be the oscillations, and for given springs the larger the mass m the slower will be the oscillations. This is confirmed by the fact that $f = (1/\pi)\sqrt{K/(2m)}$. The motion of mass m is illustrated in Fig. 5.6 as a function of time.

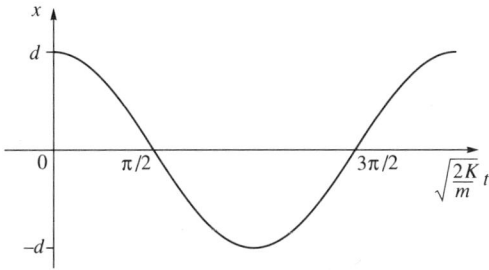

Fig. 5.6 Vibration of mass-spring system

Example 2. The *L–C* circuit

The circuit shown in Fig. 5.7(a) contains an inductance of magnitude L, a capacitor of capacitance C with an initial change Q, and a switch S which is in the open position. When the switch S is closed, as in Fig. 5.7(b), the capacitor discharges, producing a circulating current i and a charge q on the capacitor at time t. In this state the potential difference across the inductance is $L(di/dt)$, and that across the capacitor is q/C.

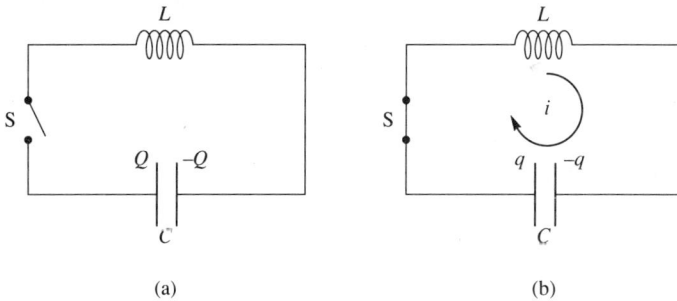

(a)   (b)

Fig. 5.7 The L–C circuit (a) Open position, (b) Closed position

Applying Kirchhoff's voltage law to the circuit, and equating to zero the sum of the potential differences around the circuit as indicated by the arrow, we obtain

$$L\frac{di}{dt} + \frac{q}{C} = 0,$$

By definition $i = dq/dt$, so in terms of q the equation governing the time variation of the charge is

$$L\frac{d^2q}{dt^2} + \frac{q}{C} = 0,$$

or

$$\frac{d^2q}{dt^2} + \frac{1}{LC}q = 0. \tag{15}$$

Identifying this result with (4) shows $\omega_0^2 = 1/LC$, so the general solution is

$$q = b \sin\left(\sqrt{\frac{1}{LC}}\, t + \varepsilon\right), \tag{16}$$

where b and ε are to be determined by the initial conditions corresponding to the manner in which the discharge is started.

The initial charge on the capacitor is Q, so if $t = 0$ is taken to be the time at which the switch S is closed, it follows that $q = Q$ when $t = 0$. The current will be zero the instant the switch is closed, so as $i = dq/dt$ it follows that $dq/dt = 0$ at $t = 0$. The initial conditions are thus $q = Q$ and $dq/dt = 0$ at $t = 0$. Arguing exactly as in Example 1, it then follows that the charge on the capacitor in Fig. 5.7(b) at time t is given by

$$q = Q \cos\left(\sqrt{\frac{1}{LC}}\, t\right), \tag{17}$$

Should it be required, the current which flows in the circuit follows from this by using the fact that

$$i = \frac{dq}{dt} = -Q\sqrt{\frac{1}{LC}} \sin\left(\frac{1}{\sqrt{LC}}\, t\right). \tag{18}$$

Comparing Examples 1 and 2 shows the forms of solution to be identical. At time t the displacement x corresponds to the charge q on the capacitor, and the angular frequency $\omega_0 = \sqrt{2K/m}$ of the mechanical system corresponds to the angular frequency $\omega_0 = \sqrt{1/(LC)}$ in the electrical circuit. ∎

It is analogies such as this between mechanical devices and electrical circuits which are exploited in electrical **analog computers**. By a suitable choice of electrical elements, together with scaling, an electrical circuit can be made analogous to a dynamical system. It is usually difficult to vary parameters in a mechanical system (mass, inertia and stiffness, etc.), though it is easy to vary them in an electrical system (inductance, capacitance and resistance, etc.). Thus electrical analog computers provide a relatively quick and easy way of examining the dynamical behavior of mechanical systems over a wide range of parameter values. Such methods are generally used to find the parameter values necessary in a mechanical system in order for it to provide optimum behavior in some suitably defined sense.

(ii) Free damped oscillations

Free damped oscillations arise when some dissipative mechanism removes energy from an oscillating physical system to which no energy is added. As a result of this process the oscillations will decay in magnitude, and eventually the system will be reduced to its equilibrium state. In the physical world such mechanisms are always present in the form of friction, air resistance, hysteresis effects in springs which dissipate energy as heat, and in electrical resistance, etc. However when these effects are small, and the operating time for the system concerned is suitably limited, it is sometimes possible to neglect the effects of damping. The results of the previous subsection may then be used to describe the system. If, however, such dissipative effects cannot be neglected, the equation describing the oscillations must be modified to take damping into account.

The basic differential equation governing free damped oscillations is

$$\frac{d^2y}{dt^2} + 2a\frac{dy}{dt} + \omega_0^2 y = 0, \tag{19}$$

in which $a > 0$ and $\omega_0 > 0$ are real. The term $2a(dy/dt)$ occurring in (19) represents a **dissipative** effect, or **damping**, and the following examples illustrate some physical situations which give rise to such a form of equation.

Example 3. Free damped oscillations of a mass-spring system

To make Ex. 1 more realistic, let us suppose that the combined effects of resistance between the mass m and the table, air resistance and hysteresis effects in the springs are to produce a force opposing the motion proportional to dx/dt. Experiment shows this to be a reasonable approximation, and if the constant of proportionality is μ, the resisting force will be $\mu(dx/dt)$. When this additional force is taken into account in the equation of motion which led to (10) the result obtained is

$$m\frac{d^2x}{dt^2} = -\mu\frac{dx}{dt} - 2Kx,$$

where the negative sign in the additional term arises because the force it represents *opposes* the motion. Thus the equation of motion including this dissipative effect becomes

$$\frac{d^2x}{dt^2} + \frac{\mu}{m}\frac{dx}{dt} + \frac{2K}{m}x = 0. \tag{20}$$

This is an equation of the form given in (19) in which $2a = \mu/m$ and $\omega_0 = \sqrt{2K/m}$. We shall see later the way in which this additional term modifies the resulting motion and frequency of oscillation when the mass m is displaced as before and then released. ∎

Example 4. Damped simple pendulum

The **simple pendulum** is a mass m suspended from the end of a light inextensible string or rod fastened rigidly at its other end, but allowed to swing in a vertical plane containing its point of suspension. This system becomes a **damped simple pendulum** when the effect of air resistance on the mass is taken into account. Figure 5.8 illustrates such a pendulum of length l, fastened at O, with a mass m at P. The angle between OP and the vertical at time t is θ.

To derive the equation of motion we will use the result from mechanics that for a rigid body rotating about a fixed axis O, the time rate of change of the moment of momentum (angular momentum) about O is equal to the sum of the moments acting about the axis through O.

The pendulum bob P swings in an arc of radius l about O, so the air resistance F acting on the bob will be tangential to the arc at P in a direction opposing the motion. If the air resistance is

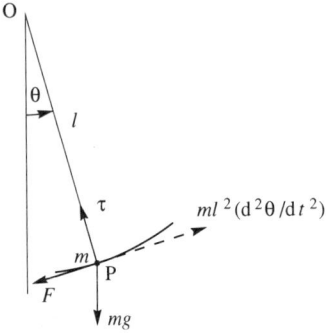

Fig. 5.8 Damped simple pendulum

assumed to be proportional to the linear speed of the bob, which is $l(d\theta/dt)$, it follows that $F = kl(d\theta/dt)$, where k is a suitable constant of proportionality. The weight mg of the bob acts vertically downwards through P, while the tension τ in the rod at P is directed from P to O as shown in Fig. 5.8.

The moment M_F produced by F about the axis through O is $M_F = lF = kl^2(d\theta/dt)$, while the moment M_W produced by the weight mg of the bob about that same axis is $M_W = mg\,l\sin\theta$, where the axis through O is normal to the plane of motion.

The moment of momentum (angular momentum) of the bob about the axis through O is $ml^2(d\theta/dt)$, so its time rate of change is $ml^2(d^2\theta/dt^2)$. Equating this to the sum of the moments acting about the axis through O gives

$$ml^2\frac{d^2\theta}{dt^2} = -kl^2\frac{d\theta}{dt} - mgl\sin\theta,$$

or

$$\frac{d^2\theta}{dt^2} + \frac{k}{m}\frac{d\theta}{dt} + \frac{g}{l}\sin\theta = 0. \tag{21}$$

This is a nonlinear differential equation because of the presence of the term $\sin\theta$. If, however, the maximum angle of swing is small, we may approximate $\sin\theta$ by θ to obtain the linear second order constant coefficient differential equation

$$\frac{d^2\theta}{dt^2} + \frac{k}{m}\frac{d\theta}{dt} + \frac{g}{l}\theta = 0. \tag{22}$$

This is an equation of the form given in (19), in which $2a = k/m$ and $\omega_0 = \sqrt{g/l}$.   ■

Example 5. Torsional oscillations of a crane

Many cranes used on construction sites comprise a vertical support column of square cross-section built up from a lattice of girders on each side, with a long counterweighted jib of similar construction at the top of the column. A typical example of such a structure is shown in Fig. 5.9(a) in which the height of the jib above the ground is h.

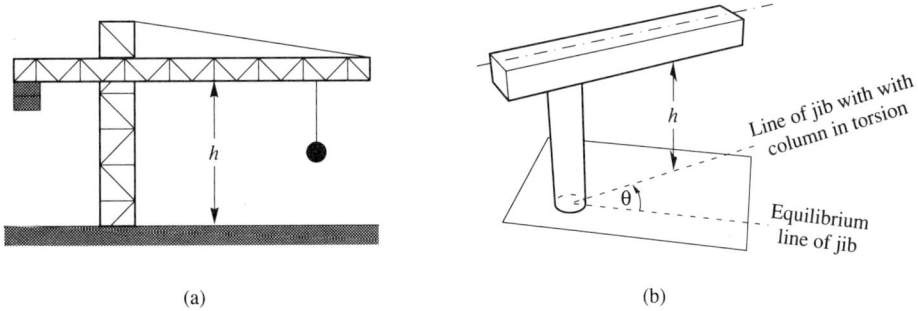

Fig. 5.9 (a) Typical crane (b) Idealized model of the crane with the jib twisted through an angle θ relative to its equilibrium position

When the base of the crane is rotated to a new position, torsional oscillations are set up about the axis of the support column which observation shows persist for several cycles before the jib comes to rest in its new position.

An idealization of this situation is shown in Fig. 5.9(b), in which the support column is replaced by a vertical rod of length h, and the jib and counterweight by a horizontal bar with the appropriate moment of inertia I about the axis of the column. The angular displacement (twist) of the idealized jib about the column relative to its equilibrium position is θ.

For small angles of twist the support column can be considered to exert a restoring moment $\mu\theta$ about its axis, where μ is a suitable constant depending on the crane. The moment about the column caused by the damping effect due to wind resistance may be taken to be proportional to $d\theta/dt$, so the moment due to this effect is $\kappa(d\theta/dt)$, where κ is a suitable constant.

Equating the total moment acting about the column to the rate of change of moment of momentum $I(d^2\theta/dt^2)$, brings us to the equation of motion

$$I\frac{d^2\theta}{dt^2} = -\kappa\frac{d\theta}{dt} - \mu\theta,$$

or,

$$\frac{d^2\theta}{dt^2} + \frac{\kappa}{I}\frac{d\theta}{dt} + \frac{\mu}{I}\theta = 0. \tag{23}$$

This is an equation of the form given in (19), in which $2a = \kappa/I$ and $\omega_0 = \sqrt{\mu/I}$. If wind resistance is neglected $\kappa = 0$, and the time period T of the crane's natural undamped oscillation is seen to be $T = 2\pi\sqrt{I/\mu}$. We shall see later how a nonzero value of κ will modify this period and also cause the oscillations to decay with time. ∎

Example 6. The *R–L–C* circuit

The circuit shown in Fig. 5.10(a) is similar to that shown in Fig. 5.7(a) of Ex. 2, except that it has a resistor of resistance R included in the circuit. As before, the capacitor has an initial charge Q when the switch S is open, and when S is closed a current $i = dq/dt$ flows at time t as shown in Fig. 5.10(b), when the charge on the capacitor is q.

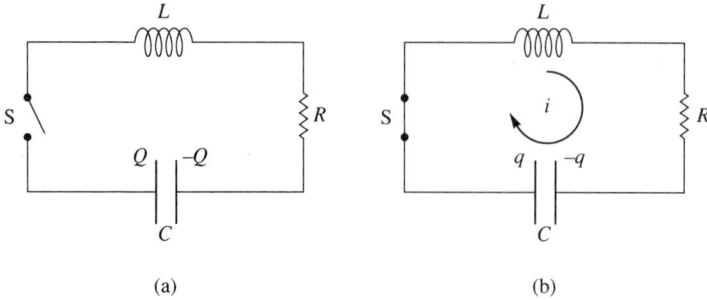

Fig. 5.10 The R–L–C circuit (a) with S open (b) with S closed

The derivation of the equation governing the time variation of the charge q follows as in Ex. 2, except that now there is a potential drop $V = iR$ across the resistor. The equation for i is thus

$$L\frac{di}{dt} + Ri + \frac{q}{C} = 0,$$

which because $i = dq/dt$ is equivalent to

$$\frac{d^2q}{dt^2} + \frac{R}{L}\frac{dq}{dt} + \frac{q}{LC} = 0. \tag{24}$$

This also is an equation of the form (19), in which $2a = R/L$ and $\omega_0 = \sqrt{1/(LC)}$. As in Ex. 2, the initial conditions are again $q = Q$ and $dq/dt = 0$ when $t = 0$.

Table 5.3 lists the most important electrical quantities together with the appropriate units for their measurement.

Table 5.3 Electrical quantities and units

Electrical quantity	Unit
Voltage or emf	volt
Current	ampere
Charge	coulomb
Resistance	ohm
Inductance	henry
Capacitance	farad

It is now time for us to consider the forms taken by the solution of (19). Referring to Sec. 5.2, we see that the nature of the solution depends on the discriminant $a^2 - \omega_0^2$; that is, on whether (a) $a > \omega_0$, (b) $a = \omega_0$ or (c) $a < \omega_0$.

Case (a) $a > \omega_0$ (overdamped case)

The solutions λ_1 and λ_2 of the characteristic equation are real distinct negative quantities, with

$$\lambda_1 = -a - \sqrt{a^2 - \omega_0^2} \quad \text{and} \quad \lambda_2 = -a + \sqrt{a^2 - \omega_0^2}. \tag{25}$$

It follows from Sec. 5.2 that the general solution of (19) is the linear combination

$$y = A \exp(\lambda_1 t) + B \exp(\lambda_2 t), \tag{26}$$

with A and B arbitrary constants. As both λ_1 and λ_2 are real and negative the solution is nonoscillatory. This is called the **overdamped** case, and although for some initial conditions an overshoot of the initial value $y(0)$ is possible, the solution will eventually decay to zero as t becomes arbitrarily large. Typical behavior of overdamped solutions is shown in Fig. 5.11, in which solution (A) is a simple decay and solution (B) has an initial overshoot followed by decay.

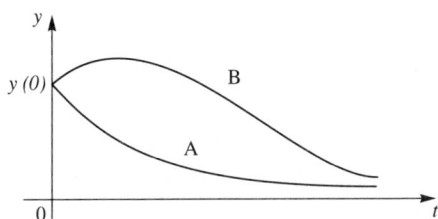

Fig. 5.11 Typical overdamped solutions, with (A) exhibiting pure decay and (B) an initial overshoot followed by decay

Case (b) $a = \omega_0$ (critically damped case)

The solutions λ_1 and λ_2 of the characteristic equation are equal and negative, so that $\lambda_1 = \lambda_2 = \lambda$, with

$$\lambda = -a. \tag{27}$$

It follows from Sec. 5.2 that the general solution of (19) is

$$y = (At + B)e^{-at}, \tag{28}$$

with A and B arbitrary constants.

This is called the **critically damped** case, and once again the solution is nonoscillatory and eventually decays to zero as t becomes arbitrarily large. Depending on the initial conditions, overshoots may or may not occur. The qualitative behavior of solutions in this case is very similar to those illustrated in Fig. 5.11.

Case (c) $a < \omega_0$ (underdamped case)

The solutions λ_1 and λ_2 of the characteristic equation are complex conjugates with negative real parts, for

$$\lambda_1 = -a - i\sqrt{\omega_0^2 - a^2} \quad \text{and} \quad \lambda_2 = -a + i\sqrt{\omega_0^2 - a^2}. \tag{29}$$

It follows from Sec. 5.2 that the general solution of (19) is

$$y = e^{-at}(A \cos \sqrt{\omega_0^2 - a^2}\, t + B \sin \sqrt{\omega_0^2 - a^2}\, t), \tag{30}$$

with A and B arbitrary real constants

The form of argument used to derive (4) from (2) shows that (30) may be re-expressed in the more convenient form

$$y = b e^{-at} \sin(\omega t + \varepsilon), \tag{31}$$

in which b and ε are arbitrary constants and

$$\omega = \sqrt{\omega_0^2 - a^2}. \tag{32}$$

Thus in this case the solution is oscillatory, with ω the angular frequency of the damped oscillation, b the amplitude of the oscillation at $t = 0$, and ε the phase angle. As before, in the form of solution given in (31), b and ε represent the two arbitrary real constants introduced when (19) is integrated. This, the most interesting of the three forms of solution of (19), is called the **underdamped** case. The solution is oscillatory because of the sine function in (31), but not periodic in the sense defined in (5), because the solution decays with time. Defining the **period** T of a damped oscillation $y(t)$ to be twice the time difference between successive zeros of $y(t)$, it is easily seen that $T = 2\pi/\omega$, so that

$$T = \frac{2\pi}{\sqrt{\omega_0^2 - a^2}}. \tag{33}$$

Comparison of (7) and (33) shows how the damping term characterized by the parameter a in (19) modifies the time period of the oscillation through its modification of the angular frequency ω in (32), and hence of y itself in (31). Typical behavior of damped oscillations is shown in Fig. 5.12.

The successive extrema (maxima and minima) of the damped oscillations in (31) occur when $dy/dt = 0$. Thus since

$$\frac{dy}{dt} = b e^{-at}[-a \sin(\omega t + \varepsilon) + \omega \cos(\omega t + \varepsilon)],$$

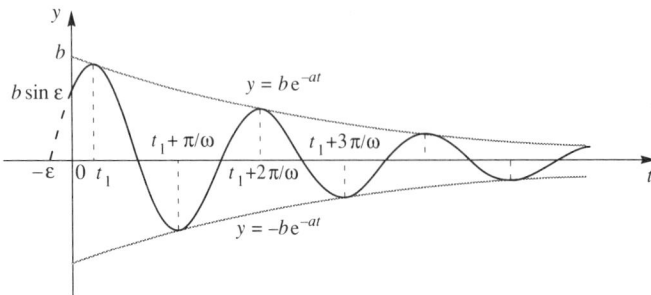

Fig. 5.12 Damped oscillations

the extrema must occur when

$$\tan(\omega t + \varepsilon) = \omega/a. \tag{34}$$

If (34) is first true when $t = t_1$, it follows from the periodicity of the tangent function that it will subsequently be satisfied at the times $t_2 = t_1 + (\pi/\omega)$, $t_3 = t_1 + (2\pi/\omega)$, $t_4 = t_1 + (3\pi/\omega)$, ... Denoting the magnitudes of the successive extrema at $t_1, t_2, \dots, t_n$, ... by $y_1, y_2, \dots, y_n, \dots$, respectively, it follows that extrema of the same kind (maxima or minima) occur at $y_1, y_3, y_5, \dots$, and extrema of the other kind (minima or maxima) occur at $y_2, y_4, y_6, \dots$, each separated by a time period $T = 2\pi/\omega$. Thus from (31) it follows that

$$\frac{y_n}{y_{n+2}} = \frac{b\exp(-at_n)\sin(\omega t_n + \varepsilon)}{b\exp[-a(t_n + T)]\sin[\omega(t_n + T) + \varepsilon]} = e^{aT}, \tag{35}$$

because $T = 2\pi/\omega$, so $\sin[\omega(t_n + T) + \varepsilon] = \sin(\omega t_n + 2\pi + \varepsilon) = \sin(\omega_n t + \varepsilon)$. As this ratio is independent of n, it follows that the ratio of the magnitudes of successive maxima, or successive minima, is always given by e^{aT}.

This result is used to describe the decay rate of damped oscillations in terms of a parameter λ called the **logarithmic decrement**, which is defined as

$$\lambda = \ln\left(\frac{y_n}{y_{n+2}}\right). \tag{36}$$

Using result (35) we see the logarithmic decrement of the solution (31) is

$$\lambda = aT = \frac{2\pi a}{\sqrt{\omega_0^2 - a^2}}. \tag{37}$$

The logarithmic decrement λ is dimensionless (a real number), so (37) shows the parameter $1/a$ has the dimensions of time. In fact inspection of (31) shows that in the time interval $1/a$ the amplitude of the oscillations decreases to $1/e$ of its initial value.

This last result suggests yet another way of describing the decay rate of oscillations. Defining the dimensionless quantity Q (a real number) by

$$Q = \frac{\omega_0}{2a}, \tag{38}$$

we see that Q is determined by the two parameters ω_0 and a which together characterize the behavior of the oscillations. In engineering applications this number is called the **Q-value** of the system (Q for quality). The Q-value is large for lightly damped systems (small a) and small for strongly damped systems (large a). When the damping is light the Q-value has a simple interpretation, for then $\omega \approx \omega_0$, so

$Q/\pi \approx$ the number of free oscillations performed by the system while the amplitude of the oscillations decays to $1/e$ of its initial value.

The Q-value for a mechanical system comprising a child sitting passively on a swing which has been set in motion and then left to come to rest is approximately 60. The

Q-value for an FM radio receiver is approximately 400, while the Q-value for a resonating microwave cavity is typically 2×10^5.

Result (38) may be used to replace the parameters a and ω_0 used to describe damped oscillations so far by Q and ω_0. Combining (32) and (38) gives

$$\omega = \omega_0 \left(1 - \frac{1}{4Q^2} \right)^{1/2}, \tag{39}$$

while combining (19) and (38) gives

$$\frac{d^2 y}{dt^2} + \frac{\omega_0}{Q} \frac{dy}{dt} + \omega_0^2 y = 0. \tag{40}$$

(iii) Forced oscillations and resonance

When either (1) or (19) is modified by the inclusion of a periodic nonhomogeneous term the resulting oscillations are said to be **forced**. This name is used because the non-homogeneous term represents an imposed (**forced**) oscillation to which the system represented by the differential equation must respond. The nonhomogeneous term is usually called the **input** to the oscillating system, or sometimes the **forcing function** of the system.

The most important forcing function used in the study of oscillatory behaviour involves setting the nonhomogeneous term equal to $F \sin \Omega t$, where F and Ω are positive constants. Such a forcing function is often used to approximate other periodic inputs, and it also serves as a standard input when evaluating system response.

First of all we shall consider the effect of such a forcing function when the oscillations are **undamped**. The governing differential equation is then

$$\frac{d^2 y}{dt^2} + \omega_0^2 y = F \sin \Omega t. \tag{41}$$

Using the method of undetermined coefficients (or otherwise), it is easily shown that provided $\Omega \neq \omega_0$, the general solution of (41) may be written

$$y(t) = b \sin (\omega_0 t + \varepsilon) + \frac{F}{\omega_0^2} \left(\frac{\omega_0^2}{\omega_0^2 - \Omega^2} \right) \sin \Omega t, \tag{42}$$

where b and ε are arbitrary constants.

This solution is the linear superposition of two sinusoidal functions with different amplitudes, angular frequencies and phases. To interpret these, notice that the first term of (42) is the complementary function y_c of (41). It describes the free oscillations performed by the system with frequency $\omega_0/2\pi$, amplitude b and phase angle ε. The second term in (42) is the particular integral y_p of (41). This term describes the forced oscillations performed by the system with the forcing frequency $\Omega/2\pi$ and amplitude $F/(\omega_0^2 - \Omega^2)$. The amplitude of the forced oscillations is determined solely by F, Ω and ω_0, and it is independent of the initial conditions for (41). Thus the response of a system described by (41) is the linear superposition of the free oscillations of the system and oscillations at the forcing frequency.

Inspection of the particular integral y_p in (42) shows that in terms of the dimensionless frequency $\bar{\Omega} = \Omega/\omega_0$, the amplification factor r forming the multiplier of the dimensionless amplitude (F/ω_0^2) in (42) is

$$r(\bar{\Omega}) = \frac{1}{1 - \bar{\Omega}^2}.$$ (43)

This is shown graphically in Fig. 5.13, from which it will be seen that $r \to \infty$ as $\Omega \to \omega_0$. The excitation of oscillations of a system by means of a forcing frequency tuned to give the maximum value of the amplitude factor is called **resonance**. In the case of (41), when resonance occurs the amplitude of each successive oscillation grows without bound. However we shall see later that this behavior is modified when damping is present.

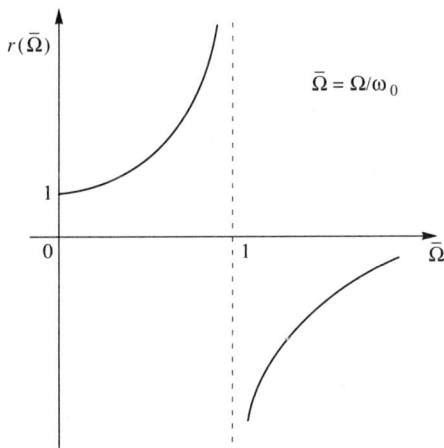

Fig. 5.13 Amplification factor $r(\bar{\Omega})$

To see why undamped oscillations at the resonance frequency grow, notice first that when $\Omega = \omega_0$, the governing equation (41) becomes

$$\frac{d^2 y}{dt^2} + \omega_0^2 y = F \sin \omega_0 t.$$ (44)

Then, using the methods of Sec. 5.2, it is a straightforward matter to show that (44) has the general solution

$$y(t) = b \sin(\omega_0 t + \varepsilon) - \frac{F}{2\omega_0} t \sin \omega_0 t,$$ (45)

with b and ε arbitrary constants. It is the presence of the factor t in the particular integral

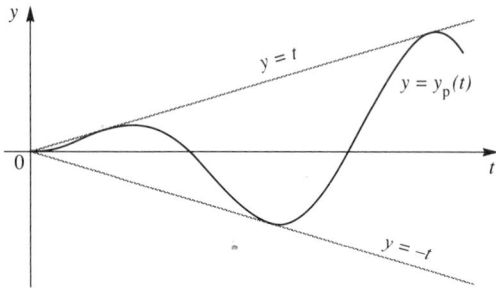

Fig. 5.14   Growth of $y_p(t)$ with time

in (45) (the last term) which causes the unbounded linear growth of the amplitude of y_p with time. This is illustrated in Fig. 5.14.

A special case of (42) arises when the amplitudes of the two sinusoids are both equal, say to k, and Ω is close to ω_0, so that

$$y(t) = k[\sin(\omega_0 t + \varepsilon) + \sin \Omega t]. \tag{46}$$

Using the trigonometric identity

$$\sin C + \sin D = 2 \sin\left(\frac{C+D}{2}\right)\cos\left(\frac{C-D}{2}\right),$$

result (46) may be rewritten

$$y(t) = 2k \sin\left(\frac{(\omega_0 + \Omega)t + \varepsilon}{2}\right)\cos\left(\frac{(\omega_0 - \Omega)t + \varepsilon}{2}\right). \tag{47}$$

When ω_0 and Ω are large, this represents a sinusoidal oscillation with frequency $f_1 = (1/4\pi)(\omega_0 + \Omega)$ and amplitude $2k$, multiplied (**modulated**) by a low frequency cosine oscillation of unit amplitude and frequency $f_2 = (1/4\pi)|\omega_0 - \Omega|$.

The effect is to produce an oscillation with the approximate frequency $(1/2\pi)\omega_0$, whose amplitude varies like

$$\left|\cos\left(\frac{(\omega_0 - \Omega)t + \varepsilon}{2}\right)\right|.$$

In physics this phenomenon is known as the production of **beats**. The frequency f_1 is called the **fundamental frequency**, and the frequency f_2 with which the amplitude varies the **beat frequency**.

Beats arise, for example, when two equal intensity sound waves with slightly different frequencies are superimposed. The effect can be observed experimentally by exciting two identical tuning forks simultaneously, one of which has had the mass of a prong increased by the addition of a small piece of modeling clay.

Figure 5.15 shows a typical example of the phenomenon of beats in which the fundamental frequency $f_1 = 1/T_1$ and the beat frequency $f_2 = 1/T_2$. Provided Ω is close to

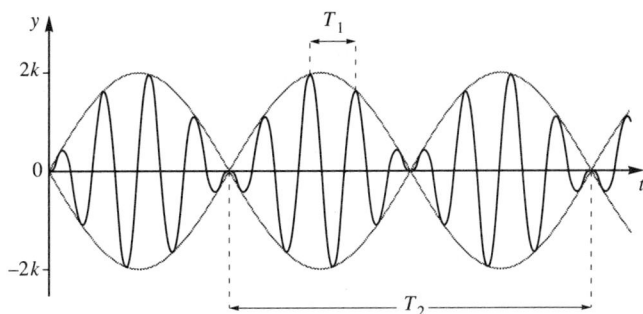

Fig. 5.15 Beats between waves of similar frequencies

ω_0 this effect remains qualitatively true even when the sinuosoidal functions in (42) have slightly different amplitudes.

We now turn our attention to the case of **damped forced** oscillations. These are described by the second order linear constant coefficient differential equation

$$\frac{d^2y}{dt^2} + 2a\frac{dy}{dt} + \omega_0^2 y = F\sin\Omega t, \quad \text{with } a>0. \tag{48}$$

Proceeding as before and using the method of undetermined coefficients (or otherwise), the general solution of (48) is found to be

$$y(t) = be^{-at}\sin(\omega t + c) + \frac{F\sin(\Omega t - \beta)}{[(\omega_0^2-\Omega^2)^2 + 4a^2\Omega^2]^{1/2}}, \tag{49}$$

where as before $\omega=(\omega_0^2-a^2)^{1/2}$, while β is such that

$$\sin\beta = \frac{2a\Omega}{[(\omega_0^2-\Omega^2)^2 + 4a^2\Omega^2]^{1/2}} \quad \text{and} \quad \cos\beta = \frac{\omega_0^2-\Omega^2}{[(\omega_0^2-\Omega^2)^2 + 4a^2\Omega^2]^{1/2}}$$

or, equivalently,

$$\tan\beta = \frac{2a\Omega}{\omega_0^2-\Omega^2}. \tag{50}$$

The first term in (49) is the complementary function y_c of (48), while the second is the particular integral y_p. Notice that the complementary function decays with time $(a>0)$, leaving only the particular integral after a sufficiently long period of time. As the constants b and ε in y_c are determined by the initial conditions for (48), it follows that after a sufficiently long period of time the solution of (48) given in (49) 'forgets' how it started. In this sense, the solution y_p given by

$$y_p(t) = \frac{F\sin(\Omega t - \beta)}{[(\omega_0^2-\Omega^2)^2 + 4a^2\Omega^2]^{1/2}} \tag{51}$$

is often called the **steady state** solution of (48). It is, of course, only steady state in the sense that this is the solution which remains after the effects due to the initial conditions have become negligible, and not in the sense that it is independent of time. For the same reason, the complementary function y_c is often called the **transient** solution of (48).

When (51) is written in the form

$$y_p(t) = \frac{F}{\omega_0^2} A(\Omega) \sin(\Omega t - \beta),$$ (52)

the **amplitude factor**

$$A(\Omega) = \frac{\omega_0^2}{[(\omega_0^2 - \Omega^2)^2 + 4a^2\Omega^2]^{1/2}}$$ (53)

is dimensionless, with $A(0) = 1$. Equating $dA/d\Omega$ to zero to identify the extremum of $A(\Omega)$ shows that $A(\Omega)$ attains its maximum value when $\Omega = \Omega_R$, with

$$\Omega_R = (\omega_0^2 - 2a^2)^{1/2}.$$ (54)

The angular frequency Ω_R corresponds to resonance of the system represented by (48) in the presence of damping. As with the system represented by (41), the amplitude of the oscillations still grows as $\Omega \to \Omega_R$, but now, because of damping, the amplitude factor remains finite, with

$$A(\Omega_R) = \frac{\omega_0^2}{2a(\omega_0^2 - a^2)^{1/2}}.$$ (55)

Examination of (54) shows resonance can only occur when the damping is small enough to satisfy the condition $\omega_0^2 > 2a^2$, for only then will Ω_R be real. If the damping is such that $\omega_0^2 - 2a^2 < 0$, but $\omega_0^2 - a^2 > 0$, the solution of (48) is still oscillatory, but under these circumstances $A(\Omega)$ merely decreases monotonically from $A(0) = 1$ to zero on the interval $0 \le \Omega < \infty$. Finally, if $\omega_0^2 - a^2 \le 0$, the solution of (48) is nonoscillatory, because then the system is either critically damped or overdamped.

The variation of the amplitude factor with angular frequency and damping is best seen in terms of the dimensionless angular frequency $\bar{\Omega} = \Omega/\omega_0$ and the dimensionless damping parameter $\bar{a} = a/\omega_0$ which engineers usually denote by ζ. In terms of these parameters, (53) becomes

$$A(\bar{\Omega}) = \frac{1}{[(1 - \bar{\Omega}^2)^2 + 4\bar{a}^2\bar{\Omega}^2]^{1/2}}.$$ (56)

The behavior of $A(\bar{\Omega})$ for some representative values of $\bar{a}$ is shown in Fig. 5.16. The chain dotted line represents the resonance angular frequency $\Omega_R = \omega_0$ in the undamped case, while the dotted line represents the locus of the maxima of $A(\bar{\Omega})$ as a function of $\bar{a}$. Its equation is given parametrically in terms of $\bar{a}$ by rewriting (54) and (55) as

$$\bar{\Omega}_R = (1 - 2\bar{a}^2)^{1/2}, \quad A(\bar{\Omega}_R) = \frac{1}{2\bar{a}(1 - \bar{a}^2)^{1/2}}.$$ (57)

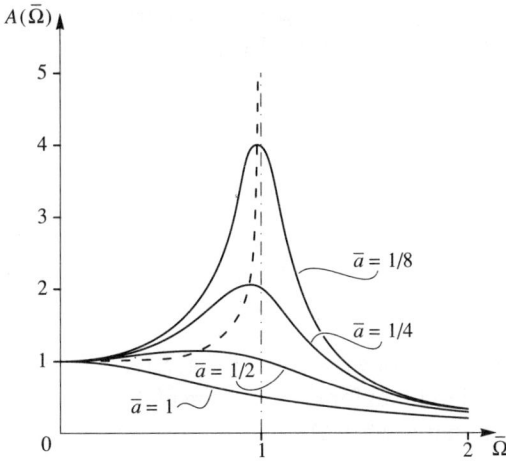

Fig. 5.16 Variation of amplitude factor $A(\bar{\Omega})$ with $\bar{a}$

The phase lag β is determined by (50) in terms of $\bar{a}$ and $\bar{\Omega}$ as

$$\beta = \text{arc} \tan\left(\frac{2\bar{a}\bar{\Omega}}{1-\bar{\Omega}^2}\right). \tag{58}$$

The variation of β with the dimensionless frequency $\bar{\Omega}$ is shown in Fig. 5.17 for some representative value of $\bar{a}$. Notice that $0 < \beta < \pi/2$ when $\bar{\Omega} < 1$, that is when $\Omega < \omega_0$, and $\pi/2 < \beta < \pi$ when $\bar{\Omega} > 1$, that is when $\Omega > \omega_0$. The phase lag $\beta = \pi/2$ when $\bar{\Omega} = 1$, that is when $\Omega = \omega_0$. In the limiting case of zero damping ($\bar{a} = 0$), the phase lag β is seen to experience a discontinuous change of magnitude π when Ω passes through the resonance frequency. In such a case the response is exactly **in phase** ($\beta = 0$), for $0 < \Omega < \omega_0$, and exactly **out of phase** ($\beta = \pi$) for $\Omega > \omega_0$.

Sometimes a steady state phenomenon of interest is related to the quantity dy_p/dt, rather than to y_p itself. This happens, for example, in the R–L–C circuit (see Ex. 6) in which y_p is identified with the steady state variation of the charge q, for then the steady

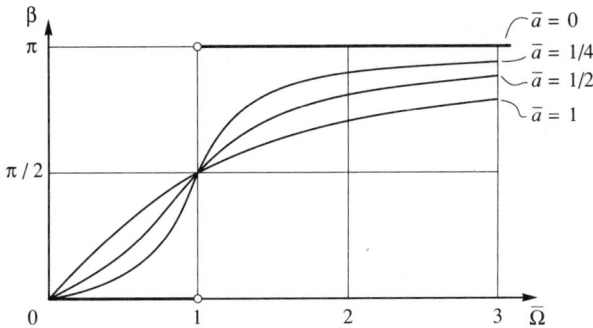

Fig. 5.17 Phase lag β for given values of $\bar{a}$

state current variation $i = dq/dt$ is described by dy_p/dt. It might also happen that y_p is identified with a displacement, when dy_p/dt would be identified with a velocity.

Differentiating (51) with respect to t gives

$$\frac{dy_p}{dt} = \left(\frac{F}{\omega_0}\right) B(\bar{\Omega}) \cos(\Omega t - \beta), \tag{59}$$

where $B(\bar{\Omega})$ is the dimensionless derivative scale factor

$$B(\bar{\Omega}) = \frac{\bar{\Omega}}{[(1 - \bar{\Omega}^2)^2 + 4\bar{a}^2\bar{\Omega}^2]^{1/2}}. \tag{60}$$

The behavior of $B(\bar{\Omega})$ for some representative values of $\bar{a}$ is shown in Fig. 5.18. Unlike the variation of the amplitude factor in Fig. 5.16, the curves for the variation of the derivative scale factor all exhibit resonance when $\bar{\Omega} = 1$; that is when $\Omega = \omega_0$, the angular frequency of resonance in the undamped case.

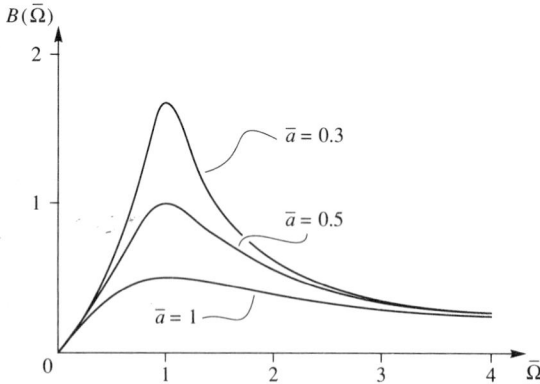

Fig. 5.18 Variation of derivative factor $B(\bar{\Omega})$ with $\bar{a}$

As damping in a mechanical system dissipates energy, it is necessary for power to be supplied to such a system if steady state oscillations are to be maintained. To determine the average power P necessary to sustain these oscillations we shall consider the forced oscillations of the damped mass-spring system of Ex. 3.

The equation of motion for forced oscillations at the angular frequency Ω follows from (20) by including a suitable forcing function. It is the equation

$$m\frac{d^2x}{dt^2} + \mu\frac{dx}{dt} + 2Kx = F_0 \sin \Omega t, \tag{61}$$

in which F_0 is the amplitude of the driving force. The steady state response $x_p(t)$ of (61), corresponding to (51), is

$$x_p(t) = \frac{F_0 \sin(\Omega t - \beta)}{m[\omega_0^2 - \Omega^2)^2 + 4a^2\Omega^2]^{1/2}}, \tag{62}$$

with $a = \mu/(2m)$, $\omega_0^2 = 2K/m$, $\beta = \text{arc tan}[2a\Omega/(\omega_0^2 - \Omega^2)]$ and $x_p(t)$ the displacement of the mass m from equilibrium at time t. Thus if W is the work done, the rate of doing work in the steady state condition is

$$\frac{dW}{dt} = F_0 \sin \Omega t \left(\frac{dx_p}{dt} \right).$$

One oscillation at the forcing angular frequency Ω occurs in the time $T = 2\pi/\Omega$, so the work done during a single cycle is

$$W = \int_0^{2\pi/\Omega} F_0 \sin \Omega t \left(\frac{dx_p}{dt} \right) dt. \tag{63}$$

Computing dx_p/dt from (62), combining the result with (63) and then integrating, shows

$$W = \frac{2\pi a F_0^2 \Omega}{m[\omega_0^2 - \Omega^2)^2 + 4a^2\Omega^2]}. \tag{64}$$

Hence as the average power P necessary to sustain the steady state oscillation during the time T is W/T, we have

$$P = \frac{a F_0^2 \Omega^2}{m[(\omega_0^2 - \Omega^2)^2 + 4a^2\Omega^2]}. \tag{65}$$

The average power P is seen to attain its maximum value P_{max} when $\Omega = \omega_0$, the angular frequency for resonance of the free oscillations of the undamped system. Thus although the value of the damping parameter influences the average power P at any given forcing angular frequency Ω, it does not influence the angular frequency at which P_{max} occurs. A typical example of the variation of P with Ω is shown in Fig. 5.19.

Let the points A and B on the curve in Fig. 5.19 be those at which the power P falls to $\frac{1}{2}P_{max}$ Then if A and B occur at the respective angular frequencies ω_1 and ω_2, the frequency

$$f_b = \frac{\omega_2 - \omega_1}{2\pi} \tag{66}$$

is called the **bandwidth** of the mechanical system.

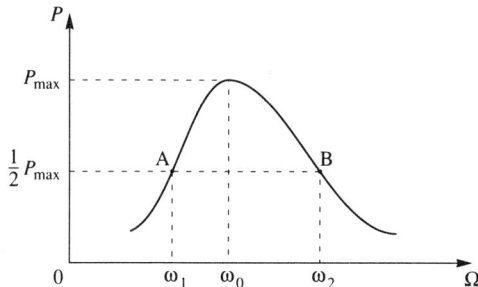

Fig. 5.19 Typical power variation curve

Knowledge of f_b provides a measure of the selectivity (response) of the system to driving frequencies in the vicinity of the peak power frequency $f_0 = \omega_0/(2\pi)$. When $\omega_0/(\omega_2 - \omega_1)$ is large the system is highly selective; that is, it is tuned to respond mainly to driving frequencies close to f_0. If, however, $\omega_0/(\omega_2 - \omega_1)$ is small, the system is non-selective and responds to a wide range of driving frequencies about f_0.

We remark in passing that the reason why a group of walkers crossing a light flexible suspension bridge should break step, is to avoid the risk of exciting a resonance which could lead to dangerously large oscillations of the structure.

Complex method for determination of $y_p(t)$. Reactance and impedance

When the forcing function is a sinusoid, an alternative and convenient method for the determination of the particular integral $y_p(t)$ involves the use of Euler's formula. The method amounts to a complex form of the method of undetermined coefficients, and it is most easily illustrated by finding the particular integral of

$$y'' + 4y' + 5y = 4\sin 3t. \tag{67}$$

The basis of the method is the fact that from Euler's formula we know $4\sin 3t = \text{Im}(4e^{3it})$, so (67) may also be written as

$$y'' + 4y' + 5y = \text{Im}(4e^{3it}). \tag{68}$$

This suggests that instead of finding $y_p(t)$ for (67), we find the complex particular integral $y_p^*(t)$ of the differential equation

$$y'' + 4y' + 5y = 4e^{3it}, \tag{69}$$

and then recover $y_p(t)$ from $y_p^*(t)$ by taking its imaginary part, in agreement with the right-hand side of (68). This process is justified because the differential equation is linear, so $\text{Im}\,[y_p^*(t)]$ must be the particular integral of (67), and although we were not seeking it, $\text{Re}\,[y_p^*(t)]$ must be the particular integral of

$$y'' + 4y' + 5y = 4\cos 3t, \tag{70}$$

because $4\cos 3t = \text{Re}(4e^{3it})$.

Let us now use this argument on (69) by setting

$$y_p^*(t) = Ae^{3it}, \tag{71}$$

where A is a complex constant to be determined. Differentiating $y_p^*(t)$ with respect to t and regarding i as a constant gives

$$y_p^{*\prime} = 3iAe^{3it} \quad \text{and} \quad y_p^{*\prime\prime} = -9Ae^{3it}.$$

Substituting these results into (69) we obtain

$$-9Ae^{3it} + 12iAe^{3it} + 5Ae^{3it} = 4e^{3it},$$

which on simplification shows

$$A = -\tfrac{1}{10}(1+3i). \tag{72}$$

Combining (71) and (72) we find

$$y_p^*(t) = -\tfrac{1}{10}(1+3i)e^{3it}, \tag{73}$$

so that

$$y_p(t) = \text{Im}[y_p^*(t)] = -\tfrac{1}{10}(3\cos 3t + \sin 3t). \tag{74}$$

This result is, of course, the one which would have been obtained by the method of undetermined coefficients starting from the expression

$$y_p(t) = A\cos 3t + B\sin 3t,$$

with A, B undetermined coefficients.

Had the particular integral been required for (70), this would have followed from the same analysis, but with

$$y_p(t) = \text{Re}[y_p^*(t)] = \tfrac{1}{10}(3\sin 3t - \cos 3t).$$

Apart from its convenience, this approach leads naturally to the concept of the **complex impedance** of the R–L–C circuit discussed in Ex. 6, and to the related concepts of **reactance** and **impedance**. If a sinusoidally varying voltage $E_0 \sin \omega t$ is imposed on the circuit shown in Fig. 5.20, then the arguments which led to (24) lead instead to

$$L\frac{d^2q}{dt^2} + R\frac{dq}{dt} + \frac{1}{C}q = E_0 \sin \omega t \tag{75}$$

as the equation for the charge q at time t. This may be transformed into the more useful equation for the current i by differentiation and use of the result $i = dq/dt$, to obtain

$$L\frac{d^2i}{dt^2} + R\frac{di}{dt} + \frac{1}{C}i = E_0 \omega \cos \omega t. \tag{76}$$

If the complex method is now used to determine the particular integral $i_p(t)$ of (76), the approach must start from the equation,

$$L\frac{d^2i}{dt^2} + R\frac{di}{dt} + \frac{1}{C}i = E_0 \omega e^{i\omega t}, \tag{77}$$

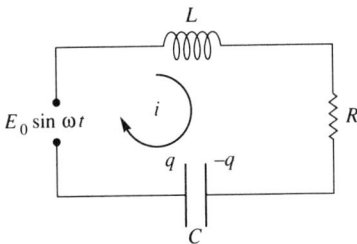

Fig. 5.20 Forced R–L–C circuit

for which a complex particular integral $i_p^*(t)$ is to be sought of the form

$$i_p^*(t) = Ae^{i\omega t}. \tag{78}$$

Arguing as before, and substituting (78) into (77), gives

$$\left(-\omega^2 L + i\omega R + \frac{1}{C} \right) Ae^{i\omega t} = E_0\omega e^{i\omega t},$$

from which it follows that

$$A = \frac{E_0\omega}{\left(\frac{1}{C} - \omega^2 L \right) + i\omega R}. \tag{79}$$

It is conventional to write A in the form

$$A = \frac{E_0}{iZ}, \tag{80}$$

when it follows from (79) that

$$Z = R + i\left(\omega L - \frac{1}{\omega C} \right). \tag{81}$$

The quantity Z in (81) is called the **complex impedance** of the R–L–C circuit. Setting

$$S = \omega L - \frac{1}{\omega C}, \tag{82}$$

which is called the **reactance** of the R–L–C circuit, enables (81) to be written

$$Z = R + iS. \tag{83}$$

Thus $\mathrm{Re}(Z) = R$ is the resistance, and $\mathrm{Im}(Z) = S$ is the reactance of the circuit. The modulus of Z is called the **impedance** of the R–L–C circuit, so in terms of the resistance and the reactance, the **impedance**

$$|Z| = \sqrt{R^2 + S^2}. \tag{84}$$

The relationship between these quantities in the complex plane is shown in Fig. 5.21, in which $\theta = \arg Z$ is determined by the two expressions

$$\sin \theta = \frac{S}{|Z|} = \frac{S}{\sqrt{R^2 + S^2}} \quad \text{and} \quad \cos \theta = \frac{R}{|z|} = \frac{R}{\sqrt{R^2 + S^2}} \tag{85}$$

and, of course, θ depends on ω through the reactance S.

In terms of these quantities (78) may be written

$$i_p^*(t) = -i\frac{E_0}{|Z|} \exp\left[i(\omega t - \theta) \right], \tag{86}$$

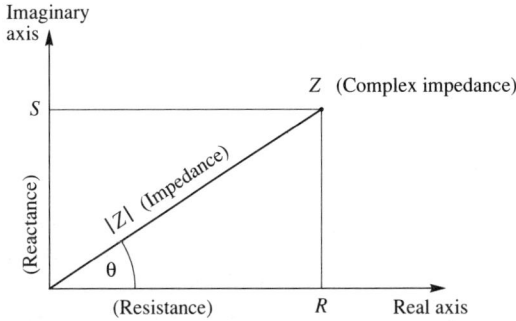

Fig. 5.21 Complex impedance of an R–L–C circuit

so as the current $i_p(t) = \text{Re}[i_p^*(t)]$, it follows that

$$i_p(t) = \frac{E_0}{\sqrt{R^2 + S^2}} \sin(\omega t - \theta). \tag{87}$$

The amplitude I_0 of $i_p(t)$ is thus

$$I_0 = \frac{E_0}{\sqrt{R^2 + S^2}}, \tag{88}$$

so the impedance $\sqrt{R^2 + S^2}$ of the R–L–C circuit is seen to be equal to the ratio E_0/I_0. It should be remembered that, like the reactance S, the impedance $\sqrt{R^2 + S^2}$, the amplitude I_0, and θ all depend on ω.

The transient solution (complimentary function) of (75) will always decay to zero because all circuits have some resistance, and $R > 0$, so eventually the current in the R–L–C circuit will be described by (87).

A comparison of the equations of motion of the mechanical systems considered in this Section with the governing equation (75) for the forced R–L–C circuit establishes the equivalence between the mechanical and electrical parameters in each case. Table 5.4 is typical of such comparisons and is based on consideration of the longitudinal oscillations

Table 5.4

Mechanical system		Electrical system
Longitudinal oscillations	Torsional oscillations	
Mass m	Moment of inertia I	Inductance I
Damping constant μ	Damping constant K	Resistance R
Spring constant κ	Torsional constant τ	Reciprocal of capacitance $1/C$
Applied force $F(t)$	Applied moment $M(t)$	Applied voltage $E(t)$
Displacement x	Angular rotation θ	Charge q
Velocity $v = dx/dt$	Rate of rotation $d\theta/dt$	Current $i = dq/dt$

of Ex. 3 and the torsional oscillations of Ex. 5. Equivalences of this type are used when simulating mechanical oscillations by electrical means.

Problems for Section 5.7

1 The balance wheel of a watch has moment of inertia I about its axis. If the restoring moment produced by the spring is $k\theta$, where θ is the angular displacement from the equilibrium position, derive the equation of motion and hence find the period T of the oscillations.

2 Two identical elastic strings of natural length $3l/4$ and elastic constant k are fastened to a mass m, while their other ends are fastened to two fixed points A and B a distance $2l$ apart, as shown in Fig. 5.22(a). If the mass m is displaced laterally by a small amount and then released, show that to terms of order x/l the equation of motion is

$$\frac{d^2x}{dt^2} + \frac{k}{2m}x = 0,$$

and hence the period T of the oscillations is

$$T = \pi\sqrt{8m/k},$$

where x is the lateral displacement of the mass at time t as shown in Fig. 5.22(b).

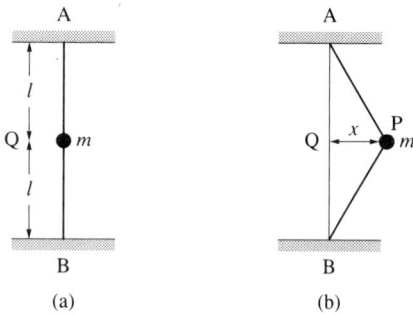

(a)   (b)

Fig. 5.22

3 Masses m_1 and m_2 resting on a smooth horizontal table are connected by a spring with elastic constant k. If the spring is stretched and both masses are then released from rest, use the fact that the center of mass of the system must remain fixed to show the equation of motion is

$$\frac{d^2x}{dt^2} + k\frac{(m_1 + m_2)}{m_1 m_2}x = 0,$$

where x is the displacement of either mass from its equilibrium position at time t.

4 Two particles of mass m are joined by identical springs with elastic constant k to a mass M, as shown in Fig. 5.23, and the mass-spring system rests on a smooth horizontal table. Derive the equation of motion for the system if its oscillations are started by pulling the two masses m apart a small distance and then releasing them from rest.

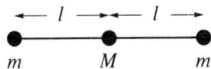

Fig. 5.23

5 An elastic string of unstretched length l is fastened at its top point (Fig. 5.24(a)), and when loaded with a mass m extends by an amount d in its equilibrium position (Fig. 5.24(b)). If the displacement of the mass-spring system from its equilibrium position is x at time t (Fig. 5.24(c)), and the air resistance acting on the mass is proportional to its speed with constant of proportionality k, show that provided the string never becomes slack, the equation of motion is

$$\frac{d^2x}{dt^2} + \frac{k}{m}\frac{dx}{dt} + \frac{g}{d}x = 0.$$

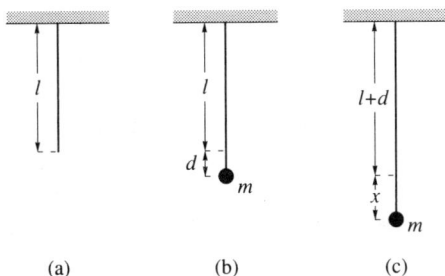

Fig. 5.24

6 Derive the equation of motion for the free oscillations of the mass m in the mass-spring system shown in Fig. 5.25, in which the mass is attached to a damping mechanism known as a dashpot. A **dashpot** is an oil-filled cylinder in which the motion of a piston provides a resisting force proportional to the speed of the piston. It may be assumed that x is the displacement of the mass from its equilibrium position at time t, the constant of proportionality for the dashpot is μ and the spring constant is k.

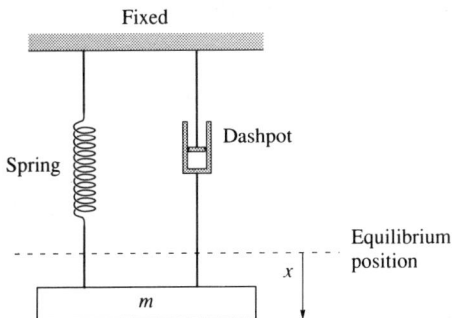

Fig. 5.25

7 The hydrometer of mass m and area of cross-section A shown in Fig. 5.26 floats in a liquid of density ρ. Find the equation of motion if, when the hydrometer is displaced vertically by a small amount x, the resistance to motion is proportional to the speed of displacement, with constant of proportionality k.

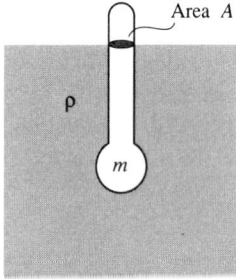

Fig. 5.26

8 The manometer tube shown in Fig. 5.27 of cross-sectional area A is filled with a liquid of density ρ, the total length of the column being l. If the displacement of the top of the right-hand column is x at time t, find its equation of motion on the assumption that the resistance to motion is proportional to the speed, with constant of proportionality equal to k. What is the time period of the free oscillations of the column if the resistance to motion is neglected?

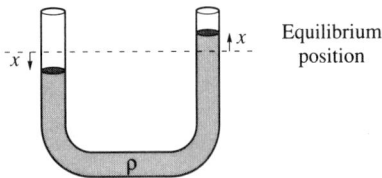

Fig. 5.27

9 An ice cube of mass m in a hemispherical bowl slides from side to side about the lowest point of the bowl. If the motion is planar, R and θ are as shown in Fig. 5.28, g is the acceleration due to gravity, the angle θ remains small and the frictional resistance is proportional to the speed of the ice cube with proportionality constant k, show the equation of motion is

$$\frac{d^2\theta}{dt^2} + k\frac{d\theta}{dt} + \frac{g}{l}\theta = 0.$$

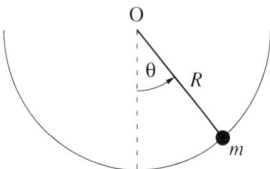

Fig.5.28

10 A horizontal plate pivots about a vertical axis through point O as shown in Fig. 5.29(a). The moment of inertia of the plate about the axis is I. An elastic string fixed to point P on the plate distant l from O has its other end fixed to a point Q outside the plate, but in its plane, and distant d from P. If the initial tension in the string is T, determine the equation of motion for small oscillations of the system, if at time t the angle between OP and its equilibrium direction is θ as shown in Fig. 5.29(b), and hence find the period of free oscillations of the system. What would be the form of the equation of motion if the system is immersed in a viscous fluid which produces a moment opposing rotation proportional to $d\theta/dt$?

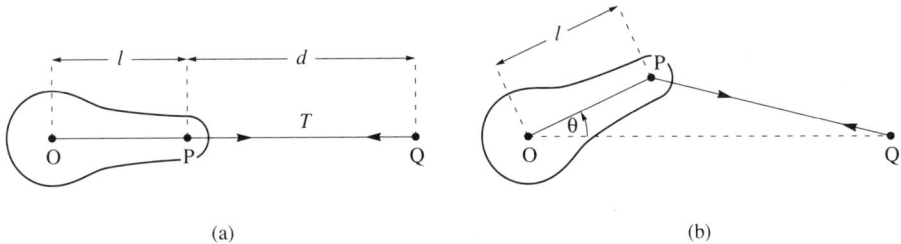

(a)   (b)

Fig. 5.29

11 Derive solution (31) for the free underdamped oscillations of (19).
12 Express solution (31) for the free underdamped oscillations of (19) in terms of ω_0 and Q. Find the form taken by this solution for large Q.
13 Derive the undamped forced solution (42) of (41).
14 Derive the undamped resonance solution (45) of (44).
15 Derive the 'beat' solution (47) from expression (46). Draw a good graph of the solution for the case $k=1$, $\varepsilon=0$, $\omega_0=3$, $\Omega=2$ for $0\le t\le 4\pi$.
16 Derive the damped forced solution (49) of (48).

Find the steady state solution for each of the following problems.
17 $y''+3y=\cos 2t$.
18 $2y''+5y=\sin t$.
19 $9y''+6y'+37y=3\sin t$.
20 $y''+6y'+13y=\cos t$.
21 $y''+4y'+5y=\sin 2t$.
22 $y''+y'+y=2\sin\frac{1}{2}t$.
23 $y''+25y=3\sin 4t$.
24 $y''+4y=4\cos 6t$.
25 $4y''+4y'+37y=3\sin 4t-\sin t$.
26 $y''+2y'+5y=4\cos 3t$.
27 $9y''+6y'+37y=2\sin t+3\cos t$.
28 $16y''+8y'+17y=5\sin 4t$.

Solve the following initial value problems.
29 $y''+16y=0$, $y(0)=1$, $y'(0)=1$.
30 $y''+4y=0$, $y(0)=0$, $y'(0)=4$.

31 $y'' + 2y' + 5y = 0,$ $y(0) = 0, \ y'(0) = 6.$
32 $y'' + 4y' + 6y = 0,$ $y(0) = 3, \ y'(0) = 0.$
33 $y'' + y = \cos 2t,$ $y(0) = 0, \ y'(0) = 1.$
34 $y'' + 9y = 2 \sin t,$ $y(0) = 1, \ y'(0) = -1.$
35 $y'' + y = 6 \sin t,$ $y(0) = 1, \ y'(0) = 1.$
36 $y'' + y = \cos t,$ $y(0) = 2, \ y'(0) = 0.$
37 $y'' + 2y' + 2y = \cos 3t,$ $y(0) = 0, \ y'(0) = 0.$
38 $3y'' + 2y' + y = \sin 2t,$ $y(0) = 0, \ y'(0) = 0.$

Use the complex method to determine the particular integral for each of the following problems.
39 $y'' + 9y = 2 \sin t.$
40 $y'' + y = 4 \cos t.$
41 $3y'' + 2y' + y = \sin 2t.$
42 $y'' + 4y' + 13y = 2 \cos 2t.$

5.8   Reduction to the normal form $u'' - f(x)u = 0$

When working with linear second order differential equations with variable coefficients it is sometimes advantageous to make a change of variable which removes the first derivative term. That is, to change variables so that the differential equation

$$y'' + p(x)y' + q(x)y = 0, \tag{1}$$

is reduced to the normal form

$$u'' - f(x)u = 0. \tag{2}$$

This may be accomplished by using the following result.

Theorem 5.8 (Reduction of a linear second order equation to normal form)

Let y be a solution of

$$y'' + p(x)y' + q(x)y = 0.$$

Set $y = uv$, and define v by the expression

$$v = \exp[-\tfrac{1}{2}\int p\,dx].$$

Then the differential equation for u becomes

$$u'' - f(x)u = 0,$$

where

$$f(x) = \tfrac{1}{2}p' + \tfrac{1}{4}p^2 - q.$$

Proof

The proof of this theorem is simple. Setting $y=uv$ in (1) and grouping terms in u' we obtain

$$u''v+(2v'+pv)u'+(v''+pv'+qv)u=0. \tag{3}$$

The first derivative term in u' arising in (3) will vanish if

$$2v'+pv=0,$$

which is when

$$v=\exp[-\tfrac{1}{2}\textstyle\int p\,dx].$$

This establishes the first part of the theorem.

We have

$$v'=-\tfrac{1}{2}pv,$$

so differentiating this with respect to x to find v'' and substituting for v' it follows that

$$v''=-\tfrac{1}{2}p'v+\tfrac{1}{4}p^2v.$$

Using these results to eliminate v' and v'' from the last term in (3), and then cancelling v, leads to the differential equation

$$u''-(\tfrac{1}{2}p'+\tfrac{1}{4}p^2-q)u=0,$$

which completes the proof. $\qquad\qquad\qquad\qquad\qquad\qquad\qquad\qquad\square$

Usually the reduction to normal form is a preliminary stage in the examination of differential equation (1), and other methods are necessary to understand the behavior of the differential equation for u. Sometimes the equation for u can be integrated in terms of elementary functions and then, as $y=uv$, the general solution of (1) can be found. This is illustrated in the next example.

Example 1. Solution after reduction to normal form

Reduce to normal form the differential equation

$$x^2y''+xy'+(x^2-\tfrac{1}{4})y=0.$$

Find two linearly independent solutions of the normal form, and hence find the general solution for y.

Solution

Dividing by x^2 ($x\neq0$) and identifying the resulting differential equation with (1) shows that

$$p=\frac{1}{x}\quad\text{and}\quad q=1-\frac{1}{4x^2}.$$

It follows directly from Theorem 5.8 that to reduce the differential equation to normal form we must first set

$$v = \exp\left(-\frac{1}{2}\int\frac{1}{x}dx\right) = x^{-1/2}.$$

Using the expressions for p and q together with the fact that $p' = -1/x^2$ we find that

$$f(x) = \frac{-1}{2x^2} + \frac{1}{4x^2} - 1 + \frac{1}{4x^2} = -1.$$

Thus, from Theorem 5.8, u satisfies the differential equation

$$u'' + u = 0.$$

Two linearly independent solutions for u are

$$u_1(x) = \cos x \quad \text{and} \quad u_2(x) = \sin x.$$

Now as $y = uv$ the corresponding linearly independent solutions of the original differential equation are

$$y_1(x) = x^{-1/2}\cos x \quad \text{and} \quad y_2(x) = x^{-1/2}\sin x.$$

The general solution of the differential equation is thus

$$y = x^{-1/2}(c_1\cos x + c_2\sin x). \qquad\blacksquare$$

The next example is one in which the equation for u in normal form cannot be integrated in terms of elementary functions.

Example 2

Reduce to normal form the differential equation

$$xy'' - 2xy' - 4y = 0.$$

Solution

Dividing by $x(x \neq 0)$ and identifying the result with (1) shows that

$$p = -2 \quad \text{and} \quad q = -4/x.$$

It follows directly from Theorem 5.8 that

$$v = e^x \quad \text{and} \quad f(x) = 1 + \frac{4}{x},$$

so u satisfies the equation in normal form

$$u'' - \left(1 + \frac{4}{x}\right)u = 0.$$

This equation cannot be integrated in terms of elementary functions, but inspection suggests that it might have a simple approximate solution when $4/x$ is small, that is when x is large. The determination of such approximate solutions for u, and hence for y, when x is large will be discussed later in the context of *asymptotic solutions*. ∎

Problems for Section 5.8

Reduce each of the following differential equations to normal form, and use this form to find the general solution of the original differential equation,

1 $y'' - 4y' + 3y = 0$.
2 $y'' + y' - 6y = 0$.
3 $y'' + 2y' + y = 0$.
4 $y'' + 6y' + 9y = 0$.
5 $y'' + 4y' + 9y = 0$.
6 $y'' - 4y' + 4y = 0$.
7 $y'' + ay' + by = 0$ when a and b are real numbers and
   (a) $\frac{1}{4}a^2 - b = -n^2 < 0$, $a \neq 0$,   (b) $\frac{1}{4}a^2 - b = 0$, $a \neq 0$,
   (c) $\frac{1}{4}a^2 - b > 0$, $b > 0$.
8 $xy'' + 2y' + 9xy = 0$.
9 $y'' + 2 \sin xy' + \cos x(1 - \cos x)y = 0$.

Reduce each of the following differential equations to normal form, find v, but do not attempt to solve the differential equation for u.

10 $y'' - 2xy' - 4y = 0$.
11 $y'' + \cos xy' + (1 + \sin^2 x)y = 0$.
12 $x^2 y'' + xy' + (x^2 - 4)y = 0$.
13 $x^2 y'' - 2y' + 4xy = 0$.
14 $xy'' - 4xy' - 2y = 0$.

5.9 The Green's function

In this section we reconsider the form of the general solution of the nonhomogeneous second order equation

$$a(x)y'' + b(x)y' + c(x)y = f(x) \tag{1}$$

first encountered in Sec. 5.5. Our purpose will be to outline a more advanced approach to the solution of (1). In this we shall show how the general solution may be expressed in terms of an integral involving a special function called the **Green's function**[5] for the

---

[5] GEORGE GREEN (1793–1841), a self-educated English mathematical physicist who was born in Nottingham. He made important contributions to the study of electricity and magnetism, but as this work was published privately in 1828 it remained largely unknown until William Thompson (Lord Kelvin) arranged for it to be reprinted in 1846. This work reflected the influence of the French school of mathematicians on Green's research, and it contained what is now known as Green's theorem. In the meantime this, and others of his theorems, had been rediscovered by Lord Kelvin, K. F. Gauss and others. Green also made significant contributions to the theory of the reflection and refraction of light and sound waves, and by the time of his death had been elected to fellowship of Caius College, Cambridge.

homogeneous equation

$$a(x)y'' + b(x)y' + c(x)y = 0 \tag{2}$$

associated with (1), when $a(x)$, $b(x)$ and $c(x)$ are continuous functions.

In particular we shall see how, when $a(x) \neq 0$ in the interval in which the solution is defined, the Green's function may be constructed by using any two linearly independent solutions ϕ_1 and ϕ_2 of (2).

We start from result (9) of Sec. 4.2 which enables the first indefinite integral in (21) of Sec. 5.5 to be written as

$$\int \frac{\phi_2 f}{aW} dx = \int_{x_0}^{x} \frac{\phi_2(t) f(t)}{a(t) W(t)} dt + k_1, \tag{3}$$

where k_1 is an arbitrary constant, x_0 is any point at which ϕ_1 and ϕ_2 are defined, and $W(t) = \phi_1(t)\phi_2'(t) - \phi_2(t)\phi_1'(t)$ is the Wronskian of ϕ_1 and ϕ_2. Grouping the first and third terms of (21) in Sec. 5.5 we obtain

$$(c_1 - k_1)\phi_1(x) - \phi_1(x) \int_{x_0}^{x} \frac{\phi_2(t) f(t)}{a(t) W(t)} dt. \tag{4}$$

The corresponding result obtained by grouping the second and fourth terms is

$$(c_2 + k_2) + \phi_2(x) \int_{x_0}^{x} \frac{\phi_1(t) f(t)}{a(t) W(t)} dt, \tag{5}$$

where k_2 is the arbitrary constant associated with the second indefinite integral.

Replacing $(c_1 - k_1)$ and $(c_2 + k_2)$ by c_1 and c_2, respectively (arithmetic is not performed on arbitrary constants), adding (4) and (5), and taking functions of x under the definite integral sign (they behave like constants as x is a limit) enables the general solution (21) of Sec. 5.5 to be written

$$y = c_1\phi_1(x) + c_2\phi_2(x) - \int_{x_0}^{x} \frac{[\phi_1(x)\phi_2(t) - \phi_2(x)\phi_1(t)]}{a(t) W(t)} f(t) dt. \tag{6}$$

If we now define the **Green's function** $G(x, t)$ for (2) to be

$$G(x, t) = -\frac{1}{a(t)} \frac{\begin{vmatrix} \phi_1(x) & \phi_2(x) \\ \phi_1(t) & \phi_2(t) \end{vmatrix}}{\begin{vmatrix} \phi_1(t) & \phi_2(t) \\ \phi_1'(t) & \phi_2'(t) \end{vmatrix}}, \qquad \text{for } x_0 \leq t \leq x, \tag{7}$$

the general solution of (1) given in (6) is expressible in terms of $G(x, t)$ and becomes

$$y = c_1\phi_1(x) + c_2\phi_2(x) + \int_{x_0}^{x} G(x, t) f(t) dt. \tag{8}$$

The first two terms of (8) represent the complementary function y_c of (1), while the term

$$\tilde{y}_p = \int_{x_0}^{x} G(x, t) f(t) dt \tag{9}$$

represents a particular solution of (1). We draw attention here to the fact that (9) is not necessarily the particular integral y_p, since $\tilde{y}_p$ may contain terms from the complementary function. It should be clearly understood that the general solution of (1) is always representable as the sum of the complementary function y_c and a particular solution $\tilde{y}_p$ (as in (8)), but that $\tilde{y}_p$ only becomes the particular integral y_p (which is unique) when it contains no terms from y_c.

Observe that once the Green's function $G(x, t)$ has been found for (2), its use in (8) then enables the general solution of (1) to be obtained for any function $f(x)$. This offers some advantage over the other methods we have discussed so far should solutions be required to the same differential equation with different nonhomogeneous terms $f(x)$. However, the main reason for the importance of the Green's function is due to the part it plays in the solution of two-point boundary value problems and integral equations.

It follows from (7) and the properties of determinants (Theorem 3.14) that

$$G(x, x) = 0, \tag{10}$$

and from (9) by differentiation[6] that

$$\tilde{y}_p'(x) = G(x, x) f(x) + \int_{x_0}^{x} \left[\frac{\partial}{\partial x} G(x, t) \right] f(t) dt,$$

so that

$$\tilde{y}_p'(x_0) = G(x_0, x_0) f(x_0) = 0. \tag{11}$$

Thus the unique particular solution $\tilde{y}_p$ in (9) satisfies the conditions

$$\tilde{y}_p(x_0) = \tilde{y}_p'(x_0) = 0. \tag{12}$$

Example 1. The Green's function

Find the Green's function for

$$y'' + y = 0,$$

and use it to obtain the general solution of

$$y'' + y = 3 \sin x.$$

---

[6] Using the general result from elementary calculus (**Leibniz' Rule**)

$$\frac{d}{dx} \int_{h(x)}^{k(x)} f(x, t) dt = \left(\frac{dk}{dx} \right) f(x, k) - \left(\frac{dh}{dx} \right) f(x, h) + \int_{h(x)}^{k(x)} \frac{\partial}{\partial x} f(x, t) dt,$$

and making the identifications $h \equiv 0$, $k \equiv x$ and $f(x, t) \equiv G(x, t) f(t)$.

Solution

Here $a(x)=1$, $b(x)=0$, $c(x)=1$, and $f(x)=3\sin x$, while the characteristic equation is

$$\lambda^2 + 1 = 0,$$

with the roots $\lambda = \pm i$. Thus two linearly independent solutions are $\phi_1 = \cos x$ and $\phi_2 = \sin x$, which are valid for all real x. Substituting into (7) gives

$$G(x,t) = -\frac{\begin{vmatrix} \cos x & \sin x \\ \cos t & \sin t \end{vmatrix}}{\begin{vmatrix} \cos t & \sin t \\ -\sin t & \cos t \end{vmatrix}} = \frac{\sin x \cos t - \cos x \sin t}{\cos^2 t + \sin^2 t},$$

so that

$$G(x,t) = \sin(x-t) \qquad \text{for } x_0 \le t \le x.$$

The general solution of the nonhomogeneous equation now follows from (8) in the form

$$y = c_1 \cos x + c_2 \sin x + 3 \int_{x_0}^{x} \sin(x-t)\sin t\, dt, \qquad \text{for all } x.$$

Performing the integration gives

$$y = c_1 \cos x + c_2 \sin x + \tfrac{3}{4}\sin x - \tfrac{3}{4}\sin(2x_0 - x) + \tfrac{3}{2}(x_0 - x)\cos x,$$

for all x and arbitrary x_0. Making the convenient choice $x_0 = 0$, this becomes

$$y = c_1 \cos x + c_2 \sin x + \tfrac{3}{2}\sin x - \tfrac{3}{2}x \cos x, \qquad \text{for all } x,$$

when $\tilde{y}_p$ is seen to be

$$\tilde{y}_p = \tfrac{3}{2}\sin x - \tfrac{3}{2}x \cos x.$$

If the term $\tfrac{3}{2}\sin x$ in $\tilde{y}_p$ is absorbed into the complementary function, the term that remains must be the particular integral, and so in this case

$$y_p = -\tfrac{3}{2}x \cos x.$$

This is the result that would have been obtained had the method of undetermined coefficients or variation of parameters been used. ∎

When two-point boundary value problems are considered for the nonhomogeneous equation (1), the definition of the Green's function given in (7) must be modified. Let us seek the solution of the boundary value problem for (1) on the interval $[a, b]$ in which the solution is subject to the **homogeneous boundary conditions**[7]

$$\alpha_1 y(a) + \beta_1 y'(a) = 0 \quad \text{and} \quad \alpha_2 y(b) + \beta_2 y'(b) = 0, \tag{13}$$

where α_1, α_2, β_1 and β_2 are given constants ($\alpha_1^2 + \beta_1^2 > 0$, $\alpha_2^2 + \beta_2^2 > 0$).

---

[7] The results which follow are also true for **mixed homogeneous boundary conditions** such as

$\alpha_{11}y(a) + \beta_{11}y'(a) + \alpha_{12}y(b) + \beta_{12}y'(b) = 0 \quad \text{and} \quad \alpha_{21}y(a) + \beta_{21}y'(a) + \alpha_{22}y(b) + \beta_{22}y'(b) = 0.$

We shall first arrive at our result using intuitive arguments, and then prove that this result is indeed the required solution. It will be necessary to make use of two linearly independent solutions ϕ_1 and ϕ_2 of the homogeneous equation (2), chosen such that ϕ_1 satisfies the boundary condition in (13) corresponding to the left-hand end point of $[a, b]$, while ϕ_2 satisfies the one at the right-hand end point. Thus ϕ_1 and ϕ_2 will satisfy the respective boundary conditions

$$\alpha_1 \phi_1(a) + \beta_1 \phi_1'(a) = 0 \quad \text{and} \quad \alpha_2 \phi_2(b) + \beta_2 \phi_2'(b) = 0. \tag{14}$$

To arrive at the solution at any interior point x of $[a, b]$, let us again appeal to arguments based on the variation of parameters solution in Sec. 5.5, in which the solution itself is written

$$y(x) = \phi_1(x) u_1(x) + \phi_2(x) u_2(x), \tag{15}$$

Now the point x divides $[a,b]$ into the two subintervals $[a, x]$ and $[x, b]$, and ϕ_1 has been chosen such that it satisfies the boundary condition at the left-hand end point of $[a, x]$, while ϕ_2 satisfies the one at the right-hand end point of $[x, b]$. Inspecting results (18) of Sec. 5.5 shows the function ϕ_1 only occurs in u_2'; so let us integrate u_2' from a to x and use the resulting expression for u_2 in (15). As ϕ_2 only occurs in u_1' and satisfies the boundary condition at b, let us integrate u_1' from b to x and use the resulting expression for u_1 in (15). When this is done we obtain

$$y(x) = \phi_2(x) \int_a^x \frac{\phi_1(t) f(t)}{a(t) W(t)} \, dt - \phi_1(x) \int_b^x \frac{\phi_2(t) f(t)}{a(t) W(t)} dt.$$

Reversing the limits in the second integral, and taking functions of x under the integral signs (they behave like constants as x is a limit) gives

$$y(x) = \int_a^x \frac{\phi_1(t) \phi_2(x) f(t)}{a(t) W(t)} \, dt + \int_x^b \frac{\phi_1(x) \phi_2(t) f(t)}{a(t) W(t)} dt, \tag{16}$$

or

$$y(x) = \int_a^b G(x, t) f(t) dt, \tag{17}$$

where

$$G(x, t) = \begin{cases} \dfrac{\phi_1(x) \phi_2(t)}{a(t) W(t)}, & \text{for } a \leq x \leq t \text{ (equivalently, } x \leq t \leq b) \\[3mm] \dfrac{\phi_1(t) \phi_2(x)}{a(t) W(t)}, & \text{for } t \leq x \leq b \text{ (equivalently, } a \leq t \leq x) \end{cases} \tag{18}$$

The function $G(x, t)$ defined in (18) is called the **Green's function for the two-point boundary value problem** associated with (1) and (13). In general, expressions of the form on the right-hand side of (17) are called **integral operators**.

It only remains for us to prove that (16) does indeed provide the solution to our problem. This is accomplished by showing that (16) satisfies the boundary conditions for y, and that it also satisfies (1). Differentiating (16) by means of Leibniz' theorem (footnote 6) to find $y'(x)$ gives

$$y'(x) = \int_a^x \frac{\phi_1(t)\phi_2'(x)f(t)}{a(t)W(t)}dt + \int_x^b \frac{\phi_1'(x)\phi_2(t)f(t)}{a(t)W(t)}dt. \tag{19}$$

Forming $\alpha_1 y(a) + \beta_1 y'(a)$ from (16) and (19) leads to the result

$$\alpha_1 y(a) + \beta_1 y'(a) = \int_a^b [\alpha_1\phi_1(a) + \beta_1\phi_1'(a)]\frac{\phi_2(t)f(t)}{a(t)W(t)}dt,$$

which is zero because $\alpha_1\phi_1(a) + \beta_1\phi_1'(a) = 0$. Thus the boundary condition at the left-hand end point is satisfied. The proof that the boundary condition at the right-hand end point is also satisfied follows in similar fashion.

To show (16) satisfies (1) we also need to determine $y''(x)$. This follows by differentiating (19), when we obtain

$$y''(x) = \frac{f(x)}{a(x)} + \int_a^x \frac{\phi_1(t)\phi_2''(x)f(t)}{a(t)W(t)}dt + \int_x^b \frac{\phi_1''(x)\phi_2(t)f(t)}{a(t)W(t)}dt. \tag{20}$$

Substituting for y, y' and y'' in the left-hand side of (1) gives

$$f(x) + \int_a^x [a(x)\phi_1'' + b(x)\phi_1' + c(x)\phi_1]\frac{\phi_2(t)f(t)}{a(t)W(t)}dt$$

$$+ \int_x^b [a(x)\phi_2'' + b(x)\phi_2' + c(x)\phi_2]\frac{\phi_2(t)f(t)}{a(t)W(t)}dt.$$

This is simply $f(x)$, because ϕ_1 and ϕ_2 are solutions of the homogeneous equation (2), so that the expressions in square brackets are both identically zero. Thus when (16) is inserted into the left-hand side of (1) it generates $f(x)$, which is the nonhomogeneous term on the right. The result is an identity in x, and our assertion that (16) is the required solution is proved.

Notice that $G(x, t)$ is a continuous function, but that its partial derivative with respect to t is discontinuous across $t = x$. These properties, and others possessed by Green's functions, play an important part in more advanced accounts of the representation of solutions in terms of integral operators.

Example 2. Boundary value problems using a Green's function

Find the Green's function associated with the two-point boundary value problem

$$y'' - k^2 y = 0, \quad \text{with} \quad y(0) = y(1) = 0.$$

Use the Green's function to solve the nonhomogeneous equation

$$y'' - k^2 y = f(x),$$

subject to the same boundary conditions.

Solution

The general solution of the homogeneous equation is

$$y = c_1 \sinh kx + c_2 \cosh kx,$$

and in this case $a = 0$ and $b = 1$. As $\phi_1(x)$ must satisfy the same left-hand boundary condition as y we require $\phi_1(0) = 0$, and so we set

$$\phi_1(x) = \sinh kx.$$

The function $\phi_2(x)$ must satisfy the same right-hand boundary condition as y, and so as $\phi_2(1) = 0$ we set

$$\phi_2(x) = c_1 \sinh kx + c_2 \cosh kx$$

and determine c_1 and c_2. Setting $\phi_2(1) = 0$ gives

$$0 = c_1 \sinh k + c_2 \cosh k,$$

and thus

$$\phi_2(x) = \frac{c_1}{\cosh k} (\sinh kx \cosh k - \cosh kx \sinh k).$$

$$= \frac{c_1}{\cosh k} \sinh k(x - 1).$$

Discarding the constant factor $c_1/\cosh k$, since ϕ_2 is indeterminate up to a multiplicative constant, we see that we may set

$$\phi_2(x) = \sinh k(x - 1).$$

Substituting into (18) then shows

$$G(x,t) = \begin{cases} \dfrac{\sinh kx \sinh k(t-1)}{k \sinh k}, & 0 \le x \le t \\[2mm] \dfrac{\sinh kt \sinh k(x-1)}{k \sinh k}, & t \le x \le 1, \end{cases}$$

where we have used the fact that $a(t) = 1$ and

$$W(t) = \begin{vmatrix} \sinh kx & \sinh k(x-1) \\ k \cosh kx & k \cosh k(x-1) \end{vmatrix} = k \sinh k.$$

The solution of the second part of the problem follows by evaluating

$$y(x) = \int_0^1 G(x,t) f(t) dt, \qquad 0 \le x \le 1.$$

This produces the solution

$$y = \frac{\sinh k(x-1)}{k \sinh k} \int_0^x \sinh kt f(t) dt + \frac{\sinh kx}{k \sinh k} \int_x^1 \sinh k(t-1) f(t) dt,$$

for $0 \le x \le 1$.   ∎

Problems for Section 5.9

Find the Green's function associated with the homogeneous form of each of the following second order equations, and use it to determine the general solution of the corresponding non-homogeneous equation.

1 $y'' + 2y' - 8y = 2e^{3x}$.

2 $y'' - 9y = e^{3x}$.

3 $y'' + 2y' + y = e^{2x}$.

4 $y'' + y = \cos x$.

5 $xy'' - 3y' = 4x - 6$.

6 $x^2 y'' - 2xy' + 2y = 2x - 8$.

[*Hint*: See Cauchy–Euler equations in Problems for Sec. 5.5]

Find the Green's function for each of the following two-point boundary value problems, and use it to determine the solution of the corresponding nonhomogeneous equation.

7 $y'' = \sin x$, with $y(0) = y\left(\dfrac{\pi}{2}\right) = 0$.

8 $y'' + 4y = f(x)$, with $y(0) = y'(1) = 0$.

9 $y'' = 0$, with $u(0) = 0$ and $u(1) + ku'(1) = 0$.

10 $y'' = x$, with $u(0) = u'(1) = 0$.

11 $y'' + 4y = 1$, with $y(0) = y'(\pi) = 0$.

12 $\dfrac{d}{dx}[(1 - x^2)y'] = 0$, with $y(0) = y'(1) = 0$.

Harder problems

13 Show that except for $t = x$ the Green's function $G(x, t)$ defined in (18) satisfies the homogeneous equation

$$a(x)G'' + b(x)G' + c(x)G = 0,$$

where a prime indicates differentiation with respect to x for any fixed t. Show also that $G'(x)$ is discontinuous when $t = x$, and such that

$$G'(x + 0) - G'(x - 0) = 1/a(x),$$

where $x + 0$ signifies the limit as x is approached from above and $x - 0$ is the limit as it is approached from below. [*Hint*: Use the integral representation of $G(x, t)$ in (6).]

14 Let the nth order differential equation

$$a_0(x)y^{(n)} + a_1(x)y^{(n-1)} + \ldots + a_n(x)y = f(x),$$

be such that $a_0(x), \ldots, a_n(x)$ and $f(x)$ are continuous functions on some interval on which $a_0(x) \neq 0$, and denote any n linearly independent solutions of the homogeneous form of the equation by $\phi_1, \phi_2, \ldots, \phi_n$. Show that

$$y = c_1\phi_1 + c_2\phi_2 + \ldots + c_n\phi_n + \int_{x_0}^{x} G(x, t)f(t)dt$$

is the general solution of the nonhomogeneous differential equation, where $G(x, t)$ is the Green's function defined by

$$G(x,t) = \frac{(-1)^{n-1}}{a_0(t)W(t)} \begin{vmatrix} \phi_1(x) & \phi_2(x) & \cdots & \phi_n(x) \\ \phi_1(t) & \phi_2(t) & \cdots & \phi_n(t) \\ \phi_1'(t) & \phi_2'(t) & \cdots & \phi_n'(t) \\ \vdots & \vdots & & \vdots \\ \phi_1^{(n-2)}(t) & \phi_2^{(n-2)}(t) & \cdots & \phi_n^{(n-2)}(t) \end{vmatrix},$$

with $W(t)$ the Wronskian of $\phi_1, \phi_2, \ldots, \phi_n$. [*Hint*: Use Theorem 3.18 to show the integral term is a particular solution.]

15 Show that the Green's function defined in Prob. 14 is the solution of the homogeneous differential equation

$$a_0(x)G^{(n)} + a_1(x)G^{(n-1)} + \ldots + a_n(x)G = 0,$$

and that it satisfies the initial conditions

$$G(x_0) = G'(x_0) = \ldots = G^{(n-2)}(x_0) = 0 \quad \text{and} \quad G^{(n-1)}(x_0) = 1/a_0(x_0),$$

for any fixed x_0 in the interval in which the Green's function is defined.

Chapter 6
Systems of Linear
Differential Equations

Physical problems which give rise to ordinary differential equations also give rise to systems of simultaneous differential equations. The natural way to study such systems is by expressing them in terms of matrices, and then using the linear algebra of matrices to develop a theory of matrix differential equations. This is carried out in Sec. 6.1 for first order constant coefficient homogeneous systems, and in Sec. 6.2 for the corresponding nonhomogeneous case. Special attention is paid to the occurrence of repeated eigenvalues of the coefficient matrix and to the way in which they alter the structure of the solution. In Sec. 6.3 only brief consideration is given to second order linear constant coefficient systems which describe coupled oscillating systems, because these may always be transformed into a first order system. However, from the practical point of view, this can only be done at the expense of doubling the size of the matrices involved.

Often it is important to study the qualitative properties of the family of solutions of a system of differential equations, rather than obtaining the exact solution to specific problems. These ideas are explored in Sec. 6.4, in which the concept of the phase plane is introduced and developed, along with that of the stability of systems. Properties of both linear and nonlinear systems are considered and related through a number of theorems of fundamental importance to the development of the subject.

The numerical solution of systems, both linear and nonlinear, is considered in Sec. 6.5 in which the Runge–Kutta method introduced in Sec. 4.8 for the numerical solution of a single first order equation is extended to a simultaneous first order system of two equations.

478

6.1 First order linear homogeneous systems of differential equations

The simplest simultaneous differential equations are first order linear homogeneous equations with constant coefficients. A typical example of such a set of equations, which we shall call a **homogeneous system** of differential equations, is

$$\dot{x}_1 - \dot{x}_2 = 4x_1 - 7x_2,$$
$$2\dot{x}_1 - 3\dot{x}_2 = 2x_1 - 3x_2, \tag{1}$$

in which the dot denotes differentiation with respect to t. It is a simple matter to check that the general solution of this system is

$$x_1 = 3c_1 e^{-2t} + 2c_2 e^t,$$
$$x_2 = 2c_1 e^{-2t} + c_2 e^t, \tag{2}$$

where c_1 and c_2 are arbitrary real constants.

Equations like (1) might, for example, arise in connection with a chemical reaction. In such a case x_1 and x_2 would each represent the amount of a reactant present at time t. We shall return to this system and the details of its solution later on in this section.

A system of differential equations can arise in many different ways, one of which is when seeking the solution of a single higher order differential equation. To see how this happens consider the second order differential equation

$$\frac{d^2 x}{dt^2} + 6\frac{dx}{dt} + 9x = 0. \tag{3}$$

The introduction of the two new variables x_1 and x_2, defined as

$$x_1 = x \quad \text{and} \quad x_2 = \frac{dx_1}{dt}, \tag{4}$$

has the effect of reducing (3) to the system of two simultaneous first order differential equations

$$\dot{x}_1 = x_2,$$
$$\dot{x}_2 = -9x_1 - 6x_2. \tag{5}$$

We know from Theorem 5.4 that the general solution of (3) is

$$x = c_1 e^{-3t} + c_2 t e^{-3t}, \tag{6}$$

where c_1 and c_2 are arbitrary constants. Thus using (6) in (4) we find that the general solution of the system (5) must be

$$x_1 = c_1 e^{-3t} + c_2 t e^{-3t},$$
$$x_2 = -(3c_1 - c_2)e^{-3t} - 3c_2 t e^{-3t}. \tag{7}$$

Before considering matrix methods of solution let us take note of the difference in structure of the solutions shown in (2) and (7) as this will be of help to us later. Whereas the solution (2) of (1) only contains exponentials, the solution (7) of (5) (equivalently (6) of (3)) contains both exponentials and the product of t with an exponential.

The extra factor t in (6) arises because the characteristic equation of (3) is

$$\lambda^2 + 6\lambda + 9 = 0,$$

and this has the double root $\lambda = -3$. Thus a basis for solutions of (3), and hence of (5), is e^{-3t} and te^{-3t}. This suggests that we must expect to see reflected in the matrix arguments which are to follow a change in the form of a solution when an analogous situation arises. The way in which this happens will become clear later.

We now turn to the main purpose of this chapter, which is to develop matrix methods for finding the general solution of systems like (1), and then to extend them to nonhomogeneous systems and to the solution of simultaneous second order differential equations. Only real solutions will concern us here, though when deriving them it will sometimes be necessary to work with complex expressions at some intermediate stage.

All matrix methods for dealing with systems like (1) must begin by expressing them in matrix form. To illustrate how this is accomplished let us consider the system in (1). Define the vector $\mathbf{x}$ and the coefficient matrices $\mathbf{C}$ and $\mathbf{D}$ as

$$\mathbf{x} = \begin{bmatrix} x_1 \\ x_2 \end{bmatrix}, \quad \mathbf{C} = \begin{bmatrix} 1 & -1 \\ 2 & -3 \end{bmatrix} \quad \text{and} \quad \mathbf{D} = \begin{bmatrix} 4 & -7 \\ 2 & -3 \end{bmatrix}.$$

Then using the definition of differentiation of a matrix given in (24) of Sec. 3.3, system (1) may be written as the **homogeneous matrix differential equation**

$$\mathbf{C\dot{x}} = \mathbf{Dx}. \tag{8}$$

A simpler but equivalent matrix differential equation may be derived from (8) on the assumption that $\mathbf{C}^{-1}$ exists. When it does, pre-multiplication of (8) by $\mathbf{C}^{-1}$ has the effect of reducing it to what is called the **normal form** (**standard form**) for such a matrix differential equation, for it becomes

$$\mathbf{\dot{x}} = \mathbf{Ax}, \tag{9}$$

where $\mathbf{A} = \mathbf{C}^{-1}\mathbf{D}$.

For the system in (1) we find

$$\mathbf{C}^{-1} = \begin{bmatrix} 3 & -1 \\ 2 & -1 \end{bmatrix}, \quad \text{so} \quad \mathbf{A} = \mathbf{C}^{-1}\mathbf{D} = \begin{bmatrix} 10 & -18 \\ 6 & -11 \end{bmatrix},$$

which shows (1) has the normal form

$$\begin{bmatrix} \dot{x}_1 \\ \dot{x}_2 \end{bmatrix} = \begin{bmatrix} 10 & -18 \\ 6 & -11 \end{bmatrix} \begin{bmatrix} x_1 \\ x_2 \end{bmatrix}.$$

It now remains for us to discover how to solve this matrix differential equation in order to obtain (2), which in matrix form becomes

$$\mathbf{x} = c_1 \begin{bmatrix} 3e^{-2t} \\ 2e^{-2t} \end{bmatrix} + c_2 \begin{bmatrix} 2e^t \\ e^t \end{bmatrix}. \tag{10}$$

Rather than working with this specific example, let us consider the general matrix differential equation in normal form

$$\dot{\mathbf{x}} = \mathbf{A}\mathbf{x}, \tag{11}$$

in which $\mathbf{x}$ is a column vector with elements $x_1, x_2, \ldots, x_n$ and $\mathbf{A}$ is an $n \times n$ matrix with constant coefficients. Working with this normal form involves no loss of generality, for we have seen that a matrix differential equation like (8) may always be reduced to normal form provided $\mathbf{C}$ is nonsingular.

We now define a **solution** of a matrix differential equation to be any column vector $\mathbf{x}$ which when substituted into the equation reduces it to an identity. This definition is an obvious generalization of the definition of a solution of an ordinary differential equation.

Case (i) A has n linearly independent eigenvectors

Throughout what is to follow we need to use the important result that if $\mathbf{x}_1$ and $\mathbf{x}_2$ are any two solutions of the homogeneous system (11), then so also is

$$\mathbf{x} = c_1 \mathbf{x}_1 + c_1 \mathbf{x}_2, \tag{12}$$

where c_1 and c_2 are arbitrary constants. Expressed in words, this result asserts that *a linear combination of any two solutions of a homogeneous system is itself a solution*, and it is a direct consequence of (25) of Sec. 3.3 and the fact that

$$\dot{\mathbf{x}}_1 = \mathbf{A}\mathbf{x}_1 \quad \text{and} \quad \dot{\mathbf{x}}_2 = \mathbf{A}\mathbf{x}_2.$$

An examination of solution (10) of (1) suggests that in the general case (11) we should seek a solution of the form

$$\mathbf{x} = \mathbf{a}e^{\lambda t}, \tag{13}$$

where $\mathbf{a}$ is a column vector with n constant elements which together with λ must be determined from the matrix differential equation itself.

Differentiation of (13) gives

$$\dot{\mathbf{x}} = \lambda \mathbf{a}e^{\lambda t}, \tag{14}$$

and combining (11) and (14) we obtain

$$\lambda \mathbf{a}e^{\lambda t} = \mathbf{A}\mathbf{a}e^{\lambda t}.$$

Cancellation of the nonzero factor $e^{\lambda t}$ then shows that the vector $\mathbf{a}$ and the number λ are related through the matrix equation

$$(A - \lambda I)a = 0, \tag{15}$$

where I is the $n \times n$ unit matrix.

Equation (15) is simply the eigenvalue problem for A which was considered in Sec. 3.10. It was shown there that a will only be nontrivial (not the zero vector) if λ is an eigenvalue of A.

Summarizing the conclusions of Sec. 3.10, we know there will be n eigenvalues $\lambda_1, \lambda_2, \ldots, \lambda_n$ of the characteristic determinant

$$\det(A - \lambda I) = 0, \tag{16}$$

though these may not all be distinct. Furthermore, as A is a real matrix, the eigenvalues will either be real or, if complex, will occur in complex conjugate pairs.

For any given eigenvalue λ_i there will be a corresponding eigenvector a_i satisfying

$$(A - \lambda_i I)a_i = 0, \tag{17}$$

though when the eigenvalue λ_i is repeated $r(>1)$ times there may be fewer than r linearly independent eigenvectors associated with it. In this section we consider only the case in which A has n linearly independent eigenvectors.

Returning now to the form of solution in (13), we see there are n linearly independent solutions, each of the form

$$x_i = a_i \exp(\lambda_i t), \qquad i = 1, 2, \ldots, n, \tag{18}$$

with the λ_i the eigenvalues of the coefficient matrix A and the a_i the corresponding eigenvectors.

It then follows at once from (18) and the linearity property in (12) that another solution of the matrix differential equation (11) must be

$$x = c_1 a_1 \exp(\lambda_1 t) + c_2 a_2 \exp(\lambda_2 t) + \ldots + c_n a_n \exp(\lambda_n t), \tag{19}$$

where $c_1, c_2, \ldots, c_n$ are arbitrary real constants. Expression (19) must in fact be the general solution of (11), because the system is linear, each term of (19) is a solution, and (19) is an arbitrary linear combination of n linearly independent solutions.

By analogy with the first order linear homogeneous equations considered in Chapter 4, we shall call (19) the **complementary function** of the homogeneous matrix differential equation (11). We have established the following fundamental result.

Theorem 6.1 (General solution of a homogeneous system)

Let A be an $n \times n$ matrix with real constant coefficients. Suppose, in addition, that A has n linearly independent eigenvectors $a_1, a_2, \ldots, a_n$, corresponding to the eigenvalues $\lambda_1, \lambda_2, \ldots, \lambda_n$, which need not all be distinct. Then the general solution of the homogeneous matrix differential equation

$$\dot{x} = Ax$$

is

$$\mathbf{x} = c_1 \mathbf{a}_1 \exp(\lambda_1 t) + c_2 \mathbf{a}_2 \exp(\lambda_2 t) + \dots + c_n \mathbf{a}_n \exp(\lambda_n t),$$

where $c_1, c_2, \dots, c_n$ are arbitrary constants.   □

Example 1. Two real distinct eigenvalues

Find the general solution of

$$\dot{\mathbf{x}} = \mathbf{A}\mathbf{x}, \quad \text{when} \quad \mathbf{A} = \begin{bmatrix} 10 & -18 \\ 6 & -11 \end{bmatrix}.$$

Solution

This matrix differential equation is the normal form of the example with which this section was introduced. The eigenvalues of $\mathbf{A}$ are the roots of

$$\det(\mathbf{A} - \lambda\mathbf{I}) = \lambda^2 + \lambda - 2 = 0,$$

which are $\lambda_1 = -2$ and $\lambda_2 = 1$. The eigenvectors $\mathbf{a}_1$ and $\mathbf{a}_2$ corresponding to λ_1 and λ_2 follow by solving the matrix equation

$$(\mathbf{A} - \lambda_i\mathbf{I})\mathbf{a}_i = \mathbf{0}, \quad i = 1, 2.$$

A routine calculation shows we may take

$$\mathbf{a}_1 = \begin{bmatrix} 3 \\ 2 \end{bmatrix} \quad \text{and} \quad \mathbf{a}_2 = \begin{bmatrix} 2 \\ 1 \end{bmatrix}$$

as the two linearly independent eigenvectors of $\mathbf{A}$.

Thus, from Theorem 6.1, the general solution of the matrix differential equation is seen to be

$$\mathbf{x} = c_1 \begin{bmatrix} 3 \\ 2 \end{bmatrix} e^{-2t} + c_2 \begin{bmatrix} 2 \\ 1 \end{bmatrix} e^t,$$

where c_1 and c_2 are arbitrary real constants. This is precisely the solution given in (7).   ■

Before examining further examples, it will be instructive to compare this method of solution with a more elementary one which makes no use of matrices. The starting point is the observation that the simplest solutions will be those in which x_1 and x_2 both vary in the same way with t. Thus, with the structure of the solution of a single homogeneous equation in mind, this suggests we should first seek solutions of the form $x_1 = Ae^{\lambda x}$ and $x_2 = Be^{\lambda x}$, with A and B arbitrary constants. Then, once these special solutions have been obtained, a way must be found to relate them to the general solution of the problem.

Substituting these forms for x_1 and x_2 into the simultaneous equations

$$\dot{x}_1 = 10x_1 - 18x_2$$
$$\dot{x}_2 = 6x_1 - 11x_2,$$

followed by cancellation of the nonzero factor $e^{\lambda t}$, gives rise to the two simultaneous equations

$$A\lambda = 10A - 18B$$
$$B\lambda = 6A - 11B.$$

After combining terms these equations become the homogeneous system

$$(10 - \lambda)A - 18B = 0$$

$$6A - (11 + \lambda)B = 0,$$

from which λ, A and B must be found. As the system of algebraic equations is homogeneous, a nontrivial solution for A and B can only be found it

$$\begin{vmatrix} (10 - \lambda) & -18 \\ 6 & -(11 + \lambda) \end{vmatrix} = 0,$$

which shows that λ must be a root of

$$\lambda^2 + \lambda - 2 = 0.$$

This last result is, of course, the characteristic equation of the matrix **A** arrived at previously when considering the eigenvalue problem. The roots of this equation (the eigenvalues of **A**) are $\lambda_1 = -2$ and $\lambda_2 = 1$.

Solving the homogeneous equations for A and B for each of the permissible values of λ shows that when $\lambda = \lambda_1 = -2$, the most general solution is

$$A = 3c_1, \quad B = 2c_1,$$

with c_1 an arbitrary constant. This is because homogeneous equations can only determine the *ratio* of A to B, and not their absolute values. Similarly, when $\lambda = \lambda_2 = 1$, the most general solution is seen to be

$$A = 2c_2, \quad B = c_2,$$

with c_2 an arbitrary constant.

Thus the most general special solutions of the required form corresponding to $\lambda = \lambda_1 = -2$ are

$$x_1 = 3c_1 e^{-2t}, \quad x_2 = 2c_1 e^{-2t},$$

while when $\lambda = \lambda_2 = 1$ they are

$$x_1 = 2c_2 e^t, \quad x_2 = c_2 e^t.$$

As these are the only possible forms of linearly independent solutions, and the differential equations are linear, a linear combination of these solutions must also be a solution. Thus the general solution is seen to be

$$x_1 = 3c_1 e^{-2t} + 2c_2 e^t, \quad x_2 = 2c_1 e^{-2t} + c_2 e^t,$$

where c_1, c_2 are arbitrary constants. These are the solutions already found by matrix methods.

Example 2. A repeated real eigenvalue

Find the general solution of

$$\dot{x} = Ax, \text{ when } A = \begin{bmatrix} 1 & 0 & 0 \\ 1 & 3 & 2 \\ 1 & 2 & 3 \end{bmatrix}.$$

Solution

The eigenvalue problem for this matrix A was considered in Ex. 4, Sec. 3.10, in which it was shown that its eigenvalues were $\lambda_1 = 5$, $\lambda_2 = 1$ and $\lambda_3 = 1$, so that $\lambda = 1$ is an eigenvalue with multiplicity 2. It was also shown that A has three linearly independent eigenvectors

$$a_1 = \begin{bmatrix} 0 \\ 1 \\ 1 \end{bmatrix}, \quad a_2 = \begin{bmatrix} -2 \\ 1 \\ 0 \end{bmatrix} \text{ and } a_3 = \begin{bmatrix} -2 \\ 0 \\ 1 \end{bmatrix},$$

where a_1, a_2 and a_3 correspond, respectively, to λ_1, λ_2 and λ_3.

Thus, from Theorem 6.1, the general solution of the matrix differential equation is

$$x = c_1 \begin{bmatrix} 0 \\ 1 \\ 1 \end{bmatrix} e^{5t} + c_2 \begin{bmatrix} -2 \\ 1 \\ 0 \end{bmatrix} e^{t} + c_3 \begin{bmatrix} -2 \\ 0 \\ 1 \end{bmatrix} e^{t},$$

where c_1, c_2 and c_3 are arbitrary real constants. This vector solution is equivalent to the three scalar solutions

$$x_1 = -2(c_2 + c_3)e^{t},$$
$$x_2 = c_1 e^{5t} + c_2 e^{t},$$
$$x_3 = c_1 e^{5t} + c_3 e^{t}.$$

■

Example 3. Complex conjugate eigenvalues

Find the general solution of

$$\dot{x} = Ax, \text{ when } A = \begin{bmatrix} 0 & -1 \\ 1 & 0 \end{bmatrix}.$$

Solution

The eigenvalues of A are the roots of

$$\det(A - \lambda I) = \lambda^2 + 1 = 0,$$

which are $\lambda_1 = -i$ and $\lambda_2 = i$. Eigenvectors corresponding to this pair of complex conjugate eigenvalues follow by solving (17) for $i=1$ and $i=2$. In the case $i=1$, $\lambda_1 = -i$, and (17) becomes

$$\begin{bmatrix} i & -1 \\ 1 & i \end{bmatrix}\begin{bmatrix} \alpha \\ \beta \end{bmatrix} = 0, \text{ where we have set } \mathbf{a}_1 = \begin{bmatrix} \alpha \\ \beta \end{bmatrix}.$$

Taking $\alpha = 1$, for convenience, this shows that the eigenvectors $\mathbf{a}_i$ may be taken as

$$\mathbf{a}_1 = \begin{bmatrix} 1 \\ i \end{bmatrix} \text{ and } \mathbf{a}_2 = \begin{bmatrix} 1 \\ -i \end{bmatrix}.$$

Thus, from Theorem 6.1, the general solution is

$$\mathbf{x} = C_1 \mathbf{a}_1 e^{-it} + C_2 \mathbf{a}_2 e^{it},$$

with C_1 and C_2 arbitrary complex constants. We require a real solution of the matrix differential equation, so the two terms in the general solution must be complex conjugates. Inspection of the form of $\mathbf{a}_1$ and $\mathbf{a}_2$ shows these are already complex conjugates, as are the factors e^{-it} and e^{it}, so we must set $C_2 = \bar{C}_1$ if the terms are to be complex conjugates. Writing $C_1 = c_1 + ic_2$, it then follows that

$$\mathbf{x} = C_1 \mathbf{a}_1 e^{-it} + \overline{(C_1 \mathbf{a}_1 e^{it})},$$
$$= 2\mathrm{Re}[(c_1 + ic_2)e^{-it}\mathbf{a}_1].$$

Thus the required real general solution of the matrix differential equation is

$$\mathbf{x} = \begin{bmatrix} 2(c_1 \cos t + c_2 \sin t) \\ 2(c_1 \sin t - c_2 \cos t) \end{bmatrix}.$$

If the factor 2 is absorbed into each of the arbitrary real constants c_1 and c_2 this vector solution is seen to be equivalent to the two scalar solutions

$$x_1 = c_1 \cos t + c_2 \sin t,$$
$$x_2 = c_1 \sin t - c_2 \cos t.$$ ∎

Example 4. Complex conjugate eigenvalues and real eigenvalue

Find the general solution of

$$\dot{\mathbf{x}} = \mathbf{A}\mathbf{x}, \text{ when } \mathbf{A} = \begin{bmatrix} 1 & 1 & 0 \\ 0 & 1 & 1 \\ 0 & -1 & 2 \end{bmatrix}.$$

Solution

The eigenvalues of $\mathbf{A}$ are the roots of

$$\det(\mathbf{A} - \lambda\mathbf{I}) = (1 - \lambda)(\lambda^2 - 3\lambda + 3) = 0,$$

which are $\lambda_1 = 1$, $\lambda_2 = \frac{1}{2}(3 - i\sqrt{3})$ and $\lambda_3 = \frac{1}{2}(3 + i\sqrt{3})$. We must now use the defining relationship (17) to find the eigenvectors $\mathbf{a}_i$. A simple calculation shows we may take

$$
\mathbf{a}_1 = \begin{bmatrix} 1 \\ 0 \\ 0 \end{bmatrix}, \quad \mathbf{a}_2 = \begin{bmatrix} 1 \\ \frac{1}{2}(1 - i\sqrt{3}) \\ -\frac{1}{2}(1 + i\sqrt{3}) \end{bmatrix} \quad \text{and} \quad \mathbf{a}_3 = \begin{bmatrix} 1 \\ \frac{1}{2}(1 + i\sqrt{3}) \\ -\frac{1}{2}(1 - i\sqrt{3}) \end{bmatrix}
$$

as the three linearly independent eigenvectors of $\mathbf{A}$.

It then follows from Theorem 6.1 that the general solution of the matrix differential equation is

$$
\mathbf{x} = c_1 \mathbf{a}_1 e^t + C_2 \mathbf{a}_2 \exp[\tfrac{1}{2}(3 - i\sqrt{3})t] + C_3 \mathbf{a}_3 \exp[\tfrac{1}{2}(3 + i\sqrt{3})t],
$$

where c_1 is an arbitrary real constant and C_1 and C_2 are arbitrary complex constants.

To obtain a real solution $\mathbf{x}$ the last two terms in the general solution must be complex conjugates. Inspection shows this will be so if $C_3 = \bar{C}_2$. Thus, setting $C_2 = c_1 + i c_2$, we find $\mathbf{x}$ becomes

$$
\mathbf{x} = c_1 \mathbf{a}_1 e^t + C_2 \mathbf{a}_2 \exp[\tfrac{1}{2}(3 - i\sqrt{3})t] + \overline{(C_2 \mathbf{a}_2 \exp[\tfrac{1}{2}(3 + i\sqrt{3})])}
$$

$$
= c_1 \mathbf{a}_1 e^t + 2 \exp(\tfrac{3}{2}t) \operatorname{Re}[(c_1 + i c_2) \exp[-(i\sqrt{3}t)/2] \mathbf{a}_2].
$$

A brief calculation shows this vector solution to be equivalent to the three scalar solutions

$$
x_1 = c_1 e^t + 2 \exp(\tfrac{3}{2}t) \left(c_1 \cos \frac{\sqrt{3}}{2} t + c_2 \sin \frac{\sqrt{3}}{2} t \right),
$$

$$
x_2 = \exp(\tfrac{3}{2}t) \left[(c_1 - \sqrt{3}c_2) \cos \frac{\sqrt{3}}{2} t + (c_2 - \sqrt{3}c_1) \sin \frac{\sqrt{3}}{2} t \right],
$$

$$
x_3 = -\exp(\tfrac{3}{2}t) \left[(c_1 - \sqrt{3}c_2) \cos \frac{\sqrt{3}}{2} t + (c_2 + \sqrt{3}c_1) \sin \frac{\sqrt{3}}{2} t \right],
$$

where now c_1, c_2 and c_3 are arbitrary real constants.　　　　　　　　　■

An alternative but equivalent approach to the solution of the general first order linear homogeneous matrix differential equation

$$
\dot{\mathbf{x}} = \mathbf{A}\mathbf{x}, \tag{20}
$$

in which $\mathbf{A}$ is an $n \times n$ matrix with real constant coefficients involves the diagonalization of $\mathbf{A}$ as discussed in Sec. 3.11.

It is known from Theorem 3.32 that the $n \times n$ matrix $\mathbf{A}$ may always be diagonalized provided it has n linearly independent eigenvectors. The diagonalization is then achieved by means of a similarity transformation involving a matrix $\mathbf{P}$ whose columns are the eigenvectors of $\mathbf{A}$.

Suppose the eigenvalues of $\mathbf{A}$ are $\lambda_1, \lambda_2, \ldots, \lambda_n$, and the corresponding eigenvectors are $\mathbf{p}_1, \mathbf{p}_2, \ldots, \mathbf{p}_n$. Then if $\mathbf{P}$ is the matrix whose ith column is $\mathbf{p}_i$, and Λ is the diagonal matrix

$$
\Lambda = \begin{bmatrix}
\lambda_1 & 0 & 0 & \cdots & 0 \\
0 & \lambda_2 & 0 & \cdots & 0 \\
0 & 0 & \lambda_3 & \cdots & 0 \\
. & . & . & \cdots & . \\
0 & 0 & 0 & \cdots & \lambda_n
\end{bmatrix},
$$

we know from Theorem 3.32 (iii) that

$$\mathbf{P}^{-1}\mathbf{AP} = \Lambda. \tag{21}$$

To make use of this result we first change the vector $\mathbf{x}$ in (20) to a new vector $\mathbf{y}$ by means of the transformation

$$\mathbf{x} = \mathbf{Py}. \tag{22}$$

Then because $\mathbf{A}$ is a matrix with constant coefficients, so also is $\mathbf{P}$, so that differentiation of (22) with respect to t gives

$$\dot{\mathbf{x}} = \mathbf{P}\dot{\mathbf{y}}. \tag{23}$$

Using (22) and (23) in (20) then gives

$$\mathbf{P}\dot{\mathbf{y}} = \mathbf{APy},$$

so that

$$\dot{\mathbf{y}} = \mathbf{P}^{-1}\mathbf{APy}. \tag{24}$$

Finally, using (21), we see that

$$\dot{\mathbf{y}} = \Lambda\mathbf{y}. \tag{25}$$

The solution of (25) is now trivial, because all the equations have been uncoupled, and (25) is equivalent to the n scalar differential equations

$$\dot{y}_i = \lambda_i y_i, \qquad i = 1, 2, \ldots, n. \tag{26}$$

Integration of (26) then gives

$$y_i = c_i \exp(\lambda_i t), \qquad i = 1, 2, \ldots, n, \tag{27}$$

where the c_i are arbitrary constants. The required solution vector $\mathbf{x}$ then follows from (22) after the solution vector $\mathbf{y}$ has been constructed, which from (27) is seen to be

$$
\mathbf{y} = \begin{bmatrix}
c_1 \exp(\lambda_1 t) \\
c_2 \exp(\lambda_2 t) \\
\vdots \\
c_n \exp(\lambda_n t)
\end{bmatrix}. \tag{28}
$$

Example 5. Solution by diagonalization

Find the general solution of

$$\dot{\mathbf{x}} = \mathbf{A}\mathbf{x}, \quad \text{when} \quad \mathbf{A} = \begin{bmatrix} 1 & 0 & 0 \\ 1 & 3 & 2 \\ 1 & 2 & 3 \end{bmatrix}.$$

Solution

This is the problem already considered in Ex. 2. In Ex. 1(i) of Sec. 3.11 the matrix $\mathbf{A}$ was shown to have the eigenvalues $\lambda_1 = 5$, $\lambda_2 = 1$ and $\lambda_3 = 1$, and the diagonalizing matrix

$$\mathbf{P} = \begin{bmatrix} 0 & -2 & -2 \\ 1 & 1 & 0 \\ 1 & 0 & 1 \end{bmatrix}$$

was such that

$$\mathbf{P}^{-1}\mathbf{A}\mathbf{P} = \Lambda = \begin{bmatrix} 5 & 0 & 0 \\ 0 & 1 & 0 \\ 0 & 0 & 1 \end{bmatrix}.$$

Setting $\mathbf{x} = \mathbf{P}\mathbf{y}$ and proceeding as indicated we find that

$$\mathbf{y} = \begin{bmatrix} c_1 e^{5t} \\ c_2 e^{t} \\ c_3 e^{t} \end{bmatrix},$$

where c_1, c_2 and c_3 are arbitrary real constants. Using this vector in $\mathbf{x} = \mathbf{P}\mathbf{y}$ then shows the general solution of the matrix differential equation to be

$$\mathbf{x} = \begin{bmatrix} 0 & -2 & -2 \\ 1 & 1 & 0 \\ 1 & 0 & 1 \end{bmatrix} \begin{bmatrix} c_1 e^{5t} \\ c_2 e^{t} \\ c_3 e^{t} \end{bmatrix}.$$

This is equivalent to the three scalar solutions

$$x_1 = -2(c_2 + c_3)e^{t},$$
$$x_2 = c_1 e^{5t} + c_2 e^{t},$$
$$x_3 = c_1 e^{5t} + c_3 e^{t},$$

which were also found in Ex. 2 above. ∎

The **initial value problem** for matrix differential equations of the type considered in this section involves finding a solution vector $\mathbf{x}$ which for $t > t_0$ satisfies

$$\dot{\mathbf{x}} = \mathbf{A}\mathbf{x}, \tag{29}$$

and at $t = t_0$ is such that $\mathbf{x}(t_0) = \mathbf{b}$, where $\mathbf{b}$ is a given column vector whose n elements $b_1, b_2, \ldots, b_n$ are all constants. The vector $\mathbf{b}$ is called the **initial condition** for matrix differential equation (29). Initial value problems are solved by first obtaining the general solution of (29), and then selecting the arbitrary constants $c_1, c_2, \ldots, c_n$ such that $x_1(t_0) = b_1, x_2(t_0) = b_2, \ldots, x_n(t_0) = b_n$.

Example 6. Initial value problem

Solve the initial value problem

$$\dot{\mathbf{x}} = \mathbf{A}\mathbf{x} \quad \text{with} \quad \mathbf{x}(0) = \begin{bmatrix} 1 \\ -1 \\ 2 \end{bmatrix}, \quad \text{given that} \quad \mathbf{A} = \begin{bmatrix} 1 & 0 & 0 \\ 1 & 3 & 2 \\ 1 & 2 & 3 \end{bmatrix}.$$

Solution

The general solution was obtained above in Ex. 5. To solve the stated initial value problem we must select c_1, c_2 and c_3 such that

$$x_1(0) = 1, \quad x_2(0) = -1 \quad \text{and} \quad x_3(0) = 2.$$

From the solution obtained in Ex. 5 we see that c_1, c_2 and c_3 must satisfy the equations

$$1 = -2c_2 - 2c_3, \quad -1 = c_1 + c_2 \quad \text{and} \quad 2 = c_1 + c_3,$$

so that $c_1 = \frac{3}{4}, c_2 = -\frac{7}{4}$ and $c_3 = \frac{5}{4}$. The solution of the initial value problem is thus

$$x_1 = e^t, \quad x_2 = \frac{1}{4}(3e^{5t} - 7e^t) \quad \text{and} \quad x_3 = \frac{1}{4}(3e^{5t} + 5e^t). \qquad \blacksquare$$

Case (ii) A has fewer than n linearly independent eigenvectors

The two methods for finding the general solution of the matrix differential equation

$$\dot{\mathbf{x}} = \mathbf{A}\mathbf{x}$$

which have just been described will both fail if the $n \times n$ matrix has fewer than n linearly independent eigenvectors: the first because there are insufficient solution vectors to form a basis for the solution space, and the second because the matrix $\mathbf{A}$ cannot be diagonalized.

To overcome this limitation, the method of solution must be modified so that the missing linearly independent solution vectors can be determined. The approach we shall adopt is suggested by consideration of the system given in (5), and its solution (7) obtained with the help of Theorem 5.4.

Expressing (5) in the normal form

$$\dot{\mathbf{x}} = \mathbf{A}\mathbf{x}, \quad \text{with} \quad \mathbf{A} = \begin{bmatrix} 0 & 1 \\ -9 & -6 \end{bmatrix}, \tag{30}$$

a routine calculation shows **A** has the double eigenvalue $\lambda = -3$, but only the single eigenvector

$$\mathbf{a}_1 = \begin{bmatrix} 1 \\ -3 \end{bmatrix}. \tag{31}$$

Thus this system is of the type we are considering in this section.

Examination of the solution given in (7) shows it may be written in the matrix form

$$\mathbf{x} = c_1 \begin{bmatrix} 1 \\ -3 \end{bmatrix} e^{-3t} + c_2 \left\{ \begin{bmatrix} 0 \\ 1 \end{bmatrix} e^{-3t} + \begin{bmatrix} 1 \\ -3 \end{bmatrix} te^{-3t} \right\}. \tag{32}$$

The first vector in (32) corresponds to the solution vector found from the single eigenvector $\mathbf{a}_1$ given above, while the bracketed vectors represent the second linearly independent solution vector we are trying to determine by matrix methods. Inspection of the bracketed vectors in (32) shows that the double root has introduced into the second solution vector of (30) a linear combination of two vectors, one scaled by e^{-3t} and the other by te^{-3t}.

This suggests that in the general case, when $\lambda = \mu$ is a real double eigenvalue of an $n \times n$ matrix **A**, we should seek the second solution vector corresponding to this double eigenvalue in the form

$$\mathbf{x} = \mathbf{b}_2 e^{\mu t} + \mathbf{b}_1 te^{\mu t}. \tag{33}$$

If (33) is to be a solution of the matrix differential equation

$$\dot{\mathbf{x}} = \mathbf{A}\mathbf{x}, \tag{34}$$

it follows by differentiating (33) and substituting it into (34) that the following expression must be an identity in t:

$$\mu\mathbf{b}_2 e^{\mu t} + \mathbf{b}_1 e^{\mu t} + \mu\mathbf{b}_1 te^{\mu t} \equiv \mathbf{A}\mathbf{b}_2 e^{\mu t} + \mathbf{A}\mathbf{b}_1 te^{\mu t}. \tag{35}$$

Cancelling the nonzero factor $e^{\mu t}$ and rearranging terms we obtain the identity

$$0 \equiv (\mathbf{A} - \mu\mathbf{I})\mathbf{b}_2 - \mathbf{b}_1 + (\mathbf{A} - \mu\mathbf{I})\mathbf{b}_1 t. \tag{36}$$

Let us now write this as

$$0 \equiv [(\mathbf{A} - \mu\mathbf{I})\mathbf{b}_2 - \mathbf{b}_1]t^0 + [(\mathbf{A} - \mu\mathbf{I})\mathbf{b}_1]t, \tag{37}$$

to indicate its dependence on t. Then this can only be true for all t (be an identity) if the matrices multiplying each power of t (both t^0 and t) are themselves zero, so we conclude that

$$(\mathbf{A} - \mu\mathbf{I})\mathbf{b}_1 = 0, \tag{38}$$

and

$$(\mathbf{A} - \mu\mathbf{I})\mathbf{b}_2 = \mathbf{b}_1. \tag{39}$$

Thus $\mathbf{b}_1$ must be the single eigenvector of $\mathbf{A}$ that can be found in the usual manner, and $\mathbf{b}_2$ then follows by solving (39) once $\mathbf{b}_1$ has been determined. With this choice of $\mathbf{b}_1$ and $\mathbf{b}_2$, (33) becomes the missing linearly independent solution vector of (34), which may be combined with the other linearly independent solution vectors to give the general solution. We have established the following general result involving a matrix $\mathbf{A}_1$ lacking a single eigenvector.

Theorem 6.2 (General solution when A has only n−1 eigenvectors)

Let the $n \times n$ matrix $\mathbf{A}_1$ have the $n-2$ real and distinct eigenvalues $\lambda_1, \lambda_2, \ldots, \lambda_{n-2}$, and the corresponding eigenvectors $\mathbf{a}_1, \mathbf{a}_2, \ldots, \mathbf{a}_{n-2}$. Suppose the two remaining eigenvalues λ_{n-1} and λ_n are equal and real, with $\lambda_{n-1} = \lambda_n = \mu$, and that to this double eigenvalue there corresponds only the single eigenvector $\mathbf{a}_{n-1}$. Then the general solution of

$$\dot{\mathbf{x}} = \mathbf{A}_1 \mathbf{x}$$

is

$$\mathbf{x} = c_1 \mathbf{a}_1 \exp(\lambda_1 t) + c_2 \mathbf{a}_2 \exp(\lambda_2 t) + \ldots + c_{n-2} \mathbf{a}_{n-2} \exp(\lambda_{n-2} t)$$
$$+ c_{n-1} \mathbf{a}_{n-1} \exp(\mu t) + c_n [\mathbf{b}_2 \exp(\mu t) + \mathbf{a}_{n-1} t \exp(\mu t)],$$

where $c_1, c_2, \ldots, c_n$ are arbitrary constants, and $\mathbf{b}_2$ is a solution of

$$(\mathbf{A}_1 - \mu \mathbf{I})\mathbf{b}_2 = \mathbf{a}_{n-1}.$$

The solution vectors

$$\mathbf{a}_1 \exp(\lambda_1 t), \quad \mathbf{a}_2 \exp(\lambda_2 t), \ldots, \quad \mathbf{a}_{n-2} \exp(\lambda_{n-2} t), \quad \mathbf{a}_{n-1} \exp(\mu t),$$
$$\mathbf{b}_2 \exp(\mu t) + \mathbf{a}_{n-1} t \exp(\mu t)$$

form a basis for the solution space of the differential equation.   □

Example 7. Matrix lacking one eigenvector

Find the general solution of

$$\dot{\mathbf{x}} = \mathbf{A}\mathbf{x}, \quad \text{where} \quad \mathbf{A} = \begin{bmatrix} 0 & 1 \\ -9 & -6 \end{bmatrix}.$$

Solution

This is the system given in (5), for which we know $\mathbf{A}$ has the double eigenvalue $\lambda = -3$, but only the single eigenvector

$$\mathbf{a}_1 = \begin{bmatrix} 1 \\ -3 \end{bmatrix}.$$

From Theorem 6.2 with the given $\mathbf{A}$, this choice for $\mathbf{a}_1$, and with $\mu = -3$, we know the general solution will be of the form

$$\mathbf{x} = c_1 \mathbf{a}_1 e^{-3t} + c_2 [\mathbf{b}_2 e^{-3t} + \mathbf{a}_1 t e^{-3t}],$$

where $\mathbf{b}_2$ is obtained by solving

$$(\mathbf{A} + 3\mathbf{I})\mathbf{b}_2 = \mathbf{a}_1.$$

Setting

$$\mathbf{b}_2 = \begin{bmatrix} \alpha \\ \beta \end{bmatrix},$$

this last result becomes

$$\begin{bmatrix} 3 & 1 \\ -9 & -3 \end{bmatrix} \begin{bmatrix} \alpha \\ \beta \end{bmatrix} = \begin{bmatrix} 1 \\ -3 \end{bmatrix},$$

so that α and β must satisfy the single condition

$$3\alpha + \beta = 1.$$

Setting $\alpha = 1$ (an arbitrary choice) gives $\beta = -2$, so that $\mathbf{b}_2$ becomes

$$\mathbf{b}_2 = \begin{bmatrix} 1 \\ -2 \end{bmatrix}.$$

Substituting for $\mathbf{a}_1$ and $\mathbf{b}_2$ to find the general solution we obtain

$$\mathbf{x} = c_1 \begin{bmatrix} 1 \\ -3 \end{bmatrix} e^{-3t} + c_2 \begin{bmatrix} 1 \\ -2 \end{bmatrix} e^{-3t} + c_2 \begin{bmatrix} 1 \\ -3 \end{bmatrix} te^{-3t}.$$

This shows that a basis for the solution space in this case comprises the vectors

$$\begin{bmatrix} 1 \\ -3 \end{bmatrix} e^{-3t} \quad \text{and} \quad \begin{bmatrix} 1 \\ -2 \end{bmatrix} e^{-3t} + \begin{bmatrix} 1 \\ -3 \end{bmatrix} te^{-3t}.$$

At first sight, the solution just obtained appears to differ from the solution given in (32), but this is simply due to the way in which the arbitrary constants have been introduced. In scalar form, the above solution becomes

$$x_1 = (c_1 + c_2)e^{-3t} + c_2 te^{-3t},$$
$$x_2 = -(3c_1 + 2c_2)e^{-3t} - 3c_2 te^{-3t},$$

If we now set $c_1 + c_2 = d_1$, this becomes

$$x_1 = d_1 e^{-3t} + c_2 te^{-3t},$$
$$x_2 = -(3d_1 - c_2) - 3c_2 te^{-3t},$$

which is precisely the form given in (7), apart from an unimportant notational change involving writing d_1 in place of c_1. This same change of arbitrary constants transforms the matrix vector solution into

$$\mathbf{x} = d_1 \begin{bmatrix} 1 \\ -3 \end{bmatrix} e^{-3t} + c_2 \left(\begin{bmatrix} 0 \\ 1 \end{bmatrix} e^{-3t} + \begin{bmatrix} 1 \\ -3 \end{bmatrix} te^{-3t} \right),$$

which is result (32) with this same notational change. ∎

Consideration of the structure of solutions given in Theorem 5.4(ii) for the case in which a repeated real root of the characteristic equation arises shows how Theorem 6.2 may be generalized. If desired, a further generalization may be made to the case in which complex conjugate roots arise. Suppose, for example, all the eigenvalues of $\mathbf{A}$ are real, with $n-3$ being distinct and the other three identical, say to μ, but with only the single eigenvector $\mathbf{a}_{n-3}$ corresponding to this triple eigenvalue. A comparison with Theorem 5.4(ii) shows that this corresponds to the case $r=3$. Thus, by analogy, we see that in addition to the solution

$$\mathbf{x}_1 = \mathbf{a}_{n-3} e^{\mu t} \tag{40}$$

corresponding to $\lambda = \mu$, which is found in the usual way, two further solutions $\mathbf{x}_2$ and $\mathbf{x}_3$ must be sought in the form

$$\mathbf{x}_2 = \mathbf{b}_2 e^{\mu t} + \mathbf{b}_1 t e^{\mu t}, \tag{41}$$

and

$$\mathbf{x}_3 = \mathbf{b}_5 e^{\mu t} + \mathbf{b}_4 t e^{\mu t} + \mathbf{b}_3 t^2 e^{\mu t}. \tag{42}$$

The vector solutions $\mathbf{x}_1$, $\mathbf{x}_2$ and $\mathbf{x}_3$ correspond to the three linearly independent solution vectors which must correspond to the triple eigenvalue.

Substitution of $\mathbf{x}_2$ and $\mathbf{x}_3$ into (34), coupled with a repetition of the previous arguments, then leads to the determination of $\mathbf{b}_1, \mathbf{b}_2, \ldots, \mathbf{b}_5$ in terms of $\mathbf{a}_{n-3}$, and thus to the next general result concerning a matrix $\mathbf{A}_2$ lacking two eigenvectors. The details are left as an exercise for the reader.

Theorem 6.3 (General solution when A has only n−2 eigenvectors)

Let the $n \times n$ matrix $\mathbf{A}_2$ have the $n-3$ real and distinct eigenvalues $\lambda_1, \lambda_2, \ldots, \lambda_{n-3}$, and the corresponding eigenvectors $\mathbf{a}_1, \mathbf{a}_2, \ldots, \mathbf{a}_{n-3}$. Suppose the three remaining eigenvalues $\lambda_{n-2}, \lambda_{n-1}$ and λ_n are equal and real, with $\lambda_{n-2} = \lambda_{n-1} = \lambda_n = \mu$, and that to this triple eigenvalue there corresponds only the single eigenvector $\mathbf{a}_{n-2}$.

Then the general solution of

$$\dot{\mathbf{x}} = \mathbf{A}_2 \mathbf{x}$$

is

$$\mathbf{x} = c_1 \mathbf{a}_1 \exp(\lambda_1 t) + c_2 \mathbf{a}_2 \exp(\lambda_2 t) + \ldots + c_{n-3} \mathbf{a}_{n-3} \exp(\lambda_{n-3} t)$$
$$+ c_{n-2} \mathbf{a}_{n-2} \exp(\mu t) + c_{n-1} [\mathbf{b}_2 \exp(\mu t) + \mathbf{a}_{n-2} t \exp(\mu t)] + c_n [\mathbf{b}_5 \exp(\mu t)$$
$$+ \mathbf{b}_4 t \exp(\mu t) + \mathbf{a}_{n-2} t^2 \exp(\mu t)],$$

where $c_1, c_2, \ldots, c_n$ are arbitrary constants, $\mathbf{b}_2$ is a solution of

$$(\mathbf{A}_2 - \mu \mathbf{I})\mathbf{b}_2 = \mathbf{a}_{n-2},$$

and $\mathbf{b}_4$ and $\mathbf{b}_5$ are solutions of

$$(\mathbf{A}_2 - \mu\mathbf{I})\mathbf{b}_4 = 2\mathbf{a}_{n-2},$$

and

$$(\mathbf{A}_2 - \mu\mathbf{I})\mathbf{b}_5 = \mathbf{b}_4.$$

The solution vectors

$$\mathbf{a}_1 \exp(\lambda_1 t), \quad \mathbf{a}_2 \exp(\lambda_2 t), \quad \ldots, \quad \mathbf{a}_{n-3} \exp(\lambda_{n-3} t), \quad \mathbf{a}_{n-2} \exp(\mu t),$$
$$\mathbf{b}_2 \exp(\mu t) + \mathbf{a}_{n-2} t \exp(\mu t), \quad \mathbf{b}_5 \exp(\mu t) + \mathbf{b}_4 t \, \exp(\mu t) + \mathbf{a}_{n-2} t^2 \exp(\mu t)$$

form a basis for the solution space of the differential equation. □

Example 8. A matrix lacking two eigenvectors

Find the general solution of

$$\dot{\mathbf{x}} = \mathbf{Ax}, \quad \text{when} \quad \mathbf{A} = \begin{bmatrix} 0 & 1 & 0 \\ 0 & 0 & 1 \\ -1 & -3 & -3 \end{bmatrix}.$$

Solution

The matrix $\mathbf{A}$ has the triple eigenvalue $\lambda = -1$, to which there corresponds only the single eigenvector

$$\mathbf{a}_1 = \begin{bmatrix} 1 \\ -1 \\ 1 \end{bmatrix}.$$

Thus $\mathbf{A}$ is lacking two eigenvectors, so that Theorem 6.3 applies. Using the notation of Theorem 6.3 we see that vector $\mathbf{b}_2$ is a solution of

$$\begin{bmatrix} 1 & 1 & 0 \\ 0 & 1 & 1 \\ -1 & -3 & -2 \end{bmatrix} \begin{bmatrix} \alpha \\ \beta \\ \gamma \end{bmatrix} = \begin{bmatrix} 1 \\ -1 \\ 1 \end{bmatrix}, \quad \text{where } \mathbf{b}_2 = \begin{bmatrix} \alpha_1 \\ \beta_1 \\ \gamma_1 \end{bmatrix}.$$

This yields the consistent set of equations

$$\alpha_1 + \beta_1 = 1, \quad \beta_1 + \gamma_1 = -1 \quad \text{and} \quad -\alpha_1 - 3\beta_1 - 2\gamma_1 = 1,$$

with a solution $\alpha_1 = 1, \beta_1 = 0, \gamma_1 = -1$, so that

$$\mathbf{b}_2 = \begin{bmatrix} 1 \\ 0 \\ -1 \end{bmatrix}.$$

Vector $\mathbf{b}_4$ is a solution of

$$
\begin{bmatrix} 1 & 1 & 0 \\ 0 & 1 & 1 \\ -1 & -3 & -2 \end{bmatrix} \begin{bmatrix} \alpha_2 \\ \beta_2 \\ \gamma_2 \end{bmatrix} = \begin{bmatrix} 2 \\ -2 \\ 2 \end{bmatrix}, \quad \text{where } \mathbf{b}_4 = \begin{bmatrix} \alpha_2 \\ \beta_2 \\ \gamma_2 \end{bmatrix}.
$$

This yields the consistent set of equations

$$
\alpha_2 + \beta_2 = 2, \quad \beta_2 + \gamma_2 = -2 \quad \text{and} \quad -\alpha_2 - 3\beta_2 - 2\gamma_2 = 2,
$$

with a solution $\alpha_2 = 2$, $\beta_2 = 0$, $\gamma_2 = -2$, so that

$$
\mathbf{b}_4 = \begin{bmatrix} 2 \\ 0 \\ -2 \end{bmatrix}.
$$

Finally, vector $\mathbf{b}_5$ is a solution of

$$
\begin{bmatrix} 1 & 1 & 0 \\ 0 & 1 & 1 \\ -1 & -3 & -2 \end{bmatrix} \begin{bmatrix} \alpha_3 \\ \beta_3 \\ \gamma_3 \end{bmatrix} = \begin{bmatrix} 2 \\ 0 \\ -2 \end{bmatrix}, \quad \text{where } \mathbf{b}_5 = \begin{bmatrix} \alpha_3 \\ \beta_3 \\ \gamma_3 \end{bmatrix}.
$$

This yields the consistent set of equations

$$
\alpha_3 + \beta_3 = 2, \quad \beta_3 + \gamma_3 = 0 \text{ and } -\alpha_3 - 3\beta_3 - 2\gamma_3 = -2, \text{ with a solution } \alpha_3 = 2, \beta_3 = 0, \gamma_3 = 0, \text{ so that}
$$

$$
\mathbf{b}_5 = \begin{bmatrix} 2 \\ 0 \\ 0 \end{bmatrix}.
$$

Using these results in Theorem 6.3 we find that the general solution of the matrix differential equation is

$$
\mathbf{x} = c_1 \begin{bmatrix} 1 \\ -1 \\ 1 \end{bmatrix} e^{-t} + c_2 \left\{ \begin{bmatrix} 1 \\ 0 \\ -1 \end{bmatrix} e^{-t} + \begin{bmatrix} 1 \\ -1 \\ 1 \end{bmatrix} te^{-t} \right\} + c_3 \left\{ \begin{bmatrix} 2 \\ 0 \\ 0 \end{bmatrix} e^{-t} + \begin{bmatrix} 2 \\ 0 \\ -2 \end{bmatrix} te^{-t} + \begin{bmatrix} 1 \\ -1 \\ 1 \end{bmatrix} t^2 e^{-t} \right\},
$$

where c_1, c_2 and c_3 are arbitrary constants.

In component form the solution becomes

$$
x_1 = (c_1 + c_2 + 2c_3)e^{-t} + (c_2 + 2c_3)te^{-t} + c_3 t^2 e^{-t},
$$
$$
x_2 = -c_1 e^{-t} - c_2 te^{-t} - c_3 t^2 e^{-t},
$$
$$
x_3 = (c_1 - c_2)e^{-t} + (c_2 - 2c_3)te^{-t} + c_3 t^2 e^{-t}.
$$

Notice that the system in this example corresponds to the one which would be obtained were the third order differential equation

$$\dddot{x} + 3\ddot{x} + 3\dot{x} + x = 0$$

to be solved by converting it to a system of equations by setting

$$x = x_1, \quad \dot{x}_1 = x_2 \quad \text{and} \quad \dot{x}_2 = x_3.$$

The constants $C_1 = c_1 + c_2 + 2c_3$, $C_2 = c_2 + 2c_3$ and $C_3 = c_3$ are arbitrary, so because $x_1 = x$, the general solution of the third order differential equation is

$$x = C_1 e^{-t} + C_2 t e^{-t} + C_3 t^2 e^{-t},$$

as would have been found had the equation been solved directly by means of Theorem 5.4(ii).

∎

We conclude this section by outlining a different method of approach to the solution of the matrix differential equation

$$\dot{x} = A_r x, \qquad r = 1, 2, \tag{43}$$

in which A_r denotes an $n \times n$ matrix short of r eigenvectors. To do this, it will be convenient to make use of the matrix partitioning notation

$$M = \{m_1 | m_2 | \ldots | m_n\}$$

introduced in Prob. 30 of Sec. 3.3, and used again in Sec. 3.11, to represent a matrix M in which the ith column is the vector m_i, for $i = 1, 2, \ldots, n$.

Using the notation and vectors introduced in Theorems 6.2 and 6.3, we define the matrices

$$Q_1 = \{a_1 | a_2 | \ldots | a_{n-1} | b_2\}, \quad \text{and} \quad Q_2 = \{a_1 | a_2 | \ldots | a_{n-2} | b_4 | b_5\}.$$

Then by using an argument similar in form to the one used in Sec. 3.11 to show that P diagonalizes A, it is possible to prove the following general result.

Theorem 6.4 (Reduction to upper triangular form)

Let A_r be either of the matrices in Theorems 6.2 or 6.3. Then

$$Q_r^{-1} A_r Q_r = T_r, \qquad r = 1, 2,$$

is an upper triangular matrix in which the elements of the leading diagonal of T_1 are $\lambda_1, \lambda_2, \ldots, \lambda_{n-2}, \mu, \mu$, and those of the leading diagonal of T_2 are $\lambda_1, \lambda_2, \ldots, \lambda_{n-3}, \mu, \mu, \mu$.

□

To use Theorem 6.4 to solve (43) we set

$$x = Q_r y \tag{44}$$

and substitute for **x** in (43). After pre-multiplication by $\mathbf{Q}_r^{-1}$ we then find that

$$\dot{\mathbf{y}} = \mathbf{T}_r\mathbf{y}, \qquad r = 1, 2. \tag{45}$$

As the system of equations represented by (45) involves an upper triangular matrix $\mathbf{T}_r$, the equations for $\dot{y}_i$ are simple in form and may be solved recursively by back substitution, starting with the last equation which is

$$\dot{\mathbf{y}}_n = \mu \mathbf{y}_n, \tag{46}$$

and then proceeding to the determination of $y_{n-1}, y_{n-2}, \dots, y_1$. To understand why this is so, all that is necessary is to notice that the general form of (46) is

$$\begin{bmatrix} \dot{y}_1 \\ \dot{y}_2 \\ \dot{y}_3 \\ \vdots \\ \dot{y}_{n-1} \\ \dot{y}_n \end{bmatrix} = \begin{bmatrix} \lambda_1 & & & & & \\ & \lambda_2 & & & & \\ & & \lambda_3 & & & \\ & & & \ddots & & \\ & & & & \mu & \\ \mathbf{0} & & & & & \mu \end{bmatrix} \begin{bmatrix} y_1 \\ y_2 \\ y_3 \\ \vdots \\ y_{n-1} \\ y_n \end{bmatrix} \tag{47}$$

Because this method of solution involves a considerable amount of routine calculation, it offers no real advantage over the first one if it is only used to solve homogeneous equations. Its real significance will become apparent in the next section when we come to solve nonhomogeneous systems. Once **y** has been constructed, **x** follows from (44) as $\mathbf{x} = \mathbf{Q}_r\mathbf{y}$.

Example 9. Solution by reduction to upper triangular form

Solve by reduction to upper triangular form the matrix differential equation of Ex. 8.

Solution

The matrix **A** is missing two eigenvectors, and so it corresponds to the case $r = 2$ in Theorem 6.4. We form the matrix

$$\mathbf{Q}_2 = \{\mathbf{a}_1 \vdots \mathbf{b}_4 \vdots \mathbf{b}_5\},$$

which from Ex. 8 is seen to be

$$\mathbf{Q}_2 = \begin{bmatrix} 1 & 2 & 2 \\ -1 & 0 & 0 \\ 1 & -2 & 0 \end{bmatrix}.$$

Then

$$\mathbf{Q}_2^{-1} = \begin{bmatrix} 0 & -1 & 0 \\ 0 & -\frac{1}{2} & -\frac{1}{2} \\ \frac{1}{2} & 1 & \frac{1}{2} \end{bmatrix},$$

and

$$T_2 = Q_2^{-1}AQ_2 = \begin{bmatrix} -1 & 2 & 0 \\ 0 & -1 & 1 \\ 0 & 0 & -1 \end{bmatrix}.$$

After the change of variable $x = Q_2 y$, the original system becomes (see (45))

$$\dot{y} = T_2 y,$$

which is equivalent to the scalar equations

$$\dot{y}_1 = -y_1 + 2y_2,$$
$$\dot{y}_2 = -y_2 + y_3,$$
$$\dot{y}_3 = -y_3.$$

Integrating the last equation we find

$$y_3 = c_3 e^{-t},$$

and using this result in the second equation then gives

$$y_2 = c_2 e^{-t} + c_3 t e^{-t}.$$

Finally, using y_3 and y_2 in the first equation, gives

$$y_1 = c_1 e^{-t} + 2c_2 t e^{-t} + c_3 t^2 e^{-t}.$$

Now $x = Q_2 y$, so that

$$x_1 = y_1 + 2y_2 + 2y_3, \quad x_2 = -y_1 \quad \text{and} \quad x_3 = y_1 - 2y_2.$$

Substituting the solutions for y_1, y_2 and y_3 just obtained we arrive at the desired solution

$$x_1 = (c_1 + 2c_2 + 2c_3)e^{-t} + 2(c_2 + c_3)t e^{-t} + c_3 t^2 e^{-t},$$
$$x_2 = -c_1 e^{-t} - 2c_2 t e^{-t} - c_3 t^2 e^{-t},$$
$$x_3 = (c_1 - 2c_2)e^{-t} + 2(c_2 - c_3)t e^{-t} + c_3 t^2 e^{-t}.$$

Replacing the arbitrary constant $2c_2$ by c_2 brings this form of solution into the one found in Ex. 8. ∎

Problems for Section 6.1

Express the following systems of differential equations in the matrix form

$$\dot{C}x = Dx.$$

1 $\dot{x}_1 - 3\dot{x}_2 = 2x_1 - x_2$
 $2\dot{x}_1 + \dot{x}_2 = 4x_1 - 3x_2.$

2 $\dot{x}_1 + \dot{x}_2 + 2\dot{x}_3 = x_1 - x_2 + x_3$
 $\dot{x}_1 - \dot{x}_2 + 3\dot{x}_3 = x_1 + 2x_2 - x_3$
 $\dot{x}_1 + \dot{x}_3 \qquad = 4x_1 + 2x_2 + x_3.$

3 $2\dot{x}_1 + 4\dot{x}_2 + 7\dot{x}_3 = 2x_1 + 3x_2 - x_3$
$\quad \dot{x}_2 - \dot{x}_3 \qquad\;\; = x_1 - x_2$
$\quad \dot{x}_1 - \dot{x}_2 \qquad\;\; = x_1 + 4x_3.$

4 $\dot{x}_1 - 2\dot{x}_2 + 3\dot{x}_3 = 3x_1 + x_2 - x_3$
$\quad \dot{x}_2 + 4\dot{x}_3 \qquad\; = x_1 + 2x_2 + x_3$
$\quad 7\dot{x}_3 \qquad\qquad = 2x_1 + 4x_2 - x_3.$

By setting $x_1 = x$, $x_2 = \dot{x}_1$ and, if necessary, $x_3 = \dot{x}_2$, express the following higher order differential equations in the matrix form

$$\dot{\mathbf{x}} = \mathbf{A}\mathbf{x}.$$

5 $\ddot{x} + 7\dot{x} - x = 0.$
6 $3\ddot{x} + 8\dot{x} + 6x = 0.$
7 $4\ddot{x} + 2\dot{x} - 3x = 0.$
8 $9\ddot{x} - 2x = 0.$
9 $2\dddot{x} + 3\ddot{x} - \dot{x} + x = 0.$
10 $3\dddot{x} - 9\ddot{x} + 3\dot{x} + 4x = 0.$

Express the following systems of differential equations in the normal form.

$$\dot{\mathbf{x}} = \mathbf{A}\mathbf{x}.$$

11 $-2\dot{x}_1 + \dot{x}_2 = x_1 + x_2$
$\quad \frac{3}{2}\dot{x}_1 - \frac{1}{2}\dot{x}_2 = 2x_2.$

12 $\dot{x}_1 + 3\dot{x}_2 \;\; = x_1 + x_2$
$\quad 2\dot{x}_1 - 4\dot{x}_2 = 2x_1 - x_2.$

13 $\dot{x}_1 + 2\dot{x}_2 \qquad\;\; = x_1 + x_3$
$\quad 2\dot{x}_1 + \dot{x}_2 - \dot{x}_3 = x_2 + x_3$
$\quad 3\dot{x}_1 + \dot{x}_2 + \dot{x}_3 = x_1 + x_2.$

14 $2\dot{x}_1 + 4\dot{x}_3 \qquad\quad = x_1$
$\quad -\dot{x}_1 + 3\dot{x}_2 + \dot{x}_3 = x_2$
$\quad \dot{x}_2 + 2\dot{x}_3 \qquad\quad = x_3.$

Find the general solution of the following systems of differential equations and, where appropriate, the solution of the associated initial value problem.

15 $\dot{x}_1 = x_1 + 6x_2$
$\quad \dot{x}_2 = 5x_1 + 2x_2,$ with $x_1(0) = 2$, $x_2(0) = -1.$

16 $\dot{x}_1 = x_1 + 3x_2$
$\quad \dot{x}_2 = -x_2.$

17 $\dot{x}_1 = 5x_1$
$\quad \dot{x}_2 = x_1 - 9x_2,$ with $x_1(0) = 1$, $x_2(0) = 2.$

18 $\dot{x}_1 = 5x_1 + 4x_2 + 2x_3$
$\quad \dot{x}_2 = 4x_1 + 5x_2 + 2x_3$
$\quad \dot{x}_3 = 2x_1 + 2x_2 + 2x_3.$

19 $\dot{x}_1 = 2x_1 - \frac{5}{2}x_2 + x_3$
$\dot{x}_2 = 2x_1 - 2x_2 + 2x_3$
$\dot{x}_3 = 0.$

20 $\dot{x}_1 = x_1 + x_3$
$\dot{x}_2 = x_3$
$\dot{x}_3 = -x_2.$

21 $\dot{x}_1 = \frac{9}{5}x_1 - \frac{2}{5}x_2 - x_3$
$\dot{x}_2 = \frac{8}{5}x_1 + \frac{1}{5}x_2 - 2x_3$ with $x_1(0) = 0, x_2(0) = 1, x_3(0) = -1.$
$\dot{x}_3 = -\frac{4}{5}x_1 + \frac{2}{5}x_2 + 2x_3.$

22 $\dot{x}_1 = 4x_1$
$\dot{x}_2 = x_2 + 3x_3$
$\dot{x}_3 = 3x_2 + x_3.$

Solve by means of diagonalization the following systems of differential equations.

23 The system in Prob. 15.

24 The system in Prob. 16.

25 The system in Prob. 17.

26 The system in Prob. 18.

27 The system in Prob. 22.

28 Starting from the vector solutions (41) and (42), derive the conditions to be satisfied by $\mathbf{b}_2, \mathbf{b}_4$ and $\mathbf{b}_5$ in Theorem 6.3.

Find the general solution of the following systems of differential equations by means of Theorems 6.2 or 6.3.

29 $\dot{x}_1 = 4x_1$
$\dot{x}_2 = x_1 + 4x_2.$

30 $\dot{x}_1 = x_2$
$\dot{x}_2 = -x_1 - 2x_2.$

31 $\dot{x}_1 = 5x_1 + 10x_2 + 7x_3$
$\dot{x}_2 = -3x_2 - 3x_3$
$\dot{x}_3 = 3x_2 + 3x_3.$

32 $\dot{x}_1 = x_1 + x_2 + 2x_3$
$\dot{x}_2 = x_2 + 3x_3$
$\dot{x}_3 = x_3.$

33 Show by reducing the differential equation to a system that the general solution of

$$\dddot{x} + \ddot{x} + 3\dot{x} - 5x = 0$$

is

$$x = c_1 e^t + e^{-t}(c_2 \sin 2t + c_3 \cos 2t).$$

34 Show by reducing the differential equation to a system that the general solution of

$$\dddot{x} - 6\ddot{x} + 12\dot{x} - 8x = 0$$

is

$$x = (c_1 + c_2 t + c_3 t^2)e^{2t}.$$

35 Show by reducing the differential equation to a system that the general solution of

$$\ddot{x} - 8x = 0$$

is

$$x = c_1 e^{2t} + e^{-t}(c_2 \sin\sqrt{3}t + c_3 \cos\sqrt{3}t).$$

Find the general solution of the following systems of differential equations by reducing them to upper triangular form by means of Theorem 6.4.

33 The system in Prob. 29.

34 The system in Prob. 30.

35 The system in Prob. 31.

36 The system in Prob. 32.

Harder problems

The following set of related problems interprets the solution of the initial value problem

$$\dot{\mathbf{x}} = \mathbf{A}\mathbf{x}, \quad \mathbf{x}(t_0) = \mathbf{x}_0, \quad t \geq t_0,$$

when the solution is written in the form

$$\mathbf{x} = \exp\left(\int_{t_0}^{t} \mathbf{A} ds\right)\mathbf{x}_0,$$

or equivalently as

$$\mathbf{x} = \exp[(t - t_0)\mathbf{A}]\mathbf{x}_0.$$

Here, and in what follows, $\mathbf{A}$ is an $n \times n$ matrix with constant elements and $\mathbf{x}$ is an n element column vector. The expression $\exp[(t - t_0)\mathbf{A}]$ will be seen to represent an $n \times n$ matrix, so from the structure of the solution the columns of $\exp[(t - t_0)\mathbf{A}]$ must constitute a basis for the solution space of the system.

The motivation for seeking a solution in this form is provided by the solution of the single homogeneous constant coefficient differential equation

$$\dot{x} = Ax, \quad x(t_0) = x_0, \quad t \geq t_0,$$

which may be written as

$$x = \exp\left(\int_{t_0}^{t} A ds\right)x_0,$$

or as

$$x = \{\exp[(t - t_0)A]\}x_0.$$

When $n = 1$ the matrix solution reduces to this result. A rigorous justification of this approach would involve a study of the convergence of the series which are involved, but these considerations will be omitted.

37 Define the **exponential matrix function** $e^{\mathbf{A}}$ as

$$e^{\mathbf{A}} = \mathbf{I} + \frac{1}{1!}\mathbf{A} + \frac{1}{2!}\mathbf{A}^2 + \frac{1}{3!}\mathbf{A}^3 + \cdots,$$

where $e^{0} = \mathbf{I}$. Then each element of the $n \times n$ matrix $e^{\mathbf{A}}$ is the sum of an infinite series, all of which will be assumed to be convergent. Show that if $\mathbf{A}$ and $\mathbf{B}$ commute $(\mathbf{AB} = \mathbf{BA})$, then

$$\mathbf{B}e^{\mathbf{A}} = e^{\mathbf{A}}\mathbf{B}.$$

38 Show that if

$$\mathbf{A} = \begin{bmatrix} 0 & -k \\ k & 0 \end{bmatrix} \quad \text{and} \quad \mathbf{B} = \begin{bmatrix} a & 0 \\ 0 & b \end{bmatrix},$$

then

$$e^{t\mathbf{A}} = \begin{bmatrix} \cos kt & -\sin kt \\ \sin kt & \cos kt \end{bmatrix} \quad \text{and} \quad e^{t\mathbf{B}} = \begin{bmatrix} e^{at} & 0 \\ 0 & e^{bt} \end{bmatrix}.$$

39 Show that

$$\frac{d}{dt}(e^{t\mathbf{A}}) = \mathbf{A}e^{t\mathbf{A}}.$$

40 Set $\mathbf{M}(t) = e^{t\mathbf{A}}e^{-t\mathbf{A}}$. By making use of the result $e^{0} = \mathbf{I}$ show that $\mathbf{M}(t) = \mathbf{I}$, and hence that $e^{t\mathbf{A}}$ is nonsingular. Use this result to prove that $(e^{t\mathbf{A}})^{-1} = e^{-t\mathbf{A}}$.

41 Let the $n \times n$ constant element matrices $\mathbf{A}$ and $\mathbf{B}$ commute, and set

$$\mathbf{M}(t) = \exp[t(\mathbf{A} + \mathbf{B})]\exp(-t\mathbf{B})\exp(-t\mathbf{A}).$$

Show by considering $\mathbf{M}(0)$ and $d\mathbf{M}/dt$, and by using the result of Prob. 37, that

$$\exp[t(\mathbf{A} + \mathbf{B})]\exp(-t\mathbf{B})\exp(-t\mathbf{A}) = \mathbf{I}.$$

Hence deduce that

$$\exp(\mathbf{A} + \mathbf{B}) = \exp(\mathbf{A})\exp(\mathbf{B}).$$

Give reasons why this result is untrue when $\mathbf{A}$ and $\mathbf{B}$ do not commute.

42 Show that

$$\int e^{-t\mathbf{A}}\,dt = -\mathbf{A}^{-1}e^{-t\mathbf{A}} = -e^{-t\mathbf{A}}\mathbf{A}^{-1},$$

to within an arbitrary constant matrix.

43 Define

$$\mathbf{B}(t) = \int_{t_0}^{t} \mathbf{A}\,ds,$$

and show that

$$\mathbf{B}(t) = (t - t_0)\mathbf{A}.$$

Hence prove that

$$\frac{d\mathbf{B}^m}{dt} = m\mathbf{A}\mathbf{B}^{m-1}, \qquad m = 1, 2, \ldots.$$

44 Take as the starting approximation to the solution of

$$\dot{\mathbf{x}} = \mathbf{A}\mathbf{x}, \quad \mathbf{x}(t_0) = \mathbf{x}_0, \qquad t \geq t_0$$

in the Picard iteration scheme (Sec. 4.7) the vector $\mathbf{x}_0$. The next approximation $\mathbf{x}^{(1)}$ is thus the solution of

$$\frac{d\mathbf{x}^{(1)}}{dt} = \mathbf{A}\mathbf{x}_0, \quad \mathbf{x}^{(1)}(t_0) = \mathbf{x}_0,$$

so that

$$\mathbf{x}^{(1)} = \mathbf{x}_0 + \int_{t_0}^{t} \mathbf{A}\mathbf{x}_0 \, dt = (\mathbf{I} + \mathbf{B})\mathbf{x}_0, \text{ where } \mathbf{B}(t) = \int_{t_0}^{t} \mathbf{A} \, ds.$$

Letting $\mathbf{x}^{(m)}$ be the mth approximation satisfying

$$\frac{d\mathbf{x}^{(m)}}{dt} = \mathbf{A}\mathbf{x}^{(m-1)}, \quad \mathbf{x}^{(m)}(t_0) = \mathbf{x}_0,$$

show by induction that

$$\mathbf{x}^{(m)} = \left(\mathbf{I} + \frac{1}{1!}\mathbf{B} + \frac{1}{2!}\mathbf{B}^2 + \frac{1}{3!}\mathbf{B}^3 + \ldots + \frac{1}{m!}\mathbf{B}^m\right)\mathbf{x}_0$$

45 By allowing $m \to \infty$ in the approximate solution $\mathbf{x}^{(m)}$ in Prob. 44, deduce that the solution to the initial value problem

$$\dot{\mathbf{x}} = \mathbf{A}\mathbf{x}, \quad \mathbf{x}(t_0) = \mathbf{x}_0, \qquad t \geq t_0$$

is

$$\mathbf{x} = e^{\mathbf{B}}\mathbf{x}_0 = \exp\left\{\int_{t_0}^{t} \mathbf{A} \, ds\right\}\mathbf{x}_0$$

or, equivalently,

$$\mathbf{x} = \exp[(t - t_0)\mathbf{A}]\mathbf{x}_0, \qquad \text{for } t \geq t_0.$$

46 Explain why it follows from the last result of Prob. 45 that the columns of $\exp[(t - t_0)\mathbf{A}]$ constitute a basis for the solution space.

47 Use the form of solution given in Prob. 45 to show that if

$$\mathbf{A} = \begin{bmatrix} 0 & 3 \\ -3 & 0 \end{bmatrix} \text{ and } \mathbf{x}_0 = \begin{bmatrix} 7 \\ -4 \end{bmatrix},$$

the solution to the initial value problem

$$\dot{\mathbf{x}} = \mathbf{A}\mathbf{x}, \quad \mathbf{x}(2) = \mathbf{x}_0, \qquad t \geq 2$$

is

$$x_1 = 7\cos 3(t - 2) - 4\sin 3(t - 2)$$
$$x_2 = -7\sin 3(t - 2) - 4\cos 3(t - 2), \qquad \text{for } t \geq 2.$$

48 Let

$$A = \begin{bmatrix} a & -k \\ k & a \end{bmatrix}.$$

Write $A = B + C$, where

$$B = \begin{bmatrix} a & 0 \\ 0 & a \end{bmatrix}, \quad C = \begin{bmatrix} 0 & -k \\ k & 0 \end{bmatrix},$$

and show that B and C commute. Use this decomposition of A to solve the initial value problem

$$\dot{x} = Ax, \quad \text{with} \quad x(0) = \begin{bmatrix} \alpha \\ \beta \end{bmatrix}$$

by using the form of solution given in Prob. 45. [*Hint*: Make use of the results of Prob. 41].

49 Explain why the method of solution used in Prob. 48 is not applicable to the system

$$\dot{x}_1 = 4x_1 - 5x_2$$
$$\dot{x}_2 = 5x_1 + 3x_2.$$

50 Let A be diagonalizable by P with

$$P^{-1}AP = \Lambda,$$

where Λ is a diagonal matrix whose elements $\lambda_1, \lambda_2, \ldots, \lambda_n$ in the leading diagonal are the eigenvalues of A. Show that

$$A^n = P\Lambda^n P^{-1},$$

and hence that

$$e^{tA} = Pe^{t\Lambda}P^{-1}.$$

Prove that

$$\exp(t\Lambda) = \begin{bmatrix} \exp(\lambda_1 t) & & & 0 \\ & \exp(\lambda_2 t) & & \\ & & \ddots & \\ 0 & & & \exp(\lambda_n t) \end{bmatrix},$$

and hence that

$$\exp(t A) = P \begin{bmatrix} \exp(\lambda_1 t) & & & 0 \\ & \exp(\lambda_2 t) & & \\ & & \ddots & \\ 0 & & & \exp(\lambda_n t) \end{bmatrix} P^{-1}.$$

51 Show by using the result of Prob. 50 that when **A** is diagonalizable by **P**, the solution to the initial value problem

$$\dot{x} = Ax, \quad x(t_0) = x_0, \quad t \geq t_0,$$

may be written as

$$
x = P
\begin{bmatrix}
\exp(\lambda_1 t) & & & \mathbf{0} \\
& \exp(\lambda_2 t) & & \\
& & \ddots & \\
\mathbf{0} & & & \exp(\lambda_n t)
\end{bmatrix}
P^{-1} x_0.
$$

Use this method to solve the initial value problem

$$\dot{x} = Ax, \quad x(0) = \begin{bmatrix} 1 \\ 0 \\ 1 \end{bmatrix}, \quad t \geq 0$$

given

$$A = \begin{bmatrix} 1 & 0 & 0 \\ 1 & 3 & 2 \\ 1 & 2 & 3 \end{bmatrix}.$$

6.2 First order linear nonhomogeneous systems of differential equations

The simultaneous differential equations

$$
\begin{aligned}
\dot{x}_1 &= a_{11}x_1 + a_{12}x_2 + \ldots + a_{1n}x_n + f_1(t) \\
\dot{x}_2 &= a_{21}x_1 + a_{22}x_2 + \ldots + a_{2n}x_n + f_2(t) \\
&\;\;\;\cdot \qquad\quad\; \cdot \qquad\quad \ldots \qquad\;\; \cdot \\
\dot{x}_n &= a_{n1}x_1 + a_{n2}x_2 + \ldots + a_{nn}x_n + f_n(t),
\end{aligned}
\tag{1}
$$

in which the coefficients a_{ij} are constants and at least one of the functions $f_1(t), f_2(t), \ldots, f_n(t)$ is nonzero, is called a first order linear **nonhomogeneous** system of constant coefficient differential equations. This system is a direct extension of the type studied in Sec. 6.1, and the nonzero function $f_i(t)$ in the ith equation is the **nonhomogeneous term** in that equation.

System (1) may be written as the matrix differential equation

$$\dot{x} = Ax + f,$$

(2)

by setting

$$
\mathbf{x} = \begin{bmatrix} x_1 \\ x_2 \\ \vdots \\ x_n \end{bmatrix}, \qquad
\mathbf{A} = \begin{bmatrix} a_{11} & a_{12} & \cdots & a_{1n} \\ a_{21} & a_{22} & \cdots & a_{2n} \\ \cdot & \cdot & \cdots & \cdot \\ a_{n1} & a_{n2} & \cdots & a_{nn} \end{bmatrix} \qquad \text{and} \quad
\mathbf{f} = \begin{bmatrix} f_1(t) \\ f_2(t) \\ \vdots \\ f_n(t) \end{bmatrix}.
\tag{3}
$$

This is called the **normal form (standard form)** of the nonhomogeneous system. The matrix differential equation

$$
\mathbf{C}\dot{\mathbf{x}} = \mathbf{D}\mathbf{x} + \mathbf{F}
\tag{4}
$$

may always be put into the normal form (2) by pre-multiplication by $\mathbf{C}^{-1}$, provided det $\mathbf{C} \neq 0$.

For example, the system

$$
\begin{aligned}
3\dot{x}_1 + \dot{x}_2 &= 13x_1 + 11x_2 + 3e^{2t} + t \\
5\dot{x}_1 + 2\dot{x}_2 &= 21x_1 + 18x_2 + 5e^{2t} + 2t,
\end{aligned}
\tag{5}
$$

is of the form (4), with

$$
\mathbf{C} = \begin{bmatrix} 3 & 1 \\ 5 & 2 \end{bmatrix}, \quad
\mathbf{D} = \begin{bmatrix} 13 & 11 \\ 21 & 18 \end{bmatrix} \quad \text{and} \quad
\mathbf{F} = \begin{bmatrix} 3e^{2t} + t \\ 5e^{2t} + 2t \end{bmatrix}.
$$

In this case

$$
\mathbf{C}^{-1} = \begin{bmatrix} 2 & -1 \\ -5 & 3 \end{bmatrix}, \quad \text{so} \quad
\mathbf{A} = \mathbf{C}^{-1}\mathbf{D} = \begin{bmatrix} 5 & 4 \\ -2 & -1 \end{bmatrix}, \quad
\mathbf{f} = \mathbf{C}^{-1}\mathbf{F} = \begin{bmatrix} e^{2t} \\ t \end{bmatrix}.
\tag{6}
$$

Thus when expressed in the normal form

$$
\dot{\mathbf{x}} = \mathbf{A}\mathbf{x} + \mathbf{f},
$$

system (5) becomes

$$
\begin{aligned}
\dot{x}_1 &= 5x_1 + 4x_2 + e^{2t} \\
\dot{x}_2 &= -2x_1 - x_2 + t.
\end{aligned}
\tag{7}
$$

The nonhomogeneous term in the first equation in (7) is e^{2t}, and in the second it is t.

We now examine the structure of the solution of the nonhomogeneous matrix differential equation (2). Let $\mathbf{x}_1$ and $\mathbf{x}_2$ be any two solutions of (2), so that

$$
\begin{aligned}
\dot{\mathbf{x}}_1 &= \mathbf{A}\mathbf{x}_1 + \mathbf{f} \\
\dot{\mathbf{x}}_2 &= \mathbf{A}\mathbf{x}_2 + \mathbf{f}.
\end{aligned}
$$

Then setting $\mathbf{w} = \mathbf{x}_1 - \mathbf{x}_2$, it follows by subtracting these equations that

$$
\dot{\mathbf{w}} = \mathbf{A}\mathbf{w}.
\tag{8}
$$

Thus any two solutions of the nonhomogeneous equation (2) can differ only by a solution of the homogeneous equation (8) associated with (2). By analogy with the scalar case, the homogeneous matrix differential equation

$$\dot{\mathbf{x}} = \mathbf{A}\mathbf{x} \tag{9}$$

associated with (2) is called the **reduced** matrix differential equation associated with (2).

If $\tilde{\mathbf{x}}_p$ is any solution of (2), it follows at once that the general solution may be written

$$\mathbf{x} = \mathbf{x}_c + \tilde{\mathbf{x}}_p, \tag{10}$$

where $\mathbf{x}_c$ is the general solution of (9). As in previous chapters, $\mathbf{x}_c$ is called the **complementary function** of (12), and $\tilde{\mathbf{x}}_p$ a **particular integral**.

It is convenient to make $\tilde{\mathbf{x}}_p$ unique and to denote it by $\mathbf{x}_p$. This may be achieved by subtracting from any particular integral $\tilde{\mathbf{x}}_p$ all terms which are linearly dependent on $\mathbf{x}_c$, and then writing the **general solution** of (2) as

$$\mathbf{x} = \mathbf{x}_c + \mathbf{x}_p. \tag{11}$$

This is permissible because the terms which are subtracted from $\tilde{\mathbf{x}}_p$ are already contained in $\mathbf{x}_c$, which is an arbitrary linear combination of all n funtions forming a basis for (2).

An **initial value problem** for (2) involves finding the solution of

$$\dot{\mathbf{x}} = \mathbf{A}\mathbf{x} + \mathbf{f}, \quad \mathbf{x}(t_0) = \mathbf{x}_0, \qquad t \geq t_0, \tag{12}$$

where the initial condition $\mathbf{x}_0$ is some given constant vector. Once the general solution (11) has been found, the solution to the initial value problem (12) follows by finding the n arbitrary constants in $\mathbf{x}_c$ such that

$$\mathbf{x}_0 = \mathbf{x}_c(t_0) + \mathbf{x}_p(t_0). \tag{13}$$

The method for obtaining the general solution of (2) depends on whether $\mathbf{A}$ is diagonalizable or only reducible to upper triangular form.

(i) A is diagonalizable

Suppose, using the notation of Theorem 3.32(iii), that $\mathbf{A}$ may be diagonalized to Λ by means of the matrix $\mathbf{P}$, so that

$$\mathbf{P}^{-1}\mathbf{A}\mathbf{P} = \Lambda, \tag{14}$$

Then setting

$$\mathbf{x} = \mathbf{P}\mathbf{y} \tag{15}$$

in (2) and pre-multiplying by $\mathbf{P}^{-1}$, it follows that

$$\dot{\mathbf{y}} = \Lambda\mathbf{y} + \mathbf{P}^{-1}\mathbf{f}. \tag{16}$$

The solution of this system is now straightforward, for as in (26) of Sec. 6.1 the equations have been uncoupled. The matrix differential equation (16) is equivalent to the n scalar nonhomogeneous equations

$$\dot{y}_i = \lambda_i y_i + g_i(t), \qquad i = 1, 2, \ldots, n, \tag{17}$$

where λ_i is the ith entry in the leading diagonal of Λ, and $g_i(t)$ is the ith entry in the column vector $\mathbf{P}^{-1}\mathbf{f}$.

Integration of (17) using the integrating factor $\exp(-\lambda_i t)$ gives

$$y_i = c_i \exp(\lambda_i t) + \exp(\lambda_i t) \int \exp(-\lambda_i t) g_i(t) \, dt, \qquad i = 1, 2, \ldots, n, \tag{18}$$

with c_i an arbitrary constant. Once the vector $\mathbf{y}$ has been constructed by means of (18), the general solution $\mathbf{x}$ of (2) then follows from the result

$$\mathbf{x} = \mathbf{P}\mathbf{y}.$$

In (18) the term $c_i \exp(\lambda_i t)$ is the complementary function, while $\int \exp(-\lambda_i t) g_i(t) \, dt$ is a particular integral of (12). The following example is typical of the solution of a nonhomogeneous system by means of diagonalization.

Example 1. Diagonalizable system

Solve the initial value problem

$$\dot{x}_1 = 5x_1 + 4x_2 + e^{2t}$$
$$\dot{x}_2 = -2x_1 - x_2 + t, \qquad \text{with} \quad x_1(0) = 2, x_2(0) = -1.$$

Solution

We have

$$\mathbf{A} = \begin{bmatrix} 5 & 4 \\ -2 & -1 \end{bmatrix}, \quad \mathbf{f} = \begin{bmatrix} e^{2t} \\ t \end{bmatrix} \quad \text{and} \quad \mathbf{x}(0) = \mathbf{x}_0 = \begin{bmatrix} 2 \\ -1 \end{bmatrix}.$$

The eigenvalues of $\mathbf{A}$ are $\lambda_1 = 3$, $\lambda_2 = 1$, and the diagonalizing matrix is

$$\mathbf{P} = \begin{bmatrix} 2 & 1 \\ -1 & -1 \end{bmatrix}, \quad \text{so } \mathbf{P}^{-1} = \begin{bmatrix} 1 & 1 \\ -1 & -2 \end{bmatrix}.$$

Thus

$$\Lambda = \mathbf{P}^{-1}\mathbf{A}\mathbf{P} = \begin{bmatrix} 3 & 0 \\ 0 & 1 \end{bmatrix}, \quad \mathbf{P}^{-1}\mathbf{f} = \begin{bmatrix} e^{2t} + t \\ -e^{2t} - 2t \end{bmatrix},$$

so that (16) becomes

$$\dot{y}_1 = 3y_1 + e^{2t} + t$$
$$\dot{y}_2 = y_2 - e^{2t} - 2t.$$

Integration of these equations gives

$$y_1 = c_1 e^{3t} - e^{2t} - \tfrac{1}{3}t - \tfrac{1}{9}$$
$$y_2 = c_2 e^t - e^{2t} + 2t + 2.$$

The general solution of the original system is

$$\mathbf{x} = \mathbf{Py},$$

with $\mathbf{y}$ the column vector with the elements y_1 and y_2; so it follows that

$$\begin{bmatrix} x_1 \\ x_2 \end{bmatrix} = \begin{bmatrix} 2 & 1 \\ -1 & -1 \end{bmatrix} \begin{bmatrix} y_1 \\ y_2 \end{bmatrix}, \quad \text{or} \quad x_1 = 2y_1 + y_2, \quad x_2 = -y_1 - y_2.$$

Substituting for y_1 and y_2 then gives

$$x_1 = 2c_1 e^{3t} + c_2 e^t - 3e^{2t} + \tfrac{4}{3}t + \tfrac{16}{9}, \quad x_2 = -c_1 e^{3t} - c_2 e^t + 2e^{2t} - \tfrac{5}{3}t - \tfrac{17}{9}.$$

If $\mathbf{x} = \mathbf{x}_c + \mathbf{x}_p$, with $\mathbf{x}_c$ the complementary function and $\mathbf{x}_p$ the particular integral, we see that

$$\mathbf{x}_c = \begin{bmatrix} 2c_1 e^{3t} + c_2 e^t \\ -c_1 e^{3t} - c_2 e^t \end{bmatrix} \quad \text{and} \quad \mathbf{x}_p = \begin{bmatrix} -3e^{2t} + \tfrac{4}{3}t + \tfrac{16}{9} \\ 2e^{2t} - \tfrac{5}{3}t - \tfrac{17}{9} \end{bmatrix}.$$

The solution to the initial value problem follows from these results by requiring $x_1(0) = 2$, $x_2(0) = -1$, which leads to the results $c_1 = 19/9$, $c_2 = -1$. The required solution is thus

$$x_1 = \tfrac{38}{9}e^{3t} - e^t - 3e^{2t} + \tfrac{4}{3}t + \tfrac{16}{9}$$
$$x_2 = -\tfrac{19}{9}e^{3t} + e^t + 2e^{2t} - \tfrac{5}{3}t - \tfrac{17}{9}. \qquad \blacksquare$$

(ii) A is reducible to upper triangular form

If the $n \times n$ matrix $\mathbf{A}$ has only $r < n$ linearly independent eigenvectors it cannot be diagonalized, but it can be reduced to upper triangular form by the matrix $\mathbf{Q}_r$ in Theorem 6.4.

Let $\mathbf{A}_r$ and $\mathbf{Q}_r$ be either of the corresponding pairs of $n \times n$ matrices in Theorem 6.4, and let $\mathbf{x}$ satisfy the nonhomogeneous matrix differential equation

$$\dot{\mathbf{x}} = \mathbf{Ax} + \mathbf{f}, \qquad r = 1, 2. \tag{20}$$

Setting

$$\mathbf{x} = \mathbf{Q}_r \mathbf{y}, \tag{21}$$

substituting (21) into (20) and pre-multiplying by $\mathbf{Q}_r^{-1}$ it then follows, as in (45) of Sec. 6.1, that

$$\dot{\mathbf{y}} = \mathbf{Q}_r^{-1} \mathbf{A}_r \mathbf{Q}_r \mathbf{y} + \mathbf{g}, \tag{22}$$

with $\mathbf{g} = \mathbf{Q}_r^{-1} \mathbf{f}$ and g_i the ith element of $\mathbf{g}$.

In terms of the upper triangular matrix $\mathbf{T}_r = \mathbf{Q}_r^{-1} \mathbf{AQ}_r$ defined in Theorem 6.4 this becomes

$$\dot{\mathbf{y}} = \mathbf{T}_r \mathbf{y} + \mathbf{g}, \qquad r = 1, 2. \tag{23}$$

Then, as in Sec. 6.1, the structure of (23) enables the y_r to be solved recursively by back substitution, starting with the last equation which is

$$\dot{y}_n = \mu y_n + g_n, \tag{24}$$

and then proceeding to the determination of $y_{n-1}, y_{n-2}, \ldots, y_1$.

To see how this comes about, it is only necessary to notice that the matrix differential equation (23) is of the form

$$
\begin{bmatrix} \dot{y}_1 \\ \dot{y}_2 \\ \dot{y}_3 \\ \vdots \\ \dot{y}_n \end{bmatrix}
=
\begin{bmatrix} \lambda_1 & & & & \\ & \lambda_2 & & & \\ & & \lambda_3 & & \\ & & & \ddots & \\ 0 & & & & \mu \end{bmatrix}
\begin{bmatrix} y_1 \\ y_2 \\ y_3 \\ \vdots \\ y_n \end{bmatrix}
+
\begin{bmatrix} g_1 \\ g_2 \\ g_3 \\ \vdots \\ g_n \end{bmatrix}.
\tag{25}
$$

Once $y_1, y_2, \ldots, y_n$ have been obtained, the vector $\mathbf{y}$ can be constructed and $\mathbf{x}$ then follows from (21) as

$$\mathbf{x} = \mathbf{Q}_r \mathbf{y}, \qquad r = 1, 2. \tag{26}$$

Example 2. A is only reducible to an upper triangular matrix

Solve the system of equations

$$\dot{x}_1 = x_2$$
$$\dot{x}_2 = x_3$$
$$\dot{x}_3 = -x_1 - 3x_2 - 3x_3 + e^{3t}$$

subject to the initial conditions $x_1(0) = 1$, $x_2(0) = x_3(0) = 0$.

Solution

In this case

$$
\mathbf{A} = \begin{bmatrix} 0 & 1 & 0 \\ 0 & 0 & 1 \\ -1 & -3 & -3 \end{bmatrix} \quad \text{and} \quad \mathbf{f} = \begin{bmatrix} 0 \\ 0 \\ e^{3t} \end{bmatrix}.
$$

Matrix $\mathbf{A}$ has already been examined in Ex. 8, Sec. 6.1, where it was shown to be associated with the matrix

$$
\mathbf{Q}_2 = \begin{bmatrix} 1 & 2 & 2 \\ -1 & 0 & 0 \\ 1 & -2 & 0 \end{bmatrix}, \quad \text{with the inverse} \quad \mathbf{Q}_2^{-1} = \begin{bmatrix} 0 & -1 & 0 \\ 0 & -\frac{1}{2} & -\frac{1}{2} \\ \frac{1}{2} & 1 & \frac{1}{2} \end{bmatrix}.
$$

It follows that

$$T_2 = Q_2^{-1} A Q_2 = \begin{bmatrix} -1 & 2 & 0 \\ 0 & -1 & 1 \\ 0 & 0 & -1 \end{bmatrix} \quad \text{and} \quad g = Q_2^{-1} f = \begin{bmatrix} 0 \\ -\frac{1}{2} e^{3t} \\ \frac{1}{2} e^{3t} \end{bmatrix}.$$

Thus (23) becomes the system

$$\dot{y}_1 = -y_1 + 2y_2$$
$$\dot{y}_2 = -y_2 + y_3 - \tfrac{1}{2} e^{3t}$$
$$\dot{y}_3 = -y_3 + \tfrac{1}{2} e^{3t}.$$

The last equation has the solution

$$y_3 = c_3 e^{-t} + \tfrac{1}{8} e^{3t}.$$

Using y_3 in the second equation leads to the solution

$$y_2 = c_2 e^{-t} + c_3 t e^{-t} - \tfrac{3}{32} e^{3t}$$

and, finally, using y_2 in the first equation we arrive at the solution

$$y_1 = c_1 e^{-t} + 2c_2 t e^{-t} + c_3 t^2 e^{-t} - \tfrac{3}{64} e^{3t}.$$

Constructing y from y_1, y_2 and y_3, the general solution x then follows from $x = Q_2 y$ as

$$x_1 = (c_1 + 2c_2 + 2c_3) e^{-t} + 2(c_2 + c_3) t e^{-t} + c_3 t^2 e^{-t} + \tfrac{1}{64} e^{3t}$$
$$x_2 = -c_1 e^{-t} - 2c_2 t e^{-t} - c_3 t^2 e^{-t} + \tfrac{3}{64} e^{3t}$$
$$x_3 = (c_1 - 2c_2) e^{-t} + 2(c_2 - c_3) t e^{-t} + c_3 t^2 e^{-t} + \tfrac{9}{64} e^{3t}.$$

The initial conditions $x_1(0) = 1$, $x_2(0) = x_3(0) = 0$ are satisfied if $c_1 = \tfrac{3}{64}$, $c_2 = \tfrac{3}{32}$ and $c_3 = \tfrac{3}{8}$, and thus the required solution to the initial value problem is

$$x_1 = \tfrac{63}{64} e^{-t} + \tfrac{15}{16} t e^{-t} + \tfrac{3}{8} t^2 e^{-t} + \tfrac{1}{64} e^{3t}$$
$$x_2 = -\tfrac{3}{64} e^{-t} - \tfrac{3}{16} t e^{-t} - \tfrac{3}{8} t^2 e^{-t} + \tfrac{3}{64} e^{3t}$$
$$x_3 = -\tfrac{9}{64} e^{-t} - \tfrac{9}{16} t e^{-t} + \tfrac{3}{8} t^2 e^{-t} + \tfrac{9}{64} e^{3t}.$$

We may use this example to illustrate the structure of the solution first mentioned in (11). If the general solution is written in the form $x = x_c + x_p$, with x_c the complementary function and x_p the particular integral, it follows that in this case

$$x_c = \begin{bmatrix} (c_1 + 2c_2 + 2c_3) e^{-t} + 2(c_2 + c_3) t e^{-t} + c_3 t^2 e^{-t} \\ -c_1 e^{-t} - 2c_2 t e^{-t} - c_3 t^2 e^{-t} \\ (c_1 - 2c_2) e^{-t} + 2(c_2 - c_3) t e^{-t} + c_3 t^2 e^{-t} \end{bmatrix}, \quad x_p = \begin{bmatrix} \tfrac{1}{64} e^{3t} \\ \tfrac{3}{64} e^{3t} \\ \tfrac{9}{64} e^{3t} \end{bmatrix}.$$

By virtue of their definitions, x_c and x_p satisfy the matrix differential equations

$$\dot{x}_c = A x_c \quad \text{and} \quad \dot{x}_p = A x_p + f,$$

with A and f the matrices already defined. ∎

Problems for Section 6.2

Find the general solution of each of the following systems of equations and hence, where appropriate, the solution of the associated initial value problem.

1 $\dot{x}_1 = 2x_1 + x_2$
$\dot{x}_2 = 3x_1 + 2.$ $x_1(0) = 1,$ $x_2(0) = -1.$

2 $\dot{x}_1 = x_1 + 6x_2 + 11$
$\dot{x}_2 = 5x_1 + 2x_2.$

3 $\dot{x}_1 = 10x_1 - 8x_2 + e^t$
$\dot{x}_2 = 6x_1 - 11x_2,$ $x_1(0) = x_2(0) = 0.$

4 $3\dot{x}_1 = x_1 - 2x_2 + 9t$
$3\dot{x}_2 = -4x_1 - x_2 + 27,$ $x_1(0) = 3,$ $x_2(0) = 0.$

5 $\dot{x}_1 = x_2 + 4$
$\dot{x}_2 = -4x_1 - 4x_2 - 1$

6 $\dot{x}_1 = x_1 - 3x_2 + 2x_3 + t$
$\dot{x}_2 =\quad -x_2 + t + 1$
$\dot{x}_3 =\quad -x_2 - 2x_3 + 2,$ $x_1(0) = x_2(0) = x_3(0) = 0.$

7 $\dot{x}_1 = -x_1 - x_3 + 1$
$\dot{x}_2 = -x_1 - 2x_2 + x_3 - 1$
$\dot{x}_3 = x_1 + x_2 - 3x_3 + 3,$ $x_1(0) = -1,$ $x_2(0) = x_3(0) = 0.$

8 $\dot{x}_1 = -x_1 + 2x_2 + 2x_3 + 1$
$\dot{x}_2 =\quad -2x_2 + 2x_3 + 1$
$\dot{x}_3 =\quad -x_2 - 5x_3 + 1,$
$x_1(0) = \frac{13}{3},$ $x_2(0) = \frac{19}{12},$ $x_3(0) = \frac{1}{12}.$

9 $\dot{x}_1 = x_2$
$\dot{x}_2 = x_3 + 4$
$\dot{x}_3 = -x_1 - 3x_2 - 3x_3 + 2.$

10 $\dot{x}_1 = 2x_1$
$\dot{x}_2 = x_3 + \sin t$
$\dot{x}_3 = -9x_2 - 6x_3.$

11 By considering the rank of C, $\{C|D\}$ and $\{C|D|F\}$, determine the implication for the solution vector x if $\det C = 0$ and

(a) $C\dot{x} = Dx,$ (b) $C\dot{x} = Dx + F.$

12 Suppose, if possible, that x_1 and x_2 are two different solutions of the initial value problem

$$\dot{x} = Ax + f, \quad x(t_0) = x_0, \quad t \geq t_0.$$

Show by considering the differential equation satisfied by $w = x_1 - x_2$ that $w \equiv 0$ for $t \geq t_0$, so the solution is unique.

Harder problems

The following problems extend the exponential matrix function introduced in the problem set of Sec. 6.1 to the solution of nonhomogeneous initial value problems of the form

$$\dot{x} = Ax + f, \quad x(t_0) = x_0, \quad t \geq t_0.$$

13 Show that

$$\frac{d}{dt}(e^{-tA}x) = -e^{-tA}Ax + e^{-tA}\dot{x},$$

and hence that

$$\dot{x} = Ax + f$$

may be written in the form

$$\frac{d}{dt}(e^{-tA}x) = e^{-tA}f.$$

14 Integrate the final result of Prob. 13 to show that the solution of the initial value problem

$$\dot{x} = Ax + f, \quad x(t_0) = x_0, \quad t \geq t_0,$$

is

$$x(t) = \exp[(t-t_0)A]x_0 + \int_{t_0}^{t} \exp[(t-\tau)A] f(\tau)\,d\tau, \quad t \geq t_0.$$

15 Show that if $f(t) = C = a$ constant matrix and $t_0 = 0$, then after using the result of Prob. 42, Sec. 6.1, the result of Prob. 14 reduces to

$$x(t) = e^{tA}(x_0 + A^{-1}C) - A^{-1}C.$$

16 Solve Prob. 7 by the method of Prob. 15, making use of the result of Prob. 50 of Sec. 6.1 to evaluate e^{tA}.

17 Use the method of Prob. 14, together with the result of Prob. 50 of Sec. 6.1, to solve

$$\dot{x}_1 = 10x_1 - 18x_2 + \sin t$$
$$\dot{x}_2 = 6x_1 - 11x_2 + 2t,$$

with $x_1(0) = x_2(0) = 0$.

18 If A and U are square constant coefficient matrices which commute (i.e. $AU = UA$), show that

(i) the product AU^{-1} commutes provided U is nonsingular,
(ii) if

$$V = e^{tA}Ue^{-tA},$$

then $V \equiv U$ for all t.
[*Hint*: Calculate dV/dt].
Use the result of Prob. 14 with $t_0 = 0$ to show that if $f(t) = e^{pt}C$, where $p \neq 0$ is a scalar constant and C is a column vector with constant elements, integration by parts coupled with the

assumption that $p\mathbf{I} - \mathbf{A}$ is nonsingular leads to the result

$$\mathbf{x}(t) = e^{t\mathbf{A}}\mathbf{x}_0 + e^{t\mathbf{A}}(p\mathbf{I} - \mathbf{A})^{-1}[e^{pt}e^{-\mathbf{A}t} - \mathbf{I}]\mathbf{C}.$$

Use the results of (i) and (ii) to show that

$$\mathbf{x}(t) = e^{t\mathbf{A}}[\mathbf{x}_0 - (p\mathbf{I} - \mathbf{A})^{-1}\mathbf{C}] + e^{pt}(p\mathbf{I} - \mathbf{A})^{-1}\mathbf{C}.$$

6.3 Second order linear systems of differential equations

This section is concerned with the solution of simultaneous linear constant coefficient second order differential equations. The main reason for interest in such systems of equations is that they include equations which may be used to model the coupled oscillations which occur in many physical systems.

For simplicity, the only case to be considered in detail will be that of systems without first derivative terms. It will be recalled from Sec. 5.6 that when related to physical systems first order derivative terms describe damping, and thus the dissipation of energy. Consequently the physical systems discussed here will all be *undamped*. The omission of damping in a first account of coupled oscillations is not serious, because apart from the decay of the solution, undamped systems exhibit all the essential features of the more general case in which damping is present. For the sake of completeness, a method by which the solution may be obtained when first derivative terms are present will be outlined at the end of the section.

The following simple examples illustrate some of the ways in which equations of this type may arise, and also the analogy that exists between mechanical and electrical systems. Figure 6.1(a) represents a mass-spring system in which masses m_1 and m_2 are connected to springs with stiffnesses k_1 and k_2. The system is assumed to be in equilibrium on a smooth horizontal table, with the left-hand end fastened rigidly to the table. In Fig. 6.1(b) the system is shown undergoing coupled oscillations along the line of

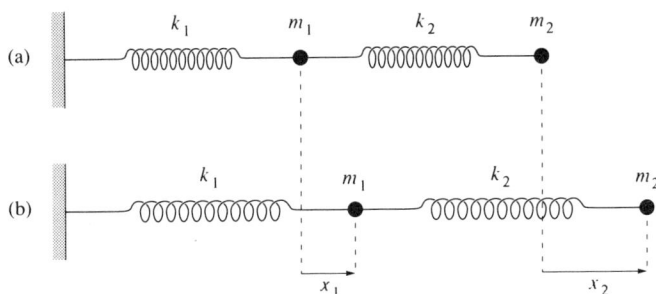

Fig. 6.1 Spring coupled oscillating masses: (a) Equilibrium configuration, (b) Displaced positions when in oscillation

the springs, with x_1 and x_2 the respective displacements of m_1 and m_2 from their equilibrium positions. The oscillations are assumed to be started by displacing m_1 and m_2 from their equilibrium positions and then releasing them, either from rest or with given initial velocities.

The equations of motion of the system are easily seen to be

$$m_1 \ddot{x}_1 + k_1 x_1 - k_2(x_2 - x_1) = 0 \tag{1}$$

$$m_2 \ddot{x}_2 + k_2(x_2 - x_1) = 0. \tag{2}$$

A direct electrical analog for this mechanical system is shown in Fig. 6.2, in which i_1, and i_2 are currents, L_1, L_2 are inductances and C_1, C_2 capacitances. Kirchhoff's current law has already been applied in the diagram, so after using the voltage law we obtain

$$L_1 \frac{di_1}{dt} + \frac{q_1}{C_1} + \frac{q_1 - q_2}{C_2} = 0, \tag{3}$$

$$L_2 \frac{di_2}{dt} + L_1 \frac{di_1}{dt} + \frac{q_1}{C_1} = 0, \tag{4}$$

where q_1, q_2 are the charges. The analogy between (1), (2) and (3), (4) becomes apparent once $L_1 \, di_1/dt$ in (4) is eliminated by means of (3) and the results $i_1 = dq_1/dt$, $i_2 = dq_2/dt$ are used, for then (3) and (4) become

$$L_1 \ddot{q}_1 + \frac{q_1}{C_1} + \frac{q_1 - q_2}{C_2} = 0 \tag{5}$$

$$L_2 \ddot{q}_2 + \frac{q_2 - q_1}{C_2} = 0. \tag{6}$$

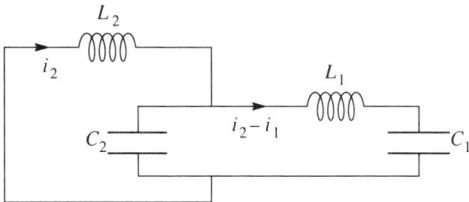

Fig. 6.2   Coupled LC circuits

There is thus a direct correspondence between mass m and inductance L, and spring stiffness k and the reciprocal of capacitance $1/C$.

A slightly more complicated mechanical system is illustrated in Fig. 6.3. In this system the three masses m_1, m_2 and m_3 rest on a smooth horizontal table and are joined by springs with stiffnesses k_1, k_2, k_3 and k_4, with the left-hand end of the system fastened rigidly to the table and the right-hand end forced to move sinusoidally along the line of the springs, with the displacement at time t being given by $X(t) = a \sin \Omega t$. The equilibrium configuration is shown in Fig. 6.3(a), and the oscillating system in Fig. 6.3(b),

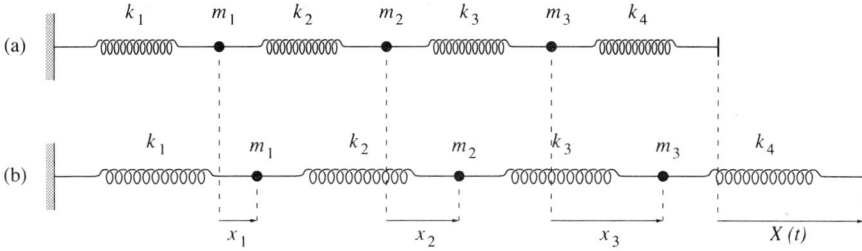

Fig. 6.3 Forced oscillations of a mass-spring system: (a) Equilibrium configuration, (b) Displaced positions when in forced oscillation

with the displacements x_1, x_2, x_3 and x_4 again being measured from the equilibrium positions.

The equations of motion for this mass-spring system are

$$m_1 \ddot{x}_1 = -k_1 x_1 + k_2 (x_2 - x_1) \tag{7}$$

$$m_2 \ddot{x}_2 = -k_2 (x_2 - x_1) + k_3 (x_3 - x_2) \tag{8}$$

$$m_3 \ddot{x}_3 = -k_3 (x_3 - x_2) + k_4 (X - x_3). \tag{9}$$

When written as a matrix differential equation (1), (2) becomes the homogeneous equation

$$\mathbf{M}\ddot{\mathbf{x}} = \mathbf{K}\mathbf{x}, \tag{10}$$

with

$$\mathbf{M} = \begin{bmatrix} m_1 & 0 \\ 0 & m_2 \end{bmatrix}, \quad \mathbf{K} = \begin{bmatrix} -(k_1 + k_2) & k_2 \\ k_2 & -k_2 \end{bmatrix} \quad \text{and} \quad \mathbf{x} = \begin{bmatrix} x_1 \\ x_2 \end{bmatrix}. \tag{11}$$

Similarly, when (7), (8) and (9) are written as a matrix differential equation they become the nonhomogeneous equation

$$\mathbf{M}\ddot{\mathbf{x}} = \mathbf{K}\mathbf{x} + \mathbf{F}, \tag{12}$$

with

$$\mathbf{M} = \begin{bmatrix} m_1 & 0 & 0 \\ 0 & m_2 & 0 \\ 0 & 0 & m_3 \end{bmatrix}, \quad \mathbf{K} = \begin{bmatrix} -(k_1 + k_2) & k_2 & 0 \\ k_2 & -(k_2 + k_3) & k_3 \\ 0 & k_3 & -(k_3 + k_4) \end{bmatrix},$$

$$\mathbf{x} = \begin{bmatrix} x_1 \\ x_2 \\ x_3 \end{bmatrix} \quad \text{and} \quad \mathbf{F} = \begin{bmatrix} 0 \\ 0 \\ k_4 a \sin \Omega t \end{bmatrix}. \tag{13}$$

In all such systems the matrix $\mathbf{M}$ is nonsingular, so these matrix differential equations may always be simplified by pre-multiplication by $\mathbf{M}^{-1}$ and brought into the **normal**

form

$$\ddot{\mathbf{x}} = \mathbf{A}\mathbf{x} + \mathbf{f}. \tag{14}$$

In what follows, (14) will represent a general system in which $\mathbf{x}$ is a column vector with elements $x_1, x_2, \ldots, x_n$, $\mathbf{A}$ is an $n \times n$ matrix with constant coefficients, and $\mathbf{f}$ is a column vector with elements $f_1(t), f_2(t), \ldots, f_n(t)$ which are given. The matrix differential equation in (14) is **homogeneous** when $\mathbf{f} \equiv 0$, and **nonhomogeneous** when at least one element of $\mathbf{f}$ is nonzero.

The **initial value problem** we shall be considering involves finding a vector $\mathbf{x}$ which satisfies the matrix differential equation

$$\ddot{\mathbf{x}} = \mathbf{A}\mathbf{x} + \mathbf{f}, \tag{15}$$

and is such that

$$\mathbf{x}(t_0) = \mathbf{x}_0, \quad \dot{\mathbf{x}}(t_0) = \mathbf{x}_1, \qquad t \geq t_0, \tag{16}$$

with $\mathbf{x}_0$, $\mathbf{x}_1$ given constant vectors.

In terms of the mechanical systems just described, the specification of $\mathbf{x}_0$ determines the initial positions of the masses at time t_0, and the specification of $\mathbf{x}_1$ their initial velocities at that time. In the electrical problem, the specification of $\mathbf{x}_0$ determines the charges at time t_0 and $\mathbf{x}_1$ the initial currents which flow.

The linearity of (14) means that if $\mathbf{x}_1$ and $\mathbf{x}_2$ are any two solutions of the associated homogeneous equation

$$\ddot{\mathbf{x}} = \mathbf{A}\mathbf{x}, \tag{17}$$

then $\mathbf{x} = c_1 \mathbf{x}_1 + c_2 \mathbf{x}_2$ is also a solution, with c_1 and c_2 arbitrary constants. Also, if $\mathbf{x}_c$ is a solution of the homogeneous equation (17) and $\tilde{\mathbf{x}}_p$ is any solution of the nonhomogeneous equation

$$\ddot{\mathbf{x}} = \mathbf{A}\mathbf{x} + \mathbf{f}, \tag{18}$$

then as in Sec. 6.2 the general solution of (18) is always expressible in the form

$$\mathbf{x} = \mathbf{x}_c + \tilde{\mathbf{x}}_p. \tag{19}$$

The solution $\mathbf{x}_c$ is again called the **complementary function**, and $\tilde{\mathbf{x}}_p$ is a **particular integral**. If all terms linearly dependent upon $\mathbf{x}_c$ are subtracted from $\mathbf{x}_p$ the resulting function $\mathbf{x}_p$ is unique, and will be called *the* **particular integral** of (18).

The determination of the general solution of the homogeneous equation

$$\ddot{\mathbf{x}} = \mathbf{A}\mathbf{x} \tag{20}$$

follows along the same lines as those discussed in Sec. 6.1. We seek a solution of the form

$$\mathbf{x} = \mathbf{a}e^{\omega t}, \tag{21}$$

with ω a parameter and $\mathbf{a}$ a constant vector, both of which are to be determined. After substituting (21) into (20) and canceling the nonzero factor $e^{\omega t}$ we obtain

$$(\mathbf{A} - \omega^2 \mathbf{I})\mathbf{a} = \mathbf{0}. \tag{22}$$

This shows that ω^2 is an eigenvalue of $\mathbf{A}$, and $\mathbf{a}$ is the corresponding eigenvector. With the exception of degenerate cases, there will always be n distinct eigenvalues λ_1, $\lambda_2, \ldots, \lambda_n$ of $\mathbf{A}$, and n corresponding linearly independent eigenvectors $\mathbf{p}_1, \mathbf{p}_2, \ldots, \mathbf{p}_n$. If we now introduce the matrix $\mathbf{P}$ whose ith column is the eigenvector $\mathbf{p}_i$ of $\mathbf{A}$ corresponding to the eigenvalue λ_i, it follows as in Sec. 6.1 that by setting

$$\mathbf{x} = \mathbf{P}\mathbf{y}, \tag{23}$$

equation (20) becomes the following matrix equation for $\mathbf{y}$

$$\ddot{\mathbf{y}} = \mathbf{P}^{-1}\mathbf{A}\mathbf{P}\mathbf{y}. \tag{24}$$

However, $\mathbf{P}^{-1}\mathbf{A}\mathbf{P}$ is the diagonal matrix

$$\mathbf{P}^{-1}\mathbf{A}\mathbf{P} = \mathbf{\Lambda} = \begin{bmatrix} \lambda_1 & & & 0 \\ & \lambda_2 & & \\ & & \ddots & \\ 0 & & & \lambda_n \end{bmatrix}, \tag{25}$$

so that (24) becomes

$$\ddot{\mathbf{y}} = \mathbf{\Lambda}\mathbf{y}. \tag{26}$$

The original system of equations has now been uncoupled, because the ith equation in (26) is just the linear second order differential equation

$$\ddot{y}_i = \lambda_i y_i, \qquad i = 1, 2, \ldots, n. \tag{27}$$

The solution vector $\mathbf{x}$ is recovered from $\mathbf{y}$ by means of (23).

Had the nonhomogeneous system (14) been considered, this same argument would have reduced it to

$$\ddot{\mathbf{y}} = \mathbf{\Lambda}\mathbf{y} + \mathbf{g}, \tag{28}$$

with $\mathbf{g} = \mathbf{P}^{-1}\mathbf{f}$ a vector with the known elements $g_1(t), g_2(t), \ldots, g_n(t)$. The ith equation in (28) is

$$\ddot{y}_i = \lambda_i y_i + g_i, \qquad i = 1, 2, \ldots, n, \tag{29}$$

which is simply the nonhomogeneous form of (27). Here again, the solution vector $\mathbf{x}$ may be recovered by means of (23) once $\mathbf{y}$ has been determined.

Inspection of (27) and (29) shows that the nature of the complementary function corresponding to (20) will be determined by the eigenvalues λ_i of $\mathbf{A}$. Positive λ_i will introduce hyperbolic functions into the complementary function, and negative λ_i will introduce sinusoidal functions into it. If, however, complex conjugate λ_i arise, they will introduce sinusoidal functions with exponential factors into the complementary function.

Of these three possibilities, only the situation in which all the λ_i are negative has special physical significance, because it is this case which corresponds to undamped coupled oscillations. The comparison of the cases in which the λ_i are either all positive or all negative is not without interest, so we shall undertake a brief examination of the nature of the solution when all the λ_i are positive, prior to considering the oscillatory case. Because it has no special physical significance, the case of complex λ_i will be omitted, but the analysis is straightforward and involves ideas similar to these used in Examples 3 and 4 of Sec. 6.1.

(i) Eigenvalues all positive

When $\lambda_i = \omega_i^2 > 0$, $i = 1, 2, \ldots, n$ the general solution of (27) may be written in the form

$$y_i(t) = c_i \cosh(\omega_i t + \alpha_i), \tag{30}$$

with c_i and α_i arbitrary constants.

If $\mathbf{p}_i$ is the ith column of $\mathbf{P}$, it follows from (23) and (30) that the general solution of (20) may be written in the form

$$\mathbf{x}(t) \equiv \mathbf{x}_c(t) = \sum_{i=1}^{n} c_i \mathbf{p}_i \cosh(\omega_i t + \alpha_i). \tag{31}$$

This result shows that the general solution of (20) is an arbitrary linear combination of the n different forms of behavior

$$\mathbf{x}_i(t) = \mathbf{p}_i \cosh(\omega_i t + \alpha_i), \qquad i = 1, 2, \ldots, n, \tag{32}$$

in which each element $x_1, x_2, \ldots, x_n$ of $\mathbf{x}$ has the same time dependent behavior; namely $\cosh(\omega_i t + \alpha_i)$. Result (32) is thus the form of the special solution vector $\mathbf{x}$ sought in (21) when $\lambda_i > 0$, for $i = 1, 2, \ldots, n$.

If the nonhomogeneous equation (14) is involved, its solution will be of the form

$$\mathbf{x} = \mathbf{x}_c + \tilde{\mathbf{x}}_p. \tag{33}$$

The complementary function $\mathbf{x}_c$ in (33) was given in (31), while (23) and (24) show that $\tilde{\mathbf{x}}_p$ is a particular integral of the form

$$\tilde{\mathbf{x}}_p(t) = \sum_{i=1}^{n} \mathbf{p}_i k_i(t), \tag{34}$$

where $k_i(t)$ are particular integrals of (14), for $i = 1, 2, \ldots, n$.

Example 1 Positive eigenvalues

Find the general solution of

$$\ddot{\mathbf{x}} = \mathbf{A}\mathbf{x} + \mathbf{f},$$

given

$$\mathbf{A}=\begin{bmatrix} -2 & 3 \\ -6 & 7 \end{bmatrix}, \quad \mathbf{x}=\begin{bmatrix} x_1 \\ x_2 \end{bmatrix} \quad \text{and} \quad \mathbf{f}=\begin{bmatrix} 1 \\ t^2 \end{bmatrix}.$$

Solution

The eigenvalues of $\mathbf{A}$ are $\lambda_1=1$, $\lambda_2=4$ and the corresponding eigenvectors are

$$\mathbf{p}_1=\begin{bmatrix} 1 \\ 1 \end{bmatrix} \quad \text{and} \quad \mathbf{p}_2=\begin{bmatrix} 1 \\ 2 \end{bmatrix}.$$

Thus the diagonalizing matrix $\mathbf{P}$ becomes

$$\mathbf{P}=\begin{bmatrix} 1 & 1 \\ 1 & 2 \end{bmatrix}, \quad \text{so} \quad \mathbf{P}^{-1}=\begin{bmatrix} 2 & -1 \\ -1 & 1 \end{bmatrix} \quad \text{and} \quad \mathbf{\Lambda}=\begin{bmatrix} 1 & 0 \\ 0 & 4 \end{bmatrix}.$$

Setting $\mathbf{x}=\mathbf{P}\mathbf{y}$, the original system reduces to

$$\begin{bmatrix} \ddot{y}_1 \\ \ddot{y}_2 \end{bmatrix}=\begin{bmatrix} 1 & 0 \\ 0 & 4 \end{bmatrix} \begin{bmatrix} y_1 \\ y_2 \end{bmatrix}+\begin{bmatrix} 2-t^2 \\ t^2-1 \end{bmatrix}.$$

Thus the system uncouples to give

$$\ddot{y}_1=y_1+2-t^2, \quad \ddot{y}_2=4y_2+t^2-1.$$

These equations have the solutions

$$y_1=c_1\cosh(t+\alpha_1)+t^2,$$
$$y_2=c_2\cosh(2t+\alpha_2)-\tfrac{1}{4}t^2+\tfrac{1}{8},$$

with c_1, c_2, α_1 and α_2 arbitrary constants.
The general solution of the original problem now follows by using the result $\mathbf{x}=\mathbf{P}\mathbf{y}$ to give

$$x_1=c_1\cosh(t+\alpha_1)+c_2\cosh(2t+\alpha_2)+\tfrac{3}{4}t^2+\tfrac{1}{8},$$
$$x_2=c_1\cosh(t+\alpha_1)+2c_2\cosh(2t+\alpha_2)+\tfrac{1}{2}t^2+\tfrac{1}{4}.$$

Had an initial value problem at $t=t_0$ been involved, the constants c_1, c_2, α_1 and α_2 would need to be determined by means of the initial conditions for $x(t_0)$ and $\dot{x}(t_0)$. ∎

(ii) Eigenvalues all negative – coupled oscillations

We now consider the main topic of this section, which is the solution of the non-homogeneous second order matrix differential equation

$$\ddot{\mathbf{x}}=\mathbf{A}\mathbf{x}+\mathbf{f}, \tag{35}$$

in the case when all the eigenvalues λ_i of the constant coefficient matrix $\mathbf{A}$ are negative. A matrix differential equation of the form (35) describes the forced oscillations of a coupled oscillatory system when the nonhomogeneous term $\mathbf{f}$ contains sinusoidal forcing terms. Related to (35) is the problem of determining the solution of the associated homogeneous matrix differential equation

$$\ddot{\mathbf{x}} = \mathbf{A}\mathbf{x}, \tag{36}$$

which describes the coupled natural oscillations associated with (35). These two problems will be considered separately.

(a) Homogeneous case – coupled natural oscillations

We will now derive the general solution of the homogeneous matrix differential equation in (36). As all the eigenvalues λ_i of $\mathbf{A}$ are negative we first set $\lambda_i = -\omega_i^2$, $i = 1, 2, \ldots, n$, where the ω_i are all real. Equation (27) then becomes $\ddot{y}_i = -\omega_i^2 y_i$, so its general solution is

$$y_i(t) = c_i \cos(\omega_i t + \beta_i), \tag{37}$$

where c_i and β_i are arbitrary constants.

If $\mathbf{p}_i$ is the ith column of the diagonalizing matrix $\mathbf{P}$ considered earlier, it follows at once from (23) and (37) that the general solution of (36), that is the complementary function $\mathbf{x}_c$ of (35), may be written as

$$\mathbf{x}(t) = \mathbf{x}_c(\mathbf{t}) = \sum_{i=1}^{n} c_i \mathbf{p}_i \cos(\omega_i t + \beta_i). \tag{38}$$

Thus the complementary function is representable as an arbitrary linear combination of the n different forms of oscillatory behavior

$$\mathbf{x}^{(i)}(t) = \mathbf{p}_i \cos(\omega_i t + \beta_i), \qquad i = 1, 2, \ldots, n. \tag{39}$$

It is for this reason that in this case the solution vector $\mathbf{x}_c(t)$ is said to describe the **coupled natural oscillations** governed by the system of differential equations represented by (36).

Each of the special oscillatory solutions in (39) is called a **normal mode** of (36), and ω_i is called the **normal mode angular frequency**, or sometimes the **natural angular frequency** of the mode. It is usual to arrange the normal mode angular frequencies in order of magnitude, and to number them so that

$$0 < \omega_1 < \omega_2 < \ldots < \omega_n.$$

When this is done, the normal mode corresponding to ω_r is called the rth **normal mode** of (36). The quantities $f_i = \omega_i/2\pi$ are called the **natural frequencies of oscillation**, or just the **natural frequencies** of the solution.

The general solution of (36) given in (38) represents the natural oscillations which can take place in the system, because as $\mathbf{f} \equiv \mathbf{0}$ there are no forcing terms present to induce any other form of behavior. The normal modes $\mathbf{x}^{(i)}(t)$, $i = 1, 2, \ldots, n$, are the special solutions originally sought in (21). Result (38) shows that every possible solution of (36) is representable as a suitable linear combination of the n normal modes of the system.

The structure of each special solution represented by a normal mode is particularly simple, because (39) shows that each element of the vector $\mathbf{x}^{(i)}$ oscillates with the *same* angular frequency ω_i. In terms of the mass-spring systems considered earlier, this means that in a normal mode each mass performs simple harmonic motion about its equilibrium position with the *same* angular frequency. The motion of any two masses in a particular mode will be said to be **in phase** when their displacements are in the same sense, and to be **180° out of phase**, or to be **phase-opposed**, when their displacements are opposite in sense. Thus if, for example, in a two mass system the vectors $\mathbf{p}_1$ and $\mathbf{p}_2$ are

$$\mathbf{p}_1 = \begin{bmatrix} 1 \\ 2 \end{bmatrix} \quad \text{and} \quad \mathbf{p}_2 = \begin{bmatrix} 2 \\ -3 \end{bmatrix},$$

the masses in the mode associated with $\mathbf{p}_1$ are in phase, while those in the mode associated with $\mathbf{p}_2$ are 180° out of phase.

The reason for these descriptions of the phase becomes apparent when the following argument is considered. Let $\chi_1(t)$ and $\chi_2(t)$ be the displacements of any two masses in a mass-spring system which is oscillating in a particular mode with angular frequency ω. Suppose the displacements are given by

$$\chi_1(t) = A^2 \cos(\omega t + \phi) \quad \text{and} \quad \chi_2(t) = B^2 \cos(\omega t + \phi),$$

where A and B are any two nonzero real numbers. Then both $\chi_1(t)$ and $\chi_2(t)$ are expressed in the standard form for simple harmonic motion in which the amplitude is positive, and as they both have the same phase angle ϕ, they oscillate in phase.

If, however, $\chi_1(t)$ and $\chi_2(t)$ are such that

$$\chi_1(t) = A^2 \cos(\omega t + \phi) \quad \text{and} \quad \chi_2(t) = -B^2 \cos(\omega t + \phi)$$

the displacements are opposite in sign. Rewriting $\chi_2(t)$ in the standard form for simple harmonic motion in which the amplitude is positive, we arrive at the equivalent expression

$$\chi_2(t) = B^2 \cos(\omega t + \phi + \pi).$$

The phase angle of $\chi_2(t)$ now differs from that of $\chi_1(t)$ by π radians, and so the oscillation represented by $\chi_2(t)$ is 180° out of phase relative to the one represented by $\chi_1(t)$. To make the phase angle unique, the convention is adopted that a phase angle ϕ is always chosen to lie in the interval

$$-\pi < \phi \le \pi.$$

The number of **degrees of freedom** of any physical system giving rise to coupled oscillations is the smallest number of variables (coordinates or charges etc.) necessary to specify the condition or state of the system at any instant of time. Thus the system in Fig. 6.1 has two degrees of freedom, while the one in Fig. 6.3 has three. The number of degrees of freedom is equal to the number n of equations represented by (36).

The matrix $\mathbf{P}$ introduced in (23) to diagonalize $\mathbf{A}$ is often called the **modal matrix** of the system represented by (36). This is because the ith column of $\mathbf{P}$ is the vector $\mathbf{p}_i$ in the ith normal mode given in (39).

Example 2. Coupled natural oscillations

Find the normal mode angular frequencies, the normal modes and the general solution of the system of equations

$$\ddot{x}_1 = -\tfrac{5}{3}p^2 x_1 + \tfrac{1}{3}p^2 x_2$$
$$\ddot{x}_2 = p^2 x_1 - p^2 x_2.$$

Solve and interpret the initial value problem for this system when $x_1(0) = 1$, $x_2(0) = 0$, $\dot{x}_1(0) = 0$ and $\dot{x}_2(0) = 0$.

Solution

This example describes the coupled oscillations of the mass-spring system illustrated in Fig. 6.1 when $m_1 = 3m_2$, $k_1 = 4k_2$ and $p^2 = k_2/m_2$. The coefficient matrix is

$$\mathbf{A} = \begin{bmatrix} -\tfrac{5}{3}p^2 & \tfrac{1}{3}p^2 \\ p^2 & -p^2 \end{bmatrix},$$

so the characteristic equation $\det(\mathbf{A} - \lambda\mathbf{I}) = 0$ becomes

$$3\lambda^2 + 8p^2\lambda + 4p^4 = 0,$$

with the roots $\lambda_1 = -\tfrac{2}{3}p^2$ and $\lambda_2 = -2p^2$. As $\lambda_1 = -\omega_1^2$ and $\lambda_2 = -\omega_2^2$, the normal mode angular frequencies are

$$\omega_1 = \sqrt{\tfrac{2}{3}}\,p \quad \text{and} \quad \omega_2 = \sqrt{2}\,p.$$

The eigenvectors $\mathbf{p}_1$ and $\mathbf{p}_2$ corresponding to λ_1 and λ_2 are

$$\mathbf{p}_1 = \begin{bmatrix} 1 \\ 3 \end{bmatrix} \quad \text{and} \quad \mathbf{p}_2 = \begin{bmatrix} 1 \\ -1 \end{bmatrix},$$

so the normal modes are

$$\mathbf{x}^{(1)}(t) = \begin{bmatrix} 1 \\ 3 \end{bmatrix} \cos(\sqrt{\tfrac{2}{3}}\,pt + \beta_1) \quad \text{and} \quad \mathbf{x}^{(2)}(t) = \begin{bmatrix} 1 \\ -1 \end{bmatrix} \cos(\sqrt{2}\,pt + \beta_2).$$

Notice that in the first normal mode the masses oscillate *in phase* with an amplitude ratio of $\tfrac{1}{3}$, whereas in the second mode the masses oscillate *180° out of phase* with an amplitude ratio of -1.

The general solution of the matrix differential equation is

$$\mathbf{x}(t) = c_1 \begin{bmatrix} 1 \\ 3 \end{bmatrix} \cos(\sqrt{\tfrac{2}{3}}\,pt + \alpha_1) + c_2 \begin{bmatrix} 1 \\ -1 \end{bmatrix} \cos(\sqrt{2}\,pt + \alpha_2).$$

This result shows that

$$x_1(t) = c_1 \cos(\sqrt{\tfrac{2}{3}}\,pt + \alpha_1) + c_2 \cos(\sqrt{2}\,pt + \alpha_2)$$
$$x_2(t) = 3c_1 \cos(\sqrt{\tfrac{2}{3}}\,pt + \alpha_1) - c_2 \cos(\sqrt{2}\,pt + \alpha_2).$$

To solve the initial value problem it is convenient to rewrite $x_1(t)$ and $x_2(t)$ as

$$x_1(t) = P\cos(\sqrt{\tfrac{2}{3}}\,pt) + Q\sin(\sqrt{\tfrac{2}{3}}\,pt) + R\cos(\sqrt{2}\,pt) + S\sin(\sqrt{2}\,pt)$$
$$x_2(t) = 3P\cos(\sqrt{\tfrac{2}{3}}\,pt) + 3Q\sin(\sqrt{\tfrac{2}{3}}\,pt) - R\cos(\sqrt{2}\,pt) - S\sin(\sqrt{2}\,pt),$$

where $P = c_1 \cos \alpha_1$, $Q = -c_1 \sin \alpha_1$, $R = c_2 \cos \alpha_2$, $S = -c_2 \sin \alpha_2$. The first two initial conditions $x_1(0) = 1$, $x_2(0) = 0$ then give

$$P + R = 1, \quad 3P - R = 0 \quad \text{so} \quad P = \tfrac{1}{4}, \quad R = \tfrac{3}{4},$$

while the remaining two conditions $\dot{x}_1(0) = \dot{x}_2(0) = 0$ give

$$\sqrt{\tfrac{2}{3}}Q + \sqrt{2}S = 0, \quad 3\sqrt{\tfrac{2}{3}}Q - \sqrt{2}S = 0 \quad \text{so} \quad Q = S = 0.$$

Thus the solution to the initial value problem is

$$x_1(t) = \tfrac{1}{4}\cos(\sqrt{\tfrac{2}{3}}pt) + \tfrac{3}{4}\cos(\sqrt{2}pt), \quad x_2(t) = \tfrac{3}{4}\cos(\sqrt{\tfrac{2}{3}}pt) - \tfrac{3}{4}\cos(\sqrt{2}pt).$$

In physical terms, these initial conditions correspond to starting the system from rest with mass m_1 displaced one unit in the positive direction from its equilibrium position, and with mass m_2 in its equilibrium position. This situation is illustrated in Fig. 6.4.

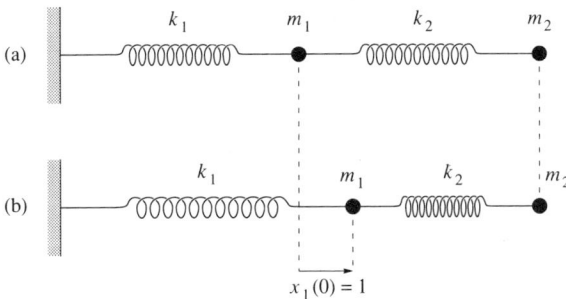

Fig. 6.4   Initial configuration for mass-spring system: (a) Equilibrium configuration, (b) Initial configuration starting from rest

Before proceeding to study the nonhomogeneous matrix differential equation (35), we must give brief consideration to a degeneracy which occurs in the homogeneous case when an eigenvalue of **A** is zero. The occurrence of a zero eigenvalue, that is a zero angular frequency for a normal mode, indicates a degenerate mode of behavior which is nonoscillatory. In such circumstances the system has one more degree of freedom than necessary in order to describe its purely oscillatory behavior. A re-examination of the formulation of the problem will enable the number of degrees of freedom, and hence the number of equations, to be reduced by one if only oscillatory phenomena are to be described. This situation is illustrated and interpreted by means of the following example.

Example 3. Degenerate case

Find the normal modes and general solution of the vibrating mass-spring system illustrated in Fig. 6.5, when the vibrations take place on a smooth table along the line of the springs, and the system is not fastened to the table in any way. The springs are of equal length l and stiffnesses k, and the particles have masses m, m and $2m$, as indicated.

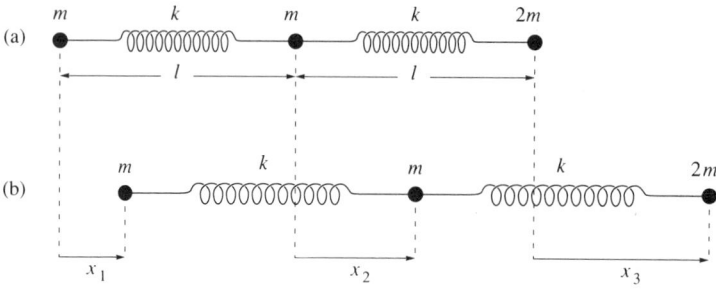

Fig. 6.5 Vibrating mass-spring system: (a) Equilibrium configuration, (b) Displaced positions when oscillating

Solution

The equations of motion are

$$m\ddot{x}_1 = k(x_2 - x_1)$$
$$m\ddot{x}_2 = -k(x_2 - x_1) + k(x_3 - x_2)$$
$$2m\ddot{x}_3 = -k(x_3 - x_2).$$

Thus the matrix differential equation to be solved is

$$\ddot{\mathbf{x}} = \mathbf{A}\mathbf{x},$$

with

$$\mathbf{x} = \begin{bmatrix} x_1 \\ x_2 \\ x_3 \end{bmatrix} \quad \text{and} \quad \mathbf{A} = \begin{bmatrix} -p^2 & p^2 & 0 \\ p^2 & -2p^2 & p^2 \\ 0 & \frac{1}{2}p^2 & -\frac{1}{2}p^2 \end{bmatrix},$$

where we have set $p^2 = k/m$. The characteristic equation $\det(\mathbf{A} - \lambda\mathbf{I}) = 0$ becomes

$$\lambda(\lambda^2 + \tfrac{7}{2}p^2\lambda + 2p^4) = 0,$$

showing that the eigenvalues are

$$\lambda_1 = 0, \quad \lambda_2 = -\tfrac{1}{4}(7 - \sqrt{17})p^2, \quad \lambda_3 = -\tfrac{1}{4}(7 + \sqrt{17})p^2.$$

Thus the angular frequencies of the normal modes are

$$\omega_1 = 0, \quad \omega_2 = \tfrac{1}{2}(7 - \sqrt{17})^{1/2}p \quad \text{and} \quad \omega_3 = \tfrac{1}{2}(7 + \sqrt{17})^{1/2}p,$$

and so the first mode is degenerate.

The eigenvectors $\mathbf{p}_1$, $\mathbf{p}_2$ and $\mathbf{p}_3$ corresponding to λ_1, λ_2 and λ_3 are

$$\mathbf{p}_1 = \begin{bmatrix} 1 \\ 1 \\ 1 \end{bmatrix}, \quad \mathbf{p}_2 = \begin{bmatrix} 1 \\ \left(\dfrac{\sqrt{17} - 3}{4}\right) \\ -\left(\dfrac{\sqrt{17} + 1}{8}\right) \end{bmatrix} \quad \text{and} \quad \mathbf{p}_3 = \begin{bmatrix} 1 \\ -\left(\dfrac{\sqrt{17} + 3}{4}\right) \\ \left(\dfrac{\sqrt{17} - 1}{8}\right) \end{bmatrix}.$$

When $\omega_1 = 0 (\lambda_1 = 0)$, (27) has the solution

$$y_1 = at + b,$$

where a and b are arbitrary real constants, and so the corresponding degenerate normal is

$$\mathbf{x}^{(1)}(t) = \mathbf{p}_1(at + b).$$

Thus the set of normal modes of the vibrating system is

$$\mathbf{x}^{(1)}(t) = \mathbf{p}_1(at + b), \quad \mathbf{x}^{(2)}(t) = \mathbf{p}_2 \cos(\omega_2 t + \beta_2) \quad \text{and} \quad \mathbf{x}^{(3)}(t) = \mathbf{p}_3 \cos(\omega_3 t + \beta_3),$$

with β_2 and β_3 arbitrary constants. The general solution of the matrix differential equation, which is an arbitrary linear combination of these normal modes, is thus

$$\mathbf{x}(t) = c_1 \mathbf{p}_1(at + b) + c_2 \mathbf{p}_2 \cos(\omega_2 t + \beta_2) + c_3 \mathbf{p}_3 \cos(\omega_3 t + \beta_3),$$

with c_1, c_2 and c_3 arbitrary constants.

The degenerate first mode, corresponding to the zero eigenvalue, represents a **rigid body translation** because in this mode, although moving, each mass is stationary relative to the other two. The presence or absence of this degenerate mode will not, of course, influence the purely oscillatory characteristics of the system.

The degeneracy can be removed by using the fact that, as no external forces act upon the system, its center of mass remains fixed. This condition introduces a linear constraint between x_1, x_2 and x_3, which enables the number of coordinates to be reduced from three to two. ∎

The last natural oscillation problem to be considered involves the free longitudinal oscillations of a system of N equal masses coupled by identical springs, with the system being fastened at the left-hand end. It is a somewhat harder problem than the previous ones we have considered, but it has been included because it is instructive mathematically and the solution is informative about properties of natural oscillations in general. It shows, for example, the interlacing of the nodes of different modes which is a property shared by all natural oscillation problems. In addition to this, by allowing the number N of coupled masses to become large, it shows how a discrete system may be used to approximate a continuous one. In this case the continuous system is an elastic rod oscillating longitudinally with its left-hand end fixed.

Example 4. Connected *N* mass-spring system

Find the natural frequencies and normal modes of the coupled mass-spring system illustrated in Fig. 6.6. The system comprises N equal masses m oscillating in line on a smooth horizontal table with a spring of stiffness k between each mass, the system being clamped at the left-hand end.

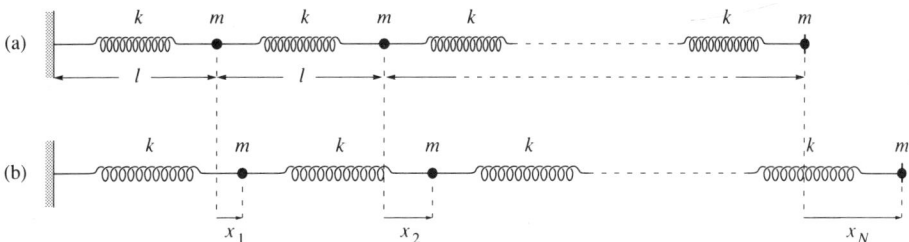

Fig. 6.6 Coupled mass-spring system: (a) Equilibrium configuration, (b) Displaced positions when in oscillation

Solution

As no external forces act on the masses, the equations of motion of the first N-1 masses take the form

$$m\ddot{x}_r = k(x_{r+1} - x_r) - k(x_r - x_{r-1}) \qquad \text{for } r = 1, 2, \ldots, N-1,$$

with the equation of motion of the Nth mass on the right being

$$m\ddot{x}_N = -k(x_N - x_{N-1}).$$

Let the displacement of the left-hand end of the system be denoted by x_0; then as the end is clamped the left-hand boundary condition is

$$x_0 = 0.$$

After division by k these equations of motion may be written in the matrix form

$$(m/k)\ddot{\mathbf{x}} + \mathbf{A}\mathbf{x} = \mathbf{0},$$

where

$$\mathbf{x} = \begin{bmatrix} x_1 \\ x_2 \\ \vdots \\ x_n \end{bmatrix} \quad \text{and} \quad \mathbf{A} = \begin{bmatrix} 2 & -1 & 0 & 0 & 0 & \cdots & 0 & 0 \\ -1 & 2 & -1 & 0 & 0 & \cdots & 0 & 0 \\ 0 & -1 & 2 & -1 & 0 & \cdots & 0 & 0 \\ & \cdot & \cdot & \cdot & \cdot & \cdots & & \cdot \\ 0 & 0 & 0 & 0 & 0 & \cdots & 2 & -1 \\ 0 & 0 & 0 & 0 & 0 & \cdots & -1 & 1 \end{bmatrix},$$

The matrix $\mathbf{A}$ is called a tri-diagonal matrix because all entries outside the leading diagonal and the diagonals immediately above and below it are zeros.

To find the natural frequencies and normal modes of this system we must seek solutions $\mathbf{x}(t)$ of the form

$$\mathbf{x}(t) = \mathbf{a}\sin(\omega t + \Phi),$$

where $\mathbf{a}$ is a constant column vector to be determined along with the frequency ω, and Φ is a constant phase angle. Substituting this expression into the matrix equation and setting $\lambda = m\omega^2/k$ we obtain

$$(\mathbf{A} - \lambda\mathbf{I})\mathbf{a} = \mathbf{0},$$

which is equivalent to

$$\begin{bmatrix} 2-\lambda & -1 & 0 & 0 & \cdots & 0 & 0 \\ -1 & 2-\lambda & -1 & 0 & \cdots & 0 & 0 \\ 0 & -1 & 2-\lambda & -1 & \cdots & 0 & 0 \\ & & & \cdots & & & \cdot \\ & \cdot & & \cdots & 2-\lambda & -1 \\ 0 & 0 & 0 & 0 & \cdots & -1 & 1-\lambda \end{bmatrix} \begin{bmatrix} a_1 \\ a_2 \\ a_3 \\ \vdots \\ \vdots \\ a_N \end{bmatrix} = \begin{bmatrix} 0 \\ 0 \\ 0 \\ \vdots \\ 0 \\ 0 \end{bmatrix}$$

The eigenvalues λ_s of this equation will determine the natural frequencies $\omega_s = (\lambda_s k/m)^{1/2}$ of the system, and the corresponding eigenvectors $\mathbf{a}^{(s)}$ the normal modes of the system. To determine ω_s and $\mathbf{a}^{(s)}$ we shall need to make use of a result obtained in Sec. 1.9.

Let us consider the homogeneous matrix system involving ω and $a_1, a_2, \ldots, a_N$ to be the difference equation

$$-a_{r-1} + (2-\lambda)a_r - a_{r+1} = 0 \qquad (r = 1, 2, \ldots, N),$$

which is possible if appropriate conditions are imposed at the ends of the system. The left-hand end is fixed with $x_0 = 0$, so $a_0 = 0$, and hence the difference equation holds when $r = 1$. The difference equation is also true when $r = N$ provided we introduce a fictitious displacement x_{N+1} such that $x_{N+1} = x_N$, which in turn implies the corresponding condition $a_{N+1} = a_N$. Thus the conditions to be satisfied at the ends of the system are

$$a_0 = 0 \qquad \text{and} \qquad a_{N+1} = a_N.$$

Now from Sec. 1.9 we know that the general solution of a difference equation involving an oscillatory solution has the form

$$a_r = A\cos r\theta + B\sin r\theta.$$

Substituting this into the difference equation, expanding the sines and cosines where necessary and grouping terms we obtain

$$[(2-\lambda)\cos r\theta - 2\cos r\theta\cos\theta]A + [(2-\lambda)\sin r\theta - 2\sin r\theta\cos\theta]B = 0.$$

This result must be true for each θ corresponding to a natural frequency, but this can only be true if the coefficients of A and B both vanish. It follows from this that

$$2 - \lambda = 2\cos\theta,$$

which is equivalent to

$$\lambda = 2\sin^2\frac{\theta}{2}.$$

The left-hand end condition $a_0 = 0$ will be satisfied by the expression for a_r if $A = 0$, so a_r must be of the form

$$a_r = B\sin r\theta.$$

If the right-hand end condition $a_N = a_{N+1}$ is to be satisfied it follows from this expression for a_r that

$$\sin(N+1)\theta = \sin N\theta.$$

Using the trigonometric identity $\sin(x+y) - \sin(x-y) = 2\cos x\sin y$, with $x+y = (N+1)\theta$ and $x-y = N\theta$, the above result simplifies to

$$\cos\frac{(2N+1)\theta}{2}\sin\frac{\theta}{2} = 0.$$

This is the condition determining the permissible values of θ, and hence of ω.

The zeros of $\sin\theta/2$ occur when $\theta = \theta_s = 2s\pi$, but we see from the form of a_r that these correspond to the **trivial solution** $\mathbf{a} = \mathbf{0}$, and so they must be disregarded. The zeros of $\cos\dfrac{(2N+1)\theta}{2}$

occur when

$$\theta = \theta_s = \frac{(2s-1)\pi}{2N+1}, \qquad \text{for } s = 1, 2, \ldots.$$

The λ_s corresponding to these θ_s are thus

$$\lambda_s = 2\sin^2\frac{\theta_s}{2} = 4\sin^2\left[\frac{(2s-1)\pi}{2(2N+1)}\right],$$

where we may now restrict s to $s = 1, 2, \ldots, N$ because of the periodicity of the sine function. The corresponding natural frequencies are

$$\omega_s = (\lambda_s k/m)^{1/2}, \qquad s = 1, 2, \ldots, N.$$

The rth element $a_r^{(s)}$ of the sth eigenvector $\mathbf{a}^{(s)}$ now follows from the expression for a_r as

$$a_r^{(s)} = \sin\left[\frac{(2s-1)r\pi}{(2N+1)}\right].$$

Here, for convenience, we have set $B = 1$ as it is an arbitrary scale factor for the eigenvector. Thus the sth normal mode is

$$\mathbf{x}^{(s)}(t) = \begin{bmatrix} \sin\left[\dfrac{(2s-1)\pi}{(2N+1)}\right] \\ \sin\left[\dfrac{(2s-1)2\pi}{(2N+1)}\right] \\ \vdots \\ \sin\left[\dfrac{(2s-1)N\pi}{(2N+1)}\right] \end{bmatrix} \sin(\omega_s t + \Phi),$$

for $s = 1, 2, \ldots, N$. If required, any general motion of the system can, of course, be represented as the linear combination of normal modes

$$\mathbf{x}(t) = c_1\mathbf{x}^{(1)}(t) + c_2\mathbf{x}^{(2)}(t) + \ldots + c_N\mathbf{x}^{(N)}(t),$$

with $c_1, c_2, \ldots, c_N$ suitably chosen constants.

The possible modes of longitudinal oscillation are illustrated in Fig. 6.7 for the case $N = 6$. In this figure the displacements of the successive masses are shown vertically with the number of the displaced mass being plotted horizontally.

The diagram illustrates some fundamental properties of all oscillatory systems possessing a tridiagonal matrix $\mathbf{A}$. These are, that the jth mode crosses the axis $(j-1)$ times between the ends of the system, while the **nodes** (positions of zero displacement) of the jth mode interlace those of the adjacent modes, that is the $(j-1)$th and the $(j+1)$th modes. The lines joining the positions of the masses in any given mode may be interpreted as the spring displacements between the masses, so the **nodes** of a mode occur where these lines cross the horizontal axis.

This oscillatory system provides a discrete approximation to a uniform continuous rod which is clamped at the left and then allowed to oscillate longitudinally. The representation of an intrinsically continuous situation by a discrete approximation is frequently used in the analysis of

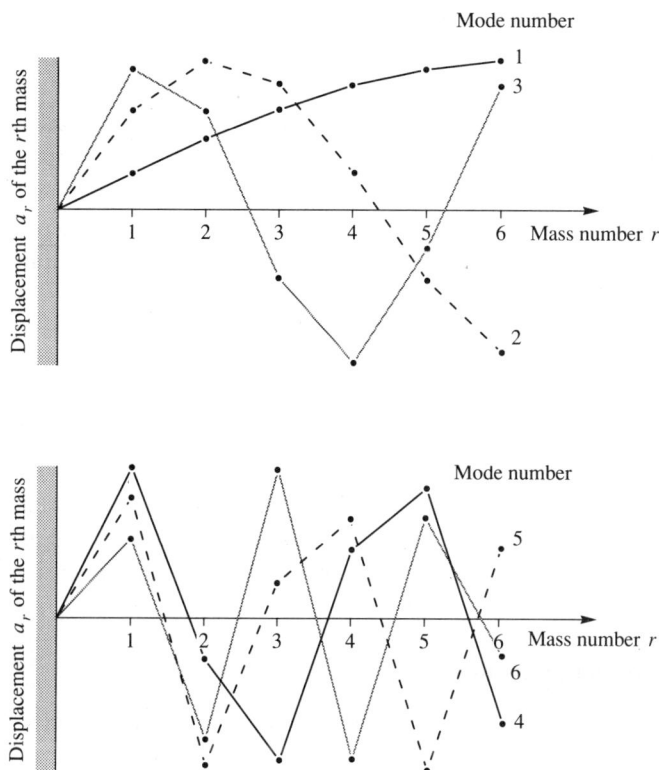

Fig. 6.7 Oscillatory modes for the mass-spring system when $N = 6$

engineering problems, though unlike here the resulting difference equations must usually be solved numerically. ∎

(b) Nonhomogeneous case – forced oscillations

It is now time to consider the nonhomogeneous matrix differential equation

$$\ddot{\mathbf{x}} = \mathbf{A}\mathbf{x} + \mathbf{f}, \tag{40}$$

in the case when all the eigenvalues λ of the constant coefficient matrix $\mathbf{A}$ are negative.

The argument which led to (28) in this section is still valid, but now $\lambda_i = -\omega_i^2$, $i = 1, 2, \ldots, n$, so the uncoupled equations corresponding to (29) become

$$\ddot{y}_i + \omega_i^2 y_i = g_i, \qquad i = 1, 2, \ldots, n, \tag{41}$$

with $g_i(t)$ the ith element of $\mathbf{g} = \mathbf{P}^{-1}\mathbf{f}$.

Once the equations in (41) have been solved for y_i, the vector $\mathbf{y}$ may be constructed, and the solution vector $\mathbf{x}$ then follows by means of the result (see (23))

$$\mathbf{x} = \mathbf{P}\mathbf{y}. \tag{42}$$

Equations (41) have solutions of the form

$$y_i(t) = y_{ci}(t) + y_{pi}(t), \tag{43}$$

where $y_{ci}(t)$ is the complimentary function associated with (41), and $y_{pi}(t)$ is the particular integral. Unlike the equations considered in Sec. 5.7, equations (41) are now undamped, so in the solutions (43) to (41) the complementary function $y_c(t)$ will not decay with time to leave only the particular integral $y_{pi}(t)$. However, for ease of reference, the terminology introduced in Sec. 5.7 will still be used, and $y_{pi}(t)$ will again be called the **steady-state solution** of (40).

The only two cases of nonhomogeneous equations of interest to us here will be when $\mathbf{f}$ is a constant vector, and when its elements are sinusoidal functions with given frequencies. In both cases the general solution follows straightforwardly by the indicated matrix method once the eigenvalues of $\mathbf{A}$ and the diagonalizing matrix $\mathbf{P}$ have been found. If, however, only the steady-state solution is required, this may be obtained more simply by applying the method of undetermined coefficients to the original system of equations, without finding either the eigenvalues of $\mathbf{A}$ or the matrix $\mathbf{P}$. This approach will be illustrated in the examples which follow, each of which also exhibits important features of forced coupled oscillations.

Example 5. Constant nonhomogeneous term

Reformulate the problem of Ex. 3 as a problem without rigid body translation so that it involves only two degrees of freedom, and hence find its general solution.

Solution

In the absence of a rigid body translation, the center of mass of the system in Ex. 3 must remain fixed. In what follows we shall use its location as a reference point from which to measure the displacements of the three masses. If the center of mass of the equilibrium configuration in Fig. 6.5(a) is located at a distance L to the right of the left-hand mass m, then taking moments about this mass gives

$$ml + 2m.2l = 4mL, \quad \text{so that } L = 5l/4.$$

Taking moments about this (fixed) center of mass in the vibratory configuration illustrated in Fig. 6.5(b) gives

$$(\tfrac{5}{4}l - x_1)m + (\tfrac{1}{4}l - x_2)m = (\tfrac{1}{4}l + x_3)2m,$$

which shows the constraint between x_1, x_2 and x_3 to be

$$x_1 + x_2 + 2x_3 = l.$$

Using this result to express x_3 in terms of x_1 and x_2 converts the first two homogeneous equations of motion in Ex. 3 into the pair of nonhomogeneous equations

$$\ddot{x}_1 = -p^2 x_1 + p^2 x_2,$$
$$\ddot{x}_2 = \tfrac{1}{2}p^2 x_1 - \tfrac{5}{2}p^2 x_2 + \tfrac{1}{2}p^2 l.$$

The third equation of motion is seen to be a linear combination of the first two, and so is redundant.

These two equations are equivalent to the matrix differential equation with a constant nonhomogeneous term

$$\ddot{\mathbf{x}} = \mathbf{A}\mathbf{x} + \mathbf{f},$$

where

$$\mathbf{x} = \begin{bmatrix} x_1 \\ x_2 \end{bmatrix}, \quad \mathbf{A} = \begin{bmatrix} -p^2 & p^2 \\ \frac{1}{2}p^2 & -\frac{5}{2}p^2 \end{bmatrix} \quad \text{and} \quad \mathbf{f} = \begin{bmatrix} 0 \\ \frac{1}{2}p^2 l \end{bmatrix}.$$

The eigenvalues of $\mathbf{A}$ follow from $\det(\mathbf{A} - \lambda\mathbf{I}) = 0$ as

$$\lambda_1 = -\tfrac{1}{4}(7 - \sqrt{17})p^2, \quad \lambda_2 = -\tfrac{1}{4}(7 + \sqrt{17})p^2.$$

These are seen to be the same as the two eigenvalues associated with vibrational modes in Ex. 3, though with different numbering. The angular frequencies of the two vibrational normal modes in this system are thus the same as before. The eigenvectors $\mathbf{p}_1$ and $\mathbf{p}_2$ corresponding to λ_1 and λ_2 are

$$\mathbf{p}_1 = \begin{bmatrix} 1 \\ \left(\dfrac{\sqrt{17} - 3}{4}\right) \end{bmatrix}, \quad \mathbf{p}_2 = \begin{bmatrix} 1 \\ -\left(\dfrac{\sqrt{17} + 3}{4}\right) \end{bmatrix},$$

and so the vibrational normal modes of the system are

$$\mathbf{x}^{(1)}(t) = \mathbf{p}_1 \cos(\omega_1 t + \beta_1), \quad \mathbf{x}^{(2)}(t) = \mathbf{p}_2 \cos(\omega_2 t + \beta_2),$$

where $\omega_1^2 = -\lambda_1$, $\omega_2^2 = -\lambda_2$. These normal modes are also the same as the vibrational modes found in Ex. 3.

The complementary function is thus

$$\mathbf{x}_c(t) = c_1 \mathbf{x}^{(1)}(t) + c_2 \mathbf{x}^{(2)}(t),$$

with c_1 and c_2 arbitrary constants. Although the general solution now follows straightforwardly by matrix methods by means of the matrix

$$\mathbf{P} = \begin{bmatrix} 1 & 1 \\ \left(\dfrac{\sqrt{17} - 3}{4}\right) & -\left(\dfrac{\sqrt{17} + 3}{4}\right) \end{bmatrix},$$

the calculation involved is tedious. It is simpler to find the particular integral $\mathbf{x}_p(t)$ as follows.

Inspection of the original system of differential equations shows suitable forms for the particular integrals to be

$$x_{p1}(t) = K_1 \quad \text{and} \quad x_{p2}(t) = K_2,$$

with K_1 and K_2 constants. Substitution into the differential equations then gives

$$-p^2 K_1 + p^2 K_2 = 0 \quad \text{and} \quad \tfrac{1}{2}p^2 K_1 - \tfrac{5}{2}p^2 K_2 + \tfrac{1}{2}p^2 l = 0.$$

Thus $K_1 = K_2 = l/4$, and so it follows that

$$\mathbf{x}_p(t) = \begin{bmatrix} l/4 \\ l/4 \end{bmatrix}.$$

The elements of the general solution $\mathbf{x}(t)=\mathbf{x}_c(t)+\mathbf{x}_p(t)$ are seen to be

$$x_1(t)=c_1\cos(\omega_1 t+\beta_1)+c_2\cos(\omega_2 t+\beta_2)+l/4,$$

$$x_2(t)=c_1\left(\frac{\sqrt{17}-3}{4}\right)\cos(\omega_1 t+\beta_1)-c_2\left(\frac{\sqrt{17}+3}{4}\right)\cos(\omega_2 t+\beta_2)+l/4,$$

$$x_3(t)=\tfrac{1}{2}(l-x_1-x_2).$$

This example illustrates the fact that when the nonhomogeneous term $\mathbf{f}$ in (40) is a constant vector, the general solution is obtained by the addition of a suitable constant vector to the complementary function. ∎

Example 6. Forced oscillations and the tuning of a vibration absorber

The equilibrium configuration of a coupled mass-spring system resting on a smooth horizontal table is shown in Fig. 6.8(a). Forced vibrations of the system are caused by driving the left-hand end of the left spring in such a way that its displacement along the line of the springs at time t is $X(t)=A\sin\Omega t$, with A and Ω positive constants. The situation is illustrated in Fig. 6.8(b). Establish the equations of motion of the system, determine their general solution when $M=3m$, $k_M=4k_m$ and $p^2=k_m/m$, and examine its general properties in terms of Ω and p.

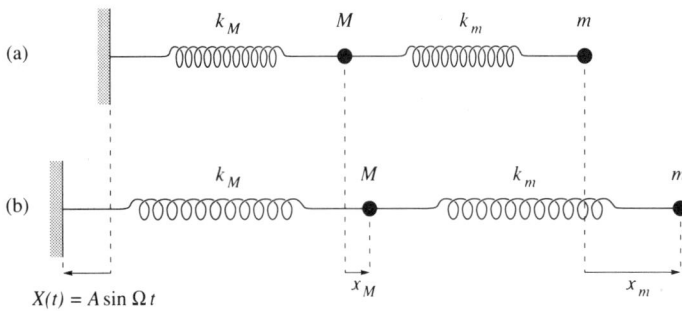

Fig. 6.8 Forced oscillations: (a) Equilibrium configuration, (b) Vibrating configuration

Solution

The equations of motion are seen to be

$$M\ddot{x}_M=-k_M(x_M-X)+k_m(x_m-x_M)$$

$$m\ddot{x}_m=-k_m(x_m-x_M).$$

In matrix form these become

$$\ddot{\mathbf{x}}=\mathbf{A}\mathbf{x}+\mathbf{f},$$

with

$$
\mathbf{x} = \begin{bmatrix} x_M \\ x_m \end{bmatrix}, \quad
\mathbf{A} = \begin{bmatrix} -\left(\dfrac{k_m + k_M}{M}\right) & \dfrac{k_m}{M} \\[2ex] \dfrac{k_m}{m} & -\dfrac{k_m}{m} \end{bmatrix}
\quad \text{and} \quad
\mathbf{f} = \begin{bmatrix} \dfrac{k_M A}{M} \sin \Omega t \\[2ex] 0 \end{bmatrix}.
$$

With $M = 3m$, $k_M = 4k_m$ and $p^2 = k_m/m$, matrices $\mathbf{A}$ and $\mathbf{f}$ become

$$
\mathbf{A} = \begin{bmatrix} -\tfrac{5}{3}p^2 & \tfrac{1}{3}p^2 \\[1ex] p^2 & -p^2 \end{bmatrix}
\quad \text{and} \quad
\mathbf{f} = \begin{bmatrix} \tfrac{4}{3}p^2 A \sin \Omega t \\[1ex] 0 \end{bmatrix}.
$$

The normal mode angular frequencies and the normal modes themselves are thus the same as in Ex. 2. It then follows from the work of that example that the diagonalizing matrix is

$$
\mathbf{P} = \begin{bmatrix} 1 & 1 \\ 3 & -1 \end{bmatrix}, \quad \text{so} \quad
\mathbf{P}^{-1} = \begin{bmatrix} \tfrac{1}{4} & \tfrac{1}{4} \\[1ex] \tfrac{3}{4} & -\tfrac{1}{4} \end{bmatrix}, \quad
\mathbf{\Lambda} = \mathbf{P}^{-1}\mathbf{A}\mathbf{P} = \begin{bmatrix} -\tfrac{2}{3}p^2 & 0 \\[1ex] 0 & -2p^2 \end{bmatrix} \quad \text{and}
$$

$$
\mathbf{P}^{-1}\mathbf{f} = \begin{bmatrix} \tfrac{1}{12}p^2 A \sin \Omega t \\[1ex] \tfrac{1}{4}p^2 A \sin \Omega t \end{bmatrix}.
$$

The equations corresponding to (41) become

$$
\ddot{y}_1 + \tfrac{2}{3}p^2 y_1 = \tfrac{1}{12}p^2 A \sin \Omega t \quad \text{and} \quad \ddot{y}_2 + 2p^2 y_2 = \tfrac{1}{4}p^2 A \sin \Omega t.
$$

These have the general solutions

$$
y_1(t) = c_1 \cos\left(\sqrt{\tfrac{2}{3}}\, pt + \alpha_1\right) + \frac{p^2 A \sin \Omega t}{4(2p^2 - 3\Omega^2)},
$$

$$
y_2(t) = c_2 \cos\left(\sqrt{2}\, pt + \alpha_2\right) + \frac{p^2 A \sin \Omega t}{4(2p^2 - \Omega^2)}.
$$

Finally, the solution vector $\mathbf{x}$ follows from the vector $\mathbf{y}$ with elements $y_1(t)$ and $y_2(t)$ by means of the result

$$
\mathbf{x} = \mathbf{P}\mathbf{y}.
$$

A routine calculation shows

$$
x_M(t) = c_1 \cos\left(\sqrt{\tfrac{2}{3}}\, pt + \alpha_1\right) + \frac{p^2(p^2 - \Omega^2)}{(2p^2 - 3\Omega^2)(2p^2 - \Omega^2)} A \sin \Omega t
$$

$$
x_m(t) = c_2 \cos\left(\sqrt{2}\, pt + \alpha_2\right) + \frac{p^4}{(2p^2 - 3\Omega^2)(2p^2 - \Omega^2)} A \sin \Omega t.
$$

The terms involving the arbitrary constants represent the complementary functions, while the others are the particular integrals. Solutions $x_M(t)$ and $x_m(t)$ become unbounded as $\Omega \to \sqrt{\tfrac{2}{3}}\,p$ and as $\Omega \to \sqrt{2}\,p$, corresponding to the existence of two **resonances** in the system. Had damping been present, then although the solutions would have become large in the neighborhood of these resonant angular frequencies, they would have remained finite.

It is interesting to notice that when $\Omega = p$, the particular integral (steady-state solution) in $x_M(t)$ vanishes. This means that when this condition is satisfied, the forcing function causes no vibration of the particle of mass $M = 3m$. This **tuning condition** for the system, involving the selection of m and k_m such that $\Omega = \sqrt{(k_m/m)}$, is used in the design of vibration absorbers whose function is to absorb unwanted vibrations. The mass M could, for example, be a piece of machinery with a periodic out of balance force in a horizontal direction, with angular frequency Ω. When tuned, the mass m attached to it by a spring would then vibrate horizontally, leaving the machine itself free from vibration.

The sensitivity of the tuning condition may be illustrated graphically. Let the steady-state solutions $x_{pM}(t)$ and $x_{pm}(t)$ be written in the form

$$x_{pM}(t) = F_M(\Omega)\sin\Omega t, \quad x_{pm}(t) = F_m(\Omega)\sin\Omega t.$$

Then Fig. 6.9(a) shows the variation of $|F_M(\Omega)|$ with Ω, and Fig. 6.9(b) the variation of $|F_m(\Omega)|$ with Ω. When expressed in this form, $|F_M(\Omega)|$ and $|F_m(\Omega)|$ represent the amplitudes of the oscillations performed by x_M and x_m, respectively. The two resonant angular frequencies occurring at the forcing frequencies $\Omega_1 = \sqrt{\frac{2}{3}}p$ and $\Omega_2 = \sqrt{2}p$ are shown as asymptotes to the curves in each diagram.

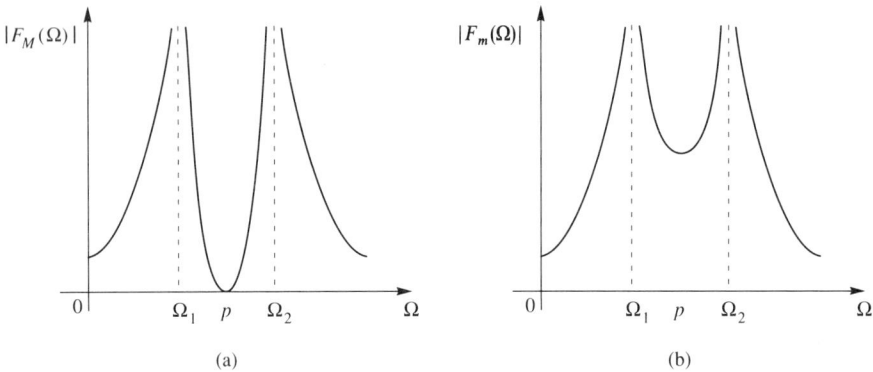

(a)   (b)

Fig. 6.9 Variation of amplitude of oscillations with Ω: (a) Amplitude of x_M, (b) Amplitude of x_m

If only the particular integrals occurring in $x_M(t)$ and $x_m(t)$ had been required they could have been obtained directly from the original equations of motion by seeking particular solutions of the form

$$x_{pM}(t) = P\sin\Omega t \quad \text{and} \quad x_{pm}(t) = Q\sin\Omega t.$$

Notice that as there is no damping in the system (no first derivative terms), terms depending on $\cos\Omega t$ may be omitted from $x_{pM}(t)$ and $x_{pm}(t)$. Substitution of these forms of solution into the original pair of differential equations, followed by the solution of the resulting simultaneous equations for P and Q, leads to the particular integrals already found and used to construct Figs 6.9. ∎

Example 7. Two forcing functions

Establish the equations of motion for the forced mass-spring system illustrated in Fig. 6.10, in which the left-hand end of the system is driven in such a way that its displacement along the line of the

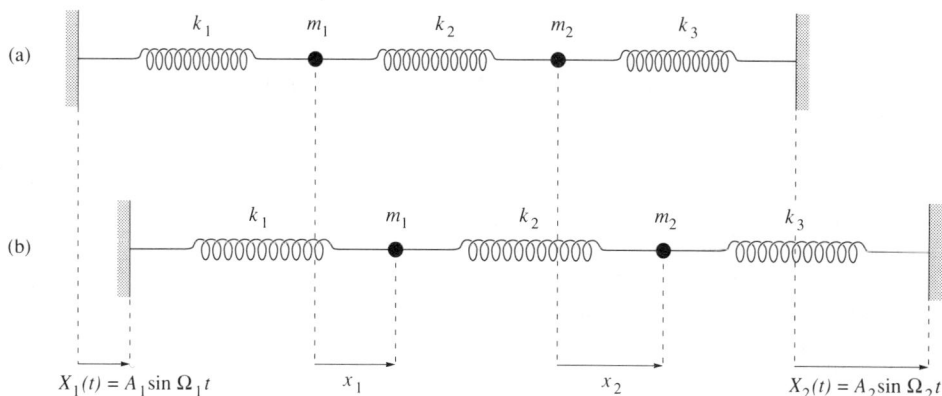

Fig. 6.10 Mass-spring system driven at each end: (a) Equilibrium configuration, (b) Vibrating configuration

springs at time t is $X_1(t) = A_1 \sin \Omega_1 t$, while the right-hand end is driven so that its corresponding displacement at time t is $X_2(t) = A_2 \sin \Omega_2 t$, where A_1, A_2 and $\Omega_1 \neq \Omega_2$ are positive constants. Find the general solution in the case that $k_3 = k_1$, $k_2 = 4k_1$, $m_2 = m_1$ and $p^2 = k_1/m_1$.

Solution

The equations of motion are seen to be

$$m_1 \ddot{x}_1 = -k_1(x_1 - X_1) + k_2(x_2 - x_1)$$
$$m_2 \ddot{x}_2 = -k_2(x_2 - x_1) + k_3(X_2 - x_2).$$

In matrix form these become

$$\ddot{\mathbf{x}} = \mathbf{A}\mathbf{x} + \mathbf{f},$$

with

$$\mathbf{x} = \begin{bmatrix} x_1 \\ x_2 \end{bmatrix}, \quad \mathbf{A} = \begin{bmatrix} -\dfrac{(k_1+k_2)}{m_1} & \dfrac{k_2}{m_1} \\[2ex] \dfrac{k_2}{m_2} & -\dfrac{(k_2+k_3)}{m_2} \end{bmatrix} \quad \text{and} \quad \mathbf{f} = \begin{bmatrix} \dfrac{k_1 A_1}{m_1} \sin \Omega_1 t \\[2ex] \dfrac{k_3 A_2}{m_2} \sin \Omega_2 t \end{bmatrix}.$$

With $k_3 = k_1$, $k_2 = 4k_1$, $m_2 = m_1$ and $p^2 = k_1/m_1$, matrices $\mathbf{A}$ and $\mathbf{f}$ become

$$\mathbf{A} = \begin{bmatrix} -5p^2 & 4p^2 \\ 4p^2 & -5p^2 \end{bmatrix} \quad \text{and} \quad \mathbf{f} = \begin{bmatrix} p^2 A_1 \sin \Omega_1 t \\ p^2 A_2 \sin \Omega_2 t \end{bmatrix}.$$

The eigenvalues of $\mathbf{A}$ given by $\det(\mathbf{A} - \lambda\mathbf{I}) = 0$ are $\lambda_1 = -p^2$ and $\lambda_2 = -9p^2$, so the angular frequencies of the normal modes are $\omega_1 = p$ and $\omega_2 = 3p$. The eigenvectors of $\mathbf{A}$ corresponding to λ_1 and λ_2 are, respectively,

$$\mathbf{p}_1 = \begin{bmatrix} 1 \\ 1 \end{bmatrix} \quad \text{and} \quad \mathbf{p}_2 = \begin{bmatrix} 1 \\ -1 \end{bmatrix}.$$

Thus the diagonalizing matrix $\mathbf{P}$ is

$$\mathbf{P}=\begin{bmatrix} 1 & 1 \\ 1 & -1 \end{bmatrix}, \quad \text{so} \quad \mathbf{P}^{-1}=\begin{bmatrix} \tfrac{1}{2} & \tfrac{1}{2} \\ \tfrac{1}{2} & -\tfrac{1}{2} \end{bmatrix}, \quad \mathbf{\Lambda}=\begin{bmatrix} -p^2 & 0 \\ 0 & -9p^2 \end{bmatrix} \quad \text{and}$$

$$\mathbf{P}^{-1}\mathbf{f}=\begin{bmatrix} \tfrac{1}{2}p^2(A_1 \sin \Omega_1 t + A_2 \sin \Omega_2 t) \\ \tfrac{1}{2}p^2(A_1 \sin \Omega_1 t - A_2 \sin \Omega_2 t) \end{bmatrix}.$$

The equations corresponding to (41) become

$$\ddot{y}_1 + p^2 y_1 = \tfrac{1}{2}p^2(A_1 \sin \Omega_1 t + A_2 \sin \Omega_2 t)$$
$$\ddot{y}_2 + 9p^2 y_2 = \tfrac{1}{2}p^2(A_1 \sin \Omega_1 t - A_2 \sin \Omega_2 t).$$

These have the general solutions

$$y_1(t)=c_1 \cos(pt+\alpha_1)+\frac{p^2}{2}\left(\frac{A_1 \sin \Omega_1 t}{(p^2-\Omega_1^2)}+\frac{A_2 \sin \Omega_2 t}{(p^2-\Omega_2^2)}\right)$$

$$y_2(t)=c_2 \cos(pt+\alpha_2)+\frac{p^2}{2}\left(\frac{A_1 \sin \Omega_1 t}{(9p^2-\Omega_1^2)}-\frac{A_2 \sin \Omega_2 t}{(9p^2-\Omega_2^2)}\right),$$

where the terms involving the arbitrary constants represent the complementary functions while the others are the particular integrals (steady-state solutions) and $\Omega_1 \neq \Omega_2$.

The required general solution follows from the fact that

$$\mathbf{x}=\mathbf{P}\mathbf{y},$$

so that

$$x_1(t)=y_1(t)+y_2(t) \quad \text{and} \quad x_2(t)=y_1(t)-y_2(t).$$

Inspection of the forms of $x_1(t)$ and $x_2(t)$ shows the presence of resonances when either of the driving frequencies equals the normal mode angular frequencies p or $3p$.   ∎

(iii) Second order matrix differential equations with first derivative term

Coupled oscillatory systems with damping, and more general systems which do not necessarily describe such phenomena, are characterized by matrix differential equations of the form

$$\mathbf{K}\ddot{\mathbf{x}}=\mathbf{L}\dot{\mathbf{x}}+\mathbf{M}\mathbf{x}+\mathbf{f}. \tag{44}$$

As with all linear systems, the solution of (44) may always be written as

$$\mathbf{x}(t)=\mathbf{x}_c(t)+\mathbf{x}_p(t), \tag{45}$$

where $\mathbf{x}_p(t)$ is the particular integral and $\mathbf{x}_c(t)$, the complementary function, is the solution of the associated homogeneous equation

$$\mathbf{K}\ddot{\mathbf{x}}=\mathbf{L}\dot{\mathbf{x}}+\mathbf{M}\mathbf{x}. \tag{46}$$

An attempt to follow the previous approach by seeking solutions of (46) of the form

$$\mathbf{x} = \mathbf{a}e^{\lambda t} \tag{47}$$

shows, after substitution into (46), that λ and $\mathbf{a}$ are related by the result

$$(\lambda^2 \mathbf{K} - \lambda \mathbf{L} - \mathbf{M})\mathbf{a} = \mathbf{0}. \tag{48}$$

Thus for nontrivial solutions to exist, λ must be a solution of

$$\det(\lambda^2 \mathbf{K} - \lambda \mathbf{L} - \mathbf{M}) = 0. \tag{49}$$

Results (48) and (49) are a generalization of the eigenvalue problem already examined. It is clear from (49) that if $\mathbf{x}$ is an n element vector, the polynomial equation represented by (49) will be of degree $2n$, so in general its roots must be found by numerical methods. There will be a vector $\mathbf{a}$ corresponding to each root of (49), and these vectors $\mathbf{a}$ must be obtained from (48) once the roots have been found. We shall not pursue this approach further.

It will suffice to say that an alternative method which may always be used involves reducing (44) to an equivalent first order system by introducing the first order derivatives of the elements of $\mathbf{x}$ as new variables. The methods of Sec. 6.2 may then be used to determine the solution of the first order system, and hence to obtain the solution vector $\mathbf{x}(t)$ of (44). When the solution vector $\mathbf{x}$ in (44) has n elements this method also involves finding the roots of a polynomial equation of degree $2n$, but once this has been accomplished the task of determining the normal modes is straightforward.

To illustrate this method we again consider the mass-spring system of Ex. 5 which is illustrated in Fig. 6.7. Let us suppose that now the mass m is subjected to viscous damping proportional to velocity. Then the governing equations of motion become

$$M\ddot{x}_M = -k_M(x_M - X) + k_m(x_m - x_M)$$
$$m\ddot{x}_m = -k_m(x_m - x_M) - \mu\dot{x}_m,$$

where μ is the viscous damping constant.

Setting $x_1 = x_M$, $x_2 = x_m$, $x_3 = \dot{x}_1$, and $x_4 = \dot{x}_2$ reduces this pair of coupled second order equations to a first order system which may be written as the matrix differential equation

$$\dot{\mathbf{x}} = \mathbf{A}\mathbf{x} + \mathbf{f},$$

where

$$\mathbf{x} = \begin{bmatrix} x_1 \\ x_2 \\ x_3 \\ x_4 \end{bmatrix}, \quad \mathbf{A} = \begin{bmatrix} 0 & 0 & 1 & 0 \\ 0 & 0 & 0 & 1 \\ -\left(\dfrac{k_m + k_M}{M}\right) & \dfrac{k_m}{M} & 0 & 0 \\ \dfrac{k_m}{m} & -\dfrac{k_m}{m} & 0 & -\dfrac{\mu}{m} \end{bmatrix} \quad \text{and} \quad \mathbf{f} = \begin{bmatrix} 0 \\ 0 \\ \dfrac{k_M}{M}X \\ 0 \end{bmatrix}.$$

A routine calculation shows that the characteristic equation $\det(\mathbf{A}-\lambda\mathbf{I})=0$ for this first order system is the quartic $\lambda^4+\dfrac{\mu}{m}\lambda^3+\left(\dfrac{k_m}{m}+\dfrac{k_m+k_M}{M}\right)\lambda^2+\left(\dfrac{k_m+k_M}{M}\right)\left(\dfrac{\mu}{m}\right)\lambda+\dfrac{k_m k_M}{mM}=0.$

In general the roots of this equation can only be found by numerical methods.

It should be emphasized that the method of reduction of a coupled second order system of equations to a first order one is *not* unique. For example, in the situation just discussed, setting $x_1=x_M$, $x_2=x_m$, $x_3=\dot{x}_1+\dot{x}_2$ and $x_4=\dot{x}_1-\dot{x}_2$ would have led to a different first order system, but to the same solutions for $x_M(t)$ and $x_m(t)$.

Problems for Section 6.3

1 Find the general solution of

$$\ddot{\mathbf{x}}=\mathbf{A}\mathbf{x},$$

when

$$\mathbf{A}=\begin{bmatrix} 2 & 1 & 1 \\ 2 & 3 & 2 \\ 1 & 1 & 2 \end{bmatrix}.$$

2 Find the general solution of

$$\ddot{\mathbf{x}}=\mathbf{A}\mathbf{x},$$

when

$$\mathbf{A}=\begin{bmatrix} 6 & -2 & 2 \\ -2 & 5 & 0 \\ 2 & 0 & 7 \end{bmatrix}.$$

3 Find the natural angular frequencies and normal modes of

$$\ddot{x}_1+2n^2x_1-n^2x_2=0$$
$$\ddot{x}_2-n^2x_1+2n^2x_2=0.$$

Solve the initial value problem for this system when

$$x_1(0)=x_2(0)=0, \quad \dot{x}_1(0)=X, \quad \dot{x}_2(0)=2X.$$

4 Find the natural angular frequencies and normal modes of

$$3\ddot{x}_1=-20x_1+4x_2$$
$$\ddot{x}_2=4x_1-4x_2.$$

Solve the initial value problem for this system when

$$x_1(0)=x_2(0)=1 \quad \text{and} \quad \dot{x}_1(0)=\dot{x}_2(0)=0.$$

5 Find the natural angular frequencies, normal modes and general solution of

$$2\ddot{x}_1 = -3x_1 + x_2 + x_3$$
$$\ddot{x}_2 = -3x_1 - 7x_2 - 3x_3$$
$$2\ddot{x}_3 = 7x_1 + 11x_2 + 3x_3.$$

6 When two LC circuits are in sufficiently close proximity, the magnetic field in one inductance influences the current in the other, and conversely. This magnetic coupling of circuits is called **mutual inductance**, and the voltage drop caused by a changing charge q is $M(dq/dt)$. The circuits illustrated in Fig. 6.11 are coupled in this manner, and the equations governing the charges q_1 and q_2 which cause the currents i_1 and i_2 to flow as indicated are

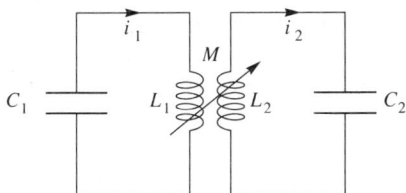

$$L_1\ddot{q}_1 + M\ddot{q}_2 + \frac{q_1}{C_1} = 0$$

$$M\ddot{q}_1 + L_2\ddot{q}_2 + \frac{q_2}{C_2} = 0.$$

Fig. 6.11 LC circuits with mutual inductance

Find both the natural angular frequencies ω and the natural frequencies of oscillation f of the charges in the circuits associated with such oscillations when $M = L_1$, $L_2 = 2L_1$, $C_2 = 2C_1$ and $p^2 = 1/(L_1C_1)$.

7 A point P with Cartesian coordinates (x, y) moves in the (x, y)-plane in such a way that

$$\ddot{x} = -10p^2x - 6p^2y$$
$$\ddot{y} = -6p^2x - 10p^2y,$$

where p is real. By determining the normal modes find the lines in the (x, y)-plane along which P can execute simple harmonic motion, and the angular frequencies with which such motion takes place.

8 A point P with Cartesian coordinates (x, y) moves in the (x, y)-plane in such a way that

$$\ddot{x} = -ap^2x - 2bp^2y$$
$$\ddot{y} = -bp^2x - ap^2y$$

with p real and $a > \sqrt{2b} > 0$. By determining the normal modes, find the lines in the (x, y)-plane along which P can execute simple harmonic motion, and the angular frequencies with which such motion takes place.

9 Three equal masses m are fastened symmetrically along a light elastic string with tension T and length $4l$ as shown in Fig. 6.12. The ends A and B of the string are fixed. When vibrating, the masses all move in a plane containing the equilibrium position of the string, with the lateral displacements x_1, x_2 and x_3 of the masses all being small. It follows by applying Newton's second law to each mass in the direction of its displacement that the equations of motion of the system are

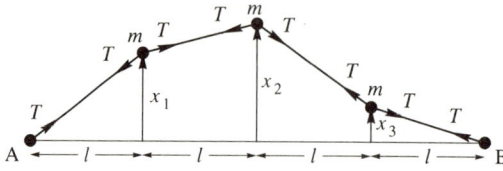

$$m\ddot{x}_1 = -\frac{2T}{l}x_1 + \frac{T}{l}x_2$$

$$m\ddot{x}_2 = \frac{T}{l}x_1 - \frac{2T}{l}x_2 + \frac{T}{l}x_3$$

$$m\ddot{x}_3 = \frac{T}{l}x_2 - \frac{2T}{l}x_3.$$

Fig. 6.12

Find the natural frequencies of vibration of the system and its corresponding three normal modes. Indicate by means of sketches the type of displacements experienced by the masses in each normal mode.

10 Two simple pendulums of length l whose bobs are each of mass m arc fastened to a horizontal beam a distance l apart. The bobs are connected by a spring with natural length l and elastic constant k. If the horizontal displacements x_1 and x_2 of the bobs from their equilibrium positions are small, then the equations of motion of the coupled pendulums shown in Fig. 6.13 are

$$m\ddot{x}_1 = -m\omega_0^2 x_1 + k(x_2 - x_1)$$
$$m\ddot{x}_2 = -m\omega_0^2 x_2 - k(x_2 - x_1),$$

where $\omega_0 = \sqrt{g/l}$ is the natural angular frequency of oscillation of the pendulums.

Find the natural angular frequencies of oscillation of the system, indicate by means of sketches the form of motion of the pendulums in each of the two normal modes, and obtain the solution of the initial value problem in which $x_1(0) = a$, $x_2(0) = 0$ and $\dot{x}_1(0) = \dot{x}_2(0) = 0$.

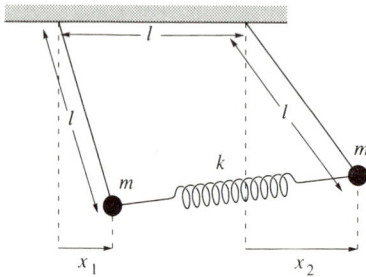

Fig. 6.13

11 Solve the initial value problem for the system

$$\ddot{x}_1 = -2\omega^2 x_1 + \omega^2 x_2 + K \sin \Omega t$$
$$\ddot{x}_2 = \omega^2 x_1 - 2\omega^2 x_2$$

when ω is real and $x_1(0) = x_2(0) = 0$ and $\dot{x}_1(0) = \dot{x}_2(0) = 0$.

12 The equations

$$3\ddot{q}_1 + 5p^2 q_1 - p^2 q_2 = 0$$
$$\ddot{q}_2 + p^2 q_2 - p^2 q_1 = E_0 \sin \Omega t$$

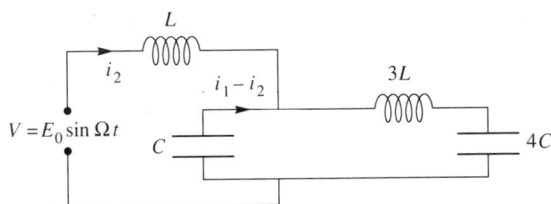

Fig. 6.14 Forced oscillations of an LC circuit

describe the charges q_1 and q_2 giving rise to the currents i_1 and i_2 flowing in the circuit shown in Fig. 6.14 when a voltage $V = E_0 \sin \Omega t$ is applied as shown, where $p^2 = 1/(LC)$. Solve the initial value problem for this system when $q_1(0) = q_2(0) = 0$ and $\dot{q}_1(0) = \dot{q}_2(0) = 0$.

13 Obtain the steady-state solution of Ex. 5 by the method of undetermined coefficients indicated at the end of the example.

14 Use the method of undetermined coefficients indicated at the end of Ex. 5 to show that the amplitudes a_M and a_m of masses M and m, respectively in the general problem formulated at the start of the example are

$$a_M = \frac{(k_m - m\Omega^2)k_M A}{(k_M + k_m - M\Omega^2)(k_m - m\Omega^2) - k_m^2}$$

$$a_m = \frac{k_m k_M A}{(k_M + k_m - m\Omega^2)(k_m - m\Omega^2) - k_m^2}.$$

Deduce that the **tuning condition** for the general system to operate as a vibration absorber is $\Omega = \sqrt{k_m/m}$.

15 By setting $x_3 = \dot{x}_1$ and $x_4 = \dot{x}_2$, reduce the following second order system containing both first and second order terms to a first order system, and hence find its general solution

$$\ddot{x}_1 = \dot{x}_2 + t$$
$$\ddot{x}_2 = \dot{x}_1 + 1.$$

Harder problems

The two questions which follow illustrate how the behavior of an oscillatory process whose natural oscillations are governed by the equation

$$\ddot{y} + 4\dot{y} + 5y = 0$$

is modified if a forcing function on $F(t) = -kx$ is introduced, where $\dot{x} = y$ and k is a constant. The process is then governed by the pair of coupled equations

$$\dot{x} = y$$
$$\ddot{y} + 4\dot{y} + 5y = -kx.$$

These equations may be reduced to a first order system by setting $x = x_1$, $y = x_2$ and $\dot{y} = x_3$ to obtain

$$\dot{x}_1 = x_2$$
$$\dot{x}_2 = x_3$$
$$\dot{x}_3 = -kx_1 - 5x_2 - 4x_3.$$

16 Solve the system above, and hence find $y(=x_2)$, when $k=2$. Comment on the nature of the solution.

17 Solve the system above, and hence find $y(=x_2)$, when $k=6$. Comment on the nature of the solution.

18 Show that if in Ex. 4, in addition to the left-hand end of the mass-spring system being clamped, the Nth mass at the right-hand end is also clamped, the eigenvalues become

$$\lambda_s = 4 \sin^2 \left(\frac{s\pi}{2N} \right) \qquad (s=1, 2, \ldots, N-1),$$

while the displacement $a_r^{(s)}$ of the rth mass in the sth mode is given by

$$a_r^{(s)} = \sin \left(\frac{rs\pi}{N} \right).$$

The next two questions develop further the use of the matrix exponential function introduced in the harder problems of Sections 6.1 and 6.2.

19 Show that the system of differential equations

$$\ddot{x}_1 - \ddot{x}_3 - 3x_1 + x_2 + 2x_3 = 0$$
$$\ddot{x}_1 + \ddot{x}_2 + \ddot{x}_3 - 2x_1 - 3x_2 - 6x_3 = 0$$
$$2\ddot{x}_1 - \ddot{x}_2 + \ddot{x}_3 + 3x_1 + x_2 + 2x_3 = 0$$

may be written in the form

$$\ddot{\mathbf{x}} = \mathbf{A}\mathbf{x},$$

where

$$\mathbf{x} = \begin{bmatrix} x_1 \\ x_2 \\ x_3 \end{bmatrix} \quad \text{and} \quad \mathbf{A} = \begin{bmatrix} 1 & 0 & 0 \\ 3 & 2 & 4 \\ -2 & 1 & 2 \end{bmatrix}.$$

Verify that the general solution is

$$\mathbf{x}(t) = [\exp(t\mathbf{B})]\mathbf{K} + [\exp(-t\mathbf{B})]\mathbf{L},$$

where $\mathbf{K}$ and $\mathbf{L}$ are arbitrary constant three element column vectors, and $\mathbf{B}$ is a 3×3 matrix such that $\mathbf{B}^2 = \mathbf{A}$.
Find $\mathbf{B}$ by diagonalizing $\mathbf{A}$.

20 If $\mathbf{P}$ is a square matrix, show that both $\mathbf{P}$ and $\mathbf{P}^2$ commute under multiplication with $\exp(-\frac{1}{2}t\mathbf{P})$ where t is a scalar.
Consider the matrix differential equation

$$\ddot{\mathbf{x}} + \mathbf{P}\dot{\mathbf{x}} + [\tfrac{1}{4}\mathbf{P}^2 + \mathbf{I}]\mathbf{x} = \mathbf{F}(t),$$

where $\mathbf{x} = [x_1, x_2, x_3]^T$, $\mathbf{P}$ is a 3×3 matrix with constant coefficients, $\mathbf{I}$ is the 3×3 unit matrix and $\mathbf{F}(t)$ is a three element column vector. Show that under the substitution

$$\mathbf{x}(t) = \exp(-\tfrac{1}{2}t\mathbf{P})\mathbf{y}(t),$$

the matrix differential equation simplifies to

$$\ddot{\mathbf{y}} + \mathbf{y} = \exp(\tfrac{1}{2}t\mathbf{P})\mathbf{F}(t).$$

Hence find the general solution of the system when

$$P = \begin{bmatrix} 0 & 1 & 1 \\ 0 & 0 & 1 \\ 0 & 0 & 0 \end{bmatrix} \quad \text{and} \quad F(t) = \begin{bmatrix} \frac{1}{2}t \\ 1 \\ 0 \end{bmatrix},$$

and thus the solution corresponding to the initial conditions

$$x(0) = \begin{bmatrix} 0 \\ 0 \\ 0 \end{bmatrix} \quad \text{and} \quad \dot{x}(0) = \begin{bmatrix} 1 \\ 0 \\ 0 \end{bmatrix}.$$

[*Hint*: Show that $P^n \equiv 0$ for integers $n > 3$, and hence calculate $\exp(\frac{1}{2}tP)$].

6.4 Qualitative theory: the phase plane and stability

Mathematical preliminaries, some physical background and the phase plane

Analytical solutions cannot usually be obtained to nonlinear differential equations, and even though such solutions may be constructed for linear differential equations with variable coefficients, their complexity often makes them difficult to interpret and use. Thus, in order to understand the general properties of solutions of both linear and nonlinear differential equations, it is helpful to find ways of obtaining general information about solutions directly from the differential equations themselves. Such information about solutions is called **qualitative** information, and it is often very useful when dealing with engineering problems. In general, attempts to obtain qualitative information solely by numerical methods are usually time-consuming; so other methods must be employed.

The types of differential equation to be considered here are second order equations of the form

$$\frac{d^2x}{dt^2} = f\left(x, \frac{dx}{dt}\right), \tag{1}$$

in which f is a function of x and dx/dt only, and first order systems of the form

$$\frac{dx}{dt} = X(x, y)$$

$$\frac{dy}{dt} = Y(x, y), \tag{2}$$

in which X and Y are functions of x and y only. In applications the independent variable t is usually the time. Notice that the form of the equations considered here is such that the time t only appears implicitly in their right-hand sides through x and dx/dt in (1), and x and y in (2). Differential equations of this type are called **autonomous** or **time-invariant** equations. For reasons which will become apparent later, system (2) is often called a **dynamical system**. Related differential equations in which t appears explicitly in the right-hand side are called **nonautonomous**, or **forced** equations.

The definitions of autonomous and nonautonomous may be formulated more precisely as follows. The second order differential equation

$$\frac{d^2x}{dt^2} = F\left(t, x, \frac{dx}{dt}\right) \tag{3}$$

is **autonomous** if

$$\frac{\partial F}{\partial t} \equiv 0,$$

and it is **nonautonomous** if

$$\frac{\partial F}{\partial t} \not\equiv 0.$$

Analogously, the system

$$\frac{dx}{dt} = R(t, x, y)$$

$$\frac{dy}{dt} = S(t, x, y), \tag{4}$$

will be **autonomous** if

$$\frac{\partial R}{\partial t} \equiv 0 \quad \text{and} \quad \frac{\partial S}{\partial t} \equiv 0,$$

and it will be **nonautonomous** if either (or both) of these conditions is violated.

The **van der Pol** equation

$$\frac{d^2x}{dt^2} + \varepsilon(x^2 - 1)\frac{dx}{dt} + x = 0, \qquad \varepsilon \geq 0 \tag{5}$$

is an example of an autonomous nonlinear second order differential equation which arose early in the study of nonlinear electrical oscillations. If $\varepsilon = 0$ (5) reduces to the equation of simple harmonic motion, but if $\varepsilon > 0$ the second term provides positive nonlinear damping (dissipation) when $x > 1$, and negative nonlinear damping (energy input) when $x < 1$.

The **Duffing** equation

$$\frac{d^2x}{dt^2} + a\frac{dx}{dt} + bx + cx^3 = P(t), \tag{6}$$

in which $a \geq 0$, b and c are constants and $P(t)$ is an arbitrary periodic function, provides an example of a nonautonomous nonlinear second order differential equation. This equation is of interest in mechanics, and it was one of the first to be studied in connection with the oscillations of a nonlinearly elastic spring with linear viscous damping. The equation becomes autonomous when the external forcing term is removed $(P(t) \equiv 0)$.

As a final example we remark that the system

$$\frac{dx}{dt} = x + 3y - 2t$$

$$\frac{dy}{dt} = 2x + y + 4 \sin t, \tag{7}$$

is a typical nonautonomous linear system.

Let us now consider the qualitative behavior of solutions of the general second order autonomous equation given in (1). Equations of motion in one dimension are of this form, so the coordinate values x_0 for which

$$f(x_0, 0) = 0 \tag{8}$$

must be equilibrium points for the mechanical system which has (1) as its equation of motion. This follows because (8) implies that the velocity $dx/dt = 0$, and (8) together with (1) then shows the acceleration $d^2x/dt^2 = 0$.

It proves convenient to study the behavior of solutions of (1) in the displacement–velocity plane, that is in the $(x, dx/dt)$-plane. This is called the **phase plane**[1] and usually we set $y = dx/dt$, so that with this understanding the phase plane becomes the (x, y)-plane. It follows from (8) and the definition of y that, in the phase plane, the equilibrium of the dynamical system associated with (1) occurs at the points $(x_0, 0)$. These are called **equilibrium points** or **critical points** of (1). The only case to be considered here will be the one in which the equilibrium points are isolated. This means that a suitably small circle centered on each equilibrium point contains no other equilibrium point.

When a solution of (1) is represented in the phase plane, the time t appears as a parameter in $x(t)$ and $y(t)$, and as t increases so a curve is traced in the phase plane. These solution curves in the phase plane are called **trajectories** or **paths** (also **orbits**), and the sense in which time increases along them is usually shown by an arrow. As $y = dx/dt$, it follows that time increases in the positive x-direction along a trajectory when $y > 0$ and in the negative x-direction when $y < 0$.

We shall see that the behavior of the trajectories in the neighborhood of the equilibrium points holds the key to the general behavior of the solutions of (1). A diagram showing the equilibrium points together with representative trajectories of an autonomous equation is called its **phase portrait**.

---

[1] The name comes from mechanics in which the one-dimensional motion of a mass m at position x may be determined from the equation of motion once x and the momentum $p = m\dot{x}$ are known. The (x, p)-plane in which the motion is represented as an oriented curve is called the *phase plane*.

For example, a trajectory in the form of a simple closed loop about an equilibrium point must correspond to a periodic motion, whereas trajectories spiraling round and entering an equilibrium point as $t \to +\infty$ must correspond to oscillatory behavior which decays to an equilibrium position. It is usual to refer to the physical problem giving rise to (1) as a **system** and to $(x(t), y(t))$ as the **state** of the system at time t.

Before considering an example, let us first notice that as $y = dx/dt$, it follows from the chain rule that

$$\frac{d^2 x}{dt^2} = \frac{dy}{dt} = \frac{dx}{dt}\frac{dy}{dx} = y\frac{dy}{dx}. \tag{9}$$

Using this in (1) we obtain the first order differential equation

$$y\frac{dy}{dx} = f(x, y), \tag{10}$$

which is usually nonlinear. Thus a **direction field** is defined throughout the phase plane (see Sec. 4.2) by the differential equation

$$\frac{dy}{dx} = \frac{f(x, y)}{y}. \tag{11}$$

The method of isoclines may be used to construct this direction field. The angle arc $\tan(dy/dx)$ made by a tangent to this field at (x, y) will be well defined at all points other than equilibrium points, at which the right-hand of (11) becomes indeterminate. Alternatively, when f is simple, the equation of the trajectories can often be found quite easily by integration of (11).

As (1) is second order, its general solution must contain two arbitrary integration constants. However the differential equation in (11) determining the trajectories in the phase plane is only first order, and so its general solution contains only one arbitrary constant of integration. This constant determines a complete trajectory for all time, but not the point on it at which the motion along it starts. We shall see that this initial point is determined by the other arbitrary constant in the general solution of (1). Thus the other arbitrary constant fixes the time origin for motion along a particular trajectory.

A given motion of a dynamical system corresponds to a single fixed trajectory in the phase plane, with time increasing along the trajectory as the motion evolves. It is for this reason that autonomous equations are called **time-invariant** equations. By analogy with the behavior of a fluid, motion along trajectories in the phase plane is sometimes referred to as a **flow**.

Example 1. Phase portrait for the simple pendulum

An undamped simple pendulum made by fastening a particle of mass m to the end of a light but rigid rod of length l pivoted at 0 is shown in Fig. 6.15(a) in which it is performing oscillations about A.

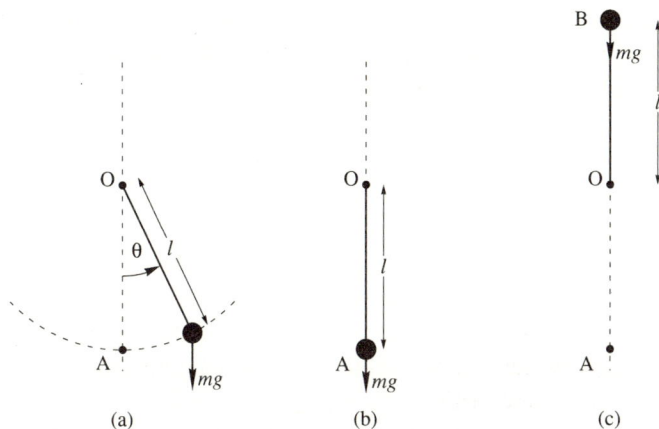

Fig. 6.15 (a) Oscillations about A, (b) Stable equilibrium position, (c) Unstable equilibrium position-inverted pendulum

The equation of motion for any angle of swing θ is

$$ml^2\ddot{\theta} + mgl\sin\theta = 0, \tag{12}$$

where a dot denotes differentiation with respect to time t. Proceeding as in (9) and setting $\xi = \dot{\theta}$, it follows that $\ddot{\theta} = \xi(d\xi/d\theta)$, so (12) becomes

$$ml^2\,\xi\frac{d\xi}{d\theta} + mgl\sin\theta = 0 \tag{13}$$

or, equivalently,

$$ml^2\,\xi\,d\xi + mgl\sin\theta\,d\theta = 0. \tag{14}$$

Integrating this exact differential equation and returning to the variables θ and $\dot{\theta}$ gives the result

$$\tfrac{1}{2}ml^2\dot{\theta}^2 - mgl\cos\theta = C, \tag{15}$$

where C is an integration constant. In terms of mechanics, this result asserts the *conservation of energy* in the system, because the first term in (15) is the kinetic and the second term together with its sign the potential energy of the system. Conservation of energy occurs because no damping term is present.

Initial conditions for (12) involve the specification of θ and $\dot{\theta}$ at time t_0 with, say, $\theta(t_0) = \theta_0$ and $\dot{\theta}(t_0) = \dot{\theta}_0$. The state of the system at time t is thus $(\theta(t), \dot{\theta}(t))$ and this evolves from the initial state $(\theta_0, \dot{\theta}_0)$ according to (12) or, equivalently, its first integral (15). ∎

Intuitively, a system is in a **stable** state when a suitably small disturbance of that state at a given time only alters the state at all subsequent times by a correspondingly small amount. If an arbitrarily small disturbance to the state of a system has the effect that it produces a large change in the state at all subsequent times, then the system is said to be in an **unstable** state. In brief, the system is said to be stable in the first case and unstable in the second.

We know from physics, and also from Chapter 5, that if the equilibrium state in Fig. 6.15(b) is slightly disturbed the pendulum will perform small oscillations about the point A for all time, so this is a stable state in the sense that although the amplitude may vary it remains bounded. This is often called **neutral stability**. The configuration in Fig. 6.15(c) is called the *inverted pendulum*. This equilibrium state is unstable, because any slight disturbance will cause the pendulum to swing down, and thereafter make large oscillations.

The indeterminacy of θ up to a multiple of 2π means that the equilibrium configuration in Fig. 6.15(b) corresponds to the infinite set of equilibrium points $\theta = 2n\pi$, $\dot{\theta} = 0$ in the $(\theta, \dot{\theta})$-phase plane, with $n = 0, \pm 1, \pm 2, \ldots$. Similarly, the equilibrium configuration in Fig. 6.15(c) corresponds to the infinite set of equilibrium points $\theta = (2n+1)\pi$, $\dot{\theta} = 0$ in the $(\theta, \dot{\theta})$ phase plane, with $n = 0, \pm 1, \pm 2, \ldots$.

A little thought shows that the phase portrait for the undamped simple pendulum which follows from (15) is of the form shown in Fig. 6.16. The periodic nonlinear oscillations which can be performed by the pendulum are represented by the roughly elliptical trajectories which surround each stable equilibrium point in the $(\theta, \dot{\theta})$-phase plane. Each of the trajectories corresponds to a different period of oscillation. The wavy curves above and below the dashed curves joining the unstable equilibrium points correspond to a whirling type of motion. This motion will be clockwise when $\dot{\theta} < 0$ and counterclockwise when $\dot{\theta} > 0$. Each of the dashed curves joining equilibrium points is a special trajectory called a **separatrix**, because motion on either side of a separatrix is quite different. Closed loops formed by separatrices must not be confused with a possible periodic motion, because such loops contain equilibrium points. It is characteristic of separatrices that they either enter or leave equilibrium points tangent to a radial line through the point.

To construct the phase portrait of Fig. 6.16 using a direction field it is necessary to find the isoclines of the first order equation in (13). For simplicity we first make (13) dimensionless by dividing by ml^2 and then setting $\xi = (g/l)^{1/2} X$, to obtain

$$\frac{dX}{d\theta} = -\frac{\sin \theta}{X}. \tag{16}$$

The phase plane is now the (θ, X)-plane, and the equation of the isoclines of (16) follow by

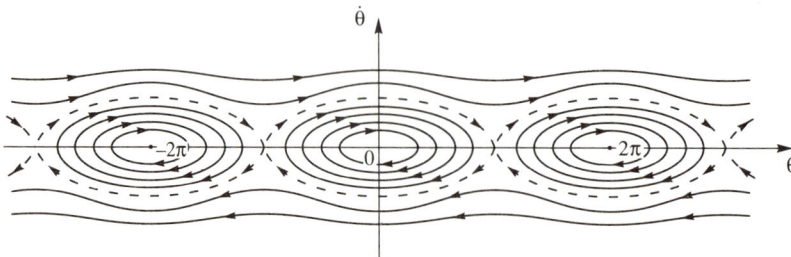

Fig. 6.16   Phase portrait of undamped simple pendulum

setting $dX/d\theta = k$ (const.) to obtain

$$X = -\left(\frac{1}{k}\right)\sin\theta. \tag{17}$$

Representative isoclines are shown in Fig. 6.17 together with the lines tangent to the trajectories which intersect them. The inclination of these tangent lines to the θ axis is arc tan k. An examination of the isoclines passing through an equilibrium point demonstrates the indeterminacy of the slope of trajectories passing through it. In Fig. 6.17 the isoclines have been used to construct an approximate trajectory.

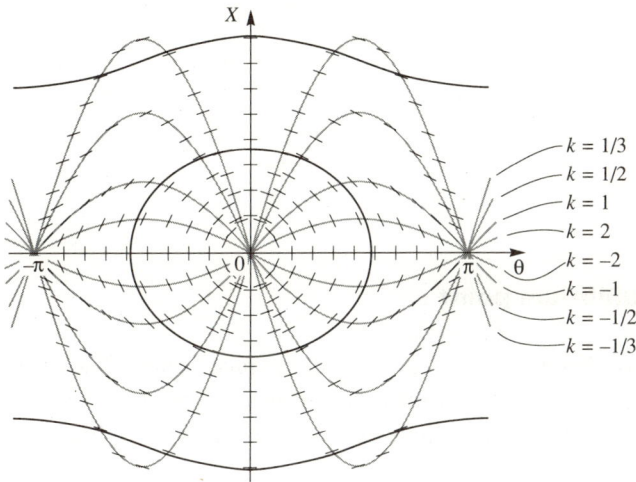

Fig. 6.17 Isoclines and approximate trajectories of $\dfrac{dX}{d\theta} = -\dfrac{\sin\theta}{X}$

The related concepts of stability and instability introduced in the last example are two of the most important qualitative features of solutions of differential equations. They enter into most aspects of engineering and science and, in particular, are fundamental to control theory. It is useful to make these ideas a little more precise by using the phase plane and defining stability and instability in terms of equilibrium points and trajectories.

Equilibrium points (critical points) are classified as being **stable, asymptotically stable** or **unstable** according to the following criteria.

Stable equilibrium point

Let an equilibrium point of (1) be located at point P in the phase plane. Then P is said to be **stable** if all trajectories which at time t_0 are close enough to P remain close to it as $t \rightarrow +\infty$. To interpret this definition graphically, let $Q_1(t_0)$, $Q_2(t_0)$, . . . be points at time

t_0 lying on the trajectories $\gamma_1, \gamma_2, \ldots$ of (1) shown in Fig. 6.18. Then $Q_1, Q_2, \ldots$ will all lie in a suitably small disk of radius ε centered on P, and for $t \geq t_0$ the trajectories $\gamma_1, \gamma_2, \ldots$ will all lie within another a suitably small disk or radius $\delta(>\varepsilon)$ centered on P.

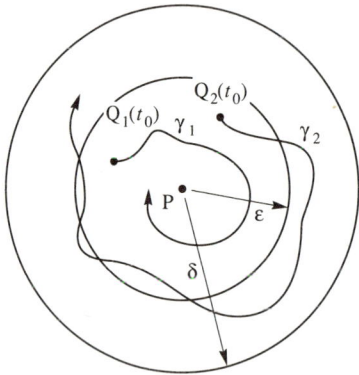

Fig. 6.18   Stable equilibrium point of (1)

Asymptotically stable equilibrium point

Let a stable equilibrium point of (1) be located at point P in the phase plane. Then P is said to be **asymptotically stable** if, in addition to stability, P has the property that all trajectories starting in a suitably small disk of radius δ about P at some time t_0 ultimately approach P as $t \to +\infty$.

A graphical interpretation of asymptotic stability is illustrated in Fig. 6.19(a) in which

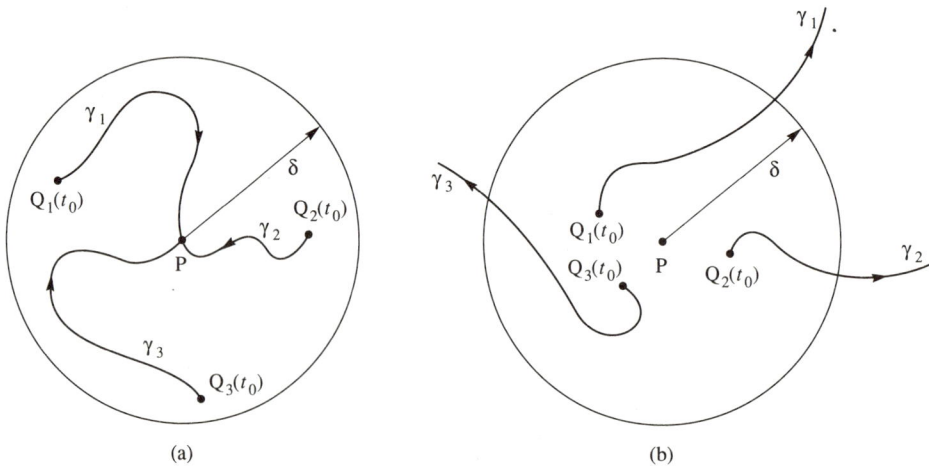

(a)   (b)

Fig. 6.19 (a) Asymptotically stable equilibrium point of (1) at P, (b) Unstable equilibrium point of (1) at P

the trajectories $\gamma_1, \gamma_2, \ldots$ starting at points $Q_1(t_0), Q_2(t_0), \ldots$ at time t_0 within a disk of radius δ about P all eventually approach the equilibrium point P as $t \to +\infty$. This disk is called a **domain of attraction** of P.

Unstable equilibrium point

An equilibrium point P of (1) which is not stable is said to be **unstable**, or **repelling**. A geometrical interpretation of an unstable equilibrium point of (1) at P is illustrated in Fig. 6.19(b) in which all trajectories $\gamma_1, \gamma_2, \ldots$ starting at $Q_1(t_0), Q_2(t_0), \ldots$ at time t_0 diverge from P as $t \to +\infty$.

The definition of stablility given here was first formulated by A. M. Liapunov[2], and for this reason it is often called *stability in the sense of Liapunov*. Other definitions of stability are possible, but this is the most straightforward one.

The above definitions show, for example, that the finite oscillations of the undamped simple pendulum considered in Ex. 1 are stable, but not asymptotically stable. This follows from the fact that trajectories suitably close to the equilibrium points $(2n\pi, 0)$, $n = 0, \pm 1, \pm 2, \ldots$, and encircling them as shown in Fig. 6.16, are all simple closed curves. They satisfy the conditions for a stable equilibrium point, but not those for an asymptotically stable one.

These same definitions show the equilibrium points $((2n+1)\pi, 0)$, $n = 0, \pm 1, \pm 2, \ldots$, in Fig. 6.16 to be unstable. This is so because at some time t_0 trajectories exist close to these points which move arbitrarily far away from them as $t \to \pm\infty$.

Unlike the undamped simple pendulum, autonomous nonlinear oscillations can arise which have only a single period. When this happens the oscillations are represented in the phase plane by a single closed trajectory Γ encircling a stable equilibrium point. Initial conditions not on the closed trajectory Γ lead to a trajectory in the phase plane which spirals either in or out until it converges to Γ. The final limiting trajectory Γ encircling an equilibrium point is called a **limit cycle**, and the convergence of trajectories to Γ shows it represents a solution which is stable, but not asymptotically stable.

This happens, for example, with the van der Pol equation (5). This limit cycle for this equation is shown in Fig. 6.20 for the case $\varepsilon = 1$. Figure 6.21 shows the convergence of two trajectories to the van der Pol limit cycle for the case $\varepsilon = 1$; one from outside the limit cycle

---

[2] ALEXANDER MIKHAILOVICH LIAPUNOV (1857–1918), a distinguished Russian mathematician whose work in differential equations forms the basis of modern stability theory. His main results first appeared in the form of a monograph published in Russia in 1892, and later in a French translation in 1907.

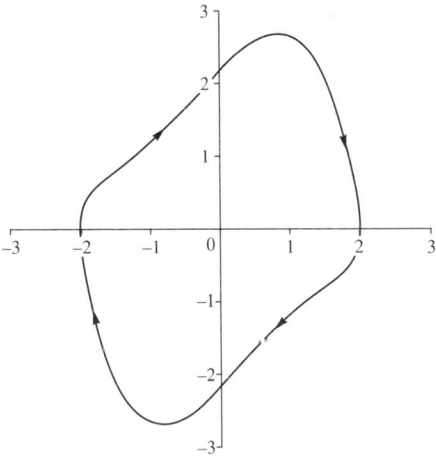

Fig. 6.20   Limit cycle for the van der Pol equation when $\varepsilon=1$

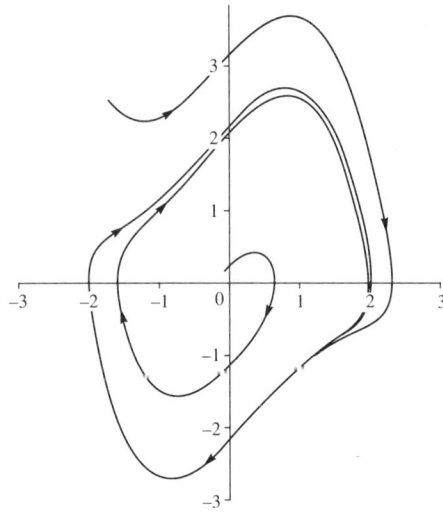

Fig. 6.21   Convergence to the limit cycle of trajectories interior and exterior to it when $\varepsilon=1$

corresponding to the initial conditions $x=-2$, $\dot{x}=3$, and the other from inside corresponding to the initial conditions $x=-0.1$, $\dot{x}=0.1$.

Differential equation (12) in Ex. 1 which gave rise to the phase portrait in Fig. 6.16 was nonlinear, though simple enough to integrate once in order that its trajectories in the phase plane could be determined analytically. Even when this analytical approach to the determination of phase portraits is not possible, it is still desirable to find the approximate pattern of trajectories close to equilibrium points because this provides information about the solution close to such points.

An approximate description of trajectories in the neighborhood of an equilibrium point $(x_0, 0)$ of (1) may be found by approximating the function $f(x, y)$ in (1) by the linear expression

$$f(x, y)=\alpha(x-x_0)+\beta y, \tag{18}$$

where

$$\alpha=\left(\frac{\partial f}{\partial x}\right)_{(x_0, 0)}, \quad \beta=\left(\frac{\partial f}{\partial y}\right)_{(x_0, 0)} \tag{19}$$

and an expression of the form $[h]_{(x_0, 0)}$ signifies that the function h is to be evaluated at the point $(x_0, 0)$.

Approximation (18) follows by expanding $f(x, y)$ in (1) in a Taylor series about the equilibrium point $(x_0, 0)$, using the fact that $f(x_0, 0)=0$, and then neglecting terms involving partial derivatives of order greater than one.[3] This form of approximation is called the **linearization** of $f(x, y)$ about $(x_0, 0)$.

To see how this works we now use this approach first to approximate the behavior of (12) in Ex. 1 close to the stable equilibrium point at $(0, 0)$, and then close to the unstable equilibrium point at $(\pi, 0)$ in Fig. 6.16. Writing (12) as

$$\ddot{\theta} = -\left(\frac{g}{l}\right) \sin \theta, \tag{20}$$

and for convenience setting $\dot{\theta} = \xi$, it follows from (20) that

$$f(\theta, \xi) = -\left(\frac{g}{l}\right) \sin \theta.$$

Now,

$$\frac{\partial f}{\partial \theta} = -\left(\frac{g}{l}\right) \cos \theta, \quad \frac{\partial f}{\partial \xi} \equiv 0,$$

so expanding $f(\theta, \xi)$ in a Taylor series about the equilibrium point $(\theta_0, 0)$, using the result that $f(\theta_0, 0) = 0$ and neglecting terms involving partial derivatives of order greater than one, we arrive at the linear approximation

$$f(\theta, \xi) = -(\theta - \theta_0)\left(\frac{g}{l}\right) \cos \theta_0.$$

At the stable equilibrium point $(0, 0)$ this yields the linear approximation

$$f(\theta, \xi) = -\left(\frac{g}{l}\right)\theta, \tag{21}$$

while at the unstable equilibrium point $(\pi, 0)$ it gives

$$f(\theta, \xi) = \left(\frac{g}{l}\right)(\theta - \pi). \tag{22}$$

Close to $(0, 0)$ the linearized form of (20) is thus

$$\ddot{\theta} + \left(\frac{g}{l}\right)\theta = 0, \tag{23}$$

while close to $(\pi, 0)$ it becomes

$$\ddot{\theta} - \left(\frac{g}{l}\right)(\theta - \pi) = 0. \tag{24}$$

---

[3] A Taylor series which is truncated after the nth order partial derivatives is called a **Taylor polynomial of degree n**, so linearization uses a Taylor polynomial of degree one.

The angle $\Theta = \theta - \pi$ is the displacement of the pendulum from its inverted unstable equilibrium position in Fig. 6.15(c), and in terms of Θ (24) becomes

$$\ddot{\Theta} - \left(\frac{g}{l}\right)\Theta = 0. \tag{25}$$

The phase portraits of the linearized form of (12) about equilibrium points $(0, 0)$ and $(\pi, 0)$ now follow from (23) and (25). The derivation of the phase portraits for these types of equation is given in the two examples which follow.

Example 2. Phase portrait for simple harmonic motion

One of the simplest examples of the use of the phase plane is in connection with the equation of simple harmonic motion

$$\frac{d^2 x}{dt^2} + \alpha^2 x = 0.$$

In terms of $y = dx/dt$ and (9) the differential equation becomes

$$y\frac{dy}{dx} + \alpha^2 x = 0,$$

so its single equilibrium point is at the origin. Integration shows x and y to be related by

$$\alpha^2 x^2 + y^2 = c.$$

The trajectories represented by this equation for different values of $c > 0$ (corresponding to different initial conditions) are ellipses in the phase plane centered on the origin. Thus, for example, periodic motion corresponding to $x(0) = 0$, $\dot{x}(0) = 1$ is the ellipse

$$\alpha^2 x^2 + y^2 = 1.$$

The origin corresponds to the stable equilibrium state $x(0) = \dot{x}(0) = 0$. Some typical trajectories are shown in Fig. 6.22. An equilibrium point encircled by simple closed trajectories in the manner shown in Fig. 6.22 is called a **center**.

To identify these results with the ones appropriate to (23) it is only necessary to set $\alpha^2 = g/l$, $x = \theta$ and $y = \dot{\theta}$. Notice that the phase portrait in Fig. 6.22 closely resembles the trajectories in Fig. 6.16 which surround the origin. However oscillations of all amplitudes represented in Fig. 6.22 and governed by (23) are **isochronous** (of equal period), whereas the corresponding ones in Fig. 6.16 governed by (12) are not.

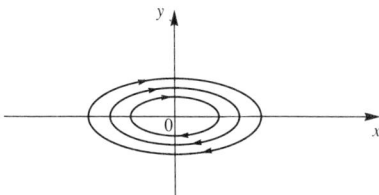

Fig. 6.22   Phase portrait for $\ddot{x} + \alpha^2 x = 0$; a center

The general time dependent solution of

$$\ddot{x} + \alpha^2 x = 0$$

is, of course,

$$x(t) = A\cos(\alpha t + \varepsilon),$$

where the amplitude $A > 0$ and phase ε are arbitrary constants. This familiar result shows explicitly the isochronous nature of the oscillations governed by this differential equation, because their period $1/\alpha$ is independent of the amplitude A. ∎

Example 3. Phase portrait of $\ddot{x} - \alpha^2 x = 0$

The equation of motion (25) for small displacements of the inverted pendulum is of the form

$$\frac{d^2 x}{dt^2} - \alpha^2 x = 0.$$

Setting $y = dx/dt$ and again using (9) this becomes

$$y\frac{dy}{dx} - \alpha^2 x = 0,$$

so here again the origin is the only equilibrium point. Integration shows the trajectories about the origin forming the phase portrait to be the family of hyperbolas

$$y^2 - \alpha^2 x^2 = c.$$

The phase portrait itself is shown in Fig. 6.23, and in that figure the asymptotes shown as dashed lines passing through the equilibrium point at the origin form the separatrices. An equilibrium point of the type occurring here is called a **saddle point**.

To identify these results with the ones appropriate to (25) it is only necessary to set $\alpha^2 = g/l$, $x = \Theta$ and $y = \dot{\Theta}$. Notice the similarity between the phase portrait in Fig. 6.23 and the pattern of trajectories in the neighborhood of the equilibrium point $(\pi, 0)$ in Fig. 6.16. Inspection of Fig. 6.23 shows that with the sole exception of initial states lying on the asymptote with a negative gradient (pushing the pendulum back into unstable equilibrium from a slightly displaced position), all other

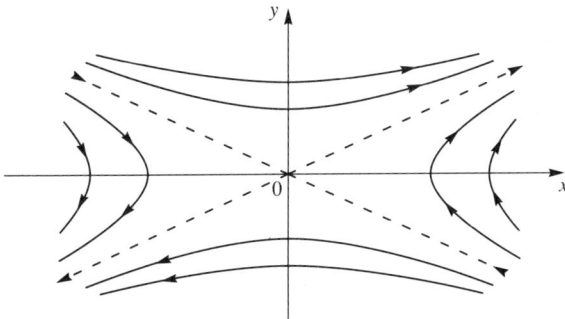

Fig. 6.23   Phase portrait for $x - \alpha^2 x = 0$; a saddle point

disturbances will lead to instability. Thus a saddle point corresponds to an unstable equilibrium position.

The general time-dependent solution of

$$\ddot{x} - \alpha^2 x = 0$$

is simply

$$x(t) = c_1 e^{-\alpha t} + c_2 e^{\alpha t}.$$

This shows explicitly that unless the initial conditions are such that $c_2 = 0$, a disturbance will cause $x(t)$ to grow exponentially. In terms of the inverted pendulum problem this means a disturbance in Θ will grow exponentially with time for as long as the approximation of $\sin \Theta$ by Θ remains valid.

∎

Examples 2 and 3 have shown how the approximate local behavior of the simple pendulum close to equilibrium points may be deduced from (20) by linearizing the function $f(\theta, \xi)$ at an equilibrium point $(x_0, 0)$. These results taken together with (18) suggest that when examining the equilibrium point $(x_0, 0)$ of

$$\frac{d^2 x}{dt^2} = f(x, y), \tag{26}$$

the equation should be linearized about $(x_0, 0)$ and the origin shifted to the point $(x_0, 0)^4$. Thus, to study the linearized behavior of (1), it is sufficient that the phase portraits associated with the linear differential equation

$$\frac{d^2 x}{dt^2} = \alpha x + \beta y, \quad \text{with} \quad y = \frac{dx}{dt} \tag{27}$$

should be investigated.

As this linear differential equation has constant coefficients, it is a routine matter to determine its general solution $x(t)$, and then to use this to obtain $\dot{x}(t)$. The phase portraits then follow parametrically in terms of t by considering the points $(x(t), \dot{x}(t))$ in the phase plane. There are a number of cases to be examined, and each will now be considered separately. The forms of $x(t)$ and $\dot{x}(t)$ follow directly from Sec 5.2, and so will merely be quoted.

(i) $\alpha = -m^2 < 0, \beta = 0$: a center

$$x(t) = A \cos(mt + \varepsilon)$$
$$y(t) = \dot{x}(t) = -mA \sin(mt + \varepsilon).$$

The trajectories described by $(x(t), \dot{x}(t))$ form a family of confocal ellipses encircling the only equilibrium point located at the origin. In this case the equilibrium point is called a

---

[4] This is accomplished by setting $X = x - x_0$, $Y = y$ and working in the (X, Y)-plane.

center and it is characteristic of stable oscillatory (sometimes called **neutrally stable**) behavior.

Care must always be exercised when linearizing differential equations. If in the case of a center the function $f(x, y)$ from which the right-hand side of (27) is constructed depends nonlinearly on y, the situation usually requires more than a linear approximation. In such a case it does not always follow that the stability of the linearized problem implies stability of the nonlinear one. The phase portrait of a center is shown in Fig. 6.24.

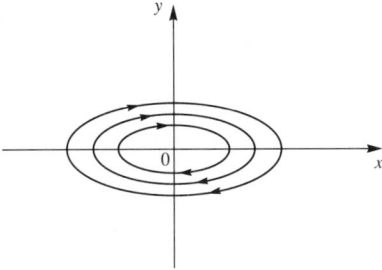

Fig. 6.24   A center; $\alpha < 0$, $\beta = 0$

(ii) $\alpha + \beta^2/4 = -m^2 < 0$, $\beta \neq 0$: a spiral

$$x(t) = A \exp\left(\tfrac{1}{2}\beta t\right) \cos(mt + \varepsilon)$$

$$y(t) = \dot{x}(t) = A \exp\left(\tfrac{1}{2}\beta t\right)\left[\tfrac{1}{2}\beta \cos(mt + \varepsilon) - m \sin(mt + \varepsilon)\right].$$

The trajectories are best constructed by setting $\theta = mt + \varepsilon$ and redefining A, so that in terms of θ as a parameter

$$x(\theta) = B \exp\left(\beta\theta/2m\right) \cos\theta$$

$$y(\theta) = \dot{x}(\theta) = B \exp\left(\beta\theta/2m\right)\left[\tfrac{1}{2}\beta \cos\theta - m \sin\theta\right].$$

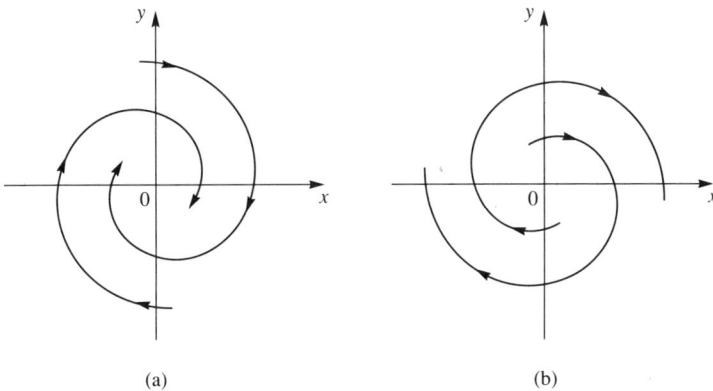

(a)   (b)

Fig. 6.25   (a) Stable attractive spiral point: $\alpha + \beta^2/4 < 0$, $\beta < 0$;   (b) Unstable spiral point: $\alpha + \beta^2/4 < 0$, $\beta > 0$

Here the coefficient B identifies a particular trajectory and θ the points on it. The trajectories are spirals centered on the single equilibrium point at the origin. The origin is called a **stable attractive spiral point** when $\beta < 0$, for then all trajectories are attracted to the origin as $t \to +\infty$. It follows from our earlier definitions that such a point is asymptotically stable. Conversely, the origin is called an **unstable spiral point** (sometimes a **repelling spiral point**) when $\beta > 0$, for then all the trajectories move away from the origin as $t \to +\infty$. A stable attractive spiral point is shown in Fig. 6.25(a), and an unstable one in Fig. 6.25(b). It follows from the earlier definitions that a stable attractive spiral point is asymptotically stable.

(iii) $\alpha + \beta^2/4 > 0$, $\alpha < 0$: improper node

$$x(t) = c_1 \exp(\lambda_1 t) + c_2 \exp(\lambda_2 t)$$
$$y(t) = \dot{x}(t) = c_1 \lambda_1 \exp(\lambda_1 t) + c_2 \lambda_2 \exp(\lambda_2 t),$$

with $\lambda_1 = \frac{1}{2}[\beta + (\beta^2 + 4\alpha)^{1/2}]$, $\lambda_2 = \frac{1}{2}[\beta - (\beta^2 + 4\alpha)^{1/2}]$.

At the origin, which is the only equilibrium point, all trajectories are tangent to one or other of the degenerate solutions $y = \lambda_1 x$ and $y = \lambda_2 x$, the first corresponding to $c_2 = 0$ and the second to $c_1 = 0$. If $\beta < 0$, so that then $\lambda_2 < \lambda_1 < 0$, the origin is stable and attractive, and thus asymptotically stable. For $\beta < 0$ and large positive values of t, all trajectories apart from the degenerate one $y = \lambda_2 x$ approach along the line $y = \lambda_1 x$. The direction $y = \lambda_2 x$ is called the **exceptional direction**.

When $\beta > 0$, so $\lambda_1 > \lambda_2 > 0$, the origin repels trajectories, and so is unstable. In the unstable case, with the exception of $y = \lambda_1 x$, all trajectories in the neighborhood of the origin leave along the line $y = \lambda_2 x$, so the origin corresponds to the values of t which are large and negative. The exceptional direction in this case is $y = \lambda_1 x$. Equilibrium points of this type are called **improper nodes with an exceptional direction**. Figure 6.26(a) shows a **stable attractive improper node** and Fig. 6.26(b) an **unstable improper node**, each with an

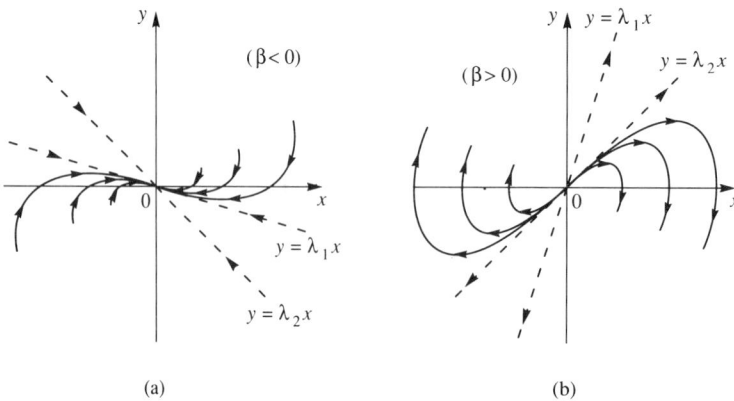

(a)   (b)

Fig. 6.26 (a) Stable attractive improper node with an exceptional direction: $\alpha + \beta^2/4 < 0$, $\alpha < 0$, $\beta < 0$, (b) Unstable improper node with an exceptional direction: $\alpha + \beta^2/4 < 0$, $\alpha < 0$, $\beta > 0$.

exceptional direction. It follows from our earlier definitions that the improper node in Fig. 6.26(a) is asymptotically stable.

A degenerate situation arises when $\alpha + \beta^2/4 = 0$ and $\beta \neq 0$, for then

$$x(t) = (c_1 t + c_2) \exp\left(\tfrac{1}{2}\beta t\right)$$
$$y(t) = \dot{x}(t) = \left[(c_1 + \tfrac{1}{2}c_2\beta) + \tfrac{1}{2}c_1\beta t\right] \exp\left(\tfrac{1}{2}\beta t\right).$$

At the origin all trajectories are tangent to the line $y = \tfrac{1}{2}\beta x$, which corresponds to the trajectory for which $c_1 = 0$, and there is no exceptional direction because the two degenerate solutions in Figs 6.26 now coincide. It follows that when $\beta < 0$ the origin is a **stable attractive improper node with no exceptional direction**, and thus is asymptotically stable. It is an **unstable improper node** when $\beta > 0$. The phase portraits in these cases are shown in Figs 6.27(a, b).

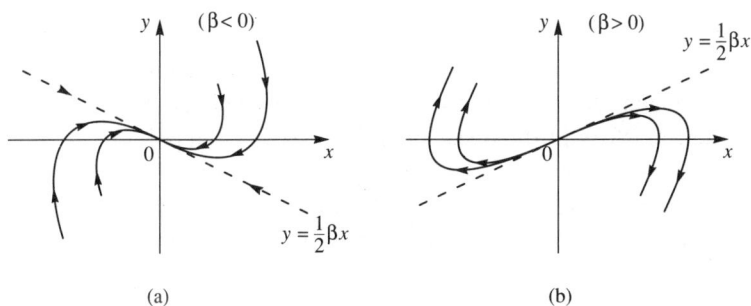

(a) (b)

Fig. 6.27 (a) Stable attractive improper node with no exceptional direction: $\alpha + \beta^2/4 = 0$, $\beta < 0$, (b) Unstable improper node with no exceptional direction: $\alpha + \beta^2/4 = 0$, $\beta > 0$.

(iv) $\alpha + \beta^2/4 > 0$, $\alpha > 0$: a saddle point

$$x(t) = c_1 \exp(\lambda_1 t) + c_2 \exp(\lambda_2 t)$$
$$y(t) = \dot{x}(t) = c_1\lambda_1 \exp(\lambda_1 t) + c_2\lambda_2 \exp(\lambda_2 t),$$

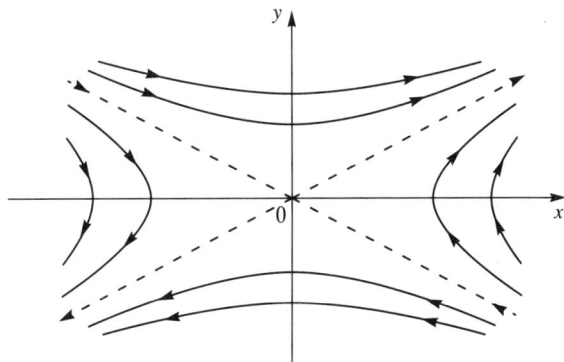

Fig. 6.28 Saddle point: $\alpha + \beta^2/4 > 0$, $\alpha > 0$

with $\lambda_1 = \frac{1}{2}[\beta + (\beta^2 + 4\alpha)^{1/2}]$, $\lambda_2 = \frac{1}{2}[\beta - (\beta^2 + 4\alpha)^{1/2}]$. It follows that because $\alpha > 0$, λ_1 and λ_2 will be of opposite sign. The only trajectories passing through the equilibrium point at the origin are $y = \lambda_1 x$ and $y = \lambda_2 x$, the first of which corresponds to $c_2 = 0$ and the second to $c_1 = 0$. The remaining trajectories form a family of hyperbolas with these two lines (the separatrices) as asymptotes. It is a direct consequence of the earlier definitions and the pattern of the trajectories shown in Fig. 6.28 that an equilibrium point of this type, called a **saddle point**, is unstable.

(v) $\alpha = 0$, $\beta \neq 0$ and $\alpha = \beta = 0$: straight line degeneracies

If $\alpha = 0$, $\beta \neq 0$ it follows that

$$x(t) = c_1 e^{\beta t} + c_2$$
$$y(t) = \dot{x}(t) = c_1 \beta e^{\beta t}.$$

The trajectories in the phase plane have now degenerated to the family of parallel straight lines $y = \beta(x - c_2)$ which are inclined to the x-axis at the angle arc tan β. When $\beta < 0$, then as $t \to +\infty$ the trajectory in the phase plane approaches the point $(c_2, 0)$ for arbitrary real c_2, and so all points on the x-axis are equilibrium points. The previous definitions imply that points on the x-axis are stable equilibrium points. If $\beta > 0$, it is clear that these points are unstable equilibrium points. These two cases are illustrated in Fig. 6.29.

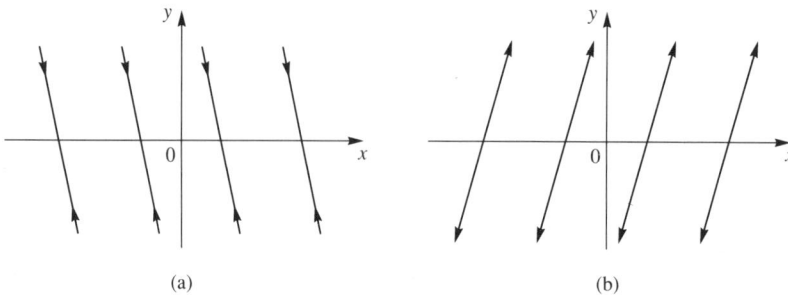

Fig. 6.29   (a) Stable equilibrium points on x-axis: $\alpha = 0$, $\beta < 0$, (b) Unstable equilibrium points on x-axis: $\alpha = 0$, $\beta > 0$

Finally, if $\alpha = \beta = 0$, then

$$x(t) = c_1 t + c_2$$
$$y(t) = \dot{x}(t) = c_1,$$

and the trajectories again degenerate to a family of parallel straight lines, but this time they are parallel to the x-axis. Inspection of the form of the solution shows it to be unstable, because the trajectories involve a uniform translation to the right in the upper

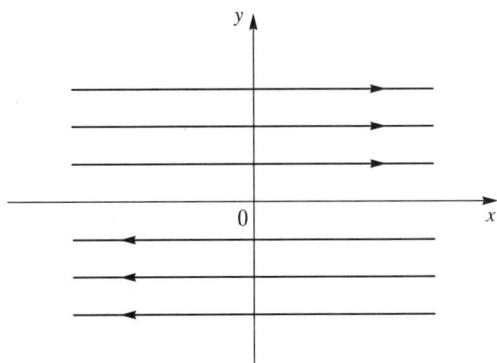

Fig. 6.30 Points on the x-axis are unstable equilibrium points: $\alpha = \beta = 0$

half of the phase plane, and a corresponding one to the left in the lower half of the phase plane (Fig. 6.30). Points on the x-axis are unstable equilibrium points.

It sometimes happens that, close to an equilibrium point of a nonlinear function $f(x, y)$ occurring in (26), the linear approximation to $f(x, y)$ vanishes identically. The linearized form of (26) then reduces to (27) with $\alpha = \beta = 0$. When this occurs, then although the equilibrium point of the linearized problem is unstable, it is not possible to infer from this that the corresponding result is also true for the nonlinear differential equation. To resolve the matter, either a better approximation must be used (say a Taylor polynomial of degree two) and its consequences examined, or a more advanced test for stability must be employed.

The results of Sections (i) to (v) are summarized in Fig. 6.31, which shows how the nature of the equilibrium points of the linear constant coefficient second order differential equation

$$\frac{d^2x}{dt^2} - \beta\frac{dx}{dt} - \alpha x = 0. \tag{28}$$

depend on the coefficients α and β.

Inspection of Fig. 6.31 shows why an equilibrium point of the nonlinear differential equation

$$\frac{d^2x}{dt^2} = f(x, y),$$

at which the linearized form degenerates to

$$\frac{d^2x}{dt^2} = 0,$$

may either be stable or unstable. The origin in the (α, β)-plane is the point at which different stability domains meet, so a higher order approximation to $f(x, y)$ may show its

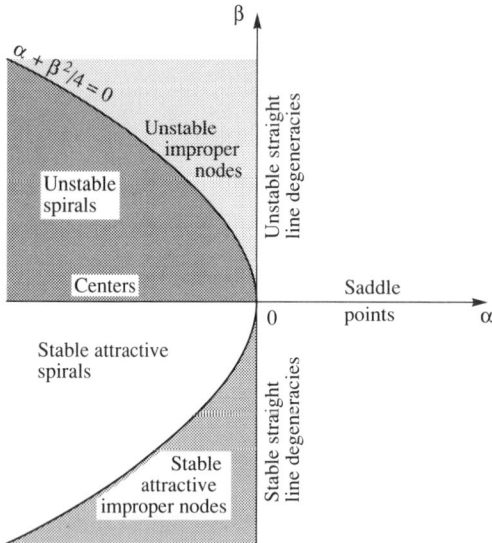

Fig. 6.31   Dependence of equilibrium points of (28) on α and β

equilibrium point to belong to any one of them. The diagram also shows why a center which in the linear approximation is stable might, in the full nonlinear case, be asymptotically stable or even unstable.

Example 4. A phase portrait

Locate and identify the nature of the equilibrium points of

$$\frac{d^2x}{dt^2} = k^2 x(x-2),$$

and use them to deduce the phase portrait for this equation.

Solution

Setting $y = dx/dt$, and using (9), the differential equation determining the trajectories is seen to be

$$\frac{dy}{dx} = \frac{k^2 x(x-2)}{y}.$$

The equilibrium points are the points in the (x, y)-phase plane at which dy/dx is indeterminate, and this occurs at $(0, 0)$ and $(2, 0)$.

Returning now to the original differential equation and comparing it with (26) we find

$$f(x, y) = k^2 x(x-2),$$

and so

$$\frac{\partial f}{\partial x} = 2k^2(x-1), \quad \frac{\partial f}{\partial y} \equiv 0.$$

Approximating $f(x, y)$ about $(x_0, 0)$ by its Taylor polynominal of degree one then gives

$$f(x, y) = f(x_0, 0) + (x - x_0)\left(\frac{\partial f}{\partial x}\right)_{(x_0, 0)} + y\left(\frac{\partial f}{\partial y}\right)_{(x_0, 0)}$$

$$= (x - x_0)2k^2(x_0 - 1).$$

Thus at $(0, 0)$ the linear approximation to $f(x, y)$ is

$$f(x, y) = -2k^2 x,$$

while at $(2, 0)$ it is

$$f(x, y) = 2k^2(x - 2).$$

Hence close to $(0, 0)$ the original differential equation is of the form

$$\frac{d^2 x}{dt^2} = -2k^2 x,$$

while close to $(2, 0)$ it is of the form

$$\frac{d^2 x}{dt^2} = 2k^2(x - 2)$$

or, setting $X = x - 2$, to

$$\frac{d^2 X}{dt^2} = 2k^2 X.$$

It then follows from (i) that $(0, 0)$ is a center and from (iv) that $(2, 0)$ is a saddle point. The local pattern of the trajectories about these points is shown in Fig. 6.32(a).

The full phase portrait is shown in Fig. 6.32(b). It follows directly from Fig. 6.32(b) and the behavior of dy/dx throughout the phase plane (or from isoclines), because we have seen that

$$\frac{dy}{dx} = \frac{k^2 x(x-2)}{y}.$$

For example, $dy/dx > 0$ when $x > 2$ or $x < 0$ and $y > 0$, with corresponding results elsewhere in the phase plane.

In fact this differential equation is integrable, and the exact equation of the trajectories is

$$\int y \, dy = \int k^2 x(x-2) \, dx,$$

or

$$\tfrac{1}{2} y^2 = \frac{k^2 x^2}{3}(x-3) + c.$$

The integration constant c is the parameter which determines individual trajectories in Fig. 6.32(b). ∎

(a)

(b)

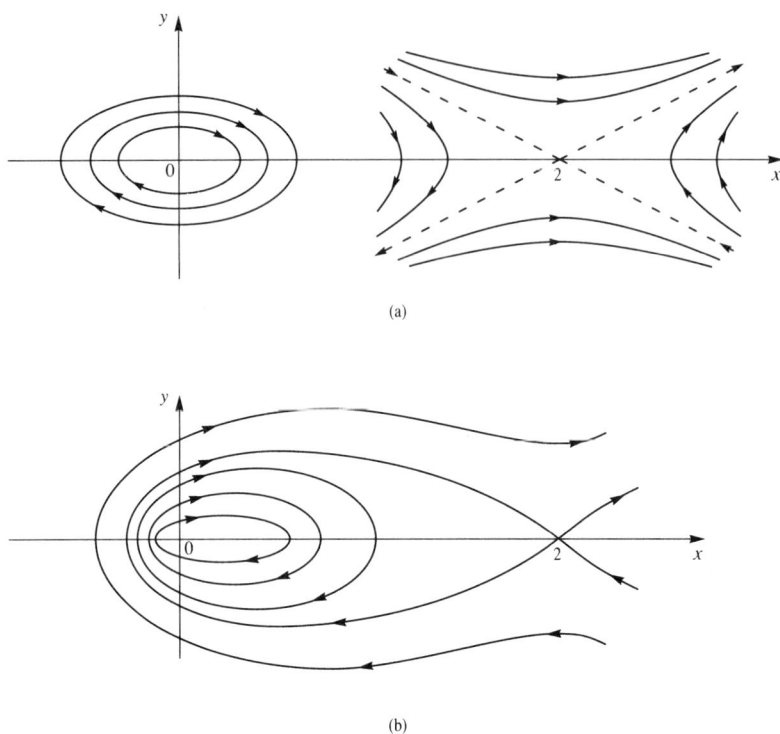

Fig. 6.32 (a) Local pattern of trajectories, (b) Full phase portrait

Sometimes the phase plane approach provides a simple representation for solutions of linear autonomous differential equations containing terms which are piecewise continuous. An example of this type is provided by a simple oscillating mass-spring system resting on a rough horizontal table, to which one end of the spring is fixed. The restoring force produced by the spring is assumed to obey Hooke's law, and the resistance is considered to be due to Coulomb friction.

The type of friction called **Coulomb friction**, or **sliding friction**, is a constant frictional force of magnitude $F_0 > 0$, independent of the velocity of the sliding mass, which always acts to oppose motion. When the mass is at rest, the Coulomb frictional force equals and opposes a force applied to induce motion until the magnitude of that force exceeds F_0, and thereafter the magnitude of the Coulomb frictional force remains equal to F_0. Thus in this mass-spring system the oscillatory mass may come to rest with the spring displaced from its natural length, provided the displacement is not sufficient to produce a restoring force with magnitude greater than F_0. If x is the displacement and $\dot{x}$ the velocity of the mass m at the end of the spring, the Coulomb frictional force F_c is modeled by

$$F_c = -F_0 \, \mathrm{sgn}(\dot{x}),$$

where sgn is the **signum function** defined as

$$sgn(\dot{x}) = \begin{cases} 1 & \text{when } \dot{x} > 0 \\ -1 & \text{when } \dot{x} < 0. \end{cases}$$

The equation of motion for the system thus becomes

$$m\ddot{x} = -\kappa x - F_0 \, sgn(\dot{x}),$$

where $\kappa > 0$ is the spring constant.

The following example uses the phase plane representation for solutions of this autonomous differential equation to determine the number of oscillations performed by the system before coming to rest, and the eventual rest position of the mass itself.

Example 5. Oscillations with Coulomb friction–a phase plane approach

A mass-spring system with Coulomb friction is governed by the piecewise continuous autonomous differential equation

$$\ddot{x} = -k^2 x - F \, sgn(\dot{x}).$$

Use the phase plane to examine the motion of the system, given that the initial conditions are $x(0) = x_0$, $\dot{x}(0) = y_0$.

Solution

Proceeding as usual and setting $y = dx/dt$, so that

$$\frac{d^2 x}{dt^2} = y \frac{dy}{dx},$$

the equation of motion becomes

$$y \frac{dy}{dx} = -k^2 x - F \qquad \text{for } y > 0,$$

and

$$y \frac{dy}{dx} = -k^2 x + F \qquad \text{for } y < 0,$$

Integration of these two first order nonlinear differential equations then gives

$$k^2 x^2 + y^2 + 2Fx = c_1 \qquad \text{for } y > 0,$$

and

$$k^2 x^2 + y^2 - 2Fx = c_2 \qquad \text{for } y < 0.$$

The trajectories represented by the first equation are semi-ellipses in the upper half of the (x, y)-phase plane, and those represented by the second equation are semi-ellipses in the lower half of the phase plane. The subsequent discussion will be simplified if we now set $X = kx$ and work in the (X, y)-phase plane, for then the semi-elliptic trajectories become the semicircles

$$X^2 + y^2 + 2(F/k)X = c_1 \qquad \text{for } y > 0,$$

and

$$X^2 + y^2 - 2(F/k)X = c_2 \qquad \text{for } y < 0.$$

In the upper half of the (X, y)-phase plane these are seen to be semicircles centered on $(-(F/k), 0)$ with radius $\sqrt{(F/k)^2 + c_1}$, while in the lower half of the phase plane they are semicircles centered on $((F/k), 0)$ with radius $\sqrt{(F/k)^2 + c_2}$. Thus the motion of the system will be represented by a trajectory described clockwise in the (X, y)-phase plane, starting from the point $(x_0/k, y_0)$ in that plane, and pieced together from the semicircles just described. The trajectory must be continuous across the X-axis, as the mass does not jump to a different position each time it comes to rest, and the trajectory will terminate on the X-axis when it reaches a point in the interval $(-F/k, F/k)$. This corresponds to the magnitude of the restoring force in the spring finally becoming less than the Coulomb frictional force.

Figure 6.33 shows the trajectory constructed in this manner when the initial point in the (X, y)-phase plane is $(7, 0)$, $F = 3$ and $k = 2$. Motion is seen to terminate after one cycle when the trajectory reaches the point $(1, 0)$ in the (X, y)-phase plane. As $X = kx$ and $k = 2$, it follows that the initial position in the (x, y)-phase plane is $(7/2, 0)$, while the equilibrium position is $(\frac{1}{2}, 0)$. The fact that $x = \frac{1}{2} > 0$ in the equilibrium position shows the spring comes to rest in tension.

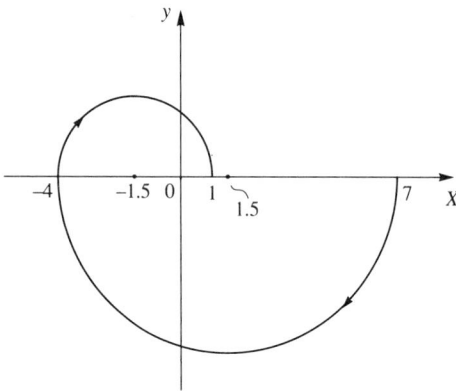

Fig. 6.33 Phase portrait for mass-spring system with Coulomb friction when $F = 3$, $k = 2$, $x_0 = 7/2$ and $y_0 = 0$. Centers of semicircles located at $(-3/2, 0)$, $(3/2, 0)$

A theorem which offers a little better understanding of when linearization provides accurate information about the local behavior of the nonlinear system from which it is derived will be given later, after a brief discussion of the system in (2). In the meantime we quote two theorems of direct relevance to the matters just discussed, the proofs of which are beyond this first account of the subject.

The first is the **Routh–Hurwitz test** for the asymptotic stability of the general solution of

$$\frac{d^n x}{dt^n} + a_1 \frac{d^{n-1} x}{dt^{n-1}} + a_2 \frac{d^{n-2} x}{dt^{n-2}} + \ldots + a_n x = 0, \tag{29}$$

in which the coefficients $a_1, a_2, \ldots, a_n$ are real constants.

In this context, the **asymptotic stability** of the solution of an nth order linear constant coefficient differential equation means that the solution decays to zero exponentially as $t \to \infty$. This property may be expressed differently as follows. Let the characteristic polynomial of (29) be

$$P(\lambda) \equiv \lambda^n + a_1 \lambda^{n-1} + a_2 \lambda^{n-2} + \ldots + a_n.$$

Then if $\lambda_1, \lambda_2, \ldots, \lambda_n$ are the zeros of $P(\lambda)$, the general solution of (29)

$$x(t) = c_1 \exp(\lambda_1 t) + c_2 \exp(\lambda_2 t) + \ldots + c_n \exp(\lambda_n t) \tag{30}$$

will be asymptotically stable if

$$\operatorname{Re} \lambda_i < 0, \qquad i = 1, 2, \ldots, n.$$

Theorem 6.5 (Routh–Hurwitz test for asymptotic stability)

Form the n numbers

$$\Delta_1 = a_1, \quad \Delta_2 = \begin{vmatrix} a_1 & 1 \\ a_3 & a_2 \end{vmatrix} \quad \Delta_3 = \begin{vmatrix} a_1 & 1 & 0 \\ a_3 & a_2 & a_1 \\ a_5 & a_4 & a_3 \end{vmatrix},$$

$$\ldots \Delta_n = \begin{vmatrix} a_1 & 1 & 0 & 0 & 0 & \ldots & 0 & 0 \\ a_3 & a_2 & a_1 & 1 & 0 & \ldots & 0 & 0 \\ a_5 & a_4 & a_3 & a_2 & a_1 & \ldots & 0 & 0 \\ . & . & . & . & . & . & . & . \\ a_{2n-1} & a_{2n-2} & a_{2n-3} & a_{2n-4} & a_{2n-5} & \ldots & a_{n+1} & a_n \end{vmatrix},$$

and set $a_r = 0$ for $r > n$. Then a necessary and sufficient condition for the zeros of $P(\lambda)$ to have negative real parts, and thus for the general solution (30) of the differential equation (29) to be asymptotically stable, is that

$$\Delta_i > 0, \qquad i = 1, 2, \ldots, n. \qquad \square$$

It is instructive to compare the implications of the Routh–Hurwitz test for the case $n = 2$, with those of Fig. 6.31. Identifying (28) with (29) when $n = 2$ shows $a_1 = -\beta$,

$a_2 = -\alpha$, so that

$$\Delta_1 = -\beta, \quad \Delta_2 = \begin{vmatrix} -\beta & 1 \\ 0 & -\alpha \end{vmatrix} = \alpha\beta.$$

The asymptotic stability conditions $\Delta_1 > 0, \Delta_2 > 0$ imply the inequalities $\alpha < 0, \beta < 0$. These last two conditions on α and β define the interior of the third quadrant in Fig. 6.31 as the region for asymptotic stability, which is in agreement with the behavior described in that figure. The test does not distinguish between the two different types of stability which occur in the third quadrant. Although the test provides no new information when $n = 2$, it is often extremely useful when $n > 2$.

It should be noticed that the Routh–Hurwitz test provides no information about the behavior of the general solution of (30) of (29) if any of the Δ_i vanish or are negative. This is illustrated by the results of Examples 2 and 3. The solution in Ex. 2 is purely oscillatory and thus is stable, though not asymptotically stable, while the solution in Ex. 3 is unstable, yet in both cases $\Delta_1 = \Delta_2 = 0$.

The other result of interest here is the **Liénard–Levinson–Smith** theorem, which provides information about when a nonlinear second order differential equation possesses a limit cycle.

Theorem 6.6 (Liénard–Levinson–Smith theorem on limit cycles)

Consider the differential equation

$$\frac{d^2 x}{dt^2} + p(x)\frac{dx}{dt} + q(x) = 0, \tag{31}$$

and define the associated functions $P(x) = \displaystyle\int_0^x p(u)\,du$ and $Q(x) = \displaystyle\int_0^x q(u)\,du$. Let $p(x)$, $q(x)$, $P(x)$ and $Q(x)$ have the following properties:

1 $p(x)$ is an even continuous function for all x;
2 $q(x)$ is an odd function with a continuous derivative for all x, and is such that $q(x) > 0$ for $x > 0$;
3 $Q(x) \to +\infty$ as $x \to +\infty$;
4 $P(x)$ has precisely one positive zero $x = x_0$, and is such that $P(x) < 0$ for $0 < x < x_0$, $P(x) > 0$ for $x > x_0$ and $P(x) \to +\infty$ as $x \to +\infty$.

Then differential equation (31) has a unique stable limit cycle in the phase plane.

$\square$

This theorem means that any differential equation which satisfies its conditions will have a periodic solution, and that the period of the solution will be uniquely determined. It means, in addition, that initial conditions not on this unique limit cycle will be associated with trajectories which converge to it as $t \to +\infty$.

Example 6. Uniqueness of limit cycle for the van der Pol equation

Prove the uniqueness of the limit cycle of the van der Pol equation

$$\frac{d^2 x}{dt^2} + \varepsilon(x^2 - 1)\frac{dx}{dt} + x = 0, \qquad \varepsilon \geq 0.$$

Solution

In the notation of Theorem 6.6, $p(x) = \varepsilon(x^2 - 1)$, $q(x) = x$, so

$$P(x) = \int_0^x \varepsilon(u^2 - 1)\,du = \varepsilon x\left(\frac{x^2}{3} - 1\right),$$

and

$$Q(x) = \int_0^x u\,du = \frac{x^2}{2}.$$

Inspection shows conditions (1), (2) and (3) of Theorem 6.6 to be satisfied. $P(x)$ has the single positive zero $x_0 = \sqrt{3}$, so as $P(x) < 0$ for $0 < x < \sqrt{3}$ and $P(x) > 0$ for $x > \sqrt{3}$, while $P(x) \to +\infty$ as $x \to +\infty$, it follows that condition (4) is also satisfied. All the conditions of Theorem 6.6 are satisfied, so the limit cycle of the van der Pol equation is unique. As the van der Pol equation only contains the single parameter ε, different values of ε will correspond to different unique limit cycles, each with a different period.

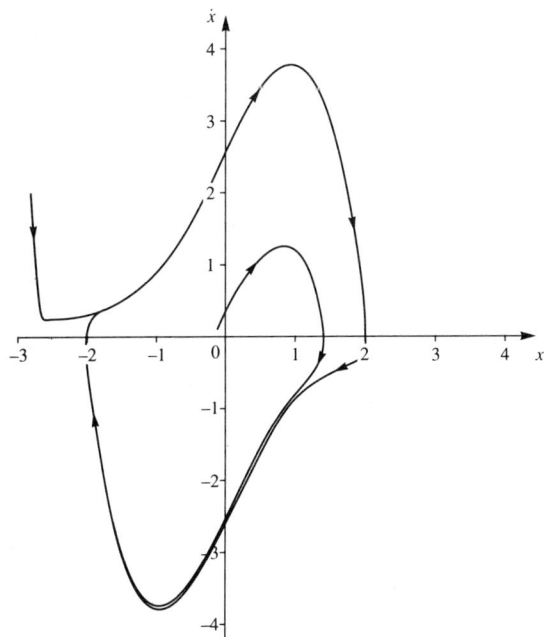

Fig. 6.34 Limit cycle and convergent trajectories for the van der Pol equation when $\varepsilon = 2$

The change in the limit cycle due to a change in ε, and the convergence to the limit cycle of trajectories with initial conditions inside and outside the limit cycle, may be seen by comparing Fig. 6.21 in which $\varepsilon = 1$, and Fig. 6.34 in which $\varepsilon = 2$.

■

First order systems – the generalized phase plane

We now turn our attention to autonomous systems of the form given in (2), that is, to systems like

$$\frac{dx}{dt} = X(x, y)$$

$$\frac{dy}{dt} = Y(x, y). \tag{32}$$

In such systems $y \neq dx/dt$ (in general), so the (x, y)-plane in which trajectories will be drawn is no longer the phase plane used for second order equations. This (x, y)-plane will be referred to as the **generalized phase plane**, and it will only reduce to the ordinary phase plane when a second order equation is expressed in the form of system (32) by setting $y = dx/dt$.

The direction along a trajectory which corresponds to increasing time will again be indicated by an arrow. However now, because $y \neq dx/dt$, it will no longer follow that time will increase in the positive x-direction when $y > 0$ and in the negative x-direction when $y < 0$, as it does in the ordinary phase plane.

A result which is essential for what is to follow, but which will not be proved, is the basic existence and uniqueness theorem for system (32). This asserts that if X, Y in (32), together with their first order partial derivatives with respect to x and y are continuous in a region of the (x, y)-plane containing the point (x_0, y_0), there exists a unique solution $x(t)$, $y(t)$ of (32) in that region such that $x(t_0) = x_0$ and $y(t_0) = y_0$.

The curve traced in the (x, y)-plane by the point $(x(t), y(t))$ as t increases is the trajectory of interest to us, and the time $t = t_0$ is the initial time at which this trajectory originates from the point (x_0, y_0).

The time may always be eliminated from an autonomous system like (32) by dividing the two equations to obtain the single first order equation

$$\frac{dy}{dx} = \frac{Y(x, y)}{X(x, y)}. \tag{33}$$

The **equilibrium points (critical points)** of (32) (equivalently (33)) occur when

$$X(x, y) = 0 \quad \text{and} \quad Y(x, y) = 0, \tag{34}$$

because then the direction field associated with (33) by dy/dx becomes indeterminate.

We now use system (32) to prove an important property of autonomous systems. This result includes the second order equation (1), as it may always be rewritten as a system.

Suppose $x(t)$, $y(t)$ are solutions of (32) for $t_0 \le t \le t_1$, and let a dot denote differentiation with respect to t. Then for any real α, if $\chi(t) = x(t + \alpha)$, $\eta(t) = y(t + \alpha)$, it follows from the chain rule that

$$\dot{\chi}(t) = \dot{x}(t + \alpha), \qquad \dot{\eta}(t) = \dot{y}(t + \alpha).$$

Thus

$$\dot{\chi}(t) = X[x(t + \alpha), y(t + \alpha)] = X(\chi, \eta)$$

and

$$\dot{\eta}(t) = Y[x(t + \alpha), y(t + \alpha)] = Y(\chi, \eta),$$

for $t_0 - \alpha \le t \le t_1 - \alpha$, so that $\chi(t)$ and $\eta(t)$ are also solutions of system (32).

This shows that the solution $\chi(t)$, $\eta(t)$ follows from $x(t)$, $y(t)$ by translating these solutions in the positive t-direction by an amount α. The implication of this result is that of the two arbitrary constants (parameters) entering into the general solution of the autonomous system (32), one identifies a particular trajectory while the other merely shifts the time origin along it.

When applied to Theorem 6.6, this result means that although periodic solutions corresponding to a limit cycle are not unique (though the period of the limit cycle is unique), one periodic solution can always be obtained from another by a shift of the time origin. That is to say, the different periodic solutions simply start at different points in the limit cycle.

The only discussion of system (32) to be offered here will be on the basis of the linearization of $X(x, y)$ and $Y(x, y)$ about any one of its equilibrium points (x_0, y_0). As $X(x_0, y_0) = Y(x_0, y_0) = 0$, it follows that the Taylor polynomials of degree one used to approximate $X(x, y)$ and $Y(x, y)$ close to (x_0, y_0) will be

$$X(x, y) = a(x - x_0) + b(y - y_0)$$
$$Y(x, y) = c(x - x_0) + d(y - y_0),$$

where

$$a = \left(\frac{\partial X}{\partial x} \right)_{(x_0, y_0)}, \quad b = \left(\frac{\partial X}{\partial y} \right)_{(x_0, y_0)}, \quad c = \left(\frac{\partial Y}{\partial x} \right)_{(x_0, y_0)} \quad \text{and} \quad d = \left(\frac{\partial Y}{\partial y} \right)_{(x_0, y_0)}.$$

Thus by shifting the origin to (x_0, y_0) by writing $\bar{x} = x - x_0$, $\bar{y} = y - y_0$, and then dropping the bars, it will be sufficient to consider the phase portraits associated with the approximating linear system

$$\frac{dx}{dt} = ax + by$$

$$\frac{dy}{dt} = cx + dy, \tag{35}$$

which has its equilibrium point at the origin.

In Sec. 6.1 we have already seen how to solve system (35) when it is written in the matrix form

$$\dot{\mathbf{x}} = \mathbf{A}\mathbf{x}, \tag{36}$$

where

$$\mathbf{x} = \begin{bmatrix} x \\ y \end{bmatrix} \quad \text{and} \quad \mathbf{A} = \begin{bmatrix} a & b \\ c & d \end{bmatrix}. \tag{37}$$

Henceforth we shall assume $\det \mathbf{A} \neq 0$, for if this is not true the two equations in (35) will be linearly dependent.

The solutions $x(t)$, $y(t)$ of (36) have been shown in Sec. 6.1 to depend on the eigenvalues λ_1, λ_2 of $\mathbf{A}$, and although the elements of $\mathbf{A}$ are real, its eigenvalues and eigenvectors may be real or complex. In what is to follow it will be helpful to express the eigenvalues in terms of two real numbers derived from the matrix $\mathbf{A}$. This may be accomplished by noticing that the characteristic equation of $\mathbf{A}$

$$\det(\mathbf{A} - \lambda\mathbf{I}) = \lambda^2 - (a+d)\lambda + (ad - bc),$$

may be written in the form

$$\det(\mathbf{A} - \lambda\mathbf{I}) = \lambda^2(\operatorname{tr}\mathbf{A})\lambda + \det\mathbf{A}, \tag{38}$$

where

$$\operatorname{tr}\mathbf{A} = a + d \quad \text{and} \quad \det\mathbf{A} = ad - bc.$$

The eigenvalues of $\mathbf{A}$ are the zeros of (38) so, in terms of $\operatorname{tr}\mathbf{A}$ and $\det\mathbf{A}$, we have

$$\lambda_1 = \tfrac{1}{2}[\operatorname{tr}\mathbf{A} - \sqrt{(\operatorname{tr}\mathbf{A})^2 - 4\det\mathbf{A}}], \quad \lambda_2 = \tfrac{1}{2}[\operatorname{tr}\mathbf{A} + \sqrt{(\operatorname{tr}\mathbf{A})^2 - 4\det\mathbf{A}}] \tag{39}$$

Using the results of Sec. 6.1, it is a straightforward matter to arrive at the phase portrait associated with the equilibrium point of (36) at the origin in the generalized phase plane in terms of $\operatorname{tr}\mathbf{A}$ and $\det\mathbf{A}$. We shall see that, as with second order equations, there are only four essentially different types of equilibrium point; namely, centers, nodes, spirals and saddle points. However, whereas second order equations only have improper nodes, at which all trajectories become tangent to one of two straight lines through the equilibrium point (these may coincide), a system may also have a proper node at which trajectories enter or leave along all possible directions.

To illustrate how the nature of equilibrium points may be identified, let us suppose eigenvalues λ_1, λ_2 of $\mathbf{A}$ in (36) to be real. If the eigenvalues are distinct, there will be two linearly independent eigenvectors $\mathbf{a}_1$, $\mathbf{a}_2$ to which they correspond, and the solution will be

$$\mathbf{x} = c_1\mathbf{a}_1\exp(\lambda_1 t) + c_2\mathbf{a}_2\exp(\lambda_2 t)$$

Writing $\mathbf{a}_1 = [a_{11}, a_{12}]^{\mathsf{T}}$, $\mathbf{a}_2 = [a_{21}, a_{22}]^{\mathsf{T}}$, and recalling that $\mathbf{x} = [x, y]^{\mathsf{T}}$, it follows that when $c_2 = 0$, $x = c_1 a_{11}\exp(\lambda_1 t)$ and $y = c_1 a_{21}\exp(\lambda_1 t)$. Thus there is a degenerate straight line trajectory $y = (a_{12}/a_{11})x$ through the origin of the generalized phase plane. Displacements along the trajectory are seen to be obtained by regarding $\mathbf{a}_1$ as a vector with components (a_{11}, a_{12}) in the generalized phase plane, and then scaling it by the quantity $c_1 \exp(\lambda_1 t)$.

Similarly, when $c_1 = 0$ there is another degenerate straight line trajectory $y = (a_{22}/a_{21})x$ through the origin. Displacements along this trajectory correspond to scaling the vector $\mathbf{a}_2$ by the quantity $c_2 \exp(\lambda_2 t)$. The two degenerate trajectories are distinct, because $\mathbf{a}_1$ and $\mathbf{a}_2$ are linearly independent.

It is not difficult to see that when the eigenvalues both have the same sign, the equilibrium point will be an **improper node** with an exceptional direction. This improper node will be stable and attractive when the eigenvalues are negative, and unstable when they are positive. If the eigenvalues have opposite signs the equilibrium point will be a **saddle point** with the directions determined by $\mathbf{a}_1$ and $\mathbf{a}_2$ as its asymptotes (separatrices).

If the eigenvalues of $\mathbf{A}$ are real and equal there may either be two linearly independent eigenvectors corresponding to the double eigenvalue, or merely a single eigenvector. It follows from the previous argument that when two eigenvectors exist the equilibrium point is an improper node with an exceptional direction, but when there is only one eigenvector the equilibrium point is an improper node with no exceptional direction.

The case of a **proper node** arises when $\mathbf{A}$ is diagonal and of the form

$$\mathbf{A} = \begin{bmatrix} \lambda_0 & 0 \\ 0 & \lambda_0 \end{bmatrix},$$

with λ_0 real. Then $\mathbf{A}$ has a double real eigenvalue λ_0 and two linearly independent eigenvectors

$$\mathbf{a}_1 = \begin{bmatrix} 1 \\ 0 \end{bmatrix}, \quad \mathbf{a}_2 = \begin{bmatrix} 0 \\ 1 \end{bmatrix},$$

so the solution becomes

$$\mathbf{x} = c_1 \mathbf{a}_1 \exp(\lambda_0 t) + c_2 \mathbf{a}_2 \exp(\lambda_0 t).$$

Thus $x = c_1 \exp(\lambda_0 t)$, $y = c_2 \exp(\lambda_0 t)$ and the trajectories are given by

$$y = (c_2/c_1)x,$$

which describes a star-like pattern of radial trajectories passing through the origin. This proper node will be stable and attractive when $\lambda_0 < 0$, and unstable when $\lambda_0 > 0$.

The identification of spiral points and centers follows in similar fashion by considering the case in which the eigenvalues are complex. **Spirals** arise when the complex eigenvalues have nonzero real parts, and **centers** when they are purely imaginary.

These results are summarized in Fig. 6.35 in terms of the parameters tr $\mathbf{A}$ and det $\mathbf{A}$. The individual cases are described in greater detail in Table 6.1, with the corresponding phase portraits (Figs 6.36(a)–(e)).

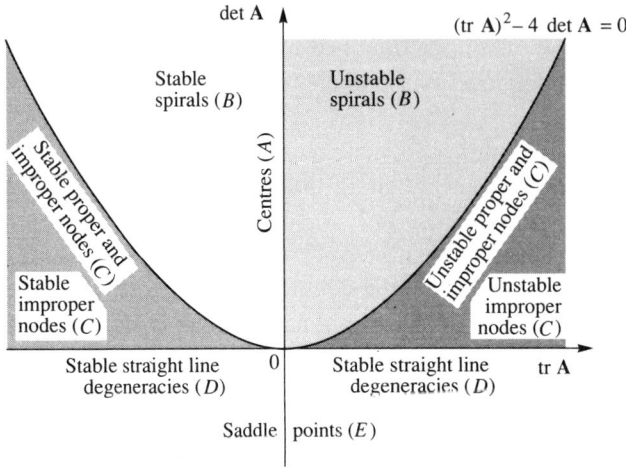

Fig. 6.35 Dependence of equilibrium points of $\dot{\mathbf{x}} = \mathbf{A}\mathbf{x}$ on tr $\mathbf{A}$ and det $\mathbf{A}$

Table 6.1 Types of phase portrait associated with $\dot{\mathbf{x}} = \mathbf{A}\mathbf{x}$ at $(0,0)$

(A) λ_1, λ_2 purely imaginary; tr $\mathbf{A} = 0$, det $\mathbf{A} > 0$; a center.

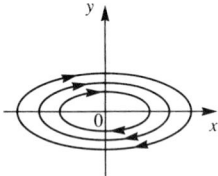

Fig. 6.36a

(B) λ_1, λ_2 complex conjugates with nonzero real parts; a spiral.
$(\text{tr }\mathbf{A})^2 - 4\det \mathbf{A} < 0, \qquad \text{tr }\mathbf{A} \neq 0.$

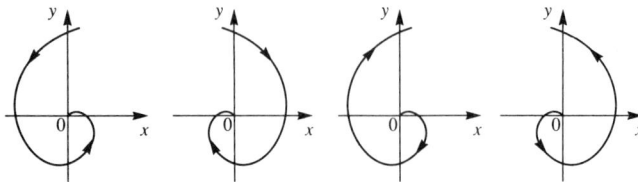

Stable attractive spirals   Unstable spirals
Re $\lambda_1 < 0$, Re $\lambda_2 < 0$ Re $\lambda_1 > 0$, Re $\lambda_2 > 0$

Fig. 6.36b

Table 6.1 (*continued*)

(C) (i) $\lambda_1 = \lambda_2$ (real); $b = c = 0$, $(\operatorname{tr}\mathbf{A})^2 - 4\det\mathbf{A} = 0$, $\det\mathbf{A} > 0$; proper node:

(ii) $\lambda_1 = \lambda_2$ (real); $b \neq 0, c \neq 0$, $(\operatorname{tr}\mathbf{A})^2 - 4\det\mathbf{A} = 0$, $\det\mathbf{A} > 0$; improper node with an exceptional direction if $\mathbf{A}$ has only one eigenvalue and two linearly independent eigenvectors, and an improper node without an exceptional direction if it has only one eigenvector.

λ_1, λ_2 real, distinct and of the same sign; $(\operatorname{tr}\mathbf{A})^2 - 4\det\mathbf{A} > 0$, $\det\mathbf{A} > 0$; improper node with an exceptional direction.

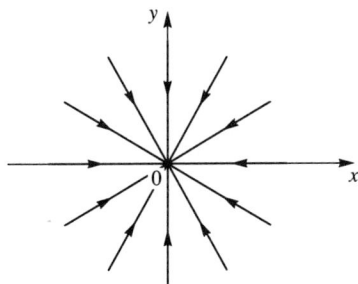

Stable attractive proper node
$\operatorname{Re}\lambda_1 = \operatorname{Re}\lambda_2 < 0$

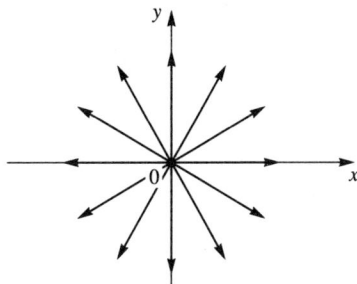

Unstable proper node
$\operatorname{Re}\lambda_1 = \operatorname{Re}\lambda_2 > 0$

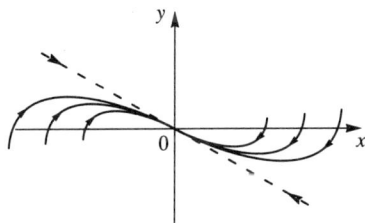

Stable attractive improper
node without an exceptional
direction

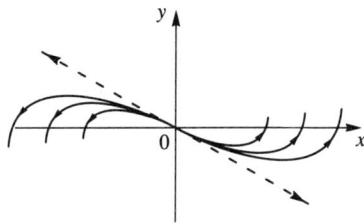

Unstable improper node without an
exceptional direction
$\operatorname{Re}\lambda_1 > 0, \quad \operatorname{Re}\lambda_2 > 0$

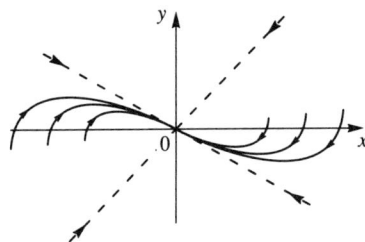

Stable attractive improper node
with an exceptional direction
$\operatorname{Re}\lambda_1 < 0, \quad \operatorname{Re}\lambda_2 < 0$

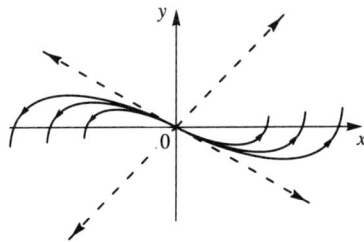

Unstable improper node with an
exceptional direction
$\operatorname{Re}\lambda_1 > 0, \quad \operatorname{Re}\lambda_2 > 0$

Fig. 6.36c

Table 6.1 (*continued*)

(D) $\lambda_1 \neq 0$ (real), $\lambda_2 = 0$; det $\mathbf{A} = 0$; parallel straight line degeneracies.

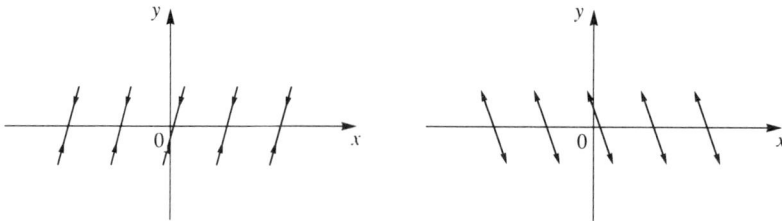

Parallel straight line degeneracies with stable equilibrium points on x-axis Re $\lambda_1 < 0$

Parallel straight line degeneracies with unstable equilibrium points on x-axis with Re $\lambda_1 > 0$

Fig. 6.36d

(E) λ_1, λ_2 real with opposite signs; det $\mathbf{A} < 0$; saddle points

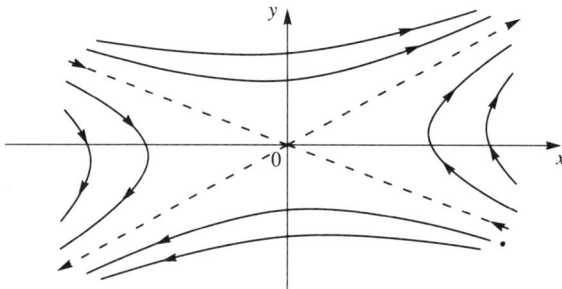

Saddle point (unstable)

Fig. 6.36e

Example 7. A typical linear system

Locate and identify the nature of the equilibrium point of the system

$$\frac{dx}{dt} = x + 6y + 11, \quad \frac{dy}{dt} = 5x + 2y - 1.$$

Solution

Identifying this system with (32) gives $X(x, y) = x + 6y + 11$ and $Y(x, y) = 5x + 2y - 1$. The equilibrium point of the system is given by solving the simultaneous equations $X(x, y) = 0$, $Y(x, y) = 0$, which have as their solution $x = 1$, $y = -2$. Thus the equilibrium point in the generalized (x, y)-phase plane is $(1, -2)$.

The standard form considered in (35) has the equilibrium point located at the origin, so to reduce the present problem to that form it is necessary to shift the origin in the generalized (x, y)-phase plane to the point $(1, -2)$. This is accomplished by means of the linear transformation $\bar{x} = x - 1$, $\bar{y} = y + 2$, and the result is the system

$$\frac{d\bar{x}}{dt} = \bar{x} + 6\bar{y}, \quad \frac{d\bar{y}}{dt} = 5\bar{x} + 2\bar{y}.$$

The generalized phase plane is now the $(\bar{x}, \bar{y})$-plane, so we must now consider the system

$$\dot{\mathbf{x}} = \mathbf{A}\mathbf{x}, \quad \text{where} \quad \mathbf{x} = \begin{bmatrix} \bar{x} \\ \bar{y} \end{bmatrix} \quad \text{and} \quad \mathbf{A} = \begin{bmatrix} 1 & 6 \\ 5 & 2 \end{bmatrix}.$$

The eigenvalues of $\mathbf{A}$ are 7 and -4, so it follows directly from (E) in Table 6.1 that $(1, -2)$ is a saddle point. ∎

Example 8. Phase portrait of a linearized system

Locate the equilibrium points of the system

$$\frac{dx}{dt} = x - x^2 + 2y + 3, \quad \frac{dy}{dt} = -x + 2y.$$

Linearize the system about each of these points and identify the nature of the equilibrium point belonging to each such system.

Solution

Identifying the system with (32) gives $X(x, y) = x - x^2 + 2y + 3$ and $Y(x, y) = -x + 2y$. Solving these equations simultaneously to find the equilibrium points of the nonlinear system shows them to be $(-1, -\frac{1}{2})$ and $(3, \frac{3}{2})$.

Using Taylor polynomials of degree one to approximate $X(x, y)$ and $Y(x, y)$ close to these points it follows from (33) and (34) that in the neighborhood of $(-1, -\frac{1}{2})$

$$X = 3(x+1) + 2(y+\tfrac{1}{2}), \quad Y = -(x+1) + 2(y+\tfrac{1}{2}).$$

Thus the linearized system of equations close to $(-1, -\frac{1}{2})$ is

$$\frac{dx}{dt} = 3(x+1) + 2(y+\tfrac{1}{2}), \quad \frac{dy}{dt} = -(x+1) + 2(y+\tfrac{1}{2}). \tag{40}$$

Similarly, in the neighborhood of $(3, \frac{3}{2})$,

$$X = -5(x-3) + 2(y-\tfrac{3}{2}), \quad Y = -(x-3) + 2(y-\tfrac{3}{2}),$$

so there the linearized system of equations is

$$\frac{dx}{dt} = -5(x-3) + 2(y-\tfrac{3}{2}), \quad \frac{dy}{dt} = -(x-3) + 2(y-\tfrac{3}{2}). \tag{41}$$

To bring system (40) into the standard form (35) we set

$\bar{x} = x + 1$, $\bar{y} = y + \frac{1}{2}$ and obtain

$$\dot{\mathbf{x}} = \mathbf{A}\mathbf{x}, \quad \text{where} \quad \mathbf{x} = \begin{bmatrix} \bar{x} \\ \bar{y} \end{bmatrix} \quad \text{and} \quad \mathbf{A} = \begin{bmatrix} 3 & 2 \\ -1 & 2 \end{bmatrix}.$$

The eigenvalues of $\mathbf{A}$ are

$$\lambda_1 = \tfrac{1}{2}[5 + i\sqrt{7}], \quad \lambda_2 = \tfrac{1}{2}[5 - i\sqrt{7}],$$

so it follows from (B) in Table 6.1 that $(-1, -\frac{1}{2})$ is an unstable spiral point of (40).

To bring system (41) into standard form we set $\bar{x} = x - 3$, $\bar{y} = y - \frac{3}{2}$ to obtain

$$\dot{\mathbf{x}} = \mathbf{A}\mathbf{x}, \quad \text{where} \quad \mathbf{x} = \begin{bmatrix} \bar{x} \\ \bar{y} \end{bmatrix} \quad \text{and} \quad \mathbf{A} = \begin{bmatrix} -5 & 2 \\ 1 & -2 \end{bmatrix}.$$

The eigenvalues of $\mathbf{A}$ are now

$$\lambda_1 = \tfrac{1}{2}[3 + \sqrt{41}], \quad \lambda_2 = \tfrac{1}{2}[3 - \sqrt{41}],$$

so it follows from (E) in Table 6.1 that $(3, \frac{3}{2})$ is a saddle point of (41).

Thus linearization of the original nonlinear system indicates that it can be expected to possess an unstable spiral point at $(-1, -\frac{1}{2})$ and a saddle point at $(3, \frac{3}{2})$. The justification for these conclusions is to be found in the next theorem which we quote but do not prove. ∎

Theorem 6.7 (Linearization theorem)

Consider a nonlinear system of the form

$$\frac{dx}{dt} = ax + by + P(x, y)$$

$$\frac{dy}{dt} = cx + dy + Q(x, y), \tag{42}$$

in which $ad - bc \neq 0$, $P(0, 0) = Q(0, 0) = 0$ and $(0, 0)$ is an isolated equilibrium point. Let $P(x, y)$, $Q(x, y)$ be continuous functions with continuous first order partial derivatives, and suppose also that

$$\lim_{(x, y) \to (0, 0)} \left[\frac{P(x, y)}{\sqrt{x^2 + y^2}} \right] = 0 \quad \text{and} \quad \lim_{(x, y) \to (0, 0)} \left[\frac{Q(x, y)}{\sqrt{x^2 + y^2}} \right] = 0.$$

Then, with only one exception, the nature of the equilibrium point $(0, 0)$ of the nonlinear system (42) will be the same as that of the linearized system

$$\frac{dx}{dt} = ax + by$$

$$\frac{dy}{dt} = cx + dy \tag{43}$$

at (0, 0), the properties of which are listed in Table 6.1. The exception arises when (43) has a proper node at (0, 0), for then system (42) may either have a node or a spiral point at (0, 0). □

Example 9. Application of linearization theorem

Apply Theorem 6.7 to Ex. 8.

Solution

Theorem 6.7 refers to equilibrium points located at the origin. Thus to apply the theorem to Ex. 8 the origin must be shifted to each of the equilibrium points in turn; that is first to $(-1, -\frac{1}{2})$ and then to $(3, \frac{3}{2})$. Application of the theorem in each case will then enable the nature of the equilibrium points of the nonlinear system to be identified.

The shift of the origin to $(-1, -\frac{1}{2})$ is accomplished by setting $x = \bar{x} - 1$, $y = \bar{y} - \frac{1}{2}$ in the full nonlinear system, to obtain

$$\frac{d\bar{x}}{dt} = 3\bar{x} + 2\bar{y} - \bar{x}^2$$

$$\frac{d\bar{y}}{dt} = -\bar{x} + 2\bar{y}.$$

Comparing this system with (42) in Theorem 6.7, and for convenience omitting the bars, shows that

$$P(x, y) = -x^2, \qquad Q(x, y) \equiv 0.$$

Clearly, the properties of Theorem 6.7 are all satisfied, so the nonlinear system

$$\frac{dx}{dt} = 3x + 2y - x^2$$

$$\frac{dy}{dt} = -x + 2y$$

and the linearized system

$$\frac{dx}{dt} = 3x + 2y$$

$$\frac{dy}{dt} = -x + 2y$$

must both have the same type of equilibrium point at (0, 0); namely an unstable spiral point. This proves that $(-1, -\frac{1}{2})$ is an unstable spiral point of the original nonlinear system and a similar argument shows that $(3, \frac{3}{2})$ is a saddle point of the system. ■

The final theorem which we merely quote is the **Poincaré–Bendixon Theorem**. This theorem gives *sufficient* conditions for the existence of a limit cycle of a nonlinear system.

Theorem 6.8 (Poincaré–Bendixon Theorem)

Consider the autonomous system

$$\frac{dx}{dt} = X(x, y), \qquad \frac{dy}{dt} = Y(x, y). \tag{44}$$

Let R be an annular region in the phase plane whose outer boundary C_1 and inner boundary C_2 are simple closed curves (no loops).
Suppose that
 (i) neither C_1 nor C_2 is a limit cycle of (44);
 (ii) R contains no limit points of (44);
 (iii) all trajectories of (44) crossing either of the boundaries C_1 or C_2 enter R as $t \to +\infty$.
   Then if Γ is a trajectory of (44) lying entirely within R for $t > t_0$, either Γ is a limit cycle of (44) or it spirals into a limit cycle. $\square$

This theorem is illustrated in Fig. 6.37, in which the limit cycle is the curve L. It is clear that this theorem implies the existence of at least one limit point of (44) inside C_2. The theorem only ensures the existence of a limit cycle, and not its uniqueness.

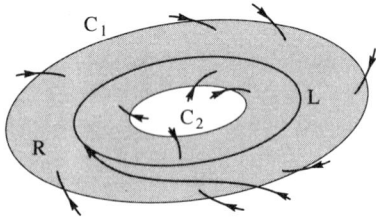

Fig. 6.37 Illustration of Poincaré–Bendixon theorem

A major difficulty when applying this theorem is to find a suitable region R. If an annular region of the form $a \le r \le b$ can be used, with its outer boundary C_1 the circle $r = b$, and its inner boundary C_2 the circle $r = a$, the subsequent argument is simplified by changing to the polar coordinates (r, θ), with $x = r \cos \theta$, $y = r \sin \theta$. For then if R contains no equilibrium points, and it can be shown that $dr/dt > 0$ on $r = a$ and $dr/dt < 0$ on $r = b$, the conditions of Theorem 6.8 are satisfied and a limit cycle must exist within R.

Example 10. Application of the Poincaré–Bendixon theorem

Prove the system

$$\frac{dx}{dt} = x - y - 2x(x^2 + y^2)$$

$$\frac{dy}{dt} = x + y - 2y(x^2 + y^2)$$

has a limit cycle.

Solution

Inspection shows the origin to be an equilibrium point. To discover if the system has any other equilibrium points, and to simplify the subsequent argument, we change to polar coordinates by setting $x = r \cos \theta$, $y = r \sin \theta$. It then follows that

$$\frac{\partial r}{\partial x} = \cos \theta, \quad \frac{\partial r}{\partial y} = \sin \theta, \quad \frac{\partial \theta}{\partial x} = -\frac{\sin \theta}{r}, \quad \frac{\partial \theta}{\partial y} = \frac{\cos \theta}{r},$$

and using the chain rule

$$\frac{dr}{dt} = \frac{\partial r}{\partial x}\frac{dx}{dt} + \frac{\partial r}{\partial y}\frac{dy}{dt}, \quad \frac{d\theta}{dt} = \frac{\partial \theta}{\partial x}\frac{dx}{dt} + \frac{\partial \theta}{\partial y}\frac{dy}{dt}$$

shows the original system to be equivalent to

$$\frac{dr}{dt} = r(1 - 2r^2), \quad \frac{d\theta}{dt} = 1.$$

Clearly this transformed system has no equilibrium points when $r > 0$, so we have established that the origin is the only equilibrium point of the system. Now $dr/dt = \frac{1}{4} > 0$ on the circle $r = \frac{1}{2}$ and $dr/dt = -1 < 0$ on the circle $r = 1$. Let us now identify region R of Theorem 6.8 with the annulus $\frac{1}{2} \leq r \leq 1$, with its outer boundary C_1 the circle $r = 1$, and its inner boundary C_2 the circle $r = \frac{1}{2}$. It then follows that R contains no equilibrium points, and that all trajectories crossing C_1 and C_2 enter R as $t \to +\infty$. Thus the conditions of Theorem 6.8 are satisfied, so the system must have a limit cycle within the region R.

Integration of the expressions for dr/dt and $d\theta/dt$ gives

$$r = (2 + a e^{-2t})^{-1/2} \quad \text{and} \quad \theta = t + t_0.$$

So, as $t \to +\infty$, so $r \to 1/\sqrt{2}$ for any value of the integration constant a. Thus the limit cycle is the circle $r = 1/\sqrt{2}$. The spiral nature of a trajectory as it approaches the limit cycle is apparent from these results. ∎

Problems for Section 6.4

1 Classify the following equations and systems as autonomous or nonautonomous, and state whether they are linear or nonlinear:

(a) $\dfrac{d^2 x}{dt^2} + 9x = 0$

(b) $\dfrac{d^2 x}{dt^2} + t\dfrac{dx}{dt} + x = 0$

(c) $\dfrac{d^2 x}{dt} + 4(x^4 - 1)\dfrac{dx}{dt} + x^2 = 0$

(d) $\dfrac{dx}{dt} = x + y + 1, \dfrac{dy}{dt} = x - y - 3$

(e) $\dfrac{dx}{dt} = 3x + y + \sin y, \quad \dfrac{dy}{dt} = x - 3y + \cos t$

(f) $\dfrac{dx}{dt} = \sinh(x + y), \quad \dfrac{dy}{dt} = \cosh(x - 2y + 7).$

Determine the equilibrium points of the following autonomous equations.

2 (a) $\dfrac{d^2x}{dt^2} + 4x = 0$

(b) $\dfrac{d^2x}{dt^2} - x + 2 = 0$.

3 (a) $\dfrac{d^2x}{dt^2} + 4\sin x = 0$

(b) $\dfrac{d^2x}{dt^2} + \sin 4x = 0$.

4 (a) $\dfrac{d^2x}{dt^2} + \sinh(3x - 1) = 0$

(b) $\dfrac{d^2x}{dt^2} + x\dfrac{dx}{dt} + 2x - 1 = 0$.

5 (a) $\dfrac{d^2x}{dt^2} - (4 - x^2) = 0$

(b) $\dfrac{d^2x}{dt^2} + k\dfrac{dx}{dt} + a^2\sin x = 0$, $k > 0$.

6 $\dfrac{d^2x}{dt^2} + a\dfrac{dx}{dt} + bx + cx^3 = 0$, with a, b, c arbitrary.

Solve the following equations. Sketch some representative trajectories in the phase plane and use arrows to indicate the direction in which time increases along a trajectory.

7 $\dfrac{d^2x}{dt^2} - 4x = 0$

8 $\dfrac{d^2x}{dt^2} + 4\dfrac{dx}{dt} + 3x = 0$

9 $\dfrac{d^2x}{dt^2} + 3\dfrac{dx}{dt} + 3x = 0$

10 $\dfrac{d^2x}{dt^2} - 3\dfrac{dx}{dt} + 3x = 0$

11 A rod of length l and mass m is freely hinged to the end of a shaft which rotates about a vertical axis at a constant rate Ω, as shown in Fig. 6.38.

The equation of motion for the system is

$$\frac{d^2\theta}{dt^2} + \left(\frac{3g}{2l} - \Omega^2\cos\theta\right)\sin\theta = 0.$$

Locate the equilibrium points in the $(\theta, \dot{\theta})$-phase plane, and find the equation of the trajectories in this plane.

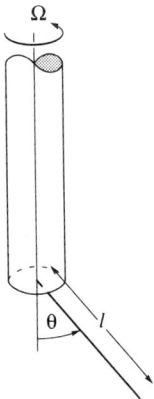

Fig. 6.38   Rod hinged to a vertical rotating shaft

12 Find the equation of the isoclines for the van der Pol equation (5) when $\varepsilon=1$. Show some representative isoclines together with their associated direction field lines and verify the trajectories in Fig. 6.21.

13 Use the method of isoclines to construct the phase portrait of Ex. 2 when $\alpha=1$.

14 Use the method of isoclines to construct the phase portrait of Ex. 3 when $\alpha=1$.

15 Repeat Prob. 14 in the case when $\alpha=2$.

16 Integrate (16) and use the result to construct the separatrices for Fig. 6.17, along with some representative trajectories.

17 Find the approximate equation of motion for the rod in Prob. 11 and determine the natural frequency of the oscillations when (a) θ is small and (b) $\theta \approx \theta_0$, where $\theta_0 = \arccos(3y/2l\Omega^2)$.

18 Integrate

$$\frac{d^2x}{dt^2} = \alpha x + \beta y \quad \text{when} \quad \beta \neq 0 \quad \text{and} \quad \alpha + \beta^2/4 = -m^2 < 0$$

to obtain the parametric representation in terms of t of the spiral trajectory given in Fig. 6.25.

19 Integrate

$$\frac{d^2x}{dt^2} = \alpha x + \beta y \quad \text{when} \quad \beta \neq 0 \quad \text{and} \quad \alpha + \beta^2/4 = 0$$

to obtain parametric representation in terms of t of the improper node with no exceptional direction given in Fig. 6.27.

In each of the following problems locate the equilibrium points, linearize the equation about each of the points to identify their nature, and use the results to sketch the phase portrait for the equation.

20 $\dfrac{d^2x}{dt^2} = \dfrac{k^2(x^2+4x+3)}{y}$.

21 $\dfrac{d^2x}{dt^2} = -a^2\sin x - k\dfrac{dx}{dt}$, with $4a^2 \leq k^2$.

22 $\dfrac{d^2x}{dt^2} = -a^2\sin x - k\dfrac{dx}{dt}$, with $4a^2 > k^2$.

23 $\dfrac{d^2x}{dt^2} = -a^2\sin x - k\dfrac{dx}{dt}\left|\dfrac{dx}{dt}\right|$.

24 Show $(0, 0)$, $(\alpha, 0)$ and $(\pi, 0)$ are equilibrium points of the differential equation

$$\frac{d^2x}{dt^2} = K(\cos x - \lambda)\sin x \qquad (K>0, |\lambda|<1)$$

in the (x, y)-phase plane, where $y = dx/dt$ and $\alpha = \arccos\lambda$. Linearize the differential equation about each of these points and thus show $(0,0)$ and $(\pi,0)$ are saddle points while $(\alpha, 0)$ is a center. Use these results to sketch the phase portrait of the nonlinear differential equation in the interval $0 \leq x \leq \pi$, and then extend it to the entire phase plane.

Each of the following problems gives values of k, F and initial conditions in the (x, y)-phase plane for a mass-spring system with Coulomb friction, in which the displacement x of a unit mass is governed by the piecewise continuous differential equation

$$\ddot{x} = -k^2 x - F \operatorname{sgn}(\dot{x}).$$

Draw the (X, y)-phase plane diagram for each problem, where $X = kx$. Use the diagram to determine the cycle in which the mass reaches equilibrium and the location of equilibrium position in the (x, y)-phase plane.

25 $k = 5$, $F = 3$ and the initial condition in the (x, y)-phase plane is $x_0 = -0.12$, $y_0 = 8$.

26 $k = 1$, $F = 1$ and the initial condition in the (x, y)-phase plane is $x_0 = 8$, $y_0 = 0$.

27 $k = 1$, $F = 1$ and the initial condition in the (x, y)-phase plane is $x_0 = 7.5$, $y_0 = 0$.

28 $k = 1$, $F = 0.5$ and the initial condition in the (x, y)-phase plane is $x_0 = 8$, $y_0 = 0$.

29 $k = 0.5$, $F = 0.5$ and the initial condition in the (x, y)-phase plane is $x_0 = -2$, $y_0 = 7$.

30 $k = \frac{1}{2}$, $F = \frac{1}{2}$ and the initial condition in the (x, y)-phase plane is $x_0 = -2$, $y_0 = 6.5$.

31 Find the equations of the trajectories in the phase plane for the differential equation

$$\frac{d^2 x}{dt^2} + k \operatorname{sgn}(x) = 0$$

when $k > 0$ is a constant. Sketch the phase portrait and use it to show the solutions are oscillatory. Consider periodic motion in which $x = A$, $dx/dt = 0$ when $t = 0$, so A is the amplitude of the oscillation. Find the time at which x first returns to zero. Use this time together with symmetry arguments to deduce the period of the oscillations. Are the oscillations isochronous or nonisochronous?

Use the Routh–Hurwitz test to determine which of the following differential equations has an asymptotically stable general solution.

32 (a) $y'' - 3y' - 28y = 0$ (b) $y'' + 6y' + 9y = 0$
(c) $y''' + 6y'' = 0$.

33 (a) $y^{(iv)} - 8y'' + 16y = 0$ (b) $y''' + 6y'' + 11y' + 6y = 0$
(c) $y^{(iv)} + 6y''' + 8y'' + 8y' + 3y = 0$.

34 (a) $y''' - 8y = 0$ (b) $y''' - 3y'' + 3y' - y = 0$
(c) $y''' + 4y'' + 7y' + 6y = 0$.

35 Use the Routh–Hurwitz theorem to show the general solution of

$$\frac{d^4 x}{dt^4} + a \frac{d^3 x}{dt^3} + b \frac{d^2 x}{dt^2} + cx = 0$$

cannot be asymptotically stable.

36 What can be deduced from the Routh–Hurwitz theorem about the roots of the cubic

$$\lambda^3 + a\lambda^2 + b = 0 \qquad (a, b) \text{ real,}$$

and what is the implication for the general solution of

$$\frac{d^3 x}{dt^3} + a \frac{d^2 x}{dt^2} + bx = 0?$$

Use Theorem 6.6 to determine which of the following differential equations has a limit cycle, and if one does not exist give reasons why this is so.

37 $\dfrac{d^2 x}{dt^2} + \varepsilon(x^2 - 1)\dfrac{dx}{dt} + x(1 + x^2) = 0$, $\varepsilon > 0$.

38 $\dfrac{d^2 x}{dt^2} + (5x^4 - 15x^2 + 6)\dfrac{dx}{dt} + x = 0$.

39 $\dfrac{d^2x}{dt^2}+(k-\cos x)\dfrac{dx}{dt}+3x=0.$

40 $\dfrac{d^2x}{dt^2}+4(x^2-1)\dfrac{dy}{dt}+x-\sinh x=0.$

Identify the nature of the singular point located at the origin in each of the following systems.

41 $\dfrac{dx}{dt}=x-2y,$ $\qquad \dfrac{dy}{dt}=4x-3y.$

42 $\dfrac{dx}{dt}=x-y,$ $\qquad \dfrac{dy}{dt}=2x-y.$

43 $\dfrac{dx}{dt}=3x-4y,$ $\qquad \dfrac{dy}{dt}=x-2y.$

44 $\dfrac{dx}{dt}=2x+3y,$ $\qquad \dfrac{dy}{dt}=x+4y.$

45 $\dfrac{dx}{dt}=2x,$ $\qquad \dfrac{dy}{dt}=x+y.$

46 $\dfrac{dx}{dt}=-2x-5y,$ $\qquad \dfrac{dy}{dt}=2x+2y.$

47 $\dfrac{dx}{dt}=3x-2y,$ $\qquad \dfrac{dy}{dt}=4x-y.$

48 $\dfrac{dx}{dt}=3x+4y,$ $\qquad \dfrac{dy}{dt}=2x+y.$

49 $\dfrac{dx}{dt}=3x+y,$ $\qquad \dfrac{dy}{dt}=-x+y.$

50 $\dfrac{dx}{dt}=10x-18y,$ $\qquad \dfrac{dy}{dt}=6x-11y.$

51 Locate the equilibrium points of the system

$$\dfrac{dx}{dt}=-ay-by^3, \quad \dfrac{dy}{dt}=x$$

in the cases (a) $a>0$, $b>0$ (b) $a<0$, $b<0$ (c) $b<0<a$. Linearize the system about these equilibrium points and identify their nature in cases (a), (b) and (c).

52 The **predator–prey** model uses the nonlinear system of equations

$$\dfrac{dx}{dt}=ax-cxy, \quad \dfrac{dy}{dt}=-by+dxy,$$

where a, b, c and d are nonnegative constants. The variable $x>0$ represents the prey population and $y>0$ the predator population. Locate the equilibrium points in the phase plane, linearize the equations about them to determine their nature, and sketch the phase portrait of the system. Interpret the phase portrait when the predators are foxes and the prey rabbits.

53 Show the nonlinear system

$$\dfrac{dx}{dt}=y-y^3, \quad \dfrac{dy}{dt}=x-x^3$$

has nine equilibrium points. Identify the nature of those in the first quadrant by linearization and determine the nature of the others by symmetry arguments.

54 Use Theorem 6.7 to justify the linearization used in Prob. 17 to find the natural frequency of the oscillations in cases (a) and (b).

55 Find the equilibrium points of the system

$$\frac{dx}{dt} = y, \quad \frac{dy}{dt} = 2x^3 + x^2 - x.$$

Identify their nature by means of linearization, using Theorem 6.7 to justify this process.

Use Theorem 6.7 to justify the linearization of the following systems about the equilibrium point at origin, and thus identify the nature of each such point.

56 $\dfrac{dx}{dt} = 2y - 3x,$ $\dfrac{dy}{dt} = x - 4y + x^2 y.$

57 $\dfrac{dx}{dt} = y - y^3,$ $\dfrac{dy}{dt} = y - 2x + x^2.$

58 $\dfrac{dx}{dt} = x + x^3,$ $\dfrac{dy}{dt} = y - x^2 + y^2.$

59 $\dfrac{dx}{dt} = x + 3y + xy,$ $\dfrac{dy}{dt} = -6x - 5y + x^3.$

60 $\dfrac{dx}{dt} = 3x + y - 3x^2 y,$ $\dfrac{dy}{dt} = -x + y + y^3.$

Use Theorem 6.8 to prove the existence of a limit cycle for each of the following systems. In each case find the equation of the trajectories in polar form with t as a parameter, and hence determine the limit cycle.

61 $\dfrac{dx}{dt} = x - y - x^3 - xy^2,$ $\dfrac{dy}{dt} = x + y - y^3 - x^2 y.$

62 $\dfrac{dx}{dt} = 9(x - y) - x(x^2 + y^2),$ $\dfrac{dy}{dt} = 9(x + y) - y(x^2 + y^2).$

63 $\dfrac{dx}{dt} = y + \dfrac{x}{\sqrt{x^2 + y^2}}[1 - (x^2 + y^2)]$ $\dfrac{dy}{dt} = -x + \dfrac{y}{\sqrt{x^2 + y^2}}[1 - (x^2 + y^2)].$

64 $\dfrac{dx}{dt} = 4y + x(1 - x^2 - y^2)$ $\dfrac{dy}{dt} = -4x + y(1 - x^2 - y^2).$

Harder problems

When linearization about an equilibrium point cannot be applied to a nonlinear equation or system, the phase portrait near that point will usually differ from the patterns studied so far, and is often more intricate. In each of the following problems either use the method of isoclines to sketch the phase portrait when none is given, or to verify the portrait if one is present.

65 $\dfrac{dy}{dt} = y^2 - x^2 - a^2,$ $\dfrac{dx}{dt} = 2xy,$ $a \neq 0.$

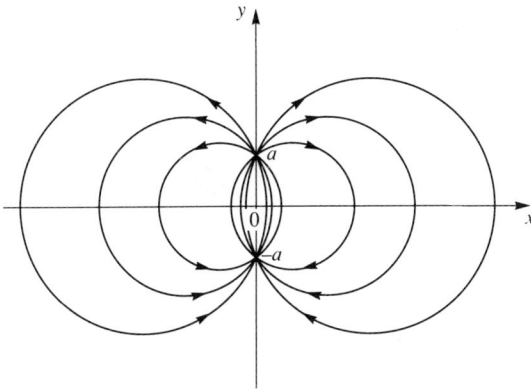

Fig. 6.39 Flow from an unstable node at $(0, a)$ to a stable attractive node at $(0, -a)$

66 $\dfrac{dy}{dt} = y^2 - x^2, \quad \dfrac{dx}{dt} = 2xy.$

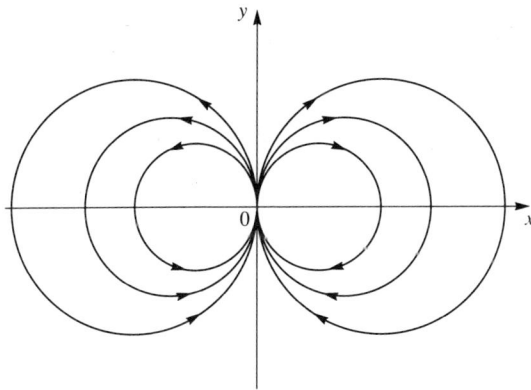

Fig. 6.40 Confluence of two nodes to produce a dipole

It can be shown for a nonlinear system, that apart from the phase portraits for a center and a spiral, all other phase portraits around isolated equilibrium points are made up of sectors of the type shown in Fig. 6.41. These are **hyperbolic sectors**, where trajectories never reach an equilibrium

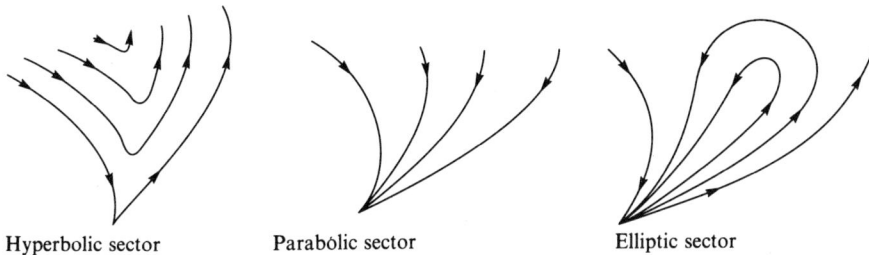

Hyperbolic sector   Parabolic sector   Elliptic sector

Fig. 6.41

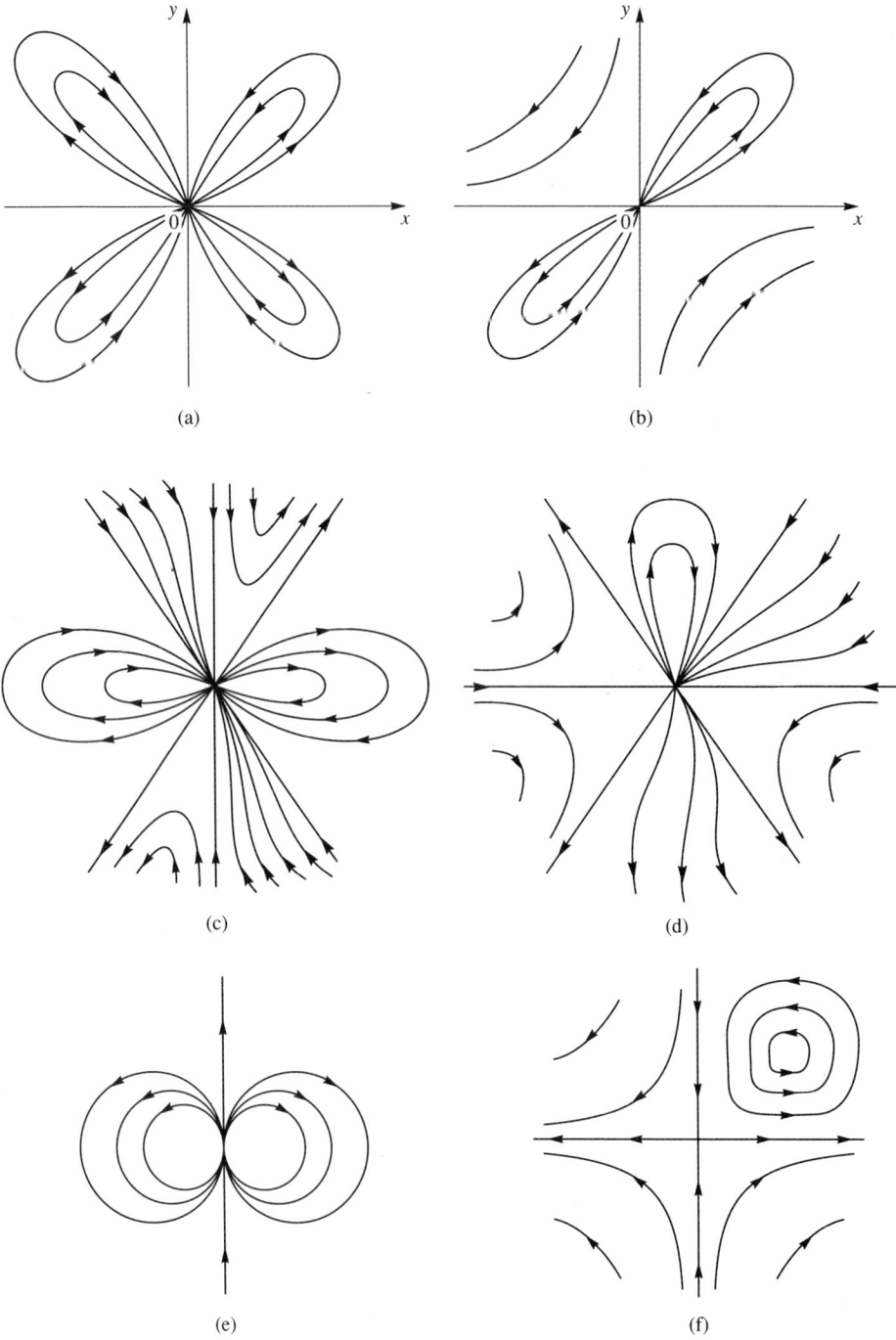

(a)

(b)

(c)

(d)

(e)

(f)

Fig. 6.42

point, **parabolic sectors**, where a trajectory either begins or ends at an equilibrium point, and **elliptic sectors**, where all trajectories begin and end at an equilibrium point.

Equilibrium points with only parabolic sectors surrounding them are **nodes,** while those with only hyperbolic sectors are called **cross points** (a **saddle point** is a cross point with four sectors). An equilibrium point with only elliptic sectors around it is called a **rose**. The dipole in Fig. 6.40 is a rose with two sectors. The trajectories separating one sector from another are the **separatrices** of the phase portrait.

67 Analyze the phase portraits shown in Fig. 6.42 into sectors of different types and identify the separatrices.

6.5 Numerical solution of systems by the Runge–Kutta method

The Runge–Kutta method introduced in Sec. 4.8 generalizes immediately to *systems* of first order differential equations. We now give the form taken by the algorithm when two simultaneous first order differential equations are involved.

Runge–Kutta algorithm for a first order system

Let the differential equations to be integrated be

$$\frac{dy}{dx} = f(x, y, z), \qquad \frac{dz}{dx} = g(x, y, z),$$

subject to the initial conditions

$$y(x_0) = y_0 \quad \text{and} \quad z(x_0) = z_0.$$

Set $x_n = x_0 + nh$, $y_n = y(x_n)$ and $z_n = z(x_n)$, where h is the step length. The algorithm then determines the values of y_{n+1} and z_{n+1} at x_{n+1} from the values of y_n and z_n at x_n by means of the following steps.

Step 1
Compute

$$k_{1n} = hf(x_n, y_n, z_n)$$
$$k_{2n} = hf(x + \tfrac{1}{2}h, y_n + \tfrac{1}{2}k_{1n}, z_n + \tfrac{1}{2}K_{1n})$$
$$k_{3n} = hf(x + \tfrac{1}{2}h, y_n + \tfrac{1}{2}k_{2n}, z_n + \tfrac{1}{2}K_{2n})$$
$$k_{4n} = hf(x_n + h, y_n + k_{3n}, z_n + K_{3n})$$

$$K_{1n} = hg(x_n, y_n, z_n)$$
$$K_{2n} = hg(x_n + \tfrac{1}{2}h, y_n + \tfrac{1}{2}k_{1n}, z_n + \tfrac{1}{2}K_{1n})$$
$$K_{3n} = hg(x_n + \tfrac{1}{2}h, y_n + \tfrac{1}{2}k_{2n}, z_n + \tfrac{1}{2}K_{2n})$$
$$K_{4n} = hg(x_n + h, y_n + k_{3n}, z_n + K_{3n}).$$

Step 2
Compute

$$k_n = \tfrac{1}{6}(k_{1n} + 2k_{2n} + 2k_{3n} + k_{4n}) \quad \text{and} \quad K_n = \tfrac{1}{6}(K_{1n} + 2K_{2n} + 2K_{3n} + K_{4n}).$$

Step 3

The numerical estimates of y_{n+1} and z_{n+1} at x_{n+1} are the given by

$$y_{n+1} = y_n + k_n \quad \text{and} \quad z_{n+1} = z_n + K_n.$$

Example 1. Runge–Kutta calculation for a system

Use the Runge–Kutta method with an integration step $h=0.1$ to determine y_1 to y_5 and z_1 to z_5, given that

$$\frac{dy}{dx} = y + z^2, \quad \frac{dz}{dx} = 2x + y^2 - z,$$

subject to the initial conditions $x_0 = 1$, $y_0 = 1$, $z_0 = -2$ (i.e. $y(1) = 1$, $z(1) = -2$).

Solution

n	x_n	k_{1n}	K_{1n}	k_{2n}	K_{2n}
0	1	0.500000	0.500000	0.431250	0.541250
1	1.1	0.357187	0.570075	0.299681	0.605730
2	1.2	0.247443	0.624187	0.215527	0.647197
3	1.3	0.199193	0.659531	0.205335	0.676236
4	1.4	0.235804	0.697609	0.291442	0.724858
5	1.5	0.386160	0.782223	0.513652	0.851506

n	x_n	k_{3n}	K_{3n}	k_{4n}	K_{4n}
0	1	0.420636	0.530712	0.357944	0.568750
1	1.1	0.292632	0.594797	0.247606	0.622673
2	1.2	0.212671	0.640170	0.198867	0.658072
3	1.3	0.205828	0.676656	0.235416	0.696058
4	1.4	0.296430	0.736201	0.386514	0.780347
5	1.5	0.531084	0.882164	0.726601	1.002599

n	x_n	k_n	K_n	y_n	z_n
0	1	0.426953	0.535446	1.000000	-2.000000
1	1.1	0.298236	0.598967	1.426953	-1.464554
2	1.2	0.217117	0.642832	1.725189	-0.865588
3	1.3	0.209489	0.676895	1.942307	-0.222756
4	1.4	0.299677	0.733346	2.151796	0.454140
5	1.5	0.533705	0.875361	2.451473	1.187486

Problems for Section 6.5

Use the Runge–Kutta method in the following problems, together with the given step length h and initial conditions, to determine y_1, y_2, and z_1 and z_2 if hand calculation is involved, or determine y_1 to y_4 and z_1 to z_4 if a programmable computer is to be used.

1 $\dfrac{dy}{dx} = \cosh(y+z), \qquad \dfrac{dz}{dx} = y + 2\sinh(y-z)$

   $y(0) = 0, \qquad z(0) = 0.5, \quad h = 0.1.$

2 $\dfrac{dy}{dx} = y + 0.5z + \sin x, \qquad \dfrac{dz}{dx} = x + z\sin y$

   $y(0) = 1.5, \qquad z(0) = 0.5, \qquad h = 0.2.$

3 $\dfrac{dy}{dx} = y + z, \qquad \dfrac{dz}{dx} = 3x + y^2 - 2z$

   $y(0) = 2, \qquad z(0) = 1, \qquad h = 0.2.$

4 $\dfrac{dy}{dx} = y + z, \qquad \dfrac{dz}{dx} = x + y - \sin z$

   $y(0.5) = 2, \qquad z(0.5) = 0, \qquad h = 0.1.$

5 $\dfrac{dy}{dx} = y^2 + z, \qquad \dfrac{dz}{dx} = x + y^2 - z$

   $y(0) = 1, \qquad z(0) = -1, \qquad h = 0.2.$

6 $\dfrac{dy}{dx} = 1 - y^2 + \sin z, \qquad \dfrac{dz}{dx} = 2x - y - z$

   $y(-0.5) = 0, \qquad z(-0.5) = -1, \qquad h = 0.2.$

Chapter 7
Laplace Transform and
z-Transform

The Laplace transform $Y(s)$ of a function $y(t)$ is defined as

$$Y(s) = \int_0^\infty e^{-st} y(t)\, dt$$

whenever the integral exists. The importance of the Laplace transform is that it provides a powerful method for solving initial value problems for linear differential equations.

In its simplest form the approach is to transform a differential equation for $y(t)$ and, as a result, to obtain an algebraic equation for the Laplace transform of the solution $Y(s)$. If the transformed equation is then solved for $Y(s)$, the required solution $y(t)$ of the original differential equation is recovered by identifying $Y(s)$ with the function $y(t)$ having $Y(s)$ as its Laplace transform. This last step is called inverting the Laplace transform, and the task is simplified by using a table of transform pairs in which commonly occurring functions $f(t)$ are paired with their Laplace transform $F(s)$, and also by developing useful rules obeyed by the Laplace transform operation.

The approach to the solution of an initial value problem by the Laplace transform differs significantly from the approach discussed in previous chapters. Instead of first finding the general solution of the differential equation, and then matching the aribtrary constants in the general solution to the initial conditions, the Laplace transform incorporates the initial conditions at the time differential equation is transformed. Thus, when the transformed solution $Y(s)$ is inverted, the solution of the initial value problem is obtained immediately.

Most of the differential equations and systems considered in this chapter have constant coefficients, though some consideration is also given to variable coefficient differential equations, special boundary value problems and simple integral equations.

Various applications of the Laplace transform are discussed at length and then its discrete analog, the z-transform, is introduced and developed. The z-transform applies to discrete data in the form of a sequence, often obtained as a result of sampling a continuous function $y(t)$ at equispaced intervals of its argument. Its applications are numerous and include control theory and data transmission systems.

The fundamentals of the Laplace transform and of inversion by means of transform pairs are discussed in Sec. 7.1, while operational properties of the transform independent of the functions involved are discussed in Sec. 7.2. Applications of the Laplace transform to a wide variety of problems are considered in Sec. 7.3. Finally, the z-transform is introduced in Sec. 7.4 and applied in Sec. 7.5 to the problems related to control theory and to linear difference equations.

594

7.1 The Laplace transform – introductory ideas

Let $f(t)$ be a given function defined for $t \geq 0$. Then if the improper integral of the second kind

$$F(s) = \int_0^\infty e^{-st} f(t) \, dt \tag{1}$$

exists, where s is a parameter, the function $F(s)$ defined by (1) is called the **Laplace transform** of the function $f(t)$. It is usual to denote the Laplace transform of $f(t)$ by $\mathscr{L}\{f\}$, so that

$$F(s) = \mathscr{L}\{f(t)\} = \int_0^\infty e^{-st} f(t) \, dt. \tag{2}$$

In this chapter s will be taken to be real, but in a more general approach it is allowed to be complex.

The operation of integration in (1) which defines $F(s)$ in terms of $f(t)$ is called the **Laplace transformation**. The term *transformation* is used because when the integral in (1) exists it transforms a function of t, usually the time, into a function of s (the transform parameter). The reason for interest in what at this stage must seem a rather strange procedure will become clear later when we come to discuss the solution of initial value problems for ordinary differential equations.

In what is to follow it will be necessary to invert a Laplace transformation. This involves determining the function $f(t)$ which has as its Laplace transformation a given function $F(s)$. This inverse relationship between $F(s)$ and $f(t)$ is expressed by saying that $f(t)$ is the **inverse Laplace transform** of $F(s)$ or, more simply, the **inverse** of $F(s)$. This is written symbolically as

$$f(t) = \mathscr{L}^{-1}\{F\}. \tag{3}$$

Various different notations are in current use to signify transformed variables. The one employed here uses lower case letters to denote the original functions of t and the corresponding upper case letters (capitals) to denote their Laplace transforms. Thus, for example, $\mathscr{L}\{f\} = F(s)$, $\mathscr{L}\{y\} = Y(s)$ and $\mathscr{L}\{\omega\} = \Omega(s)$. Let us now derive some typical Laplace transforms by direct application of the definition in (1).

Example 1

Given $f(t) = 1$ for $t \geq 0$, find $F(s)$.

Solution

We first recall the definition of an improper integral of the second kind given in (25) of Sec. 1.8. Applying this to $f(t)$ gives

$$F(s) = \mathscr{L}\{1\} = \int_0^\infty e^{-st} \, dt = \lim_{R \to \infty} \int_0^R e^{-st} \, dt = \lim_{R \to \infty} \left(-\frac{1}{s} e^{-st} \right) \Big|_0^R = \frac{1}{s},$$

provided $s > 0$, for only then will the integral converge. This has established the following Laplace transform

$$\mathcal{L}\{1\} = \frac{1}{s}, \quad \text{for } s > 0,$$ ■

Example 2

Given

$$f(t) = \begin{cases} 0, & 0 \le t < a \\ h, & a \le t \le b \quad (h = \text{const}) \\ 0, & t > b, \end{cases}$$

find $F(s)$.

Solution

The function $f(t)$ is discontinuous at $t = a$ and $t = b$, and as it vanishes outside the interval $a \le t \le b$ the question of convergence of the integral does not arise, for we have

$$F(s) = \mathcal{L}\{f\} = \int_0^\infty e^{-st} f(t)\, dt = \int_a^b h e^{-st}\, dt = \frac{h}{s}(e^{-as} - e^{-bs}).$$

This is the required Laplace transform. ■

Notice that in this last example $f(t)$ describes a pulse of amplitude h which starts at $t = a$ and stops at $t = b$, as shown in Fig. 7.1(a). In an electric circuit this corresponds to switching on a constant voltage of magnitude h at time $t = a$ and then switching it off again at time $t = b$. Similarly, in a mechanical system, this could correspond to the sudden application of a constant force of magnitude h at time $t = a$ and its sudden removal at time $t = b$.

Keeping $a > 0$ fixed, and allowing $b \to \infty$, this function becomes what is called a **step function**. In this case it is of amplitude h and it is applied at time $t = a$. Again using an electrical analogy, this could represent switching on a constant voltage of magnitude h at

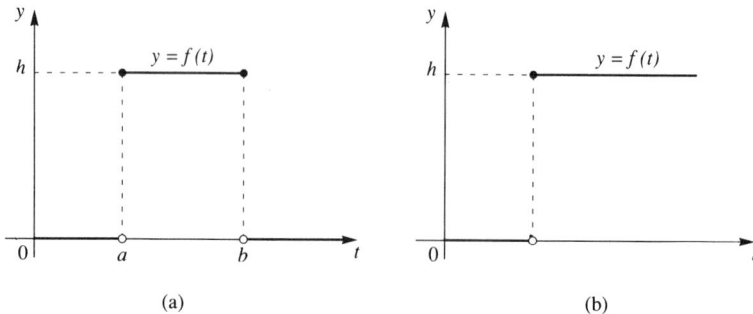

(a) (b)

Fig. 7.1 (a) Pulse of duration $b - a$, (b) Step function ($b \to \infty$)

time $t=a$ and leaving it on. It follows from the above result that the Laplace transform of such a step function is

$$F(s)=\mathscr{L}\{f\}=\frac{he^{-as}}{s}, \qquad \text{for } s>0,$$

where the condition on s is necessary for convergence. If we set $a=0$ and $h=1$ in this last expression we recover the result of Ex. 1, which may be regarded as a *unit step function* applied at time $t=0$.

Example 3

Given $f(t)=t$ for $t\geq0$, find $F(s)$.

Solution

$$F(s)=\mathscr{L}\{t\}=\int_0^\infty te^{-st}\,dt=\lim_{R\to\infty}\int_0^R te^{-st}\,dt=\frac{1}{s^2}, \qquad \text{for } s>0.$$

Thus we have shown

$$\mathscr{L}\{t\}=\frac{1}{s^2}, \qquad \text{for } s>0. \qquad\qquad ■$$

Example 4

Given $f(t)=e^{at}$ for $t\geq0$, and $a=$ constant, find $F(s)$.

Solution

$$F(s)=\mathscr{L}\{\exp(at)\}=\int_0^\infty \exp(-st)\exp(at)\,dt=\lim_{R\to\infty}\int_0^R \exp[-(s-a)t]\,dt$$

$$=\lim_{R\to\infty}\left[\left(\frac{1}{a-s}\right)\exp[-(s-a)t]\right]\Big|_0^R=\frac{1}{s-a}, \qquad \text{for } s>a.$$

Thus we have established that

$$\mathscr{L}\{e^{at}\}=\frac{1}{s-a}, \qquad \text{for } s>a. \qquad\qquad ■$$

Now we need to define a piecewise continuous function $f(t)$ on the finite interval $a\leq t\leq b$. Let $a=t_0<t_1<\ldots<t_{n+1}=b$ be any *finite* number of points in $[a, b]$. The function $f(t)$ will be said to be **piecewise continuous** on $a\leq t\leq b$ if it is continuous in each interval $t_i<t<t_{i+1}$, but it experiences *finite discontinuities* (jumps) when crossing $t_1, t_2, \ldots, t_n$. Clearly a continuous function on $[a, b]$ is a special case of a piecewise continuous function on $[a, b]$. A function will be said to be **piecewise continuous** on the interval $0\leq t<\infty$ if it is piecewise continuous on every finite interval in $0\leq t<\infty$.

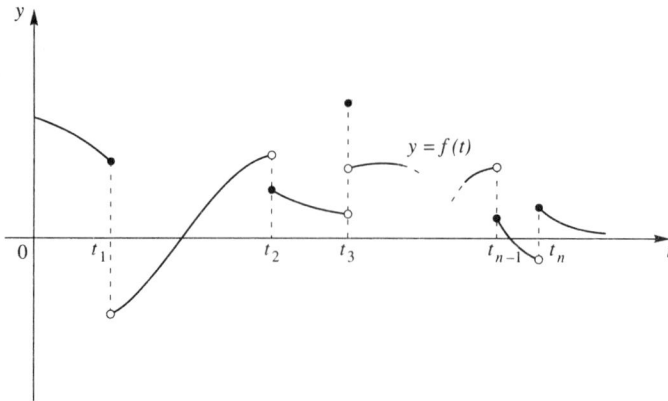

Fig. 7.2   Piecewise continuous function $y=f(t)$ with finite discontinuities at $t_1<t_2<\ldots<t_n$

The approach used in Ex. 2 to find the Laplace transform of a pulse (a particular piecewise continuous function) extends in an obvious manner to any piecewise continuous function. Thus if the function behaves as in Fig. 7.2, in which a dot represents a point belonging to the graph, and a circle a point missing from it, the Laplace transform $F(s)$ is computed as[1]

$$F(s)=\int_0^{t_1} e^{-st}f(t)\,dt+\int_0^{t_2} e^{-st}f(t)\,dt+\ldots+\int_{t_{n-1}}^{t_n} e^{-st}f(t)\,dt+\int_{t_n}^{\infty} e^{-st}f(t)\,dt. \qquad (4)$$

We have seen that a function $f(t)$ for which the integral in (1) exists defines the Laplace transform $F(s)$ of $f(t)$ uniquely. Now we must consider whether, given a specific Laplace transform $F(s)$, there is a unique function $f(t)$ with $F(s)$ as its Laplace transform. The essential result we need to use is that it can be shown that if two different functions $f_1(t)$ and $f_2(t)$ defined for $t\geq0$ have the same Laplace transform $F(s)$, then they are either identical or, at most, they only differ from one another at isolated points (cf. Footnote 1). Furthermore if $f_1(t)$ and $f_2(t)$ both have the same Laplace transform, and they are continuous, then they must be identical. This means that for all practical purposes the inverse Laplace transform is unique. Examples of two functions $f_1(t)$ and $f_2(t)$ with the same Laplace transform but which differ at isolated points are shown in Figs 7.3(a), (b).

[1] The fact that a finite discontinuity of a function at a single point does not affect its definite integral over an interval containing that point justifies writing

$$\int_{t_i}^{t_{i+1}} e^{-st}f(t)\,dt \qquad (i=1, 2, \ldots, n-1)$$

in (4) in place of the more precise expression

$$\lim_{\varepsilon,\delta\to0}\int_{t_i+\varepsilon}^{t_{i+1}-\delta} e^{-st}f(t)\,dt, \qquad \varepsilon,\delta>0.$$

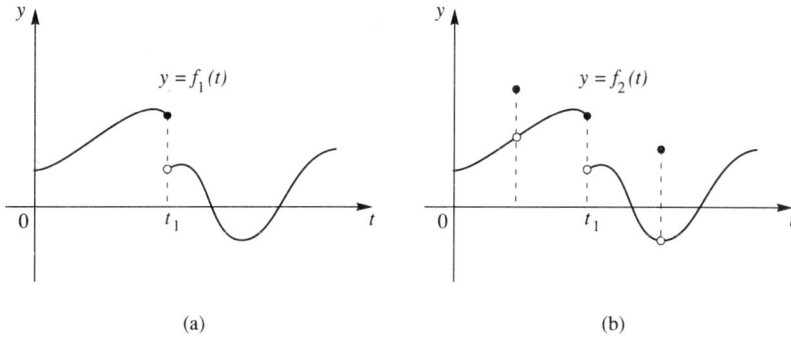

Fig. 7.3 Functions differing only at isolated points

A function $f(t)$ together with its Laplace transform $F(s)$ is called a **transform pair**. To enable the Laplace transforms of familiar functions to be determined rapidly, and inverse transforms to be determined as simply as possible, it is necessary to construct a table of transform pairs. Once this has been accomplished we shall see how the properties of the definite integral will often enable the table to be extended with very little effort. The following short table of transform pairs forms a useful starting point, and all entries in it may be obtained directly from the definition in (1).

Table 7.1 A short table of Laplace transform pairs

	$f(t)$	$F(s) = \mathcal{L}\{f(t)\}$			$f(t)$	$F(s) = \mathcal{L}\{f(t)\}$	
1	1	$1/s$	$(s>0)$	8	$\cos kt$	$\dfrac{s}{s^2+k^2}$	$(s>0)$
2	t	$1/s^2$	$(s>0)$	9	$t \sin kt$	$\dfrac{2ks}{(s^2+k^2)^2}$	$(s>0)$
3	$t^n \ (n=1, 2, 3, \ldots)$	$\dfrac{n!}{s^{n+1}}$	$(s>0)$	10	$t \cos kt$	$\dfrac{s^2-k^2}{(s^2+k^2)^2}$	$(s>0)$
4*	$t^a \ (a \text{ positive})$	$\dfrac{\Gamma(a+1)}{s^{a+1}}$	$(s>0)$	11	$e^{at} \sin kt$	$\dfrac{k}{(s-a)^2+k^2}$	$(s>a)$
5	e^{at}	$\dfrac{1}{s-a}$	$(s>a)$	12	$e^{at} \cos kt$	$\dfrac{s-a}{(s-a)^2+k^2}$	$(s>a)$
6	$t^n e^{at} \ (n=1, 2, 3, \ldots)$	$\dfrac{n!}{(s-a)^{n+1}}$	$(s>a)$	13	$\sinh kt$	$\dfrac{k}{s^2-k^2}$	$(s>|k|)$
7	$\sin kt$	$\dfrac{k}{s^2+k^2}$	$(s>0)$	14	$\cosh kt$	$\dfrac{s}{s^2-k^2}$	$(s>|k|)$

* The function $\Gamma(x)$ occurring here (with $x=a+1$), called the **gamma function**, is a generalization of the factorial function. It has the property that $\Gamma(1)=1$, $\Gamma(n+1)=n!$ when n is a positive integer, and for any $x>0$, $\Gamma(x+1)=x\Gamma(x)$. A useful result is $\Gamma(\tfrac{1}{2})=\sqrt{\pi}$ (see Sec. 8.5).

The table may either be entered with the function $f(t)$ to find the Laplace transform $F(s) = \mathscr{L}\{f(t)\}$, or with $F(s)$ to find the inverse Laplace transform $f(t) = \mathscr{L}^{-1}\{F(s)\}$. Thus, for example, from entry 3 we see that $\mathscr{L}\{t^3\} = 3!/s^4$, while $\mathscr{L}^{-1}\{1/s^5\} = t^4/(4!)$. Similarly, from entry 11 we see that

$$\mathscr{L}\{e^{-2t}\sin 3t\} = \frac{3}{(s+2)^2 + 3^2} = \frac{3}{s^2 + 4s + 13},$$

whereas after completing the square we have

$$\mathscr{L}^{-1}\left(\frac{2}{s^2 - 2s + 5}\right) = \mathscr{L}^{-1}\left(\frac{2}{(s-1)^2 + 2^2}\right) = e^t \sin 2t.$$

To discover an important way in which Table 7.1 may be extended, let the functions $f(t), g(t)$ defined for $t \geq 0$ have the respective Laplace transforms $F(s), G(s)$. Now let α, β be constants and consider $\mathscr{L}\{\alpha f + \beta g\}$. We have from (1) that

$$\mathscr{L}\{\alpha f + \beta g\} = \int_0^\infty e^{-st}[\alpha f(t) + \beta g(t)]\, dt$$

$$= \alpha \int_0^\infty e^{-st} f(t)\, dt + \beta \int_0^\infty e^{-st} g(t)\, dt$$

$$= \alpha\, \mathscr{L}\{f\} + \beta\, \mathscr{L}\{g\}.$$

Expressed in words, this important result, which expresses the *linearity* of the Laplace transformation, says that the Laplace transform of a linear combination of functions is the same combination of their individual Laplace transforms. The result forms the following theorem.

Theorem 7.1. (Linearity of the Laplace transformation)

For any two functions $f(t)$ and $g(t)$ possessing Laplace transforms and any constants α and β,

$$\mathscr{L}\{\alpha f(t) + \beta g(t)\} = \alpha\mathscr{L}\{f(t)\} + \beta\mathscr{L}\{g(t)\} \quad \text{and} \quad \mathscr{L}\{\alpha f(t)\} = \alpha\mathscr{L}\{f(t)\}. \qquad \square$$

In particular, Theorem 7.1 shows that the effect of scaling a function by the constant factor α is to scale its Laplace transform by this same factor.

Example 5

Use Theorem 7.1. and Table 7.1 to find $\mathscr{L}\{f\}$, given $f(t) = \sin kt - kt \cos kt$.

Solution

From Theorem 7.1 we have

$$\mathscr{L}\{\sin kt - kt \cos kt\} = \mathscr{L}\{\sin kt\} - k\, \mathscr{L}\{t \cos kt\}.$$

Using entries 7 and 10 of Table 7.1 gives

$$\mathscr{L}\{\sin kt - kt \cos kt\} = \frac{k}{s^2 + k^2} - \frac{k(s^2 - k^2)}{(s^2 + k^2)^2} = \frac{2k^3}{(s^2 + k^2)^2}.$$

∎

The next result to be stated is an important and immediate consequence of Theorem 7.1, and it will often be needed.

Theorem 7.2 (Linearity of the inverse Laplace transform)

Let $\mathscr{L}\{f\} = F(s)$, $\mathscr{L}\{g\} = G(s)$ and α, β be any two constants. Then the inverse Laplace transform is a linear operation with the properties that

$$\mathscr{L}^{-1}\{\alpha F(s) + \beta G(s)\} = \alpha \mathscr{L}^{-1}\{F(s)\} + \beta \mathscr{L}^{-1}\{G(s)\},$$

and

$$\mathscr{L}^{-1}\{\alpha F(s)\} = \alpha \mathscr{L}^{-1}\{F(s)\}.$$

Proof

Consider

$$h(t) = \alpha \mathscr{L}^{-1}\{F(s)\} + \beta \mathscr{L}^{-1}\{G(s)\},$$

then

$$\mathscr{L}\{h\} = \mathscr{L}\{\alpha \mathscr{L}^{-1}\{F(s)\} + \beta \mathscr{L}^{-1}\{G(s)\}\},$$

which by Theorem 7.1 becomes

$$\mathscr{L}\{h\} = \alpha \mathscr{L}\{\mathscr{L}^{-1}\{F(s)\}\} + \beta \mathscr{L}\{\mathscr{L}^{-1}\{G(s)\}\}$$

$$= \alpha F(s) + \beta G(s).$$

Taking the inverse transform and using the definition of $h(t)$ establishes the theorem.

□

In words, Theorem 7.2 says that the inverse Laplace transform of a linear combination of transforms is the same combination of their individual inverse transforms, and scaling a Laplace transform by the constant factor α scales the inverse transform by this same factor.

Theorem 7.2 is usually used in conjunction with partial fractions in order to arrive at the inverse Laplace transform of algebraic expressions involving s which are not contained in tables. This is illustrated in the following example.

Example 6. Use of Theorem 7.2 with partial fractions

Find $\mathscr{L}^{-1}\left\{\dfrac{14 + 7s - 3s^2}{s^2(s+2)}\right\}$.

Solution

Using partial fractions gives

$$\frac{14+7s-3s^2}{s^2(s+2)} = \frac{7}{s^2} - \frac{3}{s+2}.$$

Taking the inverse transform and using Theorem 7.2 we have

$$\mathcal{L}^{-1}\left\{\frac{14+7s-3s^2}{s^2(s+2)}\right\} = \mathcal{L}^{-1}\left(\frac{7}{s^2}\right) - \mathcal{L}^{-1}\left\{\frac{3}{s+2}\right\}$$

$$= 7\mathcal{L}^{-1}\left\{\frac{1}{s^2}\right\} - 3\mathcal{L}^{-1}\left\{\frac{1}{s+2}\right\} = 7t - 3e^{-2t},$$

where use has been made of entries 2 and 5 in Table 7.1. ∎

It is necessary to determine when a given function $f(t)$ defined for $t \geq 0$ has a Laplace transform, because the expression in (1) is not convergent for all functions $f(t)$. A *sufficient* condition for a Laplace transform to exist is easily formulated for functions whose magnitude grows no faster than a specific exponential rate. To arrive at such a condition we first need the notion of a function of exponential order. We shall say that a function $f(t)$ defined for $t \geq 0$ is of **exponential order** $e^{\alpha t}$ if there exist numbers α, $M > 0$ and a number $T > 0$, such that

$$e^{-\alpha t}|f(t)| \leq M, \qquad \text{for } t > T. \tag{5}$$

The function $f(t)$ in (5) may either be continuous or piecewise continuous for $t \geq 0$. This condition ensures that the absolute value of $f(t)$ grows no faster than $Me^{\alpha t}$ when $t > T$.

Example 7

Determine which of the following functions is of exponential order (i) $f(t) = \sinh 2t$ (ii) $f(t) = \exp(t^2)$ (iii) $f(t) = t^n \sin t$ with $n = 1, 2, \ldots$.

Solution

(i) $e^{-\alpha t}|f(t)| = e^{-\alpha t}\left|\dfrac{e^{2t} - e^{-2t}}{2}\right| < \dfrac{1}{2}e^{-\alpha t}e^{2t}$.

Thus if $\alpha = 2$,

$e^{-\alpha t}|f(t)| < \frac{1}{2}$ for all $t > 0$

so the function is of exponential order e^{2t} for $t > 0$, and in this case $M = \frac{1}{2}$.

(ii) $e^{-\alpha t}|f(t)| = \exp(-\alpha t)\exp(t^2) = \exp[t(t-\alpha)]$.

This function is *not* of exponential order, because $\exp[t(t-\alpha)] \to \infty$ as $t \to \infty$ for any α.

(iii) $e^{-\alpha t}|f(t)| = e^{-\alpha t}|t^n \sin t| < e^{-\alpha t} t^n$.

However $\lim\limits_{t \to \infty} e^{-\alpha t} t^n = 0$ for any $\alpha > 0$, so this function is of exponential order for any $\alpha > 0$ (and any $M > 0$). ∎

With these ideas in mind we are now in a position to formulate our next result.

Theorem 7.3 (Sufficient conditions for the existence of a Laplace transform)

Let the function $f(t)$ defined for $t \geq 0$ be piecewise continuous. Then a sufficient condition that it should have a Laplace transform defined for $s > \alpha$ is that it be of exponential order $e^{\alpha t}$ for $t > T$.

Proof

The proof is almost immediate, because by using Theorem 1.7(iii) we have

$$|\mathscr{L}\{f\}| = \left| \int_0^\infty e^{-st} f(t)\, dt \right| \leq \int_0^\infty e^{-st} |f(t)|\, dt \int_0^\infty e^{-st} M e^{\alpha t}\, dt$$

$$= M \int_0^\infty \exp[-(s-\alpha)t]\, dt = \frac{M}{s-\alpha}, \qquad \text{for } s > \alpha. \qquad \square$$

Example 8

The fact that the condition in Theorem 7.3 is only sufficient is demonstrated by considering the Laplace transform of $f(t) = e^t/\sqrt{t}$. For this function is not of exponential order, yet its Laplace transform exists and

$$\mathscr{L}\{e^t/\sqrt{t}\} = \sqrt{\frac{\pi}{s-1}}, \qquad \text{for } s > 1.$$

To establish this result, all that is necessary is to integrate $\mathscr{L}\{e^t/\sqrt{t}\,]$ by parts and use the change of variable $u = (s-1)t$ to obtain

$$\mathscr{L}\{\exp(t)/\sqrt{t}\} = \int_0^\infty t^{-1/2} \exp[-(s-1)t]\, dt = 2(s-1) \int_0^\infty t^{1/2} \exp[-(s-1)t]\, dt$$

$$= \frac{2}{(s-1)^{1/2}} \int_0^\infty u^{1/2} e^{-u}\, du = \frac{2\Gamma(3/2)}{(s-1)^{1/2}}, \qquad \text{for } s > 1,$$

where $\Gamma(x)$ is the gamma function (generalized factorial function). We shall see later that $\Gamma(3/2) = \frac{1}{2}\sqrt{\pi}$, from which the result then follows (cf. Prob. 60, Sec. 1.8). ∎

Although piecewise continuous functions of exponential order always have a Laplace transform, it does not necessarily follow that every function of s is the Laplace transform of such a function. The next theorem gives a condition which must be satisfied by the Laplace transform of every piecewise continuous function of exponential order.

Theorem 7.4 (Limiting property of the Laplace transform)

Let the function $f(t)$ defined for $t \geq 0$ be piecewise continuous and of exponential order with $\mathscr{L}\{f(t)\} = F(s)$, then

$$\lim_{s \to \infty} F(s) = 0.$$

Proof

The result follows directly from Theorem 7.3 in which it was shown that

$$|\mathscr{L}\{f\}| \leq \frac{M}{s - \alpha}, \qquad \text{for } s > \alpha.$$

Taking the limit as $s \to \infty$ it follows that

$$\lim_{s \to \infty} |F(s)| = 0,$$

and hence that

$$\lim_{s \to \infty} F(s) = 0. \qquad \square$$

Example 9

$F(s) = s$ is not the Laplace transform of any piecewise continuous function of exponential order, because $\lim_{s \to \infty} F(s) \neq 0.$ ∎

Problems for Section 7.1

Find the Laplace transforms of the following functions.

1 $f(t) = 1 + 2t$ **2** $f(t) = 1 + 3e^{-t}$

3 $f(t) = \begin{cases} h, 0 \leq t \leq a \\ 0, t > a. \end{cases}$

4 $f(t) = te^{ikt}$, k real. Equate the real and imaginary parts of the resulting equation and thus verify entries 9 and 10 in Table 7.1.

5 $f(t) = e^{at} e^{ikt}$, a and k real constants. Equate the real and imaginary parts of the resulting equations and thus verify entries 11 and 12 in Table 7.1.

6 $f(t) = \begin{cases} 1, 0 \le t < 1 \\ -1, 1 \le t \le 2 \\ 0, t > 2 \end{cases}$ **7** $f(t) = e^t \sinh 2t$

8 $f(t) = e^t \cosh 2t$ **9** $f(t) = \begin{cases} 0, 0 \le t < 1 \\ t-1, t \ge 1 \end{cases}$

10 $f(t) = \begin{cases} 0, 0 \le t < a \\ \sin(t-a), t \ge a \end{cases}$ **11** $f(t) = \begin{cases} 0, 0 \le t < 2 \\ (t-2)^2, t \ge 2 \end{cases}$

12 $f(t) = \begin{cases} 1-t, 0 \le t \le 1 \\ 0, t > 1 \end{cases}$ **13** $f(t) = \begin{cases} t, 0 \le t \le 1 \\ 0, t > 1 \end{cases}$

14 $f(t) = \begin{cases} a, 0 \le t < \alpha \\ b, \alpha \le t \le \beta \\ 0, t > \beta \end{cases}$ **15** $f(t) = \begin{cases} t, 0 \le t \le 1 \\ t-2, 1 \le t \le 2 \\ 0, t > 2 \end{cases}$

16 $f(t) = \sin at \sin bt$

Use Table 7.1 together with Theorem 7.1 to find the Laplace transforms of the following functions.

17 $f(t) = \sinh kt - \sin kt$ **18** $f(t) = \sin kt + kt \cos kt$
19 $f(t) = 1 - \cos kt$ **20** $f(t) = kt - \sin kt$
21 $f(t) = \cosh kt - \cos kt$ **22** $f(t) = e^t[\cos 3t - \tfrac{2}{3}\sin 3t]$
23 $f(t) = \tfrac{1}{2}t^2 e^t + \tfrac{1}{3}e^{-t}\sinh 3t$ **24** $f(t) = (\sin kt)^2$

Use Table 7.1 to determine the following inverse Laplace transforms, after completing the square in the denominator where necessary.

25 $\mathcal{L}^{-1}\left\{\dfrac{3}{s+4}\right\}$ **26** $\mathcal{L}^{-1}\left\{\dfrac{6}{s^2+9}\right\}$

27 $\mathcal{L}^{-1}\left\{\dfrac{2}{s^5}\right\}$ **28** $\mathcal{L}^{-1}\left\{\dfrac{3(s^2-4)}{(s^2+4)^2}\right\}$

29 $\mathcal{L}^{-1}\left\{\dfrac{3}{2}\dfrac{s}{(s^2+9)^2}\right\}$ **30** $\mathcal{L}^{-1}\left\{\dfrac{16}{s^2-4}\right\}$

31 $\mathcal{L}^{-1}\left\{\dfrac{3s-6}{s^2-4s+13}\right\}$ **32** $\mathcal{L}^{-1}\left\{\dfrac{1}{3}\left(\dfrac{s+1}{s^2+2s+2}\right)\right\}$

33 $\mathcal{L}^{-1}\left\{\dfrac{1}{s^2+6s+2s}\right\}$ **34** $\mathcal{L}^{-1}\left\{\dfrac{3}{s^2}+\dfrac{5}{(s^2+9)^2}\right\}$

35 $\mathcal{L}^{-1}\left\{\dfrac{1}{(s+1)^4}-\dfrac{1}{4}\dfrac{s}{s^2+4}\right\}$ **36** $\mathcal{L}^{-1}\left\{\dfrac{1}{(s-2)^3}+\dfrac{1}{(s+2)^3}\right\}$

37 Determine which of the following functions is of exponential order $e^{\alpha t}$ and, where appropriate, find the constant α.

(a) $e^{-3t}\cos 2t$ (b) $\sin kt$ (c) $t^{-1/2}$ (d) $(3+2t^2)/(1+t)$
(e) $t^{-1/2}$ (f) $\exp(t^3/10^7)$ (g) $t^n e^{5t}$, n any positive integer
(h) $(1/t)\sin 3t$

7.2 Operational properties of the Laplace transform

This section is concerned with the effect the Laplace transform has on various fundamental mathematical operations which need to be performed on functions. The most important of these, concerning its effect on a linear combination of functions, has already been encountered in Theorem 7.1.

Results of this type are called **operational properties** of the Laplace transform. This is because they relate to the nature of the operation performed, rather than to the specific functions involved. Such properties are of interest because the Laplace transform provides a powerful method of solving initial value problems, and operational properties often help simplify this task. Their main uses are (i) finding the Laplace transform of functions without making a direct appeal to the definition in (2), Sec. 7.1, (ii) interpreting the effect of commonly occurring factors in one member of a transform pair on the other member, (iii) expressing products of transforms in terms of the original functions of t and (iv) using operational properties to simplify the task of finding inverse transforms.

For example, by using an operational property of the type described in (ii), it is possible to deduce the Laplace transform of $e^{at} f(t)$ directly from the Laplace transform of $f(t)$. When this result is coupled with others it enables a transform such as $\mathscr{L}\{e^{-3t} t^3 \cos 4t\}$ to be deduced from the simpler transforms $\mathscr{L}\{t^3\}$ and $\mathscr{L}\{\cos 4t\}$ listed in Table 7.1. Were $\mathscr{L}\{e^{-3t} t^3 \cos 4t\}$ to be obtained directly from definition (2), Sec. 7.1, the process would be extremely tedious, because it would necessitate repeated integration by parts.

Before proceeding to the derivation of the most important operational properties that are used when solving initial value problems for linear constant coefficient differential equations and systems we first summarize the main results.

Summary of operational properties of the Laplace transform

1. **Linearity**

    Let $f(t)$, $g(t)$ have Laplace transforms $\mathscr{L}\{f(t)\}$ and $\mathscr{L}\{g(t)\}$, and let α, β be arbitrary constants. Then

    $$\mathscr{L}\{\alpha f(t) + \beta g(t)\} = \alpha \mathscr{L}\{f(t)\} + \beta \mathscr{L}\{g(t)\}.$$

2. **First shift theorem**

    Let $\mathscr{L}\{f(t)\} = F(s)$ for $s > s_0$, and take a to be any real number. Then

    $$\mathscr{L}\{e^{at} f(t)\} = F(s-a), \qquad \text{for } s > s_0 + a$$

    and, conversely,

    $$\mathscr{L}^{-1}\{F(s-a)\} = e^{at} f(t).$$

3. **Second shift theorem**

Let $\mathcal{L}\{f(t)\} = F(s)$ for $s > s_0$, and take $\tau \geq 0$ to be an arbitrary nonnegative number. Then if $\mathcal{U}(t - \tau)$ is the unit step function (see (1), Sec. 7.2),

$$\mathcal{L}\{\mathcal{U}(t - \tau) f(t - \tau)\} = e^{-\tau s} F(s), \qquad \text{for } s > s_0 \text{ and, conversely,}$$

$$\mathcal{L}^{-1}\{e^{-\tau s} F(s)\} = \mathcal{U}(t - \tau) f(t - \tau).$$

4. **Scaling theorem**

Let $\mathcal{L}\{f(t)\} = F(s)$ for $s > s_0$, and let $\lambda > 0$ be an arbitrary positive number. Then

$$\mathcal{L}\{f(\lambda t)\} = \frac{1}{\lambda} F\left(\frac{s}{\lambda}\right), \qquad \text{for } s > \lambda s_0$$

and, conversely,

$$\mathcal{L}^{-1}\left\{F\left(\frac{s}{\lambda}\right)\right\} = \lambda f(\lambda t).$$

5. **Transform of a derivative**

Let $f(t)$ be defined and continuous for $t \geq 0$ and of exponential order $e^{\alpha t}$, and let its nth derivative $f^{(n)}(t)$ be piecewise continuous on any finite interval of $t \geq 0$. Then,

$$\mathcal{L}\{f^{(n)}(t)\} = s^n \mathcal{L}\{f\} - s^{n-1} f(0) - s^{n-2} f^{(1)}(0) - \ldots - f^{(n-1)}(0), \qquad \text{for } s > \alpha.$$

In particular,

$$\mathcal{L}\{f'(t)\} = s \mathcal{L}\{f\} - f(0), \qquad \text{for } s > \alpha,$$
$$\mathcal{L}\{f''(t)\} = s^2 \mathcal{L}\{f\} - s f(0) - f'(0), \qquad \text{for } s > \alpha,$$
$$\mathcal{L}\{f'''(t)\} = s^3 \mathcal{L}\{f\} - s^2 f(0) - s f'(0) - f''(0), \qquad \text{for } s > \alpha.$$

6. **Differentiation of a transform**

Let $\mathcal{L}\{f(t)\} = F(s)$, for $s > s_0$. Then

$$\frac{d^n F(s)}{ds^n} = \mathcal{L}\{(-t)^n f(t)\}, \qquad \text{for } s > s_0$$

and, conversely,

$$\mathcal{L}^{-1}\{F^{(n)}(s)\} = (-t)^n f(t).$$

7. **Transform of an integral**

Let $\mathcal{L}\{f(t)\} = F(s)$, \qquad for $s > s_0$. Then,

$$\mathcal{L}\left\{\int_0^t f(\tau) \, d\tau\right\} = \frac{F(s)}{s}, \qquad \text{for } s > s_0$$

and, conversely,

$$\mathcal{L}^{-1}\{F(s)/s\} = \int_0^t f(\tau) \, d\tau.$$

8. **Integration of a transform**

   Let $f(t)/t$ be defined for $t \geq 0$, piecewise continuous and of exponential order such that $\mathcal{L}\{f(t)/t\} = G(s)$, for $s > s_0$. Then if $\mathcal{L}\{f(t)\} = F(s)$,

   $$\mathcal{L}\left\{\frac{f(t)}{t}\right\} = \int_s^\infty F(u)\,du, \qquad \text{for } s > s_0$$

   and, conversely,

   $$\mathcal{L}^{-1}\{G(s)\} = -\frac{\mathcal{L}^{-1}\{G'(s)\}}{t}.$$

9. **Transformation of a periodic function**

   Let the function $f(t)$ be piecewise continuous for $t \geq 0$ and periodic with period T. Then

   $$\mathcal{L}\{f(t)\} = \frac{1}{1 - e^{-sT}} \int_0^T e^{-st} f(t)\,dt.$$

10. **Convolution theorem**

    This theorem interprets the product of two Laplace transforms in terms of the convolution operation performed on the original functions of t. Let $f(t)$, $g(t)$ be piecewise continuous functions defined for $t \geq 0$ such that $\mathcal{L}\{f(t)\} = F(s)$ for $s > \alpha$, and $\mathcal{L}\{g(t)\} = G(s)$ for $s > \beta$. Then the theorem asserts that

    $$\mathcal{L}\{f * g\} = F(s)G(s), \qquad \text{for } s > s_0, \text{ where } s_0 = \max\{\alpha, \beta\} \text{ and, conversely,}$$
    $$\mathcal{L}^{-1}\{F(s)G(s)\} = f * g.$$

    Here the convolution operation $f * g$ is defined as

    $$f * g = \int_0^t f(\tau)\, g(t - \tau)\,d\tau.$$

    The convolution operation is commutative, so that

    $$f * g = g * f.$$

11. **Extended initial value theorem**

    This theorem concerns the recovery of the initial values $f(0), f'(0), \ldots, f^{(n)}(0)$ of the function $f(t)$ and its derivatives from the Laplace transform $F(s) = \mathcal{L}\{f(t)\}$. The theorem asserts that, provided $f(t)$ and the necessary derivatives exist as $t \to 0$,

    (i) $\displaystyle\lim_{s \to \infty} [s F(s)] = f(0),$

    (ii) $\displaystyle\lim_{s \to \infty} [s^2 F(s) - s f(0)] = f'(0),$

    and

    (iii) $\displaystyle\lim_{s \to \infty} [s^{n+1} F(s) - s^n f(0) - s^{n-1} f^{(1)}(0) - s^{n-2} f^{(2)}(0) - \cdots$

    $$- s f^{(n-1)}(0)] = f^{(n)}(0).$$

12. Final value theorem (see Sec. 7.3(d))

This theorem concerns the recovery of the limiting (final) value $\lim_{t \to \infty} y(t) = y(\infty)$ of the function $f(t)$ from its Laplace transform $Y(s) = \mathscr{L}\{y(t)\}$. In its general form the final value theorem asserts that

$$\lim_{s \to 0} [s\, Y(s)] = y(\infty),$$

provided the limit exists and is finite.

If only a general Laplace transform $Y(s)$ is known which is not necessarily a rational function of s, the existence of the left-hand limit is not sufficient to ensure the existence of $\lim_{t \to \infty} y(t)$. However the following restricted form of the final value theorem is true.

Let $Y(s) = \mathscr{L}\{y(t)\}$ be such that

$$Y(s) = Q(s)/P(s),$$

with

$$P(s) = a_0 s^n + a_1 s^{n-1} + \ldots + a_n, \qquad Q(s) = b_0 s^m + b_1 s^{m-1} + \ldots + b_m,$$

$0 \le m < n$ and $a_0, a_1, \ldots, a_n, b_0, b_1, \ldots b_m$ real constants. Then $\lim_{t \to \infty} y(t) = y(\infty)$ exists and is finite and

$$\lim_{s \to 0} [s\, Y(s)] = y(\infty),$$

provided all the zeros of $P(s)$, with the possible exception of one which may be located at the origin, have strictly negative real parts.

Theorem 7.5 (The first shift theorem)

Let $\mathscr{L}\{f(t)\} = F(s)$ for $s > s_0$, and take a to be an arbitrary real number. Then

$$\mathscr{L}\{e^{at} f(t)\} = F(s-a),$$

for $s > s_0 + a$ and, conversely,

$$\mathscr{L}^{-1}\{F(s-a)\} = e^{at} f(t).$$

Proof

The first result follows directly from definition (2), Sec. 7.1, because

$$\mathscr{L}\{e^{at} f(t)\} = \int_0^\infty e^{-st} e^{at} f(t)\, dt$$

$$= \int_0^\infty \exp[-(s-a)t] f(t)\, dt = F(s-a),$$

for $s - a > s_0$ or, equivalently, for $s > s_0 + a$. The converse result follows directly by taking the inverse transform. $\square$

Example 1. Use of first shift theorem

Find $\mathcal{L}\{f\}$ given (i) $f(t) = e^{-2t} t \sin 3t$, (ii) $f(t) = e^{5t} t^3$, and (iii) find $\mathcal{L}^{-1}\left\{\dfrac{4}{s^2 - 6s + 25}\right\}$.

Solution

(i) From entry 9 of Table 7.1,

$$\mathcal{L}\{t \sin 3t\} = \frac{6s}{(s^2 + 9)^2}, \qquad \text{for } s > 0.$$

Applying the first result in Theorem 7.5 with $a = -2$, which involves replacing s by $s+2$, gives

$$\mathcal{L}\{e^{-2t} t \sin 3t\} = \frac{6(s+2)}{[(s+2)^2 + 9]^2} = \frac{6s + 12}{(s^2 + 4s + 13)^2},$$

for $s > -2$.

(ii) From entry 3 of Table 7.1,

$$\mathcal{L}\{t^3\} = \frac{6}{s^4}, \qquad \text{for } s > 0.$$

Applying the first result in Theorem 7.5 with $a = 5$, which involves replacing s by $s - 5$, gives

$$\mathcal{L}\{e^{5t} t^3\} = \frac{6}{(s-5)^4}, \qquad \text{for } s > 5.$$

(iii) Completing the square in the denominator of the transform gives

$$\frac{4}{s^2 - 6s + 25} = \frac{4}{(s-3)^2 + 4^2},$$

so from the last result of Theorem 7.1 and entry 7 of Table 7.1,

$$\mathcal{L}^{-1}\left\{\frac{4}{s^2 - 6s + 25}\right\} = \mathcal{L}^{-1}\left\{\frac{4}{(s-3)^2 + 4^2}\right\} = e^{3t} \sin 4t.$$

■

Before deriving the next theorem we need to discuss some uses of the step function introduced in Sec. 7.1. The **unit step function** for $t \geq 0$ is defined as the function $\mathcal{U}(t - \tau)$, where

$$\mathcal{U}(t - \tau) = \begin{cases} 0, & 0 \leq t < \tau \\ 1, & t \geq \tau. \end{cases} \tag{1}$$

One reason for the importance of this function[2] is that when used as a multiplier of a function $f(t)$ it 'switches on' and 'switches off' $f(t)$ at given values of t. This is illustrated by the functions.

$$y(t) = \mathcal{U}(t-1)t \quad \text{and} \quad y(t) = \{\mathcal{U}(t-1) - \mathcal{U}(t-2)\}t,$$

whose graphs are shown in Figs 7.4(a, b).

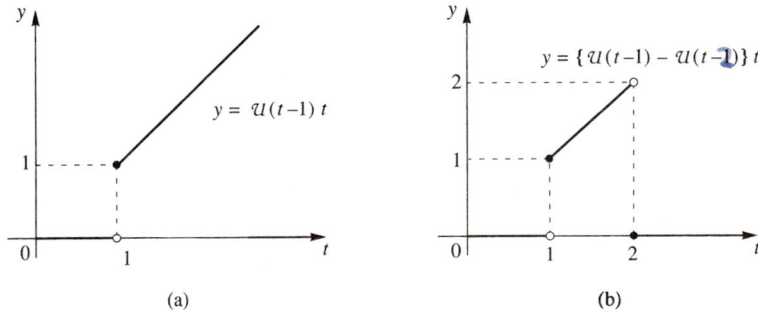

Fig. 7.4 Switching effect of the unit step function

The first function is obtained from the function $y(t) = t$ for $t \geq 0$ by 'switching it on' at $t = 1$, whereas the second function is obtained from the function $y(t) = t$ for $t \geq 0$ by 'switching it on' at $t = 1$ and then 'switching if off' again at $t = 2$.

In addition to its switching property, the unit step function may also be used to shift the graph of a function $f(t)$, defined for $t \geq 0$, to the right by an amount τ. Thus, for example, the function defined by

$$y(t) = \mathcal{U}(t-2)\exp[-\tfrac{1}{2}(t-2)]\sin 3(t-2) = \begin{cases} 0, & 0 \leq t < 2 \\ \exp[-\tfrac{1}{2}(t-2)\sin 3(t-2), & t \geq 2 \end{cases}$$

is the damped sinusoid shown for $t \geq 0$ in Fig. 7.5(a), shifted two units to the right as shown in Fig. 7.5(b). It is useful to remember that when t represents time, a shift to the right corresponds to a **time delay.**

Thus, in general, if $f(t)$ is defined for $t \geq 0$, the function

$$y(t) = \mathcal{U}(t-\tau)f(t-\tau) = \begin{cases} 0, & 0 \leq t < \tau \\ f(t-\tau), & t \geq \tau, \end{cases} \tag{2}$$

represents the function $f(t)$ shifted to the right (*delayed*) by an amount τ.

The Laplace transform of a step function of magnitude h was found as a special case of Ex. 2, Sec. 7.1. On account of its importance we now derive the result again in terms of the

---

[2] If desired, $\mathcal{U}(t-\tau)$ may be defined for all t by setting $\mathcal{U}(t-\tau) = 0$ for $t < \tau$ in (1), though this is not necessary because the Laplace transform is only applied to functions defined on the semi-infinite interval $t \geq 0$.

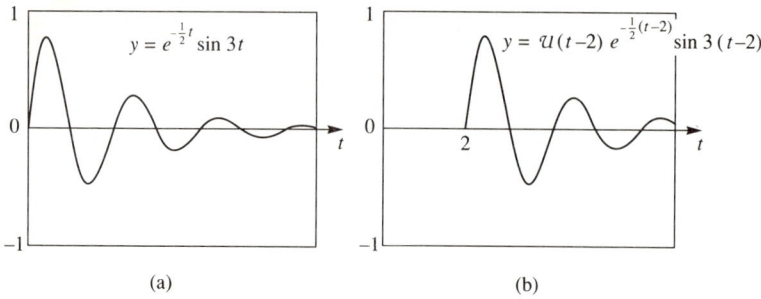

(a)   (b)

Fig. 7.5   Shifting property of the unit step function

unit step function $\mathcal{U}(t-\tau)$, with $\tau \geq 0$.

$$\mathcal{L}\{\mathcal{U}(t-\tau)\} = \int_0^\infty e^{-st}\,\mathcal{U}(t-\tau)\,dt$$

$$= \int_\tau^\infty e^{-st}\,dt$$

$$= \frac{e^{-s\tau}}{s}, \qquad \text{for } s > 0.$$

Thus we have established the important result that

$$\mathcal{L}\{\mathcal{U}(t-\tau)\} = \frac{e^{-s\tau}}{s}, \qquad \text{for } s > 0. \tag{3}$$

Example 2. Graphs using the step function

(i) Graph $y = \{\mathcal{U}\left(t - \frac{\pi}{2}\right) - \mathcal{U}(t - 2\pi)\}\sin t$,

(ii) Graph $y = \mathcal{U}(t-1)\tanh(t-1)$.

Solution

The graphs are shown in Figs 7.6 (a) and (b).

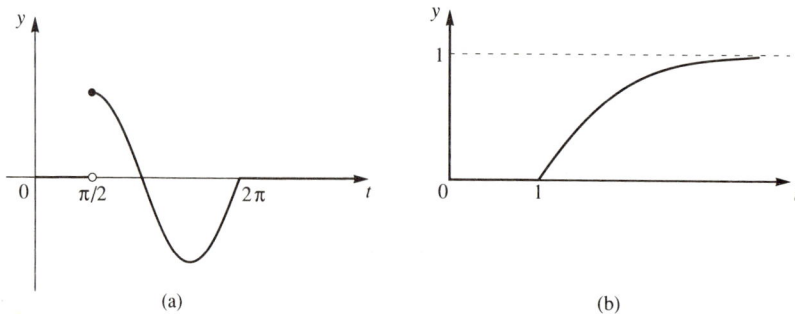

(a)   (b)

Fig. 7.6   (a) $y = \left\{\mathcal{U}\left(t - \frac{\pi}{2}\right) - \mathcal{U}(t - 2\pi)\right\}\sin t$; (b) $y = \mathcal{U}(t-1)\tanh(t-1)$ ■

With these ideas in mind we now proceed to the **second shift theorem**, sometimes known as the **time-delay theorem**.

Theorem 7.6 (Second shift theorem)

Let $\mathcal{L}\{f(t)\} = F(s)$ for $s > s_0$, and take $\tau \geq 0$ to be an arbitrary nonnegative number. Then

$$\mathcal{L}\{\mathcal{U}(t-\tau)f(t-\tau)\} = e^{-\tau s}F(s),$$

for $s > s_0$ and, conversely,

$$\mathcal{L}^{-1}\{e^{-\tau s}F(s)\} = \mathcal{U}(t-\tau)f(t-\tau).$$

Proof

By definition

$$\mathcal{L}\{\mathcal{U}(t-\tau)f(t-\tau)\} = \int_0^\infty e^{-st}\mathcal{U}(t-\tau)f(t-\tau)\,dt.$$

Using (2) and the substitution $v = t - \tau$ this becomes

$$\mathcal{L}\{\mathcal{U}(t-\tau)f(t-\tau)\} = \int_\tau^\infty e^{-st}f(t-\tau)\,dt$$

$$= \int_0^\infty \exp[-s(v+\tau)]f(v)\,dv$$

$$= e^{-s\tau}\int_0^\infty e^{-sv}f(v)\,dv$$

$$= e^{-s\tau}F(s), \quad \text{for } s > s_0,$$

which is the first result of the theorem. The converse result follows directly by taking the inverse transform. $\qquad\square$

Example 3. Applications of the second shift theorem

(i) Graph the function

$$f(t) = k[\mathcal{U}(t) - 2\mathcal{U}(t-\tau) + 2\mathcal{U}(t-2\tau) - \mathcal{U}(t-3\tau)],$$

and find $\mathcal{L}\{f\}$.

(ii) Find $\mathcal{L}\{\mathcal{U}(t-3)(t-3)^2\}$.

(iii) Find $\mathcal{L}\{\mathcal{U}(t-1)t\}$.

(iv) Find

$$\mathcal{L}^{-1}\left\{\frac{3e^{-2s}(s-4)}{s^2 - 8s + 32}\right\}.$$

Solution

(i)   The required graph is shown in Fig. 7.7

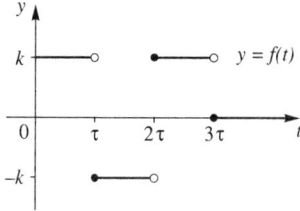

Fig. 7.7

Taking the transform of $f(t)$ gives

$$\mathscr{L}\{f\} = k[\mathscr{L}\{\mathscr{U}(t)\} - 2\mathscr{L}\{\mathscr{U}(t-\tau)\} + 2\mathscr{L}\{\mathscr{U}(t-2\tau) - \mathscr{L}\{\mathscr{U}(t-3\tau)\}],$$

so using (3) we have

$$\mathscr{L}\{f\} = k\left\{ \frac{1}{s} - \frac{2e^{-s\tau}}{s} + \frac{2e^{-2s\tau}}{s} - \frac{e^{-3s\tau}}{s} \right\}, \qquad \text{for } s > 0.$$

(ii) As $\mathscr{L}\{t^2\} = \dfrac{2}{s^3}$, for $s > 0$, it follows from the first result in Theorem 7.6 that

$$\mathscr{L}\{\mathscr{U}(t-3)(t-3)^2\} = \frac{2e^{-3s}}{s^3}, \qquad \text{for } s > 0.$$

(iii) The function to be transformed is of the form $\mathscr{U}(t-1)f(t)$ with $f(t)=t$. However, to apply Theorem 7.6, the function $f(t)$ must have the same argument as the step function, which has the argument $t-1$. So writing $t=(t-1)+1$ we see $\mathscr{U}(t-1)t = \mathscr{U}(t-1)(t-1) + \mathscr{U}(t-1)$, and thus

$$\mathscr{L}\{\mathscr{U}(t-1)t\} = \mathscr{L}\{\mathscr{U}(t-1)(t-1)\} + \mathscr{L}\{\mathscr{U}(t-1)\} = \frac{e^{-s}}{s^2} + \frac{e^{-s}}{s}, \qquad \text{for } s > 0.$$

(iv) Completing the square in the denominator of the transform gives

$$\frac{3e^{-2s}(s-4)}{s^2 - 8s + 32} = \frac{3e^{-2s}(s-4)}{(s-4)^2 + 4^2},$$

so by the converse result in Theorem 7.6 and entry 12 of Table 7.1,

$$\mathscr{L}^{-1}\left\{ \frac{3e^{-2s}(s-4)}{s^2 - 8s + 32} \right\} = \mathscr{U}(t-2)3\exp[4(t-2)]\cos 4(t-2).$$

∎

The next result determines how the transform $F(s)$ of a function $f(t)$ defined for $t \geq 0$ is modified when t in $f(t)$ is scaled by the factor λ. The result is helpful, for example, when working with mechanical or electrical analog devices in which t is the time. This is because it shows how the transform governing the response of a system must be modified when the time scale is changed.

Theorem 7.7 (Scaling theorem)

Let $\mathscr{L}\{f(t)\} = F(s)$ for $s > s_0$, and let $\lambda > 0$ be an arbitrary positive number. Then

$$\mathscr{L}\{f(\lambda t)\} = \frac{1}{\lambda}F\left(\frac{s}{\lambda}\right), \text{ for } s > \lambda s_0 \text{ and, conversely, } \mathscr{L}^{-1}\left\{F\left(\frac{s}{\lambda}\right)\right\} = \lambda f(\lambda t).$$

Proof

Using the definition of the Laplace transform, and making the substitution $u = \lambda t$, we find

$$\mathscr{L}\{f(\lambda t)\} = \int_0^\infty e^{-st} f(\lambda t)\,dt$$

$$= \frac{1}{\lambda}\int_0^\infty \exp[-su/\lambda]\,f(u)\,du$$

$$= \frac{1}{\lambda}F\left(\frac{s}{\lambda}\right),$$

where $(s/\lambda) > s_0$ or, equivalently, $s > \lambda s_0$. The converse result follows by taking the inverse transform. ☐

Example 4. Applications of the scaling theorem

(i) Find $\mathscr{L}\{t\cos 3t\}$, given that $\mathscr{L}\{t\cos t\} = (s^2 - 1)/(s^2 + 1)^2$, for $s > 0$.

(ii) If $\mathscr{L}\{f(t)\} = \dfrac{e^{-2s}(1-s)}{2s^2 + s + 4}$, find $\mathscr{L}\{f(2t)\}$.

Solution

(i) As

$$\mathscr{L}\{t\cos t\} = \frac{s^2 - 1}{(s^2 + 1)^2}, \qquad \text{for } s > 0,$$

it follows from Theorem 7.7 with $\lambda = 3$ that

$$\mathscr{L}\{3t\cos 3t\} = \frac{1}{3}\left(\frac{(s/3)^2 - 1}{[(s/3)^2 + 1]^2}\right) = 3\left[\frac{s^2 - 9}{(s^2 + 9)^2}\right], \qquad \text{for } s > 0.$$

Thus

$$\mathscr{L}\{t\cos 3t\} = \frac{s^2 - 9}{(s^2 + 9)^2}, \qquad \text{for } s > 0,$$

which agrees with entry 10 of Table 7.1.

(ii) Setting $\lambda = 2$ in Theorem 7.7 with

$$F(s) = \frac{e^{-2s}(1-s)}{2s^2 + s + 4},$$

we find

$$\mathcal{L}\{f(2t)\} = \frac{1}{2}\left(\frac{\exp[-2(s/2)][1-(s/2)]}{2(s/2)^2 + (s/2) + 4} \right) = \frac{1}{2}\frac{e^{-s}(2-s)}{(s^2 + s + 8)}.$$

∎

The value of the Laplace transform when dealing with initial value problems becomes apparent once the transformation of derivatives of $f(t)$ has been considered. The basic result which will be needed is as follows.

Theorem 7.8 (Laplace transform of the derivative $f'(t)$)

Let a function $f(t)$ which is defined and continuous for $t \geq 0$ be of exponential order $e^{\alpha t}$, and let its derivative $f'(t)$ be piecewise continuous on any finite interval in $t \geq 0$. Then the Laplace transform of the derivative $f'(t)$ exists, and is given by

$$\mathcal{L}\{f'(t)\} = s\mathcal{L}\{f\} - f(0), \qquad \text{for } s > \alpha. \tag{4}$$

Proof

Suppose first that $f'(t)$ is continuous, then

$$\mathcal{L}\{f'(t)\} = \int_0^\infty e^{-st} f'(t)\,dt = \lim_{R \to \infty} \int_0^R e^{-st} f'(t)\,dt$$

$$= \lim_{R \to \infty} \left[[e^{-st} f(t)]\big|_0^R + s \int_0^R e^{-st} f(t)\,dt \right]$$

$$= -f(0) + \lim_{R \to \infty} \left[e^{-st} f(t) + s \int_0^R e^{-st} f(t)\,dt \right].$$

Now $f(t)$ is of exponential order $e^{\alpha t}$, so the first bracketed term will vanish as $R \to \infty$ provided $s > \alpha$, while in the limit the integral becomes $\mathcal{L}\{f\}$, which is again subject to the condition $s > \alpha$. Thus we arrive at the result

$$\mathcal{L}\{f'(t)\} = s\mathcal{L}\{f\} - f(0), \qquad \text{for } s > \alpha. \qquad \square$$

The same form of argument applies when $f'(t)$ is only piecewise continuous on any finite interval in $t \geq 0$ and yields the same result, provided the interval of integration is subdivided into segments in each of which $f'(t)$ is continuous.

An immediate extension of this result to the second derivative $f''(t)$ of the function $f(t)$ in Theorem 7.8 follows by integrating by parts twice to establish that

$$\mathcal{L}\{f''\} = s^2 \mathcal{L}\{f\} - sf(0) - f'(0), \qquad \text{for } s > \alpha. \tag{5}$$

An inductive argument coupled with (4) establishes the following general result.

Theorem 7.9 (Laplace transform of the derivative $f^{(n)}(t)$)

Let a function $f(t)$ which is defined and continuous for $t \geq 0$ be of exponential order $e^{\alpha t}$, and let its nth derivative $f^{(n)}(t)$ be piecewise continuous on any finite interval of $t \geq 0$. Then the Laplace transform of the derivative $f^{(n)}(t)$ exists and is given by

$$\mathcal{L}\{f^{(n)}(t)\} = s^n \mathcal{L}\{f\} - s^{n-1} f(0) - s^{n-2} f'(0) - \ldots - f^{(n-1)}(0), \qquad \text{for } s > \alpha. \tag{6}$$

□

The main use for (6) will be in the solution of initial value problems, though on occasions it helps find the transform of a complicated function in terms of the transform of a simple related function. This application of (6) is illustrated in the next two examples.

Example 5

Use Theorem 7.9 to find $\mathcal{L}\{f\}$ given that $f(t) = te^{-3t}$ and $\mathcal{L}\{e^{at}\} = 1/(s-a)$.

Solution

As $f(t) = te^{-3t}$, $f(0) = 0$ and $f'(t) = e^{-3t} - 3te^{-3t} = e^{-3t} - 3f(t)$. Setting $n = 1$ in (6) (or using (4)) we find

$$\mathcal{L}\{f'\} = s\mathcal{L}\{f\} - f(0),$$

so as $f(0) = 0$,

$$\mathcal{L}\{e^{-3t} - 3f(t)\} = s\mathcal{L}\{f\}.$$

Thus

$$\mathcal{L}\{e^{-3t}\} - 3\mathcal{L}\{f\} = s\mathcal{L}\{f\},$$

but

$$\mathcal{L}\{e^{-3t}\} = 1/(s+3) \qquad \text{for } s > -3, \text{ and so}$$
$$\mathcal{L}\{te^{-3t}\} = 1/(s+3)^2, \qquad \text{for } s > -3,$$

which agrees with entry 6 in Table 7.1 when $n = 1$ and $a = -3$. ∎

Example 6

Use Theorem 7.9 to find $\mathcal{L}\{f\}$ given that $f(t) = \cos^2 t$.

Solution

As $f(t) = \cos^2 t$, it follows that $f'(t) = -2 \sin t \cos t$, $f''(t) = 2 - 4\cos^2 t = 2 - 4f(t)$, and so $f(0) = 1$ and $f'(0) = 0$. By setting $n = 2$ in (6) (or using (5)) we obtain

$$\mathcal{L}\{f''\} = s^2 \mathcal{L}\{f\} - sf(0) - f'(0)$$

and so using the initial values $f(0) = 1$, $f'(0) = 0$ we find that

$$\mathscr{L}\{2 - 4f(t)\} = s^2 \mathscr{L}\{f\} - s.$$

or

$$\mathscr{L}\{2\} - 4\mathscr{L}\{f\} = s^2 \mathscr{L}\{f\} - s.$$

Using the result $\mathscr{L}\{2\} = 2/s$, for $s > 0$, we obtain the result

$$\mathscr{L}\{\cos^2 t\} = \frac{2 + s^2}{s(s^2 + 4)}, \qquad \text{for } s > 0.$$ ■

Inspection of the form of (6) shows the close connection which exists between the Laplace transform of the derivative $f^{(n)}(t)$, and multiplication of the Laplace transform of $f(t)$ by s^n. This relationship forms the basis of the solution of initial value problems for linear constant coefficient differential equations by the Laplace transform. If $f(t)$ is the solution of an nth order differential equation, the quantities $f(0), f'(0), \ldots, f^{(n-1)}(0)$ are the initial values of $f(t)$, and (6) shows that they enter quite naturally when the nth derivative is transformed.

We are now in a position to consider the solution of some simple but representative initial value problems by means of the Laplace transform.

Example 7. A nonhomogeneous second order equation

Solve by means of the Laplace transform the second order equation

$$x'' + 9x = \cos 3t$$

subject to the initial conditions $x(0) = 0$, $x'(0) = 4$.

Solution

Taking the Laplace transform of the differential equation we find

$$\mathscr{L}\{x''\} + 9\mathscr{L}\{x\} = \mathscr{L}\{\cos 3t\},$$

where the linearity property of Theorem 7.1 has been used on the left-hand side. Using Theorem 7.8 with $n = 2$ (or result (5)) together with entry 8 of Table 7.1 this becomes

$$s^2 X(s) - s\,x(0) - x'(0) + 9X(s) = \frac{s}{s^2 + 9}.$$

However $x(0) = 0$ and $x'(0) = 4$, so this reduces to

$$(s^2 + 9)X(s) = 4 + \frac{s}{s^2 + 9},$$

and so

$$X(s) = \frac{4}{s^2 + 9} + \frac{s}{(s^2 + 9)^2}.$$

Here $X(s)$ is the Laplace transform of the required solution $x(t)$, so as $x(t) = \mathcal{L}^{-1}\{X(s)\}$, it follows that

$$x(t) = \mathcal{L}^{-1}\{X(s)\} = \mathcal{L}^{-1}\left\{\frac{4}{s^2+9}\right\} + \mathcal{L}^{-1}\left\{\frac{s}{(s^2+9)^2}\right\}.$$

Inspection of Table 7.1 shows

$$\mathcal{L}\{\sin 3t\} = \frac{3}{s^2+9} \quad \text{and} \quad \mathcal{L}\{t\sin 3t\} = \frac{6s}{(s^2+9)^2},$$

so that

$$\mathcal{L}^{-1}\left\{\frac{3}{s^2+9}\right\} = \sin 3t \quad \text{and} \quad \mathcal{L}^{-1}\left\{\frac{6s}{(s^2+9)^2}\right\} = t\sin 3t.$$

Using these results in the expression for $x(t)$ gives as the solution of the initial value problem

$$x(t) = \tfrac{4}{3}\sin 3t + \tfrac{1}{6}t\sin 3t, \qquad \text{for } t \geq 0. \qquad \blacksquare$$

Notice how the transform method incorporates the initial conditions at the outset. Once $X(s)$ has been found, taking the inverse transform yields the required solution immediately. This should be compared with the methods of solution discussed in the previous chapters in which the general solution is obtained first, after which the arbitrary constants must be determined from the initial conditions. It should also be observed that the occurrence of a nonhomogeneous term which is contained in the complementary function is dealt with automatically by the transform method, rather than as a special case as in previous chapters.

Example 8. A second order equation with a discontinuous nonhomogeneous term

Solve by means of the Laplace transform the second order equation

$$x'' + 2x' + 5x = \mathcal{U}(t-1) - \mathcal{U}(t-2)$$

subject to the initial conditions $x(0) = 0$ and $x' = 1$, where a prime denotes differentiation with respect to t.

Solution

Taking the Laplace transform of the equation gives

$$\mathcal{L}\{x''\} + 2\mathcal{L}\{x'\} + 5\mathcal{L}\{x\} = \mathcal{L}\{\mathcal{U}(t-1)\} - \mathcal{L}\{\mathcal{U}(t-2)\},$$

or

$$[s^2X(s) - sx(0) - x'(0)] + 2[sX(s) - x(0)] + 5X(s) = \frac{e^{-s}}{s} - \frac{e^{-2s}}{s}.$$

Making use of the initial conditions $x(0) = 0$, $x'(0) = 1$ and solving for $X(s)$ gives

$$X(s) = \frac{1}{(s^2 + 2s + 5)} + \frac{e^{-s}}{s(s^2 + 2s + 5)} - \frac{e^{-2s}}{s(s^2 + 2s + 5)}.$$

To find the inverse transform it is necessary to simplify the terms in $X(s)$ by completing the square in the denominators and using partial fractions. Routine calculations show that

$$\frac{1}{s^2 + 2s + 5} = \frac{1}{(s+1)^2 + 4},$$

and

$$\frac{1}{s(s^2 + 2s + 5)} = \frac{1}{5}\left(\frac{1}{s}\right) - \frac{1}{5}\left(\frac{s-2}{(s+1)^2 + 4}\right)$$

$$= \frac{1}{5}\left(\frac{1}{s}\right) - \frac{1}{5}\left(\frac{s+1}{(s+1)^2 + 4}\right) + \frac{3}{5}\left(\frac{1}{(s+1)^2 + 4}\right).$$

Using these results in the expression for $X(s)$ gives

$$X(s) = \frac{1}{(s+1)^2 + 4} + \frac{e^{-s}}{5}\left[\left(\frac{1}{s}\right) - \left(\frac{s+1}{(s+1)^2 + 4}\right) + 3\left(\frac{1}{(s+1)^2 + 4}\right)\right]$$

$$+ \frac{e^{-2s}}{5}\left[\left(\frac{1}{s}\right) - \left(\frac{s+1}{(s+1)^2 + 4}\right) + 3\left(\frac{1}{(s+1)^2 + 4}\right)\right].$$

The solution $x(t) = \mathscr{L}^{-1}\{X(s)\}$ follows from this result with the help of Table 7.1 and Theorems 7.5 and 7.6. To see how $x(t)$ is determined notice first that $\mathscr{L}^{-1}\{1/s\} = 1$, while from entries 7 and 8 of Table 7.1 coupled with Theorems 7.5,

$$\mathscr{L}^{-1}\left\{\frac{1}{(s+1)^2 + 4}\right\} = \frac{1}{2}e^{-t}\sin 2t, \quad \mathscr{L}^{-1}\left\{\frac{s+1}{(s+1)^2 + 4}\right\} = e^{-t}\cos 2t.$$

If these results are now used in the expression for $\mathscr{L}^{-1}\{X(s)\}$ and Theorem 7.6 is applied to take account of the exponential factors we arrive at the solution

$$x(t) = \frac{1}{2}e^{-t}\sin 2t + \frac{1}{5}\mathscr{U}(t-1)[1 - \exp[-(t-1)]\cos 2(t-1) + \frac{3}{2}\exp[-(t-1)]\sin 2(t-1)]$$

$$- \frac{1}{5}\mathscr{U}(t-2)[1 - \exp[-(t-2)]\cos 2(t-2) + \frac{3}{2}\exp[-(t-2)]\sin 2(t-2),$$

$$\text{for } t \geq 0.$$

∎

Example 9. A second order equation with a time delayed nonhomogeneous term

Solve by means of the Laplace transform

$$x'' + 3x' + 2x = f(t)$$

subject to the initial conditions $x(0) = 1$, $x'(0) = -3$, where

$$f(t) = \begin{cases} 0, & 0 \leq t < 2 \\ t - 2, & t \geq 2, \end{cases}$$

and a prime denotes differentiation with respect to t.

Solution

Notice first that $f(t)$ may be written

$$f(t) = \mathscr{U}(t-2)(t-2),$$

so the differential equation becomes

$$x'' + 3x' + 2x = \mathscr{U}(t-2)(t-2).$$

Taking the Laplace transform of this equation we obtain

$$\mathscr{L}\{x''\} + 3\mathscr{L}\{x'\} + 2\mathscr{L}\{x\} = \mathscr{L}\{\mathscr{U}(t-2)(t-2)\}.$$

Proceeding as in Ex. 6 and using Theorem 7.6 this becomes

$$[s^2 X(s) - sx(0) - x'(0)] + 3[s X(s) - x(0)] + 2X(s) = e^{-2s}\mathscr{L}\{t\}.$$

However, as $x(0) = 1$, $x'(0) = -3$ and $\mathscr{L}\{t\} = 1/s^2$, this reduces to

$$(s^2 + 3s + 2) X(s) = \frac{e^{-2s}}{s^2} + s,$$

or to

$$X(s) = \frac{e^{-2s}}{s^2(s+1)(s+2)} + \frac{s}{(s+1)(s+2)}.$$

It is now necessary to simplify these expressions so that the inverse transform $x(t) = \mathscr{L}^{-1}\{X(s)\}$ can be identified. To do this we use the partial fraction representations

$$\frac{1}{s^2(s+1)(s+2)} = \frac{1}{4}\left(\frac{1}{s}\right) + \frac{1}{2}\left(\frac{1}{s^2}\right) + \left(\frac{1}{s+1}\right) - \frac{1}{4}\left(\frac{1}{s+2}\right),$$

and

$$\frac{s}{(s+1)(s+2)} = 2\left(\frac{1}{s+2}\right) - \left(\frac{1}{s+1}\right),$$

the justification of which is left to the reader.

Using these results to simplify $X(s)$ and then taking the inverse transform gives

$$x(t) = \frac{1}{4}\mathscr{L}^{-1}\left\{\frac{e^{-2s}}{s}\right\} + \frac{1}{2}\mathscr{L}^{-1}\left\{\frac{e^{-2s}}{s^2}\right\} + \mathscr{L}^{-1}\left\{\frac{e^{-2s}}{s+1}\right\} - \frac{1}{4}\mathscr{L}^{-1}\left\{\frac{e^{-2s}}{s+2}\right\}$$

$$+ 2\mathscr{L}^{-1}\left\{\frac{1}{s+2}\right\} - \mathscr{L}^{-1}\left\{\frac{1}{s+1}\right\}.$$

Making the necessary identifications in Table 7.1 and using Theorem 7.6 we arrive at the solution

$$x(t) = \mathscr{U}(t-2)[\tfrac{1}{4} + \tfrac{1}{2}(t-2) + \exp[-(t-2)] - \tfrac{1}{4}\exp[-2(t-2)] + 2e^{-2t} - e^{-t}, \qquad \text{for } t \geq 0.$$ ∎

Example 10. A nonhomogeneous system of first order equations

Solve by means of the Laplace transform the nonhomogeneous first order system

$$4\frac{dx}{dt} - \frac{dy}{dt} + 3x = \sin t$$

$$\frac{dx}{dt} + y = \cos t,$$

subject to the initial conditions $x(0) = y(0) = 0$.

Solution

Taking the Laplace transform of the equations gives, respectively,

$$4\mathscr{L}\{x'\} - \mathscr{L}\{y'\} + 3\mathscr{L}\{x\} = \mathscr{L}\{\sin t\}$$

and

$$\mathscr{L}\{x'\} + \mathscr{L}\{y\} = \mathscr{L}\{\cos t\},$$

where the prime denotes differentiation with respect to t. Setting $\mathscr{L}\{x\} = X(s)$, $\mathscr{L}\{y\} = Y(s)$ and using (4) together with Table 7.1 these equations become

$$4[s\,X(s) - x(0)] - [s\,Y(s) - y(0)] + 3\,X(s) = \frac{1}{s^2 + 1}$$

and

$$[s\,X(s) - x(0)] + Y(s) = \frac{s}{s^2 + 1}.$$

Incorporating the initial conditions $x(0) = y(0) = 0$ leads to the simultaneous equations for $X(s)$ and $Y(s)$,

$$(4s + 3)X(s) - s\,Y(s) = \frac{1}{s^2 + 1}$$

$$s\,X(s) + Y(s) = \frac{s}{s^2 + 1}.$$

Solving for $X(s)$ gives

$$X(s) = \frac{1}{s^2 + 4s + 3} = \frac{1}{(s+1)(s+3)},$$

and using this result in the second equation we find

$$Y(s) = \frac{s}{s^2 + 1} - \frac{s}{(s+1)(s+3)}.$$

As $x(t) = \mathscr{L}^{-1}\{X(s)\}$ and $y(t) = \mathscr{L}^{-1}\{Y(s)\}$, it follows that

$$x(t) = \mathscr{L}^{-1}\left\{\frac{1}{(s+1)(s+3)}\right\},$$

and

$$y(t) = \mathscr{L}^{-1}\left\{\frac{s}{s^2+1}\right\} - \mathscr{L}^{-1}\left\{\frac{s}{(s+1)(s+3)}\right\}.$$

To simplify these inverse transforms so they can be recognized we now make use of the partial fraction representations

$$\frac{1}{(s+1)(s+3)} = \frac{1}{2}\left(\frac{1}{s+1}\right) - \frac{1}{2}\left(\frac{1}{s+3}\right)$$

and

$$\frac{s}{(s+1)(s+3)} = -\frac{1}{2}\left(\frac{1}{s+1}\right) + \frac{3}{2}\left(\frac{1}{s+3}\right),$$

the derivation of which is left as an exercise for the reader.

Using the results in the expressions for $x(t)$ and $y(t)$ gives

$$x(t) = \frac{1}{2}\mathscr{L}^{-1}\left\{\frac{1}{s+1}\right\} - \frac{1}{2}\mathscr{L}^{-1}\left\{\frac{1}{s+3}\right\},$$

$$y(t) = \mathscr{L}^{-1}\left\{\frac{s}{s^2+1}\right\} + \frac{1}{2}\mathscr{L}^{-1}\left\{\frac{1}{s+1}\right\} - \frac{3}{2}\mathscr{L}^{-1}\left\{\frac{1}{s+3}\right\}.$$

Inspection of the transform pairs in Table 7.1 shows

$$\mathscr{L}^{-1}\left\{\frac{1}{s+1}\right\} = e^{-t}, \quad \mathscr{L}^{-1}\left\{\frac{1}{s+3}\right\} = e^{-3t} \quad \text{and} \quad \mathscr{L}^{-1}\left\{\frac{s}{s^2+1}\right\} = \cos t,$$

so using these results in $x(t)$ and $y(t)$ we arrive at the required solution.

$$x(t) = \tfrac{1}{2}e^{-t} - \tfrac{1}{2}e^{-3t}, \quad y(t) - \cos t + \tfrac{1}{2}e^{-t} - \tfrac{3}{2}e^{-3t}, \quad \text{for } t \geq 0.$$

∎

Theorem 7.10 (Differentiation of a transform)

Let $\mathscr{L}\{f(t)\} = F(s)$, for $s > s_0$. Then,

$$\frac{d^n F(s)}{ds^n} = \mathscr{L}\{(-t)^n f(t)\}, \quad \text{for } s > s_0 \quad \text{and} \quad n = 1, 2, \ldots \text{ and, conversely,}$$

$$\mathscr{L}^{-1}\{F^{(n)}(s)\} = (-t)^n f(t).$$

Thus differentiation of the transform of a function $f(t)$ corresponds to multiplication of $f(t)$ by the factor $-t$.

Proof

We shall prove these results only for the case in which $f(t)$ is of exponential order $\exp(s_0 t)$, though in fact they are true for any function $f(t)$ which has a Laplace transform.

The Laplace transform

$$F(s) = \int_0^\infty e^{-st} f(t)\, dt$$

is defined for $s > s_0$ when $f(t)$ is of exponential order $\exp(s_0 t)$. Differentiation with respect to s gives

$$\frac{dF(s)}{ds} = \int_0^\infty \frac{\partial}{\partial s} [e^{-st} f(t)]\, dt$$

$$= \int_0^\infty e^{-st}(-t) f(t)\, dt = \mathcal{L}\{(-t) f(t)\}$$

for $s > s_0$. Differentiation under the integral sign with respect to s is justified by Theorem 1.10, because as $f(t)$ is of exponential order $\exp(s_0 t)$, so also is $tf(t)$. An inductive argument coupled with the fact that if $f(t)$ is of exponential order $\exp(s_0 t)$, then so also is $t^n f(t)$ for $n = 2, 3, \ldots$, establishes the first result stated in the theorem. The converse result follows directly by taking the inverse transform.

This theorem is useful because it provides a simple method for the derivation of a sequence of related Laplace transforms from a given transform, and also a convenient means for the inversion of the derivative of a transform.

Example 11. Differentiation of a transform

(i) Given that $\mathcal{L}\{e^{at} \sin kt\} = \dfrac{k}{(s-a)^2 + k^2}$ for $s > a$, find $\mathcal{L}\{t e^{at} \sin kt\}$.

(ii) Find $\mathcal{L}\{t \mathcal{L}^{-1}\{\operatorname{arc tanh}(a/s)\}\}$.

Solution

(i) From Theorem 7.10 we have

$$\mathcal{L}\{(-t)e^{at} \sin kt\} = \frac{d}{ds}\left(\frac{k}{(s-a)^2 + k^2}\right), \qquad \text{for } s > a,$$

which reduces to

$$\mathcal{L}\{t e^{at} \sin kt\} = \frac{2k(s-a)}{[(s-a)^2 + k^2]^2}, \qquad \text{for } s > a.$$

(ii) If $f(t) = \mathcal{L}^{-1}\{\operatorname{arc tanh}(a/s)\}$, then $F(s) = \mathcal{L}\{f(t)\} = \operatorname{arc tanh}(a/s)$. Hence from Theorem 7.10 (with $n = 1$), we have $\mathcal{L}\{t \mathcal{L}^{-1}\{\operatorname{arc tanh}(a/s)\}\} = -\dfrac{d}{ds} F(s) = -\dfrac{d}{ds} \operatorname{arc tanh}(a/s) = \dfrac{a}{s^2 - a^2}$, for $s > a$. ∎

Example 12. Transformation of $t^m f^{(n)}(t)$

Given $\mathcal{L}\{f(t)\} = F(s)$, find

(i) $\mathcal{L}\{tf'(t)\}$ (ii) $\mathcal{L}\{tf''(t)\}$ (iii) $\mathcal{L}\{t^2 f''(t)\}$ and   (iv) $\mathcal{L}\{t^m f^{(n)}(t)\}$.

Solution

(i) From Theorem 7.10 we have

$$\mathcal{L}\{tf'(t)\} = -\frac{d}{ds}\{\mathcal{L}\{f'(t)\}\}.$$

However by Theorem 7.8 this may be written

$$\mathcal{L}\{tf'(t)\} = -\frac{d}{ds}\{sF(s) - f(0)\},$$

and so

$$\mathcal{L}\{tf'(t)\} = -sF'(s) - F(s).$$

(ii) Using Theorems 7.10 and 7.9 we have

$$\mathcal{L}[tf''(t)] = -\frac{d}{ds}[s^2 F(s) - sf(0) - f'(0)],$$

and so

$$\mathcal{L}\{tf''(t)\} = -s^2 F'(s) - 2sF(s) + f(0).$$

(iii) Proceeding as in (ii) we have

$$\mathcal{L}\{t^2 f''(t)\} = \frac{d^2}{ds^2}[s^2 F(s) - sf(0) - f'(0)],$$

and so

$$\mathcal{L}\{t^2 f''(t)\} = s^2 F''(s) + 4sF'(s) + 2F(s).$$

(iv) In general, arguing in similar fashion, it follows that if m and n are integers,

$$\mathcal{L}\{t^m f^{(n)}(t)\} = (-1)^m \frac{d^m}{ds^m}[s^n F(s) - s^{n-1} f(0) - s^{n-2} f(0) - \ldots - f(0)^{(n-1)}].$$

∎

Theorem 7.11 (Transform of an integral)

Let $\mathcal{L}\{f(t)\} = F(s)$, for $s > s_0$. Then,

$$\mathcal{L}\left\{\int_0^t f(\tau)\,d\tau\right\} = \frac{F(s)}{s}, \qquad \text{for } s > s_0$$

and, conversely,

$$\mathcal{L}^{-1}\{F(s)/s\} = \int_0^t f(\tau)\,d\tau.$$

Thus the transform of an integral corresponds to the division of its transform by s.

Proof

To establish the first result we must start by showing $h(t) = \int_0^t f(\tau)\,d\tau$ has a Laplace transform for $s > s_0$, and then that $\mathscr{L}\{h(t)\} = F(s)/s$ for $s > s_0$.

The function $h(t)$ has a Laplace transform because $f(t)$ is of exponential order $\exp(s_0 t)$, which implies that

$$|f(t)| < Me^{\alpha t},$$

for $t > T > 0$, some $M > 0$ and $\alpha > s_0 \geq 0$. Thus, using this result, we have

$$|g(t)| = \left| \int_0^t f(\tau)\,d\tau \right| \leq \int_0^t |f(\tau)|\,d\tau < M \int_0^t e^{\alpha \tau}\,d\tau < M^* e^{\alpha t},$$

for some $M^* > 0$. This shows $|g(t)|$ is of exponential order $e^{\alpha t}$ with $\alpha > s_0 \geq 0$, and consequently that $\mathscr{L}\{g(t)\}$ exists for $s > s_0$.

To complete the proof of the first result we must now use the fact that $h(0) = 0$ and $h'(t) = f(t)$. Then from Theorem 7.8

$$F(s) = \mathscr{L}\{f(t)\} = \mathscr{L}\{h'(t)\} = s\mathscr{L}\{h(t)\} = s\mathscr{L}\left\{ \int_0^t f(\tau)\,d\tau \right\} \quad \text{for } s > s_0,$$

from which the result follows after division by s. The converse result follows by taking the inverse transform.

$\square$

Example 13. Transform of an integral

Use Theorem 7.11 to evaluate

(i) $\mathscr{L}\left\{ \int_0^t e^{a\tau} \cos k\tau\,d\tau \right\}$ and (ii) $\mathscr{L}^{-1}\{1/(s^2 + k^2)\}$,

and then to prove that if $F(s) = \mathscr{L}\{f(t)\}$,

(iii) $\mathscr{L}^{-1}\left\{ \dfrac{F(s)}{s^2} \right\} = \int_0^t \int_0^u f(\tau)\,d\tau\,du = \int_0^t (t - \tau)f(\tau)\,d\tau.$

Solution

(i) From entry 12 of Table 7.1

$$\mathscr{L}\{e^{at} \cos kt\} = \frac{s - a}{(s-a)^2 + k^2}, \quad \text{for } s > a.$$

Thus it follows directly from Theorem 7.11 that

$$\mathscr{L}\left\{ \int_0^t e^{a\tau} \cos k\tau\,d\tau \right\} = \frac{s - a}{s[(s-a)^2 + k^2]}, \quad \text{for } s > a.$$

(ii) We may write

$$\mathscr{L}^{-1}\{1/(s^2+k^2)\} = \mathscr{L}^{-1}\left\{\left(\frac{s}{s^2+k^2}\right)\left(\frac{1}{s}\right)\right\},$$

which is of the form $\mathscr{L}^{-1}\{F(s)/s\}$ with $F(s) = s/(s^2+k^2)$. However from entry 8 of Table 7.1 we know that $f(t) = \mathscr{L}^{-1}\{F(s)\} = \mathscr{L}^{-1}\{s/(s^2+k^2)\} = \cos kt$.

Thus applying the converse result given in Theorem 7.11 we obtain

$$\mathscr{L}^{-1}\{1/(s^2+k^2)\} = \mathscr{L}^{-1}\{F(s)/s\} = \int_0^t \cos k\tau \, d\tau = \frac{1}{k}\sin kt,$$

which confirms entry of Table 7.1.

(iii) If $F(s) = \mathscr{L}\{f(t)\}$, then from Theorem 7.11

$$\mathscr{L}^{-1}\left\{\frac{F(s)}{s}\right\} = \int_0^t f(\tau)\,d\tau = h(t), \text{ say.}$$

A further application of Theorem 7.11 gives

$$\mathscr{L}^{-1}\left\{\frac{\mathscr{L}\{h(t)\}}{s}\right\} = \mathscr{L}^{-1}\left\{\frac{F(s)}{s^2}\right\} = \int_0^t h(u)\,du = \int_0^t\left\{\int_0^u f(\tau)\,d\tau\right\}du = \int_0^t\int_0^u f(\tau)\,d\tau\,du.$$

The special form of double integral arising here is called an **iterated integral** because of the order in which the successive integrations are performed. When evaluating this iterated integral the first step is the determination of the inner integral, which involves integrating $f(\tau)$ with respect to τ over the interval $0 \le \tau \le u$ (with u held constant) to yield a function of u. The second step is the integration of this function of u with respect to u over the interval $0 \le u \le t$. Thus this iterated integral represents the integral of $f(\tau)$ over the shaded triangular region shown in Fig. 7.8(a).

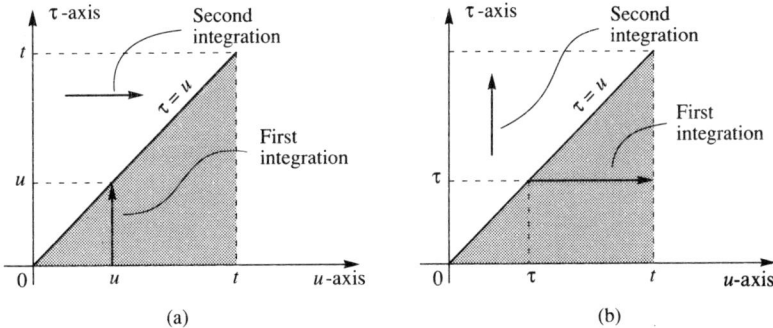

Fig. 7.8 Equivalence of iterated integrals (a) $\int_0^t\left\{\int_0^u f(\tau)\,d\tau\right\}du$ (b) $\int_0^t\left\{\int_\tau^t f(\tau)\,du\right\}d\tau$

The same integral may also be evaluated by reversing the order of the integration, so that first $f(\tau)$ is integrated with respect to u over the interval $\tau \le u \le t$ (with τ held constant), after which the resulting function of τ is integrated with respect to τ over the interval $0 \le \tau \le t$, as shown in

Fig. 7.8(b). Thus we have the result

$$\int_0^t \int_0^u f(\tau)\,d\tau\,du = \int_0^t \left\{ \int_\tau^t f(\tau)\,du \right\} d\tau$$

$$= \int_0^t \left\{ \int_\tau^t du \right\} f(\tau)\,d\tau = \int_0^t (t-\tau) f(\tau)\,d\tau.$$

We have proved that

$$\mathscr{L}^{-1}\left\{ \frac{F(s)}{s^2} \right\} = \int_0^t \int_0^u f(\tau)\,d\tau\,du = \int_0^t (t-\tau) f(\tau)\,d\tau.$$

This result will be needed in Sec. 7.3(h) when discussing Volterra integral equations. ■

Theorem 7.12 (Integration of a transform)

Let $f(t)/t$ be defined for $t \geq 0$, piecewise continuous, and of exponential order such that $\mathscr{L}\{f(t)/t\} = G(s)$, for $s > s_0$. Then if $\mathscr{L}\{f(t)\} = F(s)$,

$$\mathscr{L}\{f(t)/t\} = \int_s^\infty F(u)\,du, \qquad \text{for } s > s_0.$$

Thus integration of a transform of a function $f(t)$ corresponds to division of $f(t)$ by t. Conversely,

$$\mathscr{L}^{-1}\{G(s)\} = -\frac{\mathscr{L}^{-1}\{G'(s)\}}{t}.$$

Proof

By definition,

$$G(s) = \int_0^\infty e^{-st} \frac{f(t)}{t}\,dt, \qquad \text{for } s > s_0.$$

Thus by Theorem 7.10

$$G'(s) = \int_0^\infty e^{-st}(-t)\frac{f(t)}{t}\,dt = -\int_0^\infty e^{-st} f(t)\,dt = -F(s),$$

and so

$$\int_s^\infty F(u)\,du = -\int_s^\infty G'(u)\,du = G(s) - G(\infty).$$

However, from Theorem 7.4 it follows that $G(\infty)=0$, and thus

$$G(s) = \mathscr{L}\{f(t)/t\} = \int_s^\infty F(u)\,du, \qquad \text{for } s > s_0,$$

which was the first result to be proved.

The converse result follows from the fact that

$$\mathscr{L}^{-1}\{G(s)\} = \frac{f(t)}{t},$$

because

$$\mathscr{L}\{f(t)\} = F(s) = -G'(s),$$

so

$$f(t) = -\mathscr{L}^{-1}\{G'(s)\},$$

giving

$$\mathscr{L}^{-1}\{G(s)\} = -\frac{\mathscr{L}^{-1}\{G'(s)\}}{t}. \qquad \square$$

Example 14. Integration of transforms and inversion

Find

(i) $\mathscr{L}\left\{\dfrac{\sin kt - kt \cos kt}{t}\right\}$,

(ii) $\mathscr{L}\left\{\displaystyle\int_0^t \left(\dfrac{\sin ku - ku \cos ku}{u}\right)\right\} du$,

(iii) $\mathscr{L}^{-1}\left\{\arctan \dfrac{s}{k} + \dfrac{1}{2}\ln\left|\dfrac{k+s}{k-s}\right| - \dfrac{\pi}{2}\right\}$.

Solution

(i) From Ex. 5, Sec. 7.1,

$$\mathscr{L}\{\sin kt - kt \cos kt\} = \frac{2k^3}{(s^2 + k^2)^2}.$$

Thus from Theorem 7.12,

$$\mathscr{L}\left\{\frac{\sin kt - kt \cos kt}{t}\right\} = \int_s^\infty \frac{2k^3}{(u^2 + k^2)^2}\, du = \lim_{R \to \infty} \int_s^R \frac{2k^3}{(u^2 + k^2)^2}\, du$$

$$= \lim_{R \to \infty} \left(\frac{ku}{u^2 + k^2} + \arctan \frac{u}{k}\right)_s^R$$

$$= \frac{\pi}{2} - \arctan \frac{s}{k} - \frac{ks}{s^2 + k^2}, \quad \text{for } s > 0.$$

(ii) It follows directly from result (i) and Theorem 7.11 that,

$$\mathscr{L}\left\{\int_0^t \left(\frac{\sin ku - ku \cos ku}{u}\right) du\right\} = \frac{1}{s}\left(\frac{\pi}{2} - \arctan \frac{s}{k} - \frac{ks}{s^2 + k^2}\right), \quad \text{for } s > 0.$$

(iii) To evaluate the inverse transform involved here we need to apply the last result stated in Theorem 7.12.
Setting

$$G(s) = \arctan \frac{s}{k} + \tfrac{1}{2} \ln \left| \frac{k+s}{k-s} \right| - \frac{\pi}{2},$$

it follows by differentiation that

$$G'(s) = \frac{k}{s^2 + k^2} - \frac{k}{s^2 - k^2}.$$

From Table 7.1 we find

$$\mathscr{L}^{-1}\{G'(s)\} = \sin kt - \sinh kt,$$

so from the converse result in Theorem 7.12,

$$\mathscr{L}^{-1}\{G(s)\} = -\frac{\mathscr{L}^{-1}\{G'(s)\}}{t} = \frac{\sinh kt - \sin kt}{t}. \qquad \blacksquare$$

The next theorem concerns the transformation of periodic functions and it proves useful in many applications. In general, the function $f(t)$ is said to be periodic with period T if $f(t+T) = f(t)$ for all t. However the Laplace transform is only defined for $t \geq 0$, so it will suffice for the discussion in this chapter if $f(t+T) = f(t)$ for all $t \geq 0$. An illustration of a piecewise continuous periodic function defined for $t \geq 0$ is given in Fig. 7.9.

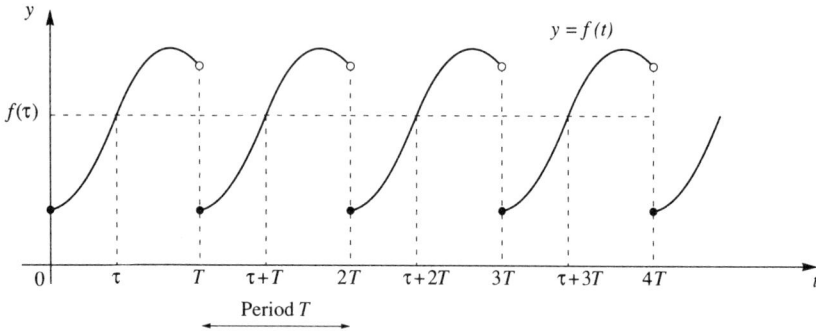

Fig. 7.9 Function $f(t)$ periodic with period T

Theorem 7.13 (Transformation of periodic functions)

Let the function $f(t)$ be piecewise continuous for $t \geq 0$ and periodic with period T. Then

$$\mathscr{L}\{f(t)\} = \frac{1}{1 - e^{-sT}} \int_0^T e^{-st} f(t) \, dt.$$

Proof

By definition

$$\mathscr{L}\{f(t)\} = \int_0^\infty e^{-st} f(t)\,dt = \int_0^T e^{-st} f(t)\,dt + \int_T^\infty e^{-st} f(t)\,dt.$$

Setting $t = \tau + T$ in the second integral gives

$$\mathscr{L}\{f(t)\} = \int_0^T e^{-st} f(t)\,dt + e^{-sT} \int_0^\infty e^{-st} f(\tau + T)\,d\tau.$$

However for $t \geq 0$, $f(\tau + T) = f(\tau)$, and thus

$$\mathscr{L}\{f(t)\} = \int_0^T e^{-st} f(t)\,dt + e^{-sT} \int_0^\infty e^{-st} f(\tau)\,d\tau.$$

Replacing the dummy variable τ by t in the last integral shows it to be $\mathscr{L}\{f(t)\}$, so after rearrangement we obtain

$$\mathscr{L}\{f(t)\} = \frac{1}{1 - e^{-sT}} \int_0^T e^{-st} f(t)\,dt. \qquad \square$$

Example 15. The transform of $f(t) = |\sin t|$

(i) Find $\mathscr{L}\{f(t)\}$ given that

$$f(t) = |\sin t|, \qquad t \geq 0.$$

(ii) Use the result of (i) to deduce $\mathscr{L}\{f(at)\}$.

Solution

(i) The graph of the periodic function to be transformed is given in Fig. 7.10, from which the period occurring in Theorem 7.13 is seen to be $T = \pi$.

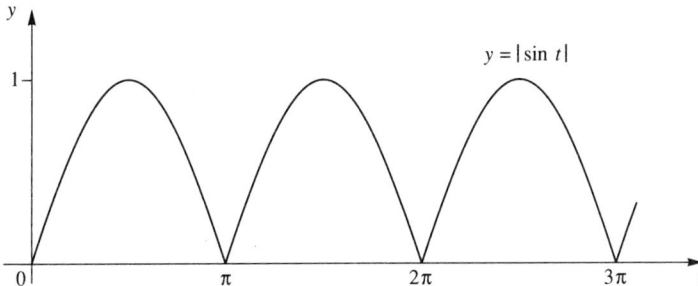

Fig. 7.10 Full wave rectification of a sine wave

We have

$$\int_0^T e^{-st} f(t)\, dt = \int_0^\pi e^{-st} \sin t\, dt$$

$$= \left\{ \frac{e^{-st}(-\cos t - s \sin t)}{s^2 + 1} \right\}\Big|_0^\pi = \frac{1 + e^{-\pi s}}{s^2 + 1}, \qquad \text{for } s > 0.$$

Thus, from Theorem 7.13,

$$\mathscr{L}\{f(t)\} = \frac{1}{1 - e^{-\pi s}}\left(\frac{1 + e^{-\pi s}}{s^2 + 1} \right)$$

$$= \frac{1}{s^2 + 1}\left(\frac{\exp(\tfrac{1}{2}\pi s) + \exp(-\tfrac{1}{2}\pi s)}{\exp(\tfrac{1}{2}\pi s) - \exp(-\tfrac{1}{2}\pi s)} \right)$$

$$= \frac{\coth(\pi s/2)}{s^2 + 1}, \qquad \text{for } s > 0.$$

The general form of the graph shown in Fig. 7.10 describes the full wave rectification of alternating current, so the function $f(t) = |\sin t|$ is often called the **full wave rectification** of a sine wave.

(ii) As

$$F(s) = \frac{\coth(\pi s/2)}{s^2 + 1}, \qquad \text{for } s > 0,$$

it follows from Theorem 7.7 with $\lambda = a$ that

$$\mathscr{L}\{f(at)\} = \mathscr{L}\{|\sin at|\} = a\,\frac{\coth(\pi s/2a)}{s^2 + a^2}, \qquad \text{for } s > 0. \qquad\blacksquare$$

The next theorem concerns the relationship between the product of two Laplace transforms and the function of t which has this product as its transform. It involves an operation called the **convolution** of two functions which is defined as follows. If $f(t)$, $g(t)$ are piecewise continuous functions defined for $t \geq 0$, then the convolution[3] of f and g denoted by $f*g$ is the integral

$$f*g = \int_0^t f(\tau) g(t - \tau)\, d\tau. \tag{7}$$

When it is necessary to indicate the variable involved in the convolution of $f(t)$ and $g(t)$ it is usual to write $(f*g)(t)$ in place of the more economical notation $f*g$ used in (7).

The convolution operation is commutative, so that

$$f*g = g*f. \tag{8}$$

This may be seen by changing the variable from τ to u in (7) by setting $u = t - \tau$, and then

---

[3] In German literature the convolution operation is known as the *Faltung* (folding) operation.

replacing the dummy variable u in the integral so obtained by τ to obtain

$$f*g = \int_0^t f(\tau) g(t-\tau) \, d\tau = -\int_t^0 f(t-u) \, g(u) \, du$$

$$= \int_0^t g(\tau) f(t-\tau) \, d\tau = g*f. \tag{9}$$

Theorem 7.14 (Convolution theorem)

Let $f(t)$, $g(t)$ be piecewise continuous functions defined for $t \geq 0$ such that $\mathscr{L}\{f(t)\} = F(s)$ for $s > \alpha$, and $\mathscr{L}\{g(t)\} = G(s)$ for $s > \beta$. Then

$$\mathscr{L}\{f*g\} = F(s)G(s), \quad \text{for } s > s_0, \text{ where } s_0 = \max\{\alpha, \beta\} \text{ and, conversely,}$$

$$\mathscr{L}^{-1}\{F(s)G(s)\} = f*g.$$

Proof

By definition

$$\mathscr{L}\{f*g\} = \int_0^\infty e^{-st} \left[\int_0^t f(\tau) g(t-\tau) \, d\tau \right] dt,$$

where the region in the (t, τ)-plane over which the integration is performed is the shaded octant shown in Fig. 7.11. Interchanging the order of integration, which is permissible because the functions f and g are of exponential order, gives

$$\mathscr{L}\{f*g\} = \int_0^\infty f(\tau) \left[\int_\tau^\infty e^{-st} f(t-\tau) \, dt \right] d\tau.$$

Now by the second shift theorem (Theorem 7.6) the inner integral is simply

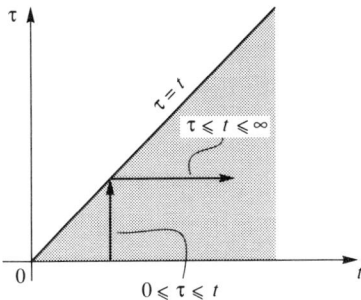

Fig. 7.11 Region of integration for convolution theorem

$e^{-st} G(s)$, and so

$$\mathcal{L}\{f*g\} = \int_0^\infty f(\tau)\, e^{-st}\, G(s)\, d\tau$$

$$= G(s) \int_0^\infty e^{-st} f(\tau)\, d\tau$$

$$= F(s)\, G(s),$$

which was to be shown.

To complete the proof it is necessary to establish the restriction on s for this last result to be valid. The exponential growth conditions imposed on f and g in the theorem imply that for some M, $N > 0$ and arbitrarily small ε, $\delta > 0$,

$$|f(t)| \le M \exp[(\alpha+\varepsilon)t] \quad \text{and} \quad |g(t)| \le N \exp[(\beta+\delta)t]$$

for large t. Thus

$$|f*g| = \left| \int_0^t f(\tau)\, g(t-\tau)\, d\tau \right|$$

$$\le \int_0^t M \exp[(\alpha+\varepsilon)t]\, N \exp[(\beta+\delta)(t-\tau)]\, d\tau$$

$$= \frac{MN\{\exp[(\alpha+\varepsilon)t] - \exp[(\beta+\delta)t]\}}{\alpha-\beta+\varepsilon-\delta} \qquad \text{for large } t.$$

It follows from this that for some K, $L > 0$,

$$|f*g| \le \begin{cases} K \exp[(\alpha+\varepsilon)t] & \text{if } \alpha \ge \beta \\ L \exp[(\beta+\delta)t] & \text{if } \alpha < \beta. \end{cases}$$

This shows $f*g$ is of exponential order $\exp(s_0 t)$ where $s_0 = \max\{\alpha, \beta\}$, and so the last part of the theorem is proved. The converse result follows directly by taking the inverse Laplace transform. $\square$

Example 16. Convolution operation and convolution theorem

(i) Find $f*g$ given $f(t) = t^2$, $g(t) = \sin t$, and use the result to find $\mathcal{L}\{t^2*\sin t\}$.

(ii) If f, g are defined as in (i) use the convolution theorem to find $\mathcal{L}\{t^2*\sin t\}$, and thus verify the result obtained by direct calculation in (i). Use $\mathcal{L}\{t^2*\sin t\}$ to find $\mathcal{L}\{t^2*\sin 3t\}$.

(iii) Find $\dfrac{d}{dt}(f*g)$, and use the result to determine $\left\{\dfrac{d}{dt}(f*g)\right\}$.

(iv) Find $\left\{ \displaystyle\int_0^t e^{3\tau}(t-\tau) \sin 2(t-\tau)\, d\tau \right\}$.

(v) Use the convolution theorem to find $\mathcal{L}^{-1}\left\{ \dfrac{1}{(s^2+1)(s^2-4)} \right\}$.

Solution

(i) $\quad f*g = \int_0^t f(\tau)g(t-\tau)\,d\tau = \int_0^t \tau^2 \sin(t-\tau)\,d\tau$

$\qquad = \int_0^t \tau^2(\sin t \cos \tau - \cos t \sin \tau)\,d\tau$

$\qquad = \sin t \int_0^t \tau^2 \cos \tau \,d\tau - \cos t \int_0^t \tau^2 \sin \tau \,d\tau$

$\qquad = \sin t \left[2\tau \cos \tau + (\tau^2 - 2)\sin \tau\right]\Big|_0^t - \cos t \left[2\tau \sin \tau - (\tau^2 - 2)\cos \tau\right]\Big|_0^t$

$\qquad = t^2 - 2 + 2 \cos t.$

Thus

$\qquad t^2 * \sin t = t^2 - 2 + 2 \cos t.$

When evaluating a convolution integral it is sometimes useful to use the fact that $f*g = g*f$, because one of these integrals may be easier to evaluate than the other. In this case there is little to choose between the equivalent convolution integrals

$\qquad f*g = \int_0^t \tau^2 \sin(t-\tau)\,d\tau \quad \text{and} \quad g*f = \int_0^t \sin \tau(t-\tau)^2 \,d\tau.$

Using Table 7.1 we find

$\qquad \mathscr{L}\{t^2 * \sin t\} = \mathscr{L}\{t^2 - 2 + 2\cos t\} = \dfrac{2}{s^3} - \dfrac{2}{s} + \dfrac{2s}{s^2+1} = \dfrac{2}{s^3(s^2+1)}, \qquad \text{for } s>0.$

Here the condition $s>0$ applies because the transform of each of the three functions t^2, -2 and $2\cos t$ is valid for $s>0$.

(ii) From Table 7.1 we have

$\qquad \mathscr{L}\{t^2\} = \dfrac{2}{s^3} \quad \text{for } s>0 \quad \text{and} \quad \mathscr{L}\{\sin t\} = \dfrac{1}{s^2+1} \quad \text{for } s>0.$

Thus by the convolution theorem

$\qquad \mathscr{L}\{t^2 * \sin t\} = \mathscr{L}\{t^2\}\mathscr{L}\{\sin t\} = \dfrac{2}{s^3(s^2+1)}.$

As each transform involved in the product is valid for $s>0$, and in the notation of Theorem 7.14 $\alpha = \beta = 0$, it follows that $s_0 = \max\{\alpha, \beta\} = 0$ and so the transform of the convolution is valid for $s>0$. This verifies the result obtained in (i) by direct calculation.

Applying Theorem 7.7 to $\mathscr{L}\{t^2 * \sin t\}$ with $\lambda = 3$ gives

$\qquad \mathscr{L}\{(3t)^2 * \sin 3t\} = \dfrac{1}{3}\dfrac{2}{(s/3)^3\left[(s/3)^2+1\right]}$

$\qquad\qquad\qquad = \dfrac{162}{s^3(s^2+9)}$

and thus

$$\mathcal{L}\{t^2 * \sin 3t\} = \frac{18}{s^3(s^2+9)}, \qquad \text{for } s > 0.$$

(iii) By definition

$$\frac{d}{dt}(f*g) = \frac{d}{dt}\int_0^t f(\tau) g(t-\tau) \, d\tau.$$

From Theorem 1.6 (differentiation under the integral sign) it follows at once that

$$\frac{d}{dt}(f*g) = \int_0^t f(\tau) \left[\frac{\partial}{\partial t} g(t-\tau) \right] d\tau + f(t) \, g(0).$$

Hence

$$\frac{d}{dt}(f*g) = g(0) f(t) + \int_0^t f(\tau) \, g'(t-\tau) \, d\tau,$$

where the prime denotes differentiation with respect to t, and thus

$$\frac{d}{dt}(f*g) = g(0) f(t) + f*g'.$$

This is the first result that was required.

Taking the Laplace transform of this result we obtain

$$\mathcal{L}\left\{ \frac{d}{dt}(f*g) \right\} = g(0) \, F(s) + \mathcal{L}\{f*g'\}.$$

Finally, using the convolution theorem on the last term together with the fact that $\mathcal{L}\{g'(t)\} = s\,G(s) - g(0)$ gives

$$\mathcal{L}\left\{ \frac{d}{dt}(f*g) \right\} = g(0) \, F(s) + F(s) \, [s\,G(s) - g(0)],$$

and so

$$\mathcal{L}\left\{ \frac{d}{dt}(f*g) \right\} = s\,F(s)\,G(s).$$

(iv) The integral involved is simply the convolution of e^{3t} and $t \sin 2t$. Thus as

$$\mathcal{L}\{e^{3t}\} = \frac{1}{s-3} \quad \text{for } s > 3 \quad \text{and} \quad \mathcal{L}\{t \sin 2t\} = \frac{4s}{(s^2+4)^2} \quad \text{for } s > 0,$$

it follows from the convolution theorem that

$$\mathcal{L}\left\{ \int_0^t e^{3\tau}(t-\tau)\sin 2(t-\tau) \, d\tau \right\} = \frac{4s}{(s-3)(s^2+4)^2} \qquad \text{for } s > 3.$$

The condition on s comes from the fact that in the notation of the convolution theorem the transform is valid for $s > s_0$, where $s_0 = \max\{\alpha, \beta\}$, and here $\alpha = 3$, $\beta = 0$.

(v) The use of the convolution theorem to find an inverse transform is based on the fact that as

$$F(s)\,G(s) = \mathscr{L}\left\{\int_0^t f(\tau)\,g(t-\tau)\,d\tau\right\}.$$

it follows directly that

$$\mathscr{L}^{-1}\{F(s)\,G(s)\} = \int_0^t f(\tau)\,g(t-\tau)\,d\tau.$$

To apply this result it is necessary to express the transform whose inverse is to be found as the product $F(s)\,G(s)$, where $F(s) = \mathscr{L}\{f(t)\}$ and $G(s) = \mathscr{L}\{g(t)\}$ are chosen so that the functions $f(t)$ and $g(t)$ are known. Once this decomposition has been accomplished the required inverse transform is obtained by evaluating $f*g$.

In this case, writing

$$\frac{1}{(s^2+1)(s^2-4)} = F(s)\,G(s),$$

the natural choice for $F(s)$ and $G(s)$ is

$$F(s) = \frac{1}{s^2+1}, \quad G(s) = \frac{1}{s^2-4}.$$

Inspection of Table 7.1 shows

$$F(s) = \mathscr{L}\{\sin t\} \quad \text{and} \quad G(s) = \mathscr{L}\{\tfrac{1}{2}\sinh 2t\},$$

so

$$f(t) = \sin t \quad \text{and} \quad g(t) = \tfrac{1}{2}\sinh 2t.$$

Thus

$$\mathscr{L}^{-1}\left\{\frac{1}{(s^2+1)(s^2-4)}\right\} = (\sin t)*(\tfrac{1}{2}\sinh 2t)$$

$$= \tfrac{1}{2}\int_0^t \sin\tau \sinh 2(t-\tau)\,d\tau$$

$$= \tfrac{1}{2}\int_0^t \sin\tau\,(\sinh 2t\cosh 2\tau - \cosh 2t\sinh 2\tau)\,d\tau$$

$$= \tfrac{1}{2}\sinh 2t\int_0^t \sin\tau\cosh 2\tau\,d\tau - \tfrac{1}{2}\cosh 2t\int_0^t \sin\tau\sinh 2t\,d\tau.$$

Evaluating these integrals and combining the results gives

$$\mathscr{L}^{-1}\left\{\frac{1}{(s^2+1)(s^2-4)}\right\} = \tfrac{1}{5}(\tfrac{1}{2}\sinh 2t - \sin t).$$

This particular result could have been obtained more easily by using partial fractions, but the purpose of the example was to illustrate the use of the convolution theorem.   ■

When the transform $F(s)$ of a function $f(t)$ has been obtained by some means, but the function $f(t)$ itself is unknown, it is often helpful to know $f(0)$, the initial value of $f(t)$. In particular, such information may be necessary when working with differential equations, in which case it may also be necessary to know $f'(0)$ and higher derivatives of $f(t)$ at $t=0$ (cf. Sec. 7.3(i) and its related problems). This information is provided by the results we call the **extended initial value theorem**. The more usual result known simply as the **initial value theorem** merely comprises result (i) of the theorem. We postpone discussion of the **final value theorem** until Sec. 7.3(d), where a restricted version will be proved suitable for applications to rational functions of s.

Theorem 7.15 (Extended initial value theorem)

Let $\mathcal{L}\{f(t)\}=F(s)$. Then provided $f(t)$ and the necessary derivatives exist as $t\to0$,

(i) $\lim\limits_{s\to\infty} [s\,F(s)]=f(0)$,

(ii) $\lim\limits_{s\to\infty} [s^2\,F(s)-s\,f(0)]=f'(0)$,

and

(iii) $\lim\limits_{s\to\infty} [s^{n+1}F(s)-s^n f(0)-s^{n-1}f^{(1)}(0)-s^{n-2}f^{(2)}(0)-\ \ldots\ -s f^{(n-1)}(0)]=f^{(n)}(0)$.

Proof

To establish (i) we must use Theorem 7.8, which may be written

$$\mathcal{L}\{f'(t)\}=s\,F(s)-f(0).$$

Now from Theorem 7.4 $\lim\limits_{s\to\infty} \mathcal{L}\{f'(t)\}=0$, so

$$\lim\limits_{s\to\infty} \mathcal{L}\{f'(t)\}=\lim\limits_{s\to\infty} [s\,F(s)-f(0)]=0,$$

and thus

$$\lim\limits_{s\to\infty} [s\,F(s)]=f(0).$$

The proofs of (ii) and (iii) are analogous apart from the use of Theorem 7.9 in place of Theorem 7.8, so we shall omit them.   □

Example 17. Extended initial value theorem

Given

$$F(s)=\frac{3s}{s^2-4s+13}$$

find $f(0)$ and $f'(0)$.

Solution

From (i) in Theorem 7.15

$$f(0) = \lim_{s \to \infty} [s F(s)] = \lim_{s \to \infty} \left(\frac{3s^2}{s^2 - 4s + 13} \right) = 3.$$

From (ii) in Theorem 7.15

$$f'(0) = \lim_{s \to \infty} [s^2 F(s) - s f(0)] = \lim_{s \to \infty} \left(\frac{3s^3}{s^2 - 4s + 13} - 3s \right)$$

$$= \lim_{s \to \infty} \left(\frac{12s^2 - 39s}{s^2 - 4s + 13} \right) = 12.$$

These results are easily verified by direct calculation because

$$\mathscr{L}^{-1}\{F(s)\} = e^{2t}(3 \cos 3t + 2 \sin 3t). \qquad \blacksquare$$

In conclusion we introduce a new concept called the **Dirac**[4] **delta function**, also sometimes called the **unit impulse** function. This so-called 'function' may be regarded as the mathematical representation of an idealized impulse;[5] that is an instantaneous transfer of a finite amount of momentum.

In an impulse the variable t represents the time, but in different physical situations the variable t may represent a distance. For example, a delta function may be used to represent a point load acting on a support beam, in which case t then represents a distance measured along the beam. We shall see that the delta function is not a function in the sense of classical analysis, but what in more advanced accounts of this subject is called a **distribution** or **generalized function**. It is, in fact, the most important member of a whole family of singularity functions which are useful in applications of mathematics.

The delta function may be introduced as the limiting form of various different approximations. For our purposes it will suffice for us to use the simplest of these approximations and to consider it as the limit as $\varepsilon \to 0 (\varepsilon > 0)$ of the rectangular pulse

$$\Delta_\varepsilon(t - t_0) = \begin{cases} 1/(2\varepsilon), & \text{for } |t - t_0| < \varepsilon \\ 0, & \text{for } |t - t_0| > \varepsilon, \end{cases} \qquad \text{for } -\infty < t < \infty.$$

The function $\Delta_\varepsilon(t - t_0)$ is called a **pre-limit delta function**.

Every function of this type bounds a unit area between its graph and the t-axis, and a sequence of such functions, centered on t_0, with $\varepsilon_1 > \varepsilon_2 > \varepsilon_3 > \ldots > 0$ is illustrated in Fig. 7.12.

---

[4] PAUL ADRIEN MAURICE DIRAC (1902–1984), an English mathematician and physicist who introduced this function in a fundamental paper on quantum mechanics presented to the Royal Society of London in 1927. In 1933 Dirac shared the Nobel prize with the German physicist ERWIN SCHRÖDINGER for their contributions to quantum mechanics.

[5] In mechanics an **impulse** is considered to occur in processes in which the transfer of momentum takes place so rapidly that only the initial and final states can be observed, and not the transition process in between. This happens, for example, when two billiard balls collide.

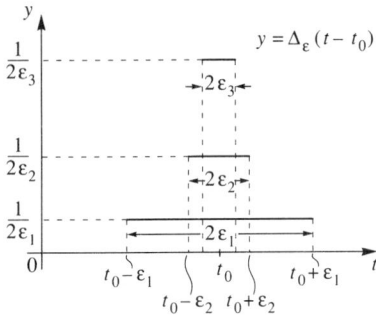

Fig. 7.12 Rectangular pulse approximations to the delta function $\delta(t-t_0)$

Hereafter we shall adopt the formal definition that the delta function centered on t_0 and denoted by $\delta(t-t_0)$ is

$$\delta(t-t_0)=\lim_{\varepsilon\to 0} \Delta_\varepsilon(t-t_0). \tag{10}$$

The infinity of the delta function at $t=t_0$ means that its defining properties must be expressed in terms of its behavior away from t_0 and its integral over the interval $-\infty<t<\infty$. Thus, formally, we arrive at the properties

(i) $\delta(t-t_0)=0,$ for $t\neq t_0$,

and (11)

(ii) $\displaystyle\int_{-\infty}^{\infty} \delta(t-t_0)\,dt=1.$

The pre-limit delta function $\Delta_\varepsilon(t-t_0)$ is an even function with respect to its argument $t-t_0$, so this is also true of $\delta(t-t_0)$, and thus it follows from (11) (ii) that

$$\int_{-\infty}^{\infty} \delta(t-t_0)\,dt = \int_{-\infty}^{\infty} \delta(t_0-t)\,dt = 1. \tag{12}$$

An important and very useful operational property of the delta function is that if $f(t)$ is defined and continuous in some neighborhood of t_0, then

$$\int_{-\infty}^{\infty} f(t)\,\delta(t-t_0)\,dt = f(t_0). \tag{13}$$

This is called the **sifting (screening) property** of the delta function, because when used in an integral it acts like a sieve to select a particular value of $f(t)$.

A purely formal justification of this result may be given as follows. Consider the approximation to $\delta(t-t_0)$ provided by $\Delta_\varepsilon(t-t_0)$ for some arbitrarily small but fixed $\varepsilon>0$.

Then from the mean value theorem for integrals and the definition of $\Delta_\varepsilon(t-t_0)$ we have

$$\int_{-\infty}^{\infty} f(t)\,\Delta_\varepsilon(t-t_0)\,\mathrm{d}t = f(\xi) \int_{t_0-\varepsilon}^{t_0+\varepsilon} \Delta_\varepsilon(t-t_0)\,\mathrm{d}t = f(\xi), \tag{14}$$

where $t_0-\varepsilon<\xi<t_0+\varepsilon$. Taking the limit as $\varepsilon\to 0$, and assuming (14) remains valid as $\varepsilon\to 0$ and $\Delta_\varepsilon(t-t_0)\to\delta(t-t_0)$, we obtain the required result

$$\int_{-\infty}^{\infty} f(t)\,\delta(t-t_0)\,\mathrm{d}t = f(t_0).$$

On account of the sifting property we may write

$$f(t)\,\delta(t-t_0) = f(t_0)\,\delta(t-t_0), \tag{15}$$

with the understanding that when either of these expressions is integrated over the interval $-\infty<t<\infty$ the result will be $f(t_0)$.

Let us now consider the integral

$$\int_{-\infty}^{t} \delta(\tau-t_0)\,\mathrm{d}\tau.$$

It follows directly from the definition of $\delta(t-t_0)$ that

$$\int_{-\infty}^{t} \delta(\tau-t_0)\,\mathrm{d}\tau = \begin{cases} 0, & \text{for } t<t_0 \\ 1, & \text{for } t\geq t_0, \end{cases}$$

but this is simply the unit step function defined in (1) and extended to $-\infty<t<\infty$ by setting $\mathcal{U}(t-t_0)=0$ for $-\infty<t<t_0$. Thus we have established that

$$\int_{-\infty}^{t} \delta(\tau-t_0)\,\mathrm{d}\tau = \mathcal{U}(t-t_0). \tag{16}$$

Assuming a justification can be provided for differentiating (16) with respect to t, a direct consequence of (16) and the fundamental theorem of calculus is the following fundamental relationship between the delta function and the unit step function

$$\frac{\mathrm{d}}{\mathrm{d}t}\mathcal{U}(t-t_0) = \delta(t-t_0). \tag{17}$$

This and other operations performed on the delta function can be justified by using the theory of distributions, so henceforth we shall use the delta function as though it were an ordinary function. It is appropriate that it should be discussed in this section because of its usefulness with the Laplace transform, and also because as it is defined via an integral it is essentially an 'operation' rather than a 'function'.

When the delta function is used with the Laplace transform it is to be understood that $t_0\geq 0$, because the Laplace transform only applies to initial value problems for which $t\geq 0$.

To obtain $\mathscr{L}\{\delta(t-t_0)\}$ for $t_0 > 0$ we make the formal assumption that

$$\mathscr{L}\{\delta(t-t_0)\} = \lim_{\varepsilon \to 0} \mathscr{L}\{\Delta_\varepsilon(t-t_0)\}. \tag{18}$$

A routine calculation shows

$$\mathscr{L}\{\Delta_\varepsilon(t-t_0)\} = \{\exp[-(t_0 - \varepsilon)s] - \exp[-(t_0 + \varepsilon)s]\}/2\varepsilon s$$

$$= \exp(-t_0 s\varepsilon)\left(\frac{\exp(\varepsilon s) - \exp(-\varepsilon s)}{2\varepsilon s}\right).$$

So from (18), after using L'Hospital's rule, we find

$$\mathscr{L}\{\delta(t-t_0)\} = \exp(-t_0 s)\lim_{\varepsilon \to 0}\left(\frac{\exp(\varepsilon s) - \exp(-\varepsilon s)}{2\varepsilon s}\right) = \exp(-t_0 s),$$

and thus

$$\mathscr{L}\{\delta(t-t_0)\} = \exp(-t_0 s). \tag{19}$$

This last result provides further evidence that $\delta(t-t_0)$ is not an ordinary function. To see this observe (19) depends continuously on t_0, so by letting $t_0 \to 0$ it may be used to define $\mathscr{L}\{\delta(t)\}$ as

$$\mathscr{L}\{\delta(t)\} = 1. \tag{20}$$

However, Theorem 7.4 asserts that if $f(t)$ is a piecewise continuous function of exponential order, then $\lim_{s \to \infty} \mathscr{L}\{f(t)\} = 0$. The theorem implies no contradiction, because although $\delta(t-t_0)$ may be regarded as a function of exponential order since it is defined as the limit of a combination of step functions, it is *not* piecewise continuous. The result in (20) is consistent with the convention used when working with the delta function in the Laplace transform that

$$\int_0^\infty \delta(t)\,dt = 1 \quad \text{and} \quad \int_0^\infty \delta(t)f(t)\,dt = f(0). \tag{21}$$

Other singularity functions which are sometimes of use are the 'derivatives' of the delta function, of which the simplest is $\delta'(t-t_0)$. To find its Laplace transform we apply the transform to the formal definition of the derivative of $\delta(t-t_0)$, and thus take as our definition

$$\mathscr{L}\{\delta'(t-t_0)\} = \mathscr{L}\left\{\lim_{h \to 0}\left[\frac{\delta(t+h-t_0) - \delta(t-t_0)}{h}\right]\right\}, \quad \text{for } t_0 > 0.$$

Assuming the transform and limiting operations may be interchanged, using (19) and again making use of L'Hospital's rule, this becomes

$$\mathscr{L}\{\delta'(t-t_0)\} = \exp(-t_0 s)\lim_{h \to 0}\left(\frac{e^{hs} - 1}{h}\right) = s\exp(-t_0 s).$$

Thus we have arrived at the formal result that

$$\mathcal{L}\{\delta'(t-t_0)\} = s\exp(-t_0 s), \quad \text{for } t_0 > 0. \tag{22}$$

This transform is a continuous function of t_0, so by letting $t\to t_0$ it may be used to define $\mathcal{L}\{\delta'(t)\}$ as

$$\mathcal{L}\{\delta'(t)\} = s. \tag{23}$$

The difference quotient in the definition of $\delta'(t-t_0)$ just used to evaluate $\mathcal{L}\{\delta'(t-t_0)\}$ may be expressed as

$$\frac{\delta(t+h-t_0)}{h} - \frac{\delta(t-t_0)}{h}.$$

This represents two delta functions of equal and opposite magnitudes located close to $t=t_0$ which coalesce in the limit as $h\to 0$. For this reason $\delta'(t-t_0)$ is called a **doublet** located at t_0.

Analogous forms of argument may be used to show that for $n=0, 1, 2, \ldots$,

$$\mathcal{L}\{\delta^{(n)}(t-t_0)\} = s^n \exp(-t_0 s) \tag{24}$$

and, correspondingly,

$$\mathcal{L}\{\delta^{(n)}(t)\} = s^n, \tag{25}$$

where $\delta^{(n)}(t-t_0) = \mathrm{d}^n[\delta(t-t_0)]/\mathrm{d}t^n$. The singularity function $\delta^{(2)}(t-t_0)$ is called a **triplet** located at $t=t_0$ and $\delta^{(3)}(t-t_0)$ a **quadruplet**, and so forth. Results (17) and (23) describe the differentiability properties and relationship between the two basic singularity functions $\mathcal{U}(t-t_0)$ and $\delta(t-t_0)$.

A simple and useful application of the doublet $\delta'(t-t_0)$ in mechanics is to a couple of unit moment applied to a beam at the point $t=t_0$. If $\delta'(t-t_0)$ is scaled by the factor M, a couple of moment M applied at the point $t=t_0$ of a beam may be represented as a 'distributed' load $M\delta'(t-t_0)$ along the beam.

The operational effect of $\delta'(t-t_0)$ when in the integrand of an integral is seen by making a formal application of integration by parts to the integral

$$\int_{-\infty}^{\infty} f(t)\,\delta'(t-t_0)\,\mathrm{d}t,$$

in which $f(t)$ is a continuous differentiable function in the neighborhood of t_0. We obtain

$$\int_{-\infty}^{\infty} f(t)\,\delta'(t-t_0)\,\mathrm{d}t = f(t)\,\delta(t-t_0)\Big|_{-\infty}^{\infty} - \int_{-\infty}^{\infty} f'(t)\,\delta(t-t_0)\,\mathrm{d}t = -f'(t_0),$$

and so we have arrived at the result

$$\int_{-\infty}^{\infty} f(t)\,\delta'(t-t_0)\,\mathrm{d}t = -f'(t_0). \tag{26}$$

This displays the special sifting property of the doublet $\delta'(t-t_0)$. By an obvious extension of this argument it is possible to give a formal proof of the general result

$$\int_{-\infty}^{\infty} f(t)\,\delta^{(n)}(t-t_0)\,dt = (-1)^n f^{(n)}(t_0), \tag{27}$$

for $n = 0, 1, 2, \ldots$, where it is understood that $\delta^{(0)}(t-t_0) = \delta(t-t_0)$.

Accepting that the delta function may in many respects be treated as an ordinary function, and so used in conjunction with such techniques as integration by substitution and integration by parts, difficulty of interpretation will still occur when equality between expressions involving delta functions and their derivatives arises. This is because the delta function has an intense singularity and so is not a function in the sense of classical analysis, and in addition the pre-limit delta function $\Delta_\varepsilon(t-t_0)$ used to define $\delta(t-t_0)$ is only one of the many possible but equivalent pre-limit delta functions which may be used for this purpose.

A suitable interpretation of equality between expressions involving delta functions in general may be found by consideration of the defining property (ii) given in (11). This suggests that equality should be interpreted as meaning *equality of the operational effect* of a delta function when acting on a function in the integrand of a definite integral. Henceforth we shall adopt this as the basis of our definition. Thus if $g(t)$ and $h(t)$ are two different expressions involving delta functions, and $f(t)$ is a continuous function of t, we shall define $g(t) = h(t)$ to mean that

$$\int_{-\infty}^{\infty} f(t)\,g(t)\,dt = \int_{-\infty}^{\infty} f(t)\,h(t)\,dt, \tag{28}$$

is true for all arbitrary continuous functions $f(t)$.

The following example illustrates how such a definition may be used in a simple and interesting but nontrivial case.

Example 18. Scaling t in the delta function

Give a formal derivation of the result

$$\delta[\lambda(t-t_0)] = \frac{1}{|\lambda|}\,\delta(t-t_0), \qquad \text{for } \lambda \neq 0.$$

Proof

Consider the integral

$$\int_{-\infty}^{\infty} f(t)\,\delta[\lambda(t-t_0)]\,dt,$$

in which $\lambda \neq 0$ and $f(t)$ is an arbitrary function of t. We shall assume that the rule for changing the variable in a definite integral can be justified for the delta function. Setting $\tau = \lambda t$, where for the moment the only restriction on λ is $\lambda \neq 0$, and using the sifting property gives

$$\int_{-\infty}^{\infty} f(t)\,\delta[\lambda(t-t_0)]\,dt = \int_{-\infty}^{\infty} \frac{1}{\lambda} f\left(\frac{\tau}{\lambda}\right)\delta(\tau-\tau_0)\,d\tau = \frac{1}{\lambda} f\left(\frac{\tau_0}{\lambda}\right) = \frac{1}{\lambda} f(t_0).$$

However, as $\delta(t-t_0)$ is an even function of its argument $t-t_0$, this last result can only be true if the factor $1/\lambda$ is replaced by $1/|\lambda|$. Now

$$\frac{1}{|\lambda|}f(t_0)=\int_{-\infty}^{\infty}\frac{1}{|\lambda|}f(t)\,\delta(t-t_0)\,dt,$$

so replacing $f(t_0)$ in the previous equation by this integral, and combining terms, gives

$$\int_{-\infty}^{\infty}f(t)\,\{\delta[\lambda(t-t_0)]-\frac{1}{|\lambda|}\,\delta(t-t_0)\}\,dt=0.$$

The function $f(t)$ was an arbitrary continuous function, so it follows from the definition of equality in (28) that

$$\delta[\lambda(t-t_0)]-\frac{1}{|\lambda|}\,\delta(t-t_0)=0,$$

which is the required result. ∎

Example 19. Oscillating system subjected to an impulsive force

The following differential equation describes the forced oscillations of an undamped oscillating mass-spring system which starts from rest and then at a subsequent time is subjected to an impulsive force in the direction of motion. Solve the differential equation and hence find the displacement $x(t)$ of the mass at time t given

$$x''+4x=\sin t+2\,\delta(t-3),$$

with $x(0)=x'(0)=0$.

Solution

Taking the Laplace transform of the equation, incorporating the initial conditions, and using the fact from (19) that $\mathscr{L}\{\delta(t-3)\}=e^{-3s}$, we find

$$s^2 X(s)+4X(s)=\frac{1}{s^2+1}+2e^{-3s}.$$

Thus

$$X(s)=\frac{1}{(s^2+1)(s^2+4)}+\frac{2e^{-3s}}{s^2+4},$$

and after the use of partial fractions this becomes

$$X(s)=\frac{1}{3}\left(\frac{1}{s^2+1}\right)-\frac{1}{3}\left(\frac{1}{s^2+4}\right)+\frac{2e^{-3s}}{s^2+4}.$$

Taking the inverse transform and applying the second shift theorem (Theorem 7.6) to the last term gives

$$x(t)=\tfrac{1}{3}\sin t-\tfrac{1}{6}\sin 2t+\mathscr{U}(t-3)\sin 2(t-3) \qquad \text{for } t\geq 0.$$

This shows, as would be expected, that until the impulse is applied to the system the response is a linear combination of the forcing oscillation $\sin t$ and a natural oscillation $\sin 2t$. The effect of the impulse at time $t=3$ is to excite a delayed natural oscillation which is then superimposed on this composite oscillation. ∎

On account of their importance in differential equations we conclude this section by discussing **Duhamel's integrals**.[6] These arise in applications of the Laplace transform in which $F(s)=\mathscr{L}\{f(t)\}$ and $G(s)=\mathscr{L}\{g(t)\}$ are known, and it is required to find $\mathscr{L}^{-1}\{s\,F(s)\,G(s)\}$.

We start from the identity

$$s\,F(s)\,G(s)=f(0)\,G(s)+\{s\,F(s)-f(0)\}\,G(s),$$

and use the fact that $\mathscr{L}\{f'(t)\}=s\,F(s)-f(0)$ to rewrite it as

$$s\,F(s)\,G(s)=f(0)\,G(s)+\mathscr{L}\{f'(t)\}\,G(s). \tag{29}$$

Taking the inverse transform and using the convolution theorem then gives

$$\mathscr{L}^{-1}\{s\,F(s)\,G(s)\}=f(0)\,g(t)+\int_0^t f'(\tau)\,g(t-\tau)\,d\tau$$

$$=f(0)\,g(t)+\int_0^t g(\tau)\,f'(t-\tau)\,d\tau, \tag{30}$$

where the second result follows from the first because $f'*g=g*f'$. Had the identity

$$s\,F(s)\,G(s)=g(0)\,F(s)+\{s\,G(s)-g(0)\}\,F(s)$$

been used, two different but equivalent expressions for $\mathscr{L}^{-1}\{s\,F(s)\,G(s)\}$ would have been obtained, corresponding to interchanging f and g in (30). These four integrals are called **Duhamel's integrals**, and for convenience we record them in the form of a theorem.

Theorem 7.16 (Duhamel's integrals)

Let $F(s)=\mathscr{L}\{f(t)\}$ and $G(s)=\mathscr{L}\{g(t)\}$, then

$$\mathscr{L}^{-1}\{s\,F(s)\,G(s)\}=\begin{cases} f(0)\,g(t)+\displaystyle\int_0^t f'(\tau)\,g(t-\tau)\,d\tau \\[2mm] f(0)\,g(t)+\displaystyle\int_0^t g(\tau)\,f'(t-\tau)\,d\tau \\[2mm] g(0)\,f(t)+\displaystyle\int_0^t g'(\tau)\,f(t-\tau)\,d\tau \\[2mm] g(0)\,f(t)+\displaystyle\int_0^t f(\tau)\,g'(t-\tau)\,d\tau. \end{cases}$$

□

[6] J. M. C. DUHAMEL (1797–1872), a French mathematician who was professor of higher algebra at the Sorbonne. In addition to his work in algebra he also made contributions to mechanics and to the study of thermal stress.

One of the most important uses of Theorem 7.16 is in connection with linear constant coefficient differential equations. It involves showing how the solution of a differential equation with zero initial conditions and an arbitrary nonhomogeneous term (forcing function) is related to the solution of the corresponding problem in which the non-homogeneous term is a unit step function. In terms of a physical system, the result shows how the response of a system to an arbitrary forcing function may be determined in terms of its response to a unit step function excitation. Indeed, it is for this reason that excitation by a step function is a standard means of assessing system response.

To obtain the desired result let the differential equation whose solution is required be

$$a_0 \frac{d^n y}{dt^n} + a_1 \frac{d^{n-1} y}{dt^{n-1}} + \ldots + a_n y = f(t), \tag{31}$$

where $y(t)$ is subject to the homogeneous initial conditions $y(0) = y^{(1)}(0) = \ldots = y^{(n-1)}(0) = 0$. Taking the Laplace transform it follows that

$$(a_0 s^n + a_1 s^{n-1} + \ldots + a_n) Y(s) = F(s), \tag{32}$$

where $Y(s) = \mathscr{L}\{y(t)\}$ and $F(s) = \mathscr{L}\{f(t)\}$.

Now let $z(t)$ be the solution of the same differential equation with homogeneous initial conditions, but with the nonhomogeneous term $\mathscr{U}(t)$, so that

$$a_0 \frac{d^n z}{dt^n} + a_1 \frac{d^{n-1} z}{dt^{n-1}} + \ldots + a_n z = \mathscr{U}(t), \tag{33}$$

with $z(0) = z^{(1)}(0) = \ldots = z^{(n-1)}(0) = 0$. Then taking the Laplace transform gives

$$(a_0 s^n + a_1 s^{n-1} + \ldots + a_n) Z(s) = \frac{1}{s}, \tag{34}$$

where $Z(s) = \mathscr{L}\{z(t)\}$.

Dividing (32) by (34) shows

$$Y(s) = s F(s) Z(s), \tag{35}$$

and thus the solution $y(t)$ of the homogeneous initial value problem for (31) is

$$y(t) = \mathscr{L}^{-1}\{s F(s) Z(s)\}. \tag{36}$$

Here $F(s)$ is the transform of the arbitrary nonhomogeneous term $f(t)$, while $Z(s)$ is the transform of the solution of the homogeneous initial value problem for (33) with the unit step function $\mathscr{U}(t)$ as its nonhomogeneous term.

Identifying $F(s)$ and $Z(s)$ in (36) with $F(s)$ and $G(s)$ in the Duhamel integrals in Theorem 7.16 it follows from the first integral that

$$y(t) = f(0) z(t) + \int_0^t f'(\tau) z(t - \tau) \, d\tau. \tag{37}$$

Alternatively, using the third integral in Theorem 7.16 together with the fact that $z(0)=0$ gives

$$y(t)=\int_0^t z'(\tau)f(t-\tau)\,d\tau. \tag{38}$$

Depending on the functions involved, sometimes one of the two equivalent results (37) and (38) is more convenient to use than the other. We have established the following important result.

Theorem 7.17 (Representation of a solution using Duhamel's integrals)

Let $y(t)$ be the solution of the constant coefficient differential equation

$$a_0\frac{d^n y}{dt^n}+a_1\frac{d^{n-1}y}{dt^{n-1}}+\ \ldots\ +a_n y=f(t),$$

with $y(0)=y^{(1)}(0)=\ \ldots\ =y^{(n-1)}(0)=0$ and $f(t)$ a piecewise continuous function. Suppose also that $z(t)$ is the solution of

$$a_0\frac{d^n z}{dt^n}+a_1\frac{d^{n-1}z}{dt^{n-1}}+\ \ldots\ +a_n z=\mathcal{U}(t),$$

with $z(0)=z^{(1)}(0)=\ \ldots\ =z^{(n-1)}(0)=0$. Then for $t>0$ the solution $y(t)$ may be represented in terms of $z(t)$ in either of the following forms

$$y(t)=f(0)z(t)+\int_0^t f'(\tau)z(t-\tau)\,d\tau,$$

or

$$y(t)=\int_0^t z'(\tau)f(t-\tau)\,d\tau. \qquad \square$$

Example 20. Application of Duhamel's integrals to a differential equation

Use Theorem 7.17 to solve the differential equation

$$y''+4y'+5y=f(t),$$

when $y(0)=y'(0)=0$ and $f(t)$ is an arbitrary piecewise continuous function. Make use of the result to obtain the solution of this differential equation when $f(t)=e^t$.

Solution

We must first determine the function $z(t)$ which satisfies the equation

$$z''+4z'+5z=\mathcal{U}(t),$$

with $z(0)=z'(0)=0$. Transforming the equation gives

$$(s^2+4s+5)\,Z(s)=\frac{1}{s},$$

and so

$$Z(s)=\frac{1}{s(s^2+4s+5)}.$$

Simplifying this expression by means of partial fractions we obtain

$$Z(s)=\frac{1}{5}\left(\frac{1}{s}\right)-\frac{1}{5}\left(\frac{s+4}{s^2+4s+5}\right),$$

and with entries 11 and 12 of Table 7.1 in mind this may be rewritten as

$$Z(s)=\frac{1}{5}\left(\frac{1}{s}\right)-\frac{1}{5}\left(\frac{s+2}{(s+2)^2+1}\right)-\frac{2}{5}\left(\frac{1}{(s+2)^2+1}\right).$$

The inverse transform now follows directly and is seen to be

$$z(t)=\tfrac{1}{5}-\tfrac{1}{5}e^{-2t}\cos t-\tfrac{2}{5}e^{-2t}\sin t,$$

or

$$z(t)=\tfrac{1}{5}-\tfrac{1}{5}e^{-2t}(\cos t+2\sin t).$$

Differentiation of this result gives

$$z'(t)=e^{-2t}\sin t,$$

so using the last representation for the solution $y(t)$ in Theorem 7.17 we obtain

$$y(t)=\int_0^t e^{-2\tau}\sin\tau f(t-\tau)\,d\tau,$$

which is the required result.

To find the solution when $f(t)=e^t$ we substitute into the above representation to obtain

$$y(t)=\int_0^t \exp(-2\tau)\sin\tau\exp(t-\tau)\,d\tau$$

$$=e^t\int_0^t e^{-3\tau}\sin\tau\,d\tau$$

$$=\tfrac{1}{10}e^t-\tfrac{1}{10}e^{-2t}(3\sin t+\cos t).$$

The same result would, of course, have been obtained had the first representation for $y(t)$ in Theorem 7.17 been used instead of the second one. ∎

Although the Laplace transform was used to determine $z(t)$ in this last example, this is not necessary, and in this case it would have been quicker to use the method of undetermined coefficients. It was, however, the Laplace transform and Duhamel's integrals which led to Theorem 7.17, the result of which is independent of the Laplace transform.

Problems for Section 7.2

In each of the following problems make use of the first translation theorem (Theorem 7.5) to find $\mathcal{L}\{f(t)\}$ when a function $f(t)$ is specified. When a function $F(s)$ is specified use the theorem, together with partial fractions where necessary, to find $\mathcal{L}^{-1}\{F(s)\}$.

1 $f(t) = e^{3t} \cos t$ **2** $f(t) = e^{2t} t \sin 2t$ **3** $f(t) = t e^{-4t}$

4 $f(t) = t^3 e^{-2t}$ **5** $f(t) = e^{-3t} \sinh 2t$ **6** $f(t) = e^{4t} \cosh 3t$

7 $f(t) = \dfrac{s^2 - 2s - 3}{(s^2 - 2s + 5)^2}$ **8** $f(t) = \dfrac{1}{s^2 + 2s + 8}$ **9** $f(t) = \dfrac{3s + 9}{s^2 + 6s - 7}$

10 $F(s) = \dfrac{2}{s^2 - 4s}$ **11** $F(s) = \dfrac{s^2 + 2s + 1}{(s - 1)(s^2 - 2s + 5)}$

12 $F(s) = \dfrac{s^4 + 4s^3 + 10s^2 + 12s + 69}{(s + 1)^3 (s^2 + 2s + 17)}$ **13** $\dfrac{s^2 - 6s + 9}{(s - 1)^2 (s^2 - 4s + 5)}$

14 $F(s) = \dfrac{3s^3 - 10s^2 + 35s - 18}{(s^2 + 9)(s^2 - 4s + 13)}$

Graph each of the following functions.

15 $y(t) = \mathcal{U}(t - 1)(4 - t^2)$ **16** $y(t) = \{\mathcal{U}(t - 1) - \mathcal{U}(t - 2)\}(1 + t^2)$

17 $y(t) = \{\mathcal{U}(t - 2) - \mathcal{U}(t - 4)\}|t - 3|$ **18** $y(t) = \mathcal{U}(t - 1) + \mathcal{U}(t - 2) + \mathcal{U}(t - 3) - 3\mathcal{U}(t - 4)$

19 $y(t) = \{\mathcal{U}(t - \pi/2) - \mathcal{U}(t - 2\pi)\} \cos t$ **20** $y(t) = \{\mathcal{U}(t - \pi/2)\} - \mathcal{U}(t - 5\pi/2)\} |\sin t|$

21 $y(t) = \mathcal{U}(t - \pi/2) \cos(t - \pi/2)$ **22** $y(t) = \mathcal{U}(t - 1) \exp[-(t - 1)]$

23 $y(t) = \{\mathcal{U}(t - 1) - \mathcal{U}(t - 4)\} \mathcal{U}(t - 1) \tanh(t - 1)$

24 $y(t) = \mathcal{U}(t - 1) \operatorname{sech}(t - 1)$

In each of the following problems make use of the second shift theorem (Theorem 7.6) to find $\mathcal{L}\{f(t)\}$ whenever a function $f(t)$ is specified, after first rewriting the form of $f(t)$ if this is necessary. Where a function $F(s)$ is given use the theorem together with partial fractions, where appropriate, to find $\mathcal{L}^{-1}\{F(s)\}$.

25 $f(t) = 3\mathcal{U}(t - 2)(t - 2)^2$ **26** $f(t) = \mathcal{U}(t - \pi/2) \cos 2(t - \pi/2)$

27 $f(t) = 2\mathcal{U}(t - 1) \cosh 3(t - 1)$ **28** $f(t) = \tfrac{1}{2}\mathcal{U}(t - \pi) \sin 2(t - \pi)$

29 $f(t) = \mathcal{U}(t - 1) t^2$ **30** $f(t) = \mathcal{U}(t - 2)(t + 1)$

31 $F(s) = \dfrac{e^{-2s}}{(s - 2)^3}$ **32** $F(s) = \dfrac{2se^{-s}}{(s^2 + 16)^2}$ **33** $F(s) = \dfrac{(s + 2)e^{-3t}}{s^2 + 4s + 20}$

34 $F(s) = \dfrac{3se^{-4s}}{s^2 - 4}$

35 $F(s) = \left(\dfrac{s+3-4(s+2)e^{-s}}{s^2+s-6}\right)e^{-s}$

36 $F(s) = \left(\dfrac{2s^2 - 2 + 2s(s^2 - 4)e^{-2s}}{s^4 - 5s^2 + 4}\right)e^{-2s}$

37 $F(s) = \left(\dfrac{3s^2 + 4s - 12}{(s^2 + 4)^2}\right)e^{-s}$

38 $F(s) = \left(\dfrac{2s^2 + 12s + 18 - s^2 e^{-s}}{s^4 + 6s^3 + 9s^2}\right)e^{-s}$

In each of the following problems make use of the scaling theorem (Theorem 7.7) to find the required transform.

39 Given $\mathcal{L}\{f(t)\} = \dfrac{s^2 + 1}{s^3 + 2s^2 + s - 1}$, find $\mathcal{L}\{f(2t)\}$.

40 Given $\mathcal{L}\{e^{2t}\cos 3t\} = \dfrac{s-2}{s^2 - 4s + 13}$, find $\mathcal{L}\{e^t \cos \tfrac{3}{2}t\}$.

41 Given $\mathcal{L}\{f(t)\} = \dfrac{2}{(s^2 + 4)(s - 1)^2}$, find $\mathcal{L}\{f(\tfrac{1}{2}t)\}$.

42 Given $\mathcal{L}\{f(t) = \dfrac{1 + se^{-s}}{s^2 + 4s - 3}$, find $\mathcal{L}\{f(\tfrac{3}{2}t)\}$.

43 Given $\mathcal{L}\{f(t)\} = \dfrac{6}{s^3(s^2 - 9)}$, find $\mathcal{L}\{f(\tfrac{1}{3}t)\}$.

44 Given $\mathcal{L}\{f(t)\} = \dfrac{2s}{(s^2 + 1)^2(s^2 - 1)}$, find $\mathcal{L}\{f(2t)\}$.

45 Given $\mathcal{L}\{f(t)\} = \dfrac{2(s-1)}{(s^2 - 2s - 5)^2}$, find $\mathcal{L}\{f(\tfrac{1}{2}t)\}$.

46 Prove from first principles that if $f(t)$ is of exponential order with exponent α then

$$\mathcal{L}\{f''(t)\} = s^2\,\mathcal{L}\{f(t)\} - sf(0) - f'(0), \qquad \text{for } s > \alpha.$$

In each of the following problems use the given function $f(t)$ together with Theorem 7.9 and the stated transform to find $\mathcal{L}\{f(t)\}$.

47 $f(t) = t^2 e^{-2t}$ and $\mathcal{L}\{e^{-2t}\} = \dfrac{1}{s+2}$.

48 $f(t) = t^3 e^t$ and $\mathcal{L}\{e^t\} = \dfrac{1}{s-1}$.

49 $f(t) = \sin^2 t$ and $\mathcal{L}\{1\} = \dfrac{1}{s}$.

50 $f(t) = \cosh^2 t$ and $\mathcal{L}\{1\} = \dfrac{1}{s}$.

51 $f(t)=t \sin 2t$ and $\mathscr{L}\{\cos 2t\}=\dfrac{s}{s^2+4}$.

52 $f(t)=t \cos 3t$ and $\mathscr{L}\{\sin 3t\}=\dfrac{3}{s^2+9}$.

In each of the following problems make use of Theorem 7.9 to find $\mathscr{L}\{f'(t)\}$ and $\mathscr{L}\{f''(t)\}$ for the given function $f(t)$.

53 $f(t)=e^{-4t}$ **54** $f(t)=t \sin t$

55 $f(t)=te^{t}$ **56** $f(t)=e^{t} \sin 2t$

57 $f(t)=t \cos 3t$ **58** $f(t)=\sinh 4t$

Use the Laplace transform to solve each of the following initial value problems.

59 $y'+3y=\sin t,\ y(0)=1.$
60 $y'-2y=e^{t},\ y(0)=3.$
61 $y''+2y'+5y=0,\quad$ with $y(0)=0,\ y'(0)=1.$
62 $y''+2y'+5y=2 \cos t,\quad$ with $y(0)=y'(0)=0.$
63 $y''+4y=2 \cos t-\sin t,\quad$ with $y(0)=0,\quad y'(0)=1.$
64 $y'''-6y''+12y'-8y=0,\quad$ with $y(0)=1,\ y'(0)=y''(0)=0.$
65 $y'''-4y'=-20-16 \sin 2t,\quad$ with $y(0)=y'(0)=y''(0)=0.$
66 $y'''+y''=16-e^{-t},\quad$ with $\quad y(0)=1,\ y'(0)=y''(0)=0.$
67 Explain why the Laplace transform cannot be used to solve the following initial value problems as stated. If a reformulation of the problem will allow the Laplace transform method to be used explain how this must be carried out.
 (i) $x''+6x'+x=\tan t,$ with $x(0)=1,\ x'(0)=0.$
 (ii) $x''+4x'+4x=t^2,$ with $x(1)=x'(1)=3.$

 (iii) $x''+16x=t \sin t+\sin\left(\dfrac{1}{3-t}\right),$ with $x(0)=2,\ x'(0)=1.$

 (iv) $x''+2x'+9x=\exp(t^2/10),$ with $x(0)=x'(0)=1.$

68 Reformulate the following initial value problem in a form suitable for solution by the Laplace transform method

$$x''+4x'+3x=4\exp[-(t-2)],\quad \text{with } x(2)=0,\quad x'(2)=2.$$

Solve the reformulated problem using the Laplace transform method and hence show that

$$x(t)=2\exp(t-2)+\exp[-2(t-2)]\,\{6 \sin(t-2)-2 \cos(t-2)\},\qquad \text{for } t>2.$$

[*Hint*: Use the method outlined in the answer to Prob. 67(ii)].

69 $x''+2x'+5x=\mathscr{U}(t-3),\quad$ with $x(0)=1,\ x'(0)=0.$
70 $x''+4x=\mathscr{U}(t-1),\quad$ with $x(0)=1,\ x'(0)=0.$
71 $x''+x=1+\mathscr{U}(t-1),\quad$ with $x(0)=1,\ x'(0)=0.$
72 $x''+x=-1+\mathscr{U}(t-1),\quad$ with $x(0)=x'(0)=0.$
73 $x''-3x'+2x=\mathscr{U}(t)-\mathscr{U}(t-2),\quad$ with $x(0)=x'(0)=0.$
74 $x''+3x'+2x=\mathscr{U}(t-2),\quad$ with $x(0)=0,\ x'(0)=1.$
75 $x''+4x=1+\mathscr{U}(t-1)\sin(t-1),\quad$ with $x(0)=0,\ x'(0)=1.$
76 $x''+3x'+2x=1+\mathscr{U}(t-1)(t-1),\ x(0)=0,\ x'(0)=1.$

Use the Laplace transform to solve the following initial value problems involving first order systems of differential equations.

77 $3x' = x - 2y + 9t$
$3y' = -4x - y + 27$, with $x(0) = 3$, $y(0) = 0$.

78 $x' = 10x - 8y + e^t$
$y' = 6x - 11y$, with $x(0) = y(0) = 0$.

79 $x' + 2x + y = \sin t$
$y' - 4x - 2y = \cos t$, with $x(0) = 2$, $y(0) = -1$,

80 $x' = y - x$
$y' = -x - 3y$, with $x(0) = -1$, $y(0) = 2$.

81 $x' = -x + 2y + 2z + 1$
$y' = -2y + 2z + 1$
$z' = -y - 5z + 1$, with $x(0) = \frac{13}{3}$, $y(0) = \frac{19}{12}$, $z(0) = \frac{1}{12}$.

82 $x' = y + z$
$y' = x + z$
$z' = x + y$, with $x(0) = 1$, $y(0) = -1$, $z(0) = 2$.

83 $x'' + y' + x = e^t$
$y'' + x' = 1$, with $x(0) = y(0) = 2$ and $x'(0) = y'(0) = 0$.

84 $y'' - x = 0$
$x'' - y = 0$, with $x(0) = x'(0) = y(0) = 0$ and $y'(0) = 1$.

In the following problems use Theorem 7.10 to obtain the indicated transform from the stated transform

85 Find $\mathscr{L}\{t \cosh kt\}$ from $\mathscr{L}\{\cosh kt\}$

86 Find $\mathscr{L}\{t^{3/2} e^t\}$ from $\mathscr{L}\{e^t/\sqrt{t}\} = \dfrac{\sqrt{\pi}}{(s-1)^{1/2}}$

87 Find $\mathscr{L}\{t^2 \sin kt\}$ from $\mathscr{L}\{\sin kt\}$
88 Find $\mathscr{L}\{t^n e^{at}\}$ from $\mathscr{L}\{e^{at}\}$

89 Find $\mathscr{L}\{t \sin kt\}$ from $\mathscr{L}\left\{\dfrac{\sin kt}{t}\right\} = \arctan \dfrac{k}{s}$

90 Given $f(t) = t^5 e^{2t}$ verifty that

  (i) $\mathscr{L}\{t f'(t)\} = s F'(s) - F(s)$ (ii) $\mathscr{L}\{t f''(t)\} = -s^2 F'(s) - 2s F(s) + f(0)$
and (iii) $\mathscr{L}\{t^2 f''(t)\} = s^2 F''(s) + 4s F'(s)$.

Use Theorem 7.11 to evaluate the following transforms and inverse transforms.

91 $\mathscr{L}\left\{\displaystyle\int_0^t \tau \cos k\tau \, d\tau\right\}$ **92** $\mathscr{L}\left\{\displaystyle\int_0^t \tau^n e^{a\tau} \, d\tau\right\}$

93 $\mathscr{L}\left\{\displaystyle\int_0^t \dfrac{\sin \tau}{\tau} \, d\tau\right\}$, given $\mathscr{L}\left\{\dfrac{\sin \tau}{\tau}\right\} = \arctan\left(\dfrac{1}{s}\right)$.

94 $\mathscr{L}\left\{\displaystyle\int_0^t \sin k\tau \sinh k\tau \, d\tau\right\}$ **95** $\mathscr{L}\left\{\displaystyle\int_0^t (\cosh k\tau - \cos k\tau) \, d\tau\right\}$

96 $\mathscr{L}\left\{\int_0^t (k\tau - \sin k\tau)\, d\tau\right.$

97 $\mathscr{L}^{-1}\left\{\dfrac{1}{s^2 - k^2}\right\}$

98 $\mathscr{L}^{-1}\left\{\dfrac{1}{(s-1)^4}\right\}$

99 $\mathscr{L}^{-1}\left\{\dfrac{e^{-4s}}{s^2}\right\}$

100 $\mathscr{L}^{-1}\left\{\dfrac{4}{s^4 - k^4}\right\}$

101 $\mathscr{L}^{-1}\left\{\dfrac{1}{(s^2 + k^2)^2}\right\}$

102 $\mathscr{L}^{-1}\left\{\dfrac{s^2 - k^2}{(s^2 + k^2)^2}\right\}$

103 Explain why it is not possible to use Theorem 7.11 to evaluate

$$\mathscr{L}^{-1}\left\{\frac{1}{s-a}\right\}, \quad \mathscr{L}^{-1}\left\{\frac{s}{s^2 + k^2}\right\} \quad \text{and} \quad \mathscr{L}^{-1}\left\{\frac{s}{s^2 - k^2}\right\}$$

in terms of transforms of piecewise continuous functions of exponential order.

104 Use Theorem 7.12 to find $\mathscr{L}\left\{\dfrac{\sin t}{t}\right\}$.

105 Use Theorem 7.12 to find $\mathscr{L}\{(\pi t)^{-1/2}\}$ given $\mathscr{L}\{2(t/\pi)^{1/2}\} = s^{-3/2}$

106 Use Theorem 7.12 to find $\mathscr{L}\{\cos kt\}$ given $\mathscr{L}\{t\cos kt\} = \dfrac{s^2 - k^2}{(s^2 + k^2)^2}$

Use Theorem 7.12 to find the following inverse transforms.

107 $\mathscr{L}^{-1}\left\{\dfrac{1}{s-a}\right\}$

108 $\mathscr{L}^{-1}\left\{\dfrac{k}{s^2 - k^2}\right\}$

109 $\mathscr{L}^{-1}\left\{\dfrac{s}{s^2 + k^2}\right\}$

110 $\mathscr{L}^{-1}\left\{\dfrac{s}{s^2 - k^2}\right\}$

Use Theorem 7.13 to transform the following periodic functions $y = f(t)$.

111

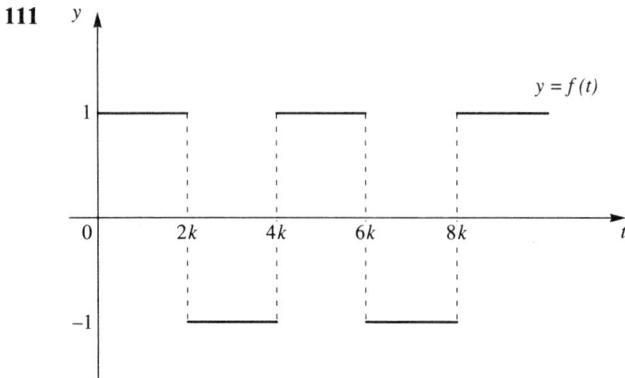

Fig. 7.13

112

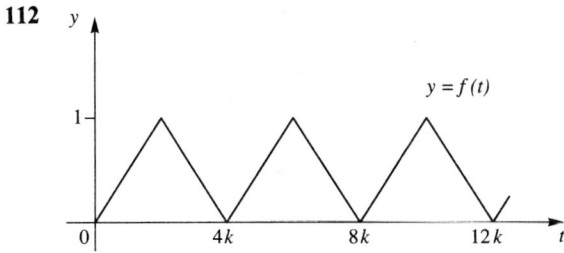

Fig. 7.14

113

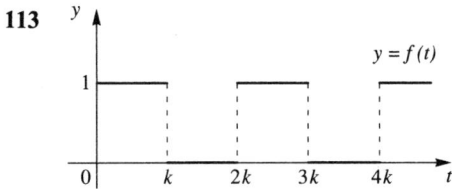

Fig. 7.15

114

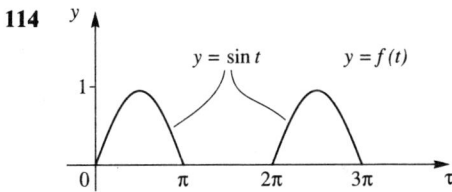

Fig. 7.16   Half-wave rectification

115

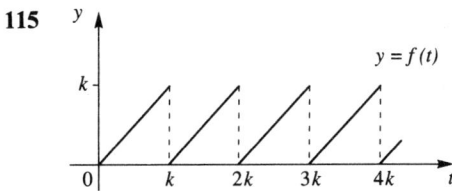

Fig. 7.17   Sawtooth function

116

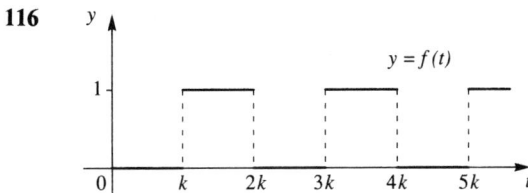

Fig. 7.18

117

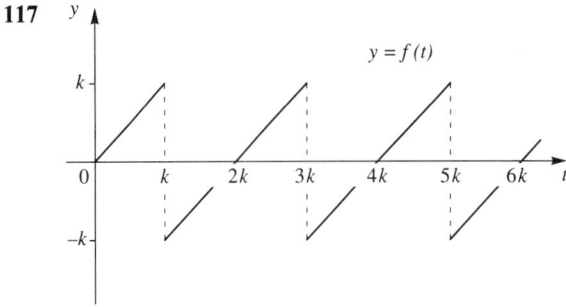

Fig. 7.19

118 Transform the periodic function $y=f'(t)$ shown in Fig. 7.20(a) and then use Theorem 7.11 to show that the transform of the periodic function $y=g(t)$ in Fig. 7.20(b) is $\mathscr{L}\{g(t)\} = \dfrac{a}{s^2}\tanh\dfrac{Ks}{2}$.

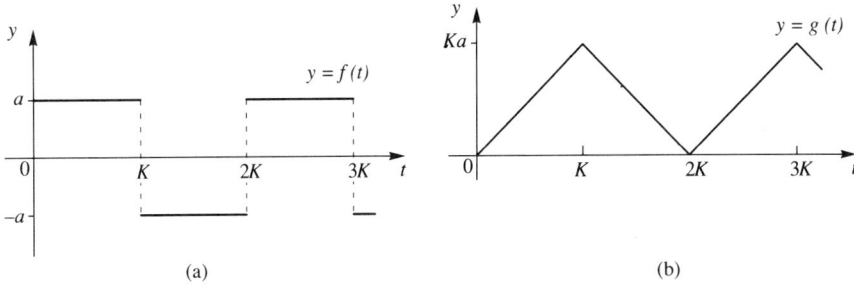

(a)   (b)

Fig. 7.20

119 Represent the staircase function shown in Fig. 7.21(a) as the difference between the two functions shown in Fig. 7.21(b), and hence find its transform.

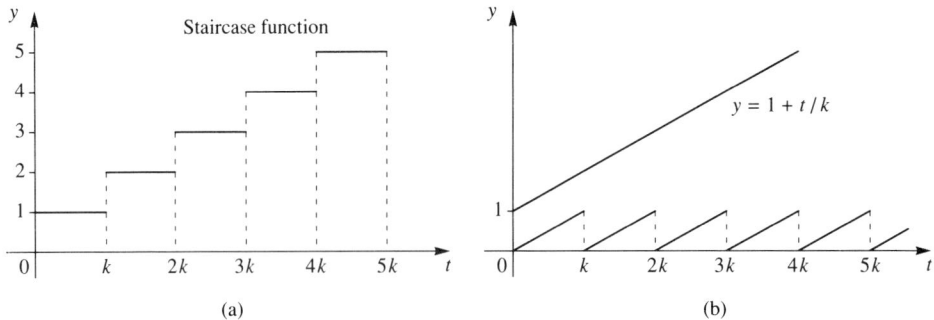

(a)   (b)

Fig. 7.21

120 Adapt the form of argument used in Prob. 119 to show that the transform of the modified staircase function $y=f(t)$ illustrated in Fig. 7.22 is $\dfrac{1}{s}\coth ks.$

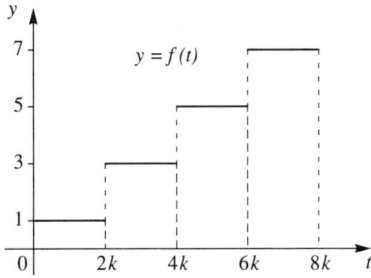

Fig. 7.22

121 If $f(t)=t^2$ and $g(t)=\mathrm{e}^{-2t}$ verify by direct calculation that $f*g=g*f$.

122 Find $t*\cos t$.

123 Find $t^{-1/2}*t^{-1/2}$, and with the aid of a change of variable deduce that $t^{-1/2}*t^{-1/2}=\text{constant}$.

124 Show $\sin kt*\sin kt=\dfrac{1}{2k}(\sin kt-kt\cos kt)$.

In the following problems use Theorem 7.14 to find the indicated transforms and inverse transforms.

125 $\mathscr{L}\{\sinh t*\cos 2t]$ **126** $\mathscr{L}\{(1-\cos t)*\cos t\}$

127 $\mathscr{L}\{\sinh kt-\sin kt)*1\}$ **128** $\mathscr{L}\{(\sin t+t\cos t)*\sinh 4t\}$

129 $\mathscr{L}^{-1}\left\{\dfrac{s}{s^4-k^4}\right\}$ **130** $\mathscr{L}^{-1}\left\{\dfrac{1}{s^2(s^2+a^2)}\right\}$

131 $\mathscr{L}^{-1}\left\{\dfrac{1}{s^2-k^2}\right\}$ **132** $\mathscr{L}^{-1}\left\{\dfrac{\mathrm{e}^{-ks}}{s^2}\right\}$

133 Given that $f(t),\ g(t)$ are differentiable and continuous functions for $0\leq t<\infty$ with $\mathscr{L}\{f(t)\}=F(s)$, and $\mathscr{L}\{g(t)\}=G(s)$, prove

$$\frac{\mathrm{d}}{\mathrm{d}t}[f(t)*g(t)]=f(t)\,g(0)+[f(t)*g'(t)]=f(0)\,g(t)+[f'(t)*g(t)],$$

and

$$\mathscr{L}\left\{\frac{\mathrm{d}}{\mathrm{d}t}[f(t)*g(t)]\right\}=sF(s)\,G(s).$$

134 Prove that

(i) $f*(g+h)=f*g+f*h,$

and

(ii) $f*1\neq f.$

In the following problems use the extended initial value theorem (Theorem 7.15) to determine $f(0)$ and $f'(0)$ directly from the stated transforms. Verify your result by finding the inverse transform.

135 $F(s)=\dfrac{1}{s^2+6s+13}$ **136** $F(s)=\dfrac{1}{s(s^2+k^2)}$

137 $F(s) = \dfrac{16(s+3)}{(s^2+4)^2}$

138 $F(s) = \dfrac{e^{-3s}}{s^2+2s+5}$

139 $F(s) = \dfrac{s^2}{(s^2+k^2)^2}$

140 $F(s) = \dfrac{s^2-1}{(s^2+1)^2}$

141 Use Theorem 7.15(iii) to deduce $f^{(n)}(0)$ given that

$$F(s) = \frac{(s+2)e^{-s}}{s^2+4s+7}.$$

Find $f(t) = \mathscr{L}^{-1}\{F(s)\}$ and hence verify your conclusion.

142 Show

$$\Delta_k(t) = \frac{k}{\pi(1+k^2 t^2)}$$

may be taken as a pre-delta function in terms of which

$$\delta(t) = \lim_{k \to \infty} \Delta_k(t).$$

Graph $\Delta_k(t)$ for some representative values of k.

143 Evaluate

$$I = \int_{-\infty}^{\infty} \frac{1+2\delta(t-1)}{1+t^2}\, dt.$$

144 Evaluate

$$I = \int_{-\infty}^{\infty} \frac{\sin^2 t}{1+2t^2}\, [\delta(t+\pi)+5\delta(t-4\pi)]\, dt.$$

145 Evaluate

$$I = \int_{-\infty}^{\infty} \frac{\sinh 3t}{1+t^2}\, [\delta(t+4)+\delta(t-4)]\, dt.$$

146 Evaluate

$$I = \int_{-\infty}^{\infty} \frac{\mathscr{U}(t-1)[1+\delta(t+1)]}{1+t^2}\, dt.$$

147 Given $f(t)$ is a continuous function, evaluate

(i) $\displaystyle\int_{-\infty}^{\infty} f(\tau)\,\delta(t-\tau)\,d\tau$ and   (ii) $\displaystyle\int_{-\infty}^{\infty} f(t)\frac{d}{dt}\,\mathscr{U}(t-t_0)\,dt.$

148 Give a formal proof that

$$f(t)\,\delta(t-t_0) = f(t_0)\,\delta(t-t_0),$$

and hence show

$$t\delta(t) = 0.$$

149 Show, by considering the integral

$$\int_{-\infty}^{\infty} f(t)\{\delta'(t-t_0)-\delta'(t_0-t)\}\,dt,$$

where $f(t)$ is an arbitrary continuous function, that $\delta'(t-t_0)$ is an odd function of its argument $t-t_0$, so that

$$\delta'(t-t_0)=-\delta'(t_0-t).$$

150 Give a formal proof that

$$\delta[(t-t_1)(t-t_2)]=\frac{\delta(t-t_1)+\delta(t-t_2)}{|t_1-t_2|}$$

[*Hint*: Use the result of Ex. 18]

151 Show by considering the integral

$$I=\int_{-\infty}^{\infty} f(t)\,t\delta'(t)\,dt$$

that

$$t\delta'(t)=-\delta(t).$$

Use this result to show $\delta'(t)$ is an odd function and interpret the sifting property of $t\delta'(t)$ when used in an integral.

152 Assuming repeated differentiation of the delta function is permissible, use the result of Prob. 151 to show that

$$t^n\delta^{(n)}(t)=(-1)^n n!\,\delta(t).$$

Interpret the sifting property of $t^n\delta^{(n)}(t)$ when used in a definite integral.

153 By writing $|\sin t|=\sin t\{2\mathscr{U}(t)-1\}$ for $-\pi\le t\le\pi$, find

$$\frac{d^2}{dt^2}|\sin t|.$$

[*Hint*: Make use of the result of Prob. 148]

154 By writing $|t|=t\{2\mathscr{U}(t)-1\}$, so

$$|t|^3=t^3\{2\mathscr{U}(t)-1\}^3=t^3\{2\mathscr{U}(t)-1\},\text{ show that}$$

(i) $\dfrac{d^2|t|}{dt^2}=2\delta(t)$ and   (ii) $\dfrac{d^4|t|^3}{dt^4}=12\delta(t).$

155 Use the identity

$$\frac{s-a}{(s-a)^2+k^2}\equiv\frac{(s-a)^2}{(s-a)^2+k^2}\cdot\frac{1}{(s-a)}$$

to show by means of the convolution theorem that

$$\mathscr{L}^{-1}\left\{\frac{s-a}{(s-a)^2+k^2}\right\}=e^{at}\cos kt.$$

156 Use the form of argument outlined in Prob. 155 to show by means of the convolution theorem that

$$\mathcal{L}^{-1}\left\{\frac{s}{s^2-k^2}\right\}=\cosh kt.$$

157 Evaluate

$$\mathcal{L}^{-1}\left\{\frac{2s^4+9s^3+12s^2+36s+16}{(s^2+4)(s+4)}\right\}.$$

158 Evaluate

$$\mathcal{L}^{-1}\left\{\frac{3s^3+4s^2-2s+1}{s^2+s-2}\right\}.$$

159 Graph the function

$$f(t)=\int_0^\infty\left\{\sum_{n=0}^\infty\delta(t-nk)\right\}dt,$$

and show

$$\mathcal{L}\{f(t)\}=\frac{1}{s(1-e^{-ks})}.$$

160 Graph the function

$$f(t)=\int_0^\infty\left\{\delta(t)+2\sum_{n=1}^\infty(-1)^n\delta(t-2nk)\right\}dt,$$

and show

$$\mathcal{L}\{f(t)\}=\frac{1}{s}\tanh ks.$$

161 Graph the function

$$f(t)=2\int_0^\infty\left\{\sum_{n=0}^\infty\delta[t-(2n+1)k]\right\}dt,$$

and show

$$\mathcal{L}\{f(t)\}=\frac{1}{s\sinh ks}.$$

162 Graph the function

$$f(t)=2\int_0^\infty\left\{\sum_{n=0}^\infty(-1)^n\delta[t-(2n+1)k]\right\}dt,$$

and show

$$\mathcal{L}\{f(t)\}=\frac{1}{s\cosh ks}.$$

163 Solve

$$x'' + 4x' + 4x = t + \delta(t - 2),$$

with $x(0) = x'(0) = 0$.

164 Solve

$$x'' + 2x' + 4x = \delta(t),$$

with $x(0) = 0$, $x'(0) = 1$.

165 Solve

$$x'' - 3x' + 2x = \delta(t) + \mathcal{U}(t - 2)$$

with $x(0) = x'(0) = 0$.

166 Solve

$$x'' + 4x = \sin 2t + \delta(t),$$

with $x(0) = x'(0) = 0$.

167 Solve

$$x' = y + \delta(t)$$
$$y' = x + 1,$$

with $x(0) = y(0) = 0$.

168 Solve

$$x' + y + 1 = 0$$
$$y' + x = \delta(t),$$

with $x(0) = 1$, $y(0) = 0$.

Use Theorem 7.17 to solve the following initial value problems after first finding $z(t)$ by any convenient means (Laplace transform or otherwise).

169 $y'' + 9y = 5 \sin 2t$, with $y(0) = y'(0) = 0$.

170 $y'' + y = 6 \sin t$, with $y(0) = y'(0) = 0$.

171 $y'' - 3y' + 2y = 2e^{-t}$, with $y(0) = y'(0) = 0$.

172 $y'' - 10y' + 41y = 17 \sin t$, with $y(0) = y'(0) = 0$.

173 $y''' + y' = e^t$, with $y(0) = y'(0) = y''(0) = 0$.

174 $y''' + y'' = 16 - e^{-t}$, with $y(0) = y'(0) = y''(0) = 0$.

175 Generalize the approach which led to Theorem 7.17 to show how the solution of the differential equation

$$a_0 \frac{d^n y}{dt^n} + a_1 \frac{d^{n-1} y}{dt^{n-1}} + \ldots + a_n y = f(t),$$

with the nonhomogeneous initial conditions $y(0) = y_0$, $y^{(1)}(0) = y_1, \ldots y^{(n-1)}(0) = y_{n-1}$ and $f(t)$ a piecewise continuous function, may be found in terms of the solution $z(t)$ of

$$a_0 \frac{d^n z}{dt^n} + a_1 \frac{d^{n-1}}{dt^{n-1}} + \ldots + a_n z = \mathcal{U}(t),$$

with $z(0) = z^{(1)}(0) = \ldots = z^{(n-1)}(0) = 0$.

7.3 Applications of the Laplace transform

This section presents a brief introduction to some of the many applications that are made of the Laplace transform. The important part played by the Laplace transform in connection with partial differential equations will be discussed in a later volume.

(a) Bending of beams

Horizontal beams in structures experience a vertical deflection due to the combined effect of their own weight and any additional load carried by the beam. If the beam is of length l, and its undeflected axis is taken to coincide with the x-axis, the downward vertical deflection $y(x)$ of the beam at position x may be shown to satisfy the differential equation

$$\frac{d^4 y}{dx^4} = \frac{w(x)}{EI}, \qquad \text{for } 0 < x < l, \tag{1}$$

where $w(x)$ is the transverse load per unit length acting on the beam and EI is the **flexural rigidity** of the beam. The two factors entering into the flexural rigidity are E, the **Young's modulus** for the material of the beam and I, the **moment of inertia** of the cross-section of the beam. It is shown in mechanics that the **bending moment** M at position x is $M = EIy''(x)$, and the **vertical shear** S is $S = EIy'''(x)$.

The boundary conditions to be applied at the ends of the beam depend on the way it is mounted.

At a **clamped** end: $\qquad\qquad\qquad y = y' = 0.$
At a **free** end: $\qquad\qquad\qquad\qquad y'' = y''' = 0.$
At a **simply supported** or **hinged** end: $\qquad y = y'' = 0.$

Three of the most commonly occurring situations are illustrated in Fig. 7.23. The cantilever beam in Fig. 7.23(a) has one end clamped and the other end free; the beam with

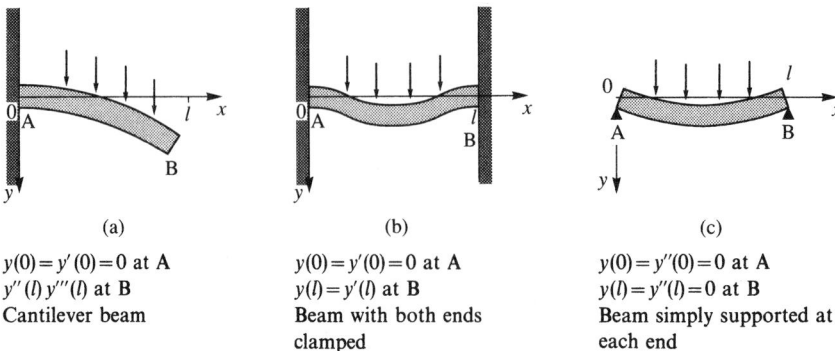

(a)	(b)	(c)
$y(0) = y'(0) = 0$ at A	$y(0) = y'(0) = 0$ at A	$y(0) = y''(0) = 0$ at A
$y''(l)\,y'''(l)$ at B	$y(l) = y'(l)$ at B	$y(l) = y''(l) = 0$ at B
Cantilever beam	Beam with both ends clamped	Beam simply supported at each end

Fig. 7.23 Typical beam mountings and boundary conditions

both ends clamped is shown in Fig. 7.23(b), while a beam which is simply supported (hinged) at each end is shown in Fig. 7.23(c).

Normally the Laplace transform is only applicable to pure initial value problems. However, the special nature of the boundary conditions in beam problems enables the transform to be used in these cases as well. To see how it must be applied we first set $y(0) = y_0$, $y'(0) = y_1$, $y''(0) = y_2$ and $y'''(0) = y_3$, and take the Laplace transform of (1) to obtain

$$s^4 Y - s^3 y_0 - s^2 y_1 - s y_2 - y_3 = \frac{W(s)}{EI}, \tag{2}$$

where $Y(s) = \mathscr{L}\{y(x)\}$ and $W(s) = \mathscr{L}\{w(x)\}$. Then, depending on the way the beam is fastened at $x = 0$, two of the four quantities y_0, y_1, y_2, y_3 will be known and so can be used in (2), leaving the other two as unknown parameters. As a result, $y(x) = \mathscr{L}^{-1}\{Y(s)\}$ can be determined with the two unspecified initial conditions as unknown parameters in the solution. These parameters are then determined by requiring the solution $y(x)$ to satisfy the two remaining boundary conditions at $x = l$. If they are required, the bending moment M and vertical shear S at position x can be found from expressions

$$M = EI y''(x), \tag{3}$$

$$S = EI y'''(x). \tag{4}$$

Example 1. Cantilever beam with concentrated load

A horizontal uniform cross-section cantilever beam of length l and constant flexural rigidity EI bends under the combined effect of its own weight M and a concentrated point load P located at $x = l/2$. Find the vertical deflection $y(x)$ of the beam, and the bending moment and vertical shear at $x = l/4$.

Solution

The deflected cantilever beam is illustrated in Fig. 7.24, with the concentrated load P applied to the midpoint of the beam.

The governing differential equation and boundary conditions are

$$\frac{d^4 y}{dx^4} = \frac{w(x)}{EI},$$

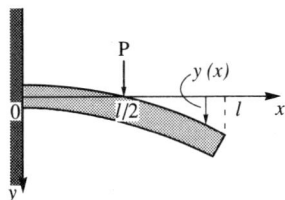

Fig. 7.24 Cantilever beam with concentrated load

with $y(0)=0$, $y'(0)=0$, $y''(l)=0$ and $y'''(l)=0$. The loading on the beam is described by the function

$$w(x)=\frac{M}{l}\{\mathcal{U}(x)-\mathcal{U}(x-l)\}+P\delta(x-\tfrac{1}{2}l),$$

in which the first term represents the distributed load per unit length due to the weight of the beam, and the second term the concentrated point load P at $x=l/2$.

Transforming the differential equation and using the fact that $y_0=y(0)=0$ and $y_1=y'(0)=0$, but $y_2=y''(0)$ and $y_3=y'''(0)$ are unknown, we obtain

$$s^4Y-sy_2-y_3=\frac{1}{EI}\left(\frac{M}{ls}-\frac{Me^{-ls}}{ls}+Pe^{-ls/2}\right),$$

and so

$$Y(s)=\frac{y_2}{s^3}+\frac{y_3}{s^4}+\frac{1}{l}\left(\frac{M}{EI}\right)\frac{1}{s^5}-\frac{1}{l}\left(\frac{M}{EI}\right)\frac{e^{-ls}}{s^5}+\left(\frac{P}{EI}\right)\frac{e^{-ls/2}}{s^4}.$$

Inverting this transform gives

$$y(x)=\frac{1}{2}y_2x^2+\frac{1}{6}y_3x^3+\frac{1}{24l}\left(\frac{M}{EI}\right)x^4-\frac{1}{24l}\left(\frac{M}{EI}\right)(x-l)^4\mathcal{U}(x-l)+\frac{1}{6}\left(\frac{P}{EI}\right)(x-\tfrac{1}{2}l)^3\mathcal{U}(x-\tfrac{1}{2}l).$$

However, as the beam only extends from $x=0$ to $x=l$, the fourth term in $y(x)$ makes no contribution in this interval because it is multiplied by the unit step function $\mathcal{U}(x-l)$. Thus in terms of the unknown parameters y_2 and y_3 we have

$$y(x)=\frac{1}{2}y_2x^2+\frac{1}{6}y_3x^3+\frac{1}{24l}\left(\frac{M}{EI}\right)x^4,\qquad\text{for }0\le x<l/2,$$

and

$$y(x)=\frac{1}{2}y_2x^2+\frac{1}{6}y_3x^3+\frac{1}{24l}\left(\frac{M}{EI}\right)x^4+\frac{1}{6}\left(\frac{P}{EI}\right)(x-\tfrac{1}{2}l)^3,\qquad\text{for }l/2<x\le l.$$

To determine y_2 and y_3 we now require the function $y(x)$ to satisfy the boundary conditions at $x=l$. Differentiation of $y(x)$ in the interval $l/2<x\le l$ gives

$$y''(x)=y_2+y_3x+\frac{1}{2l}\left(\frac{M}{EI}\right)x^2+\left(\frac{P}{EI}\right)(x-\tfrac{1}{2}l)$$

and

$$y'''(x)=y_3+\frac{1}{l}\left(\frac{M}{EI}\right)x+\left(\frac{P}{EI}\right),$$

so applying the boundary conditions $y''(l)=y'''(l)=0$ it follows that

$$y_2=\frac{l}{2}\left(\frac{M+P}{EI}\right)\quad\text{and}\quad y_3=-\left(\frac{M+P}{EI}\right).$$

The deflection $y(x)$ is thus

$$y(x)=\frac{l}{4}\left(\frac{M+P}{EI}\right)x^2-\frac{1}{6}\left(\frac{M+P}{EI}\right)x^3+\frac{1}{24l}\left(\frac{M}{EI}\right)x^4,\qquad\text{for }0\le x<l/2,$$

and

$$y(x) = \frac{l}{4}\left(\frac{M+P}{EI}\right)x^2 - \frac{1}{6}\left(\frac{M+P}{EI}\right)x^3 + \frac{1}{24l}\left(\frac{M}{EI}\right)x^4 + \frac{1}{6}\left(\frac{P}{EI}\right)(x-\tfrac{1}{2}l)^3, \quad \text{for } l/2 < x \le l.$$

The bending moment M and the vertical shear S at $x = l/4$ follow from (3) and (4) as

$$M = EIy''(l/4) = \frac{l}{32}(9M + 8P),$$

and

$$S = EIy'''(l/4) = -\tfrac{1}{4}(3M + 4P). \qquad\blacksquare$$

(b) Coupled oscillations in mechanical and electrical systems

The mathematics of coupled oscillations in any system is merely the mathematics of simultaneous differential equations describing oscillatory motion. Apart from an increase in algebraic complexity as the number of coupled oscillations increases, the method of approach to all such problems is the same. In the two examples that follow, this is illustrated in the case of a coupled mechanical oscillator and two simple electrical systems coupled by mutual induction.

Example 2. Coupled mechanical oscillations

The undamped coupled mass system illustrated in Fig. 7.25 oscillating one-dimensionally as shown has the coupled equations of motion

$$m_1 \ddot{x}_2 + k_1 x_1 - k_2(x_2 - x_1) = 0,$$
$$m_2 \ddot{x}_2 + k_2(x_2 - x_1) = 0.$$

Find the displacements $x_1(t)$ and $x_2(t)$ given that $m_1 = 3$, $m_3 = 1$, $k_1 = 4$, $k_2 = 1$ and initially $x_1(0) = 1$, $x_2(0) = 0$, $\dot{x}_1(0) = 0$ and $\dot{x}_2(0) = 0$.

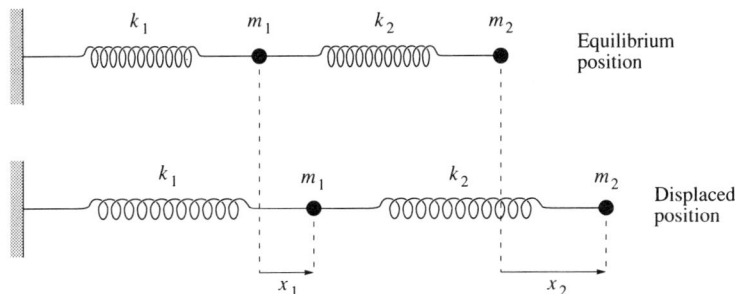

Fig. 7.25 Coupled undamped mechanical oscillator

Solution

In terms of the stated parameters the equations of motion become

$$3\ddot{x}_1 + 5x_1 - x_2 = 0 \quad \text{and} \quad \ddot{x}_2 + x_2 - x_1 = 0.$$

Transforming these equations and using the initial conditions we find

$$3s^2 X_1(s) - 3s + 5X_1(s) - X_2(s) = 0$$

and

$$s^2 X_2(s) + X_2(s) - X_1(s) = 0,$$

where $X_1(s) = \mathscr{L}\{x_1(t)\}$ and $X_2(s) = \mathscr{L}\{x_2(t)\}$.
Solving for $X_1(s)$ and $X_2(s)$ gives

$$X_1(s) = \frac{3s(s^2 + 1)}{(3s^2 + 2)(s^2 + 2)} \quad \text{and} \quad X_2(s) = \frac{3s}{(3s^2 + 2)(s^2 + 2)}.$$

After simplification by use of partial fractions these results become

$$X_1(s) = \frac{3}{4}\left(\frac{s}{3s^2 + 2}\right) + \frac{3}{4}\left(\frac{s}{s^2 + 2}\right)$$

and

$$X_2(s) = \frac{9}{4}\left(\frac{s}{3s^2 + 2}\right) - \frac{3}{4}\left(\frac{s}{s^2 + 2}\right).$$

Removing a factor 3 from the denominator of the first terms in $X_1(s)$ and $X_2(s)$ and using Table 7.1 to invert the transforms we arrive at the solution

$$x_1(t) = \tfrac{1}{4}\cos\sqrt{\tfrac{2}{3}}t + \tfrac{3}{4}\cos\sqrt{2}t \quad \text{and} \quad x_2(t) = \tfrac{3}{4}\cos\sqrt{\tfrac{2}{3}}t - \tfrac{3}{4}\cos\sqrt{2}t.$$

■

Example 3. *LC*-circuits coupled by mutual inductance

Two circuits of negligible resistance each contain a capacitance of capacity C and have coefficients of self-induction L. The coefficient of mutual inductance is M, with $L > M$. If initially one capacitance has a charge K and the other is uncharged and the current flow through the circuit is zero, find the charges on the capacitances at any time $t > 0$.

Solution

The two identical *LC*-circuits linked by a mutual inductance are represented in Fig. 7.26, with the directions of the currents i_1 and i_2 which flow being chosen as shown. Let us assume that the capacitance in the left-hand circuit is the one with the initial charge K.
An application of Kirchhoff's laws shows the governing equations to be

$$L\frac{di_1}{dt} + M\frac{di_2}{dt} + \frac{q_1}{C} = 0$$

$$L\frac{di_2}{dt} + M\frac{di_1}{dt} + \frac{q_2}{C} = 0,$$

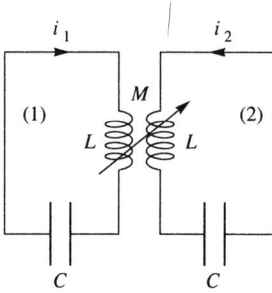

Fig. 7.26 Circuits coupled by mutual inductance

where $q_1(t)$ and $q_2(t)$ are the charges in circuits (1) and (2) at time t, and

$$\frac{dq_1}{dt} = i_1 \quad \text{and} \quad \frac{dq_2}{dt} = i_2.$$

Thus the governing equations for q_1 and q_2 are

$$LC\ddot{q}_1 + MC\ddot{q}_2 + q_1 = 0$$
$$LC\ddot{q}_2 + MC\ddot{q}_1 + q_2 = 0,$$

and the initial conditions correspond to

$$q_1(0) = K, \quad \dot{q}_1(0) = 0, \quad q_2(0) = 0 \quad \text{and} \quad \dot{q}_2(0) = 0.$$

Transforming these equations and incorporating the initial conditions gives

$$(1 + s^2 LC)Q_1(s) + s^2 MCQ_2(s) = sLCK$$

and

$$(1 + s^2 LC)Q_2(s) + s^2 MCQ_1(s) = sMCK,$$

where $Q_1(s) = \mathcal{L}\{q_1(t)\}$ and $Q_2(s) = \mathcal{L}\{q_2(t)\}$.

Rather than solving for $Q_1(s)$ and $Q_2(s)$ directly, and then inverting the transforms to find $q_1(t)$ and $q_2(t)$, the symmetry of the equations makes it easier to proceed as follows. Setting $U(s) = Q_1(s) + Q_2(s)$ and $V(s) = Q_1(s) - Q_2(s)$, it follows by adding and subtracting the equations that

$$\{1 + s^2 C(L + M)\} U(s) = sCK(L + M)$$

and

$$1 + s^2 C(1 - M)V(s) = sCK(L - M),$$

and so

$$U(s) = \frac{sCK(L + M)}{1 + s^2 C(L + M)}, \quad V(s) = \frac{sCK(L - M)}{1 + s^2 C(L - M)}.$$

Setting

$$\Omega_1^2 = \frac{1}{C(L + M)} \quad \text{and} \quad \Omega_2^2 = \frac{1}{C(L - M)}$$

simplifies the expressions for $U(s)$, $V(s)$ to

$$U(s) = \frac{sK}{s^2 + \Omega_1^2} \quad \text{and} \quad V(s) = \frac{sK}{s^2 + \Omega_2^2}.$$

Inverting these transforms gives

$$q_1(t) + q_2(t) = \mathscr{L}^{-1}\{U(s)\} = K \cos \Omega_1 t$$
$$q_1(t) - q_2(t) = \mathscr{L}^{-1}\{V(s)\} = K \cos \Omega_2 t,$$

and so

$$q_1(t) = \tfrac{1}{2}K(\cos \Omega_1 t + \cos \Omega_2 t), \quad q_2(t) = \tfrac{1}{2}K(\cos \Omega_1 t - \cos \Omega_2 t).$$

Examination of the form of these solutions shows the system to have the two fundamental frequencies of oscillation Ω_1 and Ω_2. ∎

(c) Weighting function

In control theory, the output $y_0(t)$ from a system produced by a unit delta function acting at time $t = 0$ as input $y_1(t)$ to the system when initially at rest is called the **weighting function** of the system.

Typically, the output $y_0(t)$ and input $y_1(t)$ are related by a differential equation of the form

$$a_0 \frac{d^n y_0}{dt^n} + a_1 \frac{d^{n-1} y_0}{dt^{n-1}} + \ldots + a_n y_0 = k y_1(t), \tag{5}$$

in which $k > 0$ represents the amplification factor of the input $y_1(t)$. Thus the weighting function $W(t)$ of the system described by (5) must satisfy

$$a_0 \frac{d^n W}{dt^n} + a_1 \frac{d^{n-1} W}{dt^{n-1}} + \ldots + a_n W = k\delta(t), \tag{6}$$

with $W(t) \equiv 0$ for $t < 0$. Clearly the weighting function may be discontinuous, and on occasions, it can even contain terms involving delta functions.

Example 4. Weighting function for *RC*-filter

Find the weighting function $W(t)$ of the *RC*-filter shown in Fig. 7.27, in which the input voltage $E_1(t)$ and output voltage $E_O(t)$ are related by

$$RC \frac{dE_O}{dt} + E_O = E_I.$$

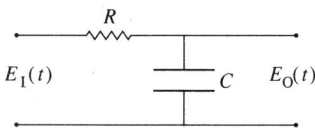

Fig. 7.27 *RC*-filter

Solution

By definition, the weighting function $W(t)$ for this RC-filter will be the solution of

$$RC\frac{\mathrm{d}W}{\mathrm{d}t} + W = \delta(t),$$

with $W(t) \equiv 0$ for $t < 0$.

Thus the initial conditions for $W(t)$ are $W(0-) = 0$ and $W'(0-) = 0$, so transforming the differential equation and using these conditions we obtain

$$sRC\,\Omega(s) + \Omega(s) = 1,$$

where $\Omega(s) = \mathscr{L}\{W(t)\}$.

Thus

$$\Omega(s) = \frac{1}{1+sRC} = \frac{1}{RC}\left(\frac{1}{s+\dfrac{1}{RC}}\right),$$

and so

$$W(t) = \mathscr{L}^{-1}\{\Omega(s)\} = \frac{1}{RC}\exp(-t/RC), \qquad \text{for } t > 0.$$

Thus we have shown that

$$W(t) = \begin{cases} \dfrac{1}{RC}\exp(-t/RC), & t > 0 \\ 0, & t < 0. \end{cases}$$

The constant $T = RC$ is called the **time constant** of the filter, and it represents the time taken for the voltage to decay to $1/e$ of its initial magnitude. ∎

Example 5. Weighting function for modified *RC*-filter

Find the weighting function $W(t)$ of the modified RC-filter shown in Fig. 7.28, in which the input voltage $E_I(t)$ and output voltage $E_O(t)$ are related by

$$RC\frac{\mathrm{d}E_O}{\mathrm{d}t} + E_O = RC\frac{\mathrm{d}E_I}{\mathrm{d}t}.$$

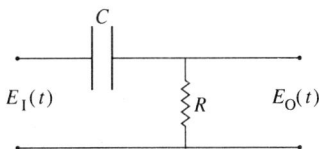

Fig. 7.28 Modified *RC*-filter

Solution

Proceeding as in the previous example, the weighting function $W(t)$ for this modified RC-filter in which the positions of the capacitance and resistor have been interchanged will be the solution of

$$RC\frac{dW}{dt} + W = RC\frac{d}{dt}\{\delta(t)\}.$$

Taking the Laplace transform of this equation and using the initial conditions $W(0-) = 0$, $W'(0-) = 0$ together with $\mathcal{L}\{\delta'(t)\} = s$ (Eq. (23), Sec. 7.2) we find

$$sRC\Omega(s) + \Omega(s) = sRC,$$

where as before $\Omega(s) = \mathcal{L}\{W(t)\}$. Thus

$$\Omega(s) = \frac{sRC}{1 + sRC} = 1 - \left(\frac{1}{RC}\right)\frac{1}{\left(s + \frac{1}{RC}\right)},$$

and so

$$W(t) = \mathcal{L}^{-1}\{\Omega(s)\} = \delta(t) - \frac{1}{RC}\exp(-t/RC), \qquad t > 0$$

Thus we have shown that

$$W(t) = \begin{cases} \delta(t) - \dfrac{1}{RC}\exp(-t/RC), & t > 0 \\ 0, & t < 0. \end{cases}$$
∎

The importance of the weighting function of a system arises from the fact that once $W(t)$ is known it can be used to express the output (response) $y_O(t)$ of the system to an arbitrary input $y_I(t)$. This result takes on a particularly simple form when the system is initially at rest so that $y_I(t) \equiv 0$ for $t > 0$, for then $y_I(0-) = 0$ together with all its derivatives at $t = 0-$.

Our derivation of this useful result will be based on the convolution theorem. Let the system in question be represented by the differential equation

$$a_0\frac{d^n y_O}{dt^n} + a_1\frac{d^{n-1} y_O}{dt^{n-1}} + \ldots + a_n y_O = k y_I(t), \tag{7}$$

with $y_O(t)$ the output, $y_I(t)$ the input and $k > 0$ the amplification factor for the input.

We will assume that the system starts from rest, so $y(0) = y^{(1)}(0) = \ldots = y^{(n-1)}(0) = 0$. Transforming (7) then leads to the result

$$P(s) Y_O(s) = k Y_I(s), \tag{8}$$

where $Y_O(s) = \mathcal{L}\{y_O(t)\}$, $Y_I(s) = \mathcal{L}\{y_I(t)\}$

and

$$P(s) = a_0 s^n + a_1 s^{n-1} + \ldots + a_n.$$

The weighting function for (7) is, by definition, the solution of

$$a_0 \frac{d^n W}{dt^n} + a_1 \frac{d^{n-1} W}{dt^{n-1}} + \ldots + a_n W = k\delta(t).$$

Thus transforming this equation and using the fact that $W(t) \equiv 0$ for $t < 0$, so that $W(0-)$ vanishes together with all its derivatives at $t = 0-$, we obtain

$$P(s)\Omega(s) = k, \quad \text{with } \Omega(s) = \mathscr{L}\{W(t)\}. \tag{9}$$

Combining (8) and (9) then shows that

$$Y_0(s) = \Omega(s) Y_1(s), \tag{10}$$

so taking the inverse transform and using the convolution theorem we find the important result

$$y_0(t) = \int_0^t W(\tau) y_1(t - \tau) \, d\tau. \tag{11}$$

The function $W(t)$ is called the **weighting function** because of this convolution integral representation for $y_0(t)$, which shows how the input $y_1(t - \tau)$ is weighted by the function $W(\tau)$ over the interval $0 \le \tau \le t$ in order to arrive at $y_0(t)$.

Expressed differently, the weighting function for a system may be regarded as a mathematical representation of its 'memory' of past events. In this sense (11) shows the extent to which the output 'now' (at time t) is determined by all past inputs y_1 over the interval $0 \le \tau \le t$. Setting $y_1(t) = \mathscr{U}(t)$ in (12), the response $y_{\mathscr{U}}(t)$ to a unit step function input is seen to be related to the response $W(t)$ to a unit delta function input by

$$y_{\mathscr{U}}(t) = \int_0^t W(\tau) \mathscr{U}(t - \tau) \, d\tau = \int_0^t W(\tau) \, d\tau. \tag{12}$$

The weighting fuction $W(t)$ may also be used to express the solution of (7) in integral form when the system starts from the arbitrary initial conditions $y_0(0) = y_0$, $y_0^{(1)}(0) = y_1, \ldots, y_0^{(n-1)}(0) = y_{n-1}$.

In such a case, after transforming (7), it follows that (8) is replaced by

$$P(s) Y_0(s) + H(s) = k Y_1(s), \tag{13}$$

where the polynomial $H(s)$ contains all the terms arising from the left-hand side of (7) when the nonhomogeneous initial conditions are included in the transformation.

Combining (10) and (13) gives

$$Y_0(s) = \Omega(s) Y_1(s) - \frac{1}{k} \Omega(s) H(s), \tag{14}$$

which is analogous to (10). Taking the inverse transform of (14) and using the convolution theorem on each term brings us to the integral representation

$$y_0(t) = \int_0^t W(\tau) y_1(t - \tau) \, d\tau - \frac{1}{k} \int_0^t W(\tau) h(t - \tau) \, d\tau, \tag{15}$$

where

$h(t) = \mathcal{L}^{-1}\{H(s)\}$ depends only on the nonhomogeneous initial conditions.

Example 6. Time response of *RC*-filter

Use the weighting function $W(t)$ found in Ex. 4 to determine the output from the filter when

(a) $E_0(0) = 0$ and $E_1(t) = \begin{cases} E, & t > 0 \\ 0, & t < 0 \end{cases}$

and

(b) $E_0(0) = k$ and $E_1(t) = \begin{cases} E \sin \omega t, & t > 0 \\ 0, & t < 0, \end{cases}$ with $E = $ constant.

Solution

(a) As the initial condition for $E_0(t)$ is homogeneous ($E_0(0) = 0$) we may use result (12) to determine $E_0(t)$ by writing $y(t) \equiv E_0(t)$ and making use of the function $W(t)$ found in Ex. 4. As a result we find the output from the filter is

$$E_0(t) = \int_0^t \frac{E}{RC} e^{-\tau/RC} \, d\tau = E(1 - e^{-t/RC}), \qquad t > 0.$$

(b) In this case the initial condition for $E_0(t)$ is nonhomogeneous ($E_0(0) = k$) so we must determine $E_0(t)$ from result (15), and to do this we need to find $h(t)$. Transforming the governing differential equation

$$RC \frac{dE_0}{dt} + E_0 = E \sin \omega t$$

gives

$$RC[sE(s) - k] + E(s) = \mathcal{L}\{E \sin \omega t\},$$

or

$$(1 + s \, RC) E(s) - kRC = \mathcal{L}\{E \sin \omega t\},$$

where $E(s) = \mathcal{L}\{E_0(t)\}$.

Identifying terms in this equation with the corresponding terms in (13) shows $P(s) \equiv 1 + sRC$, $Y_0(s) \equiv E(s)$, $H(s) \equiv -kRC$ and $Y_1(s) = \mathcal{L}\{E \sin \omega t\}$. Thus

$$h(t) = \mathcal{L}^{-1}\{H(s)\} = \mathcal{L}^{-1}\{-k \, RC\} = -kRC \, \delta(t),$$

and so (15) becomes

$$E_0(t) = \int_0^t \left(\frac{E}{RC}\right) e^{-\tau/RC} \sin \omega(t - \tau) \, d\tau + \int_0^t ke^{-\tau/RC} \delta(t - \tau) d\tau.$$

Expanding $\sin \omega(t-\tau)$ and performing the integrations gives

$$E_0(t)=\frac{E}{RC}\left(\sin \omega t \int_0^t e^{-\tau/RC}\cos \omega \tau \, d\tau - \cos \omega t \int_0^t e^{-\tau/RC}\sin \omega \tau \, d\tau\right)+\int_0^t ke^{-\tau/RC}\,\delta(t-\tau)\,d\tau$$

$$=\left(\frac{ERC}{1+(\omega RC)^2}\right)\left(\omega e^{-t/RC}+\frac{1}{RC}\sin \omega t - \omega \cos \omega t\right)+ke^{-t/RC}.$$

Thus the required solution is

$$E_0(t)=\left[k+\left(\frac{\omega ERC}{1+(\omega RC)^2}\right)\right]e^{-t/RC}+\left(\frac{ERC}{1+(\omega RC)^2}\right)\left(\frac{1}{RC}\sin \omega t - \cos \omega t\right), \qquad \text{for } t>0.$$

Should it be needed, the solution for the homogeneous initial condition follows directly from this result by setting $k=0$. If the solution is written in the form

$$E_0(t)=A[\omega e^{-t/RC}+B\sin (\omega t - \Phi)], \qquad \text{for } t>0,$$

with $A=ERC[1+(\omega RC)^2]^{-1}$, $B=[1+(\omega RC)^2]^{1/2}/RC$ and $\Phi=\arctan(\omega RC)$, a frequency dependent **phase-lag** Φ is seen to be present. Its magnitude influences the choice of the parameters R and C when designing an RC-filter for a specific purpose. ∎

(d) Final value theorem

Suppose the Laplace transform $Y_0(s)$ of the output $y_0(t)$ from a system governed by a linear differential equation is known, and that the limit

$$\lim_{t\to\infty} y_0(t)=y(\infty)$$

exists. That is, as $t\to\infty$, so $y_0(t)$ approaches a well-defined limit $y_0(\infty)$ which we shall call the **final value**. Then the final value $y_0(\infty)$ may be determined directly from $Y_0(s)$ by means of the following theorem, without the necessity of first inverting $Y_0(s)$ to find $y_0(t)$ and then proceeding to the limit. This is another example of an operational property of the Laplace transform (cf. Theorem 7.15).

Theorem 7.18 (Final value theorem)

Let $Y(s)=\mathscr{L}\{y(t)\}$ be such that

$$Y(s)=Q(s)/P(s),$$

with

$$P(s)=a_0 s^n + a_1 s^{n-1}+ \ldots +a_n, \quad Q(s)=b_0 s^m + b_1 s^{m-1}+ \ldots +b_m,$$

$0 \le m < n$ and $a_0, a_1, \ldots, a_n$, $b_0, b_1, \ldots, b_m$ real constants. Then $\lim_{t\to\infty} y(t)=y(\infty)$ exists and is finite and

$$\lim_{s\to 0}\{s\,Y(s)\}=y(\infty),$$

provided all the zeros of $P(s)$, with the possible exception of one which may be located at the origin, have strictly negative real parts.

Proof

Let $y(t)$ be the solution of an initial value problem for

$$a_0 \frac{d^n y}{dt^n} + a_1 \frac{d^{n-1}}{dt^{n-1}} + \ldots + a_n y = 0.$$

Transforming the equation gives

$$P(s)\, Y(s) - Q(s), \quad \text{or} \quad Y(s) = Q(s)/P(s),$$

where $Q(s)$ contains the transformed initial conditions for $y(t)$.

From Sec. 5.3 it follows that a basis for the solution $y(t)$ is $\exp(\lambda_1 t)$, $\exp(\lambda_2 t), \ldots$, $\exp(\lambda_n t)$, where $\lambda_1, \lambda_2, \ldots, \lambda_n$ are the zeros of the characteristic polynomial

$$P(\lambda) = a_0 \lambda^n + a_1 \lambda^{n-1} + \ldots + a_n.$$

Thus if $\mathrm{Re}\{\lambda_i\} < 0$ for $i = 1, 2, \ldots, n$ we must have $\lim_{t \to \infty} \{y(t)\} = 0$. However, if $\lambda = 0$ is a simple zero of $P(\lambda)$, and all other λ_i have strictly negative real parts, it follows that $\lim_{t \to \infty} \{y(t)\} = \text{constant} \neq 0$. For any other choice of λ_i the solution will either oscillate boundedly as $t \to \infty$ and so have no limit, or it will diverge to infinity. Thus the sufficiency of the conditions on the zeros of $P(s)$ have been established, because they are the same as those of $P(\lambda)$.

To complete the proof we now assume the zeros of $P(s)$ to satisfy the conditions of the theorem and apply Theorem 7.8 which asserts that

$$s\, Y(s) - y(0) = \int_0^\infty e^{-st} y'(t)\, dt.$$

Then

$$\lim_{s \to 0} [s\, Y(s) - y(0)] = \lim_{s \to 0} \int_0^\infty e^{-st} y'(t)\, dt,$$

so

$$\lim_{s \to 0} [s\, Y(s)] - y(0) = \int_0^\infty \lim_{s \to 0} e^{-st} y'(t)\, dt = \int_0^\infty y'(t)\, dt = y(\infty) - y(0),$$

and thus

$$\lim_{s \to 0} [s\, Y(s)] = y(\infty). \qquad \square$$

We will not justify taking the limit inside the integral in the above argument, but it will suffice to say that this is permissible because of the basis from which $y(t)$ is constructed. If $Y(s)$ is more general than the rational function considered here, a more sophisticated proof is necessary in order to justify the use of this theorem and to provide conditions under which it applies.

This theorem, together with its proof, applies equally well to the case in which $y_0(t)$ is the solution of the more general equation

$$a_0 \frac{d^n y_0}{dt^n} + a_1 \frac{d^{n-1} y_0}{dt^{n-1}} + \ldots + a_n y_0 = b_0 \frac{d^m y_1}{dt^m} + b_1 \frac{d^{m-1} y_1}{dt^{m-1}} + \ldots + b_m y_1,$$

when $0 \leq m < n$ and $y_1(t)$ is an imput which is localized in time (vanishes for some $t > t_0 > 0$). Such situations arise in control theory and they will be discussed in Sec. 7.3(e).

Example 7. Final value theorem

Where appropriate, apply the final value theorem to determine $y(\infty)$ given that

(i) $Y(s) = \dfrac{1}{s(s^2 + 4s + 5)}$,

(ii) $Y(s) = \dfrac{s+7}{s^2 + 4}$,

(iii) $Y(s) = \dfrac{1}{1 + ks}$ (k arbitrary).

Solution

(i) The denominator of $Y(s)$ has zeros at $s = 0$ and $s = -2 \pm i$, so the theorem applies and

$$y(\infty) = \lim_{s \to 0} [s \, Y(s)] = \lim_{s \to 0} \left(\frac{1}{(s^2 + 4s + 5)} \right) = \frac{1}{5}.$$

In fact

$$y(t) = \mathcal{L}^{-1}\{Y(s)\} = \tfrac{1}{5} - \tfrac{1}{5} e^{-2t} (\cos t + 2 \sin t),$$

from which the above result may be confirmed by evaluating

$$y(\infty) = \lim_{t \to \infty} y(t).$$

(ii) The denominator of $Y(s)$ has zeros at $s = \pm 2i$, so as these do not have strictly negative real parts the final value theorem does not apply.
In fact

$$y(t) = \mathcal{L}^{-1}\left\{ \frac{s+7}{s^2 + 4} \right\} = \cos 2t + \tfrac{7}{2} \sin 2t,$$

so although $y(t)$ is bounded for all $t > 0$, no limit exists as $t \to \infty$.

(iii) The denominator of $Y(s)$ has a real zero at $s = -1/k$, and this will only be strictly negative if $k > 0$. Thus when $k > 0$,

$$y(\infty) = \lim_{s \to 0} [s\, Y(s)] = \lim_{s \to 0} \frac{s}{1 + ks} = 0.$$

The final value theorem does not apply when $k < 0$.
In fact

$$y(t) = \mathcal{L}^{-1}\left\{ \frac{1}{1 + ks} \right\} = \frac{1}{k} e^{-t/k},$$

from which the above results may be confirmed. When $k < 0$ the function $y(t)$ is unbounded as $t \to \infty$. ■

(e) Control theory, transfer function and stability

In its simplest form a **feedback control system (servomechanism)** is a self-regulating device which acts to reduce to zero the difference between the system input $y_1(t)$ and the system output $y_0(t)$. A typical example of such a system is an aircraft autopilot which is required to maintain the aircraft on a specified bearing, irrespective of changes in the prevailing wind. Its function is to compare the required bearing (input) with the actual bearing (output), and then to adjust the control surfaces so that the error in the bearing $\varepsilon(t) = y_1(t) - y_0(t)$ is reduced to zero.

Schematically, the system just described may be represented as in Fig. 7.29 and it operates as follows. The required bearing $y_1(t)$ and actual bearing $y_0(t)$ are compared in an error detecting device which produces a voltage proportional to the bearing misalignment $\varepsilon(t) = y_1(t) - y_0(t)$. This error voltage is then used to actuate a servomotor connected to the appropriate control surface in such a way that the angular rotation produced will depend on the sign of $\varepsilon(t)$. There are various types of servomotor, but in the one considered here the angular rotation produced will be taken to be proportional to the applied voltage, so that the angle of rotation will return to zero when $\varepsilon(t) = 0$. For obvious reasons, the modification of $y_0(t)$ by means of the error signal $\varepsilon(t) = y_1(t) - y_0(t)$ is called **feedback**, and the path followed by $y_0(t)$ in Fig. 7.29 when being fed to the error detector is called the **feedback loop**.

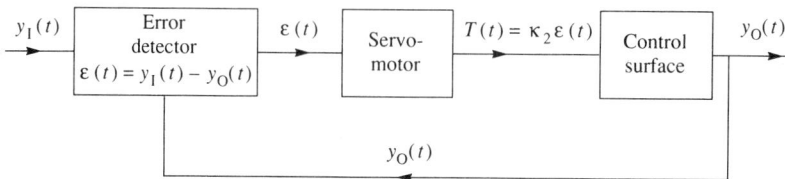

Fig. 7.29 Schematic diagram for autopilot

If the moment of inertia of the control surface about its axis is I, the torque opposing motion is $\kappa_1 \, dy_O/dt$ with $\kappa_1 > 0$ and the servomotor torque applied to the control surface is $T = \kappa_2 \varepsilon(t)$ with $\kappa_2 > 0$, the differential equation determining the variation of $y_O(t)$ in terms of $y_I(t)$ becomes

$$I \frac{d^2 y_O}{dt^2} + \kappa_1 \frac{dy_O}{dt} = \kappa_2 [y_1(t) - y_O(t)],$$

or

$$I \frac{d^2 y_O}{dt^2} + \kappa_1 \frac{dy_O}{dt} + \kappa_2 y_O = \kappa_2 y_I(t). \tag{16}$$

This simple but important type of differential equation occurs throughout control theory, and its standard form is

$$M \frac{d^2 y_O}{dt^2} + \frac{dy_O}{dt} + k y_O = k y_I(t), \tag{17}$$

in which $M > 0$ and $k > 0$.

Individual components in a control system governed by a linear differential equation are called **linear elements**. A linear element relates its output $y_O(t)$ to its input $y_I(t)$ in a linear fashion, and the simplest such element is an amplifier in which $y_O = k y_I$, with $k > 0$ the amplification factor.

A slightly more complicated but typical linear element is the RC-filter shown in Fig. 7.27, which has the governing differential equation

$$RC \frac{dE_O}{dt} + E_O = E_I.$$

In general a control system will contain a number of linear elements, some with fixed parameter values describing the system to be controlled, and others with parameters capable of variation introduced to modify the system response so that it meets specific design criteria.

We saw in Sec. 7.3(c) that a differential equation may be characterized in terms of its weighting function $W(t)$, which is its response to a unit delta function input applied when the system is at rest. If we consider an equation in the standard form (17) and set

$$\omega_n^2 = k/M \quad \text{and} \quad \zeta = \tfrac{1}{2}(kM)^{-1/2}, \tag{18}$$

so that (17) becomes

$$\frac{d^2 y_O}{dt^2} + 2\zeta\omega_n \frac{dy_O}{dt} + \omega_n^2 y_O = \omega_n^2 y_I(t), \tag{19}$$

routine calculations using the arguments of Sec. 7.3(c) show

(i) **Case $\zeta > 1$**

$$W(t) = \frac{\omega_n}{(\zeta^2 - 1)^{1/2}} \exp(-\zeta\omega_n t) \sinh[(\zeta^2 - 1)^{1/2} \omega_n t], \tag{20}$$

(ii) **Case $\zeta = 1$**

$$W(t) = \omega_n^2 t \exp(-\omega_n t),$$
(21)

(iii) **Case $\zeta < 1$**

$$W(t) = \frac{\omega_n}{(1-\zeta^2)^{1/2}} \exp(-\zeta\omega_n t) \sin[(1-\zeta)^{1/2}\omega_n t].$$
(22)

Here ω_n is called the **undamped natural frequency** and ζ the **damping factor** for the system. Case (i) with $\zeta > 1$ represents an **overdamped** system where no oscillations are possible; case (ii) with $\zeta = 1$ is the **critically damped** case again without oscillations, while case (iii) with $\zeta < 1$ is the **underdamped** case in which damped oscillations occur.

In the system described by (16) the parameters I and κ_1 are constants determined by the aircraft to be controlled, and only the amplification factor κ_2 may be selected by the designer. This will influence both ω_n and ζ, but practical limitations on the choice of κ_2 imposed by the servomotor may either cause the response to be too slow because $\zeta \gg 1$, or unwanted oscillations to arise because $\zeta \ll 1$. To overcome such problems it is usual to use as the error signal $\varepsilon(t)$ not $y_1(t) - y_0(t)$ but

$$\varepsilon(t) = y_0(t) - y_M(t),$$

where $y_M(t)$ is a modification of the output signal $y_0(t)$ produced by a linear element introduced into the feedback, so that

$$y_M(t) = L[y_0(t)].$$
(23)

Here $L[y_0(t)]$ represents a linear function of $y_0(t)$, which is usually a differential equation relating $y_M(t)$ and $y_0(t)$. Schematically such a modified feedback system takes the form shown in Fig. 7.30.

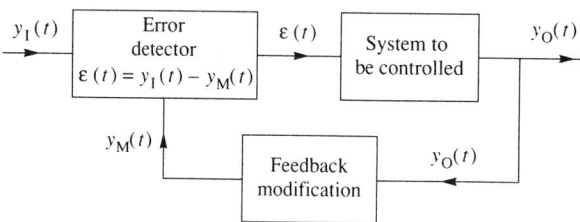

Fig. 7.30   Modified feedback control system

Let the differential equation of a typical linear element be

$$a_0 \frac{d^n y_0}{dt^n} + a_1 \frac{d^{n-1} y_0}{dt^{n-1}} + \ldots + a_n y_0 = y_1(t),$$
(24)

and suppose that $y_0(t)$ is subject to the initial conditions

$$y_0(0) = \alpha_0, \; y_1(0) = \alpha_1, \ldots, \; y_{n-1}(0) = \alpha_{n-1}.$$
(25)

Then taking the Laplace transform of (24) we find

$$P(s) Y_O(s) = Y_1(s) + R(s), \tag{26}$$

where $Y_1(s) = \mathscr{L}\{y_1(t)\}$, $Y_O(s) = \mathscr{L}\{y_O(t)\}$,

$$P(s) = a_0 s^n + a_1 s^{n-1} + \ldots + a_n, \tag{27}$$

and the polynomial $R(s)$ arises from the transformation of any nonzero initial conditions for $y_O(t)$ which may be present. If the initial conditions are homogeneous (all zero) then $R(s) \equiv 0$.

Rewriting (26), the quantities $Y_O(s)$ and $Y_1(s)$ are seen to be related by the equation

$$Y_O(s) = \left(\frac{1}{P(s)}\right) Y_1(s) + \frac{R(s)}{P(s)}. \tag{28}$$

The function

$$T(s) = \frac{1}{P(s)}, \tag{29}$$

which is the ratio if the multiplier of $Y_1(s)$ to $Y_O(s)$, is called the **transfer function** of the linear element described by (24).

The transfer function $T(s)$ may be deduced directly from (24) by observing that the polynomial $P(s)$ in (29) is derived from the left-hand side of (24) by replacing the derivatives $d^r y_O/dt^r$ by s^r and y_O by 1.

It is clear that the transfer function is, in fact, simply the Laplace transform of the response of the linear element to a unit delta function input applied at $t=0$, when $y_O(t)$ and all its derivatives are zero for $t<0$. Thus in terms of the weighting function $W(t)$ introduced in Sec. 7.3(c), we have the important relationship

$$T(s) = \mathscr{L}\{W(t)\}. \tag{30}$$

In terms of the transfer function $T(s)$, result (28) becomes

$$Y_O(s) = T(s) Y_1(s) + R(s)/P(s), \tag{31}$$

where we again stress that the term $R(s)/P(s)$ is solely due to any nonzero initial conditions which may have been imposed on $y_O(t)$.

In more complicated linear elements the relationship between the input $y_1(t)$ and output $y_O(t)$ may be of the form

$$a_0 \frac{d^n y_O}{dt^n} + a_1 \frac{d^{n-1} y_O}{dt^{n-1}} + \ldots + a_n y_O = b_0 \frac{d^m y_1}{dt^m} + b_1 \frac{d^{m-1} y_1}{dt^{m-1}} + \ldots + b_m y_1, \tag{32}$$

where $n > m$. As before, the transfer function $T(s)$ is again defined as the Laplace transform of the response of the linear element to a unit delta function input, when $y_O(t)$ and all its derivatives are zero for $t<0$. Transforming (32) and using (25) of Sec. 7.2 we find that

$$(a_0 s^n + a_1 s^{n-1} + \ldots + a_n) Y_O(s) = b_0 s^m + b_1 s^{m-1} + \ldots + b_m,$$

but $Y_O(s) = \mathcal{L}\{W(t)\} = T(s)$, so that

$$T(s) = \frac{b_0 s^m + b_1 s^{m-1} + \ldots + b_m}{a_0 s^n + a_1 s^{n-1} + \ldots + a_n}. \tag{33}$$

Thus every rational function of s may be regarded as the transfer function of a general linear element of the type characterized by (32).

Example 8. Transfer function

Find the transfer function of the linear element described by the differential equation

$$a_0 \frac{d^2 y_O}{dt^2} + a_1 \frac{dy_O}{dt} + a_2 y_O = y_I(t).$$

Give an example of a physical system with such a transfer function.

Solution

In this case the polynomial $P(s)$ in (27) is

$$P(s) = a_0 s^2 + a_1 s + a_2,$$

so from (29) the transfer function

$$T(s) = \frac{1}{a_0 s^2 + a_1 s + a_2}.$$

If, for example, the linear element had been an *RLC*-circuit, the charge $q(t)$ present in the circuit at time $t > 0$ would be related to the applied voltage $e_1(t)$ by the differential equation (see Ex. 6, in Sec. 5.7)

$$L \frac{d^2 q}{dt^2} + R \frac{dq}{dt} + \frac{1}{C} q = e_1(t).$$

Setting $y_O(t) = q(t)/C$ this becomes

$$C \left(L \frac{d^2 y_O}{dt^2} + R \frac{dy_O}{dt} + \frac{1}{C} y_O \right) = e_1(t).$$

Identifying this equation with the more general case studied in the first part of the example it follows that the transfer function relating $Y_O(s) = \mathcal{L}\{q(t)/C\}$ to $Y_1(s) = \mathcal{L}\{e_1(t)\}$ is

$$T(s) = \frac{1}{C[Ls^2 + Rs + (1/C)]}. \qquad \blacksquare$$

It is usual to describe a control system in terms of its transfer function, which is itself a function of the transfer functions of its individual linear elements. To discover how the transfer function $T(s)$ of the system as a whole is related to those of its elements let us consider Fig. 7.30.

Suppose the system is at rest for $t<0$, so that $y_O(t)$ and all its derivatives vanish at $t=0$. Then if the transfer function of the system to be controlled is $F(s)$, and the transfer function of the feedback modification system is $G(s)$, it follows that

$$Y_O(s)=F(s)E(s) \tag{34}$$

$$Y_M(s)=G(s)Y_O(s) \tag{35}$$

and

$$E(s)=\mathscr{L}\{\varepsilon(t)\}=Y_I(s)-Y_M(s), \tag{36}$$

where $Y_I(s)=\mathscr{L}\{y_I(t)\}$, $Y_O(s)=\mathscr{L}\{y_O(t)\}$ and $\varepsilon(t)=y_I(t)-y_M(t)$. The elimination of $Y_M(s)$ and $E(s)$ between these equations gives

$$Y_O(s)=\left(\frac{F(s)}{1+F(s)G(s)}\right)Y_I(s), \tag{37}$$

which shows that the transfer function of the system as a whole is

$$T(s)=\frac{F(s)}{1+F(s)G(s)}. \tag{38}$$

When, as in Fig. 7.29, the feedback variable is unmodified, $G(s)\equiv1$ and (36) simplifies to

$$T(s)=\frac{1}{1+F(s)}. \tag{39}$$

A schematic diagram equivalent to Fig. 7.30, from which (38) may be deduced, is given in Fig. 7.31. It shows the transfer functions of the linear elements involved and their relationship to the transforms of the input, output and modified feedback variables.

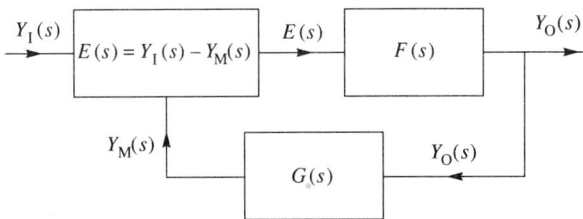

Fig. 7.31 Transfer function diagram for control system

A control system with transfer function $T(s)$ is said to be **asymptotically stable** if

$$\lim_{t\to\infty}W(t)=0. \tag{40}$$

When such a system is disturbed from rest by a unit delta function input at time $t=0$, it will eventually return to rest. This is, of course, also true of any disturbance applied to the system provided it is localized in time. If the initial disturbance grows with time the system will be said to be **unstable,** while if it oscillates boundedly for all time it will be said to be **neutrally stable** (see Sec. 6.4).

For the systems considered in this section, asymptotic stability may be deduced from the final value theorem (Theorem 7.18, Sec. 7.3(d)). It follows from that theorem that a necessary and sufficient condition for a system to be asymptotically stable is that the zeros of the denominator of its transfer function must all have strictly negative real parts.

The Routh–Hurwitz test (Theorem 6.5) is typical of the many tests which exist for asymptotic stability. Although the Routh–Hurwitz test ensures asymptotic stability, it does not guarantee that after a disturbance hãs been applied to a particular system it will return to equilibrium rapidly and without undue oscillation. This will only occur if the zeros of the denominator of the transfer function $T(s)$ lie sufficiently far to the left of the imaginary axis in the s-plane to produce an acceptably rapid exponential decay of the disturbance. More sensitive tests exist which provide qualitative information of this type, but sometimes only the precise knowledge of the location of the zeros obtained by numerical computation will suffice when designing a system.

The purpose of the transfer function $G(s)$ in the feedback loop in Fig. 7.31 is to modify the basic response of the system in a manner required by the designer. However, a wrong choice of $G(s)$ can destabilize the system, as the following example shows.

Example 9. Modification of feedback in an autopilot

The autopilot example in Fig. 7.29 led to the relationship between $y_1(t)$ and $y_0(t)$ given in (16), which corresponds to the transfer function

$$T(s)=\frac{\kappa_2}{Is^2+\kappa_1 s+\kappa_2}.$$

In terms of the transfer function diagram in Fig. 7.31 this is equivalent to setting

$$F(s)=\frac{\kappa_2}{Is^2+\kappa_1 s}\quad\text{and}\quad G(s)=1.$$

The autopilot will be asymptotically stable provided the zeros of the polynomial

$$Is^2+\kappa_1 s+\kappa_2$$

have strictly negative real parts, which is indeed the case because

$$s=\tfrac{1}{2}[-\kappa_1\pm(\kappa_1^2-4I\kappa_2)^{1/2}],$$

and I, κ_1 and κ_2 are all positive.

Now suppose the control system is modified so that $G(s)=k$. Then from (38) the new transfer function $T_M(s)$ becomes

$$T_M(s)=\frac{\kappa_2}{Is^2+\kappa_1 s+k\kappa_2},$$

and its zeros are

$$s=\tfrac{1}{2}[-\kappa_1\pm(\kappa_1^2-4kI\kappa_2)^{1/2}].$$

Provided $4kI\kappa_2 > \kappa_1^2$, this corresponds to an oscillatory behavior with angular frequency $\frac{1}{2}(4kI\kappa_2 - \kappa_1^2)^{1/2}$, which is damped by the time factor $\exp(-\frac{1}{2}\kappa_1 t)$. If $0 < k < \kappa_1^2/(4I\kappa_2)$ the system will be nonoscillatory, but still asymptotically stable. However, if $k < 0$ the zeros of the denominator of $T_M(s)$ will be real, and one will be positive, so the system will be nonoscillatory and unstable. This is hardly surprising, because by reversing the sign of k the system then acts to increase the bearing misalignment error instead of reducing it. ∎

(f) Determination of e^{tA} by the Laplace transform – systems of differential equations

In the problem sets for Sec. 6.1 and Sec. 6.2 the system of equations

$$\dot{\mathbf{x}} = \mathbf{A}\mathbf{x} + \mathbf{f}, \tag{41}$$

with the initial condition

$$\mathbf{x}(0) = \mathbf{x}_0, \tag{42}$$

was considered, with $\mathbf{x}(t)$ and $\mathbf{f}(t)$ n element column vectors and $\mathbf{A}$ a constant $n \times n$ matrix. It was shown that the system has a solution for $t > 0$ which may be written in terms of the exponential matrix e^{tA} as

$$\mathbf{x}(t) = e^{tA}\mathbf{x}_0 + \int_0^t \exp[(t - \tau)\mathbf{A}]\mathbf{f}(\tau)\,d\tau. \tag{43}$$

It was also shown that if $\mathbf{A}$ and $\mathbf{B}$ commute ($\mathbf{AB} = \mathbf{BA}$) then $e^{A+B} = e^A e^B$, so as $t\mathbf{A}$ and $-\tau\mathbf{A}$ commute the above result may be rewritten as

$$\mathbf{x}(t) = e^{tA}\mathbf{x}_0 + e^{tA}\int_0^t e^{-\tau A}\mathbf{f}(\tau)\,d\tau, \tag{44}$$

for $t > 0$.

The purpose of this section is to show how, in addition to the methods described in the problem sets for Sec. 6.1 and Sec. 6.2, the Laplace transform may also be used to determine e^{tA}.

Consider the homogeneous system

$$\dot{\mathbf{x}} = \mathbf{A}\mathbf{x} \tag{45}$$

with the solution

$$\mathbf{x}(t) = e^{tA}\mathbf{x}_0, \tag{46}$$

and take the Laplace transform of (45) to obtain

$$s\mathbf{X}(s) - \mathbf{x}_0 = \mathbf{A}\mathbf{X}(s). \tag{47}$$

Here $\mathbf{X}(s) = \mathscr{L}\{\mathbf{x}(t)\}$, with the understanding that the elements of the column vector $\mathbf{X}(s)$ are the Laplace transforms of the corresponding elements of the column vector $\mathbf{x}(t)$.

Thus

$$(s\mathbf{I} - \mathbf{A})\,\mathbf{X}(s) = \mathbf{x}_0,$$

where $\mathbf{I}$ is the $n \times n$ unit matrix, and so

$$\mathbf{X}(s) = (s\mathbf{I} - \mathbf{A})^{-1}\mathbf{x}_0, \tag{48}$$

or

$$\mathbf{X}(s) = \frac{\text{adj}\,(s\mathbf{I} - \mathbf{A})}{\det\,(s\mathbf{I} - \mathbf{A})}\,\mathbf{x}_0. \tag{49}$$

Taking the inverse transform we find

$$\mathbf{x}(t) = \mathscr{L}^{-1}\left\{\frac{\text{adj}\,(s\mathbf{I} - \mathbf{A})}{\det\,(s\mathbf{I} - \mathbf{A})}\right\}\mathbf{x}_0, \tag{50}$$

where by the inverse Laplace transform of a matrix we mean the matrix formed by taking the inverse Laplace transform of its elements.

Comparison of (46) and (50) shows

$$e^{t\mathbf{A}} = \mathscr{L}^{-1}\left\{\frac{\text{adj}\,(s\mathbf{I} - \mathbf{A})}{\det\,(s\mathbf{I} - \mathbf{A})}\right\}, \tag{51}$$

which is the required result. Once $e^{t\mathbf{A}}$ has been found, the solution of (41) subject to (42) follows directly from (44).

Example 10. Determination of $e^{t\mathbf{A}}$ and the solution of a system

Given

$$\mathbf{A} = \begin{bmatrix} 2 & -3 \\ 3 & 2 \end{bmatrix},$$

use the Laplace transform to find $e^{t\mathbf{A}}$. Use the result to solve the initial value problem

$$\dot{\mathbf{x}} = \mathbf{A}\mathbf{x} + \mathbf{f},$$

when

$$\mathbf{x}(0) = \begin{bmatrix} 2 \\ -1 \end{bmatrix} \quad \text{and} \quad \mathbf{f} = \begin{bmatrix} 1 \\ 0 \end{bmatrix}.$$

Solution

We have

$$(s\mathbf{I} - \mathbf{A}) = \begin{bmatrix} s-2 & 3 \\ -3 & s-2 \end{bmatrix}, \quad \text{so adj}\,(s\mathbf{I} - \mathbf{A}) = \begin{bmatrix} s-2 & -3 \\ 3 & s-2 \end{bmatrix},$$

and

$$\det\,(s\mathbf{I} - \mathbf{A}) = (s-2)^2 + 3^2.$$

Thus

$$(s\mathbf{I}-\mathbf{A})^{-1} = \frac{\text{adj}(s\mathbf{I}-\mathbf{A})}{\det(s\mathbf{I}-\mathbf{A})} = \begin{bmatrix} \dfrac{s-2}{(s-2)^2+3^2} & \dfrac{-3}{(s-2)^2+3^2} \\[3mm] \dfrac{3}{(s-2)^2+3^2} & \dfrac{s-2}{(s-2)^2+3^2} \end{bmatrix},$$

and so

$$e^{t\mathbf{A}} = \mathscr{L}^{-1}\{s\mathbf{I}-\mathbf{A}\}^{-1} = \begin{bmatrix} e^{2t}\cos 3t & -e^{2t}\sin 3t \\ e^{2t}\sin 3t & e^{2t}\cos 3t \end{bmatrix}.$$

Hence from (44) the solution of the required initial value problem is

$$\mathbf{x}(t) = \begin{bmatrix} x_1(t) \\ x_2(t) \end{bmatrix} = \begin{bmatrix} e^{2t}\cos 3t & -e^{2t}\sin 3t \\ e^{2t}\sin 3t & e^{2t}\cos 3t \end{bmatrix}\begin{bmatrix} 2 \\ -1 \end{bmatrix}$$

$$+ \begin{bmatrix} e^{2t}\cos 3t & -e^{2t}\sin 3t \\ e^{2t}\sin 3t & e^{2t}\cos 3t \end{bmatrix} \int_0^t \begin{bmatrix} e^{-2\tau}\cos 3\tau & e^{-2\tau}\sin 3\tau \\ -e^{-2\tau}\sin 3\tau & e^{-2\tau}\sin 3\tau \end{bmatrix}\begin{bmatrix} 1 \\ 0 \end{bmatrix} d\tau.$$

A routine calculation shows

$$x_1(t) = \tfrac{1}{13}(28e^{2t}\cos 3t + 16e^{2t}\sin 3t - 2),$$
$$x_2(t) = \tfrac{1}{13}(28e^{2t}\sin 3t - 16e^{2t}\cos 3t + 3).$$ ■

(g) Delay differential equations

In all the differential equations we have considered so far, the unknown function and its derivatives have all had the same argument. Thus, if in applications the independent variable is the time t, this means that the unknown function and all its derivatives in the differential equation are evaluated at the *same* time t. However, in many situations this is not a realistic assumption, because in practice it is their values at different times which are related.

A simple example will help here. Suppose two chemically reacting liquids are pumped through a pipe, which acts as a chemical reactor, and that the progress of the reaction is monitored at some point along the pipe. Then what is happening at the monitoring point at time t is determined by the proportions of the reactants at time $t-\tau$, where τ is the transit time of the liquid from its point of entry to the point at which it is monitored. Thus a reaction rate at time t then depends on reactant concentrations at time $t-\tau$.

We are thus led to consider differential equations of the form

$$\dot{x}(t) + kx(t-\tau) = f(t),$$

and also more general ones, in which it is essential to indicate the arguments of the functions and derivatives involved. Equations of the type shown above are called **delay**

differential equations and they belong to a more general class called **functional differential equations**.[7]

Initial value problems for these delay differential equations usually involve the specification of $x(t)$ over an interval $t_0 - \tau \leq t \leq t_0$, and this information together with the delay differential equation itself suffices to determine $x(t)$ for $t > t_0$. In this section we shall only consider the simplest such equation and illustrate by example how such problems may be solved by the Laplace transform when $t_0 = 0$ and $x(t) \equiv 0$ for $t \leq 0$.

Example 11. Delay differential equation

Solve by means of the Laplace transform

$$\frac{d}{dt} x(t) + 3x(t-2) = 1 + t,$$

given $x(t) \equiv 0$ for $t \leq 0$.

Solution

Because of the initial condition we may write

$$x(t-2) = x(t-2)\,\mathcal{U}(t-2),$$

so the delay differential equation is equivalent to

$$\frac{d}{dt} x(t) + 3x(t-2)\,\mathcal{U}(t-2) = 1 + t.$$

Taking the Laplace transform we obtain

$$s X(s) + 3e^{-2s} X(s) = \frac{1}{s} + \frac{1}{s^2},$$

and so

$$X(s) = \frac{s+1}{s^2(s + 3e^{-2s})}.$$

If we now rewrite $X(s)$ as follows, and then use the binomial theorem, we obtain

$$X(s) = \frac{s+1}{s^3}\left(1 + \frac{3e^{-2s}}{s}\right)^{-1}$$

or

$$X(s) = \frac{s+1}{s^3} \sum_{n=0}^{\infty} \frac{(-1)^n 3^n e^{-2ns}}{s^n}.$$

---

[7] Other names in use are **differential-difference equations**, **differential equations with a retarded argument** and **differential equations with deviating arguments**, the last two being commonly used in Russian literature.

Thus

$$X(s) = \sum_{n=0}^{\infty} \frac{(-1)^n 3^n e^{-2ns}}{s^{n+2}} + \sum_{n=0}^{\infty} \frac{(-1)^n 3^n e^{-2ns}}{s^{n+3}},$$

so from the second translation theorem

$$x(t) = \sum_{n=0}^{\infty} \frac{(-1)^n 3^n (t-2n)^{n+1}}{(n+1)!} \mathcal{U}(t-2n) + \sum_{n=0}^{\infty} \frac{(-1)^n 3^n (t-2n)^{n+2}}{(n+2)!} \mathcal{U}(t-2n), \qquad \text{for } t > 0.$$

Notice that for any given finite value of t, the presence of the unit step functions in the solutions means that it will only contain a finite number of terms at that time. ∎

(h) Integral equations and integro-differential equations

A **linear integral equation** for the unknown function $y(t)$ is either of the form

$$h(t) y(t) = g(t) + \int_a^b K(\tau, t) y(\tau) d\tau, \tag{52}$$

when it is called a **Fredholm integral equation**, or of the form

$$h(t) y(t) = g(t) + \int_a^t K(\tau, t) y(\tau) d\tau, \tag{53}$$

when it is called a **Volterra integral equation**.

Here a, b are constants, $h(t)$ and $g(t)$ are specified functions of t, with $g(t)$ being called the **nonhomogeneous term**, and the function $K(\tau, t)$ of the two variables τ and t, which is also specified, called the **kernel** of the integral equation. Integral equations of both types are further classified as being of the *first, second* or *third kind*, according as $h(t) \equiv 0$, $h(t) \equiv 1$, or $h(t)$ is a nonconstant function of t, respectively.

Only Volterra integral equations will be considered here because certain types are easily solved by means of the Laplace transform. They are also important because they are equivalent to initial value problems for ordinary differential equations.

In its simplest form, this connection between an initial value problem and a Volterra integral equation may be seen by considering the initial value problem

$$\frac{dy}{dt} = f(t, y), \quad \text{with } y(a) = \alpha. \tag{54}$$

Integrating with respect to t over the interval $a \leq \tau \leq t$ gives

$$y(t) = \alpha + \int_a^t f[\tau, y(\tau)] d\tau. \tag{55}$$

This is a *nonlinear* Volterra integral equation of the second kind whenever f is a nonlinear function of y, and a *linear* one when f depends linearly on y. It will be recalled that in (3) of Sec. 4.7 such an integral equation formed the basis of the Picard iterative scheme for solving an initial value problem.

As with all initial value problems transformed into Volterra integral equations, the integral equation is a complete representation of the original problem, because it incorporates both the differential equation and the initial conditions.

Let us now consider the more general problem of how to determine the Volterra integral equation corresponding to the initial value problem for the second order linear nonhomogeneous equation

$$a(t)\frac{d^2y}{dt^2}+b(t)\frac{dy}{dt}+c(t)y=f(t), \quad \text{with } y(0)=\alpha, \quad y'(0)=\beta. \tag{56}$$

Integrating with respect to t over the interval $[0, t]$, using integration by parts, and rearranging terms gives[8]

$$a(t)\frac{dy}{dt}=\beta a(0)+\alpha b(0)-\alpha a'(0)+[a'(t)-b(t)]y(t)$$

$$+\int_0^t [b'(\tau)-c(\tau)-a''(\tau)]y(\tau)d\tau+\int_0^t f(\tau)d\tau. \tag{57}$$

A further integration with respect to t over the interval $[0, t]$ gives

$$a(t)y(t)=\alpha a(0)+[\alpha b(0)+\beta a(0)-\alpha a'(0)]t+\int_0^t\int_0^u f(\tau)\,d\tau\,du$$

$$+\int_0^t [2a'(\tau)-b(\tau)]y(\tau)\,d\tau+\int_0^t\int_0^u [b'(\tau)-c(\tau)-a''(\tau)]y(\tau)d\tau. \tag{58}$$

Finally, using the result of Ex. 12(iii) of Sec. 7.2 to transform the iterated integrals we obtain the result

$$a(t)y(t)=g(t)+\int_0^t K(\tau, t)y(\tau)d\tau, \tag{59}$$

in which the kernel is

$$K(\tau, t)=2a'(\tau)-b(\tau)+(t-\tau)[b'(\tau)-c(\tau)-a''(\tau)], \tag{60}$$

and the nonhomogeneous term is

$$g(t)=\alpha a(0)+[\alpha b(0)+\beta a(0)-\alpha a'(0)]t+\int_0^t (t-\tau)f(\tau)d\tau. \tag{61}$$

Thus the initial value problem (56) has been transformed into the Volterra equation (59), with the kernel (60) and the nonhomogeneous term (61). The equivalence between (56) and (59) follows from the fact that this argument is reversible. Differentiating (59) twice with respect to t using Theorem 1.6 yields (56), and as an intermediate result (57) which determines the initial condition $y'(0)$ in terms of the integral equation. The other initial condition $y(0)$ follows directly from the integral equation itself.

---

[8] In what is to follow it will be assumed that the differential equation has no singularity at $t=0$ which renders initial data inappropriate (see Sec. 8.2).

When the coefficients $a(t)$, $b(t)$ and $c(t)$ in (56) are constants, the equivalent Volterra integral equation takes on a simpler form. It follows directly from the above results that if y is a solution of the initial value problem

$$\frac{d^2y}{dt^2}+p\frac{dy}{dt}+qy=r(t), \quad \text{with } y(0)=\alpha,\ y'(0)=\beta, \tag{62}$$

and p, q constants, then the equivalent Volterra integral equation is

$$y(t)=\alpha+(\alpha p+\beta)t+\int_0^t (t-\tau)r(\tau)\,d\tau-\int_0^t [p+(t-\tau)q]y(\tau)\,d\tau. \tag{63}$$

Example 12. Deriving a Volterra integral equation

Determine from first principles the Volterra integral equation equivalent to

$$\frac{d^2y}{dt^2}+3\frac{dy}{dt}+ty=1, \quad \text{with } y(0)=1,\ y'(0)=2.$$

Solution

Integrating over $[0, t]$ gives

$$y'(t)-y'(0)+3[y(t)-y(0)]+\int_0^t \tau\,y(\tau)\,d\tau=t,$$

which after rearrangement and use of the initial conditions becomes

$$\frac{dy}{dt}=5+t-3y-\int_0^t \tau\,y(\tau)\,d\tau.$$

A further integration gives

$$y(t)=y(0)+5t+\frac{t^2}{2}-3\int_0^t y(\tau)\,d\tau-\int_0^t\int_0^u \tau\,y(\tau)\,d\tau\,du.$$

Using the initial condition $y(0)=1$ and the result of Ex. 12(iii) of Sec. 7.2 this may be rearranged to give

$$y(t)=1+5t+\frac{t^2}{2}-\int_0^t [3+(t-\tau)\tau]y(\tau)\,d\tau.$$

In this Volterra integral equation the kernel is

$$K(\tau, t)=-[3+(t-\tau)\tau],$$

and the nonhomogeneous term is

$$g(t)=1+5t+\frac{t^2}{2}. \qquad\blacksquare$$

Example 13. Deriving an initial value problem from a Volterra equation

Find the initial value problem which corresponds to

$$y(t) = 3 + t + t^2 + \int_0^t [5t + 2(t-\tau)\tau] y(\tau) \, d\tau.$$

Solution

Setting $t = 0$ in the integral equation we find the initial condition $y(0) = 3$. Differentiating the integral equation with respect to t by means of Theorem 1.6 gives

$$\frac{dy}{dt} = 1 + 2t + \int_0^t (5 + 2\tau) y(\tau) \, d\tau + 5t \, y.$$

Setting $t = 0$ in this result gives the second initial condition $y'(0) = 1$.

Finally, a further differentiation with respect to t gives

$$\frac{d^2 y}{dt^2} = 2 + (5 + 2t) y + 5y + 5t \frac{dy}{dt}.$$

Thus the equivalent initial value problem is

$$\frac{d^2 y}{dt^2} - 5t \frac{dy}{dt} - 2(5 + t) y = 2, \quad \text{with } y(0) = 3, \ y'(0) = 1. \qquad \blacksquare$$

The simplest Volterra integral equations to solve by means of the Laplace transform are those with a kernel of convolution type. We shall say that the kernel $K(\tau, t)$ is of **convolution type** if the variables t and τ only enter into $K(\tau, t)$ as the linear combination $t - \tau$. The following are examples of kernels of convolution type;

$$K(\tau, t) = t - \tau, \quad K(\tau, t) = \sin(t - \tau) \quad \text{and} \quad K(\tau, t) = \exp[3(t - \tau)].$$

The solution of such equations by means of the Laplace transform and the convolution theorem is illustrated by the following examples.

Example 14. Linear Volterra equation with a kernel of convolution type

Solve the linear Volterra integral equation

$$y(t) = 1 + 3t + 2 \int_0^t \sin 2(t - \tau) y(\tau) \, d\tau.$$

Solution

Setting $Y(s) = \mathcal{L}\{y(t)\}$, taking the Laplace transform of the equation and using the convolution theorem on the integral term gives

$$Y(s) = \frac{1}{s} + \frac{3}{s^2} + \frac{4}{s^2 + 4} Y(s).$$

Thus

$$Y(s) = \frac{1}{s} + \frac{3}{s^2} + \frac{4}{s^3} + \frac{12}{s^4},$$

so taking the inverse transform we obtain the required solution

$$y(t) = 1 + 3t + 2t^2 + 2t^3. \qquad \blacksquare$$

Example 15. Nonlinear Volterra equation

Solve the nonlinear Volterra integral equation of the first kind

$$\int_0^t y(t-\tau)y(\tau)\,d\tau = te^{2t}.$$

Solution

Setting $Y(s) = \mathcal{L}\{y(t)\}$, taking the Laplace transform of the equation and using the convolution theorem on the integral term gives

$$[Y(s)]^2 = \frac{1}{(s-2)^2},$$

and so

$$Y(s) = \frac{\pm 1}{s-2}.$$

The solution is not unique, because taking the inverse transform shows that

$$y(t) = e^{2t} \quad \text{and} \quad y(t) = -e^{2t}$$

are both solutions. $\qquad \blacksquare$

A **linear integro-differential equation** is a linear differential equation for $y(t)$ with a nonhomogeneous term in the form of an integral containing an integrand in which the unknown function y appears linearly. The initial conditions appropriate to an integro-differential equation are those appropriate to the associated differential equation. The following example illustrates their solution by means of the Laplace transform in the specially simple case in which the integral has a convolution type kernel.

Example 16. Integro-differential equation

Solve the linear integro-differential equation

$$\frac{dy}{dt} + 2\int_0^t \exp[3(t-\tau)]\,y(\tau)\,d\tau = 1, \quad \text{with } y(0) = 3.$$

Solution

Setting $Y(s) = \mathscr{L}\{y(t)\}$, taking the Laplace transform of the equation and using the convolution theorem gives

$$s\,Y(s) - y(0) + \left(\frac{2}{s-3}\right) Y(s) = \frac{1}{s},$$

so

$$Y(s) = \frac{3(s-3)}{s^2 - 3s + 2} + \frac{s-3}{s(s^2 - 3s + 2)}.$$

Simplifying this result by means of partial fractions shows

$$Y(s) = \underbrace{\frac{6}{s-1} - \frac{3}{s-2}}_{\text{first term}} \underbrace{-\left(\frac{3}{2}\right)\frac{1}{s} + \frac{2}{s-1} - \left(\frac{1}{2}\right)\frac{1}{s-2}}_{\text{second term}},$$

or

$$Y(s) = \frac{8}{s-1} - \left(\frac{7}{2}\right)\frac{1}{s-2} - \left(\frac{3}{2}\right)\frac{1}{s}.$$

Thus, taking the inverse transform, we obtain the solution

$$y(t) = 8e^t - \tfrac{7}{2}e^{2t} - \tfrac{3}{2}.$$ ∎

(i) Variable coefficient linear differential equations

The Laplace transform may also be used to solve initial value problems for variable coefficient linear differential equations, though the task of inverting the transform to find the solution becomes much harder. This use of the Laplace transform provides another method for the determination of the Laplace transform of many of the special functions used in engineering mathematics and physics. When these special functions possess Laplace transforms, and they are solutions of a linear differential equation, it is sometimes simpler to determine their transform directly from the differential equation itself than from an explicit representation of the special function.

The situation is best illustrated by means of an example. A typical differential equation giving rise to two useful special functions is Bessel's equation (see Sec. 8.7)

$$x^2 \frac{d^2 y}{dx^2} + x \frac{dy}{dx} + (x^2 - v^2)y = 0, \tag{64}$$

in which the parameter v is an arbitrary real constant. The two linearly independent solutions possessed by this equation are denoted by $J_v(x)$ and $Y_v(x)$ and they are called, respectively, **Bessel functions of order v of the first** and **second kind.** For any fixed real v the Bessel function of the first kind $J_v(x)$ is bounded for $0 \le x < \infty$, but the Bessel function of the second kind $Y_\nu(x)$, which is bounded in every interval $0 < a \leq x < \infty$, is infinite when $x = 0$.

For the sake of simplicity, hereafter we shall confine attention to Bessel functions of the first kind of *integer order*, corresponding to $\nu = n$ with $n = 0, 1, 2, \ldots$. However, before finding $\mathscr{L}\{J_n(x)\}$ we first summarize some necessary basic properties of $J_n(x)$. It will be seen later when examining the series solutions for Bessel functions that the Bessel function of order zero $J_0(x)$ is nonvanishing at the origin and normalized so that

$$J_0(0) = 1, \tag{65}$$

whereas the functions $J_n(x)$ are such that

$$J_n(0) = 0, \quad \text{for } n = 1, 2, 3, \ldots. \tag{66}$$

The functions $J_n(x)$ and their derivatives will be shown to be related by the recursion formulas

$$J_0'(x) = -J_1(x) \tag{67}$$

and

$$2J_n'(x) = J_{n-1}(x) - J_{n+1}(x), \tag{68}$$

for $n = 1, 2, \ldots$.

Let us now use Bessel's equation (64) and the above information to determine $\mathscr{L}\{J_n(x)\}$. Setting $\nu = 0$ in (64) shows $y = J_0(x)$ satisfies the differential equation

$$x \frac{d^2 y}{dx^2} + \frac{dy}{dx} + xy = 0, \quad \text{with } y(0) = 1, \tag{69}$$

where the initial condition follows from (65).

Transforming this equation using the result of Ex. 12 of Sec. 7.2 with Theorem 7.10, and making the necessary notational changes from t to x and F to Y, gives

$$\underbrace{-s^2 Y'(s) - 2s Y(s) + y(0)}_{\mathscr{L}\{xy''\}} + \underbrace{s Y(s) - y(0)}_{\mathscr{L}\{y'\}} - \underbrace{Y'(s)}_{\mathscr{L}\{xy\}} = 0,$$

or

$$(s^2 + 1) Y'(s) + s Y(s) = 0,$$

where $Y(s) = \mathscr{L}\{y(x)\} = \mathscr{L}\{J_0(x)\}$.

Separating variables and integrating gives

$$\int \frac{dY}{Y} = -\int \frac{s}{s^2 + 1} \, ds,$$

and thus

$$\ln Y(s) = -\tfrac{1}{2} \ln(s^2 + 1) + \ln C,$$

from which we find

$$Y(s) = \frac{C}{(s^2 + 1)^{1/2}}, \tag{70}$$

where C is an arbitrary constant of integration.

To determine C we now use Theorem 7.15 (i) (the initial value theorem) together with the requirement that $y(0) = J_0(0) = 1$. Thus C must be such that

$$1 = \lim_{s \to \infty} s\, Y(s) = \lim_{s \to \infty} \left(\frac{sC}{(s^2 + 1)^{1/2}} \right) = C,$$

and so from (70)

$$Y(s) = \mathscr{L}\{J_0(x)\} = \frac{1}{(s^2 + 1)^{1/2}}. \tag{71}$$

Instead of determining $\mathscr{L}\{J_n(x)\}$ directly from (64) we shall derive the transform by making use of (71) together with the recursion formulas (67) and (68). Transforming (67) leads to

$$s\, \mathscr{L}\{J_0(x)\} - J_0(0) = -\mathscr{L}\{J_1(x)\},$$

so after using (65) and (71) we find

$$\mathscr{L}\{J_1(x)\} = 1 - \frac{s}{(s^2 + 1)^{1/2}}. \tag{72}$$

Similarly, setting $n = 1$ in (68) and transforming the recursion formula we obtain

$$2[s\, \mathscr{L}\{J_1(x)\} - J_1(0)] = \mathscr{L}\{J_0(x)\} - \mathscr{L}\{J_2(x)\},$$

Substituting for $\mathscr{L}\{J_0(x)\}$ and $\mathscr{L}\{J_1(x)\}$, and using of the fact that $J_1(0) = 0$ (see (66)), leads to the result

$$\mathscr{L}\{J_2(x)\} = \frac{[(s^2 + 1)^{1/2} - s]^2}{(s^2 + 1)^{1/2}}. \tag{73}$$

In general, by transforming (68) for arbitrary $n > 1$ and using mathematical induction it is readily established that

$$\mathscr{L}\{J_n(x)\} = \frac{[(s^2 + 1)^{1/2} - s]^n}{(s^2 + 1)^{1/2}}, \qquad \text{for } n = 0, 1, 2, \ldots . \tag{74}$$

This result, is in fact, also true when $n = v$, with v any real number such that $v > -1$.

An interesting integral relationship between $\sin x$ and $J_0(x)$ may be derived from $\mathscr{L}\{J_0(x)\}$ by means of the convolution theorem. By writing

$$\frac{1}{s^2 + 1} = \frac{1}{(s^2 + 1)^{1/2}} \cdot \frac{1}{(s^2 + 1)^{1/2}}, \tag{75}$$

and using the fact that

$$\mathscr{L}^{-1}\left\{\frac{1}{s^2+1}\right\} = \sin x \quad \text{and} \quad \mathscr{L}^{-1}\left\{\frac{1}{(s^2+1)^{1/2}}\right\} = J_0(x),$$

it follows by taking the inverse of (75) and using the convolution theorem that

$$\sin x = \int_0^x J_0(u)\, J_0(x-u)\, du. \tag{76}$$

The Laplace transform of a function may also be used to derive a representation of the function in the form of an infinite series. Let us use the result

$$\mathscr{L}\{J_0(x)\} = (s^2+1)^{-1/2}$$

which has just been established to show how a series representation of $J_0(x)$ may be found. Writing

$$\mathscr{L}\{J_0(x)\} = \frac{1}{s}\left(1+\frac{1}{s^2}\right)^{-1/2}$$

and expanding by the binomial theorem gives

$$\mathscr{L}\{J_0(x)\} = \sum_{n=0}^{\infty} \frac{1}{2}\cdot\frac{3}{2}\cdot\frac{5}{2}\cdots \frac{2n-1}{2} \frac{(-1)^n}{n!\, s^{2n+1}},$$

or

$$\mathscr{L}\{J_0(x)\} = \sum_{n=0}^{\infty} \frac{(-1)^n}{(n!)^2}\left(\frac{1}{2}\right)^{2n} \frac{(2n)!}{s^{2n+1}}.$$

Taking the inverse transform of this equation and applying it term by term to the right-hand side yields the infinite series (see (14) in Sec. 8.7)

$$J_0(x) = \sum_{n=0}^{\infty} \frac{(-1)^n x^{2n}}{2^{2n}(n!)^2}. \tag{77}$$

The following example provides two useful and nontrivial Laplace transforms, one of which is deduced from the other. It is of special interest because in the derivation of the key result the initial and final value theorems fail to provide useful information, so a different form of argument is needed to determine the constant involved. The method employed uses an approach which will be encountered again in the next chapter when discussing asymptotic expansions.

Example 17. $\mathscr{L}\{\sin\sqrt{t}\}$

Show by considering the differential equation satisfied by

$$y(t) = \sin\sqrt{t}$$

that

$$\mathscr{L}\{\sin \sqrt{t}\} = \tfrac{1}{2}\sqrt{\pi}\, s^{-3/2} \exp[-1/(4s)], \quad \text{for } s>0,$$

and deduce that

$$\mathscr{L}\left\{\frac{\cos \sqrt{t}}{\sqrt{t}}\right\} = \sqrt{\pi}\, s^{-1/2} \exp[-1/(4s)], \quad \text{for } s>0.$$

Solution

A simple calculation shows $y(t)$ satisfies the differential equation.

$$4t\frac{d^2y}{dt^2} + 2\frac{dy}{dt} + y = 0.$$

Setting $Y(s) = \mathscr{L}\{y(t)\}$ and transforming the equation with the aid of Ex. 12 (ii) gives

$$4\left[-s^2\frac{dY}{ds} - 2sY + y(0)\right] + 2[sY - y(0)] + Y = 0.$$

However $y(0)=0$, so $Y(s) = \mathscr{L}\{\sin \sqrt{t}\}$ satisfies the differential equation

$$\frac{dY}{ds} = \left(\frac{1-6s}{4s^2}\right)Y.$$

After separating variables and integrating, this yields

$$Y(s) = Cs^{-3/2}\exp[-1/(4s)],$$

where C is a constant of integration. Once C has been determined the required Laplace transform then follows from this result.

Neither the initial nor the final value theorem are useful here. The initial value theorem fails to determine C, and the final value theorem is not applicable because $\sin \sqrt{t}$ has no limit as $t \to \infty$. Consider the definition

$$\mathscr{L}\{\sin \sqrt{t}\} = \int_0^\infty e^{-st} \sin \sqrt{t}\, dt.$$

Now $|\sin \sqrt{t}| \le 1$ for all t, while for $t>0$ and large s the function e^{-st} decreases very rapidly as s increases. Thus for large s and $t>0$ the integrand $e^{-st} \sin \sqrt{t}$ only differs significantly from zero when t is very small, when we then have $\sin \sqrt{t} \simeq \sqrt{t}$.

Thus for large s it follows from entry 4 in Table 7.1 that

$$\mathscr{L}\{\sin \sqrt{t}\} \simeq \int_0^t e^{-st} t^{1/2}\, dt = \mathscr{L}\{t^{1/2}\} = \frac{\Gamma(3/2)}{s^{3/2}}.$$

However, from the remarks about the gamma function following Table 7.1 we know $\Gamma(3/2) = \tfrac{1}{2}\Gamma(1/2) = \tfrac{1}{2}\sqrt{\pi}$, so for large s

$$\mathscr{L}\{\sin \sqrt{t}\} \simeq \tfrac{1}{2}\sqrt{\pi}\, s^{-3/2}.$$

Comparing this with the form of $Y(s)$ for large s shows $C = \frac{1}{2}\sqrt{\pi}$, (because $\exp[-1/(4s)] \to 1$), and thus

$$\mathscr{L}\{\sin\sqrt{t}\} = Y(s) = \frac{1}{2}\sqrt{\pi}\, s^{-3/2}\exp[-1/(4s)], \qquad \text{for } s>0.$$

Taking $y(t) = \sin\sqrt{t}$, using the Laplace transform $Y(s)$ just deduced, and the result $\mathscr{L}\{y'(t)\} = s\,Y(s) - y(0)$ (Theorem 7.8), it follows at once that

$$\mathscr{L}\left\{\frac{\cos\sqrt{t}}{\sqrt{t}}\right\} = \sqrt{\pi}\, s^{-1/2}\exp[-1/(4s)], \qquad \text{for } s>0. \qquad \blacksquare$$

Problems for Section 7.3

Section (a): Bending of beams

1 A uniform beam of weight M, length l and flexural rigidity EI is simply supported at its ends as shown in Fig. 7.32. Find the displacement $y(x)$ due to its weight.

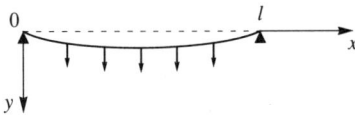

Fig. 7.32

2 Find the displacement $y(x)$ of the beam in Prob. 1 if in addition to its weight the beam also supports a concentrated load of weight $2M$ at its midpoint.

3 A uniform beam of weight M, length l and flexural rigidity EI is clamped at one end and simply supported at the other as shown in Fig. 7.33. Find the displacement $y(x)$ due to its weight.

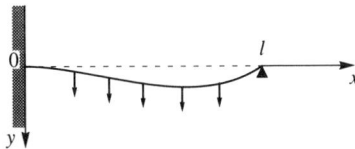

Fig. 7.33

4 A uniform beam of weight M, length l and flexural rigidity EI is clamped at its ends as shown in Fig. 7.34. Find the displacement $y(x)$ due to its own weight.

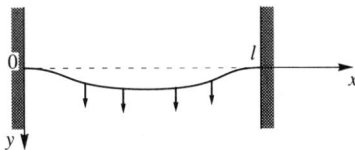

Fig. 7.34

5 A uniform beam of weight M, length l and flexural rigidity EI is clamped at one end and supported at its midpoint as shown in Fig. 7.35. Use the condition $y(\frac{1}{2}l)=0$ at the support, with the zero curvature condition $y''(l)=0$ at the free end of the beam to find the displacement $y(x)$ due to its own weight.

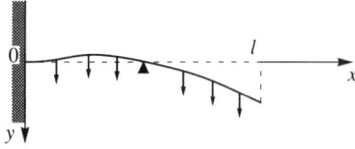

Fig. 7.35

6 The uniform beam of length l, negligible weight and flexural rigidity EI shown in Fig. 7.36 is simply supported at its ends. Find the displacement $y(x)$ if the beam supports a concentrated load of weight M at its midpoint.

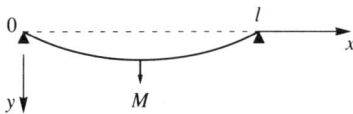

Fig. 7.36

Section (b): Coupled oscillations

7 The coupled oscillations of a mechanical system obey the equations

$$\ddot{x}_1 + 5x_1 - x_2 = 0$$

$$\ddot{x}_2 + 3x_2 - 3x_1 = 0.$$

Find $x_1(t)$ and $x_2(t)$ given that $x_1(0)=3$, $x_2(0)=0$, $\dot{x}_1(0)=0$ and $\dot{x}_2(0)=0$.

8 The coupled oscillations of a mechanical system obey the equations

$$6\ddot{x}_1 + 5x_1 - x_2 = 0$$

$$2\ddot{x}_2 + x_2 - x_1 = 0.$$

Find $x_1(t)$ and $x_2(t)$ given that $x_1(0)=2$, $x_2(0)=0$, $\dot{x}_1(0)=0$ and $\dot{x}_2(0)=0$.

9 The governing equations for the two RL-circuits with mutual inductance shown in Fig. 7.37 are

$$L\frac{di_1}{dt} + M\frac{di_2}{dt} + Ri_1 = E(t)$$

$$M\frac{di_1}{dt} + L\frac{di_2}{dt} + Ri_2 = 0.$$

Find the current $i_2(t)$ flowing in the secondary circuit at time $t>0$ if at time $t=0$ a constant voltage $E(t)=E_O$ is suddenly applied, so that

$$E(t) = \begin{cases} 0, & \text{for } t<0 \\ E_O, & \text{for } t\geq 0, \end{cases}$$

and $i_1 = i_2 = 0$ for $t<0$.

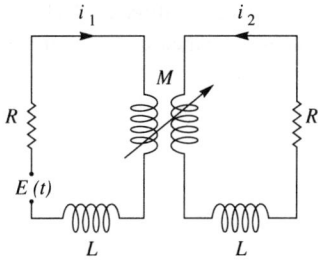

Fig. 7.37

10 Solve the system of equations

$$\ddot{x}_1 + 2x_2 = 0 \quad \text{and} \quad \ddot{x}_2 - 2x_1 = 0,$$

given $x_1(0) = 1$, $x_2(0) = 0$, $\dot{x}_1(0) = 0$ and $\dot{x}_2(0) = 0$.

11 Solve the system of equations

$$\ddot{x}_1 + 2\dot{x}_2 = 1 \quad \text{and} \quad \ddot{x}_2 - 2\dot{x}_1 = 0,$$

given $x_1(0) = x_2(0) = \dot{x}_1(0) = \dot{x}_2(0) = 0$.

12 Solve the system of equations

$$\ddot{x}_1 + 3\dot{x}_2 = 2 \quad \text{and} \quad \ddot{x}_2 - 3\dot{x}_1 = 0,$$

given that $x_1(0) = 1$, $x_2(0) = 4$, $\dot{x}_1(0) = -6$ and $\dot{x}_2(0) = 2/3$.

Section (c): Weighting function

13 Find the weighting function for the equation

$$\frac{dy}{dt} + 3y = t$$

and use it to solve the initial value problems for which (a) $y(0) = 0$ and (b) $y(0) = 2$.

14 Find the weighting function for the equation

$$\frac{dy}{dt} - 2y = 3 \cos 2t.$$

and use it to solve the initial value problems for which (a) $y(0) = 0$ and (b) $y(0) = -3$.

15 Find the weighting function for the equation

$$\frac{d^2 y}{dt^2} + 4\frac{dy}{dt} + 4y = 4y_1(t),$$

and use it to solve the initial value problem in which $y(0) = y'(0) = 0$ and $y_1(t) = \mathcal{U}(t)$.

The following three problems concern the equation

$$M\frac{d^2 y_0}{dt^2} + \frac{dy_0}{dt} + ky_0 = ky_1(t)$$

which is of considerable importance in control theory. Associated with the equation in which $M > 0$, $k > 0$ are the **undamped natural frequency** ω_n and the **damping factor** ζ defined by the expressions

$$\omega_n^2 = k/M \quad \text{and} \quad \zeta = \tfrac{1}{2}(kM)^{-1/2}.$$

16 Show that when $\zeta > 1$ the weighting function is

$$W(t) = \frac{\omega_n}{(\zeta^2 - 1)^{1/2}} \exp(-\zeta\omega_n t) \sinh[(\zeta^2 - 1)^{1/2}\omega_n t].$$

17 Show that when $\zeta = 1$ the weighting function is

$$W(t) = \omega_n^2 \exp(-\omega_n t).$$

18 Show that when $\zeta < 1$ the weighting function is

$$W(t) = \frac{\omega_n}{(1 - \zeta^2)^{1/2}} \exp(-\zeta\omega_n t) \sin(1 - \zeta^2)^{1/2}\omega_n t.$$

Section (d): Final value theorem
Where appropriate, apply the final value theorem to determine $y(\infty)$, and if no limit exists explain why.

19 (i) $Y(s) = \dfrac{1}{s(s^2 + s - 6)}$ (ii) $Y(s) = \dfrac{1}{s(4 + 3s)}$

20 (i) $Y(s) = \dfrac{2s - 3}{s^2 - 6s + 13}$ (ii) $Y(s) = \dfrac{2}{(s^2 + 4)^2}$

21 (i) $Y(s) = \dfrac{3}{s(4s^2 + s + 1)}$ (ii) $Y(s) = \dfrac{2 - s}{s(s^2 + 2s + 2)}$

22 (i) $Y(s) = \dfrac{4s}{3 - 2s}$ (ii) $Y(s) = \dfrac{s^2 - 3}{s(s^2 + 2s + 2)}$

Section (e): Transfer functions
23 In the spring-dashpot system shown in Fig. 7.38 the input $y_1(t)$ and output $y_0(t)$ are governed by the equations

$$K(y_1 - y_0) = D\frac{dy_0}{dt}.$$

Find the transfer function $T(s)$ for this system, in which for slow inputs the output is almost equal to the input, but for rapid inputs the output is slow to follow.

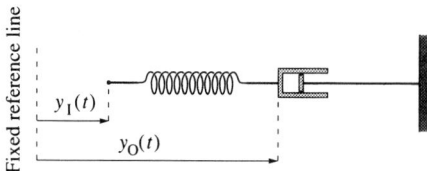

Fig. 7.38 Spring-dashpot system

24 In the spring-dashpot system shown in Fig. 7.39 the input $y_I(t)$ and output $y_O(t)$ are governed by the equation

$$Ky_O = D\frac{d}{dt}(y_I - y_O).$$

Find the transfer function $T(s)$ for this system in which a slowly changing input produces very little output, but a rapidly changing input is almost matched by the output.

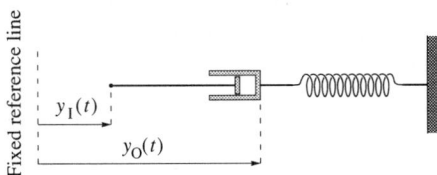

Fig. 7.39 Spring-dashpot system

25 In the hydraulic element shown in Fig. 7.40 the displacements from the neutral positions of the input, the actuating valve and the output, $x_I(t)$, $y(t)$ and $x_O(t)$, respectively, are related by the equations

$$Ky = \frac{dx_O}{dt} \quad \text{and} \quad \frac{x_I - y}{l_1} = \frac{(x_I + x_O)}{l_1 + l_2}.$$

Find the transfer function of the system.

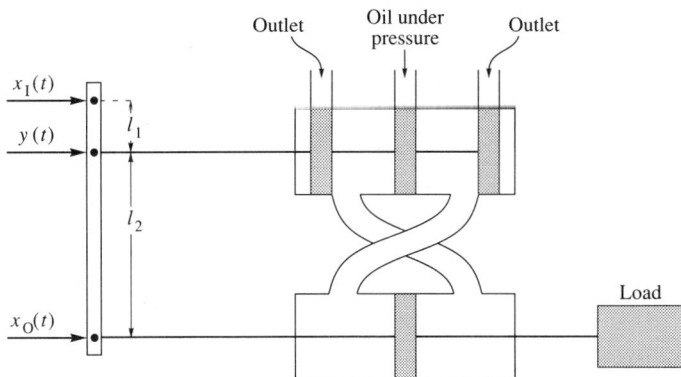

Fig. 7.40 Hydraulic system

26 In a dc motor the input voltage $e(t)$, the current $i(t)$ and the angular velocity $\omega(t)$ of the shaft of the motor are related by

$$e = Ri + K_1\omega \quad \text{and} \quad J\frac{d\omega}{dt} = K_2 i,$$

where R is the resistance of the armature and J is the moment of inertia of the armature and load. Find the transfer function $T_\omega(s)$ when the voltage $e(t)$ is regarded as the input and the angular velocity $\omega(t)$ as output. What is the transfer function $T_i(s)$ when the voltage $e(t)$ is regarded as the input and the current $i(t)$ as output?

Harder problems on transfer functions

27 The **frequency-response function** $Y(i\omega)$ for a stable linear element is defined as the output from the element (a complex quantity) when it is subjected to a sinusoidal (complex) input $y_1(t) = e^{i\omega t}$ which started at $t = -\infty$. Consequently, as the input is defined for all $t < 0$, result (11) which related input to output in terms of the weighting function $W(t)$ of the element must have its interval of integration extended to ∞ to become

$$y_0(t) = \int_0^\infty W(\tau)\, y_1(t-\tau)\, d\tau.$$

Use this result to show that a sinusoidal input $y_1(t) = Ae^{i\omega t}$ to a stable linear element will produce a sinusoidal output

$$Y(i\omega) = \int_0^\infty W(\tau)e^{-i\omega\tau}d\tau,$$

and that the amplitude of the input is multiplied by the amplification factor

$$|Y(i\omega)|$$

and its phase increased by the angle

$$\Phi = \text{arc tan}\left[\frac{1}{i}\left(\frac{Y(i\omega) - \bar{Y}(i\omega)}{Y(i\omega) + \bar{Y}(i\omega)}\right)\right],$$

where a bar denotes the complex conjugate.

28 Show that the RC-filter illustrated in Fig. 7.41 described by the equation

$$RC\frac{de_0}{dt} + e_0 = e_1$$

has the frequency-response function

$$Y(i\omega) = \frac{1}{1 + i\omega RC},$$

the amplitude amplification factor

$$|Y(i\omega)| = (1 + \omega^2 R^2 C^2)^{-1/2},$$

and the phase shift

$$\Phi = \text{arc tan}(-\omega RC).$$

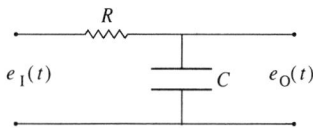

Fig. 7.41 RC-filter

29 Show that the simple servo system described by the equation

$$M \frac{d^2 y_O}{dt^2} + \frac{dy_O}{dt} + ky_O = ky_I(t)$$

has the frequency-response function

$$Y(i\omega) = \frac{\omega_n^2}{(\omega_n^2 - \omega^2) + i2\zeta\omega\omega_n},$$

the amplitude amplification factor

$$|Y(i\omega)| = \{[1 - (\omega/\omega_n)^2]^{1/2} + 4\zeta^2 (\omega/\omega_n)^2\}^{-1/2}$$

and the phase shift

$$\Phi = -\text{arc tan } \{2\zeta(\omega/\omega_n)/[1 - (\omega/\omega_n)^2]\},$$

where $\omega_n^2 = k/M$ and $\zeta = \frac{1}{2}(kM)^{-1/2}$.

30 Show that the frequency-response function of the linear element described by

$$a_0 \frac{d^n y_O}{dt^n} + a_1 \frac{d^{n-1} y_O}{dt^{n-1}} + \ldots + a_n y_O = b_0 \frac{d^m y_I}{dt^m} + b_1 \frac{d^{m-1} y_I}{dt^{m-1}} + \ldots + b_m y_I$$

is

$$Y(i\omega) = \frac{b_0 (i\omega)^m + b_1 (i\omega)^{m-1} + \ldots + b_m}{a_0 (i\omega)^n + a_1 (i\omega)^{n-1} + \ldots + a_n}.$$

Section (f): Exponential matrix e^{tA}

31 Use the Laplace transform to show that

(a) if $\mathbf{A} = \begin{bmatrix} 0 & -k \\ k & 0 \end{bmatrix}$ then $e^{tA} = \begin{bmatrix} \cos kt & -\sin kt \\ \sin kt & \cos kt \end{bmatrix}$,

(b) If $\mathbf{A} = \begin{bmatrix} a & k \\ k & a \end{bmatrix}$ then $e^{tA} = \begin{bmatrix} e^{at}\cosh kt & e^{at}\sinh kt \\ e^{at}\sinh kt & e^{at}\cosh kt \end{bmatrix}$.

32 Use the Laplace transform to show that

(a) if $\mathbf{A} = \begin{bmatrix} a & 0 & 0 \\ 0 & b & 0 \\ 0 & 0 & c \end{bmatrix}$ then $e^{tA} = \begin{bmatrix} e^{at} & 0 & 0 \\ 0 & e^{bt} & 0 \\ 0 & 0 & e^{ct} \end{bmatrix}$,

(b) if $\mathbf{A} = \begin{bmatrix} a & 0 & 0 \\ 0 & 0 & k \\ 0 & -k & 0 \end{bmatrix}$ then $e^{tA} = \begin{bmatrix} e^{at} & 0 & 0 \\ 0 & \cos kt & \sin kt \\ 0 & -\sin kt & \cos kt \end{bmatrix}$.

33 Solve the system

$$\dot{\mathbf{x}} = \mathbf{Ax} + \mathbf{f},$$

given

$$\mathbf{A} = \begin{bmatrix} 0 & 1 \\ -4 & -4 \end{bmatrix}, \quad \mathbf{x} = \begin{bmatrix} x_1 \\ x_2 \end{bmatrix}, \quad \mathbf{f} = \begin{bmatrix} 4 \\ -1 \end{bmatrix} \quad \text{and} \quad \mathbf{x}(0) = \begin{bmatrix} \frac{3}{4} \\ 0 \end{bmatrix}.$$

34 Solve the system

$$\dot{\mathbf{x}} = \mathbf{A}\mathbf{x},$$

given

$$\mathbf{A} = \begin{bmatrix} 0 & -1 & 0 \\ 1 & 0 & 0 \\ 0 & 1 & 1 \end{bmatrix}, \quad \mathbf{x} = \begin{bmatrix} x_1 \\ x_2 \\ x_3 \end{bmatrix} \quad \text{and} \quad \mathbf{x}(0) = \begin{bmatrix} 1 \\ 2 \\ 1 \end{bmatrix}.$$

Section (g): Delay differential equations

35 Solve by means of the Laplace transform

$$\frac{d}{dt}x(t) + 4x(t-1) = t,$$

given $x(t) \equiv 0$ for $t \le 0$.

36 Solve by means of the Laplace transform

$$\frac{d}{dt}x(t) + 2x(t-3) = 1,$$

given $x(t) \equiv 0$ for $t \le 0$.

37 Solve by means of the Laplace transform

$$\frac{d}{dt}x(t) + 2x(t-1) = t + k\mathcal{U}(t-\alpha),$$

given $x(t) \equiv 0$ for $t \le 0$ and $k > 0$, $\alpha > 0$ are arbitrary.

38 Solve by means of the Laplace transform

$$\frac{d}{dt}x(t) + x(t-2) = t^2,$$

given $x(t) \equiv 0$ for $t \le 0$.

Section (h): Integral equations and integro-differential equations

In each of the following problems, derive from first principles the Volterra integral equation equivalent to the given initial value problem.

39 $\dfrac{d^2 y}{dt^2} + 4y = 1$, with $y(0) = 3$, $y'(0) = -2$.

40 $\dfrac{d^2 y}{dt^2} + ty = e^t$, with $y(0) = 0$, $y'(0) = 2$.

41 $\dfrac{d^2 y}{dt^2} + 9y = 2\delta(t-1)$, with $y(0) = 1$, $y'(0) = 2$.

42 $\dfrac{d^2 y}{dt^2} + 2y = t$, with $y(0) = y'(0) = 0$.

43 $\dfrac{d^2 y}{dt^2} + \lambda^2 y = f(t)$, with $y(0) = \alpha$, $y'(0) = \beta$.

44 $\dfrac{d^2 y}{dt^2} + 3\dfrac{dy}{dt} - 2y = 1$, with $y(0) = 0$, $y'(0) = 3$.

45 $\dfrac{d^2 y}{dt^2} + 2\dfrac{dy}{dt} + 4y = t^2$, with $y(0) = y'(0) = 0$.

46 $\dfrac{d^2 y}{dt^2} + \dfrac{dy}{dt} - 3y = \sin t$, with $y(0) = 1$, $y'(0) = 0$.

47 The results of this problem may be used when transforming higher order initial value problems into equivalent Volterra integral equations. Show by means of the convolution theorem that if $F(s) = \mathscr{L}\{f(t)\}$, then

(i) $\mathscr{L}^{-1}\left\{\dfrac{F(s)}{s^2}\right\} = \displaystyle\int_0^t \int_0^u f(\tau)\, d\tau\, du = \int_0^t (t - \tau) f(\tau)\, d\tau$,

(ii) $\mathscr{L}^{-1}\left\{\dfrac{F(s)}{s^3}\right\} = \displaystyle\int_0^t \int_0^u \int_0^v f(\tau)\, d\tau\, du\, dv = \int_0^t \dfrac{(t - \tau)^2}{2} f(\tau)\, d\tau$,

and in general,

(iii) $\mathscr{L}^{-1}\left\{\dfrac{F(s)}{s^n}\right\} \displaystyle\int_0^t \int_0^{u_1} \int_0^{u_2} \cdots \int_0^{u_{n-1}} f(\tau)\, d\tau\, du_1\, du_2 \ldots du_{n-1} = \int_0^t \dfrac{(t - \tau)^{n-1}}{(n-1)!} f(\tau)\, d\tau$.

48 Use the results of Prob. 47 to find

(i) $\displaystyle\int_0^t \int_0^u \int_0^v \sin 3\tau\, d\tau$,   (ii) $\displaystyle\int_0^t \int_0^u \int_0^v e^{-3\tau} y(\tau)\, d\tau$,

(iii) $\displaystyle\int_0^t \int_0^u \int_0^v \int_0^w (1 + \tau e^{-\tau})\, y(\tau)\, d\tau\, du\, dv\, dw$.

In the two questions which follow, derive from first principles using the results of Prob. 47, or otherwise, the Volterra integral equation equivalent to the given initial value problems.

49 $\dfrac{d^3 y}{dt^3} + 2\dfrac{d^2 y}{dt^2} - \dfrac{dy}{dt} + 3y = 1$, with $y(0) = 1$, and $y''(0) = y'''(0) = 0$.

50 $\dfrac{d^3 y}{dt^3} + 2y = \sin t$, with $y(0) = 1$, $y'(0) = 3$ and $y''(0) = 2$.

In each of the following problems transform the given Volterra integral equation into an equivalent initial value problem.

51 $y(t) = \sin t + 2 \displaystyle\int_0^t \cos(t - \tau)\, y(\tau)\, d\tau$.

52 $(1 + t)\, y(t) = 1 + t - \displaystyle\int_0^t (t - 1)(1 + \tau)\, y(\tau)\, d\tau$.

53 $y(t) = 1 + t + \int_0^t [(t-\tau)(1-\tau) - 2\tau] y(\tau) d\tau.$

54 $\cos t\, y(t) = t - \int_0^t [\tau + 2\sin \tau + (t-\tau)(\tau + \cos \tau)] y(\tau) d\tau.$

Solve the following linear Volterra integral equations

55 $y(t) = 3e^{-3t} - \int_0^t \exp[-2(t-\tau)] y(\tau) d\tau.$

56 $y(t) = \sin t + \int_0^t \exp[-(t-\tau)] y(\tau) d\tau.$

57 $y(t) = \sin t + \int_0^t \cos(t-\tau) y(\tau) d\tau.$

58 Show that the solution of the linear Volterra integral equation

$$y(t) = f(t) + \lambda \int_0^t \exp[-k(t-\tau)] y(\tau) d\tau,$$

where λ, k are nonzero numbers and $f(t)$ is any function of t possessing a Laplace transform, is

$$y(t) = f(t) + \lambda \int_0^t \exp[-(k-\lambda)(t-\tau)] f(\tau) d\tau.$$

59 Use the convolution theorem to show that if $Y(s) = \mathcal{L}\{y(t)\}$, then

$$\mathcal{L}\left\{ \int_0^t \sin 2(t-\tau) \left(\frac{dy(\tau)}{d\tau} \right) d\tau \right\} = \frac{2[s\, Y(s) - y(0)]}{s^2 + 4}.$$

Put $t = 0$ in the integral equation

$$y(t) + \int_0^t \cos 2(t-\tau) y(\tau) d\tau + 2 \int_0^t \sin 2(t-\tau) \frac{dy(\tau)}{d\tau} d\tau = e^{-t} \cos t,$$

to find $y(0)$, and then solve the equation for $y(t)$.

60 Find the solution, which is finite for all time, of the nonlinear Volterra integral equation

$$y(t) = 3 - t + \tfrac{1}{9} \int_0^t y(t-\tau) y(\tau) d\tau.$$

61 Find the solution, which is finite for all time, of the nonlinear Volterra integral equation

$$y(t) = 3 \sin t + t \cos t + \tfrac{1}{8} \int_0^t y(t-\tau) y(\tau) d\tau.$$

62 Use the Laplace transform to solve the simultaneous integral equations

$$4x(t) - y(t) + 3 \int_0^t x(\tau) d\tau = 1 - \cos t$$

$$x(t) + \int_0^t y(\tau) d\tau = \sin t,$$

and hence show that

$$x(t) = \tfrac{1}{2}(e^{-t} - e^{-3t}) \quad \text{and}$$
$$y(t) = \cos t + \tfrac{1}{2}(e^{-t} - 3e^{-3t}).$$

Solve the follwing integro-differential equations

63 $\dfrac{dy}{dt} + a^2 \displaystyle\int_0^t y(\tau)\,d\tau = 3a, \quad$ with $y(0) = 2$.

64 $\dfrac{dy}{dt} + y - 2 \displaystyle\int_0^t y(\tau)\,d\tau = 4, \quad$ with $y(0) = 0$.

65 $\dfrac{dy}{dt} - y + 2 \displaystyle\int_0^t \exp[-2(t-\tau)]\,y(\tau)\,d\tau = \sin 2t, \quad$ with $y(0) = 0$.

66 $\dfrac{dy}{dt} - \displaystyle\int_0^t (t-\tau)\,y(\tau)\,d\tau = 1, \quad$ with $y(0) = 0$.

Section (i): Variable coefficient differential equations

The two problems which follow are special cases of the **Laguerre equation** with parameter n (zero or an integer) which has the form

$$x\frac{d^2 y}{dx^2} + (1-x)\frac{dy}{dx} + ny = 0.$$

This differential equation has special polynomial solutions denoted by $L_n(x)$ called **Laguerre polynomials**. These are used in mathematical physics and analysis, and also in numerical analysis in connection with numerical integration.

67 Take the Laplace transform of the equation

$$x\frac{d^2 y}{dx^2} + (1-x)\frac{dy}{dx} + 2y = 0,$$

and hence determine the transform $Y(s) = \mathscr{L}\{y(x)\}$ corresponding to the solution for which $y(0) = 1$. This is the Laplace transform of $L_2(x)$. Use the result to find $L_2(x) = \mathscr{L}^{-1}\{Y(s)\}$. Check that the solution satisfies the equation and the initial condition.

68 Take the Laplace transform of the equation

$$x\frac{d^2 y}{dx^2} + (1-x)\frac{dy}{dx} + 3y = 0,$$

and hence determine the transform $Y(s) = \mathscr{L}\{y(x)\}$ corresponding to the solution for which $y(0) = 1$. This is the Laplace transform of $L_3(x)$. Use the result to find $L_3(x) = \mathscr{L}^{-1}\{Y(s)\}$. Check that the solution satisfies the equation and the initial condition.

The two problems which follow involve the **Hermite equation** with parameter n (zero or an integer) which has the form

$$\frac{d^2 y}{dx^2} - 2x\frac{dy}{dx} + 2ny = 0.$$

The differential equation has special polynomial solutions denoted by $H_n(x)$ called **Hermite polynomials**. These are used in mathematical physics and analysis, and also in numerical analysis in connection with interpolation and numerical integration.

69 Take the Laplace transform of the equation

$$\frac{d^2 y}{dx^2} - 2x\frac{dy}{dx} + 2y = 0,$$

and by solving the resulting differential equation for $Y(s) = \mathscr{L}\{y(x)\}$, find $Y(s)$ corresponding to the solution for which $y(0) = 0$, $y'(0) = 2$. This is the Laplace transform of $H_1(x)$. Use the initial value theorem to show

$$Y(s) = 2/s^2,$$

and hence deduce that the solution of the original equation is

$$H_1(x) = 2x.$$

Verify that this solution satisfies the original equation and initial conditions.
[*Hint:* Once $Y(s)$ has been obtained, expand the exponential function and group terms in $1/s^2$ before applying the initial value theorem.]

70 Take the Laplace transform of the equation

$$\frac{d^2 y}{dx^2} - 2x\frac{dy}{dx} + 4y = 0,$$

and by solving the resulting differential equation for $Y(s) = \mathscr{L}\{y(x)\}$, find $Y(s)$ corresponding to the solution for which $y(0) = -2$, $y'(0) = 0$. This is the Laplace transform of $H_2(x)$. Use the initial value theorem to show

$$Y(s) = \frac{8}{s^3} - \frac{2}{s},$$

and hence deduce that the solution of the original equation is

$$H_2(x) = 4x^2 - 2.$$

Verify that this solution satisfies the original equation and initial conditions.
[*Hint:* Once $Y(s)$ has been obtained, expand the exponential function and group terms in $1/s$ and $1/s^3$ before applying the initial value theorem.]

Harder problems on variable coefficient equations

71 Show by taking the Laplace tranform that the **Laguerre equation** with parameter n (zero or an integer)

$$x\frac{d^2 y}{dx^2} + (1-x)\frac{dy}{dx} + ny = 0$$

has a polynomial solution $L_n(x)$ with $L_n(0) = 1$ such that

$$L_n(x) = \mathscr{L}^{-1}\left\{\frac{(s-1)^n}{s^{n+1}}\right\},$$

and deduce that

$$L_n(x) = \frac{1}{n!} e^x \frac{d^n}{dx^n} (x^n e^{-x}).$$

72 This problem involves the derivation of $\mathscr{L}\{J_1(x)\}$ directly from the governing differential equation for the Bessel function $y = J_1(x)$ which satisfies

$$x^2 \frac{d^2 y}{dx^2} + x \frac{dy}{dx} + (x^2 - 1) y = 0.$$

Take the Laplace transform of this equation and integrate the resulting differential equation for $Y(s) = \mathscr{L}\{J_1(x)\}$ to show

$$Y(s) = K + \frac{As}{(s^2 + 1)^{1/2}}.$$

Use the initial condition $J_1(0) = 0$ and the initial value $J_1'(0) = 0$ deduced from the information given in (67) and (68) in conjuction with Theorem 7.15 parts (i) and (ii) to show $1 = K = -A$, and hence that

$$\mathscr{L}\{J_1(x)\} = 1 - \frac{s}{(s^2 + 1)^{1/2}}.$$

73 This problem shows how $\mathscr{L}\{\exp(-x^2)\}$ may be deduced from the initial value problem

$$\frac{dy}{dx} + 2xy = 0, \quad \text{with} \quad y(0) = 1,$$

which has the solution $y = \exp(-x^2)$.

Take the Laplace transform of the above equation and integrate the resulting equation for $Y(s) = \mathscr{L}\{y(x)\}$ to show

$$Y(s) = \exp(s^2/4) \left\{ Y(0) - \tfrac{1}{2} \int_0^s \exp(-u^2/4) \, du \right\}.$$

Use the result obtainable from an asymptotic expansion

$$\lim_{s \to \infty} \left[\sqrt{\pi} \, s \exp(s^2) \left(1 - \frac{2}{\sqrt{\pi}} \int_0^s \exp(-u^2) \, du \right) \right] = 1$$

in conjunction with Theorem 7.15(i) to show $Y(0) = \sqrt{\pi}/2$, and hence that

$$Y(s) = \mathscr{L}\{\exp(-x^2)\} = \tfrac{1}{2} \sqrt{\pi} \exp(s^2/4) \left(1 - \frac{2}{\sqrt{\pi}} \int_0^{s/2} \exp(-u^2) \, du \right).$$

The function

$$\text{erf}\, s = \frac{2}{\sqrt{\pi}} \int_0^s \exp(-u^2) \, du$$

which is available in tabulated form and occurs in statistics, engineering and physics is called the **error function**.

The associated function

$$\operatorname{erfc} s = 1 - \operatorname{erf} s$$

is called the **complementary error function**. Thus

$$\mathscr{L}\{\exp(-x^2)\} = \tfrac{1}{2}\sqrt{\pi}\,\exp(s^2/4)\operatorname{erfc}(\tfrac{1}{2}s).$$

74 This problem shows how $\mathscr{L}\{\cos\sqrt{t}/\sqrt{t}\}$ may be deduced directly from a differential equation. The function $y(t) = \cos\sqrt{t}/\sqrt{t}$ satisfies the nonhomogeneous linear first order equation

$$2t\frac{dy}{dt} + y = -\sin\sqrt{t}.$$

Set $Y(s) = \mathscr{L}\{\cos\sqrt{t}/\sqrt{t}\}$, and with the aid of Ex. 12(i) and $\mathscr{L}\{\sin\sqrt{t}\}$ given in Ex. 17 transform the equation to obtain a first order differential equation for $Y(s)$. Solve this equation to find the general solution for $Y(s)$.

Deduce the value of the constant in the general solution by considering the form of

$$\int_0^\infty e^{-st}\frac{\cos\sqrt{t}}{\sqrt{t}}\,dt$$

for large s and comparing the result with $Y(s)$ when s is large, and hence show that

$$\mathscr{L}\left\{\frac{\cos\sqrt{t}}{\sqrt{t}}\right\} = \sqrt{\pi}\,s^{-1/2}\exp[-1/(4s)], \quad \text{for } s > 0.$$

75 Why is it not possible to deduce $\mathscr{L}\{\cos\sqrt{t}/\sqrt{t}\}$ from the differential equation

$$4t\frac{d^2y}{dt^2} + 6\frac{dy}{dt} + y = 0$$

satisfied by $y(t) = \cos\sqrt{t}/\sqrt{t}$? [*Hint*: Transform the equation to find the differential equation satisfied by $Y(s)$.]

7.4   The *z*-transform

In this section we describe the most important properties of the **z-transform**, which is the discrete variable analog of the Laplace transform. In the Laplace transform the variable $t > 0$ in the function $f(t)$ to be transformed is understood to be continuous, and it usually represents the time. However, in the z-transform, t is replaced by a quantity which increases discretely in steps of equal magnitude. The most important applications of the z-transform are encountered in the study of **linear sampled data systems** where t is the time, and each step corresponds to a time increment T, starting from $t = 0$. The steps are numbered $n = 0, 1, 2, \ldots$, with $n = 0$ corresponding to $t = 0$, $n = 1$ to $t = T$, and so on.

Discrete data problems arise from situations or processes in which either the data is inherently discrete by nature, or when for good practical reasons it is convenient to represent data depending continuously on t in a discrete form. A typical example of a physical situation giving rise to inherently discrete data is the sequence of numbers $u_0, u_1, u_2, \ldots$, generated by recording the number of vehicles passing a survey point during successive ten minute intervals, starting from a given time. A mathematical example of an inherently discrete problem is provided by a sequence of numbers $u_0, u_1, u_2, \ldots$, generated by some algorithm. A simple case in which the algorithm takes the form of an explcit formula is the sequence of numbers $u_0, u_1, u_2, \ldots$, generated by the **difference equation**[9]

$$u_{n+2} = u_{n+1} + u_n, \tag{1}$$

when u_0 and u_1 (the initial conditions) are given.

A nontrivial example of a different kind in which no algorithm is known is provided by the sequence $\pi_0, \pi_1, \pi_2, \ldots$, in which π_n denotes the number of primes less than $10n$, with $n = 0, 1, 2, \ldots$. It will be recalled that a **prime number**, or simply a **prime**, is any positive integer p greater than 1 whose only positive integer divisors are 1 and p.

The most important situation of interest to us will, however, be the one in which a function of a continuous variable t is represented in discrete form by sampling at regular intervals, starting from $t = 0$. A physical situation in which discrete data is generated artificially is provided by sampled data systems. In these systems a continuous analog signal $u(t)$ of the time t is converted to a sequence of numbers, each proportional to the signal strength at an instant of time, by using rapid instantaneous sampling of the signal at regular time intervals T. In such systems these sequences of numbers represent flows of discrete variable information and they are called **digital signals**.[10]

From now on our concern will only be with discrete data in the form of sequences of numbers $\{u_n\}_{n=0}^{\infty} = u_0, u_1, u_2, \ldots$. It will be unimportant whether the data originates in this form, or by instantaneously sampling some function $u(t)$ at successive sampling points separated by a fixed sampling interval T. As a sequence may always be regarded as having been obtained from a function by restricting the argument of the function to integral multiples of a fixed increment T (the **sampling interval**), it will be convenient to consider there always to be such a function associated with a sequence. No loss of generality is involved, because we shall only be concerned with the values of the function (sequence) at the sampling points and not at intermediate points. By making a suitable choice of origin, sampling may always be considered to start at $t = 0$. Thus the points at which sampling will occur will be $0, T, 2T, \ldots$. If the function sampled is $u(t)$, the

---

[9] When $u_0 = u_1 = 1$ this difference equation generates the **Fibonacci sequence** 1, 1, 2, 3, 5, 8, 13, 21, 34, 55, . . . , which has many uses in numerical analysis. The name of the sequence is derived from that of the author of the algebra book *Liber Abaci* published in 1202 by an Italian merchant, Leonardo of Pisa, now known to us as **Fibonacci**. The numbers in the sequence answer the following question posed in the book: 'Find the number of productive pairs of rabbits produced after n months from a single productive pair if none dies and each month every productive pair produces a new productive pair'.

[10] An example of analog to digital conversion to be found in the compact disk. Music represented by a fluctuating voltage $u(t)$ is sampled at regular time intervals and each voltage reading is converted to a numerical scale to produce a sequence $u_0, u_1, u_2, \ldots$. For convenience the numerical readings are represented by their binary codes (a sequence of 0's and 1's) and these are then stored in sequence on the disk as nonreflective (0) or reflective (1) patches capable of being detected (read) by a laser/detector system.

sequence $u_0, u_1, u_2, \ldots$ generated by sampling in this manner will be defined by

$$u_n = u(nT), \qquad n = 0, 1, 2, \ldots. \tag{2}$$

As sampling begins at $t=0$, we will always set $u_n=0$ for $n<0$.

The z-transform now to be defined is a method by which the behavior of discrete variable sequences (digital signals) may be analyzed. In the context of control theory, the z-transform determines how digital signals are affected by filters and feedback loops, and it helps determine how a system should be modified to satisfy given design criteria. Whereas the Laplace transform is based on exponential behavior using the function e^{-st} as a measure of the growth rate of a function, the z-transform is based on inverse powers of $z = e^{sT}$. In situations in which time is not involved, or when the value of the sampling time interval is unimportant, it is customary to set $T = 1$.

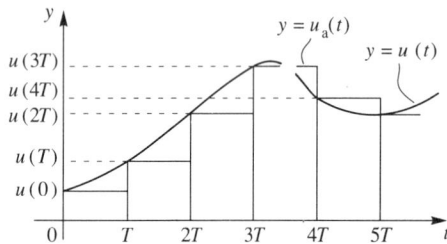

Fig. 7.42   Staircase approximation $u_a(t)$ to $u(t)$

Let us first consider the effect of applying the Laplace transform to a 'staircase' function approximation to a function $u(t)$ of the continuous variable t. In this approximation, which is illustrated in Fig. 7.42, the height u_n of the nth step is $u_n = u(nT)$. The 'staircase' approximation $u_a(t)$ to $u(t)$ is defined by

$$u_a(t) = u(nT), \qquad \text{for } nT < t < (n+1)T, \qquad n = 0, 1, 2, \ldots. \tag{3}$$

Taking the Laplace transform of $u_a(t)$ gives

$$\mathscr{L}\{u_a(t)\} = \sum_{n=0}^{\infty} \int_{nT}^{(n+1)T} e^{-st} u(nT)\,dt = \sum_{n=0}^{\infty} u(nT) \int_{nT}^{(n+1)T} e^{-st}\,dt$$

$$= \sum_{n=0}^{\infty} u(nT) \left(-\frac{1}{s} e^{-st} \right)_{nT}^{(n+1)T} = \left(\frac{1 - e^{-sT}}{s} \right) \sum_{n=0}^{\infty} u(nT) e^{-nsT}. \tag{4}$$

Notice that the series in (4) is convergent, because the function $u(t)$ from which it was derived, and to which $u_a(t)$ is an approximation, is assumed to be one for which the Laplace transform exists.

We shall set

$$\mathscr{D}\{u(nT)\} = \sum_{n=0}^{\infty} u(nT) e^{-nsT}, \tag{5}$$

and call this the **discrete Laplace transform** of the sequence $\{u(nT)\}_{n=0}^{\infty}$. Results (4) and (5) show the Laplace transform of the 'staircase' function approximation $u_a(t)$ to $u(t)$, and the discrete Laplace transform $\mathscr{D}\{u(nT)\}$, are related by the expression

$$\mathscr{L}\{u_a(t)\} = \left(\frac{1-e^{-sT}}{s}\right)\mathscr{D}\{u(nT)\}. \tag{6}$$

Now the sequence of pulses $u_s(t)$ derived from $u(t)$ by instantaneous sampling at the points $0, T, 2T, \ldots$, can be represented in the form

$$u_s(t) = \sum_{n=0}^{\infty} u(uT)\,\delta(t-nT), \tag{7}$$

from which it follows that

$$\int_0^t u_s(\tau)\,d\tau = u_a(t). \tag{8}$$

Taking the Laplace transform of (7), and using (5), gives

$$\mathscr{L}\{u_s(t)\} = \sum_{n=0}^{\infty} u(nT)e^{-nsT} = \mathscr{D}\{u(nT)\}. \tag{9}$$

Setting $z = e^{sT}$, then leads to the representation

$$\mathscr{L}\{u_s(t)\} = \sum_{n=0}^{\infty} u(nT)z^{-n} = u(0) + \frac{u(T)}{z} + \frac{u(2T)}{z^2} + \ldots. \tag{10}$$

The series on the right-hand side is called the **z-transform** of the sequence $\{u(nT)\}_{n=0}^{\infty}$. Notice that the z-transformations of $u_s(t)$ and $u(t)$ are identical, because only functional values at the sampling points are involved. Thus we need only consider the z-transform of $u(t)$ which will be denoted by $Z\{u(t)\}$. The convergence of the series in (4) implies that the series in (10) will be convergent provided $\mathscr{L}\{u(t)\}$ exists.

Definition (z-Transform)

Let $\{u(nT)\}_{n=0}^{\infty}$ be a numerical sequence whose nth term is $u_n = u(nT)$. Then the z-transform of the sequence or, equivalently, of the function $u(t)$, written $Z\{u(t)\}$ and denoted by $U(z)$ when expressed as a function of $z = e^{sT}$, is defined as

$$Z\{u(t)\} = U(z) = \sum_{n=0}^{\infty} u(nT)z^{-n},$$

whenever the series on the right-hand side converges.

The similarity of form shared by the Laplace and z-transform of $u(t)$ may be seen by comparing the Laplace transform

$$U(s) = \mathscr{L}\{u(t)\} = \int_0^{\infty} e^{-st} u(t)\,dt$$

with the z-transform when expressed in terms of e^{-sT} as

$$U(z) = Z\{u(t)\} = \sum_{n=0}^{\infty} u(nT) e^{-nsT}.$$

The most important and useful mathematical property of the z-transform is the fact that, like the Laplace transform, it is a linear operation. This, together with a related property involving the scaling of $u(nT)$, is proved in the next theorem.

Theorem 7.19 (Linearity and scaling of the z-transform)

Let α, β be arbitrary constants, and $\{u(nT)\}_{n=0}^{\infty}$, $\{v(nT)\}_{n=0}^{\infty}$ two sequences such that $Z\{u(t)\} = U(z)$ and $Z\{v(t)\} = V(z)$ are convergent. Then,

(i) the z-transform is a linear operation:

$$Z\{\alpha u(t) + \beta v(t)\} = \alpha Z\{u(t)\} + \beta Z\{v(t)\},$$

and

(ii) scaling $u(nT)$ in the z-transform is a linear operation:

$$Z\{\alpha u(t)\} = \alpha Z\{u(t)\}.$$

Proof

Result (i) follows directly from the definition of the z-transform because

$$\sum_{n=0}^{\infty} \{\alpha u(nT) + \beta v(nT)\} z^{-n} = \alpha \sum_{n=0}^{\infty} u(nT) z^{-n} + \beta \sum_{n=0}^{\infty} v(nT) z^{-n} = \alpha U(z) + \beta V(z),$$

which may be written

$$Z\{\alpha u(t) + \beta v(t)\} = \alpha Z\{u(t)\} + \beta z\{u(t)\}.$$

Result (ii) follows from (i) by setting $\beta = 0$.   □

Example 1. Some simple z-transforms

Find the z-transform in each of the following cases.

(i) The **unit pulse** $\Delta(n)$. This is the discrete analog of the delta function $\delta(t)$, and is defined as

$$\Delta(n) \equiv \{u_n\}_{n=0}^{\infty}, \quad \text{with} \quad u_n = \begin{cases} 1, & n = 0 \\ 0, & n \neq 0. \end{cases}$$

(ii) The **delayed unit pulse** $\Delta(n-k)$. This is the discrete analog of the delta function $\delta(t-a)$, and is defined as

$$\Delta(n-k) \equiv \{u_n\}_{n=0}^{\infty}, \quad \text{with} \quad u_n = \begin{cases} 1, & n = k \\ 0, & n \neq k, \text{ with } k > 0 \text{ integral.} \end{cases}$$

(iii) The **geometric sequence of pulses** is defined as the sequence

$$\{u_n\}_{n=0}^{\infty}, \quad \text{with} \quad u_n = a^n.$$

This sequence may also be considered to be derived from the function $u(t) = a^{t/T}$ by sampling with a sampling interval T.

(iv) The **unit step function**

$$\mathscr{U}(t) = \begin{cases} 0, & t < 0 \\ 1, & t \geq 0. \end{cases}$$

Solutions

(i) A direct application of the definition yields

$$U(z) = 1, \quad \text{independently of } z.$$

(ii) A further application of the definition gives

$$U(z) = \frac{1}{z^k}, \quad \text{for } z \neq 0.$$

(iii) By definition

$$U(z) = \sum_{n=0}^{\infty} a^n z^{-n} = 1 + \frac{a}{z} + \left(\frac{a}{z}\right)^2 + \dots .$$

The infinite series on the right-hand side is a geometric series with common ratio a/z and initial term unity, so its sum is

$$U(z) = \frac{1}{1 - a/z} = \frac{z}{z - a}, \quad \text{for } |z/a| > 1.$$

(iv) In this case a *function is* defined which must be sampled to obtain its z-transform. Sampling the unit step function $\mathscr{U}(t)$ with any sampling interval T leads to the sequence

$$\mathscr{U}(nT) = 1, \quad \text{for } n = 0, 1, 2, \dots .$$

Thus, by definition,

$$Z\{\mathscr{U}(nT)\} = U(z) = \sum_{n=0}^{\infty} z^{-n} = 1 + \frac{1}{z} + \frac{1}{z^2} + \dots .$$

Identification with (iii) above shows the closed form representation for $U(z)$ is

$$Z\{\mathscr{U}(t)\} = U(z) = \frac{z}{z - 1}, \quad \text{for } |z| > 1. \qquad \blacksquare$$

This example illustrates the fact that ingenuity is usually required if the z-transform of even a simple function (sequence) is to be converted into a closed form representation. The theorems which follow make this task easier. We shall discover that when a sequence may be obtained by sampling a function which has a Laplace transform, its z-transform can often be deduced directly from the Laplace transform with the help of partial fractions.

When the z-transform $U(s)$ of a sequence is known, the operation of recovering the original sequence $\{u(nT)\}_{n=0}^{\infty}$ from $U(z)$ is called finding the **inverse z-transform**. If the z-transform of the sequence $\{u(nT)\}_{n=0}^{\infty}$ exists, so that

$$Z\{u(nT)\} = U(z), \qquad (11)$$

then the inverse z-transform of $U(z)$ will be denoted by $Z^{-1}\{U(z)\}$, and we shall write

$$\{u(nT)\}_{n=0}^{\infty} = Z^{-1}\{U(z)\}. \qquad (12)$$

It must be emphasized that the inverse z-transform applied to $U(z)$ yields a sequence and *not* a function. Where a sampled function gave rise to the sequence, only its values at the sampling points can be recovered. As with the Laplace transform, the sequence $\{u(nT)\}_{n=0}^{\infty}$ and the function $U(z) = Z\{u(t)\}$ are called a **transform pair** or, to be more precise, a z-**transform pair.**

A short list of z-transform pairs is given in Table 7.2. For convenience the table also shows the corresponding Laplace transforms, the sampled value $u(nT)$ when a function $u(t)$ is involved, and the sequence value u_n when only a sequence is involved. Most of the entries can be derived by direct calculation, though some may be obtained more easily by means of the theorems given later in this section, which also enable the table to be extended.

Table 7.2 A short table of Laplace and z-transform pairs

Function $u(t)$ for $t \geq 0$ with $u(t) \equiv 0$ for $t < 0$	Sampled function $u(nT)$ or term u_n of sequence	Laplace transform $U(s) = \mathcal{L}\{u(t)\}$	z-transform $U(z) = Z\{u(t)\}$ with sampling interval T
1 $\delta(t)$	$\Delta(n) = \{u_n\}_{n=0}^{\infty}, \quad u_n = \begin{cases} 1, & n=0 \\ 0, & 0 \neq 0 \end{cases}$	1	1
2 $\delta(t-kT)$	$\Delta(n-k) = \{u_n\}_{n=0}^{\infty}, \quad u_n = \begin{cases} 1, & n=k \\ 0, & n \neq k \end{cases}$	e^{-ksT}	z^{-k}
3 $\mathcal{U}(t)$ (unit step function)	1	$\dfrac{1}{s}$	$\dfrac{z}{z-1}$
4 $a^{t/T}$	$u(nT) = a^n$	$\dfrac{1}{s-(1/T)\ln a} \quad (a>0)$	$\dfrac{z}{z-a}$
5 t	nT	$\dfrac{1}{s^2}$	$\dfrac{Tz}{(z-1)^2}$
6 t^2	$n^2 T^2$	$\dfrac{2}{s^3}$	$\dfrac{T^2 z(z+1)}{(z-1)^3}$

Table 7.2   (continued)

Function $u(t)$ for $t \geq 0$ with $u(t) \equiv 0$ for $t < 0$	Sampled function $u(nT)$ or term u_n of sequence	Laplace transform $U(s) = \mathscr{L}\{u(t)\}$	z-transform $U(z) = Z\{u(t)\}$ with sampling interval T
7 $ta^{(t/T-1)}$	nTa^{n-1}	$\dfrac{1}{(s-(1/T)\ln a)^2}$ $(a>0)$	$\dfrac{z}{(z-a)^2}$
8 e^{-at}	e^{-naT}	$\dfrac{1}{s+a}$	$\dfrac{z}{z-e^{-aT}}$
9 te^{-at}	nTe^{-naT}	$\dfrac{1}{(s+a)^2}$	$\dfrac{Tze^{-aT}}{(z-e^{-aT})^2}$
10 t^2e^{-at}	$n^2T^2e^{-naT}$	$\dfrac{2}{(s+a)^3}$	$\dfrac{T^2z(z+e^{-aT})e^{-aT}}{(z-e^{-aT})^3}$
11 $\sin kt$	$\sin knT$	$\dfrac{k}{s^2+k^2}$	$\dfrac{z\sin kT}{z^2-2z\cos kT+1}$
12 $\cos kt$	$\cos knT$	$\dfrac{s}{s^2+k^2}$	$\dfrac{z(z-\cos kT)}{z^2-2z\cos kT+1}$
13 $e^{-at}\sin kt$	$e^{-naT}\sin knT$	$\dfrac{k}{(s+a)^2+k^2}$	$\dfrac{ze^{-aT}\sin kT}{z^2-2ze^{-aT}\cos kT+e^{-2aT}}$
14 $e^{-at}\cos kt$	$e^{-naT}\cos knT$	$\dfrac{s+a}{(s+a)^2+k^2}$	$\dfrac{z(z-e^{-aT}\cos kT)}{z^2-2ze^{-aT}\cos kT+e^{-2aT}}$
15 $\sinh kt$	$\sinh nkT$	$\dfrac{k}{s^2-k^2}$	$\dfrac{z\sinh kT}{z^2-2z\cosh kT+1}$
16 $\cosh kt$	$\cosh nkT$	$\dfrac{s}{s^2-k^2}$	$\dfrac{z(z-\cosh kT)}{z^2-2z\cosh kT+1}$

Table 7.2 uses the usual convention that a transform of a function u is denoted by U. To signify which transform is involved, the variable s is shown as the argument of U when it is a Laplace transform, and the variable z when it is a z-transform. Although denoted by the same initial symbol U, the forms of the respective transforms of the same function $u(t)$ are different, depending on whether the argument is s or z. Thus in this notation $U(z)$ is not obtainable from $U(s)$ by merely replacing s by z.

Inverse z-transforms may be determined from Table 7.2 by entering it with $U(z)$ and finding the corresponding sequence $\{u(nT)\}_{n=0}$. In this process, simplification of $U(z)$ by means of partial fractions is often helpful, as are other methods which will be described later.

Example 2. A simple z-transform

Find the z-transform of the function

$$u(t) = (2 + 5t)e^{-3t}.$$

Solution

Writing

$$u(t) = 2e^{-3t} + 5te^{-3t},$$

it follows from Theorem 7.19(i) and entries 7 and 8 of Table 7.2, that

$$Z\{2e^{-3t} + 5te^{-3t}\} = \frac{2z}{z - e^{-3T}} + \frac{5Tze^{-3T}}{(z - e^{-3T})^2}$$

$$= \frac{2z^2 + z(5T - 2)e^{-3T}}{(z - e^{-3T})^2}. \qquad \blacksquare$$

Besides providing a direct listing of z-transform pairs for use as above when seeking the z-transform of a function, Table 7.2 may also be used in a variety of other ways. Some of these are illustrated by the next two examples. In the first of these the table is used to determine the z-transform of functions $u(t)$ whose Laplace transforms $U(s) = \mathscr{L}\{u(t)\}$ are known. This is accomplished by first simplifying the Laplace transform using partial fractions, and then entering Table 7.2 with each term in the partial fraction expansion to find the corresponding z-transform. The required z-transform $U(z) = Z\{u(t)\}$ follows by combining the individual z-transforms. The method is equivalent to finding $u(t) = \mathscr{L}^{-1}\{U(s)\}$, and then using Table 7.2 to determine $Z\{u(t)\}$. However the method is more powerful than this, because by allowing factors to be complex it enables complicated z-transforms to be deduced from simpler ones.

Example 3. z-transforms from Laplace transforms

(i) Find the z-transform of the function with the Laplace transform

$$U(s) = \frac{s^3 + 3s^2 + 12}{s^2(s^2 + 4)}.$$

(ii) Deduce $Z\{\cos kt\}$ from

$$\mathscr{L}\{\cos kt\} = \frac{s}{s^2 + k^2}.$$

Solutions

(i) A routine application of partial fractions gives

$$U(s) = \frac{3}{s^2} + \frac{s}{s^2 + 4}.$$

It then follows from entries 5 and 11 of Table 7.2 that the z-transform corresponding to $3/s^2$ is $3Tz(z-1)^{-2}$, while that corresponding to $s(s^2 + 4)^{-1}$ is $z(z - \cos 2T)(z^2 - 2z\cos 2T + 1)^{-1}$. The required z-transform is thus

$$U(z) = \frac{3Tz}{(z-1)^2} + \frac{z(z-\cos 2T)}{(z^2 - 2z\cos 2T + 1)}.$$

Inspection of this result shows that in fact $U(z) = Z\{3t + \cos 2t\}$.

(ii) Using partial fractions, and allowing complex numbers when factoring $s^2 + k^2$, gives

$$\mathcal{L}\{\cos kt\} = \frac{s}{s^2 + k^2} = \frac{1}{2}\left(\frac{1}{s+ik} + \frac{1}{s-ik}\right).$$

Thus entry 7 of Table 7.2 shows that

$$Z\{\cos kt\} = \frac{1}{2}\left(\frac{k}{z - e^{-ikT}} + \frac{k}{z - e^{ikT}}\right)$$

$$= \frac{z(z - \cos kT)}{z^2 - 2z\cos kT + 1}.$$

This approach is much simpler than attempting to evaluate $Z\{\cos kt\}$ directly from the definition of the z-transform. ∎

If the inverse z-transform of a complicated function $U(z)$ of z is required, it is natural to attempt to find it by first simplifying $U(z)$ by means of partial fractions, and then using Table 7.2. This approach will be successful provided the partial fraction expansion yields terms of the form found in the table, If, however, it produces terms like

$$\frac{a}{z+b} + \frac{c}{z+d} + \cdots$$

the method will fail, because these are not z-transforms of elementary functions. The problem arises because inspection of Table 7.2 shows that most z-transforms contain a factor z in their numerator. Thus when a conventional partial fraction expansion fails to produce such a factor the method must be modified to introduce one.

The simplest way of overcoming the difficulty is by expanding the function $U(z)/z$ in partial fractions (instead of $U(z)$), and then multiplying the result by z. Alternatively, if only the first few terms in the sequence $u(nT)$ (or u_n) are required, these may always be found by long division. When using this approach $U(z)$ is expanded as a power series in inverse powers of z for as many terms as necessary by means of long division. Then, because of the definition of $Z\{u(t)\}$, the terms of the sequence $u(0)$, $u(T)$, $U(2T)$, ... (equivalently $u_0, u_1, u_2, \ldots$) are simply the coefficients of $z^0, z^{-1}, z^{-2}, \ldots$, respectively.

Example 4. Inverse z-transforms

Find the inverse z-transforms of the following functions.

(i) $U(z) = \dfrac{6z}{2z - 1}$,   (ii) $U(z) = \dfrac{z}{6z^2 - 5z + 1}$,   (iii) $U(z) = \dfrac{1}{(z-1)(z-\frac{1}{2})}$,

(iv) $V(z) = \dfrac{z^2}{(z-1)(z-\frac{1}{2})}$.

Solutions

(i) Writing

$$U(z) = \frac{6z}{2z-1} = 3\left(\frac{z}{z-\frac{1}{2}}\right),$$

it follows directly from entry 4 of Table 7.2 and Theorem 7.19 (ii) (scaling) that

$$Z^{-1}\left\{\frac{6z}{2z-1}\right\} = 3\{(\tfrac{1}{2})^n\}_{n=0}^{\infty},$$

so

$$u(nT) = 3(\tfrac{1}{2})^n, \qquad \text{for } n=0, 1, 2, \ldots, \quad \text{and all } T.$$

Had long division been used it would have led to the result

$$2z-1 \overline{\left)3 + \tfrac{3}{2}z^{-1} + \tfrac{3}{4}z^{-2} + \ldots\right.}$$
$$\begin{array}{r} 6z \\ 6z - 3 \\ \hline 3 \\ 3 - \tfrac{3}{2}z^{-1} \\ \hline \tfrac{3}{2}z^{-1} \\ \tfrac{3}{2}z^{-1} - \tfrac{3}{4}z^{-2} \\ \hline \tfrac{3}{4}z^{-2} \\ \vdots \end{array}$$

Identification of the coefficients of $z^0, z^{-1}, z^{-2}, \ldots$ in the result of the long division shows $u(0)=3$, $u(T)=\tfrac{3}{2}$, $U(2T)=\tfrac{3}{4}, \ldots$, as found by the previous method.

(ii) Simplifying $U(z)$ by means of partial fractions in the usual way gives

$$U(z) = \frac{z}{6z^2 - 5z + 1} = \frac{z}{z-\frac{1}{2}} - \frac{z}{z-\frac{1}{3}}.$$

Using the linearity property of the z-transform (Theorem 7.19 (i)) together with entry 4 of Table 7.2 shows

$$Z^{-1}\left\{\frac{z}{6z^2-5z+1}\right\} = \{(\tfrac{1}{2})^n\}_{n=0}^{\infty} - \{(1/3)^n\}_{n=0}^{\infty} = \{(\tfrac{1}{2})^n - (1/3)^n\}_{n=0}^{\infty},$$

and so

$$u(nT) = (\tfrac{1}{2})^n - (\tfrac{1}{3})^n, \qquad \text{for } n=0, 1, 2, \ldots, \quad \text{and all } T.$$

(iii) In this case

$$U(z) = \frac{1}{(z-1)(z-\frac{1}{2})},$$

and simplification using partial fractions in the usual way gives

$$U(z) = \frac{2}{z-1} - \frac{2}{z-\frac{1}{2}},$$

which cannot be inverted by means of Table 7.2. Thus this is a case in which the modified approach outlined above must be used. Expanding $U(z)/z$ in partial fractions gives

$$\frac{U(z)}{z}=\frac{1}{z(z-1)(z-\frac{1}{2})}=\frac{2}{z}+\frac{2}{z-1}-\frac{4}{z-\frac{1}{2}},$$

so that

$$U(z)=2+\frac{2z}{z-1}-\frac{4z}{z-\frac{1}{2}}.$$

Using entries 1 and 4 of Table 7.2 shows

$$Z^{-1}\left\{\frac{1}{(z-1)(z-\frac{1}{2})}\right\}=2\Delta(0)+\{2\}_{n=0}^{\infty}-\{4(\tfrac{1}{2})^{n}\}_{n=0}^{\infty}.$$

Expanding the three sequences and combining corresponding terms gives

$$Z^{-1}\left\{\frac{1}{(z-1)(z-\frac{1}{2})}\right\}=\{2,\,0,\,0,\,0,\,0,\,\ldots\}+\{2,\,2,\,2,\,2,\,2,\,\ldots\}-\{4,\,2,\,1,\,\tfrac{1}{2},\,\tfrac{1}{4},\,\ldots\}$$

$$=\{0,\,0,\,1,\,\tfrac{3}{2},\,\tfrac{7}{4},\,\ldots\},$$

so that

$$u(nT)=\begin{cases}0, & \text{for } n=0,\,1\\ 2-4(\tfrac{1}{2})^{n}, & \text{for } n=2,\,3,\,\ldots, \quad \text{and all } T.\end{cases}$$

(iv) An ordinary partial fraction expansion of $V(z)$ would be of the form

$$V(z)=A+\frac{B}{z-1}+\frac{C}{z-\frac{1}{2}},$$

but as in (iii) above this is an inappropriate form for an inverse z-transform. Thus here also it is necessary to use the modified approach and to seek a partial fraction expansion of $V(z)/z$. A simple calculation establishes that

$$\frac{V(z)}{z}=\frac{z}{(z-1)(z-\frac{1}{2})}=\frac{2}{z-1}-\frac{1}{z-\frac{1}{2}},$$

so

$$V(z)=\frac{2z}{z-1}-\frac{z}{z-\frac{1}{2}}.$$

Then from entry 4 of Table 7.2 we have

$$Z^{-1}\left\{\frac{z^{2}}{(z-1)(z-\frac{1}{2})}\right\}=2-\{(\tfrac{1}{2})^{n}\}_{n=0}^{\infty}.$$

Thus in this case

$$v(nT)=2-(\tfrac{1}{2})^{n}, \quad \text{for } n=0,\,1,\,2,\,\ldots, \quad \text{and all } T. \qquad \blacksquare$$

As with the Laplace transform, properties such as linearity and scaling of the z-transform described by Theorem 7.19 are called **operational properties**. These are general properties possessed by the z-transform which are not dependent on the function $u(t)$ (or sequence $\{u_n\}_{n=0}^{\infty}$) being transformed. The most important operational properties of the z-transform are summarized below. The proofs are given later together with illustrative examples.

Summary of operational properties of the z-transform

1 Linearity and scaling of the z-transform

Let α, β be arbitrary constants, and $\{u(nT)\}_{n=0}^{\infty}$, $\{v(nT)\}_{n=0}^{\infty}$ two sequences such that $Z\{u(t)\}=U(z)$ and $Z\{v(t)\}=V(z)$ are convergent. Then,

(i) the z-transform is a linear operation:

$$Z\{\alpha u(t)+\beta v(t)\}=\alpha Z\{u(t)\}+\beta Z\{v(t)\},$$

and

(ii) scaling $u(nT)$ in the z-transform is a linear operation:

$$Z\{\alpha u(t)\}=\alpha Z\{u(t)\}.$$

2 Shifting

(i) Shifting to the right (time delay)

Let $K>0$ be an integer and T the sampling interval. Then multiplication of the z-transform of $u(t)$ by z^{-K} is equivalent in terms of t to shifting $u(t)$ to the right by an amount KT (delaying it by KT) to become $u(t-KT)$. Equivalently, if $U(z)=Z\{u(t)\}$, then

$$z^{-K}U(z)=Z\{u(t-KT)\}$$

and, conversely,

$$\{u[(n-K)T]\}_{n=0}^{\infty}=Z^{-1}\{z^{-K}U(z)\}.$$

(ii) Shifting to the left (time advance)

$$Z\{u(t+KT)\}=z^{K}U(z)-z^{K}u(0)-z^{K-1}u(T)-z^{K-2}u(2T)$$
$$-\ldots-zu[(K-1)T].$$

In particular,

$$Z\{u(t+T)\}=zU(z)-zu(0),$$
$$Z\{u(t+2T)\}=z^{2}U(z)-z^{2}u(0)-zu(T),$$
$$Z\{u(t+3T)\}=z^{3}U(z)-z^{3}u(0)-z^{2}u(T)-zu(2T).$$

3 Scaling $u(t)$ by $a^{t/T}$

Let $Z\{u(t)\} = U(t)$ and $a > 0$. Then if T is the sampling interval

$$Z\{a^{t/T}u(t)\} = U\left(\frac{z}{a}\right).$$

4 Multiplication of $u(t)$ by e^{-at}

If $Z\{u(t)\} = U(z)$, then if T is the sampling interval

$$Z\{e^{-at}u(t)\} = U(ze^{aT}).$$

5 Multiplication of $u(t)$ by t

If $Z\{u(t)\} = U(z)$, then if T is the sampling interval

$$Z\{tu(t)\} = -zT\frac{d}{dz}U(z).$$

6 Partial differentiation with respect to a parameter

Let $u(t, \alpha)$ be a function of t dependent on a parameter α, and let $Z\{u(t, \alpha)\} = U(z, \alpha)$ be its z-transform. Then,

$$Z\left\{\frac{\partial}{\partial\alpha}u(t, \alpha)\right\} = \frac{\partial}{\partial\alpha}U(z, \alpha).$$

7 Initial value theorem

If $Z\{u(t)\} = U(z)$, then

$$u(0) = \lim_{z \to \infty} U(z).$$

8 Final value theorem

Let $Z\{u(t)\} = U(z) = Q(z)/P(z)$, with $P(z)$ and $Q(z)$ polynomials in z. Then if the zeros of $P(z)$ lie within the unit circle $|z| = 1$ in the complex plane, or at $z = 1$,

$$\lim_{n \to \infty} u(nT) = \lim_{z \to 1}\left[\left(\frac{z-1}{z}\right)U(z)\right].$$

If the zeros of $P(z)$ lie within and on $|z| = 1$ with the zeros on $|z| = 1$ away from $z = 1$, $u(nT)$ will oscillate boundedly as $n \to \infty$. When any zeros of $P(z)$ lie outside $|z| = 1$, $u(nT)$ will increase without bound and may also oscillate as $n \to \infty$.

9 Discrete real variable convolution theorem

Let $u(t)$, $v(t)$ be real valued functions, with $u(t) = v(t) = 0$ for $t < 0$, such that $Z\{u(t)\} = U(z)$ and $Z\{v(t)\} = V(z)$. If T is the sampling interval and

$$U(z)V(z) = W(z) = Z\{\{w(kT)\}_{k=0}^{\infty}\},$$

then

$$U(z)V(z) = Z\left\{\left\{\sum_{n=0}^{k} u(nT)v([k-n]T)\right\}_{k=0}^{\infty}\right\},$$

and

$$w(kT) = \sum_{n=0}^{k} u(nT)v([k-n]T).$$

Throughout the remainder of this section the transform variable z will be considered to be a real variable. However in a more general approach to the z-transform it is better to allow z to be complex, and to use the powerful methods of complex analysis to determine both z-transforms and their inverses. In fact limited use of complex z has already been made in Ex. 3(ii).

The first theorem to be proved concerns the **shifting of a sequence**. This concept is best understood by analogy with the shifting of a function. A function $f(t)$ is said to undergo a shift $\xi > 0$ to the right if its graph is shifted uniformly to the right by the amount ξ. Similarly, it is said to undergo a shift $\eta > 0$ to the left if its graph is shifted uniformly to the left by an amount η. Thus $f(t - \xi)$ represents $f(t)$ shifted to the right by an amount ξ, while $f(t + \eta)$ represents $f(t)$ shifted to the left by an amount η.

If now t denotes the time, then $t - \xi$ is a **delayed time** with **time delay** ξ, and $t + \eta$ is an **advanced time** with **time advance** η. Thus if $K > 0$ is an integer and T is the sampling time interval, $f(t - KT)$ represents $f(t)$ delayed by K sampling time intervals; that is by a time delay KT. Correspondingly, $f(t + KT)$ represents $f(t)$ advanced by K sampling time intervals; that is by a time advance KT.

Now let $u(t)$ be a function which vanishes for $t < 0$, and consider the sequence

$$u_n = u(nT), \quad \text{for } n = 0, 1, 2, \ldots .$$

This function $u(t)$ is represented in Fig. 7.43(a) by the dashed line, and the sampled value by dots.

By analogy with the shifting of a function, $u[(n - K)T]$, for $n = 0, 1, 2, \ldots$, represents the *sequence* $u(nT)$ *shifted to the right* (delayed) by K sampling intervals. Similarly, $u[(n + K)T]$, for $n = 0, 1, 2 \ldots$, represents the *sequence* $u(nT)$ *shifted to the left* (advanced) by K sampling intervals.

Figure 7.43(b) shows the effect of a shift to the right (delay) by three sampling intervals (delay time $3T$). Figure 7.43(c) shows the effect of a shift to the left (a time advance), by two sampling intervals (time advance $2T$). As $u(t)$ vanishes for $t < 0$ the graph and sampled points in Fig. 7.43(b) are zero for $t < 3T$, while in Fig. 7.43(c) they are zero for $t < 0$.

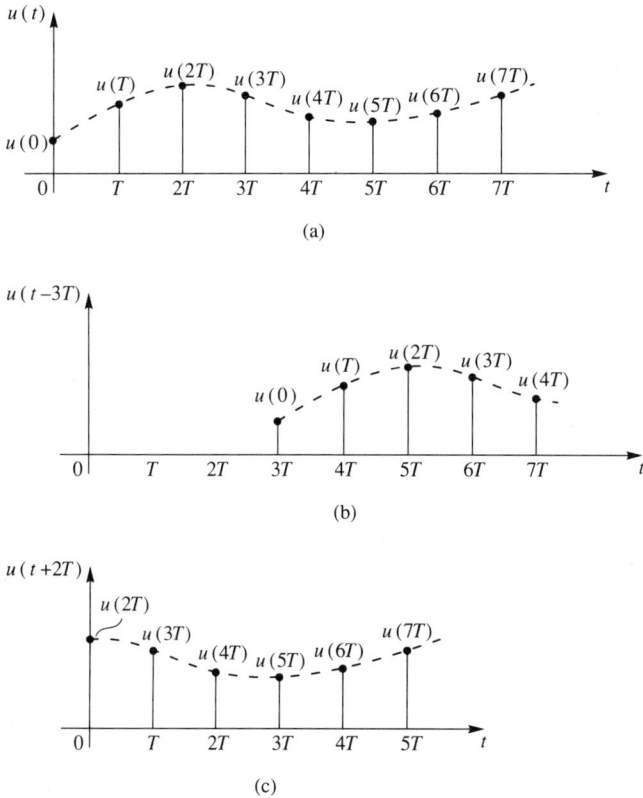

Fig. 7.43 (a) Original function and sampled values; (b) function and sampled values shifted to the right (delayed) by $3T$; (c) function and sampled values shifted to the left (advanced) by $2T$.

With these ideas in mind we now state and prove the translation theorem for the z-transform. The theorem takes different forms according as the translation is to the right (a time delay) or to the left (a time advance).

Theorem 7.20 (Shift theorem for the z-transform)

(i) Shifting to the right (time delay)

Let $K > 0$ be an integer and T the sampling interval. Then multiplication of the z-transform of $u(t)$ by z^{-K} is equivalent in terms of t to shifting $u(t)$ to the right by an amount KT (delaying it by KT) to become $u(t - KT)$. Equivalently, if $U(z) = Z\{u(t)\}$, then

$$z^{-K} U(z) = Z\{u(t - KT)\}$$

and, conversely,

$$\{u[(n - K)T]\}_{n=0}^{\infty} = Z^{-1}\{z^{-K} U(z)\}.$$

(ii) Shifting to the left (time advance)

$$Z\{u(t+KR)\} = z^K U(z) - z^K u(0) - z^{K-1} u(T) - z^{K-2} u(2T) - \ldots zu[(K-1)T].$$

In particular,

$$Z\{u(t+T)\} = zU(z) - zu(0),$$
$$Z\{u(t+2T)\} = z^2 U(z) - z^2 u(0) - zu(T),$$
$$Z\{u(t+3T)\} = z^3 U(z) - z^3 u(0) - z^2 u(T) - zu(2T).$$

Proof

(i) By definition

$$U(z) = Z\{u(t)\} = u(0) + \frac{u(T)}{z} + \frac{u(2T)}{z^2} + \cdots .$$

Thus

$$z^{-K} U(z) = \frac{u(0)}{z^K} + \frac{u(T)}{z^{K+1}} + \frac{u(2T)}{z^{K+2}} + \cdots .$$
$$= Z\{u(t-KT)\}.$$

The last result follows because the first K members of the sequence $\{u[(n-k)T]\}_{n=0}^{\infty}$ are all zero as $u(t)$ vanishes for $t < 0$.

(ii) By definition

$$Z\{u(t+KT)\} = \sum_{n=0}^{\infty} \frac{u[(n+K)T]}{z^n} = \sum_{n=0}^{\infty} \frac{u[(n+K)T]}{z^{n+K}} z^K$$

$$= z^K \left\{ \sum_{n=0}^{\infty} \frac{u(nT)}{z^n} - \sum_{n=0}^{K-1} \frac{u(nT)}{z^n} \right\}$$

$$= z^K U(z) - z^K \sum_{n=0}^{K-1} \frac{u(nT)}{z^n},$$

which is the required result. $\qquad\qquad\qquad\qquad\qquad\qquad\qquad\qquad\qquad\qquad$ □

Example 5. Shifting to right and left

(i) Use Theorem 7.20(i) to find $Z^{-1}\{U(z)\}$ given

$$U(z) = \frac{1}{(z-1)(z-\frac{1}{2})}.$$

(ii) Given $u(t) = te^{-2t}$, find $Z\{u(t+3T)\}$.

Solutions

(i) Write

$$U(z) = \frac{1}{(z-1)(z-\frac{1}{2})} = z^{-2}\left(\frac{z^2}{(z-1)(z-\frac{1}{2})}\right) = z^{-2} V(z),$$

where $V(z)$ is the function given in Ex. 4(iv). Then by Theorem 7.20(i)

$$Z\{u(t)\} = Z\{v(t-2T)\},$$

and so

$$u(nT) = v[(n-2)T], \qquad \text{for } n = 2, 3, 4, \cdots, \quad \text{with } v(-2T) = v(-T) = 0.$$

From Ex. 4(iv) we have

$$v(nT) = 2 - (\tfrac{1}{2})^n, \qquad \text{for } n = 0, 1, 2, \cdots, \quad \text{and all } T.$$

Thus $v[(n-2)T]$ follows from $v(nT)$ by replacing n by $n-2$, since the result is independent of T. Hence

$$u(nT) = v[(n-2)T] = 2 - (\tfrac{1}{2})^{n-2}$$
$$= 2 - 4(\tfrac{1}{2})^n, \qquad \text{for } n = 2, 3, 4, \cdots, \quad \text{and all } T,$$

with $u(0) = u(T) = 0$, because $v(-2T) = v(-T) = 0$.

This agrees with the result of Ex. 4(iii) in which this result was determined directly.

As T did not appear explicitly in the z-transform of $v(t)$ we could, had we wished, have set $T=1$ with the consequent simplification of $u(nT)$ to $u(n)$ and $v(nT)$ to $v(n)$.

(ii) From Theorem 7.20 (ii) we have

$$Z\{u(t+3T)\} = z^3 U(z) - z^3 u(0) - z^2 u(T) - zu(2T).$$

As $u(0) = 0$, $u(T) = Te^{-2T}$ and $u(2T) = 2Te^{-4T}$

it follows that

$$Z\{u(t+3T)\} = \frac{Tz^4 e^{-2T}}{(z - e^{-2T})^2} - z^2 Te^{-2T} - 2zTe^{-4T}.$$

∎

Theorem 7.21 (Scaling $u(t)$ by $a^{t/T}$)

Let $Z\{u(t)\} = U(t)$ and $z > 0$. Then

$$Z\{a^{t/T} u(t)\} = U\left(\frac{z}{a}\right).$$

Proof

By definition

$$Z\{a^{t/T} u(t)\} = a^0 u(0) + \frac{au(T)}{z} + \frac{a^2 u(2T)}{z^2} + \cdots = U\left(\frac{a}{z}\right).$$

□

Example 6

Use Theorem 7.21 to find $Z\{ta^{t/T}\}$.

Solution

From entry 5 of Table 7.1

$$Z\{t\} = \frac{Tz}{(z-1)^2} = U(z).$$

Thus from Theorem 7.21

$$Z\{ta^{t/T}\} = U\left(\frac{a}{z}\right) = \frac{T(a/z)}{(a/z-1)^2} = \frac{Taz}{(a-z)^2}.$$

This method is simpler than the direct approach. To see this use the definition to write

$$Z\{ta^{t/T}\} = \frac{Ta}{z} + \frac{2Ta^2}{z^2} + \frac{3Ta^3}{z^3} + \cdots$$

$$= \frac{Ta}{z}\left\{1 + 2\left(\frac{a}{z}\right) + 3\left(\frac{a}{z}\right)^2 + \cdots\right\}.$$

Recognizing that the series in the second factor is merely $\left(1 - \frac{z}{a}\right)^{-2}$ we obtain

$$U(z) = \frac{Ta}{z}\left(1 - \frac{z}{a}\right)^{-2},$$

which after simplification becomes the previous result. ∎

Theorem 7.22 (Multiplication of $u(t)$ by e^{-at})

If $Z\{u(t)\} = U(z)$,   then

$$Z\{e^{-at}u(t)\} = U(ze^{aT}).$$

Proof

By definition

$$Z\{e^{-at}u(t)\} = \sum_{n=0}^{\infty} e^{-naT}u(nT)z^{-n} = \sum_{n=0}^{\infty} u(nT)(ze^{aT})^{-n}.$$

$\square$

Example 7

Given $u(t) = t^2$ find $Z\{t^2 e^{-at}\}$.

Solution

From entry 6 of Table 7.2

$$Z\{t^2\}=\frac{T^2z(z+1)}{(z-1)^3},$$

so from Theorem 7.22

$$Z\{t^2e^{-at}\}=\frac{T^2ze^{aT}(ze^{aT}+1)}{(ze^{aT}-1)^3}=\frac{T^2z(z+e^{-aT})e^{-aT}}{(z-e^{-aT})^3}.$$

This is the simplest way to derive entry 9 of Table 7.2. ■

Theorem 7.23 (Multiplication of $u(t)$ by t)

If $Z\{u(t)\}=U(z)$, then

$$Z\{tu(t)\}=-zT\frac{d}{dz}U(z).$$

Proof

From the definition of $Z\{u(t)\}$ we have

$$-zT\frac{d}{dz}U(z)=-zT\frac{d}{dz}\left\{u(0)+\frac{u(T)}{z}+\frac{u(2T)}{z^2}+\cdots\right\}$$

$$=T\frac{u(T)}{z}+2T\frac{u(2T)}{z^2}+\cdots,$$

but the right-hand side is simply $Z\{tu(t)\}$. □

Example 8

Determine $Z\{t\sin kt\}$ from $Z\{\sin kt\}$.

Solution

From entry 10 of Table 7.2,

$$Z\{\sin kt\}=U(z)=\frac{z\sin kT}{z^2-2z\cos kT+1}.$$

Thus from Theorem 7.23,

$$Z\{t\sin kt\}=-zT\frac{d}{dz}\left\{\frac{z\sin kT}{z^2-2z\cos kT+1}\right\}$$

$$=\frac{Tz(z^2-1)\sin kT}{(z^2-2z\cos kT+1)^2}.$$

■

Theorem 7.24 (Partial differentiation with respect to a parameter)

Let $u(t, \alpha)$ be a function of t dependent on a parameter α, and let $Z\{u(t, \alpha)\} = U(z, \alpha)$ be its z-transform. Then,

$$Z\left\{\frac{\partial}{\partial\alpha} u(t, \alpha)\right\} = \frac{\partial}{\partial\alpha} U(z, \alpha).$$

Proof

The result follows directly from the definition of the z-transform because

$$Z\left\{\frac{\partial}{\partial a} u(t, \alpha)\right\} = \frac{\partial}{\partial\alpha} u(0, \alpha) + \frac{\partial}{\partial\alpha} u(T, \alpha)z^{-1} + \frac{\partial}{\partial\alpha} u(2T, \alpha)z^{-2} + \dots$$

$$= \frac{\partial}{\partial\alpha}\{u(0, \alpha) + u(T, \alpha)z^{-1} + u(2T, \alpha)z^{-2} + \dots\}$$

$$= \frac{\partial}{\partial\alpha} U(z, \alpha). \qquad\qquad \square$$

Example 9. Partial differentiation

Deduce $Z\{te^{-at}\}$ from $Z\{e^{-at}\}$ by means of Theorem 7.24.

Solution

$$Z\{e^{-at}\} = \frac{z}{z - e^{-aT}},$$

so by identifying α in Theorem 7.24 with a we have

$$Z\{-te^{-at}\} = \frac{\partial}{\partial a}\left(\frac{z}{z - e^{-aT}}\right)$$

$$= \frac{-Tze^{-aT}}{(z - e^{-aT})^2}$$

and so

$$Z\{te^{-aT}\} = \frac{Tze^{-aT}}{(z - e^{-aT})^2} \qquad\qquad \blacksquare$$

As with the Laplace transform, so also with the z-transform, initial and final value theorems may be derived enabling $u(0)$ and $\lim\limits_{n\to\infty} u(nT)$ to be deduced directly from $U(z)$.

Theorem 7.25 (Initial value theorem)

If $Z\{u(t)\} = U(z)$,   then

$$u(0) = \lim_{z\to\infty} U(z).$$

Proof

The result follows immediately from the definition of $Z\{u(t)\}$.

Example 10. Initial value theorem

Find $v(0)$ given $V(z) = \dfrac{z^2}{(z-1)(z-\frac{1}{2})}$.

Solution

Applying Theorem 7.25 we have

$$v(0) = \lim_{z \to \infty} V(z) = \lim_{z \to \infty} \frac{z^2}{(z-1)(z-\frac{1}{2})} = 1.$$

This agrees with the result of Ex. 4(iv) in which $Z^{-1}\{V(z)\}$ was found and $v(nT)$ shown to be

$$v(nT) = 2 - (\tfrac{1}{2})^n, \qquad \text{for } n = 0, 1, 2, \ldots . \qquad \blacksquare$$

Theorem 7.26 (Final value theorem)

Let $Z\{u(t)\} = U(z) = Q(z)/P(z)$, with $P(z)$ and $Q(z)$ polynomials in z. Then if the zeros of $P(z)$ lie within the unit circle $|z| = $ in the complex plane, or at $z = 1$,

$$\lim_{n \to \infty} u(nT) = \lim_{z \to 1} \left[\left(\frac{z-1}{z} \right) U(z) \right].$$

If the zeros of $P(z)$ lie within and on $|z| = 1$ with the zeros on $|z| = 1$ away from $z = 1$, $u(nT)$ will oscillate boundedly as $n \to \infty$. When any zeros of $P(z)$ lie outside $|z| = 1$, $u(nT)$ will increase without bound and may also oscillate as $n \to \infty$.

Proof

Combining (6), (9) and (10) gives

$$\mathscr{L}\{u_a(t)\} = \left(\frac{1 - e^{-sT}}{s} \right) U(z),$$

and as $z = e^{sT}$ this may be written

$$s\, \mathscr{L}\{u_a(t)\} = \left(\frac{z-1}{z} \right) U(z).$$

From the final value theorem for the Laplace transform (Theorem 7.18) we know that, subject to the conditions stated there,

$$\lim_{s \to 0} \left[s\, \mathscr{L}\{u_a(t)\} \right] = \lim_{t \to \infty} u_a(t) = \lim_{n \to \infty} u(nT).$$

However as $s \to 0$, so $z \to 1$, giving the required result

$$\lim_{n \to \infty} u(nT) = \lim_{z \to 1} \left[\left(\frac{z-1}{z} \right) U(z) \right].$$

The conditions on the location of the zeros of $P(z)$ follow from the conditions in Theorem 7.18 and the result from conformal mapping (see Volume 2) that $z = e^{sT}$ maps the left half of the z-plane onto the interior of the unit circle in the s-plane. ☐

Example 11. Final value theorem

Find $\lim_{n \to \infty} u(nT)$, given that $U(z) = \dfrac{Tze^{-aT}}{(z - e^{-aT})^2}$, with $u > 0$.

Solution

Applying Theorem 7.26 gives

$$\lim_{n \to \infty} u(nT) = \lim_{z \to 1} \left[\left(\frac{z-1}{z} \right) \frac{Tze^{-aT}}{(z - e^{-aT})^2} \right] = 0.$$

This is to be expected, because from entry 8 of Table 7.1 we see that $u(t) = te^{-at}$. The condition $a > 0$ is necessary, because if $a < 0$ the denominator of $U(z)$ will vanish for some $z > 1$ and a zero will lie outside the unit circle. ∎

The discrete summation

$$u(kT) * v(kT) = \sum_{n=0}^{k} u(nT)v([k-n]T) \tag{13}$$

is called the **discrete convolution** of the sampled functions $u(kT)$, $v(kT)$ in which T is the sampling interval. The discrete convolution is the analog of the convolution integral for functions of a continuous variable introduced in (7) of Sec. 7.2. The discrete convolution is used in the following theorem which is the z-transform analog of the convolution theorem for the Laplace transform (Theorem 7.14).

Theorem 7.27 (Discrete real variable convolution theorem)

Let $u(t)$, $v(t)$ be real valued functions, with $u(t) = v(t) = 0$ for $t < 0$, such that $Z\{u(t)\} = U(z)$ and $Z\{v(t)\} = V(z)$. If T is the sampling interval and

$$U(z)V(z) = W(z) = Z\{\{w(kT)\}_{k=0}^{\infty}\},$$

then

$$U(z)V(z) = Z\left\{ \left\{ \sum_{n=0}^{k} u(nT)v([k-n]T) \right\}_{k=0}^{\infty} \right\},$$

and

$$w(kT) = \sum_{n=0}^{k} u(nT)v([k-n]T).$$

Proof

By definition,

$$U(z) = u(0) + \frac{u(T)}{z} + \frac{u(2T)}{z^2} + \ldots ,$$

and

$$V(z) = v(0) + \frac{v(T)}{z} + \frac{v(2T)}{z^2} + \ldots .$$

Substituting these results into $W(z) = U(z)V(z)$ and grouping terms with the same inverse power of z gives

$$W(z) = u(0)v(0) + [u(0)v(T) + u(T)v(0)]z^{-1} + [u(0)v(2T) + u(T)v(t)$$

$$+ u(2T)v(0)]z^{-2} + \ldots + \left[\sum_{n=0}^{k} u(nT)v([k-n]T) \right] z^{-k} + \ldots , \qquad (14)$$

where the coefficient of z^{-k} in $W(z)$ is seen to be the discrete convolution $u(kT)*v(kT)$. Thus

$$W(z) = U(z)V(z) = Z\{\{u(kT)*v(kT)\}_{k=0}^{\infty}\},$$

which was the first result to be proved.

As, by definition,

$$W(z) = Z\{\{w(kT)\}_{k=0}^{\infty}\} = w(0) + w(T)z^{-1} + w(2T)z^{-2} + \ldots$$

$$+ w(kT)z^{-k} + \ldots , \qquad (15)$$

identification of the coefficients of z^{-k} in (14) and (15) shows $w(kT) = u(kT)*v(kT)$, which was the second result to be proved. $\qquad\qquad\qquad\qquad\qquad\qquad\qquad\qquad\Box$

Example 12. Discrete convolution

Find the discrete convolutions $w(kT) = u(kT)*v(kT)$, given that

k	$u(kT)$	$v(kT)$
0	1	4
1	2	2
2	3	1
3	2	0
4	1	0

and $u(kT) = 0$ for $k > 4$, $v(kT) = 0$ for $k > 2$.

Solution

$$w(0) = u(0)v(0) = 1\cdot4 = 4$$
$$w(T) = u(0)v(T) + u(T)v(0) = 1\cdot2 + 2\cdot4 = 10$$
$$w(2T) = u(0)v(2T) + u(T)v(T) + u(2T)v(0)$$
$$= 1\cdot1 + 2\cdot2 + 3\cdot4 = 17$$
$$w(3T) = u(0)v(3T) + u(T)v(2T) + u(2T)v(T) + u(3T)v(0)$$
$$= 1\cdot0 + 2\cdot1 + 3\cdot2 + 2\cdot4 = 16$$
$$w(4T) = u(0)v(4T) + u(T)v(3T) + u(2T)v(2T) + u(3T)v(T) + u(4T)v(0)$$
$$= 1\cdot0 + 2\cdot0 + 3\cdot1 + 2\cdot2 + 1\cdot4 = 11$$
$$w(5T) = u(0)v(5T) + u(T)v(4T) + u(2T)v(3T) + u(3T)v(2T) + u(4T)v(T) + u(5T)v(0)$$
$$= 1\cdot0 + 2\cdot0 + 3\cdot0 + 2\cdot1 + 1\cdot2 + 10\cdot4 = 4$$
$$w(6T) = u(0)v(6T) + u(T)v(5T) + u(2T)v(4T) + u(3T)v(3T) + u(4T)v(2T) + u(5T)v(1) + u(6T)v(0)$$
$$= 1\cdot0 + 2\cdot0 + 3\cdot0 + 2\cdot0 + 1\cdot1 + 0\cdot2 + 0\cdot4 = 1$$

and

$$w(nT) = 0 \quad \text{for } n > 6. \qquad \blacksquare$$

Example 13. Discrete convolution theorem

Find $W(z) = Z\left\{\left\{\sum_{n=0}^{k} u(nT)v([k-n]T)\right\}_{k=0}^{\infty}\right\}$ given $u(t) = t$ and $v(t) = e^{-at}$, and determine $w(kT)$.

Solution

From Table 7.2

$$U(z) = Z\{u(t)\} = Z\{t\} = \frac{Tz}{(z-1)^2}$$

and

$$V(z) = Z\{v(t)\} = Z\{e^{-at}\} = \frac{z}{z - e^{-at}},$$

where T is the sampling interval. Thus from the discrete convolution theorem (Theorem 7.27)

$$W(z) = U(z)V(z) = \frac{Tz^2}{(z-1)^2(z - e^{-aT})},$$

The coefficient of z^{-k} in the expansion of $W(z)$ is $w(kT)$, so

$$w(kT) = u(kT) * v(kT) = \sum_{n=0}^{k} nT \exp[-(k-n)aT]. \qquad \blacksquare$$

Problems for Section 7.4

Determine from first principles the z-transform $U(z)$ of the following sequences and functions. Where necessary take the sampling interval as T.

1 $\{u_n\}_{n=0}^{\infty}$,   given $u_n = 1/n!$

2 $\{u_n\}_{n=0}^{\infty}$,   given $u_n = \begin{cases} n, & 0 \le n \le 4 \\ 0, & n > 4 \end{cases}$.

3 $\{u_n\}_{n=0}^{\infty}$,   given $u_n = \begin{cases} 1+n^2, & \text{even } n \\ 1, & \text{odd } n \end{cases}$.

4 $\{u_n\}_{n}^{\infty}$,   given $u_n = (-1)^{n+2}\left(\dfrac{n+2}{n+1}\right)$.

5 $u(t) = e^{-at}$.

6 $u(t) = at$

7 $u(t) = \cos\left(\dfrac{\pi t}{T}\right)$.

8 $u(t) = \sin\left(\dfrac{\pi t}{2T}\right)$.

9 $u(t) = 2\,\mathscr{U}(t) + \cos\left(\dfrac{\pi t}{T}\right)$.

10 $u(t) = \mathscr{U}(t) + \sin\left(\dfrac{\pi t}{2T}\right)$.

11 Use the definition of sinh kt with $Z\{e^{kt}\}$ and $Z[e^{-kt}]$ to show

$$Z\{\sinh kt\} = \frac{z \sinh kT}{z^2 - 2z \cosh kT + 1}.$$

12 Use the definition of cosh kt with $Z\{e^{kt}\}$ and $Z\{e^{-kt}\}$ to show

$$Z\{\cosh kt\} = \frac{z(z - \cosh kT)}{z^2 - 2z \cosh kt + 1}.$$

In the following problems use Table 7.2 to determine the z-transform $U(z)$ of the given function $u(t)$.

13 $u(t) = \sin t \cos t$ **14** $u(t) = \cos^2 t$ **15** $u(t) = \sin^2 t$

16 $u(t) = \sinh t \cosh t$ **17** $u(t) = \cosh^2 t$ **18** $u(t) = \sinh^2 t$

19 Deduce the z-transform $U(z)$ of the function whose Laplace transform is

$$U(s) = \frac{2s + 7}{s^2 + 5s + 6}.$$

20 Deduce the z-transform $U(z) = Z\{\sin kt\}$ from the Laplace transform

$$\mathscr{L}\{\sin kt\} = \frac{k}{s^2 + k^2}.$$

21 Deduce the z-transform $U(z)$ of the function whose Laplace transform is

$$U(s) = \frac{2s^2 + k^2}{s(s^2 + k^2)}.$$

22 Deduce the z-transform $U(z) = Z\{\sinh kt\}$ from the Laplace transform

$$\mathscr{L}\{\sinh kt\} = \frac{k}{s^2 - k^2}.$$

Find the general term u_n in the inverse z-transforms of the following functions.

23 $U(z) = \dfrac{12z}{3z - 1}$ **24** $U(z) = \dfrac{18z - 8}{6z^2 - 5z + 1}$ **25** $U(z) = \dfrac{z^2 - z(1 + 2T)e^{-aT}}{(z - e^{-aT})^2}$

26 $U(z) = \dfrac{3z(z - \sqrt{2})}{z^2 - 2\sqrt{2}z + 4}$ **27** $U(z) = \dfrac{z}{z^2 - 3\sqrt{3}z + 9}.$

28 $U(z) = \dfrac{z^2}{z^2 - \sqrt{6}z + 3}$ **29** $U(z) = \dfrac{z^2 - 4\sqrt{2}z}{z^2 - 4\sqrt{2}z + 16}.$

In the following problems use long division to verify that the coefficients $u(0), u(T), u(2T)$ and $u(3T)$ in the expansion of $U(z)$ are as indicated.

30 $U(z) = \dfrac{Tze^{-aT}}{(z - e^{-aT})^2};$ $u(0) = 0,$ $u(T) = Te^{-aT},$ $u(2T) = 2Te^{-2aT},$ $u(3T) = 3Te^{-3aT}.$

31 $U(z) = \dfrac{T^2 z(z + e^{-aT})e^{-aT}}{(z - e^{-aT})^3};$ $u(0) = 0,$ $u(T) = T^2 e^{-aT},$

$$u(2T) = 4T^2 e^{-2aT}, \quad u(3T) = 9T^2 e^{-3aT}.$$

32 $U(z) = \dfrac{z \sinh kT}{z^2 - 2z \cosh kT + 1},$ $u(0) = 0,$ $u(T) = \sinh kT,$

$$u(2T) = \sinh 2kT,$$

$$u(3T) = \sinh 3kT.$$

33 Use Theorem 7.20(i) to find $Z^{-1}\{U(z)\}$ given

$$U(z) = \frac{1}{(z - 1)(z - 2)}.$$

34 Use Theorem 7.20 (i) to find $Z^{-1}\{U(z)\}$ given

$$U(z) = \frac{1}{(z - 2)(z - 3)}.$$

35 Given $u(t) = \cosh kt$, use Theorem 7.20(ii) to find $Z\{u(t + 2T)\}$.
36 Given $u(t) = t^2 e^{-at}$, use Theorem 7.20(ii) to find $Z\{u(t + 3T)\}$.
37 Find $Z\{e^{-2t} 3^{t/T}\}$.
38 Find $Z\{2^{t/T} \sin 3t\}$.
39 Use Theorem 7.22 to deduce $Z\{e^{-at} \sin kt\}$ from entry 11 in Table 7.2.
40 Use Theorem 7.22 to deduce $Z\{e^{-at} \cos kt\}$ from entry 12 in Table 7.2.
41 Use Theorem 7.23 to deduce $Z\{t \sinh t\}$ from $Z\{\sinh t\}$.

42 Use Theorem 7.23 to show

$$Z\{t^3\} = \frac{T^3 z(z^3 + 4z + 1)}{(z-1)^4}.$$

43 Use Theorem 7.24 to deduce $Z\{t \cos kt\}$ from $Z\{\sin kt\}$.

44 Use Theorem 7.24 to deduce $Z\{t^2 e^{-at}\}$ from $Z\{te^{-at}\}$.

45 Find $u(0)$ given

$$U(z) = \frac{z^2}{(z-\frac{1}{2})(3z-1)}.$$

46 Find $u(0)$ given

$$U(z) = \frac{(z^2-1)(2z-7)}{(3z-2)(z^2-2z+1)}.$$

Where appropriate in the following problems find $\lim_{n \to \infty} u(nT)$ from the given transform $U(z)$.

47 $U(z) = \dfrac{z^3+z+1}{z^2(z-\frac{1}{2})(z-1)}$

48 $U(z) = \dfrac{z^3+z+1}{z(z^2+\frac{1}{4})}.$

49 $U(z) = \dfrac{z^2+3z+1}{(z-\frac{1}{2})(z^2+1)}.$

50 $U(z) = \dfrac{z+3}{z^2-\sqrt{2}z+1}.$

51 Find the discrete convolutions $w(kT) = u(kT)*v(kT)$, given that

k	$u(kT)$	$v(kT)$
0	1	1
1	1	1
2	1	1
3	-1	0
4	-1	0
5	-1	0

52 Find the discrete convolutions $w(kT) = u(kT)*v(kT)$, given that

k	$u(kT)$	$v(kT)$
0	1	0
1	1	0
2	1	1
3	0	2
4	0	2

In the following problems find $W(z) = Z\left\{\left\{\sum_{n=0}^{k} u(nT)v([k-n]T)\right\}_{k=0}^{\infty}\right\}$, given functions $u(t)$, $v(t)$, and determine $w(kT)$.

53 $u(t) = e^{-at}$, $v(t) = e^{at}$

54 $u(t) = \mathcal{U}(t)$, $v(t) = \mathcal{U}(t)$.

55 $u(t) = t$, $v(t) = t$.

56 $u(t) = \mathcal{U}(t)$, $v(t) = t$.

7.5   Applications of the z-transform

This section provides a brief introduction to some of the simplest applications of the z-transform.

(a) Linear difference equations

It will be recalled from Sec. 1.9, in which an algebraic approach to linear difference equations was outlined, that a **linear constant coefficient difference equation** for u_n is an expression of the form

$$u_{n+r} + a_1 u_{n+r-1} + a_2 u_{n+r-2} + \ldots + a_r u_n = f(n), \tag{1}$$

in which $a_1, a_2, \ldots, a_r$ ($\neq 0$) are real constants, and $f(n)$ is a function of n which is independent of u_n and $n = 0, 1, 2, \ldots$. This difference equation determines u_{n+r} in terms of $u_{n+r-1}, u_{n+r-2}, \ldots, u_n$, and the sequence $\{u_n\}_{n=0}^{\infty}$ is completely determined by (1) once the **initial conditions** $u_0, u_1, \ldots, u_{r-1}$ have been specified.

The **order** of a difference equation is the difference between the largest and smallest suffixes occurring in the equation, so (1) is of order r. Difference equation (1) is said to be **nonhomogeneous** when $f(n) \neq 0$; otherwise it will be said to be **homogeneous**.

To apply the z-transform to difference equation (1) it is necessary to interpret the equation as the following relationship between shifted sequences

$$\{u_{n+r}\}_{n=0}^{\infty} + a_1 \{u_{n+r-1}\}_{n=0}^{\infty} + a_2 \{u_{n+r-2}\}_{n=0}^{\infty} + \ldots + a_n \{u_n\}_{n=0}^{\infty} = \{f(n)\}_{n=0}^{\infty}. \tag{2}$$

When doing this, as sampling is not involved, it is convenient to set the sampling interval in the z-transform equal to unity, so that $T = 1$. The initial conditions then involve the specification of the first r numbers in the sequence $\{u_n\}_{n=0}^{\infty}$, namely the numbers $u_0, u_1, \ldots, u_{r-1}$ in (2). The linearity theorem (Theorem 7.19) and the translation theorem (Theorem 7.20) may then be used to derive the z-transform of the left-hand side of (2), while Table 7.2 must be used to determine the z-transform of the right-hand side.

Hereafter we shall refer to this process as taking the **z-transform of a difference equation**. Solving the resulting equation for

$$U(z) = Z\{\{u_n\}_{n=0}^{\infty}\},$$

and then taking the inverse z-transform, leads to the required sequence $\{u_n\}_{n=0}^{\infty}$, and hence to the form of its general term u_n.

Example 1. Fibonacci sequence

Find the general term u_n in the Fibonacci sequence $\{u_n\}_{n=0}^{\infty}$ generated by the difference equation

$$u_{n+2} - u_{n+1} - u_n = 0,$$

given $u_0 = u_1 = 1$.

Solution

This is a homogeneous difference equation of order 2.
Setting

$$U(z)=Z\{\{u_n\}_{n=0}^\infty\},$$

and using Theorems 7.19 and 7.20 to obtain the z-transform of the difference equation, we obtain

$$\underbrace{z^2 U - z^2 u_0 - z u_1}_{Z\{\{u_{n+2}\}_{n=0}^\infty\}} \quad - \quad \underbrace{(zU - z u_0)}_{Z\{\{u_{n+1}\}_{n=0}^\infty\}} \quad - \quad \underbrace{U}_{Z\{\{u_n\}_{n=0}^\infty\}} \quad = 0.$$

Incorporating the initial conditions $u_0 = u_1 = 1$ and solving for $U(z)$ shows that

$$U(z)=\frac{z^2}{z^2-z-1}.$$

The required sequence $\{u_n\}_{n=0}^\infty$ follows from this result by taking the inverse z-transform. Now this is a case in which, when finding the appropriate form of partial fraction expansion, it is necessary to use the modified method discussed in the previous section. Consequently we must expand $U(z)/z$ in partial fractions by writing

$$\frac{U(z)}{z}=\frac{z}{\left(z-\frac{1}{2}-\frac{\sqrt{5}}{2}\right)\left(z-\frac{1}{2}+\frac{\sqrt{5}}{2}\right)}=\frac{A}{\left(z-\frac{1}{2}-\frac{\sqrt{5}}{2}\right)}+\frac{B}{\left(z-\frac{1}{2}+\frac{\sqrt{5}}{2}\right)}.$$

A routine calculation then shows that

$$A=\frac{\sqrt{5}+1}{2\sqrt{5}}\quad\text{and}\quad B=\frac{\sqrt{5}-1}{2\sqrt{5}},$$

so after multiplication by z we have

$$U(z)=\left(\frac{\sqrt{5}+1}{2\sqrt{5}}\right)\frac{z}{\left(z-\frac{1}{2}-\frac{\sqrt{5}}{2}\right)}+\left(\frac{\sqrt{5}-1}{2\sqrt{5}}\right)\frac{z}{\left(z-\frac{1}{2}+\frac{\sqrt{5}}{2}\right)}.$$

Inverting this result using entry 4 of Table 7.2 gives for the general term of the Fibonacci sequence

$$u_n=\left(\frac{\sqrt{5}+1}{2\sqrt{5}}\right)\left(\frac{1}{2}+\frac{\sqrt{5}}{2}\right)^n+\left(\frac{\sqrt{5}-1}{2\sqrt{5}}\right)\left(\frac{1}{2}-\frac{\sqrt{5}}{2}\right)^n.$$

When simplified this becomes

$$u_n=\frac{1}{2^{n+1}\sqrt{5}}[(1+\sqrt{5})^{n+1}-(1-\sqrt{5})^{n+1}].$$

The first few numbers in the Fibonacci sequence are tabulated below.

n	0	1	2	3	4	5	6	7	8	9	10
u_n	1	1	2	3	5	8	13	21	34	55	89

∎

Example 2. Nonhomogeneous difference equation

Find the form of the general term u_n generated by the difference equation

$$u_{n+2} - 5u_{n+1} + 6u_n = 2^n,$$

given $u_0 = 1$, and $u_1 = 0$

Solution

This is a nonhomogeneous difference equation of order 2. Setting

$$U(z) = Z\{\{u_n\}_{n=0}^\infty\},$$

and taking the z-transform of the difference equation gives

$$\underbrace{z^2 U - z^2 u_0 - zu_1}_{Z\{\{u_{n+2}\}_{n=0}^\infty\}} \quad - \quad \underbrace{5(zU - zu_0)}_{Z\{\{u_{n+1}\}_{n=0}^\infty\}} \quad + \quad \underbrace{6U}_{Z\{\{u_n\}_{n=0}^\infty\}} = \underbrace{\frac{z}{z-2}}_{Z\{2^n\}}$$

in which the z-transform of the nonhomogeneous term follows from entry 4 of Table 7.2. Incorporating the initial conditions $u_0 = 1$, $u_1 = 0$ reduces this result to

$$U(z) = \frac{z^2 - 5z}{z^2 - 5z + 6} + \frac{z}{(z-2)(z^2 - 5z + 6)},$$

and so

$$\frac{U(z)}{z} = \frac{z - 5}{(z-2)(z-3)} + \frac{1}{(z-2)^2(z-3)}.$$

Making the usual partial fraction expansion of the terms on the right-hand side leads to

$$\frac{U(z)}{z} = \left(\frac{3}{z-2} - \frac{2}{z-3} - \frac{1}{z-2} - \frac{1}{(z-2)^2} + \frac{1}{z-3} \right).$$

Combining terms and multiplying by z gives

$$U(z) = 2\left(\frac{z}{z-2} \right) - \left(\frac{z}{z-3} \right) - \frac{z}{(z-2)^2}.$$

Using entry 4 of Table 7.2 and entry 7 with $T = 1$, $a = 2$ to arrive at the inverse z-transform shows the general term u_n in the sequence is

$$u_n = 2(2^n) - 3^n - n2^{n-1},$$

or

$$u_n = 2^{n+1} - 3^n - n2^{n-1}, \qquad \text{for } n = 0, 1, 2, \ldots$$ ∎

Example 3. Simultaneous difference equations

Find the general term in each of the sequences $\{u_n\}_{n=0}^{\infty}$ and $\{v_n\}_{n=0}^{\infty}$ generated by the simultaneous difference equations

$$u_{n+1} = 2v_n + 2$$
$$v_{n+1} = 2u_n - 1,$$

given the initial conditions $u_0 = v_0 = 0$.

Solution

These are simultaneous nonhomogeneous difference equations of order 1. Setting

$$U(z) = Z\{\{u_n\}_{n=0}^{\infty}\}, \quad V(z) = Z\{\{v_n\}_{n=0}^{\infty}\}$$

and taking the z-transform of the difference equations leads to the transformed equations

$$zU = 2V + \frac{2z}{z-1}$$

$$zV - 2U - \frac{z}{z-1}.$$

Eliminating V gives

$$U(z) = \frac{2z}{z^2 - 4}.$$

Expanding $U(z)/z$ in partial fractions we obtain

$$\frac{U(z)}{z} = \frac{2}{(z-2)(z+2)} = \frac{1}{2}\left(\frac{1}{z-2}\right) - \frac{1}{2}\left(\frac{1}{z+2}\right),$$

and thus

$$U(z) = \frac{1}{2}\left(\frac{z}{z-2}\right) - \frac{1}{2}\left(\frac{z}{z+2}\right).$$

Using entry 4 of Table 7.2 to determine the inverse z-transform shows the general term u_n in sequence $\{u_n\}_{n=0}^{\infty}$ is given by

$$u_n = \tfrac{1}{2}(2)^n - \tfrac{1}{2}(-2)^n,$$

which is equivalent to

$$u_n = 2^{n-1}[1 + (-1)^{n+1}], \qquad \text{for } n = 0, 1, 2, \ldots$$

Using this expression for u_n in either of the difference equations shows that the general term v_n in the sequence $\{v_n\}_{n=0}^{\infty}$ is given by

$$v_n = 2^{n-1}[1+(-1)^n]-1, \qquad \text{for } n=0, 1, 2, \ldots.$$

Alternatively the general term v_n could have been obtained by solving the transformed equations for $V(z)$ and taking the inverse *z*-transform to arrive at v_n directly.   ∎

Example 4. Periodic solution

Find the form of the general term u_n generated by the difference equation

$$u_{n+2} - u_{n+1} + u_n = 0,$$

given

$$u_0 = 2, \quad \text{and} \quad u_1 = 3.$$

Solution

The *z*-transform of the difference equation gives

$$z^2 U - 2z^2 - 3z - zU + 2z + U = 0,$$

so

$$U(z) = \frac{2z^2 + z}{z^2 - z + 1}.$$

Inspection of Table 7.2 shows $U(z)$ to be a linear combination of entries 11 and 12 with $\cos kT = \frac{1}{2}$, corresponding to $kT = \pi/3$. To determine the precise linear combination we use the fact that $\sin kT = \sqrt{3}/2$ and set

$$\frac{2z^2 + z}{z^2 - z + 1} \equiv \frac{A(z^2 - \frac{1}{2}z)}{z^2 - z + 1} + \frac{B(\sqrt{3}/2)z}{z^2 - z + 1}.$$

Equating coefficients of z^2 and z in the numerators of this identity shows $A = 2$ and $B = 4/\sqrt{3}$, so

$$U(z) = 2\left(\frac{z^2 - \frac{1}{2}z}{z^2 - z + 1}\right) + \frac{4}{\sqrt{3}}\left(\frac{(\sqrt{3}/2)z}{z^2 - z + 1}\right),$$

and thus

$$u_n = 2\cos\frac{n\pi}{3} + \frac{4}{\sqrt{3}}\sin\frac{n\pi}{3}.   \qquad ∎$$

(b) The pulse transfer function

This section introduces the concept of a pulse transfer function, which is the discrete variable form of the transfer function for a continuous input introduced in Sec. 7.3(e). Our purpose will be to determine how a linear system described in terms of the Laplace transform by the transfer function $G(s)$ responds to a sampled data input.

We must first recall the definitions and properties of linear systems subjected to a continuous input introduced in Secs 7.3(c, e). In the case of a continuous input $y_I(t)$ and output $Y_O(t)$ (Fig. 7.44) the transfer function $G(s)$ of a linear system is defined as

$$G(s) = \frac{Y_O(s)}{Y_I(s)}, \tag{3}$$

where $Y_I(s) = \mathcal{L}\{Y_I(s)\}$ and $Y_O(s) = \mathcal{L}\{Y_O(t)\}$.

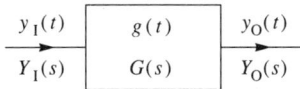

$y_I(t)$	$g(t)$	$y_O(t)$
$Y_I(s)$	$G(s)$	$Y_O(s)$

Fig. 7.44 Simple transfer function

In mathematical terms $G(s)$ is the Laplace transform of the response $g(t)$ of a system, initially at rest, to an input in the form of a unit impulse at time $t=0$, so $G(s) = \mathcal{L}\{g(t)\}$. Thus $g(t)$ is, in fact, the weighting function for the system. From (11) in Sec. 7.3 it then follows that

$$Y_O(t) = \int_0^t g(\tau) y_I(t-\tau) d\tau. \tag{4}$$

Now let the input to the system $y_I(t)$ be in the form of a train of pulses starting at $t=0$ and separated one from the other by a time interval T. If the pulse at time $t=mT$ (m an integer) has magnitude $y_I(mT)$, the input may be written in the form

$$y_I(t) = y_I(0)\delta(t) + y_I(T)\delta(t-T) + y_I(2T)\delta(t-2T) + \ldots + y_I(mT)\delta(t-mT) + \ldots \,. \tag{5}$$

Let k be any integer and t be such that $kT \le t < (k+1)T$. Then substitution of (18) into (17) followed by integration gives

$$y_O(t) = y_I(0)g(t) + y_I(T)g(t-T) + y_I(2T)g(t-2T) + \ldots + y_I(kT)g(0), \tag{6}$$

where use has been made of the sifting property of the delta function. Thus although the input $y_I(t)$ to the system is discrete, the output $y_O(t)$ is continuous, because the weighting function $g(t)$ is continuous and smooths out the pulses.

A sampled data input to the system has precisely the form given in (5), so (6) is the output from the system with transfer functions $G(s)$ when subjected to the sampled data input (5). If we now restrict consideration of the output $y_O(t)$ to the sampling times, setting $t = kT$ in (6) gives

$$y_O(kT) = y_I(0)g(kT) + y_I(T)g[(k-1)T] + y_I(2T)g[(k-2)T] \ldots + y_I(kT)g(0). \tag{7}$$

This shows that the output $y_O(kT)$ is the discrete convolution of the input and the weighting function, so we have the result

$$y_O(kT) = y_1(kT) * g(kT). \tag{8}$$

Taking the z-transform of (8) and using the discrete real variable convolution theorem (Theorem 7.27) we obtain

$$y_O(z) = y_1(z) G(z), \tag{9}$$

where

$$y_1(z) = Z\{\{y_1(kT)\}_{k=0}^\infty\}, \quad y_O(z) = Z\{\{y_O(kT)\}_{k=0}^\infty\}$$

and

$$G(z) = Z\{\{g(kT)\}_{k=0}^\infty\}.$$

Result (9) describes the relationship between the z-transforms of the input, the output and $G(z)$, which is called the **pulse transfer function**. So, as with the ordinary transfer function in (3), the pulse transfer function is expressible as

$$G(z) = \frac{y_O(z)}{y_1(z)}. \tag{10}$$

An obvious limitation of the z-transform is that although $y_O(t)$ in (6) is defined for all t, the z-transform can only determine the output $y_O(kT)$ at the sampling times. Because the application of the z-transform to the system in Fig. 7.44 automatically produces a sampled output, it has the effect of introducing a fictitious sampling device after $G(s)$ in such a way that both input and output sampling is synchronized. This is illustrated in Fig. 7.45

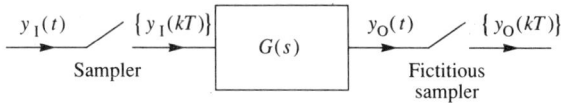

Fig. 7.45 Relationship between continuous input $y_1(t)$ and output $y_O(t)$, and the sampled input $\{y_1(kT)\}$ and sampled output $\{y_O(kT)\}$.

Example 5. Pulse transfer function

Find the pulse transfer function of the system described by the transfer function

$$G(s) = \frac{1}{(s+1)(s+3)}.$$

Solution

Using partial fractions we have

$$G(s) = \frac{1}{2}\left(\frac{1}{s+1}\right) - \frac{1}{2}\left(\frac{1}{s+3}\right).$$

Using Table 7.2 to determine the z-transforms corresponding to their Laplace transforms we find that

$$G(z) = \frac{1}{2}\left(\frac{z}{z-e^{-T}}\right) - \frac{1}{2}\left(\frac{z}{z-e^{-3T}}\right),$$

and thus

$$G(z) = \frac{z(e^{-T} - e^{-3T})}{2(z - e^{-T})(z - e^{-3T})}. \qquad \blacksquare$$

This approach extends directly to systems described by transfer functions in series. Suppose, as in Fig. 7.46(a), that the output of the first transfer function $G_1(s)$ feeds directly into the second transfer function $G_2(s)$ without further sampling. Then it follows directly from Sec. 7.3(e) that the transfer function $G_E(s)$ equivalent to the series coupling of $G_1(s)$ and $G_2(s)$ is

$$G_E(s) = G_1(s)\, G_2(s).$$

Setting $g_E(t) = \mathscr{L}^{-1}\{G_E(s)\} = \mathscr{L}^{-1}\{G_1(s)G_2(s)\}$, the pulse transfer function $G_E(z)$ for the system in Fig. 7.46(a) is seen to be

$$\frac{Y_0(z)}{Y_1(z)} = G_E(z) = Z\{g_E(t)\}. \qquad (11)$$

If, however, further sampling occurs between $G_1(s)$ and $G_2(s)$, as in Fig. 7.46(b), then when both samplers are synchronized we have

$$\frac{W(z)}{Y_1(z)} = G_1(z) \quad \text{and} \quad \frac{Y_0(z)}{W(z)} = G_2(z).$$

Thus the pulse transfer function $G_s(z)$ equivalent to $G_1(s)$ and $G_2(s)$ with intermediate sampling, as in Fig. 7.46(b), becomes

$$\frac{Y_0(z)}{Y_1(z)} = G_s(z) = G_1(z)\, G_2(z). \qquad (12)$$

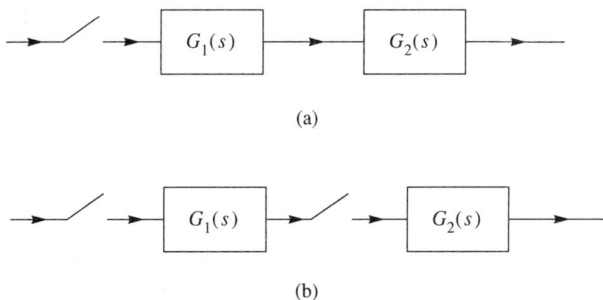

(a)

(b)

Fig. 7.46 Transfer functions in series: (a) without intermediate sampling, (b) with intermediate sampling

Notice that because of the introduction of sampling between $G_1(s)$ and $G_2(s)$ the pulse transfer functions $G_E(s)$ and $G_s(z)$ are different, so that

$$G_E(z) \neq G_s(z).$$

Example 6. Series transfer functions with intermediate sampling

Derive and compare the pulse transfer functions for the systems illustrated in Figs 7.47(a, b) in which the sampling interval is T.

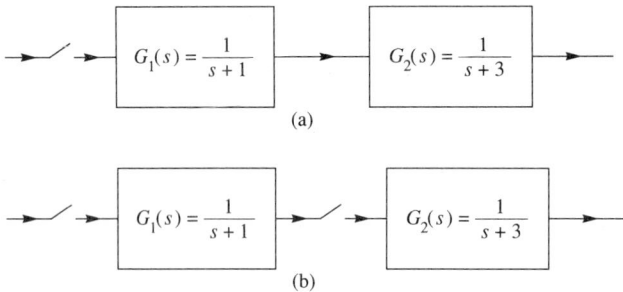

(a)

(b)

Fig. 7.47

Solutions

For the system without intermediate sampling shown in Fig. 7.36(a)

$$G_E(s) = G_1(s) G_2(s) = \frac{1}{(s+1)(s+2)}.$$

The corresponding pulse transfer function was found in Ex. 4, where it was shown that

$$G_E(z) = \frac{z(e^{-T} - e^{-3T})}{2(z - e^{-T})(z - e^{-3T})}.$$

For the system with intermediate sampling shown in Fig. 7.36(b),

$$\frac{Y_0(z)}{Y_1(z)} = G_s(z) = G_1(z) G_2(z) = \frac{z^2}{(z - e^{-T})(z - e^{-3T})}.$$

These two examples illustrate the fact that $G_E(z) \neq G_s(z)$. ∎

(c) Chebyshev polynomials

In numerical analysis, extensive use is made of a special class of polynomials called **Chebyshev polynomials.**[11] Some of the most important uses of these polynomials are to be

---

[11] PAFNUTI LIWOWICH CHEBYSHEV (1821–1894), a distinguished Russian mathematician who was Professor of Mathematics at the University of Petrograd (now Leningrad) and who made contributions to analysis and number theory. There are many variations of the spelling of his name, the most common being Chebichev and Tchebycheff.

found in approximation theory, interpolation and quadrature. It is usual to introduce Chebyshev polynomials through their defining differential equation, but here we shall adopt a different approach and start instead from a difference equation.

Consider the homogeneous second order difference equation

$$u_{n+2} - 2tu_{n+1} + u_n = 0 \tag{13}$$

in which the parameter $|t| \leq 1$, and the sequence $\{u_n\}_{n=0}^{\infty}$ generated by the difference equation when u_0 and u_1 are specified arbitrarily. Setting $U(z) = Z\{\{u_n\}_{n=0}^{\infty}\}$, and taking the z-transform of (13) we obtain

$$z^2 U - z^2 u_0 - zu_1 - 2t(zU - zu_0) + U = 0,$$

so that

$$U(z) = u_0\left(\frac{z^2}{z^2 - 2tz + 1}\right) + (u_1 - 2tu_0)\left(\frac{z}{z^2 - 2tz + 1}\right).$$

Rewriting this as

$$U(z) = u_0\left(\frac{z^2 - zt}{z^2 - 2tz + 1}\right) + \left(\frac{u_1 - tu_0}{(1 - t^2)^{1/2}}\right)\left(\frac{z(1 - t^2)^{1/2}}{z^2 - 2tz + 1}\right),$$

it is seen to be of the form

$$U(z) = A\left(\frac{z(z - t)}{z^2 - 2tz + 1}\right) + B\left(\frac{z(1 - t^2)^{1/2}}{z^2 - 2tz + 1}\right), \tag{14}$$

where we now consider A and B to be arbitrary constants (they depend on u_0 and u_1 which are themselves arbitrary).

Now as $|t| \leq 1$ we may set $t = \cos k$ in (14) and use entries 11 and 12 of Table 7.2 (with $T = 1$) to obtain the inverse z-transform of (14), and hence the general term of the sequence

$$u_n = A \cos nk + B \sin nk. \tag{15}$$

The functions $\cos nk$ and $\sin nk$ are linearly independent, so the general term of (13) is an arbitrary linear combination of these two functions, each of which is a solution of (13). Expressing u_n in terms of t by writing $k = \arccos t$ we have

$$u_n = A \cos (n \arccos t) + B \sin(n \arccos t). \tag{16}$$

The function

$$T_n(t) = \cos (n \arccos t) \tag{17}$$

occurring in (16) is called a **Chebyshev polynomial of the first kind of degree n**. That $T_n(t)$ is a polynomial in t of degree n follows from the fact that $T_n(t)$ is a particular solution of (13), so that it satisfies the relationship

$$T_{n+2}(t) - 2tT_{n+1}(t) + T_n(t) = 0, \tag{18}$$

while from (17) $T_0(t) = 1$ and $T_1(t) = t$. Using (18) repeatedly enables the generation of successive Chebyshev polynomials $T_n(t)$, starting from $T_0(t)$ and $T_1(t)$. The first eight Chebyshev polynomials $T_n(t)$ of the first kind are listed below:

$$
\begin{aligned}
T_0(t) &= 1, & T_4(t) &= 8t^4 - 8t^2 + 1, \\
T_1(t) &= t, & T_5(t) &= 16t^5 - 20t^3 + 5t, \\
T_2(t) &= 2t^2 - 1, & T_6(t) &= 32t^6 - 48t^4 + 18t^2 - 1, \\
T_3(t) &= 4t^3 - 3t, & T_7(t) &= 64t^7 - 112t^5 + 56t^3 - 7t.
\end{aligned}
\tag{19}
$$

Relationship (18) is called a recursion formula (recurrence relation) between the Chebyshev polynomials $T_n(t)$.

Differentiation of (17) shows that if $y = T_n(t)$,

$$
\frac{dy}{dt} = y' = \frac{n \sin nk}{\sin k}
$$

and

$$
\frac{d^2 y}{dt^2} = y'' = \frac{-n^2 \cos nk + n \sin nk \cot k}{\sin^2 k} = \frac{-n^2 y}{1 - t^2} + \frac{t y'}{1 - t^2},
$$

so $T_n(t)$ satisfies the differential equation

$$
(1 - t^2) y'' - t y' + n^2 y = 0. \tag{20}
$$

The other linearly independent function in (16) is $\sin(n \arccos t)$, and for convenience with the numbering of the polynomials we shall set

$$
Y_n(t) = \sin[(n+1) \arccos t], \qquad \text{for } n = 0, 1, 2, \ldots. \tag{21}
$$

The first two functions of t defined by $Y_n(t)$ are easily seen to be

$$
Y_0(t) = (1 - t^2)^{1/2} \qquad \text{and} \qquad Y_1(t) = 2t(1 - t^2)^{1/2}.
$$

Removing the factor $(1 - t^2)^{1/2}$ from $Y_n(t)$ to leave a polynomial in t we now define the new function

$$
U_n(t) = \frac{\sin[(n+1) \arccos t]}{(1 - t^2)^{1/2}} \tag{22}
$$

called a **Chebyshev polynomial of the second kind of degree n**. As $U_n(t)$ is also a solution of the difference equation (13) it must satisfy the recursion formula

$$
U_{n+2}(t) - 2t U_{n+1}(t) + U_n(t) = 0, \tag{23}
$$

with

$$
U_0(t) = 1 \qquad \text{and} \qquad U_1(t) = 2t. \tag{24}
$$

The function $y = U_n(t)$ is the other linearly independent solution of (20). The first eight Chebyshev polynomials $U_n(t)$ of the second kind calculated from (23) and (24) are listed below

$$U_0(t) = 1, \qquad\qquad U_4(t) = 16t^4 - 12t^2 + 1,$$
$$U_1(t) = 2t, \qquad\qquad U_5(t) = 32t^5 - 32t^3 + 6t,$$
$$U_2(t) = 4t^2 - 1, \qquad\quad U_6(t) = 64t^6 - 80t^4 + 24t - 1,$$
$$U_3(t) = 8t^3 - 4t, \qquad\quad U_7(t) = 128t^7 - 192t^5 + 80t^3 - 8t. \qquad (25)$$

The most important property possessed by Chebyshev polynomials is that they are **orthogonal polynomials** and, as such, belong to the general class of orthogonal functions to be discussed later. To exhibit this property we start from the two easily proved elementary definite integrals

$$\int_0^\pi \cos mk \cos nk \, dk = \begin{cases} 0 & \text{for } m \neq n \\ \frac{1}{2}\pi & \text{for } m = n \neq 0 \\ \pi & \text{for } m = n = 0, \end{cases} \qquad (26)$$

and

$$\int_0^\pi \sin nk \sin nk = \begin{cases} 0 & \text{for } m \neq n \\ \frac{1}{2}\pi & \text{for } m = n. \end{cases} \qquad (27)$$

Using (17) and setting $\cos mk = T_m(t)$, $\cos nk = T_n(t)$ and $dk = -(1-t^2)^{-1/2} \, dt$, it follows from (26) that

$$\int_{-1}^1 T_m(t) T_n(t) \frac{dt}{(1-t^2)^{1/2}} = \begin{cases} 0 & \text{for } m \neq n \\ \frac{1}{2}\pi & \text{for } m = n \neq 0 \\ \pi & \text{for } m = n = 0. \end{cases} \qquad (28)$$

This is the **orthogonality relationship** for $T_n(t)$ and it shows that the definite integral in (28) is zero unless $m = n$. The factor $W(t) = 1/(1-t^2)^{1/2}$ occurring in (28) is called the **weight function** for the polynomials $T_n(t)$.

A similar argument applied to (27) with the identifications

$$U_m(t)(1-t^2)^{1/2} = \sin([m+1]k),$$
$$U_n(t)(1-t^2)^{1/2} = \sin([n+1]k) \quad \text{and} \quad dk = -(1-t^2)^{-1/2} \, dt$$

shows that the **orthogonality relationship** for $U_n(t)$ is

$$\int_{-1}^1 U_m(t) U_n(t)(1-t^2)^{1/2} \, dt = \begin{cases} 0 & \text{for } m \neq n \\ \frac{1}{2}\pi & \text{for } m = n. \end{cases} \qquad (29)$$

Result (29) shows that the **weight function** for the polynomials $U_n(t)$ is $W(t) = (1-t^2)^{1/2}$.

Problems for Section 7.5

Section (a): Difference equations

In the following problems use the z-transform to find the general term in the sequence generated by the difference equation subject to the given initial condition.

1 $u_{n+1} = u_n + b,$ given $u_0 = a.$

2 $u_{n+1} = bu_n,$ given $u_0 = a.$

3 $u_{n+1} = bu_n + c,$ given $u_0 = a.$

4 Given $u_0 = a$ and $u_1 = b$, use the z-transform to show that the general term in the sequence generated by the difference equation

$$u_{n+2} - 2\alpha u_{n+1} + \alpha^2 u_n = 0$$

is

$$u_n = a(1-n)\alpha^n + bn\alpha^{n-1}.$$

In the following problems use the z-transform to find the general term in the sequence generated by the difference equation subject to the given initial conditions.

5 $u_{n+2} - 6u_{n+1} + 8u_n = 0,$ given $u_0 = 1,\ u_1 = 0.$

6 $u_{n+2} - 4u_{n+1} + 4u_n = 2,$ given $u_0 = 1,\ u_1 = 1.$

7 $u_{n+2} - 3u_{n+1} + 2u_n = 3^n,$ given $u_0 = 1,\ u_1 = 0.$

8 $u_{n+2} - 3u_{n+1} + 2u_n = 4^n,$ given $u_0 = 1,\ u_1 = 0.$

9 $u_{n+2} - \sqrt{2}u_{n+1} + u_n = 0,$ given $u_0 = 1,\ u_1 = 0.$

10 $u_{n+2} - \sqrt{2}u_{n+1} + u_n = 0,$ given $u_0 = 0,\ u_1 = 1.$

In the following problems use the z-transform to find the general terms in the sequences generated by the simultaneous difference equations subject to the given initial conditions.

11 $u_{n+1} = v_n + 2,\ v_{n+1} = 4u_n,$ given $u_0 = 1,\ v_0 = 0.$

12 $u_{n+1} = 3v_n + 1,\ v_{n+1} = 3u_n - 1,$ given $u_0 = v_0 = 0.$

Section (b): Pulse transfer function

In the following problems find the pulse transfer function $G(z)$ for the system described by the given transfer function $G(s)$ when the sampling interval is T. Use the result to determine the z-transform $Y_0(z)$ of the output $y_0(t)$ from the system when the time input to the system is the given function $y_1(t)$.

13 $G(s) = \dfrac{s^2 + 3s + 1}{s(s+1)(s+2)},$ $y_1(t) = \sin 2t.$

14 $G(s) = \dfrac{s+3}{(s+1)(s+4)},$ $y_1(t) = e^{-t}\cos t.$

In the following problems find the pulse transfer function for the series coupling with intermediate sampling (cf. Fig. 7.36(b)) of the given transfer functions $G_1(s)$ and $G_2(s)$ when the sampling interval is T.

15 $G_1(s) = \dfrac{1}{(s+1)^2}$ and $G_2(s) = \dfrac{s}{s^2+9}.$

16 $G_1(s) = \dfrac{1}{s^2+4s+5}$ and $G_2(s) = \dfrac{1}{s+3}.$

Chapter 8
Series Solution of Ordinary
Differential Equations

The solutions of linear variable coefficient differential equations can only be obtained in closed form in special cases. In general, it is necessary to construct the solutions of such differential equations in the form of a series. In the simplest case the solution can be developed as a Taylor series expansion about the point at which the initial data is specified. The terms in the expansion are obtained recursively by differentiating the differential equation to generate higher order derivatives, and then solving for these in terms of known lower order derivatives, starting from the initial conditions.

The Taylor series method fails if an expansion is required about a point at which the coefficient of the highest order derivative in the differential equation vanishes. This leads to the study of singular points of differential equations, and of the modified form taken by a series expansion about a regular singular point. The general method of approach used with second order differential equations is the Frobenius method. The type of series obtained when using this method depends on the roots of a quadratic equation called the indicial equation which is associated with the differential equation. Three cases arise, according as the roots are real and distinct and do not differ by an integer, they are real and distinct and do differ by an integer, or they are equal. Many of the higher functions used in applications are defined as a solution of one of these three types of differential equation. Among the simplest of these are the Legendre polynomials which are special solutions of the Legendre differential equation.

A function which is of considerable value in the development of series solutions, and also elsewhere, is the gamma function $\Gamma(x)$. The gamma function is defined in terms of an integral, and it is a generalization of factorial n to the case in which n is an arbitrary real number.

Bessel functions are specially useful in applications which involve some form of cylindrical symmetry. They are solutions of the Bessel differential equation which has a regular singular point at the origin. Like Legendre polynomials, Bessel functions have the property of being orthogonal functions, which is a concept that is developed in a systematic fashion in Chapter 9.

Special techniques called asymptotic methods are needed in order to study the behavior of solutions of differential equations or of functions defined by integrals for large values of their arguments. They take a variety of different forms, included among which is a class of methods due originally to Laplace, and an approach to oscillatory solutions called the WKBJ method.

A review of the most important ideas associated with series is contained in Sec. 8.1,

751

while the Taylor series approach is developed in Sec. 8.2. Ordinary points and singular points of a differential equation are introduced in Sec. 8.3, along with a systematic method of derivation of a series solution about an ordinary point. The series solution of the Legendre differential equation is discussed in Sec. 8.4, together with the special solutions called Legendre polynomials. These are polynomial solutions which remain finite at the singular points of the Legendre differential equation located at $x = \pm 1$, and they have valuable orthogonality properties.

The gamma function is defined and its properties are developed in Sec. 8.5. In Sec. 8.6 the Frobenius method and its extension are discussed at length, and applied to differential equations whose indicial equations have roots belonging to each of the three categories identified earlier. The Frobenius method is used again in Sec. 8.7 to develop Bessel functions.

Asymptotic expansions are introduced in Sec. 8.8 in which the Laplace and WKBJ methods are also developed for simple but useful cases. Finally, the numerical solution of initial value problems for second order differential equations by the Runge–Kutta method is described in Sec. 8.9.

8.1 Sequences, convergence and power series

The purpose of this section will be to summarize the most important properties of power series and the operations which may be performed on them. Then, in subsequent sections, these ideas will be used to solve linear variable coefficient ordinary differential equations.

As the notion of a sequence and its limit is fundamental to our discussion we begin by recalling some definitions from elementary calculus. A **sequence** is a set of numbers enumerated in a definite order. Thus the *same* set of numbers enumerated in a *different* order constitutes a *different* sequence.

Let the *i*th member of a sequence be denoted by s_i. Then the sequence s_1, s_2, $s_3, \ldots$, is called an **infinite sequence** if it contains an infinite number of terms; otherwise it is called a **finite sequence**. In the study of infinite series which is to follow, only infinite sequences are of importance.

Sequences are often represented by enumerating their elements in order, as above, by writing $s_1, s_2, \ldots$. Alternatively the condensed but explicit notation $\{s_n\}_{n=1}^N$ is used, in which the lower limit signifies the first and the upper limit the last number of the term in the sequence whose general term is s_n. The sequence will be an infinite sequence when N is infinite; otherwise it will be a finite sequence.

In the case of an infinite sequence the term with which the sequence begins is usually unimportant. Thus for convenience, unless otherwise stated, it will be assumed that a sequence begins with the term s_1. With this understanding, when using the condensed notation it is often convenient to omit the limits and to denote an infinite sequence by writing $\{s_n\}$.

The sequence $\{s_n\}$ is said to **converge** to the **limit** S if, for any $\varepsilon > 0$ (an arbitrarily small nonnegative number), there is some integer N such that

$$|s_n - S| < \varepsilon \quad \text{whenever} \quad n > N. \tag{1}$$

This definition has a useful geometrical interpretation if the terms of a sequence which converges to the limit S are represented by points on the real line. It asserts that however small we take an interval of length 2ε on the real line centered on S, there will always be an *infinite* number of points of the sequence inside the interval and a *finite* number of points outside it. Thus s_n provides an approximation to the limit S which it approaches arbitrarily closely as $n \to \infty$. This property of a limit of a sequence is illustrated diagrammatically in Fig. 8.1.

A connection between sequences and series becomes apparent, and a precise meaning can be given to the sum of infinitely many terms (numbers), if we consider the infinite

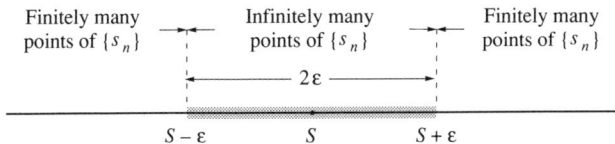

| Finitely many points of $\{s_n\}$ | Infinitely many points of $\{s_n\}$ | Finitely many points of $\{s_n\}$ |

$\longleftarrow$ 2ε $\longrightarrow$

$S - \varepsilon$ $\qquad$ S $\qquad$ $S + \varepsilon$

Fig. 8.1 The clustering of the points of the sequence $\{s_n\}$ in the interval $(S - \varepsilon, S + \varepsilon)$ containing the limit point S

series involving the real numbers $a_0, a_1, \ldots$, written

$$\sum_{n=0}^{\infty} a_n = a_0 + a_1 + \ldots, \tag{2}$$

and define its nth **partial sum** as

$$s_n = a_0 + a_1 + \ldots + a_n. \tag{3}$$

Thus s_n is the sum of all the terms in (1) up to and including the term a_n.

The infinite series (2) will be said to **converge** to the **sum** S if the sequence of partial sums $\{s_n\}$ has a limit and this limit is equal to S, so that

$$\sum_{n=0}^{\infty} a_n = \lim_{n \to \infty} s_n = S. \tag{4}$$

If the sequence $\{s_n\}$ has no limit the infinite series (1) will be said to **diverge**, or to be **divergent**. It must be stressed that although many tests for convergence are known they do *not*, as a rule, provide information about the sum of a series. A convergence test merely provides conditions under which a series is known to converge. If the actual sum of an infinite series is required it has to be determined by other means which usually involve numerical computation.

The numbers a_n comprising the terms of the infinite series (2) may be either positive or negative, and it is useful to associate with (2) the series of nonnegative terms

$$\sum_{n=0}^{\infty} |a_n| = |a_0| + |a_1| + \ldots.$$

If this series derived from (2) is convergent, the original series (2) is said to be **absolutely convergent**. It is immediately apparent that absolute convergence implies convergence, but the converse is not necessarily true. For example, the series $1 - \frac{1}{2} + \frac{1}{3} - \frac{1}{4} + \frac{1}{5} - \ldots$ is convergent to the sum $\ln 2$. However the series is not absolutely convergent because the harmonic series $1 + \frac{1}{2} + \frac{1}{3} + \frac{1}{4} + \frac{1}{5} + \ldots$ is easily shown to be divergent. Series such as this which are convergent, but not absolutely convergent, are said to be **conditionally convergent**.

The importance of absolute convergence will become apparent after these arguments have been extended to encompass power series. Suffice it to say here that a power series defines a function within an interval called its interval of convergence, and that within this interval the power series is absolutely convergent. This permits operations such as addition and differentiation which we may wish to perform on functions to be performed term by term on their power series instead. It also allows the order of terms in a power series to be interchanged without affecting its sum. Operations such as these are essential when using power series to solve ordinary differential equations and when making use of such solutions in applications.

The following examples of simple infinite series illustrate the meaning and use of the definitions given above. Expanding $(1-r)^{-1}$ by the binomial theorem we have

$$\frac{1}{1-r} = \sum_{n=0}^{\infty} r^n = 1 + r + r^2 + \ldots, \qquad \text{for } |r| < 1.$$

The infinite series on the right is, of course, the **geometric series** whose sum to infinity in the closed form $(1-r)^{-1}$ can be found by other means.

To show that this sum is in agreement with our definition of the sum of an infinite series we use the result known from elementary algebra that the nth partial sum is

$$s_n = 1 + r + r^2 + \dots + r^n = \frac{1-r^{n+1}}{1-r}.$$

The sequence $\{s_n\}$ has a limit provided $|r| < 1$, for then $\lim_{n \to \infty} r^n = 0$. Hence the series will be convergent to a sum S which from (3) is seen to be

$$S = \lim_{n \to \infty} \left(\frac{1-r^{n+1}}{1-r} \right) = \frac{1}{1-r} \qquad \text{for } |r| < 1,$$

thereby confirming the closed form expression.

As the series is convergent for $|r| < 1$, and not merely for $0 \le r < 1$, it follows directly that the geometric series is also absolutely convergent for $|r| < 1$.

An elementary example of a series which diverges, not because s_n grows without bound, but because s_n oscillates finitely as n becomes arbitrarily large, is provided by the infinite series

$$\sum_{n=0}^{\infty} (-1)^{n+1} = 1 - 1 + 1 - \dots.$$

In this case

$$s_n = 1 - 1 + 1 - \dots + (-1)^{n+1} = \begin{cases} 1, & n \text{ odd} \\ 0, & n \text{ even,} \end{cases}$$

for all n. Clearly $\{s_n\}$ has no limit[1], so the series is divergent.

For the sake of completeness, we mention in passing the four simplest and most useful tests for the convergence of series, all of which are proved in a first course on analysis. The ratio test and the nth root test are tests for absolute convergence, whereas the integral test applies only to series with positive terms, and so is simply a test for convergence. The alternating series is also only a test for convergence.

The ratio test

Let the series $\sum_{n=0}^{\infty} a_n$ be such that $a_n \ne 0$ and

$$\lim_{n \to \infty} \left| \frac{a_{n+1}}{a_n} \right| = L.$$

---

[1] The sequence $\{s_n\}$ has what are called **cluster points** or **points of accumulation** at 0 and 1. These are points such that any neighborhood of them contains infinitely many points of the sequence. In the example, $\{s_n\}$ has two cluster points so it cannot have a limit, because outside any neighborhood of a limit point there can only be a finite number of points of the sequence.

Then

   (a) the series Σa_n is absolutely convergent if $L<1$,
   (b) the series Σa_n diverges if $L>1$,
   (c) the test fails to distinguish between convergence and divergence if $L=1$, so another test must be used.

The *n*th root test

Let the series $\sum_{n=0}^{\infty} a_n$ be such that $a_n \neq 0$ and

$$\lim_{n \to \infty} |a_n|^{1/n} = L.$$

Then

   (a) the series Σa_n is absolutely convergent if $L<1$,
   (b) the series Σa_n diverges if $L>1$,
   (c) the test fails to distinguish between convergence and divergence if $L=1$, so another test must be used.

The integral test

Let $f(x)$ be a positive nonincreasing function defined for $1 \leq x < \infty$, and such that $\lim_{n \to \infty} f(x) = 0$. Then, if $a_n = f(n)$, the series of positive terms $\sum_{n=1}^{\infty} a_n$ converges, or diverges, according as the improper integral

$$\lim_{N \to \infty} \int_1^N f(x)\,dx$$

is finite, or infinite.

Alternating series test

Let the sequence of positive numbers $a_1, a_2, a_3, \ldots$ be such that (i) $a_n > a_{n+1}$ for $n = 1, 2, 3, \ldots$ and (ii) $\lim_{n \to \infty} a_n = 0$. Then the (so-called) alternating series

$$\sum_{n=1}^{\infty} (-1)^{n+1} a_n = a_1 - a_2 + a_3 - \ldots$$

formed from the sequence is convergent.

Furthermore, if the sum s of the series is approximated by the partial sum $s_N = a_1 - a_2 + a_3 - a_4 + \ldots + (-1)^{N+1} a_N$, the magnitude of the error is such that

$$|s - s_N| = \left| \sum_{n=N+1}^{\infty} (-1)^{n+1} a_n \right| < a_{N+1}.$$

Example 1. Testing for convergence

Investigate the convergence of

(a) $\displaystyle\sum_{n=0}^{\infty} \frac{(-1)^{n+1}}{(n+1)!}$,   (b) $\displaystyle\sum_{n=1}^{\infty} (-1)^n \left(\frac{nk+1}{4n+3} \right)^n$ with $k > 0$,   (c) $\displaystyle\sum_{n=1}^{\infty} \frac{1}{n^p}$ for $p > 0$,

(d) $\displaystyle\sum_{n=1}^{\infty} \frac{(-1)^{n+1}}{n}$.

Solution

(a) Setting $a_n = (-1)^{n+1}/(n+1)!$, it follows from the ratio test that

$$L = \lim_{n \to \infty} \left| \frac{a_{n+1}}{a_n} \right| = \lim_{n \to \infty} \left(\frac{(n+1)!}{(n+2)!} \right) = \lim_{n \to \infty} \left(\frac{1}{n+2} \right) = 0.$$

Thus, as $L = 0$, the series is absolutely convergent.

(b) Setting $a_n = (-1)^n [(nk+1)/(4n+3)]^n$, it follows from the nth root test that

$$L = \lim_{n \to \infty} |a_n|^{1/n} = \lim_{n \to \infty} \left[\left(\frac{nk+1}{4n+3} \right)^n \right]^{1/n} = \lim_{n \to \infty} \left(\frac{nk+1}{4n+3} \right) = \frac{k}{4}.$$

Thus the series is absolutely convergent if $0 < k < 4$, it is divergent if $k > 4$, and the test fails if $k = 4$.

(c) Setting $f(x) = 1/x^p$ we see that $f(x)$ satisfies the conditions of the integral test, so the test may be applied. Thus the series converges or diverges according as the improper integral

$$\int_1^{\infty} \frac{dx}{x^p}$$

is finite or infinite.

Routine integration shows the integral to be divergent for $0 < p < 1$ and convergent for $p > 1$. When $p = 1$

$$\int_1^{\infty} \frac{dx}{x} = \lim_{n \to \infty} \int_1^n \frac{dx}{x} = \lim_{n \to \infty} (\ln x) \Big|_1^n = \infty,$$

and hence the series is divergent for $p = 1$.

This example shows that the *harmonic series*

$$1 + \tfrac{1}{2} + \tfrac{1}{3} + \tfrac{1}{4} + \ldots$$

(corresponding to $p=1$) is divergent, whereas the series

$$1+\frac{1}{2^p}+\frac{1}{3^p}+\frac{1}{4^p}+ \ldots$$

is convergent for $p>1$.

(d) The series is convergent by the alternating series test. This follows by setting $a_n=1/n$ and noticing that $a_n>a_{n+1}$ for $n=1, 2, \ldots$, and $\lim_{n\to\infty} a_n=0$.

Suppose, for example, the sum s of the series is approximated by the partial sum

$$s_{10}=1-\tfrac{1}{2}+\tfrac{1}{3}-\tfrac{1}{4}+ \ldots -\tfrac{1}{10}=0.6456 \ldots .$$

It then follows from the last result stated in the test that the error made (the magnitude of the sum of the tail of the series) must be such that

$$|s-s_{10}|<\tfrac{1}{11}.$$

Thus

$$s_{10}-\tfrac{1}{11}<s<s_{10}+\tfrac{1}{11},$$

which provides the estimate

$$0.5547<s<0.7365.$$

This series converges very slowly to $\ln 2$, so it would not be sensible to attempt to compute $\ln 2$ by this means. In actual fact the sum of the series is $\ln 2 = 0.6931 \ldots .$ ∎

The last test we mention is a simple but very useful test for *divergence*.

*n*th term test for divergence

The series $\displaystyle\sum_{n=0}^{\infty} a_n$ is divergent if

$$\lim_{n\to\infty} a_n \neq 0.$$

Notice that although the failure of the nth term of a series to tend to zero as $n\to\infty$ will ensure its divergence, the fact that the nth term of a series tends to zero is *not* sufficient to ensure the convergence of the series. For example, the nth term of the harmonic series considered in Ex. 1(c) (corresponding to $p=1$) is $a_n=1/n$, so that $\lim_{n\to\infty} a_n=0$, yet the harmonic series is divergent.

The nth term test shows the series in Ex. 1(b) to be divergent when $k>4$, because then the magnitude of its nth term grows without bound as $n\to\infty$.

Power series, which are of prime concern to us in this chapter in connection with the solution of differential equations, are infinite series of the form

$$\sum_{n=0}^{\infty} a_n(x-x_0)^n=a_0+a_1(x-x_0)+a_2(x-x_0)^2+ \ldots , \qquad (5)$$

in which x is a real variable and a_n, x_0 are real numbers. The numbers a_n are called the **coefficients** of the power series, which is said to be *expanded* about the point x_0, or to have the *center* x_0. When x is given a specific value $x = a$ the series in (5) reduces to an ordinary infinite series of constant terms of the type just discussed which may, or may not, be convergent.

To consider the convergence of the power series in (5) we follow the previous form of argument and, as in (3), introduce its nth partial sum

$$s_n(x) = a_0 + a_1(x - x_0) + a_2(x - x_0)^2 + \ldots + a_n(x - x_0)^n, \tag{6}$$

which now depends on both n and x_0. Then for any $x = a$ the partial sums form a sequence $\{s_n(a)\}$. We shall say (5) *converges* to the sum $f(a)$ at $x = a$ if the sequence $\{s_n(a)\}$ has a limit and this limit is equal to $f(a)$.

Let f denote the **sum function** whose value for any x at which the series is convergent is determined by the sum of all the terms in (5). Then (5) may always be written as

$$f(x) = \sum_{n=0}^{\infty} a_n(x - x_0)^n = s_n(x) + R_n(x), \tag{7}$$

where

$$R_n(x) = a_{n+1}(x - x_0)^{n+1} + a_{n+2}(x - x_0)^{n+2} + \ldots \tag{8}$$

is the **remainder** after the nth partial sum, so

$$R_n(x) = f(x) - s_n(x). \tag{9}$$

Taking the absolute value of (9) when $x = a$, and using the definition of a limit in (1), shows that $\{s_n(a)\}$ will converge to $f(a)$ if for an arbitrarily small $\varepsilon > 0$ there is an integer N such that

$$|R_n(a)| = |s_n(a) - f(a)| < \varepsilon \qquad \text{when } n > N. \tag{10}$$

In general N will depend both on ε and a. Thus $|R_n(a)|$ provides a measure of the absolute error made when $f(a)$ is approximated by $s_n(a)$, and this error can be made arbitrarily small by taking n sufficiently large.

The interval on the real line containing the points x at which power series (5) is absolutely convergent, and hence convergent, is called the **interval of convergence** of the power series. By applying either the *ratio test* or the *nth root test* for absolute convergence to a power series this interval of length $2r$ may be shown to lie symmetrically about the point $x = x_0$, where $r \geq 0$. Provided the coefficients of the power series are nonvanishing, the number r, called the **radius of convergence** of the power series, is determined by using either of the expressions

$$\text{(i) } r = \lim_{n \to \infty} \left| \frac{a_n}{a_{n+1}} \right| \qquad \text{or} \qquad \text{(ii) } r = 1 \Big/ \left\{ \lim_{n \to \infty} |a_n|^{1/n} \right\}, \tag{11}$$

whenever these limits exist. It is to be understood that the radius of convergence r defined by (11)(ii) is infinite when the limit is zero, and zero when the limit is infinite. Outside the

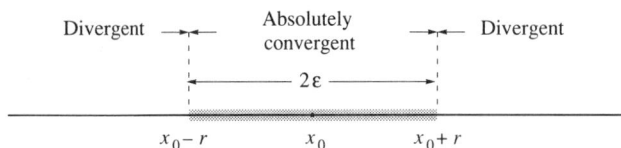

Fig. 8.2   Properties of the interval of convergence

interval of convergence the power series is divergent. A power series may, or may not, be convergent at the end points of its interval of convergence.[2] The properties of the interval of convergence are illustrated symbolically in Fig. 8.2.

Every power series has a radius of convergence, but its determination is likely to prove difficult when neither expression in (11) can be used to find it. This may happen because the limits involved are too complicated to be evaluated, or because coefficients of the power series vanish, rendering the results in (11) inapplicable without modification. However, the power series in (5) will always converge at $x = x_0$, for then the series reduces to the single term a_0. In special cases this might, of course, be the only point at which the power series converges.

A power series defines a function f, its **sum function**, which is the function to which it converges at each point x of its interval of convergence. Some elementary power series together with their radii and intervals of convergence are listed below. These results should be verified[3] by the reader. With the exception of the power series for $\ln(1 + x - a)$, which is expanded about $x = a$, all the other functions are expanded about the origin. When elementary functions exist to which these power series converge they are shown at the left.

$$\frac{1}{1-x} = \sum_{n=0}^{\infty} x^n = 1 + x + x^2 + \ldots ; \quad r = 1, \ -1 < x < 1$$

$$e^x = \sum_{n=0}^{\infty} \frac{x^n}{n!} = 1 + x + \frac{x^2}{2!} + \ldots ; \quad r = \infty, \ -\infty < x < \infty$$

$$\ln(1 + x - a) = \sum_{n=1}^{\infty} (-1)^{n+1} \frac{(x-a)^n}{n} = (x-a) - \frac{(x-a)^2}{2} + \frac{(x-a)^3}{3} - \ldots ;$$

$$r = 1, \ a - 1 < x < a + 1$$

---

[2] The full set of points at which a power series converges is called its **convergence set**, or sometimes its **domain of convergence**. This only differs from the interval of convergence by the inclusion of any end point at which the series is convergent.

[3] As even powers of x are missing in the power series for $\sin x$ the series must be modified before applying (11)(i) to find its radius of convergence. Set $u = x^2$, remove a factor x, and consider the power series in u

$$\sin x = x\left(1 - \frac{u}{3!} + \frac{u^2}{5!} - \ldots\right).$$

Similar approaches are needed when verifying the radius of convergence of the power series for $\cos x$ and arc tan x. The last result requires the application of (11)(ii) in order to determine its radius of convergence.

$$\sin x = \sum_{n=0}^{\infty} \frac{(-1)^n x^{2n+1}}{(2n+1)!} = x - \frac{x^3}{3!} + \frac{x^5}{5!} - \dots; \quad r = \infty, \ -\infty < x < \infty$$

$$\cos x = \sum_{n=0}^{\infty} \frac{(-1)^n x^{2n}}{(2n)!} = 1 - \frac{x^2}{2!} + \frac{x^4}{4!} - \dots; \quad r = \infty, \ -\infty < x < \infty$$

$$\arctan x = \sum_{n=0}^{\infty} \frac{(-1)^n x^{2n+1}}{2n+1} = x - \frac{x^3}{3} + \frac{x^5}{5} - \dots; \quad r = 1, \ -1 < x < 1$$

$$\sum_{n=1}^{\infty} (-1)^{n+1} \left(\frac{n^n}{(2n)^n + 1} \right) x^n = \left(\frac{1}{3} \right) x - \left(\frac{2^2}{(2 \cdot 2)^2 + 1} \right) x^2 + \left(\frac{3^3}{(2 \cdot 3)^3 + 1} \right) x^3 - \dots;$$

$$r = 2, \ -2 < x < 2.$$

Mathematical operations on power series

Listed below are the most important of the mathematical operations which it is permissible to perform on power series, either when using them to seek the solution of a differential equation, or when manipulating solutions obtained in the form of power series. In summary, the essential properties of power series are that, within the interval(s) of convergence,

> a power series defines a continuous function;
> a power series representation is unique;
> two power series may be added term by term;
> the terms of a power series may be rearranged;
> two power series may be multiplied term by term;
> power series may be differentiated term by term;
> power series may be integrated term by term.

More precise statements of these properties now follow.

Continuity of power series

The sum of a power series is a continuous function within its interval of convergence.

Uniqueness of power series

Let two power series expanded about $x = x_0$, each with a positive radius of convergence, have the same sum function f throughout a neighborhood of x_0. Then the two series are identical.

Thus if

$$f(x) = \sum_{n=0}^{\infty} a_n (x - x_0)^n = \sum_{n=0}^{\infty} b_n (x - x_0)^n$$

in some neighborhood of x_0, then $a_n = b_n$ for $n = 0, 1, 2, \dots$.

Addition of power series

The sum of the power series representing the functions f and g expanded about $x=x_0$ may be obtained by termwise addition of corresponding powers of $(x-x_0)$ in the two series. The result is valid for any x common to each interval of convergence, and the power series so obtained is the power series representation for the function $f+g$.

Thus if f and g are represented by power series with the positive radii of convergence r_1 and r_2, respectively, and

$$f(x)=\sum_{n=0}^{\infty} a_n(x-x_0)^n \qquad \text{for } x_0-r_1<x<x_0+r_1,$$

$$g(x)=\sum_{n=0}^{\infty} b_n(x-x_0)^n \qquad \text{for } x_0-r_2<x<x_0+r_2,$$

then

$$f(x)+g(x)=\sum_{n=0}^{\infty} (a_n+b_n)(x-x_0)^n \qquad \text{for } x_0-r<x<x_0+r,$$

where $r=\min\{r_1,r_2\}$.

Rearrangement of terms of a power series

Within its interval of convergence, the terms of a power series may be rearranged without altering its sum.

Multiplication of power series

The product of the power series representations of two functions f and g expanded about $x=x_0$ may be obtained by multiplying the two series together term by term and grouping the result to form a new power series. The power series so obtained is convergent for any x common to each of the intervals of convergence of f and g, and in this interval it is the power series representation for the function fg.

Thus if f and g have the positive radii of convergence r_1 and r_2, respectively, while

$$f(x)=\sum_{n=0}^{\infty} a_n(x-x_0)^n \qquad \text{for } x_0-r_1<x<x_0+r_1$$

and

$$g(x)=\sum_{n=0}^{\infty} b_n(x-x_0) \qquad \text{for } x_0-r_2<x<x_0+r_2,$$

then

$$f(x)\,g(x)=\sum_{n=0}^{\infty} (a_0b_n+a_1b_{n-1}+\ \ldots\ +a_nb_0)(x-x_0)^n \qquad \text{for } x_0-r<n<x_0+r,$$

where $r=\min\{r_1,r_2\}$.

Differentiation of power series

The power series representation of the derivative f' of a function f may, within the interval of convergence of the power series for f, be found by termwise differentiation of the power series.

Thus if f has the positive radius of convergence r, and

$$f(x) = \sum_{n=0}^{\infty} a_n(x-x_0)^n \qquad \text{for } x_0 - r < x < x_0 + r,$$

then

$$f'(x) = \sum_{n=1}^{\infty} na_n(x-x_0)^{n-1} \qquad \text{for } x_0 - r < x < x_0 + r.$$

Integration of power series

The power series representation of the integral of a function f may, within the interval of convergence of the power series for f, be found by termwise integration of the power series.

Thus if f has the positive radius of convergence r, and

$$f(x) = \sum_{n=0}^{\infty} a_n(x-x_0)^n \qquad \text{for } x_0 - r < x < x_0 + r,$$

then

$$\int_{x_0}^{x} f(t)\, dt = \sum_{n=0}^{\infty} a_n \int_{x_0}^{x} (t-x_0)^n\, dt = \sum_{n=0}^{\infty} \frac{a_n}{n+1}(x-x_0)^{n+1}$$

for $x_0 - r < x < x_0 + r$.

Example 2. Operations on series

Find
(i) the terms up to and including x^4 in the product fg,

(ii) $f+g$,   (iii) df/dx,   (iv) $\displaystyle\int_{1}^{x} h(t)\, dt$

and the radii and intervals of convergence of the series, given that

$$f(x) = \sum_{n=1}^{\infty} \left(\frac{1+n}{2+3n}\right)x^n, \quad g(x) = \sum_{n=1}^{\infty} \frac{(-1)^{n+1}x^n}{n^{3/2}} \quad \text{and} \quad h(x) = \sum_{n=1}^{\infty} \left(\frac{(-1)^n n^n}{1+(2n)^n}\right)(x-1)^n.$$

Solution

We first find the radii and intervals of convergence of f, g, and h. Applying (11) (i) to f with $a_n = (1+n)/(2+3n)$ gives for the radius of convergence r_f of f

$$r_f = \lim_{n \to \infty} \left[\left(\frac{1+n}{2+3n}\right)\left(\frac{5+3n}{2+n}\right) \right] = 1.$$

As f is expanded about the origin its interval of convergence is $-1<x<1$. A similar argument applied to g shows the radius of convergence r_g of g is

$$r_g = \lim_{n\to\infty} \left| \frac{(-1)^{n+1}(n+1)^{3/2}}{n^{3/2}} \frac{}{(-1)^{n+2}} \right| = \lim_{n\to\infty}\left(1+\frac{1}{n}\right)^{3/2} = 1,$$

and as g is expanded about the origin the interval of convergence is $-1<x<1$.

Applying 11(ii) to h with $a_n=(-1)^n n^n/(1+(2n)^n)$ gives for the radius of convergence r_h of h

$$r_h = 1 \Big/ \left\{ \lim_{n\to\infty} \left| \frac{(-1)^n n^n}{1+(2n)^n} \right|^{1/n} \right\} = 1/(\tfrac{1}{2}) = 2.$$

As h is expanded about the point 1 its interval of convergence is $-1<x<3$.

(i) $f(x)\,g(x)=\left(\dfrac{2}{5}x+\dfrac{3}{8}x^2+\dfrac{4}{11}x^3+\cdots\right)\left(x-\dfrac{x^2}{2^{3/2}}+\dfrac{x^3}{3^{3/2}}-\cdots\right)$

$$=\frac{2}{5}x^2+\left(-\frac{2}{5}\frac{1}{2^{3/2}}+\frac{3}{8}\right)x^3+\left(\frac{2}{5}\frac{1}{3^{3/2}}-\frac{3}{8}\frac{1}{2^{3/2}}+\frac{4}{11}\right)x^4+\cdots,$$

with the interval of convergence *common* to f and g, namely, $-1<x<1$.

(ii) $f(x)+g(x)=\sum\limits_{n=1}^{\infty}\left(\dfrac{1+n}{2+3n}+\dfrac{(-1)^n}{n^{3/2}}\right)x^n$ and its interval of convergence is the interval *common* to the intervals of convergence of f and g, namely, $-1<x<1$.

(iii) $\dfrac{df}{dx}=\sum\limits_{n=1}^{\infty}\left(\dfrac{1+n}{2+3n}\right)\dfrac{d(x^n)}{dx}=\sum\limits_{n=1}^{\infty}\left(\dfrac{n(1+n)}{2+3n}\right)x^{n-1},$

with the same interval of convergence as f, namely, $-1<x<1$.

(iv) $\displaystyle\int_1^x h(t)\,dt=\sum\limits_{n=1}^{\infty}\left(\dfrac{(-1)^n n^n}{1+(2n)^n}\right)\int_1^x (t-1)^n\,dt$

$$=\sum\limits_{n=1}^{\infty}\left(\frac{(-1)^n n^n}{(n+1)[1+(2n)^n]}\right)(x-1)^{n+1},$$

with the same interval of convergence as h, namely, $-1<x<3$. ∎

Shifting summation indices

When seeking the solution of a differential equation in the form of a power series it is necessary to add together series which do not all begin with the same power of $(x-x_0)$. In order to make effective use of the resulting summation, the terms of each series must all be brought under a single summation sign. This is accomplished by **shifting (translating)** the summation indices where necessary to bring into coincidence corresponding powers of $(x-x_0)$.

The process is most easily explained by means of examples, but first it should be remembered that the choice of symbol used for the summation index is unimportant, because

$$\sum_{m=0}^{\infty} a_m(x-x_0)^m = \sum_{n=0}^{\infty} a_n(x-x_0)^n = \sum_{r=0}^{\infty} a_r(x-x_0)^r$$

$$= a_0 + a_1(x-x_0) + a_2(x-x_0)^2 + \ldots .$$

Thus the summation symbol is a **dummy index**, in the sense that it does not appear when the series is expanded (compare this situation with the use of a **dummy variable** in a definite integral).

Example 3. Shifting a summation index

Express the series

$$\sum_{n=2}^{\infty} n(n-1)a_n x^{n-2} = 2 \cdot 1 a_2 + 3 \cdot 2 a_3 x + 4 \cdot 3 a_4 x^2 + \ldots ,$$

as a power series with its summation index starting from zero.

Solution

The index n must be reduced by 2, and this shift (translation) is accomplished by using the new index $r = n - 2$, for then $n = 2, 3, 4, \ldots$ corresponds to $r = 0, 1, 2, \ldots$. Using this change of index we find that

$$\sum_{n=2}^{\infty} n(n-1)a_n x^{n-2} = \sum_{r=0}^{\infty} (r+2)(r+1)u_{r+2} x^r,$$

and we have accomplished our objective. As the choice of symbol used for the summation index is immaterial we may, if we wish, replace r by n in the last result to obtain

$$\sum_{n=2}^{\infty} n(n-1)a_n x^{n-2} = \sum_{n=0}^{\infty} (n+2)(n+1)a_{n+2} x^n = 2 \cdot 1 a_2 + 3 \cdot 2 a_3 x + 4 \cdot 3 a_4 x^2 + \ldots .$$

∎

Example 4. Shifting suffixes to sum power series

Given that

$$f(x) = \sum_{n=2}^{\infty} n(n-1)a_n x^{n-2}, \quad g(x) = \sum_{n=1}^{\infty} na_n x^{n-1} \quad \text{and} \quad h(x) = \sum_{n=0}^{\infty} a_n x^n,$$

express

$$F(x) = (1-x^2)f(x) - (2x+1)g(x) + \lambda h(x) \qquad (\lambda \text{ constant})$$

as the sum of a finite number of terms and a power series with its summation index starting from 2.

Solution

Substituting the series into $F(x)$ and performing the indicated multiplications we find that

$$F(x) = (1-x^2)f(x) - (2x+1)g(x) + \lambda h(x)$$

$$= (1-x^2)\sum_{n=2}^{\infty} n(n-1)a_n x^{n-2} - (2x+1)\sum_{n=1}^{\infty} na_n x^{n-1} + \lambda \sum_{n=0}^{\infty} a_n x^n$$

$$= \sum_{n=2}^{\infty} n(n-1)a_n x^{n-2} - \sum_{n=2}^{\infty} n(n-1)a_n x^n - \sum_{n=1}^{\infty} 2na_n x^n - \sum_{n=1}^{\infty} na_n x^{n-1} + \sum_{n=0}^{\infty} \lambda a_n x^n.$$

Next we shift the summation indices in the first and fourth series to bring the powers of x in their general terms to the power n. This is accomplished by shifting the summation index in the first series by 2 and in the fourth series by 1, by setting $r = n-2$ and $m = n-1$, respectively, to obtain

$$F(x) = \sum_{r=0}^{\infty} (r+2)(r+1)a_{r+2} x^r - \sum_{n=2}^{\infty} n(n-1)a_n x^n - \sum_{n=1}^{\infty} 2na_n x^n - \sum_{m=0}^{\infty} (m+1)a_{m+1} x^m$$

$$+ \sum_{n=0}^{\infty} \lambda a_n x^n.$$

Replacing r and m by n, this becomes

$$F(x) = \sum_{n=0}^{\infty} (n+2)(n+1)a_{n+2} x^n - \sum_{n=2}^{\infty} n(n-1)a_n x^n - \sum_{n=1}^{\infty} 2na_n x^n - \sum_{n=0}^{\infty} (n+1)a_{n+1} x^n + \sum_{n=0}^{\infty} \lambda a_n x^n.$$

Separating out the terms corresponding to $n=0$ and $n=1$ we find

$$F(x) = 2 \cdot 1 a_2 + 3 \cdot 2 a_3 x + \sum_{n=2}^{\infty} (n+2)(n+1)a_{n+2} x^n - \sum_{n=2}^{\infty} n(n-1)a_n x^n - 2a_1 x$$

$$- \sum_{n=2}^{\infty} 2na_n x^n - a_1 - 2a_2 x - \sum_{n=2}^{\infty} (n+1)a_{n+1} x^n + \lambda a_0 + \lambda a_1 x + \sum_{n=2}^{\infty} \lambda a_n x^n.$$

Finally, combining all the summations, we arrive at the required result

$$F(x) = (2a_2 - a_1 + \lambda a_0) + [6a_3 - 2a_2 + (\lambda - 2)a_1]x + \sum_{n=2}^{\infty} [(n+2)(n+1)a_{n+2} - n(n-1)a_n - 2na_n$$

$$+ (n+1)a_{n+1} + \lambda a_n]x^n$$

$$= (2a_2 - a_1 + \lambda a_0) + [6a_3 - 2a_2 + (\lambda - 2)a_1]x$$

$$+ \sum_{n=2}^{\infty} [(n+2)(n+1)a_{n+2} + (n+1)a_{n+1} + (\lambda - n - n^2)a_n]x^n. \qquad \blacksquare$$

Up to this point a power series has been regarded as a way of defining a function (its sum function) within its interval of convergence. However it is also possible, as with a Taylor series, to generate a power series from an infinitely differentiable function f. The important question that then arises is: *when* does the sum function of the Taylor series *equal* the function f used to generate the series? We remark that a power series with a positive radius of convergence whose sum function is f is also called the **Taylor series** expansion of f or, when $x_0 = 0$, the **Maclaurin series** expansion. In this context, the nth partial sum $s_n(x)$ in (6) is called the **Taylor polynomial of degree** n associated with f.

It will be recalled that the Taylor series expansion of an infinitely differentiable function f about the point x_0 is

$$f(x) = f(x_0) + \frac{(x-x_0)}{1!} f^{(1)}(x_0) + \frac{(x-x_0)^2}{2!} f^{(2)}(x_0) + \cdots$$

$$+ \frac{(x-x_0)^n}{n!} f^{(n)}(x_0) + \cdots . \tag{12}$$

When f is not infinitely differentiable, or when a Taylor polynomial is needed to approximate f to within some required accuracy, then instead of (12) it is necessary to make use of the Taylor formula with a remainder

$$f(x) = f(x_0) + \frac{(x-x_0)}{1!} f^{(1)}(x_0) + \frac{(x-x_0)^2}{2!} f^{(2)}(x_0) + \cdots + \frac{(x-x_0)^n}{n!} f^{(n)}(x_0) + R_n(x),$$
$$\tag{13}$$

in which $R_n(x)$ is called the **remainder term**.

The sum of the first $n+1$ terms of (13) is simply the nth partial sum $s_n(x)$ of the series; so the absolute error made when $f(x)$ is represented by

$$s_n(x) = f(x) + \frac{(x-x_0)}{1!} f^{(1)}(x_0) + \frac{(x-x_0)^2}{2!} f^{(2)}(x_0) + \cdots + \frac{(x-x_0)^n}{n!} f^{(n)}(x_0), \tag{14}$$

is $|R_n(x)|$. Only if $\lim_{n \to \infty} |R_n(x)| \to 0$ will $s_n(x)$ converge to the function f used to generate the Taylor series in (12).

The remainder term $R_n(x)$ may be expressed in several different but equivalent ways, the most useful of which are as follows.

Lagrange remainder

$$R_n(x) = \frac{(x-x_0)^{n+1}}{(n+1)!} f^{(n)}(\xi), \tag{15}$$

where ξ is some number such that $x_0 < \xi < x$;

Cauchy remainder

$$R_n(x) = \frac{(x-x_0)(x-\xi)^n}{n!} f^{(n+1)}(\xi), \tag{16}$$

where ξ is some number such that $x_0 < \xi < x$;

Integral form of the remainder

$$R_n(x) = \frac{1}{n!} \int_{x_0}^{x} (x-t)^n f^{(n+1)}(t) \, dt. \tag{17}$$

Example 5. Estimation of the error associated with a Taylor polynomial

Estimate the absolute error when a Taylor polynomial is used to approximate the function $\cos x$ expanded about the point x_0.

Solution

Setting $f(x) = \cos x$, some routine calculations show that we may write

$$\frac{d^n(\cos x)}{dx^n} = f^{(n)}(x) = \cos\left(x + \frac{n\pi}{2}\right),$$

for $n = 1, 2, \ldots$. Using Taylor's formula with a remainder and expanding $\cos x$ about x_0 we arrive at the result

$$\cos x = \cos x_0 - \frac{(x - x_0)}{1!} \sin x_0 - \frac{(x - x_0)^2}{2!} \cos x_0$$

$$+ \frac{(x - x_0)^3}{3!} \sin x_0 + \ldots + \frac{(x - x_0)^n}{n!} \cos\left(x_0 + \frac{n\pi}{2}\right) + R_n(x).$$

In this case the Taylor polynomial of degree n is

$$s_n(x) = \cos x_0 - \frac{(x - x_0)}{1!} \sin x_0 - \frac{(x - x_0)^2}{2!} \cos x_0 + \ldots + \frac{(x - x_0)^n}{n!} \cos\left(x_0 + \frac{n\pi}{2}\right).$$

Making use of (15) the Lagrange remainder is seen to be

$$|R_n(x)| = \left| \frac{(x - x_0)^{n+1}}{(n+1)!} \cos\left(\xi + \frac{n\pi}{2}\right) \right|,$$

where ξ (unknown) is a number such that $x_0 < \xi < x$. As $\left|\cos\left(\xi + \frac{n\pi}{2}\right)\right| \le 1$ for all ξ, the absolute error $|R_n(x)|$ must be such that

$$|R_n(x)| \le \frac{(x - x_0)^{n+1}}{(n+1)!},$$

which is the required estimate. For any given x and x_0 this result may be used to determine the least value of n, and hence the lowest degree Taylor polynomial $s_n(x)$, for which the absolute error $|R_n(x)|$ does not exceed some predetermined value $\varepsilon > 0$ in the interval (x_0, x). ∎

Sometimes one form of the remainder term leads to a sharper estimate of the error than the others and, indeed, this happens in Example 5. Using (16) and making the factor $(x - \xi)^n$ as large as possible in the interval $x_0 < \xi < x$ we obtain the estimate

$$|R_n(x)| = \left| \frac{(x - x_0)(x - \xi)^n}{n!} \right| \le \frac{|(x - x_0)^{n+1}|}{n!},$$

which is the greater by a factor $(n+1)$ than our first estimate. The integral form of the remainder term fails to improve on the first estimate, because

$$|R_n(x)| = \left| \frac{1}{n!} \int_{x_0}^{x} (x-t)^n \cos\left(t - \frac{n\pi}{2}\right) dt \right| \le \frac{1}{n!} \int_{x_0}^{x} (x-t)^n \, dt = \frac{|(x-x_0)^{n+1}|}{(n+1)!}.$$

We are now in a position to state the precise relationship between a Taylor series and the function f used to derive it by repeated differentiation. For any specific x a Taylor series becomes an ordinary infinite series, and this series will converge if x lies within the interval of convergence of the series. As for a given x the remainder terms in (10) and (13) are the same, it follows from our earlier discussion that the sum function of a Taylor series expansion of a function f will be equal to the function f itself when x lies within the interval of convergence of the series.

A function is said to be **analytic** at $x = x_0$ if it is capable of representation in terms of a power series in $(x - x_0)$ with a positive radius of convergence. Any function which cannot be represented in this manner is said to have a **singularity** at the point x_0, or to be **nonanalytic** at x_0.

Not every infinitely differentiable function f has a Taylor series with a sum function which is equal to f at more than one point. An example of such a function with a singularity at the origin is

$$f(x) = \begin{cases} \exp(-1/x^2) & \text{for } x \neq 0 \\ 0 & \text{for } x = 0. \end{cases}$$

Attempting to represent this function by its Taylor series expanded about the origin (Maclaurin series) we find that $f(0)$ together with all its derivatives $f^{(n)}(0)$, $n = 1, 2, \ldots$ vanish, so the Taylor series is identically zero. Thus despite the function being infinitely differentiable at the origin, and its Taylor series being convergent (to zero) for all x, the sum function only represents $\exp(-1/x^2)$ at the origin.

Problems for Section 8.1

Find the coefficient a_n of the general term, the radius and the interval of convergence of the following series.

1 $x + \dfrac{x^2}{2^2} + \dfrac{x^3}{3^2} + \dfrac{x^4}{4^2} + \ldots.$

2 $x + \dfrac{x^3}{3} + \dfrac{x^5}{5} + \ldots.$

3 $1 - \frac{1}{3}(x-3) + \frac{1}{5}(x-3)^2 - \frac{1}{7}(x-3)^3 + \ldots.$

4 $x - \dfrac{x^3}{3\cdot 3} + \dfrac{x^5}{5\cdot 5!} - \dfrac{x^7}{7\cdot 7!} + \ldots.$

5 $x + \dfrac{x^2}{2^\alpha} + \dfrac{x^3}{3^\alpha} + \dfrac{x^4}{4^\alpha} + \dots$ $(\alpha > 1)$.

6 $1 + x + 2^2 x^2 + 3^3 x^3 + \dots$.

7 $1 + (x-1) + \dfrac{2!}{2^2}(x-1)^2 + \dfrac{3!}{3^3}(x-1)^3 + \dfrac{4!}{4^4}(x-1)^4 + \dots$.

8 $\dfrac{1}{4} + x + \dfrac{8^2}{2^2+3}x^2 + \dfrac{12^3}{3^3+3}x^3 + \dfrac{16^4}{4^4+3}x^4 + \dots$.

9 Given that

$$f(x) = \sum_{n=1}^{\infty} \frac{(-1)^n}{n}\left(x - \frac{1}{2}\right)^n \quad \text{and} \quad g(x) = \sum_{n=1}^{\infty} \frac{2^n}{n}\left(x - \frac{1}{2}\right)^n,$$

find the power series for $f+g$ and determine its interval of convergence.

10 Given that

$$f(x) = \sum_{n=1}^{\infty} \frac{(x-1)^n}{n^{1/2}} \quad \text{and} \quad g(x) = \sum_{n=1}^{\infty} \frac{(x-1)^n}{n^3},$$

find the power series for $f+g$ and determine its interval of convergence.

11 Making use of known series, use termwise multiplication of series to find the power series expansion about the origin up to the term in x^3 of the function $e^{3x} \ln(1+2x)$.

12 Making use of known series, use termwise multiplication of series to find the power series expansion about the origin up to the term in x^7 of the function $\sin(x^2)\sinh(2x)$.

13 By writing $\sec x = (\cos x)^{-1}$, and using the binomial theorem with the series for $\cos x$, find the power series expansion about the origin up to the term in x^4 of the function $\sec x$.

14 By writing $\operatorname{sech} x = (\cosh x)^{-1}$, and using the binomial theorem with the series for $\cosh x$, find the power series expansion about the origin up to the term in x^4 of the function $\operatorname{sech} x$. How might this result have been deduced from the result of the previous problem?

15 By writing $\tan x = \sin x \,(\cos x)^{-1}$ and making use of known series, the binomial theorem and termwise multiplication of series, find the power series expansion about the origin up to the term in x^4 of the function $\tan x$.

16 Making use of known series, use termwise multiplication of series to find the power series expansion about the origin up to the term in x^3 of the function $\sin x \exp(1 - \cos x)$.

17 Use the uniqueness property of power series to show that any power series with a sum function which is identically zero throughout a finite interval of convergence must have every coefficient equal to zero.

18 Explain why any sum function represented by a power series with a positive radius of convergence is infinitely differentiable within its interval of convergence.

19 Deduce the power series expansion about the origin for $\sec^2 x$ up to the term in x^8 from the power series expansion

$$\tan x = x + \frac{x^3}{3} + \frac{2}{15}x^5 + \frac{17}{315}x^7 + \frac{62}{2835}x^9 + \dots ,$$

for $\dfrac{-\pi}{2} < x < \dfrac{\pi}{2}$. What is its interval of convergence?

20 Deduce the power series expansion about the origin for $\sec x \tan x$ up to the term in x^7 from the power series expansion

$$\sec x = 1 + \frac{x^2}{2} + \frac{5}{24}x^4 + \frac{61}{120}x^6 + \frac{277}{8064}x^8 + \dots ,$$

for $\dfrac{-\pi}{2} < x < \dfrac{\pi}{2}$. What is its interval of convergence?

21 Deduce the power series expansion about the origin for $(\arc \sin x)/\sqrt{1-x^2}$ from the power series expansion

$$(\arc \sin x)^2 = \frac{1}{2} \sum_{n=0}^{\infty} \frac{2^{2n}(n!)^2 x^{2n+2}}{(2n+1)!\,(n+1)},$$

and find its radius and interval of convergence.

22 Deduce the power series expansion about the origin for $(\arc \sin x)^2/\sqrt{1-x^2}$ from the power series expansion

$$(\arc \sin x)^3 = x^3 + \frac{3!}{5!}3^2\left(1+\frac{1}{3^2}\right)x^5 + \frac{3!}{7!}3^2 \cdot 5^2\left(1+\frac{1}{3^2}+\frac{1}{5^2}\right)x^7 + \dots ,$$

and find the radius and interval of convergence.

23 Find the power series expansion about the origin of the derivative $f'(x)$ when $f(x)=\ln(1+x)$. Integrate the result to show

$$\ln(1+x) = x - \frac{x^2}{2} + \frac{x^3}{3} - \frac{x^4}{4} + \dots ,$$

and find the radius and interval of convergence of this series.

24 Find the power series expansion about the origin of the derivative $f'(x)$ when $f(x) = \arc \tan x$. Integrate the result to show that

$$\arc \tan x = x - \frac{x^3}{3} + \frac{x^5}{5} - \frac{x^7}{7} + \dots ,$$

and find the radius and interval of convergence of this series.

25 Find the power series expansion about the origin of the derivative $f'(x)$ when $f(x) = \arc \sin x$. Integrate the result to show that

$$\arc \sin x = x + \frac{1}{2}\frac{x^3}{3} + \frac{1}{2}\cdot\frac{3}{4}\cdot\frac{x^5}{5} + \frac{1}{2}\cdot\frac{3}{4}\cdot\frac{5}{6}\frac{x^7}{7} + \dots ,$$

and find the radius and interval of convergence of this series.

26 The **complementary error function** erf x is defined as

$$\operatorname{erf} x = \frac{2}{\sqrt{\pi}} \int_0^x \exp(-t^2)\,dt.$$

Obtain the power series expansion of erf x about the origin and deduce its interval of convergence (cf. Prob. 73, Sec. 7.3).

The quotient of two power series

When the first few terms of the **quotient of two power series** are required they may be obtained very simply as follows.

Let

$$f(x) = a_0 + a_1 x + a_2 x^2 + \ldots \qquad \text{for } -r_1 < x < r_1,$$

$$g(x) = b_0 + b_1 x + b_2 x^2 + \ldots \qquad \text{for } -r_2 < x < r_2$$

and

$$f(x)/g(x) = c_0 + c_1 x + c_2 x^2 + \ldots .$$

Then

$$\frac{f(x)}{g(x)} = \frac{a_0 + a_1 x + a_2 x^2 + \ldots}{b_0 + b_1 x + b_2 x^2 + \ldots} \equiv c_0 + c_1 x + c_2 x^2 + \ldots ,$$

so that

$$a_0 + a_1 x + a_2 x^2 + \ldots \equiv (b_0 + b_1 x + b_2 x^2 + \ldots)(c_0 + c_1 x + c_2 x^2 + \ldots).$$

Equating the coefficients of corresponding powers of x on opposite sides of this identity gives the following system of equations from which the coefficients $c_0, c_1, c_2, \ldots$, of the quotient power series may be determined sequentially:

$$a_0 = b_0 c_0$$

$$a_1 = b_0 c_1 + b_1 c_0$$

$$a_2 = b_0 c_2 + b_1 c_1 + b_2 c_0$$

$$a_3 = b_0 c_3 + b_1 c_2 + b_2 c_1 + b_3 c_0$$

$$\vdots$$

This process is valid for $-r < x < r$ with $r = \min \{r_1, r_2\}$, provided $g(x) \neq 0$ inside this interval.

A modification must be made to this form of argument when the expansion of $g(x)$ begins with the term x^m, so that $g(x) = b_0 x^m + b_1 x^{m+1} + b_2 x^{m+2} + \ldots$. All that is necessary is to write $f(x)/g(x)$ in the form

$$\frac{f(x)}{g(x)} = \left(\frac{1}{x^m} \right) \left(\frac{f(x)}{b_0 + b_1 x + b_2 x^2 + \ldots} \right),$$

and then to treat the quotient $f(x)/(b_0 + b_1 x + b_2 x^2 + \ldots)$ as before.

Apply the method described above to the following problems to determine the required number of terms in the power series expansion about the origin of the stated functions.

27 Find the power series expansion of $(1-x)/(1+2x)$ and determine its interval of convergence.

28 Expand sech $2x$ up to and including the term in x^4.

29 Find the power series expansion of $x/(1-x)^2$ and determine its interval of convergence.

30 Expand cosec x up to and including the term in x^7.

31 Re-express the following as series in which the summation index starts from zero and check your results by expanding the series

(i) $\displaystyle\sum_{n=3}^{\infty} n(n-1)x^{n-2}$ (ii) $\displaystyle x^2 \sum_{n=4}^{\infty} n(n+1)x^{n-3}$.

32 Re-express the following as series in which the summation index starts from 1 and check your results by expanding the series

(i) $\displaystyle\sum_{n=3}^{\infty} n(n+2)x^{n-1}$ (ii) $\displaystyle x \sum_{n=3}^{\infty} n(n+3)x^{n-2}$.

33 Re-express the following as the sum of a finite number of terms and a series in which the summation index starts from 1 and check your result by expanding the series

$$(1+x) \sum_{n=2}^{\infty} n(n-1)x^{n-2}.$$

34 Re-express the following as the sum of a finite number of terms and a series in which the summation index starts from 2 and check your result by expanding the series

$$(1+x^2) \sum_{n=3}^{\infty} n(n+1)x^{n-2}.$$

35 Given

$$f(x) = \sum_{n=2}^{\infty} n(n-1)a_n x^{n-2}, \quad g(x) = \sum_{n=1}^{\infty} na_n x^{n-1},$$

express

$$F(x) = (1+x^2)f(x) + xg(x)$$

as the sum of a finite number of terms and a power series with its summation index starting from 2.

36 Given

$$f(x) = \sum_{n=2}^{\infty} n(n-1)a_n x^{n-2}, \quad g(x) = \sum_{n=1}^{\infty} na_n x^{n-1},$$

express

$$F(x) = x^2 f(x) + g(x)$$

as the sum of a finite number of terms and a power series with its summation index starting from 2.

Harder problems

Bernoulli numbers

The **Bernoulli** numbers $B_0, B_1, B_2, \ldots$ are defined through the series representation

$$\frac{x}{e^x - 1} \equiv B_0 + \frac{B_1}{1!}x + \frac{B_2}{2!}x^2 + \frac{B_3}{3!}x^3 + \ldots.$$

This representation is valid for all x because

$$e^x - 1 = x\left(1 + \frac{x}{2!} + \frac{x^2}{3!} + \dots\right),$$

and the bracketed terms are never zero. Thus cross-multiplying in the defining relationship for the B_n's and canceling an x gives

$$1 = \left(1 + \frac{x}{2!} + \frac{x^2}{3!} + \dots\right)\left(B_0 + \frac{B_1}{1!}x + \frac{B_2}{2!}x^2 + \dots\right).$$

Equating the coefficients of corresponding powers of x on each side of this identity yields the following system of equations from which the Bernoulli numbers $B_0, B_1, B_2, \dots$ may be determined sequentially:

$$1 = B_0$$

$$0 = \frac{1}{2!}\frac{B_0}{0!} + \frac{1}{1!}\frac{B_1}{1!}$$

$$0 = \frac{1}{3!}\frac{B_0}{0!} + \frac{1}{2!}\frac{B_1}{1!} + \frac{1}{1!}\frac{B_2}{2!}$$

$$0 = \frac{1}{4!}\frac{B_0}{0!} + \frac{1}{3!}\frac{B_1}{1!} + \frac{1}{2!}\frac{B_2}{2!} + \frac{1}{1!}\frac{B_3}{3!} \dots$$

and, in general,

$$0 = \frac{1}{n!}\frac{B_0}{0!} + \frac{1}{(n-1)!}\frac{B_1}{1!} + \frac{1}{(n-2)!}\frac{B_2}{2!} + \dots + \frac{1}{1!}\frac{B_{n-1}}{(n-1)!}$$

for $n \geq 2$.

If this last result is multiplied by $n!$ we can make use of the binomial coefficient (cf. Prob. 24, Sec. 1.3)

$$\binom{n}{r} = \frac{n!}{r!(n-r)!}$$

to express this result in the more easily remembered form

$$\binom{n}{0}B_0 + \binom{n}{1}B_1 + \binom{n}{2}B_2 + \dots + \binom{n}{n-1}B_{n-1} = 0.$$

The first few Bernoulli numbers are

$$B_0 = 1, \qquad B_1 = -1/2, \qquad B_2 = 1/6, \qquad B_3 = 0,$$

$$B_4 = -1/30, \qquad B_5 = 0, \qquad B_6 = 1/42, \qquad B_7 = 0,$$

$$B_8 = -1/30, \qquad B_9 = 0, \qquad B_{10} = 5/66, \qquad B_{11} = 0,$$

$$B_{12} = -691/2730, \dots.$$

The relationship connecting B_{n-1} to B_0 shows that the Bernoulli numbers are all rational numbers.

37 Show that if $x = u/2$ then

$$x \coth x = \frac{u}{e^u - 1} + \frac{u}{2}.$$

Use this result together with the fact that $B_0 = 1$, $B_1 = -\frac{1}{2}$ and $B_2 = \frac{1}{6}$ to show that $B_{2n+1} = 0$ for $n = 1, 2, 3, \ldots$, and hence that

$$x \coth x = 1 + \frac{B_2}{2!}(2x)^2 + \frac{B_4}{4!}(2x)^4 + \frac{B_6}{6!}(2x)^6 + \cdots.$$

38 Justify replacing x by ix the result of Prob. 37 to obtain the power series

$$x \cot x = 1 - \frac{B_2}{2!}(2x)^2 + \frac{B_4}{4!}(2x)^4 - \cdots.$$

Use the identity

$$\tan x = \cot x - 2 \cot (2x)$$

together with the above result to show that

$$\tan x = \sum_{n=1}^{\infty} \frac{2^{2n}(2^{2n} - 1)|B_{2n}|}{(2n)!} x^{2n-1}.$$

39 Justify replacing x by ix in the series for $\tan x$ in Prob. 38 to obtain the power series

$$\tanh x = \sum_{n=1}^{\infty} \frac{2^{2n}(2^{2n} - 1)}{(2n)!} B_{2n} x^{2n-1}.$$

Use the identity

$$\operatorname{cosech} x = \coth x - \tanh \frac{x}{2}$$

together with the above result to show that

$$x \operatorname{cosech} x = 1 - \sum_{n=1}^{\infty} \frac{2(2^{2n-1} - 1)}{(2n)!} B_{2n} x^{2n}.$$

An alternative way of defining the Bernoulli number B_n is in terms of the **Bernoulli polynomial** $B_n(x)$ which occurs as the multiplier of $t^n/n!$ in the expansion

$$\frac{te^{xt}}{e^t - 1} = \sum_{n=0}^{\infty} B_n(x) \frac{t^n}{n!}.$$

The Bernoulli polynomial $B_n(x)$ is a polynomial in x of degree n, and it follows at once that the Bernoulli number $B_n = B_n(0)$, for $n = 0, 1, 2, \ldots$. The function $te^{xt}/(e^t - 1)$ is called the **generating function** for the Bernoulli polynomials. Along with Bernoulli numbers, Bernoulli polynomials find applications in numerical analysis.

40 Verify by direct expansion that

$$B_0(x) = 1, \quad B_1(x) = x - \tfrac{1}{2}, \quad B_2(x) = x^2 - x + \tfrac{1}{6},$$

$$B_3(x) = x^3 - \tfrac{3}{2} x^2 + \tfrac{1}{2}x, \quad B_4(x) = x^4 - 2x^3 + x^2 - \tfrac{1}{30}.$$

41 Show by using the generating function that

$$B_n'(x) = nB_{n-1}(x) \qquad \text{for } n = 1, 2, \ldots,$$

and hence that

$$\int_a^x B_n(t)\,dt = \frac{B_{n+1}(x) - B_{n+1}(a)}{n+1}.$$

42 Integrate the generating function and its expansion in terms of $B_n(x)$ over the interval $0 \le x \le 1$ to show that

$$\int_0^1 B_n(x)\,dx = 0 \qquad \text{for } n = 1, 2, \ldots,$$

Use this result together with $B_0(x) = 1$ and the first result of Prob. 41 to generate recursively the Bernoulli polynomials $B_0(x)$ to $B_4(x)$.

Reversion of power series

Let the function y be expressible as a power series about x_0 of the form

$$y - y_0 = a_1(x - x_0) + a_2(x - x_0)^2 + a_3(x - x_0)^3 + \ldots, \tag{A}$$

with an interval of convergence $x_0 - r < x < x_0 + r$, for some $r > 0$. A question which then arises is how the roles of x and y may be reversed so that x can be expressed as a power series about y_0 of the form

$$x - x_0 = A_1(y - y_0) + A_2(y - y_0)^2 + A_3(y - y_0)^3 + \ldots. \tag{B}$$

This process is called the **reversion** of the power series in $x - x_0$, and it corresponds to finding the power series representation of the function inverse to the one defined by the original power series. Thus if $y = f(x)$, the reversion of the power series for y corresponds to finding the power series for the inverse function $x = f^{-1}(y)$.

To determine the unknown coefficients A_1, A_2, ... in terms of the known coefficients $a_1, a_2, \ldots$, it is necessary to substitute the expression for $x - x_0$ given in (B) into (A). The required coefficients A_1, A_2, ... then follow from the equations obtained by equating the coefficients of corresponding powers of $y - y_0$ on either side of the resulting identity. As the left-hand side only contains the term $y - y_0$, this is equivalent to equating to unity the coefficient of $y - y_0$ on the right-hand side, and equating to zero the coefficients of all other powers of $y - y_0$ on the right-hand side. The coefficients A_1, A_2, ... follow recursively from the resulting equations. Thus

$$y - y_0 = a_1[A_1(y - y_0) + A_2(y - y_0)^2 + A_3(y - y_0)^3 + \ldots] + a_2[A_1(y - y_0) + A_2(y - y_0)^2 + A_3(y - y_0)^3 + \ldots]^2 + \ldots,$$

and equating coefficients of powers of $y - y_0$ gives:

Coefficients of $(y - y_0)$

$$1 = a_1 A_1,$$

Coefficients of $(y - y_0)^2$

$$0 = a_1 A_2 + a_2 A_1^2,$$
$$\vdots$$

so that

$$A_1 = \frac{1}{a_1}, \quad A_2 = \frac{-a_2 A_1^2}{a_1} = \frac{-a_2}{a_1^3}, \dots .$$

The first six coefficients A_r obtained in this manner are

$$A_1 = \frac{1}{a_1}$$

$$A_2 = \frac{-a_2}{a_1^3}$$

$$A_3 = \frac{2a_2^2 - a_1 a_3}{a_1^5}$$

$$A_4 = \frac{5a_1 a_2 a_3 - a_1^2 a_4 - 5a_2^3}{a_1^7}$$

$$A_5 = \frac{6a_1^2 a_2 a_4 + 3a_1^2 a_3^2 + 14a_2^4 - a_1^3 a_5 - 21a_1 a_2^2 a_3}{a_1^9}$$

$$A_6 = \frac{7a_1^3 a_2 a_5 + 7a_1^3 a_3 a_4 + 84a_1 a_2^3 a_3 - a_1^4 a_6 - 28a_1^2 a_2 a_3^2 - 42a_2^5 - 28a_1^2 a_2^2 a_4}{a_1^{11}}.$$

43 Using the form of argument outlined above, derive the coefficients A_3 and A_4, and then check your results against those listed above.

44 Given that

$$y = \tanh x = x - \frac{x^3}{3} + \frac{2}{15}x^5 - \frac{17}{315}x^7 + \dots ,$$

use reversion of series to show that

$$x = \operatorname{arc\ tanh} y = y + \frac{y^3}{3} + \frac{y^5}{5} + \frac{y^7}{7} + \dots .$$

45 Given that

$$y - 1 = x + \frac{x^2}{2!} + \frac{x^3}{3!} + \frac{x^4}{4!} + \frac{x^5}{5!} + \dots ,$$

use reversion of series to show that

$$x = (y - 1) - \tfrac{1}{2}(y - 1)^2 + \tfrac{1}{3}(y - 1)^3 - \tfrac{1}{4}(y - 1)^4 + \dots .$$

Interpret these two series in terms of functions of x and y.

46 Given that

$$y = \operatorname{arc\ sinh} x = x - \frac{1}{6}x^3 + \frac{3}{40}x^5 - \frac{15}{336}x^7 + \dots ,$$

use reversion of series to show that

$$x = \sinh y = y + \frac{y^3}{6} + \frac{y^5}{120} + \frac{y^7}{5040} + \dots .$$

47 Given that

$$y = \arccos x = \frac{\pi}{2} - \left(x + \frac{1}{6}x^3 + \frac{3}{40}x^5 + \frac{5}{112}x^7 + \dots \right)$$

for $|x| < 1$ and $0 < \arccos x < \pi$, use reversion of series to show that

$$x = -\left[\left(y - \frac{\pi}{2} \right) - \frac{1}{6}\left(y - \frac{\pi}{2} \right)^3 + \frac{1}{120}\left(y - \frac{\pi}{2} \right)^3 - \dots \right].$$

Is this the result you expected, and can it be simplified?

8.2 Solving differential equations by Taylor series

The **Taylor series method** of solving an initial value problem involves attempting to construct the solution in the form of a series expanded about the point x_0 at which the initial data is specified.

To understand how the process works, consider an nth order differential equation for $y(x)$ for which the initial data

$$y(x_0) = y_0, \ y^{(1)}(x_0) = y_1, \ \dots, \ y^{(n-1)}(x_0) = y_{n-1}$$

is specified at $x = x_0$ where, as usual, we use the notation $y^{(r)}(x) = d^r y/dx^r$. If the differential equation is solved for $y^{(n)}(x)$, then provided the coefficient of $y^{(n)}(x)$ does not vanish at x_0, the value of $y^{(n)}(x_0) = y_n$ (say) can be determined from this expression by making use of the initial data.

To proceed further we now differentiate the differential equation itself, thereby obtaining an equation relating $y^{(n+1)}(x)$ to $y^{(n)}(x)$, $y^{(n-1)}(x)$, $\dots$, $y^{(1)}(x)$. If, now, we again set $x = x_0$, and use the initial data together with the value of $y^{(n)}(x_0)$ that has just been determined, we arrive at $y^{(n+1)}(x_0) = y_{n+1}$ (say). Repetition of this process leads to the determination of all the derivatives of $y(x)$ at x_0. Substitution of the numbers $y_0, y_1, \dots$ into the Taylor series expansion of $y(x)$ about x_0 yields the solution in the form

$$y(x) = y_0 + y_1(x - x_0) + \frac{y_2}{2!}(x - x_0)^2 + \frac{y_3}{3!}(x - x_0)^3 + \dots .$$

In general, unless the structure of the differential equation is simple, this method is useful only for the construction of the first few terms of the Taylor series solution. This is because only in such cases can the general term of the expansion be recognized. However, despite this limitation the method finds many applications. One of the most important is its use in conjunction with numerical methods for the solution of differential equations which require numerical data at both the initial point x_0 and at some subsequent points in order to start the calculation.

Example 1. Constant coefficient equation with a nonhomogeneous term

Find the terms up to and including x^4 in the Taylor series solution of

$$y' + 2y = 3e^x \quad \text{with } y(0) = 2,$$

when expanded about the origin.

Solution

Setting $x = 0$ in the differential equation and using the initial conditions gives

$$y^{(1)}(0) = -1 = y_1.$$

Differentiating the differential equation with respect to x gives

$$y'' + 2y' = 3e^x,$$

so setting $x = 0$ and using $y^{(1)}(0) = -1$ gives

$$y^{(2)}(0) = 3 - 2y^{(1)}(0) = 5 = y_2.$$

A further differentiation followed by setting $x = 0$ and using $y^{(2)}(0) = 5$ gives

$$y^{(3)}(\sigma) = 3 - 2y^{(2)}(0) = -7$$

and, similarly,

$$y^{(4)}(0) = 3 - 2y^{(3)}(0) = 17.$$

Thus the required terms in the Taylor series solution expanded about $x_0 = 0$ are

$$y(x) = 2 + \frac{(-1)}{1!}x + \frac{5}{2!}x^2 + \frac{(-7)}{3!}x^3 + \frac{17}{4!}x^4 + \ldots,$$

$$= 2 - x + \frac{5}{2}x^2 - \frac{7}{6}x^3 + \frac{17}{24}x^4 - \ldots.$$

It is left as an exercise for the reader to verify that this initial value problem has the analytical solution

$$y(x) = e^{-2x} + e^x,$$

and that the first five terms of the series obtained from the analytical solution agree with those found above. ∎

Example 2. Variable coefficient homogeneous equation

Determine the first five nonzero terms in the Taylor series solution of

$$y'' - xy' + 3y = 0 \quad \text{with} \quad y(2) = (0), \quad y'(2) = 1,$$

when expanded about the point $x = 2$.

Solution

Rewriting the equation we have

$$y^{(2)}(x) = xy^{(1)}(x) - 3y(x),$$

so setting $x = 2$ and using the initial conditions we find

$$y^{(2)}(2) = 2.$$

Differentiating the differential equation gives

$$y^{(3)}(x) = xy^{(2)}(x) - 2y^{(1)}(x),$$

so

$$y^{(3)}(2) = 2.$$

Arguing in similar fashion we find

$$y^{(4)}(x) = xy^{(3)}(x) - y^{(2)}(x),$$

so

$$y^{(4)}(2) = 2,$$

and

$$y^{(5)}(x) = xy^{(4)}(x),$$

so

$$y^{(5)}(2) = 4.$$

The form of the Taylor series expansion about $x = 2$ is

$$y(x) = y(2) + y^{(1)}(2)(x-2) + \frac{y^{(2)}(2)}{2!}(x-2)^2 + \frac{y^{(3)}(2)}{3!}(x-2)^3$$

$$+ \frac{y^{(4)}(2)}{4!}(x-2)^4 + \frac{y^{(5)}(2)}{5!}(x-2)^5 + \ldots,$$

so as $y(2) = 0$ the first five nonzero terms are

$$y(x) = (x-2) + \frac{2}{2!}(x-2)^2 + \frac{2}{3!}(x-2)^3 + \frac{2}{4!}(x-2)^4 + \frac{4}{5!}(x-2)^5 + \ldots.$$

The fact that a Taylor series solution of a differential equation cannot always be expanded about every point x_0 may be seen by considering Bessel's equation of order zero

$$xy'' + y' + xy = 0.$$

No problem arises when a Taylor series solution of this equation is sought about any point $x_0 \neq 0$, but in this case the method fails when $x_0 = 0$. We would expect to be able to assign initial values for the solution of this equation in an arbitrary fashion, and this can indeed be done at any point other than the origin. However at $x = 0$, although the value of

$y(0)$ can be assigned arbitrarily, the differential equation constrains the solution in such a manner than $y'(0)=0$.

A better understanding of the nature of the difficulty at the origin follows by rewriting the equation as

$$y'' + \frac{1}{x}y' + y = 0.$$

This shows that the problem arises because the coefficient $1/x$ of y' has a *singularity* at the origin, and so cannot be expanded as a Taylor series about the origin. We must conclude from this that in the neighborhood of the origin a more complicated form of solution than a Taylor series is needed. The form taken by solutions close to singular points will be considered in subsequent sections.

Problems for Section 8.2

1 Determine the first five terms of the Taylor series solution of

$$y' + 4y = e^{-2x}$$

expanded about the origin, given that $y(0)=1$.

2 Determine the first five terms of the Taylor series solution of

$$y' - y = e^x$$

expanded about the origin, given that $y(0)=3$.

3 Show that for the initial value problem considered in Ex. 1

$$y^{(n)}(0) = 3[1 - 2 + 2^2 - 2^3 + \ldots + (-1)^{n-1}2^{n-1}] + (-1)^n 2^{n+1},$$

and hence that the coefficient of x^n in the Taylor series solution expanded about the origin is

$$\frac{1 + (-1)^n 2^n}{n!}.$$

4 Determine the first six terms of the Taylor series solution of

$$y' + y = e^x$$

expanded about the origin, given that $y(0)=3$.

5 Determine the first four terms of the Taylor series solution of

$$(1 + x^2)y' + 2y = 0$$

expanded about $x=1$, given that $y(1)=3$.

6 Determine the first four nonvanishing terms of the Taylor series solution of

$$y' + 3xy = 0$$

expanded about the origin, given that $y(0)=2$.

7 Determine the first three nonvanishing terms of the Taylor series solution of

$$(1+x)y' + y = \tan x$$

expanded about $x = \pi/4$, given that $y(\pi/4) = 1$.

8 Determine the first two nonzero terms of the Taylor series solution of

$$y' + (1+x^2)y = \sin x$$

expanded about $x - \pi/2$, given that $y(\pi/2) = 0$.

9 Determine the first six terms of the Taylor series solution of

$$y' - xy = \cos x$$

expanded about the origin, given that $y(0) = 1$.

10 Use the Taylor series method to show that the power series solution of

$$y'' + \lambda^2 y = 0$$

expanded about the origin, with $y(0) = \alpha$, $y'(0) = \beta$ is

$$y(x) = \alpha\left(1 - \frac{(\lambda x)^2}{2!} + \frac{(\lambda x)^4}{4!} - \frac{(\lambda x)^6}{6!} + \dots\right) + \beta\left(\lambda x - \frac{(\lambda x)^3}{3!} + \frac{(\lambda x)^5}{5!} - \dots\right),$$

and hence that

$$y(x) = \alpha\cos\lambda x + \beta\sin\lambda x.$$

11 Use the Taylor series method to show that the power series solution of

$$y'' - \lambda^2 y = 0$$

expanded about the origin, with $y(0) = \alpha$, $y'(0) = \beta$ is

$$y(x) = \alpha\left(1 + \frac{(\lambda x)^2}{2!} + \frac{(\lambda x)^4}{4!} + \frac{(\lambda x)^6}{6!} + \dots\right) + \beta\left(\lambda x + \frac{(\lambda x)^3}{3!} + \frac{(\lambda x)^5}{5!} + \dots\right),$$

and hence that

$$y(x) = \alpha\cosh\lambda x + \beta\sinh\lambda x.$$

12 Determine the first four nonzero terms of the Taylor series expansion of

$$y'' - 2y' + 2y = 0$$

expanded about the origin, given that $y(0) = y'(0) = 2$.

13 Determine the first four nonzero terms of the Taylor series solution of

$$y'' + 4y' + 5y = 0$$

expanded about the origin, given that $y(0) = 0$, $y'(0) = 1$.

14 Determine the first five terms of the Taylor series solution of

$$(1+x^2)y'' + y = 0$$

expanded about the point $x = 1$, given that $y(1) = 1$, $y'(1) = -1$.

15 Determine the first five terms of the Taylor series solution of

$$y'' - 2xy' + 2y = 0$$

expanded about the point $x = 3$, given that $y(3) = \alpha$, $y'(3) = \beta$.

16 Determine the first five nonzero terms of the Taylor series solution of

$$(1+x)y'' + y = 0$$

expanded about the origin, given that $y(0) = 2$, $y'(0) = 0$.

17 Determine the first five nonzero terms of the Taylor series solution of

$$2y'' + y' + xy = 0$$

expanded about the origin, given that $y(0) = 0$, $y'(0) = 1$.

18 Determine the first four nonzero terms of the Taylor series solution of

$$y'' + xy' + y = x^2$$

expanded about the origin, given that $y(0) = 0$, $y'(0) = 1$.

19 Determine the first four terms of the Taylor series solution of

$$xy'' + y' + y = \sin x$$

expanded about the point $x = \pi/3$, given that $y(\pi/3) = 1$, $y'(\pi/3) = \frac{1}{2}$.

20 Determine the first five nonzero terms of the Taylor series solution of

$$x^2 y'' + 2xy' + y = x^2 + x + 1$$

expanded about the point $x = 1$, given that $y(1) = y'(1) = 1$.

8.3 Solution in the neighborhood of an ordinary point

We now turn our attention to the determination of the solution of the **variable coefficient** second order linear differential equation

$$P(x)y'' + Q(x)y' + R(x)y = f(x) \tag{1}$$

in the form of a series expanded about a point x_0. It will be seen later that the type of solution obtained will depend upon whether x_0 is an ordinary point or a regular singular point of the differential equation.

Any point x_0 of (1) at which $P(x_0) \neq 0$, and about which $P(x)$, $Q(x)$, $R(x)$ and $f(x)$ are analytic functions[4] convergent in some interval $x_0 - \rho < x < x_0 + \rho$, is called an **ordinary point** of (1), also a **regular point**. All other points are called **singular points**, or **singularities**, of the differential equation.

The following existence theorem which will be stated without proof is fundamental to what is to follow in this section, because it tells us the general form taken by a series solution of (1) at an ordinary point. This is an example of what is called a *constructive* existence theorem, in the sense that as well as telling us that a solution exists, it also indicates the structure of the solution.

---

[4] Can be expanded as a Taylor series about the point x_0.

Theorem 8.1 (Existence of a power series solution at an ordinary point)

Let x_0 be an ordinary point of differential equation (1) in which $P(x)$, $Q(x)$, $R(x)$ and $f(x)$ are all analytic functions in some interval of finite length centered on x_0. Then there exists a power series solution of (1) expanded about x_0 of the form

$$y(x) = \sum_{n=0}^{\infty} a_n (x - x_0)^n,$$

which converges in an interval $x_0 - r < x < x_0 + r$ for some $r > 0$. ☐

When developing the theory of ordinary differential equations there are certain advantages in allowing both x and the solution $y(x)$ to assume complex values. We shall not develop such an approach here, but merely remark that in the complex case the radius of convergence r in Theorem 8.1 is no smaller than the smallest of the numbers comprising the radii of convergence of the analytic functions $P(x)$, $Q(x)$, $R(x)$ and $f(x)$ about the point x_0 in the complex plane, and the distance from x_0 to the closest zero (real or complex) of $P(x)$.

To obtain the precise form of solution guaranteed to exist by Theorem 8.1 in the neighborhood of an ordinary point x_0 of (1) it is first necessary to substitute the power series

$$y(x) = \sum_{n=0}^{\infty} a_n (x - x_0)^n \tag{2}$$

into (1). When doing this we make use of the results obtained by differentiation of (2)

$$y'(x) = \sum_{n=1}^{\infty} n a_n (x - x_0)^{n-1}, \tag{3}$$

and

$$y''(x) = \sum_{n=2}^{\infty} n(n - 1) a_n (x - x_0)^{n-2}, \tag{4}$$

and we also substitute into (1) the power series expansions about the point x_0 of the functions $P(x)$, $Q(x)$, $R(x)$ and $f(x)$. Then since the resulting expression is an identity in x, and so must be true for all x in the interval of convergence of the series solution (2), it follows that the coefficient of each power of $(x - x_0)$ must vanish.

Grouping together terms involving the same power of $(x - x_0)$, and equating their coefficients to zero, gives rise to **recursion relations** for the unknown coefficients $a_0, a_1, a_2,$ ... in (2). We shall see that the general coefficients may all be related to the first two coefficients a_0 and a_1, which may be regarded as the two arbitrary constants in the general solution of (1).

If equation (1) is *homogeneous* the process will generate the *complementary function*, while if it is *nonhomogeneous* it will generate the *sum* of the *complementary function* and a *particular integral.* In each case the specification of the initial conditions $y(x_0) = y_0$, $y'(x_0) = y_1$ (say) will determine a_0 and a_1, and hence the solution to the initial value problem.

This approach to the determination of a power series solution of (1) is now expressed in the form of an algorithm.

Algorithm for constructing a solution about an ordinary point

Let x_0 be an ordinary point of the differential equation

$$P(x)y'' + Q(x)y' + R(x)y = f(x).$$

Step 1. Substitute into the differential equation the expressions

$$y(x) = \sum_{n=0}^{\infty} a_n(x - x_0)^n, \quad y'(x) = \sum_{n=1}^{\infty} n a_n(x - x_0)^{n-1}$$

$$y''(x) = \sum_{n=2}^{\infty} n(n-1) a_n(x - x_0)^{n-2}$$

Step 2. Expand $P(x)$, $Q(x)$, $R(x)$ and $f(x)$ about the point x_0 as power series in $(x - x_0)$ and substitute these series into the differential equation.

Step 3. Gather together all terms involving the same power of $(x - x_0)$ to arrive at an identity of the form

$$A_0 + A_1(x - x_0) + A_2(x - x_0)^2 + \ldots \equiv 0,$$

in which each coefficient A_n involves some of the coefficients $a_0, a_1, a_2, \ldots$ and, possibly, numbers from the power series expansion of $f(x)$.

Step 4. Equate to zero each coefficient of the identity in step 3 to obtain

$$A_0 = 0, \quad A_1 = 0, \quad A_2 = 0, \ldots.$$

Step 5. Use the expressions obtained in step 4 to determine $a_2, a_3, \ldots$ in terms of a_0 and a_1 and insert these coefficients into the expression

$$y(x) = \sum_{n=0}^{\infty} a_n(x - x_0)^n,$$

to arrive at the general solution which will depend on the two arbitrary constants a_0 and a_1.

Step 6. If initial data has been given and

$$y(x_0) = y_0, \quad y'(x_0) = y_1,$$

use this in conjunction with the general solution obtained in step 5 to determine a_0 and a_1, and hence the solution of the required initial value problem.

Example 1. A second order homogeneous differential equation

Find the power series solution of

$$2y'' - 2xy' + y = 0,$$

expanded about the origin, and hence determine two linearly independent solutions of the differential equation.

Solution

It follows at once that the point $x = 0$ is an ordinary point of the differential equation, so we may set $x_0 = 0$ in Theorem 8.1 and seek a solution of the form

$$y(x) = \sum_{n=0}^{\infty} a_n x^n.$$

Setting $x_0 = 0$ in (2), (3) and (4) and substituting for y, y' and y'' as described in the algorithm leads to the result

$$\sum_{n=2}^{\infty} 2n(n-1)a_n x^{n-2} - \sum_{n=1}^{\infty} 2na_n x^n + \sum_{n=0}^{\infty} a_n x^n = 0.$$

Shifting the index in the first summation by setting $m = n - 2$ we obtain

$$\sum_{m=0}^{\infty} 2(m+2)(m+1)a_{m+2} x^m - \sum_{n=1}^{\infty} 2na_n x^n + \sum_{n=0}^{\infty} a_n x^n = 0.$$

Replacing the dummy summation index m by n, separating out the constant terms from the first and last summations and combining the remaining terms into a single summation gives

$$4a_2 + a_0 + \sum_{n=1}^{\infty} [2(n+1)(n+2)a_{n+2} - (2n-1)a_n]x^n = 0.$$

For this expression to be true for all x (to be an identity) the coefficient of each power of x must vanish, leading to the results

$$x^0: \ 2 \cdot 2 a_2 + a_0 = 0; \quad a_2 = -\frac{1}{2 \cdot 2!} a_0 \quad \text{(to show the pattern we have set } 4 = 2 \cdot 2 = 2 \cdot 2!)$$

$$x^1: \ 2 \cdot 2 \cdot 3 a_3 - a_1 = 0; \quad a_3 = \frac{1}{2 \cdot 3!} a_1$$

$$x^2: \ 2 \cdot 3 \cdot 4 a_4 - 3 a_2 = 0; \quad a_4 = \frac{3}{2 \cdot 3 \cdot 4} a_2 = -\frac{1 \cdot 3}{2^2 \cdot 3!} a_0$$

$$x^3: \quad 2 \cdot 4 \cdot 5 a_5 - 5 a_3 = 0; \quad a_5 = \frac{5}{2 \cdot 4 \cdot 5} a_3 = \frac{1 \cdot 5}{2^2 \cdot 5!} a_1$$

$$x^4: \quad 2 \cdot 5 \cdot 6 a_6 - 7 a_4 = 0; \quad a_6 = \frac{7}{2 \cdot 5 \cdot 6} a_4 = \frac{-1 \cdot 3 \cdot 7}{2^3 \cdot 6!} a_0$$

$$x^5: \quad 2.6.7 a_7 - 9 a_5 = 0; \quad a_7 = \frac{9}{2 \cdot 6 \cdot 7} a_5 = \frac{1 \cdot 5 \cdot 9}{2^3 \cdot 7!} a_1$$

$$x^6: \quad 2 \cdot 7 \cdot 8 a_8 - 11 a_6 = 0; \quad a_8 = \frac{11}{2 \cdot 7 \cdot 8} a_6 = \frac{-1 \cdot 3 \cdot 7 \cdot 11}{2^4 \cdot 8!} a_0$$

and, in general, we have the recursion formula

$$x^n: 2(n+1)(n+2) a_{n+2} - (2n-1) a_n = 0, \quad a_{n+2} = \frac{(2n-1)}{2(n+1)(n+2)} a_n,$$

$$\text{for } n = 1, 2, \ldots.$$

These results show that in this case all the coefficients of even powers of x depend on a_0, while those of the odd powers of x depend on a_1.

Inspection of the pattern of the coefficients shows that the even coefficients are given by

$$a_{2n} = -\frac{1 \cdot 3 \cdot 7 \cdot 11 \ldots (4n-5)}{2^n (2n)!} a_0, \quad n = 1, 2, 3, \ldots$$

and the odd coefficients by

$$a_{2n+1} = \frac{1 \cdot 5 \cdot 9 \cdot 13 \ldots (4n-3)}{2^n (2n+1)} a_1, \quad n = 1, 2, 3, \ldots.$$

Substituting these coefficients into the series for $y(x)$ and separating the terms depending on a_0 from those depending on a_1 we arrive at the general solution

$$y(x) = a_0 \left(1 - \frac{1}{2 \cdot 2!} x^2 - \frac{1 \cdot 3}{2^2 \cdot 3!} x^4 - \frac{1 \cdot 3 \cdot 7}{2^3 \cdot 6!} x^6 - \frac{1 \cdot 3 \cdot 7 \cdot 11}{2^4 \cdot 8!} x^8 - \ldots \right)$$

$$+ a_1 \left(x + \frac{1}{2 \cdot 3!} x^3 + \frac{1 \cdot 5}{2^2 \cdot 5!} x^5 + \frac{1 \cdot 5 \cdot 9}{2^3 \cdot 7!} x^7 + \ldots \right).$$

The rearrangement of terms in the original power series to arrive at this form of solution is justified because of the absolute convergence of the original series within its interval of convergence. The radius of convergence is, in fact, infinite, because it is easily established that each of the series in $y(x)$ has an infinite radius of convergence.

If the general solution is written as

$$y(x) = a_0 y_1(x) + a_2 y_2(x),$$

with

$$y_1(x) = 1 - \frac{1}{2 \cdot 2!} x^2 - \frac{1 \cdot 3}{2^2 \cdot 3!} x^4 - \frac{1 \cdot 3 \cdot 7}{2^3 \cdot 6!} x^6 - \ldots$$

and

$$y_2(x) = x + \frac{1}{2 \cdot 3!}x^3 + \frac{1 \cdot 5}{2^2 \cdot 5!}x^5 + \frac{1 \cdot 5 \cdot 9}{2^3 \cdot 7!} + \cdots,$$

then $y_1(x)$ and $y_2(x)$ may be taken as the two linearly independent solutions of the differential equation. Their linear independence follows directly from the fact that $y_1(x)$ is an even function and $y_2(x)$ an odd function.   ∎

Example 2. An expansion about the point 1

Find the terms up to and including $(x-1)^5$ in the power series solution of **Airy's equation**

$$y'' - xy = 0$$

expanded about the point $x = 1$.

Solution

It is clear that $x = 1$ is an ordinary point of Airy's equation so we may set $x_0 = 1$ in Theorem 8.1 and seek a solution of the form

$$y(x) = \sum_{n=0}^{\infty} a_n(x-1)^n.$$

Before substituting for y and y'' from (2) and (4) by setting $x_0 = 1$ it is necessary to express the function multiplying y as a power series expanded about the point 1. This is easily accomplished because $x = (x-1) + 1$, so the equation may be written in the form

$$y'' - (x-1)y - y = 0.$$

Substituting for y'' and y we obtain the result

$$\sum_{n=2}^{\infty} n(n-1)a_n(x-1)^{n-2} - \sum_{n=0}^{\infty} a_n(x-1)^{n+1} - \sum_{n=0}^{\infty} a_n(x-1)^n = 0.$$

Shifting the index in the first summation by setting $m = n-2$, and the index in the second summation by setting $r = n+1$, gives

$$\sum_{m=0}^{\infty} (m+2)(m+1)a_{m+2}(x-1)^m - \sum_{r=1}^{\infty} a_{r-1}(x-1)^r - \sum_{n=0}^{\infty} a_n(x-1)^n = 0.$$

Separating out the first terms from the first and last summations, replacing the dummy indices m and r by n, and combining the remaining terms under a single summation sign gives

$$1 \cdot 2a_2 - a_0 + \sum_{n=1}^{\infty} [(n+1)(n+2)a_{n+2} - a_{n-1} - a_n](x-1)^n = 0.$$

Equating the coefficient of each power of $(x-1)$ to zero now gives

$$(x-1)^0: \; 1 \cdot 2a_2 - a_0 = 0; \quad a_2 = \frac{1}{2!}a_0$$

$$(x-1)^1: \; 2 \cdot 3a_3 - a_0 - a_1 = 0; \quad a_3 = \frac{1}{3!}(a_0 + a_1)$$

$$(x-1)^2: \quad 3 \cdot 4 a_4 - a_1 - a_2 = 0; \quad a_4 = \frac{1}{3 \cdot 4}(a_1 + a_2) = \frac{1}{3 \cdot 4} a_1 + \frac{1}{4!} a_0$$

$$(x-1)^3: \quad 4 \cdot 5 a_5 - a_2 - a_3 = 0; \quad a_5 = \frac{1}{4 \cdot 5}(a_2 + a_3) = \left(\frac{1}{1 \cdot 2 \cdot 4 \cdot 5} + \frac{1}{5!}\right) a_0 + \frac{1}{5!} a_1$$

and, in general, we have the recursion formula

$$(x-1)^n: \quad (n+1)(n+2)a_{n+2} - a_{n-1} - a_n = 0; \quad a_{n+2} = \frac{1}{(n+1)(n+2)}(a_{n-1} + a_n),$$

$$\text{for } n = 1, 2, \ldots .$$

In this case the complexity of the recursion formula makes it impossible to find general expressions for the coefficients, so they must be determined sequentially, as above.

Substituting these terms into the power series and grouping terms gives for the first few terms of the solution

$$y(x) = a_0 \left[1 + \frac{1}{2!}(x-1)^2 + \frac{1}{3!}(x-1)^3 + \frac{1}{4!}(x-1)^4 + \left(\frac{1}{1 \cdot 2 \cdot 4 \cdot 5} + \frac{1}{5!}\right)(x-1)^5 + \cdots \right]$$

$$+ a_1 \left((x-1) + \frac{1}{3!}(x-1)^3 + \frac{1}{3 \cdot 4}(x-1)^4 + \frac{1}{5!}(x-1)^5 + \cdots \right).$$

The functions

$$y_1(x) = 1 + \frac{1}{2!}(x-1)^2 + \frac{1}{3!}(x-1)^3 + \frac{1}{4!}(x-1)^4 + \left(\frac{1}{1 \cdot 2 \cdot 4 \cdot 5} + \frac{1}{5!}\right)(x-5)^5 + \cdots$$

and

$$y_2(x) = (x-1) + \frac{1}{3!}(x-1)^3 + \frac{1}{3 \cdot 4}(x-1)^4 + \frac{1}{5!}(x-1)^5 + \cdots$$

may be taken as the two linearly independent solutions of the differential equation. This may be seen from the fact that $y_1(x)/y_2(x)$ is a nonconstant function of x (see Sec. 5.1). ■

Although the method of solution described in the algorithm involved a second order differential equation, after obvious modifications it may be applied equally well to a differential equation of any order. The last example illustrates this in the case of a first order equation.

Example 3. A nonhomogeneous first order equation

Find the terms up to and including the power x^5 in the series solution of

$$y' + (1+x)y = 1 + x^2$$

expanded about the origin, and use the result to find the corresponding terms in the solution satisfying the initial condition $y(0) = 3$.

Solution

This example involves modifying the algorithm and applying its method of solution to a first order equation. The point $x=0$ about which the solution is to be expanded is an ordinary point of the first order equation. This can be seen from the fact that the multiplier (unity) of the highest order derivative is nowhere zero, while the multiplier of y and the nonhomogeneous term are both analytic functions for all x. Indeed, they are already in the form of power series expanded about the origin.

Thus setting $x_0=0$ in (2) and (3) and substituting for y and y' gives

$$\sum_{n=1}^{\infty} na_n x^{n-1} + \sum_{n=0}^{\infty} a_n x^n + \sum_{n=0}^{\infty} a_n x^{n+1} = 1 + x^2.$$

Shifting the indices to bring the general terms in the first and last summations to those involving x^n we obtain

$$\sum_{m=0}^{\infty} (m+1)a_{m+1} x^m + \sum_{n=0}^{\infty} a_n x^n + \sum_{r=1}^{\infty} a_{r-1} x^r = 1 + x^2,$$

which after replacing the dummy summation indices m and r by n becomes

$$\sum_{n=0}^{\infty} (n+1)a_{n+1} x^n + \sum_{n=0}^{\infty} a_n x^n + \sum_{n=1}^{\infty} a_{n-1} x^n = 1 + x^2.$$

Grouping corresponding powers of x we arrive at the result

$$(a_1 + a_0 - 1) + (2a_2 + a_1 + a_0)x + (3a_3 + a_2 + a_1 - 1)x^2$$

$$+ \sum_{n=2}^{\infty} [(n+1)a_{n+1} + a_n + a_{n-1}]x^n = 0.$$

Equating to zero the coefficient of each power of x in this identity gives

$$x^0: \quad a_1 + a_0 - 1 = 0; \quad a_1 = 1 - a_0$$

$$x^1: \quad 2a_2 + a_1 + a_0 = 0; \quad a_2 = -\tfrac{1}{2}(a_1 + a_0) = -\tfrac{1}{2}$$

$$x_2: \quad 3a_3 + a_2 + a_1 = 0; \quad a_3 = -\tfrac{1}{3}(a_2 + a_1) = \tfrac{1}{6} + \tfrac{1}{3}a_0$$

$$x_3: \quad 4a_4 + a_3 + a_2 = 0; \quad a_4 = -\tfrac{1}{4}(a_3 + a_2) = \tfrac{1}{12} - \tfrac{1}{12}a_0$$

$$x_4: \quad 5a_5 + a_4 + a_3 = 0; \quad a_5 = -\tfrac{1}{5}(a_4 + a_3) = -\tfrac{1}{20} - \tfrac{1}{20}a_0$$

and, in general,

$$x^n: \quad (n+1)a_{n+1} + a_n + a_{n-1} = 0; \quad a_{n+1} = \frac{-1}{n+1}(a_n + a_{n-1}), \qquad \text{for } n = 2, 3, \ldots.$$

Substituting for the coefficients a_n in the expression

$$y(x) = \sum_{n=0}^{\infty} a_n x^n$$

and grouping together terms depending on a_0 we arrive at the general solution of the differential equation

$$y(x) = a_0 [1 - x + \tfrac{1}{3}x^3 - \tfrac{1}{12}x^4 - \tfrac{1}{20}x^5 + \ldots] + [x - \tfrac{1}{2}x^2 + \tfrac{1}{6}x^3 + \tfrac{1}{12}x^4 - \tfrac{1}{20}x^5 + \ldots].$$

The bracketed terms multiplied by a_0 correspond to the *complementary function*, while the terms independent of a_0 belong to the *particular integral*. The corresponding solution of the initial value problem in which $y(0) = 3$ follows from this result by matching the general solution to this condition at the origin. This shows that $a_0 = 3$. In this problem, no simple expression can be found for the general term a_n, so the terms must be determined step by step using the recursion formula for $n \geq 2$.

∎

Problems for Section 8.3

1 Find the terms up to and including x^9 in the power series solution of Airy's equation

$$y'' - xy = 0$$

expanded about the origin.

2 Find the power series solution of

$$y'' - xy' + 3y = 0$$

expanded about the origin, and show that one of its linearly independent solutions is a cubic polynomial.

3 Find the terms up to and including x^5 in the power series solution of

$$2y'' + y' + xy = 0$$

expanded about the origin.

4 Find the terms up to and including x^7 in the power series solution of

$$(1 + x^2)y'' + y = 0$$

expanded about the origin.

5 Find the power series solution of

$$y'' - 2xy' + \alpha y = 0$$

expanded about the origin.

6 Find the power series solution of

$$2y'' - 4xy' + y = 0$$

expanded about the origin.

7 Find the terms up to and including $(x - 1)^5$ in the power series solution of

$$y'' + x^2 y = 0$$

expanded about $x = 1$.

8 Find the terms up to and including $(x + 1)^5$ in the power series solution of

$$x^2 y'' + y = 0$$

expanded about $x = -1$.

9 Find the terms up to and including x^6 in the power series solution of

$$(1 - x^2)y'' - 2xy' + 12y = 0$$

expanded about the origin. Comment on the form taken by one of the linearly independent solutions.

10 Find the terms up to and including x^7 in the power series solution of

$$(1-x^2)y'' - 2xy' + 20y = 0$$

expanded about the origin. Comment on the form taken by one of the linearly independent solutions.

11 Find the terms up to and including x^6 in the power series solution of

$$y'' + xy = e^x$$

expanded about the origin.

12 Find the terms up to and including x^6 in the power series solution of

$$y'' + xy = x$$

expanded about the origin.

13 Find the terms up to and including $(x-1)^5$ in the power series solution of

$$(x+1)y' + xy = 0$$

expanded about $x = 1$.

14 Find the power series solution of

$$(x+2)y' + y = 0$$

expanded about the origin.

15 Find the power series solution of

$$(x+2)y' + y = x$$

expanded about the origin.

16 Find the terms up to and including $(x-2)^6$ in the power series solution of

$$xy' + (x+2)^2 y = 0$$

expanded about $x = 2$.

8.4 Legendre's equation and Legendre polynomials

A variable coefficient differential equation of considerable importance in engineering and physics is **Legendre's equation**[5]

$$(1-x^2)\frac{d^2 y}{dx^2} - 2x\frac{dy}{dx} + \alpha(\alpha+1)y = 0, \tag{1}$$

in which the parameter α is a real number. The equation arises naturally when seeking the solution of boundary value problems for certain types of partial differential equation

[5] ADRIEN-MARIE LEGENDRE (1752–1833), an outstanding French mathematician who was a contemporary of Laplace. He made contributions to geometry, number theory, elliptic functions and to the study of the gravitational attraction of ellipsoids which led to his interest in the differential equation which now bears his name.

where the geometry of the problem dictates the use of spherical polar coordinates with symmetry about the axis. Typical of such problems is the determination of three-dimensional potential distributions, such as an axisymmetric velocity potential in fluid mechanics or the electrostatic potential at a point in space due to a circular wire carrying a uniformly distributed electrostatic charge. Of particular importance in the above applications, and also in numerical analysis and approximation theory, is the *orthogonality property* of special solutions of Legendre's equation corresponding to integral value of α.

Using the notation of (1), Sec. 8.3, we see that in the case of Legendre's equation

$$P(x)=1-x^2, \quad Q(x)=-2x \quad \text{and} \quad R(x)=\alpha(\alpha+1).$$

As these are analytic functions for all x, and $P(x)=0$ only when $x=\pm1$, it follows that Legendre's equation has singular points at 1 and -1, with all other points being ordinary points of the equation. Let us now seek a power series solution of (1) expanded about the origin (an ordinary point). We start by setting

$$y(x)=\sum_{m=0}^{\infty} a_m x^m.$$

Substituting (2) into (1) gives

$$(1-x^2)\sum_{m=2}^{\infty} m(m-1)a_m x^{m-2}-2x\sum_{m=1}^{\infty} ma_m x^{m-1}+\alpha(\alpha+1)\sum_{m=0}^{\infty} a_m x^m=0,$$

or

$$\sum_{m=2}^{\infty} m(m-1)a_m x^{m-2}-\sum_{m=2}^{\infty} m(m-1)a_m x^m-\sum_{m=1}^{\infty} 2ma_m x^m+\sum_{m=0}^{\infty} \alpha(\alpha+1)\alpha_m x^m=0.$$

Shifting the index in the first term brings us to the result

$$\sum_{m=0}^{\infty} (m+2)(m+1)a_{m+2}x^m-\sum_{m=2}^{\infty} m(m-1)a_m x^m-\sum_{m=1}^{\infty} 2ma_m x^m$$

$$+\sum_{m=0}^{\infty} \alpha(\alpha+1)a_m x^m=0,$$

which is equivalent to

$$[2a_2+\alpha(\alpha+1)a_0]+[6a_3-2a_1+\alpha(\alpha+1)a_1]x$$
$$+\sum_{m=2}^{\infty} \{(m+1)(m+2)a_{m+2}-[m^2+m-\alpha(\alpha+1)]a_m\}x^m=0. \quad (3)$$

Equating the coefficients of each power of x to zero gives

$$x^0: \quad 2a_2+\alpha(\alpha+1)a_0=0; \quad a_2=-\frac{\alpha(\alpha+1)}{2}a_0,$$

$$x^1: \quad 6a_3-2a_1+\alpha(\alpha+1)=0; \quad a_3=\frac{2-\alpha(\alpha+1)}{6}a_1,$$

and the general recursion relation

$$x^m: \quad (m+1)(m+2)a_{m+2}-[m^2+m-\alpha(\alpha+1)]a_m=0,$$

which may be rewritten as

$$a_{m+2}=\frac{(m-\alpha)(m+\alpha+1)}{(m+1)(m+2)}a_m, \qquad m=0,1,2,\ldots. \tag{4}$$

These results show that all coefficients with even suffixes are expressible in terms of a_0, which is arbitrary, while all coefficients with odd suffixes are expressible in terms of a_1, which is also arbitrary.

The first six coefficients generated in this manner are, in terms of the arbitrary constants a_0 and a_1,

$$a_2=-\frac{\alpha(\alpha+1)}{2!}a_0$$

$$a_3=-\frac{(\alpha-1)(\alpha+2)}{3!}a_1$$

$$a_4=\frac{(\alpha-2)\alpha(\alpha+1)(\alpha+3)}{4!}a_0$$

$$a_5=\frac{(\alpha-3)(\alpha-1)(\alpha+2)(\alpha+4)}{5!}a_1 \tag{5}$$

$$a_6=-\frac{(\alpha-4)(\alpha-2)\alpha(\alpha+1)(\alpha+3)(\alpha+5)}{6!}a_0$$

$$a_7=-\frac{(\alpha-5)(\alpha-3)(\alpha-1)(\alpha+2)(\alpha+4)(\alpha+6)}{7!}a_1.$$

Inserting these coefficients into (2) and grouping terms depending on a_0 separately from those depending on a_1 gives the following power series solution of Legendre's equation with the parameter α

$$y(x)=a_0\left(1-\frac{\alpha(\alpha+1)x^2}{2!}+\frac{(\alpha-2)\alpha(\alpha+1)(\alpha+3)}{4!}x^4\right.$$

$$\left.-\frac{(\alpha-4)(\alpha-2)\alpha(\alpha+1)(\alpha+3)(\alpha+5)}{6!}x^6+\ldots\right)$$

$$+a_1\left(x-\frac{(\alpha-1)(\alpha+2)}{3!}x^3+\frac{(\alpha-3)(\alpha-1)(\alpha+2)(\alpha+4)}{5!}x^5\right.$$

$$\left.-\frac{(\alpha-5)(\alpha-3)(\alpha-1)(\alpha+2)(\alpha+4)(\alpha+6)}{7!}x^7+\ldots\right).$$

The series multiplied by a_0 is an even function and the one multiplied by a_1 is an odd function, so the solutions represented by the series

$$p_\alpha(x) = 1 - \frac{\alpha(\alpha+1)}{2!}x^2 + \frac{(\alpha-2)\alpha(\alpha+1)(\alpha+3)}{4!}x^4 - \ldots,$$ (6)

and

$$q_\alpha(x) = x - \frac{(\alpha-1)(\alpha+2)}{3!}x^3 + \frac{(\alpha-3)(\alpha-1)(\alpha+2)(\alpha+4)}{5!}x^5 - \ldots,$$ (7)

each depending on α, are linearly independent solutions of Legendre's equation. The general solution of the equation for arbitrary real α is thus

$$y(x) = a_0 p_\alpha(x) + a_1 q_\alpha(x).$$ (8)

A case of particular interest when applications are made arises when the parameter α is a nonnegative integer n. Inspection of recursion relation (4) shows that then every other coefficient beyond a_n vanishes, so we have

$$a_{n+2} = 0, \quad a_{n+4} = 0, \quad a_{n+6} = 0, \ldots.$$

Consequently, depending on whether n is even or odd, either the solution $p_n(x)$ or the solution $q_n(x)$ will reduce to a polynomial in x of degree n.

When n is even, and $n = 2m$, the general solution of Legendre's equation becomes

$$y(x) = a_0 \text{ (polynomial of degree } 2m) + a_1 q_{2m}(x), \quad m = 0, 1, 2, \ldots,$$

whereas when n is odd, and $n = 2m+1$, the general solution becomes

$$y(x) = a_0 p_{2m+1}(x) + a_1 \text{ (polynomial of degree } 2m+1), \quad m = 0, 1, 2, \ldots.$$

It can be shown that $p_{2m+1}(x)$ and $q_{2m}(x)$ diverge for $x = \pm 1$, so the special polynomial solutions of Legendre's equation are the only solutions which remain *bounded* (finite) on the interval $-1 \le x \le 1$.

As polynomial solutions may be multiplied by a constant and still remain solutions, we take advantage of this fact to normalize them in a convenient manner. Returning to recursion relation (4) we replace m by $m-2$ and rewrite it as

$$a_{m-2} = \frac{m(m-1)}{(m-n-2)(m+n-1)}a_m, \quad \text{for } m \le n.$$ (9)

This enables all the coefficients of a polynomial solution to be expressed in terms of the coefficient a_n of its highest degree term. As the coefficient a_n is now an arbitrary constant, for later convenience we choose to set

$$a_n = \frac{(2n)!}{2^n(n!)^2} = \frac{1\cdot3\cdot5\ldots(2n-1)}{n!}, \quad n = 1, 2, \ldots,$$ (10)

and $a_n = 1$ when $n = 0$.

Putting $m = n$ in (9) and using (10) gives

$$a_{n-2} = \frac{-n(n-1)}{2(2n-1)} \cdot \frac{(2n)!}{2^n(n!)^2}$$

$$= \frac{-(2n-2)!}{2^n(n-1)!(n-2)!}.$$

Repeating the calculation, but this time setting $m = n-2$ in (9) and using the above result, gives

$$a_{n-4} = \frac{-(n-2)(n-3)}{4(2n-3)} a_{n-2}$$

$$= \frac{(2n-4)!}{2^n 2!(n-2)!(n-4)!}.$$

Proceeding in this manner we arrive at the general result

$$a_{n-2m} = (-1)^m \frac{(2n-2m)!}{2^n m! \, (n-m)! \, (n-2m)!}, \tag{11}$$

for $n-2m \geq 0$.

The polynomial solutions of Legendre's equation obtained in this manner are called **Legendre polynomials**, and the polynomial solution of degree n in x is denoted by $P_n(x)$. Thus the Legendre polynomials are defined as

$$P_n(x) = \begin{cases} \displaystyle\sum_{m=0}^{n/2} (-1)^m \frac{(2n-2m)!}{2^n m! \, (n-m)! \, (n-2m)!} x^{n-2m} & (n \text{ even}) \\ \displaystyle\sum_{m=0}^{(n-1)/2} (-1)^m \frac{(2n-2m)!}{2^n m! \, (n-m)! \, (n-2m)!} x^{n-2m} & (n \text{ odd}). \end{cases} \tag{12}$$

The effect of the normalization used in (10) is to make (see Prob. 18)

$$P_n(1) = 1 \quad \text{and} \quad P_n(-1) = (-1)^n, \qquad n = 0, 1, \ldots. \tag{13}$$

Routine calculations show that the first eight Legendre polynomials are

$$\begin{aligned} P_0(x) &= 1 \\ P_1(x) &= x \\ P_2(x) &= \tfrac{1}{2}(3x^2 - 1) \\ P_3(x) &= \tfrac{1}{2}(5x^3 - 3x) \\ P_4(x) &= \tfrac{1}{8}(35x^4 - 30x^2 + 3) \\ P_5(x) &= \tfrac{1}{8}(63x^5 - 70x^3 + 15x) \\ P_6(x) &= \tfrac{1}{16}(231x^6 - 315x^4 + 105x^2 - 5) \\ P_7(x) &= \tfrac{1}{16}(429x^7 - 693x^5 + 315x^3 - 35x). \end{aligned} \tag{14}$$

Graphs of the even Legendre polynomials are shown in Fig. 8.3(a) and of the odd polynomials in Fig. 8.3(b).

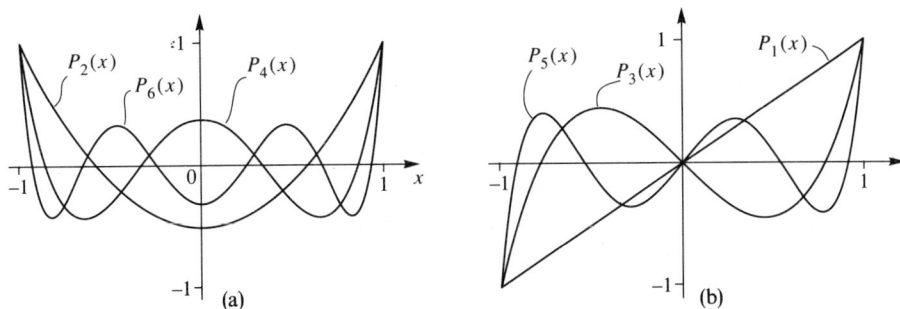

Fig. 8.3 Legendre polynomials: (a) Even polynomials, (b) Odd polynomials

Together with other special polynomials, Legendre polynomials possess the property of being **orthogonal polynomials**. The **orthogonality relation** for Legendre polynomials takes the form

$$\int_{-1}^{1} P_m(x)P_n(x)\,dx = 0 \qquad \text{for } m \neq n. \tag{15}$$

Associated with (15) is the **normalization condition**

$$\int_{-1}^{1} [P_n(x)]^2\,dx = \frac{2}{2n+1}, \qquad n = 0, 1, 2, \ldots . \tag{16}$$

Let us prove orthogonality condition (15). Suppose $P_m(x)$ and $P_n(x)$ are any two different Legendre polynomials ($m \neq n$), then

$$(1-x^2)P_m'' - 2xP_m' + m(m+1)P_m = 0 \tag{17}$$

and

$$(1-x^2)P_n'' - 2xP_n' + n(n+1)P_n = 0, \tag{18}$$

where we use a prime to signify differentiation with respect to x.

Multiplying (17) by P_n and (18) by P_m and subtracting the two results brings us to the equation

$$(1-x^2)u' - 2xu + [m(m+1)-n(n+1)]P_m P_n = 0, \tag{19}$$

where we have set $u = P_m' P_n - P_n' P_m$. Then, as $u' = P_m'' P_n - P_n'' P_m$, we may rewrite (19) as

$$\frac{d}{dx}[(1-x^2)u] + [m(m+1)-n(n+1)]P_m P_n = 0. \tag{20}$$

Integrating (20) over the interval $[-1, 1]$ gives

$$[(1-x^2)u(x)]\Big|_{-1}^{1} + [m(m+1)-n(n+1)] \int_{-1}^{1} P_m(x)P_n(x)\,dx = 0. \tag{21}$$

However, the first term vanishes, because the factor $1-x^2$ is zero at $x = \pm 1$, and as $m \neq n$

the factor $[m(m+1)-n(n+1)] \neq 0$, so we conclude that

$$\int_{-1}^{1} P_m(x) P_n(x) \, dx = 0, \qquad m \neq n,$$

as was to be proved.

The proof of the normalization condition (16) is left as an exercise for the interested reader (see Problems 15 and 17), because although not difficult the method used goes beyond the immediate needs of this section.

Expansion in terms of Legendre polynomials

Just as functions may be expanded in terms of power series, so they may also be expanded in terms of Legendre polynomials on the interval $[-1, 1]$ over which the polynomials $P_n(x)$ are defined and orthogonal.

Let $f(x)$ be a function defined on $[-1, 1]$, and set

$$f(x) = \sum_{m=0}^{\infty} a_m P_m(x), \qquad \text{for } -1 \le x \le 1. \tag{22}$$

Our task is to determine the coefficients a_m in (22) when the function $f(x)$ is known. If we multiply (22) by $P_n(x)$ and integrate the resulting expression over the interval $-1 \le x \le 1$ we obtain

$$\int_{-1}^{1} f(x) P_n(x) \, dx = \sum_{m=0}^{\infty} a_m \int_{-1}^{1} P_m(x) P_n(x) \, dx. \tag{23}$$

Result (15) shows that each integral on the right-hand side of (23) vanishes, with the exception of the one in which $m = n$. So, using (16), we find (23) reduces to

$$\int_{-1}^{1} f(x) P_n(x) \, dx = \frac{2a_n}{2n+1}, \qquad n = 0, 1, 2, \ldots,$$

and so

$$a_n = \left(\frac{2n+1}{2}\right) \int_{-1}^{1} f(x) P_n(x) \, dx, \qquad n = 0, 1, 2, \ldots. \tag{24}$$

Thus we have determined the coefficients $a_0, a_1, \ldots$ in expansion (22) in terms of a given function $f(x)$ and Legendre polynomials. We have not, of course, justified seeking an expansion of the form (22) for all functions $f(x)$. A full discussion of this problem (the **completeness** of the system of Legendre polynomials) would take us beyond this first account of the subject.

It follows immediately that as $P_n(x)$ is a polynomial in x of degree n, an arbitrary polynomial $Q(x)$ of degree n may always be expressed as a linear combination of $P_0(x)$, $P_1(x), \ldots, P_n(x)$. Thus when applying (22) and (24) to a polynomial of degree N, expansion (22) will terminate after the term involving $P_N(x)$.

Example 1. Expansion of a quadratic in terms of Legendre polynomials

Expand

$$f(x) = a + bx + cx^2$$

over the interval $[-1, 1]$ in terms of Legendre polynomials.

Solution

It follows from (14) and (24) that

$$a_0 = \frac{1}{2}\int_{-1}^{1}(a + bx + cx^2)\,dx = a + \frac{c}{3}$$

$$a_1 = \frac{3}{2}\int_{-1}^{1}(a + bx + cx^2)x\,dx = b$$

$$a_2 = \frac{5}{2}\int_{-1}^{1}(a + bx + cx^2)\frac{1}{2}(3x^2 - 1)\,dx = \frac{2c}{3},$$

and thus that

$$a + bx + cx^2 = \left(a + \frac{c}{3}\right)P_0(x) + bP_1(x) + \frac{2c}{3}P_2(x), \qquad -1 \le x \le 1. \qquad \blacksquare$$

Problems for Section 8.4

1 Verify by direct substitution that $P_6(x)$ and $P_7(x)$ are solutions of Legendre's equation corresponding to $\alpha = 6$ and $\alpha = 7$, respectively.

2 Derive the Legendre polynomials listed in (13) from definition (12).

3 Show that

$$P_8(x) = \tfrac{1}{128}(35 - 1260x^2 + 6930x^4 - 12012x^6 + 6435x^8)$$

and

$$P_9(x) = \tfrac{1}{128}(315x - 4620x^3 + 18018x^5 - 25740x^7 + 12155x^9).$$

4 Show that

$$P_n(x) = \frac{(2n)!}{2^n(n!)^2}\left(x^n - \frac{n(n-1)}{2(2n-1)}x^{n-2} + \frac{n(n-1)(n-2)(n-3)}{2 \cdot 4(2n-1)(2n-3)}x^{n-4} - \cdots\right).$$

5 Prove that the series defining $p_a(x)$ and $q_a(x)$ in (6) and (7) are absolutely convergent for $|x| < 1$.

6 Verify (14) and (15) by direct computation, using the explicit forms for $P_0(x)$, $P_1(x)$, $P_2(x)$ and $P_3(x)$ given in (14).

7 Express in terms of Legendre polynomials the function

$$f(x) = 3 + 4x^2 - x^3.$$

8 Express in terms of Legendre polynomials the function

$$f(x) = 2x + x^3 + 4x^4.$$

Harder problems

9 Consider the function

$$G(x, r) = \frac{1}{(1 - 2xr + r^2)^{1/2}}.$$

Set $u = 2xr - r^2$ and expand $(1 - u)^{-1/2}$ by the binomial theorem. Substitute for u in terms of x and r and, after expanding products, collect together all terms involving r^n. Then, by comparison with (12), show that the coefficient of r^n in this expansion is the Legendre polynomial $P_n(x)$. This proves that when the function $G(x, r)$ is expanded as a polynomial in r it may be written

$$G(x, r) = \frac{1}{(1 - 2xr + r^2)^{1/2}} = \sum_{n=0}^{\infty} P_n(x) r^n. \tag{24}$$

The function $G(x, r) = (1 - 2xr + r^2)^{-1/2}$ is called the **generating function** for Legendre polynomials.

10 This problem provides an alternative proof that the function $G(x, r)$ defined in Prob. 9 is the generating function for Legendre polynomials. Regarding x as a parameter, and using the result of Prob. 4, show first that

$$\left. \frac{\partial^n G(x, r)}{\partial r^n} \right|_{r=0} = n! P_n(x).$$

Hence show that the Taylor series expansion of $G(x, r)$ about $r = 0$ leads to result (24).

11 Show by differentiation that

$$P_n(x) = \frac{1}{2^n n!} \frac{d^n}{dx^n} [(x^2 - 1)^n] \tag{25}$$

is a solution of

$$(1 - x^2) \frac{d^2 P_n}{dx^2} - 2x \frac{dP_n}{dx} + n(n+1) P_n = 0,$$

and hence that (25) provides an alternative definition of Legendre polynomials. Expression (25) is called **Rodrigues' formula**[6] for Legendre polynomials.

[*Hint*: Set $u = (x^2 - 1)^n$ and by repeated differentiation obtain a differential equation relating $u^{(m+2)}$, $u^{(m+1)}$ and $u^{(m)}$.]

12 This problem provides an alternative derivation of Rodrigues' formula (25). Starting from the binomial expansion

$$(x^2 - 1)^n = \sum_{m=0}^{n} (-1)^m \frac{n!}{m!(n-m)!} x^{2(n-m)}$$

differentiate n times to obtain an expression for

$$\frac{d^n}{dx^n} [(x^2 - 1)^n].$$

Deduce Rodrigues' formula by comparing the expression so obtained with the result of Prob. 4.

---

[6] OLINDE RODRIGUES (1794–1851), a French mathematician who made contributions to differential geometry and to the mathematical theory of economics

13 Let P_1 and P_2 be two points in space distant r_1 and r_2 $(r_2 > r_1)$ from the origin, respectively, and let the distance between P_1 and P_2 be r as shown in Fig. 8.4, in which θ is the angle between OP_1 and OP_2. Use the familiar result from elementary trigonometry

$$r = (r_1^2 + r_2^2 - 2r_1 r_2 \cos \theta)^{1/2}$$

together with (24) to show that

$$\frac{1}{r} = \frac{1}{r_2}\left[P_0(\cos \theta) + \left(\frac{r_1}{r_2}\right) P_1(\cos \theta) + \left(\frac{r_1}{r_2}\right)^2 P_2(\cos \theta) + \ldots \right].$$

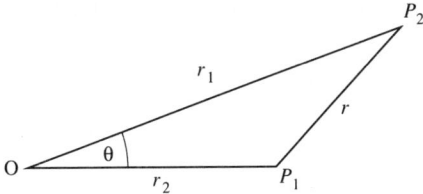

Fig. 8.4

14 Charges $\pm Q$ are located at the respective points A and B in Fig. 8.5, with $OA = OB = d$. The **electrostatic potential** ϕ at a point P distant r from the origin O and making an angle θ with OB is defined as

$$\phi = \frac{Q}{r_2} - \frac{Q}{r_1},$$

where $AP = r_1$ and $BP = r_2$.

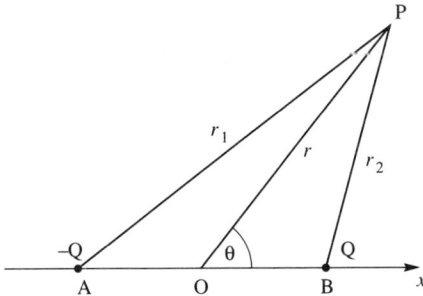

Fig. 8.5 Point charge distribution

Show by using the result of Prob. 13 that

$$\phi = \frac{2Q}{r} \sum_{m=0}^{\infty} \left(\frac{d}{r}\right)^{2m+1} P_{2m+1}(\cos \theta)\, (d < r).$$

Show that as $d \to 0$ and $Q \to \infty$ in such a way that $m = 2Qd$ remains finite, this reduces to the potential due to a **dipole**

$$\phi = \frac{m \cos \theta}{r^2}.$$

The quantity m is called the **dipole moment**.

15 Show by means of partial differentiation that the generating function $G(x, r)$ in (24) satisfies the equation

$$(1 - 2xr + r^2)\frac{\partial G}{\partial r} - (x - r)G = 0.$$

Substitute

$$G(x, r) = \sum_{n=0}^{\infty} P_n(x)r^n$$

into the equation satisfied by G and, after collecting together terms involving the same powers of r, equate the coefficient of r^n to zero to prove that

$$(n + 1)P_{n+1}(x) = (2n + 1)xP_n(x) - nP_{n-1}(x). \tag{26}$$

This **recurrence relation** for Legendre polynomials relates any three successive polynomials $P_{n+1}(x)$, $P_n(x)$ and $P_{n-1}(x)$.

16 Use recurrence relation (26) to generate $P_2(x)$ to $P_6(x)$, given that $P_0(x) = 1$ and $P_1(x) = x$.

17 Derive the recurrence relation

$$nP_n(x) = (2n - 1)xP_{n-1}(x) - (n - 1)P_{n-2}(x)$$

from result (26) in Prob. 15. Multiply this expression by $P_n(x)$ and integrate over the interval $-1 \le x \le 1$, making use of (15). Multiply (26) by $P_{n-1}(x)$ and integrate over the interval $-1 \le x \le 1$, again making use of (15).

Combine these two results to show that

$$\int_{-1}^{1} [P_n(x)]^2 \, dx = \left(\frac{2n-1}{2n+1}\right)\int_{-1}^{1} [P_{n-1}(x)]^2 \, dx.$$

Use this result to derive the **normalization condition** for Legendre polynomials

$$\int_{-1}^{1} [P_n(x)]^2 \, dx = \frac{2}{2n+1}, \qquad n = 0, 1, 2, \ldots.$$

18 Use recurrence relation (26) together with the result $P_0(x) = 1$ and $P_1(x) = x$ to prove that

$$P_n(1) = 1 \quad \text{and} \quad P_n(-1) = (-1)^n, \qquad n = 0, 1, 2, \ldots.$$

19 This problem, which provides an alternative derivation of the normalization condition for Legendre polynomials, is based on Rodrigues' formula (25). Using Rodrigues' formula prove that

$$\frac{d^n P_n(x)}{dx^n} = \frac{(2n)!}{2^n n!}.$$

Rodrigues' formula enables us to write

$$\int_{-1}^{1} [P_n(x)]^2 \, dx = \frac{1}{2^n n!}\int_{-1}^{1} P_n(x)\frac{d^n}{dx^n}[(x^2 - 1)^n] \, dx.$$

Integrate this result by parts n times to show with the help of the first result that

$$\int_{-1}^{1} [P_n(x)]^2 \, dx = \frac{(2n)!}{2^{2n}(n!)^2}\int_{-1}^{1} (1 - x^2)^n \, dx.$$

Set

$$I_n = \int_{-1}^{1} (1-x^2)^n \, dx,$$

and establish the reduction formula

$$I_n = \left(\frac{2n}{2n+1} \right) I_{n-1}.$$

Use this result to show first that

$$I_n = \frac{2^{2n}(n!)^2}{(2n+1)!} I_0,$$

and hence that

$$\int_{-1}^{1} [P_n(x)]^2 \, dx = \frac{2}{2n+1}.$$

20 Let $Q_m(x)$ be any polynomial of degree m. Then by virtue of Rodrigues' formula

$$\int_{-1}^{1} Q_m(x)P_n(x) \, dx = \frac{1}{2^n n!} \int_{-1}^{1} Q_m(x) \frac{d^n}{dx^n} [(x^2-1)^n] \, dx.$$

Integrate this result by parts n times and use it to prove that

$$\int_{-1}^{1} Q_m(x)P_n(x) \, dx = 0 \qquad \text{if } m < n,$$

and hence that

$$\int_{-1}^{1} P_m(x)P_n(x) \, dx = 0 \qquad m \neq n.$$

8.5 The gamma function $\Gamma(x)$

When working with series solutions, and elsewhere, it is often useful to generalize the notion of factorial n, namely

$$n! = 1 \cdot 2 \cdot 3 \ldots (n-1)n, \tag{1}$$

to include the case when n is any real positive number, and not just an integer. Such a generalization is provided by the **gamma function** (also called the **factorial function**) $\Gamma(x)$, defined as the improper integral

$$\Gamma(x) = \int_{0}^{\infty} e^{-t} t^{x-1} \, dt, \qquad x > 0. \tag{2}$$

The fundamental property of $\Gamma(x)$ follows from (2) by integrating by parts to obtain

$$\Gamma(x) = (-e^{-t}t^{x-1})_0^\infty + (x-1)\int_0^\infty e^{-t}t^{x-2}\,dt$$
$$= (x-1)\,\Gamma(x-1), \qquad \text{for } x-1 > 0.$$

Thus, replacing x by $x+1$, we arrive at the important general result

$$\Gamma(x+1) = x\,\Gamma(x), \qquad \text{for } x > 0. \tag{3}$$

To see that $\Gamma(x+1)$ reduces to $n!$ when $x=n$ is a positive integer, notice first that

$$\Gamma(1) = \int_0^\infty e^{-t}\,dt = 1. \tag{4}$$

Then by repeated use of (3) we find that

$$\Gamma(n+1) = n\Gamma(n) = n(n-1)\Gamma(n-1) = \ \cdots$$
$$= n(n-1)(n-2)\ldots 3\cdot2\cdot1\,\Gamma(1) = n!,$$

and we have shown that

$$\Gamma(n+1) = n!. \tag{5}$$

Thus the gamma function interpolates continuously between the successive values of $n!$ when n is not an integer. The function is readily available in tabulated form, but its graph for $x>0$ is shown in Fig. 8.6.

Repeated application of (3) establishes the useful result

$$\Gamma(a+n+1) = a(a+1)(a+2)\ldots(a+n)\Gamma(a), \tag{6}$$

for n a positive integer and $a>0$.

When written in the form

$$a(a+1)(a+2)\ldots(a+n) = \frac{\Gamma(a+n+1)}{\Gamma(a)} \tag{7}$$

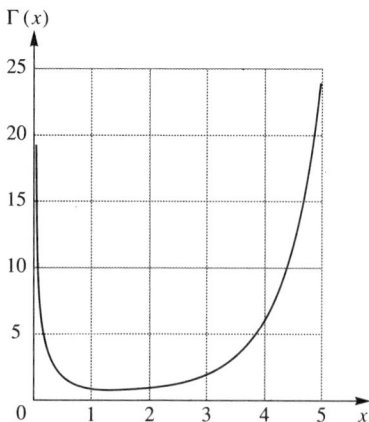

Fig. 8.6   Gamma function for positive x

it enables the product on the left-hand side to be represented in a convenient closed form in terms of a quotient of gamma functions.

More generally, as

$$a(a+d)(a+2d)(a+3d)\ldots(a+nd)$$

$$= d^{n+1}\frac{a}{d}\left(\frac{a}{d}+1\right)\left(\frac{a}{d}+2\right)\ldots\left(\frac{a}{d}+n\right),$$

by using (7) with a replaced by a/d, we arrive at the result

$$a(a+d)(a+2d)(a+3d)\ldots(a+nd) = d^{n+1}\frac{\Gamma\left(\dfrac{a}{d}+n+1\right)}{\Gamma\left(\dfrac{a}{d}\right)}, \tag{8}$$

for n a positive integer and $a>0$, $d>0$. Products of this type arise when deriving series solutions of differential equations and it is often convenient to be able to represent them in terms of gamma functions.

Two useful results we record for reference, but which are proved elsewhere (see Prob. 15 for (9) and Prob. 18 for (10)) are

$$\Gamma\left(\frac{1}{2}\right)=\sqrt{\pi}, \tag{9}$$

and

$$\Gamma(\alpha)\Gamma(1-\alpha)=\frac{\pi}{\sin \alpha\pi}, \qquad \text{for } 0<\alpha<1. \tag{10}$$

Basic properties of the gamma function

(i) $\Gamma(x)=\displaystyle\int_0^\infty e^{-t}t^{x-1}\,dt, \qquad x>0,$

(ii) $\Gamma(x+1)=x\Gamma(x), \qquad x>0,$

(iii) $\Gamma(n+1)=n!, \qquad n$ a positive integer,

(iv) $a(a+d)(a+2d)(a+3d)\ldots(a+nd)=d^{n+1}\dfrac{\Gamma\left(\dfrac{a}{d}+n+1\right)}{\Gamma\left(\dfrac{a}{d}\right)},$

 n a positive integer and $a>0$, $d>0$,

(v) $\Gamma\left(\dfrac{1}{2}\right)=\sqrt{\pi},$

(vi) $\Gamma(\alpha)\Gamma(1-\alpha)=\dfrac{\pi}{\sin \alpha\pi}, \qquad 0<\alpha<1.$

It is, in fact, possible to extend the definition of the gamma function to negative values of x, though we shall not go into the details here as they involve studying the gamma function in the complex plane. Suffice it to say that $\Gamma(x)$ is infinite when x is a negative integer, and that when x is a negative fraction (negative rational number) then $\Gamma(x)$ may be determined from property (ii) above when written in the form

$$x\Gamma(x) = \Gamma(x+1).$$

For example, to evaluate $\Gamma\left(-\dfrac{5}{2}\right)$, we set $x = -5/2$ in this last result to obtain

$$\left(-\frac{5}{2}\right)\Gamma\left(-\frac{5}{2}\right) = \Gamma\left(-\frac{3}{2}\right), \qquad \text{so } \Gamma\left(-\frac{5}{2}\right) = -\frac{2}{5}\Gamma\left(-\frac{3}{2}\right).$$

Repeating the argument with $x = -3/2$ gives

$$\left(-\frac{3}{2}\right)\Gamma\left(-\frac{3}{2}\right) = \Gamma\left(-\frac{1}{2}\right), \qquad \text{so } \Gamma\left(-\frac{3}{2}\right) = -\frac{2}{3}\Gamma\left(-\frac{1}{2}\right)$$

and, finally,

$$\left(-\frac{1}{2}\right)\Gamma\left(-\frac{1}{2}\right) = \Gamma\left(\frac{1}{2}\right), \qquad \text{so } \Gamma\left(-\frac{1}{2}\right) = -2\Gamma\left(\frac{1}{2}\right).$$

Combining these results we arrive at the result

$$\Gamma\left(-\frac{5}{2}\right) = \left(-\frac{2}{5}\right)\left(-\frac{2}{3}\right)(-2)\Gamma\left(\frac{1}{2}\right),$$

giving

$$\Gamma\left(-\frac{5}{2}\right) = -\frac{8}{15}\sqrt{\pi}.$$

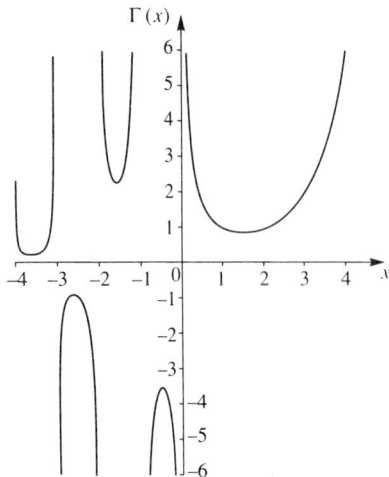

Fig. 8.7   Gamma function for positive and negative x

For reference purposes, the graph of $\Gamma(x)$ for both positive and negative x is given in Fig. 8.7.

Example 1. Expressing a product in terms of the gamma function

Use (8) to express the product

$$3 \cdot 7 \cdot 11 \cdot 15 \cdot 19 \cdot 23 \cdot 27$$

in terms of gamma functions.

Solution

Comparing the product of factors with (8) shows that $a=3$, $d=4$ and $n=6$, so it follows at once that

$$3 \cdot 7 \cdot 11 \cdot 15 \cdot 19 \cdot 23 \cdot 27 = 4^7 \frac{\Gamma(7\frac{3}{4})}{\Gamma(\frac{3}{4})}.$$

∎

Example 2. Simplifying the general term in a series

Rewrite the general term in the following series in terms of gamma functions

$$\sum_{m=1}^{\infty} \frac{1 \cdot 3 \cdot 5 \cdot 7 \dots (2m-1)}{2^{2m}} x^m$$

Solution

The product $1 \cdot 3 \cdot 5 \cdot 7 \dots (2m-1)$ in the general term of the series contains m terms, so as $a=1$ and $b=2$ it follows from (8) that

$$1 \cdot 3 \cdot 5 \cdot 7 \dots (2m-1) = 2^m \frac{\Gamma(m+\frac{1}{2})}{\Gamma(\frac{1}{2})},$$

and so the general term

$$\frac{1 \cdot 3 \cdot 5 \cdot 7 \dots (2m-1)}{2^{2m}} x^m = \frac{\Gamma(m+\frac{1}{2})}{2^m \Gamma(\frac{1}{2})} x^m.$$

∎

Problems for Section 8.5

1 Show that

$$\Gamma\left(n+\frac{1}{4}\right) = \frac{1 \cdot 5 \cdot 9 \cdot 13 \dots (4n-3)}{4^n} \Gamma\left(\frac{1}{4}\right).$$

2 Show that

$$\Gamma\left(n+\frac{1}{3}\right) = \frac{1 \cdot 4 \cdot 7 \cdot 10 \dots (3n-2)}{3^n} \Gamma\left(\frac{1}{3}\right).$$

3 Show that

$$\Gamma\left(n+\frac{2}{3}\right) = \frac{2 \cdot 5 \cdot 8 \cdot 11 \ldots (3n-1)}{3^n} \Gamma\left(\frac{2}{3}\right).$$

4 Show that

$$\Gamma\left(n+\frac{3}{4}\right) = \frac{3 \cdot 7 \cdot 11 \cdot 15 \ldots (4n-1)}{4^n} \Gamma\left(\frac{3}{4}\right).$$

In the following problems express the given products in terms of gamma functions.

5 $1 \cdot 5 \cdot 9 \cdot 13 \cdot 17 \cdot 21 \cdot 25 \cdot 29$.

6 $2 \cdot 5 \cdot 8 \cdot 11 \cdot 14 \cdot 17$.

7 $5 \cdot 7 \cdot 9 \cdot 11 \cdot 13 \cdot 15 \cdot 17 \cdot 19$.

8 $4 \cdot 7 \cdot 10 \cdot 13 \cdot 16 \cdot 19$.

In the following problems express the general term in the given series in terms of gamma functions.

9 $\displaystyle\sum_{n=1}^{\infty} \frac{2 \cdot 5 \cdot 8 \cdot 11 \ldots (3n-1)}{6^n} x^n$.

10 $\displaystyle\sum_{n=1}^{\infty} \frac{1 \cdot 3 \cdot 5 \cdot 7 \ldots (2n-1)}{8^n} x^n$.

11 $\displaystyle\sum_{n=1}^{\infty} \frac{2 \cdot 7 \cdot 12 \cdot 17 \ldots (2+5n)}{2^{n+1} n!} x^n$.

12 $\displaystyle\sum_{n=1}^{\infty} \frac{1 \cdot 4 \cdot 7 \cdot 10 \ldots (3n-2)}{3^{3n}} x^n$.

Harder problems

13 The coefficient of x^r in the expansion of $(1+x)^\alpha$, usually denoted by $\begin{pmatrix} \alpha \\ r \end{pmatrix}$ and called the binomial coefficient (see Prob. 24, Sec. 1.3), is given by

$$\begin{pmatrix} \alpha \\ r \end{pmatrix} = \frac{\alpha(\alpha-1)(\alpha-2) \ldots (\alpha-r+1)}{r!}.$$

Expand $\begin{pmatrix} \alpha \\ r \end{pmatrix}$ in terms of the gamma function.

14 Express the general term in the following series in terms of gamma functions

$$\sum_{n=1}^{\infty} \frac{(-1)^n x^{2n+\alpha}}{2^{2n} n!(\alpha+1)(\alpha+2) \ldots (\alpha+n)},$$

for any real $\alpha > 0$.

15 By writing

$$[\Gamma(\tfrac{1}{2})]^2 = \left(\int_0^\infty u^{-1/2} e^{-u} \, du\right)\left(\int_0^\infty v^{-1/2} e^{-v} \, dv\right),$$

setting $u = x^2$, $v = y^2$ and then changing to polar coordinates (r, θ) with $r^2 = x^2 + y^2$, $x = r \cos \theta$, $y = r \sin \theta$, prove that

$$\Gamma(\tfrac{1}{2}) = \sqrt{\pi}.$$

16 An alternative definition of the gamma function due to Euler is through the infinite product

$$\Gamma(x) = \lim_{n \to \infty} \frac{1 \cdot 2 \cdot 3 \ldots (n-1)}{x(x+1)(x+2) \ldots (x+n-1)} n^x.$$

Use this definition to prove that
(i) $\Gamma(x+1) = x\Gamma(x)$, $x > 0$ real,
(ii) $\Gamma(n+1) = n!$, n a positive integer,
(iii) $\Gamma(1) = 1$.

17 This problem establishes the equivalence of the definitions of $\Gamma(x)$ given in (2) and Prob. 16. Given the function

$$G(x, n) = \int_0^n \left(1 - \frac{t}{n}\right)^n t^{x-1} dt,$$

set $s = t/n$ and use integration by parts to show that

$$G(x, n) = \frac{1 \cdot 2 \cdot 3 \ldots n}{x(x+1)(x+2) \ldots (x+n)} n^x.$$

Then by showing that

$$\lim_{n \to \infty} \int_0^n \left(1 - \frac{t}{n}\right)^n t^{x-1} dt = \Gamma(x),$$

establish the equivalence of the two definitions.

18 Consider the expression

$$\Gamma(p)\Gamma(1-p) = \left(\int_0^\infty u^{p-1} e^{-u} du\right)\left(\int_0^\infty v^{-p} e^{-v} dv\right).$$

Proceeding as in Prob. 15, by first setting $u = x^2$, $v = y^2$, and changing to the polar coordinates (r, θ) with $r = x^2 + y^2$, $x = r \cos \theta$, $y = r \sin \theta$, show that

$$\Gamma(p)\Gamma(1-p) = 2 \int_0^{\pi/2} (\cot \theta)^{2p-1} d\theta.$$

Set $z = \cot^2 \theta$, and use the result

$$\int_0^\infty \frac{z^{p-1} dz}{1-z} = \frac{\pi}{\sin p\pi}$$

obtainable by an elementary complex variable method to prove that

$$\Gamma(p)\Gamma(1-p) = \frac{\pi}{\sin p\pi}.$$

The following problems concern the beta function which is useful in applications of mathematics and which is also related to the gamma function.

The **beta function** $B(p, q)$, with $p > 0$, $q > 0$, is defined by the integral

$$B(p, q) = \int_0^1 x^{p-1}(1-x)^{q-1}\,dx.$$

The conditions imposed on p and q follow because close to $x = 0$ the integrand varies like x^{p-1}, so the integral converges provided $p > 0$, while close to $x = 1$ the integral varies like $(1-x)^{q-1}$, so the integral also converges provided $q > 0$.

Setting $t = 1 - x$ in $B(p, q)$ gives

$$B(p, q) = \int_0^1 x^{p-1}(1-x)^{q-1}\,dx = \int_0^1 (1-t)^{p-1}\, t^{q-1}\,dt = B(q, p).$$

Thus the beta function is *symmetric* with respect to p and q, so

$$B(p, q) = B(q, p)$$

19 Apply integration by parts to the definition of $B(p, q)$ to prove that

$$B(p, q) = \left(\frac{q-1}{p+q-1}\right) B(p, q-1),$$

and

$$B(p, q) = \left(\frac{p-1}{p+q-1}\right) B(p-1, q).$$

20 Use the first result of Prob. 19 to prove that if $p > 0$ and $n > 0$ is an integer, then

$$B(p, n) = B(n, p) = \frac{(n-1)!}{p(p+1)(p+2)\ldots(p+n-1)}.$$

Deduce from this that if m, n are positive integers,

$$B(m, n) = B(n, m) = \frac{(m-1)!\,(n-1)!}{(m+n-1)!} = \frac{\Gamma(m)\Gamma(n)}{\Gamma(m+n)}.$$

21 Prove that
(i) $B(1, 1) = 1$ (ii) $B(\tfrac{1}{2}, \tfrac{1}{2}) = \pi$,
and

(iii) $B(m, n) = \displaystyle\int_0^\infty \frac{x^{m-1}}{(1+x)^{m+n}}\,dx = \int_0^\infty \frac{x^{n-1}}{(1+x)^{m+n}}\,dx.$

[*Hint*: In (iii) set $x = 1/(1+t)$].
22 By writing result (iii) of Prob. 21 in the form

$$B(m, n) = \int_0^1 \frac{x^{n-1}}{(1+x)^{m+n}}\,dx + \int_1^\infty \frac{x^{n-1}}{(1+x)^{m+n}}\,dx,$$

prove that

$$B(m, n) = \int_0^1 \frac{x^{m-1} + x^{n-1}}{(1+x)^{m+n}}\,dx.$$

[*Hint*: Use the substitution $x = 1/(1+t)$ in the second integral].

8.6 Frobenius' method and its extension

When working with some of the most important differential equations of applied mathematics, engineering, and physics it is often necessary to obtain a solution in the neighborhood of a singular point of the equation. This section describes how such solutions may be developed about what is called a **regular singular point** of a differential equation. To understand when the methods of this section may be applied we shall first define this concept for a general equation. Later we shall show how the definition may be simplified when the coefficients of the differential equation are polynomials, as is usually the case.

Consider a second order linear differential equation

$$p(x)y'' + q(x)y' + r(x)y = 0, \tag{1}$$

with a singular point at x_0, so that $p(x_0) = 0$. The point x_0 is said to be a **regular singular point** of (1) if

(i) $p(x_0) = 0$,

(ii) $\displaystyle\lim_{x \to x_0} \left((x - x_0) \frac{q(x)}{p(x)} \right)$ is finite, $\tag{2}$

(iii) $\displaystyle\lim_{x \to x_0} \left((x - x_0)^2 \frac{r(x)}{p(x)} \right)$ is finite.

In this definition, condition (i) identifies the location of the singular point, while conditions (ii) and (iii) must both be satisfied if x_0 is to be a regular singular point.

Any singular point at which either or both of conditions (ii) and (iii) fail to be satisfied is said to be an **irregular singular point** of (1). The study of solutions in the vicinity of irregular singular points is complicated and not of concern to us here, so henceforth attention will be confined to regular singular points.

Example 1. Legendre's equation – regular singular points

Locate and classify the singular points of

$$(1 - x^2)y'' - 2xy' + n(n+1)y = 0.$$

Solution

Using the notation of (1),

$$p(x) = 1 - x^2, \quad q(x) = -2x, \quad r(x) = n(n+1),$$

so the singular points determined by $p(x) = 0$ occur at $x = \pm 1$. To determine the nature of the singular point at 1 we must evaluate limits (ii) and (iii) of (2) with $x_0 = 1$. We obtain

$$\lim_{x \to 1} \left((x - 1) \frac{(-2x)}{(1 - x^2)} \right) = \lim_{x \to 1} \left(\frac{2x}{1 + x} \right) = 1,$$

$$\lim_{x \to 1} \left((x - 1)^2 \frac{n(n+1)}{(1 - x^2)} \right) = \lim_{x \to 1} \left(\frac{(1 - x)n(n+1)}{1 + x} \right) = 0,$$

so as both limits exist and are finite Legendre's equation has a regular singular point at 1. A similar argument shows that -1 is also a regular singular point of Legendre's equation. ∎

Example 2. Irregular singular point

Locate and identify the singular point of

$$x^2 y'' + 6y' + 3y = 0.$$

Solution

In the notation of (1)

$$p(x) = x^2, \quad q(x) = 6, \quad r(x) = 3,$$

so the singular point determined by $p(x) = 0$ occurs when $x = 0$ (there is only one singular point). Considering limits (ii) and (iii) of (1) with $x_0 = 0$ we find

$$\lim_{x \to 0} \left(x \cdot \frac{6}{x^2} \right) = \infty \quad \text{and} \quad \lim_{x \to \infty} \left(x^2 \cdot \frac{3}{x^2} \right) = 3.$$

Thus, as one of these limits is infinite, the origin is an irregular singular point of the differential equation. ∎

In certain applications it is required to know the behavior of the solution of (1) for large x. This is determined by making the transformation $x = 1/z$ in (1) and then examining the nature of the solution for small z (large x). When it is possible to expand a solution in powers of z about $z = 0$, this corresponds to developing the solution of (1) for large x as a series involving powers of $1/x$. This type of solution, called an **asymptotic solution**, has the property that the larger x becomes, the fewer are the terms needed to approximate the solution to within some given accuracy.

For the methods of this section to apply to the transformed equation it is necessary that $z = 0$ be either an ordinary point or a regular singular point of the transformed equation. If $z = 0$ is an ordinary point of the transformed equation we say that the original equation has an **ordinary point at infinity**. If, however, $z = 0$ is a regular singular point of the transformed equation, we say that the original equation has a **regular singular point at infinity**.

To determine the nature of the point at infinity in (1) set

$$x = 1/z, \tag{3}$$

from which it follows directly that

$$\frac{dy}{dx} = -z^2 \frac{dy}{dz}, \quad \frac{d^2 y}{dx^2} = z^4 \frac{d^2 y}{dz^2} + 2z^3 \frac{dy}{dz}. \tag{4}$$

Making substitutions (3) and (4) in (1) leads to an equation of the form

$$A(z) \frac{d^2 y}{dz^2} + B(z) \frac{dy}{dz} + C(z) y = 0. \tag{5}$$

Equation (1) will have an ordinary point or a regular singular point at *infinity* according as (5) has an ordinary point or a regular singular point at the *origin*.

Thus the original differential equation (1) will have an **ordinary point at infinity** if $A(0) \neq 0$ and $B(z)$, $C(z)$ are analytic functions of z at $z = 0$. Correspondingly, (1) will have a **regular singular point at infinity** if

(i) $A(0) = 0$,

(ii) $\lim\limits_{z \to 0} \left(z \dfrac{B(z)}{A(z)} \right)$ is finite, $\qquad\qquad\qquad\qquad\qquad$ (6)

(iii) $\lim\limits_{z \to 0} \left(z^2 \dfrac{C(z)}{A(z)} \right)$ is finite.

If either, or both, of conditions (ii) and (iii) in (6) fail to be satisfied we say that the original differential equation (1) has an **irregular singular point at infinity**.

Example 3. Singular points at infinity

Examine the nature of the point at infinity for the equations

(a) $y'' + y = 0$ $\qquad$ and $\qquad$ (b) $x^2 y'' + 6y' + 3y = 0$.

Solution

(a) Every finite point x_0 is an ordinary point of this equation. Making the substitutions (3), (4) the equation becomes

$$z^4 \frac{d^2 y}{dz^2} + 2z^3 \frac{dy}{dz} + y = 0,$$

which has a singular point at $z = 0$. As

$$\lim_{z \to 0} \left(z \frac{2z^3}{z^4} \right) = 2, \quad \lim_{z \to 0} \left(z^2 \cdot \frac{1}{z^4} \right) = \infty,$$

we conclude that the equation has an irregular singular point at infinity. This is hardly surprising, because the general solution is

$$y(x) = A \cos x + B \sin x$$

for all x, so no solution involving powers of $1/x$ can be expected for large x.

(b) In Ex. 2 we saw that this equation had an irregular singular point at the origin. Making the substitutions (2), (3) it becomes

$$z^2 \frac{d^2 y}{dz^2} + (2z - 6z^2) \frac{dy}{dz} + 3y = 0,$$

which has a singular point at the origin. Evaluating the limits in (6) we find

$$\lim_{z \to 0} \left[z \left(\frac{2z - 6z^2}{z^2} \right) \right] = 2 \quad \text{and} \quad \lim_{z \to 0} \left(z^2 \cdot \frac{3}{z^2} \right) = 3,$$

so the original differential equation has a regular singular point at infinity. ∎

Instead of dealing with an equation with a regular singular point of interest located at an arbitrary point x_0, it is more convenient to deal with an equation in which the singular point is located at the origin. This can always be accomplished by means of the linear transformation

$$t = x - x_0, \tag{7}$$

from which it follows that

$$\frac{dy}{dx} = \frac{dy}{dt} \quad \text{and} \quad \frac{d^2 y}{dx^2} = \frac{d^2 y}{dt^2}. \tag{8}$$

Using (7) and (8) in (1) leads to the transformed differential equation

$$p(t + x_0) \frac{d^2 y}{dt^2} + q(t + x_0) \frac{dy}{dt} + r(t + x_0) y = 0, \tag{9}$$

which now has a singular point at $t = 0$. Once the solution of (9) has been found, the solution of (8) follows from it by setting $t = x - x_0$.

Example 4. Shifting a singular point

Change the independent variable in Legendre's equation

$$(1 - x^2) \frac{d^2 y}{dx^2} - 2x \frac{dy}{dx} + n(n+1) y = 0$$

so that the singular point located at $x = -1$ is moved to the origin.

Solution

As $x_0 = -1$ it follows from (7) that $t = x + 1$. Substituting for x in Legendre's equation we obtain

$$t(2 - t) \frac{d^2 y}{dt^2} - 2(t - 1) \frac{dy}{dt} + n(n+1) y = 0,$$

which is now an equation with a regular singular point at $t = 0$. ∎

For convenience, and as no loss of generality is involved, hereafter we shall assume the regular singular point about which the solution is required to be located at the origin. Notice first that as the origin is a regular singular point at which $p(x)$, $q(x)$ and $r(x)$ are analytic functions (in most important cases they are polynomials) the structure of (1) may be simplified. Under these conditions we may always write (1), if necessary after multiplication by a suitable power of x, in the form

$$x^2 P(x) y'' + x Q(x) y' + R(x) = 0, \tag{10}$$

where

$$P(x) = P_0 + P_1 x + P_2 x^2 + \dots ,$$
$$Q(x) = Q_0 + Q_1 x + Q_2 x^2 + \dots , \tag{11}$$
$$R(x) = R_0 + R_1 x + R_x x^2 + \dots ,$$

with $P_0 \neq 0$. This follows because

(i) the origin is a singular point of (10),

(ii) $\displaystyle \lim_{x \to 0} \left(x \cdot \frac{x Q(x)}{x^2 P(x)} \right) = \lim_{x \to 0} \left(\frac{Q(x)}{P(x)} \right) = \frac{Q_0}{P_0}$ is finite ,

and

(iii) $\displaystyle \lim_{x \to 0} \left(x^2 \cdot \frac{R(x)}{x^2 P(x)} \right) = \lim_{x \to 0} \left(\frac{R(x)}{P(x)} \right) = \frac{R_0}{P_0}$ is finite ,

because $P_0 \neq 0$, and thus $x = 0$ must be a regular singular point of (10).

We now describe the method due to **Frobenius**[7] for solving differential equation (10). Motivation for his method is provided by considering the typical Cauchy–Euler equation (see Problem Section 5.5)

$$8 x^2 y'' - 2 x y' + 3 y = 0.$$

This has a regular singular point at the origin, as it is of the form given in (10) with $P_0 = 8 \neq 0$, $Q_0 = -2$, $R_0 = 3$, but its solution[8]

$$y(x) = A x^{1/2} + B x^{3/4},$$

is *not* a power series as it involves fractional powers of x.

As this solution involves fractional powers of x it is obvious that it cannot be derived by means of series (2) in Sec. 8.3, nor can it be obtained as in Sec. 8.2 because of the vanishing of the coefficients of y'' at the origin.

To overcome such difficulties, the method due to Frobenius seeks a solution of (10) in the neighborhood of the regular singular point at the origin of the form

$$y(x) = x^r \sum_{n=0}^{\infty} a_n x^n = \sum_{n=0}^{\infty} a_n x^{n+r}, \tag{12}$$

where $a_0 \neq 0$ and the index r and coefficients a_n must be determined from (10).

---

[7] GEORG FERDINAND FROBENIUS (1849–1917), a German mathematician who worked in Zurich and Berlin and whose main contributions were in group theory and analysis. He published the method described here in 1873.
[8] This is easily seen, because an algebraic solution can only be of the form $y = C x^m$. Substitution for y yields the quadratic equation for the unknown index m, $8m(m-1) - 2m + 3 = 0$ with roots $\frac{1}{2}$ and $\frac{3}{4}$, from which the stated solution then follows.

Let us see how the index r may be determined from (10) and (12). From (12) we have

$$y'(x) = \sum_{n=0}^{\infty} (n+r)a_n x^{n+r-1}, \tag{13}$$

$$y''(x) = \sum_{n=0}^{\infty} (n+r)(n+r-1)a_n x^{n+r-2}, \tag{14}$$

so substituting (12), (13), (14) into (10) gives

$$P(x)\sum_{n=0}^{\infty} (n+r)(n+r-1)a_n x^{n+r} + Q(x)\sum_{n=0}^{\infty}(n+r)a_n r^{n+r} + R(x)\sum_{n=0}^{\infty} a_n x^{n+r} = 0,$$

or

$$x^r \sum_{n=0}^{\infty} [P(x)(n+r)(n+r-1) + Q(x)(n+r) + R(x)]a_n x^n = 0. \tag{15}$$

For (12) to be a solution of (10), expression (15) must be an identity in x. Consequently the coefficient of each power of x in the series in (15) must vanish. Using (11) in (15) we find that the coefficient of the lowest power of x, corresponding to $n=0$, is

$$(P_0 r(r-1) + Q_0 r + R_0)a_0 = 0.$$

However $a_0 \neq 0$, so the equation determining the index r is seen to be the quadratic equation

$$P_0 r(r-1) + Q_0 r + R_0 = 0, \tag{16}$$

which is called the **indicial equation** associated with (10).

Let the roots of (16) (the indices) be r_1, r_2, and when these are real let $r_1 \geq r_2$. Three possibilities arise according as

(i) $r_1 \neq r_2$ and $r_1 - r_2 \neq$ an integer,
(ii) $r_1 - r_2 = N$,   a positive integer,
(iii) $r_1 = r_2$,   so the roots are equal.

We now state without proof the following fundamental theorem. It extends the method due to Frobenius, which in its original form applies only to case (i).

Theorem 8.2 (The extended Frobenius theorem)

In the differential equation

$$x^2 P(x)y'' + xQ(x)y' + R(x)y = 0 \tag{17}$$

let $P(x)$, $Q(x)$, $R(x)$ be functions with Taylor series expansions about the origin with $P(0) = P_0 \neq 0$, $Q(0) = Q_0$ and $R(0) = R_0$. Then (17) has a regular singular point at the origin.

Form the indicial equation

$$P_0 r(r-1) + Q_0 r + R_0 = 0 \tag{18}$$

and determine its roots r_1, r_2 and, if these are real, arrange them so that $r_1 \geq r_2$.

(i) When $r_1 \neq r_2$ and $r_1 - r_2$ is not equal to an integer, or when r_1 and r_2 are complex conjugates $(r_2 = \bar{r}_1)$, (17) has the two linearly independent solutions

$$y_1(x) = |x|^{r_1} \sum_{n=0}^{\infty} a_n x^n \quad \text{and} \quad y_2(x) = |x|^{r_2} \sum_{n=0}^{\infty} b_n x^n, \tag{19}$$

in the neighborhood of the origin, in which $a_0 \neq 0$, $b_0 \neq 0$ and the coefficients a_n depend on r_1 and the coefficients b_n on r_2;

(ii) When $r_1 - r_2 = N > 0$, with N an integer, (17) has the two linearly independent solutions

$$y_1(x) = |x|^{r_1} \sum_{n=0}^{\infty} a_n x^n \tag{20}$$

and

$$y_2(x) = C y_1(x) \ln|x| + |x|^{r_2} \sum_{n=0}^{\infty} b_n x^n, \tag{21}$$

in the neighborhood of the origin, where the constant C *may* be zero, $a_0 \neq 0$, $b_0 \neq 0$, and the coefficients a_n depend on r_1 and the coefficients b_n on r_2;

(iii) When $r_1 = r_2$ (17) has the two linearly independent solutions

$$y_1(x) = |x|^{r_1} \sum_{n=0}^{\infty} a_n x^n \tag{22}$$

and

$$y_2(x) = y_1(x) \ln|x| + |x|^{r_1} \sum_{n=0}^{\infty} b_n x^n, \tag{23}$$

in the neighborhood of the origin, where $a_0 \neq 0$, $b_1 \neq 0$, and the coefficients a_n and b_n depend on r_1. $\square$

Remarks about Theorem 8.2

(a) One solution of (10) is always of the form $y_1(x)$ in (19).

(b) The two solutions in case (i) are linearly independent because the nature of r_1 and r_2 means that $y_1(x)/y_2(x)$ must be a function of x.

(c) To see how to proceed in case (i) when r_1 and r_2 are complex conjugates $(r_2 = \bar{r}_1)$ see Prob. 28 at the end of this section.

(d) An attempt to find a second solution $y_2(x)$ of (10) in case (ii) by substituting $r=r_2$ $=r_1-N$ into (12) usually leads to the conclusion that a_N is infinite, so the simple Frobenius type solution is not applicable. For some problems a_N is found to be an indeterminate arbitrary constant, and then $a_{N+1}, a_{N+2}, \ldots$, can be found in terms of a_N. When this occurs, the second solution $y_2(x)$ is determined in terms of a_0 and a_N.

(e) The proof of the form taken by the second solution in both cases (ii) and (iii) is essentially the same. It is based on the integral method described in Sec. 5.6 (see Problems 29 and 30 at the end of this section).

Example 6. Case (i) $r_1 \neq r_2$ and $r_1 \quad r_2$ not an integer

Derive two linearly independent solutions of

$$2x^2 y'' - 3xy + (2-x)y = 0.$$

Solution

The equation is already in the standard form given in (10), with

$$P(x)=2, \quad Q(x)=-3 \quad \text{and} \quad R(x)=2-x.$$

so $P(0)=P_0=2$; $Q(0)=Q_0=-3$ and $R(0)=R_0=2$.

The indicial equation (16) becomes

$$2r(r-1)-3r+2=0 \quad \text{or} \quad 2r^2-5r+2=0,$$

so $r_1=2, r_2=\frac{1}{2}$, which corresponds to case (i) in Theorem 8.2.

Substituting (12), (13) and (14) into the equation and collecting together terms in x^{n+r} and terms in x^{n+r+1} gives

$$\sum_{n=0}^{\infty} [2(n+r)(n+r-1)-3(n+r)+2]a_n x^{n+r} - \sum_{n=0}^{\infty} a_n x^{n+r+1}=0.$$

Shifting the index in the last summation this becomes

$$\sum_{n=0}^{\infty} [2(n+r)(n+r-1)-3(n+r)+2]a_n x^{n+r} - \sum_{n=1}^{\infty} a_{n-1} x^{n+r}=0,$$

or

$$x^r[2r(r-1)-3r+2]a_0 + x^r \sum_{n=1}^{\infty} \{[(n+r)(2n+2r-5)+2]a_n - a_{n-1}\}x^n=0.$$

Equating to zero the coefficient of x^r, (the lowest power of x involved), in this equation merely reproduces the indicial equation. Repeating the process, but this time equating to zero the coefficient x^n in the summation yields the recursion formula

$$a_n = \frac{-a_{n-1}}{(n+r)(2n+2r-5)+2}, \qquad n=1, 2, \ldots .$$

Case $r = r_1$

Setting $r = r_1 = 2$ in the recursion formula reduces it to

$$a_n = \frac{a_{n-1}}{n(2n+3)}, \qquad n = 1, 2, \dots .$$

Thus

$$a_1 = \frac{a_0}{1 \cdot 5}, \quad a_2 = \frac{a_1}{2 \cdot 7} = \frac{a_0}{2! 5 \cdot 7},$$

$$a_3 = \frac{a_2}{3 \cdot 9} = \frac{a_0}{3! 5 \cdot 7 \cdot 9}, \quad a_4 = \frac{a_3}{4 \cdot 11} = \frac{a_0}{4! 5 \cdot 7 \cdot 9 \cdot 11}$$

and, in general,

$$a_n = \frac{a_0}{n! 5 \cdot 7 \cdot 9 \dots (2n+1)(2n+3)}, \qquad n = 1, 2, \dots .$$

Putting $a_0 = 1$, as it is an arbitrary nonzero constant, setting $r = r_1 = 2$ in (12) and substituting for a_n, one of the two linearly independent solutions is seen to be

$$y_1(x) = x^2 \left(1 + \sum_{n=1}^{\infty} \frac{x^n}{n! 5 \cdot 7 \cdot 9 \cdot \dots (2n+1)(2n+3)} \right).$$

Case $r = r_2$

Setting $r = r_2 = \frac{1}{2}$ in the recursion formula gives

$$a_n = \frac{a_{n-1}}{n(2n-3)}, \qquad n = 1, 2, \dots .$$

Thus

$$a_1 = \frac{a_0}{(-1)}, \quad a_2 = \frac{a_1}{2 \cdot 1} = \frac{a_0}{2!(-1)}$$

$$a_3 = \frac{a_2}{3 \cdot 3} = \frac{a_0}{3!(-1) \cdot 3}, \quad a_4 = \frac{a_3}{4 \cdot 5} = \frac{a_0}{4!(-1) \cdot 3 \cdot 5}$$

and, in general,

$$a_n = \frac{a_0}{n!(-1) \cdot 3 \cdot 5 \cdot 7 \dots (2n-5)(2n-3)}, \qquad n = 1, 2, \dots .$$

Setting the arbitrary constant $a_0 = 1$ as before, putting $r = r_2 = \frac{1}{2}$ in (12) and substituting for a_n gives for the other linearly independent solution

$$y_2(x) = |x|^{1/2} \left(1 - \sum_{n=1}^{\infty} \frac{x^n}{n! 1 \cdot 3 \cdot 5 \cdot 7 \dots (2n-5)(2n-3)} \right).$$

The general solution is thus

$$y(x)_2 = c_1 y_1(x) + c_2 y_2(x),$$

with c_1, c_2 arbitrary constants. Application of the ratio test to the series in the bracketed expressions in $y_1(x)$ and $y_2(x)$ shows them to be absolutely convergent for all x.

■

Example 6. Case (ii) without a logarithmic term

Find two linearly independent solutions for

$$2xy'' - (x+6)y' + 2y = 0$$

in the neighborhood of the origin.

Solution

To bring the equation into the standard form given in (10) we multiply by x to obtain

$$2x^2 y'' - x(x+6)y' + 2xy = 0,$$

from which we find

$$P(x) = 2, \quad Q(x) = -(x+6), \quad R(x) = 2x.$$

Thus as $P(0) = P_0 = 2$, $Q(0) = Q_0 = -6$, $R(0) = R_0 = 0$ the equation has a regular singularity at the origin, and its indicial equation determined by (16) is

$$r(r-4) = 0.$$

As the roots are $r_1 = 4$, $r_2 = 0$ this corresponds to case (ii) of Theorem 8.2.

Substituting for y, y' and y'' from (12), (13) and (14), shifting indices where necessary, and combining terms gives

$$2r(r-4)a_0 x^{r-1} + \sum_{n=0}^{\infty} [2(n+r+1)(n+r-3)a_{n+1} + (2-n-r)a_n]x^{n+r} = 0.$$

As usual, equating to zero the coefficient of the lowest power of x merely gives the indicial equation, but equating to zero the coefficient of x^{n+r} gives the recursion formula

$$2(n+r+1)(n+r-3)a_{n+1} + (2-n-r)a_n = 0, \quad n = 1, 2, \ldots.$$

Setting $r = r_1 = 4$ in this result shows the recursion formula for the coefficients of the first linearly independent solution $y_1(x)$ to be

$$a_{n+1} = \frac{(n+2)a_n}{2(n+1)(n+5)}, \quad n = 0, 1, \ldots.$$

Thus

$$a_1 = \frac{a_0}{5}, \quad a_2 = \frac{3a_1}{2^2 \cdot 3} = \frac{a_0}{20},$$

$$a_3 = \frac{4a_2}{2 \cdot 3 \cdot 7} = \frac{a_0}{210}, \quad a_4 = \frac{5a_3}{2 \cdot 4 \cdot 8} = \frac{a_0}{2688}$$

and, in general,

$$a_n = \frac{(n+1)}{2n(n+4)} \cdot \frac{n}{2(n-1)(n+3)} \cdot \frac{(n-1)}{2(n-2)(n+2)} \cdots \frac{2a_0}{2 \cdot 1 \cdot 5}.$$

This last result may be written

$$a_n = \frac{4!(n+1)! a_0}{2^n n! (n+4)!},$$

so the general term becomes

$$a_n = \frac{4!(n+1) a_0}{2^n (n+4)!}, \qquad n=1, 2, \ldots .$$

Setting the arbitrary constant $a_0 = 1/4!$ (for convenience), putting $r = r_1 = 4$ in (12) and using this expression for a_n we arrive at the first linearly independent solution

$$y_1(x) = \sum_{n=0}^{\infty} \frac{(x+1)x^{n+4}}{2^n(n+4)!}.$$

To obtain $y_2(x)$ it is now necessary to match the general form of the solution given in Theorem 8.2(ii) to the differential equation. Substituting for $y_2(x)$ and canceling the logarithmic terms because $y_1(x)$ is a solution we find, after multiplication by x, that

$$4Cxy_1' - (6x+8)Cy_1 + (2b_0 - 6b_1)x$$

$$+ \sum_{n=1}^{\infty} [2(n+1)(n-3)b_{n+1} + (2-n)b_n]x^{n+1} = 0.$$

Equating to zero the coefficients of x (the lowest power of x) gives

$$b_1 = \tfrac{1}{3}b_0.$$

As the terms $4Cxy_1' - (6x+8)Cy_1$ will only contribute to the terms in x^4, it follows that by equating the coefficient of x^{n+1} to zero in the summation, that b_0 to b_3 are related by

$$2(n+1)(n-3)b_{n+1} + (2-n)b_n = 0, \qquad n=1, 2, 3.$$

Thus we find

$$(n=1) \quad 2 \cdot 2(-2)b_2 + b_1 = 0, \quad \text{so } b_2 = \frac{b_1}{8} = \frac{b_0}{24}.$$

$$(n=2) \quad 2 \cdot 3(-1)b_3 = 0, \quad \text{so } b_3 = 0.$$

$$(n=3) \quad \text{leaves } b_4 \text{ arbitrary.}$$

To proceed further it is now necessary to take account of the terms of degree 4 and above in

$$4Cxy_1' - (6x+8)Cy_1 + (2b_0 - 6b_1)x + \sum_{n=1}^{\infty} [2(n+1)(n-3)b_{n+1} + (2-n)b_n]x^{n+1} = 0.$$

Equating to zero the sum of the coefficients of all terms in x^4 we find, after using our previous result to substitute for $y_1(x)$ and $y_1'(x)$, that

$$\tfrac{1}{3}C = b_3.$$

However $b_3 = 0$, so $C = 0$ and thus the logarithmic term vanishes.

Setting $C = 0$ and substituting for b_0 to b_3 reduces the second linearly independent solution to

$$y_2(x) = b_0(1 + \tfrac{1}{3}x + \tfrac{1}{24}x^2) + \sum_{n=4}^{\infty} b_n x^n,$$

where b_0 and b_4 are arbitrary and

$$2(n+1)(n-3)b_{n+1} + (2-n)b_n = 0, \qquad n = 4, 5, \dots .$$

In terms of b_4 we find that

$$(n=4) \quad 2 \cdot 5 \cdot 1 \cdot b_5 - 2b_4 = 0, \quad \text{so } b_5 = \tfrac{1}{5}b_4,$$

$$(n=5) \quad 2 \cdot 6 \cdot 2 \cdot b_6 - 3b_5 = 0, \quad \text{so } b_6 = \tfrac{1}{8}b_5 = \tfrac{1}{40}b_4,$$

$$(n=6) \quad 2 \cdot 7 \cdot 3 \cdot b_7 - 4b_6 = 0, \quad \text{so } b_7 = \tfrac{2}{21}b_6 = \tfrac{1}{420}b_4.$$

Deducing the general term b_n from the recursion formula in the way used to deduce a_n by writing

$$b_n = \frac{(3-n)}{2 \cdot n \cdot (n-4)} \cdot \frac{(4-n)}{2(n-1)(n-5)} \cdots \frac{3}{2 \cdot 6 \cdot 2} \cdot \frac{2b_4}{2 \cdot 5 \cdot 1},$$

we obtain

$$b_n = \frac{4!(n-3)}{2^{n-4}n!} b_4, \qquad n = 4, 5, \dots,$$

or, equivalently,

$$b_{n+4} = \frac{4!(n+1)}{2^n(n+4)!} b_4, \qquad n = 0, 1, 2, \dots .$$

Using this in the form of solution obtained for $y_2(x)$ so far gives

$$y_2(x) = b_0 \left(1 + \frac{1}{3}x + \frac{1}{24}x^2 \right) + b_4 4! \sum_{n=0}^{\infty} \frac{(n+1)x^{n+4}}{2^n(n+4)!}.$$

Comparing $y_1(x)$ and $y_2(x)$ shows that the last summation in $y_2(x)$ is simply a constant multiple of $y_1(x)$, and so may be absorbed into $y_1(x)$ when writing down the general solution

$$y(x) = C_1 y_1(x) + C_2 y_2(x).$$

Consequently, two linearly independent solutions are the infinite series

$$y_1(x) = \sum_{n=0}^{\infty} \frac{(n+1)x^{n+4}}{2^n(n+4)!},$$

and the quadratic polynomial

$$y_2(x) = 1 + \tfrac{1}{3}x + \tfrac{1}{24}x^2,$$

and in this case the general solution contains no logarithmic term. Both $y_1(x)$ and $y_2(x)$ are finite at the origin. ∎

Example 7. Case (ii) using the integral method

Find two linearly independent solutions for

$$x^2 y'' + xy' + (x^2 - \tfrac{1}{4})y = 0$$

in the neighborhood of the origin.

Solution

This example shows how, in special cases, the integral method of Sec. 5.6 leads very simply to a second linearly independent solution $y_2(x)$ once another linearly independent solution $y_1(x)$ has been found.

In this example the integral method enables us to avoid matching the general form of the second solution $y_2(x)$ given in Theorem 8.2(ii) to the differential equation, as was done in Ex. 6.

The equation is in the standard form given in (10) with

$$P(x)=1, \qquad Q(x)=1, \qquad R(x)=x^2 - \tfrac{1}{4},$$

so as

$$P(0) = P_0 = 1, \quad Q(0)=Q=1 \quad \text{and} \quad R(0)=R_0 = -\tfrac{1}{4}$$

the equation must have a regular singular point at the origin. The indicial equation determined by (16) becomes

$$r^2 - \tfrac{1}{4} = 0,$$

so $r_1 = \tfrac{1}{2}, r_2 = -\tfrac{1}{2}$. As $r_1 - r_2 = 1$ is an integer this corresponds to case (ii) of Theorem 8.2.

Substituting for y, y' and y'' from (12), (13) and (14), shifting indices where necessary, and combining terms in the usual way we obtain

$$(r^2 - \tfrac{1}{4})a_0 x^r + [(1+r)^2 - \tfrac{1}{4}]a_1 x^{r+1} + \sum_{n=2}^{\infty} \{[(n+r)^2 - \tfrac{1}{4}]a_n + a_{n-2}\}x^{n+r} = 0.$$

Equating to zero the coefficient of x^r simply gives the indicial equation. Equating to zero the coefficient of x^{r+1} gives

$$[(1+r)^2 - \tfrac{1}{4}]a_1 = 0,$$

but as $r= \pm\tfrac{1}{2}$ this shows that $a_1 = 0$. Finally, equating to zero the coefficient of x^{n+r} in the summation gives the general recursion formula

$$a_n = \frac{a_{n-2}}{(n+r)^2 - \tfrac{1}{4}}, \qquad n=2, 3, \ldots.$$

Setting $r=r_1 = \tfrac{1}{2}$ in this result to obtain the recursion formula for the solution $y_1(x)$ gives

$$a_n = \frac{a_{n-2}}{n(n+1)}, \qquad n=2, 3, \ldots.$$

However, as $a_1 = 0$ it follows directly that $0=a_1 = a_3 = a_5 = \ldots$, so that all the odd coefficients vanish. Thus, in terms of an arbitrary a_0,

$$a_2 = \frac{-a_0}{2 \cdot 3}, \quad a_4 = \frac{-a_2}{4 \cdot 5} = \frac{a_0}{2 \cdot 3 \cdot 4 \cdot 5} = \frac{a_0}{5!}, \quad a_6 = \frac{-a_4}{6 \cdot 7} = \frac{-a_0}{7!},$$

and, in general,

$$a_{2m} = \frac{(-1)^m a_0}{(2m+1)!}, \quad m = 0, 1, 2, \ldots.$$

Using this result in (12) with $r = r_1 = \frac{1}{2}$ we find

$$y_1(x) = a_0 x^{1/2} \sum_{m=0}^{\infty} \frac{(-1)^m x^{2m}}{(2m+1)!} = a_0 |x|^{-1/2} \sum_{m=0}^{\infty} \frac{(-1)^m x^{2m+1}}{(2m+1)!}.$$

The series in the last summation is the Maclaurin series for $\sin x$, so one of the two linearly independent solutions is (setting $a_0 = 1$),

$$y_1(x) = |x|^{-1/2} \sin x.$$

We now make use of the integral method to determine a second linearly independent solution $y_2(x)$ in terms of $y_1(x)$. Comparing the differential equation with (2), Sec. 5.6, we see that

$$a_0(x) = x^2, \quad a_1(x) = x,$$

while

$$\Phi_1(x) = y_1(x) = |x|^{-1/2} \sin x.$$

Substituting these results in (8), Sec. 5.6 gives for the second linearly independent solution

$$\begin{aligned}
y_2(x) &= B y_1(x) \int \frac{\exp[-\int (1/x) \, dx]}{[x^{-1/2} \sin x]^2} \, dx \\
&= B y_1(x) \int \operatorname{cosec}^2 x \, dx = -B y_1(x) \cot x \\
&= -B |x|^{-1/2} \cos x.
\end{aligned}$$

Thus a second linearly independent solution is

$$y_2(x) = |x|^{-1/2} \cos x.$$

The additive constant of integration has been omitted when determining $y_2(x)$. This is because it simply adds a multiple of $y_1(x)$ which may be absorbed into the first term of the general solution

$$y(x) = |x|^{-1/2}(c_1 \sin x + c_2 \cos x).$$

The differential equation in this example provides another example of case (ii) in which the logarithmic term is missing. Neither $y_1(x)$ nor $y_2(x)$ is finite at the origin. ∎

Example 8. A harder example of case (ii) with a logarithmic term

Find two linearly independent solutions for

$$xy'' + y = 0$$

in the neighborhood of the origin.

Solution

As given, the differential equation is seen to have a singular point at the origin, but it is not in the standard form shown in (10). To identify it with (10) we multiply by x to obtain

$$x^2 y'' + xy = 0.$$

This is now of the form given in (10) with

$$P(x)=1, \; Q(x)=0, \; R(x)=x,$$

so the equation must have a regular singular point at the origin. It follows that

$$P(0)=P_0 = 1, \quad Q(0)=Q_0 = 0 \quad \text{and} \quad R_0 = R(0)=0,$$

so indicial equation (16) becomes

$$r(r-1)=0,$$

showing that $r_1 = 1$ and $r_2 = 0$. As $r_1 - r_2 = 1$ is an integer this corresponds to case (ii) of Theorem 8.2.

Replacing y and y'' in the differential equation

$$xy'' + y = 0$$

by their series representations (12) and (14) we obtain

$$\sum_{n=0}^{\infty} (n+r)(n+r-1)a_n x^{n+r-1} + \sum_{n=0}^{\infty} a_n x^{n+r} = 0.$$

Shifting the summation index in the first summation to change x^{n+r-1} to x^{n+r} this becomes

$$\sum_{n=-1}^{\infty} (n+r+1)(n+r)a_{n+1} x^{n+r} + \sum_{n=0}^{\infty} a_n x^{n+r} = 0.$$

Separating out the term in the first summation corresponding to $n=-1$ and combining terms we obtain

$$r(r-1)a_0 x^{r-1} + \sum_{n=0}^{\infty} [(n+r+1)(n+r)a_{n+1} + a_n]x^{n+r} = 0.$$

Equating to zero the coefficient of x^{r-1} merely gives the indicial equation, and so provides no fresh information. However, equating to zero the coefficient of x^{n+r} for $n=0, 1, 2, \ldots$ yields the recursion formula

$$(n+r+1)(n+r)a_{n+1} + a_n = 0,$$

so that

$$a_{n+1} = \frac{-a_n}{(n+r+1)(n+r)}.$$

Setting $r=r_1 = 1$ in this expression, we see that the recursion formula for the coefficients in the solution $y_1(x)$ is

$$a_{n+1} = \frac{-a_n}{(n+1)(n+2)}, \qquad n=0, 1, 2, \ldots .$$

Thus

$$a_1 = \frac{-a_0}{1\cdot 2}, \quad a_2 = \frac{-a_1}{2\cdot 3} = \frac{a_0}{1\cdot 2\cdot 2\cdot 3}, \quad a_3 = \frac{-a_2}{3\cdot 4} = \frac{-a_0}{1\cdot 2\cdot 2\cdot 3\cdot 3\cdot 4}$$

so, in general,

$$a_n = \frac{(-1)^n a_0}{n!(n+1)!}, \quad n = 0, 1, 2, \ldots .$$

Setting the arbitrary constant $a_0 = 1$ and using this expression for a_n in (12) with $r = r_1 = 1$ yields the first linearly independent solution

$$y_1(x) = \sum_{n=0}^{\infty} \frac{(-1)^n x^{n+1}}{n!(n+1)!},$$

To determine the linearly independent solution $y_2(x)$ we must match the form of solution given in Theorem 8.2(ii) to the differential equation.

Substituting the expression

$$y_2(x) = Cy_1(x)\ln|x| + |x|^{r_2}\sum_{n=0}^{\infty} b_n x^n$$

with $r_2 = 0$ into the differential equation

$$xy'' + y = 0$$

leads to the result

$$(xy_1'' + y_1)C\ln|x| + 2Cy_1' - \frac{Cy_1}{x} + \sum_{n=2}^{\infty} n(n-1)b_n x^{n-1} + \sum_{n=0}^{\infty} b_n x^n = 0.$$

As y_1 is a solution of the differential equation the bracketed term vanishes, and after multiplication by x we obtain

$$2Cxy_1' - Cy_1 + \sum_{n=2}^{\infty} n(n-1)_n x^n + \sum_{n=0}^{\infty} b_n x^{n+1} = 0.$$

Substituting for y_1 and y_1' from the series solution obtained earlier, and shifting the summation index in the third term to change x^n to x^{n+1}, this becomes

$$2C\sum_{n=0}^{\infty} \frac{(-1)^n(n+1)x^{n+1}}{n!(n+1)!} - C\sum_{n=0}^{\infty} \frac{(-1)^n x^{n+1}}{n!(n+1)!} + \sum_{n=1}^{\infty} n(n+1)b_{n+1} x^{n+1} + \sum_{n=0}^{\infty} b_n x^{n+1} = 0,$$

or

$$C\sum_{n=0}^{\infty} \frac{(-1)^n(2n+1)x^{2n+1}}{n!(n+1)!} + \sum_{n=1}^{\infty} n(n+1)b_{n+1} x^{n+1} + \sum_{n=0}^{\infty} b_n x^{n+1} = 0.$$

Separating out the terms corresponding to $n=0$, and combining the remaining terms gives

$$(C + b_0)x + \sum_{n=1}^{\infty} \left(C\frac{(-1)^n(2n+1)}{n!(n+1)!} + n(n+1)b_{n+1} + b_n \right)x^{n+1} = 0.$$

As $y_2(x)$ is a solution of the differential equation this expression must be an identity in x, so the coefficient of each power of x must vanish.

Equating to zero the coefficients of x gives

$$b_0 = -C,$$

while equating to zero the coefficient of x^{n+1} gives

$$C\frac{(-1)^n(2n+1)}{n!(n+1)!} + n(n+1)b_{n+1} + b_n = 0, \qquad n = 1, 2, \ldots.$$

The constant C is an arbitrary scale factor so, for convenience, we set $C = 1$. Then b_{n+1} is seen to be determined from b_n by the recursion formula

$$n(n+1)b_{n+1} = \frac{(-1)^{n+1}(2n+1)}{n!(n+1)!} - b_n, \qquad n = 1, 2, \ldots.$$

The coefficients $b_2, b_3, \ldots$ are all determined by the above formula in terms of b_1. Setting $n = 1, 2, 3$ in the formula gives

$$(n=1) \quad 2b_2 = \tfrac{3}{2} - b_1, \quad \text{so} \quad b_2 = \tfrac{3}{4} - \tfrac{1}{2}b_1$$

$$(n=2) \quad 6b_3 = -\tfrac{5}{12} - b_2, \quad \text{so} \quad b_3 = -\tfrac{7}{36} + \tfrac{1}{12}b_1$$

$$(n=3) \quad 12b_4 = \tfrac{7}{144} - b_3, \quad \text{so} \quad b_4 = \tfrac{105}{5184} - \tfrac{1}{144}b_1.$$

Thus the second solution $y_2(x)$ is of the form

$$y_2(x) = y_1(x)\ln|x| - 1 + (\tfrac{3}{4} - \tfrac{1}{2}b_1)x + (-\tfrac{7}{36} + \tfrac{1}{12}b_1)x^2 + (\tfrac{105}{5184} - \tfrac{1}{144}b_1)x^3 + \cdots.$$

It only remains for us to determine b_1, and to do this we shall make use of the integral method described in Sec. 5.6. In terms of the notation used in (10) in the present section, the second linearly independent solution given in (8), Sec. 5.6, takes the form

$$y_2(x) = y_1(x)\int \frac{dx}{[y_1(x)]^2}.$$

We now develop the solution generated by this method as far as the term in x, from which we can determine b_1 by comparison with the above result in which the coefficient of x was seen to be $(\tfrac{3}{4} - \tfrac{1}{2}b_1)$. Expanding y_1 as far as the term in x^3 gives

$$y_1(x) = x - \frac{x^2}{2} + \frac{x^3}{12} + \cdots,$$

from which we find by multiplication that

$$[y_1(x)]^2 = x^2 - x^3 + \tfrac{5}{12}x^4 - \frac{x^5}{12} + \cdots.$$

Thus

$$\frac{1}{[y_1(x)]^2} = \left(\frac{1}{x^2}\right)\frac{1}{\left(1 - x + \dfrac{5}{12}x^2 - \dfrac{x^3}{12} + \cdots\right)},$$

and after long division this becomes

$$\frac{1}{[y_1(x)]^2} = \frac{1}{x^2}\left(1 + x + \frac{7}{12}x^2 + \cdots\right).$$

Consequently

$$\frac{1}{[y_1(x)]^2} = \frac{1}{x^2} + \frac{1}{x} + \frac{7}{12} + \ldots,$$

so that $\displaystyle\int \frac{dx}{[y_1(x)]^2} = -\frac{1}{x} + \ln|x| + \frac{7}{12}x + \ldots.$

Finally, after multiplication by $y_1(x)$, we see that

$$y_2(x) = y_1(x) \int \frac{dx}{[y_1(x)]^2} = y_1(x)\ln|x| - 1 + \frac{1}{2}x + \frac{1}{2}x^2 + \ldots.$$

The coefficient of x is $\frac{1}{2}$, so after comparing this result with our previous one we find

$$\tfrac{1}{2} = \tfrac{3}{4} - \tfrac{1}{2}b_1,$$

so that $b_1 = \frac{1}{2}$.

The required second linearly independent solution is thus

$$y_2(x) = y_1(x)\ln|x| - 1 + \tfrac{1}{2}x + \tfrac{1}{2}x^2 - \tfrac{11}{72}x^3 + \ldots + b_n x^n + \ldots,$$

with the coefficients b_n being given by the recursion formula

$$n(n+1)b_{n+1} = \frac{(-1)^{n+1}(2n+1)}{n!(n+1)!} - b_n, \qquad n = 1, 2, \ldots,$$

with $b_1 = \frac{1}{2}$. The solution $y_1(x)$ is the only one which is finite at the origin.   ■

Example 9. Case (iii)

Find two linearly independent solutions for

$$x(1-x)y'' + (1-5x)y' - 4y = 0$$

in the neighborhood of the origin.

Solution

To bring the equation to the standard form shown in (10) we multiply by x to obtain

$$x^2(1-x)y'' + x(1-5x)y' - 4xy = 0,$$

which corresponds to

$$P(x) = 1 - x, \; Q(x) = 1 - 5x, \; R(x) = -4x.$$

As $P(0) = P_0 = 1$, $Q(0) = Q_0 = 1$, $R(0) = R_0 = 0$, the equation has a regular singularity at the origin and the indicial equation (16) becomes

$$r^2 = 0.$$

Thus $r_1 = r_2 = 0$, and the equation corresponds to case (iii) of Theorem 8.2.

Substituting for y, y' and y'' from (12), (13) and (14), shifting indices where necessary and separating out terms leads to the result

$$r^2 a_0 x^{r-1} + \sum_{n=0}^{\infty} \{(n+r+1)^2 a_{n+1} - [(n+r)(n+r+4) + 4]a_n\}x^{n+r} = 0.$$

Equating to zero the coefficient of x^{r-1} in this identity merely reproduces the indicial equation, but equating to zero the coefficient of x^{n+r} yields the recursion formula

$$(n+r+1)^2 a_{n+1} - [(n+r)(n+r+4)+4]a_n = 0, \qquad n=0, 1, \ldots.$$

Setting $r=0$ this simplifies to

$$a_{n+1} = \frac{(n+2)^2}{(n+1)^2} a_n, \qquad n=0, 1, \ldots.$$

Thus

$$a_1 = 4a_0, \quad a_2 = (\tfrac{3}{2})^2 a_1 = (\tfrac{3}{2})^2 \cdot 4a_0, \quad a_3 = (\tfrac{4}{3})^2 a_2 = (\tfrac{4}{3})^2 (\tfrac{3}{2})^2 \cdot 4a_0$$

and, in general,

$$a_n = \left(\frac{n+1}{n}\right)^2 \left(\frac{n}{n-1}\right)^2 \left(\frac{n-1}{n-2}\right)^2 \cdots \left(\frac{3}{2}\right)^2 \frac{2^2}{1} a_0,$$

so that

$$a_n = \left(\frac{(n+1)!}{n!}\right)^2 a_0 = (n+1)^2 a_0.$$

Thus setting $r=0$ in (12) and using a_n we find that one of the two linearly independent solutions is

$$y_1(x) = \sum_{n=0}^{\infty} (n+1)^2 x^n.$$

To find a second linearly independent solution $y_2(x)$ we must match the general form of the solution given in case (iii), Theorem 8.2, to the differential equation.

We set

$$y_2(x) = y_1(x)\ln(x) + \sum_{n=1}^{\infty} b_n x^n$$

and substitute into the differential equation.

The logarithmic terms vanish because $y_1(x)$ is a solution, and after the usual manipulations we arrive at the result

$$[2x(1-x)y_1' - 4xy_1] + b_1 x + (4b_2 - 9b_1)x^2 + \sum_{n=2}^{\infty} [(n+1)^2 b_{n+1} - (n+2)^2 b_n]x^{n+1} = 0.$$

Substituting for y_1 and y_1' from the result obtained earlier and combining corresponding powers of x gives

$$(4+b_1)x + (12+4b_2 - 9b_1)x^2 + \sum_{n=2}^{\infty} [(n+1)^2 b_{n+1} - (n+2)^2 b_n + 2(n+1)(n+2)]x^{n+1} = 0.$$

This must be an identity, so equating to zero the coefficient of each power of x gives

$$4+b_1 = 0, \quad \text{so} \quad b_1 = -4$$

$$12+4b_2 - 9b_1 = 0, \quad \text{so} \quad b_2 = -12$$

and

$$(n+1)^2 b_{n+1} - (n+2)^2 b_n + 2(n+1)(n+2) = 0, \qquad n=2, 3, \ldots.$$

This recursion relation for the coefficients b_n shows that

$$b_3 = -24, \quad b_4 = -40, \quad b_5 = -60,$$

and inspection of these results suggests that

$$b_n = -2n(n+1), \qquad n=1, 2, \ldots .$$

This conjecture is easily verified by direct substitution into the recursion formula, so the second linearly independent solution is seen to be

$$y_2(x) = y_1(x)\ln|x| - 2\sum_{n=1}^{\infty} n(n+1)x^n.$$

The only solution which is finite at the origin is $y_1(x)$.   ∎

Problems for Section 8.6

Find the location and nature of all the singular points of the following equations.

1 $(x^2 - 1)y'' + 7xy' + 4y = 0.$
2 $(1 - \cos x)y'' + 3(\sin x)y' + 2y = 0.$
3 $x^4 y'' + 7xy' + (x+2)y = 0.$
4 $x(x+2)y'' + x^2 y' + 3y = 0.$
5 $(2-x)^2 y'' + 3y' + (2-x)^2 (\sin x)y = 0.$
6 $x^2 y'' + 9y' - 2y = 0.$
7 Shift the singular point located at $x = -1$ to the origin, given that

$$(x+1)(x^2 + 1)y'' + 2x^2 y' + y = 0.$$

8 Shift the singular point located at $x = 3$ to the origin, given that

$$x(x-3)y'' + 2xy' + (1+x)y = 0.$$

In the following problems find the general solution of the differential equation in the neighborhood of the origin.

9 $2x^2 y'' - xy' + (x+5)y = 0.$
10 $2xy'' + (1+x)y' + y = 0.$
11 $4xy'' + (2+x)y' + y = 0.$
12 $2xy'' + y' + xy = 0.$
13 $2x(1-x)y'' + (1-x)y' + 3y = 0.$
14 $4xy'' + 2y' + y = 0.$
15 Show that the differential equation

$$4x^3 y'' + 6x^2 y' + y = 0$$

has a regular singular point at infinity, and find the form of the solution for large positive x.

16 Show that the differential equation

$$2x^3 y'' + (3x^2 - x)y' + y = 0$$

has a regular singular point at infinity, and find the form of the solution for large positive x.

In the following problems find the general solution of the differential equation in the neighborhood of the origin.

17 $2xy'' + (3 - 2x)y' + y = 0$.

18 $(2x + x^3)y'' - y' - 6xy = 0$.

19 $xy'' - y' + 4x^2 y = 0$.

20 Find the first three terms of the two linearly independent solutions of $x^4 y'' + (2x^3 - x)y' + y = 0$.

In the following problems find one solution in the form of a series, and then use this solution in the integral method to find the leading terms in the second solution.

21 $xy'' + (1 - x)y' - y = 0$.

22 $xy'' + 2y' - y = 0$.

23 $xy'' + y' - xy = 0$.

24 Whittaker's equation is

$$y'' + \left(-\frac{1}{4} + \frac{\lambda}{x} + \frac{\frac{1}{4} - \mu^2}{x^2} \right) y = 0,$$

in which λ, μ are real, and 2μ is not an integer. Show that for $x \to 0$ the two linearly independent solutions behave as $x^{1/2 + \mu}$ and $x^{1/2 - \mu}$, and that for large positive x they behave like $x^\lambda e^{-1/2x}$ and $x^{-\lambda} e^{1/2x}$.

In the following problems find the general solution of the differential equation in the neighborhood of the origin.

25 $x^2 y'' - xy' + y = 0$.

26 $16x^2 y'' + y = 0$.

27 $4x^2 y'' + y = 0$.

Harder problems

When the indicial equation has complex conjugate roots, the general form of the solution of a second order variable coefficient linear differential equation in the neighborhood of a regular singular point located at the origin may be found as follows. Let the indicial equation have the complex conjugate roots

$$r_1 = \alpha + i\beta, \quad r_2 = \bar{r}_1 = \alpha - i\beta.$$

Then, following the Frobenius method, the complex solution corresponding to r_1 will be of the form

$$x^{\alpha + i\beta} \sum_{n=0}^{\infty} a_n x^n,$$

where the $a_n = c_n + id_n$ are now complex constants. Writing

$$x^{\alpha + i\beta} = |x|^\alpha \exp(i\beta \ln |x|),$$

it follows that

$$x^{\alpha + i\beta} = |x|^\alpha [\cos(\beta \ln |x|) + i \sin(\beta \ln |x|)].$$

Then if

$$Y(x) = y_1(x) + i y_2(x)$$

is a complex-valued solution of the differential equation, the real and imaginary parts of $Y(x)$ must also be solutions. Thus two linearly independent solutions are

$$y_1(x) = |x|^\alpha \left(\cos(\beta \ln|x|) \sum_{n=0}^\infty c_n x^n - \sin(\beta \ln|x|) \sum_{n=0}^\infty d_n x^n \right),$$

and

$$y(x) = |x|^\alpha \left(\cos(\beta \ln|x|) \sum_{n=0}^\infty d_n x^n + \sin(\beta \ln|x|) \sum_{n=0}^\infty c_n x^n \right).$$

The coefficients c_n and d_n may be found by substituting either $y_1(x)$ or $y_2(x)$ into the differential equation and equating to zero the coefficient of each power of x to make the result an identity.

28 Use the method outlined above to show that the solution of

$$x^2 y'' - 3xy' + 5y = 0$$

in the neighborhood of the regular singular point at the origin is

$$y(x) = x^2 [c_1 \cos(\ln|x|) + c_2 \sin(\ln|x|)].$$

29 Complete the details of the following outline proof of the general form taken by the second linearly independent solution in case (ii) of Theorem 8.2 when the roots r_1 and r_2 of the indicial equation differ by a positive integer N with $r_2 = r_1 - N$. Show first that if the equation

$$x^2 P(x) y'' + x Q(x) y' + R(x) y = 0$$

has a regular singular point at the origin and $r_2 = r_1 - N$ then, using the notation of Theorem 8.2,

$$2r_1 - N = (P_0 - Q_0)/P_0.$$

Let the single Frobenius type solution be

$$y_1(x) = |x|^{r_1} \sum_{n=0}^\infty a_n x^n.$$

Then, using the identifications $a_0(x) = x^2 P(x)$, $a_1(x) = x Q(x)$ in the integral method of Sec. 5.6, a second linearly independent solution is seen to be

$$y_2(x) = y_1(x) \int \frac{\exp\left(-\int \frac{Q(x)}{x P(x)} dx \right)}{[y_1(x)]^2} dx.$$

Show by division that

$$\frac{Q(x)}{x P(x)} = \frac{1}{x} \left(\frac{Q_0}{P_0} \right) \left[1 + \left(\frac{Q_1}{Q_0} - \frac{P_1}{P_0} \right) x + \dots \right],$$

and hence that

$$\exp\left(-\int \frac{Q(x)}{xP(x)}\,dx\right) = |x|^{-(Q_0/P_0)}\,e^{f(x)},$$

where $f(x)$ is a power series in x.

Justify writing

$$\frac{e^{f(x)}}{[y_1(x)]^2} = \frac{1}{a_0^2 x^{N+1}}\left(\frac{1+k_1 x+k_2 x^2+\ldots}{1+2(a_1/a_0)x+\ldots}\right)$$

$$= \frac{1}{a_0^2}\left(\frac{1}{x^{N+1}}+\frac{c_1}{x^N}+\ldots+\frac{c_N}{x}+c_{N+1}+c_{N+2}x+\ldots\right),$$

where $k_1, k_2, \ldots$, and $c_1, c_2, \ldots$, are constants. Hence show that

$$y_2(x) = \frac{1}{a_0^2}y_1(x)\left(-\frac{1}{Nx^N}-\frac{1}{(N-1)x^{N-1}}\ldots+c_N\ln|x|+c_{N+1}x+\ldots\right).$$

By setting the arbitrary constant $a_0 = 1$, and combining the product of $y_1(x)$ and the series, the second linearly independent solution is seen to be of the form

$$y_2(x) = Cy_1(x)\ln|x|+|x|^{r_2}\sum_{n=0}^{\infty}b_n x^n.$$

30 Complete the details of the following outline proof of the general form taken by the second linearly independent solution in case (iii) of Theorem 8.2 when the indicial equation has equal roots $r_1 = r_2 = s$. Show first that if the equation

$$x^2 P(x)y''+xQ(x)y'+R(x)y = 0$$

has a regular singular point at the origin and the roots of the indicial equation are equal then, using the notation of Theorem 8.2,

$$2s = (P_0 - Q_0)/P_0.$$

Let the single Frobenius type solution be

$$y_1(x) = |x|^s\sum_{n=0}^{\infty}a_n x^n.$$

Then making the identifications $a_0(x) = x^2 P(x)$, $a_1(x) = xQ(x)$ in the integral method of Sec. 5.6 a second linearly independent solution is seen to be

$$y_2(x) = y_1(x)\int \frac{\exp\left(-\int \frac{Q(x)}{xP(x)}\,dx\right)}{[y_1(x)]^2}\,dx.$$

Show by division that

$$\frac{Q(x)}{xP(x)} = \frac{1}{x}\left(\frac{Q_0}{P_0}\right)\left[1+\left(\frac{Q_1}{Q_0}-\frac{P_1}{P_0}\right)x+\ldots\right],$$

and hence that

$$\exp\left(-\int \frac{Q(x)}{xP(x)}\,dx\right) = |x|^{-(Q_0/P_0)}\,e^{f(x)},$$

where $f(x)$ is a power series in x.

Justify writing

$$\frac{e^{f(x)}}{[y_1(x)]^2} = \frac{1}{a_0^2 x}\left(\frac{1 + k_1 x + k_2 x^2 +}{1 + 2(a_1/a_0) + \ldots}\right)$$

$$= \frac{1}{a_0^2}\left(\frac{1}{x} + c_1 + c_2 x + \ldots\right),$$

where $b_1, b_2, \ldots$, and $c_1, c_2, \ldots$, are constants. Hence show that

$$y_2(x) = \frac{1}{a_0^2}\,y_1(x)\left(\ln|x| + c_1 x + \frac{c_2}{2}x^2 + \ldots\right).$$

By setting the arbitrary constant $a_0 = 1$, and combining the product of $y_1(x)$ and the series, the second linearly independent solution is seen to be of the form

$$y_2(x) = y_1(x)\ln|x| + |x|^s \sum_{n=1}^{\infty} b_n x^n.$$

8.7   Bessel functions

One of the most important and useful differential equations to arise in applications is Bessel's equation[9]

$$x^2 y'' + xy' + (x^2 - v^2)y = 0, \tag{1}$$

in which $v \geq 0$ is a specified real number. The equation has a regular singular point at the origin and its indicial equation is seen to be

$$r^2 - v^2 = 0, \tag{2}$$

so its solution in the neighborhood of the origin may be obtained from the Frobenius

---

[9] FRIEDRICH WILHELM BESSEL (1784–1846), a German astronomer who calculated the orbit of Halley's comet. In 1824 Bessel introduced the function $J_n(x)$, known as the Bessel function of the first kind of order n, through the integral representation

$$J_n(x) = \frac{1}{\pi}\int_0^\pi \cos(x\sin\theta - n\theta)\,d\theta.$$

This integral representation of the function arose as a result of his expressing an astronomical parameter called the **true anomaly** of an orbiting body in terms of a more convenient parameter called its **mean anomaly**.

method by setting

$$y(x) = \sum_{m=0}^{\infty} a_m x^{m+r}, \tag{3}$$

with $a_0 \neq 0$.

Proceeding in the usual manner, and substituting (3) into (1) gives,

$$\sum_{m=0}^{\infty} (m+r)(m+r-1)a_m x^{m+r} + \sum_{m=0}^{\infty} (m+r)a_m x^{m+r} \sum_{m=0}^{\infty} a_m x^{m+r+2}$$

$$-v^2 \sum_{m=0}^{\infty} a_m x^{m+r} = 0.$$

Shifting the index in the third term, combining the summations and separating out the terms in x^r and x^{r+1} we arrive at the result

$$(r^2 - v^2)a_0 x^r + [(r+1)^2 - v^2]a_1 x^{r+1}$$

$$+ \sum_{m=2}^{\infty} [(m+r+v)(m+r-v)a_m + a_{m-2}]x^{m+r} = 0.$$

Equating to zero the coefficients of each power of x gives

$$x^r: (r^2 - v^2)a_0 = 0 \tag{4}$$

$$x^{r+1}: [(r+1)^2 - v^2]a_1 = 0 \tag{5}$$

$$x^{m+r}: (m+r+v)(m+r-v)a_m + a_{m-2} = 0, \qquad m = 2, 3, \ldots. \tag{6}$$

Result (4) is simply the indicial equation (2) which has the roots $r_1 = v$ and $r_2 = -v$. When $r = r_1 = v$ it follows from (5) that $a_1 = 0$, and thereafter from (6) that $a_3 = 0$, $a_5 = 0, \ldots$, so that all the odd order coefficients vanish. It also follows from (6) that

$$a_{2m} = -\frac{1}{2^2 m(m+v)} a_{2m-2}, \qquad m = 1, 2, \ldots. \tag{7}$$

A standard normalization of the resulting solution is obtained by setting the arbitrary constant

$$a_0 = \frac{1}{2^v \Gamma(1+v)}, \tag{8}$$

from which we find that

$$a_2 = -\frac{a_0}{2^2(1+v)} = -\frac{1}{2^{2+v} 1! \, \Gamma(2+v)}$$

$$a_4 = -\frac{a_2}{2^2 2(2+v)} = \frac{1}{2^{4+v} 2! \, \Gamma(3+v)}$$

$$a_6 = -\frac{a_4}{2^2 3(4+v)} = -\frac{1}{2^{6+v} 3! \, \Gamma(4+v)}$$

and, in general,

$$a_{2m} = \frac{(-1)^m}{2^{2m+\nu} m! \, \Gamma(m+1+\nu)}, \qquad m = 0, 1, 2, \ldots . \tag{9}$$

Substituting this result into the series solution (3) with $r = r_1 = \nu$ gives rise to the solution of (1) known as the **Bessel function of the first kind of order ν**, denoted by $J_\nu(x)$,

$$J_\nu(x) = x^\nu \sum_{m=0}^{\infty} \frac{(-1)^m x^{2m}}{2^{2m+\nu} m! \, \Gamma(m+1+\nu)}. \tag{10}$$

A routine application of the ratio test shows the series solution (10) to be convergent for all x.

By convention, when the parameter ν assumes integral values it is denoted by n, so that as $\Gamma(m+1+n) = (m+n)!$, the expression in (10) may then be written

$$J_n(x) = x^n \sum_{m=0}^{\infty} \frac{(-1)^m x^{2m}}{2^{2m+n} m! \, (m+n)!}, \qquad n = 0, 1, 2, \ldots . \tag{11}$$

The behavior of the Bessel functions $J_0(x)$, $J_1(x)$ and $J_2(x)$ is shown in Fig. 8.8. Notice the similarity between damped cosine and sine functions and $J_0(x)$ and $J_1(x)$, respectively. It is because of such similarities that Bessel functions are classified as belonging to the class of functions called **almost periodic functions**.

A repetition of the form of argument leading to $J_\nu(x)$ (or by merely replacing ν by $-\nu$ in (10)) shows that with $r = r_2 = -\nu$, provided $r_1 - r_2 = 2\nu$ is *not* an integer the second linearly independent solution of Bessel's equation is

$$J_{-\nu}(x) = x^{-\nu} \sum_{m=0}^{\infty} \frac{(-1)^m x^{2m}}{2^{2m-\nu} m! \, \Gamma(m+1-\nu)}, \tag{12}$$

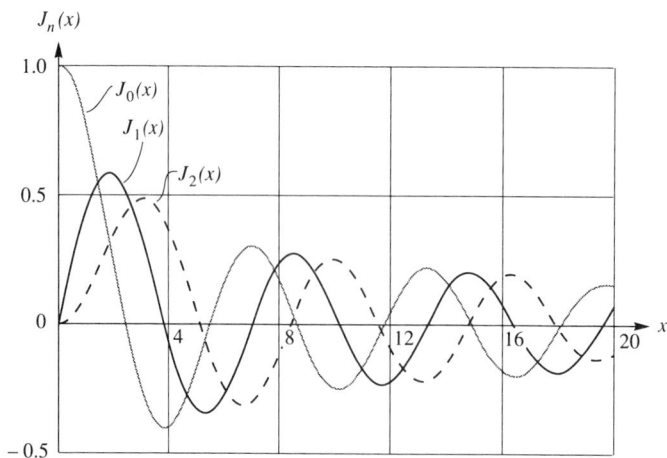

Fig. 8.8 Graphs of Bessel functions $J_0(x)$, $J_1(x)$ and $J_2(x)$

which by the ratio test is seen to converge for $|x| > 0$. Thus a **basis** (a pair of linearly independent solutions) for Bessel's equation (1) is provided by the functions $J_v(x)$ and $J_{-v}(x)$ whenever $2v$ is *not* an integer.

The failure of the functions $J_v(x)$ and $J_{-v}(x)$ to form a basis when $v = n$ (an integer) may be seen directly from (11) and (12). As the gamma function $\Gamma(m+1-n)$ is infinite at negative integer values of its argument (cf. Fig. 8.7), the first n terms of (12) vanish when $v = n$, and therefore

$$J_{-n}(x) = |x|^{-n} \sum_{m=0}^{\infty} \frac{(-1)^m x^{2m}}{2^{2m-n} m! (m-1)!}.$$

However, shifting the index to make this summation start from zero gives the result

$$J_{-n}(x) = x^{-n} \sum_{m=0}^{\infty} \frac{(-1)^{m+n} x^{2m+2n}}{2^{2m+n}(m+n)! \, m!} = (-1)^n J_n(x), \tag{13}$$

which shows that $J_{-n}(x)$ is a scalar multiple of $J_n(x)$, so $J_n(x)$ and $J_{-n}(x)$ are *not* linearly independent.

Two important series derived from (11) are

$$J_0(x) = \sum_{m=0}^{\infty} \frac{(-1)^m x^{2m}}{2^{2m}(m!)^2} = 1 - \frac{x^2}{2^2(1!)^2} + \frac{x^4}{2^4(2!)^2} - \frac{x^6}{2^6(3!)^2} + \cdots \tag{14}$$

and

$$J_1(x) = \sum_{m=0}^{\infty} \frac{(-1)^m x^{2m+1}}{2^{2m+1} m! (m+1)!} = \frac{x}{2} - \frac{x^3}{2^3 1! 2!} + \frac{x^5}{2^5 2! 3!} - \frac{x^7}{2^7 3! 4!} + \cdots . \tag{15}$$

As these series are absolutely convergent for all x they may be differentiated term by term (cf. Sec. 8.1). In particular,

$$\frac{d}{dx}[J_0(x)] = \sum_{m=1}^{\infty} \frac{(-1)^m x^{2m-1}}{2^{2m-1}(m-1)! m!} = \sum_{m=0}^{\infty} \frac{(-1)^{m+1} x^{2m+1}}{2^{2m+1} m! (m+1)!} = -J_1(x).$$

and thus we have proved the important result that

$$J_0'(x) = -J_1(x). \tag{16}$$

Other important and useful relationships between Bessel functions which may be verified directly by means of definitions (10) and (13) are

$$J_{v-1}(x) + J_{v+1}(x) = \frac{2v}{x} J_v(x), \quad \text{for } v \geq 1, \tag{17}$$

$$J_{v-1}(x) - J_{v+1}(x) = 2J_v'(x), \quad \text{for } v \geq 1, \tag{18}$$

$$x J_v'(x) = v J_v(x) - x J_{v+1}(x), \quad \text{for } v \geq 0, \tag{19}$$

$$x J_v'(x) = x J_{v-1}(x) - v J_v(x), \quad \text{for } v \geq 1, \tag{20}$$

$$\frac{d}{dx}[x^{v}J_{v}(x)] = x^{v}J_{v-1}(x), \qquad \text{for } v \geq 1, \tag{21}$$

$$\frac{d}{dx}[x^{-v}J_{v}(x)] = -x^{-v}J_{v+1}(x). \quad \text{for } v \geq 0. \tag{22}$$

Notice that (16) follows from (19) when $v = 0$, while by setting $v = 0$ in (20) we are able to determine the derivative of $J_1(x)$ in terms of $J_0(x)$ and $J_1(x)$ through the result

$$J_1'(x) = J_0(x) - \frac{1}{x}J_1(x). \tag{23}$$

Inspection of Ex. 7 of Sec. 8.6 shows the equation involved to be Bessel's equation with $v = \frac{1}{2}$. With this value of v the normalization condition determined by (8) for the coefficient a_0 in $J_{1/2}(x)$ becomes

$$a_0 = \frac{1}{2^{1/2}\Gamma(\frac{3}{2})} = \sqrt{\frac{2}{\pi}},$$

because from (3) of Sec. 8.5, $\Gamma(\frac{3}{2}) = (\frac{1}{2})\Gamma(\frac{1}{2}) = (\frac{1}{2})\sqrt{\pi}$. Thus the first solution of Bessel's equation determined in the example is seen to be

$$J_{1/2}(x) = \sqrt{\frac{2}{\pi x}} \sin x, \tag{24}$$

from which it follows at once that the second linearly independent solution determined in the example is

$$J_{-1/2}(x) = \sqrt{\frac{2}{\pi x}} \cos x. \tag{25}$$

Consequently, although in this case $r_1 = v = \frac{1}{2}$ and $r_2 = -v = -\frac{1}{2}$, so that $r_1 - r_2 = 1$, the two linearly independent solutions forming a basis for

$$x^2 y'' + xy' + (x^2 - \tfrac{1}{4})y = 0$$

are the elementary functions (24) and (25), neither of which has a logarithmic singularity at the origin. Taken together with (17), results (24) and (25) imply that $J_v(x)$ and $J_{-v}(x)$ are always elementary functions without logarithmic singularities when $v = (2m+1)/2$, $m = 0, 1, 2, \ldots$.

For reference, the first few functions generated in this manner are

$$J_{1/2}(x) = \sqrt{\frac{2}{\pi x}} \sin x, \quad J_{-1/2}(x) = \sqrt{\frac{2}{\pi x}} \cos x$$

$$J_{3/2}(x) = \sqrt{\frac{2}{\pi x}}\left(\frac{\sin x}{x} - \cos x\right), \quad J_{-3/2}(x) = -\sqrt{\frac{2}{\pi x}}\left(\sin x + \frac{\cos x}{x}\right) \tag{26}$$

$$J_{5/2}(x) = \sqrt{\frac{2}{\pi x}} \left[\left(\frac{3}{x^2} - 1 \right) \sin x - \frac{3}{x} \cos x \right],$$

$$J_{-5/2}(x) = \sqrt{\frac{2}{\pi x}} \left[\frac{3}{x} \sin x + \left(\frac{3}{x^2} - 1 \right) \cos x \right]$$

$$J_{7/2}(x) = \sqrt{\frac{2}{\pi x}} \left[\left(\frac{15}{x^3} - \frac{6}{x} \right) \sin x - \left(\frac{15}{x^2} - 1 \right) \cos x \right],$$

$$J_{-7/2}(x) = -\sqrt{\frac{2}{\pi x}} \left[\left(\frac{15}{x^2} - 1 \right) \sin x + \left(\frac{15}{x^3} - \frac{6}{x} \right) \cos x \right].$$

So far we have established that provided v is *neither* an integer *nor* a multiple of $\frac{1}{2}$ the general solution of Bessel's equation (1) is

$$y(x) = c_1 J_v(x) + c_2 J_{-v}(x). \tag{27}$$

It now only remains for us to determine a second linearly independent solution $y_2(x)$ when $v = n$ (an integer). If we consider the case $n = 0$, we see from Theorem 8.2(iii) that the second solution must be of the form

$$y_2(x) = J_0(x) \ln |x| + \sum_{m=1}^{\infty} b_n x^n. \tag{28}$$

Routine substitution of (28) into the governing equation derived from (1) with $v = 0$, namely into

$$xy'' + y' + xy = 0, \tag{29}$$

shows (cf. Ex. 9, Sec. 8.6) that

$$y_2(x) = J_0(x) \ln |x| + \sum_{m=1}^{\infty} \frac{(-1)^{m-1} h_m x^{2m}}{2^{2m} (m!)^2}, \tag{30}$$

where

$$h_m = 1 + \frac{1}{2} + \frac{1}{3} + \ldots + \frac{1}{m}. \tag{31}$$

Consequently the second solution arrived at in this manner is

$$y_2(x) = J_0(x) \ln |x| + x + \frac{x^2}{2^2} - \frac{x^4}{2^n \cdot 4^2} \left(1 + \frac{1}{2} \right) + \frac{x^6}{2^2 \cdot 4^2 \cdot 6^2} \left(1 + \frac{1}{2} + \frac{1}{3} \right) - \ldots. \tag{32}$$

If expression (32) and corresponding ones for $n = 1, 2, \ldots$ are adopted for the second linearly independent solution when v is an integer, the definition will alter according as v is integral or nonintegral. Because of this, for the sake of uniformity, a slightly different definition is adopted. The second linearly independent solution when $n = 0$, called the **Bessel function of the second kind of order zero**, and denoted by $Y_0(x)$, is defined as

$$Y_0(x) = \frac{2}{\pi} \left((\gamma - \ln 2) J_0(x) + y_2(x) \right), \tag{33}$$

where $\gamma = 0.577215 \ldots$, called **Euler's constant**, is defined as the limit

$$\gamma = \lim_{m \to \infty} \left(1 + \frac{1}{2} + \frac{1}{3} + \ldots + \frac{1}{m} - \ln m \right). \tag{34}$$

This modified definition is permissible, because we may both scale $y_2(x)$ and add to it any constant multiple of the other linearly independent solution $J_0(x)$ and it will still remain a linearly independent solution. Combining (32) and (33) we find this definition is equivalent to

$$Y_0(x) = \frac{2}{\pi} \left[\left(\gamma + \ln \left| \frac{x}{2} \right| \right) J_0(x) + \sum_{m=1}^{\infty} \frac{(-1)^{m-1} h_m x^{2m}}{2^{2m}(m!)^2} \right], \tag{35}$$

which is convergent for $|x| > 0$.

Uniformity of definition for integral and nonintegral values of v is accomplished by defining

$$Y_v(x) = \frac{J_v(x) \cos v\pi - J_{-v}(x)}{\sin v\pi} \tag{36a}$$

when v is nonintegral, and

$$Y_n(x) = \lim_{v \to n} Y_v(x), \tag{36b}$$

when v is an integer. The details will be omitted, but an application of L'Hospital's rule to (36) with $n = 0$ can be shown to yield (35). Graphs of the three most useful functions $Y_0(x)$, $Y_1(x)$ and $Y_2(x)$ are shown in Fig. 8.9. As $Y_v(x)$ represents the second linearly independent solution of Bessel's equation (1) for *all* v, the general solution of (1) may *always* be written as

$$y(x) = c_1 J_v(x) + c_2 Y_v(x). \tag{37}$$

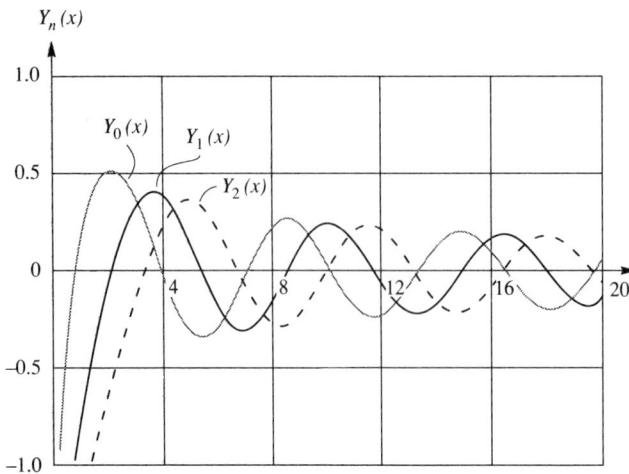

Fig. 8.9   Graph of Bessel functions $Y_0(x)$, $Y_1(x)$ and $Y_2(x)$

It is apparent from (36a) that when v is nonintegral, by redefining the arbitrary constants c_1 and c_2, (37) reduces to

$$y(x) = \tilde{c}_1 J_v(x) + \tilde{c}_2 J_{-v}(x), \tag{38}$$

which was the form of the general solution for nonintegral v given in (27).

Bessel's equation does not always arise in the standard form given in (1), but more frequently as

$$x^2 \frac{d^2 y}{dx^2} + x \frac{dy}{dx} + (k^2 x^2 - v^2)y = 0, \text{ with the general solution} \tag{39}$$

$$y(x) = c_1 J_v(kx) + c_2 Y_v(kx). \tag{40}$$

This form of the equation is obtainable from (1) by setting $x = kt$. In terms of t we have

$$\frac{dy}{dx} = \frac{1}{k} \frac{dy}{dt}, \quad \frac{d^2 y}{dx^2} = \frac{1}{k^2} \frac{d^2 y}{dt^2},$$

so (1) becomes

$$t^2 \frac{d^2 y}{dt^2} + t \frac{dy}{dt} + (k^2 t^2 - v^2)y = 0.$$

The form given in (39) follows directly by writing x in place of t.

The oscillatory nature of Bessel functions of the first kind illustrated in the graphs of $J_0(x)$, $J_1(x)$ and $J_2(x)$ in Fig. 8.8, and explicitly in the case of $J_{1/2}(x)$ and $J_{-1/2}(x)$ in (24) and (25), has important consequences which we shall examine later. It is sufficient that at this point we remark that $J_v(x)$ has an infinite number of real zeros denoted by $\mu_{v,1}, \mu_{v,2}, \mu_{v,3}, \ldots$, and that $\lim_{m \to \infty} \mu_{v,m} = \infty$. For reference purposes we record in Table 8.1 the first six zeros of $J_0(x)$, $J_1(x)$ and $J_2(x)$.

Table 8.1 Zeros $\mu_{0,m}$ of $J_0(x)$, $\mu_{1,m}$ of $J_1(x)$ and $\mu_{2,m}$ of $J_2(x)$

m	$\mu_{0,m}$	$\mu_{1,m}$	$\mu_{2,m}$
1	2.40482	0	0
2	5.52007	3.83171	5.13562
3	8.65372	7.01559	8.41724
4	11.79153	10.17347	11.61984
5	14.93091	13.32369	14.79595
6	18.07106	16.47063	17.95982

The table illustrates the important interlacing property possessed by the zeros of $J_v(x)$; namely that

$$\cdots \mu_{v+1,m} < \mu_{v,m} < \mu_{v+1,m+1} < \mu_{v,m+1} < \mu_{v+1,m+2} < \mu_{v,m+2} < \cdots.$$

Let us establish the **interlacing property** with respect to the zeros of $J_0(x)$ and $J_1(x)$. An essentially similar form of argument can be used to establish it in general, with respect to the zeros of $J_\nu(x)$ and $J_{\nu+1}(x)$ for any ν.

We start from the plausible assumptions suggested by Fig. 8.8 that:

(i) $J_0(x)$ and $J_1(x)$ have infinitely many zeros $\mu_{0,1}$, $\mu_{0,2}$, $\mu_{0,3}$, $\ldots$ and $\mu_{1,1}$, $\mu_{1,2}$, $\mu_{1,3}$, $\ldots$, respectively, where $\lim_{m\to\infty} \mu_{0,m} = \infty$ and $\lim_{m\to\infty} \mu_{1,m} = \infty$, and

(ii) both $J_0(x)$ and $J_1(x)$ oscillate about zero.

It is possible to give a rigorous justification of these assumptions, though we shall not do so here.

From (16) we have

$$J_0'(x) = -J_1(x),$$

so as $J_1(x)$ is continuous it follows that the maxima and minima of $J_0(x)$ must occur at the zeros of $J_1(x)$. In the interval between any two successive zeros of $J_1(x)$, say the interval $\mu_{1,N} < x < \mu_{1,N+1}$, the function $J_1(x)$ must be of one sign so that $J_0(x)$ must either be strictly monotonic increasing or decreasing. As $J_0(x)$ changes sign in the interval, and this can only occur once, it follows that precisely one zero $\mu_{0,N}$ of $J_0(x)$ must lie within the interval. Thus we have demonstrated that

$$\mu_{1,N} < \mu_{0,N} < \mu_{1,N+1}$$

for $N = 1, 2, 3, \ldots$, and hence the general interlacing property

$$\mu_{1,1} < \mu_{0,1} < \mu_{1,2} < \mu_{0,2} < \mu_{1,3} < \cdots.$$

Example 1. The integral representation

Show that the function

$$J_0(x) = \frac{1}{\pi} \int_0^\pi \cos(x \sin\theta)\, d\theta$$

satisfies the differential equation

$$xy'' + y' + xy = 0,$$

and so is an integral representation of the Bessel function of the first kind of order zero. Find $J_0(0)$ and $J_0'(0)$.

Solution

$$J_0'(x) = \frac{1}{\pi} \int_0^\pi \frac{\partial}{\partial x}[\cos(x\sin\theta)]\, d\theta = -\frac{1}{\pi}\int_0^\pi \sin\theta \sin(x\sin\theta)\, d\theta,$$

and

$$J_0''(x) = -\frac{1}{\pi}\int_0^\pi \sin\theta \frac{\partial}{\partial x}[\sin(x\sin\theta)]\, d\theta = -\frac{1}{\pi}\int_0^\pi \sin^2\theta \cos(x\sin\theta)\, d\theta.$$

Consequently,

$$J_0''(x) = -\frac{1}{\pi}\int_0^\pi (1-\cos^2\theta)\cos(x\sin\theta)d\theta = -J_0(x) + \frac{1}{\pi}\int_0^\pi \cos^2\theta\cos(x\sin\theta)dx.$$

Integrating by parts we find that

$$\int_0^\pi \sin\theta\sin(x\sin\theta)d\theta = -\cos\theta\sin(x\sin\theta)\bigg|_0^\pi + \int_0^\pi \cos\theta\frac{\partial}{\partial\theta}[\sin(x\sin\theta)]d\theta$$

$$= x\int_0^\pi \cos^2\theta\cos(x\sin\theta)d\theta,$$

so

$$x[J_0''(x) + J_0(x)] = -J_0(x),$$

and hence $y = J_0(x)$ is a solution of

$$xy'' + y' + xy = 0.$$

Setting $x=0$ in the integral representation for $J_0(x)$ reduces it to

$$J_0(0) = \frac{1}{\pi}\int_0^\pi 1\cdot d\theta = 1,$$

while setting $x=0$ in the integral representation for $J_0'(x)$ reduces it to

$$J_0'(0) = -\frac{1}{\pi}\int_0^\pi 0\cdot d\theta = 0. \qquad\blacksquare$$

Example 2. Integrals of Bessel functions

Evaluate $\displaystyle\int_1^2 x^{-5}J_6(x)\,dx$ and show $\displaystyle\int J_2(x)dx = \int xJ_1(x)dx - xJ_2(x).$

Solutions

(i) Integrating (22) gives

$$\int x^{-\nu}J_{\nu+1}(x)dx = -x^{-\nu}J_\nu(x)$$

so

$$\int_1^2 x^{-5}J_6(x)dx = [-x^{-5}J_5(x)]\bigg|_1^2$$
$$= J_5(1) - 2^{-5}J_5(2).$$

(ii) Using integration by parts we have

$$\int J_2(x)dx = xJ_2(x) - \int xJ_2'(x)dx.$$

Now integrating (20) with $v=2$ gives

$$\int x J_2'(x) dx = \int [x J_1(x) - 2 J_2(x)] dx,$$

so using this in the above result we find

$$\int J_2(x) dx = x J_2(x) - \int x J_1(x) dx + 2 \int J_2(x) dx,$$

and thus

$$\int J_2(x) dx = \int x J_1(x) dx - x J_2(x).$$ ∎

Modified Bessel functions

The differential equation

$$x^2 y'' + x y' - (x^2 + v^2) y = 0, \tag{41}$$

in which $v \geq 0$, is known as the **modified Bessel equation**. The equation arises in hydrodynamics, electrical engineering, acoustics, reactor physics and elsewhere. It may be derived from Bessel's equation (1) by replacing x by ix, where $i^2 = -1$, for then dy/dx becomes $-i dy/dx$ and $d^2 y/dx^2$ becomes $-d^2 y/dx^2$.

Defining

$$I_v(x) = i^{-v} J_v(ix),$$

it follows from (10) that

$$I_v(x) = x^v \sum_{m=0}^{\infty} \frac{x^{2m}}{2^{2m+v} m! \Gamma(m+1+v)}, \tag{42}$$

and from (12) that

$$I_{-v}(x) = x^{-v} \sum_{m=0}^{\infty} \frac{x^{2m}}{2^{2m-v} m! \Gamma(m+1-v)}. \tag{43}$$

The function $I_v(x)$ is called the **modified Bessel function of the first kind of order v**. Arguing as with the function $J_v(x)$, it follows directly that the general solution of (41) is

$$y(x) = c_1 I_v(x) + c_2 I_{-v}(x), \tag{44}$$

whenever v is *not* an integer.

When $v = n$ is an integer, $I_n(x)$ and $I_{-n}(x)$ are linearly dependent, because

$$I_{-n}(x) = i^n J_{-n}(ix) = i^n (-1)^n J_n(ix) = i^{-n} J_n(ix),$$

and so

$$I_{-n}(x) = I_n(x).$$

An argument similar to the one used to define $Y_v(x)$ leads to the definition of the second linearly independent solution of (41) as

$$K_v(x) = \frac{\pi}{2}\left(\frac{I_{-v}(x) - I_v(x)}{\sin v\pi}\right), \quad \text{when } v \text{ is nonintegral,} \tag{45}$$

and as

$$K_n(x) = \lim_{v \to n} K_v(x), \quad \text{for integral } n.$$

The function $K_v(x)$ is called the **modified Bessel function of the second kind of order v.** Thus the general solution of (41) may *always* be written as

$$y(x) = c_1 I_v(x) + c_2 K_v(x). \tag{46}$$

Graphs of the most frequently used functions $I_0(x)$, $I_1(x)$, $K_0(x)$ and $K_1(x)$ are shown in Fig. 8.10.

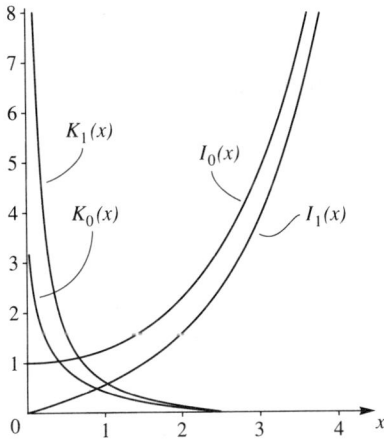

Fig. 8.10 Graphs of modified Bessel functions I_0, I_1, K_0 and K_1

An elementary change of variable establishes that the slightly more general form of modified Bessel equation

$$x^2 y'' + xy' - (k^2 x^2 + v^2)y = 0 \tag{47}$$

has the general solution

$$y(x) = c_1 I_v(kx) + c_2 I_{-v}(kx), \quad \text{if } v \text{ is not an integer,}$$

and

$$y(x) = c_1 I_v(kx) + c_2 K_v(kx), \quad \text{for } any \ v. \tag{48}$$

Differentiation of (42) with $v = 0$ establishes the useful result

$$I_0'(x) = I_1(x). \tag{49}$$

For the sake of completeness, we remark that when n is an integer, an argument similar to that used when discussing $Y_0(x)$ shows the series solution for $K_n(x)$ to be

$$K_n(x) = \frac{1}{2}\left(\frac{1}{2}x\right)^{-n} \sum_{k=1}^{n-1} \frac{(n-k-1)!}{k!}\left(-\frac{1}{4}x^2\right)^k + (-1)^{n+1}\ln\left(\frac{1}{2}x\right)I_n(x)$$

$$+ (-1)^n\frac{1}{2}\left(\frac{1}{2}x\right)^n \sum_{k=0}^{\infty} [\psi(k+1) + \psi(n+k+1)]\frac{(\frac{1}{4}x^2)^k}{k!(n+k)!}, \tag{50}$$

where

$$\psi(1) = -\gamma, \quad \psi(n) = -\gamma + \sum_{k=1}^{n-1} k^{-1}, \tag{51}$$

and

$$\gamma = 0.577215 \ldots \text{ is Euler's constant.}$$

Problems for Section 8.7

Establish the required results in the following problems by using (10) and (12).

1 $J_{v-1}(x) + J_{v+1}(x) = \dfrac{2v}{x}J_v(x), \; v \geq 1.$

2 $J_{v-1}(x) - J_{v+1}(x) = 2J_v'(x), \; v \geq 1.$

3 $xJ_v'(x) = vJ_v(x) - xJ_{v+1}(x), \; v \geq 0.$

4 $xJ_v'(x) = xJ_{v-1} - vJ_v(x), \; v \geq 1.$

5 $\dfrac{d}{dx}[x^v J_v(x)] = x^v J_{v-1}(x), \; v \geq 1.$

6 $\dfrac{d}{dx}[x^{-v} J_v(x)] = -x^{-v} J_{v+1}(x), \; v \geq 0.$

7 $2J_0''(x) = J_2(x) - J_0(x).$

8 $xJ_0''(x) = xJ_2(x) + J_0'(x).$

9 Use (18) to prove that

$$2^2 J_n'' = J_{n-2} - 2J_n + J_{n+2} \qquad \text{for } n \geq 2,$$

and then by induction that

$$2^m J_n^{(m)} = J_{n-m} - mJ_{n-m+2} + \frac{m(m-1)}{2!}J_{n-m+4} - \frac{m(m-1)(m-2)}{3!}J_{n-m+6} + \cdots + (-1)^m J_{n-m}.$$

10 Use (19) and (20) to prove that

$$\frac{d}{dx}(x^v J_v) = x^v J_{v-1} \qquad \text{and} \qquad \frac{d}{dx}(x^{-v} J_v) = -x^{-v} J_{v+1}.$$

Establish the required results in the following problems, making use of (17), (24) and (25).

11 $J_{3/2}(x) = \sqrt{\dfrac{2}{\pi x}} \left(\dfrac{\sin x}{x} - \cos x \right)$ and $J_{-3/2}(x) = -\sqrt{\dfrac{2}{\pi x}} \left(\sin x + \dfrac{\cos x}{x} \right)$.

12 $J_{5/2}(x) = \sqrt{\dfrac{2}{\pi x}} \left[\left(\dfrac{3}{x^2} - 1 \right) \sin x - \dfrac{3}{x} \cos x \right]$ and

$J_{-5/2}(x) = \sqrt{\dfrac{2}{\pi x}} \left[\dfrac{3}{x} \sin x + \left(\dfrac{3}{x^2} - 1 \right) \cos x \right]$.

13 $J_{9/2}(x) = \sqrt{\dfrac{2}{\pi x}} \left[\left(\dfrac{105}{x^4} - \dfrac{45}{x^2} + 1 \right) \sin x - \left(\dfrac{105}{x^3} - \dfrac{10}{x} \right) \cos x \right]$.

14 $J_{-9/2}(x) = \sqrt{\dfrac{2}{\pi x}} \left[\left(\dfrac{105}{x^3} - \dfrac{10}{x} \right) \sin x + \left(\dfrac{105}{x^4} - \dfrac{45}{x^2} + 1 \right) \cos x \right]$.

15 Complete the details of the calculation leading to the result

$$y_2(x) = J_0(x) \ln |x| + x + \frac{x^2}{2^2} - \frac{x^4}{2^2 \cdot 4^2} \left(1 + \frac{1}{2} \right) + \frac{x^6}{2^2 \cdot 4^2 \cdot 6^2} \left(1 + \frac{1}{2} + \frac{1}{3} \right) - \cdots$$

given in (32).

16 Show that the substitution $u = x^{1/2} y$ converts the equation

$$x^2 y'' + xy' + (x^2 - n^2) y = 0$$

into

$$u'' + \left(1 - \frac{4n^2 - 1}{4x^2} \right) u = 0.$$

Use this result to show that the linearly independent solutions of Bessel's equation of the first kind of order $\frac{1}{2}$ are proportional to $x^{-1/2} \sin x$ and $x^{-1/2} \cos x$, respectively.

17 Use the substitution $u = x^2$ to show that the general solution of

$$xy'' + y' + 4 \left(x^3 - \frac{n^2}{x} \right) y = 0$$

is

$$y = c_1 J_n(x^2) + c_2 Y_n(x^2).$$

18 Use the substitution $y = xw$ to convert the equation

$$xy'' - y' + xy = 0$$

to Bessel's equation, and hence show that its general solution is

$$y(x) = c_1 x J_1(x) + c_2 x Y_{-1}(x).$$

19 Find the general solution of

$$x^2 y'' + xy' + 9(x^2 - 1) y = 0.$$

20 Find the general solution of

$$x^2 y'' + xy' + 9(x^2 - \tfrac{4}{9}) y = 0.$$

21 Find the general solution of

$$y'' + \frac{1}{x} y' + 7\left(1 - \frac{v^2}{7x^2}\right) y = 0.$$

22 Find the general solution of

$$y'' + \frac{1}{x} y' + 2y - \frac{1}{9x^2} y = 0.$$

23 Find the general solution of

$$\frac{d^3 y}{dx^3} + \frac{1}{x}\frac{d^2 y}{dx^2} + 6\left(1 - \frac{3}{2x^2}\right)\frac{dy}{dx} = 0.$$

24 Find the general solution of

$$\frac{d^3 y}{dx^3} + \frac{1}{x}\frac{dy}{dx} + 4y - \frac{1}{4x^2} y = 0.$$

Evaluate the following integrals involving Bessel functions making use of (16) to (22) where necessary.

25 $\displaystyle\int_0^2 J_0'(x)\,dx.$

26 $\displaystyle\int x^3 J_2(x)\,dx.$

27 $\displaystyle\int x^{-2} J_3(x)\,dx.$

28 $\displaystyle\int x^{-1} J_2(x)\,dx.$

29 $\displaystyle\int x^4 J_3(x)\,dx.$

30 $\displaystyle\int J_1'(x)\,dx.$

31 $\displaystyle\int J_2'(x)\,dx.$

32 $\displaystyle\int [x^2 J_1(x) - J_0'(x)]\,dx$

33 Show that

$$\int_0^x J_{n+1}(t)\,dt = \int_0^x J_{n-1}(t)\,dt - 2J_n(x), \qquad \text{for } n > 0.$$

34 Show that

$$\int_0^x J_1(t)\,dt = 1 - J_0(x).$$

35 Basing your argument on Ex. 1, show that if

$$J_n(x) = \frac{1}{\pi}\int_0^\pi \cos(x\sin\theta - n\theta)\,d\theta,$$

then $y(x) = J_n(x)$ is a solution of

$$x^2 y'' + xy' + (x^2 - n^2) y = 0.$$

36 Given that

$$J_0(x) = \frac{1}{\pi} \int_0^\pi \cos(x \sin\theta) \, d\theta,$$

use the Maclaurin series expansion of $\cos(x \sin\theta)$ to show that

$$J_0(x) = 1 - \frac{x^2}{4} + \frac{x^4}{64} - \dots.$$

37 Rolle's theorem states that: 'If $f(x)$ is a real-valued function which is continuous on the closed interval $a \le x \le b$ and differentiable at all points of the open interval $a < x < b$ and such that $f(a) = f(b)$, then there is at least one point x_0 interior to $a \le x \le b$ at which $f'(x_0) = 0$'. Use this theorem together with the assumption that $J_0(x)$ is either convex or concave in the interval between successive zeros of $J_0(x)$ to prove the interlacing property of the zeros of $J_0(x)$ and $J_1(x)$.

Establish the required results in the following problems by using (42) and (43).

38 $I_{\nu-1}(x) - I_{\nu+1}(x) = \dfrac{2\nu}{x} I_\nu, \qquad \nu \ge 1.$

39 $I_{\nu-1}(x) + I_{\nu+1}(x) = 2I'_\nu(x), \qquad \nu \ge 1.$

40 $xI'_\nu(x) = \nu I_\nu^{(x)} + xI_{\nu+1}^{(x)}, \qquad \nu \ge 0.$

41 $xI'_\nu(x) = -\nu I_\nu(x) + xI_{\nu-1}(x), \qquad \nu \ge 1.$

42 $\dfrac{d}{dx}[x^\nu I_\nu(x)] = x^\nu I_{\nu-1}(x), \qquad \nu \ge 1.$

43 $\dfrac{d}{dx}[x^{-\nu} I_\nu(x)] = x^{-\nu} I_{\nu+1}(x), \qquad \nu \ge 0.$

Harder problems

The function $\exp[\frac{1}{2}x(t - 1/t)]$ is called the **generating function** for Bessel functions of the first kind. It has the property that $J_n(x)$ is the multiplier (coefficient) of the term in t^n in the Maclaurin series expansion of the generating function for $t \ne 0$. This may be established by first considering the product

$$\exp\left[\frac{1}{2}x\left(t - \frac{1}{t}\right)\right] = \left[1 + \frac{1}{2}xt + \frac{1}{2!}\left(\frac{1}{2}xt\right)^2 + \frac{1}{3!}\left(\frac{1}{2}xt\right)^3 + \dots\right]$$

$$\times \left[1 - \frac{1}{2}\frac{x}{t} + \frac{1}{2!}\left(\frac{x}{2t}\right)^2 - \frac{1}{3!}\left(\frac{x}{2t}\right)^3 + \dots\right],$$

and then collecting all terms in successive powers of t to arrive at the result

$$\exp\left[\frac{1}{2}x\left(t - \frac{1}{t}\right)\right] = J_0(x) + tJ_1(x) + \frac{1}{t}J_{-1}(x) + t^2 J_2(x) + \frac{1}{t^2}J_{-2}x + \dots$$

$$= \sum_{n=-\infty}^{\infty} t^n J_n(x),$$

where use has been made of (10) and (12) with $\nu = n$.

44 Use the generating function approach to show that

$$J_0(x) = \sum_{m=0}^{\infty} \frac{(-1)^m x^{2m}}{2^{2m}(m!)^2}.$$

45 Use the generating function approach to show that

$$J_1(x) = \sum_{m=0}^{\infty} \frac{(-1)^m x^{2m+1}}{2^{2m+1} m! (m+1)!}.$$

46 Differentiate the generating function with respect to x and equate the coefficient of t^n on either side of the identity to show that

$$2J_n'(x) = J_{n-1}(x) - J_{n+1}(x).$$

47 Differentiate the generating function with respect to t and equate the coefficient of t^{n-1} on either side of the identity to show that

$$J_{n-1}(x) + J_{n+1}(x) = \frac{2n}{x} J_n(x).$$

8.8 Asymptotic expansions

Many practical problems require knowledge of the way special functions and solutions of differential equations behave for large values of their argument. A typical example arises with the gamma function defined in Sec. 8.5 by the integral

$$\Gamma(x) = \int_0^{\infty} e^{-t} t^{x-1} \, dx, \tag{1}$$

for which no exact expression exists when x is nonintegral, yet the function often needs to be approximated for large x.

Approximate representations of functions for large x, with the property that the approximation gets better the larger x becomes, are called **asymptotic expansions** or **asymptotic representations** of the functions involved. They frequently take the form of **asymptotic series** involving inverse powers of x

$$a_0 + \frac{a_1}{x} + \frac{a_2}{x^2} + \dots \tag{2}$$

with $a_0, a_1, \dots$ constants and x a real variable. Such a series representation of a function $f(x)$ is usually divergent, though we shall see that if an asymptotic series is truncated after the term in x^{-n} the error made when x is large will not exceed the magnitude of the next term in the asymptotic series.

To assist in the manipulation of asymptotic expansions, and functions in general, we now describe the 'big oh' and 'little oh' notations introduced by Landau[10]. This notation

[10] EDMUND LANDAU (1877–1938), a distinguished German number theorist and analyst who worked in the University of Göttingen.

suppresses detailed information about a function involved in a limiting operation, or an approximation, and replaces it with information obtained by comparing the bahavior of the function with that of a simpler function.

We shall write

$$f(x) = O[\phi(x)].\tag{3}$$

and say $f(x)$ **is of the order of** $\phi(x)$, or just that $f(x)$ is big oh $\phi(x)$, if there exists a real positive constant A such that

$$|f(x)| \le A|\phi(x)| \quad \text{for all } x.\tag{4}$$

We shall say that $f(x)$ **is of the order of** $\phi(x)$ **in a neighborhood of** x_0, if there exists a real positive constant A and a neighborhood of x_0 such that in that neighborhood

$$|f(x)| \le A|\phi(x)| \quad \text{as } x \to x_0.\tag{5}$$

The number x_0 may be ∞, as often happens with asymptotic expansions.

If in a neighborhood of x_0 (x_0 may be infinite) the two functions $f(x)$ and $g(x)$ are such that

$$\lim_{x \to x_0} \left(\frac{f(x)}{g(x)} \right) = 0,\tag{6}$$

we shall say that $f(x)$ **is of smaller order than** $g(x)$ **in the neighborhood of** x_0, and write

$$f(x) = o[g(x)] \quad \text{as } x \to x_0.\tag{7}$$

Hereafter we adopt the usual convention and write $f(x) = O(1)$ to mean that $f(x)$ is bounded, though it may oscillate between bounds, and $f(x) = o(1)$ to mean that $f(x) \to 0$ as $x \to x_0$.

Examples of the use of this notation are:

1 $\ln(1 + x) = O(x) \quad$ as $x \to 0$,

2 $\sinh x - \left(x + \dfrac{x^3}{3!} + \dfrac{x^5}{5!} \right) = O(x^7) \quad$ as $x \to 0$,

3 $\cosh x = O(1) \quad$ as $x \to 0$,

4 $\cos x - 1 = O(x^2) \quad$ as $x \to 0$,

5 $e^{-x} = O(1) \quad$ as $x \to \infty$,

6 $\tanh x = O(1) \quad$ for all x,

7 $\sin x = x + o(x) \quad$ as $x \to 0$,

8 $\exp(-x^2) = o(1/x) \quad$ as $x \to \infty$,

9 $x^n e^{-x} = o(1) \quad$ as $x \to \infty \quad$ for $n = 1, 2, \ldots$.

It is sometimes necessary to perform algebraic operations on expressions containing the order of magnitude symbols o and O. Such operations are to be carried out according to the following rules, all of which follow directly from the basic definitions of these symbols.

1 $O[f(x)] + O[g(x)] = O[f(x) + g(x)]$ as $x \to x_0$,

2 $O[f(x)]O[g(x)] = O[f(x)g(x)]$ as $x \to x_0$,

3 $O[f(x)] o [g(x)] = o[f(x)g(x)]$ as $x \to x_0$,

4 $o[kf(x)] = o[f(x)]$ as $x \to x_0$ provided $k \neq 0$,

5 $f(x) o [g(x)] = o [f(x)g(x)]$ as $x \to x_0$,

6 $o[o(f(x)] = o[f(x)]$ as $x \to x_0$,

7 $o[f(x)] o[g(x)] = o[f(x)g(x)]$ as $x \to x_0$.

We prove only rules 2 and 5, by way of illustration. Rule 2 means that if $F(x) = O[f(x)]$ and $G(x) = O[y(x)]$ as $x \to x_0$, then $F(x)G(x) = O[f(x)g(x)]$ as $x \to x_0$. By definition we have

$$|F(x)| \leq A|f(x)| \text{and} |G(x)| \leq B|g(x)| \text{as } x \to x_0$$

for some positive A and B. Thus

$$\left|\frac{F(x)}{f(x)}\right|\left|\frac{G(x)}{g(x)}\right| = \left|\frac{F(x)G(x)}{f(x)g(x)}\right| \leq AB   \text{as } x \to x_0,$$

but this asserts that $F(x)G(x) = O[f(x)g(x)]$ as $x \to x_0$, and so

$$O[f(x)]O[g(x)] = O[f(x)g(x)]   \text{as } x \to x_0.$$

Rule 5 means that if $G(x) = o[g(x)]$ as $x \to x_0$, then $f(x)G(x) = o[f(x)g(x)]$ as $x \to x_0$. By definition we have

$$\lim_{x \to x_0}\left[\frac{G(x)}{g(x)}\right] = 0 = \lim_{x \to x_0}\left[\frac{f(x)G(x)}{f(x)g(x)}\right],$$

but this asserts that $f(x)G(x) = o[f(x)g(x)]$ as $x \to x_0$, and so

$$f(x) o [g(x)] = o[f(x)g(x)]   \text{as } x \to x_0.$$

Care must always be exercised when interpreting the equality symbol in these results which are only to be read from left to right, so in this case the equality symbol does not imply the usual symmetry of meaning. Thus it is correct to write

$$O(x^2) = O(x)   \text{as } x \to 0,$$

because

$$|x^2| = |x| \text{as } x \to 0.$$

However it is false to write

$$O(x) = O(x^2) \qquad \text{as } x \to 0,$$

because

$$|x| \not< |x^2| \qquad \text{as } x \to 0.$$

We begin our discussion of asymptotics by deriving a simple result suggested by considering the linear first order differential equation

$$y' + 2xy = 1. \tag{8}$$

This has the integrating factor $\mu = \exp(x^2)$ and the general solution

$$y(x) = \exp(-x^2) \int_a^x \exp(t^2) \, dt, \tag{9}$$

where a is an arbitrary constant.

The equation arises in various applications, among which is the propagation of radio waves. In this context x represents the distance from the transmitter. A problem which then arises is to find the approximate variation of y with x when x is large. A solution of (9) may always be obtained in the form of a series by replacing $\exp(t^2)$ by its Maclaurin series expansion and then integrating term by term. However, although the series so obtained is convergent for all x, this does not offer a practical solution to the problem. This is because of the rapid increase in the number of terms needed in the expansion to achieve a given accuracy as x increases.

Let us proceed differently by first writing

$$y(x) = \frac{\int_a^x \exp(t^2) \, dt}{\exp(x^2)},$$

and then applying L'Hospital's rule for indeterminate forms (of the type '$\infty \div \infty$') to $y(x)$. We obtain

$$\lim_{x \to \infty} y(x) = \lim_{x \to \infty} \left(\frac{\int_a^x \exp(t^2) \, dt}{\exp(x^2)} \right) = \lim_{x \to \infty} \left(\frac{\exp(x^2)}{2x \exp(x^2)} \right) = 0,$$

This result tells us that the limiting value of $y(x)$ is zero as $x \to \infty$, but it does not tell us about the behavior of $y(x)$ when x is large. To improve on this information all that is necessary is to examine the form taken by the quotient in the last expression before the limit is taken. This shows that

$$\lim_{x \to \infty} y(x) = \lim_{x \to \infty} \left(\frac{1}{2x} \right) = 0,$$

which is equivalent to the result

$$\lim_{x \to \infty} [2xy(x)] = 1.$$

Thus, for large x, the solution $y(x)$ is approximated by $1/2x$, with the result becoming more accurate, the larger x becomes. Using the order of magnitude notation this may also be written

$$y(x) = \frac{1}{2x}[1 + o(1)] \qquad \text{as } x \to \infty. \tag{10}$$

This is, in fact, the first term of the asymptotic series representation of $y(x)$ for large x. (See Ex. 2).

To proceed further we first introduce the notion of the asymptotic equivalence of two functions. We shall say that $f(x)$ is **asymptotically equivalent** to $y(x)$ as $x \to x_0$, and write

$$f(x) \sim g(x) \qquad \text{as } x \to x_0, \tag{11}$$

if

$$\lim_{x \to x_0} \left(\frac{f(x)}{g(x)} \right) = 1. \tag{12}$$

Here, the symbol $\sim$ is to be read 'is asymptotically equivalent to', and it is understood that x_0 may be ∞. Thus, in terms of the order notation, $f(x)$ and $g(x)$ are asymptotically equivalent as $x \to x_0$ if

$$f(x) = g(x)[1 + o(1)] \qquad \text{as } x \to x_0. \tag{13}$$

We must now make precise the definition of an asymptotic series before proceeding to establish some of the most important properties of such series.

A series of the form

$$a_0 + \frac{a_1}{x} + \frac{a_2}{x^2} + \cdots,$$

with $a_0, a_1, \ldots$ constants, will be called an **asymptotic series representation** of $f(x)$ for large x, or an **asymptotic expansion** of $f(x)$, if for every fixed $n = 0, 1, 2, \ldots,$

$$\lim_{x \to \infty} \left\{ \left[f(x) - \left(a_0 + \frac{a_1}{x} + \frac{a_2}{x^2} + \cdots + \frac{a_n}{x^n} \right) \right] x^n \right\} = 0. \tag{14}$$

When $f(x)$ has an asymptotic series representation we shall write

$$f(x) \sim a_0 + \frac{a_1}{x} + \frac{a_2}{x^2} + \cdots.$$

The following theorems represent the most important properties shared by asymptotic series.

Theorem 8.3 (Uniqueness of an asymptotic series)

If an asymptotic series representation of a function exists it is unique.

Proof

If possible, let $f(x)$ have the two different asymptotic representations

$$f(x) \sim a_0 + \frac{a_1}{x} + \frac{a_2}{x^2} + \dots \quad \text{and} \quad f(x) \sim b_0 + \frac{b_1}{x} + \frac{b_2}{x^2} + \dots.$$

Set

$$A_n(x) = a_0 + \frac{a_1}{x} + \frac{a_2}{x^2} + \dots + \frac{a_n}{x^n} \quad \text{and} \quad B_n(x) = b_0 + \frac{b_1}{x} + \frac{b_2}{x^2} + \dots + \frac{b_n}{x^n}.$$

Then, keeping n fixed, as these are asymptotic series we have

$$\lim_{x \to \infty} \{[A_n(x) - B_n(x)]x^n\} = \lim_{x \to \infty} [(a_0 - b_0)x^n + (a_1 - b_1)x^{n-1}$$

$$+ \dots + (a_n - b_n)] = 0,$$

which is only possible if

$$a_0 = b_0, a_1 = b_1, \dots, a_n = b_n.$$

This is true for any fixed n, and so it is true in general, which proves the theorem.

$\square$

It should be appreciated that although, when it exists, the asymptotic series representation of a function is unique, different functions may have the same asymptotic series representation. The easiest way to demonstrate this is by recognizing that functions $g(x)$ exist with an asymptotic series representation in which every coefficient a_n is zero. This happens, for example, with $g(x) = e^{-x}$, because for large x

$$e^{-x} \sim 0 + \frac{0}{x} + \frac{0}{x^2} + \dots.$$

Thus, if $f(x)$ has an asymptotic series representation, this will be the same as the representation of $f(x) + g(x)$, yet the functions $f(x)$ and $f(x) + g(x)$ will be different.

Theorem 8.4 (Asymptotic series may be added term by term)

If $f(x)$ and $g(x)$ have the asymptotic series representations

$$f(x) \sim a_0 + \frac{a_1}{x} + \frac{a_2}{x^2} + \dots, \text{ and } g(x) = b_0 + \frac{b_1}{x} + \frac{b_2}{x^2} + \dots,$$

for large x, and α, β are constants, the function $\alpha f(x) + \beta g(x)$ has the asymptotic series representation

$$\alpha f(x) + \beta g(x) \sim \alpha a_0 + \beta b_0 + \frac{\alpha a_1 + \beta b_1}{x} + \frac{\alpha a_2 + \beta b_2}{x^2} + \dots \qquad \text{for large } x.$$

Proof

The proof of this result is trivial and follows directly from the definition given in (14).

$\square$

Theorem 8.5 (Asymptotic series may be multiplied)

If $f(x)$ and $g(x)$ have the asymptotic series representations for large x given in Theorem 8.4, the function $f(x)g(x)$ has the asymptotic series representation

$$f(x)g(x) \sim c_0 + \frac{c_1}{x} + \frac{c_2}{x^2} + \dots \text{ for large } x,$$

where

$$c_n = a_0 b_n + a_1 b_{n-1} + a_2 b_{n-1} + \dots + a_n b_0.$$

Proof

Let

$$F_n(x) = a_0 + \frac{a_1}{x} + \dots + \frac{a_n}{x^n} \qquad \text{and} \qquad G_n(x) = b_0 + \frac{b_1}{x} + \dots + \frac{b_n}{x^n}.$$

Then, keeping n fixed, as $x \to \infty$, it follows that

$$f(x) - F_n(x) = \text{o}(x^{-n}) \qquad \text{and} \qquad g(x) - G_n(x) = \text{o}(x^{-n}).$$

Thus if

$$c_n = a_0 b_n + a_1 b_{n-1} + \dots + a_n b_0,$$

it must follow that

$$f(x)g(x) = F_n(x)G_n(x) + \text{o}(x^{-n}) \qquad \text{for each fixed } n.$$

As this is true for every fixed n the theorem is proved.

$\square$

Theorem 8.6 (Asymptotic series may be integrated term by term)

Let a function $f(x)$ which is continuous for large x have the asymptotic series representation

$$f(x) \sim \frac{a_2}{x^2} + \frac{a_3}{x^3} + \dots.$$

Then, for large x, this asymptotic series may be integrated term by term to give

$$\int_x^\infty f(t)\,dt \sim \frac{a_2}{x} + \frac{a_3}{2x^2} + \frac{a_4}{3x^3} + \dots + \frac{a_n}{(n-1)x^{n-1}} + \dots.$$

Proof

Set

$$F_n(x) = \frac{a_2}{x^2} + \frac{a_3}{x^3} + \dots + \frac{a_n}{x^n}.$$

Then, keeping n fixed, as $f(x)$ is continuous, for any given $\varepsilon > 0$ there exists a number $R > 0$ such that

$$|f(x) - F_n(x)| < \varepsilon x^{-n} \qquad \text{for } x \geq R.$$

Integration of this inequality gives,

$$\left| \int_x^\infty f(t)\,dt - \int_x^\infty F_n(t)\,dt \right| < \varepsilon \int_x^\infty \frac{dt}{t^n} = \frac{\varepsilon}{(n-1)x^{n-1}},$$

where use has been made of the elementary integral inequality ((13) Section 1.8).

$$\left| \int_a^b h(t)\,dt \right| \leq \int_a^b |h(t)|\,dt.$$

However, as

$$\int_x^\infty F_n(t)\,dt = \frac{a_2}{x} + \frac{a_3}{2x^2} + \dots + \frac{a_n}{(n-1)x^{n-1}}$$

we have succeeded in establishing that

$$\int_x^\infty f(t)\,dt = \frac{a_2}{x} + \frac{a_3}{2x^2} + \dots + \frac{a_n}{(n-1)x^{n-1}} + o(x^{-(n-1)}),$$

and thus

$$\int_x^\infty f(t)\,dt \sim \frac{a_2}{x} + \frac{a_3}{2x^2} + \frac{a_4}{3x^3} + \dots + \frac{a_n}{(n-1)x^{n-1}} + \dots. \qquad \square$$

The situation concerning the term by term differentiation of an asymptotic series representation of a function $f(x)$ is less satisfactory, as this is only permissible if $f'(x)$ is continuous and it is known that $f'(x)$ also has an asymptotic series representation. The difficulty arises because although $f(x)$ may have an asymptotic series representation for large x, this is not necessarily true of $f'(x)$. The relevant theorem which we state, but will not prove, is as follows

Theorem 8.7 (Differentiation of an asymptotic series)

Let $f(x)$ have the asymptotic series representation

$$f(x) \sim a_0 + \frac{a_1}{x} + \frac{a_2}{x^2} + \ldots + \frac{a_n}{x^n} + \ldots$$

for large x. Then if $f'(x)$ is continuous and also has an asymptotic series representation, it may be obtained by term by term differentiation of the asymptotic series for $f(x)$ which gives

$$f'(x) \sim -\frac{a_1}{x^2} - \frac{2a_2}{x^3} - \ldots - \frac{na_n}{x^{n+1}} - \ldots,$$

for large x. $\qquad\qquad\square$

Occasionally this theorem can be used to advantage when it is known, or can be shown subsequently, that $y(x)$ satisfies the conditions of Theorem 8.7 and is also the solution of a first order differential equation. In such circumstances we may substitute the asymptotic series representations of $y(x)$ and $y'(x)$ into the differential equation. The coefficients a_0, a_1, ... in the series can then be determined as in the series solution approach described in Sec. 8.3, though this time by equating to zero the coefficients of terms in x^{-n}.

Example 1. Differentiating an asymptotic series

Find the asymptotic series representation for large x of the solution of

$$y' + 2xy = 1.$$

Solution

Here both $y(x)$ and $y'(x)$ are continuous, so assuming that $y'(x)$ has an asymptotic series representation we shall substitute for y and y' in the differential equation using

$$y(x) \sim a_0 + \frac{a_1}{x} + \frac{a_2}{x^2} + \ldots + \frac{a_n}{x^n} + \ldots$$

and

$$y'(x) \sim -\frac{a_1}{x^2} - \frac{2a_2}{x^3} - \frac{3a_3}{x^4} - \ldots - \frac{na_n}{x^{n+1}} - \ldots.$$

This brings us to the equation

$$-\frac{a_1}{x^2} - \frac{2a_3}{x^3} - \frac{3a_3}{x^4} - \ldots - \frac{na_n}{x^{n+1}} - \ldots + 2a_0 x$$

$$+ 2a_1 + \frac{2a_2}{x} + \frac{2a_3}{x^2} + \ldots + \frac{2a_n}{x^{n-1}} + \ldots = 1.$$

Equating to zero the coefficients of x^{-n} on either side of this equation to make it an identity gives

(x) $\qquad 2a_0 = 0$ $\qquad$ so $\quad a_0 = 0,$

(x^0) $\qquad 2a_1 = 1$ $\qquad$ so $\quad a_1 = \tfrac{1}{2}$

(x^{-1}) $\qquad a_2 = 0$

(x^{-2}) $\qquad -a_1 + 2a_3 = 0$ $\qquad$ so $\quad a_3 = \tfrac{1}{2}a_1 = \tfrac{1}{4}$

(x^{-3}) $\qquad -2a_2 + 2a_4 = 0$ $\qquad$ so $\quad a_4 = a_2 = 0$

(x^{-4}) $\qquad -3a_3 + 2a_5 = 0$ $\qquad$ so $\quad a_5 = \tfrac{3}{2}a_3 = \tfrac{3}{8}.$

As the solution of this equation was given in (9), we have established that provided $y'(x)$ has an asymptotic series representation, then

$$\exp(-x^2)\int_a^\infty \exp(t^2)\,dt \sim \frac{1}{2x} + \frac{1}{4x^2} + \frac{3}{8x^4} + \cdots .$$

Comparison of this result with (10), which was obtained by a different form of argument, confirms that the leading term in the expansion is the same in both cases. We conclude from this that Theorem 8.7 was, in fact, applicable. ∎

We now introduce a few of the simpler methods by which asymptotic expansions may be determined. The first of these, and certainly the simplest, merely involves repeated integration by parts. It is best illustrated by example.

Example 2. Determination of an asymptotic series using integration by parts

Find an asymptotic series representation for large x for the complementary error function

$$\text{erfc } x = \frac{2}{\sqrt{\pi}} \int_x^\infty \exp(-t^2)\,dt.$$

Solution

Writing

$$\int_x^\infty \exp(-t^2)\,dt = \int_x^\infty \frac{2t\exp(-t^2)}{2t}\,dt,$$

integration by parts gives

$$\int_x^\infty \exp(-t^2)\,dt = -\frac{\exp(-t^2)}{2t}\Bigg|_x^\infty - \frac{1}{2}\int_x^\infty \frac{\exp(-t^2)}{t^2}\,dt$$

$$= \frac{\exp(-t^2)}{2x} - \frac{1}{2}\int_x^\infty \frac{\exp(-t^2)}{t^2}\,dt.$$

Repeating this process $n+1$ times brings us to the result

$$\text{erfc } x = \frac{\exp(-x^2)}{x\sqrt{\pi}}\left(1 - \frac{1}{2x^2} + \frac{1\cdot 3}{2^2 x^4} - \frac{1\cdot 3\cdot 5}{2^3 x^6} + \ldots + (-1)^n \frac{1\cdot 3\cdot 5 \ldots (2n-1)}{2^n x^{2n}} + R_n(x)\right),$$

where

$$R_n(x) = (-1)^{n+1} \frac{1\cdot 3\cdot 5 \ldots (2n+1)}{2^{n+1} x^{2n+2}} \exp(x^2) \int_x^\infty \frac{\exp(-t^2)}{t^{2n+2}} dt.$$

Using the fact that

$$\frac{\exp(-t^2)}{t^{2n+2}} < 2t \exp(-t^2) \qquad \text{for } t > 0,$$

and replacing the integrand involved in $R_n(x)$ by $2t\exp(-t^2)$, it follows after a further integration that

$$|R_n(x)| < \frac{1\cdot 3\cdot 5 \ldots (2n+1)}{2^{n+1} x^{2n+2}} \qquad \text{as } x \to \infty.$$

Thus

$$R_n(x) = O(x^{-(2n+2)}) \qquad \text{as } x \to \infty.$$

and so we have proved that

$$\text{erfc } x \sim \frac{\exp(-x^2)}{x\sqrt{\pi}}\left(1 + \sum_{n=1}^{\infty} (-1)^n \frac{1\cdot 3\cdot 5 \ldots (2n-1)}{2^n x^{2n}}\right),$$

as $x \to \infty$.

The ratio of the magnitude of two successive terms in this asymptotic series is

$$\frac{1\cdot 3\cdot 5 \ldots (2m-1)(2m+1)}{2^{n+1} x^{2n+2}} \Bigg/ \frac{1\cdot 3\cdot 5 \ldots (2m-1)}{2^m x^{2m}} = \frac{2m+1}{2x^2}.$$

Thus, for any given x, the magnitude of the terms of the asymptotic series will decrease until the Nth term, where N is the largest integer such that $2N + 1 < 2x^2$; thereafter the magnitude of the terms will increase (it is a divergent series). Consequently an examination of $R_n(x)$ shows that, when x is sufficiently large, by truncating the asymptotic series after only a few terms it is possible to obtain an extremely accurate approximation. For example in this case, even for $x = 4$, by taking only the first term in the series to represent erfc x, so that erfc $x \sim \exp(-x^2)/(x\sqrt{\pi})$, the absolute error ε involved is such that

$$\varepsilon < \frac{e^{-16}}{4\sqrt{\pi}} \frac{1}{2\cdot 4^2} = 4\cdot 96 \times 10^{-10}. \qquad \blacksquare$$

Naturally the meaning of the term 'large x' depends on the nature of the asymptotic representation, but it is frequently a number of the order of 10. Its interpretation in general must be determined by estimating the error involved when an asymptotic series is terminated after a given term.

The Laplace method

Many problems involving asymptotic expansions can be reduced to the study of **Laplace integrals** of the form

$$y(x) = \int_a^b e^{xa(t)} f(t) \, dt. \tag{15}$$

For example, when $a = 0$, $b = \infty$ and $a(t) = t$, the integral in (15) becomes the Laplace transform of $f(t)$. In such a case, the asymptotic behavior of $y(x)$ for large x is of relevance to the initial value theorem for Laplace transforms (Theorem 7.15).

Before describing the Laplace method for deriving asymptotic expansions for certain types of integral, let us digress for a moment and discuss one way in which **integral representations** of this type can arise from the study of differential equations. For simplicity we shall only consider the somewhat simpler case in which a solution is representable in the form

$$y(x) = \int_a^b e^{-xt} f(t) \, dt, \tag{16}$$

corresponding to setting $a(t) = -t$ in (15). The class of differential equations used to illustrate the approach is of the form

$$(a_0 + a_1 x) y'' + (b_0 + b_1 x) y' + (c_0 + c_1 x) y = 0, \tag{17}$$

in which a_0, a_1, b_0, b_1, c_0 and c_1 are constants, some of which may be zero. The following theorem will be needed.

Theorem 8.8. (Properties of Laplace integrals)

Let

$$y(x) = \int_a^b e^{-xt} f(t) \, dt,$$

where the function $f(t)$ is differentiable for $a \le t \le b$.

Then

(i) $\dfrac{dy}{dx} = -\int_a^b e^{-xt} t f(t) \, dt,$

(ii) $\dfrac{d^2 y}{dx^2} = \int_a^b e^{-xt} t^2 f(t) \, dt,$

(iii) $xy(x) = f(a) e^{-ax} - f(b) e^{-bx} + \int_a^b e^{-xt} f'(t) \, dt,$

(iv) $x\dfrac{dy}{dx} = bf(b)e^{-bx} - af(a)e^{-ax} - \displaystyle\int_a^b e^{-xt}f(t)\,dt - \int_a^b e^{-xt}tf'(t)\,dt,$

(v) $x\dfrac{d^2y}{dx^2} = a^2 f(a)e^{-ax} - b^2 f(b)e^{-bx} + 2\displaystyle\int_a^b e^{-xt}tf(t)\,dt + \int_a^b e^{-xt}t^2 f'(t)\,dt.$

Proof

Results (i) and (ii) follow by differentiating under the integral sign using Theorem 1.6. Results (iii) to (v) follow by using integration by parts. We establish result (iv) by way of illustration. We have

$$x\frac{dy}{dx} = -\int_a^b xe^{-xt}tf(t)\,dt = \int_a^b tf(t)\frac{d}{dt}(e^{-xt})\,dt$$

$$= tf(t)e^{-xt}\Big|_a^b - \int_a^b e^{-xt}\frac{d}{dt}[tf(t)]\,dt,$$

from which the required result then follows. □

The general approach to be adopted when seeking a solution of (17) of the form (16) involves the following steps.

Step 1 Substitute the expressions given in Theorem 8.8 into differential equation (17) to obtain an expression of the form

$$L[a, b, f(a)e^{-ax}, f(b)e^{-bx}] + \int_a^b e^{-xt}K[t, f(t), f'(t)]\,dt = 0, \tag{18}$$

where L and K are functions of the indicated arguments.

Step 2 Solve the differential equation

$$K[t, f(t), f'(t)] = 0 \tag{19}$$

for the function $f(t)$.

Step 3 Substitute this function $f(t)$ into the expression for L and attempt to find a, b such that

$$L[a, b, f(a)e^{-ax}, f(b)e^{-bx}] = 0, \tag{20}$$

for all x.

Step 4 If such a, b can be found, their use together with $f(t)$ in

$$y(x) = \int_a^b e^{-xt}f(t)\,dt$$

provides the required integral representation of a solution of (17).

Step 5 If no *a*, *b* exist such that $L=0$ for all *x*, then (17) does not have a solution representable in the form given in (16).

Example 3. Integral representation of solutions

Find the Laplace integral representations for the two linearly independent solutions of

$$x'' - 2xy' - 4y = 0.$$

Solution

Substituting the appropriate expressions from Theorem 8.8 into this differential equation as required in Step 1 gives

$$\int_a^b e^{-xt} t^2 f(t)\,dt - 2bf(b)e^{-bx} + 2af(a)e^{-ax}$$

$$+2\int_a^b e^{-xt}f(t)\,dt + 2\int_a^b e^{-xt}tf'(t)\,dt - 4\int_a^b e^{-xt}f(t)\,dt = 0.$$

After simplification this becomes

$$2[af(a)e^{-ax} - bf(b)e^{-bx}] + \int_a^b e^{-xt}[2tf'(t) + (t^2 - 2)f(t)]\,dt = 0,$$

which is of the form given in (18). Identifying this result with (18) shows

$$L = 2[af(a)e^{-ax} - bf(b)e^{-bx}] \quad \text{and} \quad K = 2tf'(t) + (t^2 - 2)f(t).$$

Step 2 requires us to solve the differential equation $K=0$ for the function $f(t)$, which is equivalent to solving

$$2tf' + (t^2 - 2)f = 0.$$

Separating variables and integrating gives

$$\int \frac{df}{f} = -\int \left(\frac{t^2 - 2}{2t}\right) dt,$$

and thus

$$\ln f = -\frac{t^2}{4} + \ln t + \ln c,$$

or

$$f(t) = ct \exp(-t^2/4).$$

We may, without loss of generality, set the arbitrary constant $c=1$. This is because *c* merely scales the solution in (16), which is arbitrary up to a multiplicative constant factor because it is the solution of a *homogeneous* equation.

Thus the function determined in Step 2 may be taken as

$$f(t) = t\exp(-t^2/4).$$

To accomplish Step 3 we must now attempt to find the constants a and b such that when $f(t) = t\exp(-t^2/4)$ we have $L=0$ for all x. As

$$L = 2(af(a)e^{-ax} - bf(b)e^{-bx}),$$

this is equivalent to requiring a and b to be such that

$$a^2\exp(-a^2/4)\exp(-ax) - b^2\exp(-b^2/4)\exp(-bx) = 0,$$

for all x.

This is possible only if

(i) $a=0, b=\infty$ or   (ii) $a=-\infty, b=0$.

The required integral representations then follow from Step 4 by using these a, b and $f(t)$ in (16). Thus one solution is

$$y_1(x) = \int_0^\infty \exp(-xt)t\exp(-t^2/4)\,dt,$$

and another is

$$\tilde{y}_2(x) = \int_{-\infty}^0 \exp(-xt)t\exp(-t^2/4)\,dt.$$

Replacing t by $-t$ in $\tilde{y}_2(x)$, and omitting the scale factor -1 which this introduces, we find that the second solution may be written as

$$y_2(x) = \int_0^\infty \exp(xt)t\exp(-t^2/4)\,dt.$$

It is clear that $y_1(x)$ and $y_2(x)$ are linearly independent solutions.

The general solution of the differential equation

$$y'' - 2xy' - 4y = 0$$

may thus be written as the linear combination of Laplace integrals

$$y(x) = c_1 \int_0^\infty \exp(-xt)t\exp(-t^2/4)\,dt + c_2 \int_0^\infty \exp(xt)t\exp(-t^2/4)\,dt. \qquad \blacksquare$$

Having introduced a simple way in which integral representations of the type given in (16) may be obtained, let us proceed to determine asymptotic expressions for such integrals when x is large. The basic idea underlying the Laplace method for deriving asymptotic expansions is that in many integral representations the integrand only contributes significantly to the integral throughout a *small* part of the interval of integration, and throughout the rest of the interval it is vanishingly small. In the case of integrals such as (16), provided that e^{-xt} dominates $f(t)$ when x is large, the region involved must be close to $t = 0$. Thus the approximation of $f(t)$ by some simpler function for small t can be expected to lead to an asymptotic representation of $y(x)$ for large x.

This same idea applies to the more general integrand involved in (15). Often, for large x, the function $e^{xa(t)}$ has a *sharp peak* at some point $t = t_0$ corresponding to $a(t)$ attaining a maximum at the point, while the integrand is vanishingly small outside a small neighborhood containing t_0. The behavior of the integrand in this neighborhood will then determine the asymptotic behavior of $y(x)$ for large x. In fact we shall see that when $a(t)$ has suitable properties at t_0, a simple expression known as the Laplace formula determines the asymptotic representation of $y(x)$.

Example 4. The Laplace method

Find the asymptotic behavior of

$$y_1(x) = \int_0^\infty \exp(-xt)t \exp(-t^2/4)\,dt$$

for large x.

Solution

A rigorous derivation of the asymptotic result would take us somewhat beyond this account of the subject. Thus we will only present an intuitive form of argument which will suffice as a model for use in similar situations.

For a given arbitrarily small $\varepsilon > 0$ and some suitably large fixed positive x we can always find a number $\delta > 0$ such that $e^{-x\delta} < \varepsilon$. Under these conditions, the only significant contribution to our integral from the integrand $\exp(-xt)t \exp(-t^2/4)$ will come from the interval $0 \leq t \leq \delta$. In this interval we may approximate $\exp(-t^2/4)$ by the first two terms of its Maclaurin series so that,

$$y_1(x) \sim \int_0^\delta e^{-xt} t\left(1 - \frac{t^2}{4}\right)dt.$$

However, as e^{-xt} tends to zero more rapidly than any power of t as $t \to \infty$, and $e^{-x\delta} < \varepsilon$, it will make no significant difference to this last result if the upper limit of integration is extended from δ to ∞ to become

$$y_1(x) \sim \int_0^\infty e^{-xt}\left(t - \frac{t^3}{4}\right)dt.$$

Setting $xt = \tau$ this becomes

$$y_1(x) \sim \frac{1}{x^2}\int_0^\infty e^{-\tau}\tau\,d\tau - \frac{1}{4x^4}\int_0^\infty e^{-\tau}\tau^3\,d\tau,$$

or by (2) of Sec. 8.5,

$$y_1(x) \sim \frac{\Gamma(2)}{x^2} - \frac{\Gamma(4)}{4x^4}.$$

As $\Gamma(2)=1$ and $\Gamma(4)=3!=6$ the first two terms of the asymptotic series representation of $y_1(x)$ are thus

$$y_1(x) \sim \frac{1}{x^2} - \frac{3}{2x^4} + \ \ldots \ .$$

■

Clearly, had more terms been retained in the expansion of $\exp(-t^2/4)$, in the above example, correspondingly more terms would have been obtained in the asymptotic series. Thus we have succeeded in deriving an asymptotic representation for the first of the linearly independent solutions of the differential equation considered in Ex. 3. The asymptotic form of the second solution cannot be found in this manner because of the presence of the factor e^{xt} in the integrand. We shall return to this problem in Ex. 6, after discussing the WKBJ method.

We now offer a brief discussion of the celebrated **Laplace formula** for the asymptotic representation of an integral, again with only an intuitive justification for the result. To discuss the matter fully would be inappropriate because it involves the use of advanced complex analysis. The formula relates to the asymptotic representation of integrals of the form

$$y(x) = \int_a^b e^{xa(t)} f(t) \, dt,$$

and it may be stated as follows.

Theorem 8.9 (The Laplace formula)

Let $y(x)$ have the integral representation

$$y(x) = \int_a^b e^{xa(t)} f(t) \, dt, \tag{21}$$

in which the twice differentiable function $a(t)$ attains a maximum at the single point t_0 interior to the interval of integration $a \leq t \leq b$. Then for large x the function $y(x)$ has the asymptotic representation known as the **Laplace formula**

$$y(x) \sim \left(\frac{-2\pi}{xa''(t_0)} \right)^{1/2} f(t_0) \exp\left[xa(t_0) \right] \tag{22}$$

where it is assumed that $a''(t_0) < 0$.

Proof

One form of intuitive justification of this result proceeds as follows. Let $a(t)$ have a maximum at the single point t_0 interior to $a \leq t \leq b$. Then as $a'(t_0) = 0$ we may expand $a(t)$ in a Taylor series expansion about t_0 of the form

$$a(t) = a(t_0) + \tfrac{1}{2} a''(t_0)(t-t_0)^2 + O(t-t_0)^3$$

where, of course, $a''(t_0) < 0$ because t_0 is a maximum of $a(t)$.

Consequently using this expression in the integral for $y(x)$ when x is large gives

$$y(x) \sim \exp\left[xa(t_0)\right] \int_a^b f(t) \exp\left[\tfrac{1}{2}xa''(t_0)(t-t_0)^2\right]dt.$$

As the factor $\exp\left[\tfrac{1}{2}xa''(t_0)(t-t_0)^2\right]$ only differs significantly from zero when t is close to t_0 (remember $a''(t_0) < 0$), we make a further approximation by replacing $f(t)$ by $f(t_0)$ to arrive at the result

$$y(x) \sim f(t_0)\exp\left[xa(t_0)\right] \int_a^b \exp\left[\tfrac{1}{2}xa''(t_0)(t-t_0)^2\right]dt.$$

The final step involves recognizing that because of the rapid decay of the exponential integrand in this last result when t is not close to t_0 we may, without significant error, replace a by $-\infty$ and b by ∞ to obtain

$$y(x) \sim f(t_0)\exp\left[xa(t_0)\right] \int_{-\infty}^{\infty} \exp\left[\tfrac{1}{2}xa''(t_0)(t-t_0)^2\right]dt.$$

Then using the standard probability integral

$$\int_{-\infty}^{\infty} \exp\left[-\frac{1}{2}\left(\frac{t-t_0}{\sigma}\right)^2\right]dt = \sigma\sqrt{2\pi},$$

and setting $-xa''(t_0) = \sigma^{-2}$, we arrive at the stated result

$$y(x) \sim \left(\frac{-2\pi}{xa''(t_0)}\right)^{1/2} f(t_0)\exp\left[xa(t_0)\right]. \qquad \square$$

Sometimes it is necessary to transform an integrand to bring it into a form suitable for the application of the Laplace formula. The following example in which we derive Stirling's formula for the asymptotic representation of $n!$ is a case in point.

Example 5. Stirling's formula

Derive Stirling's formula

$$n! \sim n^n e^{-n}\sqrt{2\pi n},$$

where n is a positive integer.

Solution

We start from the gamma function representation of $n!$ as

$$n! = \int_0^{\infty} e^{-t}t^n dt,$$

which we rewrite as

$$n! = \int_0^{\infty} \exp\left(-t + n\ln t\right)dt.$$

Then, as the function $\ln t$ does not attain a maximum value for finite t, we set $t = n\tau$ to obtain

$$n! = n^{n+1} \int_0^\infty \exp\left[n(\ln \tau - \tau)\right] d\tau.$$

This is now of the required form given in (21), with x replaced by n, t by τ, $f(\tau) = 1$ and $a(\tau) = \ln \tau - \tau$.

The function $a(\tau)$ has a single maximum at $\tau = 1$, $a(1) = -1$ and $a''(-1) = -1$, so its expansion is of the required form

$$\ln \tau - \tau = -1 - \tfrac{1}{2}(\tau - 1)^2 - \cdots \ .$$

Substitution into the Laplace formula (22) yields the **Stirling asymptotic formula** for factorial n

$$n! \sim n^n e^{-n} \sqrt{2\pi n}.$$

This simple result makes possible the approximation of $n!$ for large n in terms of simple functions. For example, $10! = 3\,628\,800$, whereas Stirling's formula gives $10! \sim 3\,598\,695.6$, which represents an error of less than 0.8 per cent. ∎

The WKBJ method

The WKBJ method[11] is a technique for finding asymptotic solutions of a second order differential equation when it is written in the standard form

$$u'' - f(x)u = 0. \tag{23}$$

It will be recalled that in Sec. 5.13, Theorem 5.8, it was proved that the general second order differential equation

$$y'' + p(x)y' + q(x)y = 0 \tag{24}$$

may be transformed into the form given in (23) with

$$f(x) = \tfrac{1}{2}p' + \tfrac{1}{4}p^2 - q, \tag{25}$$

by setting

$$y = uv, \tag{26}$$

with

$$v = \exp\left(-\frac{1}{2}\int p\,dx\right). \tag{27}$$

A rigorous justification of the powerful and useful WKBJ method requires the use of advanced complex analysis, as does its application to many differential equations. Instead of providing such a justification, which would be inappropriate here, we offer in its place an intuitive approach to the so-called WKBJ formulas which may be used effectively with the simplest problems. Our approach will be one of successive approximation.

---

[11] The name WKBJ method derives from the initials of the surnames of the four mathematical physicists G. Wentzel, H. A. Kramers, L. Brillouin and H. Jeffreys who in the 1920s each arrived at the method independently while considering unrelated problems connected with the Schrödinger equation in quantum mechanics.

We start by observing that when $f(x)$ is a nonzero constant the solutions of (23) will either be exponential functions or trigonometric functions. This suggests that when $f(x)$ is not constant we should try to find a solution of (23) of the form

$$u = e^{\Phi(x)}. \tag{28}$$

We will exclude functions $f(x)$ which change sign as x increases and, for convenience, assume that $f(x) > 0$. If $f(x) < 0$ the same form of argument follows through with obvious modifications.

Substitution of (28) into (23) gives for the equation to be satisfied by the unknown function Φ

$$\Phi'' + (\Phi')^2 - f = 0. \tag{29}$$

Let us assume that Φ is a slowly varying function and for the moment neglect Φ'', as it will then be small relative to $(\Phi')^2$. Our first approximation Φ_0 to Φ then satisfies the equation

$$(\Phi_0')^2 - f = 0, \tag{30}$$

and so

$$\Phi_0(x) = \pm \int f^{1/2} \, dx. \tag{31}$$

It follows directly from (30) that the first approximation to the neglected term Φ'' is

$$\Phi_0'' = \pm \tfrac{1}{2} f' f^{-1/2}. \tag{32}$$

We obtain the next approximation Φ_1 to Φ by using Φ_0'' in place of Φ'' in (29), writing Φ_1' in place of Φ', and finding Φ_1 from the equation

$$(\Phi_1')^2 = f \mp \tfrac{1}{2} f' f^{-1/2}. \tag{33}$$

Thus

$$\begin{aligned}
\Phi_1' &= (f \mp \tfrac{1}{2} f' f^{-1/2})^{1/2} \\
&= \pm f^{1/2}(1 \mp f' f^{-3/2})^{1/2} \\
&= \pm f^{1/2} - \tfrac{1}{4} f' f^{-1} + \cdots,
\end{aligned}$$

provided that the remaining terms in this binomial expansion are small. Integrating (34) gives

$$\Phi_1(x) = \pm \int f^{1/2} \, dx + \ln f^{-1/4} + \cdots. \tag{35}$$

Using Φ_1 in place of Φ in (28) we finally arrive at the two asymptotic solutions of (23) for large x

$$u_+(x) \sim [f(x)]^{-1/4} \exp\left(\int f^{1/2} \, dx\right) \quad \text{and} \quad u_-(x) \sim [f(x)]^{-1/4} \exp\left(-\int f^{1/2} \, dx\right). \tag{36}$$

The two results in (36) are called the **WKBJ formulas** for the asymptotic solutions of (23) for large x. The corresponding asymptotic results for differential equation (24), with (23) as its standard form, follow from (26), (27) and (36) as

$$y_+(x) \sim u_+(x) \exp\left(-\frac{1}{2}\int pdx\right) \quad \text{and} \quad y_-(x) \sim u_-(x) \exp\left(-\int pdx\right). \tag{37}$$

Results (37) are the WKBJ formulas for the asymptotic solutions of (24) for large x.

It is useful to have a simple criterion by which to determine when the WKBJ method can be expected to yield a good asymptotic approximation. We already have the condition that $f(x)$ must not change sign, for if it does the nature of the solution changes. An examination of the derivation of Φ_1 suggests that it will be a reasonable approximation to $\Phi(x)$ provided that when arriving at (34) it is reasonable to approximate $[1 \mp \frac{1}{2}(f'/f^{3/2})]^{1/2}$ by a truncated two term binomial expansion. Thus our additional criterion for the effectiveness of the WKBJ method as described here is that

$$|\tfrac{1}{2}(f'/f^{3/2})| \ll 1 \quad \text{for large } x. \tag{38}$$

Example 6. The WKBJ–a typical example

Use the WKBJ method to determine two linearly independent asymptotic solutions of

$$y'' - 2xy' - 4y = 0.$$

Solution

In the notation of (24)

$$p(x) = -2x \quad \text{and} \quad q(x) = -4,$$

so the function $f(x)$ in (25) becomes

$$f(x) = x^2 + 3,$$

while the function $v(x)$ in (27) becomes

$$v(x) = \exp\left[-\frac{1}{2}\int(-2x)dx\right] = \exp\left(\frac{1}{2}x^2\right).$$

Thus the transformation

$$y(x) = u(x)\exp\left(\frac{1}{2}x^2\right)$$

reduces the differential equation to the standard form

$$u'' - (x^2 + 3)u = 0.$$

Before proceeding further we remark that $f(x) > 0$ for all x and criterion (38) is satisfied, because $|\frac{1}{2}f'/f^{3/2}| = O(x^{-2})$ for large x, so the WKBJ method can be expected to yield a satisfactory result.

Now, as we are only concerned with large x, we write

$$f(x) = x^2 + 3 = x^2 \left(1 + \frac{3}{x^2}\right),$$

so

$$[f(x)]^{1/2} = x\left(1 + \frac{3}{x^2}\right)^{1/2} = x\left(1 + \frac{3}{2x^2} + O(x^{-4})\right) = x + \frac{3}{2x} + O(x^{-3}),$$

and

$$[f(x)]^{-1/4} = x^{-1/2}\left(1 + \frac{3}{x^2}\right)^{-1/4} = x^{-1/2}\left(1 - \frac{3}{4x^2} + O(x^{-4})\right).$$

Consequently,

$$\int f^{1/2} dx = \int \left(x + \frac{3}{2x} + O(x^{-3})\right) dx = \frac{1}{2}x^2 + \frac{3}{2}\ln x + O(x^{-2}),$$

and so

$$\exp\left(\pm\int f^{1/2} dx\right) = \exp\left[\pm\left(\frac{1}{2}x^2 + \frac{3}{2}\ln x + O(x^{-2})\right)\right].$$

Substituting into WKBJ formulas (26) shows that, to $O(x^{-2})$, the two linearly independent asymptotic solutions $u_\pm(x)$ are

$$u_+(x) \sim x\exp(x^2/2) \quad \text{and} \quad u_-(x) \sim x^{-2}\exp(-x^2/2) \quad \text{as } x \to \infty.$$

Using (37) we find that, to $O(x^{-2})$, the two corresponding linearly independent asymptotic solutions $y_\pm(x)$ of the original differential equation are

$$y_+(x) \sim x\exp(x^2) \quad \text{and} \quad y_-(x) \sim x^{-2} \quad \text{as } x \to \infty.$$

The asymptotic solution $y_-(x)$ corresponds to the asymptotic solution for $y_1(x)$ found in Ex. 4. The expression for $y_-(x)$ is simply the leading term in the asymptotic representation of $y_1(x)$, and a more careful analysis would yield further terms. Any multiple of $y_+(x)$ and $y_-(x)$ would, of course, also be a solution because the differential equation is homogeneous; so the solutions are indeterminate up to an arbitrary multiplicative constant. It is for this reason that such constants are usually omitted.

Similarly, the expression for $y_+(x)$ represents the leading term in the asymptotic representation of the other linearly independent solution. This corresponds to the asymptotic representation of the second solution $y_2(x)$ derived in Ex. 3. As already remarked, the asymptotic result for $y_2(x)$ cannot be found from the integral representation for $y_2(x)$ by means of the arguments used in Ex. 4, because the structure of the integrand is different. Thus this example illustrates the ability of the WKBJ method to generate asymptotic solutions which cannot be obtained by simpler methods. ∎

Example 7. Asymptotic representation of $J_\nu(x)$

(i) Use the WKBJ method to show that for large x the asymptotic form of the general solution of Bessel's equation

$$x^2 y'' + xy' + (x^2 - v^2)y = 0$$

is

$$y(x) \sim C_1(v) \frac{\cos x}{\sqrt{x}} + C_2(v) \frac{\sin x}{\sqrt{x}}.$$

(ii) Starting from the integral representation

$$J_0(x) = \frac{1}{\pi} \int_0^\pi \cos(x \sin \theta) d\theta,$$

show that when x is large $J_0(x)$ has the asymptotic representation

$$J_0(x) = \left(\frac{2}{\pi x}\right)^{1/2} \left[\cos\left(x - \frac{\pi}{4}\right) + \delta(x)\right], \quad \text{where } \lim_{x \to \infty} \delta(x) = 0.$$

Solution

(i) Rewriting Bessel's equation in the standard form given in (24) we obtain

$$y'' + \frac{1}{x}y' + \left(1 - \frac{v^2}{x^2}\right)y = 0.$$

Thus in the notation of (24),

$$p(x) = \frac{1}{x} \quad \text{and} \quad q(x) = 1 - \frac{v^2}{x^2}, \quad \text{so } f(x) = (-1)\left[1 - \left(\frac{v^2 - \frac{1}{4}}{x^2}\right)\right].$$

From (26) and (27) it follows that the transformation to the standard form

$$u'' + \left[1 - \left(\frac{v^2 - \frac{1}{4}}{x^2}\right)\right]u = 0$$

is accomplished by writing

$$y(x) = u(x)/\sqrt{x}.$$

In this case, contrary to our original assumption, $f(x) < 0$. Thus complex quantities will now arise. Working to $O(x^{-2})$ it is easily seen that as

$$f(x) = (-1)\left[1 - \left(\frac{v^2 - \frac{1}{4}}{x^2}\right)\right],$$

$$f^{1/2} = i[1 + O(x^{-2})],$$

so that

$$\int f^{1/2} dx = i[x + O(x^{-1})].$$

Arguing in similar fashion we find that, to $O(x^{-2})$,

$$f^{-1/4} \sim \text{const. (complex)}.$$

Substitution into (36) then shows that for *real* solutions we must take

$$u_+(x) \sim \cos x \quad \text{and} \quad u_-(x) \sim \sin x.$$

Using (37), or the fact that $y = u\sqrt{x}$, it then follows that the asymptotic form of the general solution of Bessel's equation must be of the form

$$y(x) \sim C_1(v)\frac{\cos x}{\sqrt{x}} + C_2(v)\frac{\sin x}{\sqrt{x}}.$$

(ii) Setting $\Omega = \theta - \pi/2$ in the integral representation for $J_0(x)$ (derived in Sec. 8.7, Ex. 1) gives

$$J_0(x) = \frac{2}{\pi}\int_0^{\pi/2}\cos(x\cos\Omega)d\Omega,$$

and after the further variable charge $t = \cos\Omega$ this becomes

$$J_0(x) = \frac{2}{\pi}\int_0^1\frac{\cos xt}{\sqrt{1-t^2}}dt.$$

Finally, writing $xt = x - \tau$, this last result becomes

$$J_0(x) = \frac{1}{\pi}\left(\frac{2}{x}\right)^{1/2}\int_0^x\frac{\cos(x-\tau)}{\sqrt{\tau}\sqrt{1-\tau/2x}}d\tau.$$

Expanding $\cos(x-\tau)$ we have

$$J_0(x) = \frac{1}{\pi}\left(\frac{2}{x}\right)^{1/2}[A(x)\cos x + B(x)\sin x],$$

where

$$A(x) = \int_0^x\frac{\cos\tau}{\sqrt{\tau}\sqrt{1-\tau/2x}}d\tau, \quad B(x) = \int_0^x\frac{\sin\tau}{\sqrt{\tau}\sqrt{1-\tau/2x}}d\tau.$$

Now in the limit of large x

$$\lim_{x\to\infty}A(x) = \int_0^\infty\frac{\cos\tau}{\sqrt{\tau}}d\tau = \left(\frac{\pi}{2}\right)^{1/2}$$

$$\lim_{x\to\infty}B(x) = \int_0^\infty\frac{\sin\tau}{\sqrt{\tau}}d\tau = \left(\frac{\pi}{2}\right)^{1/2},$$

where use has been made of the fact that these definite integrals are standard results. Thus we may set

$$A(x) = \left(\frac{\pi}{2}\right)^{1/2}(1+\varepsilon_1(x)), \quad B(x) = \left(\frac{\pi}{2}\right)^{1/2}(1+\varepsilon_2(x)),$$

where

$$\lim_{x\to\infty}\varepsilon_1(x) = \lim_{x\to\infty}\varepsilon_2(x) = 0.$$

Then, by defining

$$\delta(x) = [\varepsilon_1(x)\cos x + \varepsilon_2(x)\sin x]/\sqrt{2},$$

it follows that

$$J_0(x) = \left(\frac{2}{\pi x}\right)^{1/2} \left[\cos\left(x - \frac{\pi}{4}\right) + \delta(x) \right], \quad \text{where } \lim_{x \to \infty} \delta(x) = 0.$$

This confirms both the oscillatory behavior of $J_0(x)$, and the distribution of the zeros of $J_0(x)$ as $x \to \infty$ already noted in Sec. 8.7. The zeros of $\cos(x - \pi/4)$ occur at $x = (4n + 3)(\pi/4)$, and in the numbering of the zeros used in the table in Sec. 8.7, $\lambda_{0,1}$ corresponds to $n = 0$, $\lambda_{0,2}$ to $n = 1$ and, in general, $\lambda_{0,m}$ to $n = m - 1$. Even for m as small as $m = 6$ the approximation $\lambda_{0,6} \sim 23\pi/4 - 18.0642$ is close to the true value $\lambda_{0,6} = 18.0711$. Although we shall not prove it here, it can be shown that

$$J_v(x) \sim \left(\frac{2}{\pi x}\right)^{1/2} \left[\cos\left(x - \frac{\pi}{4} - \frac{v\pi}{2}\right) A_v(x) - \sin\left(x - \frac{\pi}{4} - \frac{v\pi}{2}\right) B(v) \right],$$

where

$$A(v) = 1 - \frac{(4v^2 - 1)(4v^2 - 3^2)}{2!(8x)^2} + \frac{(4v^2 - 1^2)(4v^2 - 3^2)(4v^2 - 5^2)(4v^2 - 7^2)}{4!(8x)^4} - \cdots,$$

and

$$B(v) = \frac{4v^2 - 1^2}{8x} - \frac{(4v^2 - 1^2)(4v^2 - 3^2)(4v^2 - 5^2)}{3!(8x)^3} + \cdots.$$

■

Problems for Section 8.8

1 Show that for large x

$$\left(1 + \frac{a}{x}\right)^{-1} \sim 1 - \frac{a}{x} + \frac{a^2}{x^2} - \frac{a^3}{x^3} + \cdots.$$

2 Show that for large x

$$\sinh \frac{1}{x} \sim \frac{1}{x} + \frac{1}{3! \, x^3} + \frac{1}{5! \, x^5} + \cdots.$$

3 Show that for large x

$$\cosh \frac{1}{x} \sim 1 + \frac{1}{2! \, x^2} + \frac{1}{4! \, x^4} + \cdots.$$

4 Show that for large x

$$\exp(1/x^2) \sim 1 + \frac{1}{x^2} + \frac{1}{2! \, x^4} + \frac{1}{3! \, x^6} + \cdots.$$

5 The **exponential integral** $E_1(x)$ is defined by the expression

$$E_1(x) = \int_x^\infty \frac{e^{-t}}{t} \, dt.$$

Show that

$$y(x) = e^x E_1(x)$$

satisfies the differential equation

$$y' = y - \frac{1}{x}.$$

Assuming that the conditions of Theorem 8.7 are satisfied, substitute the asymptotic series for $y(x)$ and $y'(x)$ into this differential equation, and hence show that

$$E_1(x) \sim e^{-x} y(x) \sim e^{-x} \left(\frac{1}{x} - \frac{1}{x^2} + \frac{2!}{x^3} - \frac{3!}{x^4} + \cdots \right).$$

6 Use repeated integration by parts to show that for large x

$$E_1(x) = \int_x^\infty \frac{e^{-t}}{t} dt \sim e^{-x} \left(\frac{1}{x} - \frac{1}{x^2} + \frac{2!}{x^3} - \cdots + (-1)^{n-1} \frac{(n-1)!}{x^n} + (-1)^n n! \int_x^\infty \frac{e^{-t}}{t^{n+1}} dt \right).$$

7 The **exponential integral** $E_n(x)$ is defined by the expression

$$E_n(x) = \int_1^\infty \frac{e^{-xt}}{t^n} dt.$$

Use repeated integration by parts to show that for large x

$$E_n(x) \sim e^{-x} \left(\frac{1}{x} - \frac{n}{x^2} + \frac{n(n+1)}{x^3} - \frac{n(n+1)(n+2)}{x^4} + \cdots \right).$$

8 The **cosine integral** $\text{ci}(x)$ is defined by the expression

$$\text{ci}(x) = \int_x^\infty \frac{\cos t}{t} dt.$$

Use repeated integration by parts to find the first six terms in its asymptotic series representation for large x.

9 The **complementary sine integral** $\text{si}(x)$ is defined by the expression

$$\text{si}(x) = \int_x^\infty \frac{\sin t}{t} dt.$$

Use repeated integration by parts to find the first six terms in its asymptotic series representation for large x.

Given that $\text{si}(0) = \pi/2$, determine the asymptotic series representation of the **sine integral**

$$\text{Si}(x) = \int_0^x \frac{\sin t}{t} dt.$$

10 The **complementary Fresnel integral** $c(x)$ arising in diffraction problems is defined by the expression

$$c(x) = \int_x^\infty \cos t^2 \, dt.$$

Use repeated integration by parts to find the first four terms in its asymptotic series representation for large x. Given that $c(0) = (1/2)\sqrt{\pi/2}$, determine the asymptotic series representation of the **Fresnel integral**

$$C(x) = \int_0^x \cos t^2 \, dt.$$

11 The **complementary Fresnel integral** $s(x)$ arising in diffraction problems is defined by the expression

$$s(x) = \int_x^\infty \sin t^2 \, dt.$$

Use repeated integration by parts to find the first four terms in its asymptotic series representation for large x. Given that $s(0) = (1/2)\sqrt{\pi/2}$, determine the asymptotic series representation of the **Fresnel integral**

$$S(x) = \int_0^x \sin^2 t \, dt.$$

12 Show that the general solution of

$$xy'' + 2y' + (1-x)y = 0$$

for $x > 0$ may be written as the sum of the Laplace integrals

$$y(x) = c_1 \int_{-1}^1 e^{xt} \sqrt{\frac{1-t}{1+t}} \, dt + c_2 \int_1^\infty e^{-xt} \sqrt{\frac{t+1}{t-1}} \, dt.$$

13 Show that the general solution of

$$xy'' + \left(\frac{3}{2} - x\right)y' - y = 0$$

for $x > 0$ may be written as the sum of the Laplace integrals

$$y(x) = c_1 \int_0^\infty \frac{e^{-xt}}{\sqrt{1+t}} \, dt + c_2 \int_0^1 \frac{e^{xt}}{\sqrt{1-t}}.$$

14 Show that one solution of

$$xy'' + y' - xy = 0$$

for $x > 0$ may be expressed as the Laplace integral

$$y(x) = \int_1^\infty \frac{e^{-xt}}{\sqrt{t^2 - 1}} \, dt.$$

15 Show that one solution of

$$xy'' - xy' - 2y = 0$$

for $x > 0$ may be written as the Laplace integral

$$y(x) = \int_0^\infty \frac{te^{-xt}}{(1+t)^3} \, dt.$$

16 Show that the general solution of

$$2y'' - xy' + (x-3)y = 0$$

for $x > 0$ may be written as the sum of the Laplace integrals

$$y(x) = c_1 \int_1^\infty \exp[-(t+1)^2 + xt]\,dt + c_2 \int_{-\infty}^1 \exp[-(t+1)^2 + xt]\,dt.$$

In the following problems, use the form of argument given in Ex. 4 to find the first three terms in the asymptotic expansion for large positive x.

17 $y(x) = \int_0^\infty \dfrac{e^{-xt}}{1+t}\,dt.$

18 $y(x) = \int_0^\infty \dfrac{e^{-xt}\ln(1+t)}{1+t}\,dt.$

19 $y(x) = \int_0^1 e^{x(t-1)}t^n\,dt.$

20 This problem provides an alternative method for the derivation of **Stirling's formula**. Using the gamma function it follows that

$$n! = \int_0^\infty t^n e^{-t}\,dt.$$

Rewrite this as

$$n! = \int_0^\infty \exp(x\ln t - t)\,dt,$$

and show that the integrand has a peak at $t = x$. Expand $x\ln t - t$ about $t = x$ and hence show that

$$n! \sim \sqrt{2\pi}\,n^{n+1/2}e^{-n}\left(1 + \frac{1}{12n} + \frac{1}{288n^2} + \cdots\right)$$

for large $n > 0$.

21 Use the Laplace formula to find an asymptotic expression for

$$I_n = \int_0^\infty (2t)^n e^{-t}\,dt,$$

for large $n > 0$.

22 Use the Laplace formula to find an asymptotic expression for

$$I(x) = \int_0^\infty \frac{\exp\{-x[(t-2)^2 + 1]\}}{1+t^2}\,dt,$$

for large $x > 0$.

23 Use Laplace's formula to show that if

$$I_n = \int_a^b [p(t)]^n\,dt,$$

where $p(t)$ is a nonnegative function with a single maximum at $t = t_0$ interior to the interval of integration and n is a positive integer, then

$$I_n \sim [p(t_0)]^n \sqrt{\frac{-2\pi p(t_0)}{np''(t_0)}}, \qquad \text{for large } n > 0.$$

What can you conclude about

$$\lim_{n \to \infty} (I_n)^{1/n}?$$

24 Use the Laplace formula to find an asymptotic expression for

$$I_n = \int_0^\infty \frac{dt}{(t^2 - 8t + 19)^n},$$

for large positive integral n.

25 Use the Laplace formula to find an asymptotic expression for

$$I(x) = \int_0^\infty \frac{\exp\left[-3x \cosh\left(\frac{x}{t} - 2\right)\right]}{\ln\left[1 + \left(\frac{x}{t}\right)^2 + \left(\frac{x}{t}\right)^4\right]} \, dt,$$

for large x.

26 Modifiy the argument presented in the text to show that when WKBJ process is applied to the equation

$$u'' + f(x)u = 0,$$

where $f(x) > 0$, the **WKBJ formulas** become

$$u_+(x) \sim [f(x)]^{-1/4} \exp\left(i \int f^{1/2} \, dx\right) \quad \text{and} \quad u_-(x) \sim [f(x)]^{-1/4} \exp\left(-i \int f^{1/2} \, dx\right).$$

In the following problems use the WKBJ method to find the form of the asymptotic solutions for large x.

27 $2y'' - xy' + (x - 3)y = 0.$

28 $xy'' + 2y' + (1 - x)y = 0.$

29 $y'' - xy = 0.$

30 $xy''(\frac{3}{2} - x)y' - y = 0.$

31 $\quad y'' + \left(-\frac{1}{4} + \frac{\lambda}{x} + \frac{\frac{1}{4} - \mu^2}{x^2}\right)y = 0.$

for real λ, μ, with 2μ not an integer. The equation is known as **Whittaker's equation**.

32 $y'' + xy = 0.$

33 $y'' - 2xy' - 16y = 0.$

34 $xy'' - (2 + x)y = 0.$

35 $xy'' + (4 - x)y' + y = 0.$

8.9 Numerical solution of second order equations by the Runge–Kutta method

A second order differential equation for $y(x)$, which is either linear or nonlinear, may be written in the form

$$\frac{d^2 y}{dx^2} = g\left(x, y, \frac{dy}{dx}\right). \tag{1}$$

By setting $dy/dx = z$ this second order equation is reduced to the system

$$\frac{dy}{dx} = z, \tag{2}$$

$$\frac{dz}{dx} = g(x, y, z). \tag{3}$$

The initial conditions for (1) at $x = x_0$, which take the form $y(x_0) = y_0$, $y'(x_0) = z_0$, become the corresponding initial conditions for the system when they are written

$$y(x_0) = y_0 \quad \text{and} \quad z(x_0) = z_0. \tag{4}$$

We may now use the Runge–Kutta method for a general system described in Sec. 6.5 to solve the above system numerically, provided we make the identification $f(x, y, z) \equiv z$ in the algorithm of Sec. 6.5. As a result we arrive at the following form of the algorithm for a second order equation.

Runge–Kutta algorithm for a second order equation

This algorithm enables the numerical solution of a second order differential equation written in the standard form

$$\frac{d^2 y}{dx^2} = g\left(x, y, \frac{dy}{dx}\right)$$

to be obtained, subject to the initial conditions

$$y(x_0) = y_0 \quad \text{and} \quad y'(x_0) = z_0.$$

The steps involved in the numerical solution are as follows.

Step 1

Set $dy/dx = z$ and rewrite the equation together with its initial conditions as the system

$$\frac{dy}{dx} = z, \quad \frac{dz}{dx} = g(x, y, z),$$

subject to the initial conditions

$$y(x_0) = y_0 \quad \text{and} \quad z(x_0) = z_0.$$

Set $x_n = x_0 + nh$, $y_n = y(x_n)$ and $z_n = z(x_n)$, where h is the step length.

The values of y_{n+1} and z_{n+1} at x_{n+1} are then determined from the values of y_n and z_n at x_n as described in Steps 2 to 4.

Step 2

Compute

$$k_{1n} = h z_n \qquad\qquad K_{1n} = h g(x_n, y_n, z_n)$$
$$k_{2n} = h(z_n + \tfrac{1}{2}K_{1n}) \qquad K_{2n} = h g(x_n + \tfrac{1}{2}h, y_n + \tfrac{1}{2}k_{1n}; z_n + \tfrac{1}{2}K_{1n})$$
$$k_{3n} = h(z_n + \tfrac{1}{2}K_{2n}) \qquad K_{3n} = h g(x_n + \tfrac{1}{2}h, y_n + \tfrac{1}{2}k_{2n}, z_n + \tfrac{1}{2}K_{2n})$$
$$k_{4n} = h(z_n + K_{3n}) \qquad K_{4n} = h g(x_n + h, y_n + k_{3n}, z_n + K_{3n}).$$

Step 3

Compute

$$k_n = \tfrac{1}{6}(k_{1n} + 2k_{2n} + 2k_{3n} + k_{4n}) \quad \text{and} \quad K_n = \tfrac{1}{6}(K_{1n} + 2K_{2n} + 2K_{3n} + K_{4n}).$$

Step 4

The numerical estimates of y_{n+1} and z_{n+1} at x_{n+1} are then given by

$$y_{n+1} = y_n + k_n \quad \text{and} \quad z_{n+1} = z_n + K_n.$$

Example 1. Runge–Kutta solution of a linear second order equation

Use the Runge–Kutta method with step length $h = 0.2$ to determine y_1 to y_5 given that

$$(1 + x^2)\frac{d^2 y}{dx^2} - 2x\frac{dy}{dx} + 12y = x,$$

subject to the initial conditions

$$y(0) = 1 \quad \text{and} \quad y'(0) = 1.$$

Solution

It is first necessary to write the differential equation in the standard form (1) by dividing by $1 + x^2$. After rearranging the equation we find

$$\frac{d^2 y}{dx^2} = \left(\frac{2x}{1 + x^2}\right)\frac{dy}{dx} - \left(\frac{12}{1 + x^2}\right)y + \frac{x}{1 + x^2}.$$

Expressing the equation in this form places no restriction on x because $1 + x^2 \geq 1$ for all x.

Setting $z = dy/dx$ converts the second order equation to the system

$$\frac{dy}{dx} = z,$$

$$\frac{dz}{dx} = \left(\frac{2x}{1+x^2}\right)z - \left(\frac{12}{1+x^2}\right)y + \frac{x}{1+x^2}.$$

In terms of the initial conditions $y(0) = 1$, $y'(0) = 1$ for the original second order differential equation, the initial conditions for the system become

$$y(0) = 1 \quad \text{and} \quad z(0) = 1.$$

Thus, in the notation of the Runge–Kutta algorithm for a second order equation,

$$g(x, y, z) = \left(\frac{2x}{1+x^2}\right)z - \left(\frac{12}{1+x^2}\right)y + \frac{x}{1+x^2},$$

$x_0 = 0$, $y_0 = 1$ and $z_0 = 1$.

The details of the calculations leading to the determination of y_1 to y_5 are set out below. Although $z_n = y'(x_n)$ is not needed in the final solution it is required in all the intermediate calculations, so it is shown in what follows.

n	x_n	k_{1n}	K_{1n}	k_{2n}	K_{2n}
0	0.0	0.200000	−2.400000	−0.040000	−2.601980
1	0.2	−0.282585	−2.276162	−0.510201	−2.019444
2	0.4	−0.654842	−1.358465	−0.790688	−0.829459
3	0.6	−0.803729	−0.130315	−0.816761	0.421649
4	0.8	−0.716722	0.943545	−0.622368	1.355787
5	1.0	−0.452117	1.629692	−0.289148	1.850694

n	x_n	k_{3n}	K_{3n}	k_{4n}	K_{4n}
0	0.0	−0.060198	−2.320831	−0.264166	−2.231915
1	0.2	−0.484529	−1.754726	−0.633530	−1.343215
2	0.4	−0.737788	−0.656726	−0.786187	−0.135794
3	0.6	−0.761565	0.484007	−0.706928	0.929218
4	0.8	−0.581144	1.334228	−0.449877	1.614581
5	1.0	−0.267048	1.784204	−0.095276	1.891750

n	x_n	k_n	K_n	y_n	z_n
0	0.0	−0.044094	−2.412923	1.000000	1.000000
1	0.2	−0.484262	−1.861286	0.955906	−1.412923
2	0.4	−0.749663	−0.744438	0.471644	−3.274209
3	0.6	−0.777885	0.435036	−0.278020	−4.018647
4	0.8	−0.595604	1.323026	−1.055904	−3.583612
5	1.0	−0.276631	1.798540	−1.651508	−2.260585

■

Example 2. Runge–Kutta solution of nonlinear second order equation

Use the Runge–Kutta method with step length $h=0.2$ to determine $y(x)$ from $x=0.5$ to $x=1.7$ given that

$$\frac{d^2y}{dx^2} - \frac{2dy}{dx} + \frac{1}{2}y^2 - \frac{1}{4}\sin y = 1,$$

subject to the initial conditions

$$y(0.5)=0.6 \quad \text{and} \quad y'(0.5)=-0.8.$$

Determine the effect of step length on the accuracy of the calculation by comparing the values of $y(0.9)$, $y(1.3)$ and $y(1.7)$ computed using a step of length $h=0.2$ with the values obtained using a step of length $h=0.4$.

Solution

Setting $z=dy/dx$ converts the second order equation to the system

$$\frac{dy}{dx}=z,$$

$$\frac{dz}{dx}=1-\tfrac{1}{2}y^2+2z+\tfrac{1}{4}\sin y.$$

In terms of this system, the original initial conditions $y(0.5)=0.6$, $y'(0.5)=-0.8$ become

$$y(0.5)=0.6 \quad \text{and} \quad z(0.5)=-0.8.$$

Thus, in the notation of the Runge–Kutta algorithm for a second order equation

$$g(x, y, z) = 1-\tfrac{1}{2}y^2+2z+\tfrac{1}{4}\sin y,$$

$$x_0=0.5, \quad y_0=0.6 \quad \text{and} \quad z_0=-0.8.$$

The first table below lists details of the calculations with step length $h=0.2$, while the second table gives the corresponding details when the step length is $h=0.4$. A comparison of the value of $y(1.7)$ computed using $h=0.2$ and $h=0.4$ shows a difference of 8.31×10^{-3}.

Calculation with $h=0.2$

n	x_n	k_{1n}	K_{1n}	k_{2n}	K_{2n}
0	0.5	-0.160000	-0.127768	-0.172777	-0.147750
1	0.7	-0.190133	-0.177743	-0.207907	-0.210516
2	0.9	-0.233253	0.260458	-0.259298	-0.314640
3	1.1	-0.297955	-0.398529	-0.337808	-0.489265
4	1.3	-0.398951	-0.632413	-0.462192	-0.787303
5	1.5	-0.562119	-1.037335	-0.665853	-1.308874
6	1.7	-0.834641	-1.760227	-1.010663	-2.255332

n	x_n	k_{3n}	K_{3n}	k_{4n}	K_{4n}
0	0.5	−0.174775	−0.151363	−0.190273	−0.178001
1	0.7	−0.211185	−0.216912	−0.233515	−0.260988
2	0.9	−0.264717	−0.325884	−0.298430	−0.399563
3	1.1	−0.346882	−0.509212	−0.399797	−0.634400
4	1.3	−0.477681	−0.823422	−0.563636	−1.041182
5	1.5	−0.693007	−1.376355	−0.837390	−1.767848
6	1.7	−1.060174	−2.387911	−1.312223	−3.132172

n	x_n	k_n	K_n	y_n	z_n
0	0.5	−0.174229	−0.150666	0.600000	−0.800000
1	0.7	−0.210306	−0.215598	0.425771	0.950666
2	0.9	−0.263285	−0.323512	0.215465	−1.166264
3	1.1	−0.344522	−0.504980	−0.047820	−1.489775
4	1.3	−0.473722	−0.815841	−0.392342	−1.994756
5	1.5	−0.686205	−1.362607	−0.866065	−2.810597
6	1.7	−1.048090	−2.363147	−1.552269	−4.173203

Calculation with $h = 0.4$

n	x_n	k_{1n}	K_{1n}	k_{2n}	K_{2n}
0	0.5	−0.320000	−0.255536	−0.371107	−0.338340
1	0.9	−0.466303	−0.520506	−0.570404	−0.742605
2	1.3	−0.796580	−1.261739	−1.048927	−1.893268
3	1.7	−1.662172	−3.501069	−2.362386	−5.522304

n	x_n	k_{3n}	K_{3n}	k_{4n}	K_{4n}
0	0.5	−0.387668	−0.369421	−0.467768	−0.523480
1	0.9	−0.614824	−0.837548	−0.801322	−1.273344
2	1.3	−1.175233	−2.197180	−1.675452	−3.541248
3	1.7	−2.766633	−6.659006	−4.325774	−11.875751

n	x_n	k_n	K_n	y_n	z_n
0	0.5	−0.384220	−0.365756	0.600000	−0.800000
1	0.9	−0.606346	−0.825693	0.215780	−1.165756
2	1.3	−1.153392	−2.163981	−0.390566	−1.991449
3	1.7	−2.707664	−6.623240	−1.543958	−4.155430

■

Problems for Section 8.9

Use the Runge–Kutta method in the following problems, together with the given step length h and initial conditions, to determine y_1 and y_2 if hand calculation is involved, or y_1 to y_4 if a computer is to be used.

1 $3(1+x)\dfrac{d^2y}{dx^2} + 2x\dfrac{dy}{dx} = 2 + 6\sin x,$

  with $y(0) = 0.5,$ $y'(0) = 1,$ $h = 0.2.$

2 $2x^2\dfrac{d^2y}{dx^2} - (1+x)\dfrac{dy}{dx} + xy = \cos x,$

  with $y(1) = 0,$ $y'(1) = 2,$ $h = 0.1.$

3 $\dfrac{d^2y}{dx^2} - y\dfrac{dy}{dx} - 2y = x,$

  with $y(0) = 2,$ $y'(0) = -1,$ $h = 0.2.$

4 $\dfrac{d^2y}{dx^2} - \dfrac{dy}{dx} + 2y^2 = 4,$

  with $y(0) = 0.5,$ $y'(0) = 0,$ $h = 0.1.$

5 $\dfrac{d^2y}{dx^2} + x^2\dfrac{dy}{dx} - (1+x)y = \cos x,$

  with $y(1) = 0,$ $y'(1) = 3,$ $h = 0.2.$

6 $\dfrac{d^2y}{dx^2} - \sin 2y\dfrac{dy}{dx} + xy = 3,$

  with $y(2) = -1,$ $y'(2) = 0,$ $h = 0.1.$

Chapter 9
Fourier Series, Sturm–Liouville Problems and Orthogonal Functions

A Fourier series is a special functional series representation of an arbitrary function on an interval, which may always be taken to be $-\pi \leq x \leq \pi$, in terms of the trigonometric functions

$$\{1, \cos x, \sin x, \cos 2x, \sin 2x, \ldots, \cos rx, \sin rx, \ldots\}.$$

The idea underlying a Fourier series differs radically from the approach used when a Taylor series expansion is made. Whereas a Taylor series expansion of $f(x)$ about $x = a$ requires the function to be infinitely differentiable at that point, a Fourier series representation merely requires it to be integrable over the interval $-\pi \leq x \leq \pi$ in some suitable sense.

Thus Fourier series representations are possible when $f(x)$ is not necessarily differentiable at each point of the interval over which it is to be represented, and it may even be discontinuous.

The coefficients of a Fourier series are obtained by using the orthogonality properties of the system of trigonometric functions given above. An immediate consequence of the periodicity of the trigonometric functions used is that a Fourier series representation of a function defined for $-\pi \leq x \leq \pi$ will be periodic with period 2π, whatever the behavior of the original function $f(x)$ outside that interval.

Different forms of Fourier series arise depending on the interval involved and on whether or not the function to be represented is even or odd. The half-range cosine series is often useful when working with even functions, while the half-range sine series is useful when working with odd functions. However, when an arbitrary interval is involved, the Fourier series will usually involve both sines and cosines.

As would be expected, the convergence properties of Fourier series differ from those of power series, and a Fourier series will converge at a point at which $f(x)$ is discontinuous, irrespective of the actual value of $f(x)$ at such a point.

The applications of Fourier series are wide-ranging, but they provide a natural description of oscillatory phenomena in terms of the fundamental harmonic components involved. They are, for example, particularly useful when the response of a linear differential equation is required to some form of periodic input which is not a simple trigonometric function.

A class of two-point boundary value problems called Sturm–Liouville problems provides a natural source of orthogonal functions, to which the trigonometric functions occurring in Fourier series also belong. These orthogonal functions offer different ways of

885

representing arbitrary functions in the form of functional series. When considering these more general functions, orthogonality over an interval with respect to a weight function has to be introduced.

The chapter begins with a discussion of trigonometric series, periodic extension and convergence in Sec. 9.1. The formal development of Fourier series and a discussion of the various different forms these series can take is to be found in Sec. 9.2. Only in Sec. 9.3 is the question of the nature of the convergence of a Fourier series actually established and a Fourier theorem proved. This section also introduces the Parseval relation in its different forms.

Integration and differentiation of Fourier series is discussed in Sec. 9.4, while Sections 9.5 and 9.6 consider the uniform convergence of Fourier series and the estimation of the effect of smoothness on their rate of convergence.

A simple method for numerical harmonic analysis is discussed in Sec. 9.7 and applied to some representative examples.

The study of Sturm–Liouville problems is introduced in Sec. 9.8 along with the associated notion of orthogonality over an interval with respect to a weight function. These ideas are applied in Sec. 9.9 to expansions in terms of Bessel functions, with both continuous and discontinuous functions being considered. Finally, in Sec. 9.10, the important topic of orthogonal polynomials is introduced and then followed by a brief discussion of Legendre, Chebyshev, Laguerre and Hermite polynomials.

9.1 Trigonometric series, periodic extension and convergence

This chapter will be concerned with the representation of an arbitrary function $f(x)$ as an infinite series of the form

$$f(x) = a_0\phi_0(x) + a_1\phi_1(x) + a_2\phi_2(x) + \dots, \tag{1}$$

valid in some interval $a < x < b$ and involving a known sequence of functions $\phi_0(x)$, $\phi_1(x)$, ... and the constants a_0, a_1, ...

Representation (1) is called a **functional series**. The most familiar example of a functional series is the Taylor series representation of a function $f(x)$ expanded about the point x_0, in which $\phi_n(x) = (x - x_0)^n$, thereby reducing (1) to a power series.

From among the many sequences of functions $\{\phi_n(x)\}$ with special properties which may be used in representation (1), one of the most important involves the use of the trigonometric functions sine and cosine. Functional series of this type, which will be our main concern, are called **trigonometric series**, and they have the general form

$$\tfrac{1}{2}a_0 + a_1 \cos x + b_1 \sin x + a_2 \cos 2x + b_2 \sin 2x + \dots,$$

in which the constants $a_0, a_1, a_2, \dots, b_1, b_2, \dots$ are called the **coefficients** of the series. The factor $\tfrac{1}{2}$ multiplying the coefficient a_0 has been introduced for reasons of convenience which will become apparent later. Henceforth, as it is more concise, we will write such series using the summation notation

$$\tfrac{1}{2}a_0 + \sum_{n=1}^{\infty} (a_n \cos nx + b_n \sin nx). \tag{2}$$

The determination of the coefficients a_n, b_n in (2) in order that it represents a given function $f(x)$ will be considered in the next section and will bring us to the study of **Fourier series**.

It will be recalled from Sec. 5.7 that a function $f(x)$ is said to be **periodic** with **period** X if

$$f(x) = f(x + X), \tag{3}$$

for all x, and there is no smaller value of X for which this result is always true.

If the trigonometric series

$$\tfrac{1}{2}a_0 + \sum_{n=1}^{\infty} (a_n \cos nx + b_n \sin nx)$$

is convergent for $-\pi \le x \le \pi$, then the function defined by this series must be periodic with period 2π, because each term is periodic and 2π is the smallest period common to all the functions $\sin nx$ and $\cos nx$ for $n = 1, 2, \dots$. As the constant $a_0/2$ may be considered to have any period, it follows at once that, provided it is convergent, the trigonometric series in (2) is periodic with period 2π.

The function $f(x)$ illustrated in Fig. 9.1 is periodic with periods X, $2X$, $3X$, ..., nX, ..., so as the smallest of these is X, this is the period of $f(x)$ in the sense of our definition.

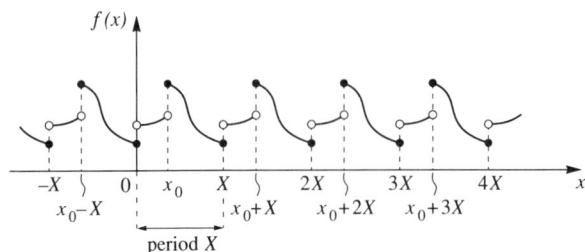

Fig. 9.1   Piecewise continuous function $f(x)$ with period X

Notice that when a function is periodic with period X, knowledge of its behavior in any interval of length X will suffice to determine its behavior for all x. In this sense the location of the origin for a periodic function is unimportant.

Thus, apart from a shift of origin, the function illustrated in Fig. 9.2 is essentially the same as the one in Fig. 9.1.

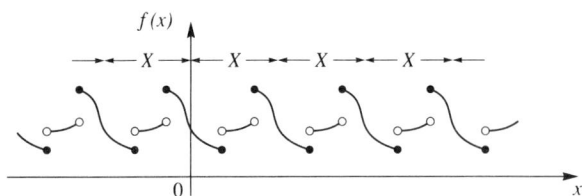

Fig. 9.2   Shift of origin with a periodic function

Notice that if a periodic function has discontinuities within its period X, then the discontinuities will themselves be periodic with period X, as illustrated in Figs 9.1 and 9.2. This consequence of periodicity will prove to be important when we come to discuss Fourier series.

We now prove a simple theorem which has useful consequences for periodic functions.

Theorem 9.1 (Shifting the origin when integrating over a period)

Let $f(x)$ be periodic with period X, and α be a real number. Then, provided the integral

$$\int_{\alpha}^{\alpha+X} f(x)\,dx$$

exists, it follows that

$$\int_{\alpha}^{\alpha+X} f(x)\,dx = \int_{0}^{X} f(x)\,dx,$$

for all real α.

Proof

Let $f(x)$ be defined for $\alpha \leq x \leq \alpha + X$, and such that the integral

$$\int_{\alpha}^{\alpha + X} f(x)\, dx$$

exists. Then if, furthermore, f is periodic with period X, it is defined for all real x and, in particular, for $0 \leq x \leq \alpha$. Thus, using an elementary property of a definite integral, we may always write

$$\int_{\alpha}^{\alpha + X} f(x)\, dx = \int_{\alpha}^{0} f(x)\, dx + \int_{0}^{X} f(x)\, dx + \int_{X}^{\alpha + X} f(x)\, dx.$$

Setting $u = x + X$ in the first integral on the right-hand side gives

$$\int_{\alpha}^{0} f(x)\, dx = \int_{\alpha + X}^{X} f(u - X)\, du.$$

However, because of the periodicity $f(u - X) = f(u)$, and so

$$\int_{\alpha}^{0} f(x)\, dx = \int_{\alpha + X}^{X} f(u)\, du = -\int_{X}^{\alpha + X} f(u)\, du = -\int_{X}^{\alpha + X} f(x)\, dx,$$

where the change of sign follows from the reversal of the limits and last result is obtained by replacing the dummy variable u by x. Consequently,

$$\int_{\alpha}^{\alpha + X} f(x)\, dx = -\int_{X}^{\alpha + X} f(x)\, dx + \int_{0}^{X} f(x)\, dx + \int_{X}^{\alpha + X} f(x)\, dx = \int_{0}^{X} f(x)\, dx,$$

and the result is proved. $\qquad\qquad\qquad\qquad\qquad\qquad\qquad\qquad\qquad\qquad\square$

Let $f(x)$ be a function defined in the interval $\alpha < x < \alpha + X$, and possibly also outside it. Then the function defined in the interval $\alpha + nX < x < \alpha + (n+1)X$, for $n = 0, \pm 1, \pm 2, \ldots$, by the expression

$$g(x - nX) = f(x)$$

is said to be a **periodic extension** of $f(x)$ onto the interval. In this context, the interval $\alpha < x < \alpha + X$ is called the **fundamental interval** for $f(x)$. Thus, for example, the graph of the periodic extension of $f(x)$ to the interval $\alpha + 2X < x < \alpha + 3X$ is obtained from the graph of $f(x)$ for $\alpha < x < \alpha + X$ by translating (shifting) it without change of shape to the new interval.

The periodic extension of the function $f(x) = x^3$ defined for all real x, from a fundamental interval chosen to be $-\pi/2 < x < 3\pi/2$, to the real line $-\infty < x < \infty$, is illustrated in Fig. 9.3. Notice the difference between the original function $f(x)$ outside the chosen fundamental interval (shown as a dashed line), and the periodic extensions (shown as full lines). Notice also that, since $f(-\pi/2) \neq f(3\pi/2)$, the graph comprising all periodic extensions is discontinuous at the ends of each individual periodic extension.

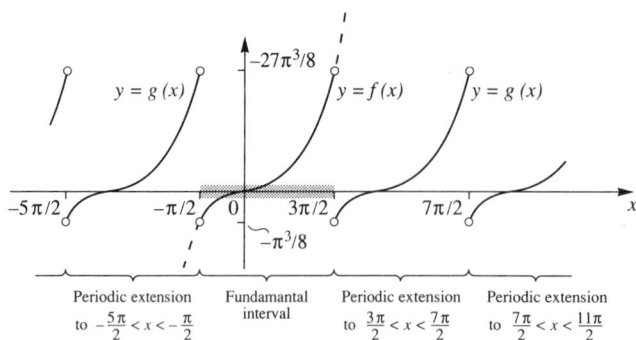

Fig. 9.3 Periodic extensions of $f(x) = x^3$ from the fundamental interval $-\pi/2 < x < 3\pi/2$

The corresponding situation in which an arbitrary continuous function $f(x)$ with the property that $f(-a) = f(a) = h$ is extended periodically outside a fundamental interval chosen to be $-a < x < a$ is shown in Fig. 9.4. When the functional values $f(\pm a) = h$ are included at the end points of the fundamental interval the graph of the periodic extension of $f(x)$ to the real line is seen to be continuous.

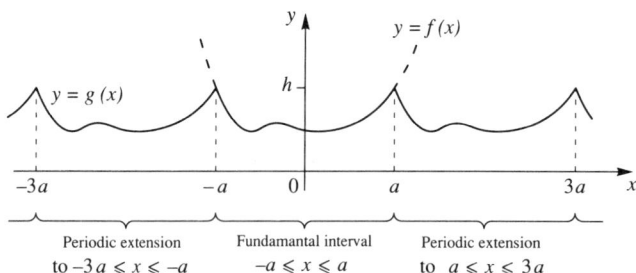

Fig. 9.4 Periodic extensions of an arbitrary continuous function $f(x)$ with fundamental interval $-a \leq x \leq a$ (period 2a) when $f(-a) = f(a) = h$

As it will be useful in what follows, let us recall the definitions of even and odd functions. The function $f(x)$ is said to be an **even function** on the interval $-L < x < L$ if

$$f(-x) = f(x), \tag{4}$$

for all x at which $f(x)$ is continuous. The graph of an even function $y = f(x)$ is *symmetric* about the y-axis.

A function $f(x)$ is said to be an **odd function** on the interval $-L < x < L$ if

$$f(-x) = -f(x), \tag{5}$$

for all x at which $f(x)$ is continuous. The graph of an odd function $y = f(x)$ is *skew-symmetric*, in the sense that the graph for negative x is obtained by first reflecting the graph for positive x in the y-axis, and then reflecting the resulting graph in the x-axis.

The following simple results, which are often useful, follow directly from the definitions of even and odd functions:

(a) the product of two *even* functions is an *even* function
(b) the product of two *odd* functions is an *even* function
(c) the product of an *even* function and an *odd* function is an *odd* function.

If $f(x)$ is both *odd* and *continuous* at the origin (5) implies that $f(0)=0$. Examples of continuous even and odd functions are, respectively, 1, x^2, cosh $3x$ and x, sin $4x$, tanh $3x$. Typical examples of discontinuous even and odd functions which jump between ± 1 are $(\cos x)/|\cos x|$ and $(\sin x)/|\sin x|$, respectively, neither of which is defined at the zeros of its denominator.

In general, a function is neither even nor odd, though it may always be decomposed into a sum of even and odd functions. To see this, let $f(x)$ for $-L < x < L$ be an arbitrary function of x. Now define $f_o(x)$ and $f_e(x)$ as

$$f_o(x) = \tfrac{1}{2}[f(x)-f(-x)] \quad \text{and} \quad f_e(x) = \tfrac{1}{2}[f(x)+f(-x)].$$

Then, clearly, $f_o(x)$ is an *odd* function and $f_e(x)$ is an *even* function on the interval $-L < x < L$, but

$$f(x) = f_o(x) + f_e(x),$$

so the result is proved.

By using a simple device based on definitions (4) and (5) it is always possible to extend an arbitrary function $f(x)$ defined only on the interval $0 \le x < L$ as an even function defined on the larger interval $-L < x < L$, or as an odd function defined on the same interval provided $f(0) = 0$. To see how this may be accomplished, let the arbitrary function $f(x)$ be defined on the interval $0 \le x < L$. Now define the function $F(x)$ on the larger interval $-L < x < L$ as

$$F(x) = \begin{cases} f(x), & 0 \le x < L \\ f(|x|), & -L < x < 0. \end{cases}$$

When so defined $F(x)$ is an even function for $-L < x < L$, but on the subinterval $0 \le x < L$ it coincides with $f(x)$. In certain types of problem such a representation is very useful, because it allows us to work with functional series using the simplifying properties of an even function $F(x)$ on the interval $-L < x < L$, and then to recover $f(x)$ from $F(x)$ by simply restricting x to the subinterval $0 \le x < L$. If the function $f(x)$ is defined at $x = L$ and $F(x)$ is then extended periodically with period $2L$ the resulting function $F(x)$ becomes an even function for all x.

Correspondingly, let $f(x)$ be an arbitrary function defined on the interval $0 \le x < L$ and such that $f(0) = 0$. Now define the function $F(x)$ on the larger interval $-L < x < L$ as

$$F(x) = \begin{cases} f(x), & 0 \le x < L \\ -f(|x|), & -L < x \le 0. \end{cases}$$

Then the function $F(x)$ is an odd function defined on the interval $-L < x < L$. We may recover $f(x)$ from $F(x)$ by again restricting x to the subinterval $0 \le x < L$. An extension of this kind allows the simplifying properties of odd functions to be used in a functional series representation, after which $f(x)$ may be recovered from $F(x)$ by restricting $F(x)$ to the subinterval $0 \leq x < L$. If we define $f(L) = 0$, then $F(x)$ may be extended periodically with period $2L$ as an odd function for all x with jump discontinuities at $\pm L$, $\pm 3L$, $\pm 5L, \ldots$.

Example 1. Even and odd extensions of functions

Extend the function $f(x) = 1 - \cos x$ defined on the interval $0 \leq x \leq \pi/2$ (a) as an even function on the interval $-\pi/2 \leq x \leq \pi/2$ and (b) as a continuous odd function on the interval $-\pi/2 \leq x \leq \pi/2$. (c) Extend the function $f(x) = 1 + \sin x$ defined on the interval $0 \leq x \leq \pi$ as an even function defined on the interval $-\pi \leq x \leq \pi$ and explain why it cannot be extended as a continuous odd function.

Solution

The function in cases (a) and (b) is such that $f(0) = 0$, so it may be extended continuously to the interval $-\pi/2 \leq x \leq \pi/2$ either as an even function or as an odd function. The function in case (c) is such that $f(0) = 1 \neq 0$, so it may only be extended continuously to the interval $-\pi \leq x \leq \pi$ as an even function. The results are shown graphically in Figs 9.5(a), (b) and (c).

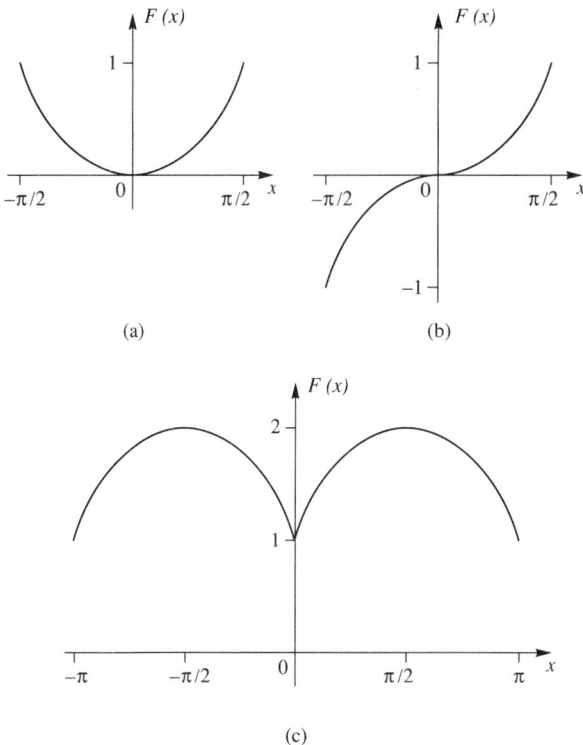

(a)   (b)

(c)

Fig. 9.5 (a) Even extension of $f(x) = 1 - \cos x$ to $-\pi/2 \leq x \leq \pi/2$ (b) Odd extension of $f(x) = 1 - \cos x$ to $-\pi/2 \leq x \leq \pi/2$ (c) Even extension of $f(x) = 1 + \sin x$ to $-\pi \leq x \leq \pi$

The next result, which we state as a theorem, will be useful when discussing Fourier series.

Theorem 9.2 (Integrating even and odd functions over $-L \leq x \leq L$)

Let $f(x)$ be either an even or an odd function for $-L \leq x \leq L$, and such that the integral

$$\int_{-L}^{L} f(x)\, dx$$

exists. Then

$$\int_{-L}^{L} f(x)\, dx = \begin{cases} 2\int_{0}^{L} f(x)\, dx & \text{if } f(x) \text{ is an even function} \\ 0 & \text{if } f(x) \text{ is an odd function.} \end{cases}$$

Proof

If the integral in the theorem exists, then

$$\int_{-L}^{L} f(x)\, dx = \int_{-L}^{0} f(x)\, dx + \int_{0}^{L} f(x)\, dx.$$

Setting $x = -u$ in the first integral on the right-hand side gives

$$\int_{-L}^{0} f(x)\, dx = -\int_{L}^{0} f(-u)\, du = \int_{0}^{L} f(-u)\, du = \int_{0}^{L} f(-x)\, dx.$$

Thus it follows that

$$\int_{-L}^{L} f(x)\, dx = \int_{0}^{L} f(-x)\, dx + \int_{0}^{L} f(x)\, dx = \int_{0}^{L} [f(-x) + f(x)]\, dx.$$

The two statements in the theorem now follow by using the fact that if $f(x)$ is an even function then $f(-x) + f(x) = 2f(x)$, while if it is an odd function $f(-x) + f(x) = 0$. $\square$

The difference between the Taylor series expansion of a function about a point x_0, and its more general representation as a functional series throughout an interval I needs to be made clear. A Taylor series expansion about x_0 is only possible for a function which is infinitely differentiable at x_0, and it represents the function as a power series in $(x - x_0)$ throughout some interval of convergence centered on x_0. The function so represented is both continuous and infinitely differentiable throughout this interval of convergence, and the series representation depends solely on the values of the function and its derivatives at the *single point* x_0.

The functional series representations we shall be developing for an arbitrary function $f(x)$ are different, in that they represent $f(x)$ over an interval I by making use of its properties, together with those of the sequence of functions involved in the representation, over the *entire* interval *I*. Because of this we will see that functional series may be used to represent both continuously differentiable functions and functions which have finitely many bounded jump discontinuities in *I*. This means we will need to discover the relationship between the functional values on either side of a jump discontinuity and the value to which the functional series representation converges at the discontinuity. It is for this reason that we speak of a functional series *representation* of a function, because a functional series representation of $f(x)$ at $x = \eta$ is not always equal to $f(\eta)$.

The most general class of functions we will need to expand over an interval *I* will be those which have a finite number of bounded jump discontinuities throughout *I*, but are continuous between each of these discontinuities. These are called **piecewise continuous** functions and they are defined formally as follows.

Let $f(x)$ be defined for $a \leq x \leq b$ and suppose that the finite number of points x_0, $x_1, \ldots, x_n$ are such that

$$a = x_0 < x_1 < x_2 < \ldots < x_n = b.$$

Then $f(x)$ is said to be **piecewise continuous** on $a \leq x \leq b$ if
(i) $f(x)$ is continuous throughout each interval

$$x_0 < x < x_1, x_1 < x < x_2, \ldots, x_{n-1} < x < x_n, \text{ and}$$

(ii) the one-sided limits[1]

$$\lim_{x \to x_k -} f(x) \quad \text{and} \quad \lim_{x \to x_k +} f(x)$$

exist and are finite, though not equal, for $k = 1, 2, \ldots, n-1$, while at x_0 the limit from the right

$$\lim_{x \to x_0 +} f(x)$$

exists and is finite and at x_n the limit from the left

$$\lim_{x \to x_n -} f(x)$$

also exists and is finite.

The graph of a typical piecewise continuous function is shown in Fig. 9.6. Notice that it is not necessary for either of the one-sided limits at an interior point of discontinuity, or the limits from the left and right at the end points, to equal the functional value at the point in question (if it is defined).

Closely related to piecewise continuous functions are piecewise smooth functions. As before, let $f(x)$ be defined for $a \leq x \leq b$ and suppose that the finite number of points

---

[1] Here, and elsewhere, we use the notation $x \to \xi -$ to signify that x tends to ξ from the left, and $x \to \xi +$ to signify that x tends to ξ from the right.

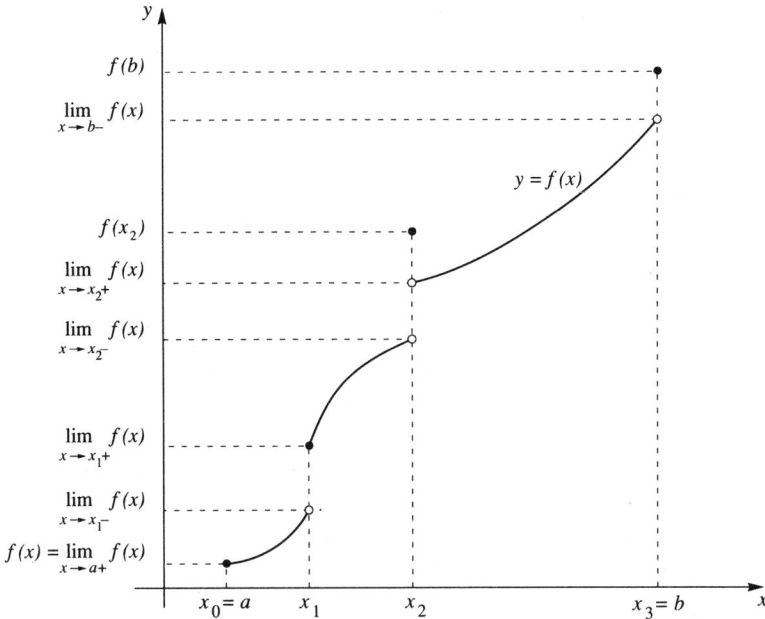

Fig. 9.6 Piecewise continuous function

$x_0, x_1, \ldots, x_n$ are such that

$$a = x_0 < x_1 < x_2 < \ldots < x_n = b.$$

Then $f(x)$ is said to be **piecewise smooth** on $a \leq x \leq b$ if
(i) $f(x)$ is continuous for $a \leq x \leq b$ and the derivative $f'(x)$ is defined throughout each subinterval

$$x_0 < x < x_1, x_1 < x < x_2, \ldots x_{n-1} < x < x_n, \text{ and}$$

(ii) $f'(x)$ has both left and right-hand derivatives

$$f'(x_k-) = \lim_{x \to x_k-} f'(x) \quad \text{and} \quad f'(x_k+) = \lim_{x \to x_k+} f'(x),$$

respectively, which are not equal, for $k = 1, 2, \ldots, n-1$, while the right-hand derivative at the left end point

$$f'(x_0+) = \lim_{x \to x_0+} f(x)$$

exists, together with the left-hand derivative at the right end point

$$f'(x_n-) = \lim_{x \to x_n-} f(x).$$

The graphical interpretation of the piecewise smooth condition is illustrated in Fig. 9.7. In the diagram the tangent lines drawn to the graph represent the left and right-hand derivatives at the points in question.

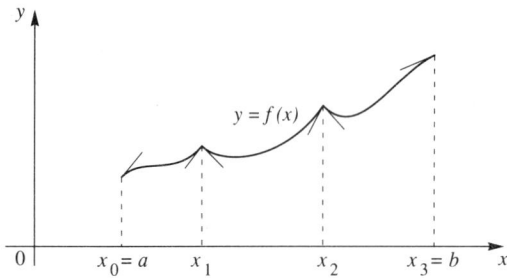

Fig. 9.7   Piecewise smooth function on $a \leq x \leq b$

When working with functional series, two very different types of convergence arise, called **pointwise convergence** and **convergence in the mean**. The first of these types of convergence involves an idea which is already familiar. We will say the functional sequence $\{f_n(x)\}$ **converges pointwise** to the **limit function** $f(x)$ for $a \leq x \leq b$ if, for every fixed value $x = x^*$ in the interval, the sequence of numbers $f_1(x^*), f_2(x^*), \ldots$ converges to the number $f(x^*)$. Expressed symbolically, this is equivalent to requiring that

$$\lim_{n \to \infty} |f(x) - f_n(x)| = 0, \tag{6}$$

for each x in the interval $a \leq x \leq b$. Thus pointwise convergence throughout an interval is a condition which must be satisfied at *each* point of the interval.

Although knowledge of the pointwise convergence property of a functional series is important, there are other properties which are equally important, and possibly even more so. These are which of the analytical properties of the partial sums of a functional series, such as continuity, differentiability and termwise integrability, are passed on to the sum function. Expressed differently, this amounts to considering a sequence of functions $\{f_n(x)\}$, which may be taken to be the partial sums of a functional series, and asking which of the analytical properties common to the function $f_n(x)$ in the sequence are shared by the limit function of the sequence

$$f(x) = \lim_{n \to \infty} f_n(x).$$

The loss of analytical properties in the limit function $f(x)$ of a sequence $\{f_n(x)\}$ which converges pointwise to $f(x)$ is illustrated in the following example.

Example 2. Loss of continuity and differentiability

Let $f_n(x) = (1 - x^2)^n$ for $-1 \leq x \leq 1$. Then each member of the sequence of functions $\{f_n(x)\}$ is both continuous and differentiable on this closed interval, and the limit function is

$$f(x) = \lim_{n \to \infty} (1 - x^2)^n.$$

As $\lim_{n \to \infty} (1 - x^2)^n = 0$ for $-1 \leq x < 0$ and $0 < x \leq 1$, while $\lim_{n \to \infty} (1 - x^2)^n = 1$ when $x = 0$, it follows that the limit function is

$$f(x) = \begin{cases} 0 & \text{for } -1 \le x < 0 \\ 1 & \text{for } x = 0 \\ 0 & \text{for } 0 < x \le 1. \end{cases}$$

Thus this sequence $\{f_n(x)\}$ converges pointwise to the limit function $f(x)$, but the limit function is discontinuous at the origin, and so is also nondifferentiable at that same point. Consequently, despite being pointwise convergent to the limit function $f(x)$ on $-1 \le x \le 1$, the limit function has lost both continuity and differentiability over the entire interval which are properties common to each of the functions $f_n(x) = (1 - x^2)^n$. Representative members of the sequence of functions $y = f_n(x)$ are illustrated in Fig. 9.8(a), and the discontinuous limit function $y = f(x)$ is shown in Fig. 9.8(b).

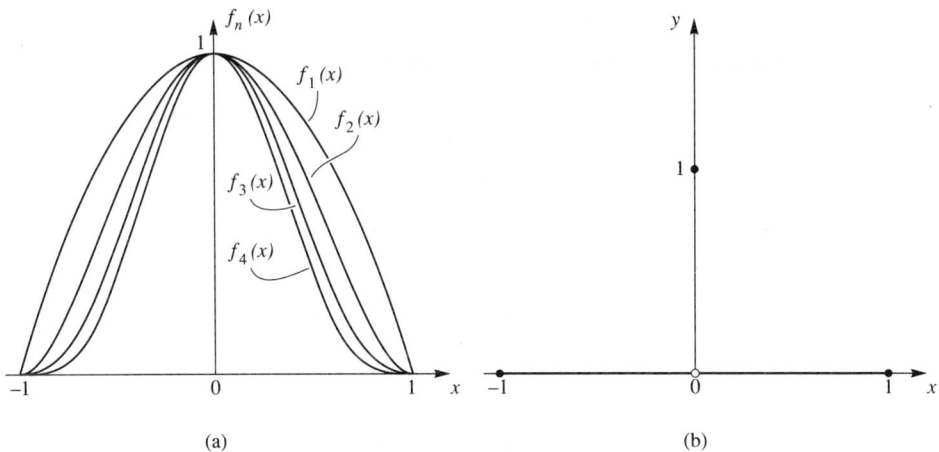

Fig. 9.8   (a) Graphs of $y = f_n(x)$ (b) Graph of $y = f(x)$ ■

We have seen that the pointwise convergence of $\{f_n(x)\}$ to the limit function $f(x)$ over an interval $a \le x \le b$ involves the convergence of the sequence of numbers $f_1(x^*)$, $f_2(x^*)$, ... to the number $f(x^*)$ at each point x^* in $[a, b]$. The other form of convergence we must consider, called **convergence in the mean**, depends on the behavior of $f_n(x)$ and $f(x)$ over the *entire* interval $a \le x \le b$. This turns out to be a natural form of convergence to consider when working with functional series representations of piecewise continuous functions. Specifically we will say that the sequence of functions $\{f_n(x)\}$ defined for $a \le x \le b$ **converges in the mean** to the function $f(x)$ if

$$\lim_{n \to \infty} \int_a^b |f(x) - f_n(x)|^2 \, dx = 0. \tag{7}$$

Expressed in words, convergence in the mean requires the integral of the square of the magnitude of the difference between $f(x)$ and $f_n(x)$ taken over the interval $a \le x \le b$ to have the limit zero as $n \to \infty$. Thus convergence in the mean is seen to be a generalization of the technique of the mean square fitting of a function to experimental data, in which the sum of the squares of the deviations of the function from the data is minimized.

To distinguish between the *pointwise convergence* and *convergence in the mean* of $f_n(x)$ to $f(x)$ as $n\to\infty$ we will signify pointwise convergence by writing

$$\lim_{n\to\infty} f_n(x)=f(x),$$

and convergence in the mean by writing

$$\lim_{n\to\infty} f_n(x)\approx f(x).$$

The next example involves one function which is both pointwise convergent and convergent in the mean to a limit function, and another which although pointwise convergent does not converge in the mean to the limit function.

Example 3. Differences between types of convergence

(i) Consider the sequence of functions $\{f_n(x)\}$ defined on the interval $0\le x\le 1$ with $f_n(x)=x^n$. This sequence is pointwise convergent to the limit function

$$f(x)=\lim_{n\to\infty} x^n,$$

which is easily seen to be the discontinuous function

$$f(x)=\begin{cases} 0 & \text{for } 0\le x<1 \\ 1 & \text{for } x=1. \end{cases}$$

This follows because for $0\le x<1$ we have $\lim_{n\to\infty} x^n=0$, while for $x=1$ we have $\lim_{n\to\infty} x^n=1$.

That this sequence is also convergent in the mean to this same limit function follows from the fact that

$$\lim_{n\to\infty}\int_0^1 |f_n(x)-f(x)|^2\,dx=\lim_{n\to\infty}\int_0^1 x^{2n}\,dx=\lim_{n\to\infty}\left(\frac{1}{2n+1}\right)=0.$$

(ii) The sequence of functions $\{f_n(x)\}$ defined for $0\le x\le 1$ with $f_n(x)=\sqrt{2nx}\,\exp(-\tfrac{1}{2}nx^2)$ converges pointwise to the limit function $f(x)\equiv 0$. However it does not converge in the mean to this limit function because

$$\lim_{n\to\infty}\int_0^1 |f_n(x)-f(x)|^2\,dx=\lim_{n\to\infty}\int_0^1 2nx\,\exp(-nx^2)\,dx=\lim_{n\to\infty}(1-e^{-n})=1,$$

and this limit is *not* zero.   ■

Uniform convergence – a deeper look at convergence

A better understanding of Fourier series, and indeed of convergence problems in general, can be achieved if the reader has some familiarity with the stronger form of convergence called **uniform convergence**. For this reason we now summarize the basic idea involved, together with some of its consequences, though as this material is not essential for what follows this subsection may be omitted at a first reading.

The basic disadvantage of pointwise convergence is that even when each of the functions $f_n(x)$ in the sequence $\{f_n(x)\}$ is continuous for $a \le x \le b$, it is not necessarily true that the limit function $f(x)$ is continuous for $a \le x \le b$. This was seen to be the case in Ex. 2 in which the limit function lost continuity at the origin. The loss of continuity in this example, which is typical, was caused by the fact that however large n may become, there were always points at which $f_n(x)$ was not close to $f(x)$. Thus, in the general case, to retain continuity in the limit, it is necessary to require that for suitably large n the difference between $f_n(x)$ and $f(x)$ remains small for all x in the interval $a \le x \le b$.

This can be accomplished in two slightly different but equivalent ways. The first of these involves taking a narrow strip of uniform width centered on the graph of $y = f(x)$ and requiring that, for suitably large n, all the graphs of $y = f_n(x)$ lie within the strip. If this can always be accomplished, no matter how narrow the strip, we say the sequence $\{f_n(x)\}$ **converges uniformly** to $f(x)$ for $a \le x \le b$.

The second, and equivalent, way of meeting the condition is to require that the magnitude of the maximum difference between $f_n(x)$ and $f(x)$ tends to zero as $n \to \infty$ for $a \le x \le b$. This provides an alternative criterion for uniform convergence.

The precise mathematical formulations of these equivalent definitions of uniform convergence are arrived at as follows. Let a sequence of functions $\{f_n(x)\}$ be defined for $a \le x \le b$, and let $\varepsilon > 0$ be an arbitrarily small pre-assigned positive number. Then the sequence of functions $\{f_n(x)\}$ will be said to converge **uniformly** to the limit function $f(x) = \lim_{n \to \infty} f_n(x)$ if it is possible to find a number $N(\varepsilon)$, independent of x and depending only on ε, such that for $n \ge N(\varepsilon)$ the graphs of all the functions $y = f_n(x)$ lie *within* a strip of width 2ε centered on the graph of $y - f(x)$. That is, for the finite number of graphs corresponding to $n = 1, 2, \ldots, N-1$, at least part of each of the graphs $y = f_1(x)$, $y = f_2(x)$, $\ldots$, $y = f_{N-1}(x)$ lies *outside* the strip, whereas the infinite number of graphs $y = f_N(x)$, $y = f_{N+1}(x)$, $\ldots$, corresponding to $n = N, N+1, \ldots$, all lie *within* the strip. Figure 9.9(a) illustrates the situation when $n < N(\varepsilon)$, while Fig. 9.9(b) corresponds to the case in which $n \ge N(\varepsilon)$.

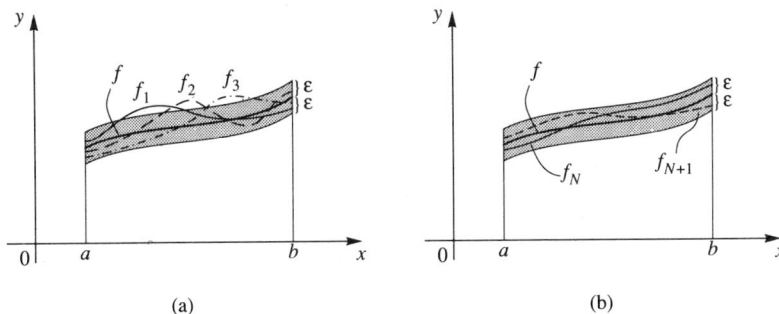

(a)   (b)

Fig. 9.9 Geometrical interpretation of uniform convergence: (a) At least part of each of the graphs of $y = f_1(x)$, $y = f_2(x)$, $\ldots$, $y = f_{N-1}(x)$ lies outside the strip; (b) The entire graph of each of $y = f_N(x)$, $y = f_{N+1}(x)$, $\ldots$, lies inside the strip.

The name *uniform convergence* derives from the way in which the graphs of $y = f_n(x)$ for $n \geq N(\varepsilon)$ are constrained to lie *uniformly* within the same strip of width 2ε. The smaller ε is taken, the narrower the strip becomes, and the larger must we take $N(\varepsilon)$. However, provided $f_n(x)$ converges uniformly to $f(x)$, the graph of each $y = f_n(x)$ for $n \geq N(\varepsilon)$ must always lie *within* the strip.

Analytically, this graphical condition means that for an $\varepsilon > 0$

$$f(x) - \varepsilon < f_n(x) < f(x) + \varepsilon \text{ or, equivalently, that } |f_n(x) - f(x)| < \varepsilon \tag{8}$$

for $n \geq N(\varepsilon)$ and all x in $a \leq x \leq b$. This is the analytical form taken by the formal definition of **uniform convergence** in terms of a *strip*.

To formulate the alternative form of the definition we need to make use of the notion of the *supremum* or *least upper bound* of a function. A number ξ is called the **supremum** or **least upper bound** of a function $h(x)$, which is defined for $a \leq x \leq b$, if

(i) ξ is an upper bound of $h(x)$ for $a \leq x \leq b$, so it is certainly true that $h(x) \leq \xi$ for x in this interval, and

(ii) ξ is the smallest (least) of all possible upper bound of $h(x)$ for $a \leq x \leq b$. Symbolically, we denote such a bound by writing either

$$\xi = \sup_{a \leq x \leq b} h(x) \quad \text{or} \quad \xi = \lub_{a \leq x \leq b} h(x),$$

where each expression has the same meaning.

Thus, for example, five upper bounds of $h(x) = 2 - x^2$ for $-3 \leq x \leq 1$ are 5, 2·4, 2, 3, and 7, but of these only 2 is the supremum, or least upper bound, so that

$$\sup_{-3 \leq x \leq 1} (2 - x^2) = 2.$$

We are now in a position to formulate the alternative form of the definition of uniform convergence. The sequence of functions $\{f_n(x)\}$ defined for $a \leq x \leq b$ will be said to **converge uniformly** to the limit function $f(x)$ if

$$\sup_{a \leq x \leq b} |f_n(x) - f(x)| \to 0 \quad \text{as } n \to \infty. \tag{9}$$

The notation of the uniform convergence of a sequence may be extended without difficulty to a functional series. All that is necessary is to identify the function $f_n(x)$ in the sequence $\{f_n(x)\}$ with the nth partial sum of the series. When this is done the series will be said to be uniformly convergent to $f(x)$ provided the sequence $\{f_n(x)\}$ converges uniformly to $f(x)$.

Example 4. Proof of uniform convergence using definition (8)

Show that the functional sequence $\{s_n(x)\}$ with

$$s_n(x) = x - \frac{x^3}{3!} + \frac{x^5}{5!} - \dots + (-1)^n \frac{x^{2n+1}}{(2n+1)!}$$

converges uniformly to the Maclaurin series for $\sin x$ for $0 \leq x \leq a$, with $a > 0$ arbitrary.

Solution

The function $s_n(x)$ is seen to be the sum of the terms of the Maclaurin series for $\sin x$ up to and including the term in x^{2x+1}, because we know that

$$\sin x = \sum_{n=0}^{\infty} (-1)^n \frac{x^{2n+1}}{(2n+1)!}.$$

To prove that $s_n(x)$ converges uniformly to this series for $\sin x$ for $0 \le x \le a$ it is necessary to show that

$$\left| s_n(x) - \sum_{n=0}^{\infty} (-1)^n \frac{x^{2n+1}}{(2n+1)!} \right|$$

is bounded by a number independent of x which tends to zero as $n \to \infty$. Now

$$\left| s_n(x) - \sum_{n=0}^{\infty} (-1)^n \frac{x^{2n+1}}{(2n+1)!} \right| = \left| \sum_{r=n+1}^{\infty} (-1)^r \frac{x^{2r+1}}{(2r+1)!} \right| \le \sum_{r=n+1}^{\infty} \frac{a^{2r+1}}{(2r+1)!},$$

where the last result follows from the triangle inequality by assigning to x its largest value in $0 \le x \le a$.

Clearly the series on the right is independent of x, and it is easily seen to be convergent by the ratio test (for all a). However, this series is the remainder after the term in a^{2n+1} of a convergent series, so it will tend to zero as $n \to \infty$, and thus the uniform convergence of $s_n(x)$ to the Maclaurin series for $\sin x$ has been proved for $0 \le x \le a$. ∎

Example 5. A case of nonuniform convergence established by using (9)

Show that the sequence of functions $\{f_n(x)\}$ with

$$f_n(x) = \frac{2nx}{1+n^2x^2}$$

does not converge uniformly to its limit function for $x \ge 0$.

Solution

The limit function is $f(x) = \lim_{n \to \infty} f_n(x) \equiv 0$, for $x \ge 0$. Differentiation of $f_n(x)$ shows that for $x \ge 0$ the maximum value occurs at $x = 1/n$, at which point the maximum value itself is $f_n(1/n) = 1$. Thus

$$\sup_{x \ge 0} \left| f_n(x) - f(x) \right| = \max_{x \ge 0} \left| f_n(x) - 0 \right| = \max_{x \ge 0} f_n(x) = 1, \quad \text{as } n \to \infty.$$

Now, from (9) we know that the sequence $\{f_n(x)\}$ can only converge uniformly to $f(x)$ if $\sup_{x \ge 0} |f_n(x) - f(x)| \to 0$ as $n \to \infty$, so as this condition is not satisfied we conclude that the sequence $\{f_n(x)\}$ does not converge uniformly to the limit function $f(x) \equiv 0$.

The failure of the uniform convergence in this example is due to the coincidence of the maximum deviation of $f_n(x)$ from $f(x)$, which is equal to unity at the point $x = 1/n$, with the maximum value of $f_n(x)$ which is itself unity. As n increases, so the location of the maximum value of $f_n(x)$ decreases

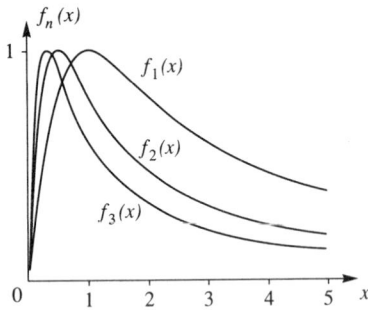

Fig. 9.10   Graph of $y = f_n(x)$

towards zero, but the magnitude of the deviation itself remains constant at unity, as shown in Fig. 9.10.

∎

The mathematical properties shared by uniformly convergent functions make it desirable to know when a function defined as the sum of a functional series is uniformly convergent. Rather than attempting to investigate this question by means of either of the above definitions, one of a variety of tests may be used. The simplest of these is called the **Weierstrass[2] *M*-test**, and it suffices for most purposes.

Theorem 9.3 (Weierstrass *M*-test for uniform convergence)

Let $\{f_n(x)\}$ be a sequence of functions defined for $a \leq x \leq b$ with the property that there are positive constants M_n such that $|f_n(x)| \leq M_n$ for all $a \leq x \leq b$. Then if the series $\sum_{n=0}^{\infty} M_n$ is convergent, the functional series $\sum_{n=0}^{\infty} f_n(x)$ is uniformly convergent (and absolutely convergent) for $a \leq x \leq b$.

Proof

Let $\varepsilon > 0$ be a given arbitrarily small number. Then, as ΣM_n is convergent, it follows that for all suitably large $N = N(\varepsilon)$

$$\sum_{n=N}^{\infty} M_n < \varepsilon.$$

For all such N, using the stated conditions of the theorem together with the triangle

---

[2] KARL THEODOR WILHELM WEIERSTRASS (1815–1897), a major German mathematician who was Professor of Mathematics at Berlin. He is regarded as one of the founders of modern analysis and, almost contemporaneously with the major English applied mathematician George Gabriel Stokes (1819–1903) who spent his working life at Cambridge, was responsible for the introduction of the notion of uniform convergence.

inequality, it follows that

$$\left|\sum_{n=N}^{\infty} f_n(x)\right| \le \sum_{n=N}^{\infty} \left|f_n(x)\right| \le \sum_{n=N}^{\infty} M_n < \varepsilon,$$

for all $a \le x \le b$.

If

$$f(x) = \sum_{n=0}^{\infty} f_n(x),$$

this result shows that

$$\left|f(x) - \sum_{n=0}^{N-1} f_n(x)\right| < \varepsilon,$$

for all $a \le x \le b$, but this is simply the condition for the uniform convergence of $f(x)$. Thus the theorem is proved. The absolute convergence of the series is an immediate consequence of the bounding of $|f_n(x)|$ by M_n. $\qquad\square$

Example 6. Proof of uniform convergence using the *M*-test

Use the Weierstrass M-test to prove that the series

$$\sum_{n=1}^{\infty} \frac{(-1)^{n+1}}{x^2 + n^2}$$

is uniformly convergent for all x.

Solution

First we identify the function $f_n(x)$ in Theorem 9.3 with the general term in the series, leading to the result

$$f_n(x) = \frac{(-1)^{n+1}}{x^2 + n^2}.$$

Then, setting $M_n = 1/n^2$, it follows that

$$|f_n(x)| = \frac{1}{x^2 + n^2} < M_n = \frac{1}{n^2}, \qquad \text{for all } x.$$

However, the series $\sum_{n=1}^{\infty} (1/n^2)$ is known to be convergent (see Ex. 1, Sec. 8.1), so it follows from the M-test that the original series is uniformly convergent for all x. $\qquad\blacksquare$

The most important properties of uniform convergence in relation to functional sequences and series may be summarized as follows.

Properties of uniform convergence

1 The uniform convergence of a sequence of continuous functions to a limit function ensures that the limit function is also continuous.

2 If the sequence of continuous functions $\{f_n(x)\}$ converges uniformly to $f(x)$ for $a \le x \le b$, then

$$\lim_{n \to \infty} \int_a^b f_n(x)\,dx = \int_a^b f(x)\,dx.$$

3 Let the sequence of continuous functions $\{f_n(x)\}$ defined for $a \le x \le b$ be such that $\sum_{n=1}^{\infty} f_n(x)$ converges uniformly. Then if the series is integrated, the operations of integration and summation may be interchanged to give

$$\int_a^b \sum_{n=1}^{\infty} f_n(x)\,dx = \sum_{n=1}^{\infty} \int_a^b f_n(x)\,dx.$$

4 Let $\{f_n(x)\}$ be a sequence of differentiable functions defined on the open interval $a < x < b$ on which they converge pointwise to the limit function $f(x)$. Then if the derivatives $f_n'(x)$ are continuous functions, and the sequence $\{f_n'(x)\}$ converges uniformly to the limit function $F(x)$, it follows that

$$f'(x) = F(x).$$

5 Let the sequence of differentiable functions $\{f_n(x)\}$ be such that the derivatives $f_n'(x)$ are continuous, and the series $\sum_{n=1}^{\infty} f_n(x)$ converges pointwise. Then if $\sum_{n=1}^{\infty} f_n'(x)$ converges uniformly, it follows that

$$\frac{d}{dx}\left[\sum_{n=1}^{\infty} f_n(x) \right] = \sum_{n=1}^{\infty} f_n'(x).$$

Thus termwise differentiation of a functional series is permissible if the series converges pointwise, the derivatives of the terms are continuous and the differentiated functional series is uniformly convergent.

To illustrate the importance of these results we remark that it is property 3 which ensures that when the Maclaurin series for $\sinh t$ is inserted into the left-hand side of the elementary result

$$\int_0^x \sinh t\,dt = \cosh x - 1,$$

we obtain the Maclaurin series for $\cosh x - 1$.

Correspondingly, it is property 5 which ensures that when the Maclaurin series for $\sinh x$ is differentiated term by term it generates the Maclaurin series for $\cosh x$, thereby verifying the elementary result

$$\frac{d}{dx}(\sinh x) = \cosh x.$$

Problems for Section 9.1

Determine the period of each of the following functions

1 $\sin x + 2 \sin 3x$ **2** $3 \sin \frac{1}{2}x + 7 \cos 2x$

3 $1 + \cos 2x + 6 \cos \frac{5}{2}x$ **4** $3-4 \sin \frac{1}{3}x + 5 \cos \frac{2}{3}x$

5 $1 + \sin \frac{1}{5}x + 6 \cos \frac{1}{2}x$ **6** $2 - \sin \frac{1}{4}x + 3 \cos 2x$

Establish the equality of the stated integrals by direct integration, thereby verifying Theorem 9.1 in the following specific cases.

7 $\displaystyle\int_{\pi/2}^{5\pi/2} \sin^2 x \, dx = \int_0^{2\pi} \sin^2 x \, dx$

8 $\displaystyle\int_{-\pi/2}^{3\pi/2} \sin^3 x \, dx = \int_0^{2\pi} \sin^3 x \, dx$

9 $\displaystyle\int_{\pi/3}^{7\pi/3} \sin^2 x \cos^2 x \, dx = \int_0^{2\pi} \sin^2 x \cos^2 x \, dx$

10 $\displaystyle\int_{\pi/4}^{9\pi/4} \cos^2 x \, dx = \int_0^{2\pi} \cos^2 x \, dx$

In each of the following problems sketch the function and some of its periodic extensions.

11 $f(x) = 1 + x^2$ for $-\pi/2 < x \leq 3\pi/2$

12 $f(x) = |x|$ for $-\frac{1}{2} \leq x \leq \frac{1}{2}$

13 $f(x) = \begin{cases} 1 & \text{for } -\pi < x \leq 0 \\ \sin x & \text{for } 0 < x \leq \pi \end{cases}$

14 $f(x) = x + |x|$ for $-1 < x \leq 1$

15 $f(x) = \begin{cases} -1 & \text{for } -1 < x < \frac{1}{2} \\ 1 & \text{for } \frac{1}{2} < x < 1 \\ 0 & \text{for } 1 < x < 2 \end{cases}$

16 $f(x) = \begin{cases} \sin x & \text{for } -\pi < x < 0 \\ \cos x & \text{for } 0 < x < \pi \end{cases}$

Where possible, extend the following functions by means of (a) an even extension and (b) an odd extension in which $f(0) = 0$.

17 $f(x) = 1 + \sin 2x$ for $0 \leq x \leq \pi$

18 $f(x) = x + \sinh x$ for $0 \leq x \leq 1$

19 $f(x) = \begin{cases} 0 & \text{for } x=0 \\ 1 & \text{for } 0<x<1 \\ 1+x^2 & \text{for } 1<x\leq2 \end{cases}$

20 $f(x) = \begin{cases} x-|x| & \text{for } 0\leq x<1 \\ x & \text{for } 1\leq x<2 \end{cases}$

Classify the following functions as being either piecewise continuous or piecewise smooth.

21 (a) $f(x) = \begin{cases} 1+\cos x & \text{for } -\pi\leq x<0 \\ 2+3x^2 & \text{for } 0\leq x\leq\pi \end{cases}$

(b) $f(x) = \begin{cases} \cosh 3x & \text{for } -1\leq x\leq1 \\ e^{3x} & \text{for } 1<x<3 \end{cases}$

(c) $f(x) = \begin{cases} x-|x| & \text{for } -2\leq x\leq0 \\ \sinh x & \text{for } 0<x\leq1 \\ (2-x)\sinh x & \text{for } 1<x\leq2 \end{cases}$

22 (a) $f(x) = \begin{cases} (1+x^2)^{-1} & \text{for } -\pi\leq x<0 \\ \cosh x & \text{for } 0\leq x\leq1 \\ \sinh x & \text{for } 1<x\leq4 \end{cases}$

(b) $f(x) = \begin{cases} 1-|x| & \text{for } -1\leq x\leq1 \\ 3\sin 2(x-1) & \text{for } 1<x\leq\pi+1 \\ (x-\pi-1)^2 & \text{for } \pi+1<x\leq3\pi \end{cases}$

(c) $f(x) = \begin{cases} \cos\left(\dfrac{1-|x|}{1+|x|}\right) & \text{for } -1\leq x\leq1 \\ \cosh(x-1) & \text{for } a<x\leq2 \end{cases}$

Harder problems

23 Show that the sequence $\{f_n(x)\}$ in which $f_n(x)=(\cos nx)/n$ converges uniformly to its limit function for all real x.

24 Show that the sequence $\{f_n(x)\}$, in which $f_n(x)=x^{2n}$, converges uniformly to its limit function for $0\leq x\leq a$, where $0<a<1$. Is the uniform convergence perserved if $a=1$?

25 Use the M-test to establish the uniform convergence of the functional series

$$\sum_{n=1}^{\infty} \frac{1+\sin nx}{3^n}$$

for all real x.

26 Prove that the sequence $\{s_n(x)\}$, in which

$$s_n(x) = \sum_{r=0}^{n} \frac{x^{2r+1}}{(2r+1)!},$$

converges uniformly to the series for $\sinh x$ for $-a\leq x\leq a$.

27 Show that the M-test fails when applied to the functional series

$$\sum_{n=1}^{\infty} \frac{(-1)^n}{3x+2n} \qquad \text{for } 0 \le x < \infty.$$

Establish the uniform convergence of the series by using definition (8) coupled with the error estimate given in the last part of the alternating series test (Sec. 8.1).

9.2 The formal development of Fourier series

We begin with a purely formal derivation of the Fourier series representation of a function $f(x)$ defined for $-\pi \le x \le \pi$. For most physical applications it will suffice if instead of allowing $f(x)$ to be a completely general function it is required to be either piecewise continuous or piecewise smooth for $-\pi \le x \le \pi$. This restriction disallows functions with extremely erratic behavior,[3] which are mainly of mathematical interest, but it includes all functions of practical interest which can be integrated using the ordinary (Riemann) integral.

The basic idea to be developed in this chapter is the way in which an arbitrary function of the type just described may be represented in the form of a trigonometric series. More precisely, we wish to set

$$f(x) = \tfrac{1}{2}a_0 + \sum_{n=1}^{\infty} (a_n \cos nx + b_n \sin nx) \tag{1}$$

in the interval $-\pi \le x \le \pi$, and then to discover how to determine the constant coefficients a_n, b_n in terms of $f(x)$.

Representation (1) of an arbitrary function $f(x)$ in terms of the sequence of functions $\{1, \cos x, \sin x, \cos 2x, \sin 2x, \ldots, \cos mx, \sin mx, \ldots\}$ is called the **Fourier series**[4] of $f(x)$, and the numbers a_n, b_n are called the **Fourier coefficients** of $f(x)$.

The use of an equality sign in (1) presupposes that (i) the class of functions which are either piecewise continuous or piecewise smooth can always be expanded in such a

---

[3] For example, it disallows the function

$$f(x) = \begin{cases} 0 & \text{if } x \text{ is irrational} \\ 1 & \text{if } x \text{ is rational,} \end{cases}$$

with x in the interval $-\pi \le x \le \pi$. This function has infinitely many finite jumps in the interval $-\pi \le x \le \pi$, and its Riemann integral is not defined, but a generalization called the **Lebesgue integral** is defined, and the Lebesgue integral of this function over the interval $-\pi \le x \le \pi$ equals zero. Functions such as this can be used to model random disturbances of finite amplitude and impulse-like behavior (noise) which affect electrical and mechanical systems.

[4] JOSEPH FOURIER (1768–1830), an outstanding French mathematical physicist whose pioneering work on heat conduction in solids published in 1822 under the title *La Théorie Analytique de la Chaleur* forms the basis of the modern theory of Fourier series and its generalizations.

manner and (ii) that the Fourier series for $f(x)$ converges to $f(x)$ for $-\pi \leq x \leq \pi$. Given that assumption (i) can be justified, assumption (ii) is reasonable whenever $f(x)$ is continuous. However, it then becomes necessary to reinterpret the meaning of the equality sign whenever $f(x)$ exhibits a discontinuity. It is for reasons of this nature that, rather than using an equality sign in (1), it is usual to indicate the connection between $f(x)$ and its Fourier series by writing

$$f(x) \sim \tfrac{1}{2} a_0 + \sum_{n=1}^{\infty} (a_n \cos nx + b_n \sin nx)$$

for $-\pi \leq x \leq \pi$. Here the relational symbol $\sim$ is to be read 'corresponds to' or 'has as its Fourier series'. This notation indicates an association between $f(x)$ and its Fourier series without making the nature of the relationship explicit. We will discover later that the symbol $\sim$ can in fact be interpreted as an equality for any x at which $f(x)$ is continuous, but that a different interpretation is needed when x is a point at which $f(x)$ is discontinuous.

Although we shall not do so here, it can be proved that assumption (i) is valid, and that all piecewise continuous or piecewise smooth functions defined for $-\pi \leq x \leq \pi$ can be represented as a Fourier series. This important property is described by saying that the sequence of trigonometric functions $\{1, \cos x, \sin x, \cos 2x, \sin 2x, \ldots, \cos mx, \sin mx, \ldots\}$ on which Fourier series are based is **complete**.

To motivate the argument, let us start by considering the simpler problem of finding the Fourier series of an arbitrary function $f(x)$ which is *continuous* for $-\pi \leq x \leq \pi$. To accomplish this we shall assume that the symbol $\sim$ can be replaced by an equality, so that we may write

$$f(x) = \tfrac{1}{2} a_0 + \sum_{n=1}^{\infty} (a_n \cos nx + b_n \sin nx), \tag{2}$$

for $-\pi \leq x \leq \pi$. Our task will then be to determine the Fourier coefficients a_n, b_n for a given function $f(x)$.

This section has been entitled a *formal* development of Fourier series because the arguments used depend on assumptions such as those just made. Only when the representation in (2) has been justified and extended to a wider class of functions, and the convergence properties of Fourier series are understood, will the arguments cease being formal and become rigorous.

The formal derivation of the expressions for the Fourier coefficients in (2) starts from the elementary definite integrals

$$\int_{-\pi}^{\pi} \sin mx \sin nx \, dx = \begin{cases} 0 & \text{for } m \neq n \\ \pi & \text{for } m = n \neq 0, \end{cases} \tag{3}$$

$$\int_{-\pi}^{\pi} \cos mx \cos nx \, dx = \begin{cases} 0 & \text{for } m \neq n \\ \pi & \text{for } m = n \neq 0 \\ 2\pi & \text{for } m = n = 0, \end{cases} \tag{4}$$

$$\int_{-\pi}^{\pi} \sin mx \cos nx \, dx = 0 \qquad \text{for all } m, n, \tag{5}$$

in which m, n are integers or zero.

The simplest way to derive these results is by using the trigonometric identities

$$\sin A \sin B = \tfrac{1}{2}[\cos(A - B) - \cos(A + B)] \tag{6}$$

$$\cos A \cos B = \tfrac{1}{2}[\cos(A + B) + \cos(A - B)] \tag{7}$$

$$\sin A \cos B = \tfrac{1}{2}[\sin(A + B) + \sin(A - B)]. \tag{8}$$

We will derive (4), because the arguments employed will be useful elsewhere. Set $A = mx$, $B = nx$, with $m \neq n$ integers. Using (7) in (4) gives

$$\int_{-\pi}^{\pi} \cos mx \cos nx \, dx = \tfrac{1}{2} \int_{-\pi}^{\pi} \cos(m + n)x \, dx + \tfrac{1}{2} \int_{-\pi}^{\pi} \cos(m - n)x \, dx$$

$$= \tfrac{1}{2} \left(\frac{\sin(m + n)x}{m + n} \right) \Bigg|_{-\pi}^{\pi} + \tfrac{1}{2} \left(\frac{\sin(m - n)x}{m - n} \right) \Bigg|_{-\pi}^{\pi} = 0,$$

which proves the first part of (4).

The value of the integral when $m = n \neq 0$ and $m = n = 0$ cannot be obtained from this last result, because in each case a zero divisor occurs on the right-hand side. When $m = n \neq 0$ the difficulty is resolved by returning to (7), setting $A = B = mx$, with $m \neq 0$, and then substituting the result into (4).

This gives

$$\cos^2 mx = \tfrac{1}{2}[\cos 2mx + 1],$$

so

$$\int_{-\pi}^{\pi} \cos^2 mx \, dx = \tfrac{1}{2} \int_{-\pi}^{\pi} \cos 2mx + \tfrac{1}{2} \int_{-\pi}^{\pi} dx$$

$$= \tfrac{1}{2} \left(\frac{\sin 2mx}{2m} \right) \Bigg|_{-\pi}^{\pi} + (\tfrac{1}{2}x) \Bigg|_{-\pi}^{\pi} = \pi.$$

The corresponding value of the integral when $m = n = 0$ follows directly from (4), which reduces to

$$\int_{-\pi}^{\pi} 1 \cdot dx = 2\pi,$$

and our derivation of the results in (4) is now complete.

The appearance, after integration, of a factor in a denominator which can vanish for some value(s) of m is of frequent occurrence when working with Fourier series. This possibility should always be considered and dealt with, when necessary, by returning to the original integral for the value(s) of m involved.

In mathematical terms the three integrals (3), (4) and (5) are the conditions which ensure that the functions of the sequence $\{1, \cos x, \sin x, \cos 2x, \sin 2x, \ldots, \cos nx,$

$\sin nx, \dots$} are **orthogonal**[5] over the interval $-\pi \leq x \leq \pi$. This simply means that when any two functions from the sequence are multiplied together and integrated over the interval $-\pi \leq x \leq \pi$ the result will be zero if the functions are different, and a positive constant if they are identical.

To proceed with our argument we now determine the coefficient a_m in (2), with m considered to be some fixed integer, or zero. To accomplish this we multiply (2) by $\cos mx$ and integrate over the interval $-\pi \leq x \leq \pi$ to obtain

$$\int_{-\pi}^{\pi} f(x) \cos mx \, dx = \int_{-\pi}^{\pi} \cos mx \left[\tfrac{1}{2} a_0 + \sum_{n=1}^{\infty} (a_n \cos nx + b_n \sin nx) \right] dx. \qquad (9)$$

Assuming term by term integration is permissible (another reason why the argument is purely formal so far) this becomes

$$\int_{-\pi}^{\pi} f(x) \cos mx \, dx = \tfrac{1}{2} a_0 \int_{-\pi}^{\pi} 1 \cdot \cos mx \, dx + a_1 \int_{-\pi}^{\pi} \cos mx \cos x \, dx$$

$$+ b_1 \int_{-\pi}^{\pi} \cos mx \sin x \, dx + a_2 \int_{-\pi}^{\pi} \cos mx \cos 2x \, dx$$

$$+ b_2 \int_{-\pi}^{\pi} \cos mx \sin 2x \, dx + \dots$$

$$+ a_m \int_{-\pi}^{\pi} \cos^2 mx \, dx$$

$$+ b_m \int_{-\pi}^{\pi} \cos mx \sin mx \, dx + \dots . \qquad (10)$$

Appeal to integrals (4), (5) shows that every integral on the right-hand side of (10) vanishes, with the exception of the one multiplying a_m which equals π. Thus it follows that

$$\int_{-\pi}^{\pi} f(x) \cos mx \, dx = \pi a_m, \qquad \text{for } m = 1, 2, \dots ,$$

while if $m = 0$ this becomes

$$\int_{-\pi}^{\pi} f(x) \, dx = \tfrac{1}{2} a_0 \int_{-\pi}^{\pi} 1 \cdot dx = \pi a_0.$$

Thus the Fourier coefficients $a_0, a_1, \dots$ are determined in terms of $f(x)$ by the integral

$$a_n = \frac{1}{\pi} \int_{-\pi}^{\pi} f(x) \cos nx \, dx, \qquad (11)$$

for $n = 0, 1, 2, \dots$. In (11) we have replaced the index m by n, simply because it is usual to

[5] This is a generalization of the notion of the orthogonality of geometrical vectors, extended to a space of infinitely many dimensions, in which each component is a function taken from the sequence, and in which the scalar product is replaced by a definite integral.

denote the general coefficient by a_n rather than by a_m. Notice that it was necessary to distinguish between m and n in (10) because m was regarded as a fixed integer, whereas the summation index n runs from 1 to ∞.

At this point it is appropriate to remark that the factor $\frac{1}{2}$ was included in the first term of (2) because of the factor 2π occurring in the integral (4) when $m=n=0$. As a result a unification of the definition of the coefficients a_n is obtained for $n=0, 1, \ldots$.

A similar form of argument leads to the definition of the Fourier coefficients b_1, $b_2, \ldots$. If (2) is multiplied by $\sin mx$ for some fixed value of m, and integrated over the interval $-\pi \le x \le \pi$, we find after appeal to (3) and (5) that

$$b_n = \frac{1}{\pi} \int_{-\pi}^{\pi} f(x) \sin nx \, dx, \tag{12}$$

for $n=1, 2, \ldots$. Combining (11) and (12) brings us to the following definition.

Definition of the Fourier series of a continuous function on $(-\pi, \pi)$

Let $f(x)$ be a continuous function defined on the interval $-\pi \le x \le \pi$. Then the **Fourier series** of $f(x)$ is defined as the trigonometric series

$$\tfrac{1}{2}a_0 + \sum_{n=1}^{\infty} (a_n \cos nx + b_n \sin nx), \tag{13}$$

in which the Fourier coefficients are given by

$$a_n = \frac{1}{\pi} \int_{-\pi}^{\pi} f(x) \cos nx \, dx, \qquad \text{for } n=0, 1, 2, \ldots$$

and $\tag{14}$

$$b_n = \frac{1}{\pi} \int_{-\pi}^{\pi} f(x) \sin nx \, dx, \qquad \text{for } n=1, 2, \ldots .$$

The expressions for the Fourier coefficients a_n and b_n in (14) are called the **Euler formulas**. It should be recalled that although this definition was derived on the assumption that the equality sign in representation (2) is valid for every continuous function, this has still to be established.

Henceforth, because it will be necessary to consider functions $f(x)$ which are either continuous or piecewise continuous, we will denote the Fourier series of $f(x)$ by writing

$$f(x) \sim \tfrac{1}{2}a_0 + \sum_{n=1}^{\infty} (a_n \cos nx + b_n \sin nx).$$

The $\sim$ sign will be interpreted as an equality when $f(x)$ is continuous, but we will leave open for the moment its meaning at any point at which $f(x)$ is discontinuous.

Following the idea introduced in Sec. 9.1, the interval $-\pi \le x \le \pi$ in which the Fourier series representation of $f(x)$ is obtained will be called the **fundamental interval**. As a

Fourier series on this interval is a trigonometric series involving functions with the common period 2π, it follows that the Fourier series representation will itself define a function which is periodic with period 2π. Thus it will replicate the behavior of $f(x)$ in the fundamental interval in each adjoining interval of length 2π. Again following Sec. 9.1, the function defined by the Fourier series in each of these intervals will be called a **periodic extension** of $f(x)$.

As the Fourier series of $f(x)$ is defined for all x it will, in particular, be defined in any interval $\alpha \le x \le \alpha + 2\pi$ of length 2π. Consequently, because of periodicity, it must also be the Fourier series of that part of the function defined by periodic extension of $f(x)$ which lies within the interval. This situation is illustrated in Fig. 9.11 in which the graph of $f(x)$ in the fundamental interval is shown as a solid line, the graphs of periodic extensions are shown as dashed lines and the function represented by the Fourier series of $f(x)$ in the interval $\alpha \le x \le \alpha + 2\pi$ is shown as a dashed line.

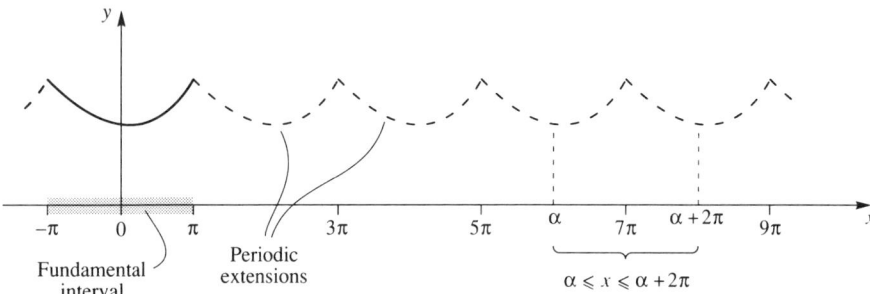

Fig. 9.11   The Fourier series of $f(x)$ in the fundamental interval $-\pi \le x \le \pi$ is also the Fourier series of the function defined by periodic extension and lying within the interval $\alpha \le x \le \alpha + 2\pi$.

Two elementary but very useful properties of Fourier series can be deduced directly from the definition of a Fourier series given above taken together with Theorem 9.2. The first of these is that if the function $f(x)$ defined on $-\pi \le x \le \pi$ is even, its Fourier series only comprises a sum of cosine terms with, possibly, an additive constant. The second is that if the function $f(x)$ defined on $-\pi \le x \le \pi$ is odd, its Fourier series only comprises a sum of sine terms.

The first result follows by noticing that if $f(x)$ is even, then $f(x) \sin nx$ is odd, and so by Theorem 9.2 the integral defining b_n in (14) must vanish for $n = 1, 2, \ldots$. The second result follows in similar fashion by noticing that if $f(x)$ is odd, then $f(x) \cos nx$ is odd, and so by Theorem 9.2 the integral defining a_n in (14) must vanish for $n = 0, 1, 2, \ldots$. A third simple but useful property of Fourier series follows by applying Theorem 9.1 to the Euler formulas (14), when either $f(x)$ is periodic with period 2π, or it is defined by periodic extension outside the fundamental interval $-\pi \le x \le \pi$. As under these circumstances the products $f(x) \cos nx$ and $f(x) \sin nx$ are periodic with period 2π, it follows immediately from Theorem 9.1 that in the Euler formulas (14) defining the Fourier coefficients a_n, b_n the interval of integration $-\pi \le x \le \pi$ may be replaced by the interval $\alpha - \pi \le x \le \alpha + \pi$, with α any real number.

The above results are sufficiently useful for them to be worth recording separately. It must, of course, be remembered that so far the arguments have been purely formal, and that they will remain so until the assumption of the convergence of the Fourier series of a continuous function $f(x)$ to the function $f(x)$ on the fundamental interval has been justified, and interpreted when $f(x)$ is discontinuous.

Elementary properties of Fourier series

1 The Fourier series representation of a function $f(x)$ defined in the fundamental interval $-\pi \leq x \leq \pi$ replicates the behavior of $f(x)$ in the fundamental interval in each adjoining interval of length 2π.

2 The Fourier series of $f(x)$ in the fundamental interval $-\pi \leq x \leq \pi$ is also the Fourier series of the function defined by periodic extension and lying within the interval $\alpha \leq x \leq \alpha + 2\pi$ for arbitrary real α.

3 If $f(x)$ is an even function defined in the interval $-\pi \leq x \leq \pi$, it follows from Theorem 9.2 that all its Fourier coefficients b_n vanish, so that $b_1 = b_2 = \ldots = 0$, and that its Fourier coefficients

$$a_n = \frac{2}{\pi} \int_0^\pi f(x) \cos nx \, dx, \qquad \text{for } n = 0, 1, \ldots .$$

4 If $f(x)$ is an odd function defined in the interval $-\pi \leq x \leq \pi$, it follows from Theorem 9.2 that all its Fourier coefficients a_n vanish, so that $a_0 = a_1 = \ldots = 0$ and that its Fourier coefficients

$$b_n = \frac{2}{\pi} \int_{-\pi}^\pi f(x) \sin nx \, dx, \qquad \text{for } n - 1, 2, \ldots .$$

5 If $f(x)$ is periodic with period 2π, or is defined outside the fundamental interval $-\pi \leq x \leq \pi$ by periodic extension, then

$$a_n = \frac{1}{\pi} \int_{-\pi}^\pi f(x) \cos nx \, dx = \frac{1}{\pi} \int_{\alpha - \pi}^{\alpha + \pi} f(x) \cos nx \, dx, \qquad \text{for } n = 0, 1, 2, \ldots ,$$

$$b_n = \frac{1}{\pi} \int_{-\pi}^\pi f(x) \sin nx \, dx = \frac{1}{\pi} \int_{\alpha - \pi}^{\alpha + \pi} f(x) \sin nx \, dx, \qquad \text{for } n = 1, 2, \ldots ,$$

for arbitrary real α.

6 The constant term $\frac{1}{2} a_0$ in the Fourier series representation (13) of $f(x)$ is

$$\frac{1}{2} a_0 = \frac{1}{2\pi} \int_{-\pi}^\pi f(x) \, dx,$$

which is seen to be the *average value* of $f(x)$ over its fundamental interval.

It should be remarked at this point that although a Fourier series is a trigonometric series, not every trigonometric series is a Fourier series. Only those trigonometric series whose coefficients a_n, b_n are determined by means of the Euler formulas (13) are Fourier series.

Thus the trigonometric series

$$\sum_{n=1}^{\infty} \left(\frac{\cos nx}{1+\ln n} + \frac{\sin nx}{3+2\ln n} \right)$$

which can be shown to be convergent for all x is *not* a Fourier series, because no function $f(x)$ exists which will generate the coefficients

$$a_n = \frac{1}{1+\ln n} \quad \text{and} \quad b_n = \frac{1}{3+2\ln n},$$

for $n = 1, 2, \ldots$.

Example 1. The Fourier series of a continuous function

Find the Fourier series representation of the function

$$f(x) = \begin{cases} 1, & \text{for } -\pi \le x \le 0 \\ 1+\sin 2x, & \text{for } 0 \le x \le \pi. \end{cases}$$

By assuming that the Fourier series of $f(x)$ converges to the function it represents whenever the function is continuous, show by substituting suitable values of x into the series that

$$\frac{1}{3} = \frac{1}{1\cdot 5} + \frac{1}{3\cdot 7} + \frac{1}{5\cdot 9} + \cdots,$$

and

$$\frac{\pi}{8} = \frac{1}{3} + \frac{1}{3\cdot 7} + \frac{1}{5\cdot 9} - \frac{1}{9\cdot 13} - \frac{1}{11\cdot 15} + \frac{1}{15\cdot 17} + \frac{1}{17\cdot 19} - \cdots.$$

Solution

The graph of $f(x)$ is shown in Fig. 9.12, from which it can be seen that the function is neither even nor odd, though the function together with its periodic extensions is everywhere continuous.

The Fourier coefficients a_n, b_n of $f(x)$ are determined by the Euler formulas given in (14). Thus it follows that the constant term

$$a_0 = \frac{1}{\pi} \int_{-\pi}^{0} 1 \cdot dx + \frac{1}{\pi} \int_{0}^{\pi} (1+\sin 2x)\, dx$$

$$= \frac{1}{\pi} \int_{-\pi}^{\pi} 1 \cdot dx + \frac{1}{\pi} \int_{0}^{\pi} \sin 2x\, dx$$

$$= \frac{1}{\pi} \cdot 2\pi + 0 = 2.$$

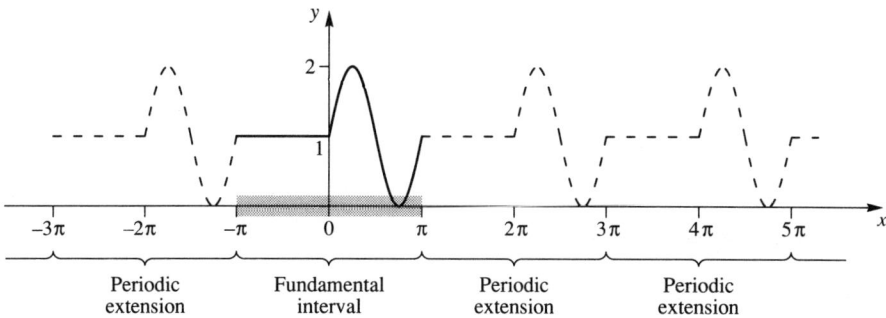

Fig. 9.12 The function $f(x)$ with its periodic extensions shown as dashed lines

Similarly, for $n \neq 0$,

$$a_n = \frac{1}{\pi} \int_{-\pi}^{0} \cos nx \, dx + \frac{1}{\pi} \int_{0}^{\pi} (1 + \sin 2x) \cos nx \, dx$$

$$= \frac{1}{\pi} \int_{-\pi}^{\pi} \cos nx \, dx + \frac{1}{\pi} \int_{0}^{\pi} \sin 2x \cos nx \, dx$$

$$= 0 + \frac{1}{\pi} \int_{0}^{\pi} \sin 2x \cos nx \, dx.$$

Using the trigonometric identity

$$\sin A \cos B = \tfrac{1}{2} [\sin(A + B) + \sin(A - B)],$$

and setting $A = 2x$, $B = nx$, we find that

$$a_n = \frac{1}{2\pi} \int_{0}^{\pi} \sin(2 + n)x \, dx + \frac{1}{2\pi} \int_{0}^{\pi} \sin(2 - n)x \, dx.$$

Integrating this last result gives

$$a_n = -\frac{1}{2\pi} \left(\frac{\cos(2+n)x}{(2+n)} \right) \Big|_{0}^{\pi} - \frac{1}{2\pi} \left(\frac{\cos(2-n)x}{(2-n)} \right) \Big|_{0}^{\pi}$$

$$= -\frac{1}{2\pi} \left(\frac{\cos(2+n)\pi - 1}{(2+n)} \right) - \frac{1}{2\pi} \left(\frac{\cos(2-n)\pi - 1}{(2-n)} \right),$$

which is true provided $n \neq 2$, because when $n = 2$ the second integration becomes invalid due to the presence of the divisor $(2 - n)$ which vanishes.

As $\cos(2 + n)\pi = (-1)^{2+n} = (-1)^n$, and $\cos(2 - n)\pi = (-1)^{2-n} = (-1)^{-n} = (-1)^n$, it follows that

$$a_n = \frac{2[1 + (-1)^{n+1}]}{\pi(2-n)(2+n)}, \qquad \text{for } n \neq 2.$$

Notice that the factor $[1 + (-1)^{n+1}]$ will be either 0 or 2, depending on whether n is even or odd.

To determine a_2 we return to the integral defining a_n and set $n=2$ to obtain

$$a_2 = \frac{1}{2\pi} \int_0^\pi \sin 4x \, dx + \frac{1}{2\pi} \int_0^\pi 0 \cdot dx.$$

However each of these integrals is zero, so

$$a_2 = 0.$$

Combining the above results we conclude that

$$a_n = \begin{cases} 2, & \text{for } n=0 \\ 0, & \text{for } n \text{ even} \\ \dfrac{4}{\pi(2-n)(2+n)}, & \text{for } n \text{ odd.} \end{cases}$$

Arguing in similar fashion we have

$$b_n = \frac{1}{\pi} \int_{-\pi}^0 \sin nx \, dx + \frac{1}{\pi} \int_0^\pi (1+\sin 2x) \sin nx \, dx$$

$$= \frac{1}{\pi} \int_{-\pi}^\pi \sin nx \, dx + \frac{1}{\pi} \int_0^\pi \sin 2x \sin nx \, dx$$

$$= 0 + \frac{1}{\pi} \int_0^\pi \sin 2x \sin nx \, dx.$$

Using the trigonometric identity

$$\sin A \sin B = \tfrac{1}{2}[\cos(A-B) - \cos(A+B)],$$

setting $A=2x$, $B=nx$ and substituting the result into the integral defining b_n gives

$$b_n = \frac{1}{2\pi} \int_0^\pi \cos(n-2)x \, dx + \frac{1}{2\pi} \int_0^\pi \cos(n+2)x \, dx$$

$$= \frac{1}{2\pi}\left(\frac{\sin(n-2)x}{(n-2)}\right)\Big|_0^\pi - \frac{1}{2\pi}\left(\frac{\sin(n+2)x}{(n+2)}\right)\Big|_0^\pi = 0,$$

provided $n \neq 2$, because when $n=2$ the first integration becomes invalid due to the presence of the divisor $n-2$ which vanishes.

To determine b_2 we return to the defining integrals above and set $n=2$. This leads to the expression

$$b_2 = \frac{1}{2\pi} \int_0^\pi 1 \cdot dx - \frac{1}{2\pi} \int_0^\pi \cos 4x \, dx$$

$$= \frac{1}{2\pi} \cdot \pi + 0 = \frac{1}{2},$$

and so

$$b_n = \begin{cases} \tfrac{1}{2}, & \text{for } n=2 \\ 0, & \text{for all other } n. \end{cases}$$

Collecting results shows that the Fourier coefficients of $f(x)$ are

$$a_n = \begin{cases} 2, & \text{for } n=0 \\ 0, & \text{for } n \text{ even} \\ \dfrac{4}{\pi(2-n)(2+n)}, & \text{for } n \text{ odd} \end{cases} \quad \text{and} \quad b_n = \begin{cases} \frac{1}{2}, & \text{for } n=2 \\ 0, & \text{for all other } n. \end{cases}$$

Now when n is an integer, $2n$ is always an even number and $2n-1$ an odd number, so we may replace n by $2n-1$ in the odd case and by $2n$ in the even case for $n = 1, 2, \ldots$, and redefine a_n in terms of the following expressions

$$a_0 = 2, \quad a_{2n-1} = \frac{4}{\pi(3-2n)(1+2n)} \quad \text{and} \quad a_{2n} = 0, \quad \text{for } n = 1, 2, \ldots.$$

Substituting the Fourier coefficients a_n, b_n into the Fourier series (13) shows that the required representation is

$$f(x) \sim 1 + \tfrac{1}{2}\sin 2x + \frac{4}{\pi} \sum_{n=1}^{\infty} \frac{\cos(2n-1)x}{(3-2n)(1+2n)}, \quad \text{for } -\pi \le x \le \pi.$$

If the Fourier series is assumed to converge to the function it represents everywhere that function is continuous, then setting x equal to a specific value, say $x = \xi$, in the series, will imply that the sum of the series is $f(\xi)$, and hence that the relational symbol $\sim$ may be replaced by an equality. As the function in question is continuous for $-\pi \le x \le \pi$, and $f(-\pi) = f(\pi)$, it follows that if this assumption is correct we may take ξ to be any real number (the function $f(x)$ together with its periodic extensions is everywhere continuous).

Setting $x = 0$ leads to the numerical result

$$1 = 1 + 0 + \frac{4}{\pi} \sum_{n=1}^{\infty} \frac{1}{(3-2n)(1+2n)},$$

which is equivalent to

$$\frac{1}{3} = \frac{1}{1 \cdot 5} + \frac{1}{3 \cdot 7} + \frac{1}{5 \cdot 9} + \cdots,$$

or to

$$\frac{1}{3} = \sum_{n=2}^{\infty} \frac{1}{(2n-3)(2n+1)},$$

which was the first result we were required to establish.

The second summation we are required to establish is arrived at by the same process, but this time by setting $x = \pi/6$. As a result, after replacing the relational symbol $\sim$ by an equality, we arrive at the expression

$$1 + \sin\frac{\pi}{3} = 1 + \frac{1}{2}\sin\frac{\pi}{3} + \frac{4}{\pi} \sum_{n=1}^{\infty} \frac{\cos(2n-1)\pi/6}{(3-2n)(1+2n)},$$

which is equivalent to

$$\frac{\pi}{8} = \frac{1}{\sin \pi/3} \sum_{n=1}^{\infty} \frac{\cos(2n-1)\pi/6}{(3-2n)(1+2n)}.$$

Now the pattern of terms in the numerator of this series is periodic, and the first few terms are listed below.

n	1	2	3	4	5	6	7	8	9	10
$\cos (2n-1)\pi/6$	$\sqrt{3}/2$	0	$-\sqrt{3}/2$	$-\sqrt{3}/2$	0	$\sqrt{3}/2$	$\sqrt{3}/2$	0	$-\sqrt{3}/2$	$-\sqrt{3}/2$

Thus as $\sin \pi/3 = \sqrt{3}/2$, the series is seen to reduce to the stated result

$$\frac{\pi}{8} = \frac{1}{3} + \frac{1}{3\cdot 7} + \frac{1}{5\cdot 9} - \frac{1}{9\cdot 13} - \frac{1}{11\cdot 15} + \frac{1}{15\cdot 19} + \frac{1}{17\cdot 21} - \cdots .$$

This example illustrates the care that must be exercised when determining Fourier coefficients should a trigonometric function be involved in the definition of $f(x)$. In deriving this particular Fourier series representation special consideration was necessary when $n=2$, because in this case the values of a_2 and b_2 were not obtainable from the general expressions for a_n and b_n.

Support for the assumption that the Fourier series representation of $f(x)$ converges to the function $f(x)$ itself wherever the function is continuous is provided by examining the behavior of the partial sums of the two related numerical series. The sum of the first 100 terms of the first series (easily seen to be convergent by the ratio test) is 0.330 833 which is approaching the exact sum of $\frac{1}{3}$ predicted on the basis of our assumption.

Furthermore, the sum of the first 100 terms of the second series (which is also convergent by the ratio test) is 0.392 724, which yields as an approximation to π the value 3.141 792. It is, in fact, necessary to sum approximately 300 terms of this series in order to determine π to an accuracy of six decimal places.

Additional evidence of the convergence of this particular Fourier series to the function $f(x)$ which it represents is provided by graphing the partial sums $s_n(x)$ of the Fourier series, defined in this case as

$$s_n(x) = 1 + \frac{1}{2}\sin 2x + \frac{4}{\pi} \sum_{m=1}^{n} \frac{\cos(2m-1)x}{(3-2m)(1+2m)}.$$

Thus the first three partial sums are

$$s_1(x) = 1 + \frac{1}{2}\sin 2x + \frac{4}{3\pi}\cos x,$$

$$s_2(x) = 1 + \frac{1}{2}\sin 2x + \frac{4}{\pi}\left[\frac{1}{3}\cos x - \frac{1}{5}\cos 3x\right],$$

$$s_3(x) = 1 + \frac{1}{2}\sin 2x + \frac{4}{\pi}\left[\frac{1}{3}\cos x - \frac{1}{5}\cos 3x - \frac{1}{21}\cos 5x\right].$$

The graph of $s_1(x)$ and $s_3(x)$ are shown in Fig. 9.13(a) and the graph of $s_5(x)$ in Fig. 9.13(b). The graph of $s_{10}(x)$ is virtually indistinguishable from the graph of $f(x)$ and so it has not been added to Fig. 9.13(b). It can be seen from the graphs that the representation improves steadily as n increases, and that even $s_5(x)$ provides a relatively good approximation to $f(x)$.

Notice the relatively slow convergence of the above numerical series, which is due to the fact that the denominator of their nth terms is only quadratic in n. This means, for example, that the second series is inconvenient for the numerical computation of π to a high accuracy; other series

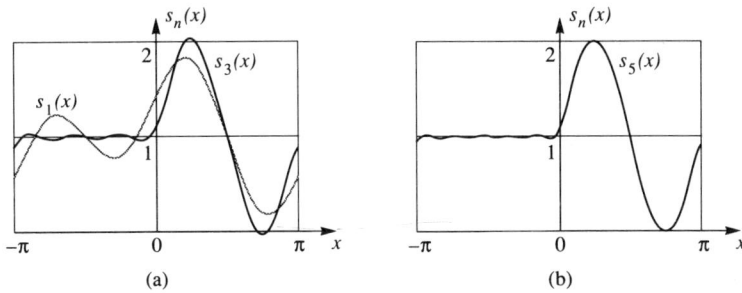

(a)   (b)

Fig. 9.13 Graphs of the approximations $s_1(x)$, $s_2(x)$ and $s_5(x)$

which converge more rapidly are available for the computation of π, as are methods for transforming series in such a way that their convergence is accelerated.

Property 2 of Fourier series listed above may be interpreted in terms of the Fourier series just obtained, namely

$$f(x) \sim 1 + \frac{1}{2}\sin 2x + \frac{4}{\pi}\sum_{n=1}^{\infty}\frac{\cos(2n-1)x}{(3-2n)(1+2n)}, \qquad \text{for } -\pi \le x \le \pi.$$

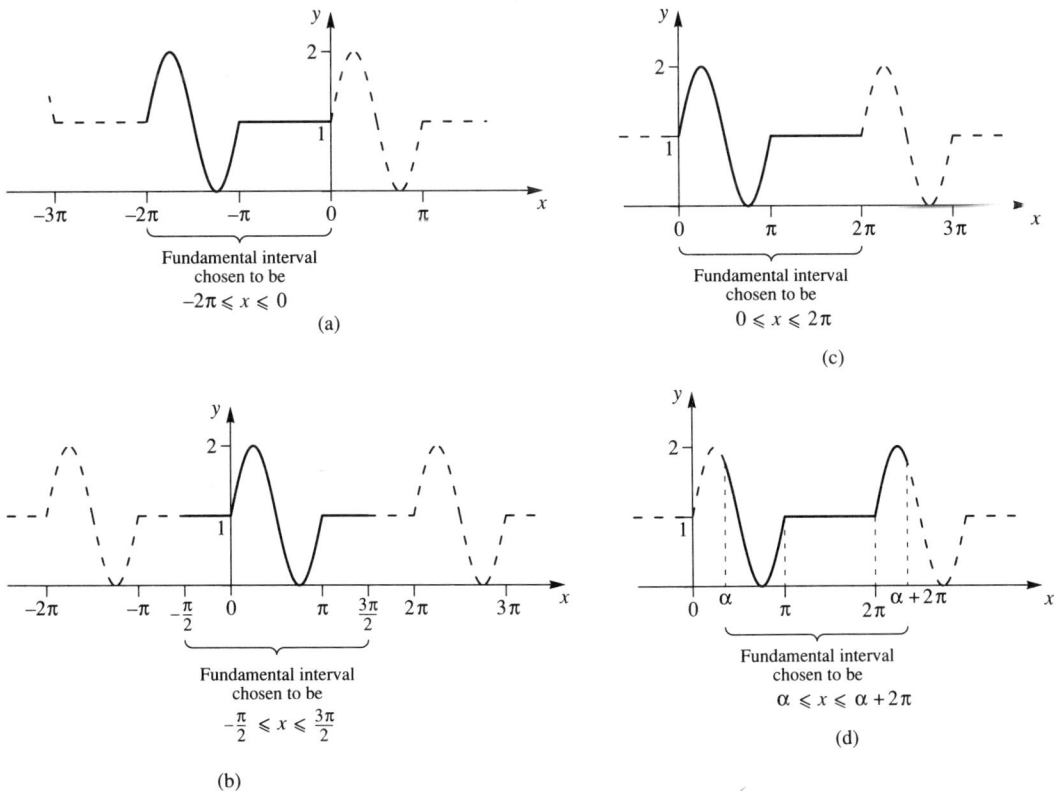

Fundamental interval
chosen to be
$-2\pi \le x \le 0$

(a)

Fundamental interval
chosen to be
$0 \le x \le 2\pi$

(c)

Fundamental interval
chosen to be
$-\frac{\pi}{2} \le x \le \frac{3\pi}{2}$

(b)

Fundamental interval
chosen to be
$\alpha \le x \le \alpha + 2\pi$

(d)

Fig. 9.14

It means, for example, that this same series represents the function defined by periodic extension in the intervals (a) $-2\pi \leq x \leq 0$, (b) $-\frac{1}{2}\pi \leq x \leq \frac{3}{2}\pi$, (c) $0 \leq x \leq 2\pi$ and, in general, $\alpha \leq x \leq \alpha + 2\pi$, with α arbitrary. The functions so represented are shown graphically in Fig. 9.14. ∎

As the integration of a piecewise continuous function presents no problems, it is immediately apparent that the definition of the Fourier series of a continuous function applies equally well to a piecewise continuous one. We now consider an example involving a piecewise continuous function which will serve to illustrate the treatment of such functions and the interpretation of the relational symbol $\sim$ in a Fourier series at points of discontinuity of $f(x)$.

Example 2. Fourier series of a discontinuous function

Find the Fourier series representation of the function

$$f(x) = \begin{cases} x + \pi & \text{for } -\pi \leq x < 0 \\ x - \pi & \text{for } 0 < x \leq \pi, \end{cases}$$

and graph its first few partial sums.

Solution

The graph of the function $f(x)$, which is an odd function because $f(-x) = -f(x)$, is shown as the full line in Fig. 9.15. Outside the fundamental interval $-\pi \leq x \leq \pi$ the periodic extensions of $f(x)$ are shown as dashed lines. Notice that $f(x)$ is discontinuous at the origin, at which point it has not been defined. Correspondingly, the periodic extensions are discontinuous at the points $x = \pm 2n\pi$ for $n = 1, 2, \ldots$ and, like $f(x)$ at the origin, they also are undefined at these points. We will see later that

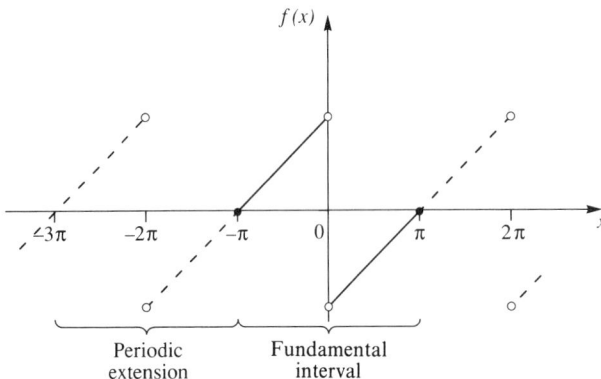

Fig. 9.15 $f(x)$ and its periodic extensions

the Fourier series representation of $f(x)$ assigns a value to $f(0)$, and hence to the value of its periodic extensions at $x = \pm 2n\pi$, independently of whether or not $f(0)$ has been defined.

As $f(x)$ is an odd function it follows from property 4 of Fourier series that its Fourier coefficients $a_n = 0$ for $n = 0, 1, \ldots$, while its coefficients b_n are given by

$$b_n = \frac{1}{\pi} \int_{-\pi}^{\pi} f(x) \sin nx \, dx$$

for $n = 1, 2, \ldots$.

Thus

$$b_n = \frac{1}{\pi} \int_{-\pi}^{0} (x + \pi) \sin nx \, dx + \frac{1}{\pi} \int_{0}^{\pi} (x - \pi) \sin nx \, dx,$$

and after combining integrals we find that

$$b_n = \int_{-\pi}^{0} \sin nx \, dx + \int_{0}^{\pi} \sin nx \, dx + \frac{1}{\pi} \int_{-\pi}^{\pi} x \sin nx \, dx.$$

Routine integration, coupled with use of the fact that $\cos n\pi = (-1)^n$, shows that

$$b_n = \frac{-2}{n}, \qquad \text{for } n = 1, 2, \ldots.$$

Hence the required Fourier series representation for $f(x)$ is

$$f(x) \sim -2 \sum_{n=1}^{\infty} \frac{\sin nx}{n}, \qquad \text{for } -\pi \leq x \leq \pi.$$

Graphs of typical partial sums of this Fourier series obtained by setting

$$s_n(x) = -2 \sum_{r=1}^{n} \frac{\sin rx}{r}$$

are shown in Fig. 9.16(a), (b), (c), to each of which has been added for purposes of comparison the graph of $f(x)$. To illustrate the convergence of $s_n(x)$ to $f(x)$ as n increases, the graph of $s_{100}(x)$ is shown in Fig. 9.16(d) for $-1 \leq x \leq 0$. The first three partial sums are

$$s_1(x) = -2 \sin x \qquad\qquad s_2(x) = -2 (\sin x + \tfrac{1}{2} \sin 2x)$$

$$s_3(x) = -2 (\sin x + \tfrac{1}{2} \sin 2x + \tfrac{1}{3} \sin 3x).$$

The graphs suggest the convergence of $s_n(x)$ to $f(x)$ as n increases at all points other than $x = 0$, at which point $f(x)$ is discontinuous. As $s_n(0) = 0$ for $n = 1, 2, \ldots$ it is clear that this Fourier series will assign to $f(x)$ at $x = 0$ the value $f(0) = 0$, despite the fact that $f(x)$ is discontinuous there and the function was not even defined at the point. Thus with this discontinuous function it appears reasonable to assume that the relational symbol $\sim$ in the Fourier series is to be interpreted as an equality wherever the function is continuous. At the origin, however, the function is discontinuous and the behavior is different. It appears that there the relational symbol $\sim$ is to be interpreted as indicating that the Fourier series converges to the mean of the values of $f(x)$ to the immediate left and right of the discontinuity.

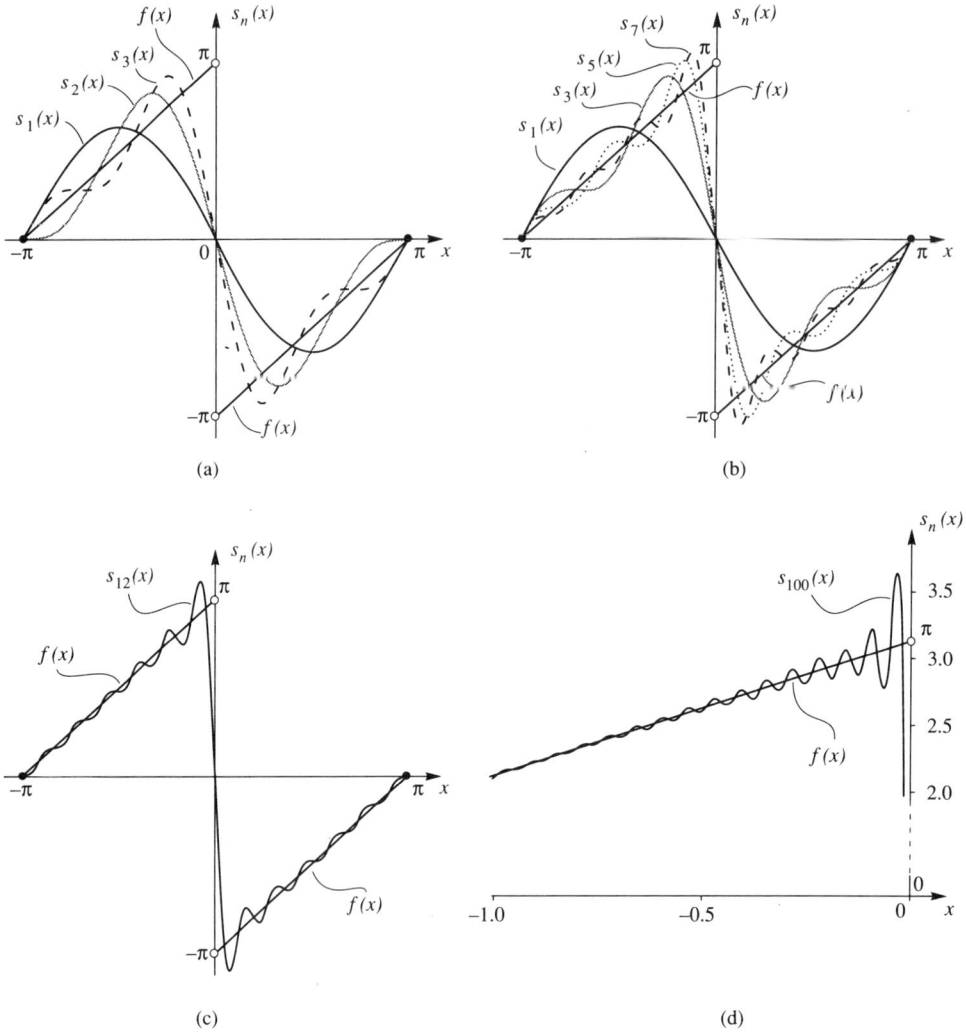

Fig. 9.16 Graphs of partial sums $s_n(x)$

Consideration of the graphical results of Examples 1 and 2 suggests the following general properties of Fourier series:

(a) the Fourier series of $f(x)$ converges to $f(x)$ at points of continuity of $f(x)$;

(b) at a point of discontinuity of $f(x)$ the Fourier series of $f(x)$ converges to the mean of the values of $f(x)$ to the immediate left and right of the discontinuity. Furthermore, the value to which the Fourier series converges at a point of discontinuity of $f(x)$ is independent of whether or not $f(x)$ is defined at that point, and of its value there if it is defined (it converges to the value zero at $x = 0$ in Fig. 9.16);

(c) there is always an overshoot of the partial sums $s_n(x)$ on either side of a jump discontinuity, and the overshoot persists as n increases. However, the larger n becomes, the narrower becomes the peak of the overshoot, and the closer it moves towards the discontinuity.[6]

Many numerical series can be summed directly by means of Fourier series, as in the previous examples, or by using Fourier series in conjunction with the Parseval relation which will be discussed in Sec. 9.3. Closed form sums of a few numerical series obtainable in this manner, which are often useful when working with Fourier series, are listed below.

Some useful numerical series

1 $\quad \dfrac{\pi}{4} = \displaystyle\sum_{n=1}^{\infty} \dfrac{(-1)^{n+1}}{n} = 1 - \dfrac{1}{2} + \dfrac{1}{3} - \dfrac{1}{4} + \cdots$

2 $\quad \dfrac{\pi^2}{12} = \displaystyle\sum_{n=1}^{\infty} \dfrac{(-1)^{n+1}}{n^2} = 1 - \dfrac{1}{2^2} + \dfrac{1}{3^2} - \dfrac{1}{4^2} + \cdots$

3 $\quad \dfrac{\pi^2}{6} = \displaystyle\sum_{n=1}^{\infty} \dfrac{1}{n^2} = 1 + \dfrac{1}{2^2} + \dfrac{1}{3^2} + \dfrac{1}{4^2} + \cdots$

4 $\quad \dfrac{\pi^2}{8} = \displaystyle\sum_{n=1}^{\infty} \dfrac{1}{(2n-1)^2} = 1 + \dfrac{1}{3^2} + \dfrac{1}{5^2} + \dfrac{1}{7^2} + \cdots$

5 $\quad \dfrac{\pi^3}{32} = \displaystyle\sum_{n=1}^{\infty} \dfrac{(-1)^{n+1}}{(2n-1)^3} = 1 - \dfrac{1}{3^3} + \dfrac{1}{5^3} - \dfrac{1}{7^3} + \cdots$

6 $\quad \dfrac{\pi^4}{96} = \displaystyle\sum_{n=1}^{\infty} \dfrac{1}{(2n-1)^4} = 1 + \dfrac{1}{3^4} + \dfrac{1}{5^4} + \dfrac{1}{7^4} + \cdots$

7 $\quad \dfrac{\pi^4}{90} = \displaystyle\sum_{n=1}^{\infty} \dfrac{1}{n^4} = 1 + \dfrac{1}{2^4} + \dfrac{1}{3^4} + \dfrac{1}{4^4} + \cdots$

8 $\quad \pi^2 = 8 + 16 \displaystyle\sum_{n=1}^{\infty} \dfrac{1}{(2n-1)^2 (2n+1)^2} = 8 + 16\left(\dfrac{1}{1\cdot 3^2} + \dfrac{1}{3^2 \cdot 5^2} + \dfrac{1}{5^2 \cdot 7^2} + \cdots \right)$

9 $\quad \dfrac{5\pi^5}{1536} = \displaystyle\sum_{n=1}^{\infty} \dfrac{(-1)^{n+1}}{(2n-1)^5} = 1 - \dfrac{1}{3^5} + \dfrac{1}{5^5} - \dfrac{1}{7^5} + \cdots$

10 $\quad \dfrac{1}{3} = \displaystyle\sum_{n=2}^{\infty} \dfrac{1}{(2n-3)(2n+1)} = \dfrac{1}{1\cdot 5} + \dfrac{1}{3\cdot 7} + \dfrac{1}{5\cdot 9} + \dfrac{1}{7\cdot 11} + \cdots$

[6] This is typical of the behavior of partial sums of Fourier series at a jump discontinuity, and Ex. 2 provides a specific illustration of what is called the **Gibbs phenomenon**. This phenomenon reflects the nonuniformity of the convergence of the partial sums of Fourier series at points where $f(x)$ has a jump discontinuity. This overshoot phenomenon is named after JOSIAH WILLARD GIBBS (1839–1903), a prominent American mathematician who while Professor of Mathematical Physics at Yale became the first person to study the effect. His most important contributions to applied mathematics were to statistical mechanics, thermodynamics and the development of vector analysis.

Fourier series on (*a, b*)

Let us now resume the formal development of the most important ideas underlying Fourier series, while keeping in mind the interpretation of the relational symbol $\sim$ suggested by the previous examples. The justification for this will follow once the convergence properties of Fourier series have been considered in sufficient detail.

It is often necessary to work with Fourier series defined either in the interval $-L \leq x \leq L$ or the interval $a \leq x \leq b$. We will now derive the form of the expansion in the interval $-L \leq x \leq L$, and then use Theorem 9.1 to extend the result to the general interval $a \leq x \leq b$.

Let a piecewise smooth or piecewise continuous function $f(x)$ be defined on the interval $-L \leq x \leq L$. Then the transformation $t = \pi x/L$ maps the interval $-L \leq x \leq L$ into the interval $-\pi \leq t \leq \pi$. Thus setting $f(x) = f(Lt/\pi) = F(t)$, it follows that $F(t)$ has the Fourier series representation in the interval $-\pi \leq t \leq \pi$

$$F(t) \sim \tfrac{1}{2}a_0 + \sum_{n=1}^{\infty} (a_n \cos nt + b_n \sin nt),$$

where

$$a_n = \frac{1}{\pi} \int_{-\pi}^{\pi} F(t) \cos nt \, dt, \qquad \text{for } n = 0, 1, \dots,$$

and

$$b_n = \frac{1}{\pi} \int_{-\pi}^{\pi} F(t) \sin nt \, dt, \qquad \text{for } n = 1, 2, \dots.$$

Changing back to the original variable x we find that the required Fourier series representation in the interval $-L \leq x \leq L$ becomes

$$f(x) \sim \frac{1}{2}a_0 + \sum_{n=1}^{\infty} \left(a_n \cos \frac{n\pi x}{L} + b_n \sin \frac{n\pi x}{L} \right), \tag{15}$$

where now

$$a_n = \frac{1}{L} \int_{-L}^{L} f(x) \cos \frac{n\pi x}{L} \, dx, \qquad \text{for } n = 0, 1, \dots, \tag{16}$$

and

$$b_n = \frac{1}{L} \int_{-L}^{L} f(x) \sin \frac{n\pi x}{L} \, dx, \qquad \text{for } n = 1, 2, \dots. \tag{17}$$

To extend Fourier series (15) and the modified Euler formulas (16) and (17) to the interval $a \leq x \leq b$ we first need to recognize that an application of Theorem 9.1 to (16) and (17) with $X = 2L$ allows us to shift the interval of integration from $[-L, L]$ to $[0, 2L]$. A further application of the theorem to the modified integrals with $\alpha = a$ and the period

$X = 2L = b - a$ gives the new result that

$$f(x) \sim \frac{1}{2} a_0 + \sum_{n=1}^{\infty} \left(a_n \cos \frac{2n\pi x}{b-a} + b_n \sin \frac{2n\pi x}{b-a} \right), \tag{18}$$

where

$$a_n = \frac{2}{b-a} \int_a^b f(x) \cos \frac{2n\pi x}{b-a} \, dx, \qquad \text{for } n = 0, 1, \ldots, \tag{19}$$

and

$$b_n = \frac{2}{b-a} \int_a^b f(x) \sin \frac{2n\pi x}{b-a} \, dx, \qquad \text{for } n = 1, 2, \ldots, \tag{20}$$

where $a \le x \le b$.

This last result comprises the next theorem.

Theorem 9.4 (The Fourier series of $f(x)$ defined on (a, b))

Let $f(x)$ be either a piecewise smooth or a piecewise continuous function defined for $a \le x \le b$. Then its Fourier series representation with $a \le x \le b$ as the fundamental interval is

$$f(x) \sim \frac{1}{2} a_0 + \sum_{n=1}^{\infty} \left(a_n \cos \frac{2n\pi x}{b-a} + b_n \sin \frac{2n\pi x}{b-a} \right),$$

where

$$a_n = \frac{?}{b-a} \int_a^b f(x) \cos \frac{2n\pi x}{b-a} \, dx, \qquad \text{for } n = 0, 1, \ldots,$$

and

$$b_n = \frac{2}{b-a} \int_a^b f(x) \sin \frac{2n\pi x}{b-a} \, dx, \qquad \text{for } n = 1, 2, \ldots.$$

Corollary

The Fourier series representation of a function $f(x)$ defined on the fundamental interval $-L \le x \le L$ is

$$f(x) \sim \frac{1}{2} a_0 + \sum_{n=1}^{\infty} \left(a_n \cos \frac{n\pi x}{L} + b_n \sin \frac{n\pi x}{L} \right),$$

where

$$a_n = \frac{1}{L} \int_{-L}^{L} f(x) \cos \frac{n\pi x}{L} \, dx, \qquad \text{for } n = 0, 1, \ldots,$$

and

$$b_n = \frac{1}{L} \int_{-L}^{L} f(x) \sin \frac{n\pi x}{L} \, dx, \qquad \text{for } n = 1, 2, \ldots .$$ □

Example 3. A Fourier series on (a, b)

Find the Fourier series of

$$f(x) = \begin{cases} 2-x, & \text{for } 0 < x \le 2 \\ 0, & \text{for } 2 < x \le 4. \end{cases}$$

Solution

The fundamental interval for this function is $0 < x \le 4$. The graph of $f(x)$ is shown as the full line in Fig. 9.17, while its periodic extensions are shown as the dashed lines.

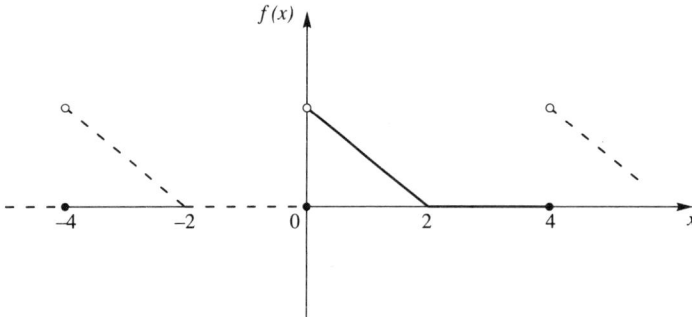

Fig. 9.17

Using the notation of Theorem 9.4 it follows that $a = 0$, $b = 4$, so that the required Fourier series representation becomes

$$f(x) \sim \frac{1}{2} a_0 + \sum_{n=1}^{\infty} \left(a_n \cos \frac{n\pi x}{2} + b_n \sin \frac{n\pi x}{2} \right)$$

in the fundamental interval $0 < x \le 4$, with

$$a_n = \frac{1}{2} \int_0^4 f(x) \cos \frac{n\pi x}{2} \, dx, \qquad \text{for } n = 0, 1, \ldots ,$$

and

$$b_n = \frac{1}{2} \int_0^4 f(x) \sin \frac{n\pi x}{2} \, dx, \qquad \text{for } n = 1, 2, \ldots .$$

Routine integration then establishes that

$$a_0 = 1, \quad a_n = \frac{2}{n^2 \pi^2} [1 - (-1)^n], \qquad \text{for } n = 1, 2, \ldots ,$$

and

$$b_n = 2/(n\pi), \qquad\qquad \text{for } n = 1, 2, \ldots .$$

The required Fourier series is thus

$$f(x) \sim \frac{1}{2} + \frac{2}{\pi} \sum_{n=1}^{\infty} \left(\frac{[1-(-1)^n]}{n^2 \pi} \cos \frac{n\pi x}{2} + \frac{1}{n} \sin \frac{n\pi x}{2} \right),$$

for $0 < x \le 4$.

Figure 9.18 shows the graph of the nth partial sum

$$s_n(x) = \frac{1}{2} + \frac{2}{\pi} \sum_{r=1}^{n} \left(\frac{[1-(-1)^r]}{r^2 \pi} \cos \frac{r\pi x}{2} + \frac{1}{r} \sin \frac{r\pi x}{2} \right)$$

corresponding to $n = 10$. The Gibbs phenomenon is again seen to be present at the jump discontinuities, which this time occur at $x = 0$ and $x = 4$. Here again, the value to which the Fourier series converges at the jump discontinuity is seen to be the mean of the values of $f(x)$ to the immediate left and right of the discontinuity, so that in this case it converges to the value 1 (remember the Fourier series is periodic, so the graph in Fig. 9.18 is periodic with period 4).

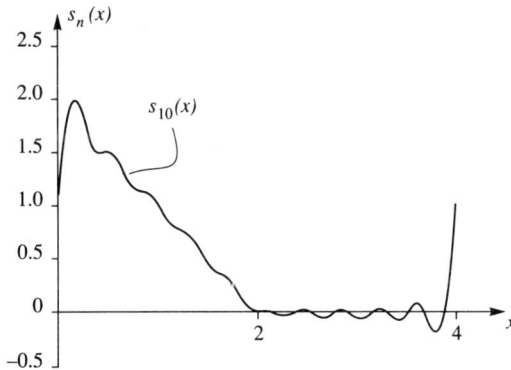

Fig. 9.18 Graph of $s_{10}(x)$ for Ex. 3

Let us assume this Fourier series converges to the value 1 at $x = 0$, as indicated by the graph in Fig. 9.18. Then if we set $x = 0$ in the right-hand side of the above Fourier series representation of $f(x)$, it follows that we must also replace the relational symbol $\sim$ by an equality and set $f(0) = 1$. When this is done we arrive at the series

$$1 = \frac{1}{2} + \frac{2}{\pi^2} \sum_{n=1}^{\infty} \frac{(1-(-1)^n)}{n^2},$$

which is equivalent to the result

$$\frac{\pi^2}{8} = \sum_{n=1}^{\infty} \frac{1}{(2n-1)^2}.$$

This illustrates one of the ways in which the Fourier series of a simple function may be used to evaluate the sum of a series. By means of a simple device it is possible to use the last result to sum a related series. Setting

$$S = 1 + \frac{1}{2^2} + \frac{1}{3^2} + \dots ,$$

it follows that

$$\frac{1}{4}S = \frac{1}{2^2} + \frac{1}{4^2} + \frac{1}{6^2} + \dots .$$

Summing this result and the series for $\pi^2/8$ gives

$$\frac{1}{8}\pi^2 + \frac{1}{4}S = 1 + \frac{1}{2^2} + \frac{1}{3^2} + \dots = S.$$

Thus $S = \pi^2/6$ and we have proved the useful result that

$$\frac{\pi^2}{6} = \sum_{n=1}^{\infty} \frac{1}{n^2}.$$

■

Fourier sine and cosine series

It is possible to expand a function $f(x)$ which is either piecewise smooth or piecewise continuous in the interval $-L \leq x \leq L$ either as a series of sines, or as a series of cosines. To see how this is accomplished let us suppose first that a sine series is required.

We define a new function $F(x)$ on $-L \leq x \leq L$ by the requirement that[7]

$$F(x) = \begin{cases} f(x), & \text{for} \quad 0 \leq x \leq L \\ -f(-x), & \text{for} \quad -L \leq x \leq 0. \end{cases} \tag{21}$$

Then clearly $F(x)$ is an *odd* function with the property that it coincides with the function $f(x)$ for $0 \leq x \leq L$ (except perhaps at the origin). A typical situation is shown in Fig. 9.19, in which the function $F(x)$ is seen to be discontinuous at the origin because $f(0+) \neq 0$.

The function $F(x)$ may now be expanded as a Fourier series on $-L \leq x \leq L$, but since it is an odd function the coefficients a_n will all vanish. It thus follows from the Corollary to Theorem 9.4 that for $-L \leq x \leq L$

$$F(x) \sim \sum_{n=1}^{\infty} b_n \sin \frac{n\pi x}{L}, \tag{22}$$

with

$$b_n = \frac{2}{L} \int_0^L f(x) \sin \frac{n\pi x}{L} dx, \tag{23}$$

for $n = 1, 2, \dots .$

---

[7] If $f(0) \neq 0$ we define $F(x)$ at the origin by the requirement that $F(0) = 0$. This makes $F(x)$ an odd function for all x in $-\pi \leq x \leq \pi$ and is in agreement with the convergence property of a Fourier series at a point of discontinuity of the function it represents.

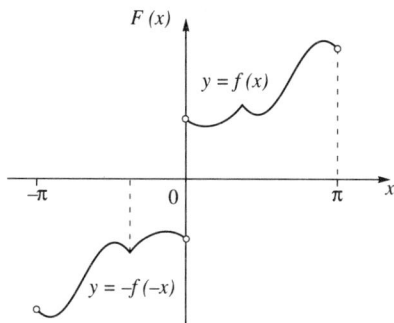

Fig. 9.19 $F(x)$ defined by extending $f(x)$ as an odd function

If, now, we confine attention to the interval $0 \le x \le L$ in which $f(x)$ and $F(x)$ coincide (except possibly at the origin), it follows from (22) that the **Fourier sine series** representation of $f(x)$ in the interval $0 \le x \le L$, also called the **half-range sine series** of $f(x)$ is

$$f(x) \sim \sum_{n=1}^{\infty} b_n \sin \frac{n\pi x}{L}, \tag{24}$$

with

$$b_n = \frac{2}{L} \int_0^L f(x) \sin \frac{n\pi x}{L} \, dx, \tag{25}$$

for $n = 1, 2, \ldots$.

If, instead of using (21), $F(x)$ is defined as an *even* function by setting

$$F(x) = \begin{cases} f(x), & \text{for} \quad 0 \le x \le L \\ f(-x), & \text{for} \quad -L \le x \le 0 \end{cases} \tag{26}$$

it will have a continuous graph of the form shown in Fig. 9.20.

An obvious extension of the form of argument just used shows that now all the coefficients b_n vanish and the Fourier series expansion of $F(x)$ on $-L \le x \le L$ becomes

$$F(x) \sim \frac{1}{2} a_0 + \sum_{n=1}^{\infty} a_n \cos \frac{n\pi x}{L},$$

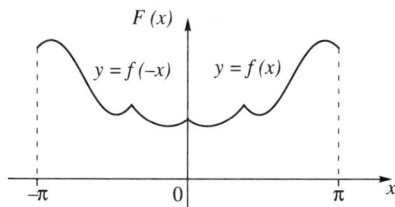

Fig. 9.20 $F(x)$ defined by extending $f(x)$ as an even function

with

$$a_n = \frac{2}{L} \int_0^L f(x) \cos \frac{n\pi x}{L} \, dx,$$

for $n = 0, 1, \ldots$.

By confining attention to the interval $0 \leq x \leq L$ in which $f(x)$ and $F(x)$ coincide it follows directly that the **Fourier cosine series** representation of $f(x)$ in the interval $0 \leq x \leq L$, also called the **half-range cosine series** of $f(x)$ is

$$f(x) \sim \frac{1}{2} a_0 + \sum_{n=1}^{\infty} a_n \cos \frac{n\pi x}{L}, \qquad (27)$$

with

$$a_n = \frac{2}{L} \int_0^L f(x) \cos \frac{n\pi x}{L} \, dx, \qquad (28)$$

for $n = 0, 1, \ldots$.   □

Example 4. Fourier sine and cosine series representations

Find the Fourier sine and cosine representations of

$$f(x) = \sin \tfrac{1}{2} x,$$

for $0 \leq x \leq \pi$.

Solution

To obtain the Fourier sine series use must be made of (24) and (25) with $L = \pi$ and $f(x) = \sin \tfrac{1}{2} x$. This leads to the result

$$f(x) \sim \sum_{n=1}^{\infty} b_n \sin nx,$$

for $0 \leq x \leq \pi$, with

$$b_n = \frac{2}{\pi} \int_0^\pi \sin \tfrac{1}{2} x \sin nx \, dx,$$

for $n = 1, 2, \ldots$.

Straightforward integration shows

$$b_n = (-1)^{n+1} \frac{8}{\pi} \left(\frac{n}{4n^2 - 1} \right),$$

so that the Fourier sine series representation of $f(x)$ becomes

$$f(x) \sim \sum_{n=1}^{\infty} (-1)^{n+1} \left(\frac{n}{4n^2 - 1} \right) \sin nx, \qquad \text{for } 0 \leq x \leq \pi.$$

Similarly, the Fourier cosine series representation follows from (27) and (28) as

$$f(x) \sim \tfrac{1}{2}a_0 + \sum_{n=1}^{\infty} a_n \cos nx,$$

for $0 \leq x \leq \pi$, with

$$a_n = \frac{2}{\pi} \int_0^\pi \sin\frac{1}{2}x \cos nx \, dx,$$

for $n = 0, 1, \ldots$.

Integration gives the results

$$a_0 = \frac{4}{\pi} \quad \text{and} \quad a_n = -\frac{4}{\pi}\left(\frac{1}{4n^2 - 1}\right),$$

for $n = 1, 2, \ldots$, and so the Fourier cosine representation of $f(x)$ becomes

$$f(x) \sim \frac{2}{\pi} - \frac{4}{\pi} \sum_{n=1}^{\infty} \left(\frac{1}{4n^2 - 1}\right) \cos nx, \qquad \text{for } 0 \leq x \leq \pi. \qquad \blacksquare$$

The complex form of a Fourier series

It is possible to re-express Fourier series in terms of the Euler formula $e^{i\theta} = \cos\theta + i\sin\theta$, and when making applications of Fourier series this often leads to a simplification of the analysis.

To arrive at the complex form of a Fourier series let us suppose that a piecewise smooth or piecewise continuous function $f(x)$ defined for $-\pi \leq x \leq \pi$ has the Fourier coefficients a_n, b_n. Then we may always write the general term of the Fourier series in the form

$$a_n \cos nx + b_n \sin nx = a_n \left(\frac{e^{inx} + e^{-inx}}{2}\right) + b_n \left(\frac{e^{inx} - e^{-inx}}{2i}\right)$$

$$= c_n e^{inx} + c_{-n} e^{-inx}, \qquad (29)$$

where

$$c_0 = \frac{1}{2}a_0 \quad \text{and} \quad c_n = \frac{a_n - ib_n}{2}, \; c_{-n} = \frac{a_n + ib_n}{2}, \qquad (30)$$

for $n = 1, 2, \ldots$.

Recalling the definitions of a_n, b_n in (14) we can thus write

$$c_n = \frac{1}{2\pi} \int_{-\pi}^{\pi} f(x)(\cos nx - i\sin nx) \, dx = \frac{1}{2\pi} \int_{-\pi}^{\pi} f(x) e^{-inx} \, dx$$

and

$$c_{-n}=\frac{1}{2\pi}\int_{-\pi}^{\pi}f(x)(\cos nx+i\sin nx)\,dx=\frac{1}{2\pi}\int_{-\pi}^{\pi}f(x)\,e^{inx}\,dx.$$

When combined, these results show that

$$c_{n}=\frac{1}{2\pi}\int_{-\pi}^{\pi}f(x)\,e^{-inx}\,dx, \tag{31}$$

for $n=0,\pm1,\pm2,\ldots$. It is also important to observe that when $f(x)$ is a real function its Fourier coefficients a_n, b_n will be real, so that the complex numbers c_{-n} and c_n defined in (31) must be complex conjugates, and hence

$$\bar{c}_{-n}=c_n. \tag{32}$$

The mth partial sum $s_m(x)$ of the Fourier series of $f(x)$ defined for $-\pi\le x\le\pi$ is

$$s_{m}(x)=\tfrac{1}{2}a_0+\sum_{n=1}^{m}(a_n\cos nx+b_n\sin nx). \tag{33}$$

Thus, in terms of (29) and (31), we have the result

$$s_{m}(x)=\sum_{n=-m}^{m}c_n\,e^{inx}. \tag{34}$$

The complex form of the Fourier series of $f(x)$ now follows by letting $m\to\infty$ to obtain

$$f(x)\sim\tfrac{1}{2}a_0+\sum_{n=1}^{\infty}(a_n\cos nx+b_n\sin nx)=\lim_{m\to\infty}\sum_{n=-m}^{m}c_n\,e^{inx}, \tag{35}$$

in which the complex numbers c_n defined by (31) are called the complex Fourier coefficients of $f(x)$.

Taken together with the above result, the arguments which were used to establish the form of the Fourier series of a function $f(x)$ defined on the interval $(-L, L)$ lead to our next theorem which generalizes (31) and (35).

Theorem 9.5 (Complex form of a Fourier series on $(-L\le x\le L)$)

Let $f(x)$ be a piecewise smooth or piecewise continuous function defined on the fundamental interval $0\le x\le L$, and by periodic extension outside it. Then for all real x the complex Fourier series representation of $f(x)$ is

$$f(x)\sim\lim_{m\to\infty}\sum_{n=-m}^{m}c_n\,e^{in\pi x/L},$$

where

$$c_{n}=\frac{1}{2L}\int_{-L}^{L}f(x)\,e^{-in\pi x/L}\,dx,$$

for $n = 0 \pm 1, \pm 2, \ldots$. The integral determining c_n may be evaluated over any interval of length $2L$. $\square$

In many applications of Fourier series involving the oscillation of physical systems, x represents the time t and the function of time $f(t)$ is periodic with period T. The angular frequency Ω of the oscillation $f(t)$ exciting the system is thus

$$\Omega = \frac{2\pi}{T} \text{ rad/unit time.} \tag{36}$$

Setting $L = \pi/\Omega$ in Theorem 9.5 and replacing x by t brings it to the form

$$f(t) \sim \lim_{m \to \infty} \sum_{n=-m}^{m} c_n e^{in\Omega t}, \tag{37}$$

where

$$c_n = \frac{\Omega}{2\pi} \int_0^{2\pi/\Omega} f(t) e^{-in\Omega t} \, dt, \tag{38}$$

for $n = 0, \pm 1, \pm 2, \ldots$.

In terms of the complex Fourier series representation of a function, the input oscillation $f(t)$, described in the time domain by a single graph of $f(t)$ against t, is characterized completely in the frequency domain in terms of $n\Omega$ by knowledge of the complex Fourier coefficients c_n, $n = 0, \pm 1, \pm 2, \ldots$. That is, once the c_n are known, $f(t)$ follows immediately from (37). The representation of $f(t)$ in terms of its complex Fourier coefficients c_n, each of which corresponds respectively to an angular frequency $n\Omega$, is called the **frequency spectrum** of $f(t)$.

As the numbers c_n are complex, two graphs are needed to display the frequency spectrum in the real plane. Writing c_n in modulus and argument form as

$$c_n = r_n \exp(i\theta_n), \qquad \text{for } n = 0, \pm 1, \pm 2, \ldots . \tag{39}$$

we will call $r_n = |c_n|$ the **amplitude** and $\theta_n = \arg c_n$ the **phase** of the component of the complex Fourier series with angular frequency $n\Omega$.

The graph of r_n against $n\Omega$ is called the **amplitude spectrum**, and the graph of θ_n against $n\Omega$ the **phase spectrum** of $f(t)$. These graphs are *discrete graphs*, because the coefficients c_n are only defined for $n = 0, \pm 1, \pm 2, \ldots$. Since $\bar{c}_{-n} = c_n$, it suffices to graph the amplitude and phase spectra for $n = 0, 1, 2, \ldots$.

Example 5. Complex Fourier series and frequency spectrum

Find the complex Fourier series representation of the function $f(t)$ with period 2 defined by

$$f(t) = \cosh t, \qquad \text{for } -1 \le t \le 1,$$

and determine its frequency spectrum.

Solution

The function is smooth, so the above representation may be used. Setting $\Omega = \pi$, because the period $T = 2$ and $\Omega = 2\pi/T$, we obtain

$$f(t) \sim \lim_{m \to \infty} \sum_{n=-m}^{\infty} c_n e^{in\pi t},$$

where

$$c_n = \tfrac{1}{2} \int_0^2 \cosh t \, e^{-in\pi t} \, dt,$$

for $n = 0, \pm 1, \pm 2, \ldots$.

Using the fact that $\cosh t = (e^t + e^{-t})/2$ we have

$$c_n = \frac{1}{4} \int_0^2 \{\exp[(1-n\pi)t] + \exp[-(1+in\pi)t]\} \, dt$$

$$= \frac{1}{4} \left[\left(\frac{\exp[(1-n\pi)t]}{1-in\pi} \right) - \left(\frac{\exp[-(1+in\pi)t]}{1+in\pi} \right) \right]\Big|_0^2$$

$$= \frac{1}{4} \left(\frac{e^2 e^{-i2n\pi} - 1}{1-in\pi} - \frac{e^{-2} e^{-i2n\pi} - 1}{1+in\pi} \right).$$

However $e^{-i2n\pi} = \cos 2n\pi - i \sin 2n\pi = 1$, so

$$c_n = \frac{1}{4} \left(\frac{e^2 - 1}{1-in\pi} - \frac{e^{-2} - 1}{1+in\pi} \right)$$

$$= \frac{1}{4} \left(\frac{(e^2 - 1 + in\pi \, e^2 - in\pi) - (e^{-2} - 1 - in\pi e^{-2} + in\pi)}{1+n^2\pi^2} \right)$$

$$\frac{1}{2(1+n^2\pi^2)} \left[\left(\frac{e^2 - e^{-2}}{2} \right) + in\pi \left(\frac{e^2 + e^{-2}}{2} \right) - in\pi \right] = \frac{1}{2(1+n^2\pi^2)} [\sinh 2 + in\pi(\cosh 2 - 1)].$$

The complex Fourier series representation of $f(t)$ is thus

$$f(t) \sim \lim_{m \to \infty} \sum_{n=-m}^{m} \left[\frac{\sinh 2 + in\pi(\cosh 2 - 1)}{2(1+n^2\pi^2)} \right] e^{in\pi t}.$$

To determine the frequency spectrum we set

$$c_n = r_n \exp(i\theta_n)$$

and compute r_n and θ_n. A simple calculation gives

$$r_n = |c_n| = \frac{[(\sinh 2)^2 + n^2\pi^2 (\cosh 2 - 1)^2]^{1/2}}{2(1+n^2\pi^2)},$$

and

$$\theta_n = \arg c_n = \arctan \left(\frac{n\pi(\cosh 2 - 1)}{\sinh 2} \right).$$

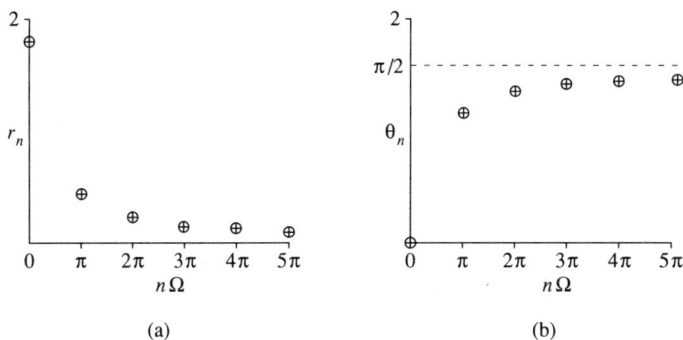

Fig. 9.21 (a) Amplitude spectrum of $f(t)$; (b) Phase spectrum of $f(t)$

Graphs of r_n and θ_n against the angular frequency $n\Omega$ (that is graphs of the amplitude and phase spectra of $f(t)$) are shown in Fig. 9.21.

If required, the real variable form of the Fourier series for $f(t)$ follows by recalling that

$$a_n = c_n + c_{-n} \qquad \text{and} \qquad b_n = i(c_n - c_{-n}),$$

which in this case yields

$$a_n = \frac{\sinh 2}{1 + n^2\pi^2} \qquad \text{and} \qquad b_n = \frac{-n\pi(\cosh 2 - 1)}{1 + n^2\pi^2}.$$

Thus the real variable Fourier series representation of the function $f(t) = \sinh t$, for $0 \le t \le 2$, is

$$f(t) \sim \frac{1}{2}\sinh 2 + \sum_{n=1}^{\infty}\left[\left(\frac{\sinh 2}{1 + n^2\pi^2}\right)\cos n\pi t - \left(\frac{n\pi(\cosh 2 - 1)}{1 + n^2\pi^2}\right)\sin n\pi t\right].$$

The approximation to $\cosh t$ provided by the partial sum $s_{10}(t)$ of this series is shown in Fig. 9.22, in which the Gibbs phenomenon can be observed at both $t = 0$ and $t = 2$. It is of interest to notice that if we assume that at $t = 0$ the Fourier series converges to the value $\frac{1}{2}(1 + \cosh 2)$ as

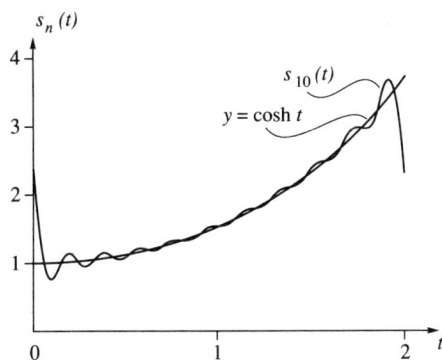

Fig. 9.22

suggested by the graph, then

$$\frac{1}{2}\left(1+\cosh 2\right)=\frac{1}{2}\sinh 2+\sum_{n=1}^{\infty}\frac{\sinh 2}{1+n^2\pi^2},$$

and so

$$\frac{1+\cosh 2}{\sinh 2}=1+2\sum_{n=1}^{\infty}\frac{1}{1+n^2\pi^2}.$$

This is yet another example of the way in which Fourier series lead to closed form expressions for sums of series. ∎

Complex Fourier series in an engineering application of differential equations

Chapter 5 showed how many physical systems of an oscillatory nature are governed by differential equations of the form

$$a_0\frac{d^m y}{dt^m}+a_1\frac{d^{m-1}y}{dt^{m-1}}+\ldots+a_m y = f(t), \tag{40}$$

in which the coefficients $a_0, a_1, \ldots, a_m$ are constants and $f(t)$ represents a time varying input which drives the system.

The general solution of (40) has the form

$$y(t) = y_c(t) + y_p(t), \tag{41}$$

with $y_c(t)$ the complementary function and $y_p(t)$ a particular integral. It will be recalled that $y_c(t)$ is a solution of the associated homogeneous equation

$$a_0\frac{d^m y_c}{dt^m}+a_1\frac{d^{m-1}y_c}{dt^{m-1}}+\ldots+a_m y_c = 0, \tag{42}$$

while the particular integral satisfies the equation

$$a_0\frac{d^m y_p}{dt^m}+a_1\frac{d^{m-1}y_p}{dt^{m-1}}+\ldots+a_m y_p = f(t), \tag{43}$$

and so is particular to the function $f(t)$.

The complementary function contains the m arbitrary constants of integration $\alpha_1, \alpha_2, \ldots, \alpha_m$ associated with the general solution of (40) and it may be written as

$$y_c(t) = \alpha_1 \exp(\lambda_1 t) + \alpha_2 \exp(\lambda_2 t) + \ldots + \alpha_m \exp(\lambda_m t), \tag{44}$$

when $\lambda_1, \lambda_2, \ldots, \lambda_m$ are distinct roots of the characteristic equation of (40),

$$a_0\lambda^m + a_1\lambda^{m-1} + \ldots + a_m = 0. \tag{45}$$

Then, for the solution of (40) to be stable for large t (for the solution to remain finite), the complementary function must decay to zero as $t \to \infty$ which is only possible if the real part of each characteristic root λ_i is negative. Thus for stability of the system described by (40) it is necessary that

$$\text{Re }\lambda_i < 0, \tag{46}$$

for $i = 1, 2, \ldots, m$. When conditions (46) are satisfied, the solution which remains after a suitably long time is the particular integral $y_p(t)$, or the so-called **steady-state** solution.

The solution $y_p(t)$ is independent of the initial conditions for (40), because they only match the constants $\alpha_1, \alpha_2, \ldots, \alpha_m$ in the general solution (41) to the required particular solution. Confining attention to equations of the type shown in (40) which possess stable solutions, let us determine the steady-state solution which results when the driving function has the form $f(t) = ce^{i\Omega t}$, with c a complex constant, Ω real and $i\Omega$ not equal to any one of the m characteristic roots $\lambda_1, \lambda_2, \ldots, \lambda_m$ (we exclude undamped resonance).

Setting $f(t) = ce^{i\Omega t}$ in (43) and using the method of undetermined coefficients shows that

$$y_p(t) = \frac{ce^{i\Omega t}}{a_0(i\Omega)^m + a_1(i\Omega)^{m-1} + \ldots + a_m}. \tag{47}$$

Referring to Sec. 7.3(3), and writing

$$T(s) = \frac{1}{a_0 s^m + a_1 s^{m-1} + \ldots + a_m}, \tag{48}$$

we recognize that $T(s)$ is the **transfer function** of the system described by (40), and that

$$y_p(t) = c\,T(i\Omega)e^{i\Omega t}. \tag{49}$$

Notice that the zeros of the denominator of the transfer function are the same as those of the characteristic equation given in (45); so a system with the transfer function (48) will be stable if conditions (46) are satisfied. For obvious reasons, the function $T(i\Omega)$ is called the **frequency response function** of the system described by (40). Thus when considering the steady-state solution of a stable system, the output response $y_p(t)$ to an input with angular frequency Ω is a periodic response with the *same* frequency. However, the amplitude and phase of the output are modified and so differ from those of the input.

To see this, notice that the amplitude and phase of the periodic input $ce^{i\Omega t}$ are $|c|$ and $\Omega + \arg c$, respectively, while the amplitude and phase of the periodic output $y_p(t)$ are $|c|\,|T(i\Omega)|$ and $\Omega + \arg c + \arg T(i\Omega)$, respectively. The **amplification factor**, or the gain, of the system $A(\Omega)$ is thus

$$A(\Omega) = \frac{|c|\,|T(i\Omega)|}{|c|} = |T(i\Omega)|, \tag{50}$$

while the **phase change** $\phi(\Omega)$ is

$$\phi(\Omega) = \arg T(i\Omega). \tag{51}$$

Thus graphs of $A(\Omega)$ and $\phi(\Omega)$ against Ω enable the steady-state response of (40) to be determined for any periodic input $f(t) = ce^{i\Omega t}$.

Continuing our intuitive approach to Fourier series, let the input $f(t)$ to (40) be a function which is periodic with period T and let us represent it by its complex Fourier series

$$f(t) = \lim_{m \to \infty} \sum_{n=-m}^{m} c_n e^{in\Omega t}, \tag{52}$$

where as before $\Omega = 2\pi/T$ is the angular frequency of $f(t)$.

We now assume that the rth derivative $d^r f/dt^r$ of the function $f(t)$ in (52) may be obtained from (52) by differentiating each term in the summation r times. Then it follows at once from our previous arguments that the periodic response forming the steady-state solution of (40) is

$$y_p(t) = \lim_{m \to \infty} \sum_{n=-m}^{m} c_n T(in\Omega) e^{in\Omega t}. \tag{53}$$

This result shows how the frequency response function influences the amplitude and phase of each Fourier frequency component of the input function. When applied to physical systems this information helps the designer select the design parameters (the coefficients $a_0, a_1, \ldots, a_m$) in order to achieve the best approximation to a desired frequency response.

If a control system is involved, its frequency response can be modified by adjusting the nature of the feedback loop. However, the equation governing the system then becomes more complicated, and (40) needs to be replaced by an equation of the form

$$a_0 \frac{d^m y}{dt^m} + a_1 \frac{d^{m-1}}{dt^{m-1}} + \ldots + a_m y = b_0 \frac{d^r f}{dt^r} + b_1 \frac{d^{r-1} f}{dt^{r-1}} + \ldots + b_r f, \tag{54}$$

where in practical systems $r < m$. In (54) the function $y(t)$ is the response of the differential equation on the left-hand side to an input represented by the right-hand side in which $f(t)$ is the driving function occurring in (40).

The transfer function for (54) is

$$T(s) = \frac{b_0 s^r + b_1 s^{r-1} + \ldots + b_r}{a_0 s^m + a_1 s^{m-1} + \ldots + a_m}, \tag{55}$$

and, as before, the corresponding frequency response function is the response of (54) to the input $f(t) = c e^{i\Omega t}$. The form of argument already used shows the frequency response function to be

$$T(i\Omega) = \frac{b_0 (i\Omega)^r + b_1 (i\Omega)^{r-1} + \ldots + b_r}{a_0 (i\Omega)^m + a_1 (i\Omega)^{m-1} + \ldots + a_m}. \tag{56}$$

Now suppose, once again, that the driving function $f(t)$ is periodic with period T, and that it is represented by the complex Fourier series given in (52). Then the steady-state response of a stable system described by (54) is still determined by (53), though now the function $T(i\Omega)$ takes the more complicated form given in (56).

As in practical systems $m > r$, it follows from (56) that

$$\lim_{\Omega \to \infty} |T(i\Omega)| = 0.$$

This shows that a very high frequency input to a stable system described by (54) produces a vanishingly small response from the system, which effectively acts as a filter.

Example 6. A typical engineering type problem

Show that the system described by the differential equation

$$\frac{d^3y}{dt^3} + 13\frac{d^2y}{dt^2} + 140\frac{dy}{dt} + 500y = 75\left(\frac{df}{dt} + 2f\right)$$

is stable. Find its frequency response function, and hence graph its amplification factor and phase change as functions of the frequency.

If the driving function $f(t)$ is periodic with period 2π, and

$$f(t) = \pi^2 - x^2, \qquad \text{for } -\pi \le x \le \pi,$$

determine the terms up to and including the multiple $3t$ in the ordinary Fourier series representation of the steady-state response $y_p(t)$.

Solution

By inspection the transfer function of the system is

$$T(s) = \frac{75(s + 2)}{s^3 + 13s^2 + 140s + 500}$$

$$= \frac{75(s + 2)}{(s + 5)(s^2 + 8s + 100)}.$$

The zeros of the denominator of $T(s)$ occur at $s = -5$ and $s = -4 \pm (i/2)\sqrt{42}$, so as their real parts are negative, conditions (46) are satisfied, thereby showing the system to be stable.

The frequency response function of the system is

$$T(i\Omega) = \frac{75(i\Omega + 2)}{(i\Omega + 5)[(i\Omega)^2 + 8(i\Omega) + 100]},$$

and graphs of the amplification factor (gain) $A(\Omega) = |T(i\Omega)|$ and the phase change (lag) $\phi(\Omega) = \arg T(i\Omega)$ as functions of Ω are shown in Figs 9.23(a), (b).

A graph of $f(t)$ and its periodic extensions is shown in Fig. 9.24. As $f(t)$ is periodic with period $T = 2\pi$, $\Omega = 2\pi/T = 1$, and the complex Fourier series representation of $f(t)$ becomes

$$f(t) = \lim_{m \to \infty} \sum_{n=-m}^{m} c_n e^{int},$$

with the coefficients c_n defined by the integral

$$c_n = \frac{1}{2\pi} \int_{-\pi}^{\pi} (\pi^2 - t^2)e^{-int}\, dt.$$

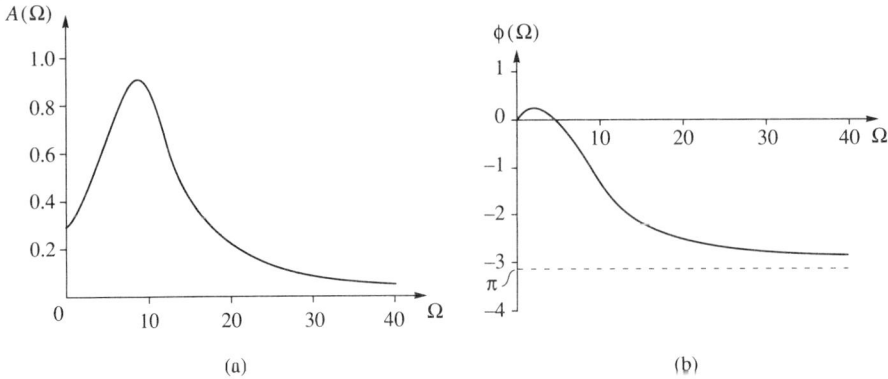

Fig. 9.23 (a) Amplification factor; (b) Phase change (lag)

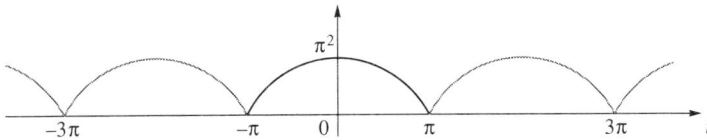

Fig. 9.24 The driving function $f(t) = \pi^2 - t^2$

Routine integration establishes that

$$c_0 = \frac{2}{3}\pi^2 \quad \text{and} \quad c_n = \frac{2(-1)^{n+1}}{n^2}, \quad \text{for } n = \pm 1, \pm 2, \dots.$$

The fact that in this the case the coefficients c_n are real can be seen from the integral defining c_n. The function $\pi^2 - t^2$ is an even function, as is $\cos nt$, while $\sin nt$ is an odd function. So writing $(\pi^2 - t^2)e^{-int} = (\pi^2 - t^2)\cos nt - i(\pi^2 - t^2)\sin nt$ we see that the imaginary part is an odd function, being the product of an even function and an odd function, so its integral from $-\pi$ to π must vanish.

To determine the required few terms in the ordinary Fourier series representation of the steady-state response $y_p(t)$ we first truncate the series in (53) by setting $m = 3$. As a result we arrive at the approximation

$$y_p(t) \approx s_3(t) = c_0 T(0) + c_1 T(i)e^{it} + c_{-1} T(-i)e^{-it} + c_2 T(2i)e^{2it} + c_{-2} T(-2i)e^{-2it}$$
$$+ c_3 T(3i)e^{3it} + c_{-3} T(-3i)e^{-3it}.$$

However, in this case the coefficients c_n are all real, and $T(s)$ is the quotient of two polynomials with real coefficients so that $T(-ni) = \bar{T}(ni)$. Thus it follows that the approximation $s_3(t)$ to $y_p(t)$ may be written

$$s_3(t) = c_0 T(0) + 2c_1 \operatorname{Re}\{T(i)e^{it}\} + 2c_2 \operatorname{Re}\{T(2i)e^{2it}\} + 2c_3 \operatorname{Re}\{T(3i)e^{3it}\}.$$

Now

$$c_0 = \frac{2}{3}\pi^2, \quad c_1 = 2, \quad c_2 = \frac{1}{2}, \quad c_3 = \frac{2}{9},$$

and

$T(0) = 3/10,$
$T(i) = 0.3255 + 0.0611i,$
$T(2i) = 0.3932 + 0.0961i,$
$T(3i) = 0.4844 + 0.0904i,$

so combining the results we obtain the approximation

$$y_p(t) \approx s_3(t) = \tfrac{1}{5}\pi^2 + 1.3018 \cos t - 0.2445 \sin t - 0.3932 \cos 2t$$
$$+ 0.0961 \sin 2t + 0.2152 \cos 3t - 0.0402 \sin 3t.$$

A graph $s_3(t)$ is shown in Fig. 9.25, to which has been added the graph of $s_{10}(t)$ for purposes of comparison. The graph of $s_{20}(t)$ is virtually indistinguishable from the graph of $s_{10}(t)$, so it has not been shown in the figure. It can be seen from the graphs that the filtering effect of the system causes the higher frequency components in $f(t)$ to have little effect on the response. This in itself provides a justification for the truncation of the series for $f(t)$.

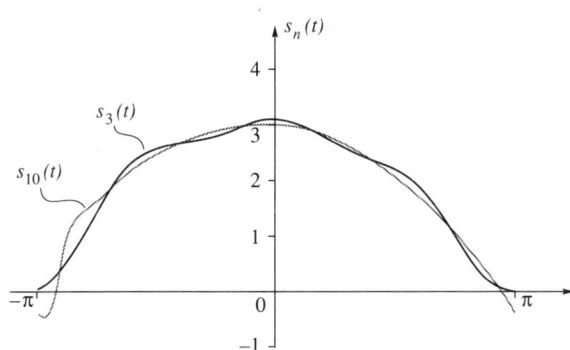

Fig. 9.25   A comparison of approximations $s_n(t)$ to $y_p(t)$ ∎

This last example demonstrates that the manual calculation of a given approximation $s_n(t)$ is straightforward, though it becomes tedious when n becomes large. Where possible, the determination of $s_n(t)$ is best carried out on a PC using mathematical sketchpad type software, such as MathCAD which was used to produce the results of Fig. 9.25 in a few minutes. In the event that such an approach is adopted, it is both simpler and quicker to determine the coefficients $c_0, c_{\pm 1}, c_{\pm 2}, \ldots, c_{\pm n}$ by numerical integration, and then to incorporate them into the approximation to $y_p(t)$ provided by the chosen $s_n(t)$. This latter approach is, of course, essential, if the driving function $f(t)$ has a complicated form or is only known numerically.

Some useful special rules

When seeking the Fourier series representation of a function $f(x)$ there are several simple results which sometimes help to reduce the amount of computation involved. They are based on the following observations:

(i) if $g(x)$ and $h(x)$ are both *even* functions, their product $f(x)=g(x)h(x)$ is an *even* function,

(ii) if $g(x)$ and $h(x)$ are both *odd* functions, their product $f(x)=g(x)h(x)$ is an *even* function,

(iii) if $g(x)$ is an *even* function and $h(x)$ an *odd* function, or conversely, their product $f(x)=g(x)h(x)$ is an *odd* function.

(iv) if k is a constant and $f(x)$ defined for $a \le x \le b$ has the Fourier series representation

$$f(x) \sim \frac{1}{2}a_0 + \sum_{n=1}^{\infty}\left(a_n \cos\frac{2n\pi x}{b-a} + b_n \sin\frac{2n\pi x}{b-a}\right),$$

then $kf(x)$ has the Fourier series representation

$$kf(x) \sim \frac{k}{2}a_0 + k\sum_{n=1}^{\infty}\left(a_n \frac{\cos 2n\pi x}{b-a} + b_n \sin\frac{2n\pi x}{b-a}\right),$$

(v) if $f(x)$ defined for $a \le x \le b$ has the Fourier series representation

$$f(x) \sim \frac{1}{2}a_0 + \sum_{n=1}^{\infty}\left(a_n \cos\frac{2n\pi x}{b-a} + b_n \sin\frac{2n\pi x}{b-a}\right),$$

then for every constant h, the function $h + f(x)$ has the Fourier series representation

$$h + f(x) \sim h + \frac{1}{2}a_0 + \sum_{n=1}^{\infty}\left(a_n \cos\frac{2n\pi x}{b-a} + b_n \sin\frac{2n\pi x}{b-a}\right).$$

Here, (i) to (iii) follow directly from the definitions of even and odd functions, while (iv) and (v) are trivial consequences of the definition of a Fourier series.

When one of results (i) to (iii) applies to a function $f(x)$, only the coefficients a_n or b_n need be determined, according as $f(x)$ is even or odd. Thus, for example, when seeking the Fourier series representation of the function $f(x)=(x^2-\pi^2)\sin x$ only the coefficients b_n need be determined, because $g(x)=x^2-\pi^2$ is an even function and $h(x)=\sin x$ is an odd function, so $f(x)$ is an odd function.

Results (iv) and (v) enable the Fourier series of the function $h + kf(x)$ for $a \le x \le b$ to be related to the Fourier series of the simpler function $f(x)$ defined on the same interval. If $h + kf(x)$ is defined on the interval $-L \le x \le L$, and in addition $f(x)$ is either even or odd, then the Fourier series of $h + kf(x)$ may be determined directly from the Fourier series for $f(x)$ for which either the coefficients a_n or the b_n are all zero, depending on whether $f(x)$ is odd or even, respectively.

Example 7. Relating Fourier series to one another

Find the Fourier series representation of the function $f(x)$ with the fundamental interval $-\pi \le x \le \pi$ shown in Fig. 9.26(a). Use the result to find the Fourier series representation of the function shown in Fig. 9.26(b).

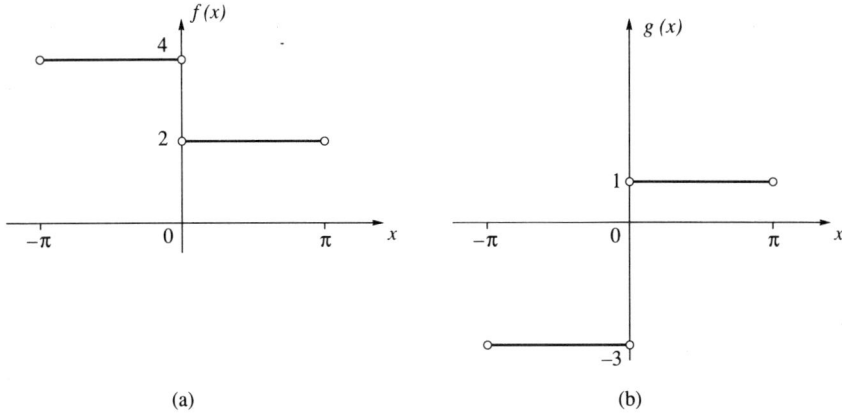

(a)   (b)

Fig. 9.26

Solution

The function $f(x)$ in Fig. 9.26 (a) given by

$$f(x) = \begin{cases} 4, & -\pi < x < 0 \\ 2, & 0 < x < \pi, \end{cases}$$

is neither even nor odd. However, the function $F(x) = f(x) - 3$ or, equivalently, the function

$$F(x) = \begin{cases} 1, & -\pi < x < 0 \\ -1, & 0 < x < \pi \end{cases}$$

shown in Fig. 9.27 is odd, and its Fourier series representation is easily seen to be given by

$$F(x) \sim -\frac{4}{\pi} \sum_{n=1}^{\infty} \frac{\sin (2n-1)x}{2n-1}, \qquad \text{for } -\pi < x < \pi.$$

This follows because $F(x)$ is odd, so all the coefficients a_n are zero, while

$$b_n = \frac{1}{\pi} \int_{-\pi}^{\pi} F(x) \sin nx \, dx = \frac{1}{\pi} \int_{-\pi}^{0} \sin nx \, dx - \frac{1}{\pi} \int_{0}^{\pi} \sin nx \, dx$$

$$= \frac{1}{\pi} \left(\frac{-\cos nx}{n} \right) \Big|_{-\pi}^{0} - \frac{1}{\pi} \left(\frac{-\cos nx}{n} \right) \Big|_{0}^{\pi}$$

$$= \frac{2}{\pi} \left(\frac{(-1)^n - 1}{n} \right), \qquad \text{for } n = 1, 2, \ldots,$$

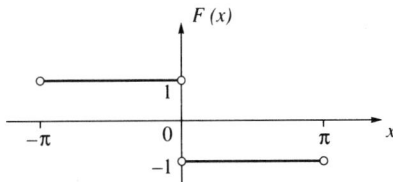

Fig. 9.27

showing that $0 = b_2 = b_4 = \ldots = b_{2n} = \ldots$, and

$$b_{2n-1} = \frac{-4}{\pi(2n-1)}, \qquad \text{for } n = 1, 2, \ldots.$$

Thus from (v), since $f(x) = F(x) + 3$, we have

$$f(x) \sim 3 - \frac{4}{\pi} \sum_{n=1}^{\infty} \frac{\sin(2n-1)x}{2n-1}.$$

To obtain the Fourier series representation of the function $g(x)$ in Fig. 9.26(b) we first recognize that

$$g(x) = \begin{cases} -3, & -\pi < x < 0 \\ 1, & 0 < x < \pi. \end{cases}$$

Thus $g(x)$ and $F(x)$ are related by

$$g(x) = -1 - 2F(x), \qquad \text{for } -\pi < x < \pi,$$

from which it follows that the Fourier series representation of $g(x)$ is

$$g(x) \sim -1 + \frac{8}{\pi} \sum_{n=1}^{\infty} \frac{\sin(2n-1)x}{2n-1}, \qquad \text{for } -\pi < x < \pi. \qquad \blacksquare$$

Table 9.1 presents a summary of the various types of Fourier series which have been considered in this section.

Table 9.1 Useful forms of Fourier series

Type of series	Series representation	Fourier coefficients
Fourier series for $f(x)$ on $-\pi \le x \le \pi$	$\frac{1}{2}a_0 + \sum_{n=1}^{\infty}(a_n \cos nx + b_n \sin nx)$	$a_0 = \frac{1}{\pi}\int_{-\pi}^{\pi} f(x)\,dx,$
		$a_n = \frac{1}{\pi}\int_{-\pi}^{\pi} f(x)\cos nx\,dx, \quad n = 1, 2, \ldots,$
		$b_n = \frac{1}{\pi}\int_{-\pi}^{\pi} f(x)\sin nx\,dx, \quad n = 1, 2, \ldots,$
		If $f(x)$ is periodic with period 2π, these integrals may be evaluated over any interval of length 2π.
Fourier series for $f(x)$ on $-L \le x \le L$	$\frac{1}{2}a_0 + \sum_{n=1}^{\infty}\left[a_n \cos\left(\frac{n\pi x}{L}\right)\right.$ $\left. + b_n \sin\left(\frac{n\pi x}{L}\right)\right]$	$a_0 = \frac{1}{L}\int_{-L}^{L} f(x)\,dx,$
		$a_n = \frac{1}{L}\int_{-L}^{L} f(x)\cos\left(\frac{n\pi x}{L}\right)dx, \quad n = 1, 2, \ldots,$
		$b_n = \frac{1}{L}\int_{-L}^{L} f(x)\sin\left(\frac{n\pi x}{L}\right)dx, \quad n = 1, 2, \ldots.$
		If $f(x)$ is periodic with period $2L$, these integrals may be evaluated over any interval of length $2L$.

Table 9.1 (continued)

Type of series	Series representation	Fourier coefficients
Fourier series for $f(x)$ on $a \leq x \leq b$	$\frac{1}{2}a_0 + \sum\limits_{n=1}^{\infty}\left[a_n\cos\left(\dfrac{2n\pi x}{b-a}\right) + b_n\sin\left(\dfrac{2n\pi x}{b-a}\right)\right]$	$a_0 = \dfrac{2}{(b-a)}\displaystyle\int_a^b f(x)\,dx,$ $a_n = \dfrac{2}{(b-a)}\displaystyle\int_a^b f(x)\cos\left(\dfrac{2n\pi x}{b-a}\right)dx, \quad n = 1, 2, \dots,$ $b_n = \dfrac{2}{(b-a)}\displaystyle\int_a^b f(x)\sin\left(\dfrac{2n\pi x}{b-a}\right)dx, \quad n = 1, 2, \dots.$
Fourier sine series for $f(x)$ on $0 \leq x \leq \pi$	$\sum\limits_{n=1}^{\infty} b_n\sin nx$	$b_n = \dfrac{2}{\pi}\displaystyle\int_0^{\pi} f(x)\sin nx\,dx, \quad n = 1, 2, \dots.$ If $f(x)$ is extended to $-\pi \leq x \leq 0$ as an odd function this series represents $f(x)$ for $-\pi \leq x \leq \pi$.
Fourier cosine series for $f(x)$ on $0 \leq x \leq \pi$	$\frac{1}{2}a_0 + \sum\limits_{n=1}^{\infty} a_n\cos nx$	$a_0 = \dfrac{2}{\pi}\displaystyle\int_0^{\pi} f(x)\,dx,$ $a_n = \dfrac{2}{\pi}\displaystyle\int_0^{\pi} f(x)\cos nx\,dx, \quad n = 1, 2, \dots.$ If $f(x)$ is extended to $-\pi \leq x \leq 0$ as an even function this series represents $f(x)$ for $-\pi \leq x \leq \pi$.
Fourier sine series for $f(x)$ on $0 \leq x < L$	$\sum\limits_{n=1}^{\infty} b_n\sin\left(\dfrac{n\pi x}{L}\right)$	$b_n = \dfrac{2}{L}\displaystyle\int_0^L f(x)\sin\left(\dfrac{n\pi x}{L}\right)dx, \quad n = 1, 2, \dots.$ If $f(x)$ is extended to $-L \leq x \leq 0$ as an odd function this series represents $f(x)$ for $-L \leq x \leq L$.
Fourier cosine series for $f(x)$ on $0 \leq x \leq L$	$\frac{1}{2}a_0 + \sum\limits_{n=1}^{\infty} a_n\cos\left(\dfrac{n\pi x}{L}\right)$	$a_0 = \dfrac{2}{L}\displaystyle\int_0^L f(x)\,dx.$ $a_n = \dfrac{2}{L}\displaystyle\int_0^L f(x)\cos\left(\dfrac{n\pi x}{L}\right)dx, \quad n = 1, 2, \dots.$ If $f(x)$ is extended to $-L \leq x \leq 0$ as an even function this series represents $f(x)$ for $-L \leq x \leq L$.
Complex Fourier series for $f(x)$ on $-\pi \leq x \leq \pi$	$\lim\limits_{m \to \infty}\sum\limits_{n=-m}^{m} c_n e^{inx}$	$c_n = \dfrac{1}{2\pi}\displaystyle\int_{-\pi}^{\pi} f(x)e^{-inx}\,dx, \quad n = 0, \pm1, \pm2, \dots.$ If $f(x)$ is periodic with period 2π these integrals may be evaluated over any interval of length 2π.
Complex Fourier series for $f(x)$ on $-L \leq x \leq L$	$\lim\limits_{m \to \infty}\sum\limits_{n=-m}^{m} c_n \exp[(in\pi x)/L]$	$c_n = \dfrac{1}{2L}\displaystyle\int_{-L}^{L} f(x)\exp[(-in\pi x)/L]\,dx,$ $n = 0 \pm 1, \pm2, \dots.$ If $f(x)$ is periodic with period $2L$ these integrals may be evaluated over any interval of length $2L$.

Problems for Section 9.2

Fourier series on $(-\pi, \pi)$ and $(0, 2\pi)$

In the following problems assume that a Fourier series converges in the manner suggested by the worked examples. Incorporate into each solution a sketch of the function to which the required Fourier series converges.

1 Show the Fourier series representation of

$$f(x)=|x|, \qquad \text{for } -\pi \leq x \leq \pi$$

is

$$f(x) \sim \frac{\pi}{2} - \frac{4}{\pi} \sum_{n=1}^{\infty} \frac{\cos(2n-1)x}{(2n-1)^2}.$$

2 Show the Fourier series representation of

$$f(x)=\begin{cases} \pi, & \text{for } -\pi \leq x < 0 \\ \pi+kx, & \text{for } 0 \leq x < \pi \end{cases}$$

is

$$f(x) \sim \left(\frac{4+k}{4}\right)\pi - \frac{2k}{\pi} \sum_{n=1}^{\infty} \frac{\cos(2n-1)x}{(2n-1)^2} + k \sum_{n=1}^{\infty} \frac{(-1)^{n+1}\sin nx}{n}.$$

3 Show that if

$$f(x)=x+x^2, \qquad \text{for } -\pi \leq x < \pi,$$

then its Fourier series representation is

$$f(x) \sim \frac{\pi^2}{3} + \sum_{n=1}^{\infty} (-1)^n \left(\frac{4\cos nx}{n^2} - \frac{2\sin nx}{n}\right).$$

4 Show that if

$$f(x)=\begin{cases} 0, & \text{for } -\pi < x \leq 0 \\ x+\pi, & \text{for } 0 < x \leq \pi, \end{cases}$$

then its Fourier series representation is

$$f(x) \sim \frac{3\pi}{4} - \frac{2}{\pi} \sum_{n=0}^{\infty} \frac{\cos(2n+1)x}{(2n+1)^2} + \sum_{n=1}^{\infty} \left(\frac{1-2(-1)^n}{n}\right)\sin nx.$$

Use the above series to sum

(i) $1+\dfrac{1}{3^2}+\dfrac{1}{5^2}+\dfrac{1}{7^2}+ \dots$ and

(ii) $1-\frac{1}{3}+\frac{1}{5}-\frac{1}{7}+\frac{1}{9}- \dots$.

5 Show that if

$$f(x)=x^2, \qquad \text{for } 0 \leq x < 2\pi,$$

then its Fourier series representation is

$$f(x) \sim \frac{4}{3}\pi^2 + 4 \sum_{n=1}^{\infty} \left(\frac{1}{n^2} \cos nx - \frac{\pi}{n} \sin nx \right).$$

Use the above series to sum

(i) $\displaystyle\sum_{n=1}^{\infty} \frac{(-1)^n}{n^2}$ and (ii) $\displaystyle\sum_{n=1}^{\infty} \frac{1}{n^2}$.

6 The function $f(x)$ defined for $-\pi \le x \le \pi$ has the graph given by the straight lines AB, BC and CD, where A is $(-\pi, \pi)$, B is $\left(-\frac{\pi}{2}, \frac{\pi}{2} \right)$, C is $\left(\frac{\pi}{2}, \frac{3\pi}{2} \right)$ and D is (π, π). Show that the function $f(x) - \pi$ is odd, and hence that $f(x)$ has the Fourier series representation

$$f(x) \sim \pi + \frac{4}{\pi} \sum_{n=0}^{\infty} (-1)^n \frac{\sin(2n+1)x}{(2n+1)^2},$$

for $-\pi \le x < \pi$.
Use the above series to sum

$$1 + \frac{1}{3^2} + \frac{1}{5^2} + \frac{1}{7^2} + \ldots .$$

7 Find the Fourier series representation of

$$f(x) = \tfrac{1}{2}(x + |x|), \qquad \text{for } -\pi \le x < \pi.$$

Use it to deduce that

$$\sum_{n=0}^{\infty} \frac{(-1)^n}{2n+1} = \frac{\pi}{4} \quad \text{and} \quad \sum_{n=0}^{\infty} \frac{1}{(2n+1)^2} = \frac{\pi^2}{8}.$$

8 Show that if

$$f(x) = \sinh(\pi - x), \qquad \text{for } 0 < x \le 2\pi,$$

then its Fourier series representation is

$$f(x) \sim \frac{2 \sinh \pi}{\pi} \sum_{n=1}^{\infty} \frac{n \sin nx}{1 + n^2}.$$

9 Show that the function

$$f(x) = \begin{cases} 0, & \text{for } -\pi < x < 0 \\ x^2, & \text{for } 0 \le x < \pi \end{cases}$$

has the Fourier series representation

$$f(x) \sim \frac{\pi^2}{6} + 2 \sum_{n=0}^{\infty} \frac{(-1)^n \cos nx}{n^2}$$

$$+ \sum_{n=0}^{\infty} \left(\frac{\pi}{2n+1} - \frac{4}{\pi(2n+1)^3} \right) \sin(2n+1)x - \sum_{n=1}^{\infty} \frac{\pi}{2n} \sin 2nx.$$

Use the series to sum

$$1+\frac{1}{2^2}+\frac{1}{3^2}+\frac{1}{4^2}+\cdots.$$

10 The function $f(x)$ is defined on the interval $-\pi \le x \le \pi$ by

$$f(x)=\begin{cases} \pi+x, & \text{for} \quad -\pi \le x < -\pi/2 \\ |x|, & \text{for} \quad -\pi/2 \le x \le \pi/2, \\ \pi-x, & \text{for} \quad \pi/2 < x \le \pi. \end{cases}$$

Show that $f(x)$ has the Fourier series representation

$$f(x) \sim \frac{\pi}{4}-\frac{2}{\pi}\sum_{n=1}^{\infty}\frac{\cos(4n-2)x}{(2n-1)^2}, \qquad \text{for} \quad -\pi \le x \le \pi.$$

11 The function

$$f(x)=\begin{cases} 1, & \text{for} \quad -\pi < x < 0 \\ x, & \text{for} \quad 0 < x \le \pi. \end{cases}$$

Show that its Fourier series representation is

$$f(x) \sim \frac{1}{2}+\frac{\pi}{4}+\sum_{n=1}^{\infty}\left[\left(\frac{1+(-1)^n}{n^2\pi}\right)\cos nx -\left(\frac{1+(-1)^n(\pi-1)}{n\pi}\right)\sin nx\right],$$

for $-\pi < x \le \pi$.

12 The function

$$f(x)=\begin{cases} 0, & \text{for} \quad -\pi < x < 0 \\ k, & \text{for} \quad 0 < x \le \pi. \end{cases}$$

Show the Fourier series representation of $f(x)$ is

$$f(x) \sim \frac{k}{2}+\frac{2k}{\pi}\sum_{n=1}^{\infty}\left(\frac{1}{2n-1}\right)\sin(2n-1)x, \qquad \text{for} \quad -\pi < x \le \pi.$$

13 The function

$$f(x)=\begin{cases} 1, & \text{for} \quad -\pi < x < 0 \\ \cos x, & \text{for} \quad 0 \le x \le \pi. \end{cases}$$

Show the Fourier series representation of $f(x)$ on the interval $-\pi < x \le \pi$ is

$$f(x) \sim \frac{1}{2}+\frac{1}{2}\cos x +\frac{2}{\pi}\sum_{n=1}^{\infty}\left(\frac{1}{4n^2-1}\right)[2n\sin 2nx-(2n+1)\sin(2n-1)x].$$

14 Show that if k is nonintegral and

$$f(x)=\cos kx, \qquad \text{for} \quad -\pi < x \le \pi,$$

then $f(x)$ has the Fourier series representation

$$f(x) \sim \frac{2k}{\pi}\sin k\pi\left(\frac{1}{2k^2}+\sum_{n=1}^{\infty}\frac{(-1)^{n+1}}{n^2-k^2}\cos nx\right),$$

for $-\pi < x \le \pi$.

Deduce that

$$\cot k = \frac{1}{k} + 2 \sum_{n=1}^{\infty} \frac{k}{k^2 - n^2\pi^2}.$$

15 Show that if k is nonintegral and

$$f(x) = \sin kx, \qquad \text{for } -\pi < x \leq \pi,$$

then $f(x)$ has the Fourier series representation

$$f(x) \sim \frac{2}{\pi} \sin k\pi \sum_{n=1}^{\infty} \frac{(-1)^n n \sin nx}{n^2 - k^2}.$$

Deduce that

$$\sec\left(\frac{k\pi}{2}\right) = \frac{2}{\pi} \sum_{n=0}^{\infty} \frac{(-1)^n (2n+1)}{(2n+1)^2 - k^2}.$$

16 By finding the Fourier series representation of

$$f(x) = \begin{cases} 0, & \text{for } -\pi < x \leq 0 \\ e^x, & \text{for } 0 < x \leq \pi, \end{cases}$$

show that

$$\frac{\pi}{2} \coth \frac{\pi}{2} = 1 + 2 \sum_{n=1}^{\infty} \frac{1}{1 + 4n^2}$$

and

$$\frac{\pi}{2} \tanh \frac{\pi}{2} = \sum_{n=0}^{\infty} \frac{1}{2n^2 + 2n + 1}$$

17 The function

$$f(x) = \begin{cases} 0, & \text{for } -\pi \leq x < 0 \\ x, & \text{for } 0 \leq x \leq \pi. \end{cases}$$

Show that its Fourier series representation is

$$f(x) \sim \frac{\pi}{4} - \frac{2}{\pi} \sum_{n=0}^{\infty} \frac{\cos(2n+1)x}{(2n+1)^2} = \sum_{n=1}^{\infty} \frac{(-1)^n}{n} \sin nx,$$

for $-\pi \leq x < \pi$.

18 Show that the Fourier series representation of

$$f(x) = \begin{cases} 0, & \text{for } -\pi \leq x < -\pi/2 \\ 1, & \text{for } -\pi/2 < x < \pi/2 \\ 0, & \text{for } \pi/2 < x \leq \pi \end{cases}$$

is

$$f(x) \sim \frac{1}{2} + \frac{2}{\pi} \sum_{n=1}^{\infty} \frac{(-1)^{n+1} \cos(2n-1)x}{(2n-1)}, \qquad \text{for } -\pi \leq x < \pi.$$

Fourier series on (a, b)

19 The function

$$f(x) = \begin{cases} -1, & \text{for } -L < x < 0 \\ 1, & \text{for } 0 < x < L. \end{cases}$$

Show that its Fourier series representation is

$$f(x) \sim \frac{4}{\pi} \sum_{n=1}^{\infty} \frac{\sin(2n-1)(\pi x/L)}{(2n-1)}, \qquad \text{for } -L < x < L.$$

20 The function

$$f(x) = \begin{cases} a, & \text{for } -L < x < 0 \\ b, & \text{for } 0 < x < L. \end{cases}$$

Find the Fourier series representation of $f(x)$ for $-L < x < L$.

21 The function

$$f(x) = \begin{cases} 1+x, & \text{for } -1 < x \le 0 \\ -1, & \text{for } 0 < x \le 1. \end{cases}$$

Show that the Fourier series representation of $f(x)$ is

$$f(x) \sim -\frac{1}{4} + \sum_{n=1}^{\infty} \left[\left(\frac{1+(-1)^{n+1}}{n^2 \pi^2} \right) \cos n\pi x - \left(\frac{2+(-1)^{n+1}}{n\pi} \right) \sin n\pi x \right],$$

for $-1 < x \le 1$.

22 The function

$$f(x) = x, \qquad \text{for } 0 < x \le 4.$$

Show that its Fourier series representation is

$$f(x) \sim \frac{8}{\pi} \sum_{n=1}^{\infty} \frac{(-1)^n \sin(n\pi x/4)}{n}, \qquad \text{for } 0 < x \le 4.$$

23 The function

$$f(x) = \tfrac{1}{2} - x, \qquad \text{for } 0 \le x < 1.$$

Show that its Fourier series representation is

$$f(x) \sim \frac{1}{\pi} \sum_{n=1}^{\infty} \frac{\sin 2n\pi x}{n}, \qquad \text{for } 0 \le x < 1.$$

24 Find the Fourier series representation of the function defined by

$$f(x) = \begin{cases} -x, & \text{for } -2 \le x < 0 \\ 0, & \text{for } 0 < x < 2. \end{cases}$$

25 The function

$$f(x) = \begin{cases} 1, & \text{for } -1 < x \le 0 \\ x, & \text{for } 0 < x \le 1. \end{cases}$$

Show that $f(x)$ has the Fourier series representation

$$f(x) \sim \frac{3}{4} + \sum_{n=1}^{\infty} \left[\left(\frac{(-1)^n - 1}{n^2 \pi^2} \right) \cos n\pi x - \frac{1}{n\pi} \sin n\pi x \right],$$

for $-1 < x \le 1$.

26 Show that the function

$$f(x) = (2 - x)^2, \qquad \text{for } 0 \le x \le 4$$

has the Fourier series representation

$$f(x) \sim \frac{4}{3} + \frac{16}{\pi^2} \sum_{n=1}^{\infty} \frac{\cos(n\pi x/2)}{n^2}, \qquad \text{for } 0 \le x \le 4.$$

27 Find the Fourier series representation of

$$f(x) = x + x^2, \qquad \text{for } -1 \le x < 1.$$

Use the resulting series to show that

$$\sum_{n=1}^{\infty} \frac{(-1)^{n+1}}{n^2} = \frac{\pi^2}{12} \quad \text{and} \quad \sum_{n=1}^{\infty} \frac{1}{n^2} = \frac{\pi^2}{6}.$$

28 The function

$$f(x) = \begin{cases} 1, & \text{for } 0 < x < \pi/4 \\ 0, & \text{for } \pi/4 < x < 3\pi/4 \\ -1, & \text{for } 3\pi/4 < x < \pi. \end{cases}$$

Show that $f(x)$ has the Fourier series representation

$$f(x) \sim \frac{4}{\pi} \sum_{n=0}^{\infty} \frac{\sin\left[(2n+1)(\pi/4)\right]}{2n+1} \cos(2n+1)x, \qquad \text{for } 0 < x < \pi.$$

29 Show that the function

$$f(x) = \begin{cases} 1 + 3x, & \text{for } 0 \le x \le \pi/3 \\ 1 + \pi, & \text{for } \pi/3 < x < 2\pi/3 \\ 1 + 3(\pi - x), & \text{for } 2\pi/3 \le x \le \pi \end{cases}$$

has the Fourier series representation

$$f(x) \sim 1 + \frac{12}{\pi} \sum_{n=1}^{\infty} \frac{\sin\left[(2n-1)(\pi/3)\right]}{(2n-1)^2} \sin(2n-1)x,$$

for $0 \le x \le \pi$.

30 The function

$$f(x) = \begin{cases} \cos(\pi x/2L), & \text{for } 0 \le x \le L \\ 0, & \text{for } L \le x < 2L. \end{cases}$$

Show that its Fourier series representation is

$$f(x) \sim \frac{1}{\pi} + \frac{2}{\pi} \sum_{n=1}^{\infty} \frac{1}{(4n^2 - 1)} \left[(-1)^{n+1} \cos \frac{n\pi x}{L} + 2n \sin \frac{n\pi x}{L} \right],$$

for $0 \le x < 2L$.

31 Show that

$$f(x) = \begin{cases} -1, & \text{for } 0 \le x < \pi/2 \\ \dfrac{4x}{\pi} - 3, & \text{for } \dfrac{\pi}{2} \le x < \pi \end{cases}$$

has the Fourier series representation

$$f(x) \sim -\frac{1}{2} + \frac{8}{\pi^2} \sum_{n=1}^{\infty} \left(\frac{(-1)^n - \cos\left(\dfrac{n\pi}{2}\right)}{n^2} \right) \cos nx,$$

for $0 \le x < \pi$.

32 The function

$$f(x) = 1 - e^{-|x|}, \qquad \text{for } -1 \le x \le 1.$$

Show by obtaining its Fourier series representation that

$$\sum_{n=1}^{\infty} \left(\frac{e + (-1)^{n+1}}{1 + n^2 \pi^2} \right) = \frac{1}{2}.$$

33 Show the function

$$f(x) = \begin{cases} 1, & \text{for } -L \le x \le 0 \\ e^{-x}, & \text{for } 0 < x < L \end{cases}$$

has the Fourier series representation

$$f(x) \sim \frac{1}{2} + \left(\frac{1 - e^{-L}}{2L} \right) + \sum_{n=1}^{\infty} \left(\frac{1 + (-1)^{n+1} e^{-L}}{n^2 \pi^2 + L^2} \right) \left(L \cos \frac{n\pi x}{L} + n\pi \sin \frac{n\pi x}{L} \right)$$

$$+ \frac{2}{\pi} \sum_{n=1}^{\infty} \frac{\sin \left[(2n+1)(\pi x/L) \right]}{(2n+1)},$$

for $-L \le x < L$.

34 Find the Fourier series representation of the function

$$f(x) = \begin{cases} \pi + \cos \pi x, & \text{for } 0 \le x < 1 \\ \pi, & \text{for } 1 < x \le 2. \end{cases}$$

Fourier sine and cosine series

35 Show that the Fourier sine series representation of

$$f(x) = \begin{cases} 0, & \text{for } 0 \le x < \pi/2 \\ k, & \text{for } \pi/2 < x < \pi. \end{cases}$$

is

$$f(x) \sim \frac{2k}{\pi} \sum_{n=1}^{\infty} \left(\frac{\cos \frac{n\pi}{2} + (-1)^{n+1}}{n} \right) \sin nx, \qquad \text{for } 0 \le x < \pi.$$

36 Show that the Fourier cosine series representation of

$$f(x) = \begin{cases} 0, & \text{for } 0 \le x < \pi/2 \\ k, & \text{for } \pi/2 < x < \pi \end{cases}$$

is

$$f(x) \sim \frac{k}{2} - \frac{2k}{\pi} \sum_{n=1}^{\infty} \frac{(-1)^{n+1}}{(2n-1)} \cos(2n-1)x, \qquad \text{for } 0 \le x < \pi.$$

37 The function

$$f(x) = x, \qquad \text{for } 0 \le x < L.$$

Extend $f(x)$ to the interval $(-L, L)$ (a) as an odd function and (b) as an even function and find their respective Fourier sine and cosine series representations. Comment on the respective rates of convergence of these two Fourier series.

38 Show by finding the Fourier sine series representation of

$$f(x) = \cos \pi x, \qquad \text{for } 0 < x < 1,$$

that

$$f(x) \sim \frac{8}{\pi} \sum_{n=1}^{\infty} \frac{n}{4n^2 - 1} \sin 2n\pi x, \qquad \text{for } 0 < x < 1.$$

What function does this series represent in the interval $(k, k+1)$ where k is an integer? Use the series to show that

$$\sum_{n=0}^{\infty} \frac{(-1)^n (2n+1)}{4(2n+1)^2 - 1} = \frac{\pi}{8\sqrt{2}}.$$

39 The function

$$f(x) = \sin x, \qquad \text{for } 0 \le x < \pi.$$

Expand $f(x)$ in a Fourier cosine series, and hence show that

$$f(x) \sim \frac{2}{\pi} - \frac{4}{\pi} \sum_{n=1}^{\infty} \frac{\cos 2nx}{(4n^2 - 1)}, \qquad \text{for } 0 \le x < \pi.$$

40 The function

$$f(x) = \cos x, \qquad \text{for } 0 \le x < \pi.$$

Show that its Fourier sine series representation is

$$f(x) \sim \frac{8}{\pi} \sum_{n=1}^{\infty} \frac{n \sin 2nx}{4n^2 - 1}, \qquad \text{for } 0 \le x < \pi.$$

41 The function

$$f(x) = e^{-\lambda \pi x}, \qquad \text{for } 0 \le x < 1,$$

where λ is a constant. Find the Fourier sine series representation of $f(x)$ and use the result to show that

$$\sum_{n=1}^{\infty} \frac{(-1)^{n+1}(2n-1)}{\lambda^2 + (2n-1)^2} = \frac{\pi}{4} \operatorname{sech}\left(\frac{\lambda \pi}{2}\right).$$

42 The function

$$f(x) = x \sin \pi x, \qquad \text{for } 0 \le x \le 1.$$

Show that its respective Fourier sine and cosine series representations are

$$f(x) \sim \frac{1}{2} \sin \pi x - \frac{16}{\pi^2} \sum_{n=1}^{\infty} \frac{n}{(4n^2 - 1)^2} \sin 2n\pi x, \qquad \text{for } 0 \le x \le 1$$

and

$$f(x) \sim \frac{1}{\pi}\left(1 - \frac{1}{2}\cos \pi n + 2 \sum_{n=2}^{\infty} \frac{(-1)^{n-1}}{(n^2-1)} \cos n\pi x\right), \qquad \text{for } 0 \le x \le 1.$$

43 The function

$$f(x) = x(\pi - x), \qquad \text{for } 0 \le x \le \pi.$$

Show that its Fourier sine series representation is

$$f(x) \sim \frac{4}{\pi} \sum_{n=1}^{\infty} \left(\frac{1 + (-1)^{n+1}}{n^3}\right) \sin nx, \qquad \text{for } 0 \le x \le \pi.$$

44 The function

$$f(x) = x(\pi - x), \qquad \text{for } 0 \le x \le \pi.$$

Show that its Fourier cosine series representation is

$$f(x) \sim \frac{\pi^2}{6} - 2 \sum_{n=1}^{\infty} \left(\frac{1 + (-1)^n}{n^2}\right) \cos nx, \qquad \text{for } 0 \le x \le \pi.$$

Complex Fourier series

45 The function

$$f(x) = e^{ikx}, \qquad \text{for } -L < x \le L,$$

where k is not an integral multiple of π/L. Show that the complex Fourier series representation of $f(x)$ is

$$f(x) \sim \lim_{m \to \infty} \sin kL \sum_{n=-m}^{m} \frac{(-1)^n}{(kL - n\pi)} \exp(in\pi x/L),$$

for $-L < x \le L$.

By taking real and imaginary parts in this representation show that

$$\cos kx \sim \sin kL \left[\frac{1}{kL} + 2kL \sum_{n=1}^{\infty} \frac{(-1)^n}{(k^2L^2 - n^2\pi^2)} \cos\left(\frac{n\pi x}{L}\right) \right]$$

and

$$\sin kx \sim 2\pi \sin kL \sum_{n=1}^{\infty} \frac{(-1)^n n}{(k^2L^2 - n^2\pi^2)} \sin\left(\frac{n\pi x}{L}\right),$$

for $-L < x \leq L$.

46 Show that when k is not an integral multiple of π/L the complex Fourier series representation of

$$f(x) = \sin kx, \qquad \text{for } -L < x \leq L$$

is

$$f(x) \sim \lim_{m \to \infty} \sum_{n=-m}^{m} c_n \exp(in\pi x/L),$$

where

$$c_n = \bar{c}_{-n} = i\pi \sin kL \left(\frac{(-1)^{n+1} n}{k^2L^2 - n^2\pi^2} \right).$$

What form does the representation take if $k = N\pi/L$ where N is an integer?

47 Show that if

$$f(x) = e^x, \qquad \text{for } -\pi < x < \pi,$$

then the complex Fourier series representation of $f(x)$ is

$$f(x) \sim \frac{\sinh \pi}{\pi} \lim_{m \to \infty} \sum_{n=-m}^{m} \frac{(-1)^n}{(1 - in)} e^{inx}, \qquad \text{for } -\pi < x < \pi.$$

Display the complex frequency spectra by graphing the modulus and phase of the complex Fourier coefficients of $f(x)$ as functions of n.

48 The function

$$f(x) = \begin{cases} 0, & \text{for } 0 \leq x < \pi/(2\Omega) \\ K \sin \Omega x, & \text{for } \pi/(2\Omega) \leq x < \pi/\Omega \\ 0, & \text{for } \pi/\Omega \leq x \leq 2\pi/\Omega. \end{cases}$$

Show that the complex Fourier series representation of $f(x)$ is

$$f(x) \sim \lim_{m \to \infty} \sum_{n=-m}^{m} c_n e^{in\Omega x},$$

where

$$c_n = \frac{-K}{2\pi(n^2 - 1)} (ine^{-in\pi/2} + e^{-in\pi}), \qquad \text{for } n = 0, \pm 2, \pm 3, \pm 4,$$

and $c_1 = \bar{c}_{-1} = -K \left(\frac{1}{8} + \frac{i}{4\pi} \right).$

49 Find the complex Fourier series representation of

$$f(x) = \begin{cases} x^2 - 1, & \text{for } -\pi < x < 0 \\ x^2 + 1, & \text{for } 0 < x < \pi. \end{cases}$$

50 The function

$$f(x) = e^{ax}, \qquad \text{for } -L < x < L.$$

Find the complex Fourier series representation of $f(x)$ and use it to deduce that the function $g(x) = e^{-x}$, for $-1 < x < 1$, has the Fourier series representation

$$f(x) \sim \sinh 1 \left(1 - 2 \sum_{n=1}^{\infty} (-1)^{n+1} \frac{\cos nx}{1 + n^2\pi^2} \right) + 2\pi \sinh 1 \sum_{n=1}^{\infty} (-1)^n \frac{n \sin nx}{1 + n^2\pi^2},$$

$$\text{for } -1 < x < 1.$$

Relating and combining Fourier series

The function

$$f(x) = \begin{cases} 0, & \text{for } -L < x < 0 \\ 1, & \text{for } \quad 0 < x < L \end{cases}$$

has the Fourier series representation

$$f(x) \sim \frac{1}{2} + \frac{2}{\pi} \sum_{n=0}^{\infty} \frac{\sin[(2n+1)(\pi x/L)]}{(2n+1)}, \qquad \text{for } -L < x < L,$$

and the function

$$g(x) = x^2, \qquad \text{for } -L < x < L$$

has the Fourier series representation

$$g(x) \sim \frac{L^2}{2} + \frac{4L^2}{\pi^2} \sum_{n=1}^{\infty} \frac{(-1)^n}{n^2} \cos \frac{n\pi x}{L}, \qquad \text{for } -L \le x \le L.$$

In each of the following problems find the Fourier series representation of the function $F(x)$ in terms of the above Fourier series and any other functions that may be necessary.

51 $F(x) = \begin{cases} -\frac{1}{2}, & \text{for } -\pi < x < 0 \\ \quad 3, & \text{for } \quad 0 < x < \pi. \end{cases}$

52 $F(x) = \begin{cases} \quad 1, & \text{for } -6 < x < 0 \\ -2, & \text{for } \quad 0 < x < 6. \end{cases}$

53 $F(x) = \begin{cases} 2x^2, & \text{for } -3 < x < 0 \\ 2x^2 + 3, & \text{for } \quad 0 < x < 3. \end{cases}$

54 $F(x) = \begin{cases} 1 - x^2, & \text{for } -1 < x < 0 \\ 2 - x^2, & \text{for } \quad 0 < x < 1. \end{cases}$

55 $F(x) = \begin{cases} 1 + \sin x + \frac{1}{3}\sin 2x, & \text{for } -\pi < x < 0 \\ 3 + \sin x + \frac{1}{3}\sin 2x, & \text{for } 0 < x < \pi. \end{cases}$

56 $F(x) = \begin{cases} -1 + \cos 2x, & \text{for } -\pi < x < 0 \\ 3 + \cos 2x, & \text{for } 0 < x < \pi. \end{cases}$

Longer problems

In each of the following problems described by a differential equation, find the frequency response function and graph its amplification factor $A(\Omega)$ and phase change $\phi(\Omega)$ as functions of the frequency Ω. Determine the frequency response function $T(mi)$ for $m = 0, 1, \ldots, 4$.

57 $\dfrac{d^3 y}{dt^3} + 15\dfrac{d^2 y}{dt^2} + 79\dfrac{dy}{dt} + 129 y = 20\dfrac{d^2 f}{dt^2} + 80\dfrac{df}{dt} + 120 f.$

58 $2\dfrac{d^3 y}{dt^3} + 21\dfrac{d^2 y}{dt^2} + 120\dfrac{dy}{dt} + 55 y = 30\dfrac{d^2 f}{dt^2} - 4 \cdot 5 f.$

59 $\dfrac{d^3 y}{dt^3} + 15\dfrac{d^2 y}{dt^2} + 79\dfrac{dy}{dt} + 129 y = 65\dfrac{df}{dt} - 195 f$

60 $\dfrac{d^3 y}{dt^3} + 14\dfrac{d^2 y}{dt^2} + 92\dfrac{dy}{dt} + 208 y = 20\dfrac{d^2 t}{dt^2} + 50\dfrac{df}{dt} + 80 f.$

9.3 Convergence of Fourier series and related results

This section provides the formal justification of the convergence properties of Fourier series, which up to this point have merely been assumed on the basis of the behavior of some typical partial sums. The importance of Fourier series, and the extensive use made of them, provides the justification for this more careful than usual study of their behavior. The detailed arguments given here may be omitted at a first reading, but all readers should study the main result of this section contained in Theorem 9.7.

The starting point for our examination of the convergence properties of Fourier series will be the result of Prob. 7, Sec. 1.7, which established the identity

$$\frac{1}{2} + \sum_{m=1}^{n} \cos mx = \frac{\sin\left(n + \frac{1}{2}\right)x}{2\sin\frac{1}{2}x}. \tag{1}$$

Integration of this identity yields the results

$$\frac{1}{\pi}\int_{-\pi}^{0} \frac{\sin\left(n + \frac{1}{2}\right)x}{2\sin\frac{1}{2}x}\,dx = \frac{1}{\pi}\int_{0}^{\pi} \frac{\sin\left(n + \frac{1}{2}\right)x}{2\sin\frac{1}{2}x}\,dx = \frac{1}{2}, \tag{2}$$

which are integrals we will need later.

For any given n, the partial sum $s_n(x)$ of the Fourier series representation of $f(x)$ on the fundamental interval $-\pi \leq x \leq \pi$ is defined as

$$s_n(x) = \tfrac{1}{2}a_0 + \sum_{m=1}^{n} (a_m \cos mx + b_m \sin mx), \tag{3}$$

which we shall refer to as a **Fourier series trignometric polynomial** of **order** n. When the definitions of a_m, b_m are incorporated into (3) it becomes

$$s_n(x) = \frac{1}{2\pi} \int_{-\pi}^{\pi} f(\tau)\,d\tau + \frac{1}{\pi}\sum_{m=1}^{n}\left(\int_{-\pi}^{\pi} f(\tau)\cos mx \cos m\tau\,d\tau + \int_{-\pi}^{\pi} f(\tau)\sin mx \sin m\tau\,d\tau \right), \tag{4}$$

where the functions $\cos mx$, $\sin mx$ have been taken under the integral signs because they are functions of x, and so are not involved in the integration with respect to the dummy variable τ.

Combining the two integrals, taking the summation sign under the integral sign and using the trigonometric identity $\cos mx \cos m\tau + \sin mx \sin m\tau = \cos m(x-\tau)$, we arrive at the result

$$s_n(x) = \frac{1}{\pi}\int_{-\pi}^{\pi} f(\tau)\left[\tfrac{1}{2} + \sum_{m=1}^{n} \cos m(x-\tau)\right]d\tau. \tag{5}$$

Appeal to the identity in (1) then shows that

$$s_n(x) = \frac{1}{\pi}\int_{-\pi}^{\pi} f(\tau)\frac{\sin\left[(n+\tfrac{1}{2})(x-\tau)\right]}{2\sin\tfrac{1}{2}(x-\tau)}\,d\tau. \tag{6}$$

This result is known as the **Dirichlet integral representation** for the partial sum $s_n(x)$, and the expression

$$D_n(x) = \frac{1}{2\pi}\frac{\sin(n+\tfrac{1}{2})(x-\tau)}{\sin\tfrac{1}{2}(x-\tau)} \tag{7}$$

is then called the **Dirichlet kernel**.

Representing $f(\tau)$ outside its fundamental interval by its periodic extensions and making the substitution $t = x - \tau$ reduces result (6) to

$$s_n(x) = \frac{1}{2\pi}\int_{x-\pi}^{x+\pi} f(x-t)\frac{\sin(n+\tfrac{1}{2})t}{\sin\tfrac{1}{2}t}\,dt. \tag{8}$$

However, the periodic extension of f outside its fundamental interval allows us to use Theorem 9.1 to shift the interval of integration in (8) by an amount x from $x-\pi \leq t \leq x+\pi$ to $-\pi \leq t \leq \pi$. When this is done we obtain the following integral representation for the partial sum

$$s_n(x) = \frac{1}{2\pi}\int_{-\pi}^{\pi} f(x-t)\frac{\sin(n+\tfrac{1}{2})t}{\sin\tfrac{1}{2}t}\,dt. \tag{9}$$

To simplify the arguments which follow, let us now assume that $f(x)$ is such that

$$\int_{-\pi}^{\pi} [f(x)]^2 \, dx < \infty.$$

This condition[8] is satisfied by most functions of practical interest, and when it is true we say $f(x)$ is **square integrable** over the interval $-\pi \leq x \leq \pi$. Thus when $f(x)$ is such a function the integral

$$\int_{-\pi}^{\pi} [f(x) - s_n(x)]^2 \, dx$$

will exist and be finite. Expanding the integrand we obtain the identity

$$\int_{-\pi}^{\pi} [f(x) - s_n(x)]^2 \, dx = \int_{-\pi}^{\pi} [f(x)]^2 \, dx - 2 \int_{-\pi}^{\pi} f(x) s_n(x) \, dx + \int_{-\pi}^{\pi} [s_n(x)]^2 \, dx. \tag{10}$$

It is an immediate consequence of the definition of $s_n(x)$, the definitions a_n, b_n in (14) of Sec. 9.2 and the orthogonality properties of the sine and cosine functions recorded in (3), (4) and (5) of Sec. 9.2, that

$$\int_{-\pi}^{\pi} [s_n(x)]^2 \, dx = \int_{-\pi}^{\pi} f(x) s_n(x) \, dx = \pi \left[\tfrac{1}{2} a_0^2 + \sum_{m=1}^{n} (a_m^2 + b_m^2) \right].$$

Thus identity (10) reduces to

$$\int_{-\pi}^{\pi} [f(x) - s_n(x)]^2 \, dx = \int_{-\pi}^{\pi} [f(x)]^2 \, dx - \pi \left[\tfrac{1}{2} a_0^2 + \sum_{m=1}^{n} (a_m^2 + b_m^2) \right]. \tag{11}$$

Now $[f(x) - s_n(x)]^2 \geq 0$, so the left-hand side of (11) is nonnegative, and thus

$$\frac{1}{2} a_0^2 + \sum_{m=1}^{n} (a_m^2 + b_m^2) \leq \frac{1}{\pi} \int_{-\pi}^{\pi} [f(x)]^2 \, dx, \tag{12}$$

for all n. As the integral on the right-hand side of (12) is finite by hypothesis and since n was arbitrary, inequality (10) establishes that the infinite series

$$\tfrac{1}{2} a_0^2 + \sum_{n=1}^{\infty} (a_n^2 + b_n^2) < \infty,$$

and so is *convergent*. The convergence of a series implies that its nth term tends to zero as

---

[8] In more advanced accounts employing the ideas of functional analysis, functions $f(x)$ which may be complex and which satisfy the condition

$$\left[\int_a^b |f(x)|^p \, dx \right]^{1/p} < \infty, \qquad \text{for } 1 \leq p,$$

are said to belong to the class of L^p functions on $[a, b]$. Thus a square integrable function is an L^2 function.

$n \to \infty$, and thus we conclude that

$$\lim_{n \to \infty} a_n = 0 \quad \text{and} \quad \lim_{n \to \infty} b_n = 0. \tag{13}$$

These results established first by Riemann, and subsequently generalized by Lebesgue[9], form the next result.

Theorem 9.6 (Riemann–Lebesgue lemma)

Let $f(x)$ defined for $-\pi \leq x \leq \pi$ be such that

$$\int_{-\pi}^{\pi} [f(x)]^2 \, dx < \infty.$$

Then

$$\lim_{n \to \infty} \int_{-\pi}^{\pi} f(x) \cos nx \, dx = 0 \quad \text{and} \quad \lim_{n \to \infty} \int_{-\pi}^{\pi} f(x) \sin nx \, dx = 0.$$

Equivalently, if a_n and b_n are the Fourier coefficients of $f(x)$, then

$$\lim_{n \to \infty} a_n = 0 \quad \text{and} \quad \lim_{n \to \infty} b_n = 0. \qquad \square$$

One further result is needed before we will be able to establish the fundamental convergence theorem for Fourier series. It will be deduced directly from the Riemann–Lebesgue lemma.

Let $-\pi \leq a < b \leq \pi$ and consider an arbitrary function $f(x)$ which is square integrable over the interval $-\pi \leq x \leq \pi$. Define $F(x)$, $G(x)$ and $H(x)$ by

$$F(x) = \begin{cases} f(x), & -\pi \leq x \leq a \\ 0, & a < x \leq \pi, \end{cases}$$

$$G(x) = \begin{cases} 0, & -\pi \leq x < a \\ f(x), & a \leq x \leq b \\ 0, & b < x \leq \pi, \end{cases}$$

$$H(x) = \begin{cases} 0, & -\pi \leq x \leq b \\ f(x), & b \leq x \leq \pi, \end{cases}$$

then

$$\int_{-\pi}^{\pi} f(x) \cos nx \, dx = \int_{-\pi}^{\pi} F(x) \cos nx \, dx + \int_{-\pi}^{\pi} G(x) \cos nx \, dx + \int_{-\pi}^{\pi} H(x) \cos nx \, dx.$$

---

[9] HENRI LEBESGUE (1875–1941) an outstanding French mathematician who when a Professor at the Collège de France used the ideas of set theory to generalize the notion of the Riemann integral.

Taking the limit of this result as $n\to\infty$ and applying the Riemann–Lebesgue lemma to each integral on the right shows that each term tends to zero in the limit and, in particular, that

$$\lim_{n\to\infty}\int_{-\pi}^{\pi}G(x)\cos nx\,dx=0,$$

which is equivalent to

$$\lim_{n\to\infty}\int_{a}^{b}f(x)\cos nx\,dx=0.\tag{14}$$

A similar argument shows that

$$\lim_{n\to\infty}\int_{a}^{b}f(x)\sin ndx\,dx=0.\tag{15}$$

As $f(x)$ was arbitrary, replacing $f(x)$ by $f(x)\sin\frac{1}{2}x$ in (14) and by $f(x)\cos\frac{1}{2}x$ in (15) and adding the integrals bring us to the result

$$\lim_{n\to\infty}\int_{a}^{b}f(x)\sin(n+\tfrac{1}{2})x\,dx=0.\tag{16}$$

We are now able to proceed to the proof of the fundamental Fourier theorem on the convergence properties of Fourier series.

Let the function $f(x)$ defined for $-\pi\le x\le\pi$ have the following properties:
(i) it is square integrable on $-\pi\le x\le\pi$,
(ii) it is piecewise continuous with finitely many segments,
(iii) it has bounded left and right-hand derivatives at each of its discontinuities,
(iv) it is continued outside its fundamental interval as a periodic extension.

Suppose a discontinuity occurs at $x=x_0$, then it follows from (8) that

$$s_n(x_0)=\frac{1}{2\pi}\int_{-\pi}^{\pi}f(x_0-x)\frac{\sin(n+\tfrac{1}{2})x}{\sin\frac{1}{2}x}\,dx.\tag{17}$$

Denote the left and right-hand limits of $f(x)$ at $x=x_0$ by $f(x_{0-})$ and $f(x_{0+})$, respectively, so that

$$f(x_{0-})=\lim_{x\to x_0-}f(x)\quad\text{and}\quad f(x_{0+})=\lim_{x\to x_0+}f(x).$$

Multiplying the first and second integrals in (2) by $f(x_{0-})$ and $f(x_{0+})$, respectively, shows that

$$\frac{1}{2}f(x_{0-})=\frac{1}{\pi}\int_{-\pi}^{\pi}f(x_{0-})\frac{\sin(n+\tfrac{1}{2})x}{2\sin\frac{1}{2}x}\,dx,\tag{18}$$

and

$$\frac{1}{2}f(x_{0+})=\frac{1}{\pi}\int_{0}^{\pi}f(x_{0+})\frac{\sin(n+\tfrac{1}{2})x}{2\sin\frac{1}{2}x}\,dx.\tag{19}$$

Thus we have from (17), (18) and (19) that

$$s_n(x_0)-\frac{1}{2}[f(x_{0-})+f(x_{0+})]=\frac{1}{\pi}\int_{-\pi}^{0}[f(x_0-x)-f(x_{0-})]\frac{\sin(n+\frac{1}{2})x}{\sin\frac{1}{2}}dx$$

$$+\frac{1}{\pi}\int_{0}^{\pi}[f(x_0-x)-f(x_{0+})]\frac{\sin(n+\frac{1}{2})x}{\sin\frac{1}{2}x}dx. \tag{20}$$

Defining

$$\Phi(x)=\left(\frac{f(x_0-x)-f(x_{0-})}{x}\right)\left(\frac{\frac{1}{2}x}{\sin\frac{1}{2}x}\right)\qquad\text{for }-\pi\le x<x_0,$$

and

$$\Psi(x)=\left(\frac{f(x_0-x)-f(x_{0+})}{x}\right)\left(\frac{\frac{1}{2}x}{\sin\frac{1}{2}x}\right)\qquad\text{for }x_0<x\le\pi,$$

we see that $\Phi(x)$ and $\Psi(x)$ are both finite except, possibly, as $x\to x_0$. To determine their behavior as $x\to x_0$, notice that the first factor in $\Phi(x)$ tends to the left-hand derivative of $f(x)$ at x_0, while the first factor in $\Psi(x)$ tends to the right-hand derivative of $f(x)$ at x_0, both of which exist by hypothesis and are finite. In each case the second factor tends to the same finite values as $x\to x_0$. Consequently both integrals in (20) are well defined.

An application of result (16) to the first integral in (20) with $f(x)$ replaced by $\Phi(x)$, $a=-\pi$ and $b=0$, and to the second integral with $f(x)$ replaced by $\Psi(x)$, $a=0$ and $b=\pi$ brings us to the fundamental result that

$$\lim_{n\to\infty}\{s_n(x_0)-\frac{1}{2}[f(x_{0-})+f(x_{0+})]\}=0. \tag{21}$$

Thus if x_0 is a point of discontinuity of $f(x)$, at $x=x_0$ its Fourier series will converge to the average of the values of $f(x)$ to the immediate left and right of the discontinuity. However, if ξ is a point of continuity[10] of $f(x)$, at $x=\xi$ the Fourier series will converge to the value of $f(\xi)$. We have thus provided the justification for the intuitive approach to Fourier series adopted in Sec. 9.3.

Theorem 9.7 (The Fourier theorem)

Let $f(x)$ defined for $-\pi\le x\le\pi$ with Fourier coefficients a_n, b_n be such that:
  (i) it is square integrable on $-\pi\le x\le\pi$,
 (ii) it is piecewise continuous with finitely many segments,
(iii) it has bounded left and right-hand derivatives at each of its discontinuities,
(iv) it is continued outside its fundamental interval as a periodic extension.

---

[10] Notice that if $f(x_{0-})=f(x_{0+})=L$, the Fourier series for $f(x)$ will converge to the value L irrespective of whether or not $f(x)$ is defined at x_0 and, if it is, of the value $f(x_0)$. Thus if $f(x)$ and $F(x)$ are two functions which differ one from the other only at isolated points their Fourier series will be identical. It is in this sense that the Fourier series representation of a function is *unique*.

Then if x_0 is a point at which $f(x)$ is discontinuous and ξ is one at which it is continuous, the Fourier series of $f(x)$ has the property that

$$\tfrac{1}{2}a_0 + \sum_{n=1}^{\infty} (a_n \cos nx + b_n \sin nx) = \begin{cases} \tfrac{1}{2}[f(x_{0-}) + f(x_{0+})] \text{ when } x = x_0 \\ f(\xi) \text{ when } x = \xi. \end{cases} \qquad \square$$

This fundamental theorem shows how the relational symbol $\sim$ used in the definition of a Fourier series is to be interpreted in terms of the functional values assumed by $f(x)$. It also justifies all the conclusions reached so far on an intuitive basis.

We remark in passing that the convergence of Fourier series may be based on the requirement that $s_n(x)$ converges in the mean to $f(x)$ (cf. (7) in Sec. 9.1), rather than on the requirement that it converges pointwise to $f(x)$, as in the above argument. To ensure convergence in the mean, the only condition which need be imposed on $f(x)$ is that

$$\int_{-\pi}^{\pi} [f(x)]^2 \, dx < \infty,$$

which applies directly to (10).

However if, as above, this condition is imposed on $f(x)$ together with the extra conditions in Theorem 9.7, then it follows immediately that $s_n(x)$ converges both pointwise and in the mean to $f(x)$. Convergence in the mean is often a more natural requirement when approximations are involved, implying as it does that $s_n(x)$ approximates $f(x)$ in a mean square sense.

Parseval's relation

An immediate consequence of Theorem 9.7 combined with (9) is that for every function $f(x)$ satisfying the conditions of Theorem 9.7 it follows that

$$\frac{1}{2}a_0^2 + \sum_{n=1}^{\infty} (a_n^2 + b_n^2) = \frac{1}{\pi} \int_{-\pi}^{\pi} [f(x)]^2 \, dx. \qquad (22)$$

This result is called the **Parseval relation**. It is the strong form of **Bessel's inequality**

$$\frac{1}{2}a_0^2 + \sum_{n=1}^{\infty} (a_n^2 + b_n^2) \leq \frac{1}{\pi} \int_{-\pi}^{\pi} [f(x)]^2 \, dx, \qquad (23)$$

which may be deduced from (10) without assuming the pointwise convergence of $s_n(x)$ to $f(x)$ as $n \to \infty$, by simply observing that the integral on the left is nonnegative, and so

$$\int_{-\pi}^{\pi} [f(x)]^2 \, dx - \pi \left[\tfrac{1}{2}a_0^2 + \sum_{n=1}^{\infty} (a_n^2 + b_n^2) \right] \geq 0.$$

Setting $t = Lx/\pi$ and $g(t) = f(\pi t/L)$, and then replacing t by x, it follows at once from (23) that for a function $g(x)$ with fundamental period $2L$ and Fourier coefficients a_n, b_n the

Parseval relation takes the form

$$\frac{1}{L}\int_{-L}^{L}[g(x)]^2\,dx = \tfrac{1}{2}a_0^2 + \sum_{n=1}^{\infty}(a_n^2 + b_n^2). \tag{24}$$

The forms taken by the Parseval relation for different types of Fourier series are listed in Table 9.2.

Table 9.2   Useful forms of the Parseval relation

Type of series	Parseval's relation
Fourier series for $f(x)$ on $-\pi \le x \le \pi$	$\dfrac{1}{\pi}\displaystyle\int_{-\pi}^{\pi}[f(x)]^2\,dx = \tfrac{1}{2}a_0^2 + \sum_{n=1}^{\infty}(a_n^2 + b_n^2)$
Fourier series for $f(x)$ on $-L \le x \le L$	$\dfrac{1}{L}\displaystyle\int_{-L}^{L}[f(x)]^2\,dx = \tfrac{1}{2}a_0^2 + \sum_{n=1}^{\infty}(a_n^2 + b_n^2)$
Fourier series for $f(x)$ on $a \le x \le b$	$\dfrac{2}{(b-a)}\displaystyle\int_{a}^{b}[f(x)]^2\,dx = \tfrac{1}{2}a_0^2 + \sum_{n=1}^{\infty}(a_n^2 + b_n^2)$
Fourier sine series for $f(x)$ on $0 \le x \le \pi$	$\dfrac{2}{\pi}\displaystyle\int_{0}^{\pi}[f(x)]^2\,dx = \sum_{n=1}^{\infty}b_n^2$
Fourier cosine series for $f(x)$ on $0 \le x \le \pi$	$\dfrac{2}{\pi}\displaystyle\int_{0}^{\pi}[f(x)]^2\,dx = \tfrac{1}{2}a_0^2 + \sum_{n=1}^{\infty}a_n^2$
Fourier sine series for $f(x)$ on $0 \le x \le L$	$\dfrac{2}{L}\displaystyle\int_{0}^{L}[f(x)]^2\,dx = \sum_{n=1}^{\infty}b_n^2$
Fourier cosine series for $f(x)$ on $0 \le x \le L$	$\dfrac{2}{L}\displaystyle\int_{0}^{L}[f(x)]^2\,dx = \tfrac{1}{2}a_0^2 + \sum_{n=1}^{\infty}a_n^2$
Complex Fourier series for $f(x)$ on $-\pi \le x \le \pi$	$\dfrac{1}{2\pi}\displaystyle\int_{-\pi}^{\pi}|f(x)|^2\,dx = \lim_{m\to\infty}\sum_{n=-m}^{m}|c_n|^2$
Complex Fourier series for $f(x)$ on $-L \le x \le L$	$\dfrac{1}{2L}\displaystyle\int_{-L}^{L}|f(x)|^2\,dx = \lim_{m\to\infty}\sum_{n=-m}^{m}|c_n|^2$

Example 1. An application of the Parseval relation

By considering the function $f(x) = \sin x$ with the fundamental interval $0 \le x \le \pi$, and using Parseval's relation, find a series representation for $\pi^2/16$ and comment on its rate of convergence.

Solution

The fundamental interval for the function $f(x) = \sin x$ is defined as $0 \le x \le \pi$, and on this interval the function is seen by inspection to satisfy the conditions of Theorem 9.7. As

$$\frac{2}{\pi} \int_0^\pi \sin^2 x \, dx = 1,$$

it follows from the Parseval relation in the form given in (24) with $L = \pi$ that

$$1 = \frac{1}{2} a_0^2 + \sum_{n=1}^\infty (a_n^2 + b_n^2),$$

where a_n, b_n are the Fourier coefficients of $f(x) = \sin x$ defined on the interval $0 \le x \le \pi$. Thus from Theorem 9.4 with $a = 0$, $b = \pi$ and $f(x) = \sin x$ we see that

$$a_0 = \frac{2}{\pi} \int_0^\pi \sin x \, dx = \frac{4}{\pi},$$

$$a_n = \frac{2}{\pi} \int_0^\pi \sin x \cos 2nx \, dx = \frac{1}{\pi} \int_0^\pi [\sin(2n+1)x - \sin(2n-1)x] \, dx$$

$$= \frac{-4}{\pi} \frac{1}{(2n-1)(2n+1)}, \qquad n = 1, 2, \ldots,$$

$$b_n = \frac{2}{\pi} \int_0^\pi \sin x \sin 2nx \, dx = \frac{1}{\pi} \int_0^\pi [\cos(2n-1)x - \cos(2n+1)x] \, dx = 0, \qquad n = 1, 2, \ldots.$$

Thus substituting the Fourier coefficients into the Parseval relation and multiplying by $\pi^2/16$ gives

$$\frac{\pi^2}{16} = \frac{1}{2} + \sum_{n=1}^\infty \frac{1}{(2n-1)^2 (2n+1)^2}.$$

This series converges fairly rapidly as its nth term is of order $1/(16n^4)$. Its sums to 5, 10 and 20 terms yield as respective approximations to π the values 3.14127, 3.1415 and 3.14159.

The product theorem for Fourier series

The Parseval relation can be generalized to the case in which a product of two functions is involved. To accomplish this let the functions $f(x)$, $g(x)$ defined for $-\pi \le x \le \pi$ satisfy the conditions of Theorem 9.7 and have the respective Fourier coefficients a_n, b_n and A_n, B_n. Then

$$f(x) + g(x) \sim \tfrac{1}{2}(a_0 + A_0) + \sum_{n=1}^\infty [(a_n + A_n)\cos nx + (b_n + B_n)\sin nx],$$

and

$$f(x) - g(x) \sim \tfrac{1}{2}(a_0 - A_0) + \sum_{n=1}^\infty [(a_n - A_n)\cos nx + (b_n - B_n)\sin nx].$$

The Parseval relation applies to both $f(x) + g(x)$ and $f(x) - g(x)$ so that

$$\frac{1}{\pi} \int_{-\pi}^{\pi} [f(x) + g(x)]^2 \, dx = \tfrac{1}{2}(a_0 + A_0)^2 + \sum_{n=1}^{\infty} [(a_n + A_n)^2 + (b_n + B_n)^2],$$

and

$$\frac{1}{\pi} \int_{-\pi}^{\pi} [f(x) - g(x)]^2 \, dx = \tfrac{1}{2}(a_0 - A_0)^2 + \sum_{n=1}^{\infty} [(a_n - A_n)^2 + (b_n - B_n)^2].$$

Subtracting these results and dividing by 4 yields the desired generalization

$$\frac{1}{\pi} \int_{-\pi}^{\pi} f(x) g(x) \, dx = \tfrac{1}{2} a_0 A_0 + \sum_{n=1}^{\infty} (a_n A_n + b_n B_n). \tag{25}$$

Extending this result in an obvious manner to functions $f(x)$ and $g(x)$ defined on the fundamental interval $a \leq x \leq b$ we obtain our next theorem.

Theorem 9.8 (Product theorem for Fourier series)

Let the two functions $f(x)$ and $g(x)$ defined on the fundamental interval $a \leq x \leq b$ satisfy conditions analogous to those in Theorem 9.7 and have the respective Fourier coefficients a_n, b_n and A_n, B_n. Then

$$\frac{2}{b-a} \int_{a}^{b} f(x) g(x) \, dx = \tfrac{1}{2} a_0 A_0 + \sum_{n=1}^{\infty} (a_n A_n + b_n B_n). \qquad \square$$

Best mean square trigonometric polynomial approximation

Let us show that the approximation to $f(x)$ provided by the partial sum $s_n(x)$ of its Fourier series is the best possible trigonometric polynomial approximation of order n to $f(x)$ in the least squares sense. Suppose the function $f(x)$ is defined for $-\pi \leq x \leq \pi$, and that

$$s_n(x) = \tfrac{1}{2} a_0 + \sum_{m=1}^{n} (a_m \cos mx + b_n \sin mx)$$

with a_n, b_n the Fourier coefficients of $f(x)$, while

$$\phi_n(x) = \tfrac{1}{2} A_0 + \sum_{m=1}^{n} (A_m \cos mx + B_m \sin mx)$$

is some other approximation with the coefficients A_n, B_n. Then to establish the stated result we need to show that the square error

$$e_n = \int_{-\pi}^{\pi} [f(x) - \phi_n(x)]^2 \, dx \tag{26}$$

will be minimized when $\phi_n(x) \equiv s_n(x)$, which will be when $A_0 = a_0$, $A_m = a_m$, and $B_m = b_m$ for $m = 1, 2, \ldots, n$.

Expanding the integrand in (26) using the definitions of the Fourier coefficients a_m, b_m and the orthogonality properties of sines and cosines we find that

$$
e_n = \int_{-\pi}^{\pi} [f(x)]^2 \, dx - 2 \int_{-\pi}^{\pi} \phi_n(x) f(x) \, dx + \int_{-\pi}^{\pi} [\phi_n(x)]^2 \, dx
$$

$$
= \int_{-\pi}^{\pi} [f(x)]^2 \, dx - \pi \left[a_0 A_0 + 2 \sum_{m=1}^{n} (a_m A_m + b_m B_m) \right]
$$

$$
+ \pi \left[\tfrac{1}{2} A_0^2 + \sum_{m=1}^{n} (A_m^2 + B_m^2) \right]
$$

$$
= \int_{-\pi}^{\pi} [f(x)]^2 \, dx - \pi \left\{ \tfrac{1}{2}(A_0 - a_0)^2 + \sum_{m=1}^{n} [(A_m - a_m)^2 + (B_m - b_m)^2] \right\}
$$

$$
- \pi \left[\tfrac{1}{2} a_0^2 + \sum_{m=1}^{n} (a_m^2 + b_m^2) \right].
$$

Clearly, the square error e_n will attain its least value when the second term on the right-hand side vanishes. This will occur when $A_0 = a_0$, $A_m = a_m$ and $B_m = b_m$ for $m = 1, 2, \ldots, n$, which completes the proof of our original assertion and establishes the following Theorem.

Theorem 9.9 (Best mean square trigonometric approximation)

Let the function $f(x)$ defined on the interval $-\pi \le x \le \pi$ have the Fourier coefficients a_n, b_n. Then the partial sum

$$
s_n(x) = \frac{1}{2} a_0 + \sum_{m=1}^{n} (a_m \cos mx + b_m \sin mx)
$$

of the Fourier series of $f(x)$ provides the best approximation to $f(x)$ in the least square sense of all possible trigonometric polynomials of order n. □

Completeness

Before leaving this section, a few general remarks should be offered about the type of function $f(x)$ which can be represented by a Fourier series. For a Fourier series to be useful, it is necessary to know that it can, in some suitable sense, represent every function belonging to a class of interest.

The most important class of functions to arise in most applications, and the one of concern to us here, is the class of functions which satisfy conditions (i) to (iii) of Theorem 9.7. When all functions belonging to a class of interest can be represented by a Fourier series, the trigonometric system

$$
\{1, \sin x, \cos x, \sin 2x, \cos 2x, \ldots\}
$$

used when developing the Fourier series is said to be **complete** with respect to that class of functions. We now indicate why the trigonometric system is complete with respect to the class of functions satisfying conditions (i) to (iii) of Theorem 9.7.

In point of fact, the system is complete with respect to a much wider class of functions, but to discuss these (Lebesgue integrable) functions would require the use of mathematical techniques which are beyond the scope of this book.

Expressed differently, the completeness of the trigonometric system with respect to a specific class of functions means there is *no* function belonging to that class, other than the function which is identically zero, which when represented as a Fourier series will have all of its Fourier coefficients zero.

Recalling the notion of the *orthogonality* of functions over an interval introduced at the start of Sec. 9.2, this means there must be no function of interest which is orthogonal to *every* function in the trigonometric system. It turns out that the most useful sense in which a Fourier series can be said to represent a function $f(x)$ over the interval $-\pi \le x \le \pi$ is through convergence in the mean (mean square convergence) to $f(x)$.

Accordingly, the trigonometric system is said to be complete with respect to a given class of functions $f(x)$ if for every function in that class the mean square error e_n defined in (26) tends to zero as $n \to \infty$, so that

$$\lim_{n \to \infty} \int_{-\pi}^{\pi} [f(x) - s_n(x)]^2 \, dx = 0. \tag{27}$$

Now we have seen that if a function $f(x)$ satisfies conditions (i) to (iii) of Theorem 9.7, its Fourier series will converge pointwise to $f(x)$ everywhere in the interval $-\pi \le x \le \pi$, except for isolated points at which $f(x)$ is discontinuous. However, as $n \to \infty$, it is possible to show that the contribution made by these points to the mean square error e_n in (26) through $s_n(x)$ becomes increasingly localized and, in the limit, will vanish (compare this situation with Ex. 3, Sec. 9.1).

Thus, when this fact is taken in conjunction with Theorems 9.7 and 9.9, we may conclude that the trigonometric system

$$\{1, \cos x, \sin x, \cos 2x, \sin 2x, \ldots\}$$

is **complete** over the interval $-\pi \le x \le \pi$ with respect to functions $f(x)$ satisfying conditions (i) to (iii) of Theorem 9.7.

Now the Parseval relation

$$\frac{1}{\pi} \int_{-\pi}^{\pi} [f(x)]^2 \, dx = \frac{1}{2} a_0^2 + \sum_{n=1}^{\infty} \left(a_n^2 + b_n^2 \right), \tag{28}$$

where a_n, b_n are the Fourier coefficients of $f(x)$, has been shown to be a direct consequence of Theorem 9.7. Consequently the completeness of the trigonometric system with respect to functions $f(x)$ satisfying conditions (i) to (iii) of Theorem 9.7 implies the Parseval relation. Conversely, if the class of functions $f(x)$ is such that (28) is true, then the trigonometric system must be complete with respect to that class of functions. For this reason the Parseval relation (28) is often called the **completeness condition**.

The significance of Bessel's inequality in relation to the Parseval relation needs to be understood. Bessel's inequality

$$\frac{1}{\pi} \int_{-\pi}^{\pi} [f(x)]^2 \, dx \geq \frac{1}{2} a_0^2 + \sum_{n=1}^{\infty} (a_n^2 + b_n^2)$$

is true for any function $f(x)$ for which the Fourier coefficients can be determined, whether or not the system of functions used to compute them is complete. On the other hand, as we have already seen, the Parseval relation is true only for a system of functions which is complete with respect to the class of functions to which $f(x)$ belongs. The difference is easily illustrated by means of a simple example.

Consider the trigonometric system

$$\{1, \cos x, \cos 2x, \sin 2x, \cos 3x, \sin 3x, \ldots\}$$

which has been obtained by deleting the function $\sin x$ from the complete trigonometric system used to develop Fourier series. Thus the deleted system is *incomplete* because of the omission of the single function $\sin x$. Consequently, as $\sin x$ is orthogonal to *every* other function in this incomplete system, every coefficient in its Fourier series representation based on this system will vanish. So, with respect to the incomplete system, Bessel's inequality with $f(x) = \sin x$ reduces to the obvious result

$$\frac{1}{\pi} \int_{-\pi}^{\pi} (\sin x)^2 \, dx \geq 0.$$

However, as $\sin x$ is its own Fourier series with respect to the complete system, the Parseval relation with $f(x) = \sin x$ becomes

$$\frac{1}{\pi} \int_{-\pi}^{\pi} (\sin x)^2 \, dx = 1,$$

which we know to be true by direct computation.

Problems for Section 9.3

1 Derive the Fourier series representation of

$$f(x) = \pi + x, \qquad \text{for } -\pi < x < \pi,$$

and use the Parseval relation to show that

$$\sum_{n=1}^{\infty} \frac{1}{n^2} = \frac{\pi^2}{6}.$$

2 Show that the Fourier series representation of

$$f(x) = |x|, \qquad \text{for } -\pi < x < \pi$$

is

$$f(x) \sim \frac{\pi}{2} - \frac{4}{\pi} \sum_{n=1}^{\infty} \frac{\cos(2n-1)x}{(2n-1)^2}, \qquad \text{for } -\pi < x < \pi.$$

Use the Parseval relation to sum the series

$$\sum_{n=1}^{\infty} \frac{1}{(2n-1)^4}.$$

3 Show by constructing the Fourier series representation of

$$f(x) = \begin{cases} 0, & \text{for} \quad \pi \le x < 0 \\ x, & \text{for} \quad 0 \le x < \pi, \end{cases}$$

using the Parseval relation and the fact that

$$\sum_{n=1}^{\infty} \frac{1}{n^2} = \frac{\pi^2}{6},$$

that

$$\sum_{n=0}^{\infty} \frac{1}{(2n+1)^4} = \frac{\pi^4}{96}$$

4 Show that the Fourier series representation of

$$f(x) = \begin{cases} \sin x, & \text{for} \quad 0 \le x < \pi \\ 0, & \text{for} \quad \pi < x \le 2\pi \end{cases}$$

is

$$f(x) \sim \frac{1}{\pi} + \frac{1}{2}\sin x - \frac{2}{\pi} \sum_{n=1}^{\infty} \frac{\cos 2nx}{(2n-1)(2n+1)}.$$

Apply the Parseval relation to the above series to show that

$$\frac{\pi^2}{8} = 1 + 2 \sum_{n=1}^{\infty} \frac{1}{(2n-1)^2 (2n+1)^2}.$$

5 By using the product theorem for Fourier series, and considering the Fourier series representation of

$$f(x) = \begin{cases} 1+x, & \text{for} \quad -1 < x < 0 \\ -1, & \text{for} \quad 0 < x < 1 \end{cases}$$

and

$$g(x) = \begin{cases} 1, & \text{for} \quad -1 < x < 0. \\ x, & \text{for} \quad 0 < x < 1 \end{cases}$$

given in Problems 21 and 25 for Sec. 9.2, show that

$$\sum_{n=1}^{\infty} \left(\frac{1-(-1)^n}{n^4} \right) = \frac{\pi^4}{48}.$$

6 Explain on the basis of the product theorem for Fourier series why the integral of the product of an even function $f(x)$ and an odd function $g(x)$, both defined for $-L \leq x \leq L$, must vanish.

9.4 Integration and differentiation of Fourier series

In this section we show that the Fourier series of a piecewise continuous function $f(x)$ defined for $L \leq x \leq L$ may always be integrated term by term to give a series representation of the function

$$F(x) = \int_c^x f(t)\,dt, \tag{1}$$

for $-L \leq c < x \leq L$, though the resulting series is not necessarily a Fourier series.

We will also establish sufficient conditions under which the Fourier series of $f(x)$ may be differentiated term by term to yield the Fourier series of the function $f'(x)$.

Integration of Fourier series

Let $f(t)$ be a piecewise continuous function defined for $-L \leq t \leq L$ and $g(t)$ be the function

$$g(t) = \begin{cases} 0, & -L \leq t < c \\ 1, & c \leq t \leq x \\ 0, & x < t \leq L. \end{cases} \tag{2}$$

Replacing the dummy variable x in Theorem 9.8 (the product theorem) by t and identifying the functions in the theorem with those above we find that

$$\frac{1}{L}\int_{-L}^{L} f(t)\,g(t)\,dt = \frac{1}{L}\int_c^x f(t)\,dt = \frac{1}{2}a_0 A_0 + \sum_{n=1}^{\infty}(a_n A_n + b_n B_n). \tag{3}$$

Now the Fourier coefficients A_n, B_n of $g(t)$ are

$$A_0 = \frac{1}{L}\int_c^x 1\,dt = \frac{(x-c)}{L}, \tag{4}$$

$$A_n = \frac{1}{L}\int_c^x \cos\frac{n\pi t}{L}\,dt, \quad B_n = \frac{1}{L}\int_c^x \sin\frac{n\pi t}{L}\,dt,$$

for $n = 1, 2, \ldots$,

Thus substituting for the A_n, B_n in (3) we find that

$$\int_c^x f(t)\,dt = \frac{1}{2}a_0(x-c) + \sum_{n=1}^{\infty}\left(a_n \int_c^x \cos\frac{n\pi t}{L}\,dt + b_n \int_c^x \sin\frac{n\pi t}{L}\,dt\right), \tag{5}$$

which is the required result, because the right-hand side is simply the result obtained by integrating the Fourier series of $f(x)$ term by term. Result (5) is not necessarily a Fourier series, because when $a_0 \neq 0$ it contains a term which is linear in x. We have thus proved the following result.

Theorem 9.10 (Integration of Fourier series)

Let the piecewise continuous function $f(x)$ defined on the interval $-L \le x \le L$ have the Fourier series

$$f(x) \sim \frac{1}{2}a_0 + \sum_{n=1}^{\infty} \left(a_n \cos \frac{n\pi x}{L} + b_n \sin \frac{n\pi x}{L} \right).$$

Then

$$\int_c^x f(t)\,dt = \frac{1}{2}a_0(x-c) + \sum_{n=1}^{\infty} \left(a_n \int_c^x \cos \frac{n\pi t}{L}\,dt + b_n \int_c^x \sin \frac{n\pi t}{L}\,dt \right),$$

for $-L \le c \le x \le L$. □

Example 1. Integrating a Fourier series

Find the Fourier series of $f(x) = x^2$, $-L \le x \le L$, and then use Theorem 9.10 to show that

$$x^3 - L^2 x = \frac{12L^3}{\pi^3} \sum_{n=1}^{\infty} \frac{(-1)^n \sin(n\pi x/L)}{n^3}, \qquad \text{for } -L \le x \le L.$$

Solution

The function $f(x) = x^2$ is an even function which is everywhere continuous on $-L \le x \le L$. The Fourier coefficients for $f(x)$ are thus

$$a_0 = \frac{2}{L} \int_0^L x^2\,dx = \frac{2L^2}{3},$$

$$a_n = \frac{2}{L} \int_0^L x^2 \cos \frac{n\pi x}{L}\,dx = (-1)^n \frac{4L^2}{\pi^2 n^2}, \quad \text{and} \quad b_n = 0, \qquad \text{for } n = 1, 2, \ldots .$$

Thus the Fourier series of $f(x) = x^2$ is

$$x^2 = \frac{L^2}{3} + \frac{4L^2}{\pi^2} \sum_{n=1}^{\infty} \frac{(-1)^n \cos(n\pi x/L)}{n^2}, \qquad \text{for } -L \le x \le L.$$

Use of the equality sign in the Fourier series is justified by virtue of Theorem 9.7, because $f(x)$ is everywhere continuous and $f(-L) = f(L)$. Replacing x by the dummy variable t, and using Theorem 9.10 to justify integrating the above expression from 0 to x, we find that

$$\int_0^x t^2\,dt = \int_0^x \frac{L^2}{3}\,dt + \frac{4L^2}{\pi^2} \sum_{n=1}^{\infty} \int_0^x \frac{(-1)^n \cos(n\pi t/L)}{n^2}\,dt,$$

and so

$$\frac{x^3}{3} = \frac{L^2 x}{3} + \frac{4L^3}{\pi^3} \sum_{n=1}^{\infty} \frac{(-1)^n \sin(n\pi x/L)}{n^3}, \qquad \text{for } -L \le x \le L.$$

This series representation of $x^3/3$ is *not* a Fourier series because of the presence of the term $L^2 x/3$ on the right-hand side. However, if the terms in x are combined on the left of the equality, as in the next result, the series on the right is, indeed, a Fourier series. To be precise, it is the Fourier series of $x^3 - L^2 x$. This fact can easily be verified by direct calculation.

$$x^3 - L^2 x = \frac{12L^3}{\pi^3} \sum_{n=1}^{\infty} \frac{(-1)^n \sin(n\pi x/L)}{n^3}, \qquad \text{for } -L \le x \le L.$$

∎

Differentiation of Fourier series

Let $f(x)$ together with its derivative $f'(x)$ be continuous for $-L \le x \le L$ with $f(-L) = f(L)$, and let the Fourier coefficients of $f(x)$ and $f'(x)$ be, respectively, a_n, b_n and A_n, B_n. Then from the Euler formulas we know that

$$A_n = \frac{1}{L} \int_{-L}^{L} f'(x) \cos \frac{n\pi x}{L} \, dn, \qquad n = 0, 1, \ldots$$

and

$$B_n = \frac{1}{L} \int_{-L}^{L} f'(x) \sin \frac{n\pi x}{L} \, dx, \qquad n = 1, 2, \ldots.$$

Integrating by parts we have

$$A_0 = \frac{1}{L} \int_{-L}^{L} f'(x) \, dx = \frac{1}{L} [f(L) - f(-L)] = 0, \tag{6}$$

because $f(-L) = f(L)$, and

$$A_n = \frac{1}{L} \left(f(x) \cos \frac{n\pi x}{L} \right) \Big|_{-L}^{L} + \frac{n\pi}{L} \int_{-L}^{L} f(x) \sin \frac{n\pi x}{L} \, dx$$

$$= \frac{(-1)^n}{L} [f(L) - f(-L)] + \frac{n\pi}{L} b_n$$

$$= \frac{n\pi}{L} b_n, \tag{7}$$

for $n = 1, 2, \ldots$, where once again use has been made of the condition $f(-L) = f(L)$.

A similar argument shows that

$$B_n = \frac{1}{L} \left(f(x) \sin \frac{n\pi x}{L} \right) \Big|_{-L}^{L} - \frac{n\pi}{L} \int_{-L}^{L} f(x) \cos \frac{n\pi x}{L} \, dx,$$

and hence that

$$B_n = -(n\pi/L)a_n, \quad \text{for } n = 1, 2, \ldots. \tag{8}$$

Substituting for A_n, B_n in the Fourier series for $f'(x)$ using (7) and (8) gives

$$f'(x) = \sum_{n=1}^{\infty} \left(A_n \cos \frac{n\pi x}{L} + B_n \sin \frac{n\pi x}{L} \right)$$

$$= \sum_{n=1}^{\infty} \frac{n\pi}{L} \left(b_n \cos \frac{n\pi x}{L} - a_n \sin \frac{n\pi x}{L} \right),$$

and so

$$f'(x) = \sum_{n=1}^{\infty} \frac{d}{dx} \left(a_n \cos \frac{n\pi x}{L} + b_n \sin \frac{n\pi x}{L} \right), \tag{9}$$

which is the result obtained by term by term differentiation of the Fourier series of $f(x)$.

A more careful argument proceeding along similar lines shows this result still to be true if $f'(x)$ is only piecewise continuous. In this case, however, the differentiated Fourier series will not converge to $f'(x)$ at any point at which $f'(x)$ is discontinuous. This forms the basis of the next result.

Theorem 9.11 (Differentiation of Fourier series)

Let $f(x)$ be a continuous function with fundamental interval $-L \leq x \leq L$ and such that $f(L) = f(-L)$. Suppose also that $f'(x)$ is piecewise continuous on the fundamental interval. Then at every point at which $f''(x)$ exists, the Fourier series for $f(x)$ may be differentiated term by term to yield a Fourier series which converges to $f'(x)$. Thus if

$$f(x) = \frac{1}{2}a_0 + \sum_{n=1}^{\infty} \left(a_0 \cos \frac{n\pi x}{L} + b_n \sin \frac{n\pi x}{L} \right), \quad \text{for } -L \leq x \leq L,$$

then

$$f'(x) = \sum_{n=1}^{\infty} \frac{d}{dx} \left(a_n \cos \frac{n\pi x}{L} + b_n \sin \frac{n\pi x}{L} \right), \quad \text{for } -L \leq x \leq L.$$

Furthermore, when $f''(x)$ does not exist at a point $x = \xi$, but the limiting values $f''(\xi)$ and $f''(\xi_+)$ both exist, differentiation of the series at $x = \xi$ is valid in the sense that the Fourier series for $f'(x)$ converges to the value

$$\tfrac{1}{2}[f'(\xi_-) + f'(\xi_+)]$$

at $x = \xi$. □

Example 2. Differentiation of Fourier series

Establish the Fourier series representations of

$$f(x) = \begin{cases} -\sin x, & -\pi \le x \le 0 \\ \sin x, & 0 < x \le \pi \end{cases} \quad \text{and} \quad g(x) = \begin{cases} -\cos x, & -\pi < x < 0 \\ \cos x, & 0 < x < \pi, \end{cases}$$

and comment on their differentiability.

Solution

The function $f(x)$ is an even function so its Fourier coefficients $b_n = 0$ for $n = 1, 2, \ldots$, while

$$a_0 = \frac{1}{\pi} \int_{-\pi}^{0} -\sin x \, dx + \frac{1}{\pi} \int_{0}^{\pi} \sin dx = \frac{4}{\pi},$$

and

$$a_n = \frac{1}{\pi} \int_{-\pi}^{0} -\sin x \cos nx \, dx + \frac{1}{\pi} \int_{0}^{\pi} \sin x \cos nx \, dx, \qquad n = 1, 2, \ldots.$$

Routine integration then shows that

$$a_{2n-1} = 0 \quad \text{and} \quad a_{2n} = \frac{-4}{\pi(4n^2 - 1)}, \qquad n = 1, 2, \ldots.$$

Thus $f(x)$ has the Fourier series representation

$$f(x) = \frac{2}{\pi} - \frac{4}{\pi} \sum_{n=1}^{\infty} \frac{\cos 2nx}{(4n^2 - 1)}, \qquad \text{for } -\pi \le x \le \pi.$$

The equality sign in the Fourier series is permitted by Theorem 9.7, since $f(x)$ is everywhere continuous.

The function $g(x)$ is an odd function, so its Fourier coefficients $a_n = 0$ for $n = 0, 1, \ldots$, while

$$b_n = \frac{1}{\pi} \int_{-\pi}^{0} -\cos x \sin nx \, dx + \frac{1}{\pi} \int_{0}^{\pi} \cos x \sin nx \, dx.$$

This time routine integration brings us to the result

$$b_{2n-1} = 0 \quad \text{and} \quad b_{2n} = \frac{8n}{\pi(4n^2 - 1)}, \qquad n = 1, 2, \ldots.$$

Thus $g(x)$ has the Fourier series representation

$$g(x) \sim \frac{8}{\pi} \sum_{n=1}^{\infty} \frac{n \sin 2nx}{(4n^2 - 1)}, \qquad \text{for } -\pi < x < \pi.$$

The function $f(x)$ satisfies the conditions of Theorem 9.11 and so may be differentiated term by term to obtain the Fourier series of $f'(x)$. Thus differentiating the Fourier series for $f(x)$ we find that

$$f'(x) = \frac{8}{\pi} \sum_{n=1}^{\infty} \frac{n \sin 2nx}{4n^2 - 1}, \qquad \text{for } -\pi \le x \le \pi,$$

which is precisely the Fourier series for $g(x)$, as would be expected.

However, the function $g(x)$ does not satisfy the conditions of Theorem 9.11, since $g(-\pi) \neq g(\pi)$ and $g(x)$ is discontinuous at the origin, so its Fourier series cannot be differentiated term by term to find the Fourier series of $-f(x)$. Indeed, it was because $g(x)$ is discontinuous that the relational symbol $\sim$ was used in place of an equality in its Fourier series representation above.

If the conditions of Theorem 9.11 are ignored and $g(x)$ is differentiated term by term the series which results is

$$\frac{16}{\pi} \sum_{n=1}^{\infty} \frac{n^2 \cos 2nx}{4n^2 - 1}.$$

This series is divergent because its nth term tends to $(4/\pi) \cos 2nx$ as $n \to \infty$, and this does not vanish.

■

Problems for Section 9.4

1 Show the Fourier series representation of

$$f(x) = x(\pi - x), \qquad \text{for } 0 \leq x \leq \pi$$

is

$$f(x) \sim \frac{\pi^2}{6} - \sum_{n=1}^{\infty} \frac{\cos 2nx}{n^2}, \qquad \text{for } 0 \leq x \leq \pi.$$

Justify using Theorem 9.10 to find the Fourier series representation of

$$g(x) = x(\pi - 2x)(\pi - x), \qquad \text{for } 0 \leq x \leq \pi.$$

2 Show that if a piecewise continuous function $f(x)$ has the Fourier series representation

$$f(x) \sim \tfrac{1}{2} a_0 + \sum_{n=1}^{\infty} (a_n \cos nx + b_n \sin nx), \qquad \text{for } -\pi \leq x \leq \pi, \text{ then}$$

$$\int_{-\pi}^{x} f(t)\, dt = \frac{1}{2}(x + \pi)a_0 + \sum_{n=1}^{\infty} \frac{1}{n} \{a_n \sin nx - b_n [\cos nx - (-1)^n]\}.$$

Find the Fourier series representation of $f(x) = x$, for $-\pi \leq x \leq \pi$, and hence the Fourier series representation of $f(x) = x^2$, for $-\pi \leq x \leq \pi$.

3 The function

$$f(x) = \tfrac{1}{2}(\pi - x), \qquad \text{for } -\pi \leq x \leq \pi.$$

Show that $f(x)$ has the Fourier series representation

$$f(x) \sim \frac{\pi}{2} + \sum_{n=1}^{\infty} \frac{(-1)^n \sin nx}{n}, \qquad \text{for } -\pi \leq x \leq \pi.$$

By considering

$$\int_{-\pi}^{x} f(t)\, dt,$$

show that

$$g(x) = \tfrac{1}{4}(3\pi - x)(\pi + x), \qquad \text{for } -\pi \leq x \leq \pi$$

has the Fourier series representation

$$g(x) \sim \frac{\pi^2}{6} - \sum_{n=1}^{\infty} \frac{(-1)^n \cos nx}{n^2}, \qquad \text{for } -\pi \leq x \leq \pi.$$

4 Find the Fourier series representation of

$$f(x) = x, \qquad \text{for } 0 \leq x \leq 2\pi.$$

Use this series to derive the Fourier series representation of

$$g(x) = 3x^2 - 6\pi x + 2\pi^2, \qquad \text{for } 0 \leq x \leq 2\pi.$$

5 Derive the Fourier series representation of

$$f(x) = x(\pi - x), \qquad \text{for } 0 \leq x \leq \pi.$$

Justify applying Theorem 9.11 to this Fourier series, and hence show that the Fourier series representation of

$$g(x) = \pi - 2x, \qquad \text{for } 0 \leq x < \pi$$

is

$$g(x) \sim 2 \sum_{n=1}^{\infty} \frac{\sin 2nx}{n}, \qquad \text{for } 0 \leq x < \pi.$$

6 Show that the Fourier series representation of

$$f(x) = |\sin x|, \qquad \text{for } -\pi \leq x \leq \pi$$

is

$$f(x) \sim \frac{2}{\pi} - \frac{4}{\pi} \sum_{n=1}^{\infty} \frac{\cos 2nx}{4n^2 - 1}, \qquad \text{for } -\pi \leq x \leq \pi.$$

What is the function $g(x)$ defined for $-\pi \leq x \leq \pi$ by the Fourier series representation

$$g(x) \sim \frac{8}{\pi} \sum_{n=1}^{\infty} \left(\frac{n}{4n^2 - 1} \right) \cos 2nx.$$

7 The function

$$f(x) = \begin{cases} 0, & \text{for } -\pi \leq x \leq 0 \\ \sin x, & \text{for } 0 \leq x \leq \pi. \end{cases}$$

Explain why Theorem 9.11 may be applied to the Fourier series representation of $f(x)$ for $-\pi \leq x \leq \pi$ to find the Fourier series representation of

$$g(x) = \begin{cases} 0, & \text{for } -\pi \leq x \leq 0 \\ \cos x, & \text{for } 0 \leq x \leq \pi, \end{cases}$$

and hence find its form.

Why cannot the same result be obtained by applying Theorem 9.11 to the Fourier series representation of the function

$$h(x) = \begin{cases} 1, & \text{for } -\pi \leq x \leq 0 \\ \sin x, & \text{for } 0 \leq x \leq \pi. \end{cases}$$

8 The function

$$f(x)=|x|, \qquad \text{for } -2 \leq x \leq 2.$$

Show that its Fourier series representation is

$$f(x) \sim 1 - \frac{8}{\pi^2} \sum_{n=0}^{\infty} \frac{1}{(2n+1)^2} \cos (2n+1) \frac{\pi x}{2}.$$

Justify applying Theorem 9.11 to this Fourier series and hence deduce the Fourier series representation of

$$g(x)=\begin{cases} -1, & \text{for } -2 < x < 0 \\ 0, & \text{for } \quad x=0 \\ 1, & \text{for } \quad 0 < x < 2. \end{cases}$$

9 Find the Fourier series representation of

$$f(x)=\begin{cases} x+1, & \text{for } -1 \leq x \leq 0 \\ -x+1, & \text{for } \quad 0 \leq x \leq 1. \end{cases}$$

Use the result to deduce the Fourier series representation of

$$g(x)=\begin{cases} 1, & \text{for } -1 \leq x \leq 0 \\ -1, & \text{for } \quad 0 \leq x \leq 1. \end{cases}$$

9.5 Conditions for the uniform convergence of a Fourier series

A direct proof of the general convergence properties of Fourier series has already been given in Sections 9.3 and 9.4, without reference to uniform convergence. Nevertheless, the importance of uniform convergence is such that it is desirable to establish precisely when a Fourier series converges uniformly to the function it represents. Such conditions are given in this section, the main result of which is Theorem 9.12. Readers should familiarize themselves with this theorem, even if they choose to omit the details of its proof at a first reading. Let us suppose that $f(x)$ is continuous and piecewise smooth in the interval $-\pi \leq x \leq \pi$ with $f(-\pi)=f(\pi)$, and that the Fourier coefficients of $f(x)$ and $f'(x)$ are, respectively, a_n, b_n and A_n, B_n.

Then integrating by parts the Euler formula for a_n gives

$$a_n = \frac{1}{\pi} \int_{-\pi}^{\pi} f(x) \cos nx \, dx$$

$$= \left[\frac{1}{\pi} \left(f(x) \frac{\sin nx}{n} \right) \Big|_{-\pi}^{\pi} - \frac{1}{n\pi} \int_{-\pi}^{\pi} f'(x) \sin nx \, dx \right]$$

$$= -\frac{B_n}{n}, \qquad \text{for } n = 1, 2, \dots, \tag{1}$$

where the terms outside the integral vanish because $\sin n\pi = \sin(-n\pi) = 0$. Similarly, integrating by parts the Euler formula for b_n gives

$$
\begin{aligned}
b_n &= \frac{1}{\pi} \int_{-\pi}^{\pi} f(x) \sin nx \, dx \\
&= \left[\frac{1}{\pi} \left(-f(x) \frac{\cos nx}{n} \right) \Big|_{-\pi}^{\pi} + \frac{1}{n\pi} \int_{-\pi}^{\pi} f'(x) \cos nx \, dx \right] \\
&= \frac{A_n}{n}, \qquad \text{for } n = 1, 2, \ldots \,,
\end{aligned}
\tag{2}
$$

where this time the terms outside the integral vanish because $f(-\pi) = f(\pi)$.

Now it follows from the elementary algebraic identity

$$
\left(|k| - \frac{1}{n} \right)^2 \equiv |k|^2 - \frac{2|k|}{n} + \frac{1}{n^2} \geq 0
$$

that

$$
\frac{|k|}{n} \leq \frac{1}{2} \left(|k|^2 + \frac{1}{n^2} \right).
\tag{3}
$$

Identifying k in (3) first with A_n and then with B_n shows that

$$
\frac{|A_n|}{n} \leq \frac{1}{2} \left(|A|^2 + \frac{1}{n^2} \right)
$$

and

$$
\frac{|B_n|}{n} \leq \frac{1}{2} \left(|B_n|^2 + \frac{1}{n^2} \right),
$$

which because of (1) and (2) is equivalent to

$$
|a_n| \leq \frac{1}{2} \left(|A_n|^2 + \frac{1}{n^2} \right) \quad \text{and} \quad |b_n| \leq \frac{1}{2} \left(|B_n|^2 + \frac{1}{n^2} \right).
\tag{4}
$$

The magnitude of the combination of terms $a_n \cos nx + b_n \sin nx$ in the Fourier series for $f(x)$ may thus be overestimated by

$$
|a_n \cos nx + b_n \sin nx| \leq |a_n \cos nx| + |b_n \sin nx|
$$

$$
\leq |a_n| + |b_n| \leq \frac{1}{2}(A_n^2 + B_n^2) + \frac{1}{n^2},
\tag{5}
$$

for $-\pi \leq x \leq \pi$. If we now set

$$
M_n = \frac{1}{2}(A_n^2 + B_n^2) + \frac{1}{n^2},
$$

it follows from the Weierstrass M-test (Theorem 9.3) that the Fourier series for $f(x)$ will converge uniformly to $f(x)$ for $-\pi \le x \le \pi$ provided the series ΣM_n is convergent.

This is easily seen to be the case. It was shown in Ex. 3, Sec. 9.2 that $\sum_{n=1}^{\infty} 1/n^2 = \pi^2/6$, so it only remains to prove that $\Sigma(A_n^2 + B_n^2)$ is convergent. Applying Bessel's inequality given in (23) of Sec. 9.3 to the Fourier series for $f'(x)$ yields the result that

$$\frac{1}{2} A_0^2 + \sum_{n=1}^{\infty} (A_n^2 + B_n^2) \le \frac{1}{\pi} \int_{-\pi}^{\pi} [f'(x)]^2 \, dx. \tag{6}$$

As $f'(x)$ is piecewise continuous, the integral is finite, so the series of positive terms is bounded above and thus is convergent. Our proof of the uniform convergence of the Fourier series of $f(x)$ to the function $f(x)$ itself for $-\pi \le x \le \pi$ is complete and we have established the next result.

Theorem 9.12 (Conditions for the uniform convergence of a Fourier series)

Let $f(x)$ be continuous and piecewise smooth on the interval $-\pi \le x \le \pi$, with $f(-\pi) = f(\pi)$. Then Fourier series of $f(x)$ converges uniformly to $f(x)$ for $-\pi \le x \le \pi$.

$\square$

9.6 Estimating the Fourier coefficients – smoothness and its effect on convergence

It is useful to have a measure of the speed with which the Fourier series of a function $f(x)$ defined on $-L \le x \le L$ converges, and such a measure is provided by estimating the magnitudes of the Fourier coefficients a_n, b_n in terms of n. This follows because by using the triangle inequality (Theorem 1.2) and the fact that $\left| \cos \frac{n\pi x}{L} \right| \le 1$, $\left| \sin \frac{n\pi x}{L} \right| \le 1$, the magnitude of the Fourier series of $f(x)$ can be overestimated as follows

$$\left| \frac{1}{2} a_0 + \sum_{n=1}^{\infty} \left(a_n \cos \frac{n\pi x}{L} + b_n \sin \frac{n\pi x}{L} \right) \right| \le \frac{1}{2} |a_0| + \sum_{n=1}^{\infty} \left| a_0 \cos \frac{n\pi x}{L} + b_n \sin \frac{n\pi x}{L} \right|$$

$$\le \frac{1}{2} |a_0| + \sum_{n=1}^{\infty} \left(\left| a_n \cos \frac{n\pi x}{L} \right| + \left| b_n \sin \frac{n\pi x}{L} \right| \right)$$

$$\le \frac{1}{2} |a_0| + \sum_{n=1}^{\infty} (|a_n| + |b_n|). \tag{1}$$

Thus the speed of convergence of the Fourier series of $f(x)$ cannot be slower than that of the series of constant terms on the right-hand side of (1).

In general, estimates of a_n, b_n are not available, but they can be found quite easily for periodic functions $f(x)$ which are continuous for all x. We shall now obtain such estimates for a function $f(x)$ which has continuous derivatives up to order p, but only a piecewise continuous derivative of order $p + 1$.

Let $f(x)$ be a continuous function defined for $-L \leq x \leq L$ and such that $f(-L) = f(L)$. In addition, let the first p derivatives $f^{(1)}(x), f^{(2)}(x), \ldots, f^{(p)}(x)$ exist and be continuous with the left and right-hand derivatives at the end points satisfying the conditions $f^{(1)}(-L) = f^{(1)}(L)$, $f^{(2)}(-L) = f^{(2)}(L), \ldots, f^{(p)}(-L) = f^{(p)}(L)$, but let the derivative $f^{(p+1)}(x)$ be merely piecewise continuous. Let the absolute value of the derivative $f^{(p+1)}(x)$ where it is continuous, and of its left and right-hand derivatives where it is discontinuous, be bounded by the positive constant K, so that denoting these derivatives by $f^{(p+1)}(x)$, it follows that

$$|f^{(p+1)}(x)| \leq K, \qquad \text{for } -L \leq x \leq L. \tag{2}$$

The Fourier coefficient a_n of $f(x)$ is defined by the Euler integral

$$a_n = \frac{1}{L} \int_{-L}^{L} f(x) \cos \frac{n\pi x}{L} \, dx, \tag{3}$$

so integrating by parts gives

$$a_n = \left[\frac{1}{n\pi} \left(f(x) \sin \frac{n\pi x}{L} \right) \bigg|_{-L}^{L} - \frac{1}{n\pi} \int_{-L}^{L} f^{(1)}(x) \sin \frac{n\pi x}{L} \, dx \right].$$

The terms outside the integral vanish because $\sin n\pi = \sin(-n\pi) = 0$, showing that

$$a_n = -\frac{1}{n\pi} \int_{-L}^{L} f^{(1)}(x) \sin \frac{n\pi x}{L} \, dx. \tag{4}$$

A further integration by parts gives

$$a_n = \left[\frac{L}{n\pi} \left(f^{(1)}(x) \cos \frac{n\pi x}{L} \right) \bigg|_{-L}^{L} - \frac{L}{n^2 \pi^2} \int_{-L}^{L} f^{(2)}(x) \cos \frac{n\pi x}{L} \, dx \right].$$

This time the terms outside the integral vanish because $f(-L) = f(L)$, showing that

$$a_n = -\frac{L}{n^2 \pi^2} \int_{-L}^{L} f^{(2)}(x) \cos \frac{n\pi x}{L} \, dx. \tag{5}$$

Repetition of integration by parts until the derivative $f^{(p+1)}(x)$ is obtained then gives

$$a_n = \pm \left(\frac{L}{n\pi} \right)^{p+1} \frac{1}{L} \int_{-L}^{L} f^{(p+1)}(x) \begin{Bmatrix} \cos \dfrac{n\pi x}{L} \\[2mm] \sin \dfrac{n\pi x}{L} \end{Bmatrix} dx, \tag{6}$$

where the choice of sign and trigonometric function in parentheses depends on the number of integrations required to arrive at result (6).

Taking the absolute value of (6), using Theorem 1.7(iii), and the bound on $f^{(p+1)}(x)$ given in (2), we see that

$$|a_n| = \left(\frac{L}{n\pi}\right)^{p+1} \frac{1}{L} \left| \int_{-L}^{L} f^{(p+1)}(x) \left\{ \begin{array}{c} \cos \dfrac{n\pi x}{L} \\[2mm] \sin \dfrac{n\pi x}{L} \end{array} \right\} dx \right|$$

$$\leq \left(\frac{L}{n\pi}\right)^{p+1} \frac{1}{L} \int_{-L}^{L} \left| f^{(p+1)}(x) \right| \left| \left\{ \begin{array}{c} \cos \dfrac{n\pi x}{L} \\[2mm] \sin \dfrac{n\pi x}{L} \end{array} \right\} \right| dx$$

$$= \left(\frac{L}{n\pi}\right)^{p+1} \frac{1}{L} \int_{-L}^{L} K \cdot 1 dx,$$

and thus

$$|a_n| \leq 2K \left(\frac{L}{n\pi}\right)^{p+1}, \qquad \text{for } n = 1, 2, \ldots . \tag{7}$$

An essentially similar argument shows that

$$|b_n| \leq 2K \left(\frac{L}{n\pi}\right)^{p+1}, \qquad \text{for } n = 1, 2, \ldots . \tag{8}$$

Using estimates (7), (8) in (1) we obtain the estimate

$$\left| \frac{1}{2} a_0 + \sum_{n=1}^{\infty} \left(a_n \cos \frac{n\pi x}{L} + b_n \sin \frac{n\pi x}{L} \right) \right| \leq \frac{1}{2} |a_0| + \frac{4KL^{p+1}}{\pi^{p+1}} \sum_{n=1}^{\infty} \frac{1}{n^{p+1}}.$$

Thus the convergence of the Fourier series of a function satisfying the stated conditions cannot be slower than the convergence of the series $\sum_{n=1}^{\infty} (1/n)^{p+1}$. This has established our next result.

Theorem 9.13 (Estimation of Fourier coefficients and convergence)

Let $f(x)$ be a continuous function defined for $-L \leq x \leq L$ and such that $f(-L) = f(L)$. In addition, let the first p derivatives $f^{(1)}(x), f^{(2)}(x), \ldots, f^{(p)}(x)$ exist and be continuous with the left and right-hand derivatives at the end points satisfying the conditions $f^{(1)}(-L)$ $= f^{(1)}(L)$, $f^{(2)}(-L) = f^{(2)}(L), \ldots, f^{(p)}(-L) = f^{(p)}(L)$, but let the derivative $f^{(p+1)}(x)$ be merely piecewise continuous. Let the absolute value of the $(p + 1)$th derivative of $f(x)$ where it is continuous, and of its left and right-hand derivatives where it is discontinuous, be bounded by the positive constant K, so that denoting all of these derivatives by $f^{(p+1)}(x)$, it follows that

$$|f^{(p+1)}(x)| \leq K, \qquad \text{for } -L \leq x \leq L.$$

Then the Fourier coefficients a_n, b_n of $f(x)$ satisfy the inequalities

$$|a_n| \leq 2K \left(\frac{L}{n\pi}\right)^{p+1}, \quad |b_n| \leq 2K \left(\frac{L}{n\pi}\right)^{p+1},$$

for $n = 1, 2, \ldots$. Furthermore, the Fourier series itself converges at a rate which is no slower than the series

$$\sum_{n=1}^{\infty} (1/n)^{p+1}. \qquad \qquad \qquad \square$$

It must be appreciated that the estimates of a_n, b_n and of the rate of convergence provided by Theorem 9.13 are conservative. In general, the rate at which a_n, b_n tend to zero as $n \to \infty$, and the corresponding rate at which the Fourier series converges, are usually faster. This situation is well illustrated by Ex. 1 of Sec. 9.4.

The function $f(x) = x^3 - L^2 x$ defined for $-L \leq x \leq L$ is continuous with $f(-L) = f(L)$, and so satisfies the conditions of Theorem 9.13. It is easily seen that $f^{(1)}(-L) = f^{(1)}(L) = 2L^2$, but that $f^{(2)}(x)$ is discontinuous at $x = \pm L$, so in the notation of Theorem 9.13 it follows that $p = 1$.

As $f(x)$ is an even function $b_n = 0$, for $n = 1, 2, \ldots$, and

$$K = \max_{-L \leq x \leq L} |f^{(2)}(x)| = \max_{-L \leq x \leq L} |6x| = 6L.$$

Thus the estimate of a_n provided by Theorem 9.13 becomes

$$|a_n| \leq \frac{12L^3}{n^2 \pi^2}, \qquad \text{for } n = 1, 2, \ldots .$$

Examination of the Fourier series for $f(x)$ obtained in the example shows that in fact

$$|a_n| = \frac{12L^3}{n^3 \pi^3},$$

so the actual rate of convergence is considerably faster than that indicated by the estimate.

9.7 Numerical harmonic analysis

The Fourier series of a function $f(x)$ defined on the interval $a < x < b$ is also called the **harmonic decomposition** of $f(x)$. The name derives from the fact that, when represented by a Fourier series, $f(x)$ is decomposed into periodic components which are multiples of the **fundamental frequency** (lowest frequency) $2\pi/(b - a)$. Each periodic component is called a **harmonic**, and the component with frequency $2n\pi/(b - a)$ is called the **nth harmonic**.

If $f(x)$ has a complicated analytical form, or if it is only known experimentally either in graphical form or at discrete points, its harmonic decomposition can only be carried out numerically. In such cases the decomposition is approximate and usually involves determining only the first few harmonics. This approach is called **numerical harmonic analysis**, and it is usually carried out by computer using a standard subroutine.

Instead of describing this sophisticated method we will develop a simple approach based on the trapezoidal integration rule which can be implemented on a programmable pocket calculator, or on a PC if one is available.

Let us suppose that a function $g(x)$ defined for $a \leq x \leq b$ is known at N equally spaced points separated by intervals of length $h = (b - a)/N$. Then denoting by $g_0, g_1, \ldots, g_N$ these $N + 1$ values of $g(x)$, it follows that

$$g_r = g(a + rh), \qquad \text{for } r = 0, 1, \ldots, N. \tag{1}$$

The trapezoidal rule for approximating the integral of $g(x)$ over the interval $a \leq x \leq b$ is

$$\int_a^b g(x)\mathrm{d}x = h(\tfrac{1}{2}g_0 + g_1 + \ldots + g_{N-1} + \tfrac{1}{2}g_N)$$

or, as $h = (b-a)/N$,

$$\int_a^b g(x)\mathrm{d}x = \left(\frac{b-a}{N}\right)(\tfrac{1}{2}g_0 + g_1 + \ldots + g_{N-1} + \tfrac{1}{2}g_N). \tag{2}$$

We know from Theorem 9.4 that the Fourier coefficients a_n, b_n of a function $f(x)$ are defined by the integrals

$$a_n = \left(\frac{2}{b-a}\right)\int_a^b f(x)\cos\frac{2n\pi x}{b-a}\,\mathrm{d}x, \qquad \text{for } n=0, 1, \ldots, \tag{3}$$

and

$$b_n = \left(\frac{2}{b-a}\right)\int_a^b f(x)\sin\frac{2n\pi x}{b-a}\,\mathrm{d}x, \qquad \text{for } n=1, 2, \ldots. \tag{4}$$

Setting

$$f_r = f(a+rh), \qquad \text{for } r=0, 1, \ldots, N, \tag{5}$$

and applying trapezoidal rule (2) to (3) and (4) after making the respective identifications

$$g(x)=f(x)\cos\frac{2n\pi x}{b-a} \qquad \text{and} \qquad g(x)=f(x)\sin\frac{2n\pi x}{b-a},$$

we find that the numerical approximations to a_n and b_n are

$$a_n = \frac{2}{N}\left(\frac{1}{2}f(a)\cos\frac{2n\pi a}{b-a} + \sum_{r=1}^{N-1} f(a+rh)\cos\frac{2n\pi(a+rh)}{b-a} + \frac{1}{2}f(b)\cos\frac{2n\pi b}{b-a}\right), \tag{6}$$

and

$$b_n = \frac{2}{N}\left(\frac{1}{2}f(a)\sin\frac{2n\pi a}{b-a} + \sum_{r=1}^{N-1} f(a+rh)\sin\frac{2n\pi(a+rh)}{b-a} + \frac{1}{2}f(b)\sin\frac{2n\pi b}{b-a}\right). \tag{7}$$

Since the Fourier series will be periodic with period $b - a$, it is usual to make the functional values $f_0, f_1, \ldots, f_{N-1}, f_N$ themselves periodic with this same period. This is accomplished by using the N values $f_0, f_1, \ldots, f_{N-1}$ and replacing the last value f_N by f_0. The $N + 1$ values to be used thus become

$$f_0, f_1, f_2, \ldots, f_{N-2}, f_{N-1}, f_0.$$

As

$$\cos \frac{2n\pi b}{b-a} = \cos\left(\frac{2n\pi(b-a) + 2n\pi a}{b-a}\right) = \cos \frac{2n\pi a}{b-a},$$

and

$$\sin \frac{2n\pi b}{b-a} = \sin\left(\frac{2n\pi(b-a) + 2n\pi a}{b-a}\right) = \sin \frac{2n\pi a}{b-a},$$

it follows that the first and last terms in (6) and (7) may be combined to give

$$a_n = \frac{2}{N} \sum_{r=0}^{N-1} f(a + rh) \cos \frac{2n\pi(a + rh)}{b-a}, \qquad \text{for } n = 0, 1, \ldots, \tag{8}$$

and

$$b_n = \frac{2}{N} \sum_{r=0}^{N-1} f(a + rh) \sin \frac{2n\pi(a + rh)}{b-a}, \qquad \text{for } n = 1, 2, \ldots. \tag{9}$$

These are the results to be used for the approximate determination of the Fourier coefficients.

Now sampling $f(x)$ at N equispaced intervals can only give reasonably reliable information about harmonics up to order $[N/2]$, where $[N/2]$ denotes the integral part of $N/2$. Thus this form of approximate harmonic analysis based on N intervals leads to the approximation of the function $f(x)$ on $a \le x \le b$ by

$$f(x) = \frac{1}{2} a_0 + \sum_{r=1}^{[N/2]} \left(a_r \cos \frac{2\pi nx}{b-a} + b_r \sin \frac{2\pi nx}{b-a} \right).$$

When using the simple numerical method in cases where only the first few harmonics are significant, the approximation to $f(x)$ is often terminated at $r = 3$ or $r = 4$, rather than at $r = [N/2]$.

Example 1. Numerical harmonic analysis

Perform a numerical harmonic analysis on the following data defined on the interval $0 \le x \le 2\pi$, and compare the numerical approximation to $f(x)$ which is obtained with the data.

r	0	1	2	3	4	5	6	7	8	9	10	11	12
f_r	3.282	1.877	0.989	1.266	0.692	−0.966	0.038	0.472	0.832	−0.306	1.787	3.194	3.282

Solution

We make the identifications $a = 0$, $b = 2\pi$, $N = 12$, $h = \pi/6$ and then use (8) and (9) to calculate a_n, b_n up to the sixth harmonic ($[N/2] = 6$). Table 9.3 gives the results.

Table 9.3

n	a_n	b_n
0	2.193	0
1	1.448	-0.103
2	0.417	-0.077
3	0.332	-0.721
4	-0.027	0.112
5	-0.158	0.168
6	0.347	0

The corresponding approximation to $f(x)$ is thus given by using these values of a_n, b_n in

$$f(x) = \tfrac{1}{2} a_0 + \sum_{r=1}^{6} (a_r \cos rx + b_r \sin rx), \qquad \text{for } 0 \le x \le 2\pi.$$

The graph of the approximation terminated after the third harmonic is shown in Fig. 9.28 on which the original discrete data is shown as crosses.

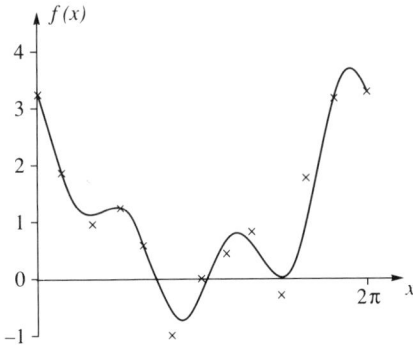

Fig. 9.28 $f(x) = \tfrac{1}{2}a_0 + \sum_{r=1}^{3} (a_r \cos r_x + b_r \sin rx)$

■

Example 2. Numerical harmonic analysis of $f(x) = |\sin x|$

Using the approximation with $N = 12$, find numerically the harmonics up to and including the sixth of the function

$$f(x) = |\sin x|, \qquad \text{for } -\pi \le x \le \pi.$$

Compare the results with the analytical values of these Fourier components and graph both $f(x)$ and its approximation $F(x)$ based on the numerically determined coefficients.

Solution

As the function $f(x)$ is even it follows that its Fourier coefficients b_n all vanish. It will suffice for us to consider only the interval $0 \le x \le \pi$. Using the numerical approximation (8) with $N = 12$ for this half-interval will increase the accuracy without increasing the effort, because it will be equivalent to using $N = 24$ with the full interval.

Thus we set $N = 12$, $a = 0$, $b = \pi$ and $h = \pi/12$ in (8), and calculate the coefficients using the result

$$a_n = \frac{1}{6} \sum_{r=0}^{11} \sin\left(\frac{r\pi}{12}\right) \cos 2nrh.$$

It must, of course, be remembered that when using the half-interval $0 \le x \le \pi$ the Fourier series (18) takes the form

$$f(x) = \tfrac{1}{2}a_0 + \sum_{n=1}^{\infty} a_n \cos 2nx,$$

in which only even multiples of the fundamental frequency arise.

Thus our approximation $F(x)$ to $f(x)$ using numerically determined coefficients up to the sixth harmonic will be

$$F(x) = \tfrac{1}{2}a_0 + a_1 \cos 2x + a_2 \cos 4x + a_3 \cos 6x.$$

We see from this that the calculation of the coefficients a_n need only proceed as far as $n = 3$.

Table 9.4

n	Approximate	Exact
0	1.2660	1.2732
1	0	0
2	−0.4318	−0.4244
3	0	0
4	−0.0926	−0.0849
5	0	0
6	−0.0447	−0.0202

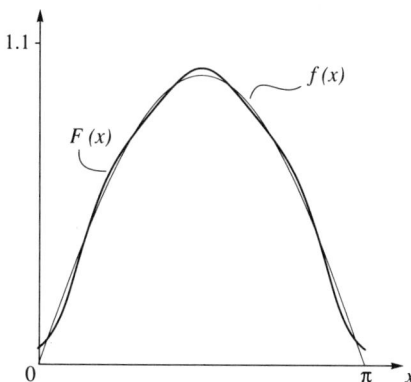

Fig. 9.29 Graphs of $f(x)$ and the numerically obtained approximation $F(x)$

The numerically determined coefficients a_0, a_1, a_2 and a_3 are given in Table 9.4, in which they are also compared with the exact results obtained in the usual way

$$a_n = \begin{cases} 4/\pi, & \text{for } n=0 \\[2mm] \dfrac{-4}{\pi(n-1)(n+1)}, & \text{for } n \text{ even} \\[2mm] 0, & \text{for } n \text{ odd.} \end{cases}$$

A comparison of the graphs of $f(x)$ and $F(x)$ is given in Fig. 9.29 for $0 \le x \le \pi$. The graphs in the full interval $-\pi \le x \le \pi$ are obtainable by recalling that $f(x)$ is an even function and reflecting the graphs of both $f(x)$ and $F(x)$ in the y-axis.

∎

Problems for Section 9.7

In the following problems data is provided at points distant $h = \pi/6$ apart over the interval $0 \le x \le 2\pi$. Carry out a numerical harmonic analysis of the data in each case and determine the Fourier coefficients a_0 to a_3 and b_1 to b_3.

	r	0	1	2	3	4	5	6	7	8	9	10	11	12
1	f_r	0.902	0.555	-1.185	-2.381	-1.033	-0.189	-0.406	-0.373	-0.518	0.075	0.356	1.019	0.902
2	f_r	2.556	1.753	0.603	-0.4	0.053	0.631	1.774	1.726	2.282	3.168	3.659	3.641	2.556
3	f_r	2.551	1.509	0.591	-0.268	0.321	0.716	1.722	1.57	1.717	2.586	3.982	3.394	2.551
4	f_r	2.702	2.212	1.835	0.626	-1.108	0.791	0.743	0.686	0.034	1.237	2.678	3.127	2.702
5	f_r	0.86	1.183	2.881	4.131	3.099	1.505	-0.345	0.114	1.853	1.683	0.881	0.771	0.86
6	f_r	2.551	1.509	0.591	-0.268	0.321	0.712	1.722	1.57	1.717	2.586	3.982	3.394	2.551
7	f_r	2.516	1.535	0.786	0.991	2.331	2.199	0.888	0.1	-0.023	1.425	2.879	3.588	2·156
8	f_r	2.437	0.963	0.503	1.084	1.807	2.122	1.03	-0.152	-0.085	1.03	2.585	3.287	2.437

9.8 Representation of functions using orthogonal systems – Sturm–Liouville problems

A Fourier series provides a special case of the representation of an arbitrary function $f(x)$ over some interval $a \le x \le b$ in the form of a functional series

$$f(x) \sim \sum_{n=1}^{\infty} c_n \phi_n(x). \tag{1}$$

The essential feature of such a series is that its constant coefficients c_n are determined in terms of $f(x)$ by using a sequence of functions $\{\phi_n(x)\}_{n=1}^{\infty}$ which possess is the property of being *mutually orthogonal* over the interval $a \le x \le b$.

Let us now give a formal definition of the concept of *orthogonality*. The sequence of functions $\{\phi_n(x)\}_{n=1}^{\infty}$ defined for $a \le x \le b$ is said to form an **orthogonal system** of functions over the interval $a \le x \le b$ if

$$\int_a^b \phi_m(x)\,\phi_n(x)\,dx = 0, \qquad \text{for } m \ne n. \tag{2}$$

Associated with each function in this system is a positive number $\| \phi_n \|$ called the **norm** of $\phi_n(x)$ defined by the integral

$$\| \phi_n \|^2 = \int_a^b [\phi_n(x)]^2\,dx. \tag{3}$$

We will see later that the norm plays an essential part in the determination of the coefficients c_n in (1).

Definitions (2) and (3) will be recognized as an immediate extension of the results expressed in (3) to (5) of Sec. 9.2 which relate to the orthogonal system of trigonometric functions

$$\{1, \cos x, \sin x, \cos 2x, \sin 2x, \ldots\},$$

used in the development of Fourier series.

More general than orthogonality, though equally important, is the concept of orthogonality with respect to a **weight function** $\rho(x)$, which we now define. The system of functions $\{\phi_n(x)\}_{n=1}^{\infty}$ defined for $a \le x \le b$ is said to be **orthogonal with respect to the weight function** $\rho(x)$ over the interval $a \le x \le b$ if $\rho(x) > 0$ and

$$\int_a^b \phi_m(x)\phi_n(x)\rho(x)\,dx = 0, \qquad \text{for } m \ne n. \tag{4}$$

In such a system the norm $\| \phi_n \|$ of $\phi_n(x)$ is defined by the following integral involving the weight function

$$\| \phi_n \|^2 = \int_a^b [\phi_n(x)]^2\,\rho(x)\,dx. \tag{5}$$

It is often convenient to *normalize* a system of orthogonal functions $\{\phi_n(x)\}_{n=1}^{\infty}$, or one which is orthogonal with respect to a weight function, so that the norm of each function is unity. Such systems are called **orthonormal systems** when the functions are orthogonal, and **orthonormal systems with respect to a weight function** when a weight function $\rho(x)$ is involved.

In either case the normalization is accomplished by working with the system of functions $\{\Phi(x)\}_{n=1}^{\infty}$, instead of the system of functions $\{\phi_n(x)\}_{n=1}^{\infty}$, where $\Phi_n(x) = \phi_n(x) / \| \phi_n \|$, for $n = 1, 2, \ldots$.

A simple example of an orthonormal system of functions over the interval $-\pi \leq x \leq \pi$ is provided by

$$\left\{ \frac{1}{\sqrt{2\pi}}, \frac{\cos x}{\sqrt{\pi}}, \frac{\sin x}{\sqrt{\pi}}, \frac{\cos 2x}{\sqrt{\pi}}, \frac{\sin 2x}{\sqrt{\pi}}, \dots \right\}. \tag{6}$$

This particular system has been obtained by normalizing the orthogonal system of trigonometric functions used in connection with Fourier series, and defined over the interval $-\pi \leq x \leq \pi$, for which

$$\|1\| = \left(\int_{-\pi}^{\pi} 1 \, dx \right)^{1/2} = \sqrt{2\pi}, \quad \|\cos nx\| = \left(\int_{-\pi}^{\pi} \cos^2 nx \, dx \right)^{1/2} = \sqrt{\pi} \text{ and}$$

$$\|\sin nx\| = \left(\int_{-\pi}^{\pi} \sin^2 nx \, dx \right)^{1/2} = \sqrt{\pi}, \qquad \text{for } n = 1, 2, \dots.$$

In a manner analogous to the formal derivation of the Fourier series representation of an arbitrary function $f(x)$ defined for $a \leq x \leq b$, it is possible to derive the coefficients c_n in the general functional series representation

$$f(x) \sim c_1 \phi_1(x) + c_2 \phi_2(x) + \dots, \tag{7}$$

using functions $\phi_n(x)$ which are orthogonal over the interval $a \leq x \leq b$. Let us proceed formally as we did with Fourier series and regard the relational symbol $\sim$ as an equality. Then multiplication of (7) by $\phi_n(x)$ followed by integration over the interval $a \leq x \leq b$ and use of the orthogonality property of the sequence of functions $\{\phi_n(x)\}_{n=1}^{\infty}$ shows that

$$c_n = \frac{1}{\|\phi_n\|^2} \int_a^b f(x) \phi_n(x) \, dx, \qquad \text{for } n = 1, 2, \dots, \tag{8}$$

where

$$\|\phi_n\|^2 = \int_a^b \{\phi_n(x)\}^2 \, dx. \tag{9}$$

Thus the square of the norm of $\phi_n(x)$ is seen to enter quite naturally into the definition of c_n given in (8).

Correspondingly, if the sequence of functions defined over the interval $a \leq x \leq b$ is orthogonal with respect to the weight function $\rho(x)$, it follows that

$$c_n = \frac{1}{\|\phi_n\|^2} \int_a^b f(x) \phi_n(x) \rho(x) \, dx, \qquad \text{for } n = 1, 2, \dots, \tag{10}$$

where now

$$\|\phi_n\|^2 = \int_a^b [\phi_n(x)]^2 \rho(x) \, dx. \tag{11}$$

Although the proof is beyond this introductory account, it can be shown that each of the systems of functions to be introduced later may be used to represent any piecewise continuous function $f(x)$ defined for $a \le x \le b$ in the form of the associated functional series (7).

In the main, the study of the convergence properties of functional series involving orthogonal functions is based on the concept of *mean square convergence*, which was introduced briefly in connection with Fourier series. The relational symbol $\sim$ in (7) can be shown to have essentially the same meaning as when used with Fourier series. In particular, if $f(x)$ is continuous for $a \le x \le b$, the symbol $\sim$ may be interpreted as an equality except, possibly, at the end points of the interval. The possible exclusion of the end points is mentioned, because if the system of functions $\{\phi_n(x)\}_{n=1}^{\infty}$ vanishes at an end point, the series in (7) cannot possibly represent a function $f(x)$ which is nonzero at that point. We will return to this matter later.

The systems of functions $\{\phi_n(x)\}_{n=1}^{\infty}$ of concern to us here are generated as solutions of a **two-point boundary value problem** for a linear homogeneous second order differential equation of special type involving a parameter λ which is independent of x. The differential equation itself is called a *Sturm–Liouville equation*. The combination of the equation and its boundary values is called a *Sturm–Liouville problem* or sometimes a *Sturm–Liouville system*.[11, 12] It is possible to deduce from such a system all the fundamental properties of the functions involved. The following simple but typical example of such a problem will serve to introduce all the essential ideas.

Consider the differential equation

$$\frac{\mathrm{d}^2 y}{\mathrm{d}x^2} + \lambda y = 0 \tag{12}$$

with parameter λ, subject to the two-point boundary conditions

$$y(0) = 0 \quad \text{and} \quad y(\pi) = 0. \tag{13}$$

This is called a two-point boundary value problem because the conditions on $y(x)$ are specified at the two distinct (boundary) points $x = 0$ and $x = \pi$.

The general solution of (12) has the form

$$y(x) = A \cos \sqrt{\lambda}\, x + B \sin \sqrt{\lambda}\, x, \tag{14}$$

where so far λ is an arbitrary parameter. However, (14) can only be the solution of the two-point boundary value problem if $y(x)$ also satisfies the boundary conditions (13). This

---

[11] JACQUES CHARLES FRANÇOIS STURM (1803–1855), an important Swiss-born mathematician who studied in Switzerland and then moved to the Sorbonne in Paris where he was appointed to the chair of mechanics previously held by Poisson. His major contributions were to the study of heat flow in solids, to algebra and to ordinary differential equations whose solutions he studied in order to determine the distribution of the zeros of a function between given limits.

[12] JOSEPH LIOUVILLE (1809–1882), an important French mathematician who was a friend of Sturm and who was professor of mathematics at the Collège de France. His important contributions were to algebra, number theory, differential geometry, complex analysis and to special functions defined by ordinary differential equations.

will be the case if

$$[y(0) = 0] \quad 0 = A, \quad \text{and}$$
$$[y(\pi) = 0] \quad 0 = B \sin \sqrt{\lambda}\, \pi.$$

Thus either $B = 0$, in which case as $A = 0$ the solution $y(x)$ becomes identically zero and so uninteresting (a **trivial solution**), or

$$\sin \sqrt{\lambda}\, \pi = 0, \tag{15}$$

which is only possible if $\lambda = n^2$, for $n = 1, 2, \ldots$. If we set $\sqrt{\lambda_n} = n$, it follows that an infinite system of nontrivial solutions $y_n(x)$ of (12) subject to (13) exists of the form

$$y_n(x) = \sin nx, \qquad \text{for } n = 1, 2, \ldots . \tag{16}$$

For convenience, in (16) the arbitrary constant B in (14) has been set equal to unity. This is permissible because differential equation (12) is both linear and homogeneous, so as $A = 0$ the solution $y_n(x)$ may be scaled in any way we wish.

The numbers

$$\lambda_n = n^2, \quad \text{with } n = 1, 2, \ldots, \tag{17}$$

are called the **eigenvalues** (also the **characteristic values**) of the Sturm–Liouville problem (12), (13), and the corresponding solutions

$$y_n(x) = \sin nx, \qquad \text{for } n = 1, 2, \ldots, \tag{18}$$

are called the **eigenfunctions** (also the **characteristic functions**) of the problem.

Let us now deduce the orthogonality of the eigenfunctions (18) over the interval $0 \le x \le \pi$ from their defining Sturm–Liouville problem (12) and (13). Take $y_m(x)$, $y_n(x)$ to be two different solutions of (12) and (13) corresponding, respectively, to $\lambda_m = m^2$ and $\lambda_n = n^2$, with $m \ne n$. Then

$$\frac{d^2 y_m}{dx^2} + m^2 y_m = 0 \tag{19}$$

and

$$\frac{d^2 y_n}{dx^2} + n^2 y_n = 0, \tag{20}$$

where

$$y_m(0) = y_m(\pi) = 0 \qquad \text{and} \qquad y_n(0) = y_m(\pi) = 0. \tag{21}$$

Multiply (19) by $y_n(x)$ and (20) by $y_m(x)$ and integrate over the interval $0 \le x \le \pi$ to obtain

$$\int_0^\pi y_n \frac{d^2 y_m}{dx^2}\, dx + m^2 \int_0^\pi y_n y_m\, dx = 0, \tag{22}$$

$$\int_0^\pi y_m \frac{d^2 y_n}{dx^2} dx + n^2 \int_0^\pi y_n y_m \, dx = 0. \tag{23}$$

Applying integration by parts to the first integral in (22) and to the corresponding integral in (23) gives

$$\left(y_n \frac{dy_m}{dx} \right)\Big|_0^\pi - \int_0^\pi \frac{dy_n}{dx} \frac{dy_m}{dx} dx + m^2 \int_0^\pi y_m y_n \, dx = 0, \tag{24}$$

and

$$\left(y_m \frac{dy_n}{dx} \right)\Big|_0^\pi - \int_0^\pi \frac{dy_n}{dx} \frac{dy_m}{dx} dx + n^2 \int_0^\pi y_m y_n \, dx = 0. \tag{25}$$

The first terms in (24) and (25) vanish by virtue of the two-point boundary conditions (21), so subtracting (25) from (24) gives

$$(m^2 - n^2) \int_0^\pi y_m(x) \, y_n(x) \, dx = 0. \tag{26}$$

However, by hypothesis $m \neq n$, so we conclude that

$$\int_0^\pi y_m(x) \, y_n(x) \, dx = \int_0^\pi \sin mx \sin nx \, dx = 0, \tag{27}$$

for $m \neq n$.

We have proved that the system of eigenfunctions

$$\{\sin x, \sin 2x, \ldots,\}$$

defined over the interval $0 \leq x \leq \pi$ forms an orthogonal system. A direct calculation shows that on this interval

$$\| \sin nx \|^2 = \int_0^\pi \sin^2 nx \, dx = \frac{\pi}{2}, \qquad \text{for } n = 1, 2, \ldots.$$

If, now, we identify the system of functions $\{\phi_n(x)\}$ in (7) with the system $\{y_n(x)\}$ in which $y_n(x) = \sin nx$, it follows that the representation of $f(x)$ given by (7) takes the form

$$f(x) \sim \sum_{n=1}^{\infty} c_n \sin nx, \qquad \text{for } 0 \leq x \leq \pi,$$

where

$$c_n = \frac{2}{\pi} \int_0^\pi f(x) \sin nx \, dx, \qquad \text{for } n = 1, 2, \ldots.$$

This is, of course, the Fourier sine series representation of $f(x)$ over the interval $0 \leq x \leq \pi$. However, in this section the Fourier sine series has been derived from the Sturm–Liouville problem given in (12) and (13), rather than directly as in Sec. 9.2. Other representations arise when the orthogonal functions $\phi_n(x)$ used in (7) occur as solutions of different Sturm–Liouville problems.

Using this example as a model we now generalize the notion of a Sturm–Liouville problem. Let $f(x)$, $p(x)$, $\rho(x)$ and $r(x)$ be continuous functions over the interval $a \leq x \leq b$, on which $\rho(x) > 0$ and $p(x) > 0$. We now define the **general Sturm–Liouville problem** as finding nontrivial solutions of the linear homogeneous second order differential equation

$$\frac{d}{dx}\left(r(x)\frac{dy}{dx}\right) + [p(x) + \lambda\rho(x)]y = 0, \tag{28}$$

in which λ is a parameter independent of x, and $y(x)$ satisfies one of the following two-point boundary conditions:

Homogeneous conditions on $y(x)$

$$y(a) = y(b) = 0, \tag{29}$$

Homogeneous conditions on $y'(x)$

$$y'(a) = y'(b) = 0, \tag{30}$$

Mixed homogeneous conditions

$$\alpha_1 y(a) + \beta_1 y'(a) = 0 \quad \text{and} \quad \alpha_2 y(b) + \beta_2 y'(b) = 0, \tag{31}$$

with α_1, β_1 and α_2, β_2 pairs of real constants, not both of which vanish in either boundary condition.

Listed below are some of the most important properties of the eigenvalues λ_n and eigenfunctions $\phi_n(x)$ of a Sturm–Liouville problem defined by the Sturm–Liouville equation (28) and one of the boundary conditions (29), (30) or (31).

Properties of Sturm–Liouville systems

1 The eigenvalues λ_n are real and can be arranged such that

$$\lambda_1 < \lambda_2 < \lambda_3 < \ldots,$$

with

$$\lim_{n \to \infty} \lambda_n = \infty.$$

2 Each eigenvalue λ_n has associated with it an eigenfunction $\phi_n(x)$ defined for $a \leq x \leq b$ which is uniquely defined up to an arbitrary multiplicative constant.

3 The eigenfunction $\phi_n(x)$ of (28) corresponding to the eigenvalue λ_n has precisely $(n-1)$ zeros in the interval $a < x < b$ (an *open* interval).

4 The eigenfunctions $\phi_m(x)$ and $\phi_n(x)$ corresponding, respectively, to the distinct eigenvalues λ_m and λ_n $(\lambda_m \neq \lambda_n)$ are orthogonal with respect to the weight function $\rho(x)$ over the interval $a \leq x \leq b$, so that

$$\int_a^b \phi_m(x)\rho_n(x)\rho(x)\,\mathrm{d}x = 0, \qquad \text{for } m \neq n.$$

5 A function $f(x)$ which is continuous and twice differentiable on the interval $a \leq x \leq b$, and which satisfies the boundary conditions of the Sturm–Liouville problem, may be expanded in a uniformly convergent series, often called an **eigenfunction expansion**,

$$f(x) = \sum_{n=1}^{\infty} c_n \phi_n(x), \qquad \text{for } a \leq x \leq b,$$

where

$$c_n = \frac{1}{\|\phi_n\|^2} \int_a^b f(x)\,\phi_n(x)\,\rho(x)\,\mathrm{d}x,$$

with

$$\|\phi_n\|^2 = \int_a^b [\phi_n(x)]^2\,\rho(x)\,\mathrm{d}x.$$

6 The coefficients c_n satisfy the generalized Parseval relation

$$\frac{2}{(b-a)} \int_a^b [f(x)]^2\,\rho(x)\,\mathrm{d}x = \sum_{n=1}^{\infty} c_n^2.$$

We will only prove the part of property 1 in which it is asserted that the eigenvalues are real, and the mutual orthogonality of the eigenfunctions which comprises property 4. A satisfactory justification of the other properties is beyond the scope of this introductory account. We will prove property 4 first, because our proof of the reality of the eigenvalues will make use of it.

Proof of property 4

Let λ_m and λ_n be distinct eigenvalues $(\lambda_m \neq \lambda_n)$ to which there correspond the respective eigenfunctions ϕ_m and ϕ_n. Then it follows that

$$\frac{\mathrm{d}}{\mathrm{d}x}\left(r\frac{\mathrm{d}\phi_m}{\mathrm{d}x}\right) + (p + \lambda_m\rho)\,\phi_m = 0, \qquad (32)$$

and

$$\frac{\mathrm{d}}{\mathrm{d}x}\left(r\frac{\mathrm{d}\phi_n}{\mathrm{d}x}\right) + (p + \lambda_n\rho)\,\phi_n = 0, \qquad (33)$$

where ϕ_m and ϕ_n satisfy one of the boundary conditions (29), (30) or (31).

Multiplying (32) by ϕ_n and (33) by ϕ_m and integrating over the interval $a \leq x \leq b$ gives

$$\int_a^b \phi_n \frac{d}{dx}\left(r\frac{d\phi_m}{dx}\right)dx + \int_a^b p\,\phi_m\phi_n\,dx + \lambda_m \int_a^b \phi_m\phi_n\rho\,dx = 0 \tag{34}$$

and

$$\int_a^b \phi_m \frac{d}{dx}\left(r\frac{d\phi_n}{dx}\right)dx + \int_a^b p\,\phi_m\phi_n\,dx + \lambda_n \int_a^b \phi_m\phi_n\rho\,dx = 0. \tag{35}$$

Applying integration by parts to the first integral in each of these equations brings us to the results

$$(r\phi_n\phi_m')\Big|_a^b - \int_a^b r\phi_m'\,\phi_n'\,dx + \int_a^b p\phi_m\phi_n\,dx + \lambda_m \int_a^b \phi_m\phi_n\rho\,dx = 0 \tag{36}$$

and

$$(r\phi_m\phi_n')\Big|_a^b - \int_a^b r\phi_m'\,\phi_n'\,dx + \int_a^b p\phi_m\phi_n\,dx + \lambda_n \int_a^b \phi_m\phi_n\rho\,dx = 0. \tag{37}$$

Subtracting (36) and (37) gives

$$[r(\phi_n\phi_m' - \phi_m\phi_n')]\Big|_a^b + (\lambda_m - \lambda_n) \int_a^b \phi_m\phi_n\rho\,dx = 0. \tag{38}$$

It is immediately obvious that the first group of terms in (38) vanishes when either boundary condition (29) or (30) is imposed.

If boundary condition (31) is used it follows directly that provided $\beta_1 \neq 0$, $\beta_2 \neq 0$,

$$\phi_m'(a) = -(\alpha_1/\beta_1)\phi_m(a) \quad \text{and} \quad \phi_n'(a) = -(\alpha_1/\beta_1)\phi_n(a),$$

and

$$\phi_m'(b) = -(\alpha_2/\beta_2)\phi_m(b) \quad \text{and} \quad \phi_n'(b) = -(\alpha_2/\beta_2)\phi_n(b),$$

which again causes the first group of terms in (38) to vanish. An obvious modification of the argument shows these groups of terms still vanish if $\beta_1 = 0$ or $\beta_2 = 0$.

Thus for any one of the boundary conditions (29), (30) or (31) it follows from (38) that

$$(\lambda_m - \lambda_n) \int_a^b \phi_m(x)\,\phi_n(x)\,\rho(x)\,dx = 0. \tag{39}$$

However, by hypothesis $\lambda_m \neq \lambda_n$, so we have proved that

$$\int_a^b \phi_m(x)\,\phi_n(x)\,\rho(x)\,dx = 0, \quad \text{for } m \neq n,$$

and property 4 has been established.

Proof that the eigenvalues λ_n are real

The proof will be by contradiction. Suppose, if possible, that the eigenvalues are complex, and corresponding to the eigenvalue

$$\lambda_n = \alpha + i\beta,$$

there corresponds a complex eigenfunction

$$\phi_n = \Phi + i\Psi.$$

Then, as λ_n and ϕ_n satisfy (28) subject to one of its boundary conditions,

$$\frac{d}{dx}\left(r\frac{d}{dx}(\Phi + i\Psi)\right) + [p + (\alpha + i\beta)(\Phi + i\Psi)] = 0.$$

Taking the complex conjugate of this equation gives

$$\frac{d}{dx}\left(r\frac{d}{dx}(\Phi - i\Psi)\right) + [p + (\alpha - i\beta)(\Phi - i\Psi)] = 0,$$

which shows that $\bar{\lambda}_n = \alpha - i\beta$ is an eigenvalue corresponding to the eigenfunction $\bar{\phi}_n = \Phi - i\Psi$. Now, by hypothesis, $\lambda_n \neq \bar{\lambda}_n$, so by property 4 the eigenfunctions ϕ_n and $\bar{\phi}_n$ must be orthogonal with respect to the weight function $\rho(x)$.
Thus

$$\int_a^b (\Phi + \Psi)(\Phi - i\Psi)\rho\,dx = 0,$$

which is equivalent to

$$\|\phi_n\|^2 = \int_a^b (\Phi^2 + \Psi^2)\rho\,dx = 0.$$

However this is impossible, because as ϕ_n is an eigenfunction its norm is a nonnegative number. We thus conclude that the eigenvalues λ_n must all be real.

Problems for Section 9.8

1 Show that the representation of an arbitrary function $f(x)$ defined over the interval $0 \leq x \leq L$ in terms of the eigenfunctions of

$$y'' + \lambda y = 0,$$

subject to the boundary conditions $y'(0) = 0$ and $y'(L) = 0$, is the Fourier cosine series.

2 Find the eigenvalues and eigenfunctions of

$$y'' + 9\lambda y = 0,$$

subject to the boundary conditions $y(0) = 0$ and $y(L) = 0$.

3 Find the eigenvalues and eigenfunctions of

$$y'' + \lambda y = 0,$$

subject to the boundary conditions $y(-\pi) = 0$, $y'(\pi) = 0$.

4 Find the eigenvalues and eigenfunctions of

$$y'' + \lambda y = 0,$$

subject to the boundary conditions $y'(-\pi) = 0$, $y(\pi) = 0$.

5 Find the eigenvalues and eigenfunctions of

$$y'' + \lambda y = 0,$$

subject to the boundary conditions $y(0) = 0$, $y(\pi) - y'(\pi) = 0$.

6 Show that the representation of an arbitrary function $f(x)$ defined over the interval $-\pi \le x \le \pi$ in terms of the eigenfunctions of

$$y'' + \lambda y = 0,$$

subject to the boundary conditions $y(0) = y(2\pi)$ and $y'(0) = y'(2\pi)$ is the ordinary Fourier series representation over the interval $0 < x < 2\pi$.

9.9 Expansions in terms of Bessel functions

The need to expand an arbitrary function $f(x)$ in terms of Bessel functions usually arises as a result of separating variables in a partial differential equation in which cylindrical polar coordinates are involved. Such problems arise in many different physical situations, ranging from the vibration of circular membranes through to problems involving electrostatic fields in cavities and steady-state temperature distributions in plates.

The result of applying the method of separation of variables in such problems is to give rise to *Bessel's equation* in the form

$$x^2 y'' + xy' + (k^2 x^2 - v^2)\, y = 0, \tag{1}$$

in which v is a fixed number and k^2 is a parameter to be determined from the given boundary conditions.

When expressed in this form Bessel's equation becomes a *Sturm–Liouville equation*, because it can be written

$$\frac{\mathrm{d}}{\mathrm{d}x}(xy') + \left(k^2 x - \frac{v^2}{x}\right) y = 0. \tag{2}$$

Inspection of (2) shows that in the notation of (28) of Sec. 9.7

$$r(x) = x, \, p(x) = -\frac{v^2}{x}, \lambda = k^2 \quad \text{and} \quad \rho(x) = x, \tag{3}$$

so k^2 is the eigenvalue which appears in the equation.

We already know from Sec. 8.7 that the general solution of (2) (equivalently (1)) in terms of k is

$$y(x) = AJ_v(kx) + BY_v(kx), \tag{4}$$

where $J_v(x)$ and $Y_v(x)$ are Bessel functions of order v of the first and second kind, respectively, and A, B are arbitrary constants.

Where $v \neq 0$, typical homogeneous boundary conditions for the solution $y(x)$ which arise in physical problems are

$$y(0) = 0 \quad \text{and} \quad y(a) = 0. \tag{5}$$

These are seen to belong to the type given in (29) of Sec. 9.8. Consequently, as (2) and (5) constitute a Sturm–Liouville problem, the solutions will possess the properties of a Sturm–Liouville system listed in Sec. 9.8.

To proceed further we need to determine the eigenvalues and eigenfunctions. Applying the boundary condition $y(0) = 0$ to (4) shows we must set $B = 0$, because $Y_v(x)$ is infinite at $x = 0$, though A remains arbitrary since $J_v(0) = 0$ provided $v \neq 0$. The general solution (4) thus reduces to

$$y(x) = AJ_v(kx), \tag{6}$$

which must also satisfy the second boundary condition $y(a) = 0$. As a result we see that k must be such that

$$J_v(ka) = 0. \tag{7}$$

It follows from this that ka may be any one of the infinity of zeros of $J_v(x)$, which we denote by

$$\mu_{v,1} < \mu_{v,2} < \mu_{v,3} < \cdots, \tag{8}$$

with the understanding that $\mu_{v,j}$ is the jth zero of $J_v(x)$. The numerical values of some of these zeros are listed in Sec. 8.7 for integral values of v.

In the event that $v = 0$, the homogeneous boundary conditions to be imposed on (2), and hence on (4), are usually of the form

$$y'(0) = 0 \quad \text{and} \quad y(a) = 0. \tag{9}$$

These boundary conditions are a special case of boundary conditions (31) in Sec. 9.7. Thus, when $v = 0$ and boundary conditions (9) apply, ka must be any one of the infinity of zeros of $J_0(x)$

$$\mu_{0,1} < \mu_{0,2} < \mu_{0,3} < \cdots, \tag{10}$$

the first few of which are also listed in Sec. 8.7. When $v \neq 0$ and boundary conditions (5) apply, then as $ka = \mu_{v,n}$, the *eigenvalues* λ_n are seen to be given by

$$\lambda_n = (\mu_{v,n}/a)^2, \quad \text{for } n = 1, 2, \dots, \tag{11}$$

while the corresponding *eigenfunctions* $y_n(x)$ which have the general form given in (6) become

$$y_n(x) = J_v\left(\frac{\mu_{v,n}x}{a}\right), \qquad \text{for } n = 1, 2, \dots . \tag{12}$$

Similarly, when $v = 0$ and boundary conditions (9) apply, the *eigenvalues* are given by

$$\lambda_n = (\mu_{0,n}/a)^2, \qquad \text{for } n = 1, 2, \dots , \tag{13}$$

and the corresponding *eigenfunctions* by

$$y_n(x) = J_0\left(\frac{\mu_{0,n}x}{a}\right), \qquad \text{for } n = 1, 2, \dots . \tag{14}$$

As the eigenfunctions $y_n(x)$ are solutions of a Sturm–Liouville problem defined over the interval $0 \le x \le a$ with the weight function $\rho(x) = x$, it follows from property 4 of Sec. 9.8 that the orthogonality condition for these Bessel functions takes the form

$$\int_0^a J_v\left(\frac{\mu_{v,m}x}{a}\right)J_v\left(\frac{\mu_{v,n}x}{a}\right)x\,dx = 0, \qquad \text{for } m \ne n. \tag{15}$$

It is instructive to notice the special nature of this condition. The orthogonality exhibited by Bessel functions involves the weight function x and it occurs between Bessel functions of the *same* order v, but with *different* zeros $\mu_{v,r}$ in their arguments.

The **Fourier–Bessel** representation of a function $f(x)$ given by property 5 of Sec. 9.8, so called because it is the analog of the trigonometric Fourier series representation of $f(x)$, becomes

$$f(x) = \sum_{n=1}^{\infty} c_n J_v\left(\frac{\mu_{v,n}x}{a}\right), \tag{16}$$

where

$$c_n = \frac{1}{\|y_n\|^2}\int_0^a f(x) J_v\left(\frac{\mu_{v,n}x}{a}\right)x\,dx, \tag{17}$$

and

$$\|y_n\|^2 = \int_0^a \left[J_v\left(\frac{\mu_{v,n}x}{a}\right)\right]^2 x\,dx = \frac{a^2}{2}[J_{v+1}(\mu_{v,n})]^2, \qquad \text{for } n = 1, 2, \dots . \tag{18}$$

The evaluation of the definite integral defining $\|y_n\|^2$ to obtain the explicit expression given in (18) can be carried out in several different ways, and it is left as an exercise for the reader. A straightforward method is outlined in Prob. 5 at the end of this section.

Example 1. Fourier–Bessel expansion of a continuous function

Expand the function $f(x) = a^2 - x^2$ in the interval $0 \le x \le a$ in terms of the eigenfunctions of the differential equation

$$xy'' + y' + k^2 xy = 0,$$

subject to the boundary conditions $y'(0) = 0$, $y(a) = 0$.

Solution

Comparing the equation with (1), and taking into account boundary conditions (9), we recognize that this is a Sturm–Liouville equation involving the Bessel function $J_0(kx)$. Thus it follows from (13) that the eigenvalues are

$$\lambda_n = (\mu_{0,n}/a)^2,$$

where $\mu_{0,n}$ is the nth zero of $J_0(x)$, and the corresponding eigenfunctions are

$$y_n(x) = J_0\left(\frac{\mu_{0,n}x}{a}\right), \qquad \text{for } n = 1, 2, \ldots.$$

Then we know from (16) to (18) that

$$f(x) = \sum_{n=1}^{\infty} c_n J_0\left(\frac{\mu_{0,n}x}{a}\right),$$

where

$$c_n = \frac{1}{\|y_n\|^2} \int_0^a (a^2 - x^2)x\, J_0\left(\frac{\mu_{0,n}x}{a}\right) dx,$$

with

$$\|y_n\|^2 = \int_0^a \left[J_0\left(\frac{\mu_{0,n}x}{a}\right)\right]^2 x\, dx = \frac{a^2}{2}[J_1(\mu_{0,n})]^2.$$

To complete the expansion we need to evaluate the integral defining c_n. As

$$\int_0^a (a^2 - x^2) J_0\left(\frac{\mu_{0,n}x}{a}\right)x\, dx = a^2 \int_0^a x\, J_0\left(\frac{\mu_{0,n}x}{a}\right) dx - \int_0^a x^3 J_0\left(\frac{\mu_{0,n}x}{a}\right) dx,$$

we shall consider each integral on the right separately.
Setting $t = \mu_{0,n}x/a$, the first integral becomes

$$a^2 \int_0^a x\, J_0\left(\frac{\mu_{0,n}x}{a}\right) dx = \left(\frac{a^4}{\mu_{0,n}^2}\right) \int_0^{\mu_{0,n}} t\, J_0(t)\, dt.$$

However, setting $v = 1$ in (21) of Sec. 8.7 and integrating gives

$$\int_0^{\mu_{0,n}} t\, J_0(t)\, dt = [t J_1(t)]\Big|_0^{\mu_{0,n}} = \mu_{0,n} J_1(\mu_{0,n}),$$

so

$$a^2 \int_0^a x\, J_0\left(\frac{\mu_{0,n}x}{a}\right) dx = \frac{a^4}{\mu_{0,n}} J_1(\mu_{0,n}).$$

Making the same variable change in the second integral brings it into the form

$$\int_0^a x^3 J_0\left(\frac{\mu_{0,n}x}{a}\right) dx = \left(\frac{a^4}{\mu_{0,n}^4}\right) \int_0^{\mu_{0,n}} t^3 J_0(t) dt.$$

Now

$$\int_0^{\mu_{0,n}} t^3 J_0(t) dt = \int_0^{\mu_{0,n}} t^2 [t J_0(t)] dt,$$

but from (21) in Sec. 8.7 with $v = 1$ we see that

$$t J_0(t) = \frac{d}{dt} [t J_1(t)],$$

so using integration by parts we find that

$$\int_0^{\mu_{0,n}} t^3 J_0(t) dt = [t^3 J_1(t)] \Big|_0^{\mu_{0,n}} - 2 \int_0^{\mu_{0,n}} t^2 J_1(t) dt$$

$$= \mu_{0,n}^3 J_1(\mu_{0,n}) - \int_0^{\mu_{0,n}} t^2 J_1(t) dt.$$

Again appealing to (21) in Sec. 8.7, but this time with $v = 2$, and then integrating, shows that

$$\int t^2 J_1(t) dt = t^2 J_2(t) + \text{const.}$$

Consequently,

$$\int_0^{\mu_{0,n}} t^3 J_0(t) dt = \mu_{0,n}^3 J_1(\mu_{0,n}) - 2 [t^2 J_2(t)] \Big|_0^{\mu_{0,n}}$$

$$= \mu_{0,n}^3 J_1(\mu_{0,n}) - 2\mu_{0,n}^2 J_2(\mu_{0,n}),$$

and hence

$$\int_0^a x^3 J_0 \left(\frac{\mu_{0,n} x}{a} \right) dx = \frac{a^4}{\mu^2} [\mu_{0,n} J_1(\mu_{0,n}) - 2 J_2(\mu_{0,n})].$$

Finally, combining results gives,

$$c_n = \frac{4 a^2 J^2(\mu_{0,n})}{\mu_{0,n}^2 [J_1(\mu_{0,n})]^2}, \qquad \text{for } n = 1, 2, \ldots .$$

Inserting this expression for c_n into the Fourier–Bessel expansion gives

$$a^2 - x^2 = 4 a^2 \sum_{n=1}^{\infty} \frac{J_2(\mu_{0,n})}{\mu_{0,n}^2 [J_1(\mu_{0,n})]^2} J_0 \left(\frac{\mu_{0,n} x}{a} \right),$$

for $0 \leq x \leq a$.

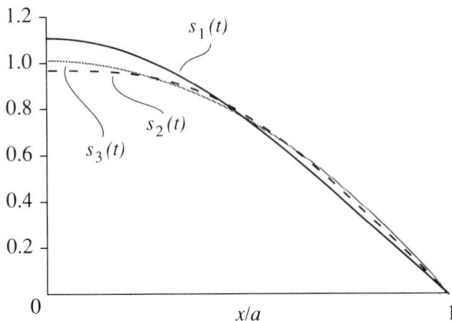

Fig. 9.30 Fourier–Bessel approximation $s_4(x)$ to $f(x) = a^2 - x^2$

The convergence of this Fourier–Bessel series to $f(x) = a^2 - x^2$ is shown in Fig. 9.30, in which the partial sum $s_n(x)$ denotes the sum of the first n terms of the above infinite series. The partial sum $s_3(x)$ provides an excellent approximation over most of the interval, except close to $x/a = 0$ where it overshoots the exact value by less than 2 per cent. When drawn on the same scale, the partial sums from $s_4(x)$ onwards are indistinguishable from $f(x)$. ∎

As a final example, we illustrate the behavior of a Fourier–Bessel expansion of a discontinuous function.

Example 2. Fourier–Bessel expansion of a discontinuous function

Expand the function

$$f(x) = \begin{cases} 1, & \text{for } 0 \le x < h \\ 0, & \text{for } h < x \le 1 \end{cases}$$

in terms of the eigenfunctions of the differential equation

$$xy'' + y' + k^2 xy = 0,$$

subject to the boundary conditions $y'(0) = 0$, $y(1) = 0$.

Solution

As with the previous example, the expansion will involve the Bessel function $J_0(kx)$, but this time as the interval involved is $0 \le x \le 1$ the eigenvalues are

$$\lambda_n = \mu_{0,n}^2, \qquad \text{for } n = 1, 2, \ldots ,$$

and the eigenfunctions are

$$y_n(x) = J_0(\mu_{0,n}x).$$

Thus the Fourier–Bessel expansion becomes

$$f(x) \sim \sum_{n=1}^{\infty} c_n J_0(\mu_{0,n}x),$$

with

$$c_n = \frac{1}{\|y_n\|^2} \int_0^1 f(x)\, x J_0(\mu_{0,n}x)\mathrm{d}x = \frac{1}{\|y_n\|^2} \int_0^h x J_0(\mu_{0,n}x)\mathrm{d}x,$$

and

$$\|y_n\|^2 = \int_0^1 [J_0(\mu_{0,n})]^2 x\, \mathrm{d}x = \tfrac{1}{2}[J_1(\mu_{0,n}x)]^2.$$

Now, setting $t = \mu_{0,n}x$, we have

$$\int_0^h x J_0(\mu_{0,n})\mathrm{d}x = \frac{1}{(\mu_{0,n})^2} \int_0^{\mu_{0,n}h} t J_0(t)\, \mathrm{d}t$$

$$= \left(\frac{h}{\mu_{0,n}}\right) J_1(\mu_{0,n}h).$$

Thus

$$c_n = \left(\frac{2h}{\mu_{0,n}}\right)\frac{J_1(\mu_{0,n}h)}{[J_1(\mu_{0,n})]^2},$$

and so

$$f(x) \sim 2h \sum_{n=1}^{\infty} \frac{J_1(\mu_{0,n}h)}{[J_1(\mu_{0,n})]^2} J_0(\mu_{0,n}x), \qquad \text{for } 0 < x < 1.$$

The behavior of this series is indicated in Fig. 9.31, in which the partial sum approximation $s_{10}(x)$ representing the sum of the first ten terms of the series is shown for the case in which $h = 0.5$. We have used the relational symbol $\sim$ in place of the equality sign because the convergence properties of Fourier–Bessel series need interpretation when discontinuous functions are involved. However, it is seen that the behavior of $s_{10}(x)$ closely resembles that of an ordinary Fourier series at a discontinuity.

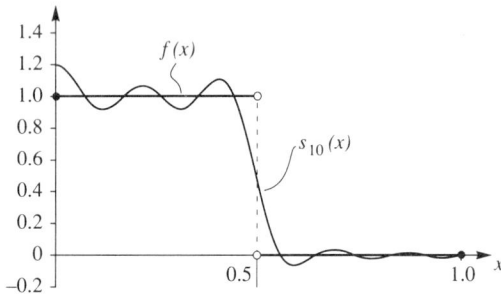

Fig. 9.31 The Fourier–Bessel approximation $s_{10}(x)$ to the discontinuous function $f(x)$ in which $h = 0.5$

Problems for Section 9.9

1 Expand the function $f(x) = x^2$ over the interval $0 \le x \le 1$ in terms of the eigenfunctions of

$$x^2 y'' + xy' + (k^2 x^2 - 4)y = 0,$$

subject to the boundary conditions $y(0) = 0$, $y(1) = 0$.

2 Expand the function $f(x) = x$ over the interval $0 \le x \le 1$ in terms of the eigenfunctions of

$$x^2 y'' + xy' + (k^2 x^2 - 1)\, y = 0,$$

subject to the boundary conditions $y(0) = 0$, $y(1) = 0$.

3 Expand the function $f(x) = x^{1/2}$ over the interval $0 \le x \le 1$ in terms of the eigenfunctions of

$$4x^2 y'' + 4xy' + (4k^2 x^2 - 1)\, y = 0,$$

subject to the boundary conditions $y(0) = 0$, $y(\pi) = 0$.

4 Expand the discontinuous function

$$f(x) = \begin{cases} 0, & \text{for} \quad 0 \le x < a \\ 1, & \text{for} \quad a < x < b \\ 0, & \text{for} \quad b < x \le 1 \end{cases}$$

in terms of the eigenfunctions of the differential equation

$$xy'' + y' + k^2 xy = 0,$$

subject to the boundary conditions $y'(0) = 0$, $y(1) = 0$, and hence show that

$$f(x) \sim 2 \sum_{n=1}^{\infty} \frac{[bJ_1(\mu_{0,n}b) - aJ_1(\mu_{0,n}a)]}{\mu_{0,n}[J_1(\mu_{0,n})]^2} J_0(\mu_{0,n}x)$$

for $0 \le x \le 1$.

5 Consider the two Sturm–Liouville equations

$$\frac{d}{dx}\left(x\frac{dU_n}{dx}\right) + \left(\mu_{v,n}^2 x - \frac{v^2}{x}\right)U_v = 0,$$

and

$$\frac{d}{dx}\left(x\frac{dV_n}{dx}\right) + \left(\alpha^2 x - \frac{v^2}{x}\right)V_v = 0,$$

defined on the interval $0 \le x \le 1$, where $U_v = J_v(\mu_{v,n}x)$ and $V_v = J_v(\alpha x)$. Multiply the first by V_v and the second by U_v; integrate the resulting equations over the interval $0 \le x \le 1$ and subtract the results to show that

$$\int_0^1 J_v(\mu_{v,n}x)\, J_v(\alpha x)\, x\, dx = \frac{\mu_{v,n} J_v(\alpha)J_v'(\mu_{v,n})}{\alpha^2 - \mu_{v,n}^2}.$$

Letting $\alpha \to \mu_{v,n}$, use L'Hospital's rule together with a justification of the fact that $J_v'(\mu_{v,n}) = -J_{v+1}(\mu_{v,n})$ to show

$$\int_0^1 [J_v(\mu_{v,n}x)]^2\, x\, dx = \frac{[J_{v+1}(\mu_{v,n})]^2}{2},$$

and hence that

$$\int_0^a \left[J_v\left(\frac{\mu_{v,n}x}{a}\right)\right]^2 x\, dx = \frac{a^2}{2}[J_{v+1}(\mu_{v,n})]^2.$$

6 Discuss, without evaluating the integrals involved, the expansion of a function $f(x)$ defined in the interval $a \le x \le b$ $(b > a > 0)$ in terms of the eigenvalues of

$$xy'' + y' + k^2 xy = 0,$$

subject to the boundary conditions $y(a) = 0$, $y(b) = 0$.

7 Consider the two Sturm–Liouville equations

$$\frac{d}{dx}\left(x\frac{dJ_v(\lambda x)}{dx}\right) + (\lambda^2 x - \frac{v^2}{x})J_v(\lambda x) = 0$$

and

$$\frac{d}{dx}\left(x\frac{dJ_v(\mu x)}{dx}\right) + (\mu^2 x - \frac{v^2}{x})J_v(\mu x) = 0.$$

Multiply the first by $J_v(\mu x)$ and the second by $J_v(\lambda x)$; integrate the resulting equations over the interval $0 \leq x \leq a$ and subtract the results to show that

$$\int_0^a x J_\nu(\lambda x)\, J_\nu(\mu x)\, \mathrm{d}x = \left(\frac{a}{\lambda^2 - \mu^2}\right) [\mu J_\nu(\lambda a) J_\nu'(\mu a) - \lambda J_\nu(\mu a) J_\nu'(\lambda a)].$$

Now let λ and μ be two different positive roots of the transcendental equation

$$h J_\nu(\xi a) + k \xi J_\nu'(\xi a) = 0,$$

with h, k constants. Use this condition to show the functions $J_\nu(\lambda x)$ and $J_\nu(\mu x)$ are orthogonal over the interval $0 \leq x \leq a$ with weight function x, so that

$$\int_0^a x J_\nu(\lambda x)\, J_\nu(\mu x)\, \mathrm{d}x = 0, \qquad \text{for } \lambda \neq \mu.$$

Let $\mu \to \lambda$ and use L'Hospital's rule to show that

$$\int_0^a x [J_\nu(\lambda x)]^2\, \mathrm{d}x = \frac{a^2}{2}\left\{ [J_\nu'(\lambda a)]^2 - J_n(\lambda a) J_n''(\lambda a) - \frac{1}{\lambda a} J_n(\lambda a) J_n'(\lambda a) \right\}.$$

9.10 Orthogonal polynomials

Orthogonal polynomials represent a special class of orthogonal functions which are comparatively simple to use and yet have wide-ranging applications. Although most occur naturally in applications to physical problems, some are also of importance in numerical analysis. Chebyshev polynomials provide an essential basis from which to develop approximation theory, while Legendre and Laguerre polynomials, like Hermite polynomials, occur in physical problems and also in the development of numerical integration techniques.

Legendre polynomials

An example of orthogonal polynomials has already been encountered in Sec. 8.4, in which Legendre polynomials were discussed along with many of their properties. It will be recalled that *Legendre polynomials* are polynomial solutions to the Legendre differential equation

$$(1 - x^2)\frac{\mathrm{d}^2 y}{\mathrm{d}x^2} - 2x \frac{\mathrm{d}y}{\mathrm{d}x} + \alpha(\alpha + 1)\, y = 0, \tag{1}$$

defined over the interval $-1 \leq x \leq 1$, when α is any one of the integers $0, 1, 2, \ldots$.

The polynomial solution $P_n(x)$ of (1) corresponding to $\alpha = n$ satisfies a Sturm–Liouville equation, because (1) can be written in the form

$$\frac{\mathrm{d}}{\mathrm{d}x}\left((1 - x^2)\frac{\mathrm{d}P_n}{\mathrm{d}x} \right) + n(n + 1)\, P_n = 0. \tag{2}$$

The boundary conditions imposed on the solutions of (2) are that they remain bounded as $x \to \pm 1$ from within the interval $-1 \leq x \leq 1$.

In the notation of Sec. 9.8 we see that for Legendre's equation

$$r(x) = 1 - x^2, \quad p(x) \equiv 0, \quad \rho(x) \equiv 1 \quad \text{and} \quad \lambda = n(n+1). \tag{3}$$

Equation (2) is an example of a singular Sturm–Liouville equation, because $r(x)$ vanishes at the end points of the interval $-1 \leq x \leq 1$ over which $P_n(x)$ is defined. The orthogonality of Legendre polynomials was established in Sec. 8.4, using arguments which were essentially the same as those used in Sec. 9.8 when discussing Sturm–Liouville problems.

For the sake of completeness we now summarize the essential properties of Legendre polynomials which have already been established.

Basic properties of the Legendre polynomials $P_n(x)$

Definitions of $P_n(x)$

The Legendre polynomial $P_n(x)$ with n an integer is defined over the interval $-1 \leq x \leq 1$ by either of the following explicit expressions:

$$P_n(x) = \frac{1}{2^n n!} \frac{d^n}{dx^n} [(x^2 - 1)^n] \; (\textit{Rodrigues' formula}) \tag{4}$$

$$P_n(x) = \begin{cases} \displaystyle\sum_{m=1}^{n/2} (-1)^m \frac{(2n-2m)!}{2^n m! (n-m)! (n-2m)!} x^{n-2m} \; (n \text{ even}) \\[4mm] \displaystyle\sum_{n=1}^{(n-1)/2} (-1)^m \frac{(2n-2m)!}{2^n m! (n-m)! (n-2m)!} x^{n-2m} \; (n \text{ odd}). \end{cases} \tag{5}$$

Recursion formula

$$(n+1) P_{n+1}(x) = (2n+1)x P_n(x) - n P_{n-1}(x). \tag{6}$$

The first eight Legendre polynomials

$$\begin{aligned} &P_0(x) = 1, && P_1(x) = x, \\ &P_2(x) = \tfrac{1}{2}(3x^2 - 1), && P_3(x) = \tfrac{1}{2}(5x^3 - 3x), \\ &P_4(x) = \tfrac{1}{8}(35x^4 - 30x^2 + 3), && P_5(x) = \tfrac{1}{8}(63x^5 - 70x^3 + 15x), \\ &P_6(x) = \tfrac{1}{16}(231x^6 - 315x^4 + 105x^2 - 5), && P_7(x) \tfrac{1}{16}(429x^7 - 693x^5 + 315x^3 - 35x). \end{aligned} \tag{7}$$

Orthogonality property

$$\int_{-1}^{1} P_m(x) P_n(x) \, dx = \begin{cases} 0, & \text{for } m \neq n \\[2mm] \dfrac{2}{2n+1}, & \text{for } m = n. \end{cases} \tag{8}$$

Useful special cases

$$P_n(1) = 1, \quad P_n(-1) = (-1)^n,$$

$$\frac{d^n}{dx^n} P_n(x) = \frac{(2n)!}{2^n n!}, \tag{9}$$

$$\int_{-1}^{1} [P_n(x)]^2 \, dx = \left(\frac{2n-1}{2n+1}\right) \int_{-1}^{1} [P_{n-1}(x)]^2 \, dx.$$

Graphs of $P_1(x)$ to $P_7(x)$

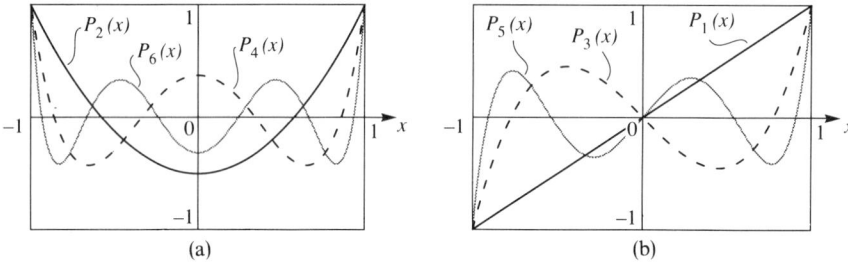

Fig. 9.32   (a) Even order Legendre polynomials;   (b) Odd order Legendre polynomials

Frequently a function to be represented in terms of Legendre polynomials is defined over the interval $a \le x \le b$, instead of the interval $-1 \le x \le 1$ on which the Legendre polynomials are defined. In such circumstances the transformation

$$t = \left(\frac{2}{b-a}\right) x - \left(\frac{b+a}{b-a}\right)$$

maps the interval $a \le x \le b$ onto the interval $-1 \le t \le 1$.

Chebyshev polynomials

The Chebyshev polynomials[13] are polynomial solutions of the Chebyshev differential equation.

$$(1-x^2)\frac{d^2y}{dx^2} - x\frac{dy}{dx} + \lambda y = 0, \tag{10}$$

---

[13] PAFNUTI LIVOVICH CHEBYSHEV (1821–1894), an important Russian mathematician whose major contributions while at the University of Petrograd (Leningrad) were to approximation theory and number theory. His orthogonal polynomials were introduced in 1854 when he showed that of all the polynomials of degree n in x such that the coefficient of x^n is unity, the Chebyshev polynomial $T_n(x)$ provides the least deviation from the value zero over the interval $-1 \le x \le 1$. Due to a change in the transliteration convention, in earlier works his name is given either as TCHEBYCHEFF or TCHEBICHEF.

defined over the interval $-1 \leq x \leq 1$. The differential equation may be written in Sturm–Liouville form as

$$\frac{d}{dx}\left((1-x^2)^{1/2}\frac{dy}{dx}\right) + \frac{\lambda}{(1-x^2)^{1/2}}y = 0, \tag{11}$$

so that in the notation of Sec. 9.8,

$$r(x) = (1-x^2)^{1/2}, \quad p(x) \equiv 0 \quad \text{and} \quad \rho(x) = \frac{1}{(1-x^2)^{1/2}}. \tag{12}$$

The boundary conditions to be imposed on (11) are that the solutions remain bounded as $x \to \pm 1$ from within the interval $-1 \leq x \leq 1$.

If a series solution is sought for (10) of the form

$$y(x) = \sum_{r=1}^{\infty} a_r x^r, \tag{13}$$

it follows that

$$\sum_{r=1}^{\infty} [(r+2)(r+1)a_{r+2} - (r^2 - \lambda)a_r]x^r = 0, \tag{14}$$

so the recursion relation for the coefficients a_r is seen to be

$$a_{r+2} = \frac{r^2 - \lambda}{(r+1)(r+2)}a_r, \quad \text{for } r = 0, 1, \ldots. \tag{15}$$

The parameter λ is the eigenvalue, and choosing

$$\lambda = \lambda_n = n^2, \quad \text{for } n = 0, 1, \ldots, \tag{16}$$

causes the sequence of coefficients a_r to vanish after a_n. Thus the polynomial solution generated by using these nonzero a_r in (13) will be of degree n. It will involve only even powers of x when n is even, and only odd powers when n is odd. Thus, as with Legendre's equation, the solutions of (10) separate into even and odd polynomial solutions.

The coefficients of the polynomials generated in this manner will depend on the choice of a_0 and a_1, but the usual choice of $a_0^{2k} = (-1)^k$ for the polynomial $T_{2k}(x)$ and $a_1^{2k+1} = (-1)^k(2k+1)$ for the polynomial $T_{2k+1}(x)$ produces the following Chebyshev polynomials with integer coefficients (see Fig. 9.33)

$$\begin{aligned}
T_0(x) &= 1, & T_1(x) &= x, \\
T_2(x) &= 2x^2 - 1, & T_3(x) &= 4x^3 - 3x, \\
T_4(x) &= 8x^4 - 8x^2 + 1, & T_5(x) &= 16x^5 - 20x^3 + 5x, \\
T_6(x) &= 32x^6 - 48x^4 + 18x^2 - 1, & T_7(x) &= 64x^7 - 112x^5 + 56x^3 - 7x.
\end{aligned} \tag{17}$$

There is a Rodrigues' type formula which generates polynomials differing from those above only in their normalization (i.e. by a factor). For the sake of simplicity, denoting these differently normalized polynomials by the same symbol $T_n(x)$, **Rodrigues' formula** for Chebyshev polynomials takes the form

$$T_n(x) = (-1)^n \frac{(2n)!}{2^n n!} (1 - x^2)^{1/2} \frac{d^n}{dx^n} (1 - x^2)^{n - 1/2}.$$ (18)

Its justification is provided by using differentiation to show that $T_n(x)$ satisfies the differential equation

$$(1 - x^2) \frac{d^2 T_n}{dx^2} - x \frac{dT_n}{dx} + n^2 T_n = 0.$$ (19)

This task is left as an exercise for the reader.

An alternative definition of Chebyshev polynomials is provided by the expression

$$T_n(x) = \cos(n \arccos x), \qquad \text{for } -1 \le x \le 1.$$ (20)

Using the trigonometric identity

$$\cos(A + B) + \cos(A - B) = 2 \cos A \cos B,$$

and setting $A + B = (n + m) \arccos x$, $A - B = (n - m) \arccos x$, we arrive at the following **recursion relation** for Chebyshev polynomials

$$T_{n+m}(x) + T_{n-m}(x) = T_n(x) T_m(x).$$ (21)

The most useful recursion formula follows from this by setting $m = 1$, for then (21) simplifies and relates three consecutive Chebyshev polynomials in the following manner

$$T_{n+1}(x) = 2x T_n(x) - T_{n-1}(x).$$ (22)

This recursion relation can be used in conjunction with the results of (17) to generate any other Chebyshev polynomials that may be required. The proof that (20) satisfies the defining differential equation (19) has already been given in Sec. 7.5(c) where Chebyshev polynomials were discussed in the context of homogeneous variable coefficient difference equations and the z-transform. We refer the reader to that section for further information about Chebyshev polynomials, and for a brief discussion of Chebyshev polynomials of the second kind $U_n(x)$.

The orthogonality of the Chebyshev polynomials $T_n(x)$ over the interval $-1 \le x \le 1$ with respect to the weight function $(1 - x^2)^{1/2}$ follows from the Sturm–Liouville theory, though it was proved in a different way in Sec. 7.5(c), in which $\| T_n \|^2$ was also determined. Thus we have the

Orthogonality property

$$\int_{-1}^{1} T_m(x) T_n(x)(1 - x^2)^{1/2} \, dx = \begin{cases} 0, & \text{for } m \ne n \\ \pi/2, & \text{for } m = n \\ \pi, & \text{for } m = n = 0. \end{cases}$$ (23)

Graphs of $T_1(x)$ to $T_7(x)$

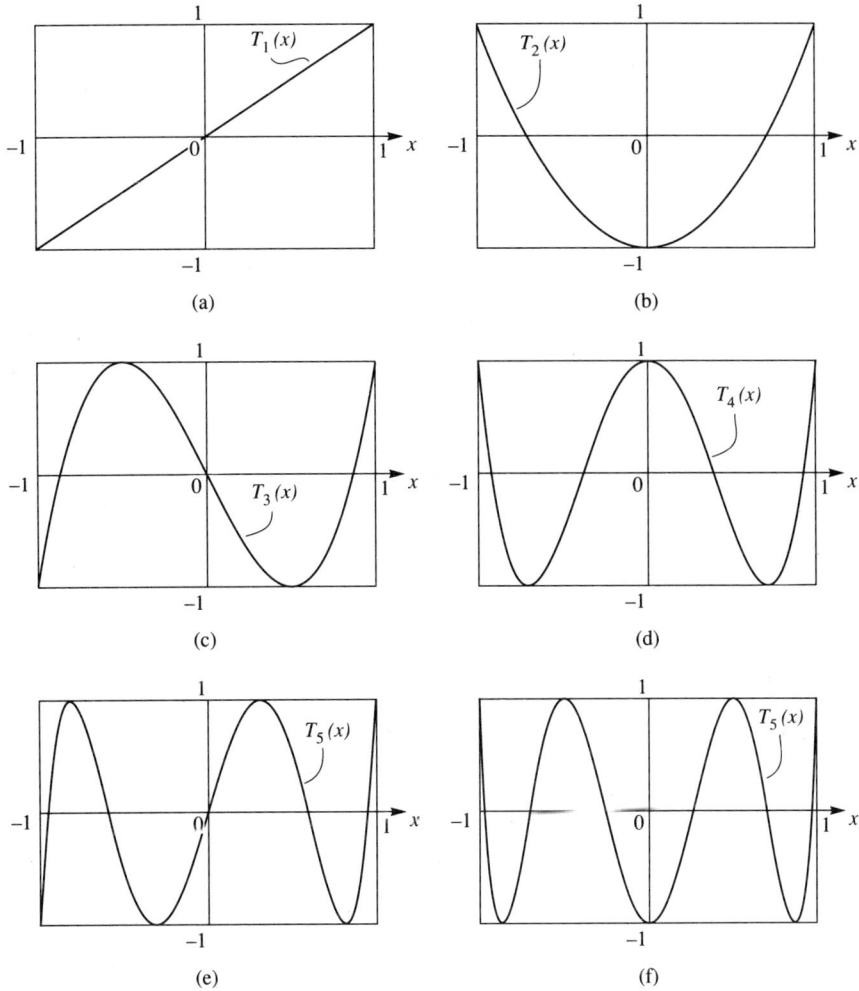

(a)

(b)

(c)

(d)

(e)

(f)

Fig 9.33 Graphs of Chebyshev polynomials

For the sake of completeness we now summarize the most important properties of $U_n(x)$ obtained in Sec. 7.5(c).

Definition of $U_n(x)$

$$U_n(x) = \frac{\sin(n \arccos x)}{(1 - t^2)^{1/2}}, \qquad \text{for } -1 \le x \le 1. \tag{24}$$

Recursion relation for $U_n(x)$

$$U_{n+1}(x) = 2x\, U_n(x) - U_{n-1}(x). \tag{25}$$

First eight Chebyshev polynomials $U_n(x)$ of the second kind

$$
\begin{aligned}
&U_0(x) = 1, &&U_4(x) = 16x^4 - 12x^2 + 1, \\
&U_1(x) = 2x, &&U_5(x) = 32x^5 - 32x^3 + 6x, \\
&U_2(x) = 4x^2 - 1, &&U_6(x) = 64x^6 - 80x^4 + 24x - 1, \\
&U_3(x) = 8x^3 - 4x, &&U_7(x) = 128x^7 - 192x^5 + 80x^3 - 8x.
\end{aligned}
\tag{26}
$$

Orthogonality property

$$
\int_{-1}^{1} U_m(x)\, U_n(x)(1 - x^2)^{1/2}\, dx = \begin{cases} 0, & \text{for } m \neq n \\ \pi/2, & \text{for } m = n \\ 0, & \text{for } m = n = 0. \end{cases}
\tag{27}
$$

If necessary, the transformation

$$
t = \left(\frac{2}{b - a}\right) x - \left(\frac{b + a}{b - a}\right)
$$

may be used to map a function defined on the interval $a \leq x \leq b$ onto one defined on the interval $-1 \leq t \leq 1$ so that Chebyshev polynomials may be used in its representation.

Laguerre and Hermite polynomials

We offer only the briefest of introductions to Laguerre and Hermite polynomials, the first of which are orthogonal with respect to the weight function e^{-x} over the interval $0 \leq x < \infty$, and the second with respect to the weight function $\exp(-x^2)$ over the interval $-\infty < x < \infty$.

Laguerre polynomials

The **Laguerre polynomials** $L_n(x)$ are polynomial solutions of the Laguerre differential equation

$$
x \frac{d^2 y}{dx^2} + (1 - x)\frac{dy}{dx} + ny = 0,
\tag{28}
$$

with $n = 0, 1, 2, \ldots$. It may be shown by differentiation and substitution into (28) that the polynomials $L_n(x)$ defined by means of the **Rodrigues' formula**[14]

$$
L_n(x) = e^x \frac{d^n}{dx^n}(x^n e^{-n}).
\tag{29}
$$

---

[14] A definition which is often used in place of the one above is

$$
L_n(x) = \frac{e^x}{n!}\frac{d^n(x^n e^{-x})}{dx^n}, \qquad \text{for } n = 1, 2, \ldots.
$$

This merely scales each polynomial by the factor $1/n!$ to make the constant term in each polynomial equal to unity.

are solutions of the Laguerre differential equation. It is apparent from (29) that $L_n(x)$ is a polynomial in x of degree n.

Repeated use of integration by parts establishes the fact that

$$\int_0^\infty x^m e^{-x} L_n(x) dx = 0, \qquad \text{for } m = 0, 1, \dots, n-1, \tag{30}$$

and

$$\int_0^\infty x^n e^{-x} L_n(x) dx = (-1)^n (n!)^2. \tag{31}$$

As $L_n(x)$ is a polynomial of degree n in x, results (30) and (31) when taken together imply the **orthogonality property** of Laguerre polynomials

$$\int_0^\infty e^{-x} L_m(x) L_n(x) dx = \begin{cases} 0, & \text{for } m \neq n \\ (n!)^2, & \text{for } m = n. \end{cases} \tag{32}$$

This shows that the Laguerre polynomials $L_n(x)$ are orthogonal with respect to the weight function e^{-x} over the interval $0 \leq x < \infty$.

It can be shown that the Laguerre polynomials $L_n(x)$ are related by the **recursion formula**

$$L_{n+1}(x) = (2n + 1 - x) L_n(x) - n^2 L_{n-1}(x). \tag{33}$$

The first eight Laguerre polynomials are

$L_0(x) = 1$
$L_1(x) = 1 - x$
$L_2(x) = 2 - 4x + x^2$
$L_3(x) = 6 - 18x + 9x^2 - x^3$
$L_4(x) = 24 - 96x + 72x^2 - 16x^3 + x^4$
$L_5(x) = 120 - 600x + 600x^2 - 200x^3 + 25x^4 - x^5$
$L_6(x) = 720 - 4320x + 5400x^2 - 2400x^3 + 450x^4 - 36x^5 + x^6$
$L_7(x) = 5040 - 35\,280x + 52\,920x^2 - 29\,400x^3 + 7350x^4 - 882x^5 + 49x^6 - x^7.$

When needed, further Laguerre polynomials may be generated by using the above results in conjunction with (33).

Hermite polynomials

The **Hermite polynomials** $H_n(x)$ are polynomial solutions of the Hermite differential equation

$$\frac{d^2 y}{dx^2} - 2x \frac{dy}{dx} + 2ny = 0, \tag{34}$$

with $n = 0, 1, 2, \dots$.

As with Laguerre polynomials, it may be shown by differentiation and substitution into (34) that the polynomials $H_n(x)$ are solutions of (34) defined by means of the

Rodrigues' formula

$$H_n(x) = (-1)^n \exp(x^2) \frac{d^n}{dx^n} \exp(-x^2). \tag{35}$$

Inspection of (35) confirms that $H_n(x)$ is a polynomial in x of degree n.

As with Laguerre polynomials, repeated use of integration by parts establishes the fact that

$$\int_{-\infty}^{\infty} \exp(-x^2) H_n(x) x^m dx = 0, \qquad \text{for } m = 0, 1, \ldots, n-1, \tag{36}$$

and

$$\int_{-\infty}^{\infty} \exp(-x^2) H_n(x) x^n dx = n! \sqrt{\pi}. \tag{37}$$

As $H_n(x)$ is a polynomial of degree n in x, results (36) and (37) when taken together imply the **orthogonality property** of Hermite polynomials

$$\int_{-\infty}^{\infty} \exp(-x^2) H_m(x) H_n(x) dx = \begin{cases} 0, & \text{for } m \neq n \\ 2^n n! \sqrt{\pi}, & \text{for } m = n. \end{cases} \tag{38}$$

This shows that the Laguerre polynomials $L_n(x)$ are orthogonal with respect to the weight function $\exp(-x^2)$ over the entire real line $-\infty < x < \infty$.

It can be shown that the Hermite polynomials $H_n(x)$ are related by the **recursion formula**

$$H_{n+1}(x) = 2x H_n(x) - 2_n H_{n-1}(x). \tag{39}$$

The first eight Hermite polynomials are

$$
\begin{aligned}
H_0(x) &= 1 \\
H_1(x) &= 2x \\
H_2(x) &= 4x^2 - 2 \\
H_3(x) &= 8x^3 - 12x \\
H_4(x) &= 16x^4 - 48x^2 + 12 \\
H_5(x) &= 32x^5 - 16x^3 + 120x \\
H_6(x) &= 64x^6 - 480x^4 + 720x^2 - 120 \\
H_7(x) &= 128x^7 - 1344x^5 + 3360x^3 - 1680x.
\end{aligned} \tag{40}
$$

When necessary, this list of Hermite polynomials may be extended by using the above results with recursion formula (39).

Problems for Section 9.10

1 Verify by differentiation that

$$\frac{d}{dx}\left((1 - x^2)\frac{dy}{dx} \right) + n(n+1)y = 0$$

is the Legendre differential equation.

2 Every polynomial

$$Q_n(x) \equiv a_0 x^n + a_1 x^{n-1} + \ldots + a_{n-1} x + a_n$$

of degree n in x may be expressed as the sum of a finite number of Legendre polynomials. This may be accomplished by expressing x^r as a linear combination of $P_0, P_1, \ldots, P_r$ for $r = 0, 1, \ldots$, and then substituting for each power of x^r in $Q_n(x)$.

The expressions for x^r up to $r = 7$ are as follows:

$$1 = P_0$$
$$x = P_1$$
$$x^2 = \tfrac{1}{3}(P_0 + 2P_2)$$
$$x^3 = \tfrac{1}{5}(3P_1 + 2P_3)$$
$$x^4 = \tfrac{1}{35}(7P_0 + 20P_2 + 8P_4)$$
$$x^5 = \tfrac{1}{63}(27P_1 + 28P_3 + 8P_5)$$
$$x^6 = \tfrac{1}{231}(33P_0 + 110P_2 + 72P_4 + 16P_6)$$
$$x^7 = \tfrac{1}{429}(143P_1 + 182P_3 + 88P_5 + 16P_7).$$

Verify the expressions for x^5 and x^7 by eliminating the appropriate powers of x from the expressions for P_1, P_3, P_5 and P_7.

3 Verify by differentiation that

$$\frac{d}{dx}\left((1-x^2)^{1/2} \frac{dy}{dx} \right) + \frac{\lambda}{(1-x^2)^{1/2}} y = 0$$

is the Chebyshev differential equation.

4 Using (13), (15) and (16), compute $T_6(x)$ by choosing a_0 so that the constant term in the resulting polynomial is unity.

5 Set $n = 2$ in (18) and verify that Rodrigues' formula generates $T_2(x)$.

6 Show by substituting into the Chebyshev differential equation the expression for $T_n(x)$ given by (18) that $T_n(x)$ is a solution.

7 Verify recursion formula (21) with $m = 2$, $n = 5$ by making use of the explicit expressions for $T_2(x)$ and $T_5(x)$ given in (17).

8 When expressed in terms of Chebyshev polynomials, the expressions for x^r up to $r = 7$ are as follows:

$$1 = T_0$$
$$x = T_1$$
$$x^2 = \tfrac{1}{2}(T_0 + T_1)$$
$$x^3 = \tfrac{1}{4}(3T_0 + T_3)$$
$$x^4 = \tfrac{1}{8}(3T_0 + 4T_2 + T_4)$$
$$x^5 = \tfrac{1}{16}(10T_1 + 5T_3 + T_5)$$
$$x^6 = \tfrac{1}{32}(10T_0 + 15T_2 + 6T_4 + T_6)$$
$$x^7 = \tfrac{1}{64}(35T_1 + 21T_3 + 7T_5 + T_7).$$

Verify the expression for x^6 by eliminating the appropriate powers of x from T_0, T_2, T_4 and T_6.

9 Show the coefficient c_n in the expansion of x^k in terms of Chebyshev polynomials

$$x^k = \sum c_n T_n(x)$$

is given by

$$c_n = \frac{2}{\pi} \int_0^\pi (\cos\theta)^k \cos n\theta \, d\theta.$$

10 Set $n = 3$ in Rodrigues' formula (28) and show that it generates $L_3(x)$.

11 Show that $L_n(x)$ defined by Rodrigues' formula (28) satisfies the recursion relation

$$L_{n+1}(x) = (2n+1-x) L_n(x) - n^2 L_{n-1}(x).$$

12 Show by substituting into the Legendre differential equation the expression for $L_n(x)$ given by (28) that $L_n(x)$ is a solution.

13 Verify (29) by direct integration in the case when $m = 2$ and $n = 3$.

14 Verify (30) by direct integration in the case when $n = 2$.

15 When expressed in terms of Laguerre polynomials, the expressions for x^r up to $r = 7$ are as follows:

$1 = L_0$

$x = L_0 - L_1$

$x^2 = 2L_0 - 4L_1 + 2L_2$

$x^3 = 6L_0 - 18L_1 + 9L_2 - L_3$

$x^4 = 24L_0 - 96L_1 + 72L_2 - 16L_3 + L_4$

$x^5 = 120L_0 - 600L_1 + 600L_2 - 200L_3 + 25L_4 - L_5$

$x^6 = 720L_0 - 4320L_1 + 5400L_2 - 3600L_3 + 430L_4 - 36L_5 + L_6$

$x^7 = 5040L_0 - 35\,280L_1 + 5920L_2 - 29\,400L_3 + 7350L_4 - 882L_5 + 294L_6 - L_7.$

Verify the expression for x^4 by eliminating the appropriate powers of x from L_0, L_1, L_2, L_3, and L_4.

16 Set $n = 3$ in (35) and show that it generates $H_3(x)$.

17 Verify (36) by direct integration in the case when $m = 1$ and $n = 3$.

18 Verify (37) by direct integration in the case when $n = 2$.

19 Verify recursion relation (39) in the case $n = 4$ by using it together with the expressions for $H_3(x)$ and $H_4(x)$ given in (40) to generate $H_5(x)$.

20 Show by substituting into the Hermite differential equation the expression for $H_n(x)$ given by (35) that $H_n(x)$ is a solution.

21 When expressed in terms of Hermite polynomials, the expressions for x^r up to $r = 7$ are as follows:

$1 = H_0$

$x = \frac{1}{2}H_1$

$x^2 = \frac{1}{4}(2H_0 + H_2)$

$x^3 = \frac{1}{8}(6H_1 + H_3)$

$x^4 = \frac{1}{16}(12H_0 + 12H_2 + H_4)$

$x^5 = \frac{1}{32}(60H_1 + 20H_3 + H_5)$

$x^6 = \frac{1}{64}(120H_0 + 180H_2 + 30H_4 + H_6)$

$x^7 = \frac{1}{128}(840H_1 + 420H_3 + 42H_5 + H_7).$

Verify the expression for x^7 by eliminating the appropriate powers of x from H_1, H_3, H_5 and H_7.

Appendices

Appendix 1
The Laplace Transform

Let $f(t)$ be a given function defined for $t \geq 0$. Then if the improper integral of the second kind

$$F(s) = \int_0^\infty e^{-st} f(t) \, dt$$

exists, where s is a parameter, the function $F(s)$ is called the Laplace transform of the function $f(t)$. It is usual to denote the Laplace transform of $f(t)$ by $\mathscr{L}\{f\}$, so that

$$F(s) = \mathscr{L}\{f(t)\} = \int_0^\infty e^{-st} f(t) \, dt.$$

Summary of operational properties of the Laplace transform

1 Linearity

Let $f(t)$, $g(t)$ have Laplace transforms $\mathscr{L}\{f(t)\}$ and $\mathscr{L}\{g(t)\}$, and let α, β be arbitrary constants. Then

$$\mathscr{L}\{\alpha f(t) + \beta g(t)\} = \alpha \mathscr{L}\{f(t)\} + \beta \mathscr{L}\{g(t)\}.$$

2 First shift theorem

Let $\mathscr{L}\{f(t)\} = F(s)$ for $s > s_0$, and take a to be any real number. Then

$$\mathscr{L}\{e^{at} f(t)\} = F(s-a), \qquad \text{for } s > s_0 + a$$

and, conversely,

$$\mathscr{L}^{-1}\{F(s-a)\} = e^{at} f(t).$$

3 Second shift theorem

Let $\mathscr{L}\{f(t)\} = F(s)$ for $s > s_0$, and take $\tau \geq 0$ to be an arbitrary nonnegative number. Then

$$\mathscr{L}\{\mathscr{U}(t-\tau) f(t-\tau)\} = e^{-\tau} F(s), \qquad \text{for } s > s_0 \text{ and, conversely,}$$
$$\mathscr{L}^{-1}\{e^{-\tau s} F(s)\} = \mathscr{U}(t-\tau) f(t-\tau).$$

4 Scaling theorem

Let $\mathscr{L}\{f(t) = F(s)$ for $s > s_0$, and let $\lambda > 0$ be an arbitrary positive number. Then

$$\mathscr{L}\{f(\lambda t)\} = \frac{1}{\lambda} F\left(\frac{s}{\lambda}\right), \quad \text{for } s > \lambda s_0$$

and, conversely,

$$\mathscr{L}^{-1}\left\{F\left(\frac{s}{\lambda}\right)\right\} = \lambda f(\lambda t).$$

5 Limiting property of the Laplace transform

Let the function $f(t)$ defined for $t \geq 0$ be piecewise continuous and of exponential order with $\mathscr{L}\{f(t)\} = F(s)$, then

$$\lim_{s \to \infty} F(s) = 0.$$

6 Transform of a derivative

Let $f(t)$ be defined and continuous for $t \geq 0$ and of exponential order $e^{\alpha t}$, and let its nth derivative $f^{(n)}(t)$ be piecewise continuous on any finite interval of $t \geq 0$. Then,

$$\mathscr{L}\{f^{(n)}(t)\} = s^n \mathscr{L}\{f\} - s^{n-1} f(0) - s^{n-2} f^{(1)}(0) - \ldots - f^{(n-1)}(0), \quad \text{for } s > \alpha.$$

In particular,

$$\mathscr{L}\{f'(t)\} = s\mathscr{L}\{f\} - f(0), \quad \text{for } s > \alpha,$$
$$\mathscr{L}\{f''(t)\} = s^2 \mathscr{L}\{f\} - sf(0) - f'(0), \quad \text{for } s > \alpha,$$
$$\mathscr{L}\{f'''(t)\} = s^3 \mathscr{L}\{f\} - s^2 f(0) - sf'(0) - f''(0), \quad \text{for } s > \alpha.$$

7 Differentiation of a transform

Let $\mathscr{L}\{f(t)\} = F(s)$, for $s > s_0$. Then

$$\frac{d^n F(s)}{ds^n} = \mathscr{L}\{(-t)^n f(t)\}, \quad \text{for } s > s_0$$

and, conversely,

$$\mathscr{L}^{-1}\{F^{(n)}(s)\} = (-t)^n f(t).$$

8 Transform of an integral

Let $\mathscr{L}\{f(t)\} = F(s)$, for $s > s_0$. Then,

$$\mathscr{L}\left\{\int_0^t f(\tau)\,d\tau\right\} = \frac{F(s)}{s}, \quad \text{for } s > s_0$$

and, conversely,

$$\mathscr{L}^{-1}\{F(s)/s\} = \int_0^t f(\tau)\,d\tau.$$

9 Integration of a transform

Let $f(t)/t$ be defined for $t \geq 0$, piecewise continuous and of exponential order such that $\mathscr{L}\{f(t)/t\} = G(s)$, for $s > s_0$. Then if $\mathscr{L}\{f(t)\} = F(s)$,

$$\mathscr{L}\left\{\frac{f(t)}{t}\right\} = \int_s^\infty F(u)\,du, \qquad \text{for } s > s_0$$

and, conversely,

$$\mathscr{L}^{-1}\{G(s)\} = -\frac{\mathscr{L}^{-1}\{G'(s)\}}{t}.$$

10 Transformation of a periodic function

Let the function $f(t)$ be piecewise continuous for $t \geq 0$ and periodic with period T. Then

$$\mathscr{L}\{f(t)\} = \frac{1}{1 - e^{-sT}} \int_0^T e^{-st} f(t)\,dt.$$

11 Convolution theorem

This theorem interprets the product of two Laplace transforms in terms of the convolution operation performed on the original functions of t. Let $f(t)$, $g(t)$ be piecewise continuous functions defined for $t \geq 0$ such that $\mathscr{L}\{f(t)\} = F(s)$ for $s > \alpha$, and $\mathscr{L}\{g(t)\} = G(s)$ for $s > \beta$. Then the theorem asserts that

$$\mathscr{L}\{f * g\} = F(s)\,G(s), \qquad \text{for } s > s_0,$$

where $s_0 = \max\{\alpha, \beta\}$.

Here the convolution operation $f * g$ is defined as

$$f * g = \int_0^t f(\tau)\,g(t - \tau)\,d\tau.$$

The convolution operation is commutative, so that

$$f * g = g * f.$$

12 Extended initial value theorem

This theorem concerns the recovery of the initial values $f(0)$, $f'(0)$, ..., $f^{(n)}(0)$ of the function $f(t)$ and its derivatives from the Laplace transform $F(s) = \mathscr{L}\{f(t)\}$. The theorem asserts that, provided $f(t)$ and the necessary derivatives exist as $t \to 0$,

(i) $\lim\limits_{s \to \infty} [s\,F(s)] = f(0)$,

(ii) $\lim\limits_{s \to \infty} [s^2\,F(s) - sf(0)] = f'(0)$,

and

(iii) $\lim\limits_{s \to \infty} [s^{n+1}\,F(s) - s^n f(0) - s^{n-1} f^{(1)}(0) - s^{n-2} f^{(2)}(0) - \ldots - sf^{(n-1)}(0)] = f^{(n)}(0)$.

13 Final value theorem

This theorem concerns the recovery of the limiting (final) value $\lim_{t \to \infty} y(t) = y(\infty)$ of the function $f(t)$ from its Laplace transform $Y(s) = \mathcal{L}\{y(t)\}$. In its general form the final value theorem asserts that

$$\lim_{s \to 0} [s\, Y(s)] = y(\infty),$$

provided the limit exists and is finite. See Sec. 7.3(d) for details.

Table A1.1 A short table of Laplace transform pairs

	$f(t)$	$F(s) = \mathcal{L}\{f(t)\}$			$f(t)$	$F(s) = \mathcal{L}\{f(t)\}$	
1	1	$1/s$	$(s>0)$	13	$\sinh kt$	$\dfrac{k}{s^2-k^2}$	$(s>\lvert k\rvert)$
2	t	$1/s^2$	$(s>0)$	14	$\cosh kt$	$\dfrac{s}{s^2-k^2}$	$(s>\lvert k\rvert)$
3	$t^n\,(n=1,2,3,\dots)$	$\dfrac{n!}{s^{n+1}}$	$(s>0)$	15	$\mathcal{U}(t-\tau)$	$e^{-s\tau}/s$	$(s>0)$
4	$t^a\ (a\ \text{positive})$	$\dfrac{\Gamma(a+1)}{s^{a+1}}$	$(s>a)$	16	$\delta(t-\tau)$	$e^{-\tau s}$	$(\tau>0)$
5	e^{at}	$\dfrac{1}{s-a}$	$(s>a)$	17	$\sin kt - kt\cos kt$	$\dfrac{2k^3}{(s^2+k^2)^2}$	$(s>0)$
6	$t^n e^{at}\,(n=1,2,3,\dots)$	$\dfrac{n!}{(s-a)^{n+1}}$	$(s>a)$	18	$(\sin kt - kt\cos kt)/t$	$\dfrac{\pi}{2}-\arctan\!\left(\dfrac{s}{12}\right)-\dfrac{ks}{s^2+k^2}$	$(s>0)$
7	$\sin kt$	$\dfrac{k}{s^2+k^2}$	$(s>0)$	19	$\lvert \sin t\rvert$	$\dfrac{\coth(\pi s/2)}{s^2+1}$	$(s>0)$
8	$\cos kt$	$\dfrac{s}{s^2+k^2}$	$(s>0)$	20	$J_0(t)$	$(s^2+1)^{-1/2}$	$(s>0)$
9	$t\sin kt$	$\dfrac{2ks}{(s^2+k^2)^2}$	$(s>0)$	21	$J_n(t)$	$\dfrac{[(s^2+1)^{1/2}-s]^n}{(s^2+1)^{1/2}}$	$n=0,1,\dots\ (s>0)$
10	$t\cos kt$	$\dfrac{s^2-k^2}{(s^2+k^2)^2}$	$(s>0)$	22	$\sin\sqrt{t}$	$(\sqrt{\pi}/2)s^{-3/2}\,e^{-1/(4s)}$	$(s>0)$
11	$e^{at}\sin kt$	$\dfrac{k}{(s-a)^2+k^2}$	$(s>a)$	23	$(\cos\sqrt{t})/\sqrt{t}$	$\sqrt{\pi}\,s^{-1/2}\,e^{-1/(4s)}$	$(s>0)$
12	$e^{at}\cos kt$	$\dfrac{s-a}{(s-a)^2+k^2}$	$(s>a)$	24	$\exp(-t^2)$	$(\sqrt{\pi}/2)\exp(s^2/4)\,\mathrm{erfc}(\tfrac{1}{2}s)$	$(s>0)$

Appendix 2
The z-Transform

Definition (z transform)

Let $\{u(nT)\}_{n=0}^{\infty}$ be a numerical sequence whose nth term is $u_n = u(nT)$. Then the z-transform of the sequence or, equivalently, of the function $u(t)$, written $Z\{u(t)\}$ and denoted by $U(z)$ when expressed as a function of $z = e^{sT}$, is defined as

$$Z\{u(t)\} = U(z) = \sum_{n=0}^{\infty} u(nT) z^{-n},$$

whenever the series on the right-hand side converges.

Summary of operational properties of the z-transform

1 Linearity and scaling of the z-transform

Let α, β be arbitrary constants, and $\{u(nT)\}_{n=0}^{\infty}$, $\{v(nT)\}_{n=0}^{\infty}$ two sequences such that $Z\{u(t)\} = U(z)$ and $Z\{v(t)\} = V(z)$ are convergent. Then,

(i) the z-transform is a linear operation:

$$Z\{\alpha u(t) + \beta v(t)\} = \alpha Z\{u(t)\} + \beta Z\{v(t)\},$$

and

(ii) scaling $u(nT)$ in the z-transform is a linear operation:

$$Z\{\alpha u(t)\} = \alpha Z\{u(t)\}.$$

2 Shift

(i) Shift to the right (time delay)
Let $K > 0$ be an integer and T the sampling interval. Then multiplication of the z-transform of $u(t)$ by z^{-K} is equivalent in terms of t to shifting $u(t)$ to the right by an amount KT (delaying it by KT) to become $u(t - KT)$. Equivalently, if $U(z) = Z\{u(t)\}$, then

$$z^{-K} U(z) = Z\{u(t - KT)\}$$

and, conversely,

$$\{u[(n-K)T]\}_{n=0}^{\infty} = Z^{-1}\{z^{-K} U(z)\}.$$

1024

(ii) Shift to the left (time advance)

$$Z\{u(t+KT)\} = z^K U(z) - z^K u(0) - z^{K-1} u(T) - z^{K-2} u(2T) - \ldots - zu[(K-1)T].$$

In particular,

$$Z\{u(t+T)\} = zU(z) - zu(0),$$
$$Z\{u(t+2T)\} = z^2 U(z) - z^2 u(0) - zu(T),$$
$$Z\{u(t+3T)\} = z^3 U(z) - z^3 u(0) - z^2 u(T) - zu(2T).$$

3 Scaling $u(t)$ by $a^{t/T}$

Let $Z\{u(t)\} = U(t)$ and $a > 0$. Then if T is the sampling interval

$$Z\{a^{t/T} u(t)\} = U\left(\frac{z}{a}\right).$$

4 Multiplication of $u(t)$ by e^{-at}

If $Z\{u(t)\} = U(z)$, then if T is the sampling interval

$$Z\{e^{-at} u(t)\} = U(ze^{aT}).$$

5 Multiplication of $u(t)$ by t

If $Z\{u(t)\} = U(z)$, then if T is the sampling interval

$$Z\{tu(t)\} = -zT\frac{\mathrm{d}}{\mathrm{d}z} U(z).$$

6 Partial differentiation with respect to a parameter

Let $u(t, \alpha)$ be a function of t dependent on a parameter α, and let $Z\{u(t, \alpha)\} = U(z, \alpha)$ be its z-transform. Then,

$$Z\left\{\frac{\partial}{\partial \alpha} u(t, \alpha)\right\} = \frac{\partial}{\partial \alpha} U(z, \alpha).$$

7 Initial value theorem

If $Z\{u(t)\} = U(z)$, then

$$u(0) = \lim_{z \to \infty} U(z).$$

8 Final value theorem

Let $Z\{u(t)\} = U(z) = Q(z)/P(z)$, with $P(z)$ and $Q(z)$ polynomials in z. Then if the zeros of $P(z)$ lie within the unit circle $|z| = 1$ in the complex plane, or at $z = 1$,

$$\lim_{n \to \infty} u(nT) = \lim_{z \to 1}\left[\left(\frac{z-1}{z}\right) U(z)\right].$$

If the zeros of $P(z)$ lie within and on $|z|=1$ with the zeros on $|z|=1$ away from $z=1$, $u(nT)$ will oscillate boundedly as $n \to \infty$. When any zeros of $P(z)$ lie outside $|z|=1$, $u(nT)$ will increase without bond and may also oscillate as $n \to \infty$.

9 Discrete real variable convolution theorem

Let $u(t)$, $v(t)$ be real valued functions, with $u(t)=v(t)=0$ for $t<0$, such that $Z\{u(t)\}=U(z)$ and $Z\{v(t)\}=V(z)$. If T is the sampling interval and

$$U(z)V(z)=W(z)=Z\{\{w(kT)\}_{k=0}^{\infty}\},$$

then

$$U(z)V(z)=Z\left\{\left\{\sum_{n=0}^{k}u(nT)v([k-n]T)\right\}_{k=0}^{\infty}\right\},$$

and

$$w(kT)=\sum_{n=0}^{k}u(nT)v([k-n]T).$$

Table A2.1 A short table comparing Laplace and z-transform pairs

Function $u(t)$ for $t \ge 0$ with $u(t) \equiv 0$ for $t<0$	Sampled function $u(nT)$ or term u_n of sequence	Laplace transform $U(s)=\mathcal{L}\{u(t)\}$	z-Transform $U(z)=Z\{u(t)\}$ with sampling interval T
1 $\delta(t)$	$\Delta(n)=\{u_n\}_{n=0}^{\infty}$, $u_n=\begin{cases}1, n=0\\0, 0 \ne 0\end{cases}$	1	1
2 $\delta(t-kT)$	$\Delta(n-k)=\{u_n\}_{n=0}^{\infty}$, $u_n=\begin{cases}1, n=k\\0, n \ne k\end{cases}$	e^{-ksT}	z^{-k}
3 $\mathcal{U}(t)$ (unit step function)	1	$\dfrac{1}{s}$	$\dfrac{z}{z-1}$
4 $a^{t/T}$	$u(nT)=a^n$	$\dfrac{1}{s-(1/T)\ln a}$ $(a>0)$	$\dfrac{z}{z-a}$
5 t	nT	$\dfrac{1}{s^2}$	$\dfrac{Tz}{(z-1)^2}$
6 t^2	$n^2 T^2$	$\dfrac{2}{s^3},$	$\dfrac{T^2 z(z+1)}{(z-1)^3}$
7 $ta^{(t/T-1)}$	nTa^{n-1}	$\dfrac{1}{[s-(1/T)\ln a]^2}$ $(a>0)$	$\dfrac{z}{(z-a)^2}$

Table A2.1 _(continued)_

Function $u(t)$ for $t \geq 0$ with $u(t) \equiv 0$ for $t < 0$	Sampled Function $u(nT)$ or term u_n of sequence	Laplace transform $U(s) = \mathscr{L}\{u(t)\}$	z-transform $U(z) = Z\{u(t)\}$ with sampling interval T
8 e^{-at}	e^{-nat}	$\dfrac{1}{s+a}$	$\dfrac{z}{z-e^{-aT}}$
9 te^{-at}	nTe^{-naT}	$\dfrac{1}{(s+a)^2}$	$\dfrac{Tze^{-aT}}{(z-e^{-aT})^2}$
10 $t^2 e^{-at}$	$n^2 T^2 e^{-naT}$	$\dfrac{2}{(s+a)^3}$	$\dfrac{T^2 z(z+e^{-aT})e^{-aT}}{(z-e^{-aT})^3}$
11 $\sin kt$	$\sin knT$	$\dfrac{k}{s^2+k^2}$	$\dfrac{z\sin kT}{z^2-2z\cos kT+1}$
12 $\cos kt$	$\cos knT$	$\dfrac{s}{s^2+k^2}$	$\dfrac{z(z-\cos kT)}{z^2-2z\cos kT+1}$
13 $e^{-at}\sin kt$	$e^{-naT}\sin knT$	$\dfrac{k}{(s+a)^2+k^2}$	$\dfrac{ze^{-aT}\sin kT}{z^2-2ze^{-aT}\cos kT+e^{-2aT}}$
14 $e^{-at}\cos kt$	$e^{-naT}\cos knT$	$\dfrac{s+a}{(s+a)^2+k^2}$	$\dfrac{z(z-e^{-aT}\cos kT)}{z^2-2ze^{-aT}\cos kT+e^{-2aT}}$
15 $\sinh kt$	$\sinh nkT$	$\dfrac{k}{s^2-k^2}$	$\dfrac{z\sinh kT}{z^2-2z\cosh kT+1}$
16 $\cosh kt$	$\cosh nkT$	$\dfrac{s}{s^2-k^2}$	$\dfrac{z(z-\cosh kT)}{z^2-2z\cosh kT+1}$

Appendix 3
The Gamma Function $\Gamma(x)$

Definition

$$\Gamma(x) = \int_0^\infty t^{x-1} e^{-t}\, dt, \qquad (x > 0).$$

Recursion relation

$\Gamma(1+x) = x\Gamma(x).$

x	$\Gamma(x)$	x	$\Gamma(x)$	x	$\Gamma(x)$	x	$\Gamma(x)$
1.00	1.000000	1.26	0.904397	1.52	0.887039	1.78	0.926227
1.01	0.994326	1.27	0.902503	1.53	0.887568	1.79	0.928767
1.02	0.988844	1.28	0.900718	1.54	0.888178	1.80	0.931384
1.03	0.983550	1.29	0.899042	1.55	0.888868	1.81	0.934076
1.04	0.978438	1.30	0.897471	1.56	0.889639	1.82	0.936845
1.05	0.973504	1.31	0.896004	1.57	0.890490	1.83	0.939690
1.06	0.968744	1.32	0.894640	1.58	0.891420	1.84	0.942612
1.07	0.964152	1.33	0.893378	1.59	0.892428	1.85	0.945611
1.08	0.959725	1.34	0.892216	1.60	0.893515	1.86	0.948687
1.09	0.955459	1.35	0.891151	1.61	0.894681	1.87	0.951840
1.10	0.951351	1.36	0.890185	1.62	0.895924	1.88	0.955071
1.11	0.947396	1.37	0.889314	1.63	0.897244	1.89	0.958379
1.12	0.943590	1.38	0.888537	1.64	0.898642	1.90	0.961766
1.13	0.939931	1.39	0.887854	1.65	0.900117	1.91	0.965231
1.14	0.936416	1.40	0.887264	1.66	0.901668	1.92	0.968774
1.15	0.933041	1.41	0.886765	1.67	0.903296	1.93	0.972397
1.16	0.929803	1.42	0.886356	1.68	0.905001	1.94	0.976099
1.17	0.926700	1.43	0.886036	1.69	0.906782	1.95	0.979881
1.18	0.923728	1.44	0.885805	1.70	0.908639	1.96	0.983743
1.19	0.920885	1.45	0.885661	1.71	0.910572	1.97	0.987685
1.20	0.918169	1.46	0.885604	1.72	0.912581	1.98	0.991708
1.21	0.915576	1.47	0.885633	1.73	0.914665	1.99	0.995813
1.22	0.913106	1.48	0.885747	1.74	0.916826	2.00	1.000000
1.23	0.910755	1.49	0.885945	1.75	0.919063		
1.24	0.908521	1.50	0.886227	1.76	0.921375		
1.25	0.906402	1.51	0.886592	1.77	0.923763		

Appendix 4
Bessel Functions $J_0(x)$, $J_1(x)$, $Y_0(x)$ and $Y_1(x)$

x	$J_0(x)$	$J_1(x)$	x	$J_0(x)$	$J_1(x)$
0.0	1.00000	0.00000	2.7	−0.14245	0.44160
0.1	0.99750	0.04994	2.8	−0.18504	0.40971
0.2	0.99002	0.09950	2.9	−0.22431	0.37543
0.3	0.97763	0.14832	3.0	−0.26005	0.33906
0.4	0.96040	0.19603	3.1	−0.29206	0.30092
0.5	0.93847	0.24227	3.2	−0.32019	0.26134
0.6	0.91200	0.28670	3.3	−0.34430	0.22066
0.7	0.88120	0.32900	3.4	−0.36430	0.17923
0.8	0.84629	0.36884	3.5	−0.38013	0.13738
0.9	0.80752	0.40595	3.6	−0.39177	0.09547
1.0	0.76520	0.44005	3.7	−0.39923	0.05383
1.1	0.71962	0.47090	3.8	−0.40256	0.01282
1.2	0.67113	0.49829	3.9	−0.40183	−0.02724
1.3	0.62009	0.52202	4.0	−0.39715	−0.06604
1.4	0.56686	0.54195	4.1	−0.38867	−0.10327
1.5	0.51183	0.55794	4.2	−0.37656	−0.13865
1.6	0.45540	0.56990	4.3	−0.36101	−0.17190
1.7	0.39798	0.57777	4.4	−0.34226	−0.20278
1.8	0.33999	0.58152	4.5	−0.32054	−0.23106
1.9	0.28182	0.58116	4.6	−0.29614	−0.25655
2.0	0.22389	0.57672	4.7	−0.26933	−0.27908
2.1	0.16661	0.56829	4.8	−0.24043	−0.29850
2.2	0.11036	0.55596	4.9	−0.20974	−0.31469
2.3	0.05554	0.53987	5.0	−0.17760	−0.32758
2.4	0.00251	0.52019	5.1	−0.14433	−0.33710
2.5	−0.04838	0.49709	5.2	−0.11029	−0.34322
2.6	−0.09680	0.47082	5.3	−0.07580	−0.34596

x	$J_0(x)$	$J_1(x)$	x	$J_0(x)$	$J_1(x)$
5.4	−0.04121	−0.34534	7.7	0.23456	0.18131
5.5	−0.00684	−0.34144	7.8	0.21541	0.20136
5.6	0.02697	−0.33433	7.9	0.19436	0.21918
5.7	0.05992	−0.32415	8.0	0.17165	0.23464
5.8	0.09170	−0.31103	8.1	0.14752	0.24761
5.9	0.12203	−0.29514	8.2	0.12222	0.25800
6.0	0.15065	−0.27668	8.3	0.09601	0.26574
6.1	0.17729	−0.25586	8.4	0.06916	0.27079
6.2	0.20175	−0.23292	8.5	0.04194	0.27312
6.3	0.22381	−0.20809	8.6	0.01462	0.27275
6.4	0.24331	−0.18164	8.7	−0.01252	0.26972
6.5	0.26009	−0.15384	8.8	−0.03923	0.26407
6.6	0.27404	−0.12498	8.9	−0.06525	0.25590
6.7	0.28506	−0.09534	9.0	−0.09033	0.24531
6.8	0.29310	−0.06522	9.1	−0.11424	0.23243
6.9	0.29810	−0.03490	9.2	−0.13675	0.21741
7.0	0.30008	−0.00468	9.3	−0.15766	0.20041
7.1	0.29905	0.02515	9.4	−0.17677	0.18163
7.2	0.29507	0.05433	9.5	−0.19393	0.16126
7.3	0.28822	0.08257	9.6	−0.20898	0.13952
7.4	0.27860	0.10963	9.7	−0.22180	0.11664
7.5	0.26634	0.13525	9.8	−0.23228	0.09284
7.6	0.25160	0.15921	9.9	−0.24034	0.06837

x	$Y_0(x)$	$Y_1(x)$	x	$Y_0(x)$	$Y_1(x)$
0.0	$-\infty$	$-\infty$	4.2	-0.09375	0.36801
0.2	-1.08111	-3.32382	4.4	-0.16334	0.32597
0.4	-0.60602	-1.78087	4.6	-0.22346	0.27375
0.6	-0.30851	-1.26039	4.8	-0.27230	0.21357
0.8	-0.08680	-0.97814	5.0	-0.30852	0.14786
1.0	0.08826	-0.78121	5.2	-0.33125	0.07919
1.2	0.22808	-0.62114	5.4	-0.34017	0.01013
1.4	0.33790	-0.47915	5.6	-0.33544	-0.05681
1.6	0.42043	-0.34758	5.8	-0.31775	-0.11923
1.8	0.47743	-0.22366	6.0	-0.28819	-0.17501
2.0	0.51038	-0.10703	6.2	-0.24831	-0.22228
2.2	0.52078	0.00149	6.4	-0.19995	-0.25956
2.4	0.51041	0.10049	6.6	-0.14523	-0.28575
2.6	0.48133	0.18836	6.8	-0.08643	-0.30019
2.8	0.43592	0.26355	7.0	-0.02595	-0.30267
3.0	0.37685	0.32467	7.2	0.03385	-0.29342
3.2	0.30705	0.37071	7.4	0.09068	-0.27311
3.4	0.22962	0.40102	7.6	0.14243	-0.24280
3.6	0.14771	0.41539	7.8	0.18723	-0.20389
3.8	0.06450	0.41411	8.0	0.22352	-0.15806
4.0	-0.01694	0.39793	8.2	0.25012	-0.10724

Answers to Odd-numbered Problems

Answers for Section 1.2

3 $2 < x < 6$

5 $-12 \leq x \leq 0$

7 $-2 \leq x < -1$ and $0 < x \leq 1$

9 $-5 < x < -2$ or $-1 < x$

11 Follows from (11) by squaring.

13 $(1+x)^n = 1 + nx + \dfrac{n(n-1)}{2!}x^2 + \dfrac{n(n-1)(n-2)}{3!}x^3 + \ldots$

If $x \geq 0$ and $n \geq 1$ all terms on RHS are positive, so $(1+n)^n \geq 1 + nx$.

15 Follows directly from Cauchy–Schwarz inequality with $b_k = 1/a_k$.

17 Indicated substitution gives

$$(S_n(x))^2 \leq \left(\sum_{k=0}^{n} c_k^2 \right) \left(\sum_{k=1}^{n+1} x^{2(k-1)} \right) = \left(\sum_{k=0}^{n} c_k^2 \right) \left(\frac{1 - x^{2(n+1)}}{1 - x^2} \right)$$

for $|x| \neq 1$, because series in x is a geometric series.

If $|x| < 1$ then $\lim_{n \to \infty} x^{2(n+1)} = 0$ and $(S_\infty(x))^2$ is finite if $\sum c_k^2$ is finite. Hence the result.

Answers for Section 1.3

9 False

11 False

13 True

21 $a_n = \dfrac{a_0}{1 \cdot 3^2 \cdot 5^2 \cdot 7^2 \ldots (2n-1)^2 (2n+1)}$

23 $a_n = \dfrac{(-1)^n a_0}{3^{\frac{1}{2}n(n+1)} 5^{\frac{1}{2}n(n+3)}}$

25 $\left(x^4 - \dfrac{1}{3x}\right)^{10} = \displaystyle\sum_{r=0}^{10} \binom{n}{r} (x^4)^{10-r} \left(\dfrac{-1}{3x}\right)^r = \displaystyle\sum_{r=0}^{10} \binom{n}{r} \left(-\dfrac{1}{3}\right)^r x^{40-5r}.$

The term independent of x occurs when $40 - 5r = 0$, so that $r = 8$. Hence the term independent of x is

$$\binom{10}{8}\left(-\dfrac{1}{3}\right)^8 = \dfrac{10!}{8!2!}\left(\dfrac{1}{3}\right)^8 = 45/3^8.$$

The term in x^{-5} corresponds to $40 - 5r = -5$, so $r = 9$. The coefficient is thus

$$\binom{10}{9}\left(-\dfrac{1}{3}\right)^9 = \dfrac{10!}{9!}\left(-\dfrac{1}{3}\right)^9 = -10/3^9.$$

Answers for Section 1.4

1 $x + \dfrac{2}{x-1} - \dfrac{1}{x+2}$

3 $\dfrac{1}{x+2} + \dfrac{2}{(x+2)^2} + \dfrac{1}{x+1}$

5 $\dfrac{x}{x^2 - 2x + 2} - \dfrac{1+x}{x^2 - x + 1} \equiv \dfrac{x}{(x-1)^2 + 1} - \dfrac{1+x}{(x-\frac{1}{2})^2 + \frac{3}{4}}$

7 $\dfrac{1}{x} - \dfrac{3}{x^2} + \dfrac{x}{x^2 + x + 1} \equiv \dfrac{1}{x} - \dfrac{3}{x^2} + \dfrac{x}{(x+\frac{1}{2})^2 + \frac{3}{4}}$

9 $\dfrac{1}{5x} - \dfrac{1}{5}\dfrac{x-2}{x^2 + 2x + 5} \equiv \dfrac{1}{5x} - \dfrac{1}{5}\dfrac{x-2}{(x+1)^2 + 4}$

Answers for Section 1.6

3 2

5 $(23 - 15i)/13$

7 $3/17$

9 0

23 Yes, No, Yes, No

25 $(-9 + i\sqrt{19})/2, \quad (-9 - i\sqrt{19})/2; \quad \left(z + \dfrac{9}{2} - i\dfrac{\sqrt{19}}{2}\right)\left(z + \dfrac{9}{2} + i\dfrac{\sqrt{19}}{2}\right)$

27 $(-3+3i)/2, -(3+3i)/2; \dfrac{1}{2}\left(2z+3-3i\right)\left(2z+3+3i\right)$

31 $-3i, 3i, -2i, 2i.$

33 $-2+i\sqrt{3}, -2-i\sqrt{3}, (-3+i\sqrt{15})/2, -(3+i\sqrt{15})/2$

Answers for Section 1.7

1 $2, \pi/3, \pi/3 + 2k\pi$

3 $4, \pi, \pi + 2k\pi$

5 $5, \alpha, \alpha + 2k\pi$ with $\alpha = \text{arc tan } (4/3)$ (acute angle)

7 $2\left(\cos\dfrac{5\pi}{6} - i\sin\dfrac{5\pi}{6}\right)$

9 $7(\cos\pi + i\sin\pi)$

11 $6\left(\cos\dfrac{\pi}{3} - i\sin\dfrac{\pi}{3}\right), \dfrac{2}{3}\left(\cos\dfrac{2\pi}{3} + i\sin\dfrac{2\pi}{3}\right)$

15 $\sin 5\theta = 5\cos^4\theta \ \sin\theta - 10\cos^2\theta\sin^3\theta + \sin^5\theta = 5\sin\theta - 20\sin^3\theta + 16\sin^5\theta, \ \cos 5\theta = \cos^5\theta$
$- 10\cos^3\theta\sin^2\theta + 5\cos\theta\sin^4\theta = 16\cos^5\theta - 20\cos^3\theta + 5\cos\theta$

17 $2^{-11/2}\left(\cos\dfrac{\pi}{12} - i\sin\dfrac{\pi}{12}\right)$

23 $8^{1/5}\exp\left[\left(\dfrac{8k-1}{20}\right)\pi i\right], \qquad (k=0, 1, 2, 3, 4).$

25 $2^{3/4}\exp\left[\left(\dfrac{1+2k}{4}\right)\pi i\right], \qquad (k=0, 1, 2, 3).$

33 $(2\sin\tfrac{1}{2}x)S = 2\sin\tfrac{1}{2}x\left[\tfrac{1}{2} + \displaystyle\sum_{r=1}^{n}\cos rx\right] = \sin\tfrac{1}{2}x + 2\displaystyle\sum_{r=1}^{n}\sin\tfrac{1}{2}x\cos rx$

$$= \sin\tfrac{1}{2}x + \sum_{r=1}^{n}[\sin(r+\tfrac{1}{2})x - \sin(r-\tfrac{1}{2})x]$$

$$= \sin\tfrac{1}{2}x + \sin\tfrac{3}{2}x - \sin\tfrac{1}{2}x + \sin\tfrac{5}{2}x - \sin\tfrac{3}{2}x + \ldots + \sin(n+\tfrac{1}{2})x - \sin(n-\tfrac{1}{2})x$$

Cancelling terms gives

$$(2\sin\tfrac{1}{2}x)S = \sin(n+\tfrac{1}{2})x,$$

from which the result then follows.

Answers for Section 1.8

1 $F(x) = x + 1 - \cos x$

3 $F(x) = \tfrac{1}{2}x^2(\ln x - \tfrac{1}{2}) + \tfrac{1}{4}$

5 $F(x) = \ln x - x + \tfrac{1}{2}x^2 - \tfrac{1}{2}e^{2x}$

7 $F(x) = \tfrac{5}{2}x^3(\tfrac{1}{2}x - 1)$

9 $F'(x) = e^{-x^2}(2xe^{-x^2} - 1)$

11 $F'(x) = \dfrac{\cos x}{2\sqrt{x}} + \dfrac{1}{x^2}\cos\left(\dfrac{1}{x^2}\right), \quad (x > 0)$

13 $F'(x) = \dfrac{1}{x} - \dfrac{1 + 2x}{x(1 + x)}$

15   Theorem 1.7(i)   Theorem 1.7(ii)

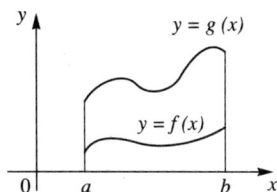

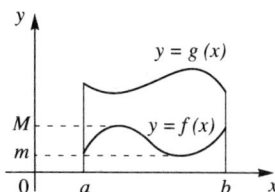

17 The theorem merely guarantees there will be a number ξ for which the result will be true, and not that ξ will be unique. In this case $\xi = -\pi/2$ and $\xi = \pi/2$, so ξ is not unique.

19 $\xi = 2/\ln 2$ is unique.

21 (a) $\xi = \frac{1}{2}(a + b)$ is unique,   (b) $\sin\xi = 0$ with $0 \le \xi \le 2\pi$, so ξ is not unique, for $\xi = 0, \pi, 2\pi$.

23 $I \le \sqrt{\dfrac{6}{5}(2 - \cos 1)} \simeq 1.32.$

25 $\pi/2$

27 Divergent

29 Divergent

31 $1/\ln 2$

33 Divergent

35 Convergent

37 Divergent; consider $\displaystyle \lim_{\varepsilon \to 0}\int_{\pi/2}^{2-\varepsilon}\frac{2\,dx}{x-2} + \lim_{\delta \to 0}\int_{2+\delta}^{3\pi/2}\frac{2\,dx}{x-2}$

39 $1/(\mu - 1)$ if $\mu > 1$, divergent if $\mu \le 1$.

41 π

43 $\frac{1}{2}\ln 2$

45 $1/\ln a$

47 $\Gamma(n + 1) = n!$ (Integral n)

49 Convergent

51 Divergent

53 Convergent

55 $F = I_0\omega/(k^2 + \omega^2), \quad G = I_0^2\omega^2/[4k(k^2 + \omega^2)].$

57 Write

$$G(x + h) = \int_{\psi(x)}^{\psi(x) + h\psi'} f(t, x + h)\,dt + \int_{\phi(x)}^{\psi(x)} f(t, x + h)\,dt + \int_{\phi(x) + h\phi'}^{\phi(x)} f(t, x + h)\,dt,$$

apply the first mean value theorem for integrals (see (15)) to the first and last terms, and use the definition of dG/dx.

59 (i) Change the value of g at a to be 0, (ii) change the value of g at b to be 0.

61 $|x^\mu f(x)| < A$, so $|f(x)| < A/x^\mu$.

Now compare $\displaystyle\int_a^\infty |f(x)|\,dx$ with $\displaystyle\int_a^\infty \frac{A}{x^\mu}\,dx$.

63 (i) Absolutely convergent, (ii) absolutely convergent, (iii) divergent, (iv) absolutely convergent.

65 Setting $f(t,\beta) = (\sin \beta t)/t$ and $g(t) = e^{-\alpha t}$ it is easily verified that the conditions of Theorem 1.9 are satisfied. We have

$$F'(\beta) = \frac{\alpha}{\alpha^2 + \beta^2},$$

so as $F(0) = 0$ it follows that

$$F(\beta) = \arctan(\beta/\alpha).$$

Thus with β fixed and $\alpha \to 0$ we have

$$\int_0^\infty \frac{\sin \beta t}{t}\,dt = \frac{\pi}{2} \qquad \text{for any } \beta > 0.$$

Answers for Section 1.9

1 $\lambda_1 = 2$, $\lambda_2 = 3$; $u_n = A \cdot 2^n + B \cdot 3^n$; $u_n = 3 \cdot 2^n - 2 \cdot 3^n$ \qquad for $n = 0, 1, 2, \ldots$.

3 $\lambda_1 = \lambda_2 = -1$; $u_n = A(-1)^n + Bn(-1)^{n-1}$; $u_n = (-1)^n + 3n(-1)^{n-1}$ \qquad for $n = 0, 1, 2, \ldots$.

5 $\lambda_1 = 2^{1/2}\,e^{3\pi i/4}$, $\lambda_2 = 2^{1/2}\,e^{-3\pi i/4}$; $u_n = 2^{n/2}\left(A\cos\dfrac{3n\pi}{4} + B\sin\dfrac{3n\pi}{4}\right)$;

$$u_n = 2^{n/2}\left(\cos\frac{3n\pi}{4} + (1+\sqrt{2})\sin\frac{3n\pi}{4}\right) \qquad \text{for } n = 0, 1, 2, \ldots.$$

7 $\lambda_1 = 2$, $\lambda_2 = -3$; $u_n = A \cdot 2^n + B(-3)^n + 4^n/14$;

$$u_n = \left(\frac{1}{10}\right)2^n - \left(\frac{6}{35}\right)(-3)^n + 4^n/14 \qquad \text{for } n = 0, 1, 2, \ldots.$$

9 $\lambda_1 = e^{i\pi/2}$, $\lambda_2 = e^{-i\pi/2}$; $u_n = A\cos\dfrac{n\pi}{2} + B\sin\dfrac{n\pi}{2} + \dfrac{\sin n}{2(1+\cos 2)} + \dfrac{\sin(n-2)}{2(1+\cos 2)}$; initial conditions

are satisfied if in u_n above

$$A = \frac{\sin 2}{2(1+\cos 2)}, \qquad B = 1.$$

Answers for Section 2.3

1 Self-checking

3 $(a^2 + b^2)^{1/2}$, $(a^2 + b^2)^{1/2}$, $(a^2 + 4b^2)^{1/2}$, $(16a^2 + 9b^2)^{1/2}$

5 (i) $\overrightarrow{OP} = \frac{1}{2}\overrightarrow{OA}$ \quad (ii) $\overrightarrow{CD} = -3\overrightarrow{AD}$ \quad (iii) $\overrightarrow{OP} = -n\overrightarrow{OA}$ \quad and \quad $\overrightarrow{PA} = (n+1)\overrightarrow{OA}$

7 $\overrightarrow{OP}_1, \overrightarrow{OP}_2, \ldots, \overrightarrow{OP}_n$ are all coplanar and of equal modulus. By symmetry their addition forms a regular n-sided polygon which is closed.

9 Use the fact that the three vectors are represented by three edges of a tetrahedron all meeting at a vertex and directed away from it, all sides of the tetrahedron being equal in length.

11 Use the result that if v_b, v_w are the velocities of the boat and the water relative to the earth, then $v_{bw} = v_b - v_w$ is the velocity of the boat relative to the water, so $v_b = v_{bw} + v_w$. Boat speed $2\sqrt{10}$ m/s; direction at angle arc tan $\frac{1}{3}$ east of north.

13 Use the result that if v_w and v_c are the velocities of the wind and the cyclist, respectively, then $v_{wc} = v_w - v_c$ is the velocity of the wind relative to the cyclist, so $v_w = v_{cw} + v_c$, wind speed $5\sqrt{2}$ m/s; direction southeast.

Answers for Section 2.4

1 (a) 1, 1, 10, $\|\overrightarrow{PQ}\| = \sqrt{102}$ (b) 2, 0, 4, $\|\overrightarrow{PQ}\| = \sqrt{20}$

    (c) 2, 6, 9, $\|\overrightarrow{PQ}\| = 11$ (d) 0, 0, -1, $\|\overrightarrow{PQ}\| = 1$.

3 $\sqrt{14}, \sqrt{2}, \sqrt{14}, \sqrt{41}$ **5** $3\mathbf{i} + 4\mathbf{j} + 5\mathbf{k}$, $3\mathbf{i} + 4\mathbf{j} + 5\mathbf{k}$

7 $-6\mathbf{i} + 9\mathbf{j}$, $8\mathbf{i} - 6\mathbf{j} + 10\mathbf{k}$, $8\mathbf{i} - 6\mathbf{j} + 10\mathbf{k}$

9 $\dfrac{1}{\sqrt{14}}(\mathbf{i} + 2\mathbf{j} + 3\mathbf{k})$, $\dfrac{1}{\sqrt{2}}(-\mathbf{i} + \mathbf{j})$, $\dfrac{1}{\sqrt{14}}(3\mathbf{i} + \mathbf{j} + 2\mathbf{k})$, $\dfrac{1}{\sqrt{41}}(-2\mathbf{i} + 6\mathbf{j} - \mathbf{k})$

11 $\sqrt{21}$ units, $\sqrt{11}$ units, $\sqrt{3}$ units

13 (a) $-4\mathbf{j}$ (b) $\dfrac{9}{\sqrt{2}}(\mathbf{i} + \mathbf{j})$ (c) $\dfrac{5}{2}\left(\sqrt{3}\mathbf{i} - \mathbf{j}\right)$

15 $\dfrac{1}{\sqrt{14}}(12\mathbf{i} + 15\mathbf{j} - 3\mathbf{k})$ units, $\sqrt{27}$ units

19 (a) $\left(\sqrt{2}, -\dfrac{\pi}{4}, -2\right)$ (b) $\left(\sqrt{6}, -\dfrac{\pi}{4}, \text{arc cos}\left(-\dfrac{2}{\sqrt{6}}\right)\right)$

21 $\left(\sqrt{13}, \dfrac{\pi}{3}, \text{arc cos}\dfrac{3}{\sqrt{13}}\right)$

Answers for Section 2.5

1 (i) 6   (ii) 35.3°   (iii) 2   (iv) $\sqrt{6}$

3 (i) -4 (ii) 131.8°   (iii) $4/\sqrt{6}$ (iv) $4/\sqrt{6}$

15 (i) $(-72/\sqrt{418})$J (Obtuse angle between force and displacement vectors. Thus work must be done by displaced object against force, and hence the negative sign). (ii) $(108/\sqrt{11})$J.

17 (a) $-1, 2, -5;$ $l = -\dfrac{1}{\sqrt{30}},$ $m = \dfrac{2}{\sqrt{30}},$ $n = \dfrac{-5}{\sqrt{30}};$

$\alpha = 100.5°$ $\beta = 68.6°$ $\gamma = 155.9°$

(b) $2, -1, 2;$ $l = \tfrac{2}{3},$ $m = -\tfrac{1}{3},$ $n = \tfrac{2}{3};$

$\alpha = 48.2°$ $\beta = 109.5°$ $\gamma = 98.2°$

(c) $0, -2, -4;$ $l = 0,$ $m = \dfrac{-1}{\sqrt{5}},$ $n = \dfrac{-2}{\sqrt{5}};$

$\alpha = 90°$ $\beta = 116.6°$ $\gamma = 153.4°$

(d) $2, 1, -3;$ $l = \dfrac{2}{\sqrt{14}},$ $m = \dfrac{1}{\sqrt{14}},$ $n = \dfrac{-3}{\sqrt{14}};$

$\alpha = 57.7°,$ $\beta = 74.5°,$ $\gamma = 143.3°$

19 (a) $-\dfrac{3}{2}\mathbf{i} + \dfrac{3}{2}\mathbf{j} + \left(2 - \dfrac{3}{\sqrt{2}}\right)\mathbf{k}$ (b) $\dfrac{7}{2}\mathbf{i} + \dfrac{7}{2}\mathbf{j} + \left(1 + \dfrac{5}{\sqrt{2}}\right)\mathbf{k}$

Answers for Section 2.6

1 $3\mathbf{i} - 2\mathbf{j} + \mathbf{k},$ $-3\mathbf{j} - 3\mathbf{k},$ $3\mathbf{j} + 3\mathbf{k}$

3 $2\sqrt{14}$ sq. units **5** $\tfrac{1}{2}\sqrt{53}$ sq. units.

7 $3\mathbf{i} - 5\mathbf{j} - 2\mathbf{k},$ $3\mathbf{i} - 5\mathbf{j} - 2\mathbf{k}$ **9** $9, 9$

11 $2\mathbf{i} - \mathbf{j} + \mathbf{k},$ $-\mathbf{i} - \mathbf{j} + \mathbf{k}$ **13** $\dfrac{\pm 1}{\sqrt{21}}(2\mathbf{i} - \mathbf{j} - 4\mathbf{k})$

19 (a) Orthogonal (b) orthogonal (c) linearly dependent (d) orthogonal (e) linearly dependent.

21 $-4\mathbf{i} + 7\mathbf{j} + 2\mathbf{k}$ units

23 $11\mathbf{i} - 7\mathbf{j} + 2\mathbf{k}$ units.

Answers for Section 2.7

1 3 **3** -1 **5** -6

7 4 **9** 1 **11** 0; the vectors are coplanar.

13 $\lambda \neq 3.$ **15** (i) Linearly dependent (ii) to (v) linearly independent.

23 $-\mathbf{j} - \mathbf{k}, -\mathbf{i}$ **25** $-2\mathbf{i} + 2\mathbf{j} - 4\mathbf{k}, \mathbf{0}$

27 Each is of the form

$\|\mathbf{a}\|^2 \|\mathbf{b}\|^2 (1 - \cos^2\theta)$ with θ the acute angle between $\mathbf{a}$ and $\mathbf{b}$.

29 $\mathbf{a} = -\dfrac{1}{\sqrt{2}}\hat{\mathbf{n}} + \hat{\mathbf{n}} \times \left(\dfrac{1}{\sqrt{2}}\mathbf{i} + \dfrac{1}{\sqrt{2}}\mathbf{j} - \dfrac{3}{\sqrt{2}}\mathbf{k}\right),$ with $\hat{\mathbf{n}} = \dfrac{1}{\sqrt{2}}(\mathbf{i} - \mathbf{j}).$

Answers for Section 2.8

1 $\pm\dfrac{1}{\sqrt{21}}(\mathbf{i} + 4\mathbf{j} + 2\mathbf{k})$ **3** $\mathbf{r} = \mathbf{i} + 2\mathbf{j} + \mathbf{k} + \lambda(2\mathbf{i} + 3\mathbf{k})$

5 $\mathbf{r} = \mathbf{i} - \mathbf{j} - 2\mathbf{k} + \lambda(\mathbf{i} + \mathbf{k})$ **7** $\mathbf{r} = 2\mathbf{i} - 3\mathbf{j} + \mathbf{k} + \lambda(-\mathbf{i} + 5\mathbf{j} - 4\mathbf{k})$

9 $r=i+j+k+\lambda(i+j+2k)$

11 (i) $r=-2i-j+k+\lambda(3i+2j-k)$

(ii) $r=\dfrac{1}{2}i-\dfrac{4}{3}j-\dfrac{3}{2}k+\lambda\left(i-\dfrac{2}{3}j-\dfrac{5}{2}k\right)$

13 (i) Intersect at $i+2j+3k$; (ii) do not intersect; (iii) intersect at $2i-j+k$

15 $1/\sqrt{2}$ **17** $\sqrt{2}$

19 $n\cdot r=3$, $x+3y-4z=3$ **21** $n\cdot r=0$, $3x-3y+z=0$

23 $x-y=1$ **25** (i) arc cos $\left(\dfrac{1}{6}\right)=80.4°$; (ii) arc cos $\sqrt{\dfrac{6}{11}}=42.4°$; (iii) arc cos $\sqrt{\dfrac{3}{5}}=39.2°$

27 $2/\sqrt{14}$ **29** $r=i+\lambda(i-j-k)$ **31** $r=2i-j+\lambda(-3i+j+2k)$

Answers for Section 2.9

1 j, k; 2   **3** $i+2j+7k$; 1   **5** 1, i; 2

7 Infinite dimensional   **9** Infinite dimensional

11 No; there is no closure when λ is irrational

13 No; no closure under addition   **15** Yes; infinite dimensional.

17 Yes; $(1, 0, 0, \ldots , 0), (0, 0, 1, \ldots , 0), \ldots , (0, 0, 0, \ldots , 1)$ $((n-1)$ vectors in all)

19 No; no closure under addition   **21** No; no closure under addition

23 Yes; $(1, 0, \ldots , 0), (0, 1, 0, \ldots , 0), \ldots , (0, 0, \ldots , 1)$.

25 Dimension 3; third vector is sum of first two.

Answers for Section 3.2

1 $\begin{bmatrix} -1 & 2 & 4 \\ 2 & 1 & 6 \\ 5 & -1 & -3 \end{bmatrix}$ **3** $\begin{bmatrix} 4 & -5 & 5 \\ 4 & -1 & 6 \\ -2 & -2 & 3 \end{bmatrix}$ **5** Not defined

7 $\begin{bmatrix} 0 & -3 & 2 \\ 3 & 0 & 5 \\ -2 & -5 & 0 \end{bmatrix}$ **9** Not defined   **11** $\begin{bmatrix} -1 & 5 & 2 \\ -1 & 1 & 1 \\ 7 & 4 & -3 \end{bmatrix}$

13 $S = \begin{bmatrix} -2 & 3/2 & 5/2 \\ 3/2 & 1 & 1 \\ 5/2 & 1 & -3 \end{bmatrix}$, $K = \begin{bmatrix} 0 & 3/2 & -3/2 \\ -3/2 & 0 & 1 \\ 3/2 & -1 & 0 \end{bmatrix}$

15 Yes. $\begin{bmatrix} 1 & 0 & 0 \\ 0 & 0 & 0 \\ 0 & 0 & 0 \end{bmatrix}$, $\begin{bmatrix} 0 & 0 & 0 \\ 0 & 1 & 0 \\ 0 & 0 & 0 \end{bmatrix}$, $\begin{bmatrix} 0 & 0 & 0 \\ 0 & 0 & 0 \\ 0 & 0 & 1 \end{bmatrix}$, $\begin{bmatrix} 0 & 1 & 0 \\ 1 & 0 & 0 \\ 0 & 0 & 0 \end{bmatrix}$, $\begin{bmatrix} 0 & 0 & 1 \\ 0 & 0 & 0 \\ 1 & 0 & 0 \end{bmatrix}$, $\begin{bmatrix} 0 & 0 & 0 \\ 0 & 0 & 1 \\ 0 & 1 & 0 \end{bmatrix}$, 6

17 Yes, provided none of the arbitrary numbers in B_{pq} is zero.

19 $\frac{1}{2}n(n+1)$.

$$\begin{bmatrix} 1 & 0 & 0 \\ 0 & 0 & 0 \\ 0 & 0 & 0 \end{bmatrix}, \begin{bmatrix} 0 & 0 & 0 \\ 1 & 0 & 0 \\ 0 & 0 & 0 \end{bmatrix}, \begin{bmatrix} 0 & 0 & 0 \\ 0 & 1 & 0 \\ 0 & 0 & 0 \end{bmatrix}, \begin{bmatrix} 0 & 0 & 0 \\ 0 & 0 & 0 \\ 1 & 0 & 0 \end{bmatrix}, \begin{bmatrix} 0 & 0 & 0 \\ 0 & 0 & 0 \\ 0 & 1 & 0 \end{bmatrix}, \begin{bmatrix} 0 & 0 & 0 \\ 0 & 0 & 0 \\ 0 & 0 & 1 \end{bmatrix}.$$

23 $A = \begin{bmatrix} 0 & 1 & 1 & 0 & 1 \\ 0 & 0 & 1 & 0 & 1 \\ 0 & 0 & 0 & 1 & 0 \\ 1 & 1 & 0 & 0 & 0 \\ 0 & 0 & 1 & 1 & 0 \end{bmatrix}$

25

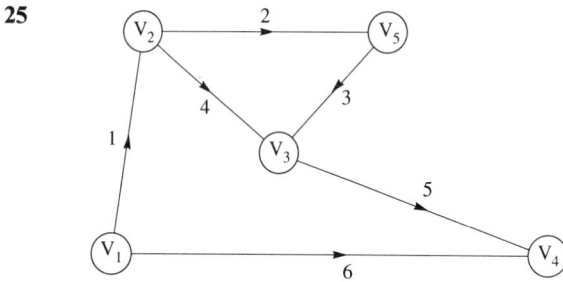

27 (Fig. 3.2) $A = \begin{bmatrix} 0 & 1 & 1 & 1 & 1 \\ 1 & 0 & 1 & 1 & 1 \\ 1 & 1 & 0 & 1 & 1 \\ 1 & 1 & 1 & 0 & 1 \\ 1 & 1 & 1 & 1 & 0 \end{bmatrix}$,   (Fig. 3.3) $A = \begin{bmatrix} 0 & 1 & 1 & 1 \\ 1 & 0 & 1 & 1 \\ 1 & 1 & 0 & 1 \\ 1 & 1 & 1 & 0 \end{bmatrix}$

Answers for Section 3.3

1 3   **3** $[1 \quad 2 \quad 3]$ **5** Not defined   **7** $\begin{bmatrix} 11 \\ 8 \\ 1 \end{bmatrix}$

9 Not defined   **11** $\begin{bmatrix} 1 & 6 & 2 \\ 5 & 2 & 11 \\ 6 & 0 & 11 \end{bmatrix}$ **13** 50   **15** $\begin{bmatrix} 0 & -\mu \\ 1/\mu & 0 \end{bmatrix}$ $(\mu \neq 0)$

17 $(ABC)^\mathsf{T} = (A(BC))^\mathsf{T} = (BC)^\mathsf{T}A^\mathsf{T} = C^\mathsf{T}B^\mathsf{T}A^\mathsf{T}$; $(A^m)^\mathsf{T} = (AA \ldots A)^\mathsf{T} = A^\mathsf{T}A^\mathsf{T} \ldots A^\mathsf{T} = (A^\mathsf{T})^m$

19 $Ax = b$, $A = \begin{bmatrix} 3 & 2 & 1 \\ 4 & -1 & 1 \\ 1 & -6 & 7 \end{bmatrix}$, $x = \begin{bmatrix} x_1 \\ x_2 \\ x_3 \end{bmatrix}$, $b = \begin{bmatrix} 7 \\ 1 \\ 4 \end{bmatrix}$

21 (i) Identity transformation; (ii) $2\times$ magnification in the x_1 and x_2 directions; (iii) shift of origin to $(-1, -3)$ and a $3\times$ magnification in the x_1 and x_2 directions; (iv) reflection in line $x_2 = x_1$; (v) scaling by λ in the x_1 direction and by μ in the x_2 direction; (vi) shift of origin to $(-3, 2)$ and a reflection in the x_2 axis.

27 $\|\mathbf{y}\|^2 = \mathbf{y}^T\mathbf{y} = (\mathbf{Ax})^T\mathbf{Ax} = \mathbf{x}^T\mathbf{A}^T\mathbf{Ax} = \mathbf{x}^T\mathbf{Ix} = \mathbf{x}^T\mathbf{x} = \|\mathbf{x}\|^2$.

$\mathbf{y} = \mathbf{Ax} + \mathbf{b}$ also preserves length because $\mathbf{b}$ merely repeats a translation.

29 Set $P^2 = [\tilde{p}_{ij}]$, then $\tilde{p}_{ij} = \mathbf{r}_i\mathbf{c}_j$ with $\mathbf{r}_i$ the ith row of $\mathbf{P}$ and $\mathbf{c}_j$ the jth column of $\mathbf{P}$. Sum of elements in ith row of P^2 is

$$\sum_{j=1}^{n} \tilde{p}_{ij} = \sum_{j=1}^{n} \mathbf{r}_i\mathbf{c}_j = \mathbf{r}_i \sum_{j=1}^{n} \mathbf{c}_j = \mathbf{r}_i\mathbf{l}$$

where $\mathbf{l} = \begin{bmatrix} 1 \\ 1 \\ \vdots \\ 1 \end{bmatrix}$. Thus $\sum_{j=1}^{n} \tilde{p}_{ij} = p_{i1} + p_{i2} + \ldots + p_{in} = 1$.

This is true for $i = 1, 2, \ldots, n$, so as $p_{ij} \geq 0$ this implies $\tilde{p}_{ij} \geq 0$ and we see P^2 is also a transition matrix. Result for P^n follows by induction.

31 Result (a) follows by using direct computation to show that $\mathbf{A}^3 = 5^2\mathbf{A}$, and then using induction to establish the stated property. Result (b) follows in similar fashion after showing that

$$\mathbf{A}^2 = \begin{bmatrix} 9 & 0 & 12 \\ 0 & 25 & 0 \\ 12 & 0 & 16 \end{bmatrix} \quad \text{and then that } \mathbf{A}^4 = 5^2\mathbf{A}^2.$$

33 Length $l = [(x_2 - x_1)^2 + (y_2 - y_1)^2]^{1/2}$ and $p = \cos\theta = (x_2 - x_1)/l$, $q = \sin\theta = (y_2 - y_1)/l$. The displacements d_1 and d_2 of the left and right-hand ends of the rod along its length are, respectively, $\alpha_1 p + \beta_1 q$ and $\alpha_2 p + \beta_2 q$. Thus the extension of the rod along its length is

$$d = \alpha_2 p + \beta_2 q - \alpha_1 p - \beta_1 q,$$

so the force F in the rod is

$$F = \frac{EA}{l}(\alpha_2 p + \beta_2 q - \alpha_1 p - \beta_1 q).$$

As the rod is in equilibrium, the sum of all horizontal and vertical forces must be zero, so resolving F in the x and y directions at the ends of the rod gives

$$F_{2x} = -F_{1x} = pF \quad \text{and} \quad F_{2y} = -F_{1y} = qF,$$

from which the required result then follows.

Answers for Section 3.4

1 $x_1 = 3$, $x_2 = 1$, $x_3 = -1$ **3** $x_1 = 2$, $x_2 = -2$, $x_3 = 3$
5 Only trial solution   **7** $x_1 = 2k$, $x_2 = -k$, $x_3 = k$ (k arbitrary)
9 $x_1 = k$, $x_2 = 2k - 1$, $x_3 = k + 2$ (k arbitrary)
11 $x_1 = k$, $x_2 = 1 - 2k$, $x_3 = 3k$, $x_4 = 1 - k$ (k arbitrary)

13 $x_1 = 2 - \frac{1}{3}k$, $x_2 = 3 + \frac{4}{3}k - l$, $x_3 = k$, $x_4 = l$ (k arbitrary)

15 No solution unless $p = 0$, $q = -3$, then $x_1 = 1$, $x_2 = 2$, $x_3 = -1$, $x_4 = -2$

19 Equivalent resistance $\frac{7}{12} R$.

21 $x_1 = x_5 - 150$, $x_2 = 850 - x_4 - x_5$, $x_3 = x_4 - 250$ with x_4 and x_5 arbitrary, but subject to the constraints $x_4 \geq 250$, $x_5 \geq 150$ and $x_4 + x_5 \leq 850$ as the number of calls in each line must be nonnegative.

Answers for Section 3.5

1 Linearly independent   **3** Linearly independent

5 Linearly dependent (vector 2 = 2 vector 3 + vector 1)

7 7   **9** 2 ($R_3 = 2R_2 - R_1$)

11 3   **13** (i) m (ii) r (iii) r (iv) m (v) r (vi) r

15 $A_E = \begin{bmatrix} 1 & 0 & 0 \\ 0 & 1 & 0 \\ 0 & 0 & 1 \end{bmatrix}$

(i) rank $A = 3$ (ii) $[1,0,0]$, $[0,1,0]$, $[0,0,1]$ (iii) $\begin{bmatrix} 1 \\ 0 \\ 0 \end{bmatrix}$, $\begin{bmatrix} 0 \\ 1 \\ 0 \end{bmatrix}$, $\begin{bmatrix} 0 \\ 0 \\ 1 \end{bmatrix}$

17 $A_E = \begin{bmatrix} 1 & 0 & 3 & 0 & 0 \\ 0 & 1 & -1 & 0 & 0 \\ 0 & 0 & 0 & 1 & 0 \\ 0 & 0 & 0 & 0 & 0 \end{bmatrix}$

(i) rank $A = 3$ (ii) $[1,0,3,0,0]$, $[0,1,-1,0,0]$, $[0,0,0,1,0]$

(iii) $\begin{bmatrix} 1 \\ 0 \\ 0 \\ 0 \end{bmatrix}$, $\begin{bmatrix} 0 \\ 1 \\ 0 \\ 0 \end{bmatrix}$, $\begin{bmatrix} 0 \\ 0 \\ 1 \\ 0 \end{bmatrix}$

19 $A_E = \begin{bmatrix} 1 & 0 & 0 & 0 & 1 \\ 0 & 1 & 0 & 0 & 0 \\ 0 & 0 & 1 & 0 & 0 \\ 0 & 0 & 0 & 1 & 0 \end{bmatrix}$

(i) rank $A = 4$ (ii) $[1, 0, 0, 0, 1]$, $[0, 1, 0, 0, 0]$, $[0, 0, 1, 0, 0]$, $[0, 0, 0, 1, 0]$

(iii) $\begin{bmatrix} 1 \\ 0 \\ 0 \\ 0 \end{bmatrix}$, $\begin{bmatrix} 0 \\ 1 \\ 0 \\ 0 \end{bmatrix}$, $\begin{bmatrix} 0 \\ 0 \\ 1 \\ 0 \end{bmatrix}$, $\begin{bmatrix} 0 \\ 0 \\ 0 \\ 1 \end{bmatrix}$

Answers for Section 3.7

1 (i) -16 (ii) 0   (iii) 0   (iv) -72 (v) 26   (vi) 1   (vii) $x^2(x^2-1)$
3 (i) -12 (ii) 3
5 $+bfg$, $+cdh$, $-bdk$; cofactors ek–fh, bg–ah, ae–bd; minors dk–gf, bk–hc
7 -35 (transpose)   **9** -35 (replace C_2 by C_2+C_3)
11 -35 (replace C_1 by C_1-C_2)   **13** 525 (remove factor 3 from C_2 and factor -5 from C_3)

23
$$\begin{vmatrix} 2x & \cos x & \sinh x \\ \cos x & \cosh x & \sinh x \\ 1 & e^x & e^{-x} \end{vmatrix} + \begin{vmatrix} x^2 & \sin x & \cosh x \\ -\sin x & \sinh x & \cosh x \\ 1 & e^x & e^{-x} \end{vmatrix}$$

$$+ \begin{vmatrix} x^2 & \sin x & \cosh x \\ \cos x & \sinh x & \sinh x \\ 0 & e^x & -e^{-x} \end{vmatrix}$$

27
$$\begin{vmatrix} k & 1 & 1 & \cdots & 1 \\ 1 & k & 1 & \cdots & 1 \\ . & . & . & \cdots & . \\ 1 & 1 & 1 & \cdots & k \end{vmatrix} = \begin{vmatrix} k+n-1 & 1 & 1 & \cdots & 1 \\ k+n-1 & k & 1 & \cdots & 1 \\ . & . & . & \cdots & . \\ k+n-1 & 1 & 1 & \cdots & k \end{vmatrix}$$

$$= (k+n-1) \begin{vmatrix} 1 & 1 & 1 & \cdots & 1 \\ 1 & k & 1 & \cdots & 1 \\ . & . & . & \cdots & . \\ 1 & 1 & 1 & \cdots & k \end{vmatrix}$$

$$= (k+n-1) \begin{vmatrix} 1 & 1 & 1 & \cdots & 1 \\ 0 & k-1 & 0 & \cdots & 0 \\ 0 & 0 & k-1 & \cdots & 0 \\ . & . & . & \cdots & . \\ 0 & 0 & 0 & \cdots & k-1 \end{vmatrix} = (k+n-1)(k-1)^{n-1}$$

Answers for Section 3.8

1 $x_1=1$, $x_2=3$ **3** $x_1=x_2=0$ **5** $x_1=1$, $x_2=-1$, $x_3=2$
7 $D=D_1=D_2=D_3=0$; $x_1=11-7k$, $x_2=2k-3$, $x_3=k$ (k arbitrary)
9 $x_1=1$, $x_2=-1$, $x_3=2$, $x_4=-1$

Answers for Section 3.9

1 $A^{-1} = \dfrac{1}{\det A} \begin{bmatrix} a_{22} & -a_{12} \\ -a_{21} & a_{11} \end{bmatrix}$ **3** $\begin{bmatrix} \frac{2}{5} & \frac{3}{10} \\ \frac{1}{5} & -\frac{1}{10} \end{bmatrix}$ **5** $\begin{bmatrix} \frac{3}{2} & -2 \\ -\frac{1}{2} & 1 \end{bmatrix}$

7 $\begin{bmatrix} \cos\theta & 0 & \sin\theta \\ 0 & 1 & 0 \\ -\sin\theta & 0 & \cos\theta \end{bmatrix}$ **9** $\begin{bmatrix} \frac{3}{8} & -\frac{1}{4} & \frac{1}{4} \\ \frac{1}{8} & \frac{1}{4} & -\frac{1}{4} \\ -\frac{3}{8} & \frac{1}{4} & \frac{3}{4} \end{bmatrix}$

11 det $\mathbf{A}=0$; no inverse   **13** $x_1=-1,\ x_2=1,\ x_3=-2$

15 $x_1 = 10y_1 + 15y_2 - 2y_3$
$x_2 = 6y_1 + 9y_2 - y_3$
$x_3 = 5y_1 + 8y_2 - y_3$

25 $\mathbf{A}^{-1} = \dfrac{-1}{21}\begin{bmatrix} 21 & -9 & -3 \\ 0 & -3 & 6 \\ -14 & 5 & -3 \end{bmatrix}$ **27** $\mathbf{A}$ is singular; no inverse

29 $\mathbf{A}^{-1} = \begin{bmatrix} 1 & 2 & 0 & 0 \\ 0 & 0 & 1 & 2 \\ -1 & 0 & 1 & 0 \\ 0 & 3 & 0 & 0 \end{bmatrix}$ **31** $\mathbf{B} = \begin{bmatrix} 11 & -\frac{1}{2} & 29 \\ -10 & 1 & -28 \\ -7 & -\frac{1}{2} & -19 \end{bmatrix}$

33 Follows directly by solving Equations (A) for v_1 and i_1 in terms of v_2 and i_2

35 $\mathbf{x}_1 = \mathbf{T}_1\mathbf{x}_2,\ \ \mathbf{x}_2 = \mathbf{T}_2\mathbf{x}_3,\dots,\ \ \mathbf{x}_n = \mathbf{T}_n\mathbf{x}_{n+1}$ so $\mathbf{x}_1 = \mathbf{T}_1\mathbf{T}_2\dots\mathbf{T}_n\mathbf{x}_{n+1}$. All $\mathbf{T}_k$ are nonsingular
(Prob. 34), so $\mathbf{x}_{n+1} = \mathbf{T}_n^{-1}\mathbf{T}_{n-1}^{-1}\dots\mathbf{T}_2^{-1}\mathbf{T}_1^{-1}\mathbf{x}_1$.

37 $(\mathbf{B}-2\mathbf{I})^{-1} = \begin{bmatrix} 1 & 2 & 3 \\ 2 & 5 & 7 \\ -2 & -4 & -5 \end{bmatrix}$.

$\mathbf{A}(\mathbf{B}-2\mathbf{A})\mathbf{x} = \mathbf{A}\mathbf{C}$ so $(\mathbf{B}-2\mathbf{A})\mathbf{x}=\mathbf{C}$ and $\mathbf{x}=(\mathbf{B}-2\mathbf{A})^{-1}\mathbf{C}$ giving

$\mathbf{x} = \begin{bmatrix} 10 \\ 24 \\ -17 \end{bmatrix}$.

Answers for Section 3.10

1 $\lambda_1 = 7,\quad \lambda_2 = -2,\quad \rho = 7,\quad \mathbf{x}_1 = \begin{bmatrix} \alpha \\ 0 \end{bmatrix},\quad \mathbf{x}_2 = \begin{bmatrix} 0 \\ \beta \end{bmatrix},\qquad \alpha,\ \beta \neq 0$ arbitrary

$\hat{\mathbf{x}}_1 = \begin{bmatrix} 1 \\ 0 \end{bmatrix},\quad \hat{\mathbf{x}}_2 = \begin{bmatrix} 0 \\ 1 \end{bmatrix},\qquad$ basis $\mathbf{x}_1,\ \mathbf{x}_2$

3 $\lambda_1 = 5,\quad \lambda_2 = -9,\quad \rho = 9,\quad \mathbf{x}_1 = \alpha\begin{bmatrix} 14 \\ 1 \end{bmatrix},\quad \mathbf{x}_2 = \begin{bmatrix} 0 \\ \beta \end{bmatrix},\qquad \alpha,\ \beta \neq 0$ arbitrary

$\hat{\mathbf{x}}_1 = \dfrac{1}{\sqrt{197}}\begin{bmatrix} 14 \\ 1 \end{bmatrix},\quad \hat{\mathbf{x}}_2 = \begin{bmatrix} 0 \\ 1 \end{bmatrix},\qquad$ basis $\mathbf{x}_1,\ \mathbf{x}_2$

5 $\lambda_1 = \dfrac{3-\sqrt{13}}{2}$, $\lambda_2 = \dfrac{3+\sqrt{13}}{2}$, $\rho = \dfrac{3+\sqrt{13}}{2}$, $\mathbf{x}_1 = \alpha \begin{bmatrix} 1 \\ \frac{1}{6}(1-\sqrt{13}) \end{bmatrix}$,

$\mathbf{x}_2 = \begin{bmatrix} 1 \\ \frac{1}{6}(1+\sqrt{13}) \end{bmatrix}$, $\alpha, \beta \neq 0$ arbitrary. $\hat{\mathbf{x}}_1 = 3\left(\dfrac{2}{20-\sqrt{13}}\right)^{1/2} \begin{bmatrix} 1 \\ \frac{1}{6}(1-\sqrt{13}) \end{bmatrix}$

$\hat{\mathbf{x}}_2 = 3\left(\dfrac{2}{20+\sqrt{13}}\right)^{1/2} \begin{bmatrix} 1 \\ \frac{1}{6}(1+\sqrt{13}) \end{bmatrix}$, basis $\mathbf{x}_1, \mathbf{x}_2$

7 $\lambda_1 = 0$, $\lambda_2 = -2\sqrt{2}$, $\lambda_3 = 2\sqrt{2}$, $\rho = 2\sqrt{2}$, $\mathbf{x}_1 = \alpha \begin{bmatrix} 2 \\ -3 \\ 8 \end{bmatrix}$, $\mathbf{x}_2 = \beta \begin{bmatrix} 2-\sqrt{2} \\ 1 \\ 0 \end{bmatrix}$

$\mathbf{x}_3 = \gamma \begin{bmatrix} 2+2\sqrt{2} \\ 1 \\ 0 \end{bmatrix}$, $\alpha, \beta, \gamma \neq 0$ arbitrary, $\hat{\mathbf{x}}_1 = \dfrac{1}{\sqrt{77}} \begin{bmatrix} 2 \\ -3 \\ 8 \end{bmatrix}$,

$\hat{\mathbf{x}}_2 = \dfrac{1}{\sqrt{7-4\cdot 2^{1/2}}} \begin{bmatrix} 2-\sqrt{2} \\ 1 \\ 0 \end{bmatrix}$, $\hat{\mathbf{x}}_3 = \dfrac{1}{\sqrt{13+8\cdot 2^{1/2}}} \begin{bmatrix} 2+2\sqrt{2} \\ 1 \\ 0 \end{bmatrix}$; basis $\mathbf{x}_1, \mathbf{x}_2, \mathbf{x}_3$

9 $\lambda_1 = \lambda_2 = \lambda_3 = 4$ (algebraic multiplicity 3), $\rho = 4$, $\mathbf{x}_1 = \begin{bmatrix} \alpha \\ 0 \\ 0 \end{bmatrix}$, $\alpha \neq 0$ arbitrary ($\lambda = 4$ has geometric

multiplicity 1); basis $\mathbf{x}_1$

11 $\lambda_1 = -2$, $\lambda_2 = \lambda_3 = 4$ (algebraic multiplicity 2), $\rho = 4$, $\mathbf{x}_1 = \begin{bmatrix} 0 \\ \alpha \\ -\alpha \end{bmatrix}$,

$\mathbf{x}_2 = \begin{bmatrix} \beta \\ 0 \\ 0 \end{bmatrix}$, $\mathbf{x}_3 = \begin{bmatrix} 0 \\ \gamma \\ \gamma \end{bmatrix}$, $\alpha, \beta, \gamma \neq 0$ arbitrary ($\lambda = 4$ has algebraic multiplicity 2)

$\hat{\mathbf{x}}_1 = \dfrac{1}{\sqrt{2}} \begin{bmatrix} 0 \\ 1 \\ -1 \end{bmatrix}$, $\hat{\mathbf{x}}_2 = \begin{bmatrix} 1 \\ 0 \\ 0 \end{bmatrix}$, $\hat{\mathbf{x}}_3 = \dfrac{1}{\sqrt{2}} \begin{bmatrix} 0 \\ 1 \\ 1 \end{bmatrix}$; basis $\mathbf{x}_1, \mathbf{x}_2, \mathbf{x}_3$

15 Eigenvalues of A, $\lambda_1 = 3$, $\lambda_2 = 4$. Eigenvectors of A and $R(A)$ $\mathbf{x}_1 = \begin{bmatrix} -3\alpha \\ \alpha \end{bmatrix}$,

$\mathbf{x}_2 = \begin{bmatrix} 0 \\ \beta \end{bmatrix}$, $\alpha, \beta \neq 0$ arbitrary, $R(\lambda_1) = 5$, $R(\lambda_2) = 3$.

17 Eigenvalues of A, $\lambda_1 = -1$, $\lambda_2 = 1$. Eigenvectors of A and $R(A)$ $\mathbf{x}_1 = \begin{bmatrix} \alpha \\ -2\alpha \end{bmatrix}$,

$\mathbf{x}_2 = \begin{bmatrix} \beta \\ 0 \end{bmatrix}$, $\alpha, \beta \neq 0$ arbitrary, $R(\lambda_1) = 3$, $R(\lambda_2) = -1$

21 Eigenvalues -2, 1, 3.

25 Eigenvalues 3, 6 and 9. C_1 centered at 5, radius $\rho_1 = 2$; C_2 centered at 6, radius $\rho_2 = 4$; C_3 centered at 7, radius $\rho_3 = 2$. Eigenvalues and circles all lie to right of the imaginary axis.

27 Eigenvalues at 1, 2 and 3. C_1 at 1, $\rho_1 = 1$; C_2 at 2, $\rho_2 = 2$; C_3 at 3, $\rho_3 = 4$. Eigenvalues all lie to right of the imaginary axis, but union of circles is intersected by the imaginary axis.

29 (i) Not orthogonal (ii) orthogonal (iii) orthogonal

33 The eigenvalues are all located off the imaginary axis (y-axis)

37 $\mathbf{k}_1 = \begin{bmatrix} 1 \\ 2 \end{bmatrix}$ $\mathbf{k}_2 = \begin{bmatrix} -\frac{6}{5} \\ \frac{3}{5} \end{bmatrix}$, $\mathbf{Q} = \begin{bmatrix} \dfrac{1}{\sqrt{5}} & \dfrac{-2}{\sqrt{5}} \\ \dfrac{2}{\sqrt{5}} & \dfrac{1}{\sqrt{5}} \end{bmatrix}$

39 $\mathbf{k}_1 = \begin{bmatrix} 1 \\ 1 \\ 1 \end{bmatrix}$, $\mathbf{k}_2 = \begin{bmatrix} -\frac{1}{3} \\ \frac{2}{3} \\ -\frac{1}{3} \end{bmatrix}$, $\mathbf{k}_3 = \begin{bmatrix} -1 \\ 0 \\ 1 \end{bmatrix}$, $\mathbf{Q} = \begin{bmatrix} \dfrac{1}{\sqrt{3}} & \dfrac{1}{\sqrt{6}} & \dfrac{1}{\sqrt{2}} \\ \dfrac{1}{\sqrt{3}} & \sqrt{\dfrac{2}{3}} & 0 \\ \dfrac{1}{\sqrt{3}} & \dfrac{-1}{\sqrt{6}} & \dfrac{1}{\sqrt{2}} \end{bmatrix}$

41 $\mathbf{k}_1 = \begin{bmatrix} 1 \\ -1 \\ 2 \end{bmatrix}$, $\mathbf{k}_2 = \begin{bmatrix} \frac{3}{2} \\ \frac{3}{2} \\ 0 \end{bmatrix}$. Basis for a two-dimensional vector space.

43 Subtract row 1 from rows 2 to n and then add columns 2 to n to column 1 to show $\det|\mathbf{A}|$ is upper triangular with $\det|\mathbf{A}| = [a+(n-1)b]\,(a-b)^{n-1}$, and hence $\det|\mathbf{A}-\lambda\mathbf{I}| = 0 = [a-\lambda+(n-1)b]\,(a-\lambda-b)^{n-1}$, so $\lambda_1 = a-(n-1)b$, $\lambda_r = a-b$, $r = 2, 3, \ldots, n$.

$x_1 = [1, 1, 1, \ldots, 1]^T$, $x_2 = [1, -1, 0, \ldots, 0]^T, \ldots, \quad x_n = [1, 0, 0, \ldots, 0]^T$.

Answers for Section 3.11

1 Eigenvalues -3, -1, $\mathbf{P} = \begin{bmatrix} -1 & 1 \\ 1 & 1 \end{bmatrix}$

3 Eigenvalues 3, 6, 9, $\mathbf{P} = \begin{bmatrix} 2 & 2 & -1 \\ 2 & -1 & 2 \\ -1 & 2 & 2 \end{bmatrix}$

5 Eigenvalues 2, 1, 1 but only the two eigenvectors $\begin{bmatrix} 0 \\ 0 \\ 1 \end{bmatrix}$, $\begin{bmatrix} 1 \\ 2 \\ -1 \end{bmatrix}$ so not diagonalizable

7 Eigenvalues 5, i, $-$i, $\mathbf{P} = \begin{bmatrix} 1 & 0 & 0 \\ 0 & 1 & 1 \\ 0 & -i & i \end{bmatrix}$

9 $\mathbf{A}^{-1} = \frac{1}{10}\begin{bmatrix} 3 & -1 \\ -2 & 4 \end{bmatrix}$, $\mathbf{A}^3 = \begin{bmatrix} 86 & 39 \\ 78 & 47 \end{bmatrix}$

11 $\mathbf{A}^{-1} = \frac{1}{18}\begin{bmatrix} 3 & 6 \\ 1 & -4 \end{bmatrix}$, $\mathbf{A}^3 = \begin{bmatrix} 94 & 114 \\ 19 & -39 \end{bmatrix}$

13 $\mathbf{A}^{-1} = \begin{bmatrix} -1 & 0 & 2 \\ -1 & 1 & 1 \\ 1 & 0 & -1 \end{bmatrix}$, $\mathbf{A}^4 = \begin{bmatrix} 17 & 0 & 24 \\ 8 & 1 & 12 \\ 12 & 0 & 17 \end{bmatrix}$

15 Write the Cayley–Hamilton theorem in the form

$$\mathbf{A}[(-1)^n\mathbf{A}^{n-1} + c_{n-1}\mathbf{A}^{n-2} + \ldots + c_1\mathbf{I}] + c_0\mathbf{I} = 0.$$

Now $\mathbf{A} \operatorname{adj} \mathbf{A} = (\det \mathbf{A})\mathbf{I}$, but $\det \mathbf{A} = P_n(0) = c_0$, so $c_0\mathbf{I} = \mathbf{A} \operatorname{adj} \mathbf{A}$ and the result follows at once.

17 Eigenvalues are 2, -1.

$$\mathbf{A}^r = \left(\frac{2^r - (-1)^r}{3}\right)\mathbf{A} + \left(\frac{2^r + 2(-1)^r}{3}\right)\mathbf{I}, \text{ so}$$

$$\mathbf{A}^{100} = \left(\frac{2^{100} - 1}{3}\right)\mathbf{A} + \left(\frac{2^{100} + 2}{3}\right)\mathbf{I}.$$

19 Eigenvalues are 1, 0, -1.

$$\mathbf{A}^r = \left(\frac{1 + (-1)^r}{2}\right)\mathbf{A}^2 + \left(\frac{1 - (-1)^r}{2}\right)\mathbf{A}.$$

The final result follows from the properties of the coefficients of $\mathbf{A}^2$ and $\mathbf{A}$, according as r is even or odd.

21 Eigenvalues are 5, 0, -5.

$$\mathbf{A}^r = 5^{r-2}\left(\frac{1 + (-1)^r}{2}\right)\mathbf{A}^2 + 5^{r-1}\left(\frac{1 - (-1)^r}{2}\right)\mathbf{A}.$$

The final results follow from the properties of the coefficients of $\mathbf{A}^2$ and $\mathbf{A}$, according as r is even or odd.

23 Eigenvalues are 2, 2, 2, so a triple eigenvalue is involved.

$$\mathbf{A}^r = r(r-1)2^{r-3}\mathbf{A}^2 + r(2-r)2^{r-1}\mathbf{A} + (2 - 3r + r^2)2^{r-1}\mathbf{I}.$$

25 Multiply by $\mathbf{A} - \mathbf{I}$ to obtain

$$\mathbf{I} = (\mathbf{A} - \mathbf{I})(b_0\mathbf{I} + b_1\mathbf{A} + b_2\mathbf{A}^2),$$

express $\mathbf{A}^3$ in terms of $\mathbf{I}$, $\mathbf{A}$ and $\mathbf{A}^2$ by the Cayley–Hamilton theorem and then determine the coefficients b_0, b_1 and b_2 by equating the coefficients of corresponding powers of $\mathbf{A}$ to make this an identity. Then use the result that

$$\det|\mathbf{A} - \mathbf{I}| = P_3(1) = c_0 + c_1 + c_2 + c_3.$$

27 Multiply by $A-I$ and then proceed as in Problem 25.

29 Eigenvalues are $\frac{1}{4}, \frac{3}{4}$, and the result for A^r follows in the usual manner. After rearranging terms we find

$$\sum_{r=0}^{\infty} A^r = \left\{ \sum_{r=0}^{\infty} \tfrac{1}{2} \left[-\left(\tfrac{3}{4}\right)^r + 3\left(\tfrac{1}{4}\right)^r \right] \right\} I + 2 \left\{ \sum_{r=0}^{\infty} \left[\left(\tfrac{3}{4}\right)^r - \left(\tfrac{1}{4}\right)^r \right] \right\} A.$$

Using the sum of a geometric series shows

$$\sum_{r=0}^{\infty} \left(\tfrac{3}{4}\right)^r = 4 \quad \text{and} \quad \sum_{r=0}^{\infty} \left(\tfrac{1}{4}\right)^r = \tfrac{4}{3},$$

from which it follows that

$$\sum_{r=0}^{\infty} A^r = 0I + \frac{2 \cdot 8}{3} A = \frac{16}{3} A.$$

Rearrangement of terms is permissible because the series involved are absolutely convergent.

Answers for Section 3.12

1 $A = \begin{bmatrix} 5 & 3 & -1 \\ 1 & 0 & -6 \\ 1 & 0 & 7 \end{bmatrix}$ **3** $A = \begin{bmatrix} 4 & 2 & 4 \\ 2 & 9 & 3 \\ 4 & 3 & -5 \end{bmatrix}$ **5** $A = \begin{bmatrix} 1 & 2 & -3 \\ 2 & -2 & 0.5 \\ -3 & 0.5 & -1 \end{bmatrix}$

7 $\Phi = z_1^2 + 2z_2^2 + 4z_3^2$; positive definite

9 $\Phi = 2z_1^2 - z_2^2 + 4z_3^2$; indefinite

$$x_1 = \frac{1}{\sqrt{2}} z_1 - \frac{1}{\sqrt{2}} z_3, \quad x_2 = z_2, \quad x_3 = \frac{1}{\sqrt{2}} z_1 + \frac{1}{\sqrt{2}} z_3;$$

$$z_1 = \frac{1}{\sqrt{2}} x_1 + \frac{1}{\sqrt{2}} x_3, \quad z_2 = x_2, \quad z_3 = -\frac{1}{\sqrt{2}} x_1 + \frac{1}{\sqrt{2}} x_3$$

11 $\Phi = -(z_1^2 + 2z_2^2 + 3z_3^2)$; negative definite

$$x_1 = -\tfrac{2}{3} z_1 + \tfrac{1}{3} z_2 + \tfrac{2}{3} z_3, \quad x_2 = \tfrac{1}{3} z_1 - \tfrac{2}{3} z_2 + \tfrac{2}{3} z_3, \quad x_3 = -\tfrac{2}{3} z_1 - \tfrac{2}{3} z_2 - \tfrac{1}{3} z_3;$$

$$z_1 = -\tfrac{2}{3} x_1 + \tfrac{1}{3} x_2 - \tfrac{2}{3} x_3, \quad z_2 = \tfrac{1}{3} x_1 - \tfrac{2}{3} x_2 - \tfrac{2}{3} x_3, \quad z_3 = \tfrac{2}{3} x_1 + \tfrac{2}{3} x_2 - \tfrac{1}{3} x_3$$

13 $\begin{bmatrix} \dfrac{1}{\sqrt{2}} & \dfrac{1}{\sqrt{2}} & 0 \\ 0 & 0 & 1 \\ \dfrac{-1}{\sqrt{2}} & \dfrac{1}{\sqrt{2}} & 0 \end{bmatrix}$ and $\begin{bmatrix} \dfrac{1}{\sqrt{2}} & 0 & \dfrac{1}{\sqrt{2}} \\ 0 & 1 & 0 \\ -\dfrac{1}{\sqrt{2}} & 0 & \dfrac{1}{\sqrt{2}} \end{bmatrix}$ **15** $A = \begin{bmatrix} 1 & 0 & 0 \\ 0 & \dfrac{3}{2} & \dfrac{-1}{2} \\ 0 & -\dfrac{1}{2} & \dfrac{3}{2} \end{bmatrix}$

17 Indefinite   **19** Positive definite

21 The argument proceeds as in Ex. 6, with the Hessian $\mathbf{H}$ being constructed from the second derivative terms. Φ, and thus $\mathbf{H}$, will be (i) negative definite corresponding to a local maximum of f if $\Delta_1 < 0$, $\Delta_2 > 0$, $\Delta_3 < 0$ (ii) positive definite corresponding to a local minimum if $\Delta_1 > 0$, $\Delta_2 > 0$, $\Delta_3 > 0$ and (iii) indefinite corresponding to a saddle point if neither of the above conditions is true.

23 (i) As $\mathbf{A}$ is symmetric, its eigenvalues are all real and so lie on the real axis in the complex plane. All the Gerschgorin circles coincide, with their common center at $z = p$ and all have the common radius $\rho = (n-1)q$. As $p, q > 0$, the eigenvalues will all be positive if $p - (n-1)q > 0$, which is a *sufficient* condition for the positive definiteness of $\mathbf{x}^{\mathsf{T}}\mathbf{A}\mathbf{x}$.

(ii) Subtract row 1 from rows 2 to n and then add columns 2 to n to column 1 to bring $|\mathbf{A}|$ to upper triangular form. It then follows that $\det|\mathbf{A}| = [p + (n-1)q](p-q)^{n-1}$, so the eigenvalues of $\mathbf{A}$ determined by $\det|\mathbf{A} - \lambda\mathbf{I}| = 0$ are $\lambda_1 = p + (n-1)q$ and $\lambda_r = p - q$ for $r = 2, 3, \ldots, n$.

(a) the *necessary and sufficient* conditions for positive definiteness $\lambda_i > 0$, $i = 1, 2, \ldots, n$ are satisfied if $p > q \ (> 0)$.

(b) the *necessary and sufficient* conditions for indefiniteness that not all the λ_i are of the same sign are satisfied if $q > p (> 0)$.

(iii) The quadratic form $\mathbf{x}^{\mathsf{T}}\mathbf{A}\mathbf{x}$ can never be negative definite because $\lambda_1 > 0$ for all $p, q > 0$.

Answers for Section 3.13

1 $L = \begin{bmatrix} 1 & 0 & 0 \\ 4 & 1 & 0 \\ 1 & -3 & 1 \end{bmatrix}$, $\quad U = \begin{bmatrix} 2 & -1 & 3 \\ 0 & 3 & 1 \\ 0 & 0 & -2 \end{bmatrix}$.

3 $L = \begin{bmatrix} 1 & 0 & 0 \\ 3 & 1 & 0 \\ 2 & -1 & 1 \end{bmatrix}$, $\quad U = \begin{bmatrix} 2 & -1 & 1 \\ 0 & 2 & 1 \\ 0 & 0 & 3 \end{bmatrix}$; $\quad x_1 = 1, \ x_2 = -1, \ x_3 = 2$

5 $L = \begin{bmatrix} 1 & 0 & 0 & 0 \\ 1 & 1 & 0 & 0 \\ -1 & 0 & 1 & 0 \\ 2 & 1 & 1 & 1 \end{bmatrix}$, $\quad U = \begin{bmatrix} -1 & 0 & 1 & 2 \\ 0 & 1 & 2 & 1 \\ 0 & 0 & 2 & 3 \\ 0 & 0 & 0 & 1 \end{bmatrix}$; $\quad x_1 = 1, \ x_2 = -1, \ x_3 = 1, \ x_4 = 1$

7 $L = \begin{bmatrix} 1 & 0 & 0 & 0 \\ 1 & 1 & 0 & 0 \\ 0 & \frac{1}{2} & 1 & 0 \\ 0 & 0 & -2 & 1 \end{bmatrix}$, $\quad U = \begin{bmatrix} 1 & 0 & 0 & 1 \\ 0 & 2 & 1 & -1 \\ 0 & 0 & -\frac{1}{2} & \frac{3}{2} \\ 0 & 0 & 0 & 5 \end{bmatrix}$; $\quad x_1 = 1, \ x_2 = -1, \ x_3 = 2, \ x_4 = -2$.

9 Not positive definite because q_{22} is complex.

11 Positive definite; $\quad Q = \begin{bmatrix} \sqrt{2} & 0 & 0 \\ \dfrac{1}{\sqrt{2}} & \dfrac{1}{\sqrt{2}} & 0 \\ 0 & \sqrt{2} & \sqrt{2} \end{bmatrix}$.

13 $L = \begin{bmatrix} 1 & 0 & 0 \\ -1 & 1 & 0 \\ 1 & 0 & 1 \end{bmatrix}$, $D = \begin{bmatrix} -1 & 0 & 0 \\ 0 & 2 & 0 \\ 0 & 0 & 3 \end{bmatrix}$.

Answers for Section 4.1

1 First order; general solution

3 First order; particular solution

5 Second order; general solution

7 Second order; particular solution

9 $y = -\frac{1}{9}\sin 3x + Ax + B.$

11 $y = \frac{2}{9}\cosh 3x + \sinh x + Ax + B$

13 $y = \frac{1}{3}x^3 - x^2 + x + 1$

15 $y = \frac{1}{12}x^4 + \frac{4}{\pi^2}\sin\left(\frac{\pi x}{2}\right) + \frac{5}{3}x - \left(\frac{48 + 21\pi^2}{12\pi^2}\right)$

17 $y = \frac{x^4}{24} + \frac{x^3}{6} - x^2 + \frac{5}{3}x - \frac{3}{4}$

19 $y' = ky$

21 $(y'')^2 = [1 + (y')^2]^{3/2}$ (Use the result from elementary calculus that the radius of curvature ρ of the curve $y = f(x)$ is $\rho = [1 + (f')^2]^{3/2}/f''$).

23 $\dfrac{dy}{dx} = ky(L - y)$ with $k > 0$ a constant of proportionality. This equation is called the **logistic equation** or the **equation of inhibited growth**. It is useful when studying population growth, the rate of learning, the spread of information, and the acceptance of newly marketed products.

25 $\dfrac{d}{dt}[(M - mt)v] = F - (M - mt)g - kv^\alpha$

27 $\dfrac{dM}{dt} = -\lambda_1 M + k_2\lambda_2 M_2\exp(-\lambda_2 t) + k_3\lambda_3 M_3\exp(-\lambda_3 t).$

Answers for Section 4.2

1 Isoclines $x = \pm\sqrt{2k}$; exact solution $y = \dfrac{x^3}{6} + A$

3 Isoclines $x = (k + 1)/k$; exact solution $y = \ln|x - 1| + A$

5 Isoclines $x = \text{arc tanh } k$; exact solution $y = \ln(\cosh x) + A$

7 Isoclines $y = \frac{1}{2}(k - x)$ **9** Isoclines $y = \pm x\sqrt{3}/\sqrt{k}$ $(k \geq 0)$

11 Isoclines $y = \frac{1}{3}kx^2$ **13** Isoclines $y = x(x - k)$

15 Isoclines $u = 4 - 2k$; the terminal speed $U = 4$, and $u = 0.9U$ when $t = \ln 100 \approx 4.61$

17 Isoclines $M = 3e^{-t} - k$; Maximum poisoning at $t = \ln(3/2) \approx 0.41$.

19 Isoclines $k = \cos\left(\dfrac{\pi}{2}t^2\right)$

21 Make the identification $f(x, y, c) \equiv y - cx - f(x)$ and then use the fact that the envelope of $f(x, y, c) = 0$, with c as a parameter, is obtained by eliminating c between $f = 0$ and $\partial f/\partial c = 0$. The singular solution is the rectangular hyperbola $2xy = a^2$.

Answers for Section 4.3

1 $x^2 + y^2 = c^2$ **3** $\ln|xy| + x - y = c$ **5** $y^3 = x^3 + c$

7 $y = c \exp(x^2/2) - 1$ **9** $(1 + y^2)(1 + x^2) = c, \quad y \neq 0$

11 $y \tan x + \sec x + y^2 = c$ **13** $2 \exp(y^2/2) = e^{1/2}[1 + \exp(x)]$

15 $1 + y^2 = 2(1 - x^2)$ **17** $y = 1/(\ln|x^2 - 1| + 2)$

19 $2e^x + e^{-2y} = 3e$ **21** $y = -\log_{10}(40 - 3.10^x)$

23 $\dfrac{2}{\sqrt{3}} \arctan\left(\dfrac{2y - x}{x\sqrt{3}}\right) = \ln\left|\dfrac{c}{x}\right|$ **25** $\dfrac{2}{\sqrt{5}} \operatorname{arc\,tanh}\left(\dfrac{2y + x}{x\sqrt{5}}\right) = \ln\left|\dfrac{c}{x}\right|$

27 $y = x/(c - \ln|x|)$.   **29** $2\ln|x + y - 1| + 3x + y = c$

Answers for Section 4.4

1 $(2xy + 3)dx + (x^2 - 1)dy = 0$

3 $(\cos x \cosh y - \sin x \sinh y)dx + (\sin x \sinh y + \cos x \cosh y)dy = 0$

5 $\left(2xe^{x/y} + \dfrac{x^2}{y}e^{x/y}\right)dx - \dfrac{x^3}{y^2}e^{x/y}dy = 0$

7 $\left(\dfrac{1}{y} - \dfrac{2y}{x^2}\right)dx + \left(\dfrac{2}{x} - \dfrac{x}{y^2}\right)dy = 0$

9 Exact; $\dfrac{x^2}{y^3} - \dfrac{1}{y} = c$

11 Exact; $x \cosh y + y \sinh x = c$

13 Exact; $xe^{x+y} + y = c$

15 Not exact; $\mu = \dfrac{1}{y^3}; \quad \dfrac{\sin xy}{y} + \dfrac{x^2}{y^2} = c$

17 Exact; $x^2 - y^2 = cy^3$

19 Exact; $\dfrac{x^3}{3} + xy^2 + x^2 = c.$

21 Not exact; $\mu = 1/x; \quad y = -2x/(x^2 + c)$

23 Not exact; $\mu = e^x; \quad e^x(x \sin y + y \cos y - \sin y) = c.$

25 Not exact; $\mu = \dfrac{1}{y^2}; \quad \dfrac{x^2}{y^2} + \dfrac{1}{y} = 4.$

27 Not exact; $\mu = 1/\sin y; \quad x = 2\sqrt{3} \sin y.$

29 Taking the total differential of $\mu_2/\mu_1 = c$ gives

$$\left(\mu_2 \frac{\partial \mu_1}{\partial x} - \mu_1 \frac{\partial \mu_2}{\partial x}\right)dx + \left(\mu_2 \frac{\partial \mu_1}{\partial y} - \mu_1 \frac{\partial \mu_2}{\partial y}\right)dy = 0. \tag{A}$$

As μ_1 and μ_2 are integrating factors of the original equation,

$$P\frac{\partial \mu_i}{\partial y} - Q\frac{\partial \mu_i}{\partial x} + \mu_i\left(\frac{\partial P}{\partial y} - \frac{\partial Q}{\partial x}\right) = 0$$

for $i = 1, 2$. Thus

$$\mu_2 P\left(\frac{\partial \mu_1}{\partial y} - Q\frac{\partial \mu_1}{\partial x}\right) - \mu_1\left(P\frac{\partial \mu_2}{\partial y} - Q\frac{\partial \mu_2}{\partial x}\right) = 0$$

which is equivalent to

$$\left(\mu_2\frac{\partial \mu_1}{\partial y} - \mu_1\frac{\partial \mu_2}{\partial y}\right)P = \left(\mu_2\frac{\partial \mu_1}{\partial x} - \mu_1\frac{\partial \mu_2}{\partial x}\right)Q,$$

which reduces (A) to the original equation

$$P(x, y)dx + Q(x, y)dy = 0.$$

Answers for Section 4.5

1 Result follows by adding $c_1 y_1' + c_1 p(x)y_1 = 0$ and $c_2 y_2' + c_2 p(x)y_2 = 0$, to get $(c_1 y_1 + c_2 y_2)' + p(x)(c_1 y_1 + c_2 y_2) = 0$.

3 Let the antiderivative of $p(x)$ be $\int p(x)dx + A$, with A an arbitrary constant. Then $\mu = \exp(A)\cdot\exp[\int p(x)dx]$, and as the integrating factor multiplies both sides of the equation the multiplicative factor e^A is immaterial. Thus, for simplicity, without loss of generality we may set $A = 0$, and hence $e^A = 1$.

5 $y = ce^{-kx} + xe^{-kx}$.

7 $y = \dfrac{c}{x} + \dfrac{1}{4}x^3$.

9 $y = c(x-2) + x(x-2) + (x-2)\ln|x-2|$.

11 $y = c\cos x + \sin x$.

13 $y = \dfrac{c}{x^n} + \dfrac{3}{x^{n-1}}$.

15 $y = c(\ln|x|)^2 - \ln|x|$.

17 $y = 2e^{2x} - e^x$.

19 $y = \frac{1}{2}(x\sqrt{1-x^2} + \arcsin x)\sqrt{\dfrac{1+x}{1-x}}$.

21 $y = x^2\{1 + \exp[(1+x)/x]\}$.

23 $x = ce^{-y/2} + y - 2$.

25 $x = \dfrac{\pi}{3}\left(\dfrac{1}{\cos y - y}\right)$.

27 $x = \dfrac{1}{6}y^4 + \dfrac{5}{6y^2}$.

29 $y = \dfrac{16}{4x - 1 + ce^{-4x}}$.

31 $y = x^4\frac{1}{2}\ln|x| + c)^2$.

33 Set $y=uv$, then $y'+p(x)y=q(x)y^{\alpha}$ becomes $u'v+uv'+p(x)uv=q(x)u^{\alpha}v^{\alpha}$. If we require $[u'+p(x)u]v=0$, then unless $v=0$, which gives the trivial solution $y\equiv0$, $v'=q(x)u^{\alpha-1}v^{\alpha}$. Thus v follows by integration once $u=c\exp[-\int p(x)dx]$ has been found.

35 $y=\dfrac{4}{x(4x-3)}$.

37 $y=\dfrac{2}{2+2\ln|x|-x}$.

39 $a^2x^3=ce^{ay}-a(y+1)-1$.

41 $u=4(1-e^{-t/2})$.

43 $M=4e^{-t}-3e^{-2t}$.

45 $i=\dfrac{V_0}{R}-\left(\dfrac{V_0}{R}-i_0\right)e^{-(R/L)t}$.

47 If $dy+[p(x)y-q(x)]dx=0$, then $P=p(x)y-q(x)$ and $Q=1$ so that $g(x)=\dfrac{1}{Q}\left(\dfrac{\partial P}{\partial y}-\dfrac{\partial Q}{\partial x}\right)=p(x)$.

The integrating factor is $\mu=\exp[\int g(x)dx]=\exp[\int p(x)dx]$, and so

$$\exp[\int p(x)dx]dy+\exp[\int p(x)dx][p(x)y-q(x)]dx=0$$

must be exact. If the solution is $F(x,y)=c$ then

$$\frac{\partial F}{\partial x}=\exp[\int p(x)dx][p(x)y-q(x)]\quad\text{and}\quad\frac{\partial F}{\partial y}=\exp[\int p(x)dx].$$

Now,

$$F=\int\frac{\partial F}{\partial x}dx+f(y)=\int\exp[\int p(x)]\,p(x)y\,dx-\int\exp[\int p(x)dx]q(x)dx+f(y)$$

or

$$F=\exp[\int p(x)dx]y-\int\exp[\int p(x)dx]q(x)\,dx+f(y),\tag{A}$$

with f an arbitrary function of y. However, we also know that

$$F=\int\frac{\partial F}{\partial y}dy+g(x)=\int\exp[\int p(x)dx]\,dy=\exp[\int p(x)dx]y+g(x),\tag{B}$$

where $g(x)$ is an arbitrary function of x. The two expressions (A) and (B) for F must be identical, so $g(x)=-\int\exp[\int p(x)dx]q(x)dx$ and $f(y)\equiv0$. Thus $F=c=\exp[\int p(x)dx]y-\int\exp[\int p(x)dx]q(x)dx$, from which (9) follows after multiplication by $\exp[-\int p(x)dx]$ and a rearrangement of terms.

49 It follows from (9) that in general

$$y=c\exp[-\int p(x)dx]+y_p,\text{ and so}$$
$$y=c_1\exp[-\int p(x)dx]+y_p\quad\text{and}\quad y_2=c_2\exp[-\int p(x)dx]+y_p.$$

Thus $y-y_1=(c-c_1)\exp[-\int p(x)dx]$ and $y_2-y_1(c_2-c_1)\exp[-\int p(x)dx]$, so that

$$\frac{y-y_1}{y_2-y_1}=\left(\frac{c-c_1}{c_2-c_1}\right)=k.$$

51 $y_1 = \dfrac{1}{2}\left(x + \dfrac{1}{x}\right)$,   for $1 \leq x \leq 2$, $y_2 = x\left(\ln x + \dfrac{5}{8} - \ln 2\right)$,   for $x \geq 2$. No.

53 $y_1 = 1 + \exp(-\tfrac{1}{2}x^2)$,   for $0 \leq x \leq 1$, $y_2 = x - 1 + \exp(1 - x) + \exp(\tfrac{1}{2} - x)$,   for $x \geq 1$. Yes.

55 $y = \dfrac{cx + 4x^3 - x^5}{4x^4}$

57 $y = \dfrac{cxe^x + x^2}{ce^x - x}$

Answers for Section 4.6

1 Rectangular hyperbolas $x^2 - y^2 = a$.

3 Confocal ellipses $2x^2 + y^2 = a^2$.

5 Parabolas $y^2 = 2ax + a$.

7 $(x^2 + y^2)^2 = a^2(x^2 - y^2)$.

9 $(x^2 + y^2)^2 = a(y^2 + 2x^2)$.

11 $y^2 = a(x - y\sqrt{3})$.

13 $r = a \sin \theta$.

15 $r = a \sin^2 \theta$, $\quad a > 0$.

17 Implicit differentiation of $u = \text{const.}$ gives

$$\frac{\partial u}{\partial x} + \frac{\partial u}{\partial y}\frac{dy}{dx} = 0, \text{ showing the gradient of the curves is}$$

$$\frac{dy}{dx} = -\frac{\partial u}{\partial x}\bigg/\frac{\partial u}{\partial y}. \text{ Thus the gradient of the orthogonal trajectories must be } \frac{dy}{dx} = \frac{\partial u}{\partial y}\bigg/\frac{\partial u}{\partial x}.$$

Substituting for $\partial u/\partial x$ and $\partial u/\partial y$ from the stated equations and rearranging gives

$$\frac{\partial v}{\partial x} + \frac{\partial v}{\partial y}\frac{dy}{dx} = 0,$$

which is the same as the result of differentiating $v = \text{const.}$ implicitly, so the result is proved.

Answers for Section 4.7

1 The equation only has the trivial solution $y \equiv 0$ which does not satisfy the given initial condition.

3 $y \equiv 1$ is a solution and another is

$$y_a(x) = \begin{cases} 1 & \text{for } -\infty < x \leq a, \\ 1 + (x - a)^2, & \text{for } x \geq a \text{ with } a \geq 0. \end{cases}$$

5 The result follows directly from Theorem 4.3.

7 $y \equiv 0$ is a solution and so also is

$$y = \begin{cases} \tfrac{1}{4}(x + \tfrac{1}{3}x^3)^2 & \text{for } x \geq 0 \\ -\tfrac{1}{4}(x + \tfrac{1}{3}x^3)^2 & \text{for } x < 0. \end{cases}$$

The result also follows from Theorem 4.3 because $f(x, y) = (1 + x^2)\sqrt{|y|}$ is not Lipschitz continuous in y in any region containing the origin.

9 $|f| \leq 1/2$ in R so condition (i) of Theorem 4.3 is satisfied. Let k and l be real constants, and by considering $|f(x, y_2) - f(x, y_1)|$ with $(x, y_1) = (x, kx^2)$ and $(x, y_2) = (x, lx^2)$ show f is not Lipschitz continuous at $(0, 0)$. It then follows that condition (ii) of Theorem 4.3 is not satisfied, so a solution exists but it is not necessarily unique.

11 If $\partial f/\partial y$ is bounded in R then $|\partial f/\partial y| \leq m$. Let (x, y_1) and (x, y_2) be any two points in R. Then by the mean value theorem

$$|f(x, y_2) - f(x, y_1)| = |y_2 - y_1| \left. \left| \frac{\partial f}{\partial y} \right| \right|_{(x, \bar{y})}$$

with $\bar{y}$ between y_1 and y_2. This yields the required Lipschitz condition when we use the result $|\partial f/\partial y| \leq m$.

13 $y_1(x) = 1 + \dfrac{x^2}{2}$, $\quad y_2(x) = 1 + \dfrac{x^2}{2} + \dfrac{x^4}{8}$,

$y_3(x) = 1 + \dfrac{x^2}{2} + \dfrac{x^4}{8} + \dfrac{x^6}{48}$, $\quad y_4(x) = 1 + \dfrac{x^2}{2} + \dfrac{x^4}{8} + \dfrac{x^6}{48} + \dfrac{x^8}{384}$.

15 (a) $y_3(x) = 1 + x + \dfrac{x^2}{2!} + \dfrac{x^3}{3!}$; (b) $y_3(x) = 1 + 2x + \dfrac{x^2}{2} - \sin x$; (c) $y_3(x) = 1 + \dfrac{x^2}{2} + \sinh x$.

Answers for Section 4.8

1

n	x_n	k_{1n}	k_{2n}	k_{3n}	k_{4n}
0	0.0	0.000000	0.010233	0.010284	0.021064
1	0.1	0.021065	0.032389	0.032559	0.044550
2	0.2	0.044551	0.057229	0.057550	0.071135
3	0.3	0.071135	0.085743	0.086280	0.102396
4	0.4	0.102394	0.120437	0.121376	0.142660
5	0.5	0.142669	0.168798	0.170908	0.207088

n	x_n	k_n	y_n
0	0.0	0.010350	1.000000
1	0.1	0.032586	1.010350
2	0.2	0.057541	1.042935
3	0.3	0.086263	1.100476
4	0.4	0.121447	1.186739
5	0.5	0.171528	1.308186

3

n	x_n	k_{1n}	k_{2n}	k_{3n}	k_{4n}
0	0.5	1.872544	−0.899703	2.095474	−1.192129
1	0.8	0.053340	−0.158427	0.127249	−0.556408
2	1.1	0.028403	−0.145906	0.075159	−0.485316
3	1.4	0.004126	−0.152294	0.047326	−0.504590
4	1.7	0.002524	−0.169642	0.049814	−0.611583
5	2.0	0.023026	−0.183461	0.059258	−0.658733

n	x_n	k_n	y_n
0	0.5	0.511993	1.000000
1	0.8	−0.094237	1.511993
2	1.1	−0.099734	1.417755
3	1.4	−0.118400	1.318021
4	1.7	−0.141453	1.199621
5	2.0	−0.147352	1.058168

5

n	x_n	k_{1n}	k_{2n}	k_{3n}	k_{4n}
0	0.0	0.704638	0.717856	0.717605	0.709868
1	0.4	0.709990	0.705120	0.705086	0.725855
2	0.8	0.726244	0.769789	0.771745	0.796506
3	1.2	0.796424	0.799908	0.799910	0.800000
4	1.6	0.800000	0.800000	0.800000	0.800000
5	2.0	0.800000	0.800000	0.800000	0.800000

n	x_n	k_n	y_n
0	0.0	0.714238	−1.000000
1	0.4	0.709376	−0.285762
2	0.8	0.767636	0.423614
3	1.2	0.799343	1.191250
4	1.6	0.800000	1.990593
5	2.0	0.800000	2.790593

7

n	x_n	k_{1n}	k_{2n}	k_{3n}	k_{4n}
0	0.0	−0.178336	−0.186009	−0.182386	−0.036151
1	0.2	−0.103410	−0.073422	−0.107936	−0.095818
2	0.4	−0.119747	−0.166666	−0.149326	−0.197504
3	0.6	−0.196779	−0.182013	−0.182068	−0.106115
4	0.8	−0.107823	−0.023539	−0.031063	0.035685
5	1.0	0.035577	0.084156	0.082541	0.119943

n	x_n	k_n	y_n
0	0.0	−0.158546	0.500000
1	0.2	−0.093657	0.341454
2	0.4	−0.158206	0.247797
3	0.6	−0.171843	0.089591
4	0.8	−0.030224	−0.082252
5	1.0	0.081486	−0.112475

Answers for Section 5.1

11 $y = c_1 + c_2 e^{x/3}$.

13 $y = c_1 + c_2 \operatorname{arc sinh} x$.

15 $y = c_1 + c_2 \operatorname{arc sin} x + (\operatorname{arc sin} x)^2$.

17 $y = c_1 + c_2 x^2 + x^3$.

19 $y = -1 + \frac{3}{2} e^x + \frac{1}{2} (\cos x - \sin x)$.

21 $y = \frac{1}{4} \left(\frac{\pi}{2} - 1 \right) + \frac{1}{2} x + \frac{1}{4} \cot x$.

23 $y = x - 2 - 4 \operatorname{arc tan} (x/2)$.

25 $y = \frac{1}{3}(x^6 - 3x + 2)$.

27 $y = \frac{5}{6} + \frac{1}{5} x - \frac{1}{30} x^6$.

29 $\frac{2}{3} \ln(3y^2 + 5) = x + \ln 4$.

31 $y = 2(x + 7)^2$.

33 (i) $y = c_1 \sin kx + c_2 \cos kx$, (ii) $y = c_1 e^{kx} + c_2 e^{-kx}$.

35 Linearly dependent: the three functions are related by $\cosh^2 3x - \sinh^2 3x = 1$.

37 Linearly independent: $x/|x|$ is ± 1 according as $x > 0$ or $x < 0$, and ± 1 is not expressible as a linear combination of x and $|x|$.

41 (a) $y = \frac{c_1}{x} + c_2 \frac{\ln x}{x}$; (b) $y = c_1 x + c_2 x^2$; (c) $y = c_1 (1 - e^x) + \frac{c_2}{1 + e^x}$.

Answers for Section 5.2

1 $y = c_1 e^x + c_2 e^{2x}$.

3 $y = e^{-2x}(c_1 \cos 3x + c_2 \sin 3x)$.

5 $y = e^{2x}(c_1 e^{\sqrt{3}x} + c_2 e^{-\sqrt{3}x})$.

7 $y - c_1 + c_2 e^{3x}$.

9 $y = c_1 e^{\sqrt{|k|}x} + c_2 e^{-\sqrt{|k|}x}$ for $k < 0$ and $y = c_1 \cos \sqrt{k}x + c_2 \sin \sqrt{k}x$ for $k > 0$.

11 $y = \frac{1}{2}(e^x + e^{3x})$.

13 $y = \frac{1}{3}(e^{3(1-x)} - 1)$.

15 $y = \frac{1}{4}(e^{2(x-2)} - e^{-2(x-2)}) = \frac{1}{2} \sinh 2(x - 2)$.

17 $y = 2xe^{-x}$.

19 $y'' + 2y' - 4y = 0$.

21 $y'' + 16y = 0$.

23 $y'' - 9y = 0$.

25 $y'' - 3y' + 5y = 0$.

27 $y = c \sin \pi x$ (c an arbitrary constant).

29 $y = \{\exp[(2 + \sqrt{2})x] - \exp[(2 - \sqrt{2})x]\}/[\exp(2 + \sqrt{2}) - \exp(2 - \sqrt{2})]$

31 $y = -\cos 2x$.

33 $k = n\pi/a$, with $n = 0, 1, 2, \ldots$; $y = c \cos\left(\dfrac{n\pi x}{a}\right)$ for $n \neq 0$ and $y \equiv 0$ when $n = 0$.

35 $k = \dfrac{(2n-1)\pi}{2a}$, with $n = 1, 2, \ldots$; $y = c \sin\left(\dfrac{(2n-1)\pi x}{2a}\right)$.

37 $y = 2e^{3x}$.

39 $y = 3e^{-2x}$.

Answers for Section 5.3

1 Linearly independent   **3** Linearly dependent   **5** Linearly dependent

7 $y = c_1 e^x + c_2 e^{2x}$, e^x, e^{2x} for all x.

9 $y = c_1 e^x + c_2 x e^x + c_3 e^{-2x}$; e^x, $x e^x$, e^{-2x} for all x.

11 $y = c_1 e^{3x} \cos 3x + c_2 e^{3x} \sin 3x$; $e^{3x} \cos 3x$, $e^{3x} \sin 3x$ for all x.

13 $y = c_1 e^{2x} + c_2 x e^{2x} + c_3 x^2 e^{2x}$; e^{2x}, $x e^{2x}$, $x^2 e^{2x}$ for all x.

15 $y = c_1 \cos 2x + c_2 x \cos 3x + c_3 \sin 3x + c_4 x \sin 3x$; $\cos 3x$, $x \cos 3x$, $\sin 3x$, $x \sin 3x$ for all x.

17 $y = c_1 + c_2 e^x + c_3 x e^x + c_4 x^2 e^x$; 1, e^x, $x e^x$, $x^2 e^x$ for all x.

19 $y = e^x - 3 \sin x$

21 $y = -\frac{1}{3}e^x + \frac{1}{12}e^{-2x} + \frac{1}{4}e^{2x}$

23 $y = \cos 2x + x \sin 2x$.

25 $y''' - y' = 0$; $y = c_1 + c_2 e^{-x} + c_3 e^x$; $y = 3 - \frac{1}{2}e^{-x} - \frac{1}{2}e^x$.

Answers for Section 5.4

1 $7e^{2x}$, $e^x(2 + 5x)$, $-6 \sin 3x + 3 \cos 3x$.

3 $20e^{3x}$, $6(\sinh 2x + \cosh 2x)$, $2 + \sin x + 3 \cos x + 3 \sinh x + 3 \cosh x$.

5 $y = c_1 e^{-x} + c_2 e^{3x}$.

7 $y = c_1 e^{-2x} + c_2 x e^{-2x} + c_3 x^2 e^{-2x}$.

9 The function u satisfies the equation $(D - \lambda_1)u = 0$, and so has the solution $u = A_1 \exp(\lambda_1 x)$. The function y satisfies the linear first order equation

$$y' - \lambda_2 y = A_1 \exp(\lambda_1 x),$$

which has the integrating factor $\exp(-\lambda_2 x)$. Thus

$$\frac{d}{dx}[\exp(-\lambda_2 x)y] = A_1 \exp[(\lambda_1 - \lambda_2)x],$$

and if $\lambda_1 \neq \lambda_2$,

$$\exp(-\lambda_2 x)y = \frac{A_1 \exp[(\lambda_1 - \lambda_2)x]}{(\lambda_1 - \lambda_2)} + B,$$

leading to the solution

$$y = A \exp(\lambda_1 x) + B \exp(\lambda_2 x) \qquad (A = A_1/(\lambda_1 - \lambda_2)).$$

If $\lambda_1 = \lambda_2 = \alpha$ we have

$$\frac{d}{dx}(e^{-\alpha x}y) = A_1,$$

leading to the solution

$$y = A_1 x e^{\alpha x} + B e^{\alpha x}.$$

11 Proceed as in Prob. 9. First set $u = (D - \lambda_2)(D - \lambda_3)y$ and solve the homogeneous equation $(D - \lambda_1)u = 0$. Then with this function u set $v = (D - \lambda_3)y$ and solve the linear first order equation $v' - \lambda_2 v = u$. Finally, with this function v solve the linear first order equation $y' - \lambda_3 y = v$. Consider separately the cases (i) distinct roots, (ii) two equal roots and (iii) three equal roots.

Answers for Section 5.5

1 $y = (c_1 + c_2 x)e^{2x} + \frac{1}{8}(2x^2 + 4x + 3)$.

3 $y = c_1 e^{-x} + c_2 e^{3x} + \frac{1}{5}e^{4x}$.

5 $y = c_1 e^x + c_2 e^{2x} + \frac{1}{10}(\sin x + 3\cos x)$.

7 $y = c_1 \cos x + c_2 \sin x + \frac{1}{2}x \sin x$.

9 Use $\sinh x = \frac{1}{2}(e^x - e^{-x})$; $y = (c_1 + c_2 x)e^x + \frac{1}{2}\cos x + \dfrac{x^2}{4}e^x - \frac{1}{8}e^{-x}$.

11 $y = e^x(c_1 \cos 2x + c_2 \sin 2x) + \frac{1}{4}xe^x \sin 2x$.

13 $y = c_1 e^{-2x} + c_2 e^{4x} - \frac{1}{3}e^x + \frac{1}{5}(3\cos 2x + \sin 2x)$.

15 $y = c_1 + c_2 e^{2x} + \frac{1}{3}e^{3x} - 2$.

17 $y = c_1 + c_2 \sin 2x + c_3 \cos 2x + 5x - 2x \sin 2x$.

19 $y = c_1 e^x + e^{-\frac{1}{2}x}\left(c_2 \cos \dfrac{\sqrt{3}}{2}x + c_3 \sin \dfrac{\sqrt{3}}{2}x\right) - x^3 - 8$.

21 $y = \cos 2x + \frac{1}{3}\sin 2x + \frac{1}{3}\sin x$.

23 $y = 1 + 2x + 2x^2 - e^{2x}$.

25 $y = 2e^x - 7xe^x + 3e^{2x} - x - 4$.

27 $y = c_1 e^x + c_2 x e^x + 2x^2 e^x$.

29 $y = e^x (c_1 \cos x + c_2 \sin x) - 2x e^x \cos x$.

31 $y = c_1 e^{2x} + c_2 x e^{2x} + \frac{1}{6} x^3 e^{2x}$.

33 $y = c_1 \cos x + c_2 \sin x + x \sin x + \cos x \ln |\cos x|$.

35 $y = c_1 \cos x + c_2 \sin x + \sin x \ln \left| \tan \dfrac{x}{2} \right|$.

37 First solve $y'' - \dfrac{1}{x} y' = 0$ by reduction of order as in Problem Section 5.1; $y = c_1 + c_2 x^2 + \frac{1}{3} x^3$.

39 $y = c_1 e^x + c_2 e^{2x} + c_3 x e^{2x} + \frac{1}{2} x^2 e^{2x}$.

41 $y = c_1 e^{-x} + c_2 \cos x + c_3 \sin x + \frac{1}{4} x e^x - \frac{3}{8} e^x$.

43 $y = e^{-x} + e^{-\frac{1}{2} x} \left(\cos \dfrac{\sqrt{3}}{2} x + \dfrac{1}{\sqrt{3}} \sin \dfrac{\sqrt{3}}{2} x \right) + x - 2$.

45 $y = \dfrac{c_1}{x} + c_2 \dfrac{\ln x}{x}$.

47 $y = c_1 \cos (2 \ln x) + c_2 \sin (2 \ln x)$.

49 $y = c_1 x^2 + c_2 x^2 \ln x$.

51 $y = c_1 x^2 + c_2 x^3 + \frac{1}{2} x$.

53 $y = c_1 (x+1)^2 + c_2 (x+1)^2 \ln(x+1) + (x+1)^3$.

Answers for Section 5.6

1 $y = e^{2x} (c_1 \sin x + c_2 \cos x)$.

3 $y = c_1 + c_2 x + c_3 e^{-x}$.

5 $y = c_1 + c_2 \arctan x$.

7 $y = c_1 x^2 + c_2 \dfrac{\sin (\sqrt{3} \ln x)}{x} + c_3 \dfrac{\cos (\sqrt{3} \ln x)}{x}$.

Answers for Section 5.7

1 $I \dfrac{d^2 \theta}{dt^2} = \dfrac{k}{I} \theta; \quad T = 2\pi \sqrt{\dfrac{I}{k}}$.

3 If x_1, x_2 are displacements of m_1 and m_2 from their equilibrium positions, the stationarity of the center of mass implies $m_1 x_1 + m_2 x_2 = 0$. The force acting on m_1 is $F = -k(x_1 - x_2)$; so the equation of motion is

$$m_1 \dfrac{d^2 x_1}{dt^2} = -k(x_1 - x_2),$$

which when either x_1 or x_2 is eliminated gives the required result.

5 In equilibrium $mg = \lambda d$, where λ is the elastic constant for the string. When displaced the restoring force $F = -(mg/d)x$ while the air resistance $R = -k(\mathrm{d}x/\mathrm{d}t)$. Thus the equation of motion is

$$m\frac{\mathrm{d}^2 x}{\mathrm{d}t^2} = -k\frac{\mathrm{d}x}{\mathrm{d}t} - \frac{mg}{d}x.$$

If the string becomes slack when moving upward Hooke's law will cease to apply, thereby invalidating the above equation of motion.

7 Downward displacement x causes an upthrust of magnitude $F = -\rho g A x$ (Archimedes' principle), while the resistance to motion $R = -k(\mathrm{d}x/\mathrm{d}t)$. The equation of motion is thus

$$m\frac{\mathrm{d}^2 x}{\mathrm{d}t^2} = -k\frac{\mathrm{d}x}{\mathrm{d}t} - \rho g A x.$$

9 The derivation of the equation of motion is similar to that of the damped simple pendulum, except that the tension τ in the pendulum string is now replaced by the reaction R normal to the surface of the bowl.

11 If the moment opposing rotation due to viscosity is $M = -k(\mathrm{d}\theta/\mathrm{d}t)$, the equation of motion becomes

$$I\frac{\mathrm{d}^2 \theta}{\mathrm{d}t^2} + k\frac{\mathrm{d}\theta}{\mathrm{d}t} + \frac{Fl}{d}\theta = 0.$$

17 $y_p = -\cos 2t$.

19 $y_p = \frac{1}{410}(42\sin t - 9\cos t)$.

21 $y_p = \frac{1}{65}(\sin 2t - 8\cos 2t)$.

23 $y_c = c_1\cos 5t + c_2\sin 5t$.

25 $y_c = e^{-\frac{1}{2}t}(c_1\cos 3t + c_2\sin 3t)$.

27 $y_c = e^{\frac{1}{3}t}(c_1\cos 2t + c_2\sin 2t)$.

29 $y = \cos 4t + \frac{1}{4}\sin 4t$.

31 $y = 3e^{-t}\sin 2t$.

33 $y = \sin t + \frac{1}{3}\cos t - \frac{1}{3}\cos 2t$.

35 $y = 4\sin t + \cos t - 3t\cos t$.

37 $y = \frac{1}{85}e^{-t}(7\cos t - 11\sin t) - \frac{1}{85}(7\cos 3t - 6\sin 3t)$.

39 $y_p = \frac{1}{4}\sin t$.

41 $y_p = -\frac{1}{137}(4\cos 2t + 11\sin 2t)$.

Answers for Section 5.8

1 $v = e^{2x};\quad u'' - u = 0;\quad u_1 = e^x,\quad u_2 = e^{-x};\quad y = c_1 e^x + c_2 e^{3x}$.

3 $v = e^x;\quad u'' = 0;\quad u_1 = c_1,\quad u_2 = c_2 x;\quad y = (c_1 + c_2 x)e^x$.

5 $v = e^{-2x};\quad u'' + 5u = 0;\quad u_1 = \cos\sqrt{5}x,\quad u_2 = \sin\sqrt{5}x;$
$y = e^{-2x}(c_1\cos\sqrt{5}x + c_2\sin\sqrt{5}x)$.

7 (a) $y = c_1 e^{-\frac{1}{2}ax} \cos(nx + c_2)$, (b) $y = (c_1 + c_2 x) e^{-\frac{1}{2}ax}$,

(c) $y = c_1 \exp(\lambda_1 x) + c_2 \exp(\lambda_2 x)$ with $\lambda_1 = \frac{1}{2}(-a - \sqrt{a^2 - 4b})$, $\lambda_2 = \frac{1}{2}(-a + \sqrt{a^2 - 4b})$.

9 $v = \exp(\cos x)$; $u'' - u = 0$; $u_1 = e^{-x}$, $u_2 = e^x$; $y = c_1 \exp(\cos x - x)$

$c_2 \exp(\cos x + x)$.

11 $v = \exp(-\frac{1}{2}\sin x)$; $u'' + (2 + \frac{1}{2}\sin x - \frac{5}{4}\cos^2 x)u = 0$.

13 $v = \exp(-1/(3x^3))$; $u'' + \frac{2}{x}\left(2 - \frac{1}{x^2} - \frac{2}{x^3}\right)u = 0$.

Answers for Section 5.9

1 $G(x, t) = \frac{1}{6}\{\exp[2(x - t)] - \exp[-4(x - t)]\}$;

$y = c_1 e^{-4x} + c_2 e^{2x} + \frac{2}{7}e^{3x}$.

3 $G(x, t) = (x - t)\exp[-(x - t)]$;

$y = c_1 e^{-x} + c_2 x e^{-x} + \frac{1}{9}e^{2x}$.

5 $G(x, t) = \frac{1}{4}\frac{x^4}{t} - 1$;

$y = c_1 + c_2 x^4 - x^2 + 2x$.

7 $G(x, t) = \begin{cases} x\left(t - \dfrac{\pi}{2}\right), & 0 \le x \le t \\ t\left(x - \dfrac{\pi}{2}\right), & t \le x \le \dfrac{\pi}{2}; \end{cases}$

$y = \dfrac{2}{\pi}x - \sin x$.

9 $G(x, t) = \begin{cases} x\left(\dfrac{t}{k+1} - 1\right), & 0 \le x \le t \\ t\left(\dfrac{x}{k+1} - 1\right), & t \le x \le 1. \end{cases}$

11 $G(x, t) = \begin{cases} -\frac{1}{2}\sin 2x \cos 2t, & 0 \le x \le t \\ -\frac{1}{2}\sin 2t \cos 2x, & t \le x \le \pi \end{cases}$

$y = \frac{1}{4}(1 - \cos 2x)$.

13 Both results follow by differentiating the expressions for $G(x, t)$ given in (18), followed either by substitution into the differential equation, or by differencing G' across $t = x$.

15 Use Theorem 3.18 to evaluate the derivatives of G and then substituting into the differential equation to establish that G is a solution. The first $n - 1$ initial conditions follow by differentiation of G with respect to x and then setting $x = t = x_0$, when two rows of the resulting determinant will always be equal. The last initial condition follows because evaluating $G^{(n-1)}$ and setting $x = t = x_0$ causes the determinant to become $(-1)^{n-1} W(x_0)$.

Answers for Section 6.1

1

$$C = \begin{bmatrix} 1 & -3 \\ 2 & 1 \end{bmatrix}, \quad D = \begin{bmatrix} 2 & -1 \\ 4 & -3 \end{bmatrix}, \quad \mathbf{x} = \begin{bmatrix} x_1 \\ x_2 \end{bmatrix}.$$

3

$$C = \begin{bmatrix} 2 & 4 & 7 \\ 0 & 1 & -1 \\ 1 & 1 & 0 \end{bmatrix}, \quad D = \begin{bmatrix} 2 & 3 & -1 \\ 1 & -1 & 0 \\ 1 & 0 & 4 \end{bmatrix}, \quad \mathbf{x} = \begin{bmatrix} x_1 \\ x_2 \\ x_3 \end{bmatrix}.$$

5

$$A = \begin{bmatrix} 0 & 1 \\ 1 & -7 \end{bmatrix}, \quad \mathbf{x} = \begin{bmatrix} x_1 \\ x_2 \end{bmatrix}$$

7

$$A = \begin{bmatrix} 0 & 1 \\ \frac{3}{4} & -\frac{1}{2} \end{bmatrix}, \quad \mathbf{x} = \begin{bmatrix} x_1 \\ x_2 \end{bmatrix}$$

9

$$A = \begin{bmatrix} 0 & 1 & 0 \\ 0 & 0 & 1 \\ -\frac{1}{2} & \frac{1}{2} & -\frac{3}{2} \end{bmatrix}, \quad \mathbf{x} = \begin{bmatrix} x_1 \\ x_2 \\ x_3 \end{bmatrix}$$

11

$$A = \begin{bmatrix} 1 & 5 \\ 3 & 11 \end{bmatrix}, \quad \mathbf{x} = \begin{bmatrix} x_1 \\ x_2 \end{bmatrix}.$$

13

$$A = \begin{bmatrix} 0 & \frac{1}{2} & 0 \\ \frac{1}{2} & -\frac{1}{4} & \frac{1}{2} \\ \frac{1}{2} & -\frac{1}{4} & \frac{1}{2} \end{bmatrix}, \quad \mathbf{x} = \begin{bmatrix} x_1 \\ x_2 \\ x_3 \end{bmatrix}$$

15

$$\mathbf{x} = c_1 \begin{bmatrix} -6 \\ 5 \end{bmatrix} e^{-4t} + c_2 \begin{bmatrix} 1 \\ 1 \end{bmatrix} e^{7t}; \quad x_1 = \tfrac{1}{11}(18e^{-4t} + 4e^{7t}), \quad x_2 = \tfrac{1}{11}(4e^{7t} - 15e^{-4t}).$$

17

$$\mathbf{x} = c_1 \begin{bmatrix} 0 \\ 1 \end{bmatrix} e^{-9t} + c_2 \begin{bmatrix} 14 \\ 1 \end{bmatrix} e^{5t}; \quad x_1 = e^{5t}, \quad x_2 = \tfrac{1}{14}(27e^{-9t} + e^{5t}).$$

19

$$\mathbf{x} = c_1 \begin{bmatrix} 3 \\ 2 \\ -1 \end{bmatrix} + 2\mathrm{Re} \left((c_2 + ic_3) \begin{bmatrix} 1 \\ \frac{2}{5}(2+i) \\ 0 \end{bmatrix} e^{-it} \right)$$

or $x_1 = 3c_1 + 2c_2 \cos t + 2c_3 \sin t$
$x_2 = 2c_1 + \frac{4}{5}(2c_2 - c_3) \cos t + \frac{4}{5}(2c_3 + c_2) \sin t$
$x_3 = -c_1$.

21

$$\mathbf{x} = c_1 \begin{bmatrix} 1 \\ 2 \\ 0 \end{bmatrix} e^t + c_2 \begin{bmatrix} 2 \\ -1 \\ 2 \end{bmatrix} e^t + c_3 \begin{bmatrix} 1 \\ 2 \\ -1 \end{bmatrix} e^{2t}; \; x_1 = \tfrac{3}{5}(e^{2t} - e^t), \; x_2 = \tfrac{1}{5}(6e^{2t} - e^t),$$

$$x_3 = -\tfrac{1}{5}(3e^{2t} + 2e^t).$$

29

$$\mathbf{x} = c_1 \begin{bmatrix} 0 \\ 1 \end{bmatrix} e^{4t} + c_2 \begin{bmatrix} 1 \\ 0 \end{bmatrix} e^{4t} + c_2 \begin{bmatrix} 0 \\ 1 \end{bmatrix} t e^{4t}.$$

31 $x_1 - 3c_2 \quad \tfrac{7}{3}c_3 + c_1 e^{5t} + 3c_3 t$
$x_2 = c_3 - 5c_2 - 5c_3 t$
$x_3 = 5c_2 + \tfrac{2}{3}c_3 + 5c_3 t.$

37 $\mathbf{Be^A} = \mathbf{B}\left(\mathbf{I} + \dfrac{1}{1!}\mathbf{A} + \dfrac{1}{2!}\mathbf{A}^2 + \cdots \right)$

$$= \mathbf{B} + \dfrac{1}{1!}\mathbf{BA} + \dfrac{1}{2!}\mathbf{BA}^2 + \cdots.$$

Now if $\mathbf{A}$ and $\mathbf{B}$ commute $\mathbf{BA} = \mathbf{AB}$, $\mathbf{BA}^2 = \mathbf{ABA} = \mathbf{A}^2\mathbf{B}, \ldots, \mathbf{BA}^m = \mathbf{A}^m\mathbf{B}$, and thus

$$\mathbf{Be^A} = \left(\mathbf{I} + \dfrac{1}{1!}\mathbf{A} + \dfrac{1}{2!}\mathbf{A}^2 + \cdots \right)\mathbf{B} = \mathbf{e^A B}.$$

39 $\dfrac{d}{dt}(e^{t\mathbf{A}}) = \dfrac{d}{dt}\left(\mathbf{I} + \dfrac{t}{1!}\mathbf{A} + \dfrac{t^2}{2!}\mathbf{A}^2 + \dfrac{t^3}{3!}\mathbf{A}^3 + \cdots \right)$

$$= \mathbf{A} + \dfrac{t}{1!}\mathbf{A}^2 + \dfrac{t^2}{2!}\mathbf{A}^3 + \cdots = \mathbf{A}e^{t\mathbf{A}}.$$

The termwise differentiation may be justified by giving due consideration to the convergence of the series.

41 $d\mathbf{M}/dt = (\mathbf{A} + \mathbf{B})\exp[t(\mathbf{A} + \mathbf{B})]\exp(-t\mathbf{B})\exp(-t\mathbf{A}) - \exp[t(\mathbf{A} + \mathbf{B})]\,\mathbf{B}\exp(-t\mathbf{B})\exp(-t\mathbf{A})$
$\quad - \exp[t(\mathbf{A} + \mathbf{B})]\exp(-t\mathbf{B})\mathbf{A}\exp(-t\mathbf{A})$
Using the result of Prob. 37 shows

$$d\mathbf{M}/dt = (\mathbf{A} + \mathbf{B})\exp[t(\mathbf{A} + \mathbf{B})]\exp(-t\mathbf{B})\exp(-t\mathbf{A}) - (\mathbf{A} + \mathbf{B})$$

$$\times \exp[t(\mathbf{A} + \mathbf{B})]\exp(-t\mathbf{B})\exp(-t\mathbf{A}) = \mathbf{0}.$$

As $\mathbf{M}(0) = e^0 e^0 = \mathbf{I}$ it follows that $\mathbf{M}(t) = \mathbf{I}$, and so

$$\exp[t(\mathbf{A} + \mathbf{B})]\exp(-t\mathbf{B})\exp(-t\mathbf{A}) = \mathbf{I}.$$

Post-multiplying first by $e^{t\mathbf{A}}$, and then by $e^{t\mathbf{B}}$, gives

$$\exp[t(\mathbf{A} + \mathbf{B})] = \exp(t\mathbf{A})\exp(t\mathbf{B}).$$

Setting $t = 1$ then gives

$$\exp(\mathbf{A} + \mathbf{B}) = \exp(\mathbf{A})\exp(\mathbf{B}).$$

Commutativity is necessary because the result of Prob. 37 has been used.

43 $\quad \mathbf{B}(t) = \int_{t_0}^{t} \mathbf{A}\,ds = \left[\int_{t_0}^{t} a_{ij}\,ds\right] = [(t-t_0)a_{ij}] = (t-t_0)\mathbf{A}.$

Thus $\mathbf{B}^m = (t-t_0)^m \mathbf{A}^m$, so that

$$\frac{d}{dt}\mathbf{B}^m = m(t-t_0)^{m-1}\mathbf{A}^m = m\mathbf{A}[(t-t_0)^{m-1}\mathbf{A}^{m-1}]$$

$$= m\mathbf{A}\mathbf{B}^{m-1}, \qquad m=1, 2, \ldots .$$

45 As $m \to \infty$ so $\mathbf{x}^{(m)} \to e^{\mathbf{B}}\mathbf{x}_0$, so in the limit

$$\mathbf{x} = e^{\mathbf{B}}\mathbf{x}_0 = \exp[(t-t_0)\mathbf{A}]\mathbf{x}_0.$$

To give a proper justification of the result it is necessary to establish the convergence of the series involved in $\mathbf{x}^{(m)}$.

47 Setting $k = -3$ in the first result of Prob. 38 shows

$$e^{t\mathbf{A}} = \begin{bmatrix} \cos 3t & \sin 3t \\ -\sin 3t & \cos 3t \end{bmatrix},$$

and the solution then follows from $\mathbf{x} = e^{t\mathbf{A}}\mathbf{x}_0$.

49 Here

$$\mathbf{A} = \begin{bmatrix} 4 & -5 \\ 5 & 3 \end{bmatrix},$$

so $\mathbf{A} = \mathbf{B} + \mathbf{C}$, with

$$\mathbf{B} = \begin{bmatrix} 4 & 0 \\ 0 & 3 \end{bmatrix}$$

and

$$\mathbf{C} = \begin{bmatrix} 0 & -5 \\ 5 & 0 \end{bmatrix}.$$

Matrices $\mathbf{B}$ and $\mathbf{C}$ do not commute, so $\exp[t(\mathbf{A}+\mathbf{B})] \neq \exp(t\mathbf{A})\exp(t\mathbf{B})$.

51

$$\mathbf{A} = \begin{bmatrix} 1 & 0 & 0 \\ 1 & 3 & 2 \\ 1 & 2 & 3 \end{bmatrix}, \quad \mathbf{P} = \begin{bmatrix} 0 & -2 & -2 \\ 1 & 1 & 0 \\ 1 & 0 & 1 \end{bmatrix}, \quad \mathbf{P}^{-1} = \begin{bmatrix} \frac{1}{4} & \frac{1}{2} & \frac{1}{2} \\ -\frac{1}{4} & \frac{1}{2} & -\frac{1}{2} \\ -\frac{1}{4} & -\frac{1}{2} & \frac{1}{2} \end{bmatrix},$$

$$\mathbf{\Lambda} = \begin{bmatrix} 5 & 0 & 0 \\ 0 & 1 & 0 \\ 0 & 0 & 1 \end{bmatrix}. \quad \text{Thus } e^{t\mathbf{A}} = \mathbf{P}\begin{bmatrix} e^{5t} & 0 & 0 \\ 0 & e^{t} & 0 \\ 0 & 0 & e^{t} \end{bmatrix}\mathbf{P}^{-1}$$

and so $\mathbf{x} = \mathbf{P}\begin{bmatrix} e^{5t} & 0 & 0 \\ 0 & e^{t} & 0 \\ 0 & 0 & e^{t} \end{bmatrix}\mathbf{P}^{-1}\begin{bmatrix} 1 \\ 0 \\ 1 \end{bmatrix}.$

Thus $x_1 = e^{t}$, $x_2 = \frac{3}{4}(e^{5t}-e^{t})$, $x_3 = \frac{1}{4}(3e^{5t}+e^{t})$.

Answers for Section 6.2

1 $\lambda_1 = 3$, $\lambda_2 = -1$, $\mathbf{P} = \begin{bmatrix} 1 & 1 \\ 1 & -3 \end{bmatrix}$, $\mathbf{P}^{-1} = \begin{bmatrix} \frac{3}{4} & \frac{1}{4} \\ \frac{1}{4} & -\frac{1}{4} \end{bmatrix}$,

$\mathbf{\Lambda} = \begin{bmatrix} 3 & 0 \\ 0 & -1 \end{bmatrix}$; $x_1 = c_1 e^{3t} + c_2 e^{-t} - \frac{2}{3}$,

$x_2 = c_1 e^{3t} - 3c_2 e^{-t} + \frac{4}{3}$; $c_1 = \frac{2}{3}$, $c_2 = 1$.

3 $\lambda_1 = 1$, $\lambda_2 = -2$, $\mathbf{P} = \begin{bmatrix} 2 & 3 \\ 1 & 2 \end{bmatrix}$, $\mathbf{P}^{-1} = \begin{bmatrix} 2 & -3 \\ -1 & 2 \end{bmatrix}$,

$\mathbf{\Lambda} = \begin{bmatrix} 1 & 0 \\ 0 & -2 \end{bmatrix}$; $x_1 = 2c_1 e^t + 3c_2 e^{-2t} + 4te^t - e^t$,

$x_2 = c_1 e^t + 2c_2 e^{-2t} + 2te^t - \frac{2}{3}e^t$; $c_1 = 0$, $c_2 = \frac{1}{3}$.

5 $\lambda_1 = -2$, $\lambda_2 = -2$, $\mathbf{Q}_1 = \begin{bmatrix} 1 & 1 \\ -2 & -1 \end{bmatrix}$, $\mathbf{Q}_1^{-1} = \begin{bmatrix} -1 & -1 \\ 2 & 1 \end{bmatrix}$,

$\mathbf{T}_1 = \begin{bmatrix} -2 & 1 \\ 0 & -2 \end{bmatrix}$; $x_1 = c_1 e^{-2t} + c_2(1+t)e^{-2t} + \frac{15}{4}$,

$x_2 = -2c_1 e^{-2t} - c_2(1+2t)e^{-2t} - 4$.

7

$\lambda_1 = -1$, $\lambda_2 = -2$, $\lambda_3 = -3$, $\mathbf{P} = \begin{bmatrix} 1 & 1 & 1 \\ -1 & 0 & -1 \\ 0 & 1 & 2 \end{bmatrix}$, $\mathbf{P}^{-1} = \begin{bmatrix} \frac{1}{2} & -\frac{1}{2} & -\frac{1}{2} \\ 1 & 1 & 0 \\ -\frac{1}{2} & -\frac{1}{2} & \frac{1}{2} \end{bmatrix}$

$\mathbf{\Lambda} = \begin{bmatrix} -1 & 0 & 0 \\ 0 & -2 & 0 \\ 0 & 0 & -3 \end{bmatrix}$; $x_1 = -e^{-2t}$, $x_2 \equiv 0$, $x_3 = 1 - e^{-2t}$.

9 $\lambda_1 = -1$, $\lambda_2 = -1$, $\lambda_3 = -1$, $\mathbf{Q}_2 = \begin{bmatrix} 1 & 2 & 2 \\ -1 & 0 & 0 \\ 1 & -2 & 0 \end{bmatrix}$,

$\mathbf{Q}_2^{-1} = \begin{bmatrix} 0 & -1 & 0 \\ 0 & -\frac{1}{2} & -\frac{1}{2} \\ \frac{1}{2} & 1 & \frac{1}{2} \end{bmatrix}$, $\mathbf{T}_2 = \begin{bmatrix} -1 & 2 & 0 \\ 0 & -1 & 1 \\ 0 & 0 & -1 \end{bmatrix}$;

$x_1 = (c_1 + 2c_2 + 2c_3)e^{-t} + 2(c_2 + c_3)te^{-t} + c_3 t^2 e^{-t} + 14$,

$x_2 = -c_1 e^{-t} - 2c_2 te^{-t} - c_3 t^2 e^{-t}$,

$x_3 = (c_1 - 2c_2)e^{-t} + 2(c_2 - c_3)te^{-t} + c_3 t^2 e^{-t} - 4$.

11 (a) If rank $\mathbf{C} = r(<n)$, only r of the left-hand sides of the system of n equations will be linearly independent. If rank $\{\mathbf{C}|\mathbf{D}\} \neq r$ the system is inconsistent; so only the trivial solution $\mathbf{x} \equiv \mathbf{0}$ is possible. If rank $\{\mathbf{C}|\mathbf{D}\} = r$ the equations are consistent, but there are only r linearly independent differential equations for the n variables $x_1, x_2, \ldots, x_n$.

(b) If rank $\mathbf{C} \neq$ rank $\{\mathbf{C}|\mathbf{D}|\mathbf{F}\}$ the equations are inconsistent and no solution exists. If rank $\mathbf{C} =$ rank $\{\mathbf{C}|\mathbf{D}|\mathbf{F}\}$ the equations are consistent but there are only r linearly independent equations for the n variables $x_1, x_2, \ldots, x_n$.

13 $\dfrac{d}{dt}(e^{-tA}x) = -Ae^{-tA}\dot{x} + e^{-tA}x$, but $Ae^{-tA} = e^{-tA}A$ (Prob. 37, Sec. 6.1) so the first result then follows. The first result is simply the left-hand side of the matrix differential equation premultiplied by e^{-tA}, and so equals $e^{-tA}f$.

15 $\quad x(t) = \exp(tA)x_0 + \displaystyle\int_0^t \exp[(t-\tau)A]C\,d\tau$

$\quad\quad\quad = e^{tA}x_0 + e^{tA}\displaystyle\int_0^t e^{-\tau A}C\,d\tau$

$\quad\quad\quad = e^{tA}[x_0 - e^{-tA}A^{-1}C + A^{-1}C]$

$\quad\quad\quad = e^{tA}[x_0 + A^{-1}C] - A^{-1}C.$

17 $x_1 = \frac{12}{5}e^{-2t} - 10e^t - \frac{16}{5}\sin t - \frac{7}{5}\cos t + 18t + 9$

$\quad x_2 = \frac{8}{5}e^{-2t} - 5e^t - \frac{9}{5}\sin t - \frac{3}{5}\cos t + 10t + 4.$

Answers for Section 6.3

1

$$x(t) = c_1\begin{bmatrix} 1 \\ 0 \\ -1 \end{bmatrix}\cosh(t+\alpha_1) + c_2\begin{bmatrix} 0 \\ 1 \\ -1 \end{bmatrix}\cosh(t+\alpha_2) + c_3\begin{bmatrix} 1 \\ 2 \\ 1 \end{bmatrix}\cosh(\sqrt{5}t+\alpha_3).$$

3 Natural angular frequencies $\omega_1 = n$, $\omega_2 = \sqrt{3}n$. Normal modes:

$$x^{(1)}(t) = \begin{bmatrix} 1 \\ 1 \end{bmatrix}\cos(nt+\alpha_1), \quad x^{(2)}(t) = \begin{bmatrix} 1 \\ -1 \end{bmatrix}\cos(\sqrt{3}nt+\alpha_2).$$

$$x_1(t) = \frac{X}{2n}\left(3\sin nt - \frac{\sqrt{3}}{3}\sin\sqrt{3}nt\right)$$

$$x_2(t) = \frac{X}{2n}\left(3\sin nt + \frac{\sqrt{3}}{3}\sin\sqrt{3}nt\right).$$

5 Natural angular frequencies $\omega_1 = 1$, $\omega_2 = \sqrt{2}$, $\omega_3 = 2$.

$$\text{Normal modes: } x^{(1)}(t) = \begin{bmatrix} 1 \\ -2 \\ 3 \end{bmatrix}\cos(t+\alpha_1), \quad x^{(2)}(t) = \begin{bmatrix} 1 \\ 0 \\ -1 \end{bmatrix}\cos(\sqrt{2}t+\alpha_2),$$

$$x^{(3)}(t) = \begin{bmatrix} 0 \\ 1 \\ -1 \end{bmatrix}\cos(2t+\alpha_3).$$

General solution $x_1(t) = c_1\cos(t+\alpha_1) + c_2\cos(\sqrt{2}t+\alpha_2)$, $x_2(t) = -2c_1\cos(t+\alpha_1) + c_3\cos(2t+\alpha_3)$, $x_3(t) = 3c_1\cos(t+\alpha_1) - c_2\cos(\sqrt{2}t+\alpha_2) - c_3\cos(2t+\alpha_3)$.

7 Normal modes: $\mathbf{x}^{(1)}(t) = \begin{bmatrix} 1 \\ -1 \end{bmatrix} \cos(2pt + \alpha_1)$, $\mathbf{x}^{(2)}(t) = \begin{bmatrix} 1 \\ 1 \end{bmatrix} \cos(4pt + \alpha_2)$. First mode is SHM along $x + y = 0$ with angular frequency $2p$. Second mode is SHM along $x - y = 0$ with angular frequency $4p$.

9 Natural frequencies of vibration $f_1 = \omega_1/2\pi, f_2 = \omega_2/2\pi, f_3 = \omega_3/2\pi$ cycles per unit time, where ω_1, ω_2 and ω_3 are the angular frequencies

$$\omega_1 = \sqrt{\frac{2T}{ml}}, \quad \omega_2 = \sqrt{\frac{(2 - \sqrt{2})T}{ml}}, \quad \omega_3 = \sqrt{\frac{(2 + \sqrt{2})T}{ml}}.$$

Normal modes:

$$\mathbf{x}^{(1)}(t) = \begin{bmatrix} 1 \\ 0 \\ -1 \end{bmatrix} \cos(\omega_1 t + \alpha_1), \quad \mathbf{x}^{(2)}(t) = \begin{bmatrix} 1 \\ \sqrt{2} \\ 1 \end{bmatrix} \cos(\omega_2 t + \alpha_2)$$

$$\mathbf{x}_3(t) = \begin{bmatrix} 1 \\ -\sqrt{2} \\ 1 \end{bmatrix} \cos(\omega_3 t + \alpha_3).$$

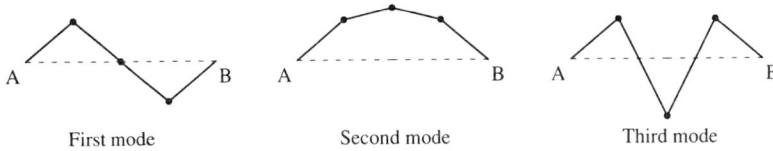

First mode   Second mode   Third mode

11 $x_1(t) = K\left(\dfrac{(2\omega^2 - \Omega^2)}{(3\omega^2 - \Omega^2)(\omega^2 - \Omega^2)} \sin \Omega t - \dfrac{\Omega \sin \omega t}{2\omega(\omega^2 - \Omega^2)} - \dfrac{\Omega \sin \sqrt{3}\omega t}{2\sqrt{3}\omega(3\omega^2 - \Omega^2)} \right)$

$x_2(t) = K\left(\dfrac{\omega^2}{(3\omega^2 - \Omega^2)(\omega^2 - \Omega^2)} \sin \Omega t - \dfrac{\Omega \sin \omega t}{2\omega(\omega^2 - \Omega^2)} + \dfrac{\Omega \sin \sqrt{3}\omega t}{2\sqrt{3}\omega(3\omega^2 - \Omega^2)} \right)$.

15 $x_1(t) = c_1 + c_3 e^t + c_4 e^{-t} - 2t$,

$x_2(t) = c_2 + c_3 e^t - c_4 e^{-t} - \frac{1}{2}t^2$,

and, although not required,

$x_3(t) = c_3 e^t - c_4 e^{-t} - 2, \quad x_4(t) = c_3 e^t + c_4 e^{-t} - t.$

17 $y = x_2 = -3c_1 \exp(-3t) + \exp(-\frac{1}{2}t)\left((c_3 - \sqrt{7}c_2)\sin\dfrac{\sqrt{7}}{2}t - (c_2 + \sqrt{7}c_3)\cos\dfrac{\sqrt{7}}{2}t \right)$

$x_1 = x = c_1 \exp(-3t) + 2\exp(-\frac{1}{2}t)\left(c_2 \cos\dfrac{\sqrt{7}}{2}t - c_3 \sin\dfrac{\sqrt{7}}{2}t \right)$,

$x_3 = \dot{x}_2 = 9c_1 \exp(-3t) + \exp(-\frac{1}{2}t)\left((3c_3 + \sqrt{7}c_2)\sin\dfrac{\sqrt{7}}{2}t - (3c_2 - \sqrt{7}c_3)\cos\dfrac{\sqrt{7}}{2}t \right).$

The solution exhibits a typical damped oscillatory behavior superimposed on a pure exponential decay.

19 Diagonalize $\mathbf{B}$ by setting $\mathbf{B} = \mathbf{P}\mathbf{\Lambda}_B\mathbf{P}^{-1}$. Then $\mathbf{B}^2 = \mathbf{A} = \mathbf{P}\mathbf{\Lambda}_B^2\mathbf{P}^{-1}$, which shows $\mathbf{\Lambda}_B^2$ is the diagonalized form of $\mathbf{A}$. Thus if the diagonal elements of $\mathbf{\Lambda}_A$ are λ_{A1}, λ_{A2} and λ_{A3}, those of $\mathbf{\Lambda}_B$ are $\sqrt{\lambda_{A1}}$, $\sqrt{\lambda_{A2}}$ and $\sqrt{\lambda_{A3}}$. Diagonalizing matrix for $\mathbf{A}$ is

$$\mathbf{P} = \begin{bmatrix} 0 & 3 & 0 \\ -2 & 11 & 2 \\ 1 & -5 & 1 \end{bmatrix}, \quad \text{when } \mathbf{\Lambda}_A = \begin{bmatrix} 0 & 0 & 0 \\ 0 & 1 & 0 \\ 0 & 0 & 4 \end{bmatrix}.$$

$$\mathbf{\Lambda}_B = \begin{bmatrix} 0 & 0 & 0 \\ 0 & 1 & 0 \\ 0 & 0 & 2 \end{bmatrix} \quad \text{and so} \quad \mathbf{B} = \tfrac{1}{72}\begin{bmatrix} 6 & 0 & 0 \\ 20 & 6 & 12 \\ -11 & 3 & 6 \end{bmatrix}.$$

Answers for Section 6.4

1 (a) Autonomous; linear   (b) nonautonomous; linear
(c) autonomous; nonlinear   (d) autonomous; linear
(e) nonautonomous; nonlinear   (f) autonomous; nonlinear
5 (a) $(-2, 0)$, $(2, 0)$ (b) $(n\pi, 0)$, $n = 0, \pm 1, \pm 2, \ldots$.
7 Hyperbolas

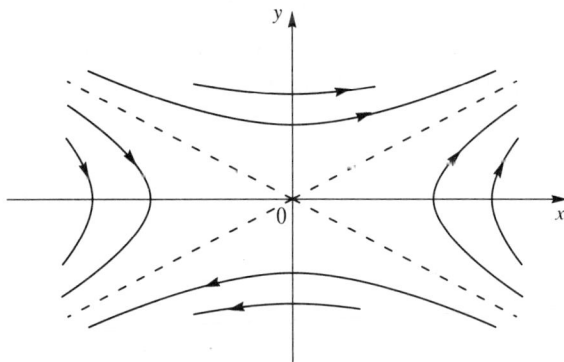

9 Spirals directed into the origin.

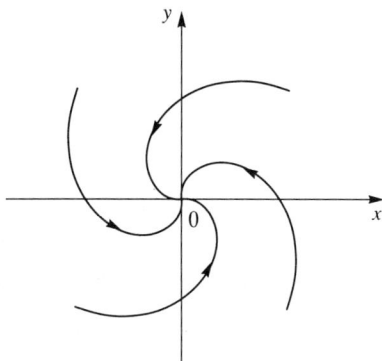

11 $(n\pi, 0)$, $n=0, \pm1, \pm2, \ldots$ and $(\text{arc } \cos(3g/2l\Omega^2), 0)$;

$$\tfrac{1}{2}\dot\theta^2 - \frac{3g}{2l}\sin\theta + \tfrac{1}{4}\Omega^2\cos^2\theta = \text{const.}$$

17 (a) $\dfrac{d^2\theta}{dt^2} + \left(\dfrac{3g}{2l} - \Omega^2\right)\theta = 0$; $\sqrt{\dfrac{3g}{2l} - \Omega^2}$;

(b) $\dfrac{d^2\Theta}{dt^2} + \Omega^2\sin^2\theta_0\,\Theta = 0$, with $\Theta = \theta - \theta_0$; $\sqrt{\Omega^2 - \left(\dfrac{3g}{2l\Omega}\right)^2}$.

21 Overdamped pendulum without an oscillatory solution. Equilibrium points on the x-axis at $x = \pm n\pi$, $n=0, 1, 2, \ldots$. Points at even multiples of π are stable nodes and points at odd multiples of π are saddle points.

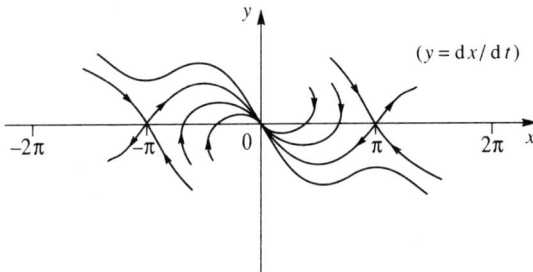

23 Equilibrium points on the x-axis at $x = \pm n\pi$, $n=0, 1, \ldots$. Points at even multiples of π are stable spirals and points at odd multiples of π are saddle points.

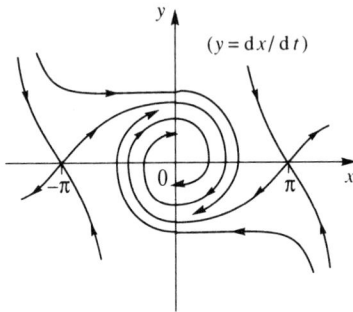

25 Equilibrium in fourth cycle at $x=0.04$, $y=0$.

27 Equilibrium in second cycle at $x=-0.5$, $y=0$.

29 Equilibrium in second cycle at $x=0$, $y=0$.

31 The trajectories are $\tfrac{1}{2}y^2 + hx = c_1$ for $x>0$ and $\tfrac{1}{2}y^2 - x = c_2$ for $x<0$. They form families of parabolas with the x-axis as their axis of symmetry. The equation of the trajectory subject to the given initial conditions is $x = A - \tfrac{1}{2}ht^2\,(x\geq0)$. Thus x first returns to zero after a time $\sqrt{2A/h}$. Symmetry shows this to be a quarter period, so the period is $4\sqrt{2A/h}$. The motion is nonisochronous, because the period depends on the amplitude.

33 (a) No, (b) Yes, (c) Yes.

35 For asymptotic stability we must have

$$a > 0, \quad \begin{vmatrix} a & 1 \\ 0 & b \end{vmatrix} > 0, \quad \begin{vmatrix} a & 1 & 0 \\ 0 & b & a \\ 0 & c & 0 \end{vmatrix} > 0,$$

$$\begin{vmatrix} a & 1 & 0 & 0 \\ 0 & b & a & 1 \\ 0 & c & 0 & b \\ 0 & 0 & 0 & c \end{vmatrix} > 0, \text{ or } a > 0, \ ab > 0, \ -a^2 c > 0, \ -a^2 c^2 > 0.$$

The last condition is impossible, so the general solution cannot be asymptotically stable.

37 Yes.

39 No if $0 < k < 0.128$, because $P(x)$ has more than one positive root; Yes if $0.128 < k < 1$; No if $k \geq 1$, because $P(x)$ has no positive roots.

41 Stable attractive spiral.   **43** Saddle point.

45 Unstable improper node with an exceptional direction.

47 Unstable spiral.

49 Improper node without an exceptional direction.

51 (a) $(0, 0)$; center   (b) $(0, 0)$; saddle point   (c) $(0, 0)$; center, $(0, \pm\sqrt{-a/b})$; saddle points.

53 Equilibrium points at all nine combinations of $x = 0$, $y = 0$, $x = \pm 1$, $y = \pm 1$. Equilibrium points at $(0, 0)$, $(\pm 1, \pm 1)$ are saddle points and those at $(0, \pm 1)$, $(\pm 1, 0)$ are centers.

55 $(0, 0)$ a center, $(-1, 0)$ a saddle point, $(0.5, 0)$ a saddle point.

57 Unstable spiral.   **59** Stable attractive spiral.

61 Consider the annular region $\frac{1}{2} \leq r \leq 2$; $dr/dt = r(1 - r^2)$, $d\theta/dt = 1$, $r = (1 + ae^{-2t})^{-1/2}$, $\theta = t + t_0$; limit cycle is circle $r = 1$.

63 Consider the annular region $\frac{1}{2} \leq r \leq 2$; $dr/dt = 1 - r^2$, $d\theta/dt = -1$, $r = (ae^{2t} - 1)/(1 + ae^{2t})$, $\theta = -t + t_0$; limit cycle is circle $r = 1$.

65 Flow comprises two families of coaxial circles all passing through the points $(0, \pm a)$ which are nodes. The node at $(0, a)$ is unstable and the one at $(0, -a)$ is stable and attractive. In terms of a flow this represents a flow from the unstable to the stable node.

67 (a) Four elliptic sectors forming a rose. Separatrices are the axes.

(b) Two elliptic and two hyperbolic sectors. Separatrices are the axes.

(c) Two elliptic sectors, two hyperbolic sectors and two parabolic sectors. Separatrices are the three radial lines.

(d) One elliptic, two parabolic and three hyperbolic sectors. Separatrices are three radial lines.

(e) Two elliptic sectors forming a dipole (rose with two sectors). Separatrix is the common tangent line.

(f) One center and three hyperbolic sectors. Separatrices are the axes. This is the full phase portrait for the predator–prey equations when x and y are allowed to be both positive (physical case) and negative (nonphysical case).

Answers for Section 6.5

1

n	x_n	k_{1n}	K_{1n}	k_{2n}	K_{2n}
0	0.0	0.112763	-0.104219	0.112986	-0.074679
1	0.1	0.114754	0.051594	0.116589	-0.028615
2	0.2	0.120104	-0.009621	0.123964	0.009469
3	0.3	0.130063	0.026173	0.136964	0.043069
4	0.4	0.147293	0.059011	0.159253	0.075226

n	x_n	k_{3n}	K_{3n}	k_{4n}	K_{4n}
0	0.0	0.113781	-0.077836	0.114709	-0.051281
1	0.1	0.117342	-0.030695	0.120068	-0.009398
2	0.2	0.124812	0.008131	0.130036	0.026337
3	0.3	0.138087	0.042414	0.147281	0.059124
4	0.4	0.161015	0.075397	0.177234	0.092312

n	x_n	k_n	K_n	y_n	z_n
0	0.0	0.113501	-0.076755	0.000000	0.500000
1	0.1	0.117114	-0.029935	0.113501	0.423245
2	0.2	0.124615	0.008653	0.230615	0.393310
3	0.3	0.137907	0.042711	0.355231	0.401962
4	0.4	0.160844	0.075428	0.493138	0.444673

3

n	x_n	k_{1n}	K_{1n}	k_{2n}	K_{2n}
0	0.0	0.200000	1.200000	0.340000	1.102000
1	0.2	0.499559	1.159471	0.665462	1.234406
2	0.4	0.894536	1.499631	1.133952	1.841861
3	0.6	1.518165	2.549064	1.924888	3.492288
4	0.8	2.672347	5.404922	3.480073	8.073531

n	x_n	k_{3n}	K_{3n}	k_{4n}	K_{4n}
0	0.0	0.344200	1.181380	0.505116	1.146503
1	0.2	0.689546	1.306911	0.898851	1.498757
2	0.4	1.192117	1.942800	1.521519	2.571740
3	0.6	2.059883	3.715989	2.673339	5.498780
4	0.8	3.827707	8.795550	5.196998	14.483573

n	x_n	k_n	K_n	y_n	z_n
0	0.0	0.345586	1.152210	2.000000	-1.000000
1	0.2	0.684738	1.290144	2.345586	0.152210
2	0.4	1.178032	1.940115	3.030324	1.442354
3	0.6	2.026841	3.744066	4.208356	3.382469
4	0.8	3.747484	8.937776	6.235197	7.126536

5

n	x_n	k_{1n}	K_{1n}	k_{2n}	K_{2n}
0	0.0	0.000000	0.400000	0.040000	0.380000
1	0.2	0.095704	0.380971	0.154258	0.383331
2	0.4	0.248652	0.416777	0.353584	0.458354
3	0.6	0.558385	0.574778	0.809187	0.730625
4	0.8	1.471927	1.203710	2.438887	1.949928

n	x_n	k_{3n}	K_{3n}	k_{4n}	K_{4n}
0	0.0	0.046080	0.390080	0.096873	0.380841
1	0.2	0.167462	0.396063	0.250514	0.417356
2	0.4	0.386290	0.482745	0.561975	0.577003
3	0.6	0.921752	0.812021	1.477506	1.209091
4	0.8	3.187579	2.549377	7.211467	5.963499

n	x_n	k_n	K_n	y_n	z_n
0	0.0	0.044839	0.386833	1.000000	-1.000000
1	0.2	0.164943	0.392852	1.044839	-0.613167
2	0.4	0.381729	0.479330	1.209782	-0.220314
3	0.6	0.916295	0.811527	1.591511	0.259015
4	0.8	3.322721	2.694303	2.507806	1.070543

Answers for Section 7.1

1 $(s+2)/s^2$

3 $\dfrac{h}{s}(1-e^{-as})$

5 $1/(s-a+ik)$

7 $2/(s^2-2s-3)$

9 e^{-s}/s^2

11 $2e^{-2s}/s^3$

13 $(1-e^{-s}-se^{-s})/s^2$

15 $(1-e^{-s})^2/s^2$

17 $2k^3/(s^4-k^4)$

19 $k^2/\{s(s^2+k^2)\}$

21 $2k^2s/(s^4-k^4)$

23 $\dfrac{1}{(s-1)^3}+\dfrac{1}{s^2+2s-8}$

25 $3e^{-4t}$

27 $\frac{1}{12}t^4$

29 $\frac{1}{4}t\sin 3t$

31 $3e^{2t}\cos 3t$

33 $\frac{1}{4}e^{-3t}\sin 4t$

35 $\frac{1}{6}t^3e^{-t}-\frac{1}{4}\cos 2t$

37 (a) $\alpha=-3$ (b) $\alpha=0$ (c) No (d) any $\alpha>0$ (e) $\alpha=k$ (f) No (g) any $\alpha>5$ (h) $\alpha=0$

Answers for Section 7.2

1 $\dfrac{s-3}{(s-3)^2+1}$ **3** $\dfrac{1}{(s+4)^2}$ **5** $\dfrac{2}{(s+3)^2-4}$

7 $te^t\cos 2t$ **9** $3e^{-3t}\cosh 4t$

11 $e^t(1+2\sin 2t)$ **13** $2te^t+e^{2t}\sin t$

15
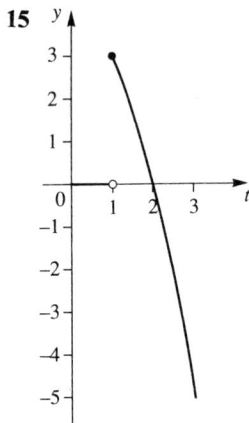

17
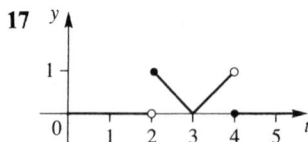

19
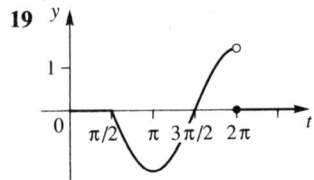

21
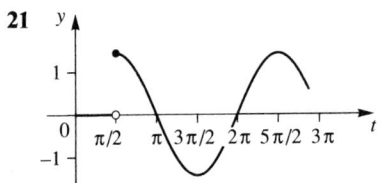

23
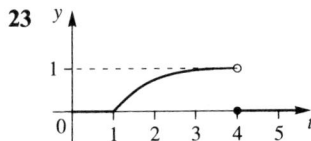

25 $F(s) = \dfrac{6e^{-2s}}{s^3}$ **27** $F(s) = \dfrac{2se^{-s}}{s^2-9}$

29 $\mathscr{L}\{\mathscr{U}(t-1)t^2\} = \mathscr{L}\{\mathscr{U}(t-1)(t-1)^2\} + 2\mathscr{L}\{\mathscr{U}(t-1)(t-1)\} + \mathscr{L}\{\mathscr{U}(t-1)\} = \dfrac{2e^{-s}}{s^3} + \dfrac{2e^{-s}}{s^2} + \dfrac{e^{-s}}{s}$

31 $f(t) = \mathscr{U}(t-2)\{\frac{1}{2}(t-2)^2 \exp[2(t-2)]\}$

33 $f(t) = \mathscr{U}(t-3)\{\exp[-2(t-3)] \cos 4(t-3)\}$

35 $f(t) = \mathscr{U}(t-1)\exp[2(t-1)] - 4\mathscr{U}(t-2)\exp[3(t-2)]$

37 $f(t) = \mathscr{U}(t-1)\{(t-1)[\sin 2(t-1) + 3 \cos 2(t-1)]\}$

39 $\mathscr{L}\{f(2t)\} = \dfrac{s^2+4}{s^3+4s^2+4s-8}$

41 $\mathscr{L}\{t\exp(\frac{1}{2}t)\sin t\} = \dfrac{1}{(s^2+1)(2s-1)^2}$

43 $\mathscr{L}\{t^3 \cosh t\} = \dfrac{2}{27s^3(s^2-1)}$

45 $\mathscr{L}\{e^t \sin t \cos t\} = \dfrac{4(2s-1)}{(4s^2-4s+5)^2}$

47 $\mathscr{L}\{t^2 e^{-2t}\} = \dfrac{2}{(s+2)^3}$, for $s > -2$

49 $\mathscr{L}\{\sin^2 t\} = \dfrac{2}{s(s^2+4)}$, for $s > 0$

51 $\mathscr{L}\{t \sin 2t\} = \dfrac{4s}{(s^2+4)^2}$, for $s > 0$

53 $\mathscr{L}\{f'\} = \dfrac{-4}{s+4}$, $\mathscr{L}\{f''\} = \dfrac{16}{s+4}$

55 $\mathscr{L}\{f'\} = \dfrac{s}{(s-1)^2}$, $\mathscr{L}\{f''\} = \dfrac{2s-1}{(s-1)^2}$

57 $\mathscr{L}\{f'\} = \dfrac{s^3-9s}{(s^2+9)^2}$, $\mathscr{L}\{f''\} = \dfrac{-27(s^2+3)}{(s^2+9)^2}$

59 $y(t) = \frac{11}{10}e^{-3t} + \frac{1}{10}(3 \sin t - \cos t)$

61 $y(t) = \frac{1}{2}e^{-t} \sin 2t$

63 $y(t) = \frac{2}{3}(\sin 2t - \cos 2t) + \frac{1}{3}(2 \cos t - \sin t)$

65 $y(t) = -2 + (\frac{5}{2} + 2t) \sin 2t + 2 \cos 2t - 5t$

67 (i) The Laplace transform of $\tan t$ does not exist due to its infinities at $\pi/2, 3\pi/2, 5\pi/2, \ldots$.

(ii) The initial conditions are applied at $t = 1$ and not at $t = 0$ as required by the Laplace transform. To use the transform method it is necessary to shift the time origin by using the variable change $\tau = t - 1$. The reformulated problem with $x(t) = x(\tau + 1) = u(\tau)$ becomes

$$u'' + 4u' + 4u = (\tau + 1)^2 \text{ with } u(0) = u'(0) = 3.$$

(iii) The function $\sin\left(\dfrac{1}{3-t}\right)$ oscillates boundedly with increasing rapidity as $|3 - t| \to 0$ and has no Laplace transform.

(iv) The function $\exp(1/10t^2)$ is not of exponential order and has no Laplace transform (cf. Ex. 7(ii)).

69 $x(t) = e^{-t}(\cos 2t + \frac{1}{2}\sin 2t) + \frac{1}{5}\mathscr{U}(t-3)[1-\exp[-(t-3)]\cos 2(t-3)$
$\quad + \frac{3}{2}\exp[-(t-3)]\sin 2(t-3)]$

71 $x(t) = 1 + \mathscr{U}(t-1)[1-\cos(t-1)]$.

73 $x(t) = 1 - 2e^t + e^{2t} - \mathscr{U}(t-2)[1-2\exp(t-2)+\exp[2(t-2)]$.

75 $x(t) = \frac{1}{2}\sin 2t + \frac{1}{4} - \frac{1}{4}\cos 2t + \frac{1}{3}\mathscr{U}(t-1)[\sin(t-1)-\frac{1}{2}\sin 2(t-1)]$.

77 $x(t) = e^t - e^{-t} - t + 3, \quad y(t) = -e^t - 2e^{-t} + 4t + 3$.

79 $x(t) = 2 - 5t + 2\sin t, \quad y(t) = 1 + 10t - 3\sin t - 2\cos t$.

81 $x(t) = 3e^{-t} - e^{-3t} + \frac{7}{3}, \quad y(t) = 2e^{-3t} - e^{-4t} + \frac{7}{12}, \quad z(t) = -e^{-3t} + e^{-4t} + \frac{1}{12}$.

83 $x(t) = 1 - t - t^2 - \frac{1}{6}t^3 + e^t, \quad y(t) = 3 + t + t^2 + \frac{1}{3}t^3 + \frac{1}{24}t^4 - e^t$.

85 $\mathscr{L}\{t\cosh kt\} = \dfrac{s^2 + k^2}{(s^2 - k^2)^2}$.

87 $\mathscr{L}\{t^2\sin kt\} = \dfrac{6s^2 - 2k^2}{(s^2 + k^2)^3}$.

89 $\mathscr{L}\{t\sin kt\} = \dfrac{2ks}{(s^2 + k^2)^2}$.

91 $\dfrac{s^2 - k^2}{s(s^2 + k^2)^2}$

93 $\dfrac{1}{s}\arctan\left(\dfrac{1}{s}\right)$

95 $\dfrac{2k^2}{s^4 - k^4}$

97 $\dfrac{1}{k}\sinh kt$

99 $\mathscr{U}(t-4)(t-4)$

101 $\dfrac{1}{2k^3}(\sin kt - kt\cos kt)$

103 In each case, when the function $F(s)$ in Theorem 7.11 is identified and represented in terms of partial fractions it is of the form $F(s) = 1 + $ proper rational functions, and from Theorem 7.4 there is no piecewise continuous function $h(t)$ of exponential order such that $\mathscr{L}\{h(t)\} = 1$.

105 $1/\sqrt{\pi t} = \sqrt{t/\pi}/t$, so $f(t) = \sqrt{t/\pi}$ and $F(s) = \frac{1}{2}s^{-3/2}$. Thus $\mathscr{L}\{1/\sqrt{\pi t}\} = \displaystyle\int_s^\infty \frac{1}{2}u^{-3/2}\,du = s^{-1/2}$.

107 e^{at} **109** $\cos kt$ **111** $\dfrac{1}{s}\tanh ks$

113 $\dfrac{1}{s(1 + e^{-ks})}$ **115** $\dfrac{1+ks}{s^2} - \dfrac{k}{s(1-e^{-ks})}$

117 $\dfrac{1}{s^2} - \dfrac{k}{s}\operatorname{cosech} ks$ **119** $\dfrac{1}{s(1-e^{-ks})}$

123 $t^{-1/2} * t^{-1/2} = \displaystyle\int_0^t \frac{1}{\sqrt{\tau(t-\tau)}}\,d\tau = \int_0^1 \frac{1}{\sqrt{u(1-u)}}\,du = \text{const}.$

125 $\dfrac{s}{(s^2-1)(s^2+4)}$ **127** $\dfrac{1}{k}\left(\dfrac{s}{s^2-k^2} + \dfrac{s}{s^2+k^2} - \dfrac{2}{s}\right)$

129 $\dfrac{1}{2k^3}(\cosh kt - \cos kt)$ **131** $\dfrac{1}{k}\sinh kt$

133 The first result follows by applying Theorem 1.6 to the definition of $f*g$, and by using the result $f*g = g*f$. The second result follows by applying Theorem 7.8 to the first result and using the convolution theorem (Theorem 7.14) to obtain $\mathcal{L}\{(f*g)'\} = g(0) F(s) + F(s)[sG(s) - g(0)]$ $= s F(s) G(s)$.

135 $f(0) = 0, f'(0) = 1$; $f(t) = \frac{1}{2}e^{-3t} \sin 2t$.

137 $f(0) = f'(0) = 0$; $f(t) = (3 + 4t) \sin 2t - 6t \cos 2t$.

139 $f(0) = 0, f'(0) = 1$; $f(t) = \dfrac{1}{2k}(\sin kt + kt \cos kt)$.

141 $f(0) = 0$,   and $f^{(n)}(0) = 0$ for $n = 1, 2, 3, \ldots$.

  $f(t) = \mathcal{U}(t - 1) \exp[-2(t - 1)] \cos\sqrt{3}(t - 1)$, so $f(t) \equiv 0$ for $0 \leq t < 1$.

143 $I = \displaystyle\int_{-\infty}^{\infty} \frac{dt}{1 + t^2} + 2 \int_{-\infty}^{\infty} \frac{\delta(t - 1)}{1 + t^2}\, dt = \pi + 1$.

145 $I = \displaystyle\int_{-\infty}^{\infty} \frac{\sinh 3t}{1 + t^2}[\delta(t + 1) + 4]\, dt + \int_{-\infty}^{\infty} \frac{\sinh 3t}{1 + t^2} \delta(t - 1)\, dt = \frac{\sinh(-3)}{2} + \frac{\sinh 3}{2} = 0$.

147 (i) $f(t)$,   (ii) $f(t_0)$, because $d[\mathcal{U}(t - t_0)]/dt = \delta(t - t_0)$.

149 Integration by parts coupled with the fact that $\delta(t - t_0) = \delta(t_0 - t)$ (the delta function is an even function) shows the integral to be zero. The result then follows by appeal to (28) and the arbitrary nature of $f(t)$.

151 $\displaystyle\int_{-\infty}^{\infty} f(t) t \delta'(t)\, dt = f(t) t \delta(t) \Big|_{-\infty}^{\infty} - \int_{-\infty}^{\infty} \{f(t) + tf'(t)\} \delta(t)\, dt = - \int_{-\infty}^{\infty} f(t)\, \delta(t)\, dt$.

Thus

$$\int_{-\infty}^{\infty} f(t)\{t\delta'(t) + \delta(t)\}\, dt = 0,$$

and as $f(t)$ is arbitrary the result follows from the definition of equality in (28). Replacing t by $-t$ in $t\delta'(t) = -\delta(t)$ gives $t\delta'(-t) = \delta(-t) = \delta(t)$, because $\delta(t)$ is an even function, so $\delta'(-t) = -\delta'(t)$. The sifting property follows from the integral above as

$$\int_{-\infty}^{\infty} f(t) t \delta'(t)\, dt = -f(0).$$

153 $d|\sin t|/dt = \cos t\, \{2\mathcal{U}(t) - 1\} + 2\sin t\delta(t) = \cos t\, \{2\mathcal{U}(t) - 1\}$. $d^2|\sin t|/dt^2 = -\sin t\, \{2\mathcal{U}(t) - 1\}$ $+ 2\cos t\delta(t) = 2\delta(t) - \sin t\{2\mathcal{U}(t) - 1\}$.

155 $\dfrac{(s - a)^2}{(s - a)^2 + k^2} = 1 - \dfrac{k^2}{(s - a)^2 + k^2}$,   so $\mathcal{L}^{-1}\left\{\dfrac{(s - a)^2}{(s - a)^2 + k^2}\right\} = \delta(t) - ke^{at}\sin kt$

  and $\mathcal{L}^{-1}\left\{\dfrac{1}{s - a}\right\} = e^{at}$. Thus $\mathcal{L}^{-1}\left\{\dfrac{s - a}{(s - a)^2 + k^2}\right\} = [\delta(t) - ke^{at}\sin kt] * e^{at} = e^{at}\cos kt$.

157 $\mathcal{L}^{-1}\left\{2s + 1 + \dfrac{3s}{s^2 + 4} - \dfrac{1}{s + 4}\right\} = 2\delta'(t) + \delta(t) + 3\cos 2t - e^{-4t}$.

159 $f(t) = \displaystyle\sum_{n=0}^{\infty} \mathcal{U}(t - nk)$,   for $t \geq 0$.

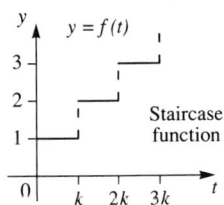

$$\mathscr{L}\{f(t)\} = \sum_{n=0}^{\infty} \frac{e^{-nks}}{s} = \frac{1}{s(1-e^{-ks})} \quad \text{(an infinite geometric series with common ratio } e^{-ks}).$$

161 $f(t) = 2 \sum_{n=0}^{\infty} \mathcal{U}[t - (2n+1)k], \quad \text{for } t \geq 0,$

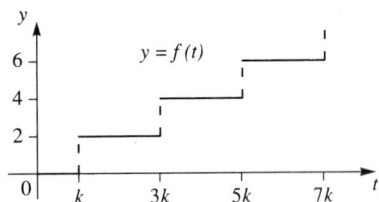

$$\mathscr{L}\{f(t)\} = 2 \sum_{n=0}^{\infty} \exp[-(2n+1)ks] = \frac{1}{s \sinh ks} \quad \text{(an infinite geometric series with common ratio}$$
e^{-2ks}).

163 $x(t) = \frac{1}{4}(t - 1 + e^{-2t} + te^{-2t}) + u(t-1)(t-1)\exp[-2(t-1)]$

165 $x(t) = e^{2t} - e^{t} + \frac{1}{2}u(t-2)\{1 + \exp[2(t-2)] - 2\exp(t-2)\}.$

167 $x(t) = \cosh t - 1, \quad y(t) = \sinh t.$

169 $y(t) = \sin 2t - \frac{2}{3}\sin 3t.$

171 $y(t) = \frac{2}{3}e^{2t} - e^{t} + \frac{1}{3}e^{-t}.$

173 $y(t) = -1 + \frac{1}{2}e^{t} - \frac{1}{2}(\sin t - \cos t).$

175 Transforming the equation gives

$$(a_0 s^{n} + a_1 s^{n-1} + \ldots + a_n)Y(s) = F(s) + H(s),$$

where $H(s)$ represents terms due to the nonhomogeneous initial conditions, for $y(t)$. As before

$$(a_0 s^{n} + a_1 s^{n-1} + \ldots + a_n)Z(s) = \frac{1}{s},$$

so

$$Y(s) = s F(s)Z(s) + s H(s)Z(s),$$

and thus

$$y(t) = \mathscr{L}^{-1}\{s F(s)Z(s)\} + \mathscr{L}^{-1}\{s H(s)Z(s)\}.$$

The first inverse transform follows from $f(t)$ and $z(t)$ as before. If $h(t) = \mathscr{L}^{-1}\{H(s)\}$ is determined, Theorem 7.16 may be used to evaluate the second inverse transform. The solution $y(t)$ then follows.

Answers for Section 7.3

1 $y(x) = \frac{1}{24}\left(\frac{M}{EIl}\right)(x^4 - 2lx^3 + l^3x)$,   for $0 \le x \le l$.

3 $y(x) = \frac{1}{48}\left(\frac{M}{EIl}\right)(2x^4 - 5lx^3 + 3l^2x^2)$,   for $0 \le x \le l$.

5 $y(x) = \frac{1}{240}\left(\frac{M}{EIl}\right)(10x - 23lx^3 + 9l^2x^2)$,   for $0 \le x \le l$.

7 $x_1(t) = \frac{3}{4}\cos\sqrt{2}t + \frac{9}{4}\cos\sqrt{6}t$
$x_2(t) = \frac{9}{4}\cos\sqrt{2}t - \frac{9}{4}\cos\sqrt{6}t$.

9 $i_2(t) = \frac{E_0}{2R}\left[\exp\left(-\frac{Rt}{L+M}\right) - \exp\left(-\frac{Rt}{L-M}\right)\right]$,   for $t > 0$.

11 $x_1(t) = \frac{1}{4}(1 - \cos 2t)$, $x_2(t) = \frac{1}{4}(2t - \sin 2t)$

13 $W(t) = e^{-3t}$ (a) $y(t) = \frac{1}{4}(e^{-3t} + 3t - 1)$

　　　　　　(b) $y(t) = \frac{1}{9}(19e^{-3t} + 3t - 1)$.

15 $W(t) = 4te^{-2t}$, $y(t) = 1 - (1 + 2t)e^{-2t}$,   for $t > 0$.

19 (i) $y(\infty)$ is not defined because denominator has positive zero $(s = 2)$
　　(ii) $y(\infty) = \frac{1}{4}$.

21 (i) $y(\infty) = 3$
　　(ii) $y(\infty) = 2$

23 $T(s) = \dfrac{K}{K + sD}$

25 $T(s) = \dfrac{l_2/l_1}{\left[1 + \left(\dfrac{l_1 + l_2}{l_1 K}\right)\right]}$.

27 $y_0(t) = A\displaystyle\int_0^\infty W(\tau)\exp[i\omega(t - \tau)]\,d\tau = Ae^{i\omega t}\int_0^\infty W(\tau)e^{-i\omega\tau}\,d\tau = Ae^{i\omega t}\,Y(i\omega)$.

Thus the amplitude of the input is multiplied by the factor $|Y(i\omega)|$ and its phase increased by the angle

$$\Phi = \arg Y(i\omega) = \arctan\left\{\frac{1}{i}\left(\frac{Y(i\omega) - \bar{Y}(i\omega)}{Y(i\omega) + \bar{Y}(i\omega)}\right)\right\}.$$

33 $e^{tA} = \mathscr{L}^{-1}\begin{vmatrix} \dfrac{s+4}{(s+2)^2} & \dfrac{1}{(s+2)^2} \\[3mm] \dfrac{-4}{(s+2)^2} & \dfrac{s}{(s+2)^2} \end{vmatrix} = \mathscr{L}^{-1}\begin{vmatrix} \dfrac{1}{s+2} + \dfrac{2}{(s+2)^2} & \dfrac{1}{(s+2)^2} \\[3mm] \dfrac{-4}{(s+2)^2} & \dfrac{1}{s+2} - \dfrac{2}{(s+2)^2} \end{vmatrix}$

$$= \begin{bmatrix} e^{-2t} + 2te^{-2t} & te^{-2t} \\ -4te^{-2t} & e^{-2t} - 2te^{-2t} \end{bmatrix}.$$

$x_1(t) = \frac{15}{4} - 3e^{-2t} - 2te^{-2t}$
$x_2(t) = 4e^{-2t} + 4te^{-2t} - 4$.

35 $x(t) = \sum\limits_{n=0}^{\infty} \dfrac{(-1)^n 4^n (t-n)^{n+2}}{(n+2)!} u(t-n), \qquad$ for $t>0$.

37 $x(t) = \sum\limits_{n=0}^{\infty} \dfrac{(-1)^n 2^n (t-n)^{n+2}}{(n+2)!} u(t-n) + k \sum\limits_{n=0}^{\infty} \dfrac{(-1)^n 2^n (t-n-\alpha)^{n+1}}{(n+1)!} u(t-n-\alpha).$

39 $y(t) = 3 - 2t \mid \frac{1}{2}t^2 - 4 \displaystyle\int_0^t (t-\tau) y(\tau) d\tau,$

41 $y(t) = 1 + 2t + 2(t-1) u(t-1) - 9 \displaystyle\int_0^t (t-\tau) y(\tau) d\tau.$

43 $y(t) = \alpha + \beta t + \displaystyle\int_0^t (t-\tau) f(\tau) d\tau - \lambda^2 \int_0^t (t-\tau) y(\tau) d\tau.$

45 $y(t) = \frac{1}{12} t^4 - \displaystyle\int_0^t \{2 + 4(t-\tau)\} y(\tau) d\tau.$

47   (i) $\mathcal{L}^{-1} \left\{ \dfrac{F(s)}{s^2} \right\} = \mathcal{L}^{-1} \left\{ \left(\dfrac{1}{s} \right) \left(\dfrac{F(s)}{s} \right) \right\} = \displaystyle\int_0^t \int_0^u f(\tau) d\tau \, du.$

However, by the convolution theorem

$$\mathcal{L}^{-1} \left\{ \frac{F(s)}{s^2} \right\} = \mathcal{L}^{-1} \left\{ \left(\frac{1}{s^2} \right) \right\} F(s) = \int_0^t (t-\tau) f(\tau) d\tau,$$

from which the result follows.

(ii) $\mathcal{L}^{-1} \left\{ \dfrac{F(s)}{s^3} \right\} = \mathcal{L}^{-1} \left\{ \left(\dfrac{1}{s^2} \right) \left(\dfrac{F(s)}{s} \right) \right\} = \displaystyle\int_0^t \int_0^u \int_0^v f(\tau) d\tau \, du \, dv.$

However, by the convolution theorem

$$\mathcal{L}^{-1} \left\{ \frac{F(s)}{s^3} \right\} = \mathcal{L}^{-1} \left\{ \left(\frac{1}{s^3} \right) \right\} F(s) = \int_0^t \frac{(t-\tau)^3}{2!} f(\tau) d\tau,$$

from which the result follows.

(iii) This result follows in similar fashion.

49 $y(t) = 1 + 2t - \dfrac{t^2}{2} + \dfrac{t^3}{6} - \displaystyle\int_0^t [2 + (t-\tau) - \frac{3}{2}(t-\tau)^2] y(\tau) d\tau.$

51 $y'' - 2y' + \frac{1}{2}y = -\frac{1}{2}\sin t,$ with $y(0)=0,$ $y'(0)=1.$
53 $y'' + 2t\, y' + (1+t) y = 0,$ with $y(0)=1,$ $y'(0)=1.$
55 $y(t) = 3(1-t) e^{-3t}.$

57 $y(t) = \dfrac{2}{\sqrt{3}} \exp(\frac{1}{2}t) \sin \dfrac{\sqrt{3}}{2} t.$

59 $y(t) = \frac{4}{3} e^{-t} + \frac{2}{3} e^{-4t} + e^{-t} (\sin t - \cos t).$

61 $[Y(s)]^2 - 8 Y(s) + 16 \dfrac{(2s^2+1)}{(s^2+1)^2} = 0,$ so $Y(s) = \dfrac{4}{s^2+1}$ or $Y(s) = 12 - \dfrac{4}{s^2+1}.$

Thus $y(t) = 4 \sin t$ or $y(t) = 12\,\delta(t) - 4 \sin t.$ Only the first solution is finite for all time.

63 $y(t) = 2\cos at + 3\sin at$.

65 $y(t) = 1 - \frac{2}{5}e^{-t} - \frac{3}{5}\cos 2t - \frac{1}{5}\sin 2t$.

67 The transformed equation is

$$s(1-s)\,Y' + (3-s)\,Y = 0$$

with the solution

$$Y(s) = \frac{C(1-s)^2}{s^3}.$$

The initial value theorem shows $C = 1$, and so

$$Y(s) = \frac{(1-s)^2}{s^3} = \frac{1}{s^3} - \frac{1}{s^2} + \frac{1}{s}.$$

Hence $L_2(x) = \mathcal{L}^{-1}\{Y(s)\} = \frac{1}{2}x^2 - 2x + 1$.

69 The transformed equation is

$$Y' + \left(\frac{s}{2} + \frac{2}{s}\right)Y = \frac{1}{s},$$

with integrating factor $s^2 \exp(s^2/4)$. Its solution is

$$Y(s) = \frac{2}{s^2} + \frac{A}{s^2}\exp(s^2/4) = \left(\frac{2+A}{s^2}\right) - \frac{A}{4}\left(1 - \frac{s^2}{8} + \frac{s^4}{96}\cdots\right)$$

The initial value theorem requires

$$0 = \lim_{s \to \infty}\,[s\,Y(s)],$$

which implies $A = 0$. Hence $Y(s)$ and $H_1(x) = \mathcal{L}^{-1}\{2/s^2\} = 2x$ follow immediately.

71 The transformed equation for $y = L_n(x)$ is

$$s(1-s)\,Y' + (1+n-s)\,Y = 0,$$

which has the solution

$$Y(s) = A\left(\frac{(1-s)^n}{s^{n+1}}\right).$$

The initial value theorem requires

$$1 = \lim_{s \to \infty}\,[s\,Y(s)] = \lim_{s \to \infty}\left(\frac{A(1-s)^n}{s^n}\right) = (-1)^n A,$$

showing $A = (-1)^n$. Hence

$$Y(s) = \frac{(s-1)^n}{s^{n+1}}, \quad \text{and} \quad \text{so } L_n(x) = \mathcal{L}^{-1}\left\{\frac{(s-1)^n}{s^{n+1}}\right\}.$$

To invert $Y(s)$ notice that

$$\mathcal{L}\{x^n e^{-x}\} = \frac{n!}{(s+1)^{n+1}},$$

but from Theorem 7.9

$$\mathscr{L}\left\{\frac{d^n}{dx^n}(x^n e^{-x})\right\}=\frac{n!\,s^n}{(s+1)^{n+1}}.$$

To convert this result to the required transform $Y(s)$ we must divide by $n!$ and replace s by $s-1$. From the first shift theorem, s will be replaced by $s-1$ if the transformed function is multiplied by e^t, so we see

$$\mathscr{L}\left\{\frac{e^x}{n!}\frac{d^n}{dx^n}(x^n e^{-x})\right\}=\frac{(s-1)^n}{s^{n+1}},$$

and thus

$$L_n(x)=\mathscr{L}^{-1}\left\{\frac{(s-1)^n}{s^{n+1}}\right\}=\frac{1}{n!}e^x\frac{d^n}{dx^n}(x^n e^{-x}).$$

73 The transformed equation is ·

$$y'-\tfrac{1}{2}sY=-\tfrac{1}{2},$$

with the integrating factor $\exp(-s^2/4)$, and hence the solution

$$Y(s)=\exp(s^2/4)\{Y(0)-\frac{1}{2}\int_0^s \exp(-u^2/4)\,du\}.$$

The variable change $u=2v$ shows

$$Y(s)=\exp(s^2/4)\{Y(0)-\int_0^{s/2} \exp(-v^2)\,dv\}$$

$$=\tfrac{1}{2}\sqrt{\pi}\exp(s^2/4)\left\{\frac{2\,Y(0)}{\sqrt{\pi}}-\frac{2}{\sqrt{\pi}}\int_0^{s/2} \exp(-v^2)\,dv\right\}.$$

Theorem 7.15(i) requires

$$1=\lim_{s\to\infty}[s\,Y(s)]=\lim_{s\to\infty}\left[\frac{1}{2}\sqrt{\pi}\,s\exp(s^2/4)\left\{\frac{2Y(0)}{\sqrt{\pi}}-\frac{2}{\sqrt{\pi}}\int_0^{s/2} \exp(-v^2)\,dv\right\}\right],$$

but replacing s by $s/2$ in the quoted limit shows the above result can only be true if $2Y(0)/\sqrt{\pi}=1$. This yields the required result.

75 The transformed equation is

$$4[-s^2\,Y'-2s\,Y+y(0)]+6[s\,Y-y(0)]+Y=0$$

and $y(t)=\cos\sqrt{t}$, so $y(0)$ is infinite.

Answers for Section 7.4

1 $U(z)=1+z^{-1}+\frac{1}{2!}z^{-2}+\frac{1}{3!}z^{-3}+\ldots+\frac{1}{n!}z^{-n}+\ldots$

3 $U(z)=1+z^{-1}+5z^{-2}+z^{-3}+17z^{-4}+z^{-5}+\ldots$

5 $U(z) = 1 + (e^{aT}z)^{-1} + (e^{2aT}z)^{-2} + (e^{3aT}z)^{-3} + \ldots$

$$= (1 - z^{-1}e^{-aT})^{-1} = \frac{z}{z - e^{-aT}}.$$

7 $U(z) = 1 - z^{-1} + z^{-2} - z^{-3} + z^{-4} - \ldots = (1 + z^{-1})^{-1} = \frac{z}{z + 1}.$

9 $U(z) = 3 + z^{-1} + 3z^{-2} + z^{-3} + 3z^{-4} + \ldots$

$$= 3(1 + z^{-2} + z^{-4} + \ldots) + z^{-1}(1 + z^{-2} + z^{-4} + \ldots)$$

$$= (3 + z^{-1})(1 + z^{-2} + z^{-4} + \ldots) = (3 + z^{-1})(1 - z^{-2})^{-1}$$

$$= \frac{z(3z + 1)}{z^2 - 1}.$$

11 The result follows directly from $\sinh kt = \frac{1}{2}(e^{kt} - e^{-kt})$ and Theorem 7.19 by using $Z\{e^{kt}\}$ and $Z\{e^{-kt}\}$.

13 Use $\sin t \cos t = \frac{1}{2}\sin 2t$ with $Z\{\sin 2t\}$ to show $U(z) = \dfrac{z \sin 2T}{2(z^2 - 2z \cos 2T + 1)}$

15 Use $\sin^2 t = \frac{1}{2}(1 - \cos 2t)$ with $Z\{1\}$ and $Z\{\cos 2t\}$ to show

$$U(z) = \frac{1}{2}\left(\frac{z}{z - 1} - \frac{z(z - \cos 2T)}{z^2 - 2z \cos 2T + 1}\right).$$

17 Use $\cosh^2 t = \frac{1}{2}1 + \cosh 2t)$ with $Z\{1\}$ and $Z\{\cosh 2t\}$ to show

$$U(z) = \frac{1}{2}\left(\frac{z}{z - 1} + \frac{z(z - \cosh 2T)}{z^2 - 2z \cosh 2T + 1}\right).$$

19 $\dfrac{2s + 7}{s^2 + 5s + 6} = \dfrac{3}{s + 2} - \dfrac{1}{s + 3}$, so $\mathscr{L}^{-1}\{U(s)\} = 3e^{-2t} - e^{-3t}$.

Thus $U(z) = \dfrac{3z}{z - e^{-2T}} - \dfrac{1}{z - e^{-3T}}$ ∴

21 $\dfrac{2s^2 + k^2}{s(s^2 + k^2)} = \dfrac{1}{s} + \dfrac{s}{s^2 + k^2}$ so $\mathscr{L}^{-1}\{U(s)\} = 1 + \cos kt$. Thus

$$U(z) = \frac{z}{z - 1} + \frac{z(z - \cos kt)}{z^2 - 2z \cos kt + 1}.$$

23 $U(z) = 4\left(\dfrac{z}{z - \frac{1}{3}}\right)$, so $u_n = 4\left(\dfrac{1}{3}\right)^n$.

25 $U(z) = \dfrac{z}{z - e^{-aT}} - \dfrac{2Tze^{-aT}}{(z - e^{-aT})^2}$, so $u_n = e^{-naT} - 2nTe^{-naT}$.

27 Identify with entry 13 of Table 7.2 by setting $e^{-2aT} = 9$ and $2e^{-aT}\cos kT = 3\sqrt{3}$, so $e^{-aT} = 3$, $kT = \pi/6$.

Thus

$$U(t) = Z\{\tfrac{2}{3}e^{-at}\sin kt\} \text{ with } a = -(1/T)\ln 3$$

and $k = \dfrac{\pi}{6T}$, so $u_n = 2{\cdot}3^{n-1}\sin\dfrac{n\pi}{3}$.

29 Identify with a linear combination of entries 13 and 14 of Table 7.2, with $e^{-2aT} = 16$ and $2e^{-aT} \cos kT = 4\sqrt{2}$, so $e^{-aT} = 4$, $kT = \pi/4$. Thus

$$U(z) \equiv AZ\{e^{-aT} \sin kt\} + BZ\{e^{-at} \cos kt\}$$

with $a = -(1/T) \ln 4$, $k = \pi/4T$. Equating the coefficients of powers of z in the numerators of this identity shows that $A = -1$, $B = 1$, and hence that

$$u_n = 4^n \left(\cos \frac{n\pi}{4} - \sin \frac{n\pi}{4} \right).$$

33 $U(z)z^{-2}V(z)$ with $V(z) = \dfrac{z^2}{(z-1)(z-2)} = \dfrac{2z}{z-2} - \dfrac{z}{z-1}$.

Thus $v(nT) = 2^{n+1} - 1$ for $n = 0, 1, 2, \ldots$, but $Z\{u(t)\} = Z\{v(t-2T)\}$, so $u(nT) = 2^{n-1} - 1$ for $n = 2, 3, \ldots$ and $u(0) = u(T) = 0$ because $v(-2T) = v(-T) = 0$.

35 $Z\{u(t+2T)\} = z^2 U(z) - z^2 u(0) - zu(T)$ but $u(0) = 1$, $u(T) = \cosh kT$, so from entry 15 in Table 7.2,

$$Z\{u(t+2T)\} = \frac{z^3(z - \cosh kT)}{z^2 - 2z \cosh kT + 1} - z^2 - z \cosh kT.$$

37 $Z\{e^{-2t}\} = \dfrac{z}{z - e^{-2T}}$, so $Z\{e^{-2t}e^{t/T}\} = \dfrac{z/3}{(z/3) - e^{-2T}}$.

41 $Z\{t \sinh t\} = \dfrac{z^3 T(2 \sinh T - 1) - zT \sinh T}{(z^2 - 2z \cosh T + 1)^2}$.

43 $Z\{t \cos kt\} = T \left(\dfrac{z^3 \cos kT - 2z^2 + 2 \cos kT}{(z^2 - 2z \cos kT + 1)^2} \right)$.

45 $u(0) = \lim\limits_{z \to \infty} u(z) = \frac{1}{3}$.

47 Permissible zero of the denominator at $z = 1$.

$$\lim_{n \to \infty} u(nT) = \lim_{z \to 1} \left[\left(\frac{z-1}{z} \right) U(z) \right] = 6.$$

49 Two zeros on unit circle $|z| = 1$ at $z = \pm i$ so limit not defined.

51 $w(0) = 1$, $w(T) = 2$, $w(3T) = 3$, $w(4T) = 1$, $w(5T) = -1$, $w(6T) = -3$, $w(7T) = -2$, $w(8T) = -1$, $w(nT) = 0$, $n > 8$.

53 $U(z) = Z\{e^{-aT}\} = \dfrac{z}{z - e^{-aT}}$, $\quad V(z) = Z\{e^{at}\} = \dfrac{z}{z - e^{aT}}$, so

$$W(z) = U(z)V(z) = \frac{z^2}{z^2 - 2z \cosh aT + 1}$$

$$w(kT) = \sum_{n=0}^{k} e^{-naT} \exp[(k-n)aT] = \sum_{n=0}^{k} \exp[(k-2n)aT].$$

55 $U(z) = V(z) = \dfrac{Tz}{(z-1)^2}$, so $W(z) = U(z) V(z) = \dfrac{T^2 z^2}{(z-1)^2}$.

$$w(kT) = T^2 \sum_{n=0}^{k} n(k-n) = T^2 \left(k \sum_{n=1}^{k} n - \sum_{n=1}^{k} n^2 \right)$$

$$= T^2 \left[\frac{k^2(1+k)}{2} - \frac{k(k+1)(2k+1)}{6} \right] = \frac{T^2}{6} k(k^2 - 1).$$

Answers for Section 7.5

1 $U = \dfrac{az}{z-1} + \dfrac{bz}{(z-1)^2}$, so $u_n = a + nb$.

3 $U = \left(\dfrac{c}{1-b} \right) \left(\dfrac{z}{z-1} \right) + \left(a - \dfrac{c}{1-b} \right) \left(\dfrac{z}{z-b} \right)$, so $u_n = \left(\dfrac{c}{1-b} \right) + \left(a - \dfrac{c}{1-b} \right) b^n$.

5 $U = \dfrac{z^2 - 6z}{(z-2)(z-4)} = \dfrac{5z}{z-2} - \dfrac{10z}{z-4}$, so $u_n = 5 \cdot 2^n - 10 \cdot 4^n$.

7 $U = \dfrac{z^2 - 3z}{(z-1)(z-2)} + \dfrac{z}{(z-1)(z-2)(z-3)} = \dfrac{5}{2} \left(\dfrac{z}{z-1} \right) - 2 \left(\dfrac{z}{z-2} \right) + \dfrac{1}{2} \left(\dfrac{z}{z-3} \right)$,

so $u_n = \dfrac{5}{2} - 2^{n+1} + \left(\dfrac{1}{2} \right) 3^n$.

9 $U = \dfrac{z^2 - \sqrt{2} z}{z^2 - \sqrt{2} z + 1}$. Identify with a linear combination of entries 11 and 12 of Table 7.2 with

$\cos kT = 1/\sqrt{2}$, so $kT = \pi/4$, and write

$$\frac{z^2 - \sqrt{2} z}{z^2 - \sqrt{2} z + 1} = A \left(\frac{z^2 - \dfrac{1}{\sqrt{2}} z}{z^2 - \sqrt{2} z + 1} \right) + B \left(\frac{\dfrac{1}{\sqrt{2}} z}{z^2 - \sqrt{2} z + 1} \right).$$

Identifying coefficients of z in the numerator shows $A = 1$, $B = -1$. Thus

$$u_n = \cos \frac{n\pi}{4} - \sin \frac{n\pi}{4}.$$

11 $zU - z = V + \dfrac{2z}{z-1}$ and $zV = 4U$, giving $U = \dfrac{z^2}{z^2 - 4} + \dfrac{2z^2}{(z-1)(z^2 - 4)}$

$$= \frac{3}{2} \left(\frac{z}{z-2} \right) + \frac{1}{6} \left(\frac{z}{z+2} \right) - \frac{2}{3} \left(\frac{z}{z-1} \right), \quad \text{so } u_n = \left(\frac{3}{2} \right) 2^n + \frac{1}{6} (-2)^n - \frac{2}{3},$$

$$v_n = u_{n+1} - 2.$$

13 $G(s) = \dfrac{1}{2s} + \dfrac{1}{s+1} - \dfrac{1}{2}\dfrac{1}{(s+2)}$,   so $G(z) = \dfrac{1}{2}\left(\dfrac{z}{z-1}\right) + \dfrac{z}{z-e^{-T}} - \dfrac{1}{2}\left(\dfrac{z}{z-e^{-2T}}\right)$.

Now $Y_1(z) = Z\{\sin zt\} = \dfrac{z\sin 2T}{z^2 - 2z\cos 2T + 1}$,   so from (9) we have

$Y_0(z) = G(z)\, Y_1(z)$.

15 $G_s(z) = G_1(z)G_2(z) = \dfrac{Te^{-T}z^2(z - \cos 3T)}{(z - e^{-T})^2\{z^2 - 2z\cos 3T + 1\}}$.

Answers for Section 8.1

1 $a_n = \dfrac{1}{n^2}$; $r = 1$; $-1 < x < 1$.

3 $a_n = (-1)^n/(2n+1)$; $r = 1$; $2 < x < 4$.

5 $a_n = 1/n^a$; $r = 1$; $-1 < x < 1$.

7 $a_n = n!/n^n$; $r = \lim\limits_{n\to\infty}\left|\dfrac{a_n}{a_{n+1}}\right| = \lim\limits_{n\to\infty}\left(1 + \dfrac{1}{n}\right)^n = e$; $1 - e < x < 1 + e$.

9 $r_f = \lim\limits_{n\to\infty}\left|-\left(\dfrac{n+1}{n}\right)\right| = 1$ and   the   interval   of   convergence   of f is $-\dfrac{1}{2} < x < \dfrac{3}{2}$;

$r_g = \lim\limits_{n\to\infty}\left(\dfrac{2^n}{n}\cdot\dfrac{n+1}{2^{n+1}}\right) = \dfrac{1}{2}$ and the interval of convergence of g is $0 < x < 1$. The sum $f(x) + g(x)$

$= \displaystyle\sum_{n=1}^{\infty}\left(\dfrac{(-1)^n + 2^n}{n}\right)\left(x - \dfrac{1}{2}\right)^n$ has as its interval of convergence the common interval $0 < x < 1$.

11 $e^{3x}\ln(1 + 2x) = 2x + 4x^3 + \tfrac{17}{3}x^3 + \dots$.

13 $\sec x = (\cos x)^{-1} = \left[1 - \left(\dfrac{x^2}{2!} - \dfrac{x^4}{4!} + \dfrac{x^6}{6!} - \dots\right)\right]^{-1} = 1 + \left(\dfrac{x^2}{2!} - \dfrac{x^4}{4!} + \dfrac{x^6}{6!} - \dots\right)$

$-\left(\dfrac{x^2}{2!} - \dfrac{x^4}{4!} + \dfrac{x^6}{6!} - \dots\right)^2 + \left(\dfrac{x^2}{2!} - \dfrac{x^4}{4!} + \dfrac{x^6}{6!} - \dots\right)^3 - \dots$

$= 1 + \dfrac{1}{2}x^2 + \dfrac{5}{24}x^4$ (up to terms in x^4).

15 $\tan x = x + \tfrac{1}{3}x^3 + \tfrac{2}{15}x^5 + \dots$.

17 Let

$$f(x) = \sum_{n=0}^{\infty} a_n(x - x_0)^n$$

be such that $f(x) = 0$ throughout the interval $x_0 - r < x < x_0 + r$, $r > 0$. Then the result is certainly true if $a_n = 0$ for all n. The uniqueness property then ensures that this must be the power series representation of the sum function.

19 $d(\tan x)/dx = \sec^2 x = 1 + x^2 + \dfrac{2}{3}x^4 + \dfrac{17}{45}x^6 + \dfrac{62}{315}x^8 + \dots$ for $\dfrac{-\pi}{2} < x < \dfrac{\pi}{2}$.

21 $\frac{1}{2}\mathrm{d}\,[(\text{arc sin }x)^2]/\mathrm{d}x = \dfrac{\text{arc sin }x}{\sqrt{1-x^2}} = x + \dfrac{2}{3}x^3 + \dfrac{2\cdot 4}{3\cdot 5}x^5 + \ldots;\quad r = 1;\quad -1 < x < 1.$

23 $f'(x) = 1 - x + x^2 - x^3 + \ldots$ (by the binomial theorem).

So $f(x) = \ln(1 + x) = \displaystyle\int_0^x \frac{1}{1+t}\,\mathrm{d}t = x - \frac{x^2}{2} + \frac{x^3}{3} - \ldots;\quad r = 1,\, -1 < x < 1.$

25 $\mathrm{d}[\text{arc sin }x]/\mathrm{d}x = 1/\sqrt{1-x^2} = 1 + \dfrac{1}{2}x^2 + \dfrac{1}{2}\cdot\dfrac{3}{4}x^4 + \dfrac{1}{2}\cdot\dfrac{3}{4}\cdot\dfrac{5}{6}x^6 + \ldots$ (by the binomial theorem).

So

$$f(x) = \text{arc sin }x = \int_0^x \frac{\mathrm{d}t}{\sqrt{1-t^2}} = 1 + \frac{1}{2}\frac{x^3}{3} + \frac{1}{2}\cdot\frac{3}{4}\frac{x^5}{5} + \frac{1}{2}\cdot\frac{3}{4}\cdot\frac{5}{6}\frac{x^7}{7} + \ldots;\quad r = 1;$$

$-1 < x < 1.$

27 $(1 - x)/(1 + 2x) = 1 + \displaystyle\sum_{n=0}^{\infty} (-1)^{n+1}2^n 3 x^n \quad$ for $-\frac{1}{2} < x < \frac{1}{2}.$

29 $x/(1 - x)^2 = \displaystyle\sum_{n=1}^{\infty} n x^n \quad$ for $-1 < x < 1.$

31 (i) $\displaystyle\sum_{m=0}^{\infty} (m + 3)(m + 2) x^{m+1}$ (ii) $\displaystyle\sum_{m=0}^{\infty} (m + 4)(m + 3) x^{m+3}.$

33 $(1 + x) \displaystyle\sum_{n=2}^{\infty} n(n - 1) x^{n-2} = \sum_{r=0}^{\infty} (r + 2)(r + 1) x^r + \sum_{m=1}^{\infty} (m + 1) m x^m$

$$= 2 + \sum_{m=1}^{\infty} [(m + 2)(m + 1) + (m + 1)m] x^m$$

$$= 2 + \sum_{m=1}^{\infty} 2(m + 1)^2 x^m.$$

35 $F(x) = \displaystyle\sum_{n=2}^{\infty} n(n - 1) a_n x^{n-2} + \sum_{n=2}^{\infty} n(n - 1) a_n x^n + \sum_{n=1}^{\infty} n a_n x^n$

$$= \sum_{m=0}^{\infty} (m + 2)(m + 1) a_{m+2} x^m + \sum_{n=2}^{\infty} n(n - 1) a_n x^n + \sum_{n=1}^{\infty} n a_n x^n$$

$$= 2a_2 + (a_1 + 6a_3)x + \sum_{n=2}^{\infty} [(n + 2)(n + 1) a_{n+2} + n^2 a_n] x^n.$$

37 $x \coth x = \dfrac{u\,\mathrm{e}^{u/2} + \mathrm{e}^{-u/2}}{2\,\mathrm{e}^{u/2} - \mathrm{e}^{-u/2}} = \dfrac{u}{2}\left(\dfrac{\mathrm{e}^u + 1}{\mathrm{e}^u - 1}\right) = \dfrac{u}{2}\left(\dfrac{2}{\mathrm{e}^u - 1} + 1\right) = \dfrac{u}{\mathrm{e}^u - 1} + \dfrac{u}{2}.$

Thus $x \coth x = \left(B_0 + B_1(2x) + \dfrac{B_2}{2!}(2x)^2 + \dfrac{B_3}{3!}(2x)^3 + \dfrac{B_4}{4!}(2x)^4 + \ldots \right) + x$

$$= B_0 + \frac{B_2}{2!}(2x)^2 + \frac{B_3}{3!}(2x)^3 + \frac{B_4}{4!}(2x)^4 + \ldots.$$

However $x \coth x$ is an even function so $B_3 = B_5 = B_7 = \ldots = 0$ from which the required result then follows.

39 The series for tanh x follows from the series for tan x by replacing x by ix because $\tanh ix = i\tan x$. The series for $x\,\operatorname{cosech} x$ then follows by multiplying the identity by x and combining terms in the series for $x\tanh(x/2)$ and the series for $x\coth x$ given in Prob. 37.

41 Differentiating the generating function with respect to x gives

$$\frac{t^2 e^{xt}}{e^t - 1} = \sum_{n=0}^{\infty} B_n'(x)\frac{t^n}{n!} = t\left(\frac{te^{xt}}{e^t - 1}\right) = \sum_{n=0}^{\infty} B_n(x)\frac{t^{n+1}}{n!}.$$

The result then follows by equating the multipliers of t^n on each side of this identity. To evaluate the integral use the result $B_n(t) = B_{n+1}'(t)/(n+1)$.

45 The original series is

$$y = 1 + x + \frac{x^2}{2!} + \frac{x^3}{3!} + \ldots = e^x.$$

Comparing the series obtained by reversion with

$$\ln(1+\theta) = \theta - \frac{\theta^2}{2} + \frac{\theta^3}{3} - \frac{\theta^4}{4} + \ldots$$

shows $x = \ln[1 + (y-1)] = \ln y$, as would be expected.

47 $y = \dfrac{\pi}{2} = -\left(x + \dfrac{1}{6}x^3 + \dfrac{3}{40}x^5 + \dfrac{5}{112}x^7 + \ldots\right),$

so

$$a_1 = -1, \quad a_2 = 0, \quad a_3 = -\tfrac{1}{6}, \quad a_4 = 0,$$

$$a_5 = \frac{-3}{40}, \quad a_6 = 0, \quad a_7 = \frac{-5}{112}, \ldots.$$

Hence $A_1 = -1$, $A_2 = 0$, $A_3 = \dfrac{1}{3!}$, $A_4 = 0$, $A_5 = \dfrac{-1}{5!}, \ldots$, which gives

$$x = -\left[\left(y - \frac{\pi}{2}\right) - \frac{1}{3!}\left(y - \frac{\pi}{2}\right)^3 + \frac{1}{5!}\left(y - \frac{\pi}{2}\right)^5 - \ldots\right].$$

The bracketed series is recognized as that for $\sin\left(y - \dfrac{\pi}{2}\right)$, but $\sin\left(y - \dfrac{\pi}{2}\right) = -\cos y$, which gives the expected result $x = \cos y$.

Answers for Section 8.2

1 $y(x) = 1 - 3x + 5x^2 - 6x^3 + \frac{17}{3}x^4 + \ldots.$

3 The expression for $y^{(n)}(0)$ follows either by inspection of the pattern of values generated by the recursion relation

$$y^{(n)}(0) = 3 - 2y^{(n-1)}(0)$$

or, rigorously, by applying mathematical induction to this result. The sum of the geometric series with common ratio $r = -2$

$$1 - 2 + 2^2 - \ldots + (-1)^{n-1} 2^{n-1} = [1 - (-2)^n]/3,$$

so $y^{(n)}(0) = 1 + (-1)^n 2^n$, from which the result then follows.

5 $y(x) = 3 - 3(x-1) - 3(x-1)^2 + \frac{7}{2}(x-1)^3 + \ldots$

7 $y(x) = 1 + \left(\dfrac{4}{4+\pi}\right)\left(x - \dfrac{\pi}{4}\right)^2 + \left(\dfrac{4\sqrt{2}(4+\pi) - 48}{3(4+\pi)^2}\right)\left(x - \dfrac{\pi}{4}\right)^3 + \ldots$

9 $y(x) = 1 + x + \dfrac{x^2}{2} + \dfrac{x^3}{6} + \dfrac{x^3}{3} + \dfrac{x^4}{8} + \ldots$

13 $y(x) = x - 2x^2 + \frac{11}{6}x^3 - x^4 + \ldots$

15 $y(x) = \alpha + \beta(x-3) + (3\beta - \alpha)(x-3)^2 + (6\beta - 2\alpha)(x-3)^3 + \left(\dfrac{57\beta - 19\alpha}{6}\right)(x-3)^4 + \ldots$

17 $y(x) = x - \frac{1}{4}x^2 + \frac{1}{24}x^3 - \frac{3}{64}x^4 + \frac{7}{640}x^5 - \ldots$

19 $y(x) = 1 + \dfrac{1}{2}\left(x - \dfrac{\pi}{3}\right) - \dfrac{3}{2\pi}\left(x - \dfrac{\pi}{3}\right)^2 + \left(\dfrac{\sqrt{3}-1}{2\pi} - \dfrac{3}{\pi^2}\right)\left(x - \dfrac{\pi}{3}\right)^3 + \ldots$

Answers for Section 8.3

1 $a_0 = 0$, $(n+1)(n+2)a_{n+2} - a_{n-1} = 0$, $\qquad n = 1, 2, \ldots$

$$y(x) = a_0\left(1 + \dfrac{1}{2\cdot3}x^3 + \dfrac{1}{2\cdot3\cdot5\cdot6}x^6 + \dfrac{1}{2\cdot3\cdot5\cdot6\cdot8\cdot9}x^9 + \ldots\right)$$

$$+ a_1\left(x + \dfrac{1}{3\cdot4}x^4 + \dfrac{1}{3\cdot4\cdot6\cdot7}x^7 + \ldots\right).$$

3 $4a_2 + a_1 = 0$, $2(n+1)(n+2)a_{n+2} + (n+1)a_{n+1} + a_{n-1} = 0$, $\qquad n = 1, 2, \ldots$

$$y(x) = a_0\left(1 - \dfrac{1}{12}x^3 + \dfrac{1}{96}x^4 - \dfrac{1}{960}x^5 + \ldots\right) + a_1\left(x - \dfrac{1}{4}x^2 + \dfrac{1}{24}x^3 - \dfrac{3}{64}x^4 + \dfrac{7}{640}x^5 - \ldots\right).$$

5 $2a_2 + \alpha a_0 = 0$, $(n+1)(n+2)a_{n+2} - (2n-\alpha)a_{n=0}$, $\qquad n = 1, 2, \ldots$

$$a_{2n} = \dfrac{-\alpha(2\cdot2-\alpha)(2\cdot4-\alpha)\ldots[2(2n-2)-\alpha]}{(2n)!}a_0,$$

$$a_{2n+1} = \dfrac{(2\cdot1-\alpha)(2\cdot3-\alpha)\ldots[2(2n-1)-\alpha]}{(2n+1)!}a_1, \qquad n = 1, 2, \ldots$$

$$y(x) = a_0\left(1 - \dfrac{\alpha}{2!}x^2 - \dfrac{\alpha(2\cdot2-\alpha)}{4!}x^4 - \dfrac{\alpha(2\cdot2-\alpha)(2\cdot4-\alpha)}{6!}x^6 + \ldots\right)$$

$$+ a_1\left(x - \dfrac{(2\cdot1-\alpha)}{3!}x^3 + \dfrac{(2\cdot1-\alpha)(2\cdot3-\alpha)}{5!}x^5 + \ldots\right).$$

7 Write $x^2 = 1 + 2(x-1) + (x-1)^2$ so the differential equation becomes

$$y'' + [1 + 2(x-1) + (x-1)^2]y = 0.$$

$$2a_2 + a_0 = 0, \ 6a_3 + a_1 + 2a_0 = 0,$$

$$(n+1)(n+2)a_{n+2} + a_n + 2a_{n-1} + a_{n-2} = 0, \qquad n = 2, 3, \ldots$$

$$y(x) = u_0[1 \ \tfrac{1}{2}(x-1)^2 - \tfrac{1}{3}(x-1)^3 - \tfrac{1}{24}(x-1)^4 + \tfrac{1}{15}(x-1)^5 + \ldots]$$

$$+ \, a_1[(x-1) - \tfrac{1}{6}(x-1)^3 - \tfrac{1}{6}(x-1)^4 - \tfrac{1}{24}(x-1)^5 + \ldots].$$

9 $a_2 + 6a_0 = 0, \ 3a_3 + 5a_1 = 0,$

$$(n+1)(n+2)a_{n-2} - (n^2 + n - 12)a_n = 0, \qquad n = 2, 3, \ldots$$

$$y(x) = a_0(1 - 6x^2 + 3x^4 + \tfrac{4}{5}x^6 + \ldots) + a_1(x - \tfrac{5}{3}x^3).$$

The second linearly independent solution reduces to a cubic.

11 $2a_2 - 1 = 0, \ (n+1)(n+2)a_{n+2} + a_{n-1} = \dfrac{1}{n!}, \qquad n = 1, 2, \ldots$

$$y(x) = a_0\left(1 - \frac{1}{2\cdot3}x^3 + \frac{1}{2\cdot3\cdot5\cdot6}x^6 - \ldots\right) + a_1\left(x - \frac{1}{3\cdot4}x^4 + \ldots\right) + \left[\frac{1}{2}x^2 + \frac{1}{2\cdot3}x^3 + \frac{1}{2\cdot3\cdot4}x^4\right.$$

$$\left. + \left(\frac{1}{4\cdot5\cdot6} - \frac{1}{2\cdot4\cdot5}\right)x^5 + \left(\frac{1}{6!} - \frac{1}{2\cdot3\cdot5\cdot6}\right)x^6 \ldots\right].$$

13 Write $x + 1 = (x-1) + 2$ and $x = (x-1) + 1$ so the differential equation becomes

$$[(x-1)+2]y' + [(x-1)+1]y = 0$$

$$2a_1 + a_0 = 0, \ 2(n+1)a_{n+1} + (n+1)a_n + a_{n-1} = 0, \ n = 1, 2, \ldots,$$

$$y(x) = a_0[1 - \tfrac{1}{2}(x-1) + \tfrac{1}{12}(x-1)^3 - \tfrac{1}{24}(x-1)^4 + \tfrac{1}{80}(x-1)^5 + \ldots].$$

15 $2a_1 + a_0 = 0, \ 2a_1 + 4a_2 - 1 = 0, \ 2a_{n+1} + a_n = 0, \qquad n = 2, 3, \ldots$

$$a_n = \frac{(-1)^n}{2^n}(1 + a_0), \qquad n = 2, 3, \ldots$$

$$y(x) = a_0(1 - \frac{1}{2}x + \frac{1}{4}x^2 - \frac{1}{8}x^3 + \ldots) + \frac{x^2}{4}\left(1 - \frac{1}{2}x + \frac{1}{4}x^2 - \frac{1}{8}x^3 + \ldots\right).$$

Compare this with the analytic solution $y = \dfrac{c}{2+x} + \dfrac{x^2}{2(2+x)}.$

Answers for Section 8.4

1 Routine calculations.

3 Routine calculations using (12) with $n = 8$ and $n = 9$.

5 Set $u = x^2$ in the series for $p_\alpha(x)$ and apply the ratio test to the resulting series in u. Remove a factor x from $q_\alpha(x)$, set $u = x^2$, and apply the ratio test to the resulting series in u.

7 $f(x) = \tfrac{13}{3}P_0 - \tfrac{3}{5}P_1 + \tfrac{8}{3}P_2 - \tfrac{2}{5}P_3, \ -1 \le x \le 1.$

9 Proceed as indicated in the problem.

11 Setting $u = (x^2 - 1)^n$, it follows that

$$(x^2 - 1)u' - 2nxu = 0.$$

Differentiate $m + 1$ times to obtain

$$(x^2 - 1)u^{(m+2)} - (2n - 2k - 2)xu^{(m+1)}$$
$$- [2n + (2n - 2) + (2n - 4) + \ldots + (2n - 2k)]u^{(m)} = 0.$$

Using the result that when $m = n$, $2n + (2n - 2) + \ldots + 2 + 0 = n(n + 1)$, reduces the equation to

$$(x^2 - 1)u^{(n+2)} + 2xu^{(n+1)} - n(n + 1)u^{(n)} = 0,$$

so $y = d^n u/dx^n$ is a solution of Legendre's equation.

Consequently

$$\frac{d^n u}{dx^n} = kP_n(x) \quad \text{with } k \text{ a constant.}$$

As $(x^2 - 1)^n = x^{2n} + \text{(lower degree terms)}$, it follows that the coefficient of x^n in $d^n u/dx^n$ is $2n(2n - 1)(2n - 2) \ldots (n + 1)$. However, from (10) the coefficient of x^n in $P_n(x)$ is $(2n)!/[2^n(n!)^2]$, so as these coefficients must be equal

$$k = \frac{2n(2n - 1) \ldots (n + 1)2^n(n!)^2}{(2n)!} = 2^n n!,$$

giving

$$\frac{1}{2^n n!} \frac{d^n[(x^2 - 1)^n]}{dx^n} = P_n(x).$$

13 $\dfrac{1}{r} = \dfrac{1}{r_2} \dfrac{1}{\left[1 - 2\left(\dfrac{r_1}{r_2}\right)\cos\theta + \left(\dfrac{r_1}{r_2}\right)^2\right]^{1/2}}$. The required result follows by setting

$x = \cos\theta$ and $r = r_1/r_2$ in (24).

15 Proceed as indicated in the problem.

17 Obtain the recurrence relation from (26) by replacing n by $n - 1$. Multiplying by $P_n(x)$, integrating, and using (15) gives

$$n \int_{-1}^{1} [P_n(x)]^2 dx = (2n - 1)\int_{-1}^{1} xP_n(x)P_{n-1}(x)dx.$$

Multiplying (26) by $P_{n-1}(x)$, integrating, and using (15) gives

$$(2n + 1)\int_{-1}^{1} xP_n(x)P_{n-1}(x)dx = n\int_{-1}^{1} [P_{n-1}(x)]^2 dx.$$

Combining these results gives

$$\int_{-1}^{1} [P_n(x)]^2 dx = \left(\frac{2n - 1}{2n + 1}\right)\int_{-1}^{1} [P_{n-1}(x)]^2 dx.$$

When used recursively this gives

$$\int_{-1}^{1} [P_n(x)]^2 dx = \left(\frac{2n - 1}{2n + 1}\right)\left(\frac{2n - 3}{2n - 1}\right)\cdots\left(\frac{3}{5}\right)\left(\frac{1}{3}\right)\int_{-1}^{1} [P_0(x)]^2 dx = \frac{2}{2n + 1}.$$

19 The leading term in $(x^2 - 1)^n$ is x^{2n}, so differentiating Rodrigues' formula n times must produce the result $(2n)!/(2^n n!)$. When integrating by parts n times to obtain the second result the terms outside the integral will always vanish because $x^2 - 1 = 0$ for $x = \pm 1$.

$$I_n = \left(\frac{2n-2}{2n+1}\right)\left(\frac{2n-2}{2n-1}\right)\left(\frac{2n-4}{2n-3}\right)\cdots\frac{2}{3}I_0$$

$$= \frac{2^n n!}{(2n+1)(2n-1)\ldots3}I_0 = \frac{2^{2n}(n!)^2}{(2n+1)2n(2n-1)\ldots3\cdot2}I_0$$

$$= \frac{2^{2n}(n!)^2}{(2n+1)!}I_0$$

where

$$I_0 = \int_{-1}^{1} dx = 2.$$

The normalization condition follows directly from these results.

Answers for Section 8.5

5 $4^8\Gamma(8\tfrac{1}{4})/\Gamma(\tfrac{1}{4})$.

7 $2^8\Gamma(10\tfrac{1}{2})/\Gamma(2\tfrac{1}{2})$.

9 $\dfrac{\Gamma(n+\tfrac{2}{3})}{2^n\Gamma(\tfrac{2}{3})}x^n$.

11 $(\tfrac{5}{2})^{n+1}\dfrac{\Gamma(\tfrac{2}{5})+n+1)}{\Gamma(n+1)\Gamma(\tfrac{2}{5})}x^n$.

13 $\dbinom{\alpha}{r} = \dfrac{\Gamma(\alpha+1)}{\Gamma(r+1)(\alpha-r+1)}$.

15 $[\Gamma(\tfrac{1}{2})]^2 = 4\int_0^\infty \exp(-x^2)\,dx\int_0^\infty \exp(-y^2)\,dy = 4\int_0^\infty\int_0^\infty \exp[-(x^2+y^2)]\,dx\,dy$

$$= 4\int_0^{\pi/2}\int_0^\infty r\exp(-r^2)\,dr\,d\theta = \pi.$$

17 $G(x, n) = n^x\int_0^1 (1-s)^n s^{x-1}\,ds.$

Integration by parts gives

$$\frac{G(x,n)}{n^x} = (1-s)^n\frac{s^x}{x}\Big|_0^1 + \frac{n}{x}\int_0^1 (1-s)^{n-1}s^x\,ds.$$

The first term on the right vanishes at each limit, so repeated integration by parts gives

$$G(x, n) = n^x\frac{1\cdot2\cdot3\ldots(n-1)}{x(x+1)(x+2)\ldots(x+n-1)}\int_0^1 s^{x+n-1}\,ds$$

$$= n^x\frac{1\cdot2\cdot3\ldots(n-1)}{x(x+1)(x+2)\ldots(x+n-1)(x+n)}.$$

The final result follows because

$$\lim_{n \to \infty} \left(1 - \frac{t}{n} \right)^n = e^{-t}.$$

19 $\quad B(p, q) = \dfrac{x^p (1-x)^{q-1}}{p} \Big|_0^1 + \left(\dfrac{q-1}{p} \right) \displaystyle\int_0^1 x^p (1-x)^{q-2} \, dx.$

As $x^p (1-x)^{q-2} = x^{p-1}(1-x)^{q-2} - x^{p-1}(1-x)^{q-1}$ it follows that

$$B(p,q) = \left(\frac{q-1}{p} \right) \int_0^1 x^{p-1}(1-x)^{q-2} \, dx - \left(\frac{q-1}{p} \right) \int_0^1 x^{p-1}(1-x)^{q-1} \, dx$$

$$= \left(\frac{q-1}{p} \right) B(p, q-1) - \left(\frac{q-1}{p} \right) B(p, q),$$

from which the first result then follows. The second result follows from the first because
$B(p, q) = B(q, p)$.

21 (i) $B(1, 1) = 1$ follows from the definition.

(ii) $B(\tfrac{1}{2}, \tfrac{1}{2}) = \pi$ follows from the definition after making the substitution $x = \sin^2 \theta$.

(iii) follows by setting $x = 1/(1+t)$, because

$$B(m, n) = \int_0^\infty \left(\frac{1}{1+t} \right)^{m-1} \left(\frac{t}{1+t} \right)^{n-1} \frac{dt}{(1+t)^2} = \int_0^\infty \frac{t^{n-1}}{(1+t)^{m+n}} \, dt,$$

which equals the first integral because $B(m, n) = B(n, m)$.

Answers for Section 8.6

1 Regular singular points at $x = \pm 1$ and a regular singular point at infinity.

3 Irregular singular point at $x = 0$ and a regular singular point at infinity.

5 Irregular singular points at $x = 2$ and at infinity.

7 $t(t^2 - 2t + 2)y'' + 2(t-1)^2 y' + y = 0.$

9 $y(x) = c_1 x^{5/2} \left\{ 1 + \displaystyle\sum_{n=1}^\infty \frac{(-1)^n x^n}{n! \, 9 \cdot 11 \cdot 13 \, \dots \, (2n+7)} \right\}$

$\quad + c_2 x^{-1} \left\{ 1 + \displaystyle\sum_{n=1}^\infty \frac{(-1)^n x^n}{n! (-5)(-3)(-1) \, \dots \, (2n-7)} \right\}.$

11 $y(x) = c_1 \displaystyle\sum_{n=0}^\infty \frac{(-1)^n n! \, x^n}{(2n)!} + c_2 x^{1/2} \displaystyle\sum_{n=0}^\infty \frac{(-1)^n x^n}{4^n n!}.$

13 $y(x) = c_1 x^{1/2} (1-x) + c_2 \left(1 + 3 \displaystyle\sum_{n=1}^\infty \frac{x^n}{(2n-3)(2n-1)} \right).$

15 Set $z = 1/x$, solve the resulting equation for z, and then write $z = 1/x$.

$$y(x) = c_1 \sum_{n=0}^\infty \frac{(-1)^n}{(2n)!} \frac{1}{x^n} + c_2 \frac{1}{\sqrt{x}} \sum_{n=0}^\infty \frac{(-1)^n}{(2n+1)} \frac{1}{x^n}.$$

17 $y(x) = c_1 x^{-1/2}(1-2x) + c_2 \displaystyle\sum_{n=0}^\infty \frac{x^n}{n!(1-4n^2)}.$

19 $y(x) = c_1 \left(1 + \sum_{n=1}^{\infty} \dfrac{(-1)^n 4^n x^{3n}}{3^n n!(3n-2)(3n-5) \ldots 7 \cdot 4} \right)$

$\qquad + c_2 \left(x^2 + \sum_{n=1}^{\infty} \dfrac{(-1)^n 4^n x^{3n+2}}{3^n n!(3n+2) \ldots 8 \cdot 5} \right).$

21 $y_1(x) = \sum_{n=0}^{\infty} \dfrac{x^n}{n!}$ (i.e. $y_1(x) = e^x$), and

$\qquad y_2(x) = y_1 \ln|x| - x - \tfrac{3}{4} x^2 + \ldots .$

23 $y_1(x) = \sum_{n=0}^{\infty} \dfrac{x^{2n}}{2^{2n}(n!)^2}$, and $y_2(x) = y_1 \ln|x| - \tfrac{1}{4} x^2 + \ldots .$

25 $y(x) = c_1 x + c_2 x \ln|x|$.

27 $y(x) = x^{1/2}[c_1 + c_2 \ln|x|]$.

Answers for Section 8.7

9 Differentiating (18) and multiplying by 2 gives

$\qquad \begin{aligned} 2^2 J_n'' &= 2(J_{n-1}' - J_{n+1}') \\ &= (J_{n-2} - J_n) - (J_n - J_{n+2}) \\ &= J_{n-2} - 2J_n + J_{n+2}. \end{aligned}$

The final result follows by induction, starting from the stated expression and showing that the result obtained by a further differentiation is the same as that obtained by replacing m by $m+1$. Then, as the result has been proved for $m=2$, it follows that it is true for all $m \geq 2$.

17 $\dfrac{dy}{dx} = 2x \dfrac{dy}{du}, \quad \dfrac{d^2y}{dx^2} = 4x^2 \dfrac{d^2y}{du^2} + 2 \dfrac{dy}{du}$, so the equation becomes

$\qquad u^2 \dfrac{d^2y}{du^2} + u \dfrac{dy}{du} + (u^2 - n^2)y = 0$. This has the solution

$\qquad y(u) = c_1 J_n(u) + c_2 Y_n(u)$, so

$\qquad y(x) = c_1 J_n(x^2) + c_2 Y_n(x^2)$.

19 $y(x) = c_1 J_3(3x) + c_2 Y_3(3x)$.

21 $y(x) = c_1 J_v(x\sqrt{7}) + c_2 Y_v(x\sqrt{7})$ which when v is nonintegral may also be written

$\qquad y(x) = c_1 J_v(x\sqrt{7}) + c_2 J_{-v}(x\sqrt{7})$.

23 Set $u = dy/dx$, then

$\qquad u = \dfrac{dy}{dx} = c_1 J_3(x\sqrt{6}) + c_2 Y_3(x\sqrt{6})$, so

$\qquad y(x) = c_1 \int J_3(x\sqrt{6})dx + c_2 \int Y_3(x\sqrt{6})dx + c$.

25 $J_1(2)$.　　　**27** $-x^{-2}J_2(x) + c$.　　　**29** $x^4 J_4(x) + c$.

31 $2x^{-1}J_0(x)+(1-4x^{-2})J_1+c.$ **33** Integrate (18).

37 Applying Rolle's theorem to $J_0(x)$ between two successive zeros shows there is at least one point x_0 interior to the interval at which $J'_0(x)=0$. As $J'_0(x)=-J_1(x)$ it follows that $J'_0(x)$ vanishes at a zero of $J_1(x)$ which must lie inside the interval. There can only be one such zero in the interval because the convexity or concavity of $J_0(x)$ implies via $J'_0(x)=-J_1(x)$ that $J_1(x)$ can only change sign once.

47
$$\frac{1}{2}x\left(1+\frac{1}{t^2}\right)\exp\left[\frac{1}{2}x\left(t-\frac{1}{t}\right)\right]=\sum_{n=-\infty}^{\infty}nt^{n-1}J_n(x),$$

which may be written

$$\tfrac{1}{2}x\sum_{n=-\infty}^{\infty}(t^n+t^{n-2})J_n(x)=\sum_{n=-\infty}^{\infty}nt^{n-1}J_n(x).$$

Equating the coefficient of t^n on either side of this identity gives the required result.

Answers for Section 8.8

9 $\mathrm{si}(x)\sim\cos x\left(\dfrac{1}{x}-\dfrac{2!}{x^3}+\dfrac{4!}{x^5}-\cdots\right)+\sin x\left(\dfrac{1}{x^2}-\dfrac{3!}{x^4}+\dfrac{5!}{x^6}-\cdots\right).$

Use $\mathrm{Si}(x)=\displaystyle\int_0^\infty\frac{\sin t}{t}dt-\int_x^\infty\frac{\sin t}{t}dt=\frac{\pi}{2}-\mathrm{si}(x).$

11 $s(x)\sim\cos x^2\left(\dfrac{1}{2x}-\dfrac{3}{8x^5}+\cdots\right)+\sin x^2\left(\dfrac{1}{4x^3}-\dfrac{15}{16x^2}+\cdots\right).$

Use $S(x)=\displaystyle\int_0^x\sin^2 t\,dt-\int_x^\infty\sin^2 t\,dt=\frac{1}{2}\sqrt{\frac{\pi}{2}}-s(x).$

17 $\displaystyle\int_0^\infty\frac{e^{-xt}}{1+t}dt\sim\frac{1}{x}-\frac{1!}{x^2}+\frac{2!}{x^3}-\cdots.$

19 $\displaystyle\int_0^\infty e^{x(t-1)}t^n dt\sim\frac{1}{x}-\frac{n}{x^2}+\frac{n(n-1)}{x^3}-\cdots.$

21 $(2t)^n e^{-t}=\exp(-t+n\ln 2+n\ln t).$ Setting $t=n\tau$ to bring the integrand to the correct form, as in Ex. 5, gives

$$I_n=2^n n^{n+1}\int_0^\infty\exp[n(\ln\tau-\tau)]d\tau.$$

The integral is of the form needed for the Laplace formula with x replaced by n and t by τ. As $a(\tau)=\ln\tau-\tau$ has a maximum at $\tau=1$, at which $a''(1)=-1$ it follows from the Laplace formula that

$$I_n\sim\sqrt{2\pi n}\,2^n n^n e^{-n}.$$

23 Write $p^n(t)=\exp[n\ln p(t)]$ and apply the Laplace formula

$$(I_n)^{1/n}=p(t_0)k^{1/n}\text{ with }k=[-2\pi p(t_0)/np''(t_0)]^{1/2}.$$

However $\lim_{n\to\infty} k^{1/n} = 1$ for any real $k > 0$, so

$$\lim_{n\to\infty}\left[\int_a^b p^n(t)\,dt\right]^{1/n} = p(t_0).$$

25 Set $x/t = \tau$ to obtain

$$I(x) = x\int_0^\infty \frac{\exp[-3x\cosh(\tau-2)]}{\tau\ln(1+\tau^2+\tau^4)}\,d\tau$$

and apply the Laplace formula with $a(\tau) = -3\cosh(\tau-2)$. The function $a(\tau)$ has a maximum at $\tau = 2$ at which $a''(2) = -3$. Setting $f(\tau) = [\tau\ln(1+\tau^2+\tau^4)]^{-1}$ shows that

$$I(x) \sim \left(\frac{\pi x}{6}\right)^{1/2}\frac{e^{-3x}}{\ln 21}.$$

27 $y = u\exp(x^2/8)$, $\quad u'' - \frac{x^2}{16}\left(1 - \frac{8}{x} + \frac{20}{x^2}\right)u = 0$

$$y_+(x) \sim \left(x + 2 + \frac{5}{x} + O(x^{-2})\right)\exp\left(\frac{1}{4}x^2 - x\right),$$

$$y_-(x) \sim \left(1 + \frac{2}{x} + \frac{5}{x^2} + O(x^{-3})\right)e^x,$$

with arbitrary multiplicative constants set equal to unity.

29 $y = u/x$, $\quad u'' - \left(1 - \frac{1}{x}\right)u = 0$.

$$y_+(x) \sim \sqrt{x}\left(\frac{1}{x} + \frac{1}{4x^2} + \frac{5}{32x^3} + O(x^{-4})\right)\exp\left(x + \frac{1}{8x}\right),$$

$$y_-(x) \sim \sqrt{x}\left(1 + \frac{1}{4x} + \frac{5}{32x^2} + O(x^{-3})\right)\exp\left[-\left(x + \frac{1}{8x}\right)\right].$$

31 The equation is in standard form with $f(x) = \frac{1}{4}\left(1 - \frac{4\lambda}{x} - \frac{1-4\mu^2}{x^2}\right)$.

$$y_+(x) \sim x^{-\lambda}\left(1 + \frac{\lambda}{x} + \frac{1-4\mu^2+10\lambda^2}{4x^2} + O(x^{-3})\right)\exp\left(\frac{1}{2}x + \frac{1-4\mu^2+4\lambda^2}{4x}\right),$$

$$y_-(x) \sim x^\lambda\left(1 + \frac{\lambda}{x} + \frac{1-4\mu^2+10\lambda^2}{4x^2} + O(x^{-3})\right)\exp\left(-\frac{1}{2} - \frac{1-4\mu^2+4\lambda^2}{4x}\right),$$

with arbitrary multiplicative constants set equal to unity.

33 $y = u\exp(x^2/2)$, $\quad u'' - x^2\left(1 + \frac{15}{x^2}\right)u = 0$.

$$y_+(x) \sim x^7\left(1 - \frac{15}{4x^2} + O(x^{-4})\right)\exp(x^2),$$

$$y_-(x) = x^{-8}\left(1 - \frac{15}{4x^2} + O(x^{-4})\right).$$

35 $y = ux^{-2}e^{x/2}$, $u'' - \dfrac{1}{4}\left(1 - \dfrac{12}{x} + \dfrac{8}{x^2}\right)u = 0.$

$$y_+(x) \sim \left(1 + \frac{3}{x} + \frac{37}{x^2} + O(x^{-3})\right)x^{-3}\exp\left(x - \frac{2}{x}\right),$$

$$y_-(x) \sim \left(1 + \frac{3}{x} + \frac{37}{x^2} + O(x^{-3})\right)x\exp(2/x).$$

Answers for Section 8.9

1

n	x_n	k_{1n}	K_{1n}	k_{2n}	K_{2n}
0	0.0	0.200000	0.133333	0.213333	0.144586
1	0.2	0.228781	0.151914	0.243973	0.155959
2	0.4	0.259868	0.157002	0.275569	0.155498
3	0.6	0.290895	0.151770	0.306072	0.145992
4	0.8	0.320052	0.138656	0.333917	0.129639

n	x_n	k_{3n}	K_{3n}	k_{4n}	K_{4n}
0	0.0	0.214459	0.144518	0.228904	0.151900
1	0.2	0.244377	0.155897	0.259961	0.156984
2	0.4	0.275418	0.155532	0.290975	0.151750
3	0.6	0.305495	0.146151	0.320126	0.138635
4	0.8	0.333016	0.129924	0.346036	0.119615

n	x_n	k_n	K_n	y_n	z_n
0	0.0	0.214081	0.143907	0.500000	1.00000
1	0.2	0.244240	0.155435	0.714081	1.143907
2	0.4	0.275469	0.155135	0.958322	1.299342
3	0.6	0.305693	0.145782	1.233791	1.454477
4	0.8	0.333326	0.129566	1.539484	1.600259

3

n	x_n	k_{1n}	K_{1n}	k_{2n}	K_{2n}
0	0.0	−0.200000	0.400000	−0.160000	0.476000
1	0.2	−0.102155	0.589737	−0.043181	0.700360
2	0.4	0.042443	0.881233	0.130566	1.072077
3	0.6	0.266407	1.423621	0.408769	1.829840
4	0.8	0.657975	2.695848	0.927560	3.798763

n	x_n	k_{3n}	K_{3n}	k_{4n}	K_{4n}
0	0.0	-0.152400	0.495392	-0.100922	0.592577
1	0.2	-0.032119	0.731061	0.044058	0.885340
2	0.4	0.149650	1.131261	0.268695	1.431031
3	0.6	0.449391	1.975175	0.661442	2.713388
4	0.8	1.037852	4.293216	1.516618	6.783084

n	x_n	k_n	K_n	y_n	z_n
0	0	-0.154287	0.489227	2.000000	-1.000000
1	0.2	-0.034783	0.722987	1.845713	-0.510773
2	0.4	0.145262	1.119824	1.810930	0.212213
3	0.6	0.440695	1.957840	1.956192	1.332037
4	0.8	1.017570	4.277148	2.396887	3.289877

5

n	x_n	k_{1n}	K_{1n}	k_{2n}	K_{2n}
0	1.0	0.600000	-0.491940	0.550806	-0.449756
1	1.2	0.508827	-0.416874	0.467140	-0.364509
2	1.4	0.434403	-0.326338	0.401769	-0.269671
3	1.6	0.378666	-0.232681	0.355398	-0.179526
4	1.8	0.340854	-0.149541	0.325900	-0.106318

n	x_n	k_{3n}	K_{3n}	k_{4n}	K_{4n}
0	1.0	0.555024	-0.465191	0.506962	-0.413343
1	1.2	0.472377	-0.382947	0.432238	-0.320963
2	1.4	0.407435	-0.290580	0.376287	-0.225245
3	1.6	0.360714	-0.201170	0.338432	-0.140298
4	1.8	0.330222	-0.126258	0.315602	-0.075791

n	x_n	k_n	K_n	y_n	z_n
0	1.0	0.553104	-0.455863	0.000000	3.000000
1	1.2	0.470016	-0.372125	0.553104	2.544137
2	1.4	0.404850	-0.278681	1.023120	2.172013
3	1.6	0.358220	-0.189062	1.427970	1.893332
4	1.8	0.328117	-0.115081	1.786190	1.704270

Answers for Section 9.1

1 2π **3** 2π **5** 20π **7** π **9** $\pi/4$

11

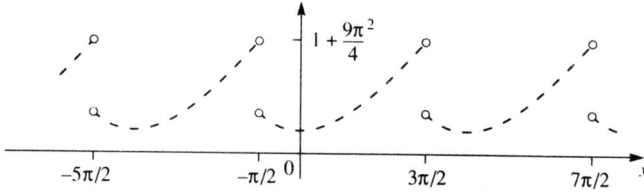

13

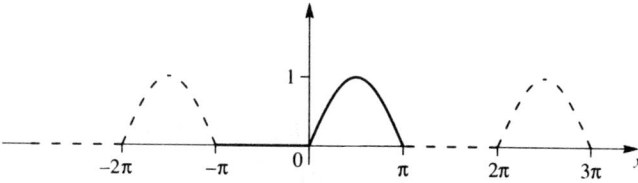

15

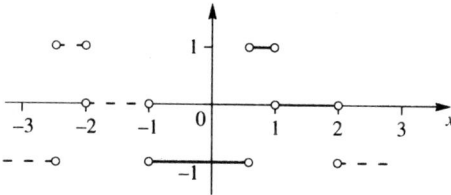

17 Only an even extension is possible because $f(0) \neq 0$.

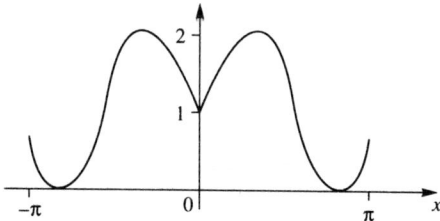

19 Both extensions are possible because $f(0) = 0$.

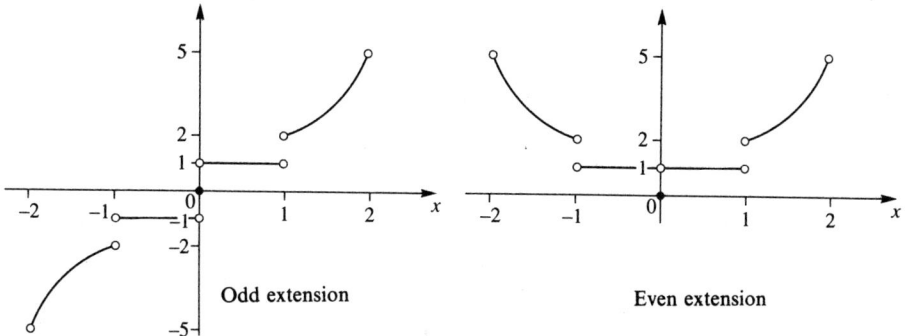

Odd extension

Even extension

21 (a) piecewise smooth   (b) piecewise continuous   (c) piecewise smooth.

23 The limit function $f(x) = \lim_{n \to \infty} f_n(x) \equiv 0$.

Now $|f_n(x) - f(x)| = \dfrac{1}{n}|\cos nx| \leq 1/n$ for all x. By taking n sufficiently large, $|f_n(x) - f(x)|$ may be made arbitrarily small for all x, thereby establishing the uniform convergence.

25 We have $|(1 + \sin x)/3^n| \leq 2/3^n$ for all real x. Setting $M_n = 2/3^n$ it follows by setting $p = 2$ in Ex. 1(c), Sec. 8.1, that $\displaystyle\sum_{n=1}^{\infty} M_n$ is convergent. Hence the functional series is uniformly convergent by the M-test for all real x.

27 The M-test fails because

$$\left|\frac{(-1)^n}{3x + 2n}\right| \leq \frac{1}{2n} \qquad \text{for } 0 \leq x < \infty,$$

so setting $M_n = 1/(2n)$ and considering $\displaystyle\sum_{n=1}^{\infty} M_n$ leads to a multiple of the harmonic series which is divergent. However, if the sum of the series is $s(x)$ and its nth partial sum is $s_n(x)$, using the error estimate provided in the alternating series test (Sec. 8.1) show that

$$|s_n(x) - s(x)| = \left|\sum_{r=n}^{\infty} \frac{(-1)^r}{3x + 2r}\right| \leq \frac{1}{3x + 2n} \leq \frac{1}{2n}$$

for $0 \leq x < \infty$. Thus, by making n suitably large, $|s_n(x) - s(x)|$ may be made arbitrarily small for $0 \leq x < \infty$. This establishes the uniform convergence.

Answers for Section 9.2

1

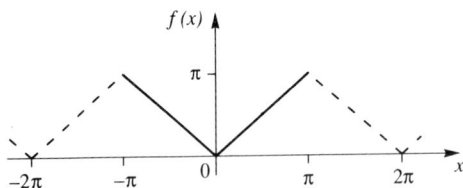

3

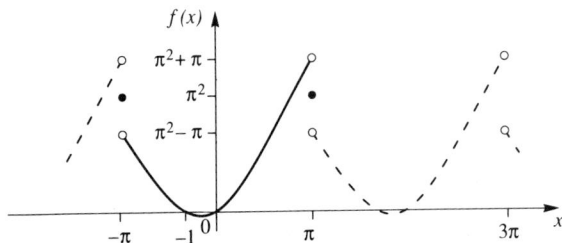

5

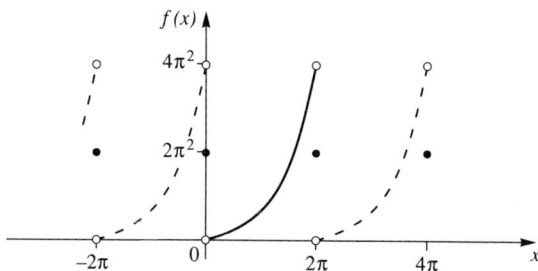

Series converges to π^2

at $x = \pi$, $\displaystyle\sum_{n=1}^{\infty} \frac{(-1)^n}{n^2} = -\frac{\pi^2}{12}$.

Series converges to $2\pi^2$ at $x = 0$, $\displaystyle\sum_{n=1}^{\infty} \frac{1}{n^2} = \frac{\pi^2}{6}$.

7

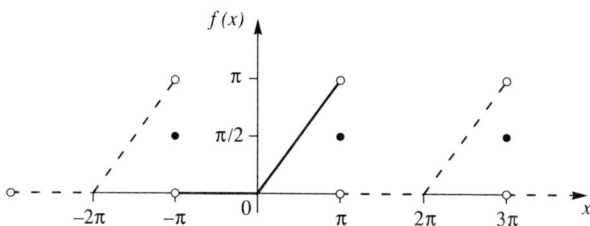

The series converges to zero at $x = 0$ from which follows the first summation. The series converges to $\pi/2$ at $x = 0$ from which follows the second summation.

9

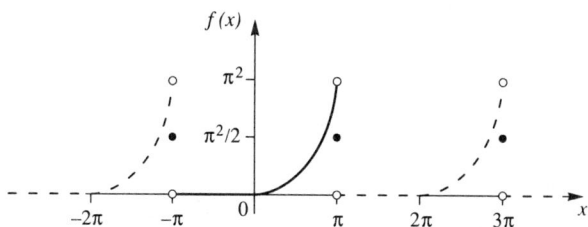

At $x = \pi$ the series converges to $\pi^2/2$, from which it follows that $\displaystyle\sum_{n=1}^{\infty} \frac{1}{n^2} = \frac{\pi^2}{6}$.

11

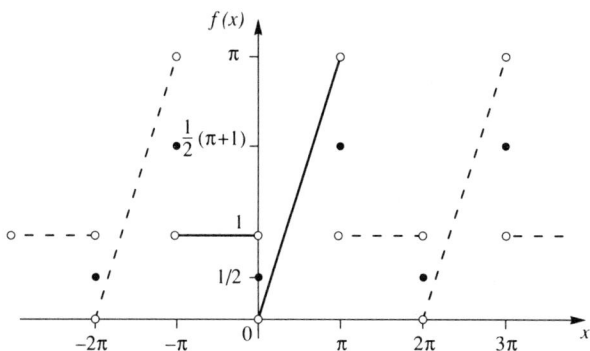

13

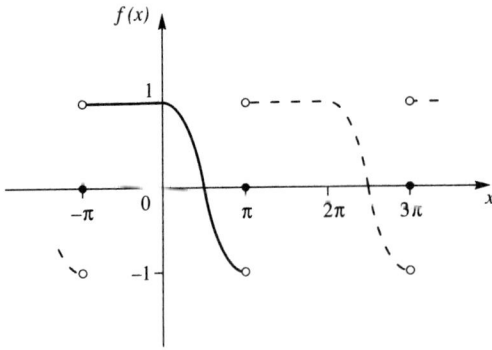

15

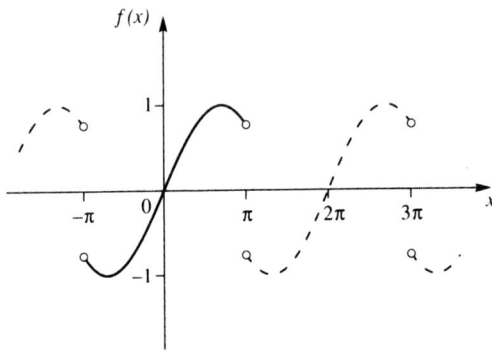

At $x = \pi/2$ the series converges to $\sin k\pi/2$. Divide by $\sin k\pi$ and write $\sin k\pi = 2 \sin k\pi/2 \cos k\pi/2$ to obtain the result.

17

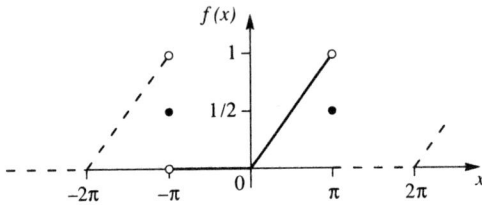

19

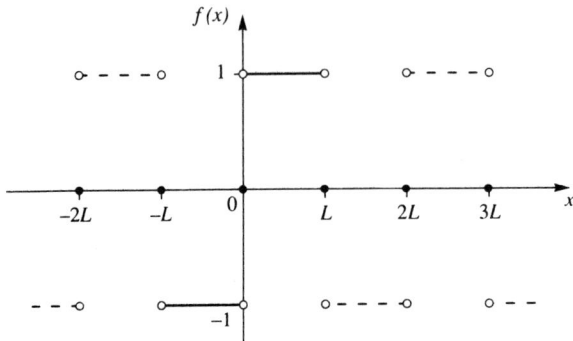

21

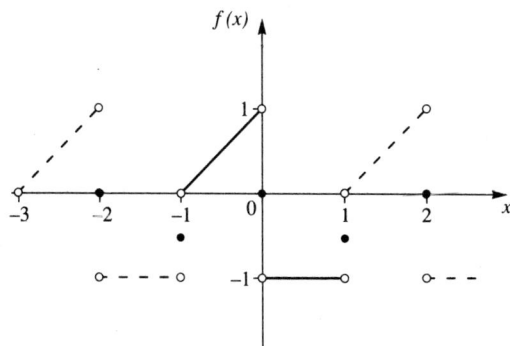

23

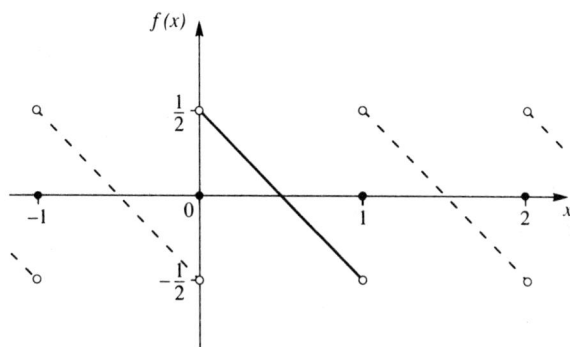

25

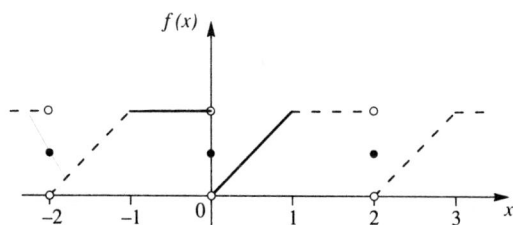

27

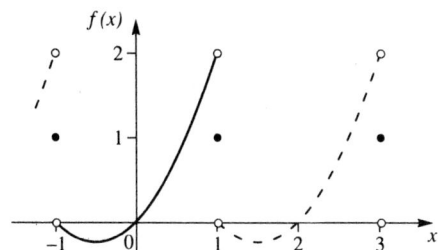

When $x = 0$ the series converges to zero leading to the first summation and when $x = 1$ the series converges to 1 (a discontinuity) leading to the second summation.

29

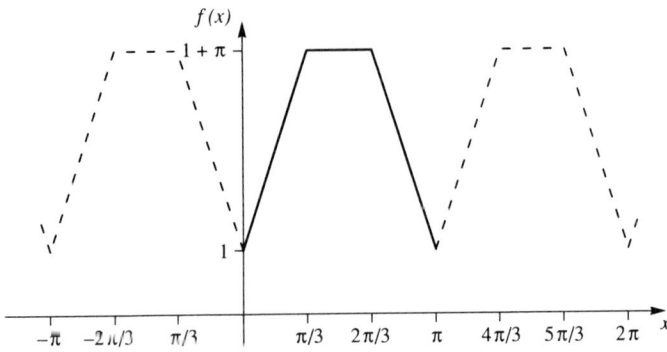

31

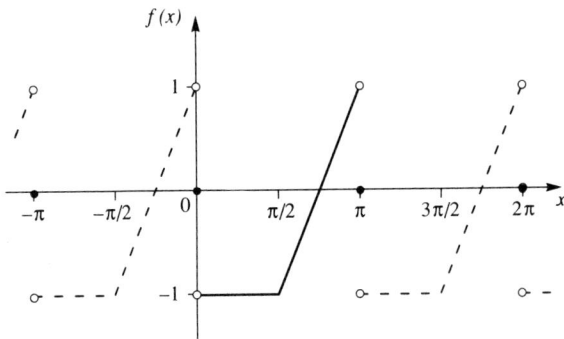

33

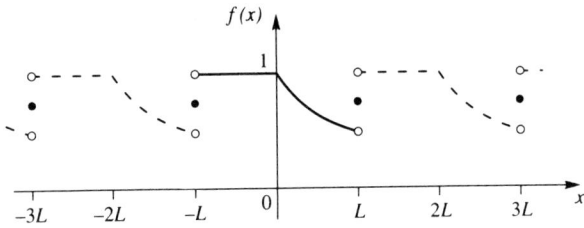

35

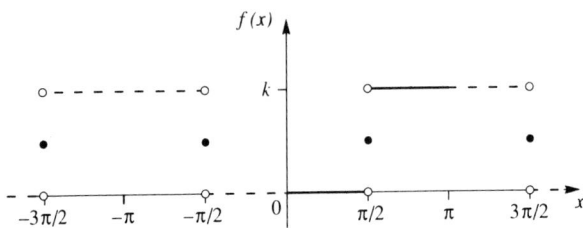

$$f(x) \sim \frac{k}{2} - \frac{2k}{\pi} \sum_{n=1}^{\infty} \frac{(-1)^{n+1}}{(2n-1)} \cos(2n-1)x, \qquad \text{for } 0 \le x < \pi.$$

37

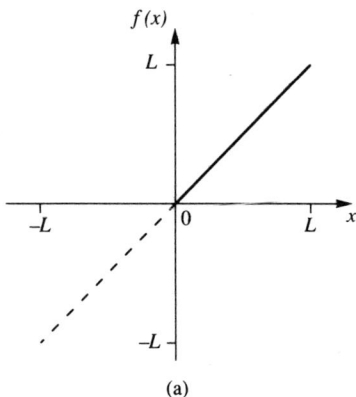

(a)

Odd extension

$$f(x) \sim \frac{2L}{\pi} \sum_{n=1}^{\infty} \frac{(-1)^{n+1}}{n} \sin \frac{n\pi x}{L},$$

for $0 \le x < L$.

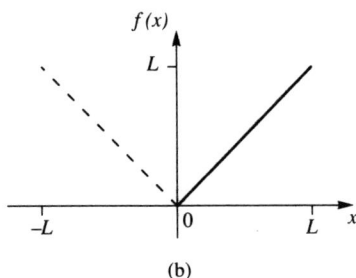

(b)

Even extension

$$f(x) \sim \frac{L}{2} - \frac{4L}{\pi^2} \sum_{n=1}^{\infty} \frac{\cos (2n-1)\frac{\pi x}{L}}{(2n-1)^2},$$

for $0 \le x < L$.

Both series converge to $f(x)$ for $0 \le x < L$, but (b) converges more rapidly than (a).

39

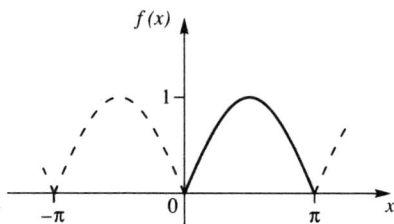

41

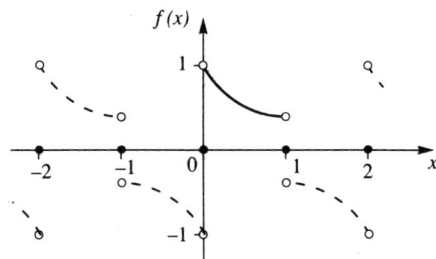

43

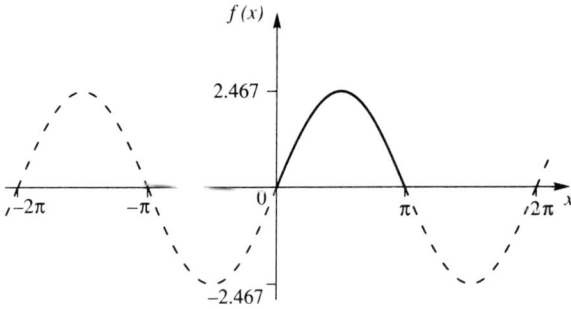

47

As $c_n = \left(\dfrac{\sinh \pi}{\pi}\right)\dfrac{(-1)^n}{1-in}$, $|c_n| = \dfrac{\sinh \pi}{\pi}\dfrac{1}{\sqrt{1+n^2}}$, and

$$\theta = \arg c_n = \begin{cases} \arctan n & (n \text{ even}) \\ \arctan n - \pi & (n \text{ odd}). \end{cases}$$

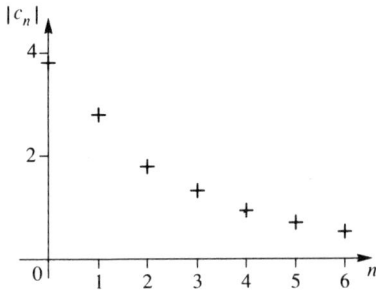

Modulus as function of n

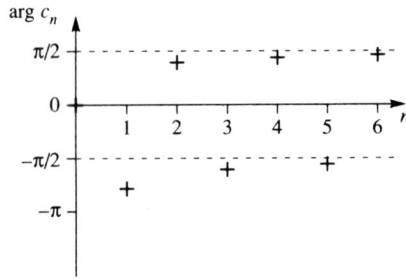

Phase as function of n

49

$$f(x) \sim \frac{4\pi^2}{3} + \lim_{m \to \infty} \sum_{n=-m}^{m}{}' \left[\left(\frac{1+(-1)^{n+1}}{in\pi}\right) + \frac{2(1+in\pi)}{n^2}\right],$$

where $\sum'$ denotes the omission of the terms corresponding to $n=0$ from the summation.

51 $F(x) = \dfrac{7}{2}f(x) - \dfrac{1}{2}$; $F(x) \sim \dfrac{7}{4} + \dfrac{7}{\pi}\displaystyle\sum_{n=0}^{\infty} \dfrac{\sin(2n+1)x}{(2n+1)}$.

53 $F(x) = 2g(x) + 2f(x)$; $F(x) \sim \dfrac{21}{2} + \dfrac{72}{\pi^2}\displaystyle\sum_{n=1}^{\infty} \dfrac{(-1)^n}{n^2}\cos\dfrac{n\pi x}{3} + \dfrac{6}{3}\displaystyle\sum_{n=0}^{\infty} \dfrac{\sin(2n+1)(\pi x/3)}{(2n+1)}$.

55 $F(x) = 1 + \sin x + \tfrac{1}{3}\sin 2x + 2f(x)$; $F(x) \sim 2 + \sin x + \tfrac{1}{3}\sin 2x + \dfrac{4}{\pi}\displaystyle\sum_{n=0}^{\infty} \dfrac{\sin(2n+1)x}{(2n+1)}$,   because

$\sin x + \tfrac{1}{3}\sin 2x$ is its own Fourier series representation.

57

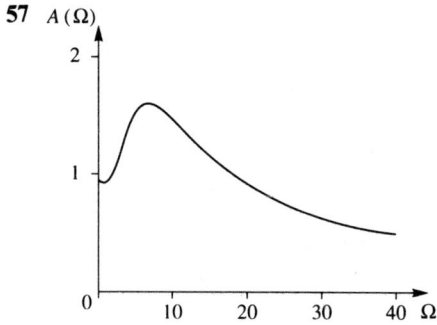

Amplitude factor $A(\Omega)$ Phase change $\phi(\Omega)$

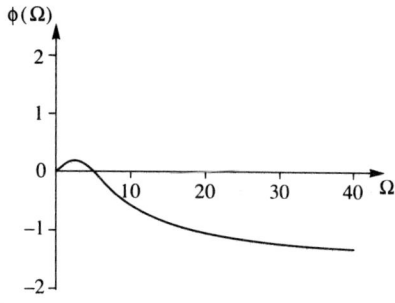

$T(0) = 0.93$ $T(i) = 0.925 + 0.069i$ $T(2i) = 0.982 + 0.185i$
$T(3i) = 1.15 + 0.253i$ $T(4i) = 1.356 + 0.196i$

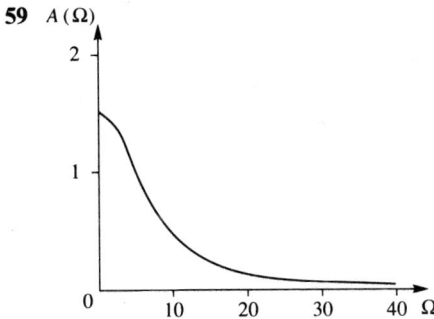

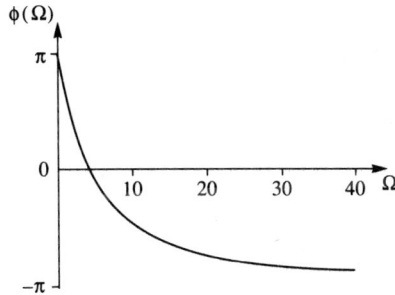

59

$T(0) = -1.512$ $T(i) = -0.899 + 1.186i$ $T(2i) = 0.222 + 1.402i$
$T(3i) = 0.954 + 0.901i$ $T(4i) = 1.15 + 0.267i$

Answers for Section 9.3

1

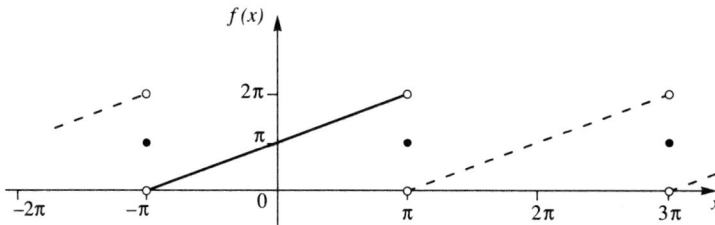

$$f(x) \sim \pi + 2\sum_{n=1}^{\infty} \frac{(-1)^{n+1}}{n}\sin x.$$

3 $$f(x) \sim \frac{\pi}{4} + \frac{1}{\pi}\sum_{n=1}^{\infty}\frac{[(-1)^n - 1]}{n^2}\cos nx - \sum_{n=1}^{\infty}\frac{(-1)^n}{n}\sin nx.$$

The Parseval relation gives

$$\frac{1}{\pi}\int_0^\pi x^2\,dx = \frac{1}{2}\left(\frac{\pi}{2}\right)^2 + \frac{4}{\pi^2}\sum_{n=1}^\infty \frac{1}{(2n-1)^4} - \sum_{n=1}^\infty \frac{1}{n^2}$$

which reduces to the required result.

5 Substitution into the product theorem gives

$$\int_{-1}^0 (1+x)\,dx + \int_0^1 -x\,dx = \frac{1}{2}\left(-\frac{1}{2}\right)\left(\frac{3}{2}\right) - \frac{1}{\pi^4}\sum_{n=1}^\infty \frac{[(-1)^n-1]^2}{n^4} + \frac{1}{\pi^2}\sum_{n=1}^\infty \left(\frac{2+(-1)^{n+1}}{n^2}\right)$$

from which the required result follows after using the known series

$$\sum_{n=1}^\infty \frac{1}{n^2} = \frac{\pi^2}{6} \quad \text{and} \quad \sum_{n=1}^\infty \frac{(-1)^{n+1}}{n^2} = \frac{\pi^2}{12}.$$

Answers for Section 9.4

1 $f(x)$ is piecewise continuous, so Theorem 9.10 applies and gives

$$\int_0^x t(\pi-t)\,dt = \int_0^x \frac{\pi^2}{6}\,dt - \sum_{n=1}^\infty \int_0^x \frac{\cos 2nt}{n^2}\,dt$$

from which it follows that

$$g(x) \sim 3\sum_{n=1}^\infty \frac{\sin 2nx}{n^3}.$$

3 $f(x)$ is piecewise continuous, so Theorem 9.10 applies and gives

$$\int_{-\pi}^x \frac{1}{2}(\pi-t)\,dt = \int_{-\pi}^x \frac{\pi}{2}\,dt + \sum_{n=1}^\infty \frac{(-1)^n}{n}\int_{-\pi}^x \sin nt\,dt,$$

from which the result follows after using the known series

$$\sum_{n=1}^\infty \frac{1}{n^2} = \frac{\pi^2}{6}.$$

5 $f(x)$ is piecewise continuous with $f(0)=f(\pi)$, so Theorem 9.11 applies and yields the stated result.

7 $f(x)$ is continuous with $f(-\pi)=f(\pi)$, so Theorem 9.11 applies and yields the stated result. $h(x)$ is only piecewise continuous, and in addition $h(-\pi)\neq h(\pi)$, each of which is sufficient to invalidate Theorem 9.11.

$$f(x) \sim \frac{1}{\pi} + \frac{1}{2}\sin x - \frac{2}{\pi}\sum_{n=1}^\infty \frac{n\sin 2nx}{(2n-1)(2n+1)}$$

$$g(x) \sim \frac{1}{2}\cos x + \frac{4}{\pi}\sum_{n=1}^\infty \frac{n\sin 2nx}{(2n-1)(2n+1)}.$$

9 $f(x) \sim \dfrac{1}{2} + \dfrac{4}{\pi^2} \displaystyle\sum_{n=1}^{\infty} \dfrac{\cos(2n-1)\pi x}{(2n-1)^2}.$

The sawtooth function $f(x)$ satisfies the conditions of Theorem 9.11, which when applied yields

$$g(x) \sim \dfrac{-4}{\pi} \sum_{n=1}^{\infty} \dfrac{\sin(2n-1)\pi x}{(2n-1)}.$$

Answers for Section 9.7

1

n	0	1	2	3
a_n	$-0 \cdot 53$	$0 \cdot 587$	$0 \cdot 75$	$0 \cdot 098$
b_n	—	$-0 \cdot 729$	$-0 \cdot 242$	$0 \cdot 363$

3

n	0	1	2	3
a_n	$3 \cdot 399$	$0 \cdot 727$	$0 \cdot 374$	$-0 \cdot 284$
b_n	—	$-1 \cdot 395$	$-0 \cdot 437$	$0 \cdot 019$

5

n	0	1	2	3
a_n	$3 \cdot 103$	$0 \cdot 15$	$-1 \cdot 312$	$0 \cdot 399$
b_n	—	$1 \cdot 027$	$-0 \cdot 032$	$-0 \cdot 107$

7

n	0	1	2	3
a_n	$3 \cdot 203$	$0 \cdot 792$	$0 \cdot 285$	$0 \cdot 045$
b_n	—	$-0 \cdot 031$	$-1 \cdot 241$	$0 \cdot 08$

Answers for Section 9.8

1 $y(x) = A \cos \sqrt{\lambda}\, x + B \sin \sqrt{\lambda}\, x$, so $y'(x) = -A\sqrt{\lambda} \sin \sqrt{\lambda}\, x + B\sqrt{\lambda} \cos \sqrt{\lambda}\, x$.

$[y'(0) = 0]$ $B = 0$ $[y'(L) = 0]$ $-A\sqrt{\lambda} \sin \sqrt{\lambda}\, L = 0$ thus $\sqrt{\lambda}\, L = n\pi$, so the eigenvalues are

$\lambda_n = \dfrac{n^2 \pi^2}{L^2}$ and the eigenfunctions are $y_n^{(x)} = \cos \dfrac{n\pi x}{L}$, $n = 0, 1, \ldots$.

3 $y(x) = A\cos\sqrt{\lambda}\,x + B\sin\sqrt{\lambda}\,x$, so $y'(x) = -A\sqrt{\lambda}\sin\sqrt{\lambda}\,x + B\sqrt{\lambda}\cos\sqrt{\lambda}\,x$.

$[y(-\pi)=0]0 = A\cos\pi\sqrt{\lambda} - B\sin\pi\sqrt{\lambda}$

$[y'(\pi)=0]0 = -A\sin\pi\sqrt{\lambda} + B\cos\pi\sqrt{\lambda}$.

Nontrivial solution if $\begin{vmatrix} \cos\pi\sqrt{\lambda} & -\sin\pi\sqrt{\lambda} \\ -\sin\pi\sqrt{\lambda} & \cos\pi\sqrt{\lambda} \end{vmatrix} = 0$

which is equivalent to $\cos 2\pi\sqrt{\lambda} = 0$.

Thus $2\pi\sqrt{\lambda} = (2n-1)\dfrac{\pi}{2}$, and so the eigenvalues are $\lambda_n = \left(\dfrac{2n-1}{4}\right)^2$, for $n = 1, 2, \ldots$. Now $A/B = \tan\ \pi\sqrt{\lambda_n}$, so setting $B = 1$ (it is arbitrary) gives for the eigenfunctions

$y_n(x) = \tan\left(\dfrac{2n-1}{4}\right)\pi\cos\left(\dfrac{2n-1}{4}\right)x + \sin\left(\dfrac{2n-1}{4}\right)x, \ n = 1, 2, \ldots$.

5 $y(x) = A\cos\sqrt{\lambda}\,x + B\sin\sqrt{\lambda}\,x$, so $y'(x) = -A\sqrt{\lambda}\sin\sqrt{\lambda}\,x + B\sqrt{\lambda}\cos\sqrt{\lambda}\,x$. $[y(0)=0]$ $A = 0$, $[y(\pi) - y'(\pi) = 0]$ $\sin\pi\sqrt{\lambda} - \sqrt{\lambda}\cos\pi\sqrt{\lambda} = 0$. Thus the numbers $\sqrt{\lambda_n}$ (the square roots of the eigenvalues) are roots of the transcendental equation $\sqrt{\lambda_n} = \tan\pi\sqrt{\lambda_n}$, $n = 1, 2, \ldots$, so the eigenfunctions are $y_n(x) = \sin\sqrt{\lambda_n}\,x$.

Answers for Section 9.9

1 $\lambda_n = \mu_{2,n}^2$, $y_n(x) = J_2(\mu_{2,n}x)$, $f(x) = x^2$ so

$x^2 = \displaystyle\sum_{n=1}^{\infty} c_n J_2(\mu_{2,n}x)$, with

$c_n = \dfrac{1}{\|y_n\|^2}\displaystyle\int_0^1 f(x)\,x\,J_2(\mu_{2,n}x)\,dx = \dfrac{1}{\|y_n\|^2}\int_0^1 x^3 J_2(\mu_{2,n}x)\,dx$.

$\|y_n\|^2 = \dfrac{1}{2}[J_3(\mu_{2,n})]^2$, $c_n = \dfrac{1}{\|y_n\|^2}\dfrac{1}{(\mu_{2,n})^4}\displaystyle\int_0^{\mu_{2,n}} t^3 J_2(t)\,dt$

$= \dfrac{2}{[J_3(\mu_{2,n})]^2}\cdot\dfrac{1}{(\mu_{2,n})^4}\cdot(\mu_{2,n})^3 J_3(\mu_{2,n}) = \dfrac{2}{\mu_{2,n}J_3(\mu_{2,n})}$.

3 $\lambda_n = n^2$, $y_n(x) = J_{1/2}(nx) = \sqrt{\dfrac{2}{n\pi x}}\sin nx$, $f(x) = x^{1/2}$ so

$x^{1/2} = \sqrt{\dfrac{2}{\pi}}\displaystyle\sum_{n=1}^{\infty} c_n \dfrac{\sin nx}{\sqrt{nx}}$, with

$c_n = \dfrac{1}{\|y_n\|^2}\displaystyle\int_0^\pi f(x)J_{1/2}(nx)\,x\,dx = \dfrac{1}{\|y_n\|^2}\sqrt{\dfrac{2}{n\pi}}\int_0^\pi x\sin nx\,dx$.

$\|y_n\|^2 = \dfrac{\pi^2}{2}[J_{3/2}(n\pi)]^2$, $c_n = \dfrac{2}{\pi^2[J_{3/2}(n\pi)]^2}\sqrt{\dfrac{2}{n\pi}}\dfrac{(-1)^{n+1}x}{n}$

$= (-1)^{n+1}\left(\dfrac{2}{n\pi}\right)^{3/2}\dfrac{1}{[J_{3/2}(n\pi)]^2}$.

5 Proceed as indicated in the problem. From (19) of Sec. 8.7 $xJ'_\nu(x) = \nu J_\nu(x) - xJ_{\nu+1}(x)$, so setting $x = \mu_{\nu,n}$ this reduces to $J'_\nu(\mu_{\nu,n}) = -J_{\nu+1}(\mu_{\nu,n})$. The last result follows by changing the variable in the integral to $t = \mu_{\nu,n}x$.

7 Proceed as indicated in the problem. Substitute the expressions

$$J'_\nu(\mu a) = -\left(\frac{h}{k\mu}\right)J_\nu(\mu a) \quad \text{and} \quad J'_\nu(\lambda a) = -\left(\frac{h}{k\lambda}\right)J_\nu(\lambda a)$$

into the right-hand side of the equation to show that it vanishes if λ, μ are different positive roots of the transcendental equation.

Answers for Section 9.10

9 $c_n = \dfrac{1}{\|T_n\|^2}\displaystyle\int_{-1}^{1} x^n T_n(x)\,\dfrac{dx}{(1-x^2)^{1/2}} = \dfrac{2}{\pi}\displaystyle\int_{-1}^{1} x^n \cos(n \arccos x)\,\dfrac{dx}{(1-x^2)^{1/2}}.$

Now set $x = \cos\theta$ to obtain the required result.

11 Using (28), express $L_n(x)$ and $L_{n+1}(x)$ in terms of $L_{n-1}(x)$, and then show that all three expressions satisfy the recursion relation.

Suggested Reading and Reference List

The following selection of titles, which is in no sense complete, has been chosen on the basis that the level of approach adopted by the authors is similar in many respects to that of this book.

Linear algebra and vector analysis

Anton, H. *Elementary Linear Algebra* 2nd edn. Wiley, New York, 1982.
Davis, H. F. and Snider, A. D. *Introduction to Vector Analysis* 4th edn. Allyn & Bacon, Boston, 1979.
O'Neil, P. V. *Introduction to Linear Algebra*. Wadsworth, California, 1979.
Strang, G. *Linear Algebra and its Applications* 3rd edn. Academic Press, New York, 1988.

Ordinary differential equations

Birkhoff, G. and Rota, G. C. *Ordinary Differential Equations* 4th edn. Wiley, New York, 1978.
Boyce, W. E. and Di Prima, R. C. *Elementary Differential Equations and Boundary Value Problems* 3rd edn. Wiley, New York, 1983.
Brauer, F. and Nobel, J. A. *Introduction to Differential Equations with Applications*. Harper & Row, New York, 1986.
Campbell, S. L. *An Introduction to Differential Equations and their Applications*. Longman, New York, 1986.
Jordan, D. W. and Smith, P. *Nonlinear Ordinary Differential Equations*. Clarendon Press, Oxford, 1977.
Ross, S. L. *Differential Equations* 3rd edn. Wiley, New York, 1984.

Mathematical modeling with differential equations

Borrelli, R. L. and Coleman, C. S. *Differential Equations: A Modeling Approach*. Prentice-Hall, New Jersey, 1987.
Giordano, F. R. and Weir, M. D. *A First Course in Mathematical Modeling*. Van Nostrand Reinhold, New York, 1985.

Special functions

Sneddon, I. N. *Special Functions of Mathematical Physics and Chemistry* 3rd edn. Longman, London, 1980.
Watson, G. N. *A Treatise on the Theory of Bessel Functions* 2nd edn. Cambridge University Press, London, 1966.

Numerical analysis

Fröberg, C.-E. *Numerical Methods.* Benjamin Cummings, Massachusetts, 1985.

Hamming, R. W. *Numerical Methods for Scientists and Engineers* 2nd edn. McGraw-Hill, New York, 1987.

Henrici, P. *Essentials of Numerical Analysis with Pocket Calculator Demonstrations.* Wiley, New York, 1982.

Ortega, J. M. and Poole, W. G., Jr. *An Introduction to Numerical Methods for Differential Equations.* Longman, New York, 1981.

Pearson, C. E. *Numerical Methods in Engineering and Science.* Van Nostrand Reinhold, New York, 1986.

Mathematical tables

Abramowitz, M. and Stegun, I. A. (eds). *Handbook of Mathematical Functions*, 10th printing with corrections. National Bureau of Standards, Washington, DC, 1972. (Also: Dover, New York, 1965.)

Index